SCIENTIFIC POULTRY PRODUCTION

A Unique Encyclopaedia

Third Revised and Enlarged Edition

Prof. P.V. SREENIVASAIAH
Head
Department of Avian Production and Management
Rajiv Gandhi College of Veterinary and Animal Sciences
PONDICHERRY : 605 009

International Book Distributing Co.
(Publishing Division)

Published by

INTERNATIONAL BOOK DISTRIBUTING CO.
(Publishing Division)
Chaman Studio Building, 2nd Floor,
Charbagh, Lucknow 226 004 U.P. (INDIA)
Tel. : Off. : 2450004, 2450007, 2459058 Fax : 0522-2458629
E-Mail : ibdco@sancharnet.in

Third Edition 2006

ISBN 81-8189-147-3

Composed & Designed at :

Panacea Computers
3rd Floor, Agarwal Sabha Bhawan, Subhash Mohal
Sadar Cantt. Lucknow-226 002
Phone : 2483312, , 9335927082
E-mail : prasgupt@rediffmail.com

Printed at:

Salasar Imaging Systems
C-7/5, Lawrence Road Industrial Area
Delhi - 110 035
Tel. : 011-27185653, 9810064311

Preface to the Third Edition

The preface to the First Edition during 1987 stated that poultry farming changed from backyard hobby to business enterprise and Veterinarians need to know more about poultry science to augment the growth of the industry. By the time Second Revised Edition was published during 1998, poultry industry and poultry science education had witnessed rapid strides of progress. Since then, the progress has further intensified; broiler and layer industries are growing at an estimated rate of 10 and 15%, respectively.

There has been an attitude change in poultry farming especially in developed countries with welfare and environmental concerns coming to the fore. In about 5-6 years, the member countries of European Union will be completely replacing the conventional cages; even alternatives for one of the most common farm practices, beak-trimming, is being thought of. The spread of Avian influenza has added an altogether different dimension to the disease control, biosecurity and basic farm technologies.

Feeding poultry has entered into a new area of enzymes, performance enhancers and other means of maximizing returns. Newer poultry products have made poultry meat, as has been till now, the most popular meat throughout the world. Development in the field of biotechnology is likely to offer opportunities for both basic research as well as applications in practical breeding; the latter has become a necessity because conventional selection processes are no more able to produce the desired/expected responses.

The Veterinary Council of India (VCI) is in the process of revising the undergraduate Veterinary syllabus. Notwithstanding the fact that two courses (instead of one earlier) are proposed, it appears inadequate. Further, it is surprising that an industry which contributes 1.2% of GDP of the country, already employing 2,000 Veterinarians and likely to continue employing a tenth of that every year, has not been properly recognized; for instance, the VCI does not stipulate that the Department of Poultry Science should be an independent Department in Veterinary colleges!

In view of the above, the Second Edition needed a thorough revision and the same is presented through this publication; suffice if I say that most of the chapters are updated keeping in view of the developments in science and art of poultry production throughout the world as well as the requirements of the undergraduates students in light of the proposed changes of syllabus by the VCI. In addition, this Edition is likely to be useful even for the postgraduate students pursuing in various fields of Poultry Science.

My sincere thanks are due to all my teachers, especially Dr P.V. Rao, Dr (Late) T.D. Mahadevan, Dr S.C. Mohapatra, Dr S. Jyotheeswara Reddy and Dr Veera Raghavan who helped open my babe eyes into the vast field of Poultry Science. I am privileged to use their teaching material for discussion in various chapters in my books. I thank the Dean, Rajiv Gandhi College of Veterinary and Animal Sciences, Pondicherry for permitting the use of computer facility and providing ample time for taking-up this revision work. I take this opportunity to thank Dr P.V.V. Satyanarayana Reddy, Dr J. Ramaprasad and Dr S. Umamaheswara Rao, Professors at College of Veterinary Science, Tirupathi for their encouragement and extending resources for writing this book.

I thank one and all at Rajiv Gandhi College of Veterinary and Animal Sciences who are directly or indirectly helpful in this endeavor; especially Dr B.B.Reddy, Professor, Department of Veterinary Physiology and Dr S. Venugopal, Assistant Professor, Department of LPM, Mr Sambasivam and Mr. Sivakumar of the Engineering section.

I am thankful to my guardian, mother, parents-in-law, brothers-in-law, brothers, sister, wife and children who have stood by me while preparing this book.

I wish to put on record my thanks to Mr Suneel Gomber, Publisher, International Book Distributing Co. who has been very keen to publish the work ever since he came to know of the revision. The elegant presentation in itself stands a testimony for his personal interest in bringing out this book.

Prof. P.V. Sreenivasaiah
Head
Department of Avian Production and Management
Rajiv Gandhi College of Veterinary and Animal Sciences
E-mail: prof.pvs@gmail.com
Website: www.prof-pvs.i8.com

Pondicherry
November 7, 2005

Contents

Preface iii

INTRODUCTION

1 Poultry industry 1

EVOLUTION, DOMESTICATION AND BREEDS

2 Evolution of birds 10

3 Domestication of poultry 19

4 Morphology and nomenclature 31

5 Breeds of different species of Poultry 49

6 Interpretation of standards 75

REPRODUCTION IN POULTRY

7 Egg - formation and structure 96

8 Male reproductive system 134

POUTRY TECHNOLOGY

9 Composition and nutritive value of eggs 147

10 Egg consumption *Vis-à-vis* coronary heart disease 164

11 Ready-to-cook chicken 194

12 Composition and nutritive value of poultry meat 211

13 Poultry meat consumption *Vis-à-vis* colon cancer 228

14 Egg quality 235

15 Poultry meat quality 257

16 Principles of food preservation 265

17 Preservation of eggs 270

18 Preservation of poultry meat 284

19 Egg products 297

20 Poultry meat products 320

21 Microbiology of eggs and egg products 345

22 Microbiology of poultry meat and poultry meat products 353

23 Non-food applications of eggs and poultry meat 360

POULTRY MANAGEMENT

24 Incubation and hatching 364

25 Climate and climatic indices 429

26 Thermoregulation 435

27 Principles of ventilation 455

28 Natural ventilation 479

29 Mechanical ventilation 494
30 Insulation 532
31 Shelter engineering 543
32 Poultry housing 560
33 Rearing systems 578
34 Brooding 601
35 Beak-trimming 615
36 Broilers and roasters 637
37 Restricted feeding 656
38 Lighting management 663
39 Judging pullets 677
40 Management of breeders and commercial layers 686
41 Litter management 702
42 Cage management 708
43 Molting : natural and induced 717
44 Poultry manure 729
45 Pest management 756
46 Carcass disposal 772
47 Environmental implications of poultry production 798
48 Farm data – recording and processing 816
49 Ducks, Turkeys, Geese, Quails and other poultry 826
50 Management requirements and specifications 835

POULTRY NUTRITION

51 Digestion, absorption and excretion 850
52 Nutrients – functions, deficiency and excess 877
53 Nutrient requirements and specifications 906
54 Feedstuffs for poultry 928
55 Feeds and feeding 950
56 Performance enhancers in poultry feeds 969
57 Ration formulation 985
58 Toxins in poultry feed 1003

POULTRY DISEASES

59 Immune system 1029
60 Profiling a disease 1051
61 Necropsy examination 1063
62 Bacterial diseases 1076

63 Fungal diseases 1105
64 Viral diseases 1111
65 Protozoan diseases 1143
66 Parasitic diseases 1160
67 Metabolic diseases 1187
68 Miscellaneous diseases and conditions 1194
69 Diagnosis of poultry diseases 1204
70 Biosecurity 1214
71 Sanitation and disinfection 1225
72 Vaccination - principles and practices 1238

POULTRY GENETICS AND BREEDING

73 Inheritance of qualitative traits 1257
74 Inheritance of quantitative traits 1288
75 Applied poultry breeding 1317
76 Biotechnology 1332

ECONOMICS OF POULTRY OPERATIONS

77 Fundamentals of economics 1349
78 Maintenance of accounts 1362
79 Agricultural marketing 1377
80 Economics of poultry operations 1395

OTHER TOPICS

81 Ethology of poultry 1404
82 Poultry welfare 1413

Appendices 1435
Abbreviations 1441
Index to tables 1448
Index to figures 1459
Index to plates 1462
Index to sites on internet 1464
Literature cited 1466
Subject index 1475

Chapter **1**

Poultry Industry

1. Global poultry

Growth of global poultry is due to 1. Technical innovations 2. Optimum temperature and lighting programs 3. Decreased mortality because of popularity of all-in all-out management, vaccination and better hygiene 4. Availability of balanced nutrition 5. Improved genetic potential and understanding of the fact that demand dictates genetic goals and 6. Government support in the form of loans, infrastructure and marketing. Globalization has helped in easy dissemination of technology as well as easy movement of material between all the countries. Global poultry development over the past five decades has been summarized in Table 1.1.

Estimated genetic gain (per annum) for the period 1998-2003 was + 30 g, + 0.43 %, + 0.44 %, + 0.17 %, - 0.13 %, + 0.75 % and – 0.03 with respect to live weight, eviscerated yield, breast meat yield, broiler livability, plant condemnations, breeder hatching eggs and food conversion ratio (FCR) to attain 2.27 kg (5 lb), respectively (*Source :* McAdam, 2004). It is now possible, at least in developed countries, to expect 320 eggs per annum with a food consumption of < 2 kg/kg egg mass and broilers weighing 1.8-2.0 kg at 5 weeks with an FCR of 1.6 to 1.8.

2. Indian Poultry Industry

2.1. Path of progress

There has been a phenomenal growth in the poultry industry in India. Most of the poultry population before 1960 consisted of "indigenous" or "Desi" birds which were mainly reared in backyard units and improved layers accounted for a meager 5% of the poultry population. During 1960's, one thirds of the total poultry population consisted of improved layers and the farmers were made aware of scientific methods of rearing, feeding and disease control.

A perceptible development of Poultry Industry in India began during 1970's and recorded rapid strides from what was a backyard rearing into a very viable enterprise. Growing demand for high-quality chicks and the need of developing strains and varieties to suit Indian conditions, necessitated the initiation of All India Coordinated Research Projects (AICRPs) separately for developing egg and meat strains. The concept of poultry for meat purpose gained momentum and many farmers ventured into poultry business with substantial increase in flock size. The contribution of improved birds increased considerably and they formed three-fourths of output of the poultry industry. Development of infrastructure for training, marketing and provision of technical input were taken on priority basis.

Simultaneously, a Central Avian Research Institute (CARI) was established during 1979 to give further fillip to poultry education, research and extension. Feed processing industries, disease diagnostic and feed analytical laboratories augmented the development of the poultry industry.

Table 1.1 Development of global poultry industry

Period	Salient features
1962 - 1972	Clear shift of rearing system from deep litter to cages. Consumer preference shifted from brown eggs to white eggs. Improved understanding of nutrition of chicken especially regarding essential amino acids. Hybridization to improve performance developed. Considerable decrease in mortality primarily due to reduced parasites and cannibalism. Intensive housing along with lighting control and equipment designing complemented the development of the poultry industry.
1976 - 1986	94% of eggs were produced by birds in cages and annual production by a layer increased from around 200 to 245 indicating a genetic gain of 4 eggs / bird / yr. Emphasis began on cage design and housing to keep the birds comfortable and at the same time to take care of welfare aspects. Automation in feeding, watering, egg collection etc. were introduced.
1986 - 1996	Egg production increased at a rate of 4 eggs / bird / yr and consumer preference partly reverted for brown eggs. Processing of eggs into products also was a feature during this period.
1996 - 2000	Egg production increased at the rate of 3 eggs / bird / yr and a hen housed production of 310 eggs per annum was recorded. Mortality reduced from 15.5 to 5.1%. Switzerland banned rearing birds in cages. Automation became much more popular for minimizing labor costs. Furnished cages were innovated and minimum cage area/bird was increased to 450 cm^2 / bird and maximum bird density was fixed at 9 hens / m^2.
2000 - 2002	Egg production increased at the rate of 2.6 eggs / hen /yr. Sweden banned beak trimming. Egg processing automation introduced.
By 2012	Increase in egg production is likely to be at a slower pace; mortality in layers is likely to reduce to about 3-4%.

Source : Elson, 2002

In the 1980's, the poultry industry recorded a further growth; but not without hurdles. The main problems cropping up during this period were high feed costs, fluctuation of prices, especially of eggs, and the exploitation of farmers by the middlemen. A need for coordination in marketing of Poultry products became more important than ever before. Thus, a National Egg Coordination Committee (NECC) came into being during 1982 to help selling eggsat farmers' prices. Grand Parent (GP) stocks were imported to meet the growing demand of egg and meat-type chicks.

In the year 2002, India accounted for 4% of the world's egg production and was ranked third in egg production behind China and the US, and 20th in chicken meat output (0.8% of world's output). During this year, 37,000 m eggs and 1.40 m tonne of carcass weight were available resulting in the *per capita* consumption of 37 eggs and 1,400 g chicken meat (Table 1.3 and 1.4). Total value of Poultry products was Rs 120,000 m. In addition, the industry has also provided 1.6 m employment which, in itself, is a remarkable feature.

Currently, broiler and layer industry are estimated to be growing at an annual rate of 10 and 5 %, respectively (See Tables 1.3 and 1.4).

Growth of poultry industry over the past six decades has been found to be as follows (Gopal *et al,* 2001) :

Table 1.2 Growth of poultry industry

Decade	% Growth
1940-49	-----
1950-59	23.53
1960-69	11.76
1970-79	35.29
1980-89	23.53
1990-99	05.89
TOTAL	100.00

It can be noticed that maximum growthof the industry occurred during the 1970's and a sudden reduction in growth rate during 1990's is attributed to the paucity of feed ingredients to and lack of marketing network for poultry products.

Egg production for the period 1993-2003 is presented in Table 1.3. A critical observation of the Relative Growth Rate (RGR) calculated clearly indicates that there was a slump in the layer industry during the 1990's and it is picking up during the new millennium. Further, layer industry (both number of layers as well as total egg production) are growing at about 5% per annum. Assuming that this would be sustained at the same rate with an additional assumption that there will be improvement of 2 eggs / bird / annum, it is predictable that, by 2013, India will have about 300 m layers producing 3,960,000 tons of eggs (about 66,000 m eggs).

Table 1.3 Number of laying hens and eggs produced

Year	Number of hens ('000)	Relative growth (%)@	Egg Production (tons)	Relative growth (%)@
1993	115,000		1,329,000	
1994	125,268	8.93	1,429,000	7.52
1995	127,836	2.05	1,496,000	4.69
1996	130,457	2.05	1,512,000	1.07
1997	133,131	2.05	1,579,000	4.43
1998	139,000	4.41	1,623,000	2.79
1999	146,000	5.04	1,683,000	3.70
2000	150,000	2.74	1,749,000	3.92
2001	160,000	6.67	1,870,000	6.47
2002	168,000	5.00	2,000,000	6.95
2003	185,000	10.12	2,200,000	10.00
Average @		4.91		5.15

@ Calculated Source : FAO

Broiler production from the year 2000 and onwards is presented in Table 1.4.

Table 1.4 Broiler production

Year	Production ('000 tons)	Relative growth (%) @
2000	1,080	
2001	1,250	15.74
2002	1,400	12.00
2003	1,600	14.29
2004 (p)	1,650	03.13
2005(f)	1,800	09.09
Average @		10.85

@ Calculated p = Preliminary f = Forecast Source : USDA, FAS

Assuming that the broiler industry continues to expand at the rate of 10% per annum with an improvement in average slaughter weight by 20 g / bird / annum, it is expected that, by the year 2013, India will be producing 5.00 m tons of broilers.

On the same lines, assuming that the population growth (2003 = 1 billion) will not exceed 2% per annum over the next decade, *per capita* availability will be about 54 eggs and 4.1 kg chicken meat by 2013.

It is in the fitness of things to note that many in the poultry industry believe that Indian broiler and layer industry will expand at a rate of 15 and 10% per annum, respectively. If these values are considered along with other additional assumptions, the *per capita* availability of eggs and chicken meat will be about 87 and 6.40 kg, respectively. However, as per a recent report (Parsai, 2005), India now ranks among the top five nations in egg production and 10th in broiler output in the world. It has a poultry population of 440.7m with an annual egg production of about 41 billion and broiler production of 1260 m at an estimated cost of Rs 150,000 m.

Further, Indian Council of Medical Research (ICMR) has recommended a minimum *per capita* consumption of 180 eggs (10.8 kg) and 4 kg of chicken meat and WHO has recommended a minimum of 183 eggs *per capita* (½ an egg per person per d). To meet these requirements, against rising population, volume of layer and broiler industry should reach four and six times the present volume, respectively; in other words, an additional 5.4 m tons of eggs (about 90,000 m eggs) and 10.6 m tons of broiler meat (about 880 m broilers) along with expected improvements in both layers and broilers.

Hence, it is unlikely that layer industry will be able to cater to the minimum requirements of eggs in the near future; whereas, broiler industry is likely to provide sufficient meat as per the ICMR recommendations by the end of the next decade itself provided the industry grows at the rate of 10 to 15% per annum.

2.2 Distribution of activities

In a study on distribution and expectation of poultry industry (Gopal *et al* 2001), representative samples of North, South, East, West and Central zones indicated that breeding and farming as well as feed manufacturing were predominant in North and South zones, commercial hatcheries and appliance manufactures were frequently found in North and South zones. Export of poultry and poultry products was a feature recorded in the East zone. Generally, East, West and Central zones recorded lower rate of poultry activity in comparison to North and South zones. South zone recorded uniform distribution of component work groups thereby rendering it an immense scope for healthy development of the industry. Approximately one fifth of the samples indicated that the main requirement for future development is advertisement / dissemination of information / trade fairs (21.53%), government support in the form of loaning export coordination etc (18.46%) and stabilized marketing infrastructure (23.07%). Expectations from the researchers were primarily 1. Development of suitable strains / varieties for variable agro-climatic conditions of rural India. 2. Availability of least cost rations for different agro-climatic zones.

2.3 Constraints

2.3.1 Escalation of feed costs

Consequent on the GATT agreement, quantitative restriction on global trade has been removed. Supplies and demand status of different commodities must be known by all concerned so that only excess is exported. For instance, during 2000-01, Karnataka state exported excess maize whereas elsewhere in the country feed manufacturers had to import maize at a higher price. Conversely, it is possible that certain items which are already available may be dumped by another trading country. Therefore, it is essential that a single window clearance system for all import and/or export requirements within a timeframe be made available, preferably on the Internet or online so that the parties concerned may reap benefits (Editorial, IVJ, 2001).

Escalation of feed prices has been a perennial problem. Inflationary trend in eggs as well as broiler meat has not been able to compensate that in feed ingredients and consequently, in compounded feed. This is explicit from the facts relating to the poultry farmers of Andhra Pradesh state presented in Table 1.5 which clearly elucidates the fact that prices of eggs, culled birds and broiler meat have been reduced whereas those of layer and broiler feed have increased (by about 28% of that during 2000). Cost of feed ingredients also has increased, on an average of about 35%.

2.3.2 Seasonal demands

Problems of escalating feed costs are accentuated by seasonal demand, especially for eggs; during summer months, demand for eggs generally records a severe slump due mainly to the popular misconception that "eggs are heat (*ushna*)". This leads to a glut of eggs in the market and a consequent nose-dive in prices. Shortage of export infrastructure like cargo space, ships, long-time storage in modified form and refrigerated transport facility further aggravates the situation. Cost involved for export activity is high compared to Western countries and hence, acts as a disincentive. Similarly, broiler meat consumption also reduces during certain festival season, depending on the region.

Quality and safety standards for both eggs and meat are yet to be strictly adhered to because they determine the viability of export in a long run. It is also necessary to promote conversion of eggs and meat into value-added products to exploit both domestic as well as overseas markets.

Table 1.5 Economics of egg and broiler production

	2000	2001	2002
Eggs, Rs/100	119	112	108
Feed, Rs/mt	4500	5000	5800
Culled bird, Rs/bird	33	30	27
Broilers, Rs/kg	29	27	25
Feed, Rs/mt	8000	9100	10300
Feed ingredients, Rs/mt	2002	2003	
Maize	4800	6000	
Jowar	3250	5250	
Soya bean cake	8500	10,000	
Deoiled GNC	7500	9500	
Rice bran	4800	7250	
Deoiled rice bran	2300	3900	

Source : Poultry planner, March 2003

2.3.3 Distribution and marketing

Poultry production units are not uniformly distributed; Southern and Northern regions account for major activities and within a region also, there are poultry pockets from where meat and eggs have to be distributed. Unlike Diary industry, marketing network has not been developed yet and hence, traders (middlemen) make most of the profits at the cost of both producers and consumers. No organization, including National Agricultural Co-operative Marketing Federation of India Limited (NAFED), NECC, State Poultry Corporations and others has been able to claim success in containing traders. If collection, handling, storage and transportation of eggs be developed in the co-operative sector (on the lines of Diary industry), not only considerable control can be exercised on prices of eggs and meat but also the availability of these products can be ensured to the entire country.

Table 1.6 Indian Broiler industry – Global competitiveness

Live broiler production drivers, 2003								
Importance	Low		Medium		High			
Factor weight	·1	2	2	3	4	5		
Factors	Land base	Industry and market structure	Labor costs	Disease status	Flock productivity	Feed cost	Overall index	Ranking
Relative Score	2	-4	2	- 2	-2	- 1	- 21	7
Broiler processing factors, 2003								
Factor weight	2	2	3	3	4	5		
Factors	Transportation logistics	Industry and market structure	Economies of scale speed	Further processing	Technology	Labor costs	Overall index	Ranking
Relative Score	- 3	- 3	- 3	- 2	-3	3	-24	7
Social factors in broiler production, 2003								
Factor weight	3	3	5	5				
Factors	Traceability	Animal welfare	Food safety	Environmental regulations	Overall index	Ranking		
Relative Score	0	3	3	2	34	4		

Domestic demand factors in broiler production, 2003						
Factor weight	3	3	5	5		
Factors	Population and growth	Age profile	GDP and growth	White meat preference	Overall index	Ranking
Relative Score	3	2	- 1	- 3	- 5	2

Competitiveness index for broilers													
Importance	Low			Medium			High			Present and future index and ranking			
Factor weight	1	2	2	3	3	3	5	5	5	2003		2012	
Factors	Government policy	Product quality	Product differentiation	Business climate	Domestic demand	Social factors	Live cost of production	Exchange rates	Processing cost	Overall index	Ranking	Overall index	Ranking
Relative Score	1	- 2	- 2	- 2	- 1	3	- 2	1	- 2	- 22	5	- 11	3

2.3.4 Foreign investments

As the reassessment of FAO, India has limited scope for attracting foreign investments because of need of 1. Greater integration - vertical and horizontal - for control of prices 2. The cost efficiencies 3. Improved distribution of resources and products and 4. Manufacturing infrastructure relating to equipment, feed etc. In fact, India is contributing only 6% of Asian output of total products although exports of egg powder doubled during 2003-04 which was primarily due to outbreak of Avian Influenza in other Asian countries (Evans, 2005).

2.4 Future competitiveness

In a major study on the future competitiveness of several countries it was predicted that many changes are required before India can hope to become competitive in the international trade (Thornton, 2004). The assessment with regards to broilers is shown in Table 1.6 which indicates that the major setbacks for production of live broilers in India are lack of marketing infrastructure, poor flock productivity, disease status and high feed cost. Similarly, broiler processing industry also is in a poor state excepting for readily available and cheap labor. Consequently, cost of production of live broilers and, in turn, processed poultry meat are costlier; this affects product quality and ultimately business climate. Although social factors are highly conducive and government policies and exchange rates are favorable, these alone cannot outweigh all the setbacks. In spite of all these, with the proper planning to solve the major problems of production andprocessing logistics, India can definitely improve its competitiveness in international market.

In view of the above, it is reasonable to state that high cost of production of eggs, lack of transportation, storage and distribution networks, paucity of egg processing industries and need for quality-control and export of eggs and their products are the major constraints to offset global competitiveness.

Augmenting the above, it is also necessary to educate the rural masses on the nutritive value of poultry products as well as the misconceptions about the consumption of eggs (especially) and meat. Suitable policies must be enforced, by both State and Union governments, in the form of incentives and support for exports, concessions in taxation (like considering poultry farming as an "industry"), long-term loans at lower interest rates from commercial banks and the like. Establishment of more vaccine production centers, feed analytical and disease diagnostic laboratories in public sector are also priority areas requiring thorough study and proper planning.

Chapter 2

Evolution of Birds

1. Theories of evolution of birds

Evolution of flight and identification of the first animal to have had such a facility has been researched, theorized and debated for decades. Hitherto there were two main theories in vogue; namely, Reptilian origin (Arboreal theory) and Dinosaur origin (Cursorial theory) (Shipman, 1997). A new theory called parental care theory has been proposed contradicting the former two by Carey (2002).

1.1 Arboreal theory (tree-to-ground model)

According to this theory, poikelothermic arboreal reptiles developed feathers for gliding, which over further sophistication, evolved into a bird. In other words, the ancestors of birds were tree-dwellers which, at the beginning, leapt from branch to branch and later on, from trees to the ground. Gliding retarded their fall on to the Earth thereby greatly reducing the impact thereon. The main contentions of this theory are :

1. Feathers, which are required mainly for flying, could have evolved only in such animals that needed aerial locomotion.
2. Aerial locomotion is required primarily for tree-living (arboreal) creatures. But if this theory were to be true, early ancestors of the birds are more likely to be four-legged which later developed membranes between their front and back legs However, flight is a high energy-requiring process and hence it is unlikely that a poikelothermic animal can generate so much energy. It is noteworthy that, anaerobically, poikelotherms can produce twice the amount of energy/kg of muscle mass than homoeotherms, but in short-lived bursts and prolonged rest; hence cannot support sustained flight. However, it is calculated that flight muscles should constitute a minimum of 16% of body weight in order to fuel flight by a poikelotherm; but *Archaeopteryx*, the first bird, had flight muscles forming only 7% of the body weight; if this concept is true, feathers appear to have been evolved exclusively for flight rather than for insulation (?).

1.2 Cursorial theory (Ground-up model)

This theory suggests that birds evolved from four-legged reptiles (Dinosaurs); scales on the fore-limbs of these creatures gradually developed into feathers that gave them upward thrust when they ran and eventually, enabled them to fly. Feathers are also likely to have been developed to conserve heat and also to swat insects from the air. The proponents of this theory argue that

1. Dinosaurs were homoeothermic as are birds. This is supported by the fact that they lived over a wide range of latitude and temperature unlike warmth-preferring reptiles.
2. Evolved feathers for insulation (?).
3. Dromeosaur, the miniature *Velociraptor*, and birds match bone-to-bone when both axial as well as appendicular skeleton are considered.
4. Homoeotherms can generate long-lasting abundant energy which can fuel flapping of wings (flight) and
5. Foot-prints of dinosaurs indicate that they were bestowed with fast locomotion and their legs appeared tucked underneath their body (as is seen in birds during flight) unlike reptiles whose legs are sprawled-out.

1.3 Parental-care theory (Carey, 2002)

This theory suggests that early ancestors of birds appear to be reptiles which, initially, were poikelothermic, laid leathery eggs, established and guarded their nests on the ground; this was followed by laying of hard-shelled eggs and development of homoeothermy to ensure constant physical environment for the developing embryo. Simultaneously, scales evolved into feathers principally for better camouflaging and insulating the parents. Subsequently, sophistication began in care for the young in form of feeding them inside the nest, forcing liquid food into the oesophagus or solid food into the mouth and the like. Number of offspring produced reduced whereas their dependence on the parent(s) increased. Egg size reduced and nesting in bushes and then, to offer better protection from predators, on small trees began.

Gradually, fore-limbs became feathered, more elongated so as to facilitate manipulation of eggs as well as landing on to the ground safely from trees; this stage was possibly followed-up by development of ability to glide and later on, to fly by flapping of wings. On the same lines, beaks appear to have developed to help construction of nest and for feeding the young ones (*cf*: popularly evolution of beaks by the birds is thought to be due to their lesser weight (*cf*: teeth) thereby facilitating flight).

Flight offered many advantages to the ancestors of birds; for instance,

1. helped to place their young high in trees and cliffs
2. allowed seasonal migration in search of new food resources thereby maximizing the food sources
3. provided a possibility of getting better nutrition due to wider area that could be covered by flight

4. better protection from the predators etc.,

2. Adaptation for flight

Attributes for flight possessed by modern-day flying birds which are mandatory for sustained flight are tabulated in Table 2.1.

Table 2.1. Mandatory requirements for sustained flight

Feathers	For formation of wing and maintain constant body temperature
Modified clavicle (Fircula or wish bone)	To accommodate considerable volume of flight muscles and stabilize shoulder joint
Bony, keeled sternum (breast-bone)	To accommodate flight muscles
Hand reduced to three fingers	First finger reducing to a small alula (bastard wing), a large second finger and a small third finger
Pygostyle (plough share bone)	By fusing of 4th to 7th coccygeal vertebrae – to support tail feathers and to act as rudder during flight
More articulating shoulder joint	To help flip the wings the position to near vertical position above the back so that with a downward thrust sufficient "lift" is created for flight
Folding fore-limbs	To tuck the feathered fore-arm close to the body for safety and protection
Respiratory system	Air-sacs to allow unidirectional air flow through the lungs providing for additional oxygen and efficient evacuation of carbon-dioxide. The additional oxygen enhances energy available for flight through endothermic metabolic pathways
	Source : Shipman, 1997

3. Paleontological evidences

In any case, to accept any of the above theories, it (they) must be supported by the following requirements :

1. Intermediary stages before flight must be functional as well as beneficiary to the concerned species and
2. There must be matching evidence(s) from fossil(s)

Unfortunately, fossil records, as far as evolution of birds are concerned, is not yet conclusive; thus, controversy persists as to the origin of flight in birds.

3.1.1 Archaeopteryx

The "first bird", called *Archaeopteryx* from Solnhofen, Germany is dated as 150 m years old; this feathered fossil is part bird and part reptile since it has feathered wings (like birds) and toothed jaw, clawed fingers and long tail (like reptiles). However, its ability for flight has been doubted because of presumably small flight muscles. Of the eight "essentials" for flight, Archaeopteryx appears to have only feathers and modified clavicle.

Structure of feathers is ideal for flight and hence, it is unlikely that they would have evolved for insulating a homoeothermic dinosaur. Therefore, some paleontologists believe that *Archaeopteryx* cannot be the ancestor for all the birds but the ancestor to the extinct enantiornithines or "opposite birds" in which foot-bones nearest to the body fuse earlier than those farther from it - which is exactly opposite to that found in modern birds.

Enantiornithines became dominant archaic land-based birds and disappeared along with dinosaurs about 65 m years ago. Their contemporaries, Ornithurines, with modern sequence of foot-bone fusion, were either aquatic, foot-propelled divers with teeth or flying land-birds. These Ornithurines, only a small group called "transitional shore birds", having similar adaptations as that of flamingos, ducks and ibises, appear to have survived the unknown that caused the mass extinction of dinosaur; these gave rise to modern birds, referred to as Neornithines, about 85 m years after the appearance of *Archaeopteryx*, about 50 m years ago.

The skeleton of *Archaeopteryx* is remarkably similar to a miniature *Velociraptor* (dromeosaur) belonging to a group of dinosaurs which are bipedal predatory creatures. However, resemblances claimed on anatomical features such as boney structure of wrist, hand and hind-limb as well as skull are under dispute for quite a long time. One school of thought believes that *Archaeopteryx* is evolved from a tree-living (arboreal) non-dinosaur species that initially leapt, then glided from perch to perch. This non-dinosaur species is presumably

1. An ancient reptile - Crocodilomorphs - a small, bipedal, tree-living crocodile or
2. Pseudosuchians - which are ancestors of dinosaurs and flying pterosaurs, which appeared about 225 m years ago? However, there is no fossil records to substantiate this theory

Cladistics, the method of classification of animals and plants on the basis of shared characteristics which are assumed to indicate common ancestry, also failed to indicate a non-theropod bird ancestor.

3.1.2 Unnamed fossils

Two recent fossil findings have tilted the balance in favor of dinosaur theory; the fossils are :

1. From Patagonia – 90 m year-old dromeosaurid dinosaur. This creature appears to be capable of tucking its arms up against its body thereby indicating the existence of bird-like folding mechanism in non-flying dinosaurs; which might have been the predecessors to creatures with flight.
2. From Madagascar – 75 m year old turkey-sized, wicked-looking, enlarged clawed archaic bird. The claws were strikingly similar to *Velociraptor*, its dinosaur ancestor (?).

3.1.3 Sino-sauropteryx

This 121 to 142 m year-old fossil was found in Liaoning province (North-Eastern China). It has a meter-long skeleton similar to *Compsognathus.* Conspicuously, it has feathers or feather-like scales along the back-bone, down the sides of the body, arms and legs; their purpose, presumably, was insulation rather than flight. However, whether the marks in the fossil are indicative of feathers or not itself is in a controversy. In any case, it is a baffling fact that *Sino-sauropteryx* has completely modern feathers similar to *Archaeopteryx*.

3.1.4 Confuciusornis sanctus

This fossil found at the same site as *Sino-sauropteryx* also dates back to 121 to 142 m years. This tooth-less opposite bird (backward-pointing first-toe and strongly curved claws to help perching) has the earliest horny beak and pygostyle (support to tail feathers).

3.1.5. Liaoningornis

This is a fossil of oldest known Ornithurine and thus differs significantly from its enantiornithine contemporary, *Confuciusornis.* This has well-developed sternum and oldest known keel for anchoring flight muscles along with a capacious and strong rib cage which is suggestive of improved respiratory system and a possibility of presence of air-sacs. This is a strong evidence of link between endothermy and the flapping flight; the supplementing the dinosaur theory.

3.1.6. Iberomesornis romerali

This is an enantiornithine fossil dating back to 120 to 130 m years found in Cuenca province (Spain). This sparrow-sized creature had a more modern shoulder than *Archaeopteryx,* and perching foot but had teeth and primitive (non-fused) pelvis and long pygostyle (10 to 15 vertebrae).

3.1.7. Eoalulavis hoyasi

This fossil, also obtained from Cuenca province (Spain), is about 115 m years old. This creature, about the size of a goldfinch, is an enantiornithine having extraordinary details of :

1. feathers on its wings and body
2. crustacean meal in its belly

3. narrow sternum with a faint keel, a robust furcula and a more modern shoulder than *Iberomesornis*
4. small, feathered first-digit – the oldest record of alula or bastard wing, which is very important for slow-flight, landing on branches and maneuverability and
5. surprisingly, fingers tipped with primitive claws.

   This creature is only 25 to 30 m years younger than *Archaeopteryx* and *Sinosauropteryx* and hence, it is likely that flight adaptations were involved early and alula might have evolved more than once since enantiornithines left no descendants.

Therefore, it is possible that enantiornithines became extinct along with dinosaurs for the same reasons viz.,

1. sudden climatic change bringing about depletion of ecological niches and/or
2. impact of a meteor.

The Avian and Mammalian survivors got an empty and ripe world for evolutionary diversification as known from Tertiary fossil record. Further, the ancestry of birds, the homoeothermic, biped, oviparous, feathered, flying creatures, possibly lies in dinosaurs.

3.1.8. Formation of feathers

Feathers are made of quill, shaft/rachis, barbs, barbules and barbicels. Differences in feathers are because of structure of barbules and barbicels. Order of formation of feathers is essentially apicobasal and therefore, the order of age of the barbs is naturally the same. In each barb, apex at the margin of the feather the first formed and the central end attached to the shaft is the last to be formed. Therefore, two time gradients can be noticed in each feather – a) from apex to base along the shaft and b) margin to centre along the barbs. Time required for formation of vane of a breast feather is about 20d from the time of plucking an old feather (Nesheim *et al.*, 1979).

Rate of growth of shaft is approximately uniform throughout the length, at least during the formation of the vane. Rate of growth of barbs decreases from apex to the base; this is important for determining the pigmentation and general pattern of feather (Nesheim *et al.*, 1979).

Most of the adult feathers have a backbone (stem, called rachis) from which barbs branch-out which in turn, give out feather's smallest unit, the barbule which is made of a single row of epithelial cells. Down (Downey) feathers do not have rachis and they are made of only barbs studded with barbules. Till now, it was generally hypothesized that the scales on dinosaurs lengthened into rachides which then became notched to form barbs and barbules (Chuong, 2002).

However, study on the newly discovered fossilized Chinese-bird-dinosaur (*Sinorthosaurs*) has revealed that evolution from scale to feather most likely followed a path in which the barbs form first and fuse to form a rachis (rather than a rachis forming first and then been sculpted into barbs and barbules) similar to down feathers which are formed before flight feathers (Chuong, 2002).

4. Salient features of Class Aves

Table 2.2 Characteristic features of Avians

System	Feature (s)
Skeletal	Presence of pneumatic bones in all birds and medullary bones in layers. Pneumatic bones reduce weight in relation to volume/surface area so as to facilitate flight. Fore-limbs are modified into wings. The 4th to 7th coccygeal vertebrae are fused to form pygostyle which supports tail feathers; the latter act as rudder during flight.
Muscular	Breast muscle forms a large proportion of body weight. For flight, they have massive muscles anchored securely on keel and shoulder along with the alula, the small first finger.
Respiratory	Air-sacs - four paired and one non-paired; all opening to lungs and communicating with pneumatic bones. They are thin-walled, dedicate structures difficult to recognize air-sacs allow unidirectional air flow and absorption of oxygen at a higher rate to fuel the endothermic reactions. Syrinx (or lower larynx) at the tracheal bifurcation into bronchi; upper larynx is only a modulator. Anatomically, both sexes have similar syrinx but, testosterone induces cocks to crow. There is no functional diaphragm.
Thermoregulatory	No sweat glands; all are homoeothermic; but, species vary over a wide range of normal body temperature (38 to 44°C).
Integument	Skin on most of the body has no secretory glands except uropygial/preen/oil gland whose secretions make the feathers water-proof. Color of skin depends on combination of pigments in upper and lower layers - carotenoids give yellow color, melanin pigments in epidermis and in both epidermis and dermis give black and darkest color, respectively, only black color in dermis gives blue/slaty-blue color, black color in dermis and yellow color in epidermis gives willow-green color and absence of pigments

	gives white color. Feathers – form 4-9% of body weight depending on sex; arranged in definite tracks (feathers tracts); a modification of scales on reptiles. Most birds have combs and wattles. The scales on the reptilian body are modified into feathers and scales are seen on the tarsus and toes of most birds.
Circulatory	4-chambered heart (unlike reptiles), blood forms 6% (adults) to 8% (chicks up to 2 weeks) of body weight, heart rate 250-350, deep body temperature 41.9°C, RBCs 2.5 - 3 x 10^6 / mm^3 depending on aged sex, nucleated RBCs, PCV 30% (young) to 40% (adult males), spleen is the reservoir of RBCs. Lymphatic system is devoid of lymph nodes; bursa of Fabricius and thymus have specific functions as far as immunity is concerned.
Excretory	Large elongated tri-lobed kidneys located tightly against the top of abdominal cavity, excretes uric acid
Digestive	Mouth has no lips and teeth, has bony mandible, the jaws are modified into beaks, tongue resembles barbed head of an arrow with point directed forwards. Crop, a pouch of oesophagus to store food. Gizzard for crushing and grinding food. Ceca – 2 blind pouches of lower small intestine, 10-15 cm long. Cloaca – common outlet for digestive, urinary and reproductive systems.
Nervous	Very small cerebral cortex (neuro-cortex) and therefore, low intelligence. Hypothalamus well-developed similar to mammals. Optic lobes particularly well-developed, eyes form larger percentage of head than mammals, can distinguish colors. Accommodation capacity of the eye is best among all species animals. Nictitating membrane is functional. Their sense of taste is poor (Chicken have 24 taste buds, Pigeons 37 and Japanese quails 62 whereas humans have 9,000 and Catfish has 100,000 taste buds) and sense of smell is absent. Sense of hearing well-developed with ears similar to reptiles and sense of touch is comparable to any other animal
Endocrine	Similar to mammals
Reproductive	Males – intra-abdominal testes, no accessory sex organs/glands. Longevity (viability) of spermatozoa is for a longer period (although they are more fragile) than that of higher species. Females – left ovary and oviduct usually functional, no placentation and hence all are oviparous. Ovary is follicular

General	Gross size of different species varies from a small humming bird to a very large ostrich; probably, no other species has such a variation in size. Members of this Class are seen in almost all climatic zones like Equatorial tropics, temperate, hot-humid and frigid regions and at all altitudes from sea level up to 5000 m above sea level. Males are homogametic and females, heterogametic. All species are feathered; in fact, birds are defined as vertebrate, oviparous, uricotelic, homoieothermic, biped, feathered, flying creatures belonging to Class *Aves*.

Chapter **3**

Domestication of Poultry

1. Domestication – general

Sequence of domestication

Table 3.1 Sequence of domestication

Species	Period	Country (ies)
Chicken	5400 BC	China – Cishan culture; but, the contribution of these birds to modern birds doubtful
	2500 to 2100 BC	From Harappan culture of Indus valley; may be main source of diffusion through the World
Geese and Mallard ducks	2500 BC	China
	1500 BC	Egypt – separately domesticated; in the West, Mallard duck was not domesticated till Middle Ages
Ring-necked pheasants	1300 BC	Greece
Turkeys	200 BC to 700 AD	Mexico
Muscovy ducks	16th Century	Columbia, Peru
Japanese quail	11th century	Japan, China, Korea
Guinea fowl	1500 AD	West African birds introduced to Europe by Portuguese explorers

Source : Crawford, 1990

1.2 Effects of domestication

1. During early stages of domestication, they started foraging on their own; therefore, seed/grass eating species preferred

2. Reproduction in captivity; as a consequence of this reproduction in poultry became less dependent on climatic and environmental factors
3. Ability to imprint which was necessary for initial domestication
4. Development of social order which facilitated rearing of large number of birds in one group.

1.3 Consequences of domestication

1. Bones larger than the wild species - is a proof of domestication (except turkeys)
2. Noticeable changes in plumage
3. Alterations to the limbs involving length, muscle attachments and joint structure
4. Changes in skin covering, muscling, fat deposition and brain size; these were accomplished at a later stage

1.4 Purpose and utilization of domestic birds

Most of the poultry species were initially used primarily for cultural needs (religion, superstition), decorative arts and entertainment. Later on, they were used as human food; for example, in ancient Rome, geese were considered sacred in the beginning and later on became a table delicacy. Similarly, in Spain, turkeys were used for cultural activities and later on as food; and Japanese quails were song birds before they were utilized for food.

2. Domestication – Chicken or domestic Fowl

2.1 Paleontological information

Information on domestication of chicken is scanty; only three fossils of *Gallus* species have been recorded. Two of them from the UK and one from Greece and Black Sea region. One of the fossils from the UK is of a coracoid bone resembling *Gallus gallus* but sufficiently different that it has been assigned to a new fossil species *gallus europaeus;* the other one is that of a radius bone closely resembling *Gallus gallus* but due to its very old age, assigned to the fossil species.

However, fossil from Greece and Black Sea region is of the tarsometatarsus bone similar to *Gallus gallus* and a coracoid bone much longer. It has been assigned to a separate species *Gallus aesculapi*.

During periods of glaciation single population of bird species across Eurasian continent fragmented into isolated population in warm southern refuges. In many species, and Eastern and a Western refuge was used. For some forest species, three subpopulations formed; Western (in Mediterranean/Middle East), Central (in India) and for Eastern Asia.

When glaciers retreated, Western and Eastern subpopulations expanded

Northward and evolved into separate species and subspecies which ultimately met depending on the length of isolation; however, these might have gone into extinction during later glaciation or ex-terminated by man. Expansion of Central Population (India) restricted because of Himalayas but persisted as modern *Gallus gallus* and related species and remained isolated because of geographic barriers.

Totally four species are known to modern Ornithology. One of these, Red Jungle Fowl (*Gallus gallus*) is undoubtedly a major contributor to domestic fowl although the role of the other three species is doubtful.

2.2 Classification

Order : *Galliformes* Sub-order : *Galli* Family : *Phasianidae*

Sub-family : *Phasianinae* Tribe : *Phasianini* Genus : *Gallus*

2.3 Wild Gallus species

There are four recognized species under the genus *Gallus* : *G. gallus* (Red Jungle Fowl, RJF), *G. sonneratti*(Grey or sonnerats), *G. lafayetti* (old name *G. stanleyi*, in Ceylon), and *G. varius* (Green, old name *G. furcatus*).

In case of *Gallus gallus*, because of geographical location, five sub-species, especially in males, have been reported. They are *G.g.gallus* (Cochin-Chinese RJF), *G.g.spadiceus* (Burmese RJF), *G.g.jabouillei* (Tonkinese RJF), *G.g.murghi* (Indian RJF) and *G.g.bankiva* (Javan RJF). These five subspecies differ in a) color of earlobes (white to red), b) shape and length of neck hackle feathers in males and c) shade of red plumage in males (golden yellow to mahogany).

2.4 Origin of modern poultry

Origin of modern poultry is not fully resolved although most agree for RJF as the only species as exclusive ancestor (Monophyletic origin). Immunogenetics also supports RJF as sole ancestor. However, some writers suggest : a) Malay fowl (*Gallus giganteus*) as a principal contributor b) Modern Asiatic breeds (Brahma, Cochin and Langshan) have different origin from Bankivoids and Malays (Polyphyletic origin).

In addition, original site of domestication is also not clear; however, it is suspected that domestication might have begun in Burma (Myanmar). There are records to show that people of Harappan culture (2500-2100 BC) of the Indus valley reared chicken and later on diffused Westward to other parts of the World at a rate of 1.5 to 3.0 km/year.

Except for Egypt, diffusion of chickens into Africa is unknown. However, India is likely to be the source of chickens keeping in view well developed trade between India and Africa.

Table 3.2 Description of wild jungle fowls

Gallus gallus	Comb is single, upright, serrated blade; a pair of wattles. Male exhibit spectacular coloring whereas females have drab color but feathers have identical morphology in both the sexes.
Gallus lafayetti	Male plumage is similar to RJF except that breast feathers are pointed and fringed. Males have a peculiar patch of bluish purple feathers on the upper breast. Plumage of females is similar to those of RJF. Some consider this as an off-shoot of RJF.
Gallus sonneratti	Male plumage is different from other species; "sealing wax" spots on some of the feathers which are actually expanded and flattened portion of rachis arranged in a series on rachis; those that are sub-terminal being white and the terminal spots being shredded and yellow. Occurs primarily in the neck-hackle, saddle and wing-coverts. Most of the body feathers are black with a white shaft and a grey border. Wing and tail feathers are black; bird appears grey. Female plumage differs from RJF in breast feathers which are white with broad black/brown borders.
Gallus varius	Most primitive of the four species. Its distinctive plumage consists of a) presence of 16 tail feathers rather than 14 in other species b) short truncate neck-hackle feathers in males (*cf* : long pointed in other species) c) male plumage is mostly glossy black, but hackle and saddle feathers are edged with bronze and yellow imparting a distinctly green coloration to the bird. In females, feathers of the back and rump are penciled (similar to Dark Cornish), upper breast feathers have dark edging, lower breast feathers are pale and rest of the plumage has irregular barring.
	Source : Crawford, 1990

Similarly, diffusion of birds eastwards has not been completely understood. However, Japanese chicken might have come from China *via* Korea during Yagoi period (300 BC to 300 AD); chicken in Pacific islands might have originated in China and India.

Recent discoveries indicate that it could even be China from where domestication began. Therefore, it is reasonable to agree that domestication was completed by about 2000 BC although some authors have reported it to be around 3200 BC. Darwin proposed 1400 BC as the time for probable introduction of chickens from China to India. Further, sites from Iran, Turkey, Syria, Greece, Romania and Ukraine also suggest the possibility of domestication prior to Indus valley civilization supporting ancient China to be the origin.

Table 3.3 Body size, reproductive traits and other features of *Gallus* species

Species →	*Gallus gallus*	*Gallus sonneratti*	*Gallus lafayetti*	*Gallus varius*
Adult weight Males, g Females, g	 800-1360 485-740	 790-1136 705-790	 790-1140 510-625	 454-795 454-795
Age at sexual maturity in captivity (years)	1-2	2	2	1-2
Clutch size	4-8	4-8	2-4	6-10
Egg size, mm	45.3 x 34.4	46.0 x 36.5	46.3 x 34.5	44.5 x 34.5
Egg shell color	White-Rosy cream	White-Rosy buff	Pinkish to buff, Brown stippled	Buffy white
Incubation period, d	19-21	20-21	20-21	21
Eclipse plumage (molting) and habitat	Most of the males molt after breeding season Prefer forest habitat and forest clearings			Molt involves only cervical feather tract of the neck; long pointed hackles of males replaced by around short feathers similar to female; this plumage persists only for a few weeks followed by complete annual molt involving all feathers tracts. Prefer sea-shore and rocky scrub land bordering cultivated land
General	Sedentary, omnivorous, behavior similar to domestic chicken with different vocalizations			

1982 Census	1072	213	432	83
Distribution	Has the largest natural range. Naturally occurs from Pakistan to China, Hainan and in East India, Burma (Myanmar), most of Indo-China, islands of Sumatra, Java, Bali.	West and South India; sympatric with RJF at boundaries of its range	Restricted to Sri Lanka	Java and on the chain of islands eastwards; sympatric with RJF in Java but occupying different habitats
				Source : Crawford, 1990

Table 3.4 Diffusion of chicken from Indus Valley

Period	Diffusion and salient features
2500-2100 BC	Chicken used for sport
1500 BC	Aryans invaded India; included chicken in their culture but not as food.
1000 -537 BC	Chicken had religious significance but forbidden as food; fighting cocks reached Persia. Persians carried westward to Mesapotamia and Asia Minor; by 700 BC, reached Greece where it was primarily used for sport and poor folk as food. Plato and Aristotle distinguished between high and low-bred chicken; egg production of high-bred chicken being poor. Two breeds were recognized from Tangare (near modern Athens) – one for fighting and other was black with crest and wattles (like anemomes; a rose-comb mutant?)
200 BC	Bearded chicken reported (muffs and beard mutant?); chicken reached Egypt and got firmly established.
100 BC	Chicken reached Romans who knew about force-feeding, hybrid vigor, caponizing, sperm competition etc. Adopted chicken as food source and employed for cock fighting, religious purposes, superstition and divination.
Dawn of Christian era - 1300 AD	Chicken probably reached Europe and then to Russia. Main diffusion through Europe occurred during the Neolithic and early Bronze age. Due to fall of Roman Empire, importance of chicken in Europe reduced and became farmyard scavengers.
1500-1600 AD	Spanish conquest brought chickens to Americas and rapidly spread to South and Central America. Later, English Dutch and French brought them to East and North America.
	Source : Crawford, 1990

3. Domestication – Turkeys

3.1 Domestic turkeys

Domestic turkeys (*Meleagris gallipavo*) were found in Mexico by Spanish who, in turn, took them to Europe. By the beginning of 16th Century, Europeans brought them to North America where they interbred with Eastern sub-population (*M.g.silvestris*) to result in a bronze bird which ultimately became foundation for nearly all domestic turkeys. *M.gallopavo* was derived from Mexican sub-species, *M.g.gallipavo*. In South-West United States, separate domestication of *M.g.merriami* seems to have been undertaken and it might have contributed to the modern domestic turkeys as well.

3.2 Classification

Order : *Galliformes* Sub-order : *Galli* Family : *Phasianidae* Sub-family : *Meleagridinae* Genus : *Meleagris* Species : 1. *M.ocellata* (Ocellated turkey; never domesticated, earlier placed under genus *Agriocharis*) 2. *M.gallipavo* : depending on geographic range and plumage, occurs as seven subspecies – *M.g.gallipavo* – Mexican, *M.g.intermedia* – Rio Grande, *M.g.merriami* – Merrianis, *M.g.mexicana* – Gould's, *M.g.silvestris* – Eastern, *M.g.meusta* – Moore's and *M.g.osceola* – Florida

3.3 Wild turkey species

Wild turkeys were similar to farmyard bronze domestic birds but slimmer in build; no massive muscling of breast and thighs unlike domestic turkeys. Eastern subspecies had bronze tips on tail feathers and upper tail coverts and hence designated as white-rumped. In plumage had metallic green tone and therefore feathers appeared black. They used to have two molts annually; one partial prenuptial molt in late winter and a second complete molt in late summer. They were omnivorous and non-migratory with an age at sexual maturity of one year, clutch size of 11 eggs, egg weight of 65 g, incubation period of 26-28 d and hatchability of more than 90%.

They were distributed over a large range in North America, Mexico, from Rio Balsas valley northward, most of US east of Rocky Mountains and South Ontario (Canada); pure subspecies is difficult to find now. However, wild species were successfully released as game birds in Canada, Hawaiian Islands, Germany, New Zealand and Prince Seal Island (near Tasmania).

3.4 Domestication and diffusion

Domesticated in Mexico during the period 200 BC to 700 AD. Introduced Caribbean by Spanish; to Cuba (1511 to 1517), to Spain (1512), to Rome (1520), to South America (Ecuador in 1587, Chile in 1646) and to Europe (1500) from where the diffused rapidly to Italy (1520), France (1538), Norway and Denmark (1550), Sweden (1552), Germany (1560), and England (1541).

Introgression of Eastern Wild Turkey (*M.g.silvestris*) into domestic Mexican bird due to crossing domestic turkeys with wild toms resulted in disappearance of bronze plumage of *M.g.silvestris* and black plumage of Mexican breeds; the body size and vigor increased; therefore, crossing became a general phenomenon; the hybrid population was called "American Bronze" which reached Europe in the early 19th century and became popular both in Europe and North America. American bronze was evolved into the modern Broad-breasted type. In spite of all these, *M.g.silvestris* was never domesticated.

In South-West US, Pueblo people independently domesticated Merriami turkey (*M.g.merriami*) by total confinement in caves during 500-700 AD; however, they were not used for food but were mainly used for feathers, spurs and bones for ornament, weapons/tools.

4. Domestication – Japanese quail

Domestication began in the 11th century in Japan, China and Korea. Japanese quail is derived from its wild counterpart and distinct from common quail (*Coturnix coturnix*). Although Zoological name of Japanese quails is often *Coturnix coturnix japonica*, most taxonomists agree that Japanese quail is a distinct species but not a subspecies of common quail (*Coturnix coturnix*); and hence, *Coturnix japonica* appears to be the correct nomenclature.

4.1 Classification

Order : *Galliformes* Family : *Phasianidae* Sub-family : *Phasianinae* Genus : *Coturnix* Species : *C. coturnix* – common quail (has six sub-species), *C. japonica* – Japanese quail, *C.coramandelica* – black-breasted quails from India, Sri Lanka and Mynmar, *C.delgorgnei* – Harlequin quail, has three sub-species from Africa and Arabia, *C.pectoralis* – pectoral quail from Australia. *C. japonica* and *C.coturnix* differ in vocalization and sympatric breeding; they do not hybridize readily and even if so, progeny will be sterile/of low fertility.

4.2 Domestication and diffusion

Domestication occurred by around the 11th century probably in Japan or from China/Korea. The birds were kept and bred for song and referred to as "song quail". By 1910, Japanese took interest in improving egg and meat production. By 1941, improved stocks were established in Korea, China and Taiwan. The entire flock in Japan was lost in World War II excepting one pen; the flock was rebuilt from the left-over pen along with those from Korea, China and Taiwan. By 1950, rebuilt stocks were sent to North America, Europe, Middle and Near East. In the US, it was initially used as game bird and later on for egg and meat production as well as a laboratory animal.

5. Domestication – Guinea fowl

Modern guinea fowl is derived from helmeted guinea fowl of Africa (*Numida meleagris*); mostly from West African species, *N.m.galeata*.

5.1 Classification

Order : *Galliformes* Family : *Numididae* Genus : *Numida* Species : *N.meleagris* Sub-species : *N.m.galeata, N.m.sabyi* – West African; *N.m.meleagris, N.m.somaliensis* – East African; *N.m.reichenowi, N.m.mitrata, N.m.marungensis, N.m.papillosa* and *N.m.coronata* – Central-South African guinea fowl

5.2 Domestication

It appears that domestication of guinea fowls occurred separately in different places. It is possible that during 1570-1300 BC, artificial hatching and rearing chicken was practiced in Egypt. By 600 BC, guinea fowls were known to Greek, Persians and Romans. Greeks called it *melanagris* meaning black-white; which later on became *meleagris*. With the fall of Roman Empire, guinea fowls disappeared from Europe but might have persisted in Greece and Italy.

During the 16th century, the Portuguese rediscovered guinea fowls in west coast of Africa and Belon during 1555 and referred to them as *poule de Guinee*. Portuguese took the birds to Europe, Americas etc. They were in Europe along with turkeys and hence, there was confusion in nomenclature.

Most of the modern guinea fowl descended from Portuguese-introduced West African sub-species *N.m.galeata*. Those in Malagasy are derived from *N.m.mitrata*, those in East Africa are from *N.m.meleagris* and *N.m.somaliensis* whereas both East and West African sub-species have contributed to guinea fowls of the Mediterranean region.

6. Domestication – Ring-necked Pheasant

Domestic stocks are likely to be derived from *Phasianus colchicus*; a subspecies from black-necked, white-winged and grey-rumped groups. Some stocks might have genes from closely related green Pheasant (*P.versicolor*). Distributed in the temperate parts of Europe, North America, Hawaii, Australia and New Zealand. No other Avian species has been so widely exploited as game bird as is Pheasant.

6.1 Classification

Order : *Galliformes* Sub-order : *Galli* Family : *Phasianidae* Sub-family : *Phasianinae* Tribe : *Phasianini* Genus : *Phasianus* Species : 1. *P.colchicus* – has 30 sub-species depending on plumage (4 black-necked, 6 white-winged, 2 kirghiz, 1 olive-rumped and 17 grey-rumped) 2. *P.versicolor* – very closely related to *P.colchicus* and hence thought to be a sub-species of the latter; has 3 sub-species. Hybrids can occur among all species and sub-species.

6.2 Domestication

Brought to Europe by around 1300 BC. Romans helped its distribution throughout their empire. By the 10th century, it was well established in France and England. By the 18th century, pheasants were seen in Europe and Asia. Chinese ring-necked Pheasant (*P.c.torquatus*) reached England in the 18th century and Europe by the 19th century. Taiwan ring-necked Pheasants (*P.c.formosanus*) were seen in Europe by 1907; therefore, European population of pheasants was highly heterogeneous. Similarly, North America was also a host to many subspecies like *P.c.bianchii, P.c.colchicus, P.c.mongolius, P.c.persius, P.c.talischensis, P.c.torquatus* and *P.c.versicolor*. In any case, Ring-necked pheasants were never fully domesticated and were kept exclusively as a game bird for the wealthy. Recently, semi-domestic stock in confinement is being tried for meat production.

7. Domestication – Mallard Ducks

Modern ducks are derived from *Anas platyrhynchos*; the first domestication occurred early in the Far East (2800 BC to 37 BC) and the second domestication occurred in Europe in the Middle Ages.

7.1 Classification

Order : *Anseriformes* Suborder : *Ansers* Family : *Anatidae* Sub-family : *Anatinae* Tribe : *Anatini* Genus : *Anas* Species : 1. *A.rubripes* – American black duck of Eastern North America, 2. *A.poecilorhyncha* – grey- or spot-billed-duck of India, SE Asia and Australia 3. *A.platyrhynchos* which has seven subspecies; they are 1. *A.p.platyrhynchos* – common Mallard; along with related species, considered as the sole progenitor of domestic ducks, 2. *A.p.conboschas* – Greenland Mallard, 3. *A.p.fulvigula* – Florida Mallard, 4. *A.p.moiculosa* – Mottled Mallard, 5. *A.p.diazi* – Mexican Mallard, 6. *A.p.wyvilliana* – Hawaiian Mallard, 7. *A.p.laysaneusis* – Laysan Mallard.

8 Domestication – Muscovy Ducks

Developed from wild Muscovy (*Cairina moschata*). Domestication undertaken in Americas during pre-Colombian times. Muscovy ducks gained entry to Europe from Africa through trans-Atlantic ships of Arab traders. By 1693, ducks reached Taiwan, to Italy from Cairo (Egypt) and Belon (France). The word muscovy is likely to be a corruption of "musk duck" because of peculiar odor emitted by old birds or because of Muscovite Company during Queen Elizabeth's reign. Muscovy ducks are native to coastal plains of Mexico, Central America and most of South America (coast of Peru west to Santa Fe of Argentina in East).

8.1 Classification

Order : *Anseriformes* Suborder : *Ansers* Family : *Anatidae* Sub-family : *Anatinae* Tribe : *Cairinini* Genus : *Cairina* Species : *C.moschata* Subspecies : Nil

9. Domestication – Goose

This species has undergone multiple domestications; very early domestication might have occurred in the Far East or Germanic tribes might have domesticated them before the great Mediterranean civilizations or domestication might have occurred in Egypt also. Wild geese have been documented during Old Kingdom (2686-1991 BC) but domestication could have occurred only by 1552-1151 BC. Geese were considered sacred and symbol of plenty by the Greeks. There are records of geese in Asia Minor during 4/5th millennium BC. Romans consumed both eggs and meat; goose fat used as medicine and goose quill as pen (4th century AD). With the fall of the Roman Empire, chickens became scavengers till Agricultural renaissance in the 19th century.

9.1 Classification

Order : *Anseriformes* Suborder : *Ansers* Family : *Anatidae* Sub-family : *Anserinae* Tribe : *Anserini* Genus : *Anser* (Grey geese) and *Branda* (Black geese); the two genera differ in configuration of V chromosome whereas the subspecies of the former (*A.a.anser* and *A.a.rubrirostris*) differ in II and V chromosomes. However, progeny of the hybrids between the two genera are usually sterile as are the backcrosses.

Table 3.5 Zoological classification of birds (Class : *Aves*)

Sub-class	Order	Family	Genus	Species	Common name
Carinatae (Flying birds)	*Anseriformes* (Ducks and Geese)	*Anseridae*	*Cairina*	*C. moschata*	Muscovy duck
			Anser	*A. anser*	Grey-legged goose
				A. albiformis	White-legged goose
		Anatidae	*Anas*	*A. platyrhynchos*	Mallard duck
				A. boscher	Wild mallard
	Galliformes (Chicken, Quail, Turkeys, Guinea fowl of African origin and Pheasants)	*Phasianidae*	*Coturnix*	*C. coturnix japonica*	Japanese quail
			Gallus	*G. domesticus*	Domestic chicken
				G. gallus	Red jungle fowl
				G. lafayetti	Ceylon jungle fowl
				G. sonneratti	Grey jungle fowl
				G. varius	Javan jungle fowl
			Phasianus	*P. colchicus*	Ring-necked pheasant
		Numitidae	*Numida*	*N. meleagris*	Ring-wattled G. fowl
				N. ptylorirhyncha	Blue-wattled G. fowl
		Meleagrididae	*Meleagris*	*M. gallipavo*	Domestic turkey
	Columbiformes (Pigeons and Doves)	*Columbiformidae*	*Columba*	*C. livea*	Rock-pigeon
			Streptopelia	*S. turtus*	Turtle dove
Ratitae (Running birds)	*Strunthiformes*	*Strunthionidae* (Large running birds)	*Struthio*	*S. camelus*	North African ostrich
				S. austrialis	South African ostrich
	Rheiformes	*Rheidae*	*Rhea*	*R. Americana abbscens*	Rhea
	Casuriformes	*Dromoiidae*	*Dromauis*	*D. novaehollandia*	Emu
		Casuriidae	*Casuarius*	*C. casuarius*	Cassowary
	Apterygiformes	*Apterigidae*	*Apterix*	*A. australis*	Kiwi
				A. owenii owenii	Kiwi

Source : Smith, 1992

Grey lag goose (*Anser anser*) of Europe and Asia are ancestors to domestic breeds of Western origin and Swan goose (*Anser cygnoides*) of Asia is ancestor to Eastern breeds; the latter are the most primitive of Grey geese whose breeding range is restricted from south-central Siberia to Kamchaka, to Central Asia and north Mongolia; and China and Japan during winter. Grey lag goose has a broader range encompassing Northern Europe and Asia.

Chapter 4

Morphology and Nomenclature

NOMENCLATURE (American Standard of Perfection, 1998)

(@ Terms generally used on waterfowl; * Source: Lever, 2000)

1. Abdomen: The under part of the body from the breast to the stern. In table dressed fowl, the breast includes the abdomen.
2. Angel wing @: Applies to defects in manner of folding of primary feathers and carriage of the primary section of the wing. Individual feathers on entire section may overlap in reverse order; a disqualification. Sometimes referred to as Angel wing.
3. Axial feather: The short feather growing between the primaries and secondaries of the wing.
4. Axillars @: The long, inner-most feathers near the juncture of the wing and body, usually three to five in number. They close the space between the secondaries and the body when the bird is in flight.
5. Back: The top section of the body from the base of the neck to the base of the tail, including the cape and saddle. In other words, the upper surface of a fowl between the shoulders and the uropygium.
6. Band @: A narrow strip or stripe on the plumage of contrasting colors.
7. Bantam: a diminutive fowl; some being distinct breeds, others being miniatures of a large breed or variety, approximately one-fourth to one-fifth their weight. Usually ornamental in character, some breeds have considerable merit as egg producers, a few as meat fowl.
8. Barring: a) Alternative transverse markings of two distinct colors on a feather, regular in shape as in Barred Plymouth Rocks, irregular in Dominiques, Hollands etc. b) The term properly applies to the alternate colored crosswise markings in Silver and Golden Penciled Hamburg females, ordinarily referred to as penicilings. c) A term used to describe defective coloration represented by transverse purple markings in blade feathers.
9. Bay: a) A light golden-brown b) Also used to designate intermingling of red and yellow in the iris caused by surface capillaries in the eyes in some varieties of fowl c) Wing-bay.
10. Beak: The horny formation projecting from the front of the head of chickens

and turkeys, consisting of upper and lower mandibles which form the forward mouth-parts.

11. Bean @: A raised, hard, bean-shaped protuberance on the tip of the upper mandible of a waterfowl.
12. Beard: a) A cluster of feathers pendent from the upper throat of some fowl, as in Crevecoeurs, Houdans, Faverolles etc.; found only in combination with muffs. b) The small tuft of long, coarse, bristly, black hairs projecting from the upper part of the breast of the adult male turkey.
13. Bib @: A contrasting color area in some breeds of ducks, covering the upper breast and lower neck.
14. Bill @ : The horny formation projecting from the front of the head of waterfowl, consisting of an upper and lower mandible which form the forward mouth-parts
15. Blade: The lower un-serrated part of a single comb. The portion to rear of the last point of the single comb is often referred to as the blade.
16. Blue: a term loosely used in referring to the general slaty color of some varieties of poultry.
17. Bluish white: A term used to describe the bluish-grey ground color of the plumage of so-called blue varieties of poultry.
18. Body: a) That portion which contains the trunk. The part exclusive of head, neck, wings, tail, thighs, shanks and toes b) When used in standard breed description i) When describing shape, it refers to the entire body as a whole ii) When describing color, it refers to the body exclusive of back and breast since these sections are described separately; more specifically, the lower sides of body, abdomen and stern.
19. Booted: Fowls that are feathered on shanks and toes and having vulture hocks are said to be booted.
20. Bow @: The upsweep of the keel on the breast, giving resemblance to the bow on a boat.
21. Bow-legged : A deformity in which the legs are farther apart at the hocks than at the feet and knee joint; a disqualification if a perceptible angle is evident at the hock when viewed from the front or back
22. Brassiness : A term descriptive of a light yellowish metallic cast commonly found in the plumage of white and parti-white varieties, and to a lesser degree in several other varieties, particularly in the hackle, wing-bow and saddle of male; a serious defect; may be hereditary or affected by exposure to sun-rays and certain items in the diet.
23. Breast: a) The forward portion of the underside of the body of fowls from the juncture of the neck and the body to a point directly between the legs b) Also applies to the flesh on both sides of the keel bone on fowls prepared for

the table.

24. Breed: An established group of individuals possessing similar characteristics, and when mated together produced offspring with those of same characteristics.
25. Buff: A medium shade of orange-yellow color with a rich golden cast; not as intense as to show reddish cast, nor so pale as to appear lemon or light yellow. The term is generally used in referring to the plumage all standard Buff varieties of poultry
26. Cap @: The area overlaying the crown or top of the skull; the plumage covering the top the head.
27. Cape: The short feathers and the juncture of the back and neck underneath the hackle and between the shoulders collectively shaped like a cape.
28. Capon: A term used to designate a castrated male fowl.
29. Carriage: a) The posture of a fowl b) Also applies to the angle of various parts or sections such as back, wing or tail, with respect to the horizontal.
30. Caruncles @: The fleshy protuberances on the naked portions of the head, face and neck of the turkey.
31. Cere @ : The bare membranes covering the bill, knob, lores and eyelids in waterfowl.
32. Cheek @ : The area on the side of the head; above the chin and below the eye, extending from the base of the bill and including the ear region.
33. Close feathered : Said of a fowl in which the feathers are held closely to the body with no perceptible angle to the body.
34. Cock : A male fowl one year old or more.
35. Cockerel : A male fowl less than one year old.
36. Comb : The fleshy protuberance on top of the head of a fowl, larger in the male than the female; usually red in color (purple in Sumatras, Birchen and Brown Red Modern Games and Silkies, Purplish-red in Sebrights.
    a) Buttercup : Consists of a single blade arising at the juncture of the head and beak rising up and slightly back to the cup-shaped crown set squarely on the centre of the skull. The rim of the cup shall bear an evenly spaced circle of points and be closed at the back. Points emerging from the centre of the cup are a serious defect.
    b) Cushion : A low, compact comb are relatively small size; it should be quite smooth, possess no depressions or no spikes and do not extend beyond the mid-point of the skull.
    c) Pea : A medium length, low comb, the top of which is marked with three low lengthwise ridges, the centre one slightly higher than the outer ones, the top of which are either undulated or marked with small

rounded serrations; this is a breed characteristic found in Ameraucanas, Brahmas, Buckeyes Cornish, Cubalayas and Sumatras. Various forms of comb are as follows :

d) Rose : A solid, broad, nearly flat on top, low fleshy comb, terminating in a well developed tapering spike, which may turn upward as in Hamburgs; is nearly horizontal as in Rose comb Leghorns; or follow the contour of the head in Wyandottes. The surface of the main part should be slightly convex and studded with small rounded protuberances. General shape varies in different breeds.

e) Silkie : An almost around, somewhatlumpy comb, inclined to be greater in width than length; covered with small corrugations on top and crossed with a narrow, transverse indentation slightly to front of the middle of comb. Sometimes, two or three small rear points hidden by crest, others without points. Generally considered to be genetically a rose comb, changed by rose comb plus crest.

f) Single : A moderately thin, fleshy formation of smooth soft surface texture, firm attracted from the beak along the top of skull with a strong base, the top portion showing five or six rather deep serrations or distinct points, the middle points being higher than the anterior or posterior, forming a semi-oval when viewed in profile. The comb always erect and much larger and thicker in male than female; may be lopped or erect in female, depending on breed. The comb is divided into three sections, the front or anterior, the middle, and that extending past the rear base of the skull, the posterior or blade.

g) Strawberry : A low set, compact comb of somewhat egg-shape with the larger portion forward and the rear extending no further than the midpoint of the skull.

h) V-shaped : A comb formed of two well-defined horn-like sections joined at their base, as in Houdans, Polish, Crevecoeurs, La Fleche and Sultans.

i) Walnut : Comb resembles one-half a walnut.

37. Condition : The state of a fowl in regard to health, including cleanliness and brightness of plumage, head parts, legs and feet.

38. Coverts : Those feathers which cover the base of the primary and secondary wing and main tail feathers.

39. Crest : An almost globular tuft of feathers on the top of the head of some fowl and waterfowl, as in Polish, Houdans, Crevecoeurs, Silkies Sultans and White Crested Ducks. Full expression is partially dependant on the "knob" which is the term most generally used to describe the bony structure and mass of tissue and follicles from which the crest grows.

40. Crop : The enlarged part of the gullet lying at the front of the body at the base of the neck and partially concealed in the cavity located between the halves of the wish-bone, it serves as a temporary storage area for ingested food.
41. Culmen @ : The top contour of the upper mandible between its point and the skull.
42. Cushion : A profuse mass of feathers over the back and base of tail of a fowl giving it a rounded effect; very pronounced in Cochin females.
43. Dewlap @ : A pendulous growth of skin under the rear of the beak or bill and extending onto the throat area. All turkeys have a form of dewlap running down the throat and front of neck, known as a throat wattle.
44. Down : a) The first soft, fine, fluffy plumule-like covering of baby chicks b) The soft, fluffy part of a feather below the web, consisting of barbs that are not hooked together; the undercolor c) Small tufts of very short barbs without a common quill, sometimes found on or between the toes and on the shanks of clean-legged fowls; a disqualification d) @ The body covering of ducklings; the additional layer of feathering that serves to keep waterfowl warm and dry.
45. Drake @ : A male duck.
46. Duck @ : Generally, any member of the *Anatinae* family; specifically, a female as distinguished from the drake or male.
47. Duck-foot : A condition where the 4th or hind toe is carried forward so as to touch or almost touch the 3rd toe instead of carried backward to touch the ground and help balance the bird; a disqualification
48. Duckling @ : A baby duck
49. Dubbed – Dubbing : A term used to describe the close trimming of the comb, wattles and earlobes; usually done in males.
50. Earlobes . The fleshy patch of bare skin below the ears, varying in size and shape with color either red, white, blue or purple depending on the breed; in all breeds, the texture should be fine and soft, the surface smooth, the outline regular and size uniform.
51. Ear-tuft : A feathered protuberance on each side the neck. The feathers grow from a slender finger-like cartilaginous appendage located slightly below the ear.
52. Eclipse molt @ : The more somber, dull plumage acquired by drakes after the breeding season.
53. Edged – Edging : a) A very narrow border or lacing of contrasting color around the entire web b) A narrow strip of contrasting color along the upper or lower web of primaries or secondaries running the length of the feather c) The white tip across the entire extreme ends of the wide main tail and

coverts of the Bronze Turkey and the same feathers of the Narragansett Turkey in which the tip ends in a light steel grey color. Also used to describe breast lacing in the male and female turkey when required.

54. Face : The skin around and below the eyes on either side of the head of a fowl; may be red, white or purple color, according to breeds; in all breeds, the texture should be fine and soft, the surface smooth.

55. Faking : A self-evident attempt to remove or conceal a disqualification of serious defect, thus creating merit which does not naturally exist; a disqualification.

56. Fawn : A light brownish tan color.

57. Feather : One of the epidermal outgrowths, which collectively forms the external covering or plumage of a fowl. A feather grows from the epidermal covering of a vascular dermal papilla, whose base is sunk in a follicle. A typical feather consists of

a) The stem or shaft which includes

i) The basal hollow and horny proximal or lower part which attaches the feather to the skin and which is known as the quill and

ii) The distal or upper part (rachis) is called the shaft and bears a series of slender somewhat obliquely directed processes on each side known as barbs. As the shaft grows out it carries with itself the pairs of barbs in succession

b) The web is formed by the barbs that bear in like manner the barbules, and these in turn the barbicels (hamuli) or little hooklets that attach onto the barbules of the next barb above uniting the whole series of processes; Color thereof being known as Surface Color

c) The fluff is that portion at the proximal or lower end of the rachis or shaft where the absence of the small hooks cause a feather to be soft and downy in character; not visible when the plumage is in natural position; the color thereof be known as undercolor

The three sections form a firm flexible composite unit which varies in shape, color and texture in the various breeds and varieties of fowl.

58. Feather-legged : A term used to designate those breeds having feather on the outer sides of shanks, and on the outer, or the outer and middle toes.

59. Finish : The term applied to fowl which indicates the completion of growth of the entire body, head parts, plumage, shanks and toes, and which must have the proper color, luster and sheen. Indicates bloom of health, ideal weight and perfection of plumage color and growth.

60. Flights : The primary feathers of the wing. The term is sometimes used to denote both primaries and secondaries, but when used in the Standard refers to primaries only.

61. Flight coverts : The stiff feathers located at the base forward of the flight feathers (primaries) and covering their base; also known as primary coverts.
62. Fluff : a) Part of a feather composed of barbs growing from the lower part of the shaft which are not hooked together as in the web, resulting in a soft downy character b) The soft, downy feathering on the inner side of the lower thighs and on the abdomen.
63. Fluted @ : Channeled, plaited, wrinkled or grooved
64. Fowl - Fowls : A collective term applying to chickens, ducks, geese and turkeys; same as poultry.
65. Frosting : A faded margin on a black lacing or spangle; a defect.
66. Gamy (Pinched) tail : A tightly folded; slim, tapering whip-like tail. a) A breed characteristic of Modern games, Cornish and to a lesser degree in Malays b) A defect in other breeds
67. Gander @ : A male goose; adult of over one year of age; young if under one year of age
68. Goose @ : Generally, any member of the subfamily *Anserinae*; specifically, a female specimen of domestic geese.
69. Ground color : The basic or predominating color of the web of a feather. In laced, penciled, barred, spangled and mottled varieties, it is the basic or predominating color on which the markings are delineated.
70. Hackle : The rear and side neck plumage of a fowl.
71. Head : In standard descriptions, the skull and face only. In general terminology, the beak, skull, face, eyes, comb and wattles.
72. Hen : A female fowl. For exhibition purposes, a female chicken or turkey one year old or more.
73. Hen feathered : A male having feathers like a female (oval instead of pointed sex feathers in hackle, saddle, wing-bow and sickles) and being identical with the female of the same variety in color and markings; as in Sebrights
74. Hock : The joint between lower thigh and shank; sometimes incorrectly referred to as the knee.
75. Keel : a) In chicken and turkeys, the lower median edge of the keel or breastbone, or more specifically, that portion of the middle sternum that resembles the keel of a boat, both as to shape and position b) @ In ducks, it is the deep, pendant fold of skin suspended from the entire underside of the body including the breast and abdomen c) @ In geese, it is the loose, pendant fold of skin suspended from the under part of the body in front of the legs.
76. Keel bone : The breast bone or sternum.
77. Knob @ : a) The horny protuberance at the juncture of the head and upper bill in African and Chinese geese b) A deformity growth on the breastbone,

usually at the front, sometimes found in chickens and turkeys, a defect c) The rounded protuberant part of the skull in crested fowl.

78. Knock-kneed : A deformity in which the legs are closer together at the hocks than at the feet. A disqualification when a perceptible angle is evident at the hock
79. Laced - Lacing : A border of contrasting color around the entire web of a feather; should be very distinct, uniform in width, and usually moderately narrow.
80. Lamella @ : Tooth-like serrations on the inner edges and roof of the bill of ducks and geese.
81. Lobe @ : Either one or two folds of skin that hang from the abdomen of many domestic geese.
82. Leaders : The single round tapering spike terminating the rear of a Rose comb; also known as spike.
83. Lores @ : The space between the eye and the upper edge of the bill.
84. Luster - Lustrous : A brilliant, glossy, luminous appearance of the feather due to the reflection of the light rays; more evident when the fowl is in perfect physical condition.
85. Main tail feathers : The straight, stiff long feathers of the tail located under and between the coverts of the female, the coverts and sickles of the male.
86. Mealy : Applies to buff or red plumage flecked with a lighter color, as if dusted with flour or meal; a defect.
87. Mossy : Confused, indistinct, irregular or disarranged color markings which destroy the desired color contrast or pattern. Also black specks appearing on the surface of parti-colored birds where White, Buff or Gold is desired
88. Mottled - Mottling : a) Plumage in which a variable percentage of feathers are tipped with white; as in Anconas, Houdans, Javas, Japanese and Belgian Bearded d'Anvers. (*cf.* Spangling in which markings may be either black or white and are located on the tip of each feather).
89. Muffs : A cluster of feathers projecting from the face below and around the sides of the eyes and extending from the beard and covering the earlobes; also known as "whiskers". Found only in combination with a beard.
90. Nuptial plumage @ : Plumage acquired by drakes following the eclipse molt; plumage of the drake prior to and during the breeding season.
91. Parti-colored : fowls having feathers of two or more colors, or shades of one color.
92. Paunch @ : Pendulous folds of flesh and skin suspended from the abdomen of geese.
93. Penciled - Penciling : Applied to several types of lines or markings on female

feathers. a) The crosswise bars on feathers of penciled Hamburgs b) The narrow concentric linear markings inside the edge on the web of the feather; sometimes multiple as in Silver Penciled or Partridge varieties, or single as in the Dark Cornish female. In all breeds, they should be characteristically narrow, uniform in width, sharply defined and continue in an unbroken line following the contour of the feather. When multiple, should also be equidistant from each other c) The fine markings on Rouen and Grey Call ducks.

94. Pinion : The distal or outer segment of the wing of a fowl.
95. Pinion feathers : The feathers growing from the pinion; i.e., the flights or primaries.
96. Pinioned @ : The surgical removal of the point of the wing at the outer joint.
97. Plumage : The collective feather covering of the entire body of a fowl, including the head, neck, wings, tail and where specified for breed, the shanks and toes.
98. Poult : The young of the domestic turkey before the sex can be determined
99. Primaries : The long, stiff feathers of the wing, growing from the pinion or outer segment, next to the secondaries, folding under and are completely hidden by the secondaries when the wing is properly folded; also known as "flight feathers" or "Pinion feathers".
100. Profile : As used in standard descriptions, refers to the side view of a fowl.
101. Pubic bones : The thin, terminal portion of the hip bones that form part of the synsacrum or pelvis. Considered important in evaluating productivity of the female fowl.
102. Pullet : For exhibition purposes, a female fowl less than one year old.
103. Pure white : An opaque white, generally applied to the unblemished white in female type feathers of white or parti-white varieties. The complement of silvery white in male sex feathers of the same varieties.
104. Quill : The basal hollow and transparent hornlike portion which attaches the feather to the fowl's body.
105. Reachy * : A term applied to Modern Game fowls (in which "reachiness" is an important point in a show bird) to denote that they stand erect.
106. Roach-back* : A convex-shape of back, and a great fault in all fowls, with the exception in Malays.
107. Rooster* : An American term for a cock.
108. Saddle : The rear of back extending to juncture of back and tail of male fowl, covered with long pointed feathers known as "saddle feather".
109. Sappiness* : A bird is said to be "sappy" when the "sap" in the feather is perceptible on the surface of the plumage. It shows in a white fowl in the form of a yellowish light primrose tint.

110. Scales : The thin horny growths completely covering the shanks and top of the toes of a fowl.
111. Scapulars : The larger feathers emerging from the region of the shoulders.
112. Secondaries : The long stiff wing feathers growing from the middle wing segments, next to the primaries. When the wing is folded, the exposed secondaries form a triangular area known as the "wing bay".
113. Self-color* : A uniform tint over feather/bird.
114. Serrated – Serration : The toothed saw-like top edge of a single comb, the points being variable in number in different breeds; those in front and rear usually being smaller than in the centre.
115. Self-color : A single uniform color throughout the plumage, as in Black and White varieties.
116. Sex feathers : a) The pointed feathers in the hackle, back, saddle, sickles and wing-bow of a male fowl, and which differ from the oval shape feathers in the same sections of the female b) The two or three top feathers in the tail of a drake or male duck which curve upward and forward and by which sex in ducks is distinguished (except in Muscovy ducks)
117. Shaft : The extension of the quill the entire length of a feather to which the barbs are attached.
118. Shafting : A color characteristic where the shaft of a feather is either lighter or darker than the color of the web. A standard requirement in some breeds; a defect in others.
119. Shank : The portion of the leg below the hock, exclusive of the foot and toes, the metatarsus; anatomically, it has tarsal bones fused into the head of the metatarsus and hence usually referred to as tarsometatarsus.
120. Shoulder : The upper side of the juncture of the wing and body.
121. Sheen : Luster, usually green or purple in the plumage; the bright, glistening effect on the plumage of certain sections.
122. Sickles : a) Main sickles – the prominent middle uppermost curving pair of long feathers in the tail of a mature male fowl b) Lesser sickles – The long curved feathers of the male chicken tail, exclusive of the top two longest main-sickles, which hang to the side of and cover most if not all the main tail. Extreme development in both number and length being a characteristic of certain breeds.
123. Side sprig : A well defined point of projection growing from the side of a single comb; a disqualification in all single-combed breeds
124. Silkie : Refers to the semiplume character of the feathers of the Silkie fowl in which the shafts are very thin and the barbs are very long, very soft and fluffy, which have no holding power and no locking arrangement; structurally similar to the fluff of a normal feather.

125. Silvery white : A lustrous white, generally applied to the lustrous white color in hackle, back, saddle and wing bow plumage of a white or parti-white male. Especially pronounced in Silver Grey Dorking.
126. Slate : A shade of grey having a bluish cast sometimes approaching black, sometimes of lighter shades.
127. Slip : A male on which the act of caponizing has not effected the complete removal of both testicles; readily distinguished from a true capon by normal development of comb and wattles.
128. Slipped wing @ : Applies to defects in manner of folding of primary feathers and carriage of the primary section of the wing. Individual feathers on entire section may overlap in reverse order; a disqualification. Sometimes referred to as Angel wing.
129. Spangle – Spangling : a) A distinct marking of contrasting color at the extremity of a feather, proximally shaped like a well-defined V with rounded end. Always is black in color and found in combination with silver or gold ground color, as in Hamburgs, or white in color and separated from bay or brown ground color by a black bar as in Sussex b) Diagonal elongated black markings on the web as in Buttercups
130. Speculum @ : A lustrous distinctly colored area of the wing of ducks
131. Spike : The single, round, tapering growth extending back from the rear of a rose comb; sometimes known as a leader.
132. Split comb : A definite division or split in the rear of the blade of a single comb; a disqualification
133. Split crest : A crest in which there is a division with feathers falling to either side; a serious defect
134. Split tail : One in which there is a distinct gap at the center of the base of the top main tail feathers, resulting from the permanent absence of a feather or feathers, or from the improper placement of main tail feathers resulting in disarrangement; a disqualification in cocks and hens
135. Split wing : One in which there is a distinct gap between the primary and secondary feathers, resulting from the permanent absence of a feather (indicated by absence of the feather follicle), and a disqualification.
136. Spur : A stiff horny projection from the rear inner side of the shanks, rounded or pointed according to age; prominent in the male, increasing greatly in size with age; sometimes multiple in character; occasionally found in females but generally rudimentary.
137. Squirrel tail : One in which any portion projects forward of the vertical from its anterior base; a disqualification except in Japanese bantams

138. Stag* : Another term for a young cock, chiefly used by Game breeders.
139. Station : A term applied to the ideal pose and symmetrical appearance, including height and reach in Modern Games.
140. Stern : The rear under part of fowl extending from the rear end of the keel bone to the ends of the pubic bones.
141. Strain : Fowl of any breed or variety that have been bred as a closed population for a number of years and which reproduce uniform characteristics with marked regularity.
142. Surface color : The color of that portion of the plumage that is exposed when the feathers are in natural position.
143. Symmetry* : Perfection in proportion.
144. Tail coverts : The feathers which cover the base of the main tail feathers in males, and the larger portion of the tail in females; curved and pointed in males, oval in females.
145. Telescope Comb : a) a pea comb with an indentation at the rear; a disqualification b) A rose comb with an inverted spike; the disqualification.
146. Tertials @ : Flight feathers on the inner or body side of the wing of ducks.
147. Top-knot* : Same as Crest.
148. Tri-colored* : A term often applied to cockerels which should be of one uniform color, when their hackles and saddles and tails are much darker, and the wing bar darker still. Chiefly in buff varieties.
149. Trio* : A cock or cockerel, and two hens or pullets.
150. Undercolor : The color of the lower or fluff portion of feathers, not visible when the feathers are in natural position.
151. Uropygium (Parson's or Pope's nose) : The fleshy and bony prominence at the posterior extremity of a fowl's body from which the tail feathers grow. It contains the free caudal vertebrae and the pygostyle (the plate of bone which forms the posterior end of the vertebral column in most fowl).
152. Uropygial gland : The oil gland. A large gland, may have two lobes, opening on the back at the base of the tail feathers, secreting an oily fluid which the fowl uses in preening it feathers. It is especially developed in waterfowl, the oil secreted aids to make the plumage shed water.
153. Variety : A subdivision of a breed. Differentiating characters include plumage color, comb type or presence of beard and muffs.
154. Vulture hock : A well-defined formation of stiff, straight and rather long feathers growing from the lower part of the lower thighs and projecting backward and downward; a disqualification in all Asiatics including their

Bantam counterparts; a standard requirement in d'Uccle, Booted and Sultans.

155. Wattles : The thin pendant growths of flesh at either side of the base of the beak and upper throat; usually much larger and longer in males than in females; usually red in color, Purple in Sumatras, Birchen and Brown red modern games and Silkies. Should be fine and soft in texture. Slightly concave in surface, regular in outline, and uniform in size.

156. Web : a) of feather - formed by the barbs that bear in like manner the barbules, and these in turn the barbicels (hamuli) or little hooklets that attach onto the barbules of the next barb above uniting the whole series of processes; Color thereof being known as Surface Color b) of feet – The stout web (membranes) between the toes of all *Anatidae.* Also a term applied to the exposed barbed parts of feathers. c) of wing – The triangular skin in front and between the joints of the wing.

157. Web-foot : A condition in fowl other than waterfowl when the web unites the toes for a greater part of their length; a disqualification

158. Wheaten : Various shades of the color of wheat. The terminal used to describe the plumage color of the females of certain varieties.

159. Wing bars : The distinct bar of color across the middle of the wing caused by the regularity of feathers and distinct color; or color and markings of the feathers known as the wing coverts.

160. Wing bay : The triangular section of the wing below the wing bar and terminating at the wing points, formed by the exposed portion of the secondaries when the wing is folded.

161. Wing bow : The surface part of the wing below the shoulder, and between the wing covert and wing front.

162. Wing-butts* : The corner or ends of the wing. The upper ends are more properly called shoulder-butts, and are thus termed by Game-fanciers. The lower, similarly, are often called lower-butts.

163. Wing coverts : The two rows of broad feathers the cover the lower portion of the secondaries.

164. Wing fronts : The extreme front portion of the wing at the shoulder, foreword of the wing bow, sometimes referred to as the "wing butts".

165. Wing points : Ends of the primaries and secondaries.

166. Wry tail : The tail of a fowl permanently carried to one side of the vertical; a disqualification

167. Yearling : An individual between January 1 and December 31 of its second year of life.

NOTE

All the figures, 4.1 to 4.25, are reproduced from THE AMERICAN STANDARD OF PERFECTION, 1998

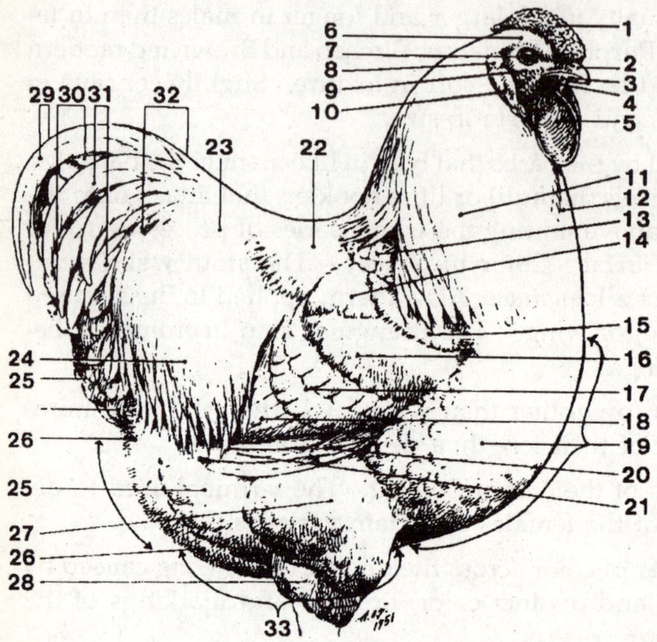

1. Comb 2. Upper mandible or beak 3. Lower mandible or beak 4. Throat 5. Wattle 6. Skull 7. Eye 8. Ear 9. Face 10. Ear-lobe 11. Hackle 12. Front of neck plumage 13. Cape 14. Shoulder 15. Wing front 16 . Wing bow 17. Wing coverts or wing bar 18. Secondaries of wing bay 19. Breast 20. Primary coverts 21. Primaries 22. Back 23. Upper saddle 24. Lower saddle 25. Rear body feathers 26. Fluff or stern 27. Lower thigh feathers 28. Hock plumage 29. main sickles 30. Main tail 31. Lesser sickles 32. Tail coverts 33. Abdomen

Fig. 4.1 Nomenclature of males

1. Skull 2. Eye 3 Ear 4. Face 5. Ear-lobe 6. Comb 7. Nostril 8. Beak 9. Wattle 10. Throat 11. Hackle 12. Front of neck plumage 13. Breast 14. Cape 15. Shoulder 16 . Wing bow 17. Wing front 18. Wing coverts or wing bar 19. Secondaries of wing bay 20. Primaries 21. Primary coverts 22. Back 23. Sweep of back 24. Cushion 25. Main tail 26. Tail coverts 27. Rear body feathers 28. Fluff or stern 29. Lower thigh plumage 30. Hock 33. Abdomen

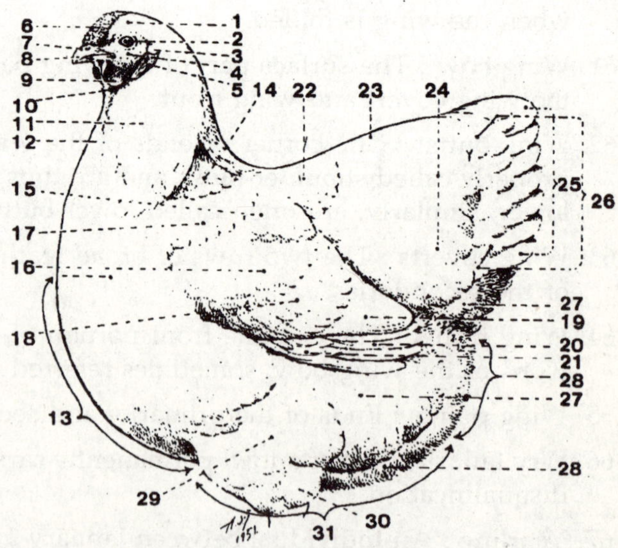

Fig. 4.2 Nomenclature of females

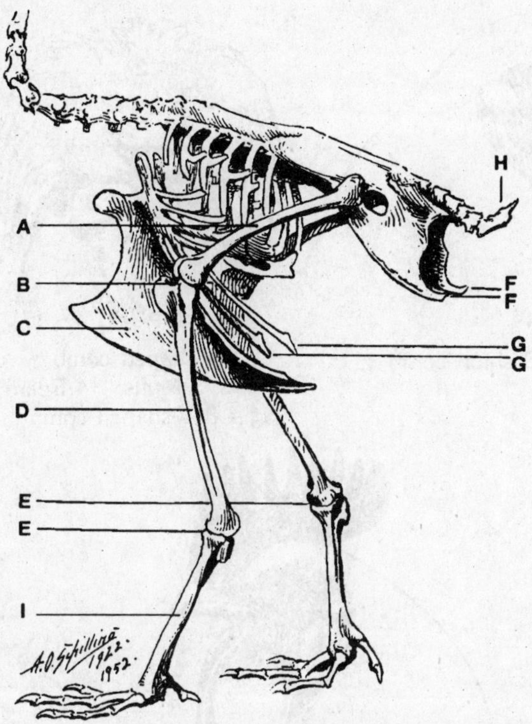

A. Thigh or upper thigh (Femur) B. Thigh joint C. Keel or breast bone (Sternum) D. Lower thigh (Tibia) E. Hock joint F. Pubic bones G. Lateral process of Sternum H. Uropigium (Coccyx) I. Shank (Tarso-metatarsus)

Fig 4.3 Skeleton of fowl

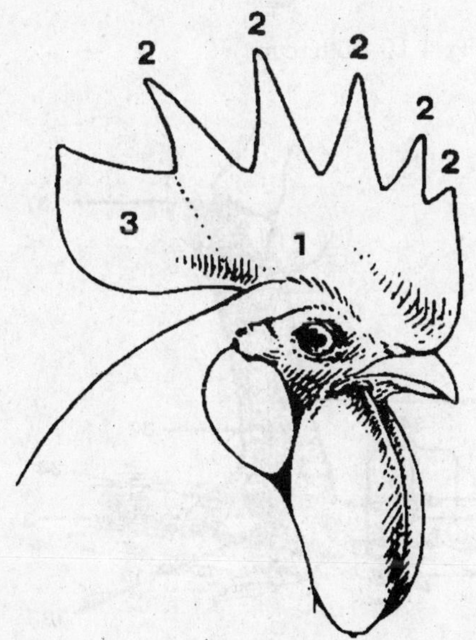

Fig 4.4 Single comb

Fig 4.5 Rose comb

1. Base 2. Points 3. Blade 1. Base 2. Cushion 3. Spike

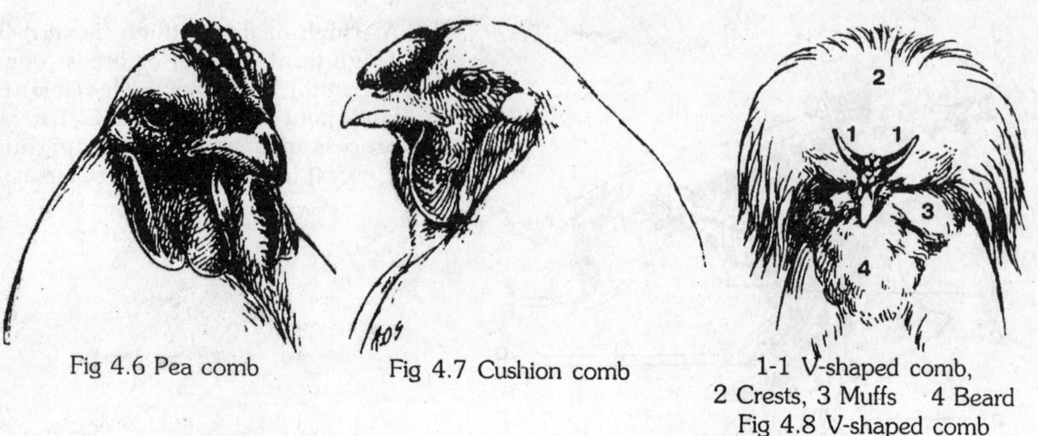

Fig 4.6 Pea comb

Fig 4.7 Cushion comb

1-1 V-shaped comb,
2 Crests, 3 Muffs 4 Beard
Fig 4.8 V-shaped comb

Fig 4.9 Strawberry comb

Fig 4.10 Buttercup

Comb patterns

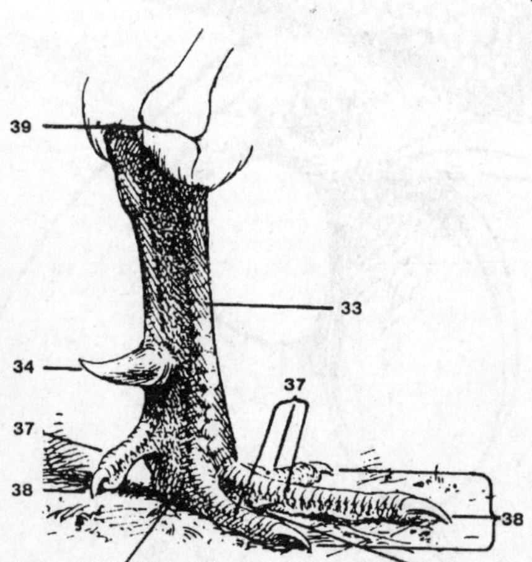

33. Shank 34. Spur 35. Foot
36. Web 37. Toes 38. Toe nails
39. Middle of hock joint
Fig 4.11 Shanks and toes (Male)

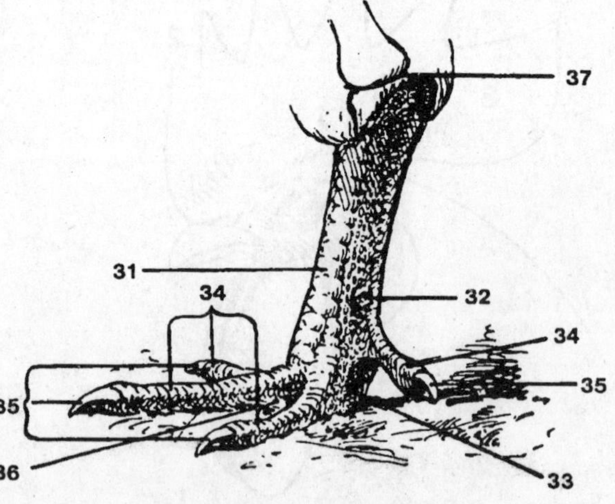

31. Shank 32. Spur 33. Foot
34. Toes 35. Toe nails
36. Web 37. Middle of hock joint
Fig 4.12 Shanks and toes (Females)

Fig 4.13 Fifth toe

A - Upper thigh B-B - Lower thigh
C-C - Shank D-D - Toes
Fig 4.14 Leg and toe fathering

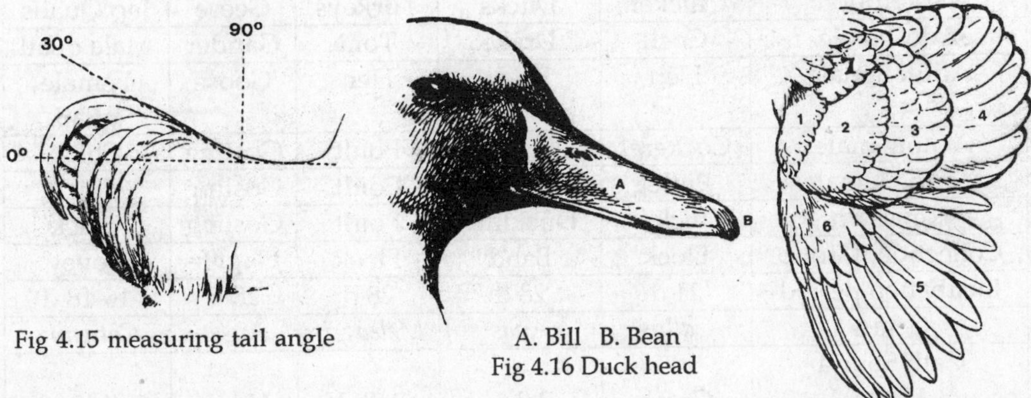

Fig 4.15 measuring tail angle

A. Bill B. Bean
Fig 4.16 Duck head

1. Front 2. Bow 3. Bar
4. Secondaries 5. Primaries
6. Primary coverts
7. Wing shoulder
Fig 4.17. Parts of a wing

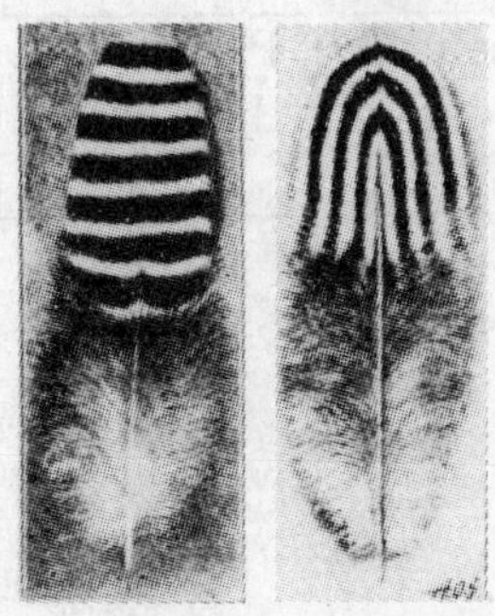

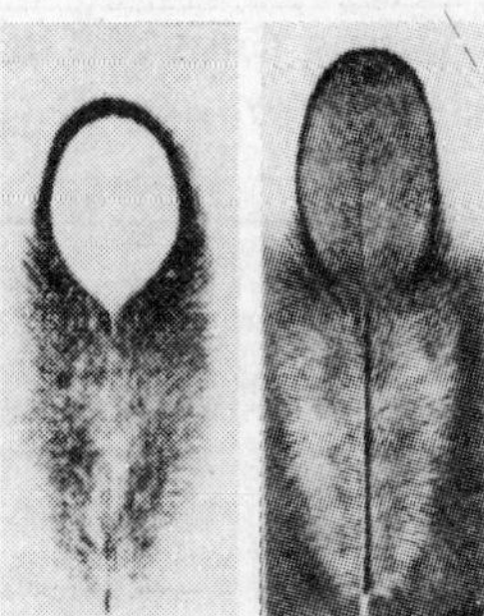

Fig 4.18 Pencilled   Fig 4.19 Laced   Fig 4.20 Barred   Fig 4.21 Striped

Plumage patterns in Chicken

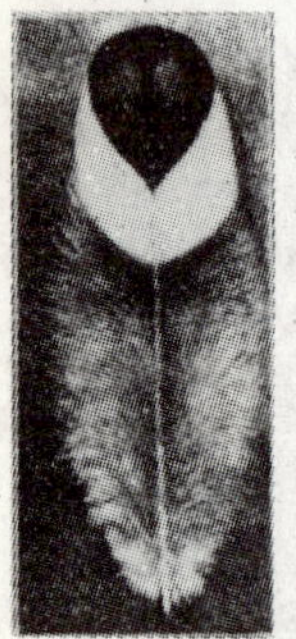
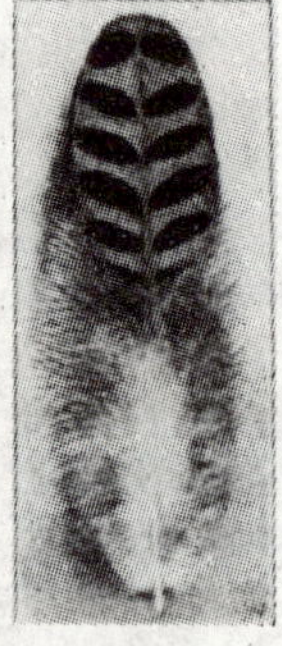

Fig 4.22 Spangled

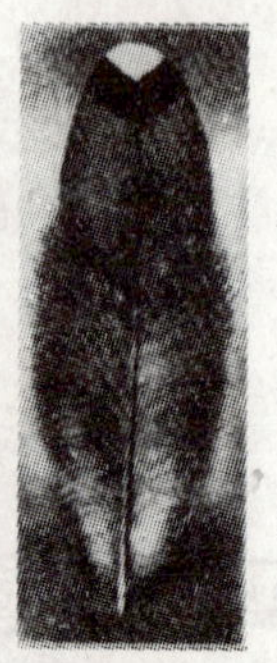

Fig 4.23 Spangle-tipped

Fig 4.24 Mottled

Fig 4.25 Stipled

Plumage patterns in Chicken (contd.)

Table 4.1 Nomenclature to denote various species of poultry

Description	Chicken	Ducks	Turkeys	Geese	Jap.Quails
Adult male	Cock	Drake	Tom	Gander	Male quail
Adult female	Hen	Duck	Hen	Goose	Female quail
Young male	Cockerel	Duckling	Poult	Gosling	Chick
Young female	Pullet	Duckling	Poult	Gosling	Chick
Newborn	Chick	Duckling	Poult	Gosling	Chick
Collection/Group	Flock	Band	Flock	Gaggle	Covey
Incubation period	21 d	28 d	28 d	28 d	16-18 d
Genus	*Gallus*	*Anas*	*Meleagris*	*Anser*	*Coturnix*
Call/Sound Males Females	 Crow Cackle	 Belch Whimper	 Gobble Gobble	 Honk Honk	 Call/Song Call/Song
Act of breeding	Mating				
Emergence of newborn	Hatching				
Term used for castrated cock is Capon. Young ones of Guinea fowl, Pigeon and Swan are referred to as Keats, Squabs and Cygnets, respectively. Incubation period of Muscovy duck eggs is 35 d.					

Chapter **5**

Breeds of Different Species of Poultry

1. Breeds and varieties of Chicken

1.1 Economic or Type classification

Type relates to a classification on the basis of the fowl's ability to produce a product of commercial value, which is virtually the purpose for which it is reared.

Table 5.1 Economic or Type classification

Type	As layer	As broody hen	As table bird	General remarks
Egg	Very good	Poor	Poor	Light body, active
Dual	Good	Good	Good	-----
Meat – Heavy	Poor	Good	Very good	Large body with plenty of flesh on the breast
Meat – Game	Poor	Good	Very good	Powerfully built
Miscellaneous	Poor	Indifferent	Very poor	Delicate build, bright plumage, small size, possess unique features intended for pleasure or fancy
Desi	Poor	Very good	Poor	-----

However, the Type classification is no more in vogue because there are possibilities of overlapping of "Type" in case of few breeds; this is particularly true in case of "Dual" purpose breeds which can be excellent egg producers or table birds. Hence, an Official or Standard classification is being currently followed.

1.2 Official or Standard classification

This classification is internationally accepted and is based on Class, Breed, Variety and Strain. Totally about 53 breeds and 176 strains have been officially classified into various Classes.

1.2.1 Class

Relates to official or standard classification with which various groups of birds can be distinguished largely on the basis of geographical regions from where they originated. Most important of various Classes are American, Asiatic, English and Mediterranean; other minor ones are Game Class, Continental Class (North European), Continental Class (Polish), Continental Class (French), Oriental Class and Miscellaneous Class.

1.2.2 Breed

Breed is a sub-division under Class, referring to a group of fowls, all members of which are descended from common ancestry, are similar in shape and conformation and breed true to type. Each breed will have its distinctive characteristics and are developed for some special purpose. Ex : Plymouth Rock, Brahma, Leghorn etc.

1.2.3 Variety

This is a sub-division of a breed; mostly differentiated by plumage color and/or special formation of comb. Ex : White Cornish, Red Cornish, Single Comb White Leghorn, Rose Comb White Leghorn etc.

1.2.4 Strain

A group of birds within a variety usually named after the breeder responsible for its development and bred for certain special characteristics (like size, maturity, egg production etc.) in view. Ex : Babcock strain of Single Comb White Leghorn.

1.3 Salient features of various Classes and breeds of Chicken

All recognized breeds and varieties of Chicken, in a chronological order of admission to Standard by the American Poultry Association, Inc. are summarized below as per Classes. The common characteristics listed are likely to vary, although to a limited extent, among some of the varieties. The detailed description of each of the varieties is beyond the scope of this book; but, the same can be obtained by the original publication titled "The American Standard of Perfection" by the American Poultry Association, 1998 Edition.

1.3.1 American Class

Table 5.2 Breeds and varieties of American Class

All breeds	Color of				Comb	Shank feathering
	Ear-lobes	Skin	Shanks	Eggs		
	Red	Yellow	Yellow [3]	Brown	Single [4]	Absent
Breeds	**Variety (ies)**					
Plymouth Rocks [1]	Barred, White, Buff, Silver-penciled, Partridge, Columbian, Blue					
Dominiques	No varieties					
Wyandottes	Silver-laced, Golden-laced, White, Black, Buff, Partridge, Silver-penciled, Columbian, Blue					
Javas	Black, Mottled					
Rhode Island Reds [2]	No plumage varieties					
Rhode Island Whites	No varieties					
Buckeyes	No varieties					
Chanteclers	White, Partridge					
Jersey Giants	Black, White					
Lamonas	White					
New Hampshires [1]	No varieties					
Hollands	Barred, White					
Delawares	No varieties					

[1] Popular breeds as female line for broiler production; [2] Popular dual purpose breed, occurs as Rose comb variety also; [3] Black in Jersey Giants [4] Rhode Island White, Dominiques and Wyandottes Rose, Buckeyes and Chanteclers Pea

1.3.2 Asiatic Class

Table 5.3 Breeds and varieties of Asiatic Class

All breeds	Color of				Comb	Shank feathering
	Ear-lobes	Skin	Shanks	Eggs		
	Red	Yellow [1]	Yellow [2]	Brown	Single [3]	Present [4]
Breeds	Varieties					
Brahmas	Light , Dark, Buff					
Cochins	Buff, Partridge, White, Black, Silver-laces, Golden-laced, Blue, Brown, Barred					
Langshans	Black, White, Blue					

[1] White in Langshans; [2] Bluish-black in Langshans; [3] Brahmas Pea; [4] Characteristic feature of the Class

1.3.3 English Class

Table 5.4 Breeds and varieties of English Class

All breeds	Color of				Comb	Shank feathering
	Ear-lobes	Skin	Shanks	Eggs		
	Red	White [3]	Yellow [4]	Brown [5]	Single [6]	Absent
Breeds	Variety(ies)					
Dorkings	White, Silver-grey, Colored, Red					
Redcaps	No varieties					
Cornish [1]	Dark, White, White-laced Red, Buff					
Orpingtons	Buff, Black, White, Blue					
Suusex	Speckled, Red, Light					
Australorps [2]	Black					

[1] Popular breed as male line for broiler production [2] Black Orpington of Australia [3] Yellow in Cornish [4] Dark-slate/Blue in Australorps [5] White in Dorkings and Redcaps [6] Pea in Cornish and Rose in Redcaps

1.3.4 Mediterranean Class

Table 5.5 Breeds and varieties of Mediterranean Class

All breeds	Color of				
	Ear-lobes	Skin	Shanks	Eggs	Shank feathering
	White	Yellow [2]	Yellow [3]	White	Absent
Breeds	Comb shape	Variety (ies)			
Leghorns [1]	Single	Dark brown, Light brown, White, Buff, Black, Silver, Red, Blood-tailed red, Colombian, , Golden duckwing			
	Rose	Dark brown, Light brown, White, Black, Buff, Silver			
Minorcas	Single	Black, White, Buff (See Plate 10)			
	Rose	Black, White			
Spanish	Single	White-faced black			
Andalusians	Single	Blue			
Anconas		Single comb, Rose comb			
Sicilian Buttercups	Buttercup				
Catalanas	Single	Buff			

[1] Popular egg producing breed; [2] White in Minorcas, Spanish, Andalusians and Catalanas; [3] Dark slate/Black in Minorca, Slaty blue in Andalusians

1.3.5 Continental Class (North European)

Table 5.6 Breeds and varieties of Continental Class (North European)

All breeds	Color of			Comb	Shanks feathering
	Skin	Eggs	Ear-lobes		
	White [1]	White [1]	White [1]	Single [2]	Absent
Breeds	Varieties			Remarks	
Barnevelders	No varieties			Black feathers with reddish-brown edging with sickles greenish black	
Hamburgs	Golden-spangled, Silver-spangled, Golden-penciled, Silver-penciled, White, Black			Ornamental	
Campines	Silver, Golden			Ornamental	
Lakenvelders	No varieties			Feathers on head, neck and tail black; rest white	
Welsummers	No varieties			Rich golden-brown feathers with greenish black tail feathers	

[1] Barnevelders and Welsummers have yellow skin, red earlobes and lay brown eggs. [2] Hamburgs Rose; Shank color yellow in case of Barnevelders and Welsummers; slate in case of Lakenvelders; Leaden blue in case of Hamburgs and Campines

1.3.6 Continental Class (Polish)

Table 5.7 Characteristics and varieties of Polish Breed

Characteristics	Varieties	
White skin, white earlobe, slaty-blue shanks, crest with or without beard and muffs, large "headgear" with V-shaped comb, lay white eggs	Non-bearded	White-crested black, Golden, Silver, White, Black-crested white, White-crested blue, Buff-laced
	Bearded (Wattle absent)	Golden, Silver, White, Buff-laced

1.3.7 Continental Class (French)

Table 5.8 Breeds and varieties of Continental Class (France)

All Breeds	Color of			Comb	Shank feathering
	Skin	Eggs	Earlobes		
	White	White [1]	White [2]	V-shaped [3]	Absent [4]
Breeds	Variety (ies)	Remarks			
Houdans	Mottled, White	Has crest, beard and five toes			
Crevecoeurs	Black	Has crest and beard			
La Fleche	Black	Has trace of crest on head			
Faverolles	Salmon, White	Has beard, muffs, feathered shanks and five toes			
All breeds are of ornamental type; shank color pinkish white (Houdans and Faverolles) to black/slate (La Fleche) or Black/dark leaden blue (Cravecoeurs). [1] Faverolles light brown; [2] Crevecoeurs and Faverolles red; [3] Faverolles Single; [4] Faverolles slightly feathered					

1.3.8 Game Class

Table 5.9 Breeds and varieties of Game Class

Breeds	Varieties
Modern games	Black-breasted red, Brown red, Golden duckwing, Silver duckwing, Birchen, Red pyle, White, Black, Wheaten
Old English games	Black-breasted red, Brown red, Golden duckwing, Silver duckwing, Red pyle, White, Black, Spangled, Blue-breasted red, Lemon blue, Blue golden duckwing, Blue silver duckwing, Self blue, Crele
Note : All breeds have red earlobes, white skin, lay white eggs and are broody; Shank color varies from black to willow green depending on the variety; Comb, wattles and earlobes are normally dubbed very close to the skin in case of males and the females have Single comb	

1.3.9 Oriental Class

Table 5.10 Breeds and varieties of Oriental Class

All Breeds	Color of				Comb shape	Shank feathering
	Skin	Earlobe	Eggs	Shanks		
	Yellow [1]	Red [1]	Brown [2]	Yellow [3]	Pea [4]	Absent

Breeds	Variety (ies)	Remarks
Sumatras	Black	has lustrous greenish-black plumage, multiple spurs, looks like wild fowl; no wattle development; long sweeping sickles
Malays	Black-breasted red, Spangled, Black, White, Red pyle, Wheaten (female)	Very tall breed; wattles very small; face and throat have no feathers
Cubalayas	Black-breasted red, White, Black	Similar to Sumatras developed in Cuba
Phoenix	Silver, Golden	Tail growth up to 6 m with black tail and lustrous greenish-black sickles and coverts
Yokohamas	White, Red shoulder	Pheasant-like body with profuse lengthy saddle feathers and long sickle and side feathers; wattles small or absent
Aseels	Black-breasted red, Dark, Spangled, White, Wheaten (female)	Face and throat have no feathers; males have chopped-off crown; wattles missing or replaced by a naked dewlap
Shamos	Black, Black-breasted red, Dark, Wheaten (female)	Plumage is scarce and leaves considerable exposed skin; wattles missing or replaced by a naked dewlap

[1] Phoenix white; [2] Sultans and Phoenix white; [3] Sultans black/dark willow, Cubalayas pinkish white, Phoenix leaden blue, Shamos yellow/black; [4] Strawberry in Malays, Single in Phoenix, Walnut in Yokohamas

1.3.10 Miscellaneous Class

Table 5.11 Breeds and varieties of Miscellaneous Class

Breeds	Variety (ies)	Remarks
Sultans	White	Has V-shaped comb, large crest, muffs, beard and feathered shanks, five toes; small wattles
Frizzles	Clean leg, Feather leg	Has frizzled feathers; shanks are yellow in White varieties and red to buff, yellow/willow in other varieties
Naked Necks	Red, White, Buff, Black	Single comb, red earlobes and yellow shanks
Araucanas	Black, Black red, Golden duckwing, Silver duckwing, White	Pea comb, red earlobes and lay blue or greenish-tinged eggs; has ear tufts; shank color willow to black; rumples; wattles very small or absent
Ameraucanas	Black, Blue, Blue wheaten, Brown red, Buff, Silver, Wheaten, White	Pea comb, red earlobes and lay blue or greenish-tinged eggs; beards and muffs but no ear tufts; shank color dark slate to black; wattles small or absent

1.4 Bantams

The word "Bantams" means miniature chickens usually weighing about one-fourth to one-fifth the weight of their large counterparts. Although they appear to be exact miniatures, they have larger head, wings, tail and feather size in proportion to their weight. Their origin is thought to be in the Orient. They are mainly ornamental in nature displayed in shows and exhibitions and hence referred to as "flower garden of the poultry world". Bantams with entirely different plumage than their large chicken counterparts are also available. Since they are small in size, their maintenance is cheaper and easier; in addition, they are adaptable to backyard rearing since they are more hardy than the regular chicken and produce slightly larger eggs in proportion to their body size. Bantams also yield fine-grained nutritious meat. Therefore, of late, Bantams are being considered a commercially viable proposition. There are about 350 kinds of Bantams named after their breed and variety with an additional "bantam" word; Ex : Buff Cochin bantams, Buff Catalana bantams etc.

2. Breeds and varieties of Ducks

Ducks, like chicken, are useful for eggs (Khaki Campbell), meat (Pekin) and exhibition (Crested). Muscovy ducks differ from the rest in their incubation period which is 35 d instead of 28 d in case of other breeds. Further, they are a different race and hence all the crosses with other breeds are sterile. The difference in size of drakes and ducks is most pronounced in Muscovy than in other breeds, Runners are popular for their gait, stance and carriage; body held virtually perpendicular to the ground. East Indies and Calls are the Bantams of the duck family.

Old drakes of Gray Runner, Mallard, Rouen and Gray call shed their showy plumage and take on a plumage resembling that of female; during fall, they once again molt and revert to male plumage.

Skin of most breeds is yellow excepting that of Aylesbury and Muscovy which have pinkish white skin.

Egg color is usually light brown with the following exceptions : Indian Runners produce white eggs; Cayuga and Black East Indie produce black eggs at the beginning of production which later on changes to blue as production progresses; Rouen produces blue eggs as well as white eggs.

2.1 Heavy-weight Class (Drake = 3.2 kg; Duck = 2.7 kg)

Table 5.12 Breeds and varieties of Ducks : Heavy-weight Class

Breeds	Variety	Sex	Color of		
			Bill	Eyes	Shanks and feet
Pekin	White	Both	Pale flesh	Deep Leaden blue	Reddish orange
Aylesbury	White	Both	Pale flesh	Deep Leaden blue	Bright, light orange
Rouen	White	Drake	Greenish yellow, black bean	Dark brown	Orange with brownish tinge
		Duck	Brownish orange dark blue blotch on upper part, black bean	Dark brown	Dusky orange
Muscovy	White	both	Pinkish flesh	Blue	Pale orange
	Black	Both	Pink shaded with horn	Brown	Black to dusky yellow
	Blue	Both	Pinkish horn	Brown	Dusky yellow
	Chocolate	Both	Pinkish horn marked with dark horn or black	Brown	Brownish yellow

2.2 Medium-weight Class (Drake > 1.8 and < 3.2 kg; Duck > 1.8 and < 2.7 kg)

Table 5.13 Breeds and varieties of Ducks : Medium-weight Class

Breeds	Variety	Sex	Color of		
			Bill	Eyes	Shanks and feet
Cayuga	Black	Both	Black	Dark brown	Black to Dusky black
Crested	White	Both	Yellow	Blue	Light orange
	Black	Both	Black	Dark brown	Black to Dusky black
Swedish	Blue	Both	Drake – greenish blue; Duck – bluish slate	Dark brown	Reddish brown with irregular grayish black markings
Buff	Buff	Both	Drake Yellow, duck brownish orange with dark bean	Brown with blue pupil	Orange yellow

2.3 Light- weight Class (Drake and Duck < 1.8 kg)

Table 5.14 Breeds and varieties of Ducks : Light-weight Class

Breeds	Variety	Sex	Color of		
			Bill	Eyes	Shanks and feet
Runner	Fawn and White	Both	Drake – yellow when young; greenish yellow in adults; black bean Duck – yellow spotted with green when young and dull green in adults	Blue or blue shaded with brown	Orange red
	White	Both	Yellow	Leaden blue	Orange
	Penciled	Drake	Yellow when young; greenish yellow in adults; black bean	Blue or blue shaded with brown	Orange red

		Duck	Yellow spotted with green when young and dull green in adults	Blue or blue shaded with brown	Orange red
	Black	Both	Black; olive tip acceptable	Brown	Black to dusky black; orange in old drakes
	Buff	Both	Drake Yellow, duck brownish orange with dark bean	Brown with blue pupil	Orange yellow
	Chocolate	Both	Blackish brown. Olive tip acceptable	Dark brown	Dark brown; orange shading in old drakes
	Cumberland blue	Both	Drake – greenish blue Duck – bluish grey	Brown	Smokey orange to grey
	Gray	Drake	Yellow to greenish yellow; dusky to black bean	Hazel-brown	Reddish orange
		Duck	Dull orange, dark bean	Dark hazel-brown	Pale reddish orange
Campbell	Khaki	Drake	Green, black bean	Dark brown	Rich dark orange
		Duck	Greenish black, black bean	Dark brown	Brown
Magpie	Black and white	Both	Yellow orange when young (spotted with green in females) and solid green in adults. Black bean	Dark grey or dark brown	Orange red spotted with grayish black when young; dark shading increasing with age especially in females
	Blue and white	Both	Same as in Black and white variety	Same as in Black and white variety	Same as in Black and white variety

2.4 Bantam Class (Drake 0.6 to 1.1 kg; Duck 0.5 to 1.0 kg)

Table 5.15 Breeds and varieties of Ducks : Bantam Class

Breeds	Variety	Sex	Color of Bill	Color of Eyes	Color of Shanks and feet
Call	Gray	Drake	Greenish yellow	Dark brown	Orange with brownish tinge
		Duck	Brownish orange	Dark brown	Dusky orange
	White	Both	Bright yellow	Blue	Bright orange
	Blue	Both	Drake – greenish blue; Duck – bluish slate	Dark brown	Reddish brown with irregular grayish black markings
	Snowy	Drake	Greenish yellow, dark bean	Brown	Purplish brown or claret
		Duck	Brownish orange, dark bean	Brown	Orange with brown tinge
	Buff	Drake	Yellow	Brown	Orange yellow
		Duck	Brownish orange, dark bean	Brown	Orange yellow
	Pastel	Drake	Apple green, dark bean	Brown	Salmon
		Duck	Brown with dark saddle and bean	Brown	Salmon
East Indies	Black	Both	Black; olive tip acceptable	Dark brown	Black to dusky black; may have orange shade in drakes
Mallard	Gray	Drake	Yellow to greenish yellow, dusky to black bean	Hazel-brown	Reddish orange
		Duck	Dull orange, dark bean	Dark hazel-brown	Pale reddish orange
	Snowy	Drake	Greenish yellow, dark bean	Brown	Purplish brown or claret
		Duck	Brownish orange, dark bean	Brown	Orange with brown tinge

2.5 Plumage patterns of Ducks

Table 5.16 Breeds and varieties of Ducks : Plumage patterns

Class	Breed	Variety	Sex	Plumage
Heavy-weight	Pekin	White	Both	Creamy white
	Aylesbury	White	Both	Web, quill and fluff of feathers in all sections pure white
	Rouen	White	Drake	Head and lower part of rump lustrous green; shoulders, tail, fronts and bows of wings and body shades of grey; streaks of wavy brown lines on shoulders, slate or slaty blue secondaries and coverts; breast, tail and primaries different shades of brown.
			Duck	Different shades of brown penciled with dark lustrous brown (neck and body), wide greenish black (back), broad, dark brown with greenish sheen (tail, wings), light brown (breast)
	Muscovy	White	Both	Web, quill and fluff of feathers in all sections pure white
		Black	Both	Black with greenish black lustrous body and back
		Blue	Both	Blue except for a patch of white in the wing-bow and bar area
		Chocolate	Both	Milk-chocolate brown except for a patch of white in the wing-bow and bar area
Medium weight	Cayuga	Black	Both	Lustrous, greenish black
	Crested	White	Both	Web, quill and fluff of feathers in all sections pure white
		Black	Both	Lustrous, greenish black
	Swedish	Blue	Both	Uniform bluish slate, with each feather of the back and upper tail cushion (especially in drakes) laced with a darker shade. Drakes typically are one to two shades darker than Ducks; front part of breast pure white forming a heart-shaped bib

Class	Breed	Variety	Sex	Plumage
	Buff	Buff	Both	Surface throughout and even shade of rich fawn-buff, with the exception of head and upper portion of neck in drake, which varies with rich fawn-buff to seal-brown, fawn-buff preferred; devoid of bluish-grey, eye streaks or white on the throat.
		Fawn and white	Both	Fawn and white head; a white line dividing the cap from cheek and base of the bill from head markings. Upper two-thirds of neck, primaries and secondaries of the wing, lower half of breast between point of breast-bone and legs and body and fluff are white; the rest fawn in color; an indistinct line of fawn runs from base of the tail to thigh.
		White	Both	Web, quill and fluff of feathers in all sections pure white
	Runner	Penciled	Drake	Bronze-green head; narrow white markings encircling eyes, dividing cap and cheek as well as head markings and bill. Neck two-thirds, primaries and secondaries of the wings (making a V-shaped marking on the wings), lower half of breast from point of breast-bone to legs and fluff are white. Rest of the body light to medium fawn
Light weight			Duck	Medium fawn head; narrow white markings encircling eyes, dividing cap and cheek as well as head markings and bill. White markings on the body same as drake with a darker shade of fawn; a lighter line of fawn running around near the edge of each feather, the border or edge a darker shade. The penciling more prominent on the back and wings.

Class	Breed	Variety	Sex	Plumage
		Black	Both	Lustrous greenish black
		Buff	Both	Surface throughout and even shade of rich fawn-buff, with the exception of head and upper portion of neck in drake, which varies with rich fawn-buff to seal-brown, fawn-buff preferred; devoid of bluish-grey, eye streaks or white on the throat.
		Chocolate	Both	Rich chocolate brown throughout
		Cumberland blue	Both	Uniform bluish slate throughout, with each feather of the back (especially in drakes) laced with a darker shade. Head, upper tail cushion and wing bows of drakes a darker shade of bluish slate
		Gray	Drake	Head and lower part of back rich lustrous green. Breast rich purplish brown and rump black with bluish green reflections. Rest of the body varying shades of grey.
			Duck	Top and rear of head, rear of neck, back, front and back of wings dark brown. Front of neck, shoulder coverts, main tail and coverts, breast and lower thighs yellowish brown penciled with dark brown. Axial feather white or creamy white
	Campbell	Khaki	Drake	Back, wings, body and breast (except lower back and tail coverts) and main tail warm khaki, rest brown bronze
			Duck	Breast, back, wings, body, tail and lower part of neck khaki; rest seal brown
	Magpie	Black and white	Both	Half of the crown, neck, under-tail cushion, body and fluff white; rest black. When viewed from above, a heart-shaped black pattern from juncture of wing bows and body through tail extending over the wing
		Blue and white	Both	Areas of black in Black and white variety replaced by blue

Class	Breed	Variety	Sex	Plumage
Bantam	Call	Gray	Drake	Back, tail, upper parts of the body and wings (excepting upper secondaries which are slaty blue and coverts which are slate with a band of white and black) grey; head and neck lustrous green; breast purplish brown or claret.
			Duck	Head and breast dark brown with the latter penciled with light brown; rest brown to light brown
		White	Both	Web, quill and fluff of feathers in all sections pure white
		Blue	Both	Uniform bluish slate, with each feather of the back and upper tail cushion (especially in drakes) laced with a darker shade. Drakes typically are one to two shades darker than Ducks; front part of breast pure white forming a heart-shaped bib
		Snowy	Drake	Neck iridescent green with white collar, back stippled grey in the front and black with white lacing at the back, primaries and secondaries grayish brown with outer webs of the latter having a band of dark brownish black overlaid with iridescent violet forming the speculum bordered at the front and back by a white bar; breast claret laced with white, tail black bordered with white, underbody and abdomen white.
			Duck	Neck fawn stippled with light brown, back stippled grey in the front and black with white lacing at the back, primaries and secondaries grayish brown with outer webs of the latter having a band of dark brownish black overlaid with iridescent violet forming the speculum bordered at the front and back by a white bar; breast light fawn, tail brownish grey bordered with white, underbody and abdomen light fawn shading to white extremities

Class	Breed	Variety	Sex	Plumage
		Buff	Drake	Head and neck rich fawn buff to seal brown; rest uniform rich fawn buff
			Duck	Uniform fawn buff
		Pastel	Drake	Neck, back, tail and wings varying shades of blue overlaid with buff and gold producing soft pastel colors with a rose or mauve sheen; breast purplish brown changing to pastel rose or ililac and blending into abdomen
			Duck	In general, blue overlaid with buff producing soft pastel colors with some suggestion of rose.; breast soft golden brown changing to light rose and blending into abdomen
	East Indies	Black	Both	Lustrous greenish black
	Mallard	Gray	Drake	Most parts different shades of grey; breast purplish brown, neck rich lustrous green, lower part of back and rump rich greenish black
			Duck	Entire plumage different shades of brown with yellowish tinge at the neck, back, tail, breast and lower part and sides of the body.
		Snowy	Drake	Neck iridescent green with white collar, back stippled grey in the front and black with white lacing at the back, primaries and secondaries grayish brown with outer webs of the latter having a band of dark brownish black overlaid with iridescent violet forming the speculum bordered at the front and back by a white bar; breast claret laced with white, tail black bordered with white, underbody and abdomen white.
			Duck	Neck fawn stippled with light brown, back stippled grey in the front and black with white lacing at the back, primaries and secondaries grayish brown with outer webs of the latter having a band of dark brownish black overlaid with iridescent violet forming the speculum bordered at the front and back by a white bar; breast light fawn, tail brownish grey bordered with white, underbody and abdomen light fawn shading to white

3. Breeds and varieties of Geese

Sebastopol is the oldest domesticated breed. Heavy breeds of geese will have broad, flat back, deep, full rounded breast, and long body. Sebastopol, Chinese, Egyptian and Canada geese are ornamental in addition to being productive; the latter is still semi-domesticated. Tufted Roman geese have a characteristic cylindrical tuft on their crown. Saddleback Pomeranian, on the other hand, is distinguishable by the dark markings on the head.

Pilgrim geese are unique for sex-linked plumage which helps sex differentiation at hatch. Egyptian geese are the smallest ofall the breeds and have a characteristic pugnacious disposition.

3.1 Heavy Class (Gander = 8.2 kg; Goose = 6.8 kg)

Table 5.17 Breeds and varieties of Geese : Heavy Class

Breeds	Variety	Sex	Color of		
			Bill	Eyes	Shanks and feet
Toulouse	Grey	Both	Pale orange with bean light horn	Dark brown or Hazel	Deep reddish orange
	Buff	Both	Light orange	Dark hazel	Orange
Embden	White	Both	Orange	Bright blue	Deep orange
African	Brown	Both	Black with black knob	Dark brown	Dark orange
	White	Both	Orange with orange knob	Light to deep leaden blue	Deep bright orange

3.2 Medium Class (Gander > 5.4 & < 8.2 kg; Goose > 4.5 & < 6.8 kg)

Table 5.18 Breeds and varieties of Geese : Medium Class

Breeds	Variety	Sex	Color of		
			Bill	Eyes	Shanks and feet
Sebastopol	White	Both	Orange	Bright blue	Deep orange
Pilgrim	Sex-linked	Gander	Orange	Bluish grey	Orange
		Goose	Orange	Hazel brown	Orange
African buff	Buff	Both	Light orange	Dark hazel	Orange
Saddleback Pomeranian	Gray	Both	Reddish pink	Blue	Orange red
	Buff	Both	Reddish pink	Blue	Orange red

3.3 Light Class (Gander < 5.4 kg; Goose < 4.5 kg)

Table 5.19 Breeds and varieties of Geese : Light Class

Breeds	Variety	Sex	Color of		
			Bill	Eyes	Shanks and feet
Chinese / China	Brown	Both	Black or dark slate with dark slate knob	Dark brown	Orange
	White	Both	Orange with orange knob	Light blue	Orange – yellow
Tufted Roman	White	Both	Pinkish to reddish orange with white bean	Blue	Orange to pinkish orange
Canada	Eastern / Common	Both	Black	Black	Black
Egyptian	Colored / Brown	Both	Reddish purple	Orange	Reddish yellow

3.4 Plumage patterns of Geese

Table 5.20 Breeds and varieties of Geese : Plumage patterns

Class	Breed	Variety	Sex	Plumage
Heavy	Toulouse	Gray	Both	Different shades of grey; tail grey and white, the ends tipped white
		Buff	Both	Different shades of buff, tail buff and white, feathers tipped pale buff
	Embden	White	Both	Pure white; traces of grey in young ones
	African	Brown	Both	Body fawn to different shades of brown; tail coverts white
		White	Both	Pure white
Medium	Sebastopol	White	Both	Pure white; traces of grey in young ones
	Pilgrim	Sex-linked	Gander	Pure white; traces of grey in young ones
			Goose	Entire body different shades of grey
	American buff	Buff	Both	Different shades of buff with tail buff and white

Class	Breed	Variety	Sex	Plumage
	Saddleback Pomeranian	Grey	Both	Tail, wings, breast, lower neck and body white except a broad band of ashy brown edged with near white beginning under secondaries, just over the shanks and extending under the abdomen to the opposite wing; rest ashy brown to grey.
		Buff	Both	Same as Grey variety with buff replacing grey
Light	Chinese / China	Brown	Both	Dark slate to dark (russet) brown with tail heavily edged with a shade approaching white.
		White	Both	Pure white
	Tufted Roman	White	Both	Pure white; traces of grey in young ones; characteristic tuft on the crown
	Canada	Eastern / Common	Both	Light grey (breast), dark grey (back and wings), black (head and neck), white (under part of the body from legs to tail)
	Egyptian	Colored / brown	Both	Black and grey body with white shoulders (with a narrow black stripe), grey breast with rich reddish brown center and under parts of the body yellowish buff penciled with black lines.

4. Turkeys

Turkeys are primarily raised for meat and most popular varieties are Broad-breasted Bronze, White Holland and Beltsville Small White. Turkey eggs have characteristic brown speckles.

Weight of adult toms recorded is 7.3 to 10.0 kg (Beltsville and Royal Palm), 10.4 to 15.0 kg (Narragansett, Black, Slate and Bourbon) and 11.4 to 16.3 kg (Bronze and White Holland); corresponding values in case of adult hens are 4.5 to 5.4 kg, 6.4 to 8.2 kg, and 7.3 to 9.1 kg, respectively.

4.1 Varieties of Turkeys

Table 5.21 Varieties of Turkeys

Variety	General description
Bronze or Broad-breasted Bronze	Plumage generally copperish bronze; tail feathers terminating in a wide edging of pure white; unexposed feathers at the breast, wings and tail black; shanks and toes dull black changing to smoky pink in adults.
Narragansett	Plumage generally black and exposed surface shades of grey (devoid of bronze); Undercover of all sections very dark slate; shanks and toes salmon in color.
White Holland or Broad White or Large White	Web, fluff and quill of feathers in all sections pure white; shanks and toes pinkish white
Black	Surface plumage is lustrous, metallic black throughout; undercover of all sections dull black; shanks and toes pink in adults and slaty black in young ones
Slate	Slaty blue plumage; shanks and toes pink in adults and deep pink in young ones
Bourbon Red	Toms have rich, dark, chestnut mahogany color with black edging at back, wings and breast; tail pure white; body and fluff shades of brownish red; shanks and toes reddish pink in adults and deep reddish horn in young ones; hens are similar to toms but do not have black edging in any section
Beltsville Small White	Web, fluff and quill of feathers in all sections pure white; shanks and toes pinkish white
Royal Palm	Toms have pre white color edged with metallic black at the neck, tail, coverts and lesser coverts, wings and breast; back is rich, metallic black over the saddle making a sharp contrast with white color of the body plumage; hens are similar to toms but black bands in plumage are less contrasting than toms.
Note : 1. Toms of all the breeds have black beard and head red in color changeable to bluish white 2. Color of eyes will be light to dark brown in Toms and hens of all varieties 3. Throat and wattles are usually red changeable to bluish white in all varieties except in Beltsville Small White (changeable to pinkish white).	

Plate 5.1 New Hampshire - Cock

Plate 5.2 New Hampshire - Hen

Plate 5.3 White Plymouth Rock-Cock

Plate 5.4 White Plymouth Rock-Hen

Plate 5.5 Barred Plymouth Rock - Cock

Plate 5.6 Barred Plymouth Rock - Hen

Plate 5.7 Rhode Island Red - Cock

Plate 5.8 Rhode Island Red - Hen

Plate 5.9 Buff Cochin - Cock

Plate 5.10 Buff Cochin - Hen

Plate 5.11 Dark Cornish - Cock

Plate 5.12 Dark Cornish - Hen

Plate 5.13 White Cornish - Cock

Plate 5.14 White Cornish - Hen

Plate 5.15 Black Australorp - Cock

Plate 5.16 Black Australorp - Hen

Plate 5.17 White Leghorn - Cock

Plate 5.18 White Leghorn - Hen

Plate 5.19 Black Minorca - Cock

Plate 5.20 Black Minorca - Hen

Plate 5.21 Black Araucanas - Cock

Plate 5.22 Black Araucanas - Hen

Plate 5.23 Black-breasted Aseel - Cock

Plate 5.24 Black-breasted Aseel - Hen

Plate 5.25 Red Naked-neck - Cock

Plate 5.26 Red Naked-neck - Hen

Plate 5.27 Black Sumatra - Cock

Plate 5.28 Black Sumatra - Hen

Plate 5.29 White Sultan - Cock

Plate 5.30 White Sultan - Hen

Plate 5. 31 Black Muscovy - Drake

Plate 5.32 Black Muscovy - Duck

Plate 5.33 Khaki Campbell - Drake

Plate 5.34 Khaki Campbell - Duck

Plate 5.35 Fawn & White Runner - Drake

Plate 5.36 Fawn & White Runner - Duck

Plate 5.37 White Sebastopol - Drake

Plate 5.38 White Sebastopol - Duck

Plate 5.39 Pilgrim - Drake

Plate 5.40 Pilgrim - Duck

Plate 5.41 Beltsville Small White - Tom

Plate 5.42 Beltsville Small White - Hen

Plate 5.43 Bronze - Tom

Plate 5.44 Bronze - Hen

Plate 5.45 White Holland - Tom

Plate 5.46 White Holland - Hen

Plate 5.47 Black - Tom

Plate 5.48 Black - Hen

CLASSIFICATION OF BREEDS (BRITISH STANDARDS)

Table 5.22 Breeds of Poultry

Hard feather	Soft feather		True Bantam
	Heavy	Light	
Asil (Rare)	Australorp	Ancona	Belgian
Belgian Game (Rare)	Barnevelder	Appenzeller	Booted (Rare)
Indian Game	Brahma	Araucana	Dutch
Ko-Shamo (Rare)	Cochin	Rumpless Araucana	Japanese
Malay (Rare)	Croad Langshan	Hamburg	Nankin (Rare)
Modern Game	Dorking	Leghorn	Pekin
Nankin-Shamo (Rare)	Faverolles	Minorca	Rosecomb
Old English Game Bantam	Frizzle	Poland	Sebright
Old English Game Carlisle	Marans	Redcap	Tuzo (Hard Feather: Rare)
Old English Game Oxford	Orpington	Scots Dumpy	
Rumpless Game (Rare)	Plymouth Rock	Scots Grey	
Shamo (Rare)	Rhode Island Red	Silkie	
Tuzo (Rare:True Bantam)	Sussex	Welsummer	
Yamato-Gunkei (Rare)	Wyandotte		

Rare			
Hard feather	Soft feather		True Bantam
	Heavy	Light	
Asil	Rhodebar (Autosexing)	Legbar (Autosexing)	Booted
Belgian Game	Wybar (Autosexing)	Welbar (Autosexing)	Nankin
Ko-Shamo Bantam	Crevecoeur	Andalusian	Tuzo (Hard Feather)
Malay	Dominique	Augsberger	
Nankin-Shamo Bantam	German Langshan	Brakel	
Rumpless Game	Houdan	Breda	
Shamo	Ixworth	Campine	
Tuzo (True Bantam)	Jersey Giant	Fayoumi	
Yamato-Gunkei	La Fleche	Friezian	
	Modern Langshan	Italiener	
	New Hampshire Red	Kraienkoppe	
	Norfolk Grey	Lakenvelder	
	North Holland Blue	Marsh Daisy	
	Orloff	Old English Pheasant Fowl	
	Transylvanian Naked Neck	Sicilian Buttercup	
	Turkens	Spanish	
		Slumtaler	
		Sultan	
		Sumatra	
		Vorwerk	
		Yokohama	

Table 5.23 Breeds of Ducks and Geese

Ducks		
Heavy	Light	Bantam
Aylesbury Blue Swedish Cayuga Muscovy Pekin Rouen Rouen Clair Saxony Silver Appleyard	Abacot Ranger Bali Campbell Crested Hook Bill Indian Runner Magpie Buff Orpington Welsh Harlequin	Black East Indian Call Crested Silver Appleyard Miniature Silver Bantam
Geese		
Heavy	Medium	Light
African Embden Toulouse	American Buff Brecon Buff Buff Back Grey Back Pomeranian	Chinese Pilgrim Roman Sebastopol Steinbacher

Table 5.24 Sitters and Non-sitters

Sitters			
Asil Australorp Barnevelder Brahma Cochin Crevecoeur Croad Langshan Dorking Faverolles	Frizzle Houdan Indian Game Ixworth Jersey Giant Jubilee Indian Game La Fleche Malay Marans	Marsh Daisy Modern Game Modern Langshan New Hampshire Red Norfolk Grey North Holland Blue Old English Game Orloff Orpington	Plymouth Rock Rhode Island Red Scots Dumpy Silkie Sultan Sumatra Game Sussex Wyandotte Yokohama
Non-sitters			
Ancona Andalusian Bresse Campine	Hamburg Lakenvelder Leghorn Minorca	Old English Pheasant Fowl Poland Redcap Scots Grey	Sicilian Buttercup Spanish Welsummer

Note : Generally speaking, Sitters and Non-Sitters divide themselves - *heavy* breeds being sitters and the *light* breeds non-sitters - the former comprising mainly American and Asiatic breeds with Indian Game, Sussex and Dorkings of British origin, while the latter are generally of Mediterranean origin. Unlike certain other countries, however, which classify three categories - *light, medium* and *heavy* - Great Britain adheres to two classes only. There are, therefore, certain exceptions to the above generalizations.
Bantams take the same classification, i.e., heavy or light, as their large prototypes.

Chapter **6**

Interpretation of Standards

1. General criteria

1.1. Weight / Size

When size and weight cannot be determined by comparison, it is advisable to actually weigh the specimens and interpret the weights in proportion to size, while at the same time preserving the ideal shape and type for the breed.

1.2. Condition and vigor

The vigor and health of a specimen is of prime importance and a necessity in the propagation and preservation of all breeds. While interpreting condition and vigor, it is necessary not only to consider the external appearance but also to assess the actual shape of the carcass and the fleshing thereof. The fading or bleaching of color from that described the standard for the beak and shanks or the pigment in yellow skin breeds is a defect unless if it is due to natural result of heavy egg production, age or seasonal changes.

In case of Wild Pattern Drakes (Rouens, Gray Calls, Mallards and Gray Runners) allowance shall be made for the normal seasonal plumage change when judging males of these breeds and varieties.

Any specimen with scaly leg sufficiently advanced to have significantly altered the shape or color of the feet or legs shall not be considered for awards. Similarly, any specimen showing evidence of a contagious or communicable disease shall be removed and ineligible to compete for any award.

1.2.1. Shape and type

Shape and type is one of the most important criteria to be considered while judging the specimens; approximately two-thirds of the total value of the "Scale of Points" is allotted to shape, except in Turkeys where the shape value becomes 62%. Specimens greatly deficient in breed type should be disqualified outright.

The terms "broad", "medium", "large", "deep" etc. used in standard descriptions should be understood to mean "relatively or comparatively" broad, medium, large, deep etc., respectively in proportion to the size and character of the breed described as well as to the two sexes within the breed.

The head is of great importance as it indicates the state of health and vitality. In all breeds where high egg production is characteristic, particular emphasis should be placed on the character of the head and eye; the head should be strong, moderately long and well filled in forward of the eyes to avoid any appearance of crow-headedness, with the skull inclined to be somewhat flat on top rather than round. The face should be clean cut, smooth and free from wrinkles, the skin fine-grain and soft in texture, the comb of the substance and size and bright red color. The over-refined, thin type of comb is not only liable to buckle or show thumb marks, but also indicates lack of constitutional vigor. The eye should be large, bright and prominent, the iris rich in color, the pupil distinctly and perfectly formed. The condition of the eye frequently indicates some form of systemic disease, including leucosis. The descriptions must be considered relatively in case of Head also, especially while considering pea-combed breeds such as Cornish, Brahma etc.

Where the back is described as "broad in entire length", the carcass should carry the desired width not only from the shoulders to the hips, but also as near as possible to base of tail; a specimen which narrows sharply from the hips is nearly as faulty as if it were narrow at the hips.

Abdominal capacity should be large enough for adequate intestinal development, which is essential for rapid digestion and assimilation of food necessary to support heavy egg production. The dimension of the abdomen can be measured by placing the thumb on the hip bone and spanning with the hand and fingers the sides of the body to the keel bone in front and back of the legs.

Heart-girth is measured by determining the width of the back and the depth of the body immediately behind the juncture of the wings and body. It is important that this portion of the body is adequate for the proper functioning of the heart, for the full development of the lungs and when of sufficient size, is conducive to health, vigor, longevity and production.

1.2.2. Color

Black or brown color in the quill of the primary or secondary of white varieties should not be construed to include stains on the quill of a feather caused by the coagulation of blood on the underside.

Brassiness on the surface plumage of white fowls is a serious defect. A few very small grayish specks in white fowls shall not prevent a specimen that is otherwise superior in type and color being judged superior to those which are less typical in shape that sound in color provided the grey specks do not appear prominently in the primary, secondary or main tail feathers.

Undercolor is also a characteristic of each color pattern; it should be considered but not granted undue emphasis.

1.2.3. Feather quality

Feather must be relatively broad, the web be of good firm texture, with a strong shaft, the barbs, barbules and barbicels closely and tightly knitted together, with the exception of a very few ornamental breeds such as Silkies, Frizzles and Sebastopol geese. Early and full feathering is associated with a good relatively broad feather of firm structure and hence full consideration should be given for this important quality. Narrow thin feathers inclined to silkiness often found in the back wings and tail coverts, are particularly undesirable.

Feathers should be broad, long, soft and fitting loosely on the body with relatively long and abundant fluff in case of Cochin. More compact feathers without fluffiness are desirable in Brahmas. Moderately broad and long feathers fitting fairly close to the body is expected in all American, English (except Cornish), French, Mediterranean, Hamburgs, Polish and Langshans. All Games, Game Bantams, Malays and Malays Bantams and Orientals should have sharp feathers, rather narrow, hard and firm with as little fluff as possible fitting very close to the body. Feathers in case of Cornish should be close fitting and hard, as in the Modern Game, but wider.

2. Faking

Faking is a deliberate attempt to deceive the judge or a prospective purchaser and it shall be a reason to disqualify that bird. In breeds where vulture hocks are a disqualification, hocks giving evidence of being plucked of long stiff feathers which form a vulture hock, is a disqualification.

Evidence of any of the following practices shall constitute faking:

1. Broken or crimped quills in sickle or main tail or saddle feathers sufficient to indicate an attempt to change the general shape of tail
2. Shaft and web giving evidence of part of feather having been cut off
3. Web of feather giving evidence of being singed
4. Scar tissue giving evidence of cutting or trimming any part of comb or earlobes of any fowl, except cocks and cockerels of Modern and Old English Game breeds, large and bantam
5. Color of plumage or condition of skin showing evidence of the use of any coloring matter in softening, deepening, intensifying or otherwise changing the natural color
6. Brittleness of feather or inflamed condition of skin in white varieties showing evidence of bleaching
7. Beak, shanks or toes showing evidence of the use of any coloring matter to change color, or the use of any device to remove foreign color
8. Face or earlobes showing evidence of the use of any coloring matter to change the color, or the use of any device to remove white or red therefrom.

3. Cutting for defects

The deduction of points for defects in any section determines the importance to be attached to such defects in relation to the total value of the section as set forth in the "Scale of points". The specific cut for any defect shall be assessed according to the severity of the defect, always within the maximum and minimum limits set forth and in any case the total not to exceed the point value of the section. In addition, individual breed or variety defects, other than those specified, shall be considered in comparable fashion to other defects; the minimum cut for any such defect shall be one fourth of 1 point and not more than the value of the section.

3.1. Weights

3.1.1. Under Standard :

a) All breeds of large chicken standard weights of Cock 2.70 kg, Cockerel 2.25 kg, Hen 2.02 kg and Pullet 1.80 kg, or weights below these :
   1. Males – two points for first 450 g and two points for each 225 g thereafter *
   2. Females – one point for each 112 g of underweight *

b) All other breeds of large chickens : two points each 450 g *

c) All turkeys, except Beltsville Small White :
   1. no cut for first 900 g
   2. two points each 450 g thereafter

d) All Bantams, one point for each 56.7 g underweight

e) All waterfowl, except Call, East Indie and Mallard ducks : same as for large chickens

f) Call, East Indie and Mallard ducks: same as for of Bantam chickens.

3.1.2. Over Standard :

a) American, English and Mediterranean Breeds :
   1. no cut for first 450 g
   2. two points for each additional 450 g *

b) All the breeds of large fowl having weight requirements :
   1. no cut for first 450 g
   2. two points for each additional 450 g *

c) Bantams :
   1. ½ point for each 28.35 g *

d) Turkeys :
   1. All varieties except Beltsville Small White :
      i. no cut for first 450 g

ii. two points for each additional 450 g

2. Beltsville Small White :
   i. Male – two points for each 450 g *
   ii. Female – two points for each 450 g *

e) Waterfowl, except Call and East Indie Ducks
   1. Ducks
      i. no cut for first 450 g
      ii. two points for each additional 450 g *
   2. Geese
      i. no cut for first 900 g
      ii. two points for each additional 450 g *

(Note: All specimens shall receive the cut for any fraction over the first under 112 g of each 450 g of overweight.

* Until disqualifying weight is reached (450 g = 1 lb, 28.35 g = 1 oz)

3.2. Breed shape

3.2.1. Comb :

a)	All Breeds	
	1. Coarse texture, oversize or undersize	½ to 1
	2. Frozen	½ to 1
b)	Rose Comb Breeds	
	1. Rough	½ to 2
	2. Irregular	½ to 2
	3. Hollow center	½ to 2
	4. Blade-shaped spike	2
	5. Improper angle of spike	1
c)	Single Comb Breeds	
	1. Points, more or less than required, each point	½
	2. Twisted	½ to 1
	3. Thumb mark	1 to 2

3.2.2. Crest :

Divided or split or twisted feathers	2

3.2.3. Eyes :

1. One eye destroyed leaving only socket	2
2. Eye showing permanent injury, but retaining form	½ to 1

3. Eyes not matching in color 2
4. Blind in one eye 2

3.2.4. Wattles :

Coarse and wrinkled ½ to 1

3.2.5. Neck (All Breeds where it is not a normal breed characteristic) :

Dewlap ½ to 1

3.2.6. Wings :

a) All breeds of fowl
   1. Less than ¾ developed 1
   2. Less than ½ developed 2
   3. Less than ¼ developed 3
   4. Primaries or secondaries missing feather or part of feather,where foreign color disqualifies - each feather 1 to 2
   5. Primaries of secondaries feather broken but not removed, where foreign color disqualifies - each feather ½ to 1
   6. Primaries of secondaries feather broken but not removed, where foreign color does not disqualify, in Buff or Parti-colored varieties - each feather ½ to 1

3.2.7. Tail :

1. Less than ¾ developed 1
2. Less than ½ developed 2
3. Less than ¼ developed 3
4. Main tail feather, absence of, where foreign color disqualifies each feather 1
5. Main tail feather, absence of, where foreign color does not disqualify each feather ½
6. Sickle, absence of, where foreign color disqualifies each feather 1
7. Sickle, absence of, where foreign color does not disqualify each feather ½
8. Pinched or "Gamy" in all breeds of chicken except Modern Games, Cornish and Malays ½ to 2
9. Split tail, the cockerel or pullet (Disqualification in Cock or Hen) 2 to 2 ½

3.2.8. Back (Any breed where Cushion is not feature of normal type)

1. Definite Cushion — 1 to 2

3.2.9. Breast :

a) All Breeds : Crooked keel or breastbone (disqualification in Turkeys and Cornish) — 2 to 5

b) Turkeys: Knob on front of keel or breastbone — 1 to 2

3.2.10. Body and Fluff (Orpington) :

Fluff within 5 cm or less of ground — 1 to 2

3.2.11. Shanks and Toes :

(Feather-leg breeds not to be penalized for a rudimentary outer toe)

a) All breeds of fowl

1. Crooked toe - each toe — ½ to 2
2. Horny, well-defined spur on females — ½
3. Missing toe nail — 1
4. Missing part of a toe, each toe — 1

b) All breeds of chicken except Sumatras - each extra spur — ½

c) Langshans, Faverolles - each feathered middle toe — ½ to 1 ½

d) Brahmas, Silkies, Booted, d'Uccles - each bare middle toe — 2

3.3. Breed of variety color

3.3.1. Bill :

All varieties of White ducks - Black in bean or bill of females — ½ to 3
(Disqualification in drakes)

3.3.2. Head - Face :

a) All Breeds where color is specified, except White-faced Black Spanish Positive enamel white in face of cock or hen — 2
(disqualification in cockerel or pullet)

b) Black in face where red is specified — 2

c) White-faced Black Spanish - Red markings directly above eyes — 1

3.3.3. Head - Eyes :

a) All Breeds - Color of iris foreign to variety, each eye — ½ to 2

b) Campines - Red iris, each eye — 1

3.3.4. Tail :(All Parti-colored varieties)

excessive white fluff at base of tail — 1

3.3.5. Shanks and toes :

a) All yellow-legged breeds, except Anconas
Dark spots on shanks, feet or toes
(no cut for slight tinge of reddish pigment) ¼ to 2

b) All white or pinkish white-legged breeds, except Houdans
Dark spots on shanks, feet or toes ½ to 2

c) White toe nail in black-legged varieties 1

3.3.6. Plumage :

a) All Breeds where luster or sheen is required – lack of, in section ¼ to 1

b) Barred Rock
1. Black feathers – each section ¼ to 2
2. Brassiness – each section 1 to 2
3. Foreign cast on surface – each section ½ to 2
4. Irregular barring – each section ¼ to 2
5. Metallic cast – each section ½ to 2

c) Black or Parti-black breeds and varieties
1. Purple barring – each section ½ to 2
2. Bronze – each section ½ to 2
(except where desired in Bronze turkeys)

d) Brown Leghorns : Grey or white in any except disqualifying sections – each section ½ to limit of section

e) Buff (solid) varieties :
1. Mealiness – each section ½ to 2
2. Light shafting – each section ½ to 1 ½
3. Black or White – each section ½ to limit of section
4. Unevenness of color ½ to 1 ½ in each section where noted
5. Slate in undercover – each section ½ to 2

f) Buff Brahma and other Buff Parti-colored breeds or varieties:
White in any except disqualifying sections – each section ½ to limit

g) Catalanas : White, except on surface where it is disqualification – each section ½ to limit

h) Laced varieties :

1.	Irregular lacing – each section	¼ to 1 ½
2.	Incomplete lacing– each section	¼ to 1 ½
3.	Too heavy lacing– each section	¼ to 1 ½
4.	Indistinct lacing– each section	¼ to 1 ½
5.	Frosting – each section	½ to 1 ½
6.	Messy centres – each section	½ to 2 ½

i) Partridge varieties : Grey or White in any except disqualifying section – each section ½ to limit of section

j) Penciled varieties : Irregular or deficient penciling – each section ¼ to 2

k) Red breeds or varieties :

1.	Bronze – each section	½ to 2
2.	Brassiness – each section	½ to 2
3.	Mealiness – each section	½ to 2
4.	Light shafting – each section	½ to 1 ½
5.	Slate in undercolor – each section	¼ to 1 ½

l) Spangled varieties ; Frosting – each section ¼ to 1 ½

m) White breeds and varieties :

1.	Brassiness – each section	1 to 2
2.	Creaminess in shaft or web (except where creamy white is specified) – each section	¼ to 1
3.	Grey specks – each section	¼ to 2

4. Disqualifications

If any of the following conditions are found the specimen becomes disqualified outright unless there is (are) doubtful evidence(s).

4.1. General disqualifications

4.1.1. All breeds and varieties :

1. Specimen lacking in breed characteristics
2. Definite indications of contagious or communicable diseases
3. Evidence of faking

4.1.2. Weights

Any bird that deviates more than 20% either up or down from the weight listed for its breed, sex and age should be disqualified. This rule applies to all large fowl, bantams, geese and turkeys (except Beltsville Small White Turkeys).

4.1.3. Shape

4.1.3.1. Comb :

a) All breeds chickens – Comb foreign to breed and variety

b) Rose-comb breeds

   1. Over-large, sufficient to obstruct sight
   2. Lopped, sufficient to obstruct sight
   3. Spike, absence of
   4. Spike, more than one
   5. Spike, inverted or telescoped

c) Pea-combed breeds

   1. Lopped below the horizontal plane of top of skull
   2. Telescoped

d) Single-comb breeds

   1. Lopped below the horizontal plane of the point where the bend occurs in all males and in females except where that lopped comb is an identified breed characteristic (No single-combed female in or near production with slightly lopped comb should be discriminated against)
   2. Split
   3. Side sprig

e) Cushion-combed breed – a spike or spikes

4.1.3.2. Beak :

All breeds of chicken and turkeys – deformed

4.1.3.3. Bill :

All breeds of ducks and geese – scoop or deformed

4.1.3.4. Eyes (All breeds)

a) Blind in both

b) Irregular pupil indicating leucosis

4.1.3.5. Crest :

a) Crested breeds

   1. Absence of crest
   2. Severely lopped crest – Two / three or more feathers falling on one side of the center line

b) Non-crested breeds – presence of crest

4.1.3.6. Beard and muffs:

1. Bearded varieties – absence of beard and muffs

2. Non-bearded varieties – presence of beard

4.1.3.7. Neck (All turkeys)
Pendulous crop

4.1.3.8. Wings :

a) All breeds
   1. Primaries or secondaries, twisted feather, except in Sebastopol geese and Frizzles
   2. Split wing
   3. Slipped wing
   4. One or more reversed main wing feathers
   5. One or more primary or secondary feathers with a split quill, except Silkies

c) All breeds chickens and turkeys – primary or secondary feathers clipped

d) All breeds waterfowl except Canada and Egyptian geese; and Muscovy, East Indie, call and Mallard ducks – primary or secondary feathers clipped
(Pinioned wings if smoothly healed are acceptable)

e) All breeds ducks and geese – Angel wings (the wing tip being inverted at the outer joint), one or both

4.1.3.9. Back :
All breeds – Crooked, roached or deformed (not to be confused with required convex Malay back or Rouen duck back or the heavy hip muscles in Cornish males)

4.1.3.10. Tail (All breeds) :

a) Main tail feathers
   1. Complete absence of all (except in Araucanas)
   2. Twisted feather, except in Sebastopol geese and Frizzles
   3. One or more reversed main feathers

b) Sickles – Twisted feather, except in Frizzles

c) Split tail in cock and hen

d) Squirrel tail, except in Japanese bantams

e) Wry tail

f) Rumpless except where it is the breed characteristic

4.1.3.11. Body :

a) All breeds – Tumor affecting health or market value of a fowl

b) All Turkeys – Crooked keel or breastbone

c) All Cornish and Aseels – Crooked keel or breastbone

(Note : Effect of full reproductive activity of the female on the shape or body outline should be considered)

4.1.3.12. Shanks and toes :

a) All breeds of fowl

1. Bow leg
2. Deformed foot or foot joints
3. Duck foot in land fowl
4. Enlarged and misshapen shank or hock
5. Knock knee
6. Web foot in land fowl

b) Four-toe breeds – more or less than four toes

c) Five-toe breeds – more or less than five toes

d) Old English, Dutch (large and bantam) and Sumatra cock – Absence of spurs

e) Clean-legged breeds : presence of, or unmistakable evidence of removal of any down, stub, feather or part of feather from shank below the hock joint, or foot or toe

f) Feather-legged breeds : Shanks not feathered down outer sides

4.1.3.13. Plumage :

All breeds except Sebrights and Campines – Hen-feathering in males

4.1.4. Color

4.1.4.1. Face :

All Mediterranean cockerels and pullets (except White-faced Black Spanish) – Positive enamel white

4.1.4.2. Ear-lobes :

All breeds where a red earlobe is specified - Positive enamel white

4.1.4.3. Bill :

All varieties of White ducks – Black in bill or bean of drakes

4.1.4.4. Shanks and toes :

All breeds – Foreign color to that described for breed and variety (Slight reddish tinge of pigment in yellow-shanked varieties not to disqualify)

4.1.4.5. Plumage :

a) All Barred, Black and Mottled varieties – Red or Yellow

b) All White varieties

1. Black in quills of primaries and secondaries
2. Foreign color, except slight grey tickling

c) All Columbian varieties including Light Brahmas – One or more solid black or brown feathers on surface of back; dark spots or mossiness in surface of back or saddle appearing in approximately 15% or more of feathers in this section, except narrow black stripes extending not over half the length of feather in saddle and near tail of male, or dark markings in cape of either sex; red feathers in plumage

5. Description of common plumage patterns

5.1. White

5.1.1. Both male and female

Web, fluff and shafts of all feathers in all sections, white

5.2. Black

5.2.1. Both male and female

Surface lustrous greenish black; undercolor (except where otherwise specified) dull black in dark-legged varieties; slate in yellow-legged varieties

5.3. Buff

5.3.1. Both male and female

Surface throughout – an even shade of golden buff; male – head, neck, hackle, back, wing-bows and saddle showing greater luster; females – some luster in hackle; undercolor – matching surface as near as possible

5.4. Silver penciled

5.4.1. Male

Head - silvery white. Neck : hackle – web of feather, lustrous greenish black with a narrow lacing of medium shade of rich, brilliant red; shaft black; front of neck – black. Back : back including saddle – web of feather lustrous greenish black with narrow lacing of silvery white, a slight shafting of silvery white permissible. Silvery white predominating on surface of upper back; saddle matching with hackle in color. Tail : main tail – web black, main tail and lesser sickles – lustrous greenish black, coverts – lustrous greenish black laced with narrow lacing of white. Wings : fronts – black, bows – silvery white, coverts – lustrous greenish black forming a distinct wing bar of this color across entire wing when folded, primaries – black with narrow edging of white on lower edge of lower webs, secondaries – lower webs black with lower half white to a point near end of feathers, terminating abruptly leaving ends of feathers black, upper webs – black, secondaries when folded forming a triangular white wing wing-bay between the wing bar and tips of secondary feathers. Breast : lustrous greenish black. Body: body – black. Fluff – black sometimes slightly tinged with grey. Legs and toes : lower thighs – black; shank and toe plumage color varies with varieties. Undercolor of all sections : slate shading lighter towards base of feathers.

5.4.2. Female

Note : Penciling in all Silver varieties should be distinct in sharp contrast to the ground color, be regular in shape, uniform in width and conform to the contour of the feather.

Each feather in the back, breast, body, wing bows and thighs should have three are more pencilings.

Head – silvery grey. Neck : hackle – black, slightly penciled with steel grey, and laced with silvery white; front of neck – penciled same as breast. Back : steel grey, with distinct black pencilings. Tail : main tail – black, except two top feathers which have black lower web and grey penciled black upper web; covers – steel grey with black pencilings. Wings : fronts, bows and coverts – steel grey with distinct black pencilings; primaries – black with steel grey diagonal pencilings on lower webs; secondaries – lower webs steel grey black pencilings extending well around tips of feathers; balance of upper webs black. Breast : steel grey with distinct black pencilings. Body: body – steel grey with distinct black pencilings. Fluff – steel grey penciled with dull black. Legs and toes : lower thighs - steel grey with distinct black pencilings; shanks and toes – varies with breed and variety. Undercolor all sections : medium slate

5.5. Partridge

5.5.1. Male

Head : web of feather lustrous rich red. Neck : hackle – web of feather lustrous greenish black with a narrow lacing of medium shade of rich brilliant red, black shaft; front of neck – black. Back : back including saddle – web of feather lustrous greenish black with narrow lacing of a medium shade of rich brilliant red, a slight shafting of rich red permissible; surface of upper back – rich brilliant red predominating; saddle – matching with hackle in color. Tail : main tail – black web; main and lesser sickles – lustrous greenish black; coverts – lustrous greenish black forming a distinct wind bar of this color across entire wing when folded; primaries – black with narrow edging of reddish bay on lower webs; secondaries – lower webs black with lower half reddish bay terminating with black at end of each feather, black upper webs, when folded form a triangular reddish bay wing-bay between the wing bar and tips of secondary feathers. Breast : lustrous greenish black. Body : black. Fluff : black slightly tinged with red. Legs and toes : lower thighs – black shanks and toes (may vary with different breeds). Undercover all sections : slate

5.5.2. Female

Note : Pencilings in all Partridge varieties should be distinct in sharp contrast to the ground color, be regular in shape, uniform in width and conform to the contour of the feather.

Each feather in the back, breast, body, wing bows and thighs should have three are more pencilings.

Head : deep reddish bay. Neck : hackle – black, slightly penciled with deep reddish bay and laced with reddish bay; front of neck – penciled same as breast. Back : deep reddish bay, with distinct black pencilings. Tail : main tail – black, except two top feathers which have black lower web and black penciled deep reddish bay upper web; coverts – deep reddish bay with distinct black pencilings. Wings : fronts, bows and coverts – deep reddish bay with distinct black pencilings; primaries – black with deep reddish bay diagonal pencilings on lower webs; secondaries – lower webs deep reddish bay with black pencilings extending well around tips of feathers; balance of upper webs black. Breast : deep reddish bay with distinct black pencilings. Body: body – deep reddish bay with distinct black pencilings. Fluff – deep reddish bay penciled with dull black. Legs and toes : lower thighs - deep reddish bay with black pencilings; shanks and toes – varies with breed and variety. Undercolor all sections : slate

5.6. Colombian

5.6.1. Male

Head - silvery white. Neck : hackle – web of feather, lustrous greenish black with a narrow lacing of silvery white, greater portion of shaft black; front of neck – white. Back : silvery white cape – black-and-white, saddle –silvery white with an elongated V-shaped black stripe increasing in width, length and density as it nears the tail coverts (the stripe should extend from near the tip of feather approximately one half to three-fourths the length of the web and allow a clean break of white between the undercolor and base of stripe. Tail : main tail – black, main tail and lesser sickles – lustrous greenish black, coverts – lustrous greenish black laced with silvery white. Wings : fronts – white, some black permissible, bows and coverts– silvery white, primaries – black with white on lower edge of lower webs, secondaries – unexposed portion of lower webs black, exposed portion white upper webs – black edged with white, the white extending around end of feather; the white ends of the upper secondaries growing progressively wider so the exposed portion in the upper row is entirely white; forming a white wing bay when wing is folded. Breast : white. Body: body – while, except under wings where it may be bluish slate. Fluff – white. Legs and toes : lower thighs – white; shank and toe plumage color varies with varieties. Undercolor of all sections : light bluish slate.

5.6.2. Female

Head : white. Neck : hackle – web of feather lustrous greenish black with a narrow lacing of silvery white, greater portion of shaft black. Back : white. Tail : main tail – black, except two top feathers which are slightly laced with white; coverts – black with narrow lacing of white. Wings : fronts, bows and coverts – white; primaries – black with white edging on lower webs; secondaries – lower webs lower portion white extending around ends and lacing upper portion of upper webs growing wider In shorter secondaries, sufficient to show a white wing bay when wing is folded; upper webs black. Breast : white. Body: white except under wings where it may be bluish slate. Fluff – white. Legs and toes : lower thighs - white; shanks and toes – vary with breed and variety. Undercolor all sections : white bluish slate

5.7. Blue

5.7.1. Male

Head : glossy black. Neck : hackle – an even shade of clear bluish slate, distinctly laced with glossy black; front of neck – same as breast. Back : back and saddle – an even shade of bluish slate, each feather distinctly laced with glossy black. Tail : main tail, lesser sickles and coverts – an even shade of bluish slate, each feather distinctly laced with glossy black. Wings : fronts and bows - even shade of clear bluish slate, each feather distinctly laced with glossy black; coverts - even shade of bluish slate, each feather having a sharply defined lacing of black; secondaries – lower webs even shade of clear bluish slate, upper webs clear bluish slate, each feather distinctly laced with black. Breast : even shade of clear bluish slate, each feather having a sharply defined lacing of black. Body : even shade of clear bluish slate, each feather having a sharply defined lacing of black. Fluff : clear bluish slate, laced with black. Legs : lower thighs - even shade of clear bluish slate, each feather having a sharply defined lacing of black. Undercolor ofall sections : bluish slate

5.7.2. Female

Head : bluish slate. Neck : hackle – an even shade of clear bluish slate, distinctly laced with glossy black; front of neck – same as breast. Back : back and saddle – an even shade of bluish slate, each feather distinctly laced with glossy black. Tail : main tail and coverts – an even shade of bluish slate, each feather distinctly laced with glossy black. Wings : fronts, bows and coverts - even shade of clear bluish slate, each feather having a sharply defined lacing of black; primaries - even shade of clear bluish slate; secondaries – an even shade of clear bluish slate, each feather distinctly laced with black. Breast : an even shade of clear bluish slate, each feather distinctly laced with black. Body : an even shade of clear bluish slate, each feather distinctly laced with black. Fluff : an even shade of clear bluish slate, laced with black. Legs : lower thighs - clear bluish slate, each feather distinctly laced with black. Undercolor of all sections : bluish slate

Note : 1. All blue varieties are considered parti-colored except self-blue 2. Blue fowl, actually of a bluish slate color genetically are black fowl in which the black pigment granules are modified in shape and distribution of surface of the feather, creating a dilution of black and causing the characteristic bluish slate color. This condition is the hybrid expression of two hereditary color factors, black and a form of white (usually with some splashing), neither of which is dominant over the other, but which are blending in character. Blue to blue will produce offspring one-half blue, the other half evenly divided in black and splashed whites; and blue to black, and blue to splashed will produce the parent types equally, while black to splashed will produce all blues.

5.8. Mottled

Note : In all sections, the white tip is to be V-shaped. Description is worded for varieties that are mottled white on a black base color. For mottled varieties of other base color, substitute that color for black in each section

5.8.1. Male

Head : lustrous greenish black, one in three tipped with white. Neck : hackle - lustrous greenish black, one in two ending with a small, sharply defined white tip; front of neck – same as breast. Back : lustrous greenish black, one in five tipped with white; saddle - lustrous greenish black, one in two ending with a small, sharply defined white tip. Tail : main tail - lustrous greenish black, ending with white tip; main and lesser sickles and coverts - lustrous greenish black, ending with white tip. Wings : fronts and bows - lustrous greenish black, one in three tipped with white; coverts - lustrous greenish black, ending with a small, sharply defined white tip; primaries and secondaries – black ending with white tip. Breast : lustrous greenish black, one in two with a white tip. Body : lustrous greenish black, one in two with a white tip. Fluff : black, slightly tinged with white. Legs : lower thighs – black slightly tipped with white. Undercolor of all sections : Dark slate

5.8.2. Female

Head : black, one in three tipped with white. Neck : hackle - lustrous greenish black, one in two tipped with white; front of neck – same as breast. Back : lustrous greenish black, one in two tipped with white. Tail : main tail - lustrous greenish black, ending with white tip; coverts - lustrous greenish black, one in three ending with white tip. Wings : fronts and bows - lustrous greenish black, one in two ending with a small sharply defined white tip; coverts - lustrous greenish black, one in two ending with a white tip; primaries and secondaries – black ending with white tip. Breast : lustrous greenish black, one in two with a small, sharply defined, white tip. Body : black, one in two ending with a small, sharply defined, white tip. Fluff : black, slightly tinged with white. Legs : lower thighs – black slightly tipped with white. Undercolor of all sections : Dark slate

6. Scale of points

6.1. Chicken

All breeds of chickens, except Crested and Bearded varieties, Modern Games and Malays, Hamburg and Rose Comb Bantams, and Japanese Bantams, the following scale of points can be employed :

Table 6.1 Scale of points : Chicken

Criterion	Total	White varieties		Other than White varieties	
		Shape	Color	Shape	Color
Symmetry	4	4		4	
Weight/Size	4	4		4	
Condition and vigor	10	10		10	
Comb	5	5		5	
Beak	3	2	1	2	1
Skull and Face	4	3	1	3	1
Eyes	4	2	2	2	2
Wattles	2	2		2	
Earlobes	4	2	2	2	2
Neck	6	3	3	1	5
Back	12	8	4	6	6
Tail	8	5	3	4	4
Wings	8	5	3	3	5
Breast	10	7	3	5	5
Body and Fluff	8	6	2	5	3
Legs and toes	8	5	3	5	3
Total	100	73	27	63	37

1. All Crested or Bearded, or both – Crest shape 8, Color 4; Beard shape 3, Color 1
2. Modern Game and Malay, Aseels and Shamos – Station 10 points
3. Japanese Bantam – Accentuated breed type 10 points
4. Hamburg and Rose Comb Bantams – substitute Comb shape 8; earlobes shape 4, Color 3
5. Buttercup and La Fleche – substitute Comb shape 10
6. Yokohamas – Tail shape 7, Color 3

In order to allow extra point value for these distinctive breed features, it is necessary to deduct proportionately from all sections in the scale of points, so that the total allowed will not exceed 100 points. For example, to allow 12 points for crest (shape 8, Color 4) deduct 1-point each from shape of Comb, wattles, earlobes, back, tail, wings, breast, body and fluff (total 8) and 1- point each from color of earlobes, neck, back and breast (total 4) making a grand total of 12.

6.2. Ducks

Table 6.2 Scale of points : Ducks

Criterion	Total	White varieties		Other than White varieties	
		Shape	Color	Shape	Color
Symmetry	4	4		4	
Weight	4	4		4	
Condition and vigor	10	10		10	
Bill	6	3	3	3	3
Eyes	4	2	2	2	2
Head	6	4	2	4	2
Neck	6	3	3	3	3
Back	12	8	4	6	6
Tail	4	2	2	2	2
Wings	8	5	3	4	4
Breast	16	12	4	10	6
Body	16	12	4	10	6
Legs and feet	4	2	2	2	2
Total	100	71	29	64	36

1. In Crested : deduct 3 points each from shape of breast and body, 2 points each from shape of head, back and wings; making a total of 12 points for shape of crest.
2. In Indian Runners : deduct 2 points from weight, 2 points each from shape of back, wings, breast and body, making a grand total of 10 to Carriage.
3. The Colored Runners : deduct 2 points from weight, two points from Shape of wings, 2 points each from Color of back, breast and body, making a grand total of 10 to Carriage.
4. The Magpie : emphasis for Color evaluation is placed on the head and back sections. Therefore, deduct 2 points each from weight, color of body and breast, 1 point each from color of bill, eyes, neck and legs, making a total of 8 points for head color and 10 points for color of back.

6.3. Geese

Table 6.3 Scale of points : Geese

Criterion	Total	White varieties		Other than White varieties	
		Shape	Color	Shape	Color
Symmetry	4	4		4	
Weight	4	4		4	
Condition and vigor	10	10		10	
Bill	6	4	2	4	2
Eyes	4	2	2	2	2
Head	8	6	2	4	4
Back	10	7	3	6	4
Tail	4	3	1	1	3
Wings	10	6	4	4	6
Breast	14	10	4	9	5
Body	13	10	3	8	5
Legs and feet	3	2	1	2	1
Total	100	75	25	64	36

6.4. Turkeys

Table 6.4 Scale of points : Turkeys

	Total	Shape	Color
Symmetry and Carriage	4	4	
Condition	4	4	
Weight	12	12	
Head	4	2	2
Eyes	4	2	2
Throat -Wattle	4	4	
Neck	5	3	2
Back and Spring ribs	12	6	6
Tail	14	6	8
Wings and shoulders	10	4	6
Breast and keel	11	6	5
Body and fluff	11	6	5
Legs and toes	5	3	2
Total	100	62	38

All illustrations (Fig 6.1 through 6.19)
Are reproduced from
THE AMERICAN STANDARD OF PERFECTION, 1998

Fig 6.1 Crow-head

Fig 6.2 Lopped Rose comb

Fig 6.3 Lopped Single comb

Fig 6.4 Coarse comb

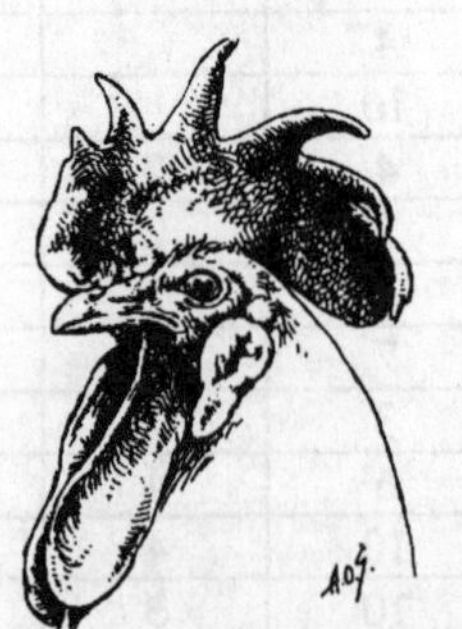
Fig 6.5 Thumb marks on comb

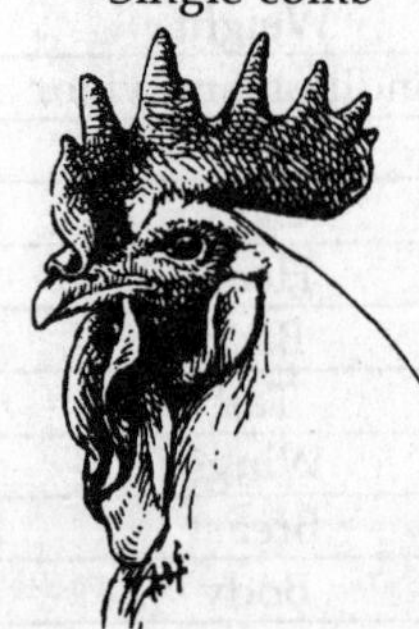
Fig 6.6 Twisted comb

Fig 6.7 Side-strips Fig 6.8 Split comb Fig 6.9 Split wing Fig 6.10 Stripped wing

Fig 6.11 Mossy

Fig 6.12 Frosting

Fig 6.13 Splashed

Fig 6.14 Shafting

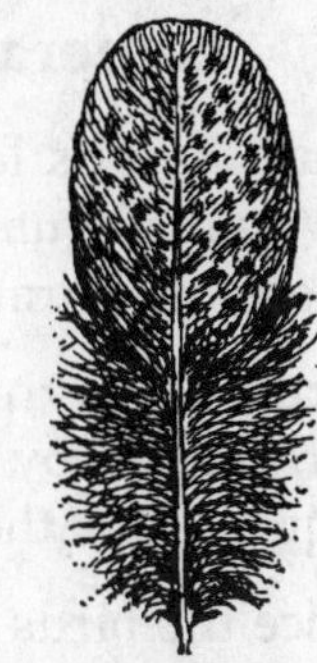

Fig 6.15 Mealy and twisted feathers

Fig 6.11 to 6.15 Feather disqualifications

Fig 6.16 Vulture hocks

Fig 6.17 Wry tail

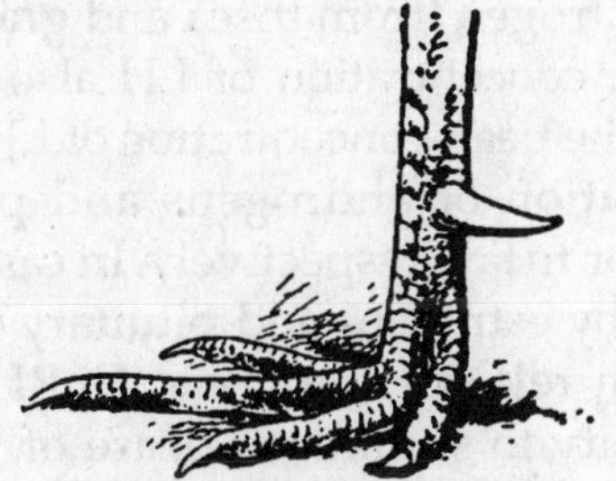

Fig 6.18 Duck foot

Fig 6.19 Scoop bill

Chapter 7

Egg – Formation and Structure

1. General

Class *Aves* is lower in the evolution ladder and therefore nature has provided several safeguards so as to ensure that the species survives and remains propitious. These are summarized below:

Birds are oviparous. They shed ova which are minute reproductive cells surrounded by yolk, albumen, shell membranes and shell; it contains all nutrients required for the complete development of the embryo, if it is fertilized.

Since the birds are lower in the evolution ladder, number of and the frequency of ovulations is higher than that in higher species, like mammals. Being oviparous, each ovum shed gets the package of other nutrients irrespective whether it is eventually fertilized or not.

The birds are uni-polytocus because many eggs are matured and ovulated singly but laid in clutches/sequences (unlike multi-polytocus animals in which many eggs are ovulated simultaneously; Nalbandov, 1970). The modern poultry breeds are capable of laying as many as 300 to 320 eggs per annum.

Within few days after fertilization accumulation of primordial germ cells occurs, usually in left gonad; by 10th day, gonads are differentiated.

2. Growth and sexual maturation (Etches, 1996)

2.1. Hormonal profiles

Of the various hormones, leutinizing hormone (LH), a glycoprotein, is of prime importance. It is required for production of androgens (from Leydig cells) in males and progesterone, androgens and estrogen (from theca and granulosa cells) in females. With increasing body weight, concentration of LH also increases in both male and female. In case of females, the peak concentration of LH is recorded 5 weeks prior to lay whereas concentration of androgens and progesterone increases several weeks and one week prior to lay, respectively. In case of females, LH is suppressed 3 to 5 weeks before lay by estrogens and pituitary becomes less sensitive to stimulation by gonadotrophin releasor hormone (GnRH). At sexual maturity, progesterone has increased ability to stimulate release of LH which is mediated by GnRH. An increased concentration of plasma androgens stimulates comb growth in both sexes.

LH – early peak at 1 week of age; pre-pubertal increase at 15 weeks of age and peak at about 3 weeks before sexual maturity. Ovarian steroidogenesis – begins at 3.5 d of incubation; plasma estradiol < 100pg/ml at 6 weeks before lay, peaks to 350pg/ml 2-3 weeks before lay, returns to basal level of 100-150pg/ml at first egg. Plasma progesterone – up to 1 week before lay a low level of 0.1-0.5ng/ml returns to basal level of 0.4-0.6ng/ml at lay (Johnson, 2000).

During molt plasma concentrations of prolactin, growth hormone, LH and ovarian steroids decrease whereas those of corticosterones, testosterone and triiodothyronine (not thyroxine) increase. Plasma thyroxine increases and it may help thermoregulation because birds lose feathers during molt (Johnson, 2000)

2.2. Secondary sexual characters

The hormones of importance are androgens and estrogen; the former is required for voice and headgear and the latter for female plumage, formation of medullary bone, development and secretory activity of oviduct, production of egg yolk precursors and mobilization of yolk from liver, nesting behavior, appetite and other related functions. Ovariectomized female (poulard) and capons will have plumage similar to males; estrogen, on the other hand, causes feminization of plumage resulting in rounded feathers and hackle and tail. In Campines and Sebright Bantams, males will have female plumage; in these breeds, a gene mutation (codes for excess aromatase production) results in conversion of androgens to estrogen; therefore, castration in these breeds leads to decreased estrogen and absence of androgens, ultimately, producing male plumage.

2.3. Alterations in sexual development

W chromosome imparts a powerful signal to differentiate into a female. Therefore, if W chromosome is present in haploid/diploid or diploid/triploid mosaics, the bird will be feminized. Feminization is because of production of aromatase early in development which results in production of estrogen and, in turn, female phenotype. Conversely, if female embryos are treated with aromatase inhibitors during first few days of incubation, will develop male phenotype and grow into a normal male but infertile because spermatids will be genetically ZW (sperm carrying W is non-functional). Removal of left ovary within first few weeks of hatching will result in a male phenotype with varying testicular development. If number of primordial germ cells is optimum, there will be normal spermatogenesis and the bird will be fertile; else, it will be infertile.

2.4. Feeding Vs reproductive performance

Restricting calorie and/or protein before sexual maturity increases age at sexual maturity and number of ovulations and decreases the number of small eggs; all being beneficial. Broiler breeders, if fed *ad libitum* will mature at 22 weeks of age with a body weight of 5 kg whereas those restricted will mature at 24 weeks with a body weight of 3 kg; the latter will produce more number of settable eggs with a higher fertility and hatchability consuming lower quantity of feed when compared to those fed *ad libidum*. Number of defective eggs also is reduced due to restricted feeding.

Details of other nutritional factors will be discussed separately under feeds and feeding.

3. Photo-periodism (Etches, 1996)

Most birds are seasonal breeders; Birds in Temperate and Northern latitudes breed under lengthening days. For reproduction, perception of light does not depend on photoreceptors in the eye; Photoreceptors in the hypothalamus are the biological transducers that convert photon energy into neural impulses which are amplified by endocrine system; the latter control ovarian and testicular functions, reproductive functions, behavior, secondary sexual characters etc.

The hypothalamic neurons produce GnRH into hypothalamic portal system to be transmitted to pituitary gonadotropes which ultimately release gonadotrophins. In females: LH and follicle stimulating hormone (FSH) are released to circulation and they bind to theca and granulosa cells of the ovarian follicles. Small ovarian follicles release androgens and estrogens and pre-ovulatory follicles release progesterone. In males, GnRH stimulates production of spermatozoa and several androgens (including testosterone)

Therefore, surgically blind birds or genetically blind birds also respond to light-dark cycle. Removal of hypothalamic receptors (tuberal hypothalamus when lesioned) makes the bird refractive to photoperiods.

3.1. Critical day length

Among domestic birds, data is available only on chicken and quails. Increase in LH depends on length of supplemental illumination between 11.5 and 13 hr (in quails) and 10.5 to 15.25 hr of light (in chicken).

3.1.1. Circadian rhythms and light perception

3.1.1.1. Pineal gland and photoperiodism (Gwinner and Han, 2000)

Melatonin in the hormone derived from Trp produced by pineal gland in a 24-hr rhythm. Concentration of melatonin is high at night and low during the day under the control of endogenous circadian oscillator.

Melatonin is also produced from a) retina, b) Harderian gland and c) gastro-intestinal tract; however, the hormone produced by these regions has similar circadian rhythm as that produced by pineal gland but, barely enters circulation and hence, has mainly local action.

Effects of melatonin include a) pineal gland as a circadian pacemaker through periodic production of melatonin (over a 24-hr cycle); eyes also contribute to circadian rhythm *via* periodic melatonin secretion by retina. However, in quail and chicken, pineal and its melatonin rhythm is almost redundant for circadian function.

In birds, light for circadian synchronization is perceived in part by the eyes from which the photic information is transmitted to supra- chiasmatic nucleus (which may be present in hypothalamus) *via* retino-hypothalamic tract. In addition, light is perceived by the pineal gland and by the encephalic photoreceptors which also play a role in synchronization. Melatonin signal does not usually seem to be involved in the regulation of hypothalamo-pituitary-gonadal axis; it may play a role in controlling seasonal phenomena such as bird-song.

Melatonin may cause direct hypothermic effects and reduce metabolic rate thereby facilitating sleep by synchronizing the physiology of birds to the external light-day cycle. Hence, chicken exhibit decreased heat tolerance during scoto-phase but body temperature is unaltered by removal of pineal gland.

3.1.1.2. Photo-sensitive phase

Internal circadian rhythm is used to identify increased light period. The period during which the bird is sensitive to light is referred to as photo-sensitive phase which is defined as the phase during which light has to be available for stimulation. All photo-periods set the circadian clock to 0 once per day at light-on which, in turn, begins to "count" the elapsed time since the photoperiod began. The photo-sensitive phase begins 11 hr after light-on and ends 2 hr later; therefore, minimum photo-period required for stimulation is more than 11 hr. However, only a small proportion of photo-sensitive phase needs to be illuminated to stimulate GnRH secretion and therefore, intermittent lighting regimes are equally effective.

A stimulatory photo-schedule must transmit two signals to the bird – 1. "Dawn" signal that allows to set circadian clock to time 0; then, the clock starts "counting" the passage of time until 11 hr from dawn or lights-on the internal oscillator signals beginning of photosensitive phase. 2. During photo-sensitive phase when exposure to light is registered by hypothalamic photo-receptors; photo-energy is used for release of hormones.

3.2. Light intensity and photo-periods

Birds never live under total darkness – some illumination in the form of stars and moon always give significant amount of light; therefore, sensitivity of birds to light intensity is relative. Ratio of light when the pen/house is illuminated to that when not illuminated will help the bird distinguish day/night. Minimumvalue of the ratio has not yet been worked-out. But, apparently, it is about 10 for the birds to recognize as "day". Effect of light intensity on hormonal levels is uncertain. The relative intensity during "day" and "night" is important for photo-stimulation itself. Rate of egg production is directly proportional to intensity between 0.2 to

5 lux; but unrelated above 5 lux; therefore, threshold of light intensity is 5 lux. 0.4 lux was not distinguished from darkness (scoto-period).

Chicken respond more effectively to orange and red lights (6640 to 7400Å); shorter wavelengths are not effective (Austic and Nesheim, 1990).

It is not known how the birds recognize artificial light in continuation/combination of sunlight or not because sunlight is several hundred times more intense than supplemental light. Hence, dim supplemental light is sufficient to provide the appropriate information regarding contrast that allows the hen to distinguish night from day.

Wild birds begin nest building, mating and egg laying during spring when the day length increases; they stop egg production and mating when the daylight is decreasing (Austic and Nesheim, 1990).

3.3. Photo-refractoriness

Inability to respond to increasing day length which is otherwise stimulatory is referred to as photo-refractoriness. This is often initiated by long days and the number of long days required for initiating photo-refractoriness depends on day-length (Johnson, 2000).

Failure to maintain gonadotrophin levels under long days is photo-refractoriness. In chicken selected for high egg production, there will be gradual decrease in egg production during 12 to 15 months of lay; anterior pituitary cannot produce sufficient gonadotrophins resulting in regression of ovary. Early onset of photo-refractoriness is common in turkeys and frequent in broiler breeder hens.

After photo-refractoriness, birds can't be photo-stimulated until they are exposed to short days for, at least 40-60 d, usually 10 to 12 weeks. Commercially, force-molting can be done by exposure to short days to produce involution of reproductive system, loss and replenishment of feathers and loss of body weight. This may reset the hypothalamus to receive and transmit photo-periodic signals.

3.4. Minimum age for photo-stimulation

Chicks (chicken and turkeys) at hatch are photo-refractory. Exposure to short days for 8 to 12 weeks will make them become photo-sensitive. In males, early sexual maturity is advantageous so that more AI can be made; on the other hand, in females, sexual maturity is delayed to 22/24 weeks to obtain good egg size.

3.5. Photoperiodism and sexual maturity

Spontaneous initiation of sexual maturity is because of decrease in the negative feedback control of steroid hormones on GnRH release. In females, concentration of LH decreases as egg production begins due to negative feedback control of gonadotrophin secretion by estrogen because small ovarian follicles begin to secrete these steroid hormones.

Photo-periodic initiation of sexual maturity is because of direct simulation of hypothalamic photo-receptors thattransmit their signal toGnRH secreting neuron of hypothalamus resulting in increased concentration of LH and FSH.

3.6. Photo-periodic response

This is a summation of photo-inhibitory and photo-stimulatory responses; and each depends upon the previous exposure to light. Long exposure to short days provides no stimulation and is required to dissipate photo-refractoriness. Short exposure to long days, on the other hand, provides a large stimulatory response with increase in concentration of LH and increased egg production. Long exposure to long days creates an inhibitory response leading to photo-refractoriness and dissipates stimulatory effects of long days.

Birds exposed to long days for long period of time when exposed to short days suddenly suffer with a severe inhibition of production of GnRH and finally cessation of reproduction. Hens transferred from 20 hours of light: 4 hours of darkness (20L : 4D) to 12L : 12D will show decreased egg production and reduced concentration of LH whereas those transferred from 4L : 20D to 12L : 12D show increased egg production and concentration of LH.

Commercially, it is recommended to use minimum photo-period to induce onset of egg production and then increase the day length to offset photo-refractoriness.

4. The Hypothalamus, Pituitary and Ovary (Etches, 1996)

4.1. The hypothalamus

Major functions of hypothalamus are 1. Perception of light and 2. Regulation of ovarian function. Secretion of gonadotrophins (LH, FSH) from anterior pituitary depends on hypothalamic peptide hormone, GnRH, the release of which, in turn, is controlled by photoreceptive centers and other centers of the hypothalamus.

Exposure to long days →increase in concentration of GnRH →increase in concentration of gonadotrophins→ovarian growth and hierarchy. Acute secretion of LH through positive feedback with progesterone from the largest ovarian follicle is responsible for pre-ovulatary release of gonadotrophins required for ovulation.

Regions of hypothalamus whose integrity is a must for normal reproduction are a) infundibular nuclear complex (which has photoreceptors), b) pre-optic region (which has GnRH producing cells) and c) supra- optic region (which is required for progesterone-induced GnRH secretion). GnRH is secreted to portal vascular system *en route* pituitary gland. Two types of GnRH (I and II) have been recognized; GnRH-I appears to be involved in tonic and acute release of gonadotrophins from anterior pituitary whereas role of GnRH-II is unknown. GnRH-I and II differ in the amino-acid residue at 5th, 7th and 8th positions; the former has tyrosine (Tyr), leucine (Leu) and glutamine (Gln) whereas the latter has histidine (His), tryptophan (Trp) and Tyr, respectively.

4.2. The anterior pituitary (Adenohypophysis)

This endocrine gland acts as a transducer of the message delivered by GnRH. In mature chicken, pulses of GnRH typically occur at intervals of 1 to 3 hr and each episode lasting for 15 to 60 min; in each pulse, concentration of LH increases up to 300% of the basal value. Concentration of FSH also increases. Both LH and FSH have α and β subunits; the former being same in both the hormones, whereas, the latter determines their specificity. Mammalian gonadotrophins are more stimulatory on birds and *vice versa*.

Levels of circulating FSH are important for establishment and maintenance of pre-ovulatory hierarchy aswell as regulation of follicle atresia.Inhibin, a hormone produced and highest levels by 4 largest pre-ovulatory follicles, decreases the circulating levels of FSH (Johnson, 2000).

Both LH and FSH regulate follicular growth and maintain large yolk-filled follicles. LH is most active simulator of steroid production from the hierarchical follicles and non-hierarchical follicles from which they are recruited. In males, growth of seminiferous tubules is also stimulated by gonadotrophins and steroid production from testes is increased by LH. In other words, LH stimulates steroidogenesis in follicular and testicular cells and initiates ovulation whereas, FSH may stimulate steroidogenesis and stimulate ovulation under experimental conditions. ACTH can also induce ovulation in hens.

4.3. The Ovary

Distribution of primordial germ cells to the ovaries becomes asymmetrical by 4th d of incubation. By 10th d, regression of right oviduct begins because of Mullerian inhibiting substance (Johnson, 2000). In *Falconiformes* and in brown Kiwi both right and left ovary and oviduct are functional. In sparrows and pigeons, 5% of the population will have both the ovaries developed (Johnson, 2000)

Number of oocytes increases from 28,000 (9th d of incubation) to 680,000 on 17th d of incubation; however, at the time of hatching when oogenesis is terminated, it reduces to 480,000. In an immature bird, the ovary has 2000 visible ova and only 200-500 of these mature in domestic species and still less in wild species (Johnson, 2000)

Removal of functional ovary in less than 20d-old chicks results in hypertrophy of the rudiment into a tissue similar to testis; even spermatogenesis occurs. But, since Wolffian duct system does not develop in genetic female, there will be no duct connection to cloaca and therefore the bird will be infertile. If testis or ovo-testis is formed, the genetic female has all sex characteristics of the male (large red comb, plumage, crowing, male copulatory behavior etc.) (Nalbandov, 1970).

If functional ovary is removed in more than 20d-old birds, the rudiment develops into a functional ovary. But, Mullerian duct develops unilaterally and therefore, there will be no oviduct of the side of rudimentary ovary. For this reason, the bird cannot lay an egg although it has a fully functional ovary (Nalbandov, 1970).

A normal female with functional ovary secrete estrogen as early as soon after hatch which prevents development of rudimentary gonad (Nalbandov, 1970).

The mature ovary weighs about 35 g with 3-4 large maturing follicles and 8-12 follicles of ever diminishing size (Leeson and Summers, 2000). It also has numerous small white follicles less than 6 mm in diameter (Johnson, 2000). In pullets before maturity, the ovarian follicles will be less than 1 mm in diameter and weigh less than 100 mg. However, in 9 days before ovulation, 18-20 g yolk material is moved into the follicle increasing its size to 20-40 mm in diameter (Nalbandov, 1970). Ovarian function depends upon neuro-endocrine information a) relayed from hypothalamus/pituitary and b) between ovarian tissues.

Arterial supply to the ovary consists of ovarian artery from left reno-lumbar artery or as a direct branch from dorsal aorta. Maximum blood flow moves towards 5 largest follicles. All veins (anterior and posterior) join to form a common vein which drains into posterior vena cava (Johnson, 2000).

Usually, left ovary is functional and is attached by mesovarian ligament at the cephalic end of left kidney (Johnson, 2000). Female pronucleus is derived from primordial germ cell migrating to the gonad. The primordial germ cells are surrounded by cells of mesodermal origin →differentiate into layer of granulosa cells →these are surrounded by theca tissue and separated by a basement membrane. The innermost cells of theca tissue (highly vascularized especially in mature follicles) produce steroids. In mature follicles, arterial supply forms a network except in a zone at the apex of the follicle called as stigma which eventually ruptures during ovulation.

In other words, surrounding the oocyte and yolk, follicle comprises concentric layers of 1. oocyte plasma membrane 2. perivitelline membrane 3. granulosa cells 4. basal lamina 5. theca interna and externa. Arteries pass through theca to basal lamina (Johnson, 2000).

Growth of follicle occurs in three phases : a) slow growth – lasting months to years, up to 60-100μm in diameter b) increasingly rapid growth – lasting several months, mainly deposition of yolk protein and c) rapid growth – 6-11d prior to ovulation (fowl, duck and pigeon), majority of yolk protein and lipids are deposited. The case of chicken, 2 g yolk protein per day is transported and the follicle grows from 8 to 37 mm in diameter and increase in volume by 3500 to 8000 times (Johnson, 2000).

Bulk of yolk is deposited in the last3-4 d prior to ovulation (Leeson and Summers, 2000). Yolk material synthesized in liver (except immunoglobulins which are not synthesized in liver) is transported *via* circulation to ovarian follicles. Yolk formation and deposition is rapid 10d prior to ovulation. This process is stimulated by estrogen. To sustain egg production, the rate of lipid synthesis in liver increases by 14 folds. Estrogen also increases rate of protein synthesis. About 65% of the yolk solids is lipoprotein complex as particles of 27 to 35 nm in diameter → VLDL (density < 1.006 g/ml) containing 12% protein, 80% lipid; the latter composed of 70-75% triacylglycerols (TAG), 20-25% phospholipids (PL) and 4% cholesterol. Palmitic and oleic acids are predominant in VLDL whereas, lecithin constitutes 77% and cephalin 18% of PL.

VLDL reaches basal lamina which acts as a filter allowing only those particles less than or equal to 40 nm in diameter. Apo-VLDL-II, a protein of VLDL, not only decreases the particle size of VLDL to 21-35 nm facilitating its entry into yolk through perforations in the vasculature of theca interna but also confers resistance to VLDL from degradation by the action of lipoprotein lipase thereby decreasing the loss of VLDL particles.

Estrogen also stimulates production of vitellogenin (a phosphoglycoprotein, which is absent in non-laying hens) in liver. Vitellogenin provides two phosphoproteins viz., phosvitin and lipovitellin to the developing yolk and serves for their transport into plasma. Uptake of vitellogenin by theoocyte is mediated by specific receptors on the surface of vitelline membrane.

Concentration of some of the proteins in the oocyte may be more than that in plasma because they are actively taken up by it. Albumin is the major plasma protein passively transferred into the oocyte. α_2-glycoprotein is actively taken up by the egg yolk. Only IgG is seen in egg yolk. Specific binding proteins from plasma also accumulate - biotin-binding (different from avidin in white), riboflavin-, thiamin-, Vitamin A(or retinol-)-, iron- and cholecalciferol-binding proteins are found in yolk.

Delivery of vitellogenin and VLDL from surface of vitelline membrane into the yolk is receptor-mediated endocytosis by vesicle formation, cleavage and rearrangement; the entire process completed within 20 min. Extensive projections on the surface of the vitelline membrane creates more surface area for receptor-mediated endocytosis. Inside the oocyte, vitellogenin and VLDL are localized to yolks spheres which are acted upon by proteolytic cathepsin D and converted to lipovitellin, triacylglycerol, cholesteroland phospholipids. At the beginning, equal quantity of protein and fat and, later on, more lipids are deposited. Deposition of yolk is terminated 24hr before ovulation (Johnson, 2000).

Role of various hormones in female reproduction has been outlined in Fig.7.1.

Vitellogenin production is controlled by gene(s); each gene producing slightly different vitellogenin and, in turn, different forms of phosvitin and lipovitellin. Spontaneous alteration in "restricted ovulator (ro)" gene → a recessive sex-linked mutation → decrease in egg production to only few eggs. This is because of non-functional apo-B/vitellogenin receptor (no or restricted number of receptors) leading to large accumulation of vitellogenin and VLDL in plasma which, in turn, results in failure of endocytosis of yolk precursors. This defect is carried usually by the heterozygous male. An autosomal recessive abnormality (rd/rd) prevents deposition of riboflavin-binding protein into both yolk and albumen; such hens cannot reproduce.

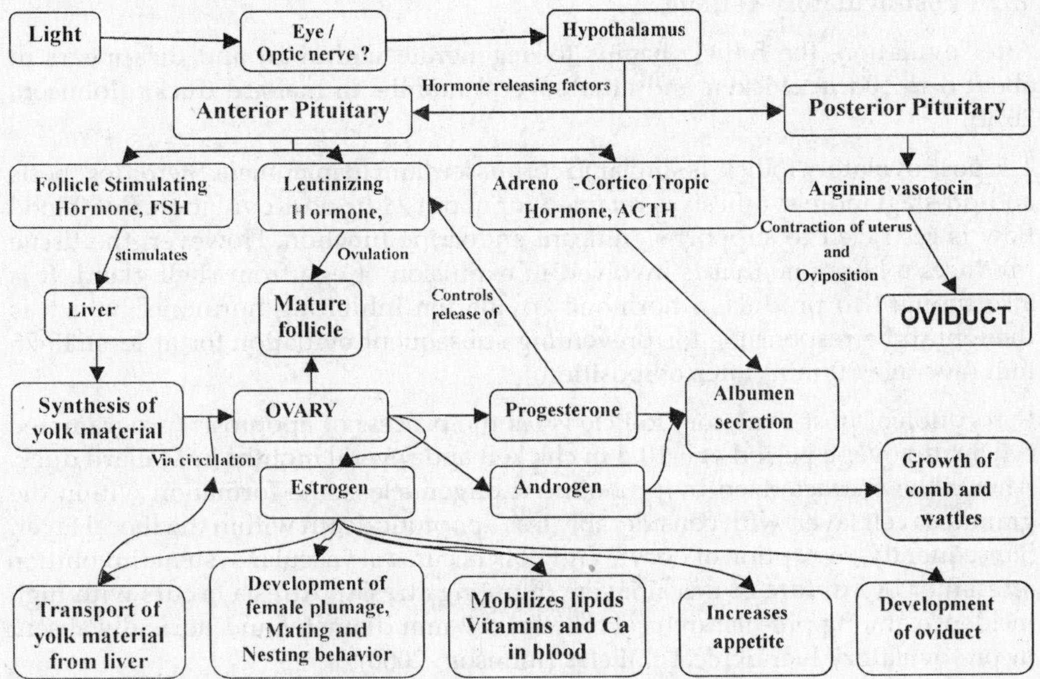

Fig 7.1 Hormones in female reproduction

6. Ovulation (Etches, 1996)

Collagen fibers are interlaced all over the surface of follicle except at stigma where they run parallel. Further, even the outer coating of connective tissue is absent at stigma facilitating ovulation. Fibroblasts in stigma produce collagenase and other proteases which weaken the junctions between cells. Fibronectin, a cell adhesion molecule, associated with collagen fibers is enzymatically degraded by cathepsins released from lysozymes of fibroblasts in thecal tissue. Expulsion of ovum is assisted by contraction of myofibrils (β-actin ?) In the follicular walls.

Ovulation at a site other than stigma results in flow of blood onto the surface of ovum producing blood spots which severely affect the aesthetic appeal when the egg is broken by the consumer.

About 2 hr preceding ovulation, protrusions of granulosa cells into the cytoplasm of the yolk are withdrawn thereby making them attached to the basement membrane. Therefore, soon after ovulation, post-ovulatory follicle contains all the endocrine and structural tissues which were present before ovulation.

Injection of corticosteroids has been found to induce ovulation or modify the timing of ovulation; therefore, it is possible that catecholaminergic innervation of ovary may be essential for maintenance of follicular hierarchy.

6.1. Post-ovulatory follicle

After ovulation, the follicle begins to degenerate within 1d and disappears in about 8 to 10d in chicken and after several months in mallard ducks (Johnson, 2000).

The post-ovulatory follicle is similar to corpus leutium in mammals. Steroidogenesis and prostaglandin synthesis is retained for about 24 hr post-ovulation. But, blood flow is restricted to support significant endocrine function. However, the tissue produces a hormone that is involved in expulsion of egg from shell gland. It is also thought to produce a hormone "ovulation inhibiting hormone" which is thought to be responsible for preventing subsequent ovulation for at least 15-75 min (average 30 min) after oviposition.

Resorption of post-ovulatory follicle is *via* the process of apoptosis (programmed cell death) over a period of 6-10 d in chicken and several months in Mallard duck. Atresia the characterized by pronounced oligonucleosome formation within the granulosa cell layer, with considerably less apoptotic death within the thecal layer. Subsequently, resorption of oocyte and yolk occurs *via* vascular system (involution atresia) or by rupture of thecal layers (bursting atresia). Atresia occurs with high incidence among pre-hierarchical follicles < 9 mm diameter and normally absent in pre-ovulatory hierarchical follicles (Johnson, 2000).

Follicles that become enlarged and yolk-filled but fail to ovulate become atretic by either resorption of the yolk material or its slow leakage into abdominal cavity. Under natural conditions, presence of atretic follicles indicates either the start of natural brooding of eggs for hatching or onset of molt (Johnson, 2000).

In Mallard ducks and pheasants, ovulated follicles remain as long as 90d after ovulation. Therefore, seasonal egg production can be calculated by counting ruptured follicles in these species (Nalbandov, 1970).

6.2. Steroidogenesis from ovarian follicles

Most follicles in the ovary do not ovulate; but participate in production of steroid hormones. As sexual maturity approaches, these hormones stimulate the deposition of calcium into the medullary bone, production of female-specific plumage, yolk-precursors in liver, growth of reproductive tract and development of secondary sexual characteristics (comb, spurs, softening and spreading of pubic bones, deposition of pigments in beak, shanks and vent, etc.).

Ovarian follicles are classified depending on diameter into small yellow follicles (5-10 mm), large white follicles (2-4 mm) and small white follicles (< 1 mm); the size of the follicles increases from F_6 to F_1 and it is F_1 that actually undergoes rupture and ovulation. Large yolk-filled follicles measure 20 to 40 mm in diameter (Johnson, 2000).

6.2.1. Hierarchical follicles

The granulosa cells produce predominantly progesterone which a) potentiates pre-ovulatory surge of LH 4-6 hr before ovulation and b) a precursor for androstenedione and testosterone synthesis by thecal layer. Theca interna produces androstenedione and theca externa produces estrogen. Steroidogenesis of granulosa and thecal cells is predominant under regulatory control of LH *via* adenylyl cyclase (cAMP); but, are insensitive to FSH (Johnson, 2000).

6.2.2. Peri-hierarchical follicles (< 8 mm in diameter) and stromal tissue

The follicles produce 3β-hydroxy dehydrogenase and no progesterone. Follicles measuring up to 2 mm in diameter produce FSH and vaso-active intestinal peptide; the latter changes ova to LH-dependent hierarchical follicles (Johnson, 2000).

6.2.3. Highest concentration of hormones

Progesterone – 6-4 hr before ovulation and coincides with pre-ovulatory surge of LH. Androgens – 10-6 hr before ovulation; secreted by 4 largest follicles. Estrogen – 6-4 hr before ovulation; secreted by 4 largest follicles; minor peak 23-18 hr before ovulation. Corticosterone – at ovulation; has a daily rhythm; regulates pre-ovulatory surge of LH (?). Prolactin – 10 hr before ovulation; has anti-steroidogenic effect; produced by anterior pituitary, increases during summer months because of increased photoperiod, lower levels recorded 6 hr before ovulation. Others – prostaglandin $F_{2\alpha}$ from F_1 follicle – at ovulation (Johnson, 2000).

6.2.4. Other hormones

Prostaglandin E_1 and prostaglandin E_2 – increased follicle plasminogen activator activity; therefore, likely to mediate enzymatic rupture of stigma. prostaglandin $F_{2\alpha}$, prostaglandin E_2, acetyl choline, oxytocin – increase contraction of smooth muscles in connective tissue wall of follicle; with proteolytic enzymes, may help ovulation (Johnson, 2000).

Thecal cells of small follicles synthesize progestins, androgen, estrogen; granulosa cells of mature follicles produce progesterone; at once yolk accumulation begins, estrogen production decreases, androgen production increases till F_3 stage of follicles (depending on size of follicles).

7. The ovulatory cycle (Etches, 1996)

The ovulatory cycle occurs in sequences (clutches) of 2 to 360 eggs and rate of egg production is directly proportional to sequence length which is defined as the interval between consecutive ovulations; length of ovulatory cycle varies from 25.12 hr to 28.53 hr when the sequence length is from 2 to 6. Last ovulation of the sequence is separated by the first ovulation of the next sequence by about 40 hr because, one full day elapses during which the ovulation does not occur. It is not possible to accurately identify the time of ovulation; however, since second and all subsequent ovulations occur only after 30-45 min of oviposition, the term used to mean laying of egg, of the preceding egg, length of ovulatory cycle is about 27 hr and therefore, in a sequence of at least 4 eggs, if 27 hr is subtracted from the time of first oviposition, time of ovulation can be obtained.

Follicular rupture (ovulation) is restricted to about 8 hr of the day beginning at about 0600 hr and ending at about 1400 hr. This period is called as "open period" which is the consequence of circadian rhythm and therefore, occurs only once during each light-dark period. Therefore, it follows that length of ovulatory cycle of hens under 14L:10D, is 24 hr; but, only a very small number of birds in a flock can achieve this during peak production.

Time of oviposition varies between species : Chicken – 9 a.m. to 3 p.m. Turkeys – 12 noon to 5 p.m. Japanese quail – 4:30 pm to 6 p.m. Chukar Partridge – 10 a.m. to 6 p.m. Guinea fowl – 9 a.m. to 5 p.m.

If 24 hr is subtracted from the interval between consecutive ovipositions within a sequence, the advance is the time of ovipositionrelative to the previous oviposition, called "individual lag" can be estimated. Cumulative lag is about 8 hr regardless of sequence length (because open period is restricted to 8 hr). Lag is longest at the beginning and at the end of sequences and minimal in the middle of sequences. These characteristics reflect the unique endocrine and neuro-endocrine regulation of ovarian function which requires synchronization of physiological systems regulating the ability to ovulate in response to pre-ovulatory LH surge with the appropriate phase of circadian rhythm which restricts the generation of the pre-ovulatory surge of LH to an 8 hr period of the day.

7.1. Release of LH

Pre-ovulatory release of LH appears to be circadian in rhythm because a) advancement or retardation of the photo-schedule causes a corresponding shift in the time of ovulation b) periodicity of the open period for LH release is equal to period of light-dark cycle c) ovipositions are not equally distributed throughout the day in the absence of light-dark cycle and therefore, hens poses exquisitely sensitive mechanisms to detect other nocturnal/diurnal signals from the physical environment (temperature, noise and feeding cycles).

Location of open period for LH release is closely linked to the transition from light-on to light-off (photo-phase to scoto-phase). Under standard 14L: 10D photo-period, the mean time of oviposition is about 15 hr after dusk. Increase the decrease of scoto-phase to 18 hr or 6 hr increases or decreases in the mean time of oviposition by 4 and 3 hr, respectively. Under short ahemeral cycles (Ahemeral cycle indicates that photo-period + scoto-period is not equal to 24 hr), precision of relationship is obscured because of broad distribution of times of oviposition; but, modal time of oviposition is 15 hr after dusk under 14L : 7D. Effect of photo-schedule on time of ovulation/oviposition is summarized in Table 7.1.

Time of oviposition is a consequence of physiological circadian rhythm which restricts pre-ovulatory surge of LH to an 8 hr period of the day. This is usually set by the transition from light to darkness; 1 hr of darkness sufficient to synchronise the timing of LH release in most hens. The relationship between light-dark transition and the modal time of lay can be modified, particularly with ahemeral light-dark cycles.

With 24hr light-dark cycles, lag the longest at the beginning and at the end of sequences and shortest in the middle of sequences. These shows that circadian rhythm control a threshold in the neuro-endocrine events that culminate in the generation of the pre-ovulatory surge of LH. Open period is a threshold in the neuro-endocrine events controlling the pre-ovulatory LH surge. However, in reality, existence of circadian rhythm does not appear to be true !.

Table 7.1 Photo-schedule *Vs* time of ovulation/oviposition

Photo-schedule	Time of ovulation/oviposition
Conventional : 14L:10D to 17L:7D	Ovulation - early morning hours of photo-phase
Long ahemeral : 14L:14D	Very narrow distribution of oviposition. Mean time – 8 hr after dusk; all ovipositions during dark completed within a short period
Ahemeral : 14L:7D	Ovipositions immediately after dusk
Conventional 24hr day 10L:14D 4L: 20D Intermittent lighting 14(0.25L:0.75D):10D	 Many ovipositions in darkness Many ovipositions in night 75% of eggs during darkness

7.2. The ovulatory sequence

With 24hr light-dark cycle, time between last ovulation of one sequence to first ovulation of the next sequence is about 40 hr. Sequences end when F_1 follicle does not mature in synchrony with the last one in the circadian threshold within the mechanisms that generate the pre-ovulatory LH surge. With 28hr light-dark cycle, follicle always mature within the low threshold phase and sequences no longer exist; ovulations occur once every 28hr. With 21 hr light-dark cycle, number of circadian cycles with ovulations decrease and hence size of sequence reduced to 3 or 4 eggs.

Follicular maturation is independent of period of circadian rhythm. Therefore, interval between consecutive eggs is independent of photo-schedules. But interval between last ovulation of the sequence and first ovulation of the next is decreased to 34 hr under a 21 hr light-dark cycle because the period of circadian rhythm is reset to equal the period of photo-schedule.

Length of ovulatory sequence is determined by rate of follicular maturation under 24 hr photo – schedules; follicle maturing once in 24 or 23 hr results in long sequences with 100% lay. With 23 hr photo-schedules, follicle maturing once in 24hr advances in every cycle until a mature follicle is not present producing a "pause"; on the other hand, follicle maturing once in 23 hr produces long sequences with 100% lay.

In most hens, follicle maturation takes 25hr and therefore, in conventional 14L:10D, maturation of ovarian follicle and openperiod for LH release gets desynchronized to produce a pause. Therefore, during last day of a sequence, there will be no ovulation.

7.3. Follicular maturation

Follicular maturation includes both growth of follicle and development of endocrine capability in the follicular tissue. Accumulation of mass in follicles is exponential at once the follicle enters into hierarchy. But follicle size is not a major determinant of maturation. Considering F_1 as ovulable stage, steroidogenesis from follicular cells shifts from estrogen to androgens between F_n and F_2 position. Follicular structure remains the same with rapid accumulation of yolk. During last few hours before ovulation, F_1 produces only progesterone. All follicles in the ovary are subjected to pre-ovulatory surges in plasma concentration of LH and steroid hormones because several other follicles would have ovulated before each of them ovulating. Therefore, changes in theca and granulosa cells transform F_2 to F_1 stage.

Theca loses ability to produce androgens 12 hr after shifting to F_1 from F_2 position. This may release hindrance for progesterone synthesis or the thecal cells may not be able to convert progesterone to androgens thereby causing sudden increase in concentration of progesterone. Increased concentration of plasma progesterone triggers secretion of GnRH in hypothalamus which, in turn, leads to secretion of LH followed by increased concentration of plasma LH; increase plasma LH stimulates further production of progesterone and *vice versa*. This positive feedback ultimately results in ovulation.

Therefore, shift in plasma progesterone concentration in response to exogenous LH is indicative of follicular maturation. Process of maturation of follicle begins as soon as the follicle enters F_1 position.

Follicular maturity can also be measured depending on the response to injection of a) GnRH/gonadotrophins; amount of pituitary extract (chicken) required for ovulation decreases as ovulation time of F_1 approaches or b) ACTH/ Corticosteroids – ovulation within 6 hr preceded by increased concentration of LH and progesterone indicates mature follicle.

In any case, it is very difficult to identify hens with rapid follicular maturation by any of the above methods. However, itcan be done by examining sequence length under less than 24hr photo-schedule; if the hen lays an egg during each cycle, it has the ability to mature a follicle within the period of photo-schedule. That means, hens laying every 23 hr in 14L:9D must be able to mature follicles within 23hr. But, egg size will decrease because rate of ovulation and egg size are negatively correlated.

8. Egg formation (Etches, 1996)

The female reproductive tract of chicken, referred to as oviduct, receives ovum within 15 min of ovulation, provides appropriate environmental for fertilization, and secretes albumen, shell membranes, shell and cuticle in concentric layers around the ovum. For oviposition, female seeks nesting place where it lays its egg. These are co-ordinated by mechanical stimulation of the tract, neuro-endocrine control of smooth muscles of the shell gland and endocrine control of calcium mobilization from medullary bones and feed.

8.1. The oviduct

Before attaining sexual maturity (pre-pubertal) the oviduct development is insignificant whereas in a laying hen the development is complete because of increased concentrations of androgens, estrogen and progesterone. The weight of the oviduct increases from 1.1 g in an immature pullet (4 months) to 77.2 g in a pullet after laying 1st egg; the corresponding length of oviduct is 9.69 cm and 67.74 cm. Hen in full molt will have 4.20 g oviduct 16.92 cm long (Nesheim *et al.*, 1979). The oviduct is suspended to the body cavity by a ligament occupying throughout the dorso-lateral part of the body cavity. In most hens, left oviduct is functional. Reproductive system of hen and vertical section of egg are shown schematically in Fig. 7.2 and 7.3.

The oviduct is seven-layered viz. 1. outer peritoneum (epithelial cells), 2. Longitudinal layer of smooth muscles, 3. Sheet of connective tissue (with blood vessels), 4. Circular layer of smooth muscles, 5. 2nd sheet of connective tissue, 6. Inner lining of mucosa (epithelial cells) and 7. Outer lining of mucosa (glandular cells interspersed with non-secretory ciliated cells). Different layers are well developed at different regions of the oviduct; for example, muscle fibers are limited in infundibulum but are abundant in shell gland and vagina. The muscular tissue is well innervated by autonomic nervous system which maintains proper contraction of various segments; the tissue is also innervated by sensory nerves. Unlike in mammals, diameter of the oviduct is almost uniform (Nalbandov, 1970).

Oviduct is innervated by both sympathetic and parasympathetic fibers. Infundibulum from aortic plexus, magnum from aortic and renal plexus, shell gland by hypogastric.n. (from aortic plexus, sympathetic) and pelvic.n. (from left pelvic plexus, parasympathetic). Both α and β adrenergic receptors are present throughout the oviduct and they control the motility of the oviduct (Johnson, 2000)

8.1.1. Infundibulum

This is a funnel-shaped upper-most portion of the oviduct, approximately 9 cm long, and is usually inactive excepting soon after ovulation. The ovum will be engulfed within 15 minutes after ovulation and the yolk stays here for an average of 18 min. Fertilization, if any, can occur here.

The sperms are stored in sperms pouches in vagina, where they remain viable for an average of 7d. Sperms move to infundibulum after the previous egg is laid and hence, after one mating or artificial insemination, fertile eggs can be expected for 7d. If the infundibulum fails to engulf the yolk, it falls into the abdominal cavity, where it is usually resorbed; such birds are referred to, though incorrectly, as "internal layers". Infundibulum has no egg forming function although secretory cells in this region (towards magnum) may contribute to extra-vitelline layer of yolk membrane and caudal part of infundibulum secretes initial albumen layer. The activity of infundibulum is controlled by ovum *per se*.

8.1.2. Magnum

This is the longest part of the oviduct (33 cm) which is highly vascular and secretes albumen around the yolk. The yolk is tumbled in the magnum for uniform deposition of white material. The duration of stay (average) is 2 hr 54 min and most of the proteins are secreted here.

Mucosa in this region is folded into 15-22 primary ridges (and secondary ridges) each 4-5 mm in height. Musculature is not extensive but is sufficient to propel the incomplete egg to the next section as well as tumble for uniform deposition of white. Tubular glands secrete 80% of egg white (ovalbumin, lysozyme, ovotransferrin and ovomucoid). The epithelial cells secrete avidin (progesterone-dependant) and, probably, ovomucin. All secretions are likely to be due to mechanical distension. Proteins are produced from endoplasmic reticulum, packaged by Golgi complex; the nascent granules coalesce to form large storage granules which collect at the apical region of the secretory cells.

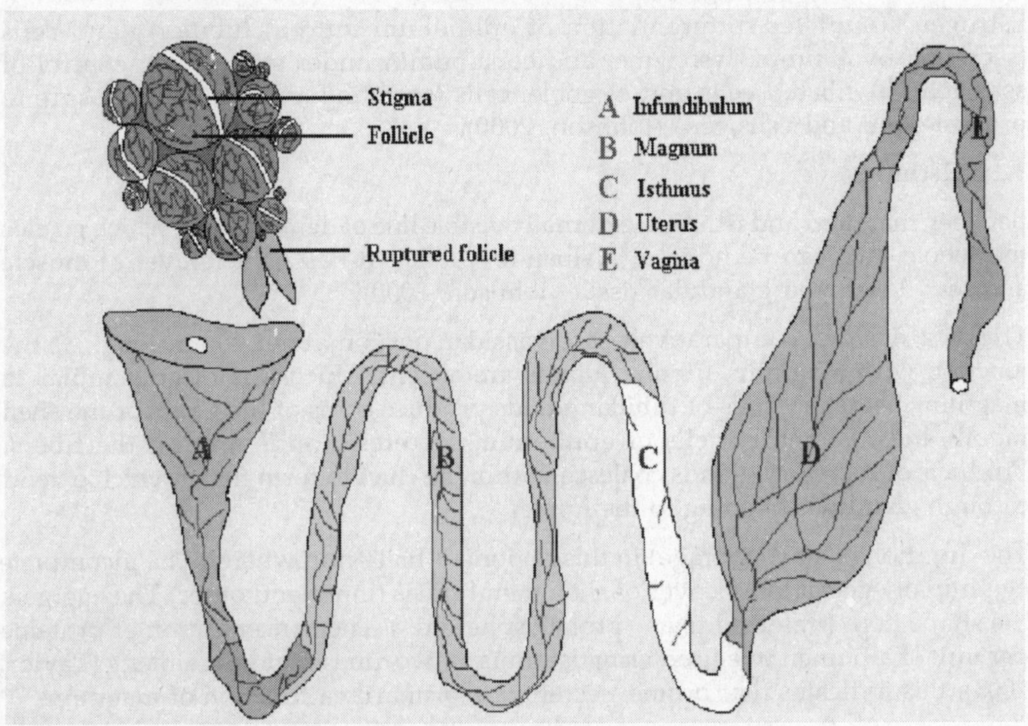

Fig. 7.2 Reproductive system of a hen

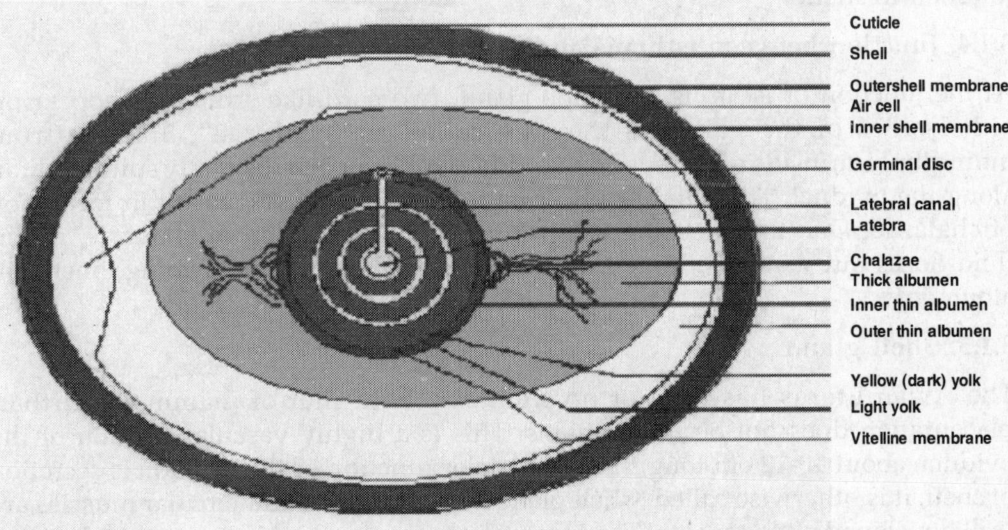

Fig. 7.3 Vertical section of egg

Estrogen stimulates differentiation of epithelium unto a) tubular gland cells (secreting ovalbumin, lysozyme, and conalbumin under stimulatory control of estrogen), b) ciliated cells and c) goblet cells (secreting avidin after exposure to progesterone and estrogen) (Johnson, 2000).

8.1.3. Isthmus

Between magnum and isthmus, externally visible line of demarcation, which girdles the whole duct can be noticed (Nalbandov, 1970). It has a thick layer of muscle and less developed glandular tissue (Johnson, 2000).

This is a narrow, comparatively less vascular portion about 10 cm long. At the junction with magnum, tubular glands are absent. Mucosa is ridged similar to magnum. Secretory cells of tubular glands.produce cores of the fibers of the shell membranes. Secretory cells of epithelium secretes mantle around the fibers. Products of secretory glands coalesce within the duct to form fibers which extend through glandular opening to the lumen.

The duration of stay (average) in this regionis 1 hr 14 min wherein the incomplete egg initially gets a loose cover of shell membranes (inner and outer) The egg gets the shape here (referred to as "prolate spheroid"). The concentration of proteins per unit of albumen in egg entering isthmus is two times that in a laid egg (Taylor, 2003); this indicates that protein secretion is primarily a function of magnum.

Initial crystals (nuclei, particularly the mammillary cores) of shell material ($CaCO_3$) are formed on the outer shell membrane in the distal part of isthmus, which are the prerequisites for subsequent calcification. This process is called as nucleation for shell formation.

8.1.4. Junction between isthmus and shell gland

At the junction of isthmus and shell gland, two cord-like structures appear on either sides of the yolk and they are called as "chalazae". Starting from infundibulum, inside surface is having ridges running in a distinctly spiral fashion along the oviduct. This might cause rotation of the yolk and result in formation of chalazae. Due to squeezing action during the formation of chalazae, a thin fluid oozes out which collects round the vitelline membrane forming inner thin albumen.

8.1.5. Shell gland

The Avian uterus has little or no similarity with that of mammals; further, placentation does not occur in avians. This is a highly vascular portion of the oviduct, about 10-12 cm long; since the major function of this portion is secretion of shell, it is otherwise called "shell gland". Longitudinal and circular muscles are well developed in this region. The incomplete egg stays here for a longest duration of 20 hr 40 min, on average, depending on photo-period.

The incomplete egg imbibes about 15 g water and exchanges several electrolytes including sodium, potassium and chlorine with the shell gland fluid; the process is referred to as "plumping". The fluid also contains carbonic anhydrase, acid phosphatase and esterase; bicarbonates and variety of additional ions (Johnson, 2000). It also gets $CaCO_3$, protein, pigment, cuticle and other components of shell. After this, it is forcibly expelled due to contraction of smooth muscles surrounding mucosa.

Mucosa has small leaf-shaped folds with ciliated and non-ciliated cells containing tubular glands. At the first 2 cm, the intrauterine junction, 8 ml of water containing electrolytes is imbibed during the first 5 hr with the help of tubular shell glands. This region may also be responsible for organization of fibers into mamillary cores which anchor crystals of $CaCO_3$ to shell membrane. Ciliated cells are secretory only in uterus.

Cuticle, the outermost acellular layer is formed by non-ciliated secretory cells after shell formation is complete. It gives the characteristic sheen (shining, bloom) to the egg.

8.1.5.1. Deposition of calcium

Calcium is transported to shell gland fluid by ciliated and/or non-ciliated cells. It is suspected that cells of tubular gland and/or epithelial cells contain carbonic anhydrase enzyme. The mechanism of calcium deposition is still not completely understood. Plasma calcium levels increase from 100 μg/ml before lay to 200-270 μg/ml during shell formation. Most of the calcium associatedwith VLDL particles and vitellogenin is unavailable. Free/Unbound/Inorganic calcium which forms only 20% of total calcium is available for deposition.

Secretory cells of tubular glands have calcium binding protein (CaBP) which help movement of calcium from shell gland to shell gland fluid. Plasma estrogen increases which causes a concomitant increase in $1\alpha 25(OH)_2D_3$ (Vitamin D_3) in kidney; the latter promotes absorption of calcium from intestines.

Throughout calcification, super-saturated calcium is maintained and concentration of ionic calcium decreases continuously in blood plasma. Ultimate source of calcium is feed. But, shell formation occurs during scoto-period when birds do not eat. Therefore, bone is important source of calcium. Storage depots in bones (in the endostial surface of femur and tibia with interconnecting spicules, developed under the stimulation of estrogen and this labile form of $Ca_3\ (PO_4)_2$ is called as "medullary bone" which maintains a steady concentration of available calcium in long bones. Intense medullary bone formation alternates with periods of severe bone depletion. Vitamin D_3 primarily and ovarian steroids regulate the synthesis of calcium binding protein (calbindin D_{28k})(Johnson, 2000).

If calcium consumption decreases, entire skeleton will be demineralized (to maintain medullary bone depots); therefore, integrity of cortical bone decreases which may precipitate as "cage layer fatigue". Bone reserves are withdrawn when plasma calcium levels decline. If calcium content of feed is = 3.6%, most of egg calcium comes from intestines. The calcium in feed is 2%, bone supplies 30-40% of egg shell calcium. On calcium-free diet, skeleton is the main source of calcium (Johnson, 2000).

Shell formation occurs during the night when the birds will not be feeding; therefore, medullary bone is the main source of calcium during darkness. Calcium in blood exists as a) non-diffusable protein-bound calcium – bound to plasma calcium binding proteins like vitellogenin and albumin, and b) diffusible ionized calcium. Estrogen increases plasma calcium levels by increasing the production of plasma calcium binding proteins (Johnson, 2000).

Plasma ionized calcium decreases of calcium-deficient diet thereby leading to reduction or cessation of lay; LH levels decrease but, LH-releasing activity is not reduced (Johnson, 2000).

Osteoclasts are involved through stimulation by parathyroid hormone (PTH) for mobilization of bone calcium. Reduced plasma calcium levels stimulate PTH and *vice versa*. Hence, during shell formation, concentration of PTH will be highest.

Osteoblasts, which increase calcium deposition are active throughout calcification but, during non-calcification, the activity is highest indicating increased bone mineralization.

Osteoclasts help resorption of calcium from bone through acid phosphatase enzyme during darkness whereas, osteoblasts help deposition of calcium into the bone through alkaline phosphatase enzyme during light. Feed consumption increases 2 hr before darkness and the feed is stored in crop to release calcium continuously during scoto-phase; if shell-grit is fed separately, hens feed on it during the last 2 hr before darkness so that calcium need of shell formation is met.

Calcium secretion from shell gland increases after about 7 hr after ovulation; reaches maximum and then comes back to basal level; it decreases 2 hr before oviposition. Stimulus for calcium secretion is ovulation but not the presence of egg. On the other hand, termination of calcification is not related to removal of egg from shell gland (Johnson, 2000).

Calcification is not due to distension or autonomic innervation. It occurs slowly at first and then increases to 300 mg/hr for about 15 hr, then again slowly during the last 2 hr before oviposition (Johnson, 2000).

On average, a commercial layer weighs 1.8 kg and assuming that it produces at the rate of 75%, totally 274 eggs are produced. Assuming that each of the eggs weighs 56.7 g, the hen has produced 1.55 kg of shell which is made of 620 g of elemental calcium (approximately 34% of the body weight or 2.7 times the calcium present in the entire skeleton (North, 1972). On the same lines, assuming 12%

protein and 11% fat in each of the eggs, the hen would have produced 1.86 kg protein and 1.71 kg fat; both approximately equal to her own body weight.

8.1.5.2. Deposition of carbonate

Bulk of carbonate comes from respiration. CO_2 and water from blood enter the cells of shell gland and combine in presence of carbonic anhydrase enzyme present in epithelial cells, tubular gland cells and shell gland fluid to form H_2CO_3, the carbonic acid, which dissociates to form HCO_3^- and H^+; HCO_3^- is further dissociated to CO_3^{2-} and H^+ by carbonic anhydrase in the shell gland fluid; the H^+ so formed diffuse to circulation causing slight fall in pH of both blood and subsequently of urine. NH_3 serves as a vehicle for excretion of H^+ and hence renal clearance of $NH4^+$ increases. pH of the shell gland fluid is not altered because of high HCO_3^- concentration and quick removal of H^+ into circulation.

8.1.5.3. Shell formation

The reaction $Ca^{2+} + CO_3^{2-} \rightarrow CaCO_3$ occurs non-enzymatically in the lumen of the shell gland on the outer shell membrane.

Due to loss of CO_2 from blood, plasma concentration of CO_2 decreases causing decreased concentration of HCO_3^- in blood (respiratory alkalosis) and urine. Although plasma CO_2 levels fall, increased H^+ in plasma triggers polypnoea (respiratory rate increases from 7 to 12/min) which ultimately reduces plasma CO_2.

It is evident that during summer, there will be polypnoea (panting) induced at 29.4°C which results in reduced plasma concentration of CO_2 and consequent on this HCO_3^- availability for shell formation reduces leading to production of thin-shelled eggs. This is compounded by the fact that the birds lack sweat glands rendering them highly susceptible to heat stress. Renal elimination of HCO_3^- to restore blood pH further accentuates the situation. Further, activity of carbonic anhydrase also reduces at higher temperature. Reduced feed consumption under high temperature does pose additional problems of calcium availability, but, the exact mechanism (s) that limit calcium homeostasis under thermal stress is(are) not completely understood.

Hence during summer, anti-stress factors such as Vitamin C and B-complex are generally given to reduce panting reflex. Increasing nutrient density of feed, feeding sodium bicarbonate (0.3% in feed) or offering oyster shells in separate feeders or offering carbonated water have also been found beneficial.

Biochemical events during shell formation are shown in Fig. 7.4.

8.1.5.4. Dynamics of shell formation

Shell formation also causes large changes in the flux of K^+, Mg^{2+} and glucose into and Na^+ and Cl^- out of the shell gland fluid; these are likely to be associated with active transport of CO_2, HCO_3^- and Ca^{2+} into the shell gland fluid using adenosine triphosphatase (ATPase) systems requiring Na^+, K^+, Cl^- and Mg^{2+}. Glucose appears to fuel the active transport (against concentration gradient) of Ca^{2+} and HCO_3^- between blood (source) and shell gland fluid (destination). Plasma levels of Na^+, K^+, Cl^-, Mg^{2+} and glucose remain unchanged indicating that these may be recirculated in the shell gland fluid.

Volume of shell gland fluid increases during the first 6 hr, maintained at that level during the next 6 hr and decreases during the last 6 hr of calcification; the exact significance of such changes is not yet understood.

Calcification is terminated about 2 hr before oviposition through an unknown mechanism associated with increased P and Mg levels in shell gland fluid; the latter is incorporated in the outer layer of shell. It is likely that levels of P and Mg along with P-containing proteins are involved in termination of shell formation

8.1.5.5. Color of shell

Pigments from both ciliated and non-ciliated cells are deposited during last 2 hr of shell formation. However, reports indicating that deposition of pigments occurs as early as 5 hr and as late ½ hr before oviposition are available.

Color of eggs is the genetically determined trait. Succinyl CoA and glycine the starters of the porphyrin formation, combine to form porphobilinogen which is enzymatically transferred to uroporphyrin; the latter is decarboxylated successively, first to yield coproporphyrinogen and subsequently to protoporphyrinogen. Auto-oxidation of protoporphyrinogen yields colorless

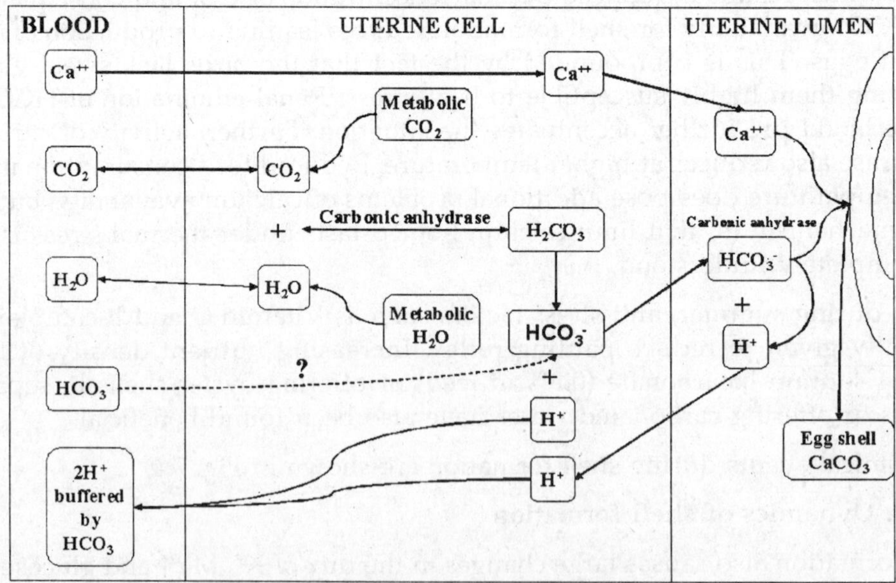

Fig. 7.4 Biochemistry of shell formation

porphyrin which, upon incorporation of double bonds (un-saturation), is converted to colored porphyrin. Oviduct is the most likely origin of egg-shell color (not the heme degradation) and this occurs in mitochondria which are most abundant in tubular gland cells of the uterus. Therefore, porphyrin produced in the oviduct is referred to as ooporphyrin. Blue colour of the egg shells (in breeds like Araucanas, Ameraucanas) is due to the pigment oocyan.

Bulk of the pigment is deposited on the cuticle and some pigmentation does occur within the calcite matrix. Tinted egg shell color in some White Leghorns is thought to be due to leakage of protoporphyrin into the lumen of uterus.

That pigmented shells have a higher strength is yet a controversial issue. However, colored eggs tend to retain more heat than white eggs and hence it is believed that this is one of the reasons why eggs of wild birds are heavily colored to tolerate absence of the broody hen for a longer time during natural incubation.

8.1.5.6. Color of yolk

Yolk color is one of the easily recognizable quality factors of an egg in spite of the fact that it has no nutritive value. Consumer preference in respect of yolk color is also highly variable. Dark-colored yolk is required while preparing pasta, mayonnaise and other products.

Color of yolk (yellow-orange) is due to presence of fat-soluble carotenoids in the lipid portion of lipoproteins; majority of them are hydroxyl compounds called xanthophylls with minor amounts of carotenes. Types and amounts of carotenoids are diet-dependent. Major xanthophylls of yolk are lutein, zaexanthin and cryptoxanthin, all derived from maize(corn), alfalfa meal, maize(corn) gluten (Stadelman and Cotterill, 1995).

Xanthophyll pigments obtained from feed cannot be utilized by the hen and thus transformed to egg virtually unchanged. The pigments are either yellow or red and their relative proportion determines the final appearance of yolk. Rate of deposition of pigments is genetically determined which plateaus and higher levels of intake. Even synthetic pigments can be used in feed (Table 7.3). Carotenoids in alfalfa, yellow maize and dried algal meal are shown in Table 7.2.

Table 7.2 Carotenoids in alfalfa, yellow maize and algal meal

Carotenoid	Alfalfa	Yellow maize	Algal meal
Cryptoxanthin	7	8	---
Lutein	46	54	86
Neoxanthin	14	---	---
Violaxanthin	16	---	4
Zeaxanthin	4	23	2
Others	13	15	8
Bioavailability of xanthophylls, %	57.49	75.51	107.88
			Source : Stademan and Cotterill, 1995

Lutein is the dominant carotenoid in yolk (62% of total carotenoids) of eggs from birds fed maize and alfalfa diets. Zeaxanthin is the next predominant carotenoid (12.9% of total carotenoids).

8.1.6. Vagina

8.1.6.1. Sperm storage (in vivo)

Sperms are stored in tubular glands located in a narrow band of about 2-5 mm wide immediately posterior to uterus and anterior to vagina. The mucosa is thrown into three circular folds with epithelial surface having both non-secretory ciliated cells and non-ciliated cells secreting an unidentified acid mucopolysaccharide. These folds are called "sperm storage tubules" have an organization similar to tubular glands. A hen has about 25,000 such tubules each of which can hold few to several hundred sperms (average of about 400) are oriented parallel to each other with their heads to the blind end of the gland.

Table 7.3 Total carotenoid content (ppm) of feed ingredients

Ingredient	Content
Algal meal	2000-4000
Clover meal	500
Crustaceans (Crab, Shrimp etc.)	80
Grass meal	200-760
Alfalfa meal	100 to 550
Corn gluten meal	100-300
Marigold meal	4275
Paprika meal	275-1650
Yeast (Phaffia sp.)	340

Source : Hunton, 1987

Hen inseminated with 1.50 x 10^8 sperms will retain about 10^7 (about 7%) sperms. Secretory cells, having numerous microvilli projecting from apical border of cells, produce lipids, glycogen and acid phosphatase; significance of either the microvilli or the secretory products is unknown.

Site of fertility is about 60 cm away and therefore, sperm release is so arranged so as to achieve maximum concentration at the infundibulum which is hospitable to sperms. Sperms may be stored in clefts and shallow tubular glands; but not as a regular feature.

Mechanism of sperm release is unknown. Sperms may be released continuously and ascend at random. Many sperms penetrate ovum. Sperm-host glands are always full or empty; therefore, filling and emptying appears to be all-or-none process. Mechanism that attracts sperms into the storage gland is not known.

Table 7.4 Summary of egg formation in oviduct

Part	Structure	Function(s)	Duration (Average)
Infundibulum	Funnel-shaped, 9 cm long	Engulf the ovulated follicle; fertilization if viable sperms are available	18 min@
Magnum	33 cm long; longest part; highly vascular	Yolk is tumbled for uniform secretion of albumen under the influence of estrogen	2 hr 54 min@
Isthmus	10 cm long; narrower and less vascular	First a loose cover of shell membranes (inner and outer); "Plumping-out" of water and electrolytes; shape of the egg (Prolate-spheroid) decided; initial crystals of shell deposited (nucleation for shell formation)	1 hr 14 min@
Shell gland	10 to 12 cm long; highly vascular;	Formation of shell; if plumping-out is incomplete, it will be completed before calcification; cuticle is also secreted.	20 hr 40 min@
Vagina	12 cm long; vascular; Sperm pouches can store sperms	No egg forming function	---
Total time required for egg formation (Average)			25 hr 6 min@
Average time gap between oviposition and subsequent ovulation			30 min@
Total time required			25 hr 36 min@
@ *Source* : Taylor, 2003			

Release of sperms may be due to mechanical distension as egg passes through the sphincter between shells gland and vagina during oviposition although sperms are found throughout the oviduct at all times of the day. Therefore, it is also possible that sperm release may be continuous or may be autonomally released by each gland. The sphincter may also block entry of non-motile sperms.

Sperms fill the utero-vaginal glands in a sequential fashion without mixing so that with successive artificial insemination (AI) sperms with latest AI is most likely to fertilize (Johnson, 2000).

8.1.6.2. Role of vagina

Vagina is the last part of the oviduct, short, S-shaped, vascular and 12 cm long. Both layers of musculature help in oviposition. Summary of egg formation in different parts of the oviduct is given in Table 7.4.

8.2. Oviposition

Except the last egg in a sequence, oviposition is initiated by pre-ovulatory endocrine events associated with ovulation. Under the influence of estrogen, pubic bones spread and vent enlarges to facilitate oviposition. With definite lighting schedule, LH release occurs 8 hr before ovulation time. With a photoperiod between 0500 hr and 1900 hr, if an egg in a sequence is laid at 1600 hr, ovulation cannot occur on that they because 8 hr before 1600 hr happens to the 0800 hr when the lights are on. Since LH release occurs during dark, the hen skips a day resulting in the minimum of one day's pause (Austic and Nesheim, 1990).

In about 40% of hens, egg rotates through 180° about 1 hr before oviposition and laid blunt-end first; further details on this rotation is not known. Normal egg is too long to turn in a horizontal plane within the pubic arch; therefore, egg must drop from its horizontal position high up between ischia to a point opposite the tips of pubic bones. Time required for rotation is 1-2 min and why eggs are formed small-end forward is not known. However, rotation of the egg so as to make broad-end forward is likely to make the muscular pressure required for expelling the egg more effective when it is applied to the small-end (Austic and Nesheim , 1990).

8.2.1. Pre-incubation inside the oviduct

After ovulation, if the released ovum is fertilized in infundibulum, the zygote will have to stay for about 25 hr inside the oviduct at the body temperature of the hen (41°C). After 5hr of fertilization, that is when the incomplete egg is leaving isthmus, it undergoes I mitotic division.

Chicken egg is describes as telolecithal (yolk at one end) and megalecithal or polylecithal (good amount of yolk). Cytoplasm is confined to a small area called blastodisc or germinal disc which contains egg nucleus or germinal vesicle at its center. The most typical cleidoic egg taken for embryological experiments is that of chicken.

At ovulation, I maturation division occurs to form secondary oocyte and a minute I polar body. Soon after ovulation, ovarian follicle is disorganized and the secondary oocyte initiates II maturation division and, at the time of engulfing into the infundibulum, it pauses at metaphase (Arora, 1992).

Many sperms enter the egg (polyspermy) but during karyogamy, nucleus from only one sperm fuses with the female pronucleus and others degenerate (Arora, 1992)

8.2.2. Cleavage / Segmentation

This is a process of converting uni-nucleated blastodisc into blastomeres. I cleavage occurs when zygote is in magnum and 3 hr after fertilization. Only blastodisc is divided and hence referred to as discomeroblastic cleavage occurring perpendicular to the axis of mitotic spindles and coinciding with axis passing through animal pole (where blastodisc is located) and vegetal pole (diametrically opposite to animal pole) (Arora, 1992).

After 1-2 hr stay in uterus 6-7 hr after fertilization), four divisions occur to result in 32 cells. However, complete separation of daughter cells doesn't occur and the cells are in continuation with yolk both peripherally and dorsally. II cleavage occurs at right angles to I and the III cleavage bisects the axes formed by the first two. The IV cleavage is radially oriented and hence centrally located daughter cells are closed on all lateral surfaces. Next 2 hr following this (8-9 hr after fertilization) 14-16 laterally closed cells but not separated from yolk (due to absence of lateral cleavage) are seen at the centre.

During the next 2 hr (10-11 hr after fertilization), 80-90 cells form the central area of the embryo and by the end of 10 hr after fertilization, underside of the embryo gets separated from the yolk due to formation of sub-germinal fluid. Subsequently, during the next 4 hr (12 to 15 hr after fertilization), rapid cell division results in formation of a sheet of epithelial cells over the germinal disc which is separated from the yolk by the sub-germinal fluid; this is referred to as blastoderm. During the remaining time in the shell gland, cells on the upper surface of the embryo multiply rapidly and become smaller. Cells on the lower surface are shed and they fall to the surface of yolk below. Therefore, a central area which is translucent (*Area pellucida*) is formed. At oviposition, a well-defined *area pellucida* surrounded concentrically by a marginal zone which is in close contact with the yolk (*area opaqua*) are seen.

Before or soon after oviposition, the embryo differentiates into 2 layers of cells referred to as gastrulation (Austic and Nesheim, 1990). Hence, on the vitelline membrane of a fresh egg (after being broken) a clear white ring (high gastrula stage), referred to as blastoderm, may be seen (Taylor, 2003). In case of an infertile egg, only the cells surrounding the ovum multiply to form a heap of cells appearing as a white spot (referred to as blastodisc) on the vitelline membrane of a fresh egg. Blastoderm and blastodisc are shown in Fig. 7.5.

Blastodisc is about 3.5 mm in diameter wherein nucleus of Pander is surrounded by yolk spheres. On the contrary, blastoderm is about 4.4 mm in diameter made of about 30,000 to 40,000 cells with the nucleus of Pander separated by a layer of sub-germinal fluid. The area having sub-germinal fluid underneath appears transparent and hence referred to as area pellucida which is surrounded by area opaqua; the latter is so named because it has yolk spheres underneath and hence appears opaque. This stage is referred to as Stage X embryo. Therefore, in a fertile egg, a raised disc which is white/less intense colored than surrounding yolk can be seen on the vitelline membrane (Leeson and Summers, 1990)

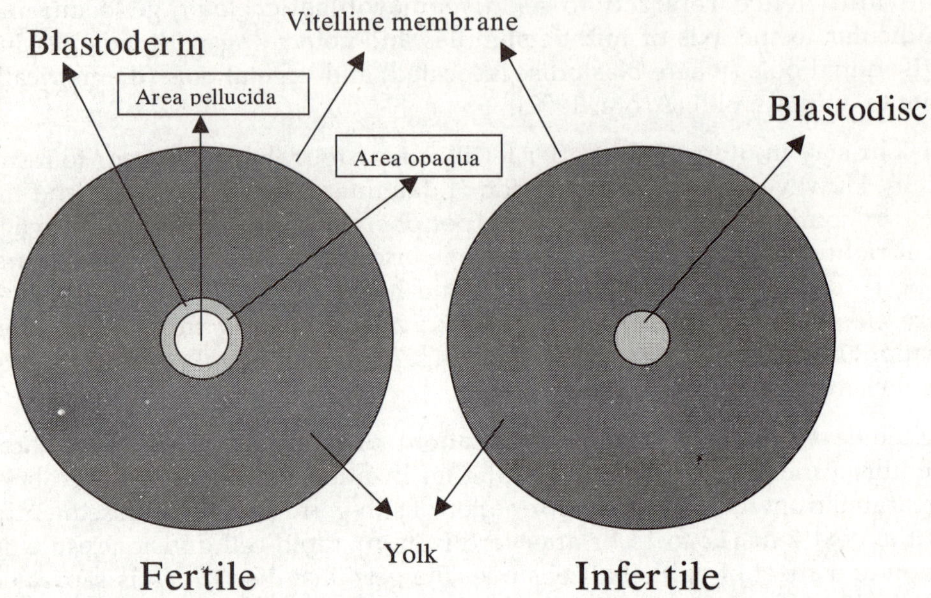

Fig. 7.5 Germinal disc on the surface of yolk in fertile and infertile eggs

8.2.3. Contraction of muscles of shell gland

Frequency and intensity of contraction of muscles increase throughout the stay of incomplete egg in shell gland. Just before oviposition, frequency and intensity of contraction increases and soon after diminishes; this is preceded by increased and decreased plasma concentration of prostaglandin E_2 and $F_{2\alpha}$ (both produced from granulosa cells of the two largest pre-ovulatory follicles in association with pre-ovulatory surge of the LH and venous drainage from follicles) and arginine vasotocin (released from posterior pituitary gland presumably due to neural reflex similar to parturition and/or possibly, ovary itself), respectively. Prostaglandin E_2 causes relaxation of vaginal sphincter whereas the other two are potent smooth muscle stimulants.

In case of first ovulation of a sequence, pre-ovulatory release of prostaglandins from the largest pre-ovulatory follicle causes marked contraction of shell gland; but, does not cause oviposition.

From second ovulation till the last ovulation, prostaglandin $F_{2\alpha}$ from the largest pre-ovulatory follicle is responsible for oviposition of the previous egg in sequence.

Last oviposition occurs without followed by ovulation and without increases in prostaglandin $F_{2\alpha}$ and venous drainage. Therefore, the terminal oviposition is associated with marginal increase in venous concentration of prostaglandin $F_{2\alpha}$ in the second largest pre-ovulatory follicle. Actual cause for release of prostaglandin $F_{2\alpha}$ is not known for the last oviposition, whereas, for others, pre-ovulatory surge of LH has been incriminated.

Corticosterone has also been suspected to help oviposition.

Oviposition is influenced by feeding schedule; normally, oviposition occurs during the feeding time. For instance, the feeding time is 8 a.m. to 4 p.m., oviposition also occurs during that time; if feeding time is shifted to 8 p.m. to 4 a.m., time of oviposition also changes to 8 p.m. to 4 a.m. (Austic and Nesheim, 1990).

In case of young flock of broiler breeders in peak production, there will be a conflict between desire to feed Vs nest. With lights-on at 5 a.m., the birds feed actively for 30-40 min followed by continuous but less active feeding until clean-up some 2-4 hr later. If desired to feed is more than that for nesting, eggs are dropped along the feeder line (referred to as floor-eggs); especially the first eggs of long clutches laid early in the day. If more than 1% of eggs are on slats around the feeders, it indicates that initial feeding time is too late in relation to time of light-on (Leeson and Summers, 2000).

8.2.4. Nesting behavior

Includes nest investigation before lay, selection of nest site, oviposition and exit from the nest. Specific pre-laying call with nest investigation and cackling soon after oviposition have been recorded. Nesting behavior differs when housed in cages/deep-litter and among hens. Nest investigation takes about 1 hr during which time the hen inspects about 40 sites (sometimes repeatedly also), lifts the legs up to the hip and slowly moves towards the selected site (exaggerated gait). After entering the nest, the hen stays for about 1-2 hr either sitting quietly or constructing a rudimentary nest by throwing pieces of nesting material to fall all around it. As egg is expelled, the hen assumes a vertical penguin-like stance; therefore, egg is dropped into the nest from a height of several cm.

In case of cages, no nest investigation can be done and oviposition is preceded by stereotyped pacing for about 1 hr, sitting on the cage floor/exhibiting stereotyped nest-building behavior without nesting material. Oviposition occurs while hen is standing/squatting and the egg falls from a height several cm from the floor.

These behavioral characteristics are likely to be associated with the plasma concentration of prostaglandins and arginine vasotocin. Search for appropriate nest is a compulsive behavior which reduces motivation to eat/drink; it is also associated with decreased motility of gizzard 2 hr before oviposition. This may be associated with both ovulation and oviposition. In other words, feed and water intake may be regulated directly or indirectly by pre-ovulatory surges in estrogen, androgens and progesterone that precede ovulation and increase in plasma concentrations of prostaglandins and arginin vasotocin that precede oviposition.

Broody hens develop a brood (incubation) patch characterized by increased circulation, edema and thickening of epidermis. Prolactin and estrogen are the hormones controlling the brooding tendency and during this period, concentration of LH, progesterone, testosterone and estradiol decrease (Johnson, 2000).

8.2.5. Egg-laying sequence (Clutch)

On reaching sexual maturity (20/22 weeks), domestic hen lays sequences with ≥ 2d pause or with high incidence of abnormal (soft-shelled, double-yolked) eggs of the first 2 weeks. It reaches peak production of ≥ 90% 6-10 weeks later; egg production gradually decreases over a period of 40-50 weeks (depending upon meat/egg type). Finally, egg production stops and the hen enters into molt. The typical hen produces 280 eggs (80%) in a 50-week production period (Johnson, 2000).

Group of eggs is called clutch (the case of non-domesticated species prior to incubation) and number of eggs laid onsuccessive days is called sequence (in case of domestic birds)(Johnson, 2000) and number of days of gap between two consecutive clutches/sequences is referred to as pause. Generally, ovulation occurs after 15 to 75 min (average 30 min) after oviposition (except for first egg in a sequence) and it is preceded 8 hr earlier by release of LH from anterior pituitary (Austic and Nesheim, 1990). Therefore, there will be a progressive shift in time of laying of egg in the day. As sequence becomes shorter. Lag between successive ovipositions increases and length of ovulation-oviposition cycle becomes > 24hr. The lag between first ovulation in a sequence (C_1) to the last ovulation of a sequence (C_t) is greater in chicken (4 to 8 hr) than in Japanese quails (1 ½ to 2 hr). Normally, release of LH is restricted to 4-11 hr period beginning at the onset of scoto-phase (open period). Chicken usually ovulates first egg in a sequence early in the photo-phase whereas, Japanese quail ovulates for its C_1 egg 8-9 hr after the onset of photo-phase (Johnson, 2000)

Size of the sequence varies between and within a species; European partridges lay in a single clutch of 12-20 eggs per annum. Pigeons lay 8 sequences/annum of 2 eggs each with an interval of 45d during fall and winter and 30-32 d during spring and early summer between sequences. Bobwhite quails have an average sequence size of 19.2 eggs during early May and 11.3 eggs during late July (Johnson, 2000).

Ovulation-oviposition cycle needs more than 24 hr (24+ hr) in case of chicken, turkey, bobwhite quail, and possibly, in Japanese quails. Pigeons need 40-44 hr to complete the cycle and Khaki Campbell need 23.5 to 24.5 hr. (Johnson, 2000).

8.2.6. Physiological basis of sequence and pause

Assuming lights-on and 6 a.m. and lights-off at 8 p.m. (14 hr of light, 14L:10D), the photosensitive phase begins at 5 p.m. and ends at 7 p.m. Follicular rupture (ovulation) is restricted to 6 a.m. to 2 p.m. (open period, 8 hr). Taking into consideration the average time required for formation of egg (25 hr 6 min) and average time lag between oviposition and subsequent ovulation (30 min), supposing, a hen ovulates for the first time at 6 a.m., first egg will be laid at 7.36 a.m. which will be followed by second ovulationat 8.06 a.m. and so on. Therefore, the time of laying of successive eggs in this example will be 7:36 a.m., 9:12 a.m., 10:48 a.m., 12:24 noon, 2 p.m. and 3:36 p.m.; the last egg being the consequence of ovulation at 2 p.m. of the preceding day. After this sequence of 6 eggs, no ovulation can occur because the open period ends at 2 p.m. and hence, the hen takes a pause of at least 1d. It can be noticed that the first and the last egg in a sequence are separated by 8 hr; in other words, cumulative lag for the sequence is 8 hr after which the "window" of time for release of LH does not correspond to development of F_1 follicles or the light program (Leeson and Summers, 2000). This cycle of sequence and pause can repeat throughout the egg-laying period.

Rate of FSH secretion is not controlled and hen's pituitary is very resistant to estrogenic inhibition (unlike mammals). Therefore, in the hen with a long sequence, the valve is opened wider for a more rapid but still steady outpouring of FSH whereas in those with short sequences, valve is set for a low but steady flow of FSH (Nalbandov, 1970). Therefore, high producing birds do produce eggs in long clutches of even 30 eggs and more. This is a consequence of intensive breeding for frequent ovulation; such birds ovulate once every 24hr and produce at a very high rate. For instance, a hen laying eggs in clutches of 30, will produce at a rate of {(100 * 30) / 31} = 96.77%. The corresponding values of % egg production with respect to clutch size of 4,6,8,10 and 12 with a pause duration of 1d is 80.00, 85.71, 88.89, 90.91 and 92.31, respectively.

With ahemeral lighting schedule of < 24hr, ovulation can be induced every 23hr also; but, the egg size will be severely reduced and hence, is not commercially practiced.

8.2.7. Position of the egg in a sequence

Eggs laid early in the day (first few eggs in a sequence) are larger because they over-stay in the oviduct for about 2 hr; therefore, they are subjected to longer pre-incubation in the oviduct. It is for this reason that higher embryo mortality has been recorded in eggs laid during the beginning of the sequence. Eggs laid later in the day (last few eggs in a sequence) will be smaller and have more shell material; therefore hatchability of these eggs also is reduced. Extent of embryonic development and hatchability have been found to show a curvilinear relationship (Taylor, 2003).

It follows by the above facts that higher the sequence size smaller will be the proportion of eggs with increased pre-incubation; therefore, % hatchability of eggs laid in long sequences will be higher than those laid in short sequences. In other words, sequence size and % hatchability are positively correlated (Leeson and Summers, 2000).

Yolk size has been found to be nearly constant in all eggs in a sequence (Austic and Nesheim , 1990) and therefore, weight differences between the eggs within a sequence are primarily due to differences in weight of albumen and/or shell.

8.3. Significance of shape of egg

The prolate spheroid shape of the eggs serves the following purposes:

1. In a nest there will be minimum air-space in between the eggs thereby reducing heat loss and allowing better use of nest-space. This also facilitates brooding because the hen can leave the eggs for some time for feed and water without much cooling of eggs; particularly during the first week of incubation.
2. The shape does not allow linear rolling of eggs; instead, they roll in a circular path around the pointed end. This would help protect the eggs from falling out of the nest in case a momentum is created due to turning the eggs by the hen or some other reasons when the hen is absent.
3. The shape of the egg is convenient causing least strain during oviposition.
4. The arrangement of calcium carbonate crystals to form the final shape offers greatest resistance from outside, especially from end to end, and least resistance from within. The long axis of the crystals are perpendicular to that of the egg in the outer layer of the shell whereas, the inner layer of the shell is amorphous so that in the event of the egg falling down, the chances of complete breakage is reduced and the chick can easily pip-out of the shell and the time of hatching.

9. Structure of egg (Etches, 1996)

Vertical section of an egg is shown in Fig. 7.3. Considering for all domestic species, yolk constitutes 32-35%, albumen forms 52-58% and shell + membranes constitute 9-14%. The geometry of an egg is shown in Fig. 7.6.

Considering V = volume (cm^3), L = length (cm), B = breadth (cm), W = weight (g) and S = surface area (cm^2), V and S of an egg can be estimated by the following formulae :

$$V = \frac{k\pi LB^2}{6} \quad \text{where } k = 0.85 \text{ to } 0.99 \text{ (Average: 0.929)} \quad \text{or } V = 0.913W$$

$$S = k\left(\frac{\pi LB^2}{6}\right)^{0.67} \quad \text{where, } k = 4.63 \text{ to } 5.07 \text{ (Average: 4.85)} \quad \text{or } S = kW^{0.67}$$

where $k = 4.558$

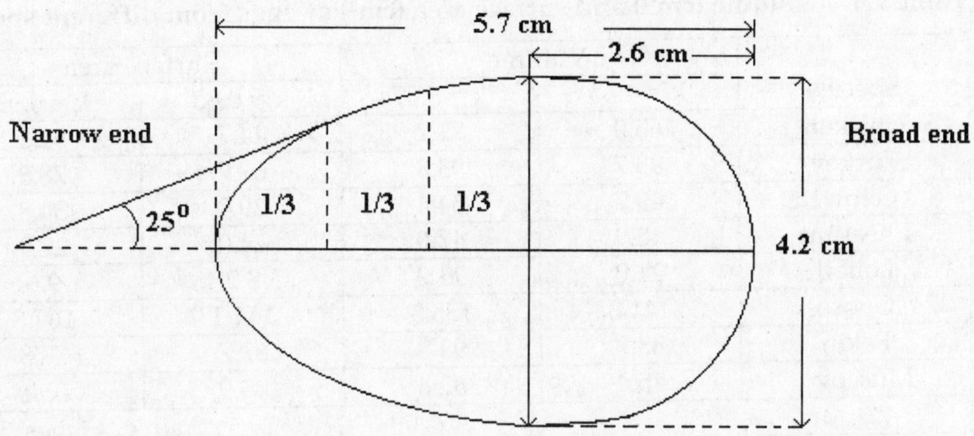

Fig. 7.6. Geometry of chicken egg

The above equations have been utilized to calculate V and S of eggs of different species of poultry shown in Table 7.6.

Structure of different parts of chicken egg is outlined below :

9.1. Cuticle

Cuticle is the outer-most acellular layer about 10μm thick, laid 30 min before oviposition. It is made of protein + polysaccharide + lipid. May be unevenly distributed over the shell surface ranging, in depth, from 0.5 to 12.8 μm. Dry weight of cuticle in a 60g egg is 12 mg; although it cannot give any additional integrity to the shell structure, it is likely to reduce microbial invasion into the contents and evaporation of water; however, cuticle-less eggs have been found to hatch equally well (Johnson, 2000).

Table 7.5. Gross composition of eggs from different species

Species	Weight (g)	Albumen (%)	Yolk (%)	Shell (%)	Length (cm)	Breadth (cm)
Chicken	58	56	32	12	5.7	4.2
Turkey	85	53	33	14	6.6	4.8
G.fowl	40	52	35	13	4.9	3.7
Pheasant	32	53	35	12	4.4	3.5
J.quail	9	58	33	9	3.2	2.4
Goose	155	56	32	12	8.7	6.1
Pekin	92	57	33	10	6.6	4.8
Muscovy	80	53	35	12	6.2	4.5
						Source : Etches, 1996

Table 7.6 Volume (cm3) and surface area (cm²) of eggs from different species

Species	Volume		Surface area	
	k = 4.63	k = 5.07	k = 0.85	k = 0.99
Chicken	65.0	71.2	44.7	52.1
Turkey	85.7	93.8	67.7	78.8
G.fowl	49.7	54.4	29.9	34.8
Pheasant	42.9	47.0	24.0	27.9
J.quail	21.0	23.0	8.2	9.6
Goose	141.8	155.3	144.1	167.8
Pekin	85.7	93.8	67.7	78.8
Muscovy	75.4	82.6	55.9	65.1

Source : Etches, 1996

9.2. Shell

9.2.1. Shell structure

a) Mammillary knob layer : Fibers of the outer shell membrane are incorporated into the shell in areas called mammillary cores; the latter made of primarily protein with carbohydrates and mucopolysaccharides, serve as nucleation sites for an outward growth of hexagonal-rhombohedral crystals of $CaCO_3$. This portion, 100μm thick, contributes most of the organic matter in shell.

In addition, calcium deposits radiate from bottom of mammillary cores to penetrate outer shell membrane during the first 5 hr of calcification. Plumping process stretches the shell membranes and hence increases the distance between the mammillary cores. Therefore, crystals growing laterally come in contact with the neighboring ones; those which grow outward may extend up to shell surface and, at some points, the crystal columns may not meet resulting in shell pores (0.3 to 0.9μm in diameter)(Johnson, 2000); not all need extend from inner shell membrane to shell surface. In fact, only about 6,000 to 10,000 of approximately 20,000 shell pores are of this type (distributed over 0.034% of shell surface), referred to as effective shell pores through which exchange of gases and water vapor is possible; the rest of the shell pores, which may begin either from inside or from shell surface, will terminate within the palisade layer making them unsuitable for respiration and water loss during incubation. Total area of shell pores in a turkey egg (surface area = 90 cm²) is 2.2 mm² (Johnson, 2000), i.e., 0.024%.

b) Palisade (spongy) layer : Organic shell matrix is a series of layers of protein + acid mucopolysaccharides on which calcification occurs. Present within the matrix are calcium-binding proteins and carbonic anhydrase enzyme. Shell matrix along with calcified crystals constitute palisade layer which forms the inner-most region of the shell has a greater density than the outer region. It is made mainly of crystalline $CaCO_3$ and its deposition starts after 5-6 hr of entry into shell gland. The crystals are located as columns, 200μm thick, directly over mammillary knobs that are perpendicular to the shell surface (Johnson, 2000).

c) Surface crystalline layer : This layer is 3-8μm thick, denser than the palisade layer and is perpendicular to shell surface.

9.2.2. Chemical composition

Table 7.7 Chemical composition of egg shell calcium

Component	g %
Water	0.5
Protein	3.2
Ash	95.3
Ca	37.7
K	0.0414
Na	0.0964
P	0.106
Fe	0.0016
Mg	0.376
Pb	< 0.000007
	Source : Sim *et al.* 2000

9.2.3. Respiration through egg shell

Number of shell pores depends on metabolic demand before functioning of lungs and it decreases with egg weight (Johnson, 2000). On average, a fertile chicken egg consumes 4.6 l oxygen and produces about 3.9 l carbon dioxide. An infertile egg, on the other hand, loses 400 mg of water per d when kept at 50% relative humidity (RH). Average diffusion loss during incubation for 117 species of birds is 15% (Johnson, 2000).

Egg shell ($CaCO_3$) has 2% organic matter with trace amounts of Mg and P; function of organic matter is not known.

Strength of egg shell is partly due to a) palisade layer and b) compact and vertically arranged $CaCO_3$ in the upper palisade layer + cuticle; both contributing equally. Initiation of a crack cannot be prevented by mere strong shell material but strong shells will have ability to prevent cracks from growing. Pressure required from inside to cause breakage (explosion) is more than 6900 Pa (100 lb/In^2) which is more than that for an automobile tyre.

Shell quality can be assessed by specific gravity, thickness, weight, weight per unit area etc. Deformation of shell under a constant mass or force required to break the shell also can be used to measure shell strength. However, none of these methods is very precise to accurately assess the shell quality.

9.3. Shell membranes

The two shell membranes are apposed to each other except at air cell. They are mats of intermeshing of fibers lying parallel to the surface of the egg. Each fiber has a protein core of 0.53 nm and 0.80 nm diameter in inner and outer shell membrane, respectively, surrounded by a glycoprotein mantle of 0.13 nm in both the membranes. Larger fibers of outer shell membrane are more tightly woven than the inner shell membrane.

Inner shell membrane is 50-70μm thick whereas outer shell membrane is 15 to 25μm in thickness. The protein network is stabilized by disulfide bridges and lysine-derived bonds. Being made of collagen 10% and glycoprotein 70-75%, it is semi-permeable allowing gases, water and crystalloids but not albumen. Shell thickness and membrane thickness are not correlated; membrane thickness reduces with age (Johnson, 2000).

9.4. Albumen

In a fresh egg, albumen is made of 57% thick albumen, 17% inner thin albumen, 23% outer thin albumen and 3% chalazae. Albumen is a major reservoir of water for the embryo and poorer source of protein than yolk (at best, may provide equal amount of protein as that of yolk). It has traces of lipid and 1/3 rd of the mineral content of the egg. Excepting niacin and riboflavin, only few vitamins are stored in albumen. Minerals also are unequally distributed between albumen and yolk. Ovomucin fibers in chalazae are twisted the opposite directions.

Major proteins in albumen are listed in Table 7.8.

9.5. Yolk membranes

Yolk membrane has 4 layers; outer two from oviduct and inner two from ovary; the innermost is called vitelline membrane and it has trilaminar structure made of phospholipids bilayer, 8 nm thick, tightly apposed to perivitelline layer. It is perforated at vegetal region (distant from blastodisc) and in the region of blastodisc, it has microvilli projecting into perivitelline space, some of which even penetrate the perivitelline layer and help as sperm-binding sites. After fertilization and few nuclear divisions, vitelline membrane forms boundary for blastomeres.

Perivitelline layer, which is likely to be secreted by granulose cells in the developing follicle, is made of loosely intermeshed network of fibers 4μm thick.

Middle continuous layer and extra (outer) vitelline layer is secreted sequentially by the oviduct soon after engulfing of the ovum. The middle layer is amorphous (50-100 nm thick) and extra-vitelline layer, thickest of all layers (6μm), is made of inter-woven fibers deposited in concentric layers around the ovum.

Table 7.8 Major proteins in albumen

Protein	Remarks
Ovalbumin	Major source of amino-acids for the embryo; decreases enzymatic activity of the egg.
Ovotransferrin	Also called conalbumin; Iron-chelator of
Ovomucoid	Serine protease inhibitor; Major protease inhibitor of the egg (primarily trypsin); modulates enzymatic degradation of albumen during embryo development.
Ovomucin (α and β)	Glycoprotein; gives gel-like quality to albumen, decreases microbial invasion, has antiviral properties (?)
Lysozyme	Lyses Gram-positive organisms
Flavoprotein	Binds riboflavin
Avidin	Binds biotin; progesterone-dependent
Ovoinhibitor, Cystatin, Ovo-macroglobulin	Protease inhibitors
	Source : Johnson, 2000

9.6. Yolk

White yolk accumulates slowly from hatch till the small follicle is recruited into the hierarchy of large follicles 5 to 7 d before ovulation. As successive layers of yolk (yellow and white alternatively) are deposited concentrically around the core, the female pro-nucleus rises to the surface of the yolk on a column of white yolk which is contiguous with the core. The summit on which the ovum is located is nucleus of Pander. Most of the yolk (95-99%) is deposited 7-9 d before ovulation (North, 1972).

Chapter **8**

Male Reproductive System

1. General

Class *Aves* is lower in the evolution ladder and therefore, analogous to those in females, nature has provided several safeguards so as to ensure that the species survives and remains propitious. For instance, sperm concentration in semen as well as longevity of spermatozoa inside the oviduct is higher in Avians than those in higher species. On average, sperms are viable in the sperm-host glands at the utero-vaginal junction, for a period of 7-14 d in chicken and 40-50 d in turkeys. Functional significance of similar glands in infundibulum (found in chicken, doubtful in turkeys) is not known (Johnson, 2000).

Sex determination in non-ratites depends on the ratio of Z chromosome to autosomes because this ratio determines whether the enzyme aromatase is produced or not; if produced, testosterone will be converted to estradiol and the embryo becomes female and *vice versa*. Mullerian inhibiting substance inhibits synthesis of aromatase and the embryo develops into a male (Kirby and Froman, 2000).

Formation and differentiation of male reproductive system during incubation is summarized in Table 8.1.

Table 8.1 Formation and differentiation during incubation

Day	Gonadal events
1	Wolffian duct begins to form
1 ½	Primordial germ cells enter circulation
2	Mesonephros begins to form
2 ½	Germinal ridge formation
3	Primordial germ cells enter germinal ridge
4	Wolffian duct opens into cloaca
5	Onset of mesonephric function; Rete cords appear
5 ½ - 6 ½	Primary sex cords appear
7	Gonadal differentiation
9	Rapid growth of interstitial tissue
10-15	Interval of maximum mesonephric function
11	Contact between testis and mesonephros limited to hilus of testis
13	First spermatozoa appear
20	Differentiation of *rete testis* begins

Source : Kirby and Froman, 2000

In chicken and turkeys, by 3 to 4 weeks and onwards, males become heavier than females. Mature turkey hen weighs approximately 70% of the mature tom turkey. Muscles will be longer with more sarcomeres in males than in females. In case of broilers, sex dimorphism is evident by 7th day of age itself which is suggestive that sex dimorphism may not be related to hormones. Bone growth continues for a longer period, the feed consumption is higher and fat deposition is more in males than females; in other words, female have to consume more feed to deposit the same mass than males.

In Japanese quail females are larger than the males because of greater mass of reproductive system (but no difference in carcass weight). Unlike chickens and turkeys, in guinea fowl, females are bigger the about 12 to 14 weeks of age with larger breast muscles and skeleton and greater deposition of fat before sexual maturity.

Caponization, the removal of testes, neither affects production characteristics (growth, FCR) nor carcass quality in chicken and turkeys; therefore, it is out of practice now.

Formation of spur in case of males is not dependent on hormones but is only a sex-limited trait. Spurs will be longer and larger especially during first year after maturity in case of males.

Amount of androgen required to decrease secretion of GnRH increases as the birds grow; and therefore, LH increases as sexual maturity approaches. Peak LH concentration is recorded after sexual maturity and concentration of androgens is analogous with spermatogenesis and testicular growth. Increased concentration of plasma LH with sexual maturity is because of loss of ability of androgens to suppress LH from pituitary. Increased concentrations of plasma androgens stimulate comb growth in both sexes.

Functions of male include a) production of semen and b) to copulate and ejaculate semen when allowed for natural mating. A sexually active rooster produces, on average, 3 billion (3×10^9) sperms/d i.e., 3.5×10^4 sperms/sec (Etches, 1996).

2. The testes (Etches, 1996)

2.1. Structure

The reproductive system of the male consists of paired testes, epididymis (ED) and highly convoluted deferent duct running along ureter.

The testes are located in the center of the body cavity attached by ligament (mesorchium, which also conducts nerves and blood vessels) to the dorsal surface of peritoneal cavity adjacent to adrenal gland and ventral to kidneys; dorsal surface is very close to the ventral surface of the spine. Both the testes are functional. At birth, their total weight is 2-4 g which increases to 25-35 g at maturity; left testis heavier by 0.5 to 3.0 g than the right one.

Each testis is surrounded by connective tissue (CT) capsule containing two types of parenchymal tissue namely anastomosing seminiferous tubules (ST) and interstitial tissue. In a sexually mature male, seminal epithelium is compartmentalized into basal and adluminal regions through tight junctions between adjacent Sertoli cells. It contains developing germ cells in distinct associations (stages) which are arranged sequentially in a helix that extends along the length of ST. The interstitial tissue contains blood vessels, nerves, lymphatic vessels, peri-tubular epithelial cells and Leydig cells (LC) (Kirby and Froman, 2000).

On the dorso-medial aspect of the testis is a series of ducts (rete testis, efferent ducts, connecting ducts and the epididymal ducts) that ultimately empties into deferent duct called as epididymis (ED), which in association with deferent ducts is referred to as excurrent ducts. Each of the distal deferent ducts straightens and then abruptly widens at the juncture with the cloaca to form receptacle of the deferent duct. Each of the deferent ducts terminates in the cloaca urodeum as a papilla immediately below the ostium of ureter. The receptacle becomes bean-shaped when engorged with semen (Kirby and Froman, 2000).

2.2. Accessory reproductive organs

The accessory reproductive organs, which are proximal to or integral part of cloaca, consist of a) para-cloacal vascular bodies - located alongside the receptacle of the deferent duct and are necessary for lymphatic tissue tumescence because they are the sites where lymph is produced by ultrafiltration of blood, b) dorsal proctodeal gland - located within the wall of proctodeum; during copulation/ massage, the non-intromittent phallus forms as tumescent lymphatic tissue which is everted through vent just before ejaculation and c) lymphatic folds (Kirby and Froman, 2000).

2.3. Hormonal profiles and spermatogenesis

Duration of photoperiod is the principal environmental factor that stimulates spermatogenesis in *Galliforms*. However, photoreceptors does not seem to be located either in retina or in pineal gland; but may be found in brain itself at medial basal hypothalamus and *lobus parolfactorius* in the ventral fore-brain. Deep encephalic photoreceptors modulate the activity of GnRH secreting neurons in the hypothalamus. GnRH causes the release of FSH and LH; the former acts on Sertoli cells potentiated by testosterone. The latter acts on Leydig cells which produce testosterone and androstenedione (a precursor of testosterone). Testosterone is required for spermatogenesis, maintenance of excurrent ducts and secondary sexual characters/behavior.

The LC being dispersed between the tubules, produce several androgens, primarily testosterone. As sexual maturity approaches, production of testosterone is stimulated by increased plasma concentrations of gonadotrophins, especially, LH. In mature males, blood levels of LH is maintained by the following negative feedback loop : increased testosterone levels → decreased secretion of GnRH → inhibits secretion of LH → decreased concentration of LH → decreased concentration of androgens → increased secretion of GnRH → increased secretion of LH.

ST → network of inter-connected ducts → empty into *rete testis* (RT). Periphery of ST is lined with spermatogonia (SG, I stage of spermatogenesis, diploid) which undergoes mitosis so as to maintain constant population of stem cells for spermatogenesis (diploid); SG will be differentiated to spermatocytes (haploid) by the Sertoli cells (SC).

SC, situated at the periphery of the tubules along with SG, may produce androgen binding protein so as to increase androgen levels in ST, estrogen and inhibin. Estrogen may have an important role in differentiation and maturation of sperms. Inhibin and estrogen may regulate secretion of FSH *via* a negative feedback loop similar to that of LH and testosterone. After proliferation of SC, there will be increased concentration of androgens which, in turn, stimulates meiosis.

Transformation of the SG begins at the periphery of the ST; at each stage, it is transported closer to the lumen of the ST. Spermatozoa (SZ) will be released to the lumen and their maturation is synchronized within the regions of each ST. Therefore, all germ cells within a region are at the same stage of differentiation whereas, different regions will be at different stages of maturation. Hence, production of SZ is continuous although spermatogenesis in each section of ST is phasic.

Ability of testes to produce sperms is dependant on proliferation of SC which occurs before sexual maturity. From birth to 10 weeks of age, increase in weight of testes is only about 60-100 mg and number of SC increases from 10^6 to 10^8. If one testis is removed within a few weeks after hatch, the other enters into compensatory growth to achieve the same testicular mass and SC as in combined testes. If hemi-castration is done after mitosis of SC is complete, no compensatory growth occurs. Therefore, number of SC is critical for spermatogenesis.

2.4. Semen

2.4.1. Volume and concentration

Number of SC and daily sperm production is a function of testicular size; large-sized males will have larger testes and produce more semen. If all factors are constant, daily production of semen is constant.

Up to 87% of daily sperm production may be collected by abdominal massage method provided that the males are mating frequently or ejaculated 5 times a week in artificial insemination (AI) program. In the absence of ejaculation, spermatozoa are stored for a few days in lower vas deferens to be ultimately reabsorbed into excurrent ducts.

Volume and concentration of sperms decrease if ejaculate is collected frequently. Therefore, frequency of collection depends on the objective; if largest number of offspring per mating is required, semen collection is made once a week and if largest number of offspring per male is required, collection is made 5 times a week. Volume and concentration of semen from different species, number of sperms Vs number of ejaculations per week and motility of sperms at different regions are given in Table 8.2, 8.3 and 8.4, respectively. However, number of fully formed SZ/g of testis/d is 92.5 x 10^6 and 80 - 120 x 10^6 in *Coturnix* and *Gallus*, respectively.

In chicken, color of semen will be pearly white with a pH of 7.0 to 7.2 and is collected when the cocks are at least 20 to 22 weeks (light breeds) or 22 to 24 weeks (heavy breeds) of age.

2.4.2. Transport and storage

Fluids from ST move SZ released into the lumen to RT. Re-absorption of the fluids takes place at RT and *vasa efferentia* resulting in concentration of SZ which then pass through ED to *vas deferens* (VD), reservoir of semen. In case of *Coturnix*, sperm concentration is enhanced by 60 folds and transport through excurrent ducts takes about 24 hr (several days in chicken). Cells of RT and ED synthesize and secrete proteins which help SZ to gain motility and fertilizability. Release of fully formed SZ into lumen of ST is called as spermiation.

SZ can reach lower VD from ED in 24 hr and most of SZ in ED would have been transported within 72 hr. 40-60% of SZ/semen is given out in the first ejaculate on abdominal massage.

At ejaculation, semen from VD flows out, on contraction of well-developed muscles in the lower region, through papillae into urodeum, lymph-engorged lateral phallic folds of proctodeum and to cloaca in that order; it exits with such a force that it seldom touches cloaca.

Copulatory organs of water-fowl are larger, although functionally similar, than gallinaceous birds; the phallus-like folds extend several cm from cloaca. The semen is transported in an open groove (*Phallus sulcus*) that spirals down the outside surface of the protruding organ while base of the phallus fills the vent.

Table 8.2 Volume and concentration of semen

	Volume, ml		Concentration *		
	Mean	Range	Mean	Range	No. of AI **
Broiler Breeder	0.350	0.10-0.90	5.7	3.0-8.0	20
Layer - Light weight	0.150	0.15-0.30	5.0	5.0-7.5	7.5
Layer - Medium weight	0.200	0.08-0.50	5.0	3.5-6.0	10
Turkey - Light weight	0.150	0.08-0.30	9.0	8.0-14.0	13.5
Turkey - Heavy weight	0.200	0.10-0.33	9.5	9.0-13.5	19
Guinea fowl	0.075	0.05-0.15	6.0	4.0-8.0	4.5
Duck - Pekin	0.230	0.10-1.00	4.0	0.02-6.0	9.2
Duck - Muscovy	1.10	1.0-1.5	1.8	----	20

* 10^9 sperms/ml ** 10^8 sperms / AI *Source* : Etches, 1996

Table 8.3 Number of sperms *Vs* number of ejaculations

Characteristic	Number of ejaculations/week		
	2	3	5
Number of sperms/wk, billion	22	25	23
Number of sperms/ejaculate, billion	6	5	4
Number of sperms collected/week, billion	11	15	20
% of sperms produced that was collected	50	60	87
Potential number of AI/week *	110	150	200

* 10^8 sperms / AI billion = 10^9 *Source* : Etches, 1996

Table 8.4 Motility of sperms at different regions

Characteristic	Source of spermatozoa (n = 30 Cocks)		
	Testis	Epidydimis	*Vas deferens*
Motility	Rare	< 50%	All
Freezing capacity *	3 / 69	5 / 69	57 / 77
% hens fertile	4	13	74
Fertile eggs by fertile hens	3 of 19	16 of 36	250 of 351
% fertile by fertile hens	16	44	71

* Fertile hens/hens inseminated *Source* : Etches, 1996

2.4.3. Secondary sex glands

Domestic birds lack secondary sex glands (*Cf* : mammals). Therefore, seminal fluid is derived entirely from testes and/or excurrent ducts; lymphatic exudates may be added to the semen in case of abdominal massage for collection of semen. Chicken, turkeys and Japanese quails have a gland in the dorsal proctodeum. This gland can be considered an accessory sex gland in case of Japanese quails because, unlike in chicken and turkeys, it contributes, under the influence of androgens, a foamy secretion to the semen. The foam has been found to increase sperm transport and might provide high oxygen levels for aerobic metabolism of the SZ thereby contributing to fertility. Removal of the foam glands leads to severe reduction in fertility.

2.5. Spermatozoa

2.5.1. Morphology

Transformation of spermatids to SZ without further cell division is referred to as spermiogenesis which consists of a) formation of acrosome and axoneme, b) loss of cytoplasm and c) replication of nucleohistones with nucleoprotamine which accompanies nuclear condensation. Spermatids condense to about 3% of their volume during spermiogenesis.

Spermatozoa are long, cylindrical cells tapered at both ends, 0.5-0.7μm at their widest point, 100 μm long and 10 μm^3 in volume whereas SG are 6 μm in diameter and 75 μm^3 in volume. SZ consists of conical acrosome (AC), head (H) having a slightly bent, cylindrical nucleus, mid-piece (MP) and tail (T) all embedded within the cell membrane.

Acrosome, a derivative of Golgi complex, contains at least one proteolytic enzyme which is required for fertilization.

Head contains nucleus, with highly condensed chromosomes, with its membrane abutting cell membrane thereby eliminating cytoplasm. Mitochondria and cytoskeleton of the cell give rise to MP and T giving motility to the SZ.

Mid-piece is attached to H by proximal centriole (PC); distal centriole (DC) forms the core of MP and is fused with the PC. DC also gives rise to 2 central fibers projecting into T; these fibers are surrounded by 9 sets of triplet tubules (presumably originated between PC and DC) extending throughout the proximal to the end (distal) piece of T. MP is surrounded by a helix of about 25-30 mitochondria for fueling the activities of SZ. The long flagellum accounts for 84% of the cell's length.

2.5.2. Metabolism

It is surprising that, *in vivo*, fertilizing capacity of SZ lasts for 7 to 14 d whereas, *in vitro*, it decreases within 15 min at body temperature. Further, from testes up to ED, SZ do not receive energy nor dispatch waste directly to blood stream.

Cock and tom semen differ in energy metabolism; the former can utilize glucose both aerobically and anaerobically whereas the latter can utilize only aerobically. Fowl SZ can utilize PL when glucose is not available; but it cannot metabolize glutamate although it is available at high concentrations. Therefore, glutamate is not an essential component in any semen diluent but it is beneficial as a preservative probably as an alternative for Cl^- ions.

Seminal fluid is a poor reservoir of glucose and hence cannot cater to the energy requirements of SZ. If lymphatic fluid is present in semen (as in case of abdominal massage), it contributes significant amounts of protein. Composition of seminal plasma is given in Table 8.5.

Table 8.5 Composition of seminal plasma *Vs* blood plasma *

Component	Seminal plasma		Blood plasma
	Chicken	Turkeys	
Glucose	0-18	--	12
Cl^-	46	23	121
Na^+	145	140	160
K^+	13	20	6
Ca^{++}	1.4	0.3	6
Glutamate	75	88	0.2
Lactate	3.7	2.4	5.5
Pyruvate	0.3	0.4	0.4
α-keto glutarate	0.4	0.2	0.1
Carnitine	3.2	1.7	0.2
Acetyl carnitine	0.5-2.0	0.5-2.0	0.1
Protein (g/l)	8	22	40

* concentrations in mM

Source : Etches, 1996

3. Mating/ AI

3.1. Natural mating

Birds are polygamous with exception of few species like pigeons, swans, etc. Neither the presence of male nor mating is a pre-requisite for ovulation. However, in flock matings, peck-order is important for optimum fertility; dominant males prevent low-ranking males from mating by establishing their territory, especially when in close proximity. Cockerels may mate up to 30 times a day, but, on more than half of those matings, there will be no transfer of semen. Frequency of mating appears to be higher during morning and afternoon.

3.1.1. Courtship

3.1.1.1. Male

Courtship in cocks/cockerels can be described over the following three steps :

a) Waltz – male drops one wing and approaches the hen with short shuffling side-steps. It may also venture a rear approach wherein it holds the comb/ neck of the hen or flaps its wings over the hen.

b) Tidbitting – circling round the hen with exaggerated high steps, pecking and scratching the ground while giving calls.

c) Cornering – male runs to a corner, stamps its feet and settles down in the litter with movements that make a nest-like depression while vocalizing to transmit courtship message to the hens. It may also exhibit dust-bathing, wing-flapping, head-shaking, feather-ruffling, tail-wagging, bill-wiping (ducks and geese), preening, strutting (turkeys) and whining.

3.1.1.2. Female

Receptive hen crouches before the male with slightly spread legs. The male mounts, grasps the neck feathers with the beak and treads on the hen's back. The hen raises the tail and moves it to one side while the male lowers its tail and slides over the rear of the hen. The cloaca is everted and cloacal contact is effected for a very brief period of < 1 sec.

3.1.1.3. Mating

Semen is ejaculated through the engorged phallic folds into everted cloaca; the male quickly retracts and slides off the female. The female assumes a characteristic stance and in about 3 to 4 sec shakes vigorously while the male may circle and waltz around the female.

Males prefer to mate with females having a phenotype that they recognize from the rearing period, but will mate with other females also. In any case, correlation courtship behavior and mating is poor.

In small flocks, females mate with the same male in the absence of morphological differences in the male

Courtship in case of turkeys is called strutting; the male drops the wings so that its tips scrape the ground while stepping forward in a deliberate motion and trumpeting a series of deep notes. Frequency of such behavior increases with sexual maturity. Mature toms in a flock tend to trumpet together.

3.2. Artificial inseminaton

3.2.1. Advantages

1. Method of choice for pedigree mating.
2. Allows unlimited number of single male matings without requiring extensive breeding equipment.
3. Preferential mating avoided.
4. Problems of trap-nesting avoided.
5. If, due to some reason(s), a male of superior qualities cannot mate, it can still contribute to next generation.
6. Very old and/or heavy males which are not able to mate can still be used for breeding.
7. In layers raised in conventional cages, fertile eggs can be obtained.
8. Incidence of sexually transmitted disease(s) can be minimized.
9. In turkeys, due to heavy body, under natural mating, on most occasions, fertility will be very low

10. In case of Guinea fowl, it is common that the males will have a harem of 3 to 4 females. Therefore, in order to broaden the genetic base, AI becomes a necessity
11. In Muscovy as well as common ducks, mating ratio required for optimum fertility under natural mating is high. Hence, AI will help obtain fertile eggs at low cost.
12. For production of inter-specific crosses

3.2.2. Disadvantages

1. Labor-consuming
2. Chances of cross-contamination of birds, especially paratyphoid infections, through AI equipment.
3. Involves handling of birds.

3.2.3. Semen collection

In chicken, turkeys, Guinea fowl, Pheasants and Quails, semen collection can be done by abdominal massage method. In case of Ducks and Geese, semen is usually collected by intercepting the flow of semen during natural mating.

3.2.3.1. Abdominal massage method

Semen is obtained by massaging the male's back and moving the hand in the direction of the bird's vent and supporting the bird by keeping its legs between the operator's legs or held by an assistant. The fingers of the hand are outspread at the start of the stroke so as to converge on the vent. A trained male would evert the vent by the strokings and the operator must quickly move his hands one above and the other below so as to squeeze the protruded vent by the thumb and index finger inward and downward at a point just above the vent to force the yield of semen. Sustained pressure at this point will cause some males to yield more semen. Unrestrained and relaxed males yield more semen than restrained and excited ones.

Abdominal massage is not identical to natural mating but still it stimulates ejaculation and flow of semen by tumescence of the phallic folds. Semen is collected by aspiration into an ampoule containing semen diluent at 15°C.

Advantages of using diluents are a) prevents cold-shock to the SZ since the vial itself will be cooler b) diluents buffer the acidic products of metabolism of SZ and c) volume of semen to be handled will be practical and convenient commercially.

Since Avian SZ can metabolize glucose/fructose, one of these is included in all diluents. TES, BES and PO_4^{--} act as buffers in the range of pH 6.8 to 7.5. Milk powder and albumen are added to help freezing.

Under commercial conditions wherein semen dilution and/or storage is/are not practiced, semen may be collected in a funnel (with tubular portion blocked with paraffin wax) or in small sugar tubes or any other suitable container and inseminated, with the help of tuberculin syringe, within 30 min of collection.

3.2.4. Semen evaluation

Semen collected is evaluated so as to assess the fertilizability of the SZ and also to fix dilution rate to ensure 10^8 sperms/AI.

Color of semen must be white to pearly white; yellow (fecal contamination) and brownish red (presence of RBCs) colors are not acceptable. Motility is assessed over a subjective scale of 1 to 5/10 with lower values for poor motility using diluted semen (1:3 or 1:4). Neat semen can be viewed under microscope for swirling mass of cells sweeping across the field. Volume is found to fix dilution rate later and if weigh of semen is fund, density can be calculated as the ratio of weight to volume; density of semen is expected to be 1 mg/μl. Sperm concentration can be assessed either by automatic counters (spermatocrit) or by transmittance (spectrometric). Integrity of SZ can be estimated by staining technique (eosin-nigrosin or eosin-aniline) to count live and dead SZ or by fluorescence technique (ethidium bromide).

3.2.5. Insemination dose

Minimum number of SZ for optimum fertility depends on several factors including age of male and female, method of collection and interval between collection and AI, technical skill etc. Theoretically, 5×10^7 SZ appear to be sufficient; but, under commercial conditions, 10^8 (chicken) to 2×10^8 (turkeys) SZ are recommended. In case of turkeys, AI a week prior to onset of lay has been found to increase fertility throughout the laying period.

Therefore, in chicken, number of AI possible can be found out by dividing the product of volume of semen (ml) and concentration (sperms/ml) by 10^8. In general, 5 to 20 AI can be done per collection and as each male can yield semen 4 to 5 times/week, 40 to 200 hens can be inseminated over a week out of semen from a cock.

Commercially, it is difficult to maintain accuracy if volume to be inseminated is < 0.05 ml. Automatic AI pipettes are available for easier handling and dispensation of diluted semen.

In view of the above, total volume of diluted semen (ml) can be calculated as follows :

Let volume of undiluted semen be x, concentration per ml in undiluted semen be y, sperms per AI be 10^8 and volume of undiluted semen per AI be 0.05; then, total number of inseminations possible = $(x.y)/10^8$ and each insemination dose is 0.05 ml; therefore, total volume of diluted semen required is $0.05 \{(x.y)/10^8\}$.

In other words, x ml of undiluted semen, containing y SZ/ml, should be diluted to a volume of $k.\{(x.y)/10^8\}$ ml which will suffice AI of $\{(x.y)/10^8\}$ hens; k is the volume of diluted semen (containing 10^8 SZ) to be inseminated to each hen. Although k is usually 0.05 ml, it can be changed depending on convenience.

Assuming that a cock yielded 0.4 ml of semen (x) with SZ concentration of 3 x 10^9/ml (y), it can be used to inseminate (0.4 x 3 x 10^9)/10^8 = 12 hens ($x.y/10^8$); since minimum dose of AI is 0.05 ml (k), total volume of diluted semen required for inseminating 12 hens is 12 x 0.05 = 0.6 ml [$k.\{(x.y)/10^8\}$]. That means, 0.4 ml of undiluted semen must be extended to 0.6 ml in order to facilitate AI of 12 hens with 10^8 sperms in an insemination dose of 0.05 ml.

3.2.6. Timing of AI

Domestic birds differ widely in timing of oviposition and hence, timing of AI also changes according to species. In case of chicken, AI done at or within 2 hr after oviposition resulted in 20 to 40% fewer fertile eggs. This is presumably because of fewer contractions of shell gland and vagina 1 to 3 hr after oviposition preventing the movement of SZ into sperm tubules.

Therefore, AI has to be done when majority of the birds have laid and fertile eggs are expectable only on second day after AI because at the time of AI most of the ova would have left infundibulum afterwhich no fertilization can occur. High fertility can be expected for 7 d (chicken) and 14 d (turkeys) after AI. It is highly recommended to repeat AI after 5 d (chicken) or 7 d (turkeys) to ensure high fertility. Under commercial conditions, some poultry breeders do perform AI twice a week.

3.2.7. Placement of semen

Cloaca is everted by securing the bird and applying a gentle pressure on the abdomen towards the vent; White Leghorns commonly evert the cloaca with little effort whereas, broiler breeders may require training and careful handling for satisfactory results. Vaginal entry is exposed (constricted opening can be seen on the left side of the bird) and the AI pipette is inserted to about 3 cm (chicken) or 6 cm (turkeys) depth. A short interval of few seconds is allowed for the vagina to return to normal position and then the semen is deposited.

It is highly desirable that the AI is performed within 30 min of collection of semen.

3.2.8. *In vitro* storage of semen

Avian semen is more fragile than that of mammals. Therefore, whenever AI is not possible to be completed within 30 min, *in vitro* storage becomes a necessity. Semen can either be held in a liquid medium or be frozen; the latter is not commercial because of fragility of Avian SZ (98 to 99% of SZ are lost due to freezing and thawing) as well as cost and availability of liquid nitrogen.

Liquid semen can be stored up to 48 hr with sufficient supply of oxygen and glucose with sufficient buffering to maintain pH and an ambient temperature of 5 to 7°C. Oxygen can be supplied in large vessels with continuous agitation/pumping of air into semen and requirement of glucose/fructose as well as pH is taken care of by the diluents used which have sufficient buffering ability.

Hence, it is possible to hold diluted semen for 24 hr (turkeys) to 48 hr (chicken) at 5°C with minimum loss of fertilizability; liquid semen is being used even for grand-parent flocks.

Chapter **9**

Composition and Nutritive Value of Eggs

"Milk is an animal product and cannot be, by any means, be included in a strict vegetarian diet. It serves the purpose of meat to a large extent. In medical language, it is classified as animal food. A layman does not consider milk to be animal food. On the other hand, eggs are regarded by layman as a flesh-food. In reality, they are not. Nowadays, sterile-eggs are also produced. The hen is not allowed to see the cock and yet it lays egg. A sterile-egg never develops into a chick. Therefore, he who can take milk should have no objection to taking sterile-eggs "

- Mahatma Gandhi

1. General

Nutritive worth of the eggs was understood in real sense after the knowledge of intermediary metabolism which described the role of several nutrients in human health was discovered. Subsequently, information on protein quality, role of vitamins and minerals etc. further strengthened importance of egg in human nutrition. In addition, unlike milk, that egg cannot be adulterated, and can be held at room temperature for a longer time added to the popularity of eggs in human diets.

It is said that there are thousand methods of cooking eggs which indicates the versatility of eggs to suit several processing manipulations. The advent of sophisticated analytical equipment like amino-acid analyzer, gas-liquid chromatography etc. provided the detailed chemical composition of eggs which clearly proved the superiority of egg in human diets. It is now generally accepted that eggs are the Nature's gift to mankind.

At the beginning, there was sufficient resistance for egg consumption in India due to the misconception that egg has life in it and hence is non-vegetarian. It is noteworthy that food obtained from vegetation alone can be called "vegetarian" and all others are naturally "non-vegetarian"; therefore, both milk and eggs are definitely non-vegetarian. Further, same as the milk cannot give birth to a calf, table eggs also can not yield a chick because both lack a zygote. It is worthwhile to mention that milk has more number of cells per unit volume than eggs.

Therefore, it is reasonable to state that if milk is considered vegetarian, then eggs are more vegetarian and if milk is considered non-vegetarian, then eggs are less non-vegetarian than milk.

Consequent on improved literacy and general awareness, there is a substantial improvement in proportion of population accepting egg in their diet. As per a sample survey conducted in 1996, Indian population consisted of vegetarians 21.1%, persons consuming only eggs, no meat 3.7%, non-vegetarians 74.2%; as income increased, vegetarians and those consuming only eggs increased and non-vegetarians decreased. Non-vegetarian households cooked 91.5% mutton, 82.6% fish, 76% chicken, 12.7% beef, 2.7% pork. As per a recent FAO (2002) report, 80% of India is non-vegetarian.

Eggs are excellent for growing children because they provide nutritional variety through high-quality protein for muscle growth in response to exercise and help balance the energy being a low energy high protein food. Those who are interested in weight loss, egg being nutrient-dense, are one of the important components of their diet; in addition, it helps control blood sugar, triacylglycerol and HDL cholesterol levels. For elderly, high-quality protein of the egg will help counter sarcopenia (age-related loss of skeletal muscle mass) because old people require more protein/kg than young adults. For pregnant women, choline is an important nutrient the development of neurons and memory centers and reduction in programmed cell-death and lifelong changes in nerve growth factors and calretinin; egg is an excellent source of choline (Evans, 2004).

It is necessary to note that, like any other dietary ingredient, eggs do have shortcomings and/or limitations. Developments in feeding and nutrition have helped not only to minimize such shortcomings but also to produce eggs enriched with specifically desired nutrient(s); such eggs are referred to variedly as "diet eggs", "designer eggs", "enriched eggs" etc.

Premium/designer/diet eggs specially developed to contain desired components such as ω-3 fatty acids, lutein, and Vitamin E as well as organic, free-range, vegetarian, GM-free and natural eggs are now available. Consumption of such eggs increases plasma levels of lutein and zeaxanthin which are required for lowering the risk of age-related macular degeneration, atherosclerosis and cataract. Eggs providing disease-fighting proteins are likely to be developed through application of biotechnology in the near future; such eggs, called immune eggs, will contain biological compounds having immuno-regulatory and anti-inflammatory properties. Egg yolk is an excellent source of iron for infants and teenage girls consuming eggs are less likely to develop breast cancer and cardiac arrest (Evans, 2004).

2. Gross composition

Proximate composition of eggs from different species is given in Table 9.1.

Table 9.1 Gross composition of eggs from different species

	Chicken	Duck	Goose	Quail	Turkey
Total egg weight, g *	56.8	79.5	163.6	10.2	89.8
Weight , g (Edible portion) **	50	70	144	9	79
Energy, MJ/100 g edible portion	0.660	0.774	0.774	0.661	0.715
g/100 g edible portion					
Water	74.57	70.83	70.43	74.35	72.5
Protein	12.14	12.81	13.87	13.05	13.68
Lipid	11.15	13.77	13.27	11.09	11.88
Carbohydrate	1.20	1.45	1.35	0.41	1.15
Ash	0.94	1.14	1.08	1.10	0.79
Fiber	Nil	Nil	Nil	Nil	Nil

* Calculated
MJ (Mega joule) = 239 kcal

** Shell = 12% of egg weight
Source : Stadelman *et al.* 1988

The gross composition of eggs from different species is more or less the same though not identical. Hence, further discussions in this Chapter primarily refer to chicken eggs and it is reasonable to assume that the same are applicable to the eggs of other species also.

When the gross composition of eggs is compared with those of most commonly consumed cereals like maize, rice and wheat (Table 9.2), it is clear that starch (carbohydrate) and fiber replace most of water and hence are high in energy and provide bulk through fiber when consumed.

3. Composition of albumen Vs yolk

Another feature that attracts special mention is the near contrasting composition of the components of the edible part of the egg i.e., albumen and yolk (Tables 9.3, 9.4, 9.5, 9.6, 9.7 and 9.8). Albumen and yolk are in the ratio of 2:1 by weight. Albumen has 90% water and contributes most of carbohydrates and more than half of protein, niacin, riboflavin of an egg. On the other hand, yolk has the entire lipid and therefore most of the energy and all the fat-soluble vitamins in the egg. In addition, most of the minerals and other vitamins are concentrated in yolk. Yolk color which is due to xanthophylls has no nutritive value to humans.

4. Proteins of albumen

Albumen has more than 40 proteins of which 7 proteins form about 90% of the total dry matter in albumen (4 g). Fresh egg has 57% thick white, 23% outer thin white, 17% inner thin white and 3% chalazae. Albumen is the major reservoir of water and poorer source of protein than yolk (equal amount of protein at the best) to the embryo. It has traces of lipid and one third the mineral content of egg. Excepting niacin and riboflavin, only few vitamins are stored in albumen. Minerals are unequally distributed between albumen and yolk. Major proteins in albumen are listed in Table 9.9.

Amino-acid composition of all albumen proteins has been determined except that of ovoinhibitor an ovoglycoprotein. Primaries amino-acid sequence and responding mRNA and DNA sequences are determined for ovalbumin, ovotransferrin, ovomucoid, lysozyme, cystatin, ovoflavoprotein and avidin. 3-D structures are proposed for lysozyme, ovalbumin and ovotransferrin and preliminary data on shape of cystatin and avidin are available (Etches, 1996).

Table 9.2 Gross composition of eggs *Vs* maize, rice and wheat

	Chicken	Maize	Rice	Wheat
Total egg weight, g *	56.8			
Weight , g (Edible portion) **	50	100	100	100
Energy, MJ/100 g edible portion	0.660	1.49	1.51	1.40
g/100 g edible portion				
Water	74.57	12	13	12
Protein	12.14	9.5	6.7	12.2
Lipid	11.15	4.3	0.7	2.3
Carbohydrate	1.20	72.9	78.9	71.8
Ash	0.94	1.30	0.70	1.70
Fiber	Nil	2.1	0.4	2.1

* Calculated
MJ (Mega joule) = 239 kcal

** Shell = 12% of egg weight
Source : Ockerman, 1978

4.1. Raw eggs and human health

Egg is a package of nutrients produced by the hen to support healthy and normal development of the embryo, in case it is fertilized. Therefore, there is no reason to expect an egg to contain constituents totally compatible to human beings; on the contrary, proteins in egg albumen (Table 9.9) impart antimicrobial properties so that the microbes gaining entry on to the inner shell membrane during the formation of air cell are prevented from gaining access to the embryo.

Most of the antimicrobials act by binding nutrients which are highly essential for microbial growth or by inhibiting proteases elaborated by the microbes in order to utilize protein present in egg (albumen in the beginning). Hence, it is possible that these anti-nutritional factors may have similar action inside the human alimentary canal, especially when raw eggs are consumed.

Table 9.3 Composition of albumen and yolk : Gross

		Albumen	Yolk	Egg contents	Membranes + shell
Energy, MJ	/100 g	0.172 - 0.213	1.456 - 1.498	0.728 -0.820	Negligible
	/egg	0.100 - 0.126	0.272 - 0.293	0.368 – 0.418	Negligible
Protein, g	%	9.4-10.9	16.0-16.4	12.0-12.9	6.4
	/egg	3.1-3.5	3.0-3.1	6.1-6.6	0.45
Lipid, g	%	Traces	30.6-34.1	11.5-12.3	Traces
	/egg	Traces	5.7-6.3	5.7-6.3	Traces
Carbohydrates, g	%	0.8	0.6-0.7	0.7-0.9	Traces
	/egg	0.3	0.11-0.13	0.14-0.17	Traces
Solids, g	%	10.7-12.4	48.9-50.8	24.7-26.3	98.4
	/egg	3.5-4.0	9.1-9.4	12.9-13.4	6.0
Water, g	%	87.6-89.3	49.2-50.1	73.7-75.3	Nil
	/egg	28.4-29.0	9.1-9.3	37.6-38.4	Nil
Ash, g	%	0.6-0.7	1.7	1.0	92.0
	/egg	0.02-0.23	0.3	0.52	6.4

MJ (Mega joule) = 239 kcal — *Source* : Etches, 1996

4.1.1. Anti-nutritional proteins

4.1.1.1. Avidin

Avidin is a glycoprotein which binds biotin; 4 biotin molecules bound to one molecule of avidin forms a stable complex which is resistant to denaturation and proteolysis. It is presumed that avidin in albumen is not totally satiated and hence if yolk and albumen are mixed, biotin of yolk also will be bound by avidin. However, biotin deficiency due to raw egg consumption is very rare because biotin is not a critical vitamin to human beings since it is widely distributed in nature. But, if raw egg albumen is consumed as a sole source of biotin, typical deficiency symptoms characterized by dermatitis in non-hairy regions, retarded growth and activity, somnolence etc., can appear which is referred to as "raw-egg disease". Biotin gets released from avidin following coagulation of albumen (Osuga and Fenny, 1974).

Table 9.4 Composition of albumen and yolk : Amino-acids, mg

	Whole egg	Yolk	Albumen
Weight of contents, g	55.1	16.7	38.4
Total protein, g	6.6	2.74	3.88
Ala	380	140	240
Arg	420	190	230
Asp	650	250	400
Cys	150	50	110
Glu	850	330	520
Gly	220	80	140
His	160	70	90
Ile	360	150	210
Leu	570	240	330
Lys	450	200	250
Met	210	60	150
Phe	350	120	230
Pro	260	110	150
Ser	500	230	270
Thr	320	140	180
Trp	110	40	70
Tyr	280	120	160
Val	430	160	270

Source : Stadelman and Cotterill, 1995

4.1.1.2. Lysozyme

It cleaves 1-4 glycosidic linkages between polymers of N-acetyl glucosamine and N-muramic acid thereby causing lysis of cell wall of gram positive organisms. It is likely to have little or no effect on human beings (Osuga and Fenny, 1974).

4.1.1.3. Ovomucoid and Ovoinhibitor

Both are sufficiently stable even under acidic conditions expected in stomach. Ovomucoid of quail eggs inhibits human trypsin whereas chicken ovomucoid and ovoinhibitor appear to have no such effect. Anaphylactic reaction encountered in some children under 1 yr of age is thought to be due to ovomucoid (Osuga and Fenny, 1974).

Table 9.5 Composition of albumen and yolk : Lipids, g

	Whole egg	Yolk	Albumen
Weight of contents, g	55.1	16.7	38.4
Total lipid, g	6.0	5.8	Nil
Saturated fatty acids (SFA)	2.01(33.50)	1.95	Nil
8:0, Caprylic	0.027(0.50)	0.027	Nil
10:0, Capric	0.082(1.37)	0.080	Nil
12:0, Lauric	0.027(0.50)	0.026	Nil
14:0, Myristic	0.022(0.37)	0.022	Nil
16:0, Palmitic	1.370(22.83)	1.310	Nil
18:0, Stearic	0.462(7.70)	0.459	Nil
20:0, Arachidic	0.022(0.37)	0.022	Nil
Monounsaturated fatty acids (MUFA)	2.5(42.17)	2.50	Nil
14:1, Myristoleic	0.005(0.08)	0.005	Nil
16:1, Palmitoleic	0.214(3.57)	0.211	Nil
18:1, Oleic	2.310(38.50)	2.250	Nil
Polyunsaturated fatty acids (PUFA)	0.73(12.17)	0.72	Nil
18:2, Linoleic	0.660(11.00)	0.650	Nil
18:3, Linolenic	0.011(0.18)	0.014	Nil
20:4, Arachidonic	0.055(0.92)	0.051	Nil
Cholesterol	0.213(3.55)	0.213	Nil
Lecithin	1.27(21.17)	1.22	Nil
Cephalin	0.253(4.22)	0.241	Nil

Value in the parentheses is calculated as % of total lipid

Source : Stadelman and Cotterill, 1995

4.1.1.4. Ovotransferrin (Conalbumin)

This protein binds 2 atoms of Fe per molecule. It can also bind other cations with the following relative stabilities : $Fe^{3+} > Cr^{3+}.Cu^{2+} > Mn^{2+}.Co^{2+}.Cd^{2+} > Zn^{2+} > Ni^{2+}$. This is very important and potent inhibitor of microbial growth; but, effect of binding cations is yet to be recorded in case of higher animals including man.

4.1.1.5. Ovoflavoprotein

Amount of this protein I an egg is genetically determined. It binds riboflavin on an equimolar basis. However, its nutritional inhibitory effects are not yet recorded in human beings.

Table 9.6 Composition of albumen and yolk : Major minerals

Element	Albumen		Yolk	
	Content	% of total egg reserve	Content	% of total egg reserve
Ca, mg	4	13	25	87
Cl_2, mg	66	70	29	30
Cu, ng	9	27	24	73
I_2, ng	1	4	24	96
Fe, mg	0.1	10	1.0	90
Mg, mg	4.2	66	2.2	34
Mn, ng	2	10	21	90
P, mg	8	7	102	93
K, mg	57	77	17	23
Na, mg	63	88	9	12
S, mg	62	69	28	31
Zn, ng	50	7	660	93

Source : Etches, 1996

Table 9.7 Composition of albumen and yolk : Trace minerals (μg/g)

Element	Albumen	Yolk
Al	0.1	2.7
As	Trace	0.2
Bo	1.2	Trace
Br	---	7.5
F	0.2	4.3
Pb	6.1	8.1
Mo	0.1	Trace
Si	21.0	31.0
Sr	Trace	Trace
Ti	0.1	0.2
V	0.2	0.1

Source : Wells and Belyavin, 1987

4.1.2. Other factors

In addition to the proteins, raw egg is likely to make a thick lining on the mucous membrane of gastro-intestinal tract, especially stomach; this might inhibit secretion of gastric and/or intestinal juices resulting in impaired digestion and/or absorption.

Egg yolk can harbor organisms like Salmonella which can cause gastro-intestinal disorder in human beings; however, such a possibility is very low if eggs are purchased from organized poultry farms.

Table 9.8 Composition of albumen and yolk : Vitamins

Component	Albumen		Yolk	
	Content	% of total egg reserve	Content	% of total egg reserve
Vitamin A, IU	--	--	260	100
Vitamin D, IU	--	--	27	100
Vitamin E, mg	--	--	87	100
Vitamin B_{12}, µg	--	--	48	100
Biotin, µg	258	24	835	76
Choline, mg	--	--	238	100
Folic acid, ng	6	19	26	81
Lonositol, mg	1.5	26	4.4	74
Niacin, ng	35	78	10	22
Vitamin B_6, ng	8	12	57	87
Vitamin B_2, ng	110	61	70	39
Vitamin, B_1, ng	4	8	48	92
Pantothenic acid, ng	90	11	730	89

Source : Etches, 1996

5. Yolk infrastructure

Yolk consists of polygonal, tightly packed yolk spheres 140 µm in diameter bound by tri-laminar membrane 7 nm thick. The yolk spheres contain yolk proteins and lipids in sub-droplets, particles, lamellar bodies and a finely dispersed aqueous phase (Etches, 1996). Composition of yolk has been schematically shown in Fig. 9.1

5.1. Sub-droplets

These are 100-2500 nm in diameter; contain 23% of yolk protein in the form of phosvitin (two molecules) and lipovitellin (two molecules). The four molecules of synthesized in liver as precursors molecules (vitellogenin) and transferred to ovary *via* circulation; these molecules enter ovum through endocytosis and cleaved into a) two phosvitins (molecular weight 28,000 and 34,000; the latter having a carbohydrate side-chain) which bind Ca, Mg and Fe; these are 11 nm granules forming about 80% of protein bound P and containing 54% serine of which 90-95% is phosphorylated and b) two lipovitellins (α and β forms; the latter having 15-20% lipid and molecular weight of 400,000) (Etches, 1996).

Table 9.9 Major proteins in albumen

Protein	Proportion (%)	Molecular weight (kDa)	Characteristics
Ovalbumin	54	45	Enzyme inhibitor, binds Fe, Mn, Zn, Cu and other trace metals
Ovotransferrin	12	76	Binds metallic ions
Ovomucoid	11	28	Inhibits trypsin
Ovomucin	3.5	110	Inhibits viral hemaggutination
Lysozyme	3.4	14.3	Lyses bacteria
Ovoglobulin G_2	4	33-49	Foaming agent
Ovoglobulin G_3	4	33-49	Foaming agent
Ovoinhibitor	1.5	49	Inhibits serine proteases
Cystatin	0.05	12.7	Inhibits thioproteases
Ovoglycoprotein	1.0	24.4	Sialoprotein
Ovoflavoprotein	0.8	32	Binds riboflavin
Ovomacroglobulin	0.5	760-900	Strongly antigenic
Avidin	0.05	16.3	Binds biotin
Others	4.2	Unknown	-----

Source : Etches, 1996

5.2. Particles

These are 27 nm in diameter containing 65% of yolk solids and 95% of yolk lipids and only 12% of yolk proteins. Yolk lipids contain VLDL (molecular weight 3,000 to 10,000 kDa; largest of the yolk macromolecules). The proteins may exist in a membrane-like form within yolk spheres as lamellar bodies (100-200 nm diameter). Apoproteins in plasma are incorporated into VLDL and transferred into yolk by receptor-mediated endocytosis. Of the apoproteins, apo-VLDL-II is the major one being identical to apovitellenin-I of plasma and transferred unchanged into yolk. Apoprotein-B of plasma is cleaved upon entry into yolk to yield apoproteins III, IV, V and VI (Etches, 1996).

5.3. Aqueous phase

Aqueous phase, mainly made of proteins (plasma albumin, plasma α_2 glycoprotein and IgG) with traces of lipid, constitutes 10% of yolk solids. IgG gives maternal immunity and minor proteins present act as carriers of vitamin molecules (Etches, 1996).

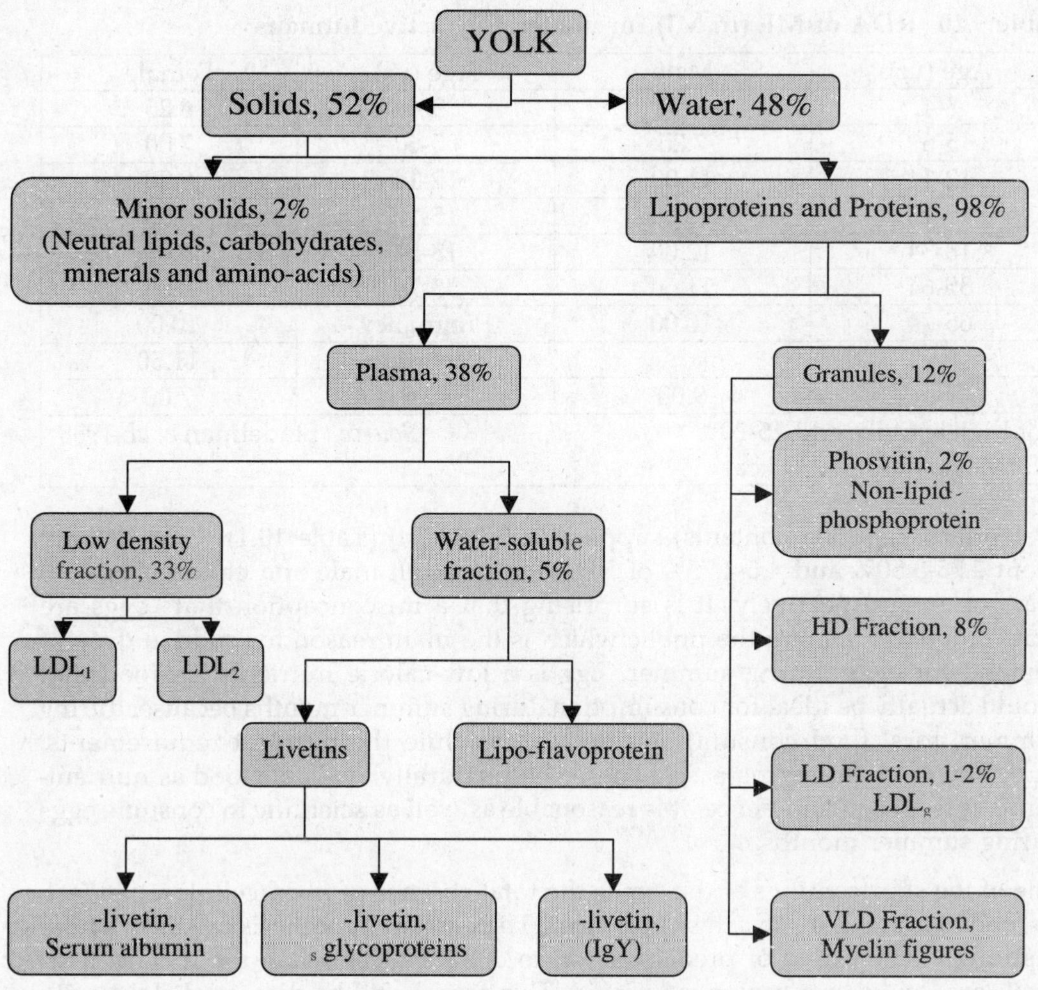

Fig. 9.1 Composition of yolk

Source : Stadelman and Cotterill, 1995

6. Recommended daily allowance (RDA) for humans

6.1. Carbohydrates, fiber and energy

Recommended daily allowance of metabolizable energy (ME) for humans is given in Table 9.10.

Table 9.10 RDA of ME (in MJ) for moderately active humans

Age (yr)	Male	Age (yr)	Female
3-4	6.50	3-4	6.25
5-6	7.25	5-6	7.00
12-14	11.00	12-14	9.00
15-17	12.00	15-17	9.00
18-34	12.00	18-54	9.00
35-64	11.50	55-74	8.00
65-74	10.00	Pregnancy	10.00
		Lactation	11.50
75 +	9.00	75 +	7.00

For high activity add 15-20%
MJ (Mega joule) = 239 kcal

Source : Stadelman *et al.* 1988

A standard egg (50g contents) supplies 0.33-0.41 MJ (Table 10.1) which is only about 2.75-3.50% and 4.5-4.75% of RDA for an adult male and children up to 6 years of age, respectively. It is surprising that a misconception that "eggs are heat" still exists among the public which is the main reason for sudden drop in demand for eggs during summer. Egg is a low-calorie nutrient-rich food that should actually be ideal for consumption during summer months because, during summer, total food consumption decreases while the nutrient requirements, especially of vitamins, increases. There is no naturally available food as nutrient-dense as is an egg and hence, it is reasonable as well as scientific to consume eggs during summer months.

One of the shortcomings of the egg is the total absence of fiber (cellulose); in fact, no food of animal origin can supply fiber. The current hypothesis on dietary fiber is strongly suggestive of predisposition to colon cancer when the dietary fiber levels are low over a prolonged period. This aspect will be discussed along with nutritive value of meat.

6.2. Proteins and amino-acids

Human beings require 6g protein per MJ of energy. That means, an average active adult man and woman requires 72 and 54 g of protein, respectively. Unlike proteins of vegetable origin, egg proteins are not only completely digestible but also are highly absorbable. Biological value (BV), the proportion of absorbed nitrogen retained in the body, of egg protein is 96 and it is the highest among all the foodstuffs; therefore, BV of egg protein is considered as 100 and is the FAO reference protein. In other words, egg proteins are completely utilized by human body. An egg can supply 10-12% of RDA of an average active adult.

High BV of egg proteins indicates that the protein contains amino-acids in required quantity, proportion and in an available form. A large egg can supply 10% of

RDA of protein (McNamara, 2000). Contribution of an egg per day (50g contents) to the RDA of some of the essential amino-acids is shown in Table 9.11.

Table 9.11 RDA of amino-acids met by one egg

Amino-acid	% RDA
Ile	54.00
Leu	49.50
Lys	46.50
Met	17.50
Phe	31.50
Thr	60.50
Trp	28.00
	Source : Stadelman and Cotterill, 1977

Considering that Trp, Met and Lys are most likely to be deficient in diets of Indians, nutrient worth of eggs needs no further emphasis. However, caution has to be exercised while feeding children under 1 yr of age with egg albumen; some of the proteins may be absorbed intact and cause anaphylactic reactions. Yolk can be fed even to children aged 3 months and egg albumen safely to children above 1 yr of age.

6.3. Fat and fatty acids

An average active adult requires 50 g of fat per d and hence an egg can supply 11% of it. Egg lipids contain unsaturated fatty acids twice as that of saturated ones; of the unsaturated fatty acids, polyunsaturated fatty acids account for about 25% and linoleic acid, the essential fatty acid forms 86% of the polyunsaturated fatty acids (Table 9.5).

Cholesterol content is one of the limitations of the egg since it is incriminated as a predisposing factor for coronary heart disease (CHD). This aspect has been highly controversial and considerable literature is available on this issue. Hence, it is discussed under a separate Chapter titled "Egg consumption *Vis a Vis* CHD"

6.4. Minerals

Mineral content (major and trace) of eggs given in tables 10.6 and 10.7 indicates that excepting Ca which is concentrated in shell, egg is a good source of all minerals required for formation of blood, bone and soft tissues. Proportion of RDA (US) met by an egg is tabulated in Table 9.12.

Table 9.12 RDA of minerals met by one egg

Mineral	%RDA
Calcium	2.7
Copper	4.0
Iodine	4.0
Iron	5.8
Magnesium	1.4
Phosphorus	10.0
Selenium	17.0*
Zinc	4.7
* McNamara, 2000	*Source* : Stadelman and Cotterill, 1995

Table 9.13 RDA of vitamins met by one egg

Vitamin	% RDA
A	11.8
D	6.3
E	3.3
B_1	3.2
B_2	10.0
Niacin	0.3
B_6	7.3
Choline	Not required
Pantothenic acid	7.2
Folic acid	4.0
Biotin	7.4
B_{12}	9.2
Source : Nesheim *et al.*, 1979 and Stadelman and Cotterill, 1977	

6.5. Vitamins

Egg is an excellent source of all vitamins except vitamin C which is absent in eggs because embryos can biosynthesize the vitamin in their kidneys. Proportion of RDA supplied by an egg (50g contents) with respect to vitamins is shown in Table 9.13.

7. Nutritive value of poultry products

In India, eggs are mainly consumed popularly as full-boiled, half-boiled and omelettes, in few of the metropolitan cities, various other products are becoming popular as luncheon foods. As many women are taking up employment, egg products industry is gaining momentum and many ready-to-eat products are being introduced into the market. Proximate composition of selected commercial products is given in Tables 9.14 and 9.15.

Table 9.14 Composition of selected egg products, g/100g

Product	Water	Protein	Lipid	Carbohydrate	Ash	Energy, MJ
Frozen						
Whole-egg	74.57	12.14	11.15	1.20	0.94	0.66
Albumen	88.07	10.14	Trace	1.23	0.56	0.21
Yolk	48.20	16.10	34.10	---	1.69	1.58
Sugared yolk	50.82	12.92	25.50	9.49	1.27	1.35
Dehydrated						
Whole-egg	4.14	45.83	41.81	4.77	3.45	2.49
Whole-egg, stabilized	1.87	48.17	43.95	2.38	3.63	2.57
Albumen, stabilized	14.62	76.92	0.04	4.17	4.25	1.47
Albumen flakes, stabilized	8.54	82.40	0.04	4.47	4.55	1.57
Yolk powder	4.65	30.52	61.28	0.39	3.16	2.87
MJ (Mega joule) = 239 kcal				*Source* : Stadelman *et al.* 1988		

Table 9.15 Composition of frozen, dried and cooked whole-eggs

Nutrient	Frozen	Dried	Hard-boiled	Poached	Fried	Omelette or Scrambled
Weight, g	50	13.27	50	50	46	64
Water, g	37.28	0.56	37.28	37.14	33.06	48.83
Energy, MJ	0.33	0.33	0.33	0.33	0.35	0.40
Protein, g	6.07	6.08	6.07	6.04	5.37	5.96
Lipid, g	5.58	5.54	5.58	5.55	6.41	7.08
Carbohydrate, g	0.60	0.64	0.60	0.60	0.53	1.37
Ash, g	0.47	0.45	0.47	0.67	0.63	0.76
MJ (Mega joule) = 239 kcal				*Source* : Stadelman *et al.* 1988		

7.1. Effect of cooking on nutritive value

Cooking brings about several changes in any product and egg is no exception. Protein coagulation begins at 62-65°C and of all the cooking methods, frying causes the most damage in the form of losses of vitamin B_1 (25%), B_2 (25%) and folic acid (31%). Hard-cooking causes a loss of 25% loss of folic acid. Frying in oil, butter or lard decreases moisture and adds fat and Na. Poaching reduces vitamin B_2 and niacin but increases Na content. Nutrient content of omelettes is highly variable due to difference in the ingredients usedfor different types of omelettes; however, moderate loss of folic acid does occur in omelettes.

Contrary to popular belief, cooking increases digestibility and coagulation of protein ensures the safety of protein, especially of albumen, to human beings.

8. Residues in eggs

Residues in eggs belong mainly to 3 categories : a) insecticides/pesticides – which gain entry through feed ingredients, b) heavy metals – which can enter poultry products through feed and water and c) veterinary drugs – which gain access through feed, water or parenteral use.

8.1. Insecticides

Tolerances of organochloride insecticides in poultry meat, fat and eggs (Table 9.16) indicates that such insecticides through feed might not, under normal circumstances, accumulate to unacceptable levels. This is because, unlike plants, animals are capable of eliminating these products, albeit to a limited extent. In any case, eggs are more prone to be contaminated than poultry meat with the same level of insecticide in feed.

Table 9.16 Tolerance of organochloride insecticides in poultry meat, fat and eggs

Insecticide	Tolerance, ppm on fat basis
Dieldrin/Aldrin	0.20
Chlordane, Endrin	0.05
Total DDT, Lindane	1.00
β-hexachloro-cyclohexane	0.10
Heptachlor (Epoxide), Heptachloro-benzene, α-hexachloro-cyclohexane	0.20
	Source : Kan, 1991

8.2. Heavy metals

Tolerance level of Cd, Pb and Hg in eggs is 0.01, 0.10 and 0.03 ppm, respectively (Kan, 1991). The entry of these heavy metals is mainly due to man-made environmental pollution. Fortunately, eggs do not pose a potential problem of heavy metals because negligible amounts of these are impregnated into eggs.

8.3. Veterinary drugs

Coccidiostats form a very important group of chemicals that are likely to be imparted into eggs and meat. Hence withdrawal periods have been stipulated to each of the coccidiostats and if followed rigorously, there should not be much danger to the consumers from these chemicals. Tolerance levels are also set for amprolium, metchlorpindol, decoquinate, ethopabate, halofuginone, lasolacid, nicarbazin, robenidine and zoaline in poultry meat. Re-circulation of these drugs through litter has posed additional problems because low level of these drugs will be passed on to the eggs (yolk).

Chloramphenicol is by far the most studied antibiotic as a residue in eggs. It can persist in yolk for fairly long time of up to 66d after administration depending on dosage. It disappears rapidly from meat but slowly from skin. However, with the advent of newer generation antibiotics, use of chloramphenicol has become a very rare possibility.

Sulfadimidine and oxolinic acid have been found as residues in egg albumen. Use of hormones has been outdated more because of their limited economic advantage; their residues are more harmful than all the advantages put together.

Chapter **10**

Egg Consumption *Vis a Vis* Coronary Heart Disease

It is interesting to note that due to propaganda against cholesterol which began during early 1970s, there was a 46% reduction in egg consumption in US but plasma cholesterol levels reduced only by 1%. Now, most industrialized countries do not recommend specific restrictions on dietary cholesterol intakes for the population (McNamara, 2000); in addition, interest in cholesterol has fallen for consumers in the same way as salt and sugar as they become tired and confused by conflicting information (Zeidler, 2000).

1. Lipoproteins (LP)

1.1. Structure

1.1.1. Cholesterol and triacylglycerol (TAG) are indispensable structural and metabolic components of all animal cells. They are hydrophobic and therefore have to be encapsulated in a shell of phospholipids (PL) and special protein called apo- lipoproteins (Apo) forming a family of lipid-protein particles or lipoproteins (LPs). Apo maintains structure of the particles and direct their metabolism by acting as recognition proteins for variety of plasma enzymes and cell membrane receptors. LPs help transport of cholesterol and TAG in aqueous environment (Frier *et al*, 1999).

Structure and functions of the four main LPs are given in Table 10.1 and site of synthesis and functions of various Apo are shown in Table 10.2.

Table 10.1 Structure and functions of lipoproteins (LPs)

LP	Main Apo	Functions
CM	B_{48}, A-I, C-II, E	Main transporter of dietary TAG; synthesized in gut after meals, not present in normal fasting plasma
VLDL	B_{100}, C-II, E	Main carrier of endogenous TAG, synthesized in liver, precursor of LDL
LDL	B_{100}	Main cholesterol carrier in blood, generated from VLDL in blood
HDL	A-I, A-II	Smallest most abundant LP; transports cholesterol from peripheral tissues to liver for excretion
		Source : Frier *et al*, 1999

Table 10.2 Apolipoproteins and their functions

Apo	Site of synthesis	Functions
A-I	Liver, intestines	Maintain HDL protein, activates lecithin-cholesterol acyl transferase (LCAT)
B_{100}	Liver	Main structural protein of VLDL and LDL; involved in transport of cholesterol and TAG; binds to LDL receptor.
B_{48}	Intestines	Main structural protein of chylomicrons(CM)
C-II	Liver	Activates lipoprotein lipase (LPL), directs CM and VLDL TAG hydrolysis
E	Liver, intestines, macrophages, glial cells	Binds to LDL receptor and also to another specific liver receptor; important in brain lipid transport

Source : Frier *et al*, 1999

1.2. LP metabolism

Generally, 100 g of TAG and 0.5 g of cholesterol flow through each day in a normal healthy individual. Fats (TAG) are acted upon by digestive enzymes to produce free cholesterol, fatty acids (FA), monoacylglycerols (MAG), and diacylglycerols (DAG). All these combine with bile salts to form water-soluble micelles which carry lipids to absorptive sites in intestines. TAG is completely absorbed whereas 50% of cholesterol is absorbed and the rest is excreted through feces (Frier *et al*, 1999).

On entry into intestinal cells (enterocytes), they are reconstituted to form TAG and cholesteryl ester; both packaged in chylomicrons (CM) and secreted into intestinal lymphatics which empty into thoracic duct from where they ultimately reach circulation (Frier *et al*, 1999).

In circulation, CM is acted upon by lipoprotein lipase (LPL) in endothelial surface of capillary beds in adipose tissue, cardiac and skeletal muscles to release TAG, some of the PL and protein shed to plasma (precursor to HDL) and a remnant (rich in cholesterol) which is quickly taken up by liver; therefore, dietary cholesterol reaches liver (Frier *et al*, 1999).

In liver, cholesterol is incorporated to hepatocyte membrane where it is oxidized to bile acids or re-packaged into endogenous TAG-rich counterpart of CM, called VLDL. During fasting, VLDL is the primary source of circulating TAG energy. The VLDL is acted upon by LPL to form intermediate density LP (IDL) which is ultimately remodeled in liver into cholesterol enriched product called LDL with a plasma half-life of 3d (Frier *et al*, 1999).

LDL is removed from circulation by high affinity LDL-receptor pathway or by others less understood scavenger mechanism which may lead to incorporation of LDL into atheromatous plaques. Therefore, increased concentration of LDL predisposes to CHD (Frier *et al*, 1999).

Cholesterol-rich HDL is synthesized in liver and intestines and also as the part of lipolysis of CM (Frier *et al*, 1999).

Defective (retarded) TAG lipolysis leads to hypertriglyceridemia, consequently, low circulating levels of HDL. Actual function of HDL is to remove cholesterol from peripheral tissues through action of plasma LCAT and transport it centripetally for hepatic excretion; therefore, HDL reduces risk of CHD (Frier *et al*, 1999).

1.3. Fatty acids and nomenclature

Fats are made of FA esterified with an alcohol, usually, glycerol to form TAG. FA acids are of two types – a) Saturated fatty acids (SFA): having no double bonds and b) Unsaturated fatty acids (UFA) : having at least one double bond; the latter, depending on number of double bonds, are grouped into i) Monounsaturated fatty acids (MUFA) and ii) Polyunsaturated fatty acids (PUFA).

CHD is related to proportions of SFA and UFA, in general, and SFA, MUFA and PUFA, in particular. SFA are primarily used for energy purposes whereas PUFA and MUFA have other important functions in addition to their role as energy sources. Fat in animals contains UFA mainly of 16 carbons and more; hence, nomenclature of UFA of length =16 carbons is given in Table 10.3.

1.4. Functions of fat and cholesterol

Fat is an essential nutrient as a source of energy, insulator to vital organs, source of essential FA, and source of fat-soluble vitamins (A, D, E and K) and part and parcel of cell membranes (especially nerve tissue). Similarly, cholesterol also is essential as a precursor of steroid hormones (sex steroids and corticosteroids) and bile acids apart from being a compulsory constituent of cell membranes. It is also true that fat may cause obesity but, not due to its intrinsic structural and/or biological property (Kritchevsky, 2000).

Myelin is 75% fat and 25% of it is cholesterol; therefore, cholesterol is important for brain cell membrane stability. Two other important fats are docosahexaenoic acid (DHA) and phosphatidyl choline (lecithin, PC). Brain is 70% fat and most of it is DHA. In healthy brain, phospholipids (PL) are usually attached to DHA as the UFA. PL protects nerves from injury and deterioration due to ingestion of toxins and endogenous free radical production. DHA deficiency can result in attention deficit disorder and hyperactivity in children. DHA supplementation is useful in schizophrenia, Alzheimer's and others. 75% of circulating cholesterol is actually synthesized in liver and not directly eaten. Cholesterol levels less than

100 mg/dl may cause death (Kaminski, 2000).

Table 10.3 Nomenclature of unsaturated fatty acids (UFA)

Common name	Structure	Position of double bond(s)		Series
		From α-carbon*	From ω-carbon**	
Palmitoleic, 16:1	$CH_3(CH_2)_5CH=CH(CH_2)_7COOH$	9	7	ω7
Oleic, 18:1	$CH_3(CH_2)_7CH=CH(CH_2)_7COOH$	9	9	ω9
Erucic, 22:1	$CH_3(CH_2)_7CH=CH(CH_2)_{11}COOH$	13	9	ω9
Nervonic, 24:1	$CH_3(CH_2)_7CH=CH(CH_2)_{13}COOH$	15	9	ω9
Linoleic, 18:2, LA	$CH_3(CH_2)_4CH=CHCH_2CH=CH(CH_2)_7COOH$	9,12	6, 9	ω6
α-Linolenic, 18:3, LNA	$CH_3CH_2CH=CHCH_2CH=CHCH_2CH=CH(CH_2)_7COOH$	9,12,15	3,6,9	ω3
γ-Linolenic, 18:3, GLN	$CH_3(CH_2)_4CH=CHCH_2CH=CHCH_2CH=CH(CH_2)_4COOH$	6,9,12	6,9,12	ω6
Arachidonic, 20:4, AA	$CH_3(CH_2)_4CH=CHCH_2CH=CHCH_2CH=CHCH_2CH=CH(CH_2)_3COOH$	5,8,11,14	6,9,12,15	ω6
Timnodonic, 20:5, EPA	$CH_3CH_2CH=CHCH_2CH=CHCH_2CH=CHCH_2CH=CHCH_2CH=CH(CH_2)_3COOH$	5,8,11,14, 17	3,6,9,12, 15	ω3
Clupanadonic, 22:5, DPA	$CH_3CH_2CH=CHCH_2CH=CHCH_2CH=CHCH_2CH=CHCH_2CH=CH(CH_2)_5COOH$	7,10,13,16, 19	3,6,9,12, 15	ω3
Cervonic, 22:6, DHA	$CH_3CH_2CH=CHCH_2CH=CHCH_2CH=CHCH_2CH=CHCH_2CH=CHCH_2CH=CH(CH_2)_2COOH$	4,7,10,13, 16,19	3,6,9,12, 15,18	ω3
* Carboxyl carbon EPA = Eicosapentaenoic acid	** Methyl carbon DPA = Dodecapentaenoic acid	All double bonds are of *cis* type DHA = Dodecahexaenoic acid		

Human body requires cholesterol and liver manufactures 3000 mg/d of cholesterol which is equivalent to that in one dozen eggs. Experts in human fat requirement state that cholesterol must not be avoided in the diet; it is the base molecule which is converted into steroidhormones like estrogen, progesterone, testosterone and cortisol. If cholesterol levels are too low, brain-and nerve-related problems may develop (Kaminski, 2000). A person weighing 70 kg has 140 g of cholesterol (0.2% of body weight) mostly in nervous tissue (Narahari, 2001)

Eicosanoids

Eicosanoids are the compounds derived from 20-carbon polyenoic (polyunsaturated) FA. They can be grouped into a) prostanoids comprising prostaglandins (PG), prostacyclines (PC) and thromboxanes (TX), b) leukotrienes (LT) and c) lipoxins (LX).

PUFA are generally anti-atherogenic. Linoleic acid (LA) can be converted to either dihomo γ-linoleic acid (DLA) or arachidonic acid (AA). The former (a ω6-series compound) prevents aggregation of platelets (anti-aggregating). The latter, another ω6 compound, can produce thromboxane A_2 (TXA_2) and/or prostacyclin

PGI_2. TXA_2 supports aggregation of platelets whereas, PGI_2 is anti-aggregating (Howell, 2000).

Metabolism of α-linolenic acid (LNA), a ω3-series compound, involves formation of eicosapentaenoic acid (EPA), docosahexaenoic acid (DHA) and finally prostacyclin PGI_3 which is anti-aggregating and TXA_3 which is inactive (Howell, 2000).

ω3 FA decrease the conversion of AA (a ω6 FA) into pro-inflammatory PGs, LTs and TXs (Kaminski, 2000).

Stress → essential FA are mobilized from cell membrane → converted to PGs, LTs and TX mediators. In case of ω6 FA, mediators are PGE_2 which exaggerates fever reaction, LT_4 which can produce inflammatory response 1000 times that of histamine is and TXA_2 which causes extreme smooth muscle contraction and hypercoagulation. In case of ω3 FA, the mediators are PGE_3 which does not exaggerate fever, TXA_3 which opposes TXA_2 and LT_5 which opposes LT_4. Therefore, Eskimos with high-fat diets have low incidence of CHD and asthma (Kaminski, 2000).

1.5. Units of measurement

Conventionally, cholesterol, LDL, HDL, TAG etc., are measured in mg/dl. However, SI unit for these estimates is mmol. Conventional units are still used; but, SI units are followed with the conventional units indicated in parenthesis. The conversion procedure is shown below with cholesterol and the same is applicable for other molecules :

Let molecular weight of a compound be x Da. Then, x g/L = 1 mole/L or 1000 mmol/L. In other words, x mg/L = 1 mmol/L. Dividing both numerator and denominator of LHS by 10, we get 0.1x mg/dl = 1 mmol/L. Therefore, if conventional mg/dl or mg% value is divided by one tenth the molecular weight of the compound, it will be converted to mmol/L. Conversely, if mmol/L value is multiplied by one tenth the molecular weight, it will be converted to mg/dl or mg%.

Molecular weight of cholesterol ($C_{27}H_{46}O$) is 386.66 Da. Therefore, to convert cholesterol (total, LDL, HDL) values in conventional mg/dl or mg% into mmol/L, it is divided by 39. Molecular weight of some of the important compounds is as follows : TAG (tristearate, $C_{57}H_{110}O_6$) : 890, Fibrinogen : 340000, Homocysteine ($HS\text{-}CH_2CH_2CH(NH_2)COOH$) : 135, Apo A-I : 28000, Apo B_{100} : 550000, LP(a) : 400000 to 700000, C-reactive protein (CRP) : 125 kD and so on..

2. Coronary heart disease

2.1. Pathophysiology

Pathophysiology of coronary disease was centered around cholesterol sometime back and even today, it is no doubt one of the factors always considered; however, endothelial injury, platelet aggregation, smooth muscle cell proliferation and migration, Chemo-attractants and eicosanoids as well as LDL, especially oxidized LDL have become equally important for development of CHD (Kritchevsky, 2000).

Atheroma/arteriosclerosis/CHD is a patchy focal disease of the arterial wall. Coronary arteries are at high risk (Boon *et al.* 1999). Fatty streak or subendothelial accumulation of lipids attracts macrophages (monocytes) into endothelial space. Macrophages take-up lipids (oxidized LDL) and become lipid-laden macrophages giving an appearance of "foam" cells. Smooth muscle cells migrate into this lesion deranging the endothelial cell function. Therefore, entry of LPs into the vessel wall becomes uncontrolled. If the plaque remains stable, a fibrous cap forms which will be calcified over a period of time leading to remodeling of the vessel wall. Consequent on this, lumen of the vessel will be compromised and gets narrowed. In certain cases, the lesion may rupture because of inflammatory process and metalloproteinase activity causing a turbulent flow, extrusion of lipids and fatty gruel; tissue factors will be exposed leading to cascade of events and intravascular thrombosis (Massie and Granger, 2005). Plaque rupture may lead to rapid growth of the lesion or occlusion of the vessel and is thought to be the cause of most acute coronary syndrome (Boon *et al.* 1999), particularly during stress (Kaminski, 2000).

A mature fibro-lipid plaque has a core of extracellular lipid surrounded by smooth muscle cells and is separated by the lumen by a cap of collagen-rich fibrous tissue. Number and state of evolution of plaques both increase with age, but the rate of progression of individual plaques, even in thesame patient, is very variable (Boon *et al.* 1999).

Receptors on macrophage surface within the atherosclerotic plaques bind and accumulate oxidized LDL. Formation of antibodies to oxidized LDL may also be important for plaque formation (Baron, 2005)

2.2. Risk factors

More than 200 risk factors for CHD have been listed of which elevated blood cholesterol, elevated blood pressure, smoking and obesity appear to be most important. Guidelines given regarding these factors are purely based on probability and averages and hence are more statistical than medical; therefore, they are for the population and hence may not suit all the individuals; more so when the individual happens to be away from the mean value on either sides (Kritchevsky, 2000).

Investigations on animal models do not accurately reproduce human pathology and epidemiological studies are often unable to distinguish between risk factors (which have causative relationship with disease) and markers (which are only linked to the causative factor). Some of the important factors of coronary arterial disease are tabulated in Table 10.4 (Boon *et al.*, 1999).

Table 10.4 Some important risk factors of CHD

Fixed	Modifiable
Age Male sex Family history	Smoking Hypertension Lipid disorders diabetes mellitus Hemostatic variables Sedentary lifestyle Obesity Dietary deficiency of antioxidants and PUFA
	Source : Boon *et al.*, 1999

Effect of risk factors is multiplicative rather than additive and it is necessary to make a difference between relative risk Vs absolute risk; the former indicates proportional increase in risk whereas the latter is the probability of actual chance of the event. For instance, a man aged 35 years, plasma cholesterol 10 mmol/L, smoking 40 cigarettes/d is relatively much more likely to die of CHD within the next decade than a non-smoking woman of same age with normal cholesterol; still, absolute risk of dying is small (Boon *et al.*, 1999).

2.2.1. Family history

CHD often runs in families because of genetic factor and/or shared environment (similar diet, smoking etc.). It is noteworthy that nearly 40% of CHD (Boon *et al.*, 1999) is familial indicating a profound influence of genetic factors. Some persons are genetically equipped to process ω3 fatty acids to EPA (precursor for anti-inflammatory PG, LT and TX) and DHA (essential component of healthy nerves and brain) (Kaminski, 2000).

2.2.2. Smoking

Smoking is the most important avoidable cause of CHD. It has a strong, consistent, dose-linked relationship with CHD. Highest risk is among the young people and risk reduces within six months of quitting (Boon *et al.*, 1999).

2.2.3. Type of fat

Excess SFA is conducive to CHD. PUFA are susceptible for oxidation into peroxy radicals which cause oxidative damage with inflammation leading to CHD, cancer etc. (Stadelman and Cotterill, 1995). Saturated fat and *trans*-FA are causally related to CHD (Kaminski, 2000).

2.2.4. ω3 and ω6 fatty acids

Phytochemicals found in soya protein and ω3-rich foods can also protect against

CHD. ω3 and ω6 FA are essential because they cannot be synthesized the body; the former are the most beneficial being the precursors for anti-inflammatory PGs, LTs and TXs whereas the latter produce mediators which are pro-inflammatory (Kaminski, 2000). Knowledge of percentage breakdown of ω3 and ω6 FA in various oils is the critical factor for making right dietary choices (Kaminski, 2000).

2.2.5. Cholesterol

Total cholesterol *per se* doesn't cause atherosclerosis (Kaminski, 2000). Dietary cholesterol *per se* doesn't contribute greatly to the circulating cholesterol; further, although blood cholesterol levels in men with CHD were significantly higher than controls, in no case did they reflect the level of cholesterol in their diet (Gertler *et al.* 1950). It is interesting that low blood cholesterol levels (< 160mg/dl) caused a) increased mortality b) 20% more cancer c) 40% more CHD no cancer deaths d) 35% more deaths from injury and e) 50% more deaths due to problems in digestive system (Kritchevsky, 2000).

2.2.6. Lipoproteins

HDL is protective against atherosclerosis. It increases by weight loss, drinking two glasses of red wine/grape juice daily, exercising, eating tofu etc. HDL increases with exercise and modest consumption of alcohol. LDL is linked with obesity, sedentary lifestyle, hyperglycemia and hyperinsulinemia, consumption of *trans*-fatty acids (margarine) and/or saturated fatty acids (fried foods). LDL will not enter lining of blood vessel unless oxidized (Kaminski, 2000). Only after oxidation, LDL can in any way be related to CHD and the oxidation of LDL can be decreased by powerful antioxidants like β-carotene, E and, C and selenium.

HDLs carry 2 hydrolytic enzymes (platelet activating factor, PAF) comprising acetyl hydrolase and paraoxonase which decrease pathological effects of PAF-mimics carried on oxidized LDL (Lands, 2000).

If hyperlipidaemia/hyperlipoproteinemiais familial, high incidence of premature CHD is recorded. Incidence of CHD has a strong positive correlation with LDL, weak positive correlation with TAG and negative correlation with HDL levels in blood (Boon *et al.*, 1999). Over-consumption of sugar, refined carbohydrate (CHO), not eggs, are strongly associated with obesity and increased LDL.

2.2.7. Diabetes mellitus

This condition increases the incidence of CHD with a tendency to diffuse coronary atheroma. Inappropriate insulin therapy (insulin resistance) also increases risk of CHD (Boon *et al.*, 1999).

2.2.8. Refined carbohydrates

Fat-free snacks contained refined CHO (sugar, flour) and hence are easily digested and absorbed resulting in hyperglycaemia. Therefore, fat-free is not calorie-free. Over consumption of refined CHO appears to be causally linked to Syndrome X characterized by obesity, hypertension, insulin resistance, hyperinsulinemia, hyperglycaemia and also hypercholesterolemia (particularly LDL). Hyperglycaemia leads to Maillard reaction with proteins to form glycosylated protein which loses its ability to guard against oxidation. Superoxide dismutase (free radical-quenching defensive protein) becomes ineffective. Protein glycosylation directly contributes to the ageing by increasing free radical activity. Unquenched free radical or reactive oxygen species (ROS) through glycosylation oxidizes LDL cholesterol and cause extensive damage to cells leading to ageing and progression of many chronic diseases like cancer. ROS can also lead to cataracts, advancing arthritis and renal insufficiency.

2.2.9. Other factors

Surprisingly, 50% of those with CHD do not have high cholesterol. Therefore, there is no reason to avoid eggs. Other factors such as TAG, LP(a), Apo A-1, Apo B, C-reactive protein (CRP) and fibrinogen; of these factors, Apo A-1 and Apo B are hereditary. CRP and fibrinogen indicate pro-clottingresponse to inflammatory stress. However, all these can be modified by lifestyle, diet and exercise modifications (Kaminski, 2000).

2.2.9.1. Homocysteine

In a healthy artery, endothelial cells form a continuous protective layer that regulates passage of substances from plasma to underlying artery wall. If endothelial cell is damaged → permeability altered → allows direct interaction between elements from blood and artery wall. Homocysteine, a powerful molecular abrasive scrapes in the layers of blood vessel → damages endothelium → increased platelet utilization and formation of atherosclerotic lesions. Deficiency of Vitamins B_{12}, folic acid and/or B_6 results in impaired conversion of homocysteine to cystine during the biosynthesis of the latter. Hence, there will be elevated levels of homocysteine which is associated with depression, multiple sclerosis, diabetes, birth defects, Alzheimer's, rheumatoid arthritis and osteoporosis. Therefore, high homocysteine represents possible danger to heart, bone, brain and nervous system (Kaminski, 2000).

2.2.9.2. Triacylglycerol

These are esterified fatty acids found primarily in the core of CM and VLDL. Increased TAG in blood, not cholesterol → impaired fibrinolyticsystem → impairs body's pro-clotting response to inflammatory stress → a possible etiology for CHD independent of LDL. High fried foods and saturated fats, excess glucose, refined CHO, fat-free products rich in sugar and flour increase TAG levels thus predisposing to CHD. Therefore, TAG levels can be used as predictors of myocardial infarction (MI) and severity of coronary disease.

2.2.9.3. Lipoprotein (a)

LPs transport fats that are not soluble in water (primarily TAG and cholesterol esters) through circulation. LP(a) is formed by LDL cholesterol attached to a protein component called Apo LP(a). LP(a) is related to CHD through atherothrombogenesis and its levels are determined genetically. High levels of LP(a) in a seemingly "healthy" patient makes the subject susceptible to CHD. Therefore, LP(a) is the most important genetic factor and a better predictor of severity of CHD. LP(a) binds to endothelial and macrophage cells, fibrinogen and fibrin thereby promoting deposition of cholesterol and other fatty waste in blood vessel walls. It also prevents clot lysis adding fibrin and other debris to atherosclerotic plaque. It may also inhibit plasminogen activity leading to suppression of smooth muscle cell growth and promotion of proliferation of muscle cells commonly seen in atherosclerotic lesions. LP(a) levels are most important indicators of high risk stroke.

2.2.9.4. Apolipoprotein A-I

This is a major component of HDL and hence higher levels of this LP indicates protection against CHD. Hence, this is the best negative predictor of a family history of CHD in young men whereas HDL cholesterol levels serve the same purpose in young women (Kaminski, 2000).

2.2.9.5. Apolipoprotein B_{100}

This is a primary substance in LDL and hence, unlike Apo A-1, it is positively related to CHD. Therefore, Apo A-1 and Apo B_{100} give important information about CHD, especially of premature atherosclerosis than plasma lipoproteins.

2.2.9.6. Ratio of Apo B_{100} to Apo A-1

This ratio is useful for a wide range of ages and cardiovascular conditions. Higher the ratio more severe the CHD; likely to be inherited to the next generation.

2.2.9.7. Fibrinogen

This globulin molecule synthesized in liver affects various blood clotting factors including aggregation of blood platelets. It has a direct effect on vascular wall and is a prominent acute-phase reactant. It plays a key role in arterial occlusion by promoting atherosclerotic plaques, thrombus formation, endothelial injury and hyper-viscosity. Hence, fibrinogen may trigger atherogenetic effect associated with elevated lipids in blood stream; a synergistic effect between ratio of total cholesterol to HDL and fibrinogen. Increased fibrinogen levels are linked to brain infarction. Smoking, obesity, inflammation, stress, oral contraceptives and ageing are associated with high fibrinogen levels; ROS are associated with all fibrinogen-raising agents (Kaminski, 2000).

2.2.9.8. C-reactive protein

CRP is associated with production of inflammatory cytokines which increase coagulation and damage to vascular endothelium thereby increasing the risk of CHD. This effect is independent of other lipid and non-lipid factors, including smoking. Previous infection with *Helicobacter pylori* or *Chlamydia pneumoniae* may cause chronic inflammation leading to raised level of CRP. Aspirin reduces inflammation thereby reducing CRP (Kaminski, 2000).

2.2.9.9. Free radicals

Free radicals attack DNA followed by mutagenesis and carcinogenesis. Body has enzyme system and antioxidants to limit free radical formation. But, their levels have to be maintained by supplying them through diet. Phytochemicals in several plants like soyabeans, garlic, cabbage, ginger, licorice, *Umbelliferae* (carrots, celery) and flax are therefore very important as sources of carotenoids, tocopherols, phenolics and flavonoids (Stadelman and Cotterill, 1995). Therefore, activity/disease associated with free radical generation and excess are involved (smoking, poorly controlled diabetes). Vitamins A, C and E, Se and others are the free radical scavengers (Kaminski, 2000).

Therefore, it is reasonable to argue that CHD risk is not due to a single dietary component but it is due to complex and interactive nature of effects of diverse dietary factors on both serum lipids and CHD risk. Solution in the form of increasing PUFA in diet with decrease in total fat, SFA and cholesterol may be too simplistic (Kritchevsky, 2000).

3. Egg lipids

For assessment of quality for human consumption, egg fat can be divided into a) PUFA, b) SFA and c) cholesterol. To evaluate the "healthiness" of a product, the levels and relationships between these groupings within any dietary fat are compared to scientifically established values (Noble and Penny, 2002).

3.1. Fatty acids and lipoproteins

About 16% of the egg fat is LA which is an essential FA for humans. LA with other more UFA (AA and DHA) constitutes about 20% of the egg fat and are very essential for health. Oleic acid, the MUFA forms about 44% of the egg fat and the rest 35% is saturated fat. Therefore, egg fat is predominantly unsaturated and compares well with good quality table margarines which contain PUFA 20-25% of total FA; even liver has twice the saturated fats and half the unsaturated fats as that in an egg (Noble and Penny, 2002).

Yolk lipid has 65.5% TAG, 28.3% PL and 5.2% cholesterol; PL are made of 73% phosphatidyl choline (lecithin), 15% phosphatidyl ethanolamine (cephalin), 5.5% lyso-lecithin, 2.5% sphingomyelin, 2.1% lyso-cephalin, 0.9% plasmalogen and 0.6% ionositol phospholipids. MUFA in Yolk lipid have been thought to be superior to

PUFA in reducing the risk of CHD; the former reduces only VLDL and LDL whereas the latter, in addition to VLDL and LDL, reduces HDL also. Further, PUFA is highly susceptible for oxidation (hence, anti-oxidants have to be supplemented) resulting in free radicals which destroy fat-soluble vitamins. Conjugated LA

Table 10.5 Fatty acid composition and content of standard hen's egg

Component	Composition	
	% of egg fat	g/60g egg
Saturated fatty acids	35	1.785
Monounsaturated fatty acids	44	2.244
Polyunsaturated fatty acids	20	1.020
Linoleic acid	16	0.816
α-linolenic	1	0.051
C20 and C22	3	0.153
Esterified cholesterol	1	0.030
Free cholesterol	4	0.240
Source : Noble and Penny, 2002		

Table 10.6 Fatty acid composition of lipid fractions of yolk

Fatty acid	% of total fatty acid			
	Crude lipid	Triacylglycerol	Lecithin	Cephalin
16:0, Palmitic	23.5	22.5	37.0	21.6
18:0, Stearic	14.0	7.5	12.4	32.5
16:1, Palmitoleic	3.8	7.3	0.6	Trace
18:1, Oleic	38.4	44.7	31.4	17.3
18:2, Linoleic	16.4	15.4	12.0	7.0
18:3, Linolenic	1.4	1.3	1.0	2.0
20:4, Arachidonic	1.3	0.5	2.7	10.2
20:5, Eicosapentaenoic	Nil	Nil	Nil	Nil
22:5, Docosapentaenoic	0.4	0.2	0.8	3.0
22:6, Docosahexaenoic	0.8	0.6	2.1	6.4
Source : Stadelman and Cotterill, 1995				

(CLA) of yolk reduces risk of CHD and cancer. (Stadelman and Cotterill, 1995).

3.2. PUFA : SFA ratio (P/S ratio)

This is widely accepted parameter for PUFA adequacy in the diet. P/S ratio recommended for a healthy diet is 0.45 and the ratio for egg fat is 0.59. Meat also has a P/S ratio lower than that of eggs (0.2 to 0.3) (Noble and Penny, 2002).

3.2.1. ω-6/ ω-3 ratio

PUFA derived from LA (the ω-6 series) and those from LNA (ω-3 series) have different metabolic functions; therefore, ω-6/ ω-3 ratio has gained importance. The recommended ratio is 6.0 whereas the egg has a ratio of 8.0 (Noble and Penny, 2002).

Table 10.7 Composition of low density lipoproteins* of egg

Component	LDL_1	LDL_2
Total lipid, %	89	86
Lipid phosphorus, %	1.0	1.14
Lipid molar N/P ratio	1.04	1.14
Cholesterol, % of lipid	3.3	3.4
Free fatty acids, % of lipid	5.0	4.5
Phospholipid – Lecithin, %	84	83
Phospholipid – Cephalin, %	13	14
Non-lipid residue	11	14
Nitrogen, % of non-lipid residue	14.8	13.7
Phosphorus, % of non-lipid residue	0.15	0.16

* Diameter 170-600Å (avr. 300 Å) *Source* : Stadelman and Cotterill, 1995

3.2.2. Cholesterol

Egg contains 250-300 mg of cholesterol which is high on absolute terms; but, when calculated per unit weight of fat, it is comparable to most of the other animal products; liver has twice the amount of cholesterol compared to eggs (Noble and Penny, 2002). Eggs are good sources of cholesterol, DHA and lecithin (Kaminiski, 2000)

3.2.3. Other protective factors

a) Carotenoid pigments in yolk are also natural anti-oxidants which not only act as anti-carcinogens but also help elimination of free radicals and reduce LDL levels thereby protecting against CHD. b) Phosvitin, the protein of yolk, is a potential natural antioxidant safer than synthetic ones. c) Anti-oxidative property of yolk lipids help protect DHA and ω3 fatty acid from rancidity. d) Taurine of egg (also present in milk and meat) prevents formation of atherosclerotic plaques thereby providing protection against CHD. e) Betaine, a methyl donor, of egg reduces the plasma homocysteine levels offering protection against CHD (refer 1.3.2.9.1 for details) f) Natural minor sterols in egg (brassicasterol, campesterol, stigmasterol and β-sitosterol) increase HDL levels and protect against CHD (Narahari, 2003).

Hence, egg is an affordable, nutrient-dense, convenient commodity which is far in excess of its potential contribution to changes in plasma cholesterol (McNamara, 2000).

4. Enriched / Designer / Diet eggs

Notwithstanding the fact that cholesterol is not the only causative factor for CHD, it is definitely advantageous if the lipid content of the egg be manipulated so that not only its nutritive value improves but also makes it more safe for human consumption. Nutrients of particular interest are a) cholesterol b) PUFA c) ω3 FA, especially EPA, DPA and DHA and d) antioxidants like Vitamins A, E, and minerals like Se etc.

Cruikshank (1934) was the first to demonstrate that dietary fat can change composition of lipids in poultry. Fat metabolism of the hen is such that any fat-like component fed is likely to find its way into yolk (Noble and Penny, 2002).

4.1. Definition

The term "designer food" describes a food tailored to contain specific concentration and proportionsof nutrients critical to good health. Other terms used are functional foods, neutraceuticals and phytochemicals sources. The concept of nutritionally modifying foods to provide health benefits beyond the nutrients they naturally contain has been referred to as "designer" or "functional" food production. Such efforts had yielded eggs that contain at least 150 mg DHA, 120 mg of ω3 FA and DHA precursor, LNA (Elswyk, 2000).

4.2. Components modified

Hen lays eggs for the sake of embryo and hence, dietary changes can cause only a few changes in composition. However, FA composition of TAG in VLDL can be altered to a certain level by diet (Etches, 1996). Total fat, protein and sugar levels cannot be altered by diet but composition of FA, minerals, vitamins and certain non-nutrient compounds like pigments and anti-oxidant can be manipulated through diet (Narahari, 2001).

4.2.1. Fats and fatty acids

Egg lipid composition is because of combination of a) *de novo* lipogenesis, b) incorporation of lipid components from diet and c) feedback inhibition of dietary long chain PUFA. Therefore, it is possible to alter FA composition ofeggs through dietary manipulation of long chain PUFA. Hens fed flax seed produced eggs rich in ω3 FA (7-12% of yolk lipids) in the order LNA > DHA > DPA > EPA. This is suggestive that hen can convert dietary LNA to EPA, DPA and DHA via desaturase and elongase enzyme systems (Sim, 2000).

It is not possible to make major changes in total fat content in eggs; but, the composition of fat is easily amenable for dietary manipulations. Diets containing fish oils (menhaden oil, at 3% levels) rich in EA and DHA resulted in decreased levels of AA and increased levels of EPA, DHA and LA. However, at 5% level, taste-panel could detect objectionable off-flavors in eggs. Hence, plant feedstuffs rich in ω-3 FA like flaxseed was incorporated at 10-30% levels to obtain enriched eggs (Gill, 2001).

The amount of fat as a proportion of total egg weight is defined by the bird in order to ensure viable embryo in case it is fertilized. Therefore, attempts to reduce egg fat resulted in decreased egg weight. Attempts to reduce SFA content were not successful but manipulation of PUFA possible and the eggs produced to satisfy the human dietary provision of ω-3 and ω-6 fatty acids are called "designer eggs"; eggs with 200 mg of ω-3 FA DHA, are being produced (Noble and Penny, 2002).

4.2.2. Cholesterol

Dietary fat, cellulose, lecithin, emulsifiers and D-thyroxin enhance yolk cholesterol levels while pectin, oat hulls, plant sterols, triparanol, azasterols and probucol have an opposite effect. Many of the cholesterol problems associated with high fat intake can be solved by adding garlic powder (3-5%) or copper compounds (250 mg/kg) (Esmail, 2003).

Shell eggs are the first to be fortified with neutraceuticals in an effort to decrease cholesterol. Neutraceuticals are food components which also provide health benefits for better well-being and disease prevention. Profiles of FA, fat-soluble vitamins and minerals (iodine, fluorine and manganese), B-vitamins of the egg can be altered easily (Zeidler, 2000). Although it is not possible to reduce cholesterol levels significantly without severely affecting egg weight and/or production, certain hypocholesterolemic ingredients/compounds can reduce LDL and/or increase HDL content of eggs whereas some others have beneficial effects when included in hen's diet; these are tabulated in Table 10.8 (Narahari, 2001).

4.2.3. Vitamins, minerals and others

Designer egg was the first commercial shell egg product enriched with ω3 and ω6 FA, and Vitamin E. Egg can be enriched with only about 2 mg/g of ω3 fatty acids and six times the content of vitamin E; however, many consume vitamin E capsules to the extent of 400-1000 IU and therefore, fortification is meaningless (Zeidler, 2000). Feed vitamin content has the greatest and widespread influence on egg vitamin content. The efficiency of vitamin transfer into eggs as a function of intake may vary from 60% (riboflavin) to 5% (thiamine and folacin) (Gill, 2001).

Egg leutein has three-fold bio-availability than that from spinach or other supplements. Se and folate contents can be easily increased through diet. Folate from eggs is more bio-available than from other sources. Through maize-soy diets, soy isoflavones can be transferred to eggs. ω-3 fatty acids are particularly necessary for mental development of children

Table 10.8 Ingredients/substances having health functions

Ingredient/Substance	Effect(s)
MUFA, Oleic acid	Increase HDL cholesterol
PUFA (linoleic and linolenic)	Reduce total serum cholesterol
ω-3 fatty acids – eicosapentaenoic acid (EPA), docosahexaenoic acid (DHA) and α-linolenic acid (LNA)	Reduce blood pressure and serum LDL; inhibit thrombosis
Antioxidants – BHA, BHT, ethoxyquin	Prevent peroxidation and rancidity
Vitamin E and selenium	Anti-oxidant, increase HDL, reduce LDL; delay ageing process
Carotenoids, β-carotene	Anti-oxidant, anti-carcinogenic
Turmeric powder	Anti-oxidant, anti-microbial
Grape seed	Reduces LDL
Linseed/oil, canola/rapeseed meal/oil	Rich in ALN, Reduce total serum cholesterol
Fish meal/oil	Rich in EPA, DHA; increase HDL, reduce LDL
Brassicas : rapeseed, cabbage, turnip etc.	Reduce LDL
Glandless cotton seed	Reduce total serum cholesterol
Allyl sulfide (from garlic)	Reduce LDL, anti-carcinogenic
Lycopene (from tomato pomace)	Reduce LDL, anti-oxidant
Organic chromium	Reduce total serum cholesterol
Zinc	Protects heart
Lumiflavin, lumichrome (in egg)	Anti-oxidant, anti-carcinogenic
Tocotrienols (in brans)	Reduce LDL
Statin (in cell membrane protein)	Reduce LDL
Aspirin	Reduce LDL and clumping of RBCs

Source : Narahari, 2001

Table 10.9 Composition of NEE *Vs* Commercial egg

Component	Quantity/100 g contents	
	Commercial egg	NEE
Total SFA, g	3.3	2.8
Total UFA, g	6.4	6.9
MUFA, g	4.4	4.4
PUFA, g	1.2	2.2
Linoleic acid (ω-6), g	1.0	1.3
ALN, g	0.03	0.4
EPA+DHA, g	0.08	0.4
ω-6 : ω-3 ratio	9.90	1.75
UFA : SFA ratio	1.94	2.46
Cholesterol, mg	400	350
Carotenoid pigments, mg	1.5	2.2
Vitamin A, I.U	1200	1700
Vitamin E, mg	2	8
Thiamine, mg	0.1	0.15
Riboflavin, mg	0.32	0.45
Niacin, mg	0.1	0.15
Folacin, µg	5	10
Vitamin B_{12}, µg	1.0	1.8
Biotin, µg	18	25
Choline, mg	510	820
Iron, mg	2.3	4.0
Zinc, mg	1.4	2.7
Selenium, µg	Traces	1.8
Chromium, µg	---	Traces

Source : Narahari, 2001

(Noble and Penny, 2002). Color of yolk can be increased (oxycarotenoids, maize, maize gluten or synthetic chemicals) or modified (deep olive color due to gossypol when cotton seed meal is incorporated in diet). Concentration of fat-soluble (A,D and E) and water-soluble (biotin, pantothenic acid) vitamins can be altered; similarly, mineral content can be increased to certain extent by modifying the diet (Etches, 1996).

4.3. Composition of designer eggs

At the outset, it is necessary to note that composition of designer eggs depends totally on the diet and hence is not constant. The diet, in turn, depends on the component(s) which is(are) targeted for enrichment or reduction. Therefore, composition of the designer eggs available in literature shows a wide range of variation.

Composition of an NEE egg in comparison to commercial egg is shown in Table 10.9.

Designer eggs rich in ω-3 FA (LA, EPA and DHA), carotenoid pigments, vitamin E, organic Se and Cr help minimizing the risk of thrombosis, platelet aggregation, TAG accumulation, and CHD through reduction in LDL and/or VLDL; organic Cr also promotes insulin production and vitamin E acts as antioxidant and retards effects of ageing (Narahari, 2003)

Table 10.10 Nutritive value of enriched and conventional eggs*

Component	Enriched	Conventional
Calories	75	75
Protein, g	6	6
Carbohydrates, g	0.6	0.6
Total fat, g	6	6
Saturated fat, g	1.5	2.2
Polyunsaturated fat, g	1.35	0.90
ω-6 fatty acids, mg	750	800
ω-3 fatty acids, mg	350	60
18:3 fatty acids, mg	250	40
22:6 fatty acids, mg	100	20
Ratio of ω-6: ω-3	2.6	13.0
Monounsaturated fat, g	2.8	2.4
Cholesterol, mg	180	210
		Source : Gill, 2001

AA, a metabolite of LA, was significantly decreased in designer egg thereby reducing the ratio of ω6 to ω3 FA from 10:1 to 1:1 and consumption of such designer eggs (Table 10.11) produced desirable changes in plasma of human beings (Table 10.12). There was an increase in ω3 FA, especially DHA, which accumulated in platelets. Therefore, egg fat can be designed to suit health-conscious consumers (Sim, 2000).

Table 10.11 Fatty acid profile of ω3-enriched eggs

		(% of yolk fatty acids)
	Designer egg	Normal egg
SFA	27.25	31.5
PUFA	28.0	20.0
ω6 Fatty acids	14.5	18.5
ω3 Fatty acids	13.6	2.0
PUFA: SFA	1.02	0.63
ω6: ω3	1.08	10.05
		Source : Sim, 2000

Table 10.12 Change in plasma of humans after consuming two ω3 PUFA-enriched eggs with their habitual diets for a period of three weeks (%)

Total cholesterol	0.3	5.75
LDL cholesterol	5.0	12.75
HDL cholesterol	8.0	- 0.50
TAG	- 37.5	- 7.25

Source : Sim, 2000

FA composition of breast milk of nursing mothers also showed significant increases in LNA, DHA, DPA and EPA after consuming two designer eggs daily for six weeks (Table 10.13) (Sim, 2000).

Subjects consuming DHA-enriched foods had significantly lower values for platelet aggregation, plasma total cholesterol and plasma LDL; there was no change in HDL and plasma TAG decreased by 30% (Horocks and Yeo, 2000).

FA composition of designer eggs resembles breast milk fat and therefore yolk oil from designer eggs (extracted by a suitable procedure) can form the base for infant food industry (Sim, 2000).

4.4. Advantages of designer eggs

Epidemiological and clinical studies have revealed that CHD will decrease even if a small quantity of ω3 FA (0.5 g/d) is consumed over long period of time. One large designer egg will give more than 600mg of ω3 PUFA and 6 mg tocopherol with PUFA: SFA = 1:1 and ω6: ω3 = 1:1.

Table 10.13 Fatty acid composition of breast milk as influenced by consumption of designer eggs for six weeks

FA	% of breast milk ω3 FA		Designer egg
	Before	After	
DHA	0.25	0.60	2.0
DPA	0.10	0.30	0.9
EPA	---	0.30	0.2
α-linolenic	1.40	2.50	9.3
Arachidonic	0.40	0.50	1.00

Source : Sim, 2000

Generally, diets rich in ω-3 type PUFA, similar to those in cold-water marine fish like salmon, cod, sardines, anchovies and others, lowered the risk of atherosclerosis and CHD. Therefore, eggs rich in ω-3 type PUFA are considered "heart-healthy" or "nutritionally enhanced" or "nutritionally enriched" and they look, cook and taste similar to conventional eggs and have similar storage qualities.

They also contain higher levels of certain vitamins and minerals.

4.5. Disadvantages

Major disadvantage of designer eggs is the low enrichment level and worldwide prohibition on making medical/physiological claims. Feeding excess fish meal can precipitate in fishy odor in eggs (Etches, 1996) and they cost 30-130% more than the conventional eggs (Gill, 2001).

High level of long chain PUFA in designer eggs and the diet used to produce such eggs renders them highly vulnerable for auto-oxidation. Hence, natural form of tocopherols are used to stabilize dietary sources of ω3 PUFA. This is particularly true in case of flax seeds which produces fish flavor during storage due to rancidity and lipid peroxidation in chicken tissues and eggs. Cholesterol oxidation is also increased by presence of long chain PUFA in the order fish oil > flax seed oil > sunflower seed oil > palm oil. Increased tocopherols reduces cholesterol oxidation regardless of FA profile and also increases vitamin E content of eggs when fed for at least eight days (Sim, 2000). Suitable antioxidants at optimum levels must be incorporated in diets so that sufficient concentration is transmitted to the eggs to obviate auto-oxidation.

4.6. Production of designer eggs

The following are the general guidelines for producing designer eggs (Narahari, 2001). However, ration formula, ingredients to be used and feeding system for the birds depends on the nutrient(s) selected for modification and other local constraints concerning the feeding and management of birds. Care has to be taken to consider restrictions concerning rearing systems and levels of inclusion of the ingredients while formulating diets for producing designer eggs. Hence, the guidelines given below have to be modified suitably.

1. Age of layers – above 30 weeks.
2. If strains producing white eggs are popular in the locality, it is better to choose strains producing brown eggs and *vice versa* to facilitate easy identification of NEE in the marketing channel.
3. System of rearing : if deep-litter or free-range, preferably, access to vegetation can be provided; however, NEE can be produced by birds reared in cages also.
4. If there is no access to vegetation, herbs or vegetable waste, the feed should have a minimum of 8-10% fiber.
5. Metabolizable energy (ME) could be 0.42MJ (100 kcal)/kg less than the commercial feed to help reduce cost of production; for birds on range, ME need not be reduced.
6. Antibiotics and antimicrobials are avoided in feed and water.
7. Carotenoid-rich ingredients like corn gluten meal or marigold meal or to-

mato pomace may be included to give color to yolk

8. First grade oil-rich fish meal or fish oil to be incorporated in the diet because they are rich in ω-3 FA, EPA and DHA or ground, full-fat linseed, glandless cotton seed or rapeseed meal which are rich in LNA which can be converted to EPA and DHA by the hen to be incorporated.
9. Sufficient anti-oxidants either natural (vitamin E, selenium) or synthetic (0.002% W/W of BHA/BHT/ethoxyquin) to be added to avoid rancidity.
10. Activated charcoal may be added to remove fishy odor.
11. Change of feed to feed to obtain NEE must be gradual (over 2 weeks) especially if ingredients like linseed, fish oil etc. are incorporated in the diet.

The eggs are subjected laboratory assay for detailed composition and also to a taste-panel assay; the composition of the eggs has to be as per the claim on the label for the eggs and taste-panel should certify the acceptability of the eggs before releasing into the market (Narahari, 2001).

LNA is of no benefit unless converted to EPA, DPA and DHA. ω3 enriched eggs can be produced by diets having a) seed oils - eggs rich in LNA but poor in DHA and DPA are produced, and b) microalgae (marine algae, *Schizocpytrium* sp. has only DHA at 7.4% of total FA) - eggs will have exclusively DHA and DPA. Therefore, to get eggs rich in EPA, the most important ω3 FA, fish meal/fish oil is a must in the diet. British nutrition foundation recommends 1.4 g DPA + DHA/ week/adult; there is no mention of LNA because of poor conversion. Conversion of LNA to long chain PUFA in humans is uncertain and therefore diet is better if it contains EPA and DHA. Freshwater fishes are not a good source of long chain PUFA as are marine fishes. Enrichment products used to fortify the hen's egg are given in Table 10.14.

Table 10.14 Enrichment products to fortify hen's egg with ω3 FA

Product	%	mg			
	Inclusion	Linolenic	EPA	DHA + DPA	Total
Flax seed	10	292	6	132	430
Can alna	17	290	--	163	454
Canola seed	15	100	4	33	137
Marine algae	2.4	--	--	163	176
Menhaden oil	3	26	30	185	241
Mackerel oil	4	23	40	251	343
Millet	68	23	--	49	103

Source : Farrell, 2000

Upon feeding of flax seed which has 42% oil of which 53% is LNA, virtually no additional EPA was deposited in egg yolk; but about 33% is likely to be converted to DPA and DHA (Farrell, 2000).

Microalgae can be used up to 4.3% level in layer diets in anticipation of increased DHA content as well as beneficial effects on egg production FCR and body weight. However, due to reduction in ω6 FA, egg weight decreased by 1g at highest level of inclusion (4.3%). Microalgae was included at 0.86, 2.57 and 4.29% level in diets to supply 165, 495 and 825 mg of DHA/hen/d. The treatments produced an increased DHA content in eggs from 26.6 mg in control to 134.4, 170.3 and 220.0 mg in the above treatments, respectively (Abril *et al.*, 2000)

Conjugated LA (CLA) isomers have anti-carcinogenic properties. In case of layers, egg and yolk weights increased with CLA being incorporated in yolk lipids. CLAs are present in beef, lamb, poultry, cheese, seafoods, butter, milk and vegetable oils with fat; meat from ruminants being the richest sources. The properties of CLAs include anti-carcinogenic, anti-atherosclerotic, antioxidant, immuno-modulative and antibacterial (Watkins *et al.*, 2000).

5. Reducing risk of CHD

Hence, the best nutritional advice for an average healthy person is moderation, variety and balance; in other words, it is most healthy to eat a little of everything and not too much of anything (Kritchevsky, 2000).

Considering that in India *per capita* consumption of eggs is around 40 and even in urban areas it may not exceed 150 to 200, consumption of eggs should not pose serious threat of elevation in incidence of CHD. However, it is definitely essential to gear up propaganda against smoking, alcoholism, excessive and/or indiscriminate consumption of saturated fats and elevated blood pressure.

Yet another serious consideration that is warranted before a conclusion is drawn on cholesterol which is characterized by a great degree of idiosyncracy is the experimental subjects employed by deriving conclusion(s). Not many experiments on human subjects, taking representatives from different types of habits, have been conducted and most experimental results are based on laboratory animals. Not only that the metabolic responses in laboratory animals might be dissimilar to that in humans, but also the dose of cholesterol given to them in relation to the body weight would be unreasonably high. Hence, the results obtained from such trials must be viewed with utmost caution.

5.1. Cholesterol

Epidemiological studies indicated no significant relationship between dietary cholesterol and CHD when multivariate analysis included dietary fat and fiber. In a larger study, dietary cholesterol at 24 mg/MJ did not contribute to the risk of either MI or fatal CHD. Across cultures, negative correlation recorded between *per capita* consumption of egg and CHD mortality indicates that independent effect of cholesterol in causing CHD is doubtful. High dietary cholesterol (more than 750 mg/d) with lower intake of fruit and vegetables and higher intake of animal products will result in lower fiber and antioxidant vitamins and Vitamin B complex and PUFA; all these are defined risk factors for CHD. Therefore, increased risk of high dietary cholesterol may not be due to dietary cholesterol but rather than the multiple nutrients which are absent from the diet. Further, epidemiological data indicate that dietary cholesterol is not a significant contributor to either elevated plasma cholesterol levels or increased CHD incidence across population. Dietary cholesterol, when consumed within physiological ranges, has only limited effect on plasma cholesterol; 15-20% of the population have hypersensitivity to dietary cholesterol and majority of population have little, if any, plasma cholesterol response to changes in diet (McNamara, 2000).

In addition to the above, it has been found that in normal mixed diets which contain above 450 mg/d of cholesterol, no significant increase in cholesterol occurs with one or two eggs per day barring certain idiosyncrasies. Besides, high cholesterol diet suppressed cholesterol synthesis to a tune of 40 to 750 mg/d indicating a high degree of variability. Therefore, just reduction in dietary cholesterol will not presumably decrease population serum cholesterol levels whereas, reduction in intake of saturated fats would do. Understandably, it follows that controllers in cholesterol alone could not reduce the population risk of CHD whereas that in association with reductions in smoking and blood pressure might be more effective.

Moreover, as on today, single foodstuff which can meet all the nutrient requirements of humans without any limitation(s) is yet to be identified and/or produced. Most of the so-called "essential nutrients" do produce harmful effects, not only when deficient in diet, but also when consumed in excess; effects of excess of vitamins A, D, pyridoxine etc. minerals like Ca, P etc., carbohydrates, proteins and fats have all been well documented. In other words, each foodstuff has one or more harmful component(s) to some or more or all people.

Therefore, it follows that excluding egg from human diet just because it contains more cholesterol is unscientific *per se*. Egg is no doubt Nature's gift to human beings containing almost all the essential nutrients in a concentrated form. Hence, it is not only nutrient-dense but also an economical foodstuff (details of nutritive value of eggs are dealt in a separate Chapter).

5.1.1. Low blood cholesterol

In probably one of the most dramatic and the findings, Swedish scientists have revealed after studying over men and women, 25,000 each, for more than 20 years reported the following in persons with low blood cholesterol (< 150 mg/dl) :

1. Risk of injury was 2.75 times higher
2. Suicides among men were 4.1 times more frequent

3. Risk of colon cancer was 16% higher than those having blood cholesterol levels > 200 mg/dl

Low cholesterol levels were also found in criminals, aggressive people and those with low self-control. It as the presumed that a decrease in cholesterol to below person's normal value may be more important than habitually low levels. Lowering blood cholesterol is beneficial to people who already have heart disease; but in normal subjects, the advantages is not too clear; although deaths due to CHD reduced, deaths due to other causes, especially colon cancer, increased. Therefore, it has been suggested to have a blood cholesterol level between 150 and 200 mg/dl (Health and Nutrition Magazine, February 1993)

5.2. Unsaturated fatty acids

Normal human diets contain primarily oleic acid which is a neutral atherogenic factor. If diet is rich in MUFA, platelet aggregation is inhibited and LDL resistant to oxidation is generated. However, if MUFA contribute > 20-25% of total calorie, then weight gain occurs due to higher calorie intake. Elaidic acid (18:1), the *trans* form of MUFA, released upon hydrogenation of polyunsaturated vegetable oils (manufacture of margarines and shortenings) is atherogenic to an extent that of SFA. When it is used to substitute short chain SFA, it decreases both LDL and HDL, especially the latter more severely than either SFA or PUFA (Kritchevsky, 2000).

Highly unsaturated fatty acids (HUFA, ω3-type) from tissue PL are released when AA, a ω6 FA is released decreasing ω6-eicosanoid (TxA_2 and PGI_2, especially former) production. Therefore, dietary ω3 FA decrease inflammation and myocardial arrhythmia thereby reducing CHD risk. Therefore, it is necessary to identify the metabolic patterns that maintain HUFA, precursors of ω6-eicosanoids, in tissue lipids (Lands, 2000).

5.3. Complex carbohydrates

Replacement of SFA with starch results in elevated levels of TAG and VLDL and reduction in HDL with little effect on LDL. Hence, for an overweight person, reduction in total energy should be the goal rather than low fat high CHO diet. Non-absorbable but water-soluble complex CHO, like that present in oat bran fiber, decrease LDL cholesterol in hypercholesterolemic subject. On the other hand, non-absorbable, water-insoluble complex CHO present in fruits, vegetables and certain cereals do reduce risk of CHD; but, such effects are not related to changes in serum lipids. There may be an interaction of CHD with intake of fiber and SFA (Kritchevsky, 2000).

5.4. Others

Fruits and vegetables are good sources of vitamins and flavonoids. Vitamins E and C as well as carotenoids scavenge free radicals which are responsible for oxidation of LDL and subsequent CHD. Vitamin E also decreases platelet aggregation. All these effects of antioxidants are independent of their beneficial effect on serum lipids.

More emphasis on anti-thrombogenic ω3 FA, dietary fiber and dietary antioxidants along with weight control and regular physical activity is important to reduce the risk of CHD. Less emphasis on single dietary constituent, such as cholesterol and its primary source eggs, is the informed approach (Kritchevsky, 2000).

Structural requirement for an anti-arrhythmatic compound appears to be a long acyl/hydrocarbon chain with at least two double bonds and free carboxyl group at one end. Hence, even all-*trans*-retinoic acid is antiarrhythmic (Leaf *et al.* 2000).

It is in the fitness of things that risk of suffering a coronary heart disease is considered analogous to a road accident (Boon *et al.*, 1999). An inexperienced driver in an old car with poor brakes, bald tyres and defective steering is much more likely to have an accident than an experienced driver in a new car. However, there is a random element which means that, occasionally, the good driver will have accident few minutes after leaving home.

5.5. Framingham Heart Study

The most influential and respected investigation of the causes of heart disease is the Framingham Heart Study. Although subjects consumed cholesterol over a wide range, there was little or no difference in the levels of cholesterol in their blood and, thus, no relationship between the amount of cholesterol eaten and levels of blood cholesterol was found (although it is interesting that women who had the highest levels of cholesterol in their blood were the ones who had eaten the least cholesterol). Next, the scientists studied intakes of saturated fats but again they could find no relation. There was still no relation when they studied total calorie intake. They then considered the possibility that something was masking the effects of diet, but no other factor made the slightest difference. After twenty-two years of research, the researchers concluded: "There is, in short, no suggestion of any relation between diet and the subsequent development of CHD in the study group."

6. Avoiding CHD

6.1. Linoleic and α-linolenic acids

ω3 FA competitively inhibit the production of TXA_2 and availability of PGI_2 thereby increasing the availability of PGI_3. Therefore, increase in dietary LNA relative to LA may reduce the risk of thrombogenecity and CHD (Kritchevsky, 2000). CLA (*cis* 9, *trans* 11 octadecadienoic acid) decreased severity of experimental atherosclerosis (Kritchevsky, 2000).

6.1.1. Dietary guidelines

1. Should not be aimed at avoiding all fats.
2. Deliberate attempt to replace ω6 with ω3 FA; Corn/Safflower to be replaced with Olive/Canola/Flax seed.
3. Food rich in EPA and DHA (fish, fish oil) to be consumed; egg yolk, especially from hens raised on feeds containing flax seed, DHA or other ω3 sources, is the only land source of EPA and DHA.
4. Avoid foods containing highly amounts of simple sugars and refined CHO (table sugar, refined flour, rice etc.) and replace with foods that registered a low glycemic index.
5. Glucose-and sucrose-containing foods, especially soft drinks, candy and other confectioneries to be excluded.
6. Potato products and all baked foods with refined flour, rice etc. to be excluded.
7. Fruits and vegetables should replace snacks.
8. Antioxidants and vitamin supplements to be included.
9. Therapeutic doses of B_6, B_{12} and folic acid are critical to patients who carry homocysteinemic gene.

Deficiency of either ω-3 or ω-6 FA causes physical and biochemical changes in humans; large intake of ω-3 FA leads to increased requirements of anti-oxidants and vitamin E, delayed clotting time, inhibition of amino acid metabolism for PG formation and immuno-suppression. On the other hand, increased intake of ω-6 FA produces increased blood viscosity, vasospasm, vasoconstriction and hastened clotting time. Therefore, the ratio between ω-6 to ω-3 FA is more important than absolute levels of each of them; a ratio of 6 : 1 causes no pharmaceutical effect although a ratio of 4:1 to 10:1 has been suggested by some for infant feeding (Gill, 2001). At least 0.5% of energy intake should be from LNA and ratio of ω6 to ω3 should be less than 4:1 (Sim, 2000)..

In a study of over 80,000 female nurses, Harvard researchers actually found that increasing cholesterol intake by 200 mg for every 1000 calories in the diet (about an egg a day) did not appreciably increase the risk for heart disease. While it's true that egg yolks have a lot of cholesterol—and, therefore may slightly affect blood cholesterol levels—eggs also contain nutrients that may help lower the risk for heart disease, including protein, vitamins B_{12} and D, riboflavin, and folate. So, when eaten in moderation, eggs can be part of a healthy diet.

There is no compelling reason to avoid eggs in diet. Only oxidized and rancid cholesterol that causes coronary plaque. Hypercholesterolemia is not a disease but only a lab finding which needs further evaluation. Egg is an affordable, nutrient-dense, convenient commodity which is far in excess of its potential contribution to changes in plasma cholesterol (McNamara, 2000).

6.2. Dietary fats

6.2.1. Saturated Fats

Saturated fats are mainly animal fats. They are found in meat, seafood, whole-milk dairy products (cheese, milk, and ice cream), poultry skin, and egg yolks. Some plant foods are also high in saturated fats, including coconut and coconut oil, palm oil, and palm kernel oil. Saturated fats raise total blood cholesterol levels more than dietary cholesterol because they tend to boost both good HDL and bad LDL cholesterol. The net effect is negative, meaning it's important to limit saturated fats.

6.2.2. Trans Fats

Trans fatty acids are fats produced by heating liquid vegetable oils in the presence of hydrogen. This process is known as hydrogenation. The more hydrogenated an oil is, the harder it will be at room temperature. For example, spreadable tub margarine is less hydrogenated and so has fewer trans fats than stick margarine. Most of the trans fats in the American diet are found in commercially prepared baked goods, margarines, snack foods, and processed foods. Commercially prepared fried foods, like French fries and onion rings, also contain a good deal of trans fat. Trans fats are even worse for cholesterol levels than saturated fats because they raise bad LDL and lower good HDL. While you should limit your intake of saturated fats, it is important to eliminate trans fats from partially hydrogenated oils from your diet.

6.2.3. Unsaturated Fats—Polyunsaturated and Monounsaturated

Unsaturated fats are found in products derived from plant sources, such as vegetable oils, nuts, and seeds. There are two main categories: polyunsaturated fats (which are found in high concentrations in sunflower, corn, and soyabean oils) and monounsaturated fats (which are found in high concentrations in canola, peanut, and olive oils). In studies in which polyunsaturated and monounsaturated fats were eaten in place of carbohydrates, these good fats decreased LDL levels and increased HDL levels. Composition of common oils and fats in comparison to that of eggs is given in Table 10.15. It is clear from the table that MUFA and PUFA composition of eggs is better than the cooking fats as well as palm oil.

7. Recommendations

Advice of American Heart Association (AHA): 1. Eat less calories 2. Eat a smaller % of calorie as fat 3. Substitute unsaturated fats for saturated fats 4. Include some polyunsaturated fats (Lands, 2000). Hence, American Dietetics Association (ADA) recommends about 12% of total calories from MUFA (i.e., 14g per Mcal). (Narahari, 2003)

AHA is no longer putting a limit on the number of eggs that could be eaten

Table 10.15 Percentage of specific types of fat in common oils and fats*

Oils	Saturated	Mono-unsaturated	Poly-unsaturated	Trans
Canola	7	58	29	0
Safflower	9	12	74	0
Sunflower	10	20	66	0
Corn	13	24	60	0
Olive	13	72	8	0
Soyabean	16	44	37	0
Peanut	17	49	32	0
Palm	50	37	10	0
Coconut	87	6	2	0
Cooking Fats				
Shortening	22	29	29	18
Lard	39	44	11	1
Butter	60	26	5	5
Egg**	35	44	20	0

* Values expressed as percent of total fat; data are from analyses at Harvard School of Public Health Lipid Laboratory and U.S.D.A. publications.
** From Table 10.5

because science did not justify the recommendation on the number of egg yolks a person may consume in a week. Research now indicates that people who didn't eat eggs had the highest rate of heart disease and where egg consumption was highest, heart disease was lowest. Although serum cholesterol increased following consumption of eggs, the ratio of LDL to HDL remained the same at 2.6. Therefore, risk of heart disease did not change due to eating of eggs (Evans, 2003).

For adults age 20 years or over, the latest guidelines from the National Cholesterol Education Program recommend the following optimal levels: a) Total cholesterol less than 200 milligrams per deciliter (mg/dl) b) HDL cholesterol levels greater than 40 mg/dl c) LDL cholesterol levels less than 100 mg/dl.

Many health agencies, including the American Dietetic Association, the American Diabetes Association, and the American Heart Association, once recommended limiting fat intake to 30% or less of total daily calories as a means of preventing disease. Today, these recommendations focus on limiting intake of saturated fat, and have relaxed a bit with regard to total fat intake because there is no good evidence for any particular "optimal" amount of total fat in a healthy diet.

The relation of fat intake to health is one of the areas that Harvard researchers have examined in detail over the last 20 years in two large studies. The Nurses' Health Study and the Health Professionals Follow-up Study have found no link between the overall percentage of calories from fat and any important health outcome, including cancer, heart disease, and weight gain. What was important in these studies was the type of fat in the diet. There are clear links between the different types of dietary fats and heart disease. Logically, most of the influence that fat intake has on heart disease is due to its effect on blood cholesterol levels.

People with diabetes, though, should probably limit themselves to no more than two or three eggs a week, as the Nurses' Health Study found that for such individuals, an egg a day might increase the risk for heart disease. Similarly, people who have difficulty controlling their blood cholesterol may also want to be cautious about eating egg yolks and choose foods made with egg whites instead.

8. Diagnosis and treatment

It is very clear that cholesterol is not the only factor and therefore a multifactorial analysis involving LDL cholesterol, TAG, HDL CRP etc. is mandatory to assess the risk of CHD. A comprehensive cardiovascular risk profile is shown in Table 10.16.

Table 10.16 Comprehensive cardiovascular risk profile

Risk factors, mg/dl				
	Risk level			Range in brackets indicates
	Low	Moderate	High	
TAG	(0-200)	200-400	400-600	Reference
Total cholesterol	(150-200)	200-240	240-300	Desirable
LDL cholesterol	(80-130)	130-160	160-210	Desirable
Apo B		(49-103)	103-160	Reference
Protein factors				
	Protective	Desirable	High-risk	
HDL cholesterol	(60-100	35-60)	0-35	Desirable
Apo A-1		(101-198)	25-101	Reference
Independent risk factors				
		Desirable	High-risk	
LP(a), mg/dl		(5-40)	40-80	Optimum
Homocysteine, µmol/L		(3-16)	16-26	Optimum
CRP, mg/dl		(0-2)	2-8	Optimum
Fibrinogen, mg/dl		(185-408)	408-630	Optimum
Ratios				
	Optimum	Moderate	High-risk	
Total cholesterol/HDL	0-3.7	3.7-5.0	5.0-9.0	
Apo B/Apo A-1	0-0.65	0.65-0.92	0.92-1.65	

Source : Kaminski, 2000

Clinical targets for healthy humans : Body mass index (BMI) – 20 to 26 kg/m^2; Glucose – 70 to 110 mg/dl; Cholesterol - < 200 mg/dl and proportion of ω3-HUFA in plasma - > 45% (Lands, 2000). However, it is very important to note that there is no true "normal" range for serum lipids; for instance, Western population normally has 20% higher cholesterol than Asian counterparts (Baron, 2005)

Notwithstanding the above, the best screening and treatment strategy for adults who do not have CHD is not clear. Several algorithms have been developed to guide the clinician in treatment decisions, but management details are not individualized. Measuring cholesterol alone is the least expensive strategy and is adequate for low risk individuals. For individuals with cholesterol > 200 mg/dl, fasting LDL and HDL values are needed.

Middle-aged men having serum cholesterol levels > 230 mg/dl, have a 10% risk of CHD before 65 years of age whereas, if the cholesterol level is about 170 mg/dl, the risk reduces to about 3%. For women, probability of death due to CHD is about a third of that in men. For each 10 mg/dl increase in cholesterol/LDL cholesterol, CHD risk increases by 10% in men; conversely, for each 5 mg/dl increase in HDL cholesterol, CHD risk decreases by 10%; former effect being smaller and the latter effect being greater in women (Baron, 2005).

Chapter 11

Ready-to-cook Chicken

Meat is defined as the flesh of animals used as food, and usually, it also includes organs such as liver, kidneys, brain and other edible tissues. The process of obtaining ready-to-cook meat from the live birds is referred to as "dressing of birds" or "preparation of ready-to-cook chicken" and it greatly determines the quality and shelf-life of the meet.

1. Flow-chart

Poultry dressing refers to slaughtering, feather removal and evisceration and conversion of poultry carcasses into a number of products is referred to as "further processing". Flow chart for preparation of ready-to-cook chicken is shown in fig. 11.1.

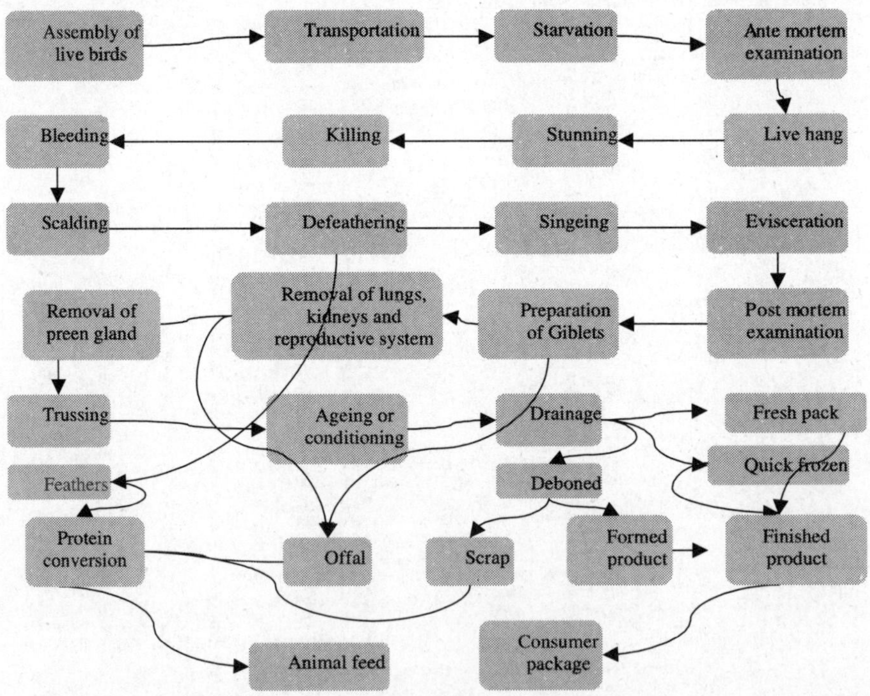

Fig. 11.1 Flow-chart for conversion of live poultry into poultry meat

1.1. Advantages

The advantages of preparation of ready-to-cook product are a) for the consumers, it will be convenient for further processing b) saves time for the consumers c) it is easy to carry d) hygienic e) has a pleasing appearance f) easy for marketing and f) can be stored for variable periods depending on the method of preservation.

2. Assembly and transportation

Birds for slaughter are procured from the site of production to the processing plant. Broilers are placed in crates. These crates are usually made of polyethylene or fiber glass to help easy cleaning. Further they are uniform in weight whether wet or dry and are thin-walled thereby requiring lesser space in the trucks. In addition to the above, they also durable and sturdy making them cost-effective. Hence, crates are most often used although other systems like cages on rollers or fork lift trucks as well as automatic bird-hauling methods are also in use. In case of turkeys, special turkey handing trailers or trucks are often used. When crates are used for turkeys, the height of the crates is usually greater than that for chickens.

In case of both chicken and turkeys, birds per crate is dictated by the size of bird, length of haul, environmental conditions (especially temperature and humidity) etc.

Ducks are usually hauled on trailers. The loading is unique as the ducks are herded onto trailers and off the trailers into holding pens where water is provided at the processing plant. Geese are handled the same as ducks when numbers are sufficient to justify the equipment. With small numbers of geese, crates meant for turkeys are used. Other species such as pheasants, quail, guinea fowl and pigeons are usually handled in crates (Stadelman *et al.*, 1988).

If the birds are to be transported, it must be done during cool parts of the day so as to reduce stress and shrinkage (weight lost during transit). Broilers are preferably caught at night for loading because, during that time it is easier to catch, the birds struggle less and settle down in crates quickly and easily. In addition, during summer months, it will be cooler during night which would minimize shrinkage during transit.

2.1. Considerations during assembly

2.1.1. Minimizing bruises

This is a major concern during assembly. Great care must be exercised during catching and loading of live birds into the crates, onto the trucks and finally into the pens. Personnel involved in catching and handling of birds for slaughter must be trained suitably to eliminate bruising of tissue or fractures during assembly. Proper maintenance of records regarding bruising and fractures will help identify personnel responsible so that suitable remedial measures can be undertaken.

2.1.1.1. Time of bruising

Bruises caused up to 12 hr prior to slaughter are pink to red. If it is more than 12 hr and less than 24hr, blue to purple color will result. If the bruise is more than 24hr old, a greenish color develops which disappears usually after 4 to 5d. Bruises heal faster on younger birds than older birds. Most bruises occur within 12 to 24hr prior to processing.

2.1.1.2. Reasons for bruising

Type of house construction, placement of equipment and other management factors such as not allowing feeders to run empty during the final 24hr or failing to have the house ready when the catching crews arrive may also resulting bruising. Ideally, all equipment, especially feeders and sulfur must be removed from floor pens before the catching crew arrives. The lights in the turned out or dimmed to prevent birds from becoming excited.

Breast-blisters can be attributed to sex, weight and litter management. Caked and wet litter predisposes to blisters and scabby hips. In heavy birds, excessive percentage of breast-blisters are recorded when they are transported in wooden crates and/or over long distances.

2.1.2. Shrinkage

Birds meant for slaughter are starved insuch a way that the digestive tract cleared without affecting the yield. It is generally agreed that the birds have to be starved for at least 4 hr to reduce fecal contamination. Shrinkage of edible tissues is detectable when feed and water are withheld in excess of 12 hr; a linear relationship between shrinkage% and duration of starvation has been reported. However, nutritive value of meat is not affected by duration of starvation.

Shrinkage of 3-4% is usually permitted. The shrinkage above the maximum level is borne by the seller/hauler.

3. Starvation

After procurement of birds, they are off-fed (starved) for 4-12 hr during which time only cool drinking water is provided.

3.1. Advantages

1. Broilers consume 120-140 g of feed per d during 6th week, spent hens also consume around 100-120 g per d, and growing birds 90-100 g per d. Feed consumed in the day of slaughter is not utilized for weight gain and hence is a waste. The cost of poultry feed can be saved by starting the birds.
2. Birds fed prior to slaughter will have feed in their gastro-intestinal tract

which will hinder the evisceration process and also increases the chances of contamination of the carcass in case the intestines are torn during evisceration.

3. Starvation helps *ante- mortem* examination of birds. If there is any latent infection or if the birds are drugged to mask any existing infection, starvation results in precipitation of the disease thereby assisting *ante-mortem* inspection.
4. Starvation also helps obtain accurate and more reliable dressing percentage. Dressing percentage is the ratio of total edible portion to the live weight times 100. Since all the birds do not consume the same quantity of feed, the denominator of the ratio will be non-uniformly increased. Therefore, not only the dressing percentage value is reduced, the reduction being non-uniform, cannot be corrected. Starvation helps emptying the gastro-intestinal tract and hence makes the dressing percentage value more accurate and reliable.
5. Starvation is presumed to help reaching the correct ultimate pH.

4. *Ante- mortem* examination

After starvation, the birds are subjected to *ante- mortem* examination and any bird showing indication(s) of sickness is not taken for further processing.

As per the specifications of Bureau of Indian Standards (BIS, IS: 6559-1972), proper *ante-mortem* inspection of all poultry is essential to ensure that they are not affected which any disease or other condition which may render the flesh unwholesome.

The birds showing any clinical evidence of a disease or condition that would result in condemnation of the carcass on *post-mortem* examination should be tagged as condemned. The birds shall be segregated from other poultry and held for separate slaughter, evisceration and *post-mortem* examination.

During *ante- mortem* examination, the following details shall be noted:

1. Evidence of cruelty to poultry by overcrowding and suffocation, exposure to extremities of temperature and transport over long distances without feed and water.
2. Disease symptoms which may affect the general health of the poultry or depreciate the meat.
3. Presence of scheduled infections and contagious diseases or symptoms which may suggest that such a disease is developing and
4. Species, sex, age and body temperature.

Condemned poultry, if not already dead, shall be killed in the casualty room only and shall not be conveyed into any department of the establishment used for edible portions.

5. Live-hang

Removal of birds from the crates or from trucks is important as loading. All care to minimize bruising and fracture have to be taken, especially in case of turkeys. In case of ducks, care must be given to avoid overcrowding in the holding pens or chutes through which they are driven. If ducks are allowed to pile up, there will be skin scratches; if piling is severe, smothered ducks will result.

The hanging of all poultry on shackles should be done so as the minimize struggling which is evidenced by wild wing flapping. It is also result in bruising of wings and bruising of birds adjacent to them on the line. In addition, birds showing excessive wing flapping are likely to yield tougher meat due to rapid depletion of glycogen from the tissues leading to lower ultimate pH at the onset of *rigor mortis*.

Use of carbon dioxide gas to immobilize broilers and turkeys in crates has been found to minimize struggling; however, it is not practiced commercially.

6. Humane slaughter (Stunning)

In most of the countries, law requires that all animals must be stunned prior to slaughter; but, it is not unusual that the birds are bled without stunning. Although, in case of large animals, there are many methods of stunning, the method of choice in case of poultry is electric shock. There are many methods of electrical stunning and most effective methods are those that insure positive contact between the head and feet of the bird with a source of electricity and the ground. The amperage used must be sufficient to produce desired immobilization and at the same time in the heart functioning. Excessive amperage can result in stoppage of heart leading to improper bleeding; in case of toms, may result in violent reaction which may lead to breakage of clavicle.

Head of the bird (hung on chutes) will be made to come in contact with a water- or saline-bath where they are subjected to 150-200 V electric shock for 1-2 sec. The other methods of electrical stunning which are not frequently used are a) knife which has an electric current in the blade b) electric plate or wands which touch the heads of the birds as the move along the conveyor. To obtain the current supply, one electric line is attached to the track and another one to the shocking device. When the circuit is completed by touching the bird's head, the current runs through the bird and stuns it. The size of electrical charge can be regulated according to the size and species of poultry being slaughtered.

Stunning has been found to accelerate rate of pH fall *post-mortem*, relax the muscles which hold the feathers and increase tenderness. However, this does reduce the amount of blood loss during bleeding; but still, the bleeding in this method will be more efficient than in other methods. Ultimate pH is also not affected by stunning.

7. Killing and bleeding

Following stunning, poultry are normally dispatched by cutting the jugular vein and a carotid artery. Decapitation results in reduced flow of blood and hence is not normally advised. Total blood loss amounts to about 4% of live weight and 35 to 50% of the total blood during the 3 min bleeding period. Blood constitutes about 10% of the total body weight of young chicken; it reduces as the birds mature. Females lose less blood than the males following the slitting of vessels. Commercially, 90 to 180 seconds are allowed for bleeding prior to scalding.

7.1. Jutka method

This method is followed by Hindus and is the most humane method because the bird is decapitated in one stroke and hence cannot feel the pain. But, brain is completely severed and hence, struggling is reduced, consequent on which, the bleeding also is less. Besides, when the head is severed, the wind-and the food-pipe are severed. Hence, when the bird struggles, contents of the crop and trachea might ooze out and contaminate the carcass.

7.2. Halal method

This method is usually followed by Muslims. The blood vessels are severed through an incision just behind the earlobes and a part of spinal cord is also severed. The birds feel pain and hence this method is more inhumane than other methods. Violent struggling and perfect bleeding follow. Since respiratory and/or digestive system are not severed, there is no possibility of contamination from the secretions of crop and/or trachea. Perfect bleeding ensures prolonged keeping quality as well as better consumer preference. Most of the commercial slaughterhouses follow this method (referred to as "Modified kosher") preceded with electrical stunning.

7.3. Kosher's method

This method is usually practiced the Jews wherein a special knife is used to pierce the upper palate (roof of the mouth) after opening the beaks to reach the cranial cavity. The knife is twisted to effect decerebration (removal of brain) and the brain is severed from the spinal cord. This is followed by severing of the blood vessels for bleeding. This method is very impractical under commercial conditions.

7.4. Ultimate pH

When the birds are bled, there will be cessation of oxygen supply to the cells which results in reduced body temperature and various other processes. In an attempt to maintain body temperature, the cells switch on to glycolysis, the anaerobic ATP-generating system. This result in breakdown of glycogen/glucose to lactic acid and thereby pH drops. When the pH reaches 5.5-5.6, the glycolytic enzymes are inhibited and the glycolysis is turned off. This pH is referred to as ultimate pH.

8. Scalding

8.1. Dry method

In this method, the feathers are removed by hand. The primary, secondary, tail and sickle-feathers are removed first because, soon after killing, nerve endings lose hold on feathers causing the feathers to become still (setting of feathers); otherwise, the skin will be abraded resulting in an unsightly appearance, contamination and reduced shell-life. Meat in this method will be absolutely tender and soft, tastier and juicy. But, it is impracticable under commercial conditions because it is time-and labor-consuming.

8.2. Waxing

The primary, secondary and tail feathers are removed and then the bird is dipped in molten wax; then, removed and placed in cold water for solidification of wax. The wax is then scraped off by using a blunt knife to get a clean bird. The wax subsequently can be molten once again, sieved to remove the feathers and reused again. The feathers becomes loose due to heat at the base resulting in expansion of feather follicles. This method also is time-consuming and costly. Although this method is no more used for chicken, it is practiced in case of ducks and geese whose feathers do not come out easily and the skin surface is slippery.

8.3. Wet method

In this case, water is maintained at a known temperature and the bled birds are dipped for a known duration. Heat expands both feather follicles and the roof of the feathers. However, coefficient of expansion of the former is greater than the latter and hence, the feathers become loose. Scalding may also act through nerve-muscle relaxing mechanism.

Earlier, boiling water was used; but, it caused melting of the subcutaneous fat and uneven accumulation resulting in blotchy appearance. In addition, it also results in partial cooking of skin and subsequent tearing and contamination.

However, by experiments, it is now specifically known as to what temperature is required for scalding different types of birds (Table 11.1). Nowadays, automatic scalding tanks available which maintain a set temperature and therefore have made the scalding process very easy and efficient.

Table 11.1 Methods of scalding

Method	Water temperature °C	°F	Duration (min)	Type of birds
Soft-scalding	50	122	Up to 2.5	Broilers and heavy fowl
Semi-scalding	53-55	126-130	1-2	Broilers and heavy fowl
Sub-scalding	59-60	138-140	0.75-1.5	Turkey, spent-hens
Hard-scalding	≥63	≥145	Variable	Mature birds, waterfowl
Source : Stadelman *et al.*, 1988				

Hard-scalding at 71.1-82.2°C for 30-60 sec produces carcasses which are slightly puffy because the flesh expands at carcass appears plump and it is easier to remove the feathers. Hard-scalding is used mostly for waterfowl because it is only satisfactory way to release feathers and the skin of waterfowl does not discolor unlike that of other species of poultry.

Sub-scalding will result in breaking down of outer layer of skin; the flesh being not affected. Defeathering will be usually easy after sub-scalding and skin color will be uniform provided the skin surface is kept moist and covered.

Generally, scalding carcasses in the range between 54.4-58.8°C should be avoided because the temperature is too hot to keep all the skin intact and too low to remove all the epidermal layer of skin. As a result, blotches and unsightly patches are formed if the skin dehydrates (Mountney and Parkhurst, 1995).

Combination of temperature and time is very important in semi-scalding because high temperatures for a long time will cause patches of skin to peel off leaving blotchy, unsightly areas when the skin becomes dry. However, skin will remain intact in this method so that the carcass can be subjected to different methods of chilling and packing. Defeathering is difficult and more hand pinning is required after semi-scalding.

Recent developments in semi-scalding include, completely enclosed scalders, blood and feathers not splashed around the room and reduced noise and odor levels. The carcasses are showered with hot water and conveyed through humidity cabinets where they are sprayed with steam at 60°C (Mountney and Parkhurst, 1995).

Soft-scalding might be acting through a nerve-muscle relaxing mechanism. Scalding time is much less critical factor than scalding temperature in feather release. Sub-scalding resulted in a complete removal of epidermal layer of the skin; therefore, skin surface must be kept moist through chilling, packaging and freezing to prevent severe discoloration and a toughening of the dry skin. High temperature scalding usually results in toughening; the effect aggravated by longer duration of scalding.

In none of the above methods, skin has to be removed because it has some protein value, yields some fat during cooking and reduces contamination of meat.

In view of the above, scalding procedure requires control of both time and temperature. Sufficient heat must be applied in order to cause relaxation of muscles in feather follicles and the duration should not be long enough to cause toughening of muscles. Either the time as is short enough so no part of epidermal layer of the skin is removed or long enough to obtain complete removal of feathers during defeathering. Scalding procedures has a direct effect on appearance of the defeathered carcass and tenderness of surface muscles.

Use of microwave energy for release of feathers has not yet been commercially viable.

9. Defeathering

Scalding temperature is of more importance than scalding time. Starvation for 8 hr increase the force required to remove feathers.

Currently, the commercial practice is to defeather the carcasses by electrical defeatherer. The carcasses are tumbled against rubber fingers of the equipment producing an effect similar to beating of the body surface and rubbing. Since the feather follicles have expanded due to scalding, feathers will be picked by the rubber fingers. Use of rubber fingers has been found to increase toughness of the muscles which is directly proportional to the weight of the fingers and the duration of defeathering. Lighter weight fingers have been developed to minimize damage to the meat tenderness.

In case of waterfowl, an additional step after the removal of rough feathers is required. Ducks and/or geese, after usual defeathering, are dragged through hot wax containing rosin. The wax is cooled to become pliable and then stripped from the carcass taking residual feathers and down with the wax.

After defeathering, body hairs referred to as filoplumes and small pin feathers remain on the carcass. Pin feathers can be removed by using a blunt knife. The carcasses is subjected to flame singeing to remove filoplumes. Singeing must be controlled because high temperatures in the flame will cook muscle under the skin similar to over-scalding leading to irreversible toughening of the muscle. Singeing has been found to reduce surface microbial load and increase flavor and palatability of meat. After singeing, the carcasses are washed and transferred to evisceration line.

10. Evisceration

Washed carcasses after singeing are hung by the feet in shackles. In case of large turkeys, a three-point suspension is sometimes used to hang the bird by the head and both feet. With provision to mechanical evisceration, two-point suspension is used.

Head, upper oesophagus and crop are removed from the front of the carcass. In incision is made through the abdominal wall under the tail. The cu is continued around the vent so that intestines are free of any connection to the skin or abdominal wall muscles. All organs of the body cavity are removed through this opening. The process of removal of viscera (the organs in the abdominal cavity) is referred to as evisceration. Since the birds do not have functional diaphragm, all the organs in the thoracic and abdominal cavity will be referred to as viscera.

To retain the identity of the viscera with the carcass and to be in a position for each inspection, the viscera are generally left handing outside the body but still attached to the carcass. The evisceration is performed by supporting the bird with one hand and inserting the fingers of the other hand through the incision in the abdomen. The three middle fingers (sometimes thelittle finger, too) extended, slide past the viscera until the heart is reached. They are then party closed in a loose grip, followed by a gentle twisting action, and the viscera are slipped out of the body and released (Mountney and Parkhurst, 1995).

11. *Post-mortem* examination

At this stage, *post-mortem* inspection of the carcass is made by a qualified and certified Veterinarian. Various conditions have been listed; but, in general, any indication (like pustules, severe hemorrhages in liver, spleen, kidney) of systemic (general) infection, makes the carcass unfit for human consumption. Specifications for *post-mortem* examination as per BIS (IS: 6559-1972) are briefly outlined in Table 11.2.

In case of laying hens, eggs, if any, from the oviduct and ova, of at least 1 cm in diameter, from ovary are collected; the latter are kept identified with each carcass until the bird is passed as wholesome by the inspector.

The oil (preen) gland situated at the base of the tail is removed by cutting deep and backward towards the tail, under it, to ensure complete removal since it gives a very bitter and disagreeable taste.

The carcass is completely washed, both inside and outside. Water is absorbed by the carcass, mostly in the membranes the body cavity and between the skin and muscle tissues during this process; the extent of moisture absorption depends on the washing procedure, duration, temperature of wash water as well as the cuts in opening the body cavity.

11.1. Preparation of giblets

When the carcass is passed by the inspector, the heart, liver and gizzard are saved as giblets. The gizzard is cut lengthwise; the contents and the inner chitinous lining are carefully removed. Gall bladder from the liver without breaking, and blood and pericardium from heart are removed. These three, the edible part of viscera (sometimes neck also included), put together constitute the giblets. The giblets are washed, wrapped in a parchment paper to be sold separately or placed in body cavity.

11.2. Inedible viscera

The intestines, proventriculus, lower oesophagus, spleen, lungs and reproductive organs are discarded. In order to get through clearing of organs from the abdominal cavity a vacuum nozzle is generally utilized. Kidneys are removed by a specially shaped vacuum nozzle and lungs can be removed with either a hand rake or a vacuum system.

After the carcass is passed by the inspector, the inedible viscera is pulled free of the carcass and dropped into a carry-off system. The giblets are trimmed and cleaned. The carcass is given thorough washing, inside and outside, to remove blood clots or other foreign material. Wind pipe is pulled and cut off closest to the thoracic cavity. The feet and hocks are then cut from the carcass. Care must be taken so as to not cut through the cartilage cap on the lower end of the tibia or leg bone. Similarly, during evisceration, it is important not to cut across major muscles because such excised muscles undergo toughening.

12. Trussing

Trussing refers to fastening or securing of wings and legs to the body with a view to give a more compact and attractive appearance to the carcass.

For trussing, a string that is approximately 4 to 5 times the length of the chicken is ideal. With the bird on its back (tail away from processor), middle of the string is placed under the tail, bringing both sides up and cross over the top of the tail. Each of the strings is wrapped around the end of each drumstick and pulled to draw the legs together, crossing strings over each other again (Plate 11.1; www.hormel.com).

The bird flipped over so that the backside is up, keeping neck away strings are pulled up over the thighs and wrapped around the upper wings, catching the tips of the wings in the loop. The string is wrapped around the wing, close to the body and then both ends are brought to the upper side. If there is a flap of skin at the neck, it is folded up and the two strings are tied over it (plates 11.2 and 11.3; www.hormel.com).

Besides, it also assists packaging by minimizing the chance of damage to the packing material. Trussing also helps conserve space in the freezer. The neck skin is drawn as far over the back as possible and the wings are locked over the back to hold the skin in place. The legs of the carcass are secured under the fold of the skin below the keel. The carcass is ready for chilling.

A chicken does not have to be trussed before it is roasted. When a chicken is trussed, it takes longer for the dark meat in the inner thigh area to reach its proper doneness when it is trussed (80°C to 82°C). When the dark meat is cooked until it reaches the appropriate temperature, the white meat will many times be too dry. If it is important that the bird keeps its shape while roasting, it is best to truss it. If it isn't important that it to keeps its shape, it is generally better not to truss the chicken, because the white and dark meat will cook more evenly.

13. Chilling or ageing or conditioning

The temperature of the carcass must be brought to 4°C or lower within 4 hr after trussing for maintaining the top quality of meat. This can be achieved by using chilling tanks in which crushed ice produced from wholesome water or block-type ice sprayed with clean water and crushed, is used. The chilling tank must maintain temperature of 2-4°C throughout.

Chilling is essential to allow *rigor mortis* to proceed with surface microbial growth under control. Usually, $^{1}/_{3}$ ice has been found to be efficient and economical. There will be some water uptake during ageing and using the polyphosphates and other chemicals have been found to be beneficial in increasing tenderness, water-holding capacity, shelf-life etc. (Sreenivasaiah, 1985). Among the gain allowed varies with the class poultry and with what disposition is to be made of the product. Higher levels of water uptake is allowed in case of broilers which are to be distributed as non-frozen product than for frozen carcass. With turkeys, the larger the bird the less water uptake, as a percentage of carcass weight, is allowed.

In the US, eviscerated washed bird is dropped into a prechiller wherein water at less than 18.3°C is regularly agitated; it also serves as a very effective washer. A second chiller with a water temperature of less than 2°C is used for final chilling. As per USDA regulations, carcasses must be chilled to at least 4.4°C internal temperature.

After chilling, the carcasses are drained for 3-5 min, graded and vacuum-packed in polyurethane packs. They are then either marketed as ready-to-cook chicken or frozen if they have to be stored for a long-time before marketing.

14. Yield

In chicken, dressing percentage, which is the ratio of total edible portion (including giblets) to live weight times 100, ranges between 70 and 75%; giblets accounting for 3-5%. The estimated % yield of inedible parts are blood 4.0, feathers 6.5, head 3.0, shanks and feet 4.5, viscera 7.5 and miscellaneous (trimmed and condemned) 2.5. However, proportion of meat skin and bone vary among different species of poultry (Tables 11.4 and 11.5).

Table 11.2 ***Post-mortem* examination guide**

Disease/condition	Remarks
Chronic Respiratory Disease	Acute nature – condemn. If acute symptoms have disappeared – may be passed after adequate cleaning
Salphingitis	Extensive lesions – condemn. Mild lesions – acceptable after thorough cleaning
Inflammation of joint and tendon sheath	Acute cases – condemn. Chronic cases with emaciation – condemn. Only thickening of joints – passed after removal of affected parts
Neurolymphomatosis	Condemn
Leukemia (visceral form)	Condemn
Tumors	Local, non-malignant in skin – may be passed after removing the affected portion. Organs, musculature involved – Condemn
Enteritis including coccidiosis	Condemned except in some chronic cases where it may be passed after removing affected organs
Hepatitis	Condemn
Presternal bursitis	Mild and Chronic may be passed. Extensive – Condemn
Abscesses	Depending on nature and degree
Hemorrhages	Condemned if due to disease/syndrome. Mechanical reasons – cleaned and passed
Abnormal odor and staining	Condemn
Emaciation	Condemn
Contaminated	Which cannot be cleaned – Condemn
Tearing, wounds, other lesions etc.	If extensive – Condemn. If less severe – acceptable after removal of damaged part. Teared liver – Condemn
Tuberculosis	Condemn
Coligranuloma	Few nodules in intestines – discard liver
Leukemia, peritonitis, ascites	Condemn
Fowl Pox	Skin and mouth types – Condemn. Oculonasal type – acceptable after removal of affected parts
Fowl cholera	Condemn
Newcastle Disease	Condemn

Table 11.3 Minimum chilling (ageing) time (in ice and water)

Species	Class	Ageing time (hr)
Chicken	Broilers	4
	Roasters	6
	Fowl	4
Turkey	Fryers	24
	Young hens	8
	young toms	8
Ducks	Roaster	4

Source : Stadelman *et al.*, 1988

Table 11.4 Yields (%) of poultry species (raw)

Species	Bone	Skin	Meat
Broilers	34	14	52
Turkey	21	12	67
Duck	28	38*	34
Goose	19	34*	47
Guinea fowl	17	19*	64
Squab	23	12	65
Pheasant	14	10	76
Quail	10	14	76

* Includes fat

Source : Stadelman *et al.*, 1988

Table 11.5 Proportion of meat, skin and bone in chicken fryer parts (%)

Part	Weight (g)*	Meat	Skin	Bone
Breast	235	71.0	7.7	19.4
Thigh	164	64.7	9.2	23.8
Leg	157	52.8	7.2	37.3
Back	97	41.3	12.3	43.1
Rib	79	44.5	6.4	46.3
Wing	129	32.4	16.5	47.7
Neck	80	26.2	22.2	48.9
Total bird**	920	56.8	9.7	31.0

* 2.5% weight lost during separation of parts

** Includes 20 g each of gizzard and liver and 9 g of heart

Source : Stadelman *et al.*, 1988

Table 11.6 Proportion of various cuts to eviscerated weight in broilers

Cut	%
Split breast with back	17.2
Keel cut breast	13.5
Keel	8.8
Wishbone cut breast	14.8
Quartered breast – anterior portion	10.6
Quartered breast – posterior portion	7.3
Split breast with shoulder	15.3
Split breast with ribs	15.3
Split breast	13.5
Thigh with back	16.6
Thigh with back portion	13.9
Thigh portion with back	11.6
Thigh with connecting back and skin	10.9
Thigh	9.1
Drumstick	8.4
Wing	7.4
Wing with breast portion	8.2
Wing segments – proximal (I joint)	3.4
Wing segments – distal (II joint)	2.8
Wing segments – tips	1.9
Front (fore) quarter	25.1
Rear quarter or whole leg with back	24.4
	Source : Stadelman *et al.*, 1988

15. Cut-up parts

In India, sale of live poultry has given way for ready-to-cook chicken and there is a trend, especially in metropolitan cities, for specific parts of ready-to-cook chicken. Therefore, in the near future, demand for specific part(s) is(are) likely to arise. In Western countries, the market has developed for the cut-up parts and each of the parts has been specifically defined. However, BIS has not yet standards for different cuts. However, relative proportion of various cuts of broiler carcasses is given in Table 11.6.

15.1. Chicken feet/paws

Chicken's foot is about 10 cm long and comprises about 5% of the lightweight; the upper part is the shank and the bottom end with the toes is called the paw, which makes up about 2.7% of the bird's weight.

It is interesting to note that the magnitude of export of frozen chicken feet/paws, especially to China, was about 280 million tons during 2001; the value was about 135 million US $. Therefore, with an estimated 1.8 million tons of broilers and about 185 million layers produced in India, it is worth exploring the possibility of export of chicken feet/paws to the neighboring countries in the coming future.

16. Grading of dressed chicken

Table 11.7 BIS Grading characteristics for dressed chicken

Characteristic	Grade - I	Grade – II
Conformation	Free of deformities that detract from its appearance or that affect the normal distribution of flesh. Slight deformities, such a slightly curved or dented breastbones and slightly curved back may be present	Slight abnormalities, such as dented, curved or crooked breastbone, crooked back, or misshapen legs or wings which do not materially affect the distribution of flesh or the appearance of the carcass or part.
Fleshing	The breast is moderately long and deep and has sufficient flesh to give it a rounded appearance with the flesh carrying well to the crest of the breastbone along its entire length.	The breast has a substantial covering of flesh carrying up to the crest of the breastbone sufficiently to prevent a thin appearance.
Fat covering	The fat is well distributed so that there is a noticeable amount of fat in the skin in the areas between the heavy feather tracts.	The fat under the skin is sufficient to prevent a distinct appearance of the flesh through the skin; specially on the breast and legs.
Defeathering	Free of pin feathers, diminutive feathers and hair which are visible to the inspector or grader.	Not more than an occasional protruding pin feather or diminutive feather shall be in evidence under a careful examination.
Cuts and tears	Free of cuts and tears on the breast and legs.	The carcass may have very few cuts and tears.
Discoloration	Discoloration due to bruising shall be free of clots. Flesh bruises and discolorations of the skin such as "blue back", are not permitted on the breast or legs of the carcass or on these individual parts and only lightly shaded discoloration permitted elsewhere.	Discoloration due to bruising shall be free of clots. Moderate areas of discoloration due to bruises in the skin or flesh and moderately shaded discoloration of the skin, such as "blue back", are permitted.
Freezer burn	may have an occasional pock mark due to drying of the inner layer of skin (derma), provided that none exceeds the area of a circle 0.5 cm in diameter.	may have a few pock marks due to drying of the inner layer of skin (derma), provided that no single area exceeds that of a circle 1.5 cm in diameter.

The USDA has set standards for both live poultry as well as ready-to-cook chicken, turkeys and geese. However, BIS has specifications for grading of dressed chicken which is given in Table 11.7. In any case, marking the carcasses with grades and/or specific demand for graded chicken are not yet a common practice in India. On most occasions, chicken are dressed and sold on the same day.

17. Processing waterfowl

17.1. Ducks

Ducks after delivery are held for 1-2 hr to settle down before slaughter. Carcasses are scalded twice; first to 60°C for 2 ½ min and then at 61.1-62.7°C. The carcasses are defeathered, dipped in wax at 90.5°C and then cooled in water at 12.7°C. Wax is removed by a wax stripping machine and the remainder by pinners. Subsequently, the ducks are eviscerated and chilled in a continuous chiller and packaged (Mountney and Parkhurst, 1995).

17.2. Geese

Geese are forced to swim through a canal to clean their feathers. They are scalded at 65.5-67.7°C. Flight and shoulder feathers are removed as soon as the carcasses our removed from the scalder and then defeathered. The carcasses then are dipped in molten wax and 104.4°C for 3-5 sec, cooled, redipped at 71.1°C for 3-5 sec. Then, dipped in cold water to harden the wax which is removed by a wax stripping machine. Subsequently, the carcasses are dipped in water at 82.2°C to tighten the skin so as to facilitate removal of pin feathers by the pinners. Carcasses are then eviscerated and inspected (Mountney and Parkhurst, 1995).

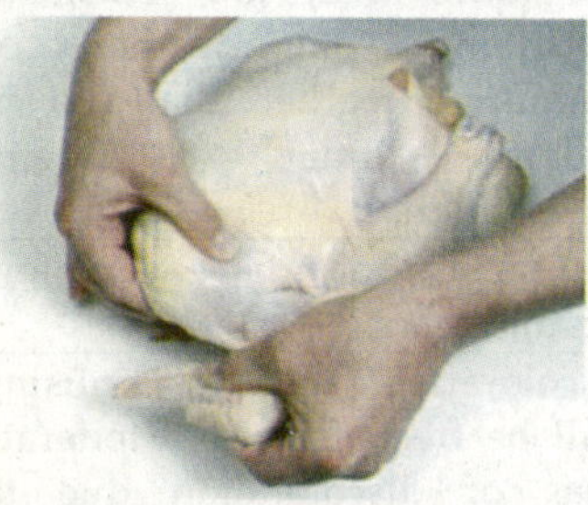

Plate 11.1 Tucking wings underneath the chicken

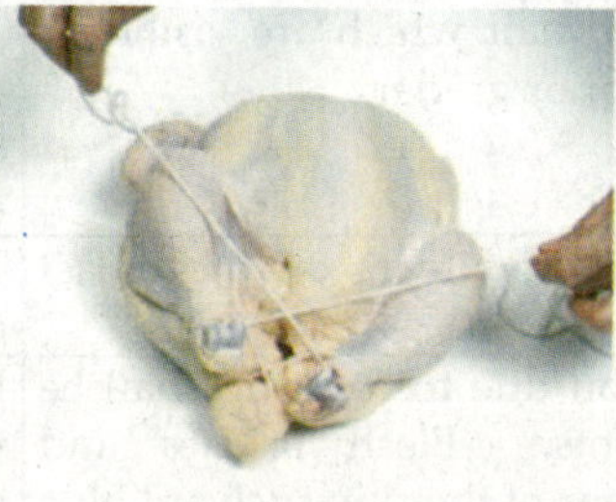

Plate 11.2 Wrapping strings

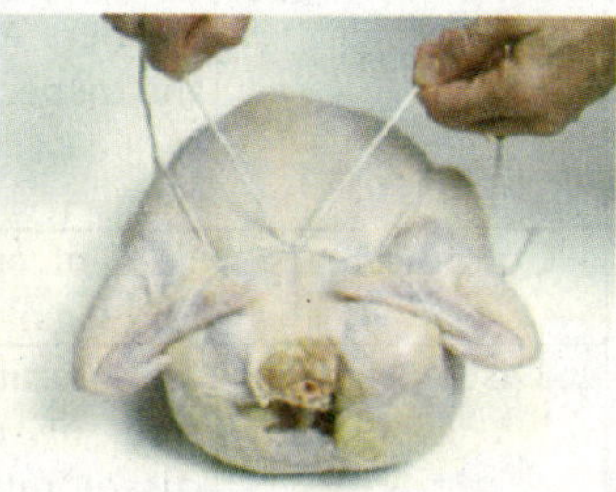

Plate 11.3 Final stages of trussing

Chapter **12**

Composition and Nutritive Value of Poultry Meat

1. Raw meat

1.1. Gross composition

Poultry meat is a highly nutritious food for human beings. Similar to eggs, poultry meat is also a low calorie food and therefore, a concentrated food highly suitable for all age groups and for all seasons. The gross chemical composition of poultry meat is highly variable depending on age and type of birds as well as their diet and environment under which they were reared and other factors. Composition of poultry meat is presented in Tables 12.1 to 12.4.

Table 12.1 Composition of meat – Gross (g / 100 g edible portion)

Nutrient	Broilers	Turkey	Goose	Duck
Water	65.99	70.40	49.66	48.50
Protein	18.60	20.42	15.86	11.49
Total lipids	15.06	8.02	33.62	39.34
Carbohydrates	Nil	Nil	Nil	Nil
Fiber	Nil	Nil	Nil	Nil
Ash	0.80	0.88	0.87	0.68
Energy, MJ	0.90	0.67	1.55	1.69
			Source : Stadelman *et al.*, 1988	

Table 12.2 Composition of meat – Vitamins (per 100 g edible portion)

Nutrient	Broilers	Turkey	Goose	Duck
A, Retinol equivalent	41	2	51	17
C, mg	1.60	Nil	2.80	NA
B_1, mg	0.06	0.06	0.20	0.08
B_2, mg	0.12	0.16	0.21	0.24
Niacin, mg	6.80	4.08	3.93	3.61
Pantothenic acid, mg	0.91	0.81	0.95	NA
B_6, mg	0.35	0.41	0.19	0.39
Folic acid, µg	6.00	8.00	13.00	4.00
B_{12}, µg	0.31	0.40	0.25	NA
NA = Data not available			*Source* : Stadelman *et al.*, 1988	

Table 12.3 Composition of meat – Minerals (mg / 100 g edible portion)

Nutrient	Broilers	Turkey	Goose	Duck
Calcium	11.00	15.00	11.00	12.00
Iron	0.90	1.43	2.40	2.50
Magnesium	20.00	22.00	15.00	18.00
Phosphorus	147.00	178.00	139.00	234.00
Potassium	189.00	266.00	209.00	308.00
Sodium	70.00	65.00	63.00	73.00
Zinc	1.31	2.20	1.36	NA
Copper	0.48	0.10	0.24	0.27
Manganese	0.02	0.02	NA	NA

NA = Data not available *Source* : Stadelman *et al.*, 1988

Table 12.4 Composition of meat – lipids (g / 100 g edible portion)

Nutrient	Broilers	Turkey	Goose	Duck
Saturated fat	4.50 (29.90)	2.37 (29.50)	13.10 (33.30)	9.35 (27.80)
Monounsaturated fat	6.73 (44.70)	3.44 (42.90)	19.43 (49.40)	19.10 (56.80)
Polyunsaturated fat	3.16 (21.00)	1.86 (23.20)	5.11 (13.00)	3.70 (11.00)
Cholesterol	0.075	0.068	0.076	0.080
PUFA/Saturated fat ratio	0.28	0.32	0.16	0.13

Value in the parentheses indicates % of total fat *Source* : Stadelman *et al.*, 1988

1.1.1. Proteins and amino-acids

Poultry meat contains more proteins as well as more essential amino acids than other meats. Poultry meat proteins are of high biological value; the biological value of poultry meat varies inversely with its collagen (fiber) content. Therefore, meat from older birds is of low biological value than that from younger birds. In general, the biological value poultry meat ranges between 65 and 85. The average composition of poultry meat with respect to amino acids and % RDA met by 100 gm of meat assuming digestibility as 80% and biological value as 75 is given in Tables 12.5 and 12.6, respectively.

1.1.2. Lipids

Among the nutrients poultry meat, lipid content is the most variable. It depends on sex, age, species, diet etc. Most of the fat poultry meat is subcutaneous. Poultry meat contains higher proportion of unsaturated fatty acids than fat from other meats. About 50% of the unsaturated fatty acids, containing 18 carbons, is linoleic acid (LA). The cholesterol content of poultry meat is quite low; chicken breast meat contains 79 mg whereas, drumstick meat has 91 mg and chicken fat has 65 mg per 100 g.

Fatty acid composition of poultry tissues and that of different species of poultry are given in Tables 12.7 and 12.8, respectively.

Table 12.5 Amino-acid composition of chicken and turkey meat (% of protein)

Amino-acid	Chicken	Turkey
Arg	6.7	6.5
Cys	1.8	1.0
His	2.0	3.0
Ile	4.1	5.0
Leu	6.6	7.6
Lys	7.5	9.0
Met	1.8	2.6
Phe	4.0	3.7
Thr	4.0	4.0
Trp	0.8	0.9
Tyr	2.5	1.5
Val	6.7	5.1

Source : Mountney and Parkhurst, 1995

Table 12.6 Amino-acid composition of poultry meat (mg/100 g)

Amino-acid	Content in 100 g	% RDA
Arg	1206	Not required
Ile	738	64
Leu	1188	64
Lys	1350	101
Met	324	17
Phe	720	40
Thr	720	86
Trp	144	34

Source : Stadelman *et al.*, 1988

1.1.3. Minerals and vitamins

Proportion of RDA supplied by 100 g of broiler meat with respect to minerals and vitamins is given in Table 12.9.

Table 12.7 Fatty acid composition of poultry tissues (% of lipid)

Component		SFA	MUFA	PUFA
Broilers –	Fresh	32	37	31
	Skin	31	46	23
	Giblets	38	31	30
	Light meat, no skin	37	32	31
	Dark meat	31	38	30
Turkeys –	Light meat	42	23	35
	Dark meat	39	26	35

Source : Stadelman *et al.*, 1988

Table 12.8 Fatty acid composition of different species of poultry (% of lipid)

	Chicken	Turkey	Duck	Goose	Pigeon
SFA	28-31	28-33	27	30	23
Oleic acid	47-51	39-51	42	57	56
Linoleic acid	14-18	13-31	24	8	17
Linolenic acid	0.7-1.0	0.8-1.3	1.4	0.4	0.7
Arachidonic acid	0.3-0.5	0.2-0.7	0.20	0.05	0.04
Iodine number	63-80	73-79	87	67	82

Source : Mountney and Parkhurst, 1995

Table 12.9 Minerals and vitamins supplied by 100 g broiler meat

Minerals	% RDA	Vitamins	% RDA
Calcium	1.38	A	4.10
Iron	5.00	C	2.67
Magnesium	5.71	B_1	4.00
Phosphorus	18.38	B_2	7.06
Zinc	8.73	Niacin	35.79
		B_6	15.91
		Folic acid	1.50
		B_{12}	10.33

Source : Stadelman *et al.*, 1988

Poultry meat is an excellent source of minerals and vitamins. It supplies good amounts of phosphorus and iron. Iron in meat is not only highly available, but also increases the availability of iron from other sources. Meat, being predominantly skeletal muscle, is very rich in potassium (the intracellular ion). Meat is probably the best natural source of niacin and is a good source of riboflavin.

The organ meat (liver) is an excellent source of iron, vitamins A, B_2 and niacin. Consumption of 100 g of organ meat, especially liver, supplies seven times the requirement of vitamin A, about 14% of the requirement of vitamin B_1, 1½ times the requirement of vitamin B_2, about ¾th the requirement of niacin and 1/3rd the requirement of vitamin C.

2. Factors affecting composition (Stadelman *et al.*, 1988)

2.1. Diet

Diet can have significant effect on composition. Feeds containing higher proportion of fat and/or high calorie-protein ratio tend to increase carcass fat. Similarly, broilers growing at high ambient temperature also tend to deposit more fat. Force-feeding geese increases weights of livers due to excessive deposition of fat. Fatty livers are also recorded in caged layers.

Fatty acid composition of turkey fat is considerably influenced by composition of dietary fat. In case of chicken, dietary fat influences the fatty acid composition of thigh meat and abdominal fat whereas, that of liver was least influenced. Therefore, fatty acid composition of chicken and turkey fat can be favorably altered by diet so as to increase P/S ratio.

Increased protein, niacin and tocopherol contents in diet were found to be reflected in the carcass composition also.

2.1.1. Sex

Fat content in males is usually lower than that in females, both in case of chicken and turkeys. Females tend to deposit more fat Mute estrogens. Male broiler meat contained more riboflavin and sodium than that of females. However, differences between sexes in other components are not consistent and significant.

2.1.2. Age

Fat deposition, niacin and thiamin increase with age in case of broilers with concomitant reduction in moisture and riboflavin. In case of turkeys, cooked edible yield increased with age.

2.1.3. Part of Carcass

2.1.3.1. Proportion of meat skin and bone

Difference in composition of meat from various species of poultry can be partly attributed to differences in yields among different species (Table 11.4) and proportion of meat, skin and bone in different parts within a species (Table 12.10).

Table 12.10 Proportion of meat, skin and bone in chicken fryer parts (%)

Part	Approximate weight (g)*	Meat	Skin	Bone
Breast	235	71.0	7.7	19.4
Thigh	164	64.7	9.2	23.8
Leg	157	52.8	7.2	37.3
Back	97	41.3	12.3	43.1
Rib	79	44.5	6.4	46.3
Wing	129	32.4	16.5	47.7
Neck	80	26.2	22.2	48.9
Total bird	990**	56.8	9.7	31.0

* 2.5% lost during separation of parts ** includes 49 g of giblets

Source : Stadelman *et al.*, 1988

2.1.3.2. Light and Dark meat

The terminologies "light" and "dark" meat refers to visual appearance. For instance, the thigh and leg muscles appear darker (reddish) whereas the breast meat is pale in color. This is mainly because of difference in amount of exercise the respective muscles are subjected to. The leg muscles have to work more; hence they contain more adipose tissue, blood (myoglobin and iron), mitochondria and energy-producing enzyme activities than breast meat. Of the two types of meat, light and dark, the former has more protein, and saturated fat whereas, fat and cholesterol are normally high in the latter. Fat in light meat is more saturated and skin has more fat, cholesterol and monounsaturated fatty acids. Organ meat (liver) has very high levels of cholesterol whereas abdominal fat and breast meat are low in cholesterol. Breast meat has more protein whereas heart and back have more fat. Other differences include higher niacin and pyridoxine and lower iron, zinc, sodium, thiamin, riboflavin and tocopherols in light meat than in dark meat.

Proximate composition of light and dark raw poultry meat is given in Table 12.11

2.1.3.3. Broiler parts

Proximate composition of broiler parts is given in Table 12.12 to elucidate the compositional differences in the carcass.

3. Cooked and processed meat

3.1. Effect of processing

During the process of preparation of ready-to-cook chicken, several steps involved have significant effect on nutritive value of raw chicken meat. For instance, scalding and plucking, chilling, low temperature distribution and storage, and freezing are additional processing steps which are likely to reduce some of the nutrients. In some of the Western countries, irradiation is approved as a processing step for raw meat; in such conditions, irradiation also affects nutritive value. Mechanical separation of meat, popularly referred to as mechanically deboned poultry meat (MDPM), also affects the nutritive value of raw meat. These effects are summarized in Table 12.13.

Table 12.11 Proximate composition of light and dark meat (per 100 g edible portion)

Component	Chicken		Turkey	
	Light	Dark	Light	Dark
Water, g	74.86	73.99	73.82	74.88
Calories, MJ	0.477	0.523	0.481	0.523
Protein, g	23.20	20.08	23.56	20.07
Fat, g	1.65	4.31	1.56	4.38
Ash, g	0.98	0.94	1.00	0.93

Source : Stadelman *et al.*, 1988

3.2. Effect of cooking

Raw chicken has little or no flavor and it develops during cooking. The principles of cooking poultry are basically the same as for cooking meats.

3.2.1. Cooking methods

3.2.1.1. Dry heat

3.2.1.1.1. Roasting

Roasting in pan over temperature of 163^0 C ensures the adequate browning of meat for good flavors and good appearance.

3.2.1.1.2. Broiling

It consists of cooking meat by direct radiant heat such as the open fire of a gas flame, live coals or electric oven. Broiling is applied to tender cuts that are at least 2.5 cm thick. Thinner cuts will be too dry if broiled. Broiling is carried out at a temperature of 176^0 C until the top side is down. Broiling is a faster method of cooking meat by dry heat than roasting.

Table 12.12 Proximate composition of broiler parts (100 g edible portion)

Part	Water (g)	Energy (MJ)	Protein (g)	Fat (g)	Ash (g)	Cholesterol (mg)
Flesh only	75.46	0.50	21.39	3.08	0.96	70
Skin	54.22	1.46	13.33	32.35	0.41	109
Giblets	74.87	0.52	17.88	4.47	0.99	262
Back with skin	58.10	1.33	14.05	28.74	0.64	79
Breast with skin	69.46	0.72	20.85	9.25	1.01	64
Leg with skin	69.91	0.78	18.14	12.12	0.85	83
Neck with skin	59.99	1.24	14.07	26.24	0.55	99
Wing with skin	66.21	0.93	18.13	15.97	0.69	77
Separable fat	28.91	2.63	3.73	67.95	0.28	58

Source : Stadelman *et al.*, 1988

3.2.1.1.3. Pan Broiling

Meat is placed in a cold girdle and heated so that meat cooks slowly. Any fat that accumulates in the pan is removed so that the meat will continue to pan broil rather than pan fry.

3.2.1.1.4. Frying

Two methods of frying are pan frying and deep frying. Too high temperature results in inside uncooked and too low temperature results in greasier product.

3.2.1.2. Moist heat

This method is used for less tender cuts, meat become tender owing to the conversion of connective tissue to gelatin.

3.2.1.2.1. Braising

In this method meat is cooked with or without the addition of water, the meat is first carefully browned on all sides by broiling, pan broiling or frying. Tomatoes and fruit juices may be added as liquids.

Table 12.13 Effect of processing on nutritive value of raw poultry meat

Processing	Effect (s)
Immersion chilling	1. Moisture pickup: maximum permissible is 12%. Most of the moisture taken-up by skin or between muscle and skin, and hence, meat moisture generally unchanged. 2. Leaching of solids, protein, minerals (ash), total nitrogen and increase in drip, volatile and cooking losses. Potassium and phosphorus levels reduced.
Air chilling	1. More weight loss than immersion chilling. 2. Less thawing drip and cook-loss
Ice-packed breast portions	More loss of potassium, phosphorus and magnesium than surface-frozen ones.
Freezing	1. Thaw-drip from 1-6%. 2. Slow freezing – higher loss of amino-acids and peptides, more drip. 3. Thawing prior to cooking – more cook-loss. 4. Longer the storage period greater is cook-loss; loss of niacin, thiamin and riboflavin particularly high if storage extends beyond 8-12 months. 5. Fluctuations in temperature – loss of thiamin
Irradiation	1. Pyridoxine and thiamin particularly sensitive. 2. Even riboflavin can be lost to a tune of 50% depending on dosage.
Mechanically separated (deboned) poultry meet (MDPM)	1. Variation in calcium content depending on deboning machine. 2. Fact expressed from skin into the meat faction and most of the skin protein (collagen) remains in bony residue. 3. Due to phospholipids portion, which is highly unsaturated, it is susceptible to oxidation

3.2.1.2.2. Stewing

Large pieces of tough meat are cooked in sufficient water until tender.

3.2.1.2.3. Pressure-cooking

This method takes less time. Pressure-cooked meats are less juicy and cooking losses are great.

3.2.1.2.4. Tandoori chicken

The cooking is done in a clay oven called a tandoori. Tender chicken, either whole or cut is used. The skin is removed from the chicken and the flesh is pricked with a fork and sprinkled with salt. Tandoori sauce is then smeared on the chicken which is then left aside for 6-8 hours.

Effects of cooking on nutritive value of meat occur due to 1. Cook-and/or drip-losses 2. Heat-lability of the meat constituents and 3. Absorption of the constituents from the cooking medium.

3.3. Cook-loss

Cooking loss averages about 26% in poultry meat. Stewing resulted in minimum cook-loss (22%), followed by frying (23%) and roasting (31%). Cooking of frozen parts might be cause of cook-loss between 37.2 and 52.7%.

The cook-out fluid contained small amounts of protein and amino-acids whereas fats remain usually unaltered. On the other hand, fat may be absorbed from skin into the meat and some portion of fat from skin, especially from species which tend to contain more fat (ducks and geese), is lost in the cook-out fluids.

Of the various components, thiamin is most heat-sensitive while pyridoxine and pantothenic acid can be lost into cook-out fluid or drip. Pyridoxine may also be lost due to heat the degradation. Depending on type of cooking, magnesium (steaming), phosphorus and sodium (roasting) may be lost significantly. Calcium is thought to migrate from bones into muscle during cooking process. Such a migration is also thought to be responsible for increase in calcium content of meat (with bone intact) during storage. If at a different cooking methods on lipid content and retention of B vitamins in chicken meat are shown in Tables 12.14 and 12.15, respectively; mineral content of cooked whole chicken (giblets) and percentage nutrient retention in cook poultry products are given in Tables 12.16 and 12.17, respectively.

It is clear from the Table 12.17 that moist heat causes more leaching and cooking under pressure causes for the losses of thiamin. Of the dry-heat methods, broiling, roasting and microwave are least destructive with the latter causing minimum loss of thiamin and possibly pyridoxine also.

Amount of fat absorbed by the meat depends on type of frying (Table 12.14). On average, 17.9 g of fat is absorbed per 100 g of coating. Effect of frying on nutrient retention appears to be similar to that following dry-heat cooking. Reading does not seem to affect significantly the nutrients of the product.

Effects of each of the methods of preservation and storage of raw meat will be discussed in a separate chapter.

4. Contaminants in meat and meat products

Contaminants can be defined as undesirable compounds that are present in relatively small amounts, as a result of human action. Classification of contaminants is shown in Fig 12.1, 12.2, 12.3 and 12.4.

Table 12.14 Effect of cooking methods on lipid content of chicken meat

Cooking method	Type of meat	Lipids, %		Apparent retention of lipids
		Raw	Cooked	
Stewing	Light meat with skin	11.07	9.97	90
	Light meat without skin	1.65	3.79	242
	Dark meat with skin	18.34	14.66	80
	Dark meat without skin	4.31	8.98	208
	Mature chicken : Flesh with skin	20.33	18.87	93
	Mature chicken : Fresh without skin	6.32	11.89	188
Roasting	Light meat with skin	11.07	10.85	98
	Light meat without skin	1.65	4.51	273
	Dark meat with skin	18.34	15.78	86
	Dark meat without skin	4.31	9.73	226
Frying	Light meat with skin	11.07	15.44	139
	Light meat without skin	1.65	5.54	336
	Dark meat with skin	18.34	27.22	146
	Dark meat without skin	4.31	11.62	270

Source : Stadelman *et al*., 1988

4.1. Environmental contaminants

4.1.1. Heavy metals

Heavy metals too are natural constituents of meat. But, the larger utilization of them by man, their concentration in the environment has increased. Thus, gradually increases in concentration of heavy metals have been recorded in animal products. Of the several heavy metals, lead, cadmium and mercury are particularly important because they are toxic and they accumulate in tissues. Tolerance levels of these poultry meat are 0.05, 0.30 and 0.50 ppm, respectively and those in poultry liver are 0.05, 1.00 and 0.05 ppm, respectively (Kan, 1991).

4.1.2. Xenobiotics

4.1.2.1. Pesticides

Most important in this category is dichlorodiphenyl trichloroethane (DDT) which is very stable and accumulates in fat tissue. Major degradation product is dichlorodiphenyl dichloroethane (DDE) which is even more persistent than DDT itself.

The other important group includes cyclodienes which are isomers of Hexachloro cyclohexane or benzehexachloride (HCH or BHC). Aldrin, Dieldrin, Heptachlor, Heptachlor epoxide, Endrin, Chlordane etc. are the popular pesticides of this category. Fungicide Hexachloro benzene (HCB) also can accumulate in body fat.

Table 12.15 Effect of cooking methods on retention of B vitamins in chicken meat (%)

Cooking Method	B_1	Niacin	B_2	B_6	Folic acid	Pantothenic acid	B_{12}
Dry heat*	73	82	90	80		77	65
Moist heat*	53	60	95	49			48
Grilled	78						
Stewed	58					67	
Fried	65						
Fried (light meat)					100		
Roasted						64	
Broiled (leg muscle)	67-80	80-81	69-92				
Broiled (breast)	60-75	78-91	67-114				
Microwave (breast) **	79		86				
Oven (breast) **	79		92				

* with skin ** tu rkey meat *Source* : Stadelman *et al.*, 1988

Table 12.16 Mineral content of cooked whole carcasses (with giblets)

Mineral, mg%	Raw	Fried	Roasted	Stewed
Calcium	11	21	15	14
Iron	1.31	1.79	1.66	1.53
Magnesium	20	21	23	19
Phosphorus	149	158	182	144
Potassium	189	190	212	163
Sodium	70	284	79	66
Zinc	1.48	1.91	2.16	1.98
Copper	0.064	0.089	0.080	0.071
Manganese	0.029	0.064	0.031	0.029

Source : Stadelman *et al.*, 1988

Pesticides are classified depending on persistence in body tissues (adipose tissue) as 1. Highly accumulating - HCB, β-HCH, most of cyclodiene group 2. Moderately accumulating - Endrin, Heptachlor etc. 3. Low accumulating - α- and γ-HCH and 4. Very low accumulating - Methoxychlor.

Table 12.17 Nutrient retention in cooked chicken products (%)

Nutrient	Dry heat		Moist heat	
	Meat only *	Meat + skin	Meat only	Meat + skin
Water	63	59	70	70
Protein	101	99	101	98
Total lipid	192	57	186	59
Ash	98	92	87	80
Calcium	97	87	94	88
Iron	77	76	66	67
Magnesium	77	76	66	67
Phosphorus	80	79	69	69
Potassium	80	78	57	58
Sodium	82	76	71	69
Zinc	105	100	104	99
Copper	95	91	93	91
Manganese	77	72	78	73
Thiamin	73	71	53	55
Riboflavin	90	89	95	93
Niacin	82	82	60	61
Cyanocobalamin	65		48	
Vitamin A	75	75	75	75
Cholesterol	95			

* cooked with skin

Source : Stadelman *et al.*, 1988

In case of poultry, the rate of accumulation in relation to the corresponding level in feed has been found to be 10-30% for highly accumulating group and 3-20% for DDT in case of meat; whereas, the figures may be far higher particularly depending on fat content which is highly variable.

Chlorinated pesticides are excreted through both feces and urine; DDT is transferred to eggs also. Half-life of DDT with respect to abdominal fat is 5-10 weeks. Under normal circumstances, the residues in meat will be well within tolerable limits; however, sufficient information seems not available in our country regarding pesticide residues in meat and eggs.

4.1.2.2. Polychlorinated biphenyls (PCBs)

Concentration of PCBs is relatively low in meat and meat products. Less-chlorinated biphenyls are more toxic and tend to accumulate to lower levels than PCBs.

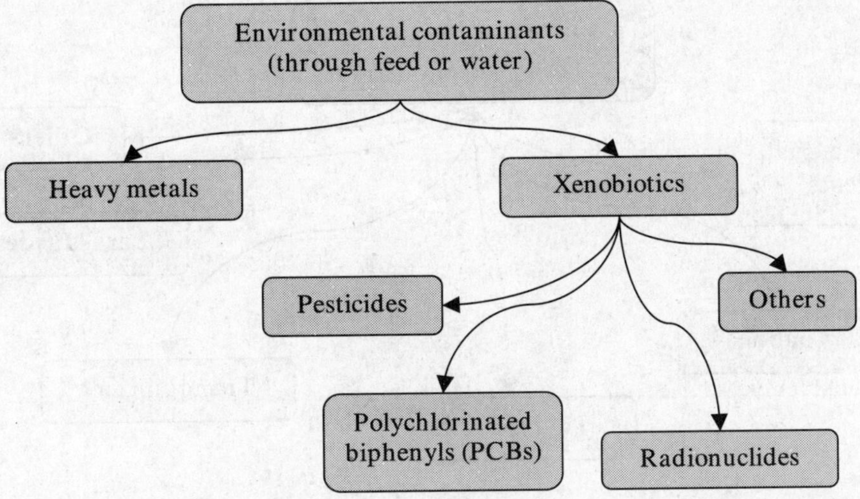

Fig. 12.1 Environmental contaminants in meat and meat products

4.1.2.3. Radionuclides

In animal tissues in natural radioactivity of 100 to 200 Becquerel/kg (i.e. one disintegration/sec or 27 pCi) which is due to presence of natural isotopes (like 40_K mainly and 14_C partially). However, additional man-induced radioactivity is not expectable in our country.

4.1.2.4. Others

Meat of quail can be poisonous if it gains access to toxic plants such as hemlock (which is not toxic to quails).

4.2. Contamination from animal production

Many chemicals are used during the animal's life till slaughter and many of them given orally are metabolized and excreted satisfactorily if the time-lapse between administration of chemicals and slaughter is long enough. However, parenterally administered compounds remain at the site application for considerable time. Thus, withdrawal times suggested for various compounds should be strictly adhered to.

4.2.1. Antibacterial agents

These are used for curative, preventive or nutritive purposes; mostly for the latter via feed or water the commonly used antibacterials are tabulated in Table 12.18 along with the withdrawal periods and effects of residues.

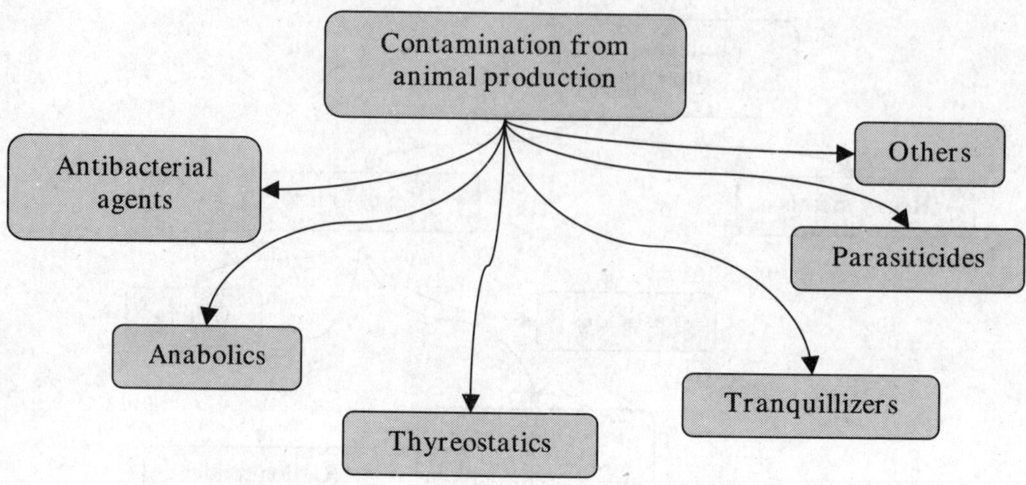

Fig. 12.2 Contaminants in meat and meat products from animal production

4.2.2. Anabolics

Use of anabolics is not common nowadays because it has not been economical. Hence, their residue, especially in poultry meat, does not pose practical problems.

4.2.3. Thyreostatics

These are substances which suppress thyroid function and thus they can cause increased growth rate due to more water retention in the muscle mass and increased filling of gastro-intestinal tract. However, these are banned because just increasing the water content of muscle is considered improper practice.

4.2.4. Tranquillizers

These are used during transportation before slaughter, especially in pigs. Since these are, if administered, are given shortly before slaughter, chance of residues is high. However, if recommended dose is strictly adhered to, the residual levels will be far below the maximum tolerance levels.

4.2.5. Parasiticides

Parasiticides are applied on and administered into the animal; but such an application generally is not done immediately before slaughter. However, if some of the ectoparasiticides (like lindane) which are fat-soluble are applied, they can penetrate and lodge in subcutaneous fat. But, these compounds are not of serious importance in poultry meat.

4.3. Contaminants from meat processing and packaging

Polycyclic aromatic hydrocarbons (PAHs)

These compounds are formed during pyrolysis of organic matter (smoked meat).

Some of the PAHs like benzpyrene are potent carcinogens. Details of these compounds will be discussed along with preservation of poultry meat.

4.3.1. Nitrosamines

Found in cured meats and will be discussed separately along with the preservation of poultry meat.

Table 12.18 Commonly used antibacterials and effects of their residues

Antibacterials	Minimum withdrawal period	Effect of residues (man)
Sulfonamides (Tolerance level 0.1-1 ppm)	2 weeks	Residual bacteriostatic effect
Nitrofurans	3 weeks	Carcinogenic
Tetracyclines	3-5d	Residual antibiotic effect
Chloramphenicol	Not available	Aplastic Anemia (grey baby syndrome)
Penicillin	Not required; quickly excreted	Allergic reactions in sensitive people; otherwise, safest antibiotic

4.3.2. Disinfectants

Contamination of meat by disinfectants is very rare and most of the disinfectants used on utensils can be removed by washing with water itself. Even if some of the surface-acting disinfectants to gain access to cut surfaces by use of utensils not properly cleaned after disinfection, their toxicity is generally low. However, those with the chlorine-releasing compounds have to be properly removed to avoid formation of compounds of toxicity by the reaction of chlorine with several meat components.

Spices added during the manufacture of meat products may form an avenue for ethylene dioxide or methyl bromide (which are used on spices) to enter meat products. Chances of such contamination to reach levels of serious public health importance are too low.

4.3.3. Migration from packaging material

Plastics which have become very popular as packaging material contain a variety of low-molecular weight compounds such as plasticizers, stabilizers, lubricants etc. Most of these are lipophilic and hence can migrate into fatty parts of the food packed. For longer time storage, polyvinylchloride films are unsuitable and films impermeable to oxygen and water vapor are required. Vinyl chloride, a carcinogenic, should not exceed levels of 0.01 ppm of products.

4.3.4. Heavy metals

Meat can be contaminated with metals, as a result of packaging, although use of metals (with an exception of canned meat) for packing meat and meat products is not a common practice. Contamination of meat (especially beef) with lead has been recorded.

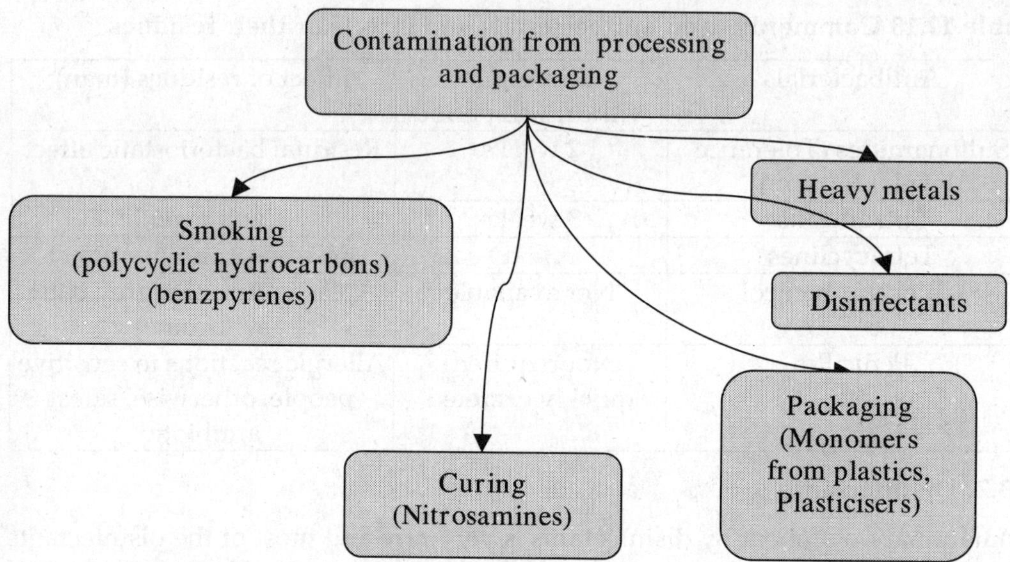

Fig. 12.3 Contaminants in meat and meat products from processing and packaging

4.4. Bio-contaminants

These are products produced by living organisms if they are present in the food and most of the products are toxins. Microbial spoilage also is included in this group

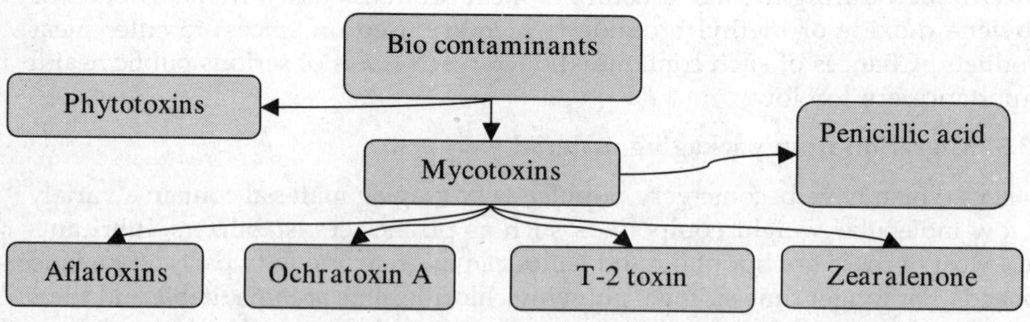

Fig. 12.4 Bio-contaminants in meat in meat products

4.4.1. Mycotoxins

Most of the mycotoxins are modified and eliminated by adult animals important ones in this group are aflatoxins, ochratoxin A, T-2 toxin, zearalenone and penicillic acid. Most of these toxins are heat-stable. Therefore, toxin-producing molds must be avoided in feed of meat animals or must be destroyed in feed by suitable methods like ammonia and formaldehyde treatment.

4.4.2. Aflatoxins

These are coumarin derivatives produced mainly by *Aspergillus flavus* and *Aspergillus parasiticus;* the former produces aflatoxins. Tolerance level of aflatoxins in respect of meat and meat products is not yet set.

4.4.3. Ochratoxin A

This toxin, produced by several species of *Aspergillus* and *Penicillium,* is strongly nephrotoxic; but, its occurrence is not very common.

4.4.4. T-2 toxin

This is produced by *Fusarium* species. It is cytotoxic causing necrosis of mucus layers of gastro-intestinal tract in chicks and possibly in humans.

4.4.5. Zearalenone

Produced by *Fusarium roseum;* this has estrogenic and teratogenic properties. Its frequency of occurrence is quite low.

4.4.6. Penicillic acid

This is a cytotoxic, cardiotoxic and carcinogenic toxin which has minimal toxicity to chicken.

4.4.7. Phytotoxins

Can enter meat through certain plants like hemlock; quail meat becomes poisonous after they eat hemlock.

Chapter **13**

Poultry Meat Consumption *vis a vis* Colon Cancer

Colorectal cancer is the fourth most common cancer in the United States. It is the only cancer that strikes men and women almost equally. In 2000, colorectal cancer will account for 10% of all new cancers and 10% of all cancer deaths in the United States. Worldwide there is a 20-fold variance in incidence rates of colorectal cancer. In general, the variance has been attributed to dietary and environmental differences. Low incidence of colorectal cancer in Africa was attributed to high dietary fiber intake.

Consumption of meat faced a minor setback during the late 1970s due to a hypothesis put forward relating it with colon cancer. Considerable research has been and is being carried out in this regard and salient observations are outlined below.

Colorectal cancer is a slow developing cancer that may be asymptomatic and therefore remain undiagnosed for several years after initial development. The cancer can develop in any of the four sections of the colon or the rectum. Adenomatous polyps are thought to be precursors of cancer development and they may provide information on the initiation phase of malignant cell proliferation. Ninety-five per cent of colorectal cancers are adenocarcinoma.

1. Etiology

Genetics, experimental, and epidemiologic studies suggest that colorectal cancer results from complex interactions between inherited susceptibility and environmental factors. It has been suggested that dietary factors may be responsible for a significant but poorly quantitated number of cancer cases; diets high in total fat, protein, calories, alcohol, and meat (both red and white) and low in calcium and folate, are associated with an increased incidence of colorectal cancer.

Efforts to identify causes and to develop effective preventive measures have led to the hypothesis that adenomatous polyps (adenomas) are precursors for the vast majority of colorectal cancers. While most of these adenomas are polypoid, flat and depressed lesions may be more prevalent than previously recognized.

Large flat and depressed lesions are more likely to be severely dysplastic. Specialized techniques may be needed to identify, biopsy, and remove such lesions. In effect, measures which reduce the incidence and prevalence of adenomas may result in a subsequent decrease in the risk of colorectal cancer. The finding of an adenoma on flexible sigmoidoscopy may warrant colonoscopy to evaluate the more proximal colon for synchronous neoplasms. Many of the intervention trials employ adenoma recurrence or disappearance as a surrogate end point. The evolution of a carcinoma from a small adenoma, however, takes many years.

Colon cancer is more a function of environment (diet) than genetic. Various food additives or contaminants can act as carcinogens and/or co-carcinogens. Certain deficiency(ies) can alter the metabolic pathways to stimulate neoplasms. Due to intake of selected macro-nutrients, metabolic abnormality might result which, in turn, can increase risk of colon cancer. However, considering the incidence of colon and breast cancers has not increased significantly over the past five decades in industrialized countries like the US and Japan, it is imperative to conclude that it is the diet rather than contaminants and/or industrialization, the primary etiology of colon cancer.

Keeping the above in view, another interesting fact was noticed that urban population had higher incidence of colon cancer than rural. Among the urban populace, the so-called "Lacto-ovo-vegetarians" or "Vegans" (those who consume milk, eggs and vegetarian diet) had lower incidence than those consuming meat also in their diet (the so-called "Non-vegetarians"). Therefore, meat was soon thought to be responsible for colon cancer and appeared are the several advertisements proclaiming that "meat causes colon cancer".

Further research of colon cancer revealed very interesting results. Colon cancer was very highly and positively correlated to total fat intake, be it saturated or unsaturated, or be it vegetable or animal origin. Consumption of refined carbohydrates with little fiber predisposed whereas high-fiber diets offered protective effect, against colon cancer.

These observations led for the development of a hypothesis that colon cancer is probably caused by unidentified initiating carcinogens whose deleterious effects are possibly extended by certain promoting agents which comprise primarily bile acids. The metabolism of gut microflora is altered by the dietary fat; which when excess can trigger these organisms to metabolize bile acids and others into promoters (co-carcinogens) and/or carcinogens.

Total dietary fiber had a significant inverse relationship to the risk of colorectal adenoma of the distal colon. Increasing intakes of total dietary fiber also had a significant inverse association with the development of incident polyps of the distal colon. Total dietary fiber had a significant inverse relation to the risk of recurrent colorectal adenomatous polyps in women but not in men.

It is well-documented that dietary fiber absorbs moisture in the gut and swells resulting in increased stool bulk. Increased bulk can cause dilution of promoters and/or carcinogens. In addition, they might modify the metabolism of tumorogenic compounds. Gut microflora of high-risk group comprised higher number of anaerobes in contrast to low-risk group. The fiber may also absorb some/many of the carcinogens to be excreted ultimately through fecal matter. Alfalfa, wheat straw and others can bind considerable amounts of bile acids whereas wheat bran, oat hulls and synthetic fibers can bind negligible amounts.

High fat diet was accompanied by 1. Increased fecal excretion of total bile acid and cholesterol metabolites and 2. Increased β-glucuronidase activity of fecal bacteria. Incidentally, many of the exogenous and endogenous substances, including those which are tumorogenic metabolites, are excreted via bile as glucuronide conjugates. It is also been found that increased activity of NAD-and NADP-dependent 7α hydroxy steroid dehydrogenase (7α HSDH) also indicates high risk of colon cancer.

It is clearly evident from Table 13.1 that the total amount of each of the bile acids excreted is approximately the same whereas the concentration of many of the bile acids were significantly higher in high-risk group than in low-risk group.

In this connection, what is particularly important to note is that high animal protein diets did not predispose was high-fat and low fiber diets with the without animal protein did. Further, animal protein intake didnot influence bile acids and neutral sterols in feces at all. However, animal protein may affect time of appearance as well as size and number of tumors.

In an effort to further delineate what specific food groups provide a protective effect, several studies investigated the food sources of dietary fiber. The main dietary contributors of fiber are fruits, vegetables, legumes and cereals/grains.

1.1. Fruit Fiber

High fruit fiber intake had a significant inverse relationship to cancer of the colon, rectum, and colon/rectum. In addition, a significant inverse trend with intake of fiber from fruits and vegetables combined has been reported. It may be noted that there is a mixture of results in the literature, and that these studies show associations - not cause and effect. There does appear to be an association between increasing fruit intake and fruit fiber, in particular, and decreased incidence of adenomatous polyps and/or colorectal cancer. More research is necessary in this area to clarify if any particular constituent of fruit is more protective than another.

Table 13.1 Dietary intake of various nutrients and fecal excretion of certain metabolites in low- and high-risk individuals

Constituents	Low-risk group	High-risk group
Dietary (g/d)		
Total protein	93	89
Total fat	110	115
Saturated fat	59	49
Other fats	51	66
Carbohydrates	320	285
Total fiber	32	14
Fecal (g/d)		
Fresh feces excreted	277	76
Fiber	26	9
Fecal dry matter	61	22
Fecal bile acids (mg/d) (on dry matter basis)		
Cholic acid	0.02 (12)	0.24 (6)
Chenodeoxycholic acid	0.13 (8)	0.03 (5)
Deoxycholic acid	1.72 (104)	3.74 (88)
Lithocholic acid	1.40 (84)	3.27 (77)
Ursodeoxycholic acid	0.08 (5)	0.13 (3)
3α 7β 12α trihydroxy 5β cholanic acid	0.04* (2)	9.12 (3)
12 ketolithocholic acid	0.06* (4)	0.13 (3)
Other bile acids	0.93* (56)	3.80 (89)
Total bile acids	4.59* (277)	11.70 (275)

* significantly different — Value in parentheses indicates total per d (mg)

Source : Reddy, 1982

1.2. Vegetable Fiber

Non-starch polysaccharide from vegetables had a significant inverse relationship with the risk of developing colorectal cancer. Vegetable, citrus and high-fiber grains together had an inverse association to the risk for colon cancer.

1.3. Cereal/grains

Given the substantial studies with null results and the remainder with varying positive and negative results there is little evidence that cereal fiber exerts a protective effect with respect to the development of colorectal adenomas and cancer.

2. Fiber Components

Fiber from food sources has been researched extensively. However the measurable components, i.e. soluble, insoluble, pectin, cellulose, hemicellulose and lignin, have not been as extensively covered in the literature. The components may hold a clue to further explain the relationship between plant foods, mainly fruits, vegetables and whole grains, and colorectal cancer.

2.1. Soluble Fiber

Soluble fiber had a significant inverse relationship with adenomatous polyps of the distal colon. Pectin had a significant inverse trend in relation to the risk of colon cancer. Specifically, it was found protective in all male subjects, in males over 67 years old, and in men with cancer of the proximal colon. Soluble fiber was also inversely associated with cancer of the proximal colon in women.

2.2. Insoluble Fiber

Increased consumption of cellulose was inversely related to the risk of development of adenomatous polyps in the distal colon in men and women, cancer of the colon (30% reduction), in general, and proximal colon (50% reduction), in particular, in women.

The remainder of the components, hemicellulose and lignin either were not measured or showed no significant trend in relation to colorectal adenoma or cancer. Given the small sampling of studies available for the components of fiber and their relationship to colorectal adenomas and cancer, it may be cautiously concluded that soluble fiber, including pectin, as well as insoluble fiber, including cellulose, exerts a protective effect. However, further research must be conducted to clarify the relationships.

3. Conclusion

The key to developing an overall conclusion about the relationship between dietary fiber and colorectal adenomas and cancer is to look at the studies in aggregate. No single study can account for all the variation in human diets, accurately measure all variables and account for all confounders. Therefore, one must be cautious when interpreting the results.

Total dietary fiber was inversely associated with colorectal adenomas and/or cancer in 73% of the studies, fruit fiber in 77% of the studies, and vegetable fiber in 66% of the studies. Cereal/grain fiber was found protective in 36% of the studies but the null hypothesis or no effect was found for cereal fiber in 45% of the studies. Therefore, based on the aggregate review of these studies, intakes of fruit fiber, vegetable fiber and total dietary fiber are inversely related to the development of colorectal adenomas and cancer. The result for cereal fiber are not consistent, therefore no conclusion can be made.

Although the studies are few, it may also be cautiously concluded that soluble and insoluble components of fiber exert a protective effect against the development of colorectal adenomas and cancer. An important point to make is that fruits, vegetables and cereal/grains are the main sources of these fiber components in the human diet.

As our knowledge of food composition continues to expand, we may find other components, e.g., antioxidants, isoflavones, carotenoids, in these foods to be anti-carcinogenic. Although additional research is warranted to further delineate the specific components of each food that may impart a protective effect upon the endothelial lining of the colon and rectum, food pattern analyses are still useful in that they allow for interaction effects of multiple qualities in whole foods. Fruit and vegetable fiber and cereal fiber are likely complementing each other in the prevention of cancer. Cereal fiber also binds bile acids reducing transit time and increasing stool bulk. Fatty acids are then converted to short-chain fatty acids which lower colonic pH and inhibit the conversion of primary to secondary bile acids. The secondary bile acids are thought to promote carcinogenesis.

3.1. Implications for practice

The above discussion clearly points out that it is a high fiber content of the diet that offers protection against colon cancer. It definitely does not follow that just because meat and eggs are devoid of fiber (for that matter any of the foods of animal origin), they predispose a person to colon cancer. Further, it is necessary to emphasize that no component(s) of meat, in general, and poultry meat, in particular, has(have) been incriminated as predisposing factor(s) of colon cancer.

It must always be remembered that neither meat and/or egg alone constitute the entire diet of any individual nor they constitute the sole sources of any/many of the nutrients. In fact, various foodstuffs are selectively mixed and consumed by the Indian population. Therefore, campaigning against food items of very high nutritive value due to lower content of one of the nutrients is highly unreasonable and therefore, unacceptable. On the contrary, the public must be made aware of the nutritive worth of these animal products and the same time must be informed about the importance of dietary fiber so that requisite amount of plant foodstuffs, especially vegetables and fruits, are also incorporated in their diets.

Moreover, as on today, single foodstuff which can meet all the nutrient requirements of humans without any limitation(s) is yet to be identified and/or produced. Most of the so called "essential nutrients" do produce harmful effects, not only when deficient in diet, but also when consumed in excess; effects of excess of vitamins A, D, pyridoxine etc. minerals like Ca, P etc., carbohydrates, proteins and fats have all been well documented. In other words, each foodstuff has one or more harmful component(s) to some or more or all people.

Therefore, it follows that excluding poultry meat from human diet just because it doesn't contain fiber is unscientific *per se*. (details of nutritive value of poultry meat are dealt in a separate Chapter).

Considering that in India *per capita* consumption of poultry meat is around 1.6 kg and even in urban areas it may not exceed 2.5 to 3.0 kg, consumption of poultry meat should not pose serious threat of colon cancer. However, it is definitely essential to gear up propaganda to promote consumption of fruits and vegetables.

Based on the results, both vegetarians and non-vegetarians, with adequate intake, will benefit from fruit and vegetable fiber. Continued advocacy of a diet high in fiber, with a dietary goal of 20-35 grams per day, is recommended.

Chapter 14

Egg Quality

In general, the term quality has been defined by Kramer (1951) as "the sum of the characteristics of a given food item which influence the acceptability or preference for that food by the consumer." The term "quality" refers to that inherent property of any product which determines the degree of excellence and those conditions with the consumer wants and for which is prepared to pay.

Egg quality is a general term that relates to various standards that are imposed on the eggs. These standards can be broken down into those used for determining the quality of the egg shell itself (exterior egg quality) and those standards which relate to the quality of the interior of the egg (interior egg quality). Some of these standards are based on subjective measures of egg quality and some are based on a more quantitative measure of egg quality. In general, exterior and interior egg quality standards are based on shell cleanliness, shell soundness, shell texture, shell shape, relative viscosity of the albumen, freedom from foreign matter in the albumen, shape and firmness of the yolk, and freedom from yolk defects. In order to classify eggs into the various grades used, an evaluation of all these items needs to be done (Koelkebeck, 2005).

Because of the readiness with which eggs spoil, the term "fresh" has become synonymous with the idea of desirable quality in eggs. As a matter of fact the actual age of an egg is quite subordinate to other factors which affect the quality (Hastings, 2003).

It is mandatory that, in the marketing channel, a high quality has to be maintained in order that the product fetches a higher price and retains a sustained demand. Supply of poor quality produce would soon result in reduced consumer preferences and decreased demand.

Yolk color, cleanliness of the shell, consistency of egg white and size of eggs are considered as "very important", in that order, by the consumers. Strength of shell and size of yolk are also considered as "important". Color of shell is generally considered "not important" (Forbes, 2002)

In case of eggs, besides marketing, is equally essential to know whether the egg is fit for incubation or not. Both the above factors necessitate the assessment of egg quality without breaking it.

1. Assessment of egg quality

Eggs are judged for quality by three sets of criteria namely, a) external appearance b) candling and c) interior equality.

1.1. External appearance

The shell must be sound in the sense that there must not be cracks, which is visible to the naked eye. Such cracks would result in the spoilage of egg by microorganisms. Besides, there will be problems in packaging and transport of eggs with unsound shell.

Eggs in the marketing channel must be clean in order to minimize the microbial invasion of the contents. Added to this, the neighboring eggs also get contaminated. Hence, is advisable not to allow unclean eggs to enter into the marketing channel. Unsound and unclean eggs are unfit for incubation.

Size of the egg is a very important criterion, both in marketing as well as in incubation. Uneven sized eggs not only pose problems in packaging and transport but also while setting them into setter trays. Therefore, only medium and uniform sized eggs are fit for incubation.

Shape of the egg also plays a vital role in the marketing channel. Very long eggs as well as very round eggs do not fit properly into the egg filler flats and hence there are liable to break during the packing and transportation. Besides, such eggs do not fit properly into the setter (incubation) tray, thereby rendering them unfit for incubation. Cracked shells allow easy entry to microbes followed by quick rotting of eggs. Eggs with cracked shells are neither fit for incubation nor for marketing.

Consumers show certain color preferences, few liking white, few brown-shelled eggs. However, this does not seem to be an important criterion in our country at present.

1.1.1. Production of eggs with good external quality (Hastings, 2003).

1. Hens that produce a good number of eggs, and at the same time an egg that is moderately large (average 56.7 g each) to be maintained.
2. Good housing, regular feeding and watering, and above all clean, dry nests.
3. Gathering of eggs, preferably once every hour from 9 am to 4 pm.
4. The confining of all broody hens as soon as discovered.
5. The rejection of all eggs found in a nest which was not visited the previous day as "doubtful".
6. The placing of all eggs, especially during summer, in the coolest spot available as soon as gathered (preferably refrigerated immediately after collection).

7. The prevention, at all times, of moisture in any form coming in contact with the egg's shell.
8. The selling of young cockerels before they begin to annoy the hens. Also the selling or confining of old male birds.
9. Cracked and dirty, as well as small eggs must not be marketed.
10. Keeping eggs away from musty cellars or bad odors.
11. Keeping the egg as cool and dry as possible while *en route* to market.
12. The marketing of all eggs as soon as possible; at least once per week and oftener, when facilities permit.
13. The use of strong, clean cases or cartons and good fillers.

1.2. Candling

Eggs were being artificially incubated long before there was electricity. Originally, to see "inside" before hatch time, eggs were held up to the flame of a candle - probably because candles provide a nice, small, intense source of light. Hence, the word "Candling" to mean determination of certain quality characteristics of internal parts of the eggs, without breaking the shell, by holding it against a light source.

A sixteen candle-power electric lamp is the most desirable. The light is enclosed in a dark box, and the eggs are held against openings about the size of a half dollar. The candler holds the egg large end forward, and gives it a quick turn in order to view all sides, and to cause the contents to whirl within the shell. To the expert this process reveals the actual condition of the egg (Hastings, 2003). Normally, expert candlers candle about one dozen eggs/min.

In Western countries, individual candling of eggs by hand is mostly replaced by automated mass scanning devices; however, it may take some more time for such developments in India.

1.2.1. Procedure (Mountney and Parkhurst, 1995)

In each hand, two eggs are held; one egg is supported by the tips of the thumb and index finger and the other, with small end pointed toward the palm, is held against the palm with the other fingers. After one egg is candled, it is shifted back in a rotating motion to the palm of the hand simultaneously bringing the second egg to the candling position. The eggs are viewed alternatively before the light. After reasonable ability to quickly rotate the eggs in both the hands is attained, candling becomes a very rapid and a mechanical process for the candler.

1.2.2. Criteria

The criteria involved in candling are 1. Shell quality 2. Air-cell characteristics and 3. Yolk characteristics.

1.2.2.1. Shell quality

While candling, an approximate assessment of the size and shape of the egg is possible. Very small and very large eggs as well as very long and very round eggs can be identified.

Besides, cleanliness of the shell can be visualized; dirty eggs are grouped roughly in three classes (Hastings, 2003):

1. Plain dirties those to which soil or dung adheres.
2. Stained eggs, those caused by contact with damp straw or other material which discolors the shell (plain dirties when washed usually show this appearance).
3. Smeared eggs, those covered with the contents of broken eggs.

For the first two classes of dirty eggs the producer is to blame. The third class originates all along the route from the nest to consumer. The percentage of dirty eggs varies with the season and weather conditions, being noticeably increased during rainy weather. In grading, about five percent of farm-grown eggs are thrown out as dirties. These dirties are sold at a loss of at least twenty per cent.

Plate 14.1 Highly porous shell (Courtesy Scott Shilala through Internet)

Added to the above, porosity of the shell can be easily appreciated. Amount of light passing through the eggs gives a direct indication of the thickness (porosity) of the shell. Cracks in the shell, which are not visible to the naked eye, can be identified by candling. Non-uniform distribution of shell material can also be detected. Very thin-shelled or cracked eggs lose their weight faster and hence deteriorate quickly; therefore, such eggs are unfit for incubation. Any defect in the shell makes the egg unsuitable for incubation

1.2.2.2. Air-cell

When the egg is formed within the hen the contents fill the shell completely. As the egg cools the contents shrink, and the two layers of membrane separate in the large end of the egg, causing the appearance of the bubble or air cell. Evaporation of water from the egg further shrinks the contents and increases the size of the air cell. The size of the air cell is commonly taken as a guide to the age of the egg. But when we consider that with the same relative humidity on a hot July day, evaporation would take place about ten times as fast as on a frosty November morning, and that differences in humidity and air currents equally great occur between localities, we see that the age of an egg, judged by this method, means simply the extent of evaporation, and proves nothing at all about the actual age (Hastings, 2003).

Even as a measure of evaporation, the size of the air cell may be deceptive, for when an egg with an air cell of considerable size is roughly handled, the air cell breaks down the side of the egg, and gives the air cell the appearance of being larger than it really is. Still rougher handling of shrunken eggs may cause the rupture of the inner membrane, allowing the air to escape into the contents of the egg. This causes a so-called watery or frothy egg (Hastings, 2003).

Notwithstanding the above, under normal conditions, size, shape and condition of the air-cell are useful in assessing the interior quality of the eggs. As the eggs becomes older, there will be increased loss of moisture and therefore increased air-cell size especially when the egg shell is thinner, temperature and air movement are higher and humidity in the environment lower. The United States Department of Agriculture (USDA) has laid down standards for weight and air-cell depth (Table 14.1).

The air-cell size is measured by an air-cell gauge. Air-cell in eggs to be incubated must be at the broad-end, firm an immobile. Any defect in the air-cell precludes the eggs from being incubated. Moving (tremulous), bubbly (many bubbles, indicates rough handing of eggs) and displaced air-cells do not come in the way of suitability of eggs for human consumption; but, eggs with any of these defects are unfit for incubation.

1.2.2.3. Yolk quality

This is identified by color, visibility and movement of yolk. Egg is twirled while candling for this purpose. Assessment of yolk quality is especially easy in white-shelled eggs. In a fresh egg, yolk is at the centre of the thick albumen and held at that position by chalazae. Therefore, it is not possible to detect the boundaries of the yolk and upon twirling; the yolk exhibits very limited movement. Color of the yolk also is faintly visible.

As the egg becomes older, liquefaction of the thick albumen occurs and results in the displacement of the yolk to the surface (because, yolk being rich in fat has a lower specific gravity than albumen), and its boundaries thus become very distinct.

Table 14.1 USDA quality and grade specifications

<table>
<tr><th rowspan="2">USDA grade</th><th colspan="2">Minimum weight</th><th rowspan="2">USDA quality</th><th rowspan="2">Air-cell depth (maximum)</th></tr>
<tr><th>Oz/dozen</th><th>g/egg</th></tr>
<tr><td>Jumbo</td><td>30</td><td>70.9</td><td>AA</td><td>0.3 cm</td></tr>
<tr><td>Extra large</td><td>27</td><td>63.8</td><td>A</td><td>0.5 cm</td></tr>
<tr><td>Large</td><td>24</td><td>56.7</td><td>B</td><td>1.0 cm</td></tr>
<tr><td>Medium</td><td>21</td><td>49.6</td><td rowspan="3">C</td><td rowspan="3">> 1.0 cm, unfit for human consumption</td></tr>
<tr><td>Small</td><td>18</td><td>42.5</td></tr>
<tr><td>Peewee</td><td>15</td><td>35.4</td></tr>
</table>

Besides, upon twirling, it moves very fast all round. Liquefaction results in creation of greater differences osmotic pressure between albumen and yolk because albumen has 88% water as against 48% in yolk. Therefore, there will be ingress of water into the yolk across the vitelline membrane. Hence, the yolk bulges which, in turn, makes the yolk flatter and heavier.

Therefore, in very old eggs, yolk is very distinct, flatter and sluggish in movement. As per USDA, yolk movement top to bottom and sidewise should not exceed 1 ½ and 1 cm, respectively.

Depending on the yolk and air-cell characteristics, albumen quality is assessed because the albumen is colorless and hence transparent.

1.2.3. Defects identifiable by candling

1.2.3.1. Shell

1.2.3.1.1. Thin-shelled egg

This may be due to calcium deficiency, increased temperature, respiratory diseases, inheritance and other physiological disturbances. Such eggs are also referred to as "glassy-shelled" eggs.

1.2.3.1.2. Soft-shelled egg

Occurs when the egg is laid before-time. The time spent in the shell gland is not sufficient enough to produce normal shell and hence, such eggs will be soft to touch.

1.2.3.1.3. Mottled shell

Eggs with this effect appear patchy during candling.

1.2.3.1.4. Body-checked egg

Egg broken in shell gland and partially repaired by calcification. A prominent ridge is seen around the waist of the egg (equatorial bulge). This may be due to stress causing release of adrenaline which, in turn, causes contraction of shell gland to end in breakage of egg.

1.2.3.1.5. Corrugated eggs

Encountered in hens suffering from infectious bronchitis. It is also seen copper deficiency because copper is required for offering rigidity to shell-membranes. Therefore, in copper deficiency, there will be excessive plumping-out and production of larger than normal eggs with this shell defect.

1.2.3.1.6. A and B eggs

Formed when one egg (A) is retained in the shell gland beyond the normal length will next egg (B) reaches shell gland and presses against it. Egg A is laid on the same day either shell-less or with wrinkled shell corresponding to the area of its contact with egg B.

1.2.3.1.7. Pimpling

Can occur only on shell surface or anywhere between shell membranes and exterior surface of egg. In the latter, a hole may be present on the shell allowing contents to leak. Pimpling is thought to be due to mass of albumen-like debris produced in magnum getting attached to outer shell membrane before the onset of shell calcification.

1.2.3.1.8. Coated eggs

Sand-paper texture of the shell due to overstay in shell gland which may be because of adrenaline released under stress or external disturbance.

1.2.3.2. Air-cell

1.2.3.2.1. Tremulous air-cell

1.2.3.2.2. Bubbly air-cell

1.2.3.3. Yolk

1.2.3.3.1. Addled (vitelline membrane broken)

1.2.3.3.2. Rots

1.2.3.3.3. Mottled

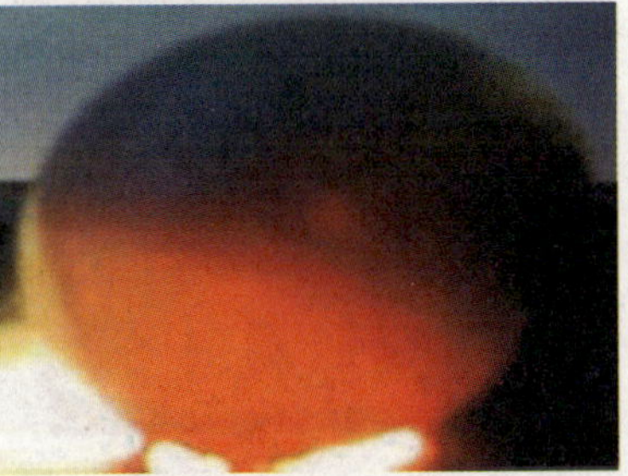

Plate 14.2 Rotten egg

1.2.3.4. Others

1.2.3.4.1. Blood spots

During the beginning of the production, there will be some bleeding during ovulation and the same will be attached to the yolk. It is likely to be inherited and hence such eggs are not referred for incubation, although blood-spotted eggs may hatch equally well. The efficiency of vitamins A and K in the feed of layers also results in increased incidence of blood spots. When feed is defective, most of the eggs produced show blood spots whereas, if the problem is due to genetic causes, only few weeks will have blood spots.

1.2.3.4.2. Meat spots

When the hemoglobin in the blood spot is reduced, it appears brown in color and referred to as meat spot. Sometimes, epithelial tissue from the oviduct may slough (denude) and appear in the albumen, which also is referred to as meat spot. The eggs with meat spots are not referred for incubation although they also hatch equally well.

1.2.3.4.3. Double-yolked eggs

Double-yolked eggs result due to multiple ovulation, followed by infundibulum engulfing both the yolks. Sometimes, when one yolk is lost into the body cavity

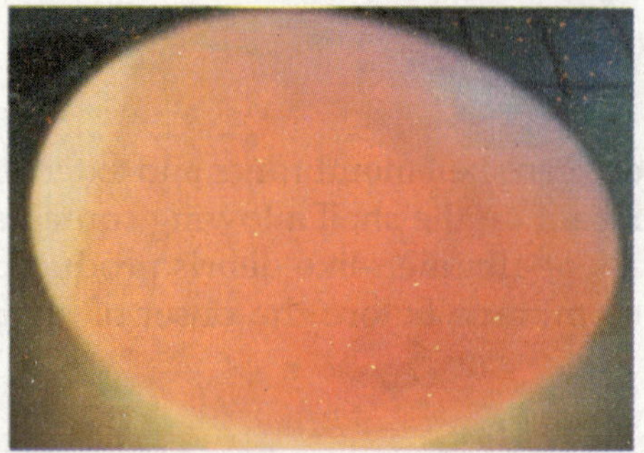
Plate 14.3 Blood-ring

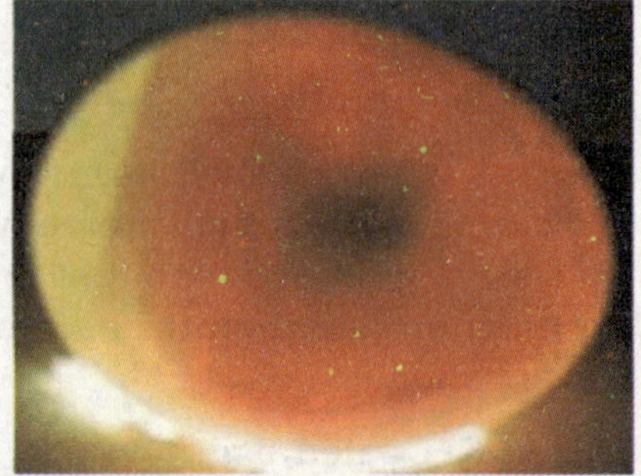
Plate 14.4 Blood-spot

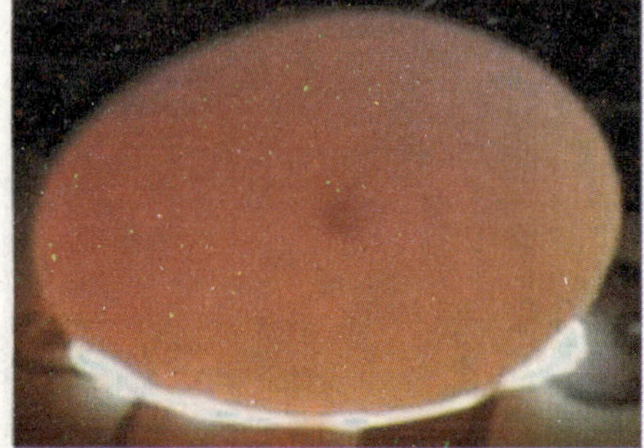
Plate 14.5 Blood-spot

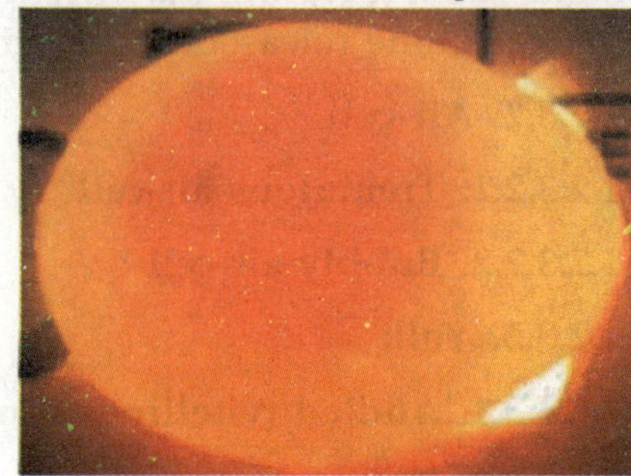
Plate 14.6 Blood-spot

for a day and is picked up by the infundibulum when the next day's yolk is released. The eggs unfit for incubation because a) insufficient space for two embryos b) insufficient nutrition and c) the egg is unsuitable for setting in the setter tray.

1.2.3.4.4. Yolk-less eggs

These are formed when sloughed-out tissue of ovary and/or oviduct; therefore, frequency of this defect is quite low.

1.2.3.4.5. Heat spot

In case of fertile eggs held at or above the temperature of20°C (68°F), the embryos grow, but soon die bearing partially developed spots. Such eggs are not accepted well by the consumers and are also unsuitable for incubation.

1.2.3.4.6. Bloody egg

If blood is present in albumen, upon candling, the egg appears bloody. This may be due to internal injury in the oviduct and subsequent bleeding. Such eggs show red-tinged albumen upon candling and are unfit for both human consumption and incubation.

1.2.3.4.7. Egg in egg

A fully formed egg, due to disturbance or some other unknown reasons, sometimes moves backwards up to the infundibulum by anti-peristalytic movements of the oviduct. One more ovulation and other processes follow resulting in egg-in-egg. Such eggs are unfit for incubation mostly because of their abnormal size, shape and difficulty for restoration of the embryos

1.2.3.4.8. Misshapen eggs

Misshapen eggs result due to several factors including respiratory diseases. Such eggs are best discarded in the marketing channel as well as for incubation.

1.2.4. Shortcomings of candling

Although candling has several advantages remunerated above, itdoes suffer from a shortcomings listed below.

1. Candling is more objective and subjective – it depends on person doing the operation and borderline quality is difficult to be differentiated.
2. Opened-out appearance need not coincide with the candle grade.
3. Candling fails to reflect the exact color of the yolk.
4. Candling cannot assess the flavor of the egg.
5. It is difficult to candle brown eggs.
6. Fine cracks in fresh eggs are not visible until moisture from the egg moves into the crack.
7. Scratch marks and weaknesses of the shell are confused with hair cracks.
8. Speckled brown eggs are frequently identified as dirty eggs.
9. Only extremes in air-space can be identified.
10. Small blood and/or meat spots attached to yolk and/or chalazae escape identification.
11. As percentage of faults in a batch increases, the candler normally concentrates on that particular fault only; in the process, other faults may escape notice.
12. As faults approach 10%, it is difficult to physically remove eggs in a short time available during candling.
13. Candlers may remove eggs with no faults (phantom faults) due to false identification which may approach as high as 2-3%.

1.3. Internal characteristics

The modern poultry industry is not satisfied with the traditional system of the handling and processing of eggs which is based on candling - visual inspection of the eggs. Currently, the operator of the conveyer does not have the opportunity to inspect 25,000 eggs per hour and to estimate the freshness, weight, and bacterial infection, presence of technical spoilage, eggshell defects without elimination of subjectivity, fatigability and destruction. That is why the problem of automation of egg quality control is rather difficult (Posudin *et al.*, 1992).

The main parameters of the egg which are inspected during its quality evaluation are: freshness of the egg, presence of the interior quality defects, weight, density and shape (egg index, asymmetry) of the egg, the state of the eggshell, the size of the air cell, albumen and yolk quality, Haugh unit, ratio of albumen weight to yolk weight, and the eggshell thickness (Posudin *et al.*, 1992).

Interior egg quality relates to the functional, aesthetic, and microbiological contamination factors of the albumen (white) and yolk. Good interior egg quality is essential to consumers who use eggs in many common baking and cooking items. As soon as the egg is laid, its quality begins to decline. As time of storage increases, the overall egg quality as measured by conventional grading standards declines. Even with eventual breakdown of interior egg quality, the chemical composition of the whole egg, albumen, and yolk does not change very much (Koelkebeck, 2005).

Along with candling, it is always advisable to determine the actual internal quality of randomly chosen eggs. The eggs are to be broke-open on a glass sheet kept on a horizontal plane.

The equipment used is slide calipers (least count 0.01 cm), spherometer (least count 0.1 mm) and screws gauge/shell-thickness measure (least count 0.01 mm).

1.3.1. Shape index

Before breaking open, the longest length and broadest width of the egg is measured by using a slide calipers. The shape index is calculated as the ratio of breadth to length times 100. Normal-shaped eggs will have a shape index ranging from 70 to 75.

After measuring length and breadth of the egg, the egg is carefully broken on to a glass sheet kept on a horizontal plane. Care has to be taken to ensure that vitelline membrane is not broken and yolk remains intact.

The following internal characteristics are assessed:

1.3.2. Albumen quality

1.3.2.1. Albumen index

The breadth and length of the thick albumen are measured by using slide calipers and height of the albumen is measured at 3 or 4 locations by using a spherometer. Albumen index is calculated as the ratio of average height of the albumen to the average of breadth and length. In case of chicken eggs, albumen index does not exceed 0.1.

In case of waterfowl, an alternate formula has been suggested to determine goose egg quality which is as follows:

$Gq = 4HW^{0.286}$, where H = height of thick albumen (mm) and W = weight of egg (g); the height of thick albumen estimated with the formula refers to an egg with an average weight of 128 g (Stanislaw and Cywa-Benko, 2002).

1.3.2.2. Haugh Unit Score (HUS)

The weight and size of egg influences unequally the height and length/width of albumen. Therefore, there is a likelihood of underestimating the albumen index of large-sized eggs and *vice versa*.

Therefore, Haugh (1937) developed a formula to give weightages to the gravitational force and weight of the egg. For a standard egg (2 oz or 56.7 g), Haugh unit score (HUS) is equal to 100 times the logarithm of the height of thick albumen (in mm) is shown below:

$$HUS = 100 \log \left[H - \frac{\sqrt{G}\left\{30\,(W)^{0.37} - 100\right\}}{100} + 1.9 \right],$$

where H = height of the thick albumen (mm), G = Gravitational force (32.2 ft/sec²), W = weight of the egg (g).

The above formula for chicken eggs can also be simplified to facilitate calculations as: $HUS = 100[\log \{H - 1.7W^{0.37} + 7.6\}]$

In cm-g-sec (CGS) system, the formula can be modified as follows:

$$HUS = 100 \log \left[H - \frac{\sqrt{G}\left\{30\,(W)^{0.37} - 100\right\}}{100} + 10.54 \right],$$

where G = 981 cm/sec².; or $HUS = 100\,[\log\{H - 9.40W^{0.37} + 41.875\}]$

The above formula is so developed that the term $\left[- \frac{\sqrt{G}\left\{30\,(W)^{0.37} - 100\right\}}{100} + 1.9 \right]$ becomes zero when W = 56.7 (standard weight for chicken eggs) and hence, HUS = 100 log H.

In addition, the coefficient of $(W)^{0.37}$ indicates that the average egg weight in chickens is one thirtieth or 3.33% of average body weight of the hen. It follows therefore, that the formula for HUS can be applicable to only chicken since both standard weight of eggs as well as egg weight as a proportion of body weight of the hen differ in different species.

Therefore, converting the HUS formula for chicken eggs into a generalized form i.e.,

$$HUS = 100 \log \left[H - \frac{\sqrt{G}\left\{P\,(W)^{0.37} - 100\right\}}{100} + K \right],$$

where, P is the ratio of body weight of hen to the egg weight and K = constant. Values of P and K have been derived for different species (Table 14.2) considering both average body weight and standard egg weight for the species in question. For all the species, HUS = 100 log H when W = standard egg weight.

Table 14.2 Modified constants for calculation of HUS for eggs of different species

Species	P	Standard weight, g	K	Egg weight as % of body weight
Chicken	30	56.7	10.54	3.33
Ducks	40	80.0	32.07	2.50
Geese	40	165.0	51.54	2.50
Japanese quail	12.5	10.0	- 22.14	8.00
Guinea fowl	40	40.0	17.73	2.50
Turkey	100	90.0	134.22	1.00
				Sreenivasaiah, 1998

To demonstrate the use of the above modified formulae, calculations are shown by using equations for chicken eggs as well as by the modified formulae and the same are shown in Table 14.3. It is evident that the new formulae are useful in obtaining accurate HUS values for different species of poultry. In fact, it is not possible to calculate HUS for turkey and goose eggs by using the formula for chicken eggs.

Table 14.3 HUS calculated by the formula for chicken eggs *Vs* the modified formulae

Species	Height, mm	Weight, g	HUS calculated by	
			Formula for chicken eggs	Modified formula
Chicken	9.2	60	92 (AA)	----
Ducks	5.0	70	20.64 (C)	90.60 (AA)
Geese	8.3	180	*	74.71 (A)
Japanese quail	6.8	36	112.31 (AA)	93.82 (AA)
Guinea fowl	4.5	12	135.78 (AA)	58.69 (A)
Turkey	8.7	95	*	72.87 (A)
Value in the parentheses indicates USDA quality				* Cannot be calculated

USDA quality depending on HUS is shown in Table 14.4.

Table 14.4 USDA quality based on HUS

Quality	HUS
AA	≥ 79
A	55 to 78
B	31 to 54
C *	< 31
	* Unfit for human consumption

1.3.2.3. pH of albumen

In a fresh egg, pH of albumen will be about 7.8 which rises to a maximum of 9.7 due mainly to loss of carbon dioxide with a concomitant loss of albumen quality (liquefaction). Thus, pH measurement itself is a good criterion for albumen quality.

1.3.2.4. Van Waggenen Chart

This comprises a set of photographic standards, each with a numerical value of 1 to 5 for broken-out eggs. The broken-out egg is given the number of the photograph which most nearly corresponds with it.

1.3.2.5. USDA chart

This comprises of a set of pictures of broken-out eggs corresponding to high, medium and low AA, A and B quality; totally nine pictures. The pictures are given numbers from 1 to 9; lowest (1) corresponding to high AA and the highest (9) corresponding to low B quality. The broken-out egg is given the number of the picture which most nearly corresponds with it.

1.3.2.6. Optical and spectroscopic methods of egg quality evaluation

These methods (Posudin *et al.*, 1992) can be considered as alternatives to visual candling of the eggs. Spectrophotometry of shell color, yolk color, blood and meat spots present the first attempt of the automation of egg quality control. The principle of blood and meat spots detection was based on the measurement of light transmission at 575 mm where hemoglobin has the absorption band.

The disadvantage of this method is the dependence of the transmitted light on the size, color of egg, on the eggshell - sickness and contamination.

The detection of the technical spoilage of the eggs on the basis of transmission spectra analysis is based on the fact that the fresh white eggs have maximum transmission at 573-580 nm, brown eggs have maximum transmission at 590-596 and 636-638 nm, eggs with blood spots have absorption maxima at 538-540 and 573-576 nm. It is very attractive to use the spectrophotometry of eggs for the nondestructive determination of the fecundated eggs.

The fact is that fecundated and nonfecundated eggs differ with the production of carbon di oxide as the result of egg metabolism. But carbon dioxide is responsible for the absorption at 426.8 nm; the measurement of the absorption of CO_2 can make it possible to determine the fecundated eggs and to select them.

The reflectance spectra of eggs can be applied to the estimation of pigmentation of the eggshell. The fluorescence characteristics of the eggs present convenient quality criteria.

Hence, the fluorescence spectroscopy is a rather promising approach for the detection of different types of spoilage and quantitative estimation of eggshell pigmentation.

1.3.2.7. Others

Other criteria include surface area of the albumen and % of thick white; but, both are difficult to assess and hence, are not much in vogue.

1.3.3. Yolk quality

1.3.3.1. Yolk index

Yolk index is the ratio of height of yolk to its diameter *in situ*; the former measured by a spherometer and the latter by a slide calipers. Yolk index of chicken eggs does not exceed 0.5.

1.3.3.2. Yolk color

Color of yolk can be directly assessed by comparing with standard colors in a yolk color rotor or a yolk color fan.

A standard colorimetric method in which quantity of acetone-extracted pigments from yolk is measured colorimetrically againsta standard β-carotene or potassium dichromate can be employed. However, this method is effective for color produced by xanthophylls pigments from corn only.

Similarly, measurement of light reflected from egg yolk surface can also replace visual or colorimetric methods.

1.3.4. Shell quality (Baerdemaeker *et al.*, 2005).

1.3.4.1. Direct methods

Different direct methods have been used; the most widely used being the compression fracture force measured during quasi-static compression, which is a measure of its material strength. Other methods include puncture tests and impact tests. All these direct methods are destructive.

1.3.4.2. Indirect methods

These include both destructive and non-destructive methods. Each of the methods measures a parameter that is related to egg shell strength. The correlation between the different methods is moderate and the choice of which method to use often depends on the application.

1.3.4.2.1. Destructive methods

1.3.4.2.1.1. Shell thickness

Measuring the egg shell thickness is a frequently used indirect method to have an indication of the egg shell strength This is determined by finding out the shell thickness by taking samples from 3 to 4 locations (broad-end, small-end, middle region etc.) and removing from underlying shell membranes. A screw-gauge or a shell-thickness measure is used for the purpose. In chicken eggs, shell thickness varies between 0.24 and 0.34 mm. Shell thickness < 0.24 mm indicates a high-risk of breakage during transportation and/or marketing.

1.3.4.2.1.2. Shell weight

Calculating the weight of egg shell as a percentage of the egg.

1.3.4.2.1.3. Shell porosity

Can be assessed by determining the breaking strength of the shell.

1.3.4.2.1.4. Specific gravity

Another method of assessing shell quality is by estimating specific gravity of eggs. Eggs are individually dipped in salt solutions of specific gravity ranging from 1.060 to 1.100 at intervals of 0.005. For 3L of water, 276, 298, 320, 342, 365, 390, 414, 438 and 462 g of salt has to be added to obtain the requisite specific gravity. Eggs must be immersed in a basin of water just prior to measurement of specific gravity. Eggs have to be immersed first in solution of lowest specific gravity and the subsequent ones serially till it floats for the first time. This method is especially accurate in case of fresh eggs which have small air-cells. In case of old eggs, specific gravity reduces due to increase air-cell size rather than reduced shell thickness

1.3.4.2.1.5. Measurement of Young's modulus

Young's modulus is the measure of material strength of the egg shell, not of structural strength. A piece of egg shell is glued to a paper clip, which is attached into a bench screw for measurement. The sample is excited by a sound wave (burst chirp-type, 1000-3000Hz) produced by a speaker just below the attached piece of shell. The vibration of the shell is measured by a laser-vibrometer; the third vibration mode is the most dominant and is used to calculate the Young's modulus. This method is relatively fast (5 sec) and gives information about the material properties itself.

1.3.4.2.2. Non-destructive methods

These methods are used to measure egg shell strength on the assumption that the indirect values are correlated with the direct values although only moderate correlations exist among the parameters.

1.3.4.2.2.1. Quasi-static compression

Quasi-static compression of the eggs between two parallel plates is one of the widely used methods. The slope of the force-deformationline provides a measure of the shell stiffness. The deformation, also known as non-destructive deformation, which can be defined as the amount an egg shell bends or deflects under an applied force, can be used to estimate the force required to fracture the egg shell.

1.3.4.2.2.2. Vibration analysis

Vibration analysis is a non-destructive method of measuring egg shell strength, useful for crack detection, as a measure of shell strength and to study the differences between genetic strains.

1.3.4.2.2.2.1. Crack detection

Under commercial circumstances, mechanical sensors can be used for egg shell crack detection. A small impactor excites the egg shell. A combination of the amplitude of the rebounds and/or the number of rebounds of the impactor is used to indicate egg shell integrity. However, this reveals only local shell quality information and the crack detectors have to test several locations (24-32 points) for each egg in order to obtain satisfactory results. Under practical, industrial conditions, the crack detection rate ranges from 70 to 85%, while the percentage of falsely rejected intact eggs was 0.3 to 1.0%.

1.3.4.2.2.2.2. Oscillation response

When the egg is subjected to non-destructive impact excitation, the shell will react with an oscillation response. Eggs with damaged shells show higher number of resonant peaks than intact eggs. For intact eggs, the impulse response will be similar on every point on the equator, whereas the eggs with a damaged shell show a different response on different locations of the equator. Only four measurements for each shell are taken and correlation between repeated measurements of the same egg was calculated. With this approach, up to 90% of the cracked eggs can be identified with false rejects below 1%. This technique is fast because any vibration typically iasts for about 10 ms.

1.3.4.2.2.2.3. Egg shells strength

Using a mass-spring-damper model, it is shown that the damping of the vibration provides extra information linking the dynamic measurements to static measurements. Further, shape of the egg is needed to link dynamic to static measurements. A multiple linear regression model was constructed using the dynamic stiffness (K_{dyn}), the damping and the shape index of the eggs as explanatory variables in order to predict the static stiffness with a high correlation of 0.90. Dynamic stiffness has been found to be related to shell breakage in practice which is the most important parameter.

Dynamic stiffness has a moderate heritability indicating a possibility of including it in selection schemes in future.

1.3.4.2.2.2.4. Shell porosity

Shell porosity can be assessed by loss of weight after storage for 2 weeks at 37.8°C.

Visual measurement of shell porosity is possible by the dye method. Egg shell is dyed in an alcoholic 0.3% (W/V) solution of methylene blue or crystal violet and then breaking it. The number of pores penetrated by the dye is compared with photographic standards and a score of 1(minimum pores) to 10 is allotted to each shell. Weight loss is closely related to dye score.

1.3.4.3. Egg shell color

Estimated by reflectance method (similar to that for yolk color)

1.3.5. Flavor

Flavor is a very important quality attribute which can be assessed only after the egg is broken. Off-odors in eggs can be due to rotting of eggs or due to certain chemical changes or chemical contaminants (Table 14.5)

Table 14.5 Off-odors and off-flavors in eggs

Material used	Effect on eggs	Remarks
Insecticides		
BHC	Taste and smell like BHC	Do not use
Malathion (liquid)	Taste and smell like cat urine	If needed, use at night
Sevin	Slightly bitter, especially in albumen	Do not spray on eggs
Packing material Moldy, dirty damp filler-flats	Musty, cardboard smell and taste	Oil keeps filler-flats damp for mold growth
Sanitizers-detergents Chlorine products (concentrated)	Slight bitter taste and chlorine smell	Manufacturer's recommendations to be followed
Destainers Citric acid	Slight off-flavor depending on duration of exposure	Manufacturer's recommendations to be followed
Fruits Apples	Bad, bitter, cardboard flavor	Eggs not to be stored near it

Material used	Effect on eggs	Remarks
Vegetables and citrus fruits	Corresponding flavor and odors	Eggs not to be stored with such items
Hen-house odors		
Manure	Picked-up if eggs are not gathered frequently	Eggs to be gathered frequently
Egg room	Mushy odors etc.	Egg rooms must be kept clean and free of odors
		Source : Mountney and Parkhurst, 1995

1.3.6. Functional properties

If the eggs are assessed for their industrial uses, differences in functional properties such as emulsifying capacity, viscosity and foaming will be of much practical significance.

1.4. Factors affecting egg quality

Several factors influence egg quality; this includes genetic factors and environmental factors. The latter group includes primarily temperature and humidity during storage and length of the storage. The effects of these factors are tabulated below :

Table 14.6 Influence of time, temperature and humidity on egg quality

	Fresh	Stored at 12.8°C			Stored at 37.8°C		
Storage period (d)	0	2	7	14	2	7	14
Moisture loss, %	0	0.08	0.13	0.53	1.05	2.94	5.20
Change in air-cell size, cm	0	0	0.3	0.3	0.3	0.9	0.9
Dye porosity score	3.5	5.9	6.75	7.8	7.9	8.7	9.6
Albumen height, mm	6.5	6.4	5.7	4.6	5.1	4.3	3.2
HUS	77.0	77.0	73.2	64.7	66.1	55.7	48.3
USDA Score	5.3 avg.	4.6 avg.	5.6 low	6.6 high	6.0 low	8.4 avg.	8.8 low
USDA Quality	A	A	A	B	A	B	B
Van Waggenen Score	2.1 avg.	2.3 avg.	2.7 low	2.9 high	2.9 high	3.7 low	4.3 avg.
Quality	A	A	A	B	B	B	C

2. Grading of eggs

Eggs are among the most difficult of food products to grade, because each egg must be considered separately and because the actual substance of the egg cannot be examined without destroying the egg (Hastings, 2003).

Grading is a classification of eggs according to characteristics that have an economic value. Usually, eggs weight is the main criterion for grading.

2.1. Advantages of grading

1. Allows the consumer to exercise preference.
2. Provides incentive to the producer to adjust his production methods.
3. Gives confidence to the consumer and reputation to the producer.
4. Facilitates the mechanical processes of packaging and distribution along with appearance.
5. Helps establishment of minimum characteristics for protection of the consumer.

2.2. Disadvantages of grading

1. Wasteful because matching supply to demand is difficult and impossible in short-term.
2. It is rigid and slow to react to market changes.
3. May inhibit attempts to stimulate and develop demand when grading is made mandatory.

2.3. Egg grading and marketing rules

2.3.1. Short title, application and commencement

These rules may be called "Table-eggs grading and marketing rules, 1968". They shall apply to table-eggs produced in India and they shall come into force on the date of their publication in the Official Gazette.

2.3.2. Definitions

In these rules, a) "Agricultural Marketing Adviser" means the Agricultural Marketing Adviser to the Government of India and includes any officer subordinate to him to whom the powers under these rules may be delegated by the Agricultural Marketing Adviser; b) "Schedule" means a schedule appended to these rules; c) "Table-eggs" means edible eggs derived as a product of poultry husbandry.

2.3.3. Grade designation

The grade designation to indicate the quality of table eggs shall be set out in Column 1 of Schedule II (Table 14.7).

2.3.4. Grade designation mark

1. The grade designation mark in the case of each table egg shall consist of a design incorporating the word "Agmark" and the grade of the egg has been approved by the Agricultural Marketing Adviser placed centrally in a circle of not less than 13 mm in diameter.
2. The grade designation mark in case of containers in which graded table eggs are packed shall consist of a label in the color as specified below :
   a) Extra large - White
   b) Large - Red

c) Medium - Blue
d) Small - Yellow

The label specifying the grade designation and bearing the design of an outline map of India with the word "Agmark" and figure of the rising sun, with the words "Produce of India", in English and Hindi.

The grade designation mark to be attached to each container of table eggs shall consist of the label specifying the grade designation and shall be in the colors specified above

2.3.5. Method of marking

1. The grade designation mark shall be marked legibly on each table egg in indelible ink on the shell by means of a rubber stamp and in a manner approved by the Agricultural Marketing Adviser.
2. The grade designation mark label shall be attached by means of a lead seal bearing the word AGMARK to each container of table eggs and shall clearly show the following particulars :
   a) Grade designation of table eggs
   b) Number of table eggs
   c) Net weight of table eggs
   d) Name of grading station
   e) Date of dispatch

SCHEDULE II

Table 14.7 Grade designations and quality of table eggs (as per BIS)

Grade	weight of egg, g per			Shell	Air-cell	Albumen	Yolk
	Egg	Dozen	Unit of ten				
A Extra large	≥ 60	≥ 715	≥ 596	Clean, unbroken and sound; shape normal	Up to 4 mm in depth; practically regular or "better"	Clear, reasonably firm	Fairly well centered, practically free from defects, outline indistinct
A Large	53-59	631-714	526-595				
A Medium	45-52	535-630	446-525				
A Small	38-44	436-534	380-445				
B Extra large	≥ 60	≥ 715	≥ 596	Clean to moderately stained and sound; shape slightly abnormal	Up to 8 mm in depth; may be free and slightly bubbly	Clear, may be slightly weak	May be slightly off-centered, outline slightly visible
B Large	53-59	631-714	526-595				
B Medium	45-52	535-630	446-525				
B Small	38-44	436-534	380-445				

Source : Indian Poultry Industry Year book, 1986

2.3.6. Method of packing

1. Only sound, clean and dry containers which is suitable for the purpose and are free from any undesirable smell and from any insect infestation or fungus contamination shall be used for packing.
2. The containers shall be securely closed and sealed in a manner approved by the Agricultural Marketing Adviser.
3. Packing material, if used, shall be clean, dry, sweet smelling and free from any taint liable to impart any objectionable flavor to the table eggs.
4. Table eggs of different grades shall be packed separately, as far as possible; and if table eggs of more grades than one or packed in one container, a layer of clean paper or clean straw shall be placed between the different grades and separate label of appropriate color shall be attached to the container giving the particulars in respect of the table eggs of each grade packed in the container.

The USDA standards for assessing quality of individual shell-eggs are summarized in Table 14.8.

Table 14.8 Summary of US Standards for quality of individual shell-eggs

Quality factor	AA quality	A quality	B quality
Shell	Clean, unbroken, practically normal	Clean, unbroken, practically normal	Clean to slightly stained *, unbroken, abnormal
Air-cell	≤ 0.3 cm in depth, unlimited movement and free or bubbly	≤ 0.5 cm in depth, unlimited movement and free or bubbly	> 0.5 cm in depth, unlimited movement and free or bubbly
White	Clear, firm	Clear, reasonably firm	Weak and watery; small blood- and meat-spots present **
Yolk	Outline slightly defined, practically free from defects	Outline fairly well defined, practically free from defects	Outline plainly visible, enlarged and flattened. Clearly visible germ development but no blood. Other serious defects

* 3% and 1.5% of surface if localized and scattered, respectively

** aggregating to ≤ 0.3 cm in diameter

Source : Mountney and Parkhurst, 1995

For eggs with dirty or broken shells, the standards of quality (USDA) provide to additional qualities namely,

1. Dirty – unbroken, adhering dirt or foreign material, prominent stains, moderate stained areas in excess of B quality.
2. Check – broken or cracked shell but membranes intact; not leaking.

Table 14.9 Summary of UK Standards for quality of individual shell-eggs

Quality factor	Class A	Class B	Class C
Cuticle	Normal, clean, undamaged	Dirty, damaged, treated	Dirty, damaged, treated
Shell	Normal, clean, undamaged	Normal, undamaged	Cracked, misshapen, rough textured or any other abnormality
Air-space	Height not exceeding 6 mm (4 mm for Extra fresh), stationary	Height not exceeding 9 mm, mobile	Height exceeding 9 mm or damaged
Albumen	Clear, limpid, of a gelatin-like consistency, free of all foreign bodies of any kind	Clear, limpid, free of all foreign bodies of any kind	Clear, no discoloration or turbidity, small foreign bodies permissible
Yolk	Visible on candling as a shadow only, without clearly discernible outline, not moving appreciably away from centre of the egg on rotation, free from all foreign bodies of any kind	Visible on candling as a shadow only, free from all foreign bodies of any kind	Distinct on candling, "sided" or "stuck", no discoloration, small foreign bodies permissible
Germ cell	Imperceptible development	Imperceptible development	Imperceptible development
Smell	Free of foreign smell	Free of foreign smell	Free of foreign smell
Wet or dry cleaning	Not permitted	Permitted	Permitted
Notes (for guidance only)		This category includes naturally dirty shelled and washed eggs	This category includes various shell faults and dry cracked eggs (membranes unbroken)

Source : Forbes, 2002

Chapter **15**

Poultry Meat Quality

1. Quality defined

Before poultry meat quality is addressed, the term quality should be clearly defined as it relates to poultry. Quality is "in the eye of the beholder."

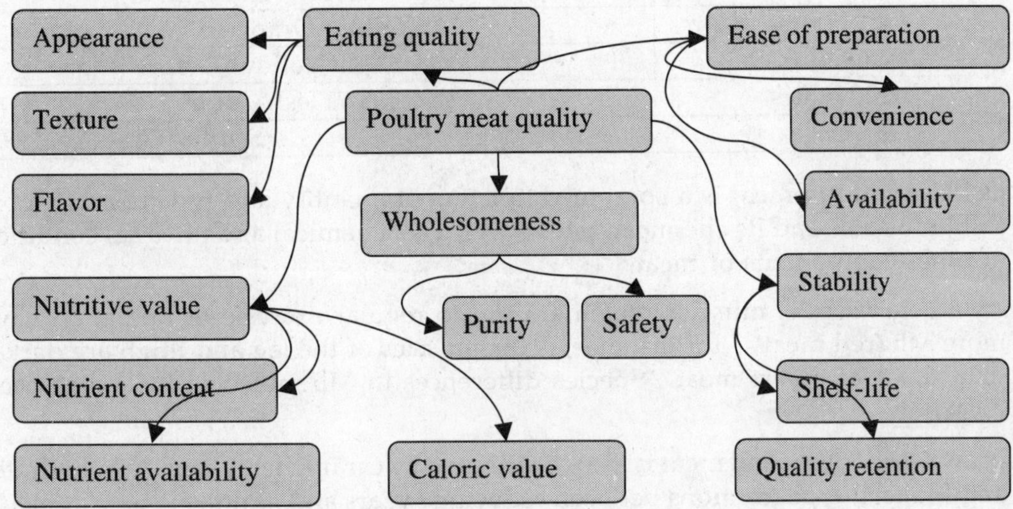

Fig. 15.1 Factors affecting poultry meat quality

Whether or not a poultry product meets the consumer's expectations depends upon the conditions surrounding various stages in the bird's development from the fertilized egg through production and processing to consumption (Northcut, 1997). Although there are a number of characteristics that determine the overall quality of meat (Figure 15.1), the following discussion will focus only on appearance, texture, and flavor.

2. Factors affecting quality

2.1. Appearance (Color)

Color of cooked or raw poultry meat is important because consumers associate it with the product's freshness, and they decide whether or not to buy the product based on their opinion of its attractiveness. Poultry is unique because it is sold with and without its skin. In addition, it is the only species known to have muscles that are dramatic extremes in color (whiteand dark meat). Breast meat is expected to have a pale pink color when it is raw, while thigh and leg meat are expected to be dark red when raw. There are times when poultry meat does not have the expected color, and this has created some special problems for the poultry industry.

Table 15.1 Color changes in a bruise over time for broiler muscle

Age of Bruise	Color of Bruise
2 minutes	Red
12 hours	Dark Red-Purple
24 hours	Light Green-Purple
36 hours	Yellow-Green-Purple
48 hours	Yellow-Green (Orange)
72 hours	Yellow-Orange
96 hours	Slight Yellow
120 hours	Normal, Flesh Color

Source : Gregory, 1992

Color of poultry meat is a combined function of quantity and type of myoglobin (Mb) molecule and its chemical state as well as of chemical and physical condition of other components of meat.

As a general rule, muscles which are put to regular exercise or operation have more Mb (red meat) – for instance, is the muscles of the leg and thigh are darker than those of breast meat. Species differences in Mb molecule have also been recorded.

Brown color of cooked meat is also contributed by caramelization of carbohydrates and Mailard-type reactions between reducing sugars and amino groups. Met Mb is the principal undesirable pigment on raw meat surface. Thus, addition of vitamin C can help re-oxidize the met Mb and avoid formation of brown color on raw meat surface. Vitamin C can also be given as a pre-slaughter injection.

Poultry meat color is affected by factors such as bird age, sex, strain, diet, intramuscular fat, meat moisture content, pre-slaughter conditions and processing variables. Color of meat depends upon the presence of the muscle pigments myoglobin and hemoglobin. Discoloration of poultry can be related to the amount of these pigments that are present in the meat, the chemical state of the pigments, or the way in which light is reflected off the meat. The discoloration can occur in an entire muscle, or it can be limited to a specific area, such as a bruise or a broken blood vessel. When an entire muscle is discolored, it is frequently the breast muscle. This occurs because breast muscle accounts for a large portion of the live weight (~5%), it is more sensitive to factors that contribute to discoloration, and being already of light appearance, small changes in color become more noticeable. Extreme environmental temperatures or stress due to live handling before processing can cause broiler and turkey breast meat to be discolored. The extent of the discoloration is related to each bird's individual response to the conditions (Northcut, 1997).

Another major cause of poultry meat discoloration is bruising. Majority of downgrading of carcasses (28 per cent) is from bruises (AMS, 1995) but is often difficult to determine the actual reason(s) for bruising. The color of the bruise, the amount of "blood" present, and the extent of the "blood clot" formation in the affected area are good indicators of the age of the injury and may give some clues as to its origin. A bruise will vary in appearance from a fresh, "bloody" red color with no clotting minutes after the injury to a normal flesh color 120 hr later

Table 15.2 Pigments in fresh, cured and cooked meat

Pigment	Formation	State of iron	Color	Remarks
Myoglobin, Mb	Reduction of Met Mb; OxyMb-O_2	++	Purplish red	---
Oxy Mb	Mb + O_2	++ or +++	Bright red	---
Meth Mb	Oxidation of Mb or Oxy Mb	+++	Brown	---
Nitroso Mb	Mb + Nitric oxide (NO)	++	Bright red (pink)	Cured meat
Nitrosomet Mb	Met Mb + NO	+++	Crimson	Cured meat
Met Mb Nitrite	Met Mb + Excess nitrite	+++	Reddish brown	Cured meat
Globin myo hemochromogen	Mb or Oxy Mb subjected to denaturation (heat); irradiation of globin hemochromogen	++	Dull red	Cooking effect
Globin myo hemichromogen	Heating or denaturation of Mb, Oxy Mb, Meth Mb or Globin myo hemochromogen	+++	Brown; sometimes grayish	Cooking effect
NO-myo hemochromogen	Heating or denaturation of nitroso Mb	++	Bright red	Cooking effect
Sulph Mb	Mb+H_2O+O_2	++	Green	One double bond of Mb saturated
Met Sulph Mb	Oxidation of sulph Mb	+++	Red	One double bond of Mb saturated
Choleglobin	H2O2 + Mb/Oxy Mb	++ or +++	Green	One double bond of Mb saturated
Nitrihemin	Large excesses nitrite and heat on nitroso met Mb	+++	Green	Globin absent
Verdohaem	Excessive heating and denaturation	+++	Green	Globin absent
Bile pigments	Severe heating and denaturation; even more than verdohaem	Absent	Yellow or colorless	

Source : Lawrie, 1979

(Table 15.1). The amount of "blood" present and the extent of clot formation are useful in distinguishing if the injury occurred during catching/transportation or during processing. Injuries that occur in the field are usually magnified by processing plant equipment or handling conditions in the plant.

Pigments found in fresh, cured and cooked meat are given below:

2.2. Texture (Tenderness)

Texture and tenderness are most important quality factors of poultry meat. Texture can be defined, as seen by the eye, as a function of the size of bundles of fibers into which the perimyseal septa of connective tissue divide the muscle longitudinally. It follows that muscles which have more number of smaller muscle fibers, is and hence "feel" like fine-grained than those muscles which have smaller number of bigger muscle fibers. Hence, muscles of males are coarser than those of females.

Tenderness is a feeling to the palate and it depends on ease of penetration by the teeth, ease with which meat breaks into pieces and amount of residue left-over after chewing; all these have relationship with texture.

Many methods have been developed to measure tenderness, but with variable success; the popular ones being the a) force required to penetrate or cut or mince the meat b) chemical methods such as estimation of concentration of connective tissue or enzymatic digestion. Tenderness depends both on quality of connective tissue as well as proteins of sarcoplasm and myofibrils. The measurement of shear-force mostly reflects the state of myofibrillar proteins.

After consumers buy a poultry product, they relate the quality of that product to its texture and flavor when they are eating it. Whether or not poultry meat is tender depends upon the rate and extent of the chemical and physical changes occurring in the muscle as it becomes meat. When an animal dies, blood stops circulating, and there is no new supply of oxygen or nutrients to the muscles. Without oxygen and nutrients, muscles run out of energy, and they contract and become stiff. This stiffening is called *rigor mortis*. Eventually, muscles become soft again, which means that they are tender when cooked (Northcut, 1997).

Anything that interferes with the setting of *rigor mortis*, or its resolution soon after, will affect meat tenderness. For example, birds that struggle before or during slaughter cause their muscles to run out of energy quicker, and *rigor mortis* sets much faster than normal. The texture of these muscles tends to be tough because energy was reduced in the live bird. A similar pattern occurs when birds are exposed to environmental stress (hot or cold temperatures) before slaughter. High pre-slaughter stunning, high scalding temperatures, longer scalding times and machine picking can also cause poultry meat to be tough.

Tenderness of portioned or boneless cuts of poultry is influenced by the time *post-mortem* of the deboning. Muscles that are deboned during early *post-mortem* still have energy available for contraction. When these muscles are removed from the carcass, they contract and become tough. To avoid this toughening, meat is usually "aged" for 6 to 24 hours before deboning; however, this is costly for the processor. When poultry is deboned early (0 to 2 hr *post-mortem*), 50 to 80 % of the meat will be tough. On the other hand, if the processor waits 6 hr before deboning, 70 to 80 % of the poultry meat will be tender. The poultry industry has recently started using post-slaughter electrical stimulation immediately after death to hasten *rigor* development of carcasses and reduce "aging" time before deboning. This is different from energy depletion in the live bird which causes meat to be tough. When electricity is applied to the dead bird, the treatment acts like a nerve impulse, and causes the muscle to contract, use up energy and enter *rigor mortis* at a faster rate. In the live bird, the same treatment causes meat to be tough; however, after death, the treatment causes tender deboned poultry meat within 2 hr postmortem instead of the four to 6 hr required with normal aging. Although electrical stimulation is still in the developmental stages, it seems that processors using it can debone carcasses right out of the chiller and save on their equipment costs, time, space and energy requirements (Northcut, 1997).

Age, breed, sex profusely influence tenderness; meat from older animals and males being "tougher" than that from younger animals and females. Intramuscular fat "dilutes" connective tissue elements and thus enhances tenderness. Ageing process adds to the tenderness whereas faster *post-mortem* pH fall makes the meat tougher.

Cooking process generally converts collagen to gelatin which has tenderizing effect whereas, the denaturation of protein that accompanies cooking toughens the meat; the overall effect depends on time and temperature of cooking; prolonged cooking at lower temperature is most preferred for optimum tenderness.

Special case has to be taken to avoid toughening processes especially cold-shortening and thaw-rigor; the former is the shortening and toughening that results during cooking of refrigerated meat which was cooled to below 15 to 19°C while it was still in pre-rigor condition; the latter is also a severe shortening during thawing and cooking but of frozen muscle, which was frozen before ATP level had fallen appreciably. Accelerated Freeze Drying (AFD) process also causes reduced tenderness due to the effects of plate temperature on sarcoplasmic and myofibrillar proteins. Ionising radiation (IR), on the other hand, has tenderizing effect presumably through its action on collagen.

Artificial tenderization has been tried since time immemorial – by beating meat, marinating it with vinegar, wine, salt, tenderizing enzymes etc. Non-toxic proteolytic enzymes produced by certain plants, fungi and bacteria are becoming popular commercially. They can be used in chill-water or pumped to the carcass or given as pre-slaughter injection; the latter being most effective at the rate of 5-

10% of tenderizing enzymes (about 1.1 mg/kg live weight) 1-30 min before slaughter. Most of these enzymes disintegrate upon cooking. Common enzymes used are plant proteases (ficin from fig, papaine from pawpaw, bromelein from pineapple) and bacterial proteases (rhozyme, hydralase) and fungal amylase. Bacterial and fungal proteolytic enzymes digest sarcolemma and plant proteases break-up mucopolysaccharide of the ground substance and then the connective tissue to convert them into amorphous mass.

2.3. Flavor

Flavor is another quality attribute that consumers use to determine the acceptability of poultry meat. Both taste and odor contribute to the flavor of poultry, and it is generally difficult to distinguish between the two during consumption (Fig. 15.2). When poultry is cooked, flavor develops from sugar and amino acid interactions, lipid and thermal oxidation and thiamin degradation. These chemical changes are not unique to poultry, but the lipids and fats in poultry are unique and combine with odor to account for the characteristic "poultry" flavor (Northcut, 1997).

Few factors during production and processing affect poultry meat flavor. This means that it is not only difficult to produce a flavor defect, but it is difficult to enhance flavor during production and processing. Age of the bird at slaughter (young or mature birds) affect the flavor of the meat. Minor effect on meat flavor are related to bird strain, diet, environmental conditions (litter, ventilation, etc.), scalding temperatures, chilling, product packaging, and storage; however, these effects are too small for consumers to notice.

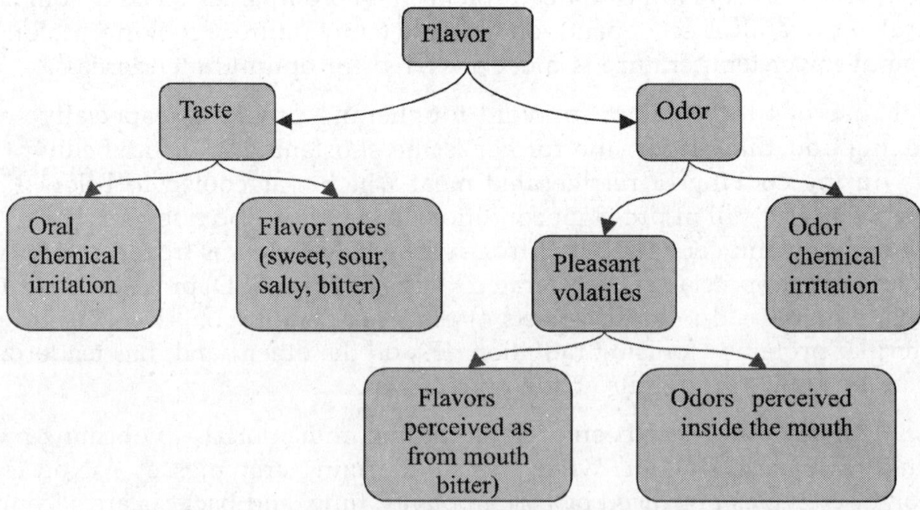

Fig. 15.2 Components of poultry meat flavor (Lawless, 1991)

Odor is the most important attribute determining the flavor; the latter is also dependent on taste, texture, temperature and pH. It is very difficult to measure odor and taste. The best way, however, is by a trained taste panel. Odor is mainly identified by the olfactory cells of the nasal mucus membrane and taste by specialized cells on tongue, soft-palate and top of the gullet. Although several chemicals have been isolated and found to have specific taste, high degree of idiosyncracy has been recorded in identification of taste by both taste panel and consumers. In any case, odor differences are more easily discerned than that of taste

Cooked meat produces an aroma which is again due to a complex phenomenon. The main chemical participating are pyrazines and those which contain S or O_2. Common volatiles from cooked meat are H_2S, NH_3, acetaldehyde, acetone, and diacetyl; traces of formic, acetic, propionic, butyric and iso-butyric acids and diamethyl sulfide are also part of flavor volatiles. Many of the major volatiles are derivatives of amino-acids upon pyrolysis, especially of cysteine, and also of glycine, glutamic acid, β-alanine, threonine, histidine, lysine, leucine, isoleucine, serine or valine. Inosine, ribose, mono nucleotides and hypoxanthine are also very important flavor precursors.

There are non-volatile flavor constituents as well in meat; for example, lactic acid, carnosine, creatine and hypoxanthine which upon heating can participate in Maillard-type reactions to produce characteristic flavors. In fact, flavor chemistry is one of the most popular fields of current research and already more than 200 chemicals have been identified with flavors. With so many chemicals, variability in flavor of meat is a very complex phenomenon.

2.4. Juiciness and water-holding capacity (WHC)

Perishability of meat in relation to this moisture content has already been discussed. WHC indicate the amount of fluid to meat can hold and hence is particularly important in comminuted meats like sausages and in frozen meats to avoid loss of drip. In other words, WHC has a direct relationship with retention of juices or juiciness of meat.

Due to drop in pH after death (formation of lactic acid from glycolysis) to about 5.5 which happens to be the isoelectric pH of meat proteins, WHC of meat falls. However, the loss of WHC is greater if pH fall is faster and *vice versa*. It is also expectable on the same lines that any factor leading to denaturation of muscle proteins results in reducing WHC. On the contrary, conditioning (ageing) has enhancing the WHC; salts of weak acids, particularly polyphosphates, enhance WHC by binding additional molecules of water.

3. Conclusion

The most important aspect of poultry meat is its eating quality - a function of the combined effects of appearance, texture and flavor. Live production affects poultry meat quality by determining the state of the animal at slaughter. Poultry processing affects meat quality by establishing the chemistry of the muscle constituents and their interactions within the muscle structure. The producer, processor, retailer and consumer all have specific expectations for the quality attributes of poultry; however, the ultimate authority will always be the consumer.

Chapter 16

Principles of Food Preservation

Processing and preservation of foods are known since time immemorial and probably were initiated when primitive men first use salt for preservation and cooked to prolong the keeping quality of meat. These were followed by sun-drying, use of ice and snow and others. However, the modern food processing has its origin in the development of canning by Nicholas Appert during 1809. Several innovative researches resulted in newer methods of food processing and technology and lot of research is still going on the various aspects of food preservation.

World War II imposed a very serious necessity of finding out methods and means to transport food to the war front with all the nutrients in a convenient form and with as little a bulk as possible. This prompted concerted efforts by scientists although the globe to develop methods of preserving and transportation of various kinds of foods, including poultry meat, for the defense personnel.

1. Basic concepts

Depending on the perishability when stored at the room temperature, the fresh foods both of animal and vegetable origin are classified (Table 16.1)

Table 16.1 Perishability of foods

Perishability	Time required for spoilage	Water content	Examples
High	1-7d	Medium to high	Animal tissues (meats, poultry, seafoods and milk), plant tissues (juicy fruit and vegetables)
Moderate	1 to several weeks	Medium	Plant tissues like root-crops, mature apples and pears; and eggs
Slight	At least one year	Low	All plant tissues comprising food-grains, dry beans, dry peas and nuts
			Source : Duckworth, 1975

It can be clearly understood that perishability of fresh foods is directly related to their water and nutrient contents which is utilized by the micro-organisms causing spoilage of food. Therefore, aims of commercial preservation are:

1. To retain wholesomeness, nutritive value and sensory quality of foods by economical methods by controlling the growth of microorganisms.
2. To minimize or arrest chemical, physical and physiological changes of an undesirable nature and
3. To obviate contamination.

1.1. Water in foods

It is obvious from the above information that water in foods has a direct bearing on their perishability.

Since water is a dipole (carrying both positive primitive charges), a portion of total water is totally bound to specific charged sites. For instance, hydroxyl group of sugars and polysaccharides, free carbonyl and amino groups hanging in a protein chain (as in the case of aspartic acid, glutamic acid, lysine etc.) and others in which water can be held by hydrogen bonding, or by ion-dipole bonds or by other strong interactions. At these sites, there will be several concentric layers of bound water whose binding strength is highest in the innermost layer. In addition to this, there is a portion of total water which has its vapor pressure depressed because it is located in small capillaries. The remaining portion of water is available as a solvent and is referred to as "free water". Different forms of water foods are indicated in Table 16.2.

Table 16.2 Water in foods

Component	g/g solids
Total water	3.00
Water bound to specific sites	0.10
Non-freezable water	0.40
Non-solvent water	0.35
Solvent water	2.15
	Source : Duckworth, 1975

It is obvious from the above Table that approximately 28-30% of the total water is bound and hence is not available to microbes since that portion of water is neither available as a solvent nor for reactions. The microbes, therefore, grow and multiply utilizing the nutrients through the solvent water acting as the medium for various reactions.

1.2. Water activity

The term "water activity, a_w" quantitates the relationship between moisture in foodstuffs and ability of the microorganisms to grow on them; the index of availability of water to support microbial growth and take part in chemical reactions is a_w it is also the ratio of vapor pressure of water above the material (P) and the vapor pressure of pure water (P_0) at the same temperature. i.e., $a_w = P \div P_0$ or is the equilibrium relative humidity (ERH) ÷ 100, where ERH is relative humidity of the environmental in which the material neither loses nor gains moisture or, in other words, the material will be at equilibrium. Fresh meat has an a_w of 0.99. Range of a_w and microbial inhibition (Table 16.3) and minimum a_w for growth and toxin production of food-poisoning microorganisms (Table 16.4) are given below:

Water activity of a food is not the same thing as its moisture content. It is a measure of that water which is not bound to the food and is therefore available to microbes and other organisms for growth. Foods can have a water activity anywhere between 0 (bone dry) and 1 (pure water). They may have the same moisture content and yet have quite different water activities and, therefore, stabilities. The converse is also true. For example, wheat with a moisture content of about 18 per cent has the same water activity (about 0.7) as rapeseed which has a moisture content of only about 8 per cent.

Water activity is the same as the equilibrium relative humidity (ERH) of the headspace over a food, but expressed as a decimal fraction rather than as a percentage. Thus a food with a water activity of 0.75 has an ERH of 75 %. For any given food the water activity increases as the moisture content increases. A moisture absorption isotherm is a graphical representation of this relationship (Steele, 1989).

Table 16.3 a_w *Vs* Microbial inhibition

Range of a_w	Organisms inhibited at the lowest a_w of the range
1.00 to 0.95	Gm negative rods, bacterial spores, some yeasts
0.95 to 0.91	most cocci, *Lactobacillus*, *Bacillus*, vegetative cells, some molds
0.91 to 0.87	Most yeasts
0.87 to 0.80	Most molds, *Staphylococcus aureus*
0.80 to 0.75	Most halophilic bacteria
0.75 to 0.65	Xerophilic molds
0.65 to 0.60	Osmophilic yeasts
0.50	All microorganisms

Source : Mossel, 1975

Table 16.4 Minimum a_w *Vs* microbial growth and toxin production

Food-poisoning organism		Minimum a_w for	
		Growth	Toxin production
Staphylococcus aureus	Aerobic growth	0.75 to 0.86	---
	Anaerobic growth	0.90	---
	Aerobic enterotoxin production		
	Type A	---	0.86 to 0.90
	Type B	---	0.93 to 0.97
	Type C	---	0.93

Food-poisoning organism		Minimum a_w for Growth	Minimum a_w for Toxin production
Campylobacter parahaemolyticus		0.937 to 0.992	---
Salmonella spp.		0.930 to 0.999	---
Clostridium botulinum	Spores - Type A	0.95	---
	Type B	0.94	---
	Type E	0.97	---
	Vegetative growth –		
	Type A	0.93 to 0.94	0.93 to 0.94
	Type B	0.93 to 0.94	0.93 to 0.94
	Type E	0.94 to 0.97	0.94 to 0.97
Clostridium perfringens		0.93 to 0.97	---
Bacillus cereus		0.955	---
Aspergillus flavus	Aflatoxin production	---	0.85
		Source : Riemann and Bryan, 1979	

Therefore, most of the food preservation processes mainly aim at lowering the a_w or availability of water to microbes and/or killing of microbes. Reducing the initial load of contamination of the food by the microbes is very cardinal to prolong the shelf-life of the food.

Below a water activity of 0.91 most bacteria, including pathogens such as *Clostridium botulinum*, cannot grow, but an exception is *Staphylococcus aureus*, which grows even a water activity of 0.86. Below 0.80 most moulds cannot be grown and below 0.60 no microbiological growth is possible. However, there remain a number of food spoilage microbes that can grow within the range 0.8 - 0.6. Even for products with an water activity below 0.6, a knowledge of water activity is important in determining the shelf life as water activity influences deterioration reactions such as oxidation, hydrolysis or non- enzymic browning. Although it is desirable to know the water activity of a food, it is not always easy to measure it, especially for quality control (Steele, 1989).

1.3. Measurement of Water Activity

Most techniques measure relative humidity of the headspace of a contained food. There have been many electronic instruments developed to carry out this measurement. These instruments have the following characteristics (Steele, 1989):

All need frequent calibration.

All depend on the equilibration of the moisture in the sample and headspace air.

This can take from minutes to several hours depending on the geometry of the system and water activity of the food.

All have limited accuracy (or hysteresis) in some regions although for some instruments this is negligible.

They give incorrect readings when organic materials such as acetic acid or glycerol are present. These errors are often undetectable.

Some of the sensors are attacked by volatile acids and this limits their life.

It is difficult to measure the water activity of more than twenty samples a day.

A method of measuring the water activity of foods that avoids many of the above problems has been developed (Steele, 1989). It has been used to measure the water activity of hundreds of equilibrated samples per day. The refractive index of an appropriate detector liquid equilibrated with the headspace of the food is measured and the water activity of the liquid read off from a calibration chart. The water activity of such a detector liquid after equilibration is equal to the water activity of the food. Because the sensor is a liquid there is no hysteresis of the water activity scale. The initial concentration of water in the detector liquid can be adjusted so that its water activity is close to that of the sample, thus minimizing equilibration time. A feature of the method is that at temperatures between 15 and 25°C the readings are relatively insensitive to temperature. This does not mean that there can be temperature differentials within the measuring system but it eliminates the need for controlled temperature rooms.

Chapter **17**

Preservation of Eggs

1. Introduction

The ancient Chinese stored eggs up to several years by immersion in a variety of such imaginative mixtures as salt and wet clay; cooked rice, salt and lime; or salt and wood ashes mixed with a tea infusion. Although the Chinese ate them with no ill effects of which we are aware, the eggs thus treated bore little similarity to fresh eggs, some exhibiting greenish-gray yolks and albumen resembling brown jelly (Curtis *et al.* on internet).

Generally, eggs can be preserved which shell or their contents may be preserved by various methods. Preservation of shell-eggs is primarily aimed at maintaining the interior quality rather than at minimizing the microbial spoilage because microbes definitely gain entry into the egg during the air-cell formation, but only the nature and number of microbes vary. Under normal handing, subsequent contamination of the contents of the egg is of lesser consequence. It is agreeable that there can be penetration of the microbes into the egg contents even during storage, transportation etc., but still preservation of shell-eggs does not specifically aim at preventing microbial spoilage.

In case of egg contents, the main aim is to prevent microbial spoilage and hence, to remove free water or to make it non-available to microbes. Into the first category falls the egg powder and accelerated freeze dried egg and to the latter category, frozen egg.

2. Washing of eggs

In most of the layer farms in our country, eggs are collected either in filler-flats are in baskets; in case of the latter, care has to be taken not to overfill the basket so that eggs at the bottom do not crack due to weight on them. Almost all the eggs are sterile at lay but microbes from dust, fecal material, old packaging material etc. soon contaminate the shell surface. Washing eggs definitely opens the mucous plugs (cuticle).

In any type of washing eggs, water used for washing should be warmer than the egg; otherwise, the contents shrink and the wash water along with contaminants will easily penetrate through the effective pores. For immersion type of washers, the water temperature should not be more than 15°C warmer than the egg temperature. If spray-washing is practiced, temperature of water could be still higher.

Iron content of water is a very important criterion because, iron can satiate conalbumin, the antibacterial protein present in albumen and excess of iron left-over will be available for bacteria for growth and multiplication. Hence, water for washing eggs should not contain more than 2 ppm (lesser the better) of iron to obviate bacterial spoilage.

Cooling of eggs during drying causes shrinkage of contents creating a negative pressure inside. Hence, even non-motile bacteria can swim into the egg through medium of water by capillary action. Therefore, eggs must be dried as soon as possible to prevent bacterial penetration through the pores of the shell.

2.1 BIS recommendations

As per BIS (IS: 6696-1972), the washing practice must be as follows: prepare the detergent solution daily with the approved detergents and maintain a proper ratio between detergent and germicide. About 200 ppm germicide is recommended. Start washing eggs in 40-50L of detergent solution. Wash all the eggs the day they are collected. Dry the eggs before they are packed; a drier may be used to hasten drying of eggs.

2.2 USDA recommendations

Cleaning eggs during washing is related to: wash water temperature, water quality characteristics (i.e. hardness, pH), detergent type and concentration, and defoamer. Replacement water in washer tanks should be added continuously to maintain a constant overflow rate, according to USDA regulations. Chlorine or quaternary ammonium sanitizing compounds may be used as part of replacement water provided they are compatible with the detergent. Only potable water may be used to wash eggs and USDA requires a certificate to this effect. Rate and extent of bacterial growth during storage iron content of wash water and hence, it is important to monitor the iron content of the wash water. USDA suggests that water with iron content in excess of 2 ppm should not be used unless deionized. Iron contamination may influence microbial growth when shell membranes are penetrated. As bacteria grow on membranes in an iron-rich environment, they can produce metabolic products which allow microorganisms to penetrate and diffuse into the albumen, providing a more favorable medium for microorganism growth, able to satisfy their iron requirements.

Most processors use wash water much hotter than the minimum 90 F. Product chosen as detergents and the detergent dispenser must be listed as approved for use on eggs in the current List of Propriety Substances and Nonfood Compounds (USDA), FSIS, Miscellaneous Publication Number 1419. The best approach for reducing microbial populations in wash water tanks, and therefore on eggs, occurs when detergents are added in amounts sufficient to maintain a pH of 11.

Alkaline cleaning formulations produce an initial pH in the wash water near 11, and pH during operation continues in the 10-11 range which is usually unfavorable for growth of most bacteria. Many shell egg processors have no idea what the pH of their wash water is. Often, those who monitor measure pH only at the start of a shift. PH may be 10 or 11 at the beginning of the shift, but recycling wash water, overflow losses and added replacement water all contribute to reduce pH levels. Detergents elevate the pH of egg washers and are dispensed, for the most part, in concentrations necessary to clean the egg shell.

Defoamers play an important role in egg washing. When defoamers are not dispensed properly, foam in the wash tanks builds up and overflows eventually. When foam spills from the tanks, it can interfere with water level detection, and affect water temperature and pH.

Washing, drying, and candling unit operations are continuous generally. Eggs detected as "dirties" at candling must not be soaked in water for cleaning, because soaking in water for as little as one to three minutes can facilitate microbial penetration through the egg's shell.

2.2.1. Rinse and dry

After washing, the hot water rinse may contain chlorine, or quaternary sanitizers which are compatible with the washing compound. Iodine sanitizing rinse may not be used as part of the replacement water.

Ambient air dries the eggs. At this point the surface temperature of the egg reaches approximately 35°C.

Shell eggs may be oiled then, provided operations are conducted in a manner to avoid contamination of the product. Processing oil that has been previously used and which has become contaminated can be filtered and heat treated at 82.2°C for three minutes prior to reuse.

2.2.2. Storage

Eggs have to be stored at 10 to 15.6°C. Researchers have found that the growth rate of *Samonella enteritidis* in eggs responds directly to the temperature at which the eggs were stored, and that holding eggs at 4.4 to 7.2°C reduced the heat resistance of *S. enteritidis.* It has been suggested that refrigeration reduces the level of microbial multiplication in shell eggs, and lowers the temperature at which the organism is killed during cooking. This, in and of itself, may be adequate justification to store eggs at 4.4 to 7.2°C.

Humidity in the storage environment is important both inmaintaining egg weight and preventing microbial growth. Storage relative humidity of 60% can cause weight loss and a corresponding increase in air cell size. However, storage in relative humidity of 80% can promote microbial growth.

3. HACCP plan for shell-egg processing plants

(Curtis *et al.*, on internet)

Egg temperature (initial and throughout processing and storage), and wash water pH and temperature play key roles in reducing microbial growth in shell eggs, and should be key in developing an HACCP plan for shell egg processing plants. Many processors are initiating their own Hazard Analysis Critical Control Points plans to get a jump on anticipated future regulatory needs. Shell eggs can acquire bacteria from every surface they come in contact.

3.1 Incoming eggs

Egg temperature at processing is very important. USDA regulations require that wash water temperature be at 32.2°C or higher, or at least 11.1°C warmer than the highest egg temperature (which ever is greater). These temperatures must be maintained throughout the cleaning cycle.

Temperature of incoming eggs will vary from season to season and from operation to operation. Initial internal egg temperatures of 16.7 to 20° C are common although pre-processing coolers are held generally between 10 and 15.6°C, egg temperatures decline only slightly. Where the processing plant is adjacent to production facilities, internal egg temperatures range generally from 31.1 to 35.6°C when they reach the processing area.

Regulations require also, that wash water be changed every four hours or more if needed, to maintain sanitary conditions. When the difference between wash water temperatures is $\geq$ 22.2°C, thermal checks and cracks increase, allowing surface microbes greater access to the interior of the egg.

Contact between wash water and eggs during processing causes internal egg temperature to increase. Although blow drying following washing causes a slight decrease, internal egg temperature rises generally throughout the process, and can continue to rise for up to 6 h after eggs are placed in a cooler.

4. Causes of deterioration of eggs

4.1 Shell-eggs

Shell-eggs deteriorate or spoil in the marketing channels due to various reasons; the important ones are described below:

4.1.1. Embryo development

Embryo development is a serious problem with fertile eggs only. These are usually the discards from the hatchery mostly due to their unsuitable size or some other defects including those of the shell. Embryo development is initiated at 20°C and hence the growth of blastoderm results in a "heat-spot" or "germ-spot" rendering them unfit for consumption. Nowadays, most of the eggs from organized poultry units come from unmated females and hence, the problem of heat-spot has become inconspicuous.

4.1.2. Shrinkage

Shrinkage is defined as the loss of moisture from the egg contents. The degree of shrinkage depends upon 1. Number of effective shell pores 2. Temperature and 3. Relative humidity of surrounding environment; the former two are positively correlated while the latter has a negative relationship. When temperature and relative humidity are constant (10°C and 80% as in cold storage), moisture loss is least affected by other factors. Size of the egg also has an inverse relationship with shrinkage because the relative surface area in case of smaller eggs is greater than that in case of larger eggs.

4.1.3. Liquefaction

Liquefaction can be regarded as the breakdown of thick white into thin white, which is more or less a continuous process. On an average, fresh eggs have 54% thick white. Liquefaction is principally temperature dependent; a linear positive correlation exists between the two.

The gelatinous structure of egg white and integrity of chalazae are believed to be due to ovomucin, a glycoprotein present in albumen within isoelectric pH of 10.2. The albumen has 10 times more ovomucin than thin albumen. Liquefaction is associated with decline in ovomucin content. Ovomucin has two components designated α (carbohydrate-rich) and β (carbohydrate-poor) ovomucins which are non-covalently bonded. β-ovomucin seems to be substantially responsible for gelatinous properties of thick egg-white gel.

It is hypothesized that the structure of thick white is due to interaction between ovomucins which are acidic and lysozyme which are cationic in nature. In fact, lysozyme co-precipitates with ovomucin. On the contrary, is also suggested that the liquefaction is due to increased interaction of ovomucin and lysozyme at a higher pH; both systems may be operating.

As pH increases from 7 to 9, more α-ovomucin-lysozyme complexes are formed and β-ovomucin is likely to inhibit for the complex formation. However, such an aggregation (complexing) does not occur in natural thinning of egg white. Thus, formation of ovomucin-lysozyme complex alone is not the cause of egg-white thinning but degradation of β-ovomucin mainly responsible for liquefaction.

It is also speculated that magnesium, due to its small size and high charge density, replaces lysozyme in ovomucin complexes and gives rise to most stable Mg-ovomucin complex in egg albumen.

Since loss of carbon dioxide from eggs does occur and it results in elevated pH of contents, it can also be hypothesized that the microfilaments of ovomucin are held together by carbon dioxide molecules thereby giving a gelatinous structure to albumen.

4.1.4. Ingress of water into the yolk

This is due to osmosis, since yolk and albumen differ in their water content; the former having lower amount (48%) than the latter (88%). This gradient particularly takes an upper hand when liquefaction is complete, since the thick albumen is more viscous and has less water content than the thin albumen. The ingress of water into the yolk is directly proportional to the amount of carbon dioxide in the thick white and also directly related to the surrounding temperature.

4.1.5. pH of egg

pH of albumen records an initial increase in alkalinity due to loss of carbon dioxide but subsequently becomes reduced due to entry of carbon dioxide from the yolk before the finally escapes through shell-pores resulting in an alkaline pH. Yolk pH progressively proceeds towards alkalinity due to loss of carbon dioxide. A fresh egg has about 0.5% carbon dioxide and pH increases from 7.6 to 9.5 due to progressive loss of carbon dioxide. An alkaline pH of a certain range is suitable for microbial growth. Hence, the microbial contamination can further bring about faster deterioration in egg quality.

4.2 Opened-out eggs

The main cause for deterioration of opened-out eggs (eggs contents) is microbial spoilage. Egg contents being highly nutritious become quickly a target for microbes for growth, multiplication and in the process, spoilage.

5. Preservation of eggs

5.1 Shell-eggs

It is obvious that shell-eggs have to be properly and adequately preserved so that their quality does not deteriorate in the marketing channels. Different methods of preservation of eggs (both with shell and without shell) are shown in Fig 17.1.

5.1.1. Wet methods

Under these methods, eggs come in contact with a solution usually prepared with water as solvent. Most of the shell-pores will be blocked during the process and hence a small pin hole has to be made before boiling the eggs stored by wet methods to avoid breakage of shell due to expansion of eggs contents as well as air-cell.

Immersion in different liquids too numerous to mention was explored, lime water

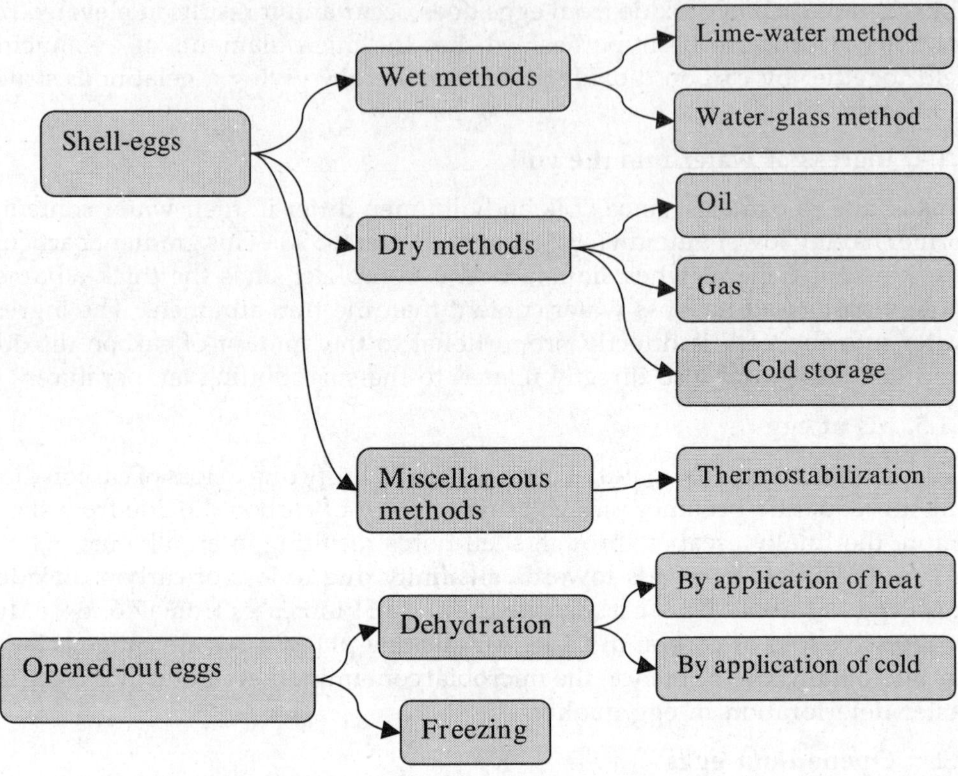

Fig 17.1 Methods of preservation of eggs

being a favorite in the 18th century. During the early 20th century, water glass was used with considerable success (Curtis *et al.*, on internet).

Essentials for any of the wet methods are a) should not induce any taste or odor b) solution should not deteriorate even after long periods of storage c) prevent gaseous exchange between egg and atmosphere and d) preserve edible quality of eggs under storage.

5.1.1.1. Lime-water method

5.1.1.1.1. Principle

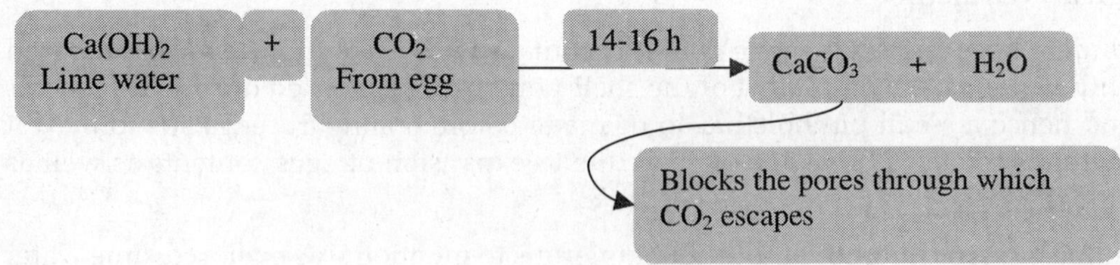

5.1.1.1.2. Procedure

1. One kg of quicklime (CaO) is mixed with 1L of water and allowed to react.
2. After the reaction is over, 120 g of salt is added to increase the specific gravity so that none of the eggs touch the bottom of the container when immersed and get broken due to the weight of eggs on top of them.
3. Further, 5-6 L of water is added and stirred thoroughly.
4. Solution filtered through muslin cloth and to the filtrate, a small quantity of slaked-lime (available on the muslin cloth) is added in order to maintain the concentration, since there will be production of water molecule replacing every molecule of lime water during the reaction resulting in dilution of solution.
5. The eggs are held in the solution for 14-16 h and later on, removed and stored at room temperature.

5.1.1.1.3. Efficacy

1. Eggs may be stored up to 3-4 weeks.
2. Albumen and yolk quality maintained even under room temperatures.

5.1.1.1.4. Advantages

1. Easily available, cheap and easy to prepare
2. Solution is alkaline and hence, surface bacteria killed
3. No residual flavor encountered
4. The chemical reaction occurs at all effective shell-pores and therefore, the method is very efficient.
5. The chemical reaction produces $CaCO_3$, which is the shell material itself but not produced by the hen.

5.1.1.1.5. Disadvantages

1. Hatching eggs cannot be preserved, since embryos die due to increased carbon dioxide tension within the egg.
2. Due to blockage of shell-pores, embryo cannot respire.
3. Thin-shelled eggs when removed from the solution will be subjected to a cooling effect. This is likely to produce anomalous expansion of water present in albumen which is closer to the shell. Since all the shell-pores are blocked through the chemical reaction, the egg shell being thin, is likely to crack.

5.1.1.2. Water-glass method

A 10% solution of sodium silicate ($NaSiO_3$) is popularly called as "water-glass". Water glass, a bacteria-resistant solution, discouraged the entrance of spoilage organisms and evaporation of water from eggs. It did not penetrate the egg shell, imparted no odor or taste to the eggs and was considered to have somewhat antiseptic properties. However, it did a rather poor job at relatively high storage temperatures. Eggs preserved in a water glass solution and stored in a cool place keep 8 to 9 months (Curtis *et al.*, on internet).

5.1.1.2.1. Principle

Water glass is a colloid solution and hence, sodium silicate molecules absorb the shell surface and physically block the effective shell-pores, without any chemical reaction unlike lime-water method.

5.1.1.2.2. Procedure

1. Water is boiled thoroughly to remove the dissolved CO_2, which otherwise form a complex with sodium silicate.
2. Calculated amount of sodium silicate is added.
3. Eggs are dipped in the cooled water-glass solution and kept overnight.

5.1.1.2.3. Efficacy

The solution is most effective at 25°C.

5.1.1.2.4. Disadvantages

1. Efficacy is less at higher temperatures.
2. The chemical is costly and not easily available.
3. The procedure is time-consuming and requires accurate weighing.
4. At -1°C, the solution becomes a mass and hence cannot be used.

5.1.2. Dry methods

Eggs will not come in contact with water in any of these methods and therefore, they are referred to as "dry methods". Dry packing in various substances ranging from bran to wood ashes was used occasionally, but costs of transporting the excess weight of the packing material far exceeded the dubious advantages.

5.1.2.1. Oil

In an attempt to seal the shell pores to prevent loss of moisture and carbon dioxide, a great variety of materials including cactus juice, soap and shellac were investigated with varying degrees of success. The only coating considered fairly efficient was oil which is still used today (Curtis *et al.*, on internet).

Of the vegetable oils, excepting coconut oil, all the colored and impart the same onto the eggs. Besides, they are unsaturated and hence are prone to rancidity, costly and are human foods. Therefore, mineral oils (by-products of petroleum industry) are generally used. The standards of egg-coating mineral oil are: specific gravity: 0.830 to 0.840, viscosity: 50 poises, colorless, pour point: (-) 1.11°C, flash point: 140°C and neutral in reaction.

5.1.2.1.1. Methods

Eggs can be dipped (400 cc/200 eggs) or sprayed (over the broad end, 400 cc/ 1000 eggs). Usually, eggs are kept broad end up in filler flats and sprayed cover ½ to ¾ of the shell surface. Bactericides and/or fungicides can be added to the oil to enhance efficacy

5.1.2.1.2. Principle

Most of the effective pores are present towards the broad end of the egg; the use of oil blocks these pores and prevents CO_2 loss. Spraying with heated oil is more beneficial.

5.1.2.1.3. Precaution

Oil should be done 5-6 h (winter) or 8-10h (summer) after lay to avoid syneresis (weeping of eggs).

5.1.2.2. Gas

5.1.2.2.1. Principle

Inert gases are used. Eggs are kept in plastic bags or special plastic retail cartons which are filled with the gas and sealed. Usually N_2:CO_2 = 94:6 is used. Higher CO_2 pressure outside the eggs precludes its loss from inside the eggs the affording protection to egg quality.

5.1.2.2.2. Disadvantages

1. Transportation is difficult 2. Costly and cumbersome and 3. Causes turbidity or cloudiness of the albumen.

5.1.2.3. Cold-storage

During the first half of the 20th century, storing eggs in refrigerated warehouses was a common practice. Preservation was later improved with the introduction of carbon dioxide into the cold storage atmosphere (Curtis *et al.*, on internet).

Cold-storage of eggs is done at the temperature of 10.0-15.6°C and a relative humidity of 75-90%. And ante-room is a must to avoid the entry of hot air causing air pockets and corners of cold-storage room which can subsequently trigger mold growth under hot humid conditions. Air circulation is also important to effect a proper loss of heat from eggs. It should be borne in mind that freezing point of albumen is (-)0.42°C, that of yolk is (-) 0.65°C and that of albumen + yolk is (-) 2.22°C. Therefore, cold-storage temperature should not be less than (-) 2.22°C. Eggs keep for a long time (up to 8-10 months) in cold-storage.

5.1.2.3.1. Disadvantages

1. After 7-8 months in cold-storage, eggs develop a metallic taste referred to as "storage taste".

2. Mold growth, if not controlled, cause whitish marks on the shell (mucous growth), referred to as "old man's beard".
3. Sweating: when the eggs are taken out of the cold-store, due to lower temperature of the egg, the atmospheric moisture condenses on the shell surface which is referred to as "sweating". Thus, eggs are to be brought through graded temperature cabins viz., 10, 15.6, 21.1°C and then to room temperature.
4. Costly, since proper maintenance of cold-storage is compulsory.

5.1.3. Miscellaneous methods

Several methods are attempted and/or advocated:

1. 21.1°C for 5-6 d; but fertile eggs may still show embryo development.
2. 100°C (boiling water) – eggs rotated to have even coagulation of albumen. But embryo growth may still occur.
3. Thermo stabilization – 54.4°C for 30 min: This is stabilization of albumen quality by application of heat. Several combinations of temperature and duration are tried to ensure a 100% kill of embryos and also to have uniform coagulation of albumen close to the shell.

Under village conditions, boiling water and cold water may be mixed (in such a way that the candle just melts; melting point of paraffin candle is approximately 54.4°C) and eggs can be dipped for 30 min.

Oiling of thermostabilized eggs increases the shell-life to 4-6 weeks at room temperature. This method was extensively practiced by housewives of the late 19th century (Curtis *et al.,* on internet).

5.1.3.1. Principle

To effect peripheral coagulation of albumen and hence inhibit CO_2 loss. Thermostabilization influences only albumen quality and yolk quality unaffected.

5.1.3.2. Effects of Thermostabilization

1. Only stabilization of the existing egg quality is possible.
2. pH of albumen increases.
3. Stabilizes albumen quality.
4. Many of the bacteria on shell surface are killed.
5. Loss of CO_2 reduced.
6. Even in hottest summer, eggs keep for 10-12 d at room temperature.

5.1.3.3. Disadvantages

1. Cake-volume is reduced
2. Requires specific temperature for efficacy

3. Requires an additional fuel for heating and maintaining temperature

Keeping the above methods for preservation of shell-eggs in view, lime-water method appears to be the most practicable one under village conditions followed by oiling and thermostabilization, in that order. The large number of eggs is to be preserved at a centralized facility, cold-storage is the method of choice.

5.2 Opened-out eggs

Eggs are classified under "most perishable" commodity. The rate of spoilage is very rapid when the eggs are opened out because of high nutritive value of the contents acting as an excellent medium for the growth of microorganisms.

During summer, there will be a temporary reduction in demand for eggs and surplus eggs may be broken and contains be preserved for future marketing.

5.2.1. Dehydration

Whole egg contents (that is, egg without shell) has about 74% moisture. This helps microbial growth. Removal of this moisture not only reduces the chances of microbial spoilage, but also reduces the bulk of the contents. However, application of heat for removal of water has some short-comings like loss of few B-complex vitamins and slight reduction in quality of protein; but, the greatest advantage is that the resultant product, the egg powder, can be stored at room temperature in cans and reconstituted (similar to milk powder) as and when required. Removal of water by application of cold is very expensive and transportation of the finished product, accelerated freeze dried (AFD) egg, which should be under low temperatures also adds to the cost of production. The main advantage of AFD method is there will be no loss in nutritive value. Hence, the AFD products are especially meant for defense personnel.

5.2.1.1. By application of heat

5.2.1.1.1. Candling and breaking

The eggs are refrigerated overnight so that candling will be easy. The eggs are candled and those with blood-and meat-spots are rejected. The eggs can be washed, if many are reasonably unclean. For washing, even cold water can be used since the eggs are going to be broken soon after washing and drying.

The eggs are broken individually although, mechanized egg-breaking machines are commonly used in Western countries. The egg contents are further assessed for flavor and presence of any other defects. Only good eggs are taken for further processing.

In case albumen and yolk have to be powdered separately, they are separated at the stage.

5.2.1.1.2. Homogenization

The next process is homogenization, wherein the contents are mixed in a low-speed homogenizer taking ample care to avoid foaming/frothing during the process. The homogenization is followed by filtration through a fine sieve to remove chalazae, vitelline membrane, bits of shell etc. The filtrate is called "egg mélange" (albumen/yolk mélange, as the case may be).

5.2.1.1.3. Desugaring

Egg is probably the only animal product containing free α-D glucose (0.48% w/w). The free aldehyde group of glucose combines with the å-amino group of lysine (a very important essential amino-acid), especially at high temperatures (as employed in the preparation of egg powder forming a Schiff's base rendering lysine non-available. This not only offsets the nutritive value of the egg severely, but also results in browning of egg powder, bitter taste, reduced solubility and off-flavor. This reaction being non-enzymatic, is referred to as non-enzymatic browning and also popular after the scientist who discovered it as Maillard reaction. Glucose also reacts with cephalin part of phospholipids in egg yolk fraction of the stored egg powder resulting in loss of palatability. Hence, glucose must be removed from the egg mélange and the process is called as "desugaring".

5.2.1.1.3.1. By bacteria

Bacteria that can be used is *Aerobacter aerogenes* which is nonpathogenic and multiplies by using glucose. However, it requires 24-30 h at a pH of 7-7.5. Besides, it gives a bad flavor and bad odor to the finished product. Hence, this method is not popularly practiced.

5.2.1.1.3.2. By yeasts

Earlier, *Sacchromyces cerevesiae* and now *Torula monoza* is used to ferment glucose. Yeasts are added at the concentration of 0.07 to 0.15%. This method also is not much practiced mainly because of yeasty flavor it imparts to the powder which is disliked by many.

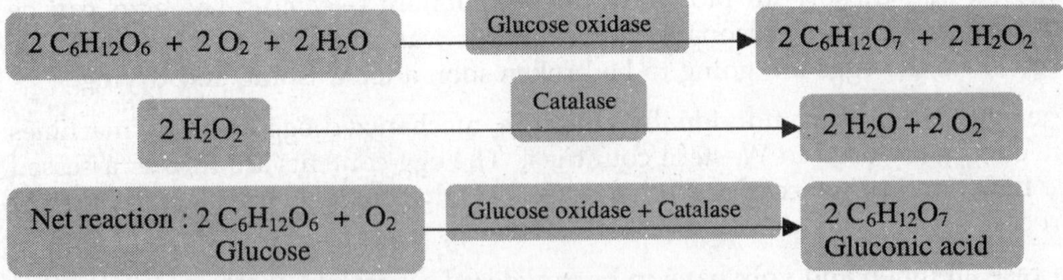

5.2.1.1.3.3. By enzymes

Glucose oxidase and catalase are added to the egg mélange and left for 6h at 30-33°C and pH of 7-7.5. The aldehyde carbon of glucose is oxidized to gluconic acid making it refractory for Maillard reaction. The reactions are as shown above.

5.2.1.1.3.4. Pasteurization

After desugaring, the egg mélange ispasteurized (a process in which egg mélange will be freed of all the pathogenic organisms) at 60°C for 4-5 min. *Salmonella* organisms for the reference for the pasteurization process. Recently, chemicals such as β-propiolactone, butadiene oxidase etc. have also been effectively used for pasteurization.

5.2.1.1.3.5. Drying

The pasteurized egg mélange is dried, usually in a spray-drier (pan-drying, plate-drying are also practiced). The egg mélange is sprinkled as a fine mist and dry-air (moisture nearly 0%) at the temperature of 121-232°C is sent from the other side. The egg mélange is instantaneously dried and the powder is collected and canned in tins.

5.2.1.1.3.6. Storage and reconstitution

The egg powder can be stored at room temperature for a very long time as long as the lid is closed as soon as the required amount of powder is taken out. The egg powder can be reconstituted with two times the quantity of water and the lid of the tin should be kept tightly closed after use.

5.2.1.1.3.7. BIS Specifications

Moisture: maximum 2%, Coliform count: maximum 100/g, Standard Plate Count: maximum 7.6×10^4/g, pH: 7.9, Protein: 45%, Lecithin and fat: 38%.

5.2.1.2. By application of cold (AFD Egg)

Due to sudden cooling, water separates and it can be removed by vacuum. This process is expensive and therefore, meant for manufacture of foods for defense personnel.

5.2.2. Freezing

In this process also nutritive value is not lost. But, major disadvantages are

1. The finished product must be stored at low temperatures throughout
2. Transportation is expensive.

Egg mélange, after desugaring, is frozen at temperature of - 20° to - 30°C and the frozen product is stored at a temperature of - 5°C or lower. Pasteurization of the egg mélange is not essential because at the temperatures of freezing and storage, all the bacteria die. However, upon thawing, yolk might show some clumps (coagulated masses) due to reasons still incompletely understood. Several anticoagulants are tried; the best being addition of papaine at a rate of 0.04% to the egg mélange before freezing.

Chapter 18

Preservation of Poultry Meat

Poultry meat preservation is principally aimed at inhibiting microbial spoilage although retention of the qualities of meat is equally important. The methods of preservation of poultry meat can be grouped as under:

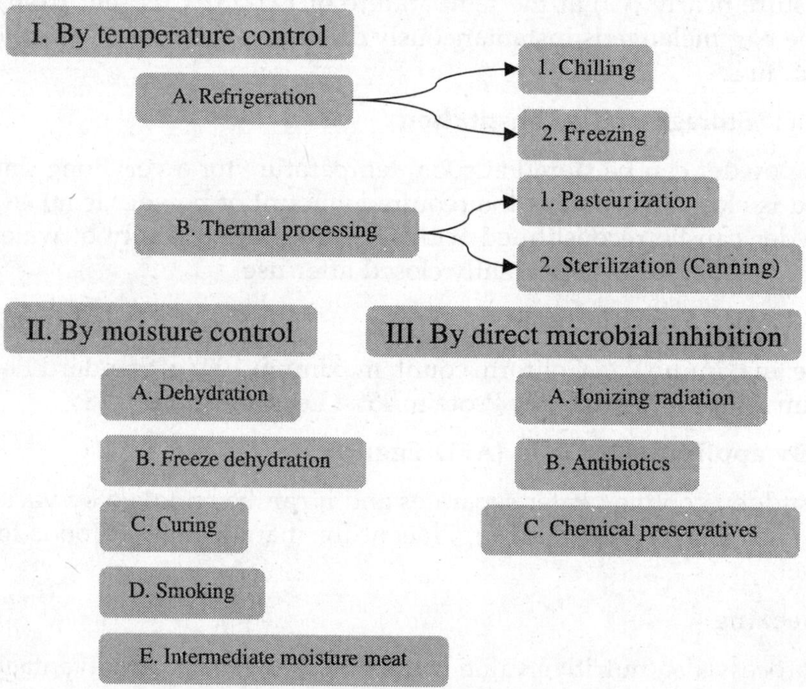

Fig 18.1 Methods of preservation of poultry meat

1. By temperature control

Chicken should be stored at a temperature outside of the temperature zone in which bacteria, that causes food-borne illness, grows quickly. The danger temperature zone is a range between 4 and 60°C.

1.1. Refrigeration

1.1.1. Chilling

Meat is kept under refrigeration temperatures (i.e. 6-8°C) preferably after prepackaging. Refrigeration definitely retards, but cannot eliminate, the microbial growth. The psychrophilic bacteria definitely take an upper hand and produce slime within 3-8 d. The duration required for slime formation is inversely related to the refrigeration temperature and initial load of microbes. Prepackaged chicken meat can be stored by this method for maximum of 7-8 d.

Raw chicken can be stored in a refrigerator for several days. If it is not going to be used within the recommended time, it should be frozen to prevent it from perishing. Leftover cooked chicken should be wrapped tightly and refrigerated as soon as possible. Chicken meat should never be left at room temperature for more than two hours. If cooked chicken is not going to be used within four days of cooking, it should be frozen.

Raw or cooked chicken can be stored safely in a refrigerator at 4°C or lower for several days. The amount of time that it can be refrigerated will depend on the freshness of the meat when purchased, the temperatures it is exposed to before refrigeration and the type of packaging used.

If the raw juices are leaking from the original package, it should be removed and the chicken placed in a bowl and covered with wax paper, foil or rewrapped tightly in plastic before placing in the refrigerator. The package should be placed on a dish with sides to prevent any meat juices from dripping on other foods. It should be kept away from other foods so they do not come in contact with the raw juices. The meat should be stored in the coldest section of the refrigerator.

Chicken giblets and ground poultry should only be stored in the refrigerator for 1 d. If necessary to store for a longer period of time, the chicken products should be frozen.

Chicken leftovers should be cooled and refrigerated as soon as possible, limiting the amount of time it is exposed to room temperatures. Cooked chicken can be stored for up to 3-4 d in a refrigerator at 4°C or less. If leftovers are not going to be used within this time, they can be frozen and stored for up to 3-4 months. Leftover stuffing should be removed from the chicken as soon as possible to minimize the possibility of bacterial growth and then stored in a covered container in the refrigerator. The stuffing can be stored for up to three days, but if it is not going to be used within that time it should be frozen. Stuffing can be kept in the freezer for up to a month.

1.1.2. Freezing

Freezing changes water in meat to ice crystals thereby rendering it non-available for microbes. In addition, water present inside the microbes also will be frozen which limits their growth severely.

Table 18.1 Rate of freezing *Vs* changes in histology of muscle (meat)

Rate of freezing (min)	Type of ice formation	Remarks
5	Many, very small, intracellular	Sarcolemma undamaged; low drip fluid
25*	4 intracellular ice columns	Unfrozen intracellular contents too thinly spread over surface between ice crystals and sarcolemma to avoid damaging the latter; more drip fluid
50	Single, relatively small intracellular ice column	Layer of unfrozen intracellular contents thick enough to separate ice from sarcolemma and protect it; low drip fluid
75**	Single, very large intracellular ice column	Unfrozen intracellular contents are too less to protect sarcolemma from damage; more drip fluid
100	Relatively small, extracellular ice crystals	Water drawn osmotically from the cell through intact sarcolemma to form small ice crystals without distorting fibers; less drip fluid
200-500	Large, extracellular ice crystals	Ice crystals large enough to damage sarcolemma; more drip fluid
>500 and < 750	Very large, extracellular ice masses	Fibers are pressed together between the ice masses and hence sarcolemma is undamaged; more drip fluid

* Slowest freezing rate at which more than one ice crystal forms within muscle cell
** Slowest freezing rate at which ice crystal forms intracellularly
Source : Lawrie, 1979

However, there will be exudation of fluid (drip) on thawing, which contains proteins, peptides, amino-acids, lactic acid, purines, B-complex vitamins and various salts; therefore, definitely reduces both eating quality as well as nutritive value.

The time taken to pass from 0 to - 5°C is usually regarded as an indicator of speed (rate) of freezing. Lower the freezing time, better the quality of meat because ice crystals will be intracellular and smooth so that sarcolemma remains undamaged. Slower freezing (taking more than 100 min) results in extracellular ice crystals, and when sufficiently large, they can distort sarcolemma and result in increased drip loss. Damage to sarcolemma can also to under freezing time of 75 min (Table 18.1).

The chicken should be frozen as quickly as possible. The quicker it freezes the better it will be when thawed. To speed up the freezing process, the package is placed on the floor or against the wall of the freezer since these are the coldest parts.

However, under commercial conditions, it is too expensive to achieve freezing rate < 100 min and therefore, the common practice is to achieve freezing in < 1-2 h.

Drawback of freezing, in addition to drip loss, is "freezer burn" which is defined as whitish or amber-colored patches seen on the surface of frozen meat caused by sublimation of ice crystals into the freezer atmosphere. This happens when vapor pressure of water on the refrigeration coil is less than that of the surface of the meat. This leads to desiccation of water from muscular tissue to the freezer coils. This can be prevented by regular defrosting of the freezing equipment, pre-chilling meat with air of high relative humidity at – 3.9°C, dipping meat in water before freezing or by proper prepackaging of meat

Frozen prepackaged meat remains edible for a very long time (even years) and in the US, frozen meat is very popular. Chilled chicken carcass when properly prepackaged may be kept in freezer chambers maintained at – 20 to - 40°C for optimum results.

In any case, freezing can be avoided as far as possible because it causes the chicken to be less tender and juicy. For freezing,the chicken should be as fresh as possible. Raw meat is removed from the package it came in and rewrapped tightly, using plastic wrap, foil or freezer paper. If storing for more than two months, double wrapping is recommended. The wrap must be pulled tightly against the entire surface of the chicken to prevent ice crystals from forming in areas that are not wrapped tight. Ice crystals form in these areas because moisture has been drawn out of the meat, causing the chicken to become tough in these areas (freezer burn). The package has to be suitably marked with contents and the date so as to know how long it has been stored in the freezer.

It is always best to freeze and store frozen food in a freezer unit, rather than a refrigerator freezer. The freezer units will maintain a temperature of -17.8°C or below, which will allow food to be stored for longer periods of time. A refrigerator freezer will generally only maintain a temperature of -12.2 to -3.9°C and is opened more often, which causes fluctuation in temperature. If meat is stored in a refrigerator freezer, it should be used within two or three months. Whole chicken stored in a freezer unit can be stored safely for up to a year, and chicken pieces can be stored up to nine months.

Table 18.2 Suggested periods for storage of poultry meat

	Refrigerator (4°C)	Freezer (-17.8°C)
Whole Chicken	Two to Three Days	Twelve Months
Chicken Parts	Two to Three Days	Nine Months
Giblets	One Day	Three Months
Ground Chicken	One Day	Three Months
Cooked Chicken	Three to Four Days	Three to Four Months
Note: If storing longer than the storage times shown above, double wrapping is suggested to help keep in moisture.		

1.2. Thermal processing

1.2.1. Criteria

Thermal processing is developed spending on certain physical principles such as:

1.2.1.1. Thermal death time (TDT)

It is defined as the time required to achieve sterility at 121°C, usually designated F_o.

1.2.1.2. Decimal Reduction value (D)

This is the time required to kill 90% of the bacterial cells at a given temperature.

1.2.1.3. Z value

Regression coefficient of log of F_o or D on temperature is referred to as Z value. In other words, it is a number of °C by which the temperature has to be raised to attain one log (tenfold) increase in death rate of bacterial cells.

The species of bacteria used as reference to evaluate heat processing are *Clostridium botulinum* and *Clostridium sporogenes*; the latter is more heat-resistant than the former and therefore is usually the reference organism. Z value for *Clostridium botulinum* is 10 whereas its F_o value is 2.8

1.2.2. Pasteurization

This is rendering meat free from pathogenic microorganisms. Usually a temperature of 75°C for 3-6 h is employed so as to attain an internal temperature of 45-55°C in the carcass. This method is not commonly practiced since there is a very wide variation in pasteurization conditions depending on several factors like type of meat, type of initial microbial load etc. Besides, there are easier and better methods than the method for preservation of poultry meat.

1.2.3. Sterilization (Canning)

In this process, the meat is processed to a point at which most micro-organisms and their spores are killed and therefore ensures a near indefinite storage life of meat inside the can, at any ambient temperature, provided it is kept sealed.

1.2.3.1. Procedure for preparation of poultry meat cans

1. Meat procurement: usually meat from older birds is used because the severe cooking process employed in canning might result in separation of meat and bones from younger birds.
2. Pre-cooking: meat is pre-cooked in boiling water for 5-10 min. Ghee can be used to fry the meat till it gets a light brown color.
3. Filtration: to remove the cook-out juices and meat-fat that has melted-out during cooking. The filtrate is called "chicken-essence" which contains amino acids, salts and vitamins. Chicken-essence is easily digestible and therefore is ideal for children and convalescents. Hence, chicken-essence is sterilized and ampouled to make it convenient for marketing.
4. Later on, steamed-meat is filled into lacquered cans (sulfur-resistant). Desired amounts of spices and gravy are added. Finally, water is added leaving about 1.5 cm of space below the brim and the mouth is half sealed so that steam can come out in a heavy jet from all from the circumference.
5. The half-sealed cans are kept in a pressure-cooker and steamed once again for 7-10 min.
6. The lid is sealed firmly while hot to avoid entry of air. This can be done by vacuum method also.
7. The sealed cans are processed in a pressure cooker at 1.057 kg/cm^2 for 30-60 min.
8. The cans are cooled under running tap-water for 30 min or by spraying.
9. The cans are suitably labeled and are ready for storage and/or transport; they can be stored at any ambient temperature.

2. By moisture control

2.1. Dehydration

By using hot-air, moisture is removed from meat. A higher surface area quickens the process of dehydration and hence meat is cut into small pieces (minces). Hot-air drying is not suitable for raw meat because of faster movement of solutes and consequent case-hardening. Therefore, cooked meat is minced and dried at temperatures below 70°C. Air-dried meat will have usually 5.6% moisture. Dehydration can also beeffected by microwave energy or by hot-plates. In any case, chicken meat is not that commonly used for dehydration by this method.

2.2. Freeze dehydration

2.2.1. Principle

The principle involved in this process is to remove water from meat by sublimation from the frozen state rather than by evaporation. At eutectic or triple point of water (0.0075°C, 4 mm of Hg pressure), water in foods exists in all three states (ice, liquid and vapor). Therefore, and less than 0.0075°C, water can be sublimed and removed by vacuum without application of heat. Heat exchange can be accelerated by incorporating plates to supply heat and hasten drying during the second phase. This process is termed "accelerated freeze drying, AFD".

The meat retains its original honeycomb texture and translocation of salts is avoided. Water from minute interstices is removed and therefore the meat proteins are undamaged. Consequently, there is little or no reduction in nutritive value of meat

However, the processes is costly and hence it is usually used for manufacturing food for defense personnel. The AFD meat can be stored at room temperatures for several months.

The main drawback of AFD meat is Maillard reaction, especially when the residual moisture is less than 2-3%. The other effects on tenderness can be easily alleviated by use of papaine in water used for rehydration.

2.3. Curing

2.3.1. Principle

The principle involved in curing is toremove moisture food by using salt. Nitrites are added to get the desired red color (Fig 18.2) of nitrosomyoglobin in the meat. Nitrates are added so that bacteria convert nitrates to nitrite which, in turn, will be available for color reactions.

Dressed chicken carcasses are kept in the curing solution for 2 to 3 days during which there will be leaching out of water so that free water availability for the microbes is minimized. Added to this, sodium nitrite destroys the food poisoning bacteria, especially *Clostridium botulinum.* After the carcasses are taken out, they are washed in cold water for one hour, packed and stored, preferably under refrigeration temperatures. Vitamin C (0.0005%) when added can accelerate the curing process. Sugar reduces the saltiness and polyphosphates, if added, enhance eating quality.

2.3.2. Curing solution

Curing solution contains salt 13.60%, Sodium nitrate 0.21%, Sodium nitrite 0.28%, Sugar 6.80% and monosodium glutamate 0.05%. The ingredients are dissolved in mineral free water to a Salinometer reading of 60° at 20°C. The Curing solution should be maintained at 1.1 to 2.2°C.

2.3.3. Shortcomings

The major drawback of the cured meats is the formation of nitrosamines (Table 18.3) which are carcinogenic. However, it is found that the consumption of cured meat, even in advanced countries, is not as high as to be major factor to cause cancer in human beings. Cured meats normally contain 10 to 50 ppb of nitrosamines whereas the BIS accepted level is as high as 200 ppm.

2.3.3.1. Nitrates, nitrites and nitrosamines

Nitrates are distributed widely in nature and are non-toxic. Vegetables contain high levels of nitrates which they obtain from soil (fertilizers) and contribute 86% of total intake of nitrates by man whereas, cured meat and daily products constitute only 10%. Human saliva also contains nitrate is converted nitrites by oral microbes; on average, 7 ppm of nitrite, amounting to 75% of normal daily exposure, is produced. The acidic pH in stomach further increases the formation of nitrosamines especially in case of smokers than non-smokers because thiocyanate (catalyst for the formation of nitrosamines from secondary amines) concentration in saliva is higher among smokers. Hence, just elimination of nitrite from food does not guarantee elimination from its exposure to man.

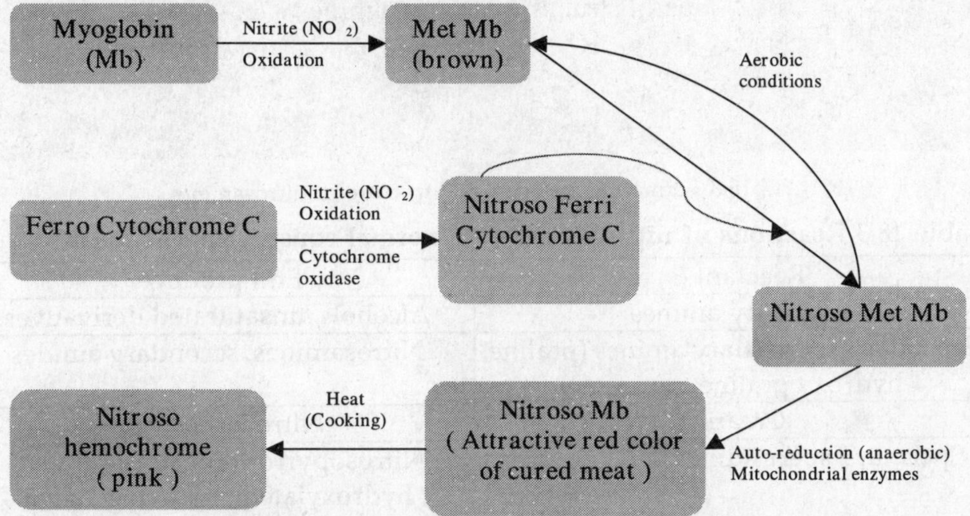

Fig 18.2 Color of cured and cooked-cured meat

High temperature, frozen storage (due to removal of water in the form of ice thereby bringing about concentration effect), lecithin, nitrophenols (present in smoked meat) accelerate the formation of nitrosamines. Microbes have a triple way of catalyzing nitrosamines formation namely by reducing nitrate to nitrite, reducing pH and non-enzymatically (?) assisting nitrosation process.

Nitrites produced by microbial action on nitrates are further reduced to nitrous acid (HNO_2) under acidic conditions. The various reactions of nitrous acid are outlined in Table 18.2 which clearly indicates that different types of N-nitrosamines are produced (Fig 18.3). Depending on the groups R_1 and R_2, the compounds are named. Of the various nitrosamines, dimethyl- and diethyl-nitrosamines are the most common and most potent carcinogens.

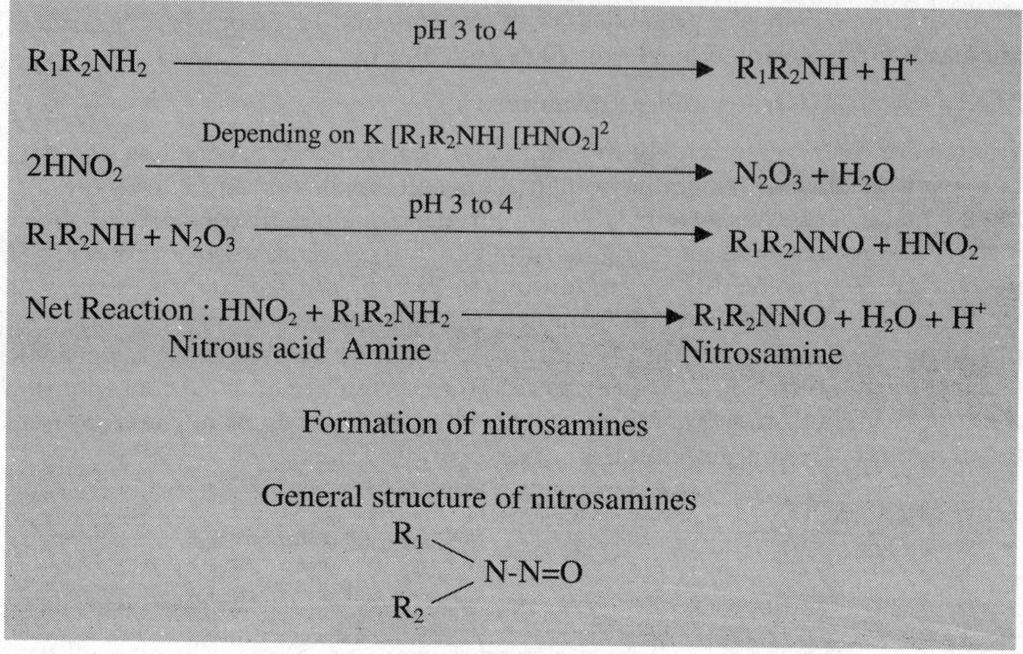

Fig 18.3 Formation and general structure of nitrosamines

Table 18.3 Reactions of nitrous acid with normal constituents of meat

Reactant (s)	End products
Primary amines	Alcohols, unsaturated derivatives
Secondary and tertiary amines (praline, hydroxy praline, sarcosine)	Nitrosamines, secondary amides
Creatine	Nitrososarcosine
Spermine, spermidine (poly- amines)	Nitrosopyrrolidine, unsaturated, hydroxylated and halogenated nitrosamines
Urea, carbamates	Nitroso-derivatives
N-acetyl tryptophan	Nitrosamines
Ornithine, lysine	Nitrosopyrrolidine, Nitrosopiperidine
Choline, lecithin	diamethyl nitrosamine
Activated methylene group	Oximes
Phenols	Nitrosophenols, alcohols
Thiols	Nitrite-derivatives
Reductans (Vit C)*	Dehydro-reductans
* React faster with nitric acid than all other amines and hence scavenge all the nitrite	

Nitrosamines are known to cause malignant lesions predominantly in liver and kidneys; and sometimes in esophagus, at least in laboratory animals (rats, minks etc.). Therefore, lower their levels in human diet better it will be.

2.4. Smoking

When wood be subjected to pyrolysis, cellulose, hemicellulose and lignin will yield several components into the smoke which can act as meat preservatives (Table 18.4). Usually, temperature of the smoke will be 43°C and hardwood and maize cobs are preferred because the yield smoke at the lower temperature. The carcasses are hung in the chamber in which the smoke is generated. Recently, liquid smoke is being used by completely removing the particle phase of smoke which has carcinogens. It also a common practice to smoke the cured meat to enhance consumers acceptance

Table 18.4 Components of smoke

Components	Function (s)
Phenols	Antioxidants, flavoring agents, bacteriostats
Alcohols	Mild bacteriostatic, not important in meat
Organic acids (1 to 10 carbons)	Coagulation of surface proteins, especially in frankfurters
Ketones	Flavor and color
Aldehydes	Antioxidants

Besides, the smoke also contains complex polynuclear hydrocarbons like benzpyrenes and benzanthracenes which are carcinogenic. Benzpyrenes and benzanthracenes are derivatives of incomplete combustion of fat. Hence, meat should not be broiled over the flames of burning fat; the fire should be quickly put off with water. Cooking of meat should be preferably done by the heat from coals. Prolonged consumption of these compounds can cause cancer of the alimentary tract, especially stomach. In home-smoked products, 3,4 benzpyrene is the common carcinogen. The minimum values (in µg/kg wet weight) of polycyclic aromatic hydrocarbons in smoke products reported are acenapthylene 137.7; flourene 20.6; phenanthracene 86.5; anthracene 19.8; pyrene 5.9; fluoranthene 4.6; 3,4 benzpyrene 1.3.

2.5. Intermediate moisture meat

Many foods contain moisture levels of about 20 to 50% by weight and hence they are very ideal for microbial growth. If the concentration of the aqueous solution present in a food be increased to a point where microbiological activity can be brought under control, several desirable features will start to manifest like

1. spoilage reduced and food is conserved
2. safety improved and health hazards diminished

3. shelf life increased and great convenience in use
4. soft and moisture texture retained.

It has been found that the final product contains 15 to 30% moisture, they will be in equilibrium with a relative humidity of 65 to 85% (in other words, aw of 0.65 to 0.85) and such foods are referred to as intermediate moisture foods (IMF). Included in this group are jams, jellies, cakes, candies, dried fruits etc. Meat contains 65 to 80% moisture and hence, it is necessary to either remove or concentrate in such a way as to leave only 15 to 30% moisture in the product.

This controlled reduction in water content can be achieved either by adsorption (drying followed by rehumidification) or by desorption (infusion of food in a solvent of higher osmotic pressure like salt and/or sugar). The processes can be accelerated by raising the temperature. Recently, humectants in the form of diols of varying carbon length (Glycerol, Sorbitol etc.) have been used to produce IMF without the removal of water. They reduce available moisture besides being antimicrobials.

Most of the bacteria, many molds and yeasts cease to grow and/or produce toxin in IMF. However, lipid oxidation and Maillard reaction, particularly the latter, are triggered. Vacuum packing, lower storage temperature (although IMF can be stored even at room temperature) and addition of anti-oxidants have been successful in containing oxidation. Addition of sulphur dioxide or sulfites has been popularly used to contain Maillard reaction. Minimizing the initial load of microbes is important in achieving longer shelf life.

3. By direct microbial inhibition

3.1. Ionising radiation

Only high energy cathode rays or soft *x*-rays from generators and γ-rays from radioactive sources like Co^{60} are useful. However, the energy level of the rays should not exceed 9 Mega-electron volt (MeV); otherwise there will be induced radioactivity in certain components of meat. 1 electron volt (1 eV) is defined as the energy gained on an electron in moving through 1 volt and 1 MeV = 1 million eV. Cathode rays are useful to treat the surface whereas, the *x*-rays and γ-rays can penetrate into the depth of the tissue. It is advisable to deliver the rays at a higher rate to minimize the chemical change in the finished product.

3.1.1. Dosage assessment

Dose of radiation is assessed, similar to Pasteurization and canning by

3.1.1.1. Radiation death time (D)

D value is defined as the dose required to kill a known bacteria *Clostridium botulinum* (in case of meat). 4.5 to 4.8 m rads is the dose required for *Clostridium botulinum.* Rad is the unit equivalent to absorption of 100 ergs/g of food and 1 m rad = 1 million rads

3.1.1.2. DM value

This is the radiation dose which reduces 90% of microbial population in meat.

3.1.1.3. DF value

This is the radiation dose which causes 90% reduction in enzyme activity; usually ranges between 5 and 70 mrads.

Depending on DM and DF values, commercial radiation is categorized as follows:

1. Radappertization (Pasteurization) – is the dose sufficient to cause commercial stability
2. Radacidation – is the dose sufficient to kill the viable, specific non-spore-forming pathogenic organisms and
3. Radonization – is the dose sufficient to enhance keeping quality by causing a substantial reduction in the number of viable specific spoilage organisms.

3.1.2. Advantages

Ionizing radiation (IR) effectively inactivates bacteria with the minimum chemical changes in the product. It can also penetrate considerable thickness of the food even with metal packing. The ionizing radiations produce ions and thereby the chemically excited molecules in the food interact among themselves in unusual ways forming free-radicals, polymers and, in presence of oxygen, peroxides. This inhibits microbial growth; and similar changes in microbes result in their death. Irradiation of meat in wet state can cause tenderization due to the shrinkage of collagen.

3.1.3. Disadvantages

Peroxide formation from lipids resulting in "wet-dog" odor, destruction of Vitamins K and B_1; Vitamin B_{12} is most radio-sensitive, loss of juiciness and induced radioactivity. However, when proper control of dose, rate and duration of radiation is ensured, the advantages outweigh the disadvantages.

3.2. Antibiotics

Usually, broad-spectrum antibiotics are used. Suitability of an antibiotic depends upon the type of spoilage organism and stability, solubility and toxicity of the antibiotic. Pre-slaughter infusion of the antibiotic is the best method although it is not always practicable. Otherwise, perfusion, spraying or dipping of carcasses in antibiotic solutions or impregnation of packing material with antibiotics can be practiced. However, due to severe restrictions on the traces of antibiotics in human foods as well as the cost and availability of antibiotics has made it impracticable to use them as meat preservatives.

3.3. Chemical preservatives

Chemical such as polyphosphates, nicotinamide, ascorbic acid, sulphur dioxide, propionic acid, benzoic acid and others can be used to increase shelf life and/or eating quality of meat. Since many other more efficient methods of meat preservation are available, chemicals are not commonly used to preserve meat.

Chapter **19**

Egg Products

The term egg products refer to processed and convenience forms of eggs for commercial, foodservice, and home use. These products can be classified as refrigerated liquid, frozen, dried, and specialty products. For many years, eggs were marketed primarily as shell eggs, but in recent years egg consumption in the form of egg products has increased.

Consumption of egg products in 1982 was 12% of the total eggs eaten or 0.86 billion pounds of liquid eggs. By 1997, the numbers increased to about 28.8% of the total egg production, or about 52.7 million cases of shell eggs. Today, the production of frozen eggs has leveled out, some growth is noted in dried egg production, and production of refrigerated liquid eggs has greatly increased.

Many new convenience forms of egg products are reaching the marketplace, both in the home and through foodservice and commercially processed foods. In fact, tremendous growth of the use of egg products has occurred in the foodservice industry, particularly in breakfast menu items, and in the utilization of hard-cooked eggs on salad bars.

Because they provide certain desirable functional attributes, eggs and egg products are widely used as ingredients in many food products.

Fueled by increasing consumer demand for more convenience food products, growth of the egg products industry is expected to continue.

1. Commercial egg products

These products include liquid, frozen and dried whole egg magma, albumen and yolk. Eggs used for preparing these products are washed, if they were not washed earlier, and given a rinse in water containing 100 to 200 ppm of available chlorine to minimize microbial contamination from shells. The eggs are broken and the contents are collected as whole egg, albumen or yolk, as the case may be. Subsequently, they are blended (homogenized), pasteurized, cooled and packaged.

1.1. Products

1.1.1. Refrigerated liquid egg products

These are widely used, especially by the foodservice industry and the commercial food industry. They are available in polyethylene coated fiber or laminated foil and paper cartons, and hermetically sealed polyethylene bags. These containers range in size from bags containing a few ounces to cartons (250g to 2.5 kg) and lacquer coated tins and plastic pails up to 18 kg. They are available as:

1. Egg whites
2. Egg yolks
3. Various blends of yolk and white

1.1.2. Frozen egg products

These are used as ingredients by food processors. Products containing egg yolk usually have salt, sugar or corn syrup added to prevent gelation or increased viscosity during freezing. They are packed in 13.5kg containers and in 1.8, 2.7, 3.6, and 4.5kg pouches or waxed or plastic cartons. The following products are popular in the US.

1. Egg white
2. Egg yolks
3. Salted yolks
4. Whole eggs
5. Salted whole egg
6. Sugared yolks
7. Whole eggs with com syrup
8. Whole eggs with added yolk (fortified)
9. Whole eggs with yolks and corn syrup
10. Yolk and white blends with or without sweeteners or salts

1.1.3. Dried Egg Products

These are used primarily as ingredients in the food industry. They are not commonly sold directly to consumers. For foodservice use, they are generally sold in various packages. The popular products under this category are:

1. Spray-dried egg white solids
2. Instant egg white solids
3. Whole egg or yolk solids
4. Stabilized (glucose-free) whole egg yolk solids
5. Free-flowing whole egg or yolk solids (with sodiumsilicoaluminate added as a free flow agent)
6. Blends of whole egg and/or yolk with carbohydrates (sugar or corn syrup) added
7. Pan Dried Albumen (used by the confectionery industry)

1.1.4. Specialty Egg Products

These are marketed to institutional and consumer users.

1. Chopped hard-cooked, peeled eggs, cryogenically frozen and used by salad bars in restaurants
2. Whole hard-cooked, peeled eggs, plain or pickled usually packed in a citric acid solution with preservative or a pickling solution
3. Frozen hard-cooked egg rolls or long eggs — albumen is cooked around a center core of egg yolk. Cryogenically frozen and used, sliced, for salads
4. Frozen quiche mixes
5. Frozen scrambled egg mix in boilable pouch
6. Scrambled egg mixes (frozen, refrigerated liquid, or dried)
7. Freeze-dried scrambled eggs (for campers)
8. Freeze-dried precooked scrambled eggs
9. Egg substitutes are refrigerated liquid or frozen egg products formulated as substitutes for whole eggs. Such products contain only egg white. The yolk is replaced with other ingredients such as non-fat dried milk, vegetable oils, emulsifiers, stabilizers, anti-oxidants, gums, artificial color, minerals and vitamins.

1.1.5. Frozen precooked products:

1. Egg patties
2. Fried eggs
3. Crepes
4. Scrambled eggs
5. Egg pizza
6. Omelettes
7. French toast
8. Quiches
9. Egg breakfast sandwiches

1.2. PROCESSING

Due to developments in processing technology, today's egg products are highly superior to those 35 to 40 years ago.

1.2.1. Holding

Eggs for processing must be of good quality to avoid excessive yolk breakage in egg breaking machines. Eggs to be processed are usually held in refrigerated storage no longer than 7 to 10 days.

1.2.2. Breaking

The initial step in making egg products is breaking the eggs and separating the yolks, whites, and shells. Eggs are processed by completely automated equipment which removes eggs from filler flats, washes and sanitizes the outside shells, and breaks and separates the eggs into whites, yolks, and mixtures of white and yolk. In recent years, technology for automated egg breaking equipment has changed dramatically, allowing as many as 180,000 eggs per hour (50 eggs per second) to be broken.

The liquid egg product is filtered, mixed, and chilled before entering further processing stages.

1.2.3. Pasteurization

Various times and temperatures are used for effective pasteurization, depending on the product (Table 19.1). The Pasteurization time-temperatures combinations are developed depending on Thermal death curves which reflect the resistance of microorganisms. These curves are regression lines and indicated time required to destroy 90% of viable cells in a product into space by temperature (D values) or the time required to kill all test organisms inoculated into a product (F values). Higher the Z value, greater is the heat resistance and it has a direct relationship with water activity. The values obtained theoretically will be confirmed by experimentation for each product. All egg products are monitored for pathogenic organisms. Pasteurized liquid egg products routinely contain less than 1000 organisms per gram. Salmonella tests are run regularly by the egg products industry and USDA. Only salmonella-negative products can be sold.

1.2.3.1. Yolk and whole egg

These are usually pasteurized in their liquid form. Liquidegg white is pasteurized when sold as a liquid or frozen product. Dehydrated egg white with glucose removed is normally pasteurized in a hot room at 54°C for 7 days.

1.2.3.2. Refrigerated liquid egg products

These are transported directly from the breaking plant to the user in insulated thermal tank trucks or in portable refrigerated vats. For long hauls, mechanical refrigeration, liquid carbon dioxide, or liquid nitrogen cooling systems may be used.

Liquid whole egg and yolk must be maintained below 4°C and egg white below 7°C. They should be used on a first in, first out, basis.

1.2.3.3. Frozen Egg Products

These are produced by filling a container with pasteurized chilled liquid egg and freezing in a blast freezer at a temperature of -23°C. When thawed, frozen whole egg becomes quite fluid and easy to handle, but frozen raw yolk has a gelatinized consistency. However, when yolk is blended with sugar, corn syrup, or salt (usually at levels of 2 to 10%) before freezing, the product will become fluid when thawed.

1.2.3.4. Dried Egg Products

Usually, these are produced by spray drying although some egg white is dried on trays to produce a flake or granular form. Before egg white is dried, glucose is removed. This produces dried egg white products with excellent storage stability. Whipping aids may be added to produce dried egg white products with good whipping properties. Non-reducing carbohydrates such as glucose-free corn syrup and sucrose are added to some products to preserve their whipping properties and to improve their storage stability.

Table 19.1 Minimum Pasteurization requirement of liquid egg products in the US

Liquid egg products	Minimum temperature (°C)	minimum time (min)
Albumen (without chemicals)	56.7 55.6	3.5 6.2
Whole egg	60.0	3.5
Salted whole egg (≥2% salt)	63.3 62.2	3.5 6.2
Sugared whole egg (2-12%)	61.1 60.0	3.5 6.2
Plain yolk	61.1 60.0	3.5 6.2
Sugared yolk (≥2%)	63.3 62.2	3.5 6.2
Salted yolk (2-12%)	63.3 62.2	3.5 6.2

Source : Stadelman *et al.*, 1988

1.2.4. Storage and handling

Proper storage and handling is critical for all egg products to prevent bacterial contamination.

1.2.4.1. Frozen egg products

These have a long shelf life when kept at less than -12°C. They should be thawed under refrigeration or under cold running water in unopened containers.

1.2.4.2. Refrigerated liquid products

These can be kept at recommended temperatures, unopened, for 2 to 6d, depending on the microbial quality of the product. Refrigerated liquid egg products with extended shelf life should be stored according to the processor's recommendations.

1.2.4.3. Egg white solids

As long as these are kept dry, are stable during storage even at room temperature. Spray dried egg white with glucose removed has an almost infinite shelf life.

1.2.4.4. Dried whole egg and yolk solids

These should be kept cool, less than 10°C, to maintain quality. Once containers of egg solids have been opened, they should be resealed tightly to prevent contamination and absorption of moisture. If dried eggs are combined with dry ingredients and held for storage, they should be sealed tightly in a closed container and stored in the refrigerator at 0° to 10°C. Reconstituted eggs should be used immediately. Preparation of egg powder is discussed separately under "preservation of eggs".

1.2.4.5. Plain non-stabilized whole egg solids

These have a shelf life of about one month at room temperature and about a year at refrigerated temperatures.

1.2.4.6. Plain stabilized whole egg solids

The shelf life increases to one year at room temperature.

1.2.4.7. Non-stabilized egg yolk solids

These have a shelf life of up to one year at room temperature and more than a year at refrigerated temperatures.

1.2.4.8. Stabilized egg yolk solids

These have a shelf life of about eight months at room temperature and over a year at refrigerated temperatures.

1.3. Functional properties

Because eggs are poly-functional, they provide many desirable attributes as a food ingredient.

Most uses of eggs as a food are associated with four functions — foam formation, coagulation, emulsification, and nutrition.

1.3.1. Foam Formation

Foaming is the incorporation of air into a food product, usually by whipping. This is often done to produce a leavening action. While many food ingredients form foams, eggs and egg products are especially good foaming agents because they produce a large foam volume which is relatively stable for cooking, and they coagulate on heating to maintain a stable foam structure.

In the preparation of angel food cake, for instance, the foaming attributes of the egg white proteins become very important. The small air globules are stabilized by the egg white proteins surrounding them. During heating, the proteins coagulate, further stabilizing the established foam.

1.3.2. Coagulation

Coagulation of egg protein is the conversion of the liquid egg to a solid or semisolid state, usually accomplished by heating. Coagulation is important in many food formulations such as custards, cakes, soufflés, and pie fillings. In many food products, coagulated egg protein binds together other ingredients such as those in meat loaves and patties. This property of the egg is difficult to duplicate with any other food ingredient.

Egg protein coagulates over a wide temperature range. The temperature of coagulation is influenced by pH, salts, other ingredients, and duration of heating. Egg white coagulates at lower temperatures (62° to 65°), egg yolk at higher temperatures (65° to 70°), and whole egg at intermediate temperatures. The addition of salt or sugar raises the coagulation point.

1.3.3. Emulsification

Emulsification is the stabilization of the suspension of one liquid in another. Egg yolk or products containing egg yolk are excellent food emulsifiers. In mayonnaise, for example, egg yolk acts as an emulsifier to keep oil suspended in vinegar. Phospholipids and certain proteins contribute to the emulsifying properties of whole eggs and yolks.

1.3.4. pH Factor

The pH of shell eggs and egg products is an important factor in quality retention, processing, and performance.

Initially, an egg contains about 30 ml of dissolved carbon dioxide (CO_2), all in the white. It exists primarily in the carbonate form. The pH of egg white is initially about 8.2 to 8.4, but as the egg ages and loses carbon dioxide through the shell, the pH eventually increases to about 9.4. Egg white is one of a few alkaline food products. Depending on the age of the egg, whole egg pH will range between 7.0 and 7.6. Egg yolk does not contain CO_2 and has a constant pH of about 6.0 which does not normally change much during storage. Probably the loss of CO_2 accounts for the increase in pH of many egg products during cooking or dehydration. Storage at refrigerated temperatures greatly slows down pH changes and minimizes quality breakdown.

1.4. Advantages of commercial egg products

1.4.1. Quality

Most egg products are virtually indistinguishable from fresh eggs in nutritional value, flavor, and most functional properties. These qualities are well retained during proper storage.

1.4.2. Safety

Egg products are pasteurized to destroy *Salmonella* and other bacteria.

1.4.3. Economy

Reduced handling cost, minimal shipping cost, and elimination of breakage result in reduced-cost formulations.

1.4.4. Minimal Storage Space

A 45 kg drum of dried egg white solids is equivalent to the whites from about 28 cases (360 large eggs per case) of shell eggs. 45 kg of dried whole egg solids are equivalent to about 10 cases of large shell eggs. A 13.5 kg can of frozen eggs is equivalent to about 22 dozen large shell eggs.

1.4.5. Stability

When properly stored according to their type, egg products will keep their quality over several months.

1.4.6. Uniformity

Egg products can be produced to definite specifications to assure the same performance in formulations time after time.

1.4.7. Convenience

Bulk quantities may be ordered and ingredients weighed and incorporated into formulas with less labor. Equipment needs are minimal, cleanup is simplified, and, except for packaging, there is no waste for disposal Doneness.

2. Separating egg white and yolk

It is often necessary to separate egg whites and yolks for many recipes. Beaten egg whites are used in many baked items and desserts such as meringues, cakes, and soufflés, providing air and volume to the dish. Yolks are required for sauces such as hollandaise and mayonnaise and for sweet items such as butter cream frosting and custards. It is important to describe proper methods for separating eggs successfully.

2.1. Traditional method

The traditional method for separating eggs is to break the egg over a bowl, splitting the shell into halves, and then passing the contents of the egg from one half of the shell to the other half. This allows the white to fall into the bowl as the yolk is transferred back and forth between the shell halves.

2.1.1. Procedure

1. It is best to use three bowls for this method: one to catch the white as the egg is being separated, one to store the separated whites, and one to store the separated yolks.
2. As the egg is being passed between the shell halves, make sure the yolk does not break and spill into the bowl containing the white.
3. Once all of the white has fallen into the bowl, the yolk can be placed into a different bowl to store the yolks.
4. It is also important to transfer the white to a different bowl before separating the next egg.
5. When separating several eggs, care should be taken not to drop any yolk into the bowl of already separated whites. If this happens, it can be difficult to remove the portion of yolk from the bowl of whites. In addition, most recipes requiring beaten egg whites will be negatively affected by the presence of any yolk.

2.1.2. Shortcomings

One negative aspect of the traditional egg separating method is the possibility that the contents of the egg may become contaminated during the procedure. When the egg is passed back and forth between the shell halves, it may become contaminated if bacteria are present on the shell. Bacteria may be present on the shell even after it is sanitized and the shell may also become contaminated from other food sources that it may come in contact with. In addition, albumen can come in contact with the operator's hands, further increasing the chances of contamination.

2.2. Needle method

A second method that may be used for separating the white and the yolk is to insert a needle into one end of the egg, creating a hole that can be enlarged by moving the needle in a circular direction. The egg white should drain through the hole, leaving the yolk behind. The egg shell can then be cracked open to remove the unbroken yolk.

2.2.1. Shortcomings

While this method may be easy to do, it can take awhile for the egg white to drain from the shell. There is also a slight possibility that this method may transfer germs existing on the shell to the contents of the egg.

2.3. Funnel method

A third method of separating the white and the yolk is the funnel method. Place a small funnel, with an opening small enough so that the yolk does not slide through but yet large enough so that the whites can slide through easily, over a container and crack the egg over the funnel. Care has to be exercised not to break the yolk in the process. The egg white slides through the funnel opening into the container, leaving the yolk behind. The yolk and the whites are then transferred to other containers. This method is more sanitary because the contents of the egg have very little contact with the outside of the shell.

2.4. Separator tool method

Perhaps the easiest and most sanitary egg separation method is with the use of an egg separating tool. It is basically a tray which is centered in a circular frame with slots around the perimeter. When an egg is broken over the separator, yolk slides into the center tray while white falls through the slots in the frame and into a container placed beneath the separator. The tools come in different styles, are usually constructed of plastic or metal, and are inexpensive.

3. Beating egg white and yolk

3.1. Egg white

Egg whites that are correctly beaten may increase in volume by up to 8 times. They should be extremely smooth and firm, but not dry, forming stiff peaks. A copper bowl is often used because slight acidity of the copper results in a chemical reaction with the egg whites which helps to stabilize them. If copper is not available, the next best choice is stainless steel. A pinch of cream of tartar per egg white can be added as a stabilizer, replacing the acidic properties of the copper. A balloon whip or large wire whisk may be used to beat the egg whites by hand. It is extremely important that the bowl and whisk be very clean and dry and that no trace of oil is present. Egg whites will not increase to the desired volume if contaminated with any trace of oil. This is also true if any yolk is present in the egg whites. Plastic bowls and utensils should never be used because plastic tends to hold some oil even after thorough cleaning.

The beaten egg whites should be folded immediately into any other ingredients used in the selected recipe. It is a good idea to have the other ingredients prepared first and then beat the egg whites as the final step so that they will not have time to break down and lose volume.

Whole eggs may be separated into whites and yolks immediately after removal from the refrigerator, but the whites should be allowed to reach room temperature before beating because this allows the egg whites to increase in volume more rapidly when beaten. However, the egg whites should not remain without refrigeration for more than 2 hours, which includes the preparation time, in order to reduce the possibility of bacterial growth. Egg whites may also be warmed more quickly by placing the bowl of whites over warm water.

3.1.1. Hand-beating

3.1.1.1. Procedure

1. Add a pinch of cream of tartar per egg white or about ¼ teaspoon for every 4 egg whites. This will help stabilize the egg whites and prevent them from losing their volume after the beating has stopped. Cream of tartar does not need to be added if a copper bowl is used.
2. Start slowly, using a circular motion at about 2 strokes per second. When the egg whites begin to foam, usually after 30 seconds or so, increase the speed to about 4 strokes per second. The idea is to get as much air into the egg whites as possible and to keep them in constant motion.
3. After 2 to 3 minutes of vigorous beating, the egg whites will increase in volume.
4. After another 2 minutes or so, the egg whites should reach their maximum volume.

3.1.2. Beating with an electric mixer

A hand held electric mixer also works well for beating egg whites, because it offers freedom of moving the spinning beaters all around the bowl, keeping the egg whites in motion. A stationary mixer, in which the beater spins as well as rotates around the circumference of the bowl, is an excellent tool and eliminates all of the manual work. Kitchen tools such as blendersand food processors should not be used for beating egg whites.

3.1.2.1. Procedure for blending with a mixer

1. When using any electric mixer, start at a slow speed and continue for about 60 seconds. As with manual beating, cream of tartar is added to the egg whites for stabilization of the beaten whites.
2. Gradually increase the speed until the egg whites reach their full volume. Operating the mixer at a high speed from the start will not allow the egg whites to reach their full volume.
3. The entire beating process usually takes no more than 3 minutes with better quality electric mixers.

3.1.3. Testing the beating process

Beaten egg whites can be tested for the proper volume, regardless if they are manually beaten or beaten with a machine. Pull some from the bowl on the end of a whisk or spoon to see if the egg whites form soft peaks. egg whites will not fall off when whisk or spoon is held upside down. If the egg whites fall off, continue beating. Don't over-beat the egg whites because they will begin to break down and will not blend properly with other ingredients.

3.2. Egg yolk

3.2.1. Ribbon Stage

Egg yolks that will be used for the preparation of desserts are often required to be blended with sugar and beaten until the mixture reaches the "ribbon" stage. This helps prevent the yolks from becoming granular when heat is applied. When some of the egg yolk and sugar mixture is lifted with a beater or spoon and falls back into the bowl forming a slowly disappearing ribbon on the surface, the ribbon stage has been reached. The egg yolks and sugar can be beaten by hand using a whisk or they can be beaten using an electric mixer.

3.2.1.1. Procedure

1. Add the required number of egg yolks (depending on the recipe) to a stainless steel or glass bowl and begin to lightly beat the yolks.
2. Add a small quantity of the sugar (the quantity depends on the recipe and the number of yolks used) and continue beating, but more vigorously. While beating the mixture, gradually add the remaining sugar. The mixture will become thicker and the color will lighten.
3. Continue beating until the ribbon stage is reached (approximately 3 minutes from start to finish when using a whisk), but do not over-beat or the egg yolks may become granular. The mixture will be thick and the color will be pale yellow. The mixture will form a ribbon on the surface when some of it falls onto the surface from a whisk, a spoon, or electric beaters.

3.2.2. Cooking and Beating

For some recipes, egg yolks may need to be cooked as they are beaten. This is true of hollandaise sauce, which is an egg yolk and butter sauce flavored with lemon juice and pepper. When preparing hollandaise sauce, make sure that everything needed are within arms reach because once the process begins, the beating of the egg yolks cannot stop.

3.2.2.1. Procedure

1. Place the egg yolks in a small saucepan (which is not heated) and add one tablespoon of cold water per yolk.
2. Whisk the egg yolks and water (off the heat) for about 45 to 60 seconds and then place the pan over low heat and continue to whisk rapidly. The yolks must heat slowly or they may become granular. If the heat is too high, the yolks will scramble. It is important to keep the egg yolks in constant motion.
3. A sink or large bowl of ice water should be nearby so that if the egg yolks start to become too warm, the pan can be plunged into the cold water to stop the cooking. When the egg yolks are smooth and have increased in volume, remove the pan from the heat. You should be able to briefly see the bottom of the pan between strokes of the whisk.

4. While whisking continuously, slowly add clarified butter or softened whole butter. (Make sure the clarified butter is not too warm.) Whisk in small quantities of butter at a time. The sauce will not thicken if too much butter is added right away. One large egg yolk is able to absorb as much as 85 to 90g of butter, but it is safer to use less than this (60g per yolk is suitable) to make sure that the yolks absorb all of the butter.
5. Using clarified butter makes the hollandaise sauce thicker whereas using whole butter, which contains milk solids and water, makes a thinner sauce. A thin sauce is appropriate for use on light recipes of fish and seafood or asparagus and a thick sauce is best for serving on eggs Benedict or steak.
6. To finish the sauce, slowly add some lemon juice while whisking and then add some salt and white or black pepper to taste.

There are several variations of hollandaise sauce that can be prepared.

All of the variations are made exactly like hollandaise except that the flavorings used to finish the sauces are different for each one.

1. Béarnaise Sauce: flavored with tarragon, shallots, pepper, and wine.
2. Choron Sauce: flavored the same as béarnaise with the addition of tomato paste.
3. Colbert Sauce: the same as béarnaise except it is also flavored with meat glaze.
4. Chantilly Sauce: whipped cream folded into the hollandaise sauce.
5. Vin Blanc Sauce: hollandaise sauce flavored with white wine fish stock.
6. Maltaise Sauce: hollandaise flavored with orange juice (in addition to the lemon juice).

4. Manufactured egg products

There are innumerable methods to use eggs in kitchen bound only by imagination. More than 1000 methods are given in literature exclusively meant for cooking of eggs emphasizing the fact that egg is one of the most versatile ingredients for culinary manipulations. However, all the methods of cooking eggs cannot be exhausted and hence, most commonly employed methods are described below:

4.1. Cooked egg

4.1.1. With shell (Boiled egg)

Using the term "boiled" when referring to cooking eggs in the shell can be misleading, because eggs referred to as "hard-boiled" or "soft-boiled" should never be cooked at a full boil for the entire length of the cooking time. Eggs cooked in the shell with heat that is too high or with a cooking time that is too lengthy, will become tough and rubbery and a dark line may form between the yolk and the white. The following steps can be used for cooking eggs in the shell:

1. Pierce the large end of the eggs with a pin or needle. This pierces the air cell, allowing the air to escape, preventing a flat spot from being formed on the large end of the egg during the cooking process. It also helps in making the eggs easier to peel after cooking.
2. Pour cold water into a saucepan and add 1½ teaspoons of salt per quart of water. (The salt may help make the peeling process easier). Make sure there is enough water in the pan so that the eggs will be completely covered. Bring the water to a boil and with a large spoon, place the eggs in a single layer on the bottom of the pan.
3. When the water returns to a boil, turn down the heat so that the water is at a low simmer and then begin timing the eggs for the desired doneness. Do not cover the pan.
4. After the required time has elapsed, run cold water over the eggs to stop the cooking process. This will help prevent discoloration of the yolk and will also assist with the peeling process. Running cold water over the eggs creates steam between the egg white and the shell which makes the shell easier to remove.
5. To peel the egg, simply roll it lightly over a hard surface which will crack the shell, making it easy to remove. It is also worth noting that very fresh eggs are more difficult to peel than older eggs.

Table 19.2 Cooking time *Vs* degree of doneness

Size	Degree of Doneness	Time Required (min)*
Medium	Soft-cooked yolk	4
	Medium-cooked yolk	6
	Hard-cooked yolk	11
Large	Soft-cooked yolk	5
	Medium-cooked yolk	7
	Hard-cooked yolk	12
Extra Large	Soft-cooked yolk	6
	Medium-cooked yolk	8
	Hard-cooked yolk	13
* For eggs taken directly from refrigerator		

Use the appropriate cooking times (Table 19.2) as a guide for the desired firmness for the yolk of each egg size (the whites will be firm). Temperature of the egg at the start of the cooking process will affect the cooking time. An egg that is at room temperature at the start of the cooking process will require about 1 minute less cooking time for each time listed above.

American egg board (1981) has recommended that eggs be placed in water at about 40°C and then the water temperature be brought to simmering boil (about

90°C) at which point the heater is put off. The eggs are left in hot water for 15 min. However, when temperature regulation is not practicable at consumer level, the eggs can be put into water at simmering point (about 80°C when small bubbles appear at the bottom of the vessel) and continue to heat water till boiling after which heater can be put off. Eggs are left in hot water for 12 min (full-boiling) or 6 min (half-boiling).

Eggs should not be placed in cold water and then boiled. This is because, upon heating, water gets heated first and hot water along with contaminants, especially from surface of shell, enters into the egg. In addition, exact duration of cooking cannot be judged.

Soon after the requisite cooking time is completed, eggs have to be cooled quickly in cold potable water. This will ensure shrinkage of cooked interior portion from shell facilitating easy removal of shell. Otherwise, the coagulated albumen upon expansion during cooking adheres tightly to shell and part of it comes out along with shell during peeling (which is referred to as "peeling effect"). The causes loss of albumen and also gives a very disagreeable look to the product.

In case of very fresh eggs, space to accommodate expansion of contents is so less (because of smaller air-cell size) that itis almost impossible to avoid peeling effect. It is also found that pH of albumen should be about 8.9 to avoid peeling effect. This pH can be expected after 1-7 d of holding. During the holding period, care has to be taken to keep the relative humidity high to minimize desiccation. In other words, eggs have to be "aged" before cooking. Ageing of eggs can be done they constantly a flow at 20°C for 16h.

Overcooking results in greenish color on the surface of the yolk due to formation of ferrous sulfide (iron from yolk and hydrogen sulfide from albumen). This can be overcome by quick cooling soon after cooking.

Hard-cooked eggs can be served whole or they may be cut and sliced. An egg slicer can be used to prepare evenly sliced sections that can be used in salads or as a garnish.

Hard-cooked peeled eggs can be packaged in preservative solution containing citric acid, sorbic acid, NaCl and polyphosphates and held at 5°C or less; under this method, cooked eggs may be preserved up to six weeks or more. Hard-cooked peeled eggs can be frozen, but very fast, by using cryogenic temperatures; otherwise, albumen separates into layers with liquid trapped in between.

4.1.2. Without shell

The best way to guard against the spread of bacteria and food-borne illness is to cook eggs thoroughly. Cooking eggs thoroughly does not mean that the eggs should be overcooked, which can make them tough and rubbery. It simply means the eggs should be cooked to a temperature that kills any bacteria that may be present. Most harmful bacteria cannot survive in a temperature of 71.1°C or greater. In fact, salmonella is killed instantly when subjected to a temperature of 71.1°C. An egg (white and yolk) requires a temperature of up to 70°C before it sets properly. The white alone requires a somewhat lower temperature before it coagulates, usually in the 60° to 65.6°C range. These temperatures are only slightly less than what is required to destroy all of the harmful bacteria that may be present, so heating eggs to 71.1°C should not cause eggs to be overcooked, unless they are held at that temperature (or higher) for an extended period.

Cooking eggs slowly with heat that is not too high should destroy harmful bacteria as well as allow for proper doneness. There are exceptions to this, such as when cooking a plain omelette. The eggs are cooked very quickly, but the heat is also much higher, which takes care of any possible bacterial contamination. Baked egg dishes can be checked using a kitchen thermometer placed in the center of the dish. The thermometer should register 160°F to ensure proper doneness.

Eggs cooked in the microwave may not cook evenly, so it is important to rotate the dish several times during the cooking process. Covering the dish during microwave cooking is recommended because steam is trapped inside the dish, which helps to cook the eggs more quickly and evenly and keeps them from drying out.

4.2. Deviled / Stuffed egg

Hard-cooked eggs can also be cut in half the long way so that the solid yolk can be removed from the two halves. The yolk is then blended with other ingredients such as mayonnaise, mustard, and seasonings and the mixture is stuffed back into the yolk cavity of the egg halves and served. This is known as stuffed eggs or deviled eggs.

4.3. Pickled eggs

Small eggs are preferred. The hard-cooked peeled eggs are placed in pickling solution consisting of vinegar or 5% food-grade acetic acid with other ingredients (spices, salt, garlic etc.) to taste. Pickling solution penetrates albumen in 1 d and yolk in about 6 d. Eggs in pickling solution remain edible for months at room temperature. Red beet juice in the pickling solution can give attractive red color. For good results, even hot pickling solution can be for onto the hard-cooked peeled eggs.

4.4. Scotch eggs

Scotch eggs are hard-cooked peeled eggs wrapped in layers of sausage and deep fat-fried. Yolk should be at the centre of the boiled eggs; otherwise, the sausage will not and properly to the surface. Recipe for sausage mix is dependent on local preferences; but, usually contain Soya protein, ground meat and seasonings. Hard-cooked peeled eggs are kept moist, then rolled in wheat flour and wrapped into formed sausage. Finally, they are again rolled in breading mix (consisting of cornflakes) and then subjected to deep-fat frying in oil at 177°C for 7 min. Such eggs have a shelf-life of 30d at 4°C.

4.5. Poaching

4.5.1. Stovetop

A poached egg cooked on the stovetop is one that is cooked in simmering water without the shell. Unlike a boiled or coddled egg that benefits from the use of an older egg, a poached egg is best when a very fresh egg is used. This is because the fresh egg, when placed into the heated water, will not spread out like an older egg, yielding better results with the shape and texture of the egg. If an older egg must be used, it can be simmered in the shell for a few seconds so that the white is just slightly congealed. When the egg is broken into the simmering water, it will not spread out as much. One tablespoon of vinegar added to the water will also help with coagulating the white to keep it from spreading too much.

The following steps can be used for poaching eggs:

1. Pierce the large end of the eggs with a pin.
2. Add cold water to a large saucepan and bring the water to a boil. With a slotted spoon, lower the eggs into the boiling water for no more than 10 seconds. This helps to slightly set the whites so that they do not spread out as much when the eggs are poached.
3. Add 1 tablespoon of vinegar to the boiling water and then turn down the heat so that the water is simmering.
4. It is best to open the egg shell as close to the surface of the simmering water as possible so that the egg will spread out less.
5. The poaching process should be timed for about 4 minutes. Remove the eggs from the water with a large slotted spoon in the order in which they entered the water. This allows the last egg to be cooked the same amount of time as the first. Wait a few seconds between the removals of each egg. Each egg can be bathed in warm water to remove any trace of the vinegar. The egg white should be firm, but the yolk should remain liquid. If there are any streamers of egg white extending out from the main portion of the egg, they can be trimmed off before serving.

4.5.2. Metal egg poaching

Another method for poaching eggs involves a metal egg poaching form which makes the process almost foolproof. The metal form is a shallow, oval, slotted container with an attached handle. The egg is broken into the form which is then lowered into simmering water. The metal form or poacher is convenient to use and creates a pleasing shape. Using fresh eggs or older eggs is less of a concern when using the metal poacher.

Regardless of how the eggs are poached, the water should be at a simmer and not a full boil; because boiling water can break albumen. If metal poaching tools are used, several can be placed in the pan at one time. If the forms are not used, several eggs can also be poached at one time, but more care must be taken when breaking the eggs over the water so that each egg has its own space in the pan.

Poached egg can be put in cold water to prepare cold-poached egg which can be reheated in fairly hot water (not very hot water to avoid toughening of albumen) before serving.

4.5.3. Microwave

If the egg is to be poached in a microwave, there are various poaching dishes that can be used, such as the cooking dish shown at the right.

Crack the egg into the dish and then poke the yolk and white with a toothpick several times to reduce the tendency for it to burst while cooking. The egg will also burst or explode if the cooking time or heat is too excessive. Cover the dish securely with a plastic food wrap.

Realizing that different microwave wattages and cooking at different altitudes will affect the cooking time, it will be necessary to experiment in order to achieve the best results, allowing a little longer or a little less time for the desired hardness. When cooking only one egg, as a general rule for average elevations of 1,000 feet, set the desired time around 40 sec at a medium cooking power for the first stage of cooking. A time of 40 sec for a medium size egg and 45 sec for a large egg will usually give it a soft consistency.

After cooking the egg for the first phase, allow it to remain covered in the cup, in the microwave for 15 sec or so and reset the microwave for 10 to 15 seconds at a medium power level.

Cooking the egg in two phases allows the egg to cook slowly, moving from a soft poached egg to a hard poached egg if desired by adding a few more seconds and letting it stand slightly longer either between or after the cooking phases. If the egg yolk or white explodes, reduce the time for the phase causing the reaction. Also, consider cooking the egg in three phases, each one with a decreased amount of time to succeed with a perfect result. Achieving the desired results can be accomplished by doing some testing and watching the egg during the final cooking stage.

Since all microwaves differ in cooking power and corresponding temperatures, test the results several different ways to determine which is best for the desired results or hardness.

4.6. Scrambled egg

Scrambling is a method of preparing an egg in which the white and the yolk are blended together and cooked in a sauté pan.

1. Butter is dropped in a metal or nonstick sauté pan that has been placed on moderately high heat. The butter should foam and bubble, but not turn brown.
2. Several eggs are stirred slightly in a bowl to blend the whites and the yolks. Salt and pepper can be added to the eggs or added when the eggs are served.
3. The eggs are then stirred, scraped from the bottom of the pan, and turned during the cooking process.
4. As the eggs begin to congeal, they begin to cook very rapidly. When 2 or 3 eggs are scrambled together, it usually takes a minute or less before the cooking process is complete. A larger quantity of eggs will require several more minutes.
5. It is best to remove the eggs from the heat source when they are still very moist, because the internal heat of the eggs will complete the cooking process. Eggs that become too dry indicate that they have been overcooked.

Another method of scrambling eggs is to mix eggs and milk in the ratio of 7:3. Eggs are beaten with salt and pepper lightly. Before butter begins to sizzle, the eggs are poured onto the pan and heat is kept low and using a wooden spatula, stirred slowly and continuously. After the egg is set, pan is removed from the fire and milk is added; stirred till eggs set into a soft, creamy mixture and served.

The scrambled eggs mixes can be packaged in suitable flexible packs and frozen for storage. Scrambled eggs are likely to turn greenish over storage the formation of ferrous sulfide which can be avoided by adding 0.1% of monosodium phosphate or citric acid to chelate iron.

4.7. Shirred-egg

Shirred eggs are prepared as a broiled or baked egg dish. Eggs are placed in small buttered dishes referred to as ramekins and broiled until the white is set, but the yolk remains liquid. The eggs may also be baked, but the cooking time is longer and the eggs may become tough and rubbery ifcooked too long. Variations of the basic shirred egg may include the addition of herbs, cheese, and/or cream.

4.8. Coddled egg

A coddled egg is cooked more slowly than a boiled egg, but basically yields the same results, except that the egg is a bit tenderer. The following steps can be used for coddling eggs:

1. Pierce the large end of the eggs with a pin. This pierces the air cell, allowing the air to escape, preventing a flat spot from being formed on the large end

of the egg during the cooking process. It also helps in making the eggs easier to peel after cooking.

2. Add cold water and 1½ teaspoons of salt per quart of water in a saucepan. Make sure there is enough water in the pan so that the eggs will be completely covered. Place the eggs in a single layer in the saucepan and leave the pan uncovered. Add cold water and 1½ teaspoons of salt per quart of water in a saucepan. Make sure there is enough water in the pan so that the eggs will be completely covered. Place the eggs in a single layer in the saucepan and leave the pan uncovered.
3. Place the pan on medium heat and bring the water to a simmer, but not to a full boil. Remove the pan from the heat and cover it. The length of time that the eggs remain in the covered pan determines the degree of firmness of the yolk. Soft yolk: 4 to 6 min, Medium yolk: 6 to 8 min, Hard yolk: 20 to 25 min

When coddling eggs, the size of the egg and its temperature at the start of the cooking process will have an affect on the cooking time. An extra large egg used directly from the refrigerator will require the full cooking time as stated above and a medium egg that has been brought up to room temperature before cooking will require only the minimum time listed.

To stop the cooking process, run cold water over the eggs. It is best to use older eggs for coddling because they peel easier. Soft-cooked coddled eggs are often served in an egg cup and eaten directly from the shell because they are difficult to peel.

Another method used for coddling eggs involves the use of a special porcelain dish with a screw top. The egg, without the shell, is placed in the dish, the cover is screwed on, and the dish is placed in a pan of heated water. When the cooking process is complete, the dish is removed from the water and is used to serve the egg.

4.9. Fried egg

Frying is another popular method of cooking eggs and it is easy to do. Butter or cooking fat is heated in the bottom of the pan. Whole eggs are cracked and opened over the pan. The eggs should be opened as close to the bottom of the pan as possible so that they maintain a pleasing shape and do not spread out too much. The eggs are cooked until the whites are firm and the yolk is runny or firm, depending on how they are desired.

There are several methods used to finish cooking the eggs. They can be left unturned and can be basted with the hot fat. A few drops of water can be added and the pan can be covered to steam cook the eggs. The eggs can also be finished by carefully turning them over using a spatula once they have firmed up on the bottom. The eggs are then cooked until the yolks are at a desired doneness, such as over easy (runny yolk), over medium (soft yolk), or over hard (firm yolk).

Fresh eggs with higher thick albumen are soft-fried in oil with only slight cooking of yolks. The cooking is done on a griddle inside a 10 cm diameter ring so as to have uniform-sized product. The fried eggs are cryogenically frozen and packaged. The product has to be thawed fast to avoid gelation of yolk.

For frying, the fat should be hot, not sizzling. The eggs are broken into a cup and saucer and slid into the fat (about 7.5 ml per egg); cooked for 2-3 min till the albumen is no more transparent. Otherwise, the broken contents can be slid into a pan has-filled with oil at 180°C. With a wooden spatula, the egg can be rolled till it cooks (about 30-60 sec).

4.10. Roasted egg

A roasted egg is prepared using two different cooking processes. It is first hard-cooked in simmering water and then it is placed in the oven and roasted in the shell. It is removed from the oven when the shell becomes brown. This method is used to prepare eggs traditionally served during the Jewish Passover.

It must be noted that, regardless of the cooking method, eggs should not be overcooked. Eggs cooked too long or at too high a temperature may become tough and rubbery and will not bevery appealing. Unless eggs are rapidly cooled, they continue cooking after removal from the heat source.

4.11. Egg omelette

Milk in scrambled eggs is replaced by the water in omelettes. Several varieties of omelettes are possible, limited only by imagination. However, there are two basic kinds of omelettes namely Savory omelette (made from beaten whole eggs and folded into three layers or into half) and Spanish omelette (generally unfolded, the top surface lightly browned under a grill).

A sweet omelette is usually made with separated egg – the albumen is beaten stiff to give a puffed (scuffle) omelette; it is always folded in half.

Regardless of the type of omelette, the omelette should be just set at still moist, the texture soft and creamy, with no part still distinctly liquid and raw when served. Fast cooking with good manipulations of the mixture is the secret behind good omelettes.

An omelette is usually made with 2 or 3 eggs and is cooked very quickly in a sauté pan. The bottom of the pan should be about 18-20 cm in diameter so that the eggs will be no more than 0.6 cm in height in the pan. A nonstick pan works very well. Salt is rubbed on the pan and wiped clean after each use. The salt treatment helps prevent eggs from sticking to the pan the next time it is used.

4.11.1. Procedure (General)

1. To prepare an omelette, crack open 2 or 3 eggs over a bowl, add salt and pepper, and stir the eggs just until the whites and yolks begin to blend. The eggs should not be vigorously beaten.
2. Place the sauté pan on high heat and thoroughly coat the bottom and sides with butter. After the butter is melted, it will begin to foam. The pan is hot enough to continue when the foaming stops, but the butter has not begun to brown.
3. When the eggs are poured into the pan, they should begin to coagulate almost immediately.
4. With a few side to side movements of the pan, distribute the eggs evenly.
5. After a few more seconds, the eggs should be cooked enough to begin forming the mass into an omelette shape.
6. Jerking the pan toward you should cause the omelette to roll over upon it as it hits the side of the pan. Continue doing this for few seconds until the omelette is folded into a pleasing form. The omelette should be tender and moist. A dry omelette indicates that it has been cooked too long. Tilt the omelette from the pan onto a plate and it is ready to serve. The actual cooking time is usually no more than 30 to 45 sec.

Additional ingredients, such as herbs, can be added to the eggs before they are cooked without changing the way the basic omelette is cooked. Other ingredients, such as cheese or finely chopped meats, can be added just before finishing the cooking process. The extra ingredients will probably require that a spatula be used to fold the omelette in the pan rather than trying to roll it over by jerking the pan. The extra ingredients can also be placed on top of the plain omelette when it is served.

4.11.2. Plain omelette

Egg is broken, seasoned with salt and pepper and about 7.5 ml of water is added. The mixture is beaten lightly with the fork. The mixture is cooked for about 1 min on uniformly greased hot pan. After the mixture settles, allowed for cooking for another 4-5 sec. It should be removed from the pan without breaking it and served. It can be frozen for further storage.

4.11.3. Filled omelette

Any plain omelette can be filled by cheese, cooked fish, chopped meat, tender onion, mushroom, shrimp, prawn, tomato, chicken meat, breadcrumbs etc... The cooking procedure is similar to plain omelette.

4.11.4. Spanish omelette

Chopped onion and tomatoes with diced potatoes and chopped pimentos are prepared. Onion is fried for 5 min till soft but not colored. Tomatoes, potatoes and pimentos are added to the cooked onions and cooked thoroughly. Eggs are beaten, seasoned and poured over vegetables. The resultant omelette is cooked slowly still light brown side on the underside and turned upside down to cook the other side too till light brown color.

4.12. Frittata

A Frittata is an Italian version of the French omelette. It is open-faced and is not folded over like a French omelette and the preparation is also a bit different. A French omelette is cooked very quickly on high heat and any other ingredients such as cheese or vegetables are placed on the omelette just before it is folded over. A frittata is cooked more slowly and any additional ingredients are stirred into the eggs and cooked at the same time. A frittata is substantial enough to be served as a lunch or dinner entrée.

4.13. French toast

This is prepared by soaking slices of bread in a blended liquid egg mixture similar to the formula for omelettes and then pan-frying the coated bread. Special bread with multiple small pores and yellow color is commercially used to get uniformly colored product.

4.14. Custards

These consist of cooked mixture of eggs, milk, sugar and flavorings. The custards can be frozen for storage and distribution.

4.14.1. Stirred custard

When custard mixture is cooked on a stove to creamy pourable consistency, it is called stirred custard. It is eaten as a sauce poured over cake or fruit.

4.14.2. Baked custard

When the custard mixture is cooked in a water bath in an oven to a firm gel-like consistency, it is called baked custard. It is a dessert itself but, can also be used as a base for toppings or sauces.

4.15. Others

As mentioned earlier, eggs are one of the most versatile cooking ingredients and hence varieties of preparations are possible such as roll-ups, Soufflé's, Quiches, Waffles, Long-eggs, salads, egg-drinks and various other snack-foods including egg yogurt.

Chapter **20**

Poultry Meat Products

Most commonly, freshly dressed chicken are used for cooking in our country. The other forms of fresh meat are chilled and frozen; the latter have to be thawed in a refrigerator. Poultry meat is also a very versatile material for preparation of various products. Some of the important products are discussed in this chapter.

1. Poultry meat types

1.1. Description

All poultry, is described and classified using a number of different criteria:

1.1.1. Kind

Describes whether the poultry is turkeys, or chickens, or ducks, etc.

1.1.2. Class

Describes how the poultry is categorized, such as a fryer or roaster.

1.1.3. Grade

Describes the quality of the bird based on USDA guidelines. Common grades for poultry are A, B, and C, with A being the best quality.

1.1.4. Style

Describes the bird as being whole, cut into sections, (halves, quarters, etc.), or cut into individual parts, (breast, leg, thigh, etc.).

1.1.5. Size or Weight

Turkeys are usually specified by individual weight.

1.1.6. Type

Describes whether meat source is fresh, frozen, hard-chilled, etc.

1.1.7. Packaging

Specifies how the meat is packaged and received at the site of further processing.

1.1.8. Temperature

An ideal temperature for fresh, frozen, or hard-chilled poultry meat for processing, shipping and storage is given.

1.1.9. Breed

Quality, flavor, size, and the ratio of meat to bone can be affected by the breed of Poultry. Hence, breed also is indicated under special circumstances.

1.2. Chicken

There are several varieties of chicken available to the consumer, such as regular chicken, and Poussin, also known as spring chicken etc... A Poussin is an extremely young, small chicken, and it provides a very mild flavor. Because it is so young, it has very little fat. Regular chickens are classified according to their age. Listed below are some of the common classifications.

1.2.1. Broiler-Fryer

is a chicken ranging in age from 6 to 8 weeks and weighing from 0.7 to 1.8 kg. Their meat is very tender and they can be prepared by any cooking method, such as broiling, braising, frying, roasting, and grilling.

1.2.2. Roaster

is a chicken ranging in age from 3 to 5 months and weighing from 1.5 to 3kg. Their meat is tender and more flavorful than that of broiler-fryer chickens. They make a good roasting chicken but can be prepared by other methods and are good in other dishes.

1.2.3. Stewing Chickens

are mature chicken, which is over 10 months old and weighing in the range of 2 to 3 kg. Their meat is very flavorful but tougher than that of the broiler-fryers and roasters. They are best used for stews and soups, or should be cooked slowly with a moist heat method such as simmering or braising.

1.2.4. Capons

are male chickens that have been castrated. They are generally under 8 months old and will weigh in the range of 2 to 4kg. Capon has more white meat but generally has a higher fat content. Their meat is the most flavored of all the chickens and it is very tender. However, with availability of broiler fryers and roasters, capon production has become a rare enterprise.

1.3. Chicken parts

There are an unlimited number of chicken products available, consisting of fresh, frozen, uncooked, fully cooked, and many heat-and-serve products. Many products are available unbreaded, breaded, seasoned, or marinated. Some products are made from formed chicken, which consists of chopped chicken pieces that are pressed and formed into specific shapes and used for products such as nuggets, patties, fillets, and tenders.

1.3.1. Whole Chickens

Available fresh, frozen, bone-in, boneless, uncooked, fully cooked, and seasoned.

1.3.1.1. Parts

Parts available are drumsticks, thighs, wings, and breasts. They are also available as legs (drumstick and thigh attached),leg quarters, breast quarters,breast halves, and poultry halves. Wingettes and drummettes are available from the wing. The chicken parts are available uncooked, fresh and frozen. Various parts are also available boneless, fully cooked or seasoned.

1.3.2. Breast

One of the most popular chicken parts available, the chicken breast, can be purchased in many different forms. They are available fresh and frozen in various cuts, such as whole breasts, breast quarters and breast halves. They can be found bone-in, boneless, skin-on, skinless, uncooked, fully cooked, breaded, and unbreaded. They are available seasoned with flavors such as lemon pepper, BBQ Mesquite and Italian. Chicken breasts are also used to make other chicken products.

1.3.3. Chicken wings

These are another very popular of chicken products, available in many forms. They can be found uncooked, fully cooked, breaded, and unbreaded. They are available marinated, barbecued or glazed, and seasoned with spicy and rotisserie flavorings. Buffalo wings are chicken wings served as a popular finger food which are generally available in varying degrees of hotness, from mild to fiery hot and in various flavors, such as BBQ, honey BBQ, Mexican BBQ, honey Dijon, oriental, and teriyaki. Wingettes and drummettes are also available from the wing.

1.3.4. Cutlets

are boneless chicken breasts or legs that have been pounded to tenderize and to provide a piece of meat that is more uniform in thickness, allowing the meat to cook more evenly. Cutlets are generally boneless and skinless, and are available breaded, unbreaded, uncooked, fully cooked, and seasoned with various flavors.

1.3.5. Thighs

is available fresh, frozen, bone-in, boneless, skin-on, skinless, uncooked, fully cooked, breaded, and unbreaded. They can also be found seasoned and marinated.

1.3.6. Drumstick

is the bottom portion of the leg below the knee joint and consists of all dark meat. Drumsticks can be found fresh or frozen and are generally available with the skins on.

1.3.7. Fillets

Fillets of meat sliced from the chicken breast, which are available uncooked, fully cooked, breaded, unbreaded, and seasoned with flavors, such as BBQ mesquite, lemon pepper and Italian.

1.3.8. Chicken breast cut into strips

are available uncooked, fully cooked, breaded, unbreaded, and seasoned with flavors, such as barbecue, garlic and herb, teriyaki, grilled, Southwestern, and fajita. Strips are found cut in various widths.

1.3.9. Tenders

are full pieces or chunks of chicken tenderloins, which are available uncooked, fully cooked, breaded, unbreaded, and seasoned with flavors, such as barbecue, buffalo, and sweet-and-sour.

1.3.10. Giblets

consists of the neck, liver, heart, and gizzard.

1.4. Types of Turkey

1.4.1. Classes of Turkey

1.4.1.1. Fryer / Roaster

A small turkey of 1.8 to 3.6kg is classified as a fryer-roaster and is usually no older than 4 months.

1.4.1.2. Young

A 4 to 8-month-old turkey is referred to as a young roaster, which has soft, smooth skin and tender meat.

1.4.1.3. Yearling

A 12-month-old turkey is called a yearling and the meat and skin are still reasonably tender, but not as tender as a young turkey.

1.4.1.4. Mature or Old

A mature turkey is 15 months or older and is not well suited for roasting because the meat is much tougher.

1.4.1.5. Whole turkeys

may be labeled hen or tom turkey, but this does not make a significant difference in the quality. The flavor and tenderness of the meat is determined by the age of the turkey at the time it is brought to market, rather thanthe gender. Most turkeys used for roasting are between 4 and 9 months old and range in weight from 3.5 to 10kg. Tom turkeys that are ready for market are usually 6.8kg or more, while hens usually weigh less than 6.8kg.

1.4.2. Types of Turkey

1.4.2.1. Fresh

Turkey meat begins to freeze at about -3.3°C. Therefore, any turkey labeled "fresh" in a food store has never been allowed to be cooled to a temperature lower than -3.3°C. A fresh turkey should be cooked no later than 2 days after it is purchased or by the "use by date" on the label, otherwise it must be frozen until it is ready to use.

1.4.2.2. Refrigerated

Fresh turkeys labeled "refrigerated" are chilled to -3.3°C; hence, they may seem a bit stiff on the surface, as though they have been frozen.

1.4.2.3. Frozen

Even though turkey meat begins to freeze at -3.3°C, in order to be considered "frozen", a turkey must be cooled to a temperature of -17.7°C or below. Commercial processing companies use a flash freezing process that quickly cools turkeys to -17.7°C or below. This ensures that when the turkey is defrosted, the meat will be at the same level of freshness as the day it was originally frozen. Listed below are additional types of frozen turkeys that may be available:

1.4.2.4. Prestuffed

Prestuffed frozen turkeys can be purchased with a variety of stuffing flavors and in a large selection of sizes. Prestuffing and freezing a turkey at home can be dangerous. Home freezers cannot freeze as quickly as commercial freezing processes. The stuffing increases the growth rate of harmful bacteria, which may multiply rapidly before the turkey has a chance to freeze properly.

1.4.2.5. Self-Basting

A self-basting turkey is injected with various ingredients, such as oil, juices, and seasonings, before it is frozen. When the turkey is roasted, the added ingredients keep the meat moist and tender as well as provide extra flavor.

1.4.2.6. Boneless Turkey Roast

Whole turkeys are available that have been deboned and packaged. The preparation, cooking, and carving are much more convenient. Another type of turkey roast consists of pieces of white or dark meat or both, that are removed from the bones, placed in a foil pan, and frozen to be cooked when needed.

1.4.2.7. Hard-Chilled

Turkeys that are hard-chilled are cooled to temperatures between -17.7°C and -3.3°C. They are not considered fresh or frozen and so they are not labeled as such. Since they are not considered frozen, the turkeys mustbe handled as though they are fresh to ensure proper food safety.

1.4.3. Turkey Parts

Turkey is often cut into parts, packaged, and sold fresh or frozen. Turkey breasts, drumsticks, thighs, and wings are often sold this way. The breast section of a turkey is quite large and it is often cut into smaller cutlets that are thinly sliced.

1.4.4. Ground Turkey

Turkey parts can be ground and used in the same way as beef cuts that are ground into hamburger. Ground turkey can be used to make patties, meatloaf, casseroles, and many similar types of dishes and it has much less fat than ground beef.

1.4.5. Specialty Turkeys

1.4.5.1. Smoked

Smoked turkeys are ready to eat because the smoking process cures and cooks the meat with indirect heat. They are available in a wide range of flavors depending on the type of fuel used for the smoking process.

1.4.5.2. Free-Range

A turkey known as "free-range" indicates that the bird was allowed to roam outdoors, which may have a positive affect on the flavor of the meat, especially if the roaming area was not too crowded.

1.4.5.3. Organic

An organic turkey refers to a bird that is allowed to eat only organic feed and is allowed to roam outdoors (free-range). In order to be classified as organic, the turkey can never receive any antibiotics. Growth hormones are also prohibited, which is the case for all commercially raised poultry.

1.4.5.4. Natural

This refers to turkeys that have limited processing with no artificial ingredients or coloring added. The term "natural" does indicate that a turkey is free-range. The meat is also likely to be a bit drier than other types of turkeys that have had additional processing.

1.4.5.5. Kosher

A kosher turkey must be raised and processed with strict guidelines under rabbinical supervision. The turkeys are free-range birds that are fed only grain, are never given any antibiotics, and are individually inspected. When they are processed, the turkeys are soaked in a salty brine solution to provide maximum tenderness and to give the meat a unique flavor.

1.4.5.6. Heritage Turkeys

Before turkeys were raised in large commercial sites and mass marketed, most people ate a variety of breeds raised on small farms. Today these various breeds are known collectively as "Heritage Turkeys", which are making a comeback. Heritage turkeys are free-range birds and include breeds such as Jersey Buff, Bourbon Red, Black Spanish, and Narragansett. The various breeds generally have a longer body, smaller breast muscles, and are bit leaner than commercially raised birds. Heritage turkeys require an additional 2 to 3 months to grow to the proper size for processing. This makes the turkeys more expensive than commercially raised birds, but most people feel the extra expense is justified because of the excellent flavor and the texture and tenderness of the meat. Most heritage turkeys are raised on small farms where they can be directly purchased or they can be conveniently purchased from a number of online sites and shipped directly to the consumer.

2. Thawing

2.1. Chicken

There are several methods that can be used for thawing fresh frozen chicken. Chicken should never be thawed out on the kitchen counter and one of the methods described below can be employed:

2.1.1. Refrigerator

2.1.1.1. Recommended Method

Thawing chicken in the refrigerator is the slowest but safest method. The temperature of the refrigerator should be maintained at 1.7°C to 4.4°C to discourage growth of harmful organisms as the chicken thaws. Leave the chicken wrapped and place on a platter or a tray to catch the drippings as it thaws. Thawing time: about 11h per kg.

2.1.1.2. Cold Water

Thawing the chicken in cold water is a faster method than thawing in the refrigerator, but the proper precautions must be taken when using this thawing method. Fill the sink with enough cold tap water to cover the chicken and place the wrapped chicken in water. The water should be changed every 30 minutes. Do not use warm water, even though it will thaw the chicken faster, because it will also cause the growth of bacteria. Do not use this sink for other purposes during thawing period and be sure the thawing water does not splash onto other preparation surfaces or food. Once the chicken is thawed, remove it from the sink and clean all utensils and surfaces affected during the thawing period with hot water and soap. Thawing time: 2 ½h per kg

2.1.1.3. Microwave

Thawing chicken in a microwave is a quick method but is not recommended because of the difficulty in determining the proper defrosting time. Defrosting times vary according to different microwaves and according to the form of chicken (whole or pieces) under thawing. Chicken should be loosely wrapped during microwaving. Generally it is best to start out by microwaving at a defrost or medium-low setting for 2 min and then letting the chicken stand for 2 min before checking progress. Turn the chicken and repeat this procedure if needed, being careful that the meat does not start to cook. Thawing largeitems in the microwave does not work well and should be avoided, if possible. See the equipment manual for defrosting times.

2.1.2. Other thawing guidelines

Be sure the chicken, particularly a large whole bird, is defrosted thoroughly to ensure proper cooking. Place a hand inside the cavity of the chicken to check for ice crystals. If there are any present, more thawing time is needed.

Thawed chicken should be cooked as soon as possible. If not using immediately, store in the refrigerator and use within 24 hr of thawing.

Remove the giblets from the cavity of a whole chicken as thawing allows.

It is safe to refreeze chicken once it has been cooked but some of its quality will be lost. While thawing, be sure that drippings do not contaminate other food or preparation surfaces

2.2. Thawing Turkey

2.2.1. Refrigerator Thawing

Placing a frozen turkey in the refrigerator is the only reliable and safe method for thawing and it is the only method of thawing that is recommended. The turkey should be kept in its original wrapper during the thawing process and it should be placed on a platter to catch any juices that may leak from the package. Thawing the bird in the refrigerator may require several days, especially if the turkey is a large size, but it is absolutely the safest method for thawing. The cool temperature of the refrigerator (usually no higher than 4.4°C) discourages the growth of harmful organisms in the turkey as it thaws. Up to 11 h defrosting time per kg should be allowed to properly thaw the turkey. The required time may vary because of the individual temperature settings of different refrigerators.

2.2.2. Cold Water Thawing

Thawing a frozen turkey in cold water is another defrosting method that has been used for years, but with increased awareness of illness duo to bacterial growth, it is generally not recommended. There is a good chance that bacteria may rapidly multiply in thawed areas near the surface of the bird while the interior may still be frozen.

If absolutely necessary, the following steps can be taken to make it as safe as possible:

1. As with thawing in the refrigerator, the bird should be kept in its original wrapper.

2. If the original wrapper has any punctures or tears in it, the turkey should be placed in another plastic bag and sealed.
3. The turkey should be placed breast side down and it should be completely covered with water.
4. The water must be changed every 30 min.

Cold water thawing is much faster than thawing in the refrigerator (usually 8 to 10 times faster), but it can be annoying having to change the water every 30 min, especially when thawing a large turkey. At least 1h defrosting time per kg should be allowed when using the cold water method.

Once the turkey is thawed, it should be cooked immediately.

2.2.3. Microwave Thawing

Some people have used a microwave oven for thawing meat and while this method may be risky for thawing red meat, it is downright unsafe for thawing poultry. While waiting for the bird to fully defrost, bacteria will grow quickly in areas where the turkey is already defrosted.

2.2.4. Additional safety recommendations

1. Never thaw turkey at room temperature. Harmful bacteria will grow rapidly creating toxins that may still be present after the turkey is cooked.
2. Frozen turkey that has been thawed should not be frozen again until after the meat is cooked in order to reduce the risk of harmful bacterial growth.
3. A frozen prestuffed turkey should not be thawed before it is cooked because bacteria can multiply rapidly in the stuffing while the turkey is defrosting.

3. Doneness

Doneness indicates satisfactory completion of cooking process.

3.1. Checking Doneness

Checking doneness is accomplished in basically the same manner regardless of the cooking method you are using. Some methods are more accurate than others and some are more suitable in regard to the cooking method and to the cut of chicken. When the chicken is done, it should be allowed to rest (a waiting period before carving) for 10 to 15 min. This allows the juices to be distributed through the meat before it is carved.

The length of time a chicken will have to cook depends on whether it is stuffed and/or trussed and on the quantity of other ingredients, such as potatoes and vegetables, added to the pan. The best way to determine if it has cooked long enough is to check for doneness. Chicken should not be overcooked; but, if undercooked, it will not be fully flavored and will not have the desired texture.

It also needs to be cooked to the proper doneness to make it safe to eat. Shown below are some common methods used to determine doneness.

3.1.1. Thermometer

Using a thermometer is the most accurate method for testing doneness of the chicken. A regular meat thermometer or an instant read thermometer can be used. A regular meat thermometer is inserted before placing the chicken in the oven or other heat source and remains there throughout the cooking time.

An instant read thermometer is used to check doneness once the chicken is cooked. The chicken is taken away from the heat source and the instant read thermometer is immediately inserted into the thickest part of the breast or thigh (it should not be touching a bone). The thermometer provides a temperature reading in approximately 15 sec.

Table 20.1 Checking doneness

Part	Temperature (°C)
Whole chicken – thigh area	79.4-82.2
Whole chicken – breast area	76.7-79.4
Chicken breast and wings	76.7-79.4
Chicken parts – dark meat	82.2
Ground chicken	76.7
Stuffed whole chicken	73.9
If the proper temperature is not reached, the chicken should be returned to the heat source for further cooking.	

3.1.2. Piercing

Another method for testing doneness is to prick the chicken with a fork or the tip of a knife and check to see if the juices that escape run clear. When pricked in the thigh or breast, the juices should run clear, with no pinkish coloring. Tilting a whole bird up, so the juices from the cavity run out, should also produce clear juices. If the juices have any pinkish coloring, the chicken is not done and should be returned to the heat source for further cooking.

3.1.3. Visual

The visual method of determining doneness is to make a small slit with the tip of a knife into the thickest part of the chicken and then pry the slit open. The meat should be opaque with no signs of pink coloring. If the meat shows any signs of not being done, the chicken should be returned to the heat source for further cooking. Visually check the skin, which should be golden brown and check the legs, which should move around easily in their joints.

3.2. Tenderness Tips

1. Avoid freezing whenever possible to eliminate additional moisture loss during thawing, which results in less tender meat.
2. Keep chicken from drying out in the refrigerator by keeping it tightly wrapped. If the chicken dries out, it will become tough.
3. Leaving the skin on the chicken, while cooking it, helps to hold in juices, increases tenderness.
4. To keep breast area of chicken from drying out during roasting, place a piece of foil over this area. Remove during last 30 min of roasting time to allow the skin to brown properly.
5. Cook chicken to the proper temperature, because undercooking the chicken will cause it to be tough and overcooking the chicken causes loss of moisture, making the chicken drier.
6. Let roasted chicken rest for 10 to 15 min before carving to allow juices to be distributed throughout the meat. Standing the chicken up with bottom end up allows more juices to run into the drier breast area.
7. Cutting meat across the grain will produce slices with shorter fibers, resulting in more tender pieces.
8. When adding cooked chicken to dishes that have a long cooking time, it is best to use dark meat because it will stay moist longer than white meat.

3.3. Cooking Tips

1. Roasting chicken on a rack, broiling and grilling are cooking methods that allow fat to drip away from the meat.
2. Poaching, steaming and microwaving are methods of cooking where no additional fat is used. All provide for less fat content in the meat when it is done.
3. When frying or browning chicken in a pan, use a nonstick skillet, which requires less added fat, or use a nonstick skillet with a fat free nonstick cooking spray to reduce the amount of fat used.
4. Reduce added fat by seasoning chicken in marinades that are low-fat or fat-free. Use ingredients such as low-fat yogurt, juices, wine, herbs, and spices.
5. Removing the skin before eating chicken eliminates about two thirds of the fat content.
6. When stewing chicken for soup, let broth cool and then discard fat that forms on top before reheating to serve.
7. When adding chicken to a recipe that calls for a measured amount, determine how much chicken is needed by following a standard of 1kg of boneless chicken equals approximately 7 cups of cubed chicken.

8. When roasting a chicken, a non-trussed chicken will cook faster and more evenly than a trussed chicken.
9. Covered chicken takes longer to cook in the oven than uncovered chicken.
10. When frying, grilling, broiling, or sautéing chicken, remove pieces as they get done to avoid overcooking while finishing other pieces. White meat and smaller pieces, such as breasts and wings, will get done faster than dark meat pieces, such as legs and thighs.
11. For a quick test of doneness when roasting a chicken, hold on to the leg, move it around, and side to side. The leg should move freely at the joint if it is done.
12. Do not overcrowd chicken pieces when cooking. Leaving space between them will allow them to brown and cook more evenly.
13. If using a marinade for basting, set some marinade aside before placing raw chicken in it to marinate. Never reuse marinade that the chicken was marinated in for basting.
14. Be sure to use a sharp knife when cutting or carving chicken. Sharp knives will make the job a lot easier, especially when having to cut in the joint areas, and will provide neatly cut slices and pieces

4. Poultry meat products

4.1. Roasting

Roasting is accomplished by cooking the chicken uncovered in a hot oven. The chicken is cleaned, seasoned, stuffed (optional), and then placed on a rack in a shallow roasting pan. The rack is not necessary, but it will improve the quality by holding the bird out of the juices and allowing the heat to reach evenly around the entire bird, providing a chicken that is golden brown on all sides. It will also allow the juices to evaporate more easily, developing a caramelized layer, which will produce a very flavorful base for making a sauce or gravy.

The chicken should be cooked in at a higher heat, for a short period of time, to crisp the skin and seal in the juices to provide moist, tender meat. Generally roasting at about 232°C for 15 to 20 min is adequate.. The heat should then be reduced to 190°C for the rest of the cooking time. Basting the bird throughout the cooking time is not necessary, unless the chicken is smaller, generally under three pounds. The smaller birds do not have the fat necessary to flavor the meat and keep it moist. Cooking temperatures and times will vary according to the size of the chicken and varying recipes.

4.1.1. Roasted whole bird

Chicken carcasses are normally stuffed at the neck end with forcemeat or other stuffing. Stuffing is also put into the body cavity. The bird then is put into a roasting tin and brush with molten fat or oil (basting). Breast region may be covered with foil and cook for 40 min per kg over, until the thickness part of the thigh, when pricked, gives out a clear liquid without any blood. The carcass can be changed to the other side half-way through the cooking process. After cooking process is over, the product can be served with traditional accompaniments.

4.1.2. Roasting in foil

This is most convenient and hygienic. No basting is required as the carcass is enclosed in a foil. A loose enclosing has to be made to allow collection of fluids oozing out during cooking process. The foil is opened during the last 15-20 min of the cooking to brown the carcass. There oven temperature should be 220°C for 60-70 min per kg of carcass weight.

4.2. Baking chicken

Baking is similar to roasting in that the chicken is cooked in a preheated oven. Roasting involves the cooking of a whole chicken whereas baking generally involves the cooking of chicken pieces. The pieces can be skinless, skin-on, boneless or bone-in.

Generally, baked chicken has a coating applied to it before baking. There are many different coatings used, such as a mixture of herbs, breadcrumbs and sauces. The chicken can also have other ingredients baked with it, such as potatoes and vegetables.

The chicken is baked in a shallow pan, which should be large enough so that the pieces do not touch, allowing the chicken to cook and brown evenly. It is common for the chicken to be turned once through the cooking time but this may depend on the recipe.

The chicken should be baked until the juices run clear when pierced. The amount of time the chicken needs to bake will depend on the amount of chicken and other ingredients being cooked at the time. If some pieces finish cooking before others, they can be removed from the oven and kept warm by placing aluminum foil loosely over them.

4.3. Braising and Stewing Chicken

Braising and stewing are very similar methods of cooking. They both use the same process of searing to enhance color and flavor, and slow cooking in liquid to produce tender, moist meat. Once the meat is browned, it is cooked in a covered pan, either on top of the stove or in the oven. It is best to use the same pan that was used to sear the meat to get all the benefit of the flavored pieces in the bottom of the pan when making a sauce. Select a heavy sauté or frying pan that is both suitable for the stove top and the oven. The pan should also have a fairly tight fitting lid to prevent desiccation.

To begin braising or stewing, chicken has to be thoroughly dried off with a paper towel to ensure even browning. Heat enough oil, or oil and butter mixture, to sufficiently cover the bottom of a heavy pan, and then place the pieces of chicken in the hot oil. To ensure even browning, do not overcrowd the pieces. Cook over a medium high heat until pieces are nicely browned on all sides. Once browned, remove pieces from the pan using tongs and then quickly sear any other desired ingredients in that same pan. Replace the chicken pieces in the pan and add the required liquid and flavorings. Heat until liquid boils, cover the pan and reduce heat to a simmer, cooking until chicken is done. To prevent the meat from becoming dry and stringy, keep the liquid at a simmer through the remaining cooking time and do not allow it to boil. The covered pan can also be placed in a low temperature oven to finish cooking, rather than on top of the stove.

The meat should be checked occasionally throughout cooking to see if liquids need replenishing. When done, the chicken and all other ingredients should be removed from the pan, so the sauce can be made. After braising, the pan drippings can be thickened with a paste mixture of butter and flour, or a mixture of cornstarch and water to make a sauce. The sauce is then served with the chicken. When stewing, thickening of the liquid may or may not be required, depending on the recipe. The liquid, chicken and other ingredients are served as one dish.

Braising and stewing are almost identical methods except stewing involves the use of more liquid and the ingredients, including the meat, are cut into more bite size pieces. Sometimes the chicken is not seared before stewing, but most often it is.

4.4. Frying Chicken

Frying chicken is a quick and simple way of cooking chicken where the cooking can be accomplished by the use of two different methods, pan-frying and deep-fat frying. Both methods work on the principle of using hot oil to cook the chicken, producing chicken with a crispy brown outside and juicy, flavorful meat inside. The chicken is generally coated with a thin layer of flour, crumbs or batter. Chicken pieces are generally used for frying, rather than a whole chicken.

4.4.1. Pan-Frying

When pan-frying, a heavy skillet with deep sides is generally used. The chicken pieces are coated with flour that has been seasoned with salt, pepper and at times, paprika. It can also be seasoned with other spices, such as chili or curry powder, etc., to create a desired flavor.

Once the chicken is coated, it should be placed on a rack to allow the pieces to dry, which may take 20 to 30 min. Allowing the pieces to dry will provide for more even browning of the chicken.

1. To fry, heat 5 or 6 tablespoons of oil in a heavy skillet over a medium-high burner until the oil is very hot. Place the chicken pieces in the hot oil, skin

side down, one piece at a time. Leave enough space between pieces so that they are not crowded. This allows the pieces to cookand brown more evenly.

2. Continue to cook over a medium heat, turning until all sides are golden brown and the meat is cooked thoroughly.
3. As the pieces finish cooking, they should be removed from the pan and placed on a paper towel to allow grease to drain.
4. The chicken can be checked for proper doneness using one of three methods. The best way to check for doneness is to use a meat thermometer. Chicken breast must reach 76.7°C and other pieces should be 82.2°C. If a meat thermometer is not available, a visual test can be performed by piercing the pieces with a fork. If thoroughly cooked, the juices should run clear. Doneness can also be checked by cutting into the thickest part of the chicken and making sure the meat is opaque all the way through.

4.4.2. Deep-fat Frying

The chicken pieces for deep-fat frying are generally coated with a crumb coating or a batter. If a crumb coating is used, the pieces should be set aside to dry for 20 to 30 min after they are coated. If coating with a batter, the pieces of chicken can be coated just before frying. When deep-fat frying, be sure the individual pieces are not too large. Large pieces of chicken are difficult to cook properly, resulting in an outside crust that is overcooked or an inside that is not cooked thoroughly.

1. To fry, fill a deep pan approximately half full of oil and heat to between 176.7°C and 190.6°C. To test the temperature, drop in a cube of bread, it should brown in 50 sec if the oil is about 185°C.
2. Using tongs or a metal spatula, place three to four pieces of chicken in the hot oil, being careful that they are not overcrowded. Cook until golden brown, approximately 6 to 10 min.
3. As the pieces of chicken get done, remove them from the oil and place them on a paper towel to allow the grease to drain.

When using either method of frying, an important part of the process is keeping the oil at the proper temperature. Once the chicken is placed in the hot oil, the temperature of the oil will drop. But, as long as the oil continues to bubble consistently, it will cook the chicken properly. If the temperature of the oil would be brought back up to between 176.7°C and 190.6°C, the outside of the chicken would burn before the inside finished cooking. To prevent this from happening, the temperature may need to be turned down a little once the chicken starts browning, but make sure the oil stays hot enough. If the oil is not hot enough, the chicken will be greasy. Controlling the temperature of the oil is an important part of producing delicious fried chicken

4.5. Grilling and Broiling

Grilling and broiling are very similar methods of cooking chicken. They both use a dry heat that quickly cooks the surface and then slowly moves to the middle of the meat. The main difference between the two methods is that grilling applies the heat to the bottom surface of the chicken, and broiling applies the heat to the top surface. Also, grilling infuses the chicken with a smoky flavor from the meat juices that drip during the grilling process. When broiling, this infusion of flavor does not occur.

Before cooking, check the size of the chicken pieces. Cut the larger pieces to smaller sizes so all pieces will cook more evenly. Flatten chicken breasts so that they are more uniform in thickness. If cooking a whole chicken, it should be butterflied for more even cooking. To butterfly, the backbone is removed, the breast bone is pressed down to flatten the chicken and skewers are used to help the bird lay flat by threading them through the breast and thigh area

When grilling or broiling, the chicken will have a tendency to dry out so it must be watched carefully during the cooking process. Coating the chicken with a little oil or marinating it will aid in preventing it from drying out while cooking. Also, temperature at which the chicken is cooked and the distance the chicken is from the heat source are both important to produce tender, juicy, properly cooked chicken.

4.5.1. Grilling

The grilling process cooks foods over a heat source, either directly, indirectly, or a combination of both. Grilling temperatures typically reach as high as 343.3°C, but any temperature above 148.8°C is suitable as a grilling temperature. The high heat of grilling sears the surface of the chicken, sealing in the natural juices and creating tender meat with a flavorful crust. The required cooking temperature and the method of grilling (direct, indirect, or a combination) depends on the size of the chicken or the pieces. It is important to cook the chicken to its proper doneness but not overcooked.

There are many different types of grills available today that can be used when grilling chicken. It is important that the grill is set up properly and reaches the appropriate temperature for the type of chicken that is being grilled to ensure that it produces in a juicy, tender finished product that is cooked to the proper doneness.

A medium heat should be used when grilling chicken. Using too high of a heat will cause some parts to cook too quickly and dry out while other parts will not be done all the way through. To check the temperature of the grill, use the palm of your hand for testing. The thicker the cut the farther away from the heat source it should be or the heat source should be at a lower temperature to prevent the outside of the cut from burning before the inside is properly cooked. Depending on the type of chicken, direct and/or indirect heating can be used. A whole chicken or a thick piece, such as a full breast, may require direct heat to seal the outside and indirect to allow the cut to cook thoroughly to the center.

4.5.1.1. Indirect Heat

Cooking with indirect heat occurs when you use an area of the grill that is not directly over the heat source. Using indirect heat slows the cooking process down, which allows the center of the cut to cook thoroughly without burning the outside. Indirect heat is good for cooking whole chickens or larger pieces, such as a full chicken breast.

4.5.1.2. Direct Heat

Cooking with direct heat occurs when you cook the meat directly over the heat source. The chicken is cooked quickly over medium or high heat coals or over burners set to medium or high heat on a gas grill. Direct heat is used when grilling food that cooks in 30 or less minutes, such as boneless chicken breasts and chicken pieces.

4.5.2. Broiling

When broiling chicken, there is no benefit from the infusion of smoked flavoring that occurs when grilling, causing the chicken to be fairly bland in taste. This can be remedied by the use of seasoning, such as a mixture of herbs, marinades or basting sauces.

Use seasoning to add flavor to the meat. Chicken pieces should be placed on the broiler rack skin side down. Do not line the broiler rack with aluminum foil because it will prevent the drippings from falling into the pan below and the drippings that remain on the foil may cause flare-ups to occur.

Similar to grilling, the distance from the heat source is important for proper cooking and producing chicken that is golden brown, juicy, tender, and thoroughly cooked but not overcooked. To check for proper cooking distance for broiling, place pieces on the broiler rack and place the rack on the broiler pan. Set the broiler pan in the oven and measure the distance from the heat source in the oven to the top of the chicken. It should be 12.5 to 15 cm away; adjust oven racks accordingly.

Remove broiler pan from the oven and preheat the broiler for 9 or 10 minutes. With chicken pieces skin side down on the broiler rack, place them in the oven to cook. As with grilling, it is necessary to watch the chicken carefully as it cooks, making sure the edges are not cooking too fast and if they are, rearrange the pieces or adjust the heat accordingly.

When one side is nicely browned, turn the chicken pieces to finish cooking.

When grilling or broiling, all the pieces will not cook at the same rate so it is necessary to remove them as they finish cooking to avoid overcooking.

Whenever possible, it is best to leave the skin on the chicken during grilling and broiling, If desired, remove the skin before serving. Leaving the skin on will provide a juicier, more tender piece of chicken and will not increase the fat content as long as it is removed before eating. If the chicken has already had the skin removed, coat it before cooking to prevent it from drying out. The chicken can be coated with cooking oil, a liquid marinade or any type of moist marinade..

The chicken must be checked for proper doneness when using either cooking method. The best way to check doneness is to use a meat thermometer. Ground chicken and chicken breast must reach 76.7°C. Whole chickens and other pieces should be 82.2°C. If a meat thermometer is not available, a visual test can be performed by piercing the pieces with a fork. If thoroughly cooked, the juices should run clear. Doneness can also be checked by cutting into the thickest part of the chicken and making sure the meat is opaque all the way through.

4.5.2.1. Grilling and broiling tips

1. To prevent dryness, leave the skin on the chicken during cooking, which helps preserves the chicken's natural moisture.
2. Be sure racks are clean and coat them with vegetable oil or a nonstick vegetable oil spray to help prevent sticking.
3. When grilling, aromatic woods, wherever possible, can be added to the preheated coals to give the chicken a distinctive flavor.
4. Place smaller pieces of chicken around the outer edges, further away from the main heat source, to allow them to cook slower.
5. Do not use a fork to turn the chicken as it cooks. The piercing causes the juices to escape.
6. To speed grilling or broiling time, partially cook the chicken in the microwave first. Microwave on high approximately 9 to 12 min per kg, or 3 to 4 min if using cut up parts. Grill or broil the microwaved pieces of chicken immediately to finish cooking.

4.6. Barbecuing

The whole chicken is cut in half, lengthways. The legs are cut in half at the joints. The halves are fixed to remain flat with skin side towards the exterior on which molten butter is basted. The halves are simmered for 2 min, basted again and grilled for 12-15 min; the process is repeated till the carcasses are cooked.

4.6.1. Difference Between Barbecue and Grilling

People often use the term barbecue when referring to foods that are grilled, but barbecuing and grilling are two different processes (Tale 20.2).

4.7. Microwave

Microwaving is a quick and convenient method for cooking chicken but the flavor of the chicken will not be enhanced as it is with other cooking methods. When cooked properly, it will provide a tender, juicy meat. The flavor can be enhanced by the use of salt, pepper and other seasonings. This is a good method for cooking chicken that is going to be used in dishes that require cooked chicken or when just a small serving is required.

Table 20.2 Barbecuing *Vs* Grilling

Barbecuing	Grilling
Barbecuing refers to foods that are cooked with a long, slow process using indirect, low-heat generated by smoldering logs or wood chips that smoke-cook the food.	Grilling refers to foods that are cooked quickly and directly over high heat.
The fuel and heat source are separate from the cooking chamber, but the cooking chamber contains enough heat to properly cook the food over a long period of time.	
The cooking chamber fills with smoke, giving the food its characteristic smoked flavor, which varies depending on the type of wood that is used for the fuel.	The high heat of grilling sears the surface of meat, sealing in the natural juices and creating a flavorful crust.
The best temperature for barbecuing is between 93° and 148.9°C. If the temperature rises above 148.9°C, it is considered grilling.	Grilling temperatures typically reach 260°C or more, but any temperature above 148.9°C is considered a grilling temperature.

When microwaving chicken, it is best to follow the manufacturer's instruction manual because different varieties of microwaves vary in cooking time. Chicken pieces cook better than whole birds in the microwave and are generally cooked at a high setting. Whole chickens should be cooked at a medium setting. Place the chicken parts in a microwave safe dish arranged so that the larger, denser pieces are placed around the outer edge, and the smaller pieces are placed in the middle. Part way through the cooking time the chicken pieces should be rearranged by turning and exchanging positions of the pieces that are cooking faster with those that are cooking more slowly. This will assist in getting most of the pieces done at approximately the same time. While cooking, the chicken should be covered with wax paper or plastic wrap, leaving one corner open for venting purposes.

4.8. Poaching

Poaching is a healthy method of cooking chicken because no fat is added. It retains the chicken's flavor, tenderness and moisture through a gentle simmering process. Poaching is an especially good method to use when cooking chicken that normally has tougher meat, such as stewing hens and chicken that has a tendency to be a little dry, such as skinless breasts. Cooking by this method provides a very flavorful chicken because it draws additional flavor from the meat and bones as it cooks.

The poached chicken can be eaten as part of a main meal or used as an ingredient for salads, sandwiches, pot pies, and other dishes. The liquid, in which the chicken is poached, can also be used. It can be used as a broth or reduced to use in making a sauce to be served with the chicken.

1. To poach, place the chicken in a sauce pan or Dutch oven with the vegetables you are using.
2. The chicken and vegetables should fit fairly tight in the pan.
3. The chicken and vegetables should be completely covered with liquid.
4. Do not use a pan that is too large, avoiding the use of too much water, which would result in a liquid that has a diluted flavor. The pan should be just large enough so that the liquid can move freely around the chicken.
5. Add other flavorings, if desired. The liquid used for poaching can be water, stock or water with the addition of other ingredients such as wine or fruit juices.
6. Bring the liquid to a boil and skim off the foam that forms on top, then reduce heat to low and partially cover; let liquid gently simmer until the chicken is finished cooking.
7. Check for doneness by piercing the thickest areas of the meat and check to see that the juices run clear.
8. When done, remove the chicken and let it sit at room temperature to rest for approximately 15 min before cutting. This allows the juices to redistribute through the meat.

When using the poaching method of cooking, it is important to bring the liquid to a boil and then lower the temperature so that the liquid is simmering very gently. The liquid should barely be moving with only a few bubbles breaking the surface occasionally, not bubbling as it does when boiling. Vigorous boiling will toughen the meat and make it stringy.

4.9. Steaming

Steaming is an ideal method for cooking boneless chicken breasts and small whole birds. It retains the flavor, tenderness and moisture through the use of steam. It is a healthy method of cooking because no additional fat is used.

1. When steaming chicken with a traditional steamer, fill the steamer pot half full of water and bring it to a full boil using a high heat.
2. For additional flavor, herbs or other flavorings can be added to the water.
3. Place the chicken in the steamer in a single layer, leaving a little room around the pieces. This allows the steam to circulate freely, cooking the meat at a more even rate.
4. Place the steamer in the pot over the boiling water and make sure no water is coming up through the holes in the steamer.
5. Cover and cook for 8 to 10 min.
6. Check for doneness and if thoroughly cooked, remove chicken from steamer and use the water for a broth to be eaten on its own or to make a sauce to serve with the chicken.

4.9.1. Steaming Tips

1. Marinate the chicken before steaming to infuse the chicken with a distinctive flavor.
2. Infuse flavor into the chicken by adding ingredients to the steaming water, such as onions, carrots, celery, and fresh gingerroot.
3. Making a few cuts through the top and bottom surface of the chicken will allow the heat to penetrate more evenly throughout the cooking process.
4. Other ingredients, such as vegetables, can be steamed with the chicken, but do not overcrowd.
5. Avoid removing the cover to the pot during the cooking process. This will allow heat and steam to escape, resulting in extended cooking times.

4.10. Sautéing

Sautéing is a cooking method that quickly cooks the chicken using a little oil and high heat. Olive oil, vegetable oil, corn oil, canola oil, and soy oil are commonly used oils. Sautéing and the searing process, used in browning the meat in the beginning steps of braising, are basically the same methods of cooking, except searing browns the meat and seals in its juices, but does not completely cook the meat. Sautéing browns the meat and seals in the juices but it also thoroughly cooks the meat. Seared chicken requires the use of another cooking method, such as braising, to finish the cooking process.

The type of chicken cooked using this method is generally skinless, boneless chicken breasts, which are generally served with a sauce made from the pan drippings. The breasts are most often pounded to form cutlets, which provides breast pieces that are more even in thickness, preventing areas of the breast from becoming overdone while trying to get thicker areas to the proper doneness.

1. To sauté, preheat pan to condition it before adding oil.

2. Add only enough oil to lightly coat the bottom of the pan. A nonstick pan or a well-seasoned pan may not require as much oil. Be sure the pan and cooking oil are at the proper temperature before adding the chicken so it will begin to cook immediately once it is placed in the pan.
3. The chicken can be lightly dredged in flour before being placed in the pan. The light flour coating helps provide a good surface color when sautéing, but is not required. Seasoning, such as salt, pepper and other herbs or rubs, should be applied before cooking to enhance the flavor of the chicken as it cooks.
4. Place the chicken pieces with the best side down so that when they are finished cooking, they will be served with the best side up.
5. Cook the first side until it is a golden brown color. Then, using tongs, turn the pieces over and finish cooking on the second side.
6. The chicken pieces should not touch each other and they should only be turned once. Turning them more than once can affect their color and flavor.
7. If necessary, adjust the heat during cooking. When properly done, the chicken can be removed from the pan and if desired, a sauce can be made from the pan drippings.

4.11. Stir-frying

Stir-frying, like sautéing, is a cooking method that quickly cooks the chicken in a little oil, using high heat. The difference between the two methods is that sautéing cooks serving size pieces of meat and stir-frying involves cooking smaller size pieces that have been cut into strips, cubes or diced pieces all similar in size. Stir-frying generally includes meat and other food ingredients cooked at different intervals, which are added together at the end of the cooking process to make one dish. They are cooked according to how quickly each food cooks, with the quickest added last. Generally the process starts by cooking the chicken and the slowest cooking vegetables first, such as carrots, cauliflower and onions. Then vegetables such as broccoli, green beans, peppers, sugar snap peas, and snow peas are added after the first ingredients have been stir-fried for several minutes. After few minutes, vegetables, such as mushrooms, celery, peas etc., are added. Any fresh herbs should be added at the very end of the cooking time. Fruits are also used in many stir-fry recipes and should be added as directed.

Stir-frying is a fast paced cooking method so it is important that everything is ready before beginning to cook. Oil, chicken, vegetables, seasonings, and sauce ingredients are pre-measured and placed within easy reaching distance. The pan used for stir-frying should be a heavy skillet with deep sides to allow the ingredients to be stirred and tossed without being spilled over the edges of the pan while cooking.

1. Start the cooking process by heating the oil in the pan, trying to coat the

entire surface, bottom and sides, with the oil, using approximately 3 ½ to 5 tablespoons of oil per kg of ingredients being cooked. Vegetable, corn and groundnut oil all work well for stir-frying.

2. When the pan is hot, add the chicken and the ingredients that require a longest cooking time, making sure all are exposed to the oil and the hottest area of the pan. Cook on high heat for several minutes, constantly stirring and tossing until the chicken is almost done. It will be just barely pink in the middle when cut into.
3. Remove the chicken, set it aside and cover lightly to keep it warm. Add the next ingredients according to the recipe and cook until just tender, stirring constantly. Add the stir-fry sauce according to the recipe and heat until hot and bubbling.
4. At the end of the process, the chicken and all other ingredients are added and cooked on high heat until the chicken is completely cooked. The other ingredients should be tender but still have a crunchy texture. The mixture should be heated thoroughly, but not overcooked. Once it is done, it should be served immediately.

4.12. Miscellaneous Products

4.12.1. Diced chicken

Skinless and boneless chicken diced up into pieces. Diced chicken is available fresh, frozen, uncooked, and fully cooked.

4.12.2. Ground chicken

Ground chicken consists of ground white meat or a combination of white and dark. It is available in bulk form or formed into unbreaded chicken patties and chicken balls.

4.12.3. Shredded chicken

Shredded chicken, also referred to as pulled chicken, is fully cooked chicken with the skin removed. It is pulled off the bones and then pulled into shreds. Shredded chicken, generally found frozen, is available in white meat only or a combination of white and dark. It is also available marinated and in various flavors.

4.12.4. Stuffed chicken breasts

Fully cooked, breaded chicken stuffed in the center. They are available with stuffing with broccoli and cheese, seasoned butter and chives (known as chicken Kiev), and Swiss cheese and ham (known as Cordon Bleaus).

4.12.5. Chicken patties and nuggets

Round patties of chicken, which are generally available fully cooked and breaded. The meat in some varieties is made from chunks of meat pressed together to form the patty and in other varieties the meat is ground and mixed with other ingredients to form the patties. Poultry patties should have at least 10% skin. Skinless light meat is chopped in a cutter along with ice and blended with emulsified skin, mechanically deboned poultry meat, water and seasoning till water is absorbed. The mixture is mechanically formed into desired shape to form patties and nuggets of specific sizes. The patties are flash-frozen, packaged and stored or battered, breaded, flash-fried, frozen, packaged stored. Patties are usually not fully cooked and hence, good manufacturing practices are mandatory.

4.12.6. Formed chicken

Formed chicken consists of chopped chicken pieces, which are pressed together to form a particular shape. The chicken is fully cooked and breaded. Formed chicken is used to make nuggets, patties, fillets, and tenders.

4.12.7. Chicken hot dogs (Frankfurters)

Hot dogs processed from the combination of white and dark chicken meat. They are lower in fat than pork or beef hot dogs. Mechanically separated poultry meat (whose temperature is always maintained below 4°C) is mixed with water (50% each of water and ice), seasonings, salt, sodium eryhthorbate, sodium nitrite and other optional ingredients. The mixture is emulsified, stuffed into casings and linked. This is cooked in fast air-circulated smoke-house using gradient heating. The final temperature on the surface must be 82°C and into the temperature should be 71-72°C. The product is showered, chilled, peeled and vacuum-packaged.

4.12.8. Chicken bologna and salami

The mix prepared for frankfurters are stuffed into fibrous casings and cooked or smoked. The product, referred to as bologna, then is chilled and held at 2-4°C before slicing.

Salami is mechanically deboned chicken, which is seasoned, cured and shaped. It is sliced and served as luncheon meat.

4.12.9. Chicken sausage

Sausages made with chicken meat free from bones, skin and tendons. The meat is ground, mixed with seasonings and re-ground. The product is filled into sausages, mechanically, and cooked. They are available in a variety of types and sizes.

4.12.10. Chicken steaks

Raw breast and thigh meat are cut from bones without skin. Several masses of meat are tenderized by kneading, seasoned and fried to obtain chicken steaks.

4.12.11. Deli chicken roast

Chicken breast roll consisting of meaty-grained chicken breast cured, seasoned and then packed into rolls and gently cooked by oven roasting, poaching or steaming. Also available smoked. The deli chicken roast is generally sliced thin and used on sandwiches.

4.12.12. Chicken ham

A product produced from chicken meat, which is cured and smoked. The size and shape will depend on how it is processed. Chicken ham is a low-fat alternative to pork ham. It can be sliced thick or thin, and served hot or cold.

4.12.13. Canned chicken products

Chicken meat can be purchased canned and is available as all white meat or all dark meat. The canned chicken is convenient for use in salads, sandwiches, and casseroles. Other canned products containing chicken include soups and stews.

4.12.14. Chicken pot pie

A baked chicken dish that has a bottom and top pastry crust which contains pieces of chicken, chunks of potatoes, other vegetables and a creamy sauce. Chicken pot pies are sold as a frozen product in most food stores, generally in a single serving size.

4.12.15. Chicken bouillon

Chicken bouillon is available in powder, crystal, or cube form. Bouillon is used as a flavoring agent for foods such as soup, stew, sauce, and gravy. It is also available with or without salt added.

Chapter **21**

Microbiology of Egg and Egg Products

1. Microbial load

Shell eggs are often referred to as "limited universe" to which the cuticle, true shell and shell membranes provide a mediating boundary to the bulk environment; albumen along with shell offers a very important barrier for the microbes. However, microbes do gain ntry into the egg contents either before oviposition or after it.

1.1. Before oviposition

Chances of upward migration and/or copulation resulting in contamination of eggs are very low because, the bacteria have to encounter parts of oviduct which are actively secreting antimicrobial factors. Even bacteria lodged in yolk also seem to account for negligible frequency of rotting of eggs indicating that the number of bacteria under such circumstances may be too low.

However, *Salmonella* organism, especially in duck eggs, cannot be ruled out from being of public health importance. Others Avian pathogens are unlikely to be of public health importance. *Salmonella* can enter to the gut through water or feed and later on, migrate to the follicles (yolk). Similarly, *Escherichia coli, Campylobacter spp*, and *enterococci* also can reach follicles and oviduct.

The risk of eggs being contaminated with harmful bacteria and causing illness is very low. The odds of becoming ill from consuming eggs is no greater than with any other perishable type of food and the risk is often less than many foods. It is estimated that only 0.005% (1 in 20,000) of eggs may be contaminated with the salmonella bacteria.

1.2. After oviposition

The egg acquires the initial load of microbes from vent and nesting material. Since cuticle will be wet at this stage, some of them, even non-motile ones, can pass through the shell pores along with the movement of moisture due to capillary action. Formation of air-cell accentuates this migration. Thin-shells also facilitate the entry of organisms. However, it is uncertain that organisms entering through this route could be main factor for rotting of eggs because only 1% nest-clean eggs rot during storage.

Soon after lay, eggs come into contact with fecal matter, litter material and dirt. Good managemental practice should aim to keep the eggs away from fecal matter by regularly changing the litter in the nest boxes and replacing it whenever it becomes soiled. Nest box with a provision for rolling of eggs out of it can definitely reduce the microbial load on the surface of eggs.

Fecal material is usually deposited on the floor and so suitable managemental precautions should be taken to minimize floor-eggs; this includes training hens to enter S. box for lay, management of litter to limit litter moisture to 30%, frequent collection of eggs etc. Longer an egg remains in nest box higher will be the microbial load. Cleanliness of the nest box is very important because air cell formation due to cooling of eggs in the nest box can result in sucking of bacteria onto inner shell membrane. This further emphasizes the need of frequent collection and fumigation of eggs.

When eggs are being collected, hands of the personnel must be clean and trays for gathering eggs and the storage room must be cleaned and disinfected. It is advisable to collect all the eggs from nest boxes before collecting floor-eggs, if any. After collection, damage to the cuticle must be avoided.

The benefits of good egg collection practices will only accrue if they are applied to every egg collected on every day for which management must be vigilant, personnel properly trained and continuously motivated.

Recently, surface disinfection of eggs by high intensity ultraviolet (UV) radiation by mounting UV-lamps on the egg grading machine in relatively close proximity to the surface of eggs has been tried. UV cannot penetrate the shell of an egg and hence, kills only micro-organisms present on the surface of the egg (Plattel, 2003).

1.3. Number of microorganisms

Average number of microorganisms on the shell surface is 10^5 (Range – few hundred to tens of millions). Dust, soil and feces, in that order, are the main sources of contamination and most of the bacteria are gram-positive type (*micrococci*). But, rotting of eggs is caused by Gram-negative bacteria which are in small numbers at the beginning (Table 16.1). Fumigation of eggs soon after gathering can promptly reduce bacterial contamination.

Molds, although of lesser importance, can also cause addling of eggs. Their hyphae penetrate the shells as dark or colored patches and then the molds grow on shell membranes. They generally cause gelling of albumen. High humidity during storage can compound the problems of molds.

2. Antimicrobial defense

2.1. Physical

2.1.1. Shell

2.1.1.1. Water loss

Shell of an egg has shell pores whose main function is to allow restoration of the embryo i.e., allow oxygen intake and carbon dioxide escape. At the same time, since egg contents, of which albumen is the outermost, has as much has 88% water, it is reasonable to expect evaporation of moisture unless relative humidity in the environment is very high. In fact, such a loss of water is essential for increasing air-cell size so that embryos can breathe and obtain sufficient space for pipping. But, in infertile (table) eggs for market, loss of water has the minimized either by increasing humidity (which causes problems of mold growth) or by increasing resistance of shells for loss of water vapor by blocking the pores.

Table 21.1 Types of microorganisms on the shell and in rotten eggs

Frequency	On the shell	In rotten eggs
Occasional	*Streptococcus, Sarcina, Aeromonas, Proteus, Serratia*	*Pseudomonas aeruginosa, Achromobacter, Cytophaga, Bacillus, Micrococcus, Streptococcus, Arthrobacter*
Infrequent (in small numbers)	*Staphylococcus, Arthrobacter, Bacillus, Pseudomonas, Achromobacter, Alcaligenes, Flavobacterium, Cytophaga, Escherichia, Aerobacter*	*Pseudomonas maltophilia, Hafnia, Citrobacter*
Frequent (common)	*Micrococcus*	*Pseudomonas fluorescens, Pseudomonas putida, Alcaligenes, Proteus, Escherichia*
		Source : Stadelman and Cotterill, 1977

2.1.1.2. Water-repellent property

Shell pores are sufficiently narrow to easily allow a normal column of water by capillary action. Thus, if egg becomes wet, the embryos may die of suffocation. Thus, the normal shell has been endowed with water-repellent property. This property is further reinforced by the cuticle which plugs shell pore canals by the varying degree and thereby offering water-resistant property to the egg shell.

Thus, if the egg is partially or completely cuticle-less, its shell will have only water-repellent property but not water-resistance. Therefore, anything that damages the cuticle, including washing of eggs by a forced jet of water, can greatly impair the resistance of the shell.

It is necessary to emphasize that along with water entering through capillary action, bacteria can migrate into the eggs.

2.1.2. Shell membranes

Shell membranes are made up of mesh work of anastomosing fibers having keratin as core and a glycoprotein mantle. Therefore, they act as filters. Surprisingly, retention of organisms on the shell membranes ultimately helps addling of eggs probably by providing an area for colonization.

2.1.3. Albumen

Albumen offers both physical and chemical barrier to microbes. Physical barrier is due to viscosity, which is a function of carbon dioxide/pH-dependent ovomucin-lysozyme interaction hindering the movement of bacteria.

The gelatinous structure of egg white and integrity of chalazae are believed to be due to ovomucin, a glycoprotein presenting albumen with an isoelectric pH of 10.2. Thick albumen has ten times more ovomucin than thin albumen. Ovomucin has two components namely α (carbohydrate-rich) and β (carbohydrate-poor) non-covalently bonded. B-ovomucin seems to be substantially responsible for gelatinous properties.

It is hypothesized that the structure of thick white is due to interaction between ovomucin which are acidic and lysozyme which is cationic in nature. In fact, lysozyme co-precipitates with ovomucin.

2.2. Chemical

Albumen is primarily responsible for chemical defence of the egg against microorganisms. Proteins present in albumen along with their properties (Table 7.7) and their anti-nutritional (antimicrobial) properties have already been discussed (Chapter 9, section 1.4.1.1).

3. Rotting of eggs

Under normal conditions, only a few organisms can grow on the shell due to low water activity. In fact, unless protected by fecal matter, soil etc., many die due to desiccation. Bacteria so localized into the eggs mainly accompanying water which enters through Very action or during air-cell formation or when a warm egg contracts while cooling. Antibacterials in wash water do not guarantee protection against all microbes.

As shown in Table 16.1, gram-positive organisms constitute most of the flora entering into the egg and onto the inner shell membrane. From then on, the gram-negative organisms, though they are fewer in number, take over primarily because they can withstand reduced nutrient levels and unfavorable environment. Augmenting to this, the nutrient requirements of gram-negative bacteria are of simple nature and of limited number.

Action of lysozyme at the junction between inner shell membrane and albumen as well as nutrient-limiting action of conalbumin play a very important role in containing the growth of gram-positive bacteria.

3.1. Egg rots by pure cultures

Under commercial conditions, rotting of eggs occurs due to a mixture of organisms. Therefore, to understand the effects of individual organisms, pure cultures were inoculated into the egg; the results are summarized in Table 16.2. It must be remembered that the list of organisms is not exhaustive and hence many other organisms are also likely to cause rotting of eggs under practical situations. Rotting with gas production has to be handled more carefully because a slight agitation or a slight tap on the shell of such eggs, referred to as "popper" or "banger", can result in "explosion" due to deranged balance in pressure within the egg leading not only to contamination of other eggs but also to obstruction to further work because of rotten egg smell.

4. Precautions to prevent contamination of egg & egg products

4.1. Cleanliness

It is important to follow the basic rules of cleanliness and proper hygiene when preparing eggs and egg products. Hands must be washed before and after egg preparation. If multiple foods are being prepared at one time, hands should be washed thoroughly after preparing each different food item to avoid spreading possible bacteria from one food to another (cross contamination) and personnel.

The risk of eggs being contaminated with harmful bacteria and causing illness is very low. The odds of becoming ill from consuming eggs is no greater than with any other perishable type of food and the risk is often less than many foods. As already indicated, only 0.005% (1 in 20,000) of eggs may be contaminated with the *salmonella*, but even with a risk this low, it is wise to cook eggs to the proper doneness to ensure safety. Proper cooking kills the *salmonella* in any of the eggs that may have it.

Cutting boards used in the kitchen are best when they are constructed of non-porous materials such as tempered glass or heavy plastic. Tempered glass cutting boards are safer to use because there is less of a problem with cracks and pores harboring bacteria as there is with wood or soft plastic surfaces. Cutting boards should be washed thoroughly after each use with hot, soapy water. A mixture of bleach and water or an antibacterial spray can also be used to kill germs. All other work surfaces, such as countertops, and all utensils used in food preparation should also be thoroughly cleaned.

Table 21.2 Egg rots produced by pure cultures

Organism	Change(s)		Type of rot	Remarks
	Yolk	Albumen		
Proteus spp.	Dark, brown	Dark, brown	Black rot – Type 2	Proteolysis, H_2S production
Aeromonas liquefaciens	Gelatinous, black	Grey, watery	Black rot – Type 1	Proteolysis, H_2S production, lecithinase produced
Certain enterobacters	Encrusted with custard-like material, sometimes flecked with olive-green pigment	---	Custard rot	Proteolysis *, H_2S production *, lecithinase produced
Serratia marcescens	Surrounded with custard-like material	Red throughout	Red rot	Proteolysis *, lecithinase and soluble pigments produced
Pseudomonas maltophilia	Gelatinous, amber-like striped olive-green, almond odor	---	Green rot	Proteolysis, H_2S production, insoluble pigments *
Pseudomonas fluorescens	Fluorescent green changing to pink white, surrounded by custard-like material	---	Pink rot	Proteolysis, lecithinase and insoluble pigments produced
Pseudomonas putida	---	Fluorescent green	Fluorescent green	Production of insoluble pigments
Pseudomonas aeruginosa	---	Fluorescent blue	Fluorescent blue	Production of insoluble pigments
Flavobacterium, Cytophaga	---	---	Yellow rot	Yellow pigment on shell membrane at the site of microbial growth; soluble pigments produced
Other enterobacters, Alcaligenes	---	---	Colorless rot	

* May not contribute to rotting

Source : Stadelman and Cotterill, 1977

Kitchen washcloths and towels that have been used on multiple surfaces can spread germs. Paper towels or other disposable cloths are to be preferably used whenever possible. Eggs that are cracked, leaking, and/or soiled should not be used.

4.2. Cross Contamination

Cross contamination is also important to guard against. Various types of foods should be kept separate from each other during storage and preparation. Ready-to-eat foods should never be kept next to raw eggs, raw meats, or raw fish. Germs from perishable food items may contaminate the ready-to-eat foods. If cutting boards are used it is a good idea to use one for meats and a different one for fruits and vegetables. Never use the same knives and utensils for preparing multiple food items unless they are washed before using them on a different item. The knife that was used to cut raw beef should not be used to chop a hard-cooked egg unless the knife has been thoroughly washed first. It is also important to wash hands often during food preparation to avoid transferring harmful bacteria from one food item to the next.

It is best not to separate egg whites and yolks by splitting open the eggshell and passing the contents between the two shell halves. The egg may become contaminated if bacteria are present on the shell. Bacteria may be present on the shell even after it is cleaned and the shell may also become contaminated from other food sources that it may come in contact with.

Do not use the two halves of the shell for removing bits of the shell from an egg mixture and never use the shell halves to measure other foods for a recipe.

Salmonella may be found not only in eggs, but in other foods such as chicken, cheese, orange juice, tomatoes, and alfalfa sprouts. It can be spread quite easily from one food to another, which is why it is important to guard against cross contamination during food preparation.

Cross contamination can occur when bacteria are transferred from one food to another, from contaminated kitchen equipment to food, or from people to food. The number of incidents of people becoming infected from salmonella in eggs has steadily declined during the past few years. This is due mainly to quality control measures on the farm, in processing facilities, and during shipping to food stores and also because of increased awareness of proper food handling procedures by food service personnel and consumers.

4.3. Proper Storage

Eggs should never be stored at room temperature, but there are recipes that require eggs to be at or near room temperature before incorporating them into the other ingredients in the recipe. Egg whites that will be beaten should be at room temperature because this helps the whites to reach their maximum volume when beaten. Approximately 30 min is required for eggs to reach room temperature after removing them from the refrigerator. Eggs should not be away from refrigeration for more than 2 hr, so the time required to allow the eggs to warm to room temperature, as well as the total preparation time of the recipe, should not exceed the 2 hr maximum. This should be considered when planning the steps required for the preparation of a recipe using eggs that must be at room temperature.

4.3.1. Refrigerator Storage

Eggs should be stored in the refrigerator in the carton they were packed in. Many refrigerators provide storage for eggs in special units in the door, but this is not the ideal place for storing eggs because the temperature fluctuates so much in the door when it is opened and closed. Eggs should be stored in the coldest part of the refrigerator where the temperature remains constant. Eggs keep best when they are stored at temperatures of no higher than 4°C. The ideal temperature range is 0.5°C to 3.3°C. When the temperature is above 4°C, harmful bacteria may grow rapidly. Although salmonella are not destroyed in temperatures below 4°C, any of the bacteria that may be present will not multiply when the temperature is below 4°C.

Eggs should be stored with the rounded end pointed up in order to keep the air cell on top and to help keep the yolk centered in the egg. Never store eggs next to strong smelling foods because eggshells are porous and will allow strong odors to be absorbed into the egg over time. This is another reason why it is a good idea to store eggs in the original protective carton.

4.3.2. Freezer Storage

For long term storage, eggs and egg products may be frozen. If stored properly, eggs will emerge from the freezer no better or worse, in terms of quality, than when they first entered the freezer. The temperature of the freezer compartment must be at -17.8°C or less and the eggs should be stored in an area of the freezer where there is the least amount of temperature change. Eggs and egg products should not be stored in the door compartment of the freezer, especially if the door is opened frequently.

Whole eggs can be beaten slightly and placed in a container with a tight seal and stored in the freezer for as long as a year. Egg whites may be stored in the freezer for up to a year in a tightly sealed container. Egg yolks may also be stored in the freezer, but sugar or salt must be added to keep the yolks from becoming too thick and gelatinous over time. Add a pinch of salt per yolk if the yolks will be used for savory dishes or about a ¼ teaspoon of sugar per yolk if the yolks will be used for sweet dishes.

Like other perishable foods that have been frozen, eggs should be defrosted in the refrigerator and should never be allowed to thaw at room temperature. Thawing foods on the countertop encourages the growth of harmful bacteria, especially on the outside edges of the food.

Chapter **22**

Microbiology of Poultry Meat and Poultry Meat Products

1. Microbial load

Many species of organisms gained entry into poultry meat either through systemic route (infection) or by contamination during handling, processing and storage. These organisms could be pathogenic (either by invading the human body or by the toxins produced by them) of non-pathogenic (spoilage organisms or non-spoilage type).

Live poultry generally have 6×10^2 to 8.1×10^3 organisms per sq cm of skin area and after processing and evisceration; it increases to 1.1×10^4 to 9.3×10^4 per sq cm. Live turkeys will generally have 7.5×10^2 to 4.1×10^4 organisms per sq cm.

Off-odors appear to the bacterial number is 3.2×10^6 to 10^8 per sq cm and slime formation is evident at 3.2×10^7 to 10^9 organisms per sq cm. More than half (50 to 60%) of the microbial load on carcasses soon after processing are chromogenic, about ¼ (20 to 25%) are *Pseudomonas,* colorless cocci, and closely related forms, and the remaining ¼ (20 to 25%) are miscellaneous group. Slime is mainly caused by *Pseudomonas* and *Alcaligenes.*

1.1 Pathogenic organisms

About 26 diseases of poultry have been reported to be transmissible to humans. However, poultry meat is the largest single source of different species of *Salmonella* organisms which cause gastro-intestinal disturbances by causing an infection. Similarly, *Streptococci* can also cause infection to humans. Organisms like *Staphylococcus* and *Clostridium* cause food poisoning by producing toxins which contaminate the food. Diseases caused by many other organisms,except probably paracolons (Arizona group), are of too low frequency.

1.2 Spoilage organisms

These normally inhabit soil, water or air and hence are more resistant than pathogens. These organisms normally invade the meat, utilize the nutrients for the growth and produce their end products, which mostly bring about undesirable changes in organoleptic quality of meat. Some of them do produce desirable changes as in the case of fermenting organisms (lactic-cultures).

2. Source of microorganisms

Water is a source of many of the spoilage organisms as well as few of the pathogenic ones. *Pseudomonas, Chromobacterium , Acinetobacter, Aeromonas, Alcaligenes, Yersinia, Micrococcus* and *Bacillus* constitute natural flora of surface water. Enterococci, *Proteus, Enterobacter, Escherichia coli* and *Salmonella* are common contaminants which can survive in water from few days to several weeks. Water may be contaminated dust, litter, feed, feathers, feet, beaks and faeces while in troughs.

Soil harbors *Acinetobacter, Actinomyces, Alcaligenes, Bacillus, Chromobacterium, Clostridium, Cytophaga, Enterobacter, Escherichia, Flavobacterium, Micrococcus* and *Moraxella;* Yeasts and molds. Spores of bacteria can survive in soil for years (for example : *Salmonella* up to 200 d).

Table 22.1 Common contaminants during production of ready-to-cook chicken

Stage	Organism (s)
Process-water	*Pseudomonas, Aeromonas,*
Ice	*Acinetobacter, Moraxella, Pseudomonas, Corynebacterium,* and Cocci
Workers	*Salmonella*
Scalding	*Clostridium, Micrococcus, Proteus, Salmonella, Staphylococcus, Streptococcus*
Defeathering	*Salmonella,* aerobes
Spray-washing	*Pseudomonas*
Cutting and handling	*Salmonella,* aerobes
Mechanical deboning	*Salmonella, Staphylococcus aureus, Clostridium perfringens, Enterococci,* Yeast and Molds
Eviscerated cut-up poultry	*Pseudomonas, Micrococcus, Achromobacter, Flavobacterium, Alcaligenes, Proteus, Bacillus, Sarcina, Streptococcus, Eberthella, Salmonella, Escherichia, Aerobacter, Streptomyces, Penicillium, Oospora, Cryptococcus,* and *Rhodotorula*

Several environmental factors like nature of soil, pH, sunshine etc. influence their survival. Hence, types of organisms for which the birds are exposed through soil and dust depend on the local agro-climatic conditions. Improperly managed litter can support growth of microbes. Mold spores and Yeasts also grow in litter and may be transmitted through dust or air. Many pathogens are shed in feces by infected birds. During transportation of birds, contamination with organisms of fecal origin does increase and during dressing of chicken (Table 22.1), the chances of cross-contamination further increases. Utensils and equipment, processed water, ice, workers, aerosols and dust, scalding, evisceration, washing, cooling, chilling (ageing), cutting and packaging can be potential source of contamination and hence at each stage proper care is essential.

3. Symptoms of spoilage

Symptoms of spoilage by invading microorganism(s) can be summarized by taking the biochemical activities of *Clostridium welchii* as a reference organism, an anaerobe, degrading meat as follows:

Table 22.2 Spoilage of meat by *Clostridium welchii*

Stage	Action
Liquefaction	Production of collagenase which hydrolyses the connective tissue between muscle bundles
Gas production	Free amino-acids acted upon by deaminases to produce H_2, CO_2, NH_3 in meat
Foul smell	Glycogen, if present, fermented to acetic and butyric acids
Increased membrane permeability	Histidine decarboxylated to histamine and CO_2
Increased penetration	Hyaluronic acid in mucopolysaccharides (brown substance) disintegrated by hyaluronidase
Toxin production	Toxic to consumer

Superficially recognizable symptoms of meat spoilage are as shown in Table 22.3.

4. Factors affecting growth of microorganisms

4.1 Temperature

The ranges listed below (Table 22.4) are not water-tight and the organisms can grow, but relatively slowly, in temperatures other than their own optima. Poultry carcasses stored at 0°C need 18 d for spoilage whereas those stored at 2.8°C spoil in 11 d and at 20°C in just 2 d. Microorganisms are viable even at temperatures below - 2°C as well as above 66°C. Whenever a microorganism have to grow at temperatures other than own optimum, its nutrient and other requirements become more specific and often increased; *Lactobacillus arabinosus* needs the amino-acids Phe, Tyr and Trp at 39°C; Phe and Tyr at 37°C and none at 26°C.

4.2 Moisture and osmotic pressure (a_w)

Fresh meat has a_w close to 0.99 and hence is highly perishable. Temperature, aw and oxygen status interact determining the growth of an organism in question. For example *Staphylococcus aureus* needs an optimum a_w of 0.995; but at 38°C, it can grow on dried meat (a_w=0.88). Similarly, aerobically it can grow at a_w of 0.86 but, anaerobically, it requires a minimum aw of 0.90. pH also interacts with these factors and the organism becomes more and more requirement-specific as one or many of the requirements are not met. Yeasts and molds tolerate lower a_w and

Table 22.3 Superficially recognizable symptoms of spoilage

O_2 status	Organism	Symptoms of spoilage
Present	Bacteria	Slime on meat surface; discoloration by destruction of meat pigments or growth of colonies of colored organisms; production of taints and off-odors; fat decomposition
Present	Yeasts	Yeast slime; discoloration; off-odors and taste; fat decomposition
Present	Molds	Surface "Stickiness" and "Whiskers"; discoloration; odors and taints; fat decomposition
Absent	Bacteria	Putrefaction accompanied by foul odor; gas production; souring
		Source : Lawrie, 1979

hence can withstand higher concentrations of salt and sugar unlike bacteria. Most of the preservation methods of meat involves binding, removing or immobilizing free water (to lower a_w) so as to obviate growth of microorganisms.

4.3 pH

Most bacteria show optimum growth at neutral pH (7.0) and do not grow well below a pH of 4.0 or above a pH of 9.0. Optimum pH for an organism is a function of several variables like temperature aw etc. High ultimate pH therefore, is undesirable. Hence, animals should not be subjected to fatigue, stress and other factors which are likely to deplete glycogen levels thereby causing high ultimate pH.

Table 22.4 Temperature (range) for growth of food poisoning microorganisms (°C)

Microorganism/Toxin	Lower limit (°C)	Upper limit (°C)
Psychrophiles	- 2 (optimum)	7 (optimum)
Mesophiles	10 (optimum)	40 (optimum)
Thermophiles	43 (optimum)	66 (optimum)
Salmonella	5.2	44.0 to 47.0
Staphylococci	6.7	45.4
Staphylococci – toxin production	10.0	37.0 to 40.0 (optimum)
Campylobacter parahaemolyticus	3.0 to 13.0	42.0 to 44.0
Clostridium botulinum (types A and B)	10.0 to 12.0	48.0 to 50.0
Clostridium perfringens	6.5	50.0
Bacillus cereus	7.0	49.0
Aspergillus flavus (aflatoxin production)	4.0 to 5.0	-
		Source : Riemann and Bryan, 1979

4.4 OR potential (E_h)

E_h is directly proportional to tissue oxygen levels. Soon after death, Eh falls due to loss of blood (O_2), cessation of respiration and due to action of surviving O_2-utilising enzyme systems. In meat, molecules which have electropositive character, like nitrite, can elevate E_h and exert an antibacterial effect. In any case, the effect of E_h is primarily to prolong lag phase and microbes soon adjust themselves to the existing E_h.

4.5 Atmosphere

When the meat is wrapped with impermeable or semi-permeable material, initially aerobes grow utilizing the available O_2. If the wrapping material is impermeable, the aerobes which cannot tolerate high CO_2 levels are taken over by those which can tolerate low O_2 levels. *Pseudomonas* and *Achromobacter* spp. are highly sensitive to CO_2 as are molds; yeasts and *Lactobacilli* are relatively resistant. Changes in meat wrapped in permeable material are similar to those without being wrapped.

5. Microbiology of preserved raw meat products

5.1 Frozen meat

Freezing and frozen storage can kill or sublethally damage the microorganisms. At - 10°C or lower, the organisms die over a period of time but above this temperature, some of them can recover. Aerobic counts reduce drastically (up to 99%) due to freezing depending on type of and time taken for freezing. Bacteria which associate themselves with the water-film around the carcass are most vulnerable to freezing process. Freezing does not kill *Salmonellae, micrococci* and *enterococci.* Organisms that survive freezing process are killed during cooking. Frozen meat does not spoil while in frozen state but it gets spoiled if thawed and kept at refrigerator temperatures after a period of time by psychrophiles. Carcasses must be fast-frozen, held at or near - 18°C and thawed just before cooking to avoid problems of pathogens like *Salmonella.*

5.2 Dehydrated meat

Heat used during dehydration or freezing stage of free-drying destroys some of the microorganisms. In addition, dried poultry meet contains less than 10% moisture and packed in water-impermeable films do not support microbial growth. If moisture content exceeds 10%, molds and possibly yeasts can grow and spoil the product. Spores of *Clostridium perfringens* survive cooking, freezing, drying and later on, germinate upon reconstitution. Cooked had dehydrated product can also be contaminated with *Salmonella, Staphylococci* and others; control of contamination during the dehydration process and initial microbial status of the raw material are of utmost importance.

5.3 Cured and smoked meat

Curing favors survival of *Micrococci* and fungi, and smoking kills most of the non-sporeforming organisms. Spoilage is mainly by molds if stored in gas-permeable films. *Staphylococcus aureus* and toxigenic molds tolerated to 10% NaCl and chances of *Clostridium perfringens* can be minimized at storage temperatures.

5.4 Irradiated meat

γ-rays from Co^{60} or Cs^{137} or fast electrons of up to 10 MeV energy can be used to irradiate packaged poultry products. The dose of 200 to 700 krads prolongs shelf-life and 500 to 700 krad causes considerable death of pathogens. Higher dosage is required on frozen carcass for the same effect. *Moraxella* is the primary flora that can cause spoilage of irradiated meat under refrigerated storage. Under anaerobic conditions, lactic acid bacteria can cause spoilage. Spores of *Clostridium perfringens* do survive pasteurizing dose of irradiation and hence may cause public health hazards, although, it is of rare occurrence. Packaging before irradiation avoids subsequent contamination and storage at all below 6°C retards growth of residual microbes, if any, and germination of spores.

5.5 Heat-processed meat

Vegetative cells on poultry carcass die during cooking but the spores, which may even come from spices used, of *Clostridium perfringens* and *Bacillus cereus* may not be destroyed. In canned meat, most of the bacteria and spores are killed. Poultry meat rolls which contain spices can contain *Clostridium perfringens*. Cooking destroys all vegetative cells and most of the spores on the surface; but those in the internal parts survive. Barbecuing has similar effects as that of cooking. Frying is lethal to all surface cells but not to spores in deeper parts. If packaging is done after cooking, further contamination occurs. The shelf-life of these products, therefore, depends on extent of post-cooking contamination.

6. Control of microbial contamination and spoilage

6.1 Sanitation

(Wallner-Pendleton *et al* on internet)

The first step in preventing bacterial contamination of poultry meat and eggs is to raise the birds in a clean and dry environment. Frequent removal of wet caked litter, cleaning of waterers, adding fresh bedding as needed, and good ventilation are very important. Providing adequate space per bird will also help maintain cleanliness and prevent many of the common diseases.

Basic sanitary husbandry is very important in broiler production as most birds are processed with the skin on. Raising birds with clean skin and feathers is very helpful in reducing surface contamination. Proper scalding to remove feathers, as well as careful evisceration, is also important. Bacterial contamination can also occur through accidental rupture and spillage of the gastrointestinal contents during evisceration. If obvious contamination occurs, trimming of that portion of the carcass is preferred. Frequent hand washing and dipping of cutting utensils in hot water will also minimize cross contamination.

Producers raising birds for their own consumption should be familiar with signs of disease, both in live and in freshly slaughtered birds. Obvious sick birds should either be treated or culled from the flock prior to slaughter. It is absolutely necessary to have a good working knowledge of the normal anatomy of the bird to detect disease in the carcass. Any change from the "normal" in the carcass should be discarded.

Even with the most careful husbandry and processing practices, the final product is never sterile. Shelf life of freshly processed refrigerated chicken is seldom greater than seven to ten days before spoilage occurs.

Spray-washing of carcasses and rinsing of equipment in water containing hypochlorite or other approved disinfectants kills many of the organisms. Chlorinated water (45 to 55 ppm Cl_2) has been found to reduce microbial comes by 50 to 90% and enhance shelf-life.

6.2 Packaging

If packaging material is O_2-impermeable, it will have best result in minimizing spoilage. Vinyledene chloride-vinyl chloride polymer is becoming increasingly popular.

6.3 CO_2

Atmospheric concentration of 10 to 25% CO_2 delays most of the aerobic organisms especially at 4°C or lower. This effect is proportional to the concentration up to 25% after which discoloration begins.

6.4 Temperature

Lower the temperature longer the shelf-life. Poultry meat has to be stored below 3°C or preferably, at – 2°C at which temperature meat does not freeze.

6.5 pH

Carcasses immersed in 0.12% lactic acid or dil. HCl or sprayed with a solution of 7.5% sorbic acid in 70: 20: 10 propylene glycol, water and glycerine or use of polyphosphates in meat has been found to enhance shelf-life. The *Lactobacilli* were mixed in chill-water at the rate of 10^8 cells/ml also extended shell-life. Antibiotics have also been tried although their use under commercial conditions is neither economical nor advisable.

Chapter **23**

Non-food Applications of Eggs and Poultry Meat

1. Eggs

Major uses of eggs, other than as human food, are as a component of pet foods, shampoos and vaccine production process. There are other uses of eggs too; the same are tabulated below:

Table 23.1 Non-food applications of eggs

Health-promoting uses	1. Egg albumen is an antidote to counteract some toxins and irritants consumed accidentally. 2. Protects mucous membrane of stomach and intestine and prevents ulcer formation. 3. Water-holding and binding properties - relieves enteritis caused by toxins and microbes. Egg white is, therefore, the most popular natural remedy for gastritis, enteritis, diarrhea, dysentery and dehydration (Narahari, 2003).
Therapeutic uses	1. Egg yolk is used to prepare a parenteral fat emulsion (Intralipid) which is used to administer fat-soluble drugs. 2. Lecithin of yolk, in combination of Vit B_{12}, helps nervous tissue development and mental ability in persons suffering from Alzheimer's disease. 3. Toxicity of diamidine and other anti-protozoal drugs is reduced when encapsulated in lecithin from yolk. 4. Yolk lipoprotein is used in media for growth of cells in biotechnology and genetic engineering. 5. Egg yolk and chalazae are excellent sources of sialic acid which has powerful anti-microbial, anti-inflammatory, and anti-viral properties; hence used in treatment of *H.pylori* and other microbial infections causing ulcers, colon cancer, gastritis and enteritis. 6. Yolk is a rich source of IgY antibodies which can be used to treat human rotavirus, *E.coli*.

Therapeutic uses	7. *Streptococcus, Pseudomonas, Staphylococcus* and *Salmonella* infections. 8. Egg white, especially chalazae, is used for application on burns, cuts, insect bites and rashes with pasting of shell membranes over affected skin to reduce inflammation and promote healing. 9. Eggs are used in Indian systems of medicine in pastes made from herbs, sesame oil and green mung bean to apply over fractures before PoP bandage is put. 10. Egg is a cheap and good medium for Ig and vaccine production. Even anti-snake venom has also been produced at a considerably less cost from eggs (Narahari, 2003).
Antibody-farming	1. Yolk antibodies used to treat cystic fibrosis. 2. Anti-gluten specific IgY is used to treat celiac disease. 3. Anti-prion antibodies used to detect prions. 4. Passive immunization with IgY antibodies to prevent dental caries by *Streptococcus mutans* (Noble and Penny, 2002).
Immuno-stimulating uses	1. Lumiflavin, lumichrome and sulphoraphane are the natural anti-oxidants present in eggs which are known to retard multiplication of cancer-inducing viruses and prevent normal cells turning cancerous. 2. Lysozyme (G_1 globulin), G_2 and G_3 globulins, ovomacroglobulin, antibody IgY and other natural antimicrobials and immuno-stimulants in egg prolong the life of AIDS patients. (Narahari, 2003).
Animal feeds	1. Pet foods for dogs and cats. Domestic animals – ground egg shell for layers after sterilization. 2. Experimental in growth studies (FAO reference protein). 3. Egg shells from processing plants are often dried and fed to laying hens as an excellent source of calcium and protein.
Fertilizers	Eggs and egg shells discarded from hatcheries and processing plants are dried and ground to be used as fertilizers
Biologicals	1. Culture media – Egg yolk is an excellent enrichment ingredient for bacteriological media (detection of *Clostridium botulinum* and *Staphylococcus*). 2. Yolk, albumen and whole egg are used in culture media in the form of coagulated, dehydrated etc. 3. Chicken embryo culture is used for isolation and culturing of several bacteria and viruses which is a pre-requisite for vaccine production.

	4. Low-density lipoproteins in yolk are used in production of monoclonal antibodies. 5. Source of immunoglobulins: hens produce many eggs with the desired immunoglobulins and hence better than other laboratory animals.
Biochemistry and medical diagnostics	1. Sialic acid isolated from chalazae and vitelline membrane – used in studies on immune system and macro-monitor the structure. 2. The egg white has numerous unique proteins which offer opportunities in new technologies. a) Lysozyme is separated from egg white using ion exchange resins and utilized as an anti-microbial agent and in other pharmaceutical compounds. b) Lysozyme – to protect ready-to-eat foods from *Clostridium botulinum* & *Listeria monocytogenes.* c) Avidin, another egg white protein, is separated along with lysozyme. Avidin-biotin technology is being used in various medical diagnostic applications such as immuno-assay, histopathology and gene probes. 3. Lecithin – as a nutrient in human diet. 4. Neutral egg oil – has many food and non-food applications. 5. Ovomucoid as a high-affinity ligand in liquid chromatography. 6. Conalbumin as a ligand of iron
Medical and pharmaceutical	1. Vaccine production – chick embryo inoculation is the most popularly employed method to manufacture several vaccines for poultry, other animals and man. The following organisms are cultured in embryonated eggs – Fowl pox, Mumps, Rabies, Laryngotracheitis, Newcastle disease, Psittacosis, Cow pox, Eastern and Western equine fever, Canine Distemper, Typhus, Influenza, Louping ill, Pigeon pox, Q fever, Rocky Mountain fever, Yellow fever, Venezuelan equine fever, Blue tongue, Infectious Bronchitis, Rift Valley fever etc. 2. Artificial insemination – was an important constituent of egg-yolk citrate diluent of liquid semen till the advent of frozen semen technology.
Manufacturing	1. Leather – tanning of leather; limited nowadays. 2. Cosmetics and shampoos – cleansing of hair, giving a natural sheen and soft feeling.

	1. Adhesives – retained albumen in egg shells can be dried and used as adhesive to fasten cork inserts in metal caps on beverage bottles
Art	1. Egg decorating in Easter season. 2. Painting – color printing and tempera painting. 3. Others – photographic plates, painter's ink, spinning synthetic fibers, dye mordants in textiles and rubber manufacture

2. Slaughter-house byproducts

Bureau of Indian Standards (BIS) has set standards for handling, storage and transport of slaughter-house by-products (ISI: 8895-1978) which is shown in Table 23.2.

Table 23.2 Handling, storage and transport of slaughter-house by-products

By-product	Utilization	Remarks
Blood	1. Human food, pharmaceuticals (plasma, albumin, fibrin). 2. Livestock feed. 3. Fertilizer and others	i) Blood for human food must be collected hygienically ii) Transported within 3 to 6 hours; otherwise preservatives to be added Blood not useful for any of the above is added with preservatives such as formalin or Lysol
Pancreas	Trypsin, Insulin, Pancreatin, Chymotrypsin	Pancreas removed within 30 min, chilled, frozen immediately; packaged and transported
Endocrine glands	Pituitrin, Thyroxin, Adrenalin	Same as for pancreas
Liver	Liver extract, glycogen, Vit B_{12} etc.	Collected within 1 to 2 hours and stored at 4^oC
Feathers *	Feather meal, Pillows etc.	----------

* Not included in ISI bulletin

Chapter **24**

Incubation and Hatching

The freshly laid Avian egg contains everything that the embryo needs for its growth and development - excepting for oxygen and heat. The oxygen diffuse is into the egg from the surrounding air through microscopic pores in the egg shell. The pores allow the carbon dioxide produced by the embryo to diffuse out of the egg but they also permit loss of water vapor from the egg. Thus, the regulation of gas exchange between the egg and its environment is closely related to its water balance; both are covered by the diffusive conductance of the egg shell until the embryo penetrates (pips) the chorioallantoic membrane are egg shell with the aid of its egg tooth. The shell diffusive conductance is a measure of the diffusibility of gas/water molecule through the pores; it depends on the shell geometry (porosity and thickness) and diffusion coefficient of the diffusive molecules. The adult bird has a key role in incubation providing not only the heat necessary for embryonic development but also the microclimate of the egg. In the poultry industry and for research purposes, the added bird is conveniently replaced by an incubator. The physiological functions of developing embryos were elucidated using artificially incubated chicken eggs (Tazawa and Whittow, 2000).

1. Natural brooding and hatching

Heat transfer from the body of the hen to the egg is very difficult to measure. Most birds develop "brood patch", a seasonal bare patch of skin, on part of the thorax and abdomen through which it directly transfers heat to the eggs. In addition to loss of feathers, there will be increase in size and numbers of blood vessels in the brood patch (Tazawa and Whittow, 2000).

The hen can adjust the rate of heat transfer by standing or leaving the egg, but also by closeness with which the bird applies its patch to the egg. In addition, the hen also responds physiologically to variations in the temperature of the egg, increasing its metabolic heat production in response to cooling of the egg. The efficiency of heat transfer diminishes with decreasing ambient temperature and clutch size. If the egg is very cold, "cold vasodilatation" occurs in the brood patch, to increase blood flow and thereby the temperature of the brood patch. Brood patch temperature varies in different Avian species from 34.9°C to 42.4°C so that the brood patch temperature is always 1.1 to 5.5°C warmer than the egg temperature in different species (Tazawa and Whittow, 2000).

The hen herself incubates the eggs by keeping the brood-patch on the top of the eggs. Most of the time, the eggs will be horizontal on the floor of the nest.

The temperature at the top of the eggs will always be higher (37.2 to 37.8°C) because it is in contact with the brood-patch; the lower side of the egg will be cooler (32.6°C) and the centre of the egg will be intermediate (35°C). At the beginning, top of the yolk containing embryo is towards the brood-patch and it rotates within the albumen-sac when egg is rotated to keep embryo near the brood-patch. Therefore, embryo will be a few mm away from the body of the hen.

As the embryo grows, the extra embryonic membranes (EEM) develop and blood flowing through them warms the embryo. Hence, the centre of the egg becomes warmer than before. Simultaneously, metabolic activity contributes heat and hence, in the later stage incubation, centre and bottom of eggs also become as warm as the portion in contact with the brood-patch. However, circulation of blood through EEM is more important than metabolic heat. Hence, the hens leave the eggs for cooling during later stages. The effect of blood flow on heat flow is more pronounced in large eggs than in smaller ones.

The main barrier to heat loss from the egg is a thin layer of air immediately adjacent to the shell — the boundary layer. If the egg is in a nest, the nest itself imposes an additional resistance to heat loss. Nests, in addition to protecting against heat loss, may also mitigate the effects of solar radiation (Tazawa and Whittow, 2000).

With commercialization of poultry industry, natural brooding has been replaced by artificial methods of incubating and hatching eggs. However, under village and backyard rearing, even now, natural brooding is being practiced.

2. Selection and care of eggs

2.1. General

Commercial hatching eggs may be collected as often as four or five times daily to ensure egg quality especially when daily temperatures are above 29.4 to 32.2°C. Keep nest eggs separate from eggs found on the floor so that disease organisms do not spread. Do not incubate dirty floor eggs with nest eggs; they may spread disease to clean eggs.

Eggs have to be checked for cleanliness soon after collection. Eggs free of clinging dirt or debris and those with a small amount of adhering dirt that can be easily removed are retained for incubation. Dirty eggs are never incubated.

Hatching eggs should not be washed. Bacteria can be forced through the porous shell and into the egg. Washing also removes the protective sealing substance from the shell, leaving it vulnerable to penetration by other bacteria.

Only eggs of average size are incubated. Excessively large eggs hatch poorly; small eggs hatch into small, unthrifty chicks. Similarly, abnormally shaped eggs hatch poorly and hence not incubated.

All cracked or thin shelled eggs are discarded because they cannot retain the moisture needed for proper chick development. Penetration by disease-causing organisms increases in cracked eggs. In addition, eggs with loose or bubbly air cells are discarded.

The following list summarizes the care required while handling hatching eggs:

1. Before setting eggs in an incubator, the eggs must be procured and/or produced from well-managed, healthy flock which is fed properly balanced diets.
2. Keep the nest full of clean, dry litter. Collect the eggs early in the morning and frequently during the day to prevent excessive chilling or heating of the eggs.
3. Do not wash eggs unless necessary. If it is necessary to wash eggs always use a damp cloth with water warmer than the egg. This causes the egg to sweat the dirt out of the pores. Never use water cooler than the egg. Also, do not soak the eggs in water. If the egg is allowed to soak in water for a period of time, the temperature difference can equalize and bacteria have a greater chance of entering through the pores.
4. Be sure eggs are dry before storing. Never place damp or wet eggs in a Styrofoam carton for storage.
5. Store the clean fertile eggs in an area which is kept at 12.8 to 18.3°C and 70-75% relative humidity. Never store eggs at temperatures at or above 23.9°C and at relative humidity lower than 40%. These conditions can decrease hatchability dramatically in a very short period of time.
6. Slant or turn the fertile eggs daily while they are being stored. Store the eggs small end down and slanted at 30-45°. Putting a wooden or thermocole piece of 5 cm x 10 cm cross-section under one end of the carton or storage container and changing it to the other end daily works well.
7. Do not store eggs for more than 10-14 d. After 14 d of storage, hatchability begins to decline significantly.
8. Just before setting the eggs, allow them to warm to room temperature (21.1-26.6 °C) and remove any cracked eggs.

2.2. Fumigation

Eggs and equipment before storage or use are fumigated. Under-fumigation does not kill the bacteria, but over-fumigation can kill the chick embryo in the egg. Therefore, use recommended amounts of chemicals at the right time for the length of time specified.

A room or cabinet large enough to hold the eggs is required. It must be relatively air tight and equipped with a small fan to circulate the gas. Calculate the inside volume of the structure by multiplying the inside length by the width and by the height.

Eggs are stacked inside the room or cabinet on wire racks, in wire baskets, or on egg flats so air can circulate among the eggs. Eggs from the cases are removed for good air circulation. Formaldehyde gas is produced by mixing 20 gram of potassium permanganate ($KMnO_4$) with 40 cc of formalin (37.5 % formaldehyde) for each m^3 of space (1 X concentration) in the fumigating structure. The ingredients are mixed in an earthenware or enamelware container with a capacity at least 10 times the total volume of the ingredients. Sometimes, even 3 X concentration is employed.

The gas within the structure is allowed for circulation for 20 min and then expelled. The temperature during fumigation should be above 21.1°C. Eggs are allowed to air out for several hours before placing them in cases.

2.3. Hatching Egg Storage

When large number of eggs have to be incubated at a time, it becomes necessary that large number of breeding birds is available to produce the desired number of eggs. Maintenance of this large breeding flock becomes very expensive which will be reflected on the cost of the day-old chick. In addition, eggs produced after the desired number of eggs for incubation is obtained have to be sold as table eggs although they are fertile.

Alternatively, a huge incubation facility may have to be created to accommodate all the eggs produced by the building flock; which a gain is very expensive. Therefore, it is necessary that eggs produced over a period of time be stored/ pooled so that the size of the breeding flock as well as incubation facility required is minimized. However, eggs produced during the earlier days will have their embryos grown partially and hence they will not hatch on the same day. This makes it mandatory that the embryo should be stored in such a way that they neither grow nor die. It has been found that at the temperature of 12.8 to 18.3°C (15.6°C, preferably), the embryos remained quiescent and hence, this temperature is referred to as "physiological zero". To avoid excessive loss of moisture, it is ideal that the relative humidity be maintained above 70%. The above two conditions are provided in a cold storage and fertile eggs can be stored ideally for 7 to 10 d (for a maximum of 15 d) without seriously affecting hatchability. All the eggs stored under the above conditions hatch on the same day; in other words, the hatching will be synchronized.

When eggs are to be stored for longer than 14d, reductions in hatchability can be minimized by "flushing" the eggs with nitrogen (certified as "pure") as soon as the eggs have been cased and before sealing the neck of the plastic bag in which the fertile eggs are stored.

Eggs must be incubated as soon as convenient. The hatchability of eggs stored for less than seven to ten days remains high with proper storage conditions. Eggs held longer experience reduced hatches. After three weeks of storage, the hatchability is near zero per cent.

If the eggs are not incubated within three or four days, turn them daily. Turning the eggs prevents the yolks from touching the shell and injuring the embryo. Store the eggs with small ends down and slanted at an angle of 30 to 45°. Large numbers of eggs can be stored on egg flats and in cases with one end of the case elevated to give the proper slant. Turn the eggs by elevating alternate ends of the case or flat each day.

The eggs should warm slowly before being placed in the incubator. The shock of warming the eggs too rapidly will cause moisture to condense on the shell. This may lead to disease problems.

2.3.1. Changes during storage (Ven, 2004)

1. Liquefaction and increase in pH of albumen. Egg quality is best at lay but highest hatchabilities are produced from eggs that have been stored for 1-2 d indicating that eggs from "suboptimal" albumen quality hatched better. Recent research also has indicated that the embryo itself initiates alterations in albumen characteristics to optimize its environment for incubation.
2. Hatchability showed a linear decrease, estimated to be 0.7% for extra day of storage; the reductions being especially conspicuous after 7d of storage.

2.3.2. Reducing the effects of storage (Ven, 2004)

Losses in hatchability due to storage of eggs can be reduced when the duration of storage for every batch of eggs is known beforehand by the following methods:

1. As far as possible, eggs from young maternal flocks should be stored rather than those from old flocks.
2. Eggs destined to be stored up to a maximum of 3d, should be stored at 18-21°C; those to be stored for a period of 4-7 d at 15-18°C; those which have to be stored beyond 7d at 10-12°C.
3. Relative humidity during storage should be 70-80%; it may be increased to 88% for eggs from old flocks or when extended storage period is necessary.
4. Incubation period should be extended by 1 hr for every d of storage.
5. Pre-warming eggs immediately prior to start incubation can reduce hatchability losses in stored eggs because, during the pre-warming period, the temperature of various components of the egg becomes homogeneous before the initiation of incubation leading to a more uniform embryonic development. Stored eggs are pre-warmed at 20-25°C for 5-18 hr just prior

incubation. However, effects are particularly apparent when the eggs were restored for prolonged period (more than 14d).

6. In case of turkey eggs, increase in the incubation temperature during the first two weeks of incubation has been beneficial. However, such a practice may be practically difficult in multistage incubation.
7. Hatching eggs are stored with pointed end up to maintain central position of the yolk (and to the embryo) during the storage. This position also helps minimizing dehydration and adhesion of membranes.
8. Pre-warming of eggs at the farm immediately after oviposition at 37.5°C for 6 hr before they are sent to cold storage was beneficial especially when the hatching eggs were stored for 14d.

2.4. Location of incubation facility

All incubators and hatchers are placed inside a building away reasonably far away from the farm premises and well protected against severe weather. It is easier to check the temperature and humidity if the incubator is kept indoors. The room must be well ventilated and have a system for maintaining comfortable temperatures.

3. Incubation periods

Allometric relationship between initial egg mass (IEM) and its incubation period (I_p, in d) for about 1100 Avian species from 19 Orders was

$I_P = 11.59\,IEM^{0.244}$; $N = 1161$, $SE = 0.005$, $r^2 = 0.700$ (Crafer *et al.*, 2005).

4. Physical conditions required for artificial incubation

The conditions described below refer to the forced-air incubators which are most common. However, still-air incubation is discussed separately.

Each of the physical conditions is briefly described below:

4.1.1. Number of days in setter and hatcher

At the beginning of development of artificial incubation and hatching, an incubator (setter)-cum-hatcher was introduced in which separate trays were provided for the first 18d and the last 3d of incubation. Most of these machines were all-in all-out type (Single stage).

However, as the industry developed into a large commercial sector, it became necessary to have separate machines for the first 18d and the last 3d of incubation because

1. Single stage incubation was not always economical and hence multistage incubation became most common

Table 24.1 Incubation period of different Avian Species

Species	Days	Species	Days
Chicken	21	Lophura (Gallopheasants)	
Turkey	28	White Tailed	24-25
Pekin and most other Domestic Duck	28	Kalij	24-25
Muscovy Duck	33-35	Silver	25
Domestic Geese	28-32	Imperial	24
Canadian and Egyptian Geese	35	Edwards	21
Pheasant, Partridge, Chukar partridge	23.5	Swinhoe	25
Coturnix quail	17	Fireback	24-25
Guinea Fowl	26	Phasianius (Game Pheasant)	
Bantam chicken	21	Black Necked	24-25
Pigeon	17	Kirghiz (monzolia)	24-25
Pheasant	23-28	Green	24-25
Grouse	25	Lophophorus (monals)	
Emu	42-53 (Av 52)	Himalayan	27-30
Ostrich	42		
Swan	35-40		
Canary	13	Crossoptilon (Eared Pheasant)	
Parakeet	19	White Eared	24
Various Game Bird Breeds Chrysolophus (Buffed Pheasant)		Brown Eared	26-27
Golden	23-24	Trogopans Satyr	22
Lady Amherst	23-24	Polyplectron (Peacock Pheasant)	
Syrmaticul (Long Tailed Pheasant)		Germains	22
Elliots	24-24	Grey Peacock	22
Mikado	27-28	Pavo (Peafowl)	
Reeves	24-25	Indian	27-28
Relnartia (Argus)		Black Shouldered	27-28
Crested Argus	25	Green	27-28
Catreus (cheer)		Quail	
Cheer	26-27	Common	15-17
		Bob White	15-17

2. In multistage incubators, it is difficult to clean meconium, shell debris and other material expected during pipping and hatching.
3. Therefore, disinfection of an incubator-cum-hatcher with different stages of

Table 24.2 Physical conditions for chicken eggs

	Incubator (setter)	Hatcher
Number of days	First 18 d	19th , 20th and 21st d
Position	broad-end up, in contact with each other	Horizontal
Temperature, °C	37.6 to 37.9	36.1 to 36.7
Relative humidity, %	55	65
Turning	At least 6 times a day	Not required
Oxygen, %	21	21
Carbon dioxide, %	0.04	0.04

embryos being present is extremely difficult.

4. In addition, frequent opening of the door leads to changes in both temperature and humidity which can adversely affect other eggs at different stages of incubation.
5. Since turning is not required during the last 3d of incubation, a separate arrangement becomes necessary which is extremely difficult both in terms of design of the machine as well as its operation.
6. Similarly, high humidity requirement than the last 3d of incubation will adversely affect other embryos which are less than 18d of age.
7. Over-hauling of the machinery will become virtually impossible under the commercial set-up if incubator-cum-hatcher is employed in a large scale enterprise.

In view of the above, it is most common nowadays to have a separate machine called "setter" for the first 18d and a "hatcher" for the last 3d of incubation.

4.1.2. Position of eggs at setting

Eggs are kept broad-end up and in contact with each other in a setter. This is primarily because hatching eggs are selected with air-cell at the broad-end so that embryos when set broad-end up can breathe easily. It has also been recorded that embryos during growth communicate between each other when the eggs are in contact with each other which can help synchronization of hatch especially when the differences in age of embryos at setting is by a few hours.

In the hatcher, the eggs with fully grown embryo are kept horizontal to help them come out of the shell easily. Further, more space is required to accommodate movements of the egg during the pipping process and hence horizontal position will be advantageous.

4.1.3. Temperature

Table 24.3 Effect of temperature on growth of chicken embryo

Temperature, °C	Remarks on growth of embryo
15 or below	No growth
25	Development not possible; some activity may be present; prolonged exposure causes death of embryo
25 to 35	Limited development; high rate of embryo abnormalities due to faulty utilization of nutrients
35	Sustains growth; but very slow growth
37.6 to 37.9	Optimum growth; Physiological optimum
38 to 39	Growth accelerated but not synchronized which might extend incubation period and reduce hatchability
Above 40	Severely restricted growth and development; high rate of abnormalities and embryo deaths

Incubator is put on a day or two before setting the eggs to ensure that proper incubation conditions can be obtained. Thermometer is placed in the incubator in such a way that the bulb is 2.5 cm above the screen floor and the bulb of the thermometer does not touch the eggs or side of the incubator. The sides and top of the incubator should fit together securely to prevent heat loss.

It is clear from the above table that embryos tolerate cold better than heat. It is for this reason that opening of doors in multistage incubators does not pose serious problems unless it is too abnormally frequent. There could be minor fluctuations of 0.5°C above and below the desired temperature, but there should not be prolonged periods of high or low temperatures. Hatching eggs can take an amazing amount of abuse because they are well protected and insulated, but they are sensitive to extreme heat. Operating the incubator at 40.6°C for 30 min will harm many embryos, but operating at 32.2°C for 3 to 4 hr will merely slow the rate of growth. Maintain a forced-air incubator at 37.8°C for best results.

It is interesting to note that during the first 10-15 d of incubation, the embryo behaves more like a poikelotherm (the evolutionary lean on reptiles) and tries to adjust its temperature depending on that of the incubator. When the embryo breaks into air-cell and pips, it begins pulmonary respiration by taking in atmospheric air into the lungs; at this stage, the embryo changes over to homoeothermy. It is for this reason that temperature of the hatcher be preferably reduced by 1.2 to 1.8°C although, embryos might tolerate temperature similar to that setter provided all the conditions like humidity, ventilation etc. are optimum.

Thermometer placement is not as critical in the forced-air incubator because of better air circulation. The sides and top of the incubator must fit tightly to prevent heat and moisture loss.

Temperature fluctuations of 1°C or less are allowed, but there should be no prolonged periods of high or low temperatures. Hatching eggs can take much abuse, but they are sensitive to high heat. Many embryos die if held at 40.6°C for 30 min. Incubating eggs at 32.2°C for 3 to 4 hr will slow development and growth.

4.1.3.1. Management during power outage (failure)/load-shedding

In spite of a power failure, the hatch can be saved. The key is to keep the eggs as warm as possible until the power returns.

This can be done by placing a large cardboard box or blankets over the top of small incubators for additional insulation. To warm the eggs, place candles in jars, light them and place the jars under the box that covers the incubator. Be careful not to put any flammable material closer than a foot from the top of the candles. The heat from the candles can easily keep the eggs above 32.2°C until the power returns.

Embryos have survived at temperatures below 32.2°C for up to 18 hr. Therefore, continue to incubate the eggs after the outage; then candle them 4 to 6 d later to check for further development or signs of life. If, after 6 d, embryos are found dead, then incubation can be terminated. Most of the time, a power outage will delay hatching by a few days and decrease the hatchability to 40-50 %.

4.1.4. Humidity

The term relative humidity (RH) refers to the amount of moisture (humidity) in the air to that of its saturation value at a given temperature. Of the several methods of measuring RH, dry- and wet-bulb thermometer is most accurate and generally practiced in setters and hatchers.

4.1.4.1. Working of dry- and wet-bulb thermometer

The dry-bulb thermometer directly records the temperature. The wet-bulb thermometer comprises a dry-bulb thermometer whose mercury bulb is covered with a wick immersed in distilled water. Water moves by capillary action through the wick and keeps the mercury bulb wet. Water from the wick evaporates, the rate of which is inversely proportional to the humidity (at a given temperature), and during this process, it takes away heat equivalent to the latent heat of vaporization of water from the bulb, causing a cooling effect. Therefore, the magnitude of difference between the dry-and wet-bulb thermometers, at a given temperature, is inversely related to the humidity. Standard charts are available from which RH can be obtained depending on the dry-and wet-bulb readings. The readings (°C) for chicken eggs are shown in Table 24.4.

Table 24.4 Dry- and wet-bulb readings (°C) for incubating chicken eggs

	Dry-bulb	Wet-bulb
Setter (Incubator)	37.8	29.4
Hatcher	36.1 to 36.7	30.6 to 31.1

Eggs during incubation lose weight due to evaporation of moisture. For ideal hatching, chicken eggs should lose 10 to 11% of their weight during the first 18 days of incubation (in setter). To effect this, RH must be 55%; any variation in RH results in improper egg weight loss, the relationship between RH and weight loss being inverse. In either case, survivability of the embryo and consequently, the hatchability, are severely affected.

In the hatcher, RH is increased by 10% to help keep the beaks of the pipping chicks wet to facilitate pipping process. In addition, the hair-like down feathers floating in the hatcher absorb moisture, become heavier and settle down; otherwise, the down feathers may lodge in the eyes of the chicks as well as the hatchery personnel.

Setter door should not be opened three days before hatching else vital moisture will be lost. The chicks need added moisture when hatching to prevent the membranes from drying. Low humidity makes it more difficult for the chick to tear the tough shell membranes and may cause the chick to stick to the shell and die.

Goose eggs hatch better if sprinkled or dipped in clean, lukewarm (37.8°C) water for ½ min each day during the last half of the incubation period. Use warm water; cool water may suck bacteria into the eggs.

An excellent method of determining proper humidity is to candle the eggs and observe the size of the air cell after 7, 14, and 18 d of incubation (Fig 24.1). As incubation progresses, the size of the air cell increases because of moisture loss. Humidity adjustments can be made after each candling operation.

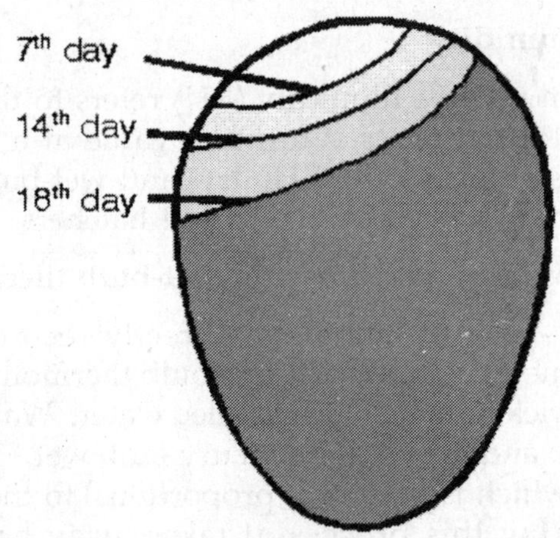

Fig. 24.1 Air-cell size during incubation

4.1.5. Turning

In forced-air incubators, the eggs are placed, small end down, in special incubating trays that mechanically turn the eggs. If using such a machine, be sure all trays operate properly.

Turn eggs at least two or three times daily during the first 18 d of incubation. Turning prevents the embryo from sticking to the shell membranes when left in one position too long. Do not turn eggs during the 3d before hatching. The embryos are moving into hatching position and should not be disturbed.

4.1.5.1. Significance of turning during incubation

During incubation, eggs lose carbon dioxide, both due to respiration of the embryo and also that of thick albumen. Therefore, thick albumen liquefies; consequently, yolk which is rich in fat (and hence low in specific gravity) gets off-centered, floats to the surface and sticks to the shell. This can result in death of the embryo, particularly during first three days of incubation. To avoid such an eventuality, turning is mandatory. Turning is also associated with:

4.1.5.1.1. Orientation of EEM

The orientation of allantoic, amniotic, albumen-and yolk-sacs is extremely important during the first week of incubation (Fig 24.2) because it determines the positioning of *area vasculosa* (a.v), a network of blood vessels on the yolk-sac membrane, to the albumen. The positioning of a.v is very important for formation of sub-embryonic fluid. Terming ensures proper relationship between EEM so that normal moment of albumen to the amniotic fluid (which the embryos swallow) is not hindered.

4.1.5.1.2. Growth of EEM

Turning is likely to stimulate growth of EEM by sheer physical movements it creates and gravity, both of which may influence cell migration. On the other hand, smaller a.v reduces area of EEM and limits restoration. Reduced chorio-allantoic membrane (CAM) also severely restricts oxygen uptake in later part of incubation and limits calcium absorption from shell. CAM is also required for formation of complete albumen-sac.

4.1.5.1.3. Water balance

Reduced a.v and consequent reductions in sub-embryonic fluid production leads to reduce allantoic fluid; volume of albumen increases and amount of water sent to yolk-sac reduces. Hence, embryo suffers from dehydration although sufficient water is present in egg. Dehydration is primarily due to the failure of transportation of water to yolk-sac rather than reduced allantoic fluid volume.

4.1.5.1.4. Nutrition

Utilization of albumen is directly proportional to size of a.v. Hence, reduced flow of albumen to amnion complicates the adverse effects of reduced av. Turning maximizes formation of albumen-sac and ensures its proper orientation with respect to yolk-sac. Albumen protein enters amniotic fluid which is swallowed by the embryo into its gut. The excess proteins moved into the yolk which later on, supports the newly-hatched chick. Turning seems to have effect on utilization of yolk also.

In view of the above, it is apparent that in case of species whose eggs have more albumen need very precise turning requirements whereas, even those species whose eggs have left albumen (like wild birds, flying birds, Japanese quails etc.) turning, even if not effected, can cause little change in hatch results; however, incubation period of the eggs with more yolk is generally more than those with more albumen.

It also follows that, turning is most critical during the first 3 to 7 d of incubation when embryo size is small enough to result in sticking to the shell whereas after 1st week, embryos will be large with considerably less movement.

Eggs have to be turned at least 6 times a day or as to the manufacturer's recommendation (usually once every hour). However, when the birds are selected and reproduced over several generations by artificial incubation, turning twice a day itself will suffice. In the hatcher, the embryo is completely development and hence turning is not required.

4.1.6. Ventilation

Proper ventilation is important for successful incubation. During embryonic development, oxygen enters the egg through pores in the shell, and carbon dioxide escapes in the reverse manner. As the embryo grows, it needs a larger supply of fresh air. The air openings of the incubator are gradually opened as incubation progresses. When increasing the air flow, be careful not to reduce humidity.

When the electric power fails, open a forced-air incubator immediately so fresh air can enter. If the incubator is left closed, the oxygen supply will be exhausted and the chicks will suffocate. Hold room temperature at 23.9°C or more.

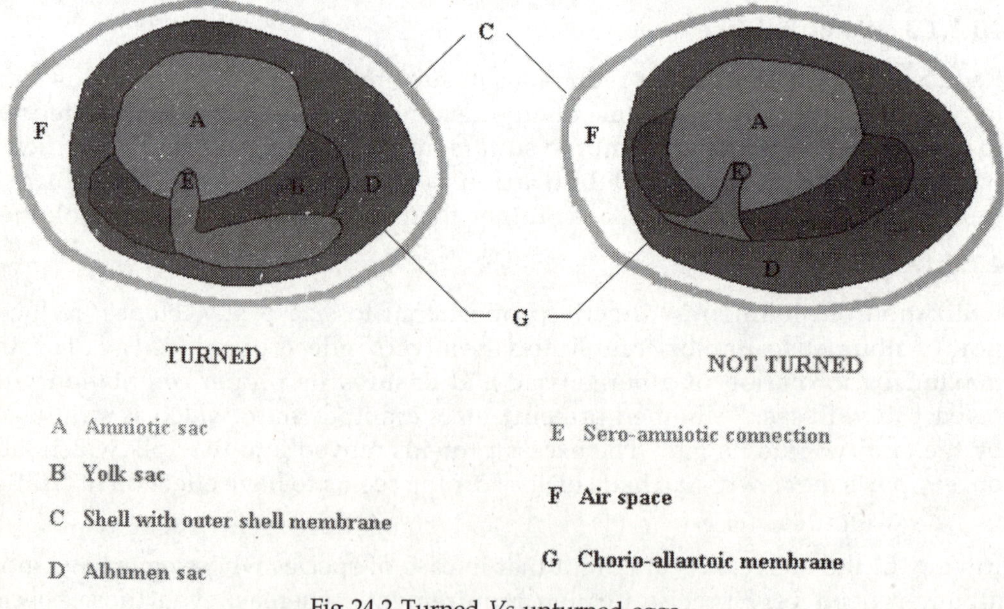

Fig 24.2 Turned *Vs* unturned eggs

4.1.7. Pulling the hatch (Salazar, 2003)

A common source of contamination and subsequent chick quality problems is the transfer of incubating embryos from the setter to the hatcher. This is when a "dangerous" mixture of broken, contaminated eggs, high humidity and wet surfaces can come together, resulting in microbial contamination.

Hatches and related equipment are the areas of a hatchery most likely to become contaminated and so every hatchery unit should be inspected after cleaning and washing. Hard-to-reach surfaces like fan blades, cooling coils and ducting are prone to build up debris and the sources of re-contamination.

Trays should be thoroughly cleaned, washed and disinfected. Most importantly, they should be completely dry before reuse. Incubating eggs should never be transferred into wet hatch trays.

Another significant factor contributing to the production of contaminated chicks is poor timing of setting and pulling. Chicks that had pulled from the hatcher too early will be processed with their navels still open. Setting schedules should be adjusted to take account for the seasonal variation in incubation time.

Hatcher units should be pulled in order of readiness. Often, the first chicks that are pulled at the beginning of the shift are still green and not quite ready, resulting in contamination and mortality.

Over-letting the chicks during spray vaccination must be avoided. It makes the chick box bottoms wet, providing an excellent growth medium for microbial growth.

Equipments such as vaccinators, conveyor belts (if any), sexing tables, chick boxes etc. should be kept in clean and in good condition.

Dehydration is a frequent problem in a poorly operated hatchery. The hatches with the most advanced hatch should be pulled first. Prolonged unnecessary holding of the chicks in the hatcher box causes dehydration. A good rule of thumb is to pull the hatch when 5-10% of the chicks are still damp behind their necks. Factors such as age of the flock, breed cross and time of the year also should be borne in mind.

Chick-holding room temperature is maintained slightly above 24°C with the relativity of 50% to prevent dehydration.

4.1.7.1. Chicks with black buttons are threads

Black buttons consist of a plug or string of remnant tissue at has to the chick's navel. They are caused by high incubation temperatures, which lead to premature hatching and closing of skin around the navel before the extra-embryonic membranes are drawn into the body cavity. Black buttons, threads, ectopic viscera and cull chicks are the consequences. This has to be controlled by proper temperature regulation in the setters.

4.1.7.2. Characteristics of top quality broiler chicks

1. Good and uniform size – hatched from eggs that weigh a minimum average of 48-50 g.
2. Alert, strong and active with high numbers of chicks eating or drinking at any given time
3. Free from bacterial infections
4. Free of defects and deformities like crossed beaks, blind chicks, lameness etc.
5. Bright, plump and fully fleshed shanks; no signs of dehydration (skin folds or prominent blood vessels)
6. No swelling or red coloration of hock joints
7. Well-healed navels; no black buttons or weepy navels
8. Abdomen soft and pliable; not distended, bloated or hardened
9. Mortality during the 1st week < 1% and not exceeding 1.5% at 14d of age

4.1.8. Others

Eggs incubated on different days must be marked with the date they were placed in the incubator. This prevents eggs from being overlooked and left in the incubator after they should have hatched.

All chicks should be out of their shells by the end of the 21st d unless proper incubation conditions were not maintained. Chicks hatched after the 22nd d are not usually healthy and vigorous. After the chicks have hatched and fluffed up, remove them from the incubator and place them under a brooder with feed and water.

When the hatch is completed, disconnect the incubator from the power supply and remove all shells, unhatched eggs, and debris. Wash the interior of the incubator with a warm detergent solution. Rinse with a sanitizing solution, and fumigate with formaldehyde as described earlier. Fumigate the incubator only after removing the eggs. After drying, the incubator is ready for reuse or storage.

Cleanliness and sanitation are the most practical and economical methods to prevent disease and produce quality chicks. Keep the hatching area and equipment clean and sanitary. Cover all windows, air intakes, and doors with screens to keep out insects and rodents. A good sanitation program has no substitute.

Physical conditions required for different species of birds in comparison to those of chicken are shown in Table 24.5.

Table 24.5 Physical conditions for incubating eggs of different species

Item	Chicken	Turkey	Duck[1]	Muscovy Duck	Goose[2]	Guinea	Peafowl
Incubation period (d)	21	28	28	35-37	28-34	28	28-30
Temperature[3] (°C, dry-bulb)	37.8	37.2	37.8	37.8	37.2	37.8	37.2
Temperature (°C, wet-bulb)	29.4-30.6	28.9-30.0	29.4-30.0	29.4-30.0	30.0-31.1	29.4-30.6	28.9-30.0
No Egg Turning After	18th d	25th d	25th d	31st d	25th d	25th d	25th d
Open Vents Additional ¼	10th d	14th d	12th d	15th d	1st d	14th d	14th d
Open Vents (if needed)	18th d	25th d	25th d	30th d	25th d	24th d	25th d

Item	Pheasant	Bobwhite Quail	Coturnix Quail	Chukar Partridge	Grouse	Pigeon
Incubation period (d)	23-28	23-24	17	23-24	25	17
Temperature[3] (°C, dry-bulb)	37.8	37.8	37.8	37.8	37.8	37.8
Temperature (°C, wet-bulb)	30.0-31.1	28.9-30.6	29.4-30.0	27.2-28.3	28.3-30.6	29.4-30.6
No Egg Turning After	21st d	20th d	15th d	20th d	22nd d	15th d
Open Vents Additional ¼	12th d	12th d	8th d	12th d	12th d	8th d
Open Vents (if needed)	20th d	20th d	14th d	20th d	21st d	14th d

[1] Duck eggs reportedly hatch better in still-air incubators than in forced-air incubators.
[2] Better hatchability may be obtained if goose eggs are sprinkled with warm water or dipped in lukewarm water for half a minute each day during the last half of the incubation period.
[3] For Forced-air incubators. Add 1.0-1.5°C to the recommended temperatures for still-air incubators.

5. Incubators

Fairly constant environmental conditions can be maintained in an incubator. Incubators are available in many different models and sizes with capacities ranging from two to thousands of eggs. The larger incubators are rooms in which environmental conditions are carefully controlled.

For incubating large numbers of eggs or setting eggs more than once weekly, separate incubating and hatching units are used. The incubators should be large enough to hold a three-week supply of eggs. The hatcher unit can be small, but large enough to hold the largest setting of eggs.

Eggs at various stages of incubation are held in the incubator. The eggs are transferred to the hatcher on the 18th d and held in the hatcher until completely hatched. Hatcher is thoroughly cleaned and disinfected after each group of eggs hatches.

There are two basic types of incubators, forced-air and still-air incubators. The size and type of incubator selected depends on needs and future plans.

5.1. Forced-draft (Forced-air)

Here outside air is forcibly sucked in, warmed and distributed uniformly by fan. Therefore, there is no temperature gradient unlike still-air incubator. Hence, temperature can be measured amidst the eggs; most of the modern-day incubators belong to this category.

Eggs are placed in stacks of trays. The capacity of these incubators is large. Most units have automatic equipment for turning the eggs and spray-mist nozzles for holding proper humidity levels. Most of the information given above refers to forced-air incubation.

5.2. Still-air

These are usually small but may hold 100 eggs or more. They do not have fans. Air exchange is made by escaping warm, stale air at the top and entering cool, fresh air near the bottom. Air circulation is limited but heat is convected to different parts of the incubator; therefore only one layer of eggs can be incubated. It follows that egg temperature will not be uniform at all points and the top of the incubator is warmest as hot air is lighter and moves to the top; in a way, it stimulates natural brooding.

Hence, temperature of the still-air incubator should be measured at the top of the eggs. Incubating temperatures in these machines must be about 1 to 1.7°C above the temperatures in forced-air incubators. Deviations which have to be made while incubating eggs in a still-air incubator are detailed separately.

In any of the incubators mentioned above, the total volume of air surrounding the eggs is too small in comparison to natural brooding and hence heat loss is not efficient. Hence, there could be a rise in temperature beyond the tolerance of embryos, especially in still-air incubators, adversely affecting incubation results. This is particularly true during the last week of incubation when they embryos are least tolerant to higher temperatures/heat stress. Therefore, it has to be noted that the temperature of the embryo is more important than that of the egg; but unfortunately, there is no convenient method to measure the temperature of the embryo.

6. Systems of incubation

6.1. Multistage incubation

The incubator/setter will have several batches of eggs; in other words, at any point of time, there are eggs at different stages of incubation. Therefore, it is very difficult to provide ideal temperature to all batches of eggs; consequently, no more improvement is probably expectable from multi-stage incubators by just hatchery managemental modifications. In addition, over-hauling of the incubation machinery is difficult because one or the other batch of eggs will always be present in the setter. The system is not labor-friendly, poses problems in efficient control of humidity and/or temperature as well as for regular sanitation procedures.

6.2. Single stage or all-in all-out incubation

Only one batch of eggs will be present in the incubator and hence is very similar to natural brooding. Therefore, modifications in temperature to maximize hatch results is possible; higher temperature can be provided initially to offset heat loss because of evaporation of moisture from the eggs at the beginning and later on, the temperature is then reduced to the average requirement; temperature is further reduced by the last part of incubation, all these simulating natural brooding. This system facilitates sanitation, effective control of humidity and/or temperature, and overhauling of the incubating equipment. Hence, durability of the equipment will increase, disease control will be efficient and hatch results will be maximum. Therefore, this system of incubation is becoming more popular especially in very large hatchery operations.

In India, forced-draft incubators and hatchers with multistage system of incubation are common.

7. Testing for fertility

7.1. Destructive method

Before collecting eggs for incubation, the males are allowed with the females at least for 10 days and fresh eggs are obtained at random and broken open to observe for the presence of blastoderm. After ascertaining optimum fertility, the eggs are collected and pooled for setting into an incubator (setter). If optimum fertility is not achieved, reasons for the same should be explored and corrected before collecting hatching eggs.

Since the eggs are broken to test fertility, the method is referred to as "destructive method".

7.2. Non-destructive method

It is necessary to test the incubated eggs for fertility. If large numbers of infertile eggs are incubated, they can be found and discarded, and the extra space used for additional eggs. This test, done by candling, will not injure the young embryos and is reliable for eliminating eggs that will not hatch. Eggs with white shells are easier to test and can be tested earlier than dark shelled eggs.

The eggs are normally tested after 4 to 7d of incubation or on 18th d when they are being transferred to the hatcher (Fig 24.3). Since the eggs are not broken for testing fertility, it is also referred to as "non-destructive method" of fertility testing. The eggs after candling can be grouped into:

7.2.1. Fertiles

Fertile eggs after 4 to 7 d of incubation clearly show development of blood vessels and clearly moving embryo at the centre appearing dark. A live embryo is spider-like in appearance, with the embryo representing a spider's body and the large blood vessels spreading out much like a spider's legs.

At 18th d, a large air-cell at the broad end with the bottom portion opaque due to the presence of developed embryo is extremely easy to identify. In fact, an entire setter tray is candled at a time (mass-candling) and those appearing translucent are remove as infertile ones.

7.2.2. Infertiles

Infertile refers to an unfertilized egg or an egg that started developing but died before growth. An infertile egg after 4 to 7 d of incubation appears as a clear egg except for a slight shadow cast by the yolk and at 18th day they are translucent with a large air-cell; many times, the vitelline membrane is broken and the contents appear uniformly yellowish.

7.2.3. Dead germs

Dead germ refers to an embryo that died after growing large enough to be seen when candled. A dead germ can be distinguished by the presence of a blood ring around the embryo. This is caused by the movement of blood away from the embryo after death. There will be no movement of the embryo and the network of blood vessels, if noticed, will not be extensive.

If not sure whether the embryo is alive or not, incubation of such eggs is continued and retested later. A second test can be on 18th d.

7.2.4. Dead-in-shells

This group comprises embryos which have developed sufficiently butare unable to come out of the shells whether they are able to pierce the shell or not. High incidence of dead-in-shells is indicative of improper hatchery management.

In addition to the above, it is always advisable to break open all the unhatched eggs in a separate facility to find out the actual reason those eggs failed to hatch so that any

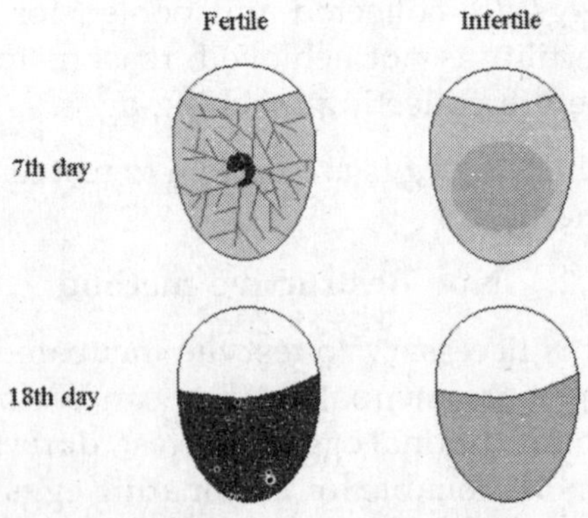

Fig 24.3 Fertility testing during incubation

improvements possible can be taken-up.

8. Factors affecting incubation results

Four factors are of major importance in incubating eggs artificially: temperature, humidity, ventilation and turning. Of these factors, temperature is the most critical. However, humidity tends to be overlooked and causes many hatching problems. Extensive research has shown that the optimum incubator temperature is 37.8°C when relative humidity is 60 %. Concentrations of oxygen should be above 20 % (optimum 21%), carbon dioxide should be below 0.5 % (optimum 0.04%), and air movement past the egg should be 0.3 m^3/min.

8.1. Temperature

Thermometer used for recording temperature must be checked for accuracy because an error of even 1°C for 21 d can seriously interfere with an embryo's growth. To check an incubator thermometer, place its bulb next to the bulb of a clinical (oral kind used to check body temperature) or a laboratory thermometer. Hold them under lukewarm water and compare the readings. Both should read the same temperature, so make adjustments for any error in the incubator thermometer. A thermometer in which the mercury column is separated must be discarded.

The longer an egg is held at a temperature above 24°C prior to setting, the shorter the incubation period. Embryo development is halted below 24°C. Increasing the incubating temperature shortens the incubation period. Hatching eggs are stored prior to incubation at around 13 to 15.5°C for waterfowl and game birds and 12.8 to 18.3°C for chicken and turkeys, with relative humidity of 70-75%. Eggs stored more than 7 days should be kept at 12.8°C. Hatching eggs should be stored at a constant temperature as fluctuating temperatures reduce hatchability.

Eggs, even when stored in a cold store, should be incubated as soon as possible to prevent reduction in hatchability. The hatchability of eggs older than 7 to 10 d decreases rapidly; at 3 weeks, it reaches 0%. If the eggs are not incubated within 3 to 4 d, they have to be turned daily. Turning the eggs prevents the yolks from touching the shell and injuring the embryo.

The hatching eggs from the cold storage should never be directly loaded into a setter because

1. The embryos present inside the setter (in case of multistage incubation) as well as those which are being loaded will be subjected to cold shock and heat shock, respectively, severely reducing hatchability.
2. Temperature of the setter reduces depending on the number of eggs loaded from cold storage which triggers solenoids put on heaters in compensation.
3. When the eggs from the cold store are loaded, there will be condensation of moisture leading to sudden drop in relative humidity of the setter.

4. Therefore, solenoids controlling water flow allow more water to flow into the setter to compensate for the reduced relative humidity.
5. Subsequently, moisture condensed on eggs evaporates resulting in sudden increase in relative humidity; since there is no provision to reduce relative humidity, all eggs will be exposed to high humidity till it is brought back to normal after fresh air circulating inside the setter.
6. Therefore, not only all the eggs incubated are affected but also the incubation machinery is put to unnecessary strain which ultimately leads to reduced hatchability and early mechanical problems.

8.2. Humidity

The relative humidity of the air within an incubator should be about 60 %. During the last 3 d (the hatching period) the relative humidity should be nearer 65-70 %. (Too much moisture in the incubator prevents normal evaporation and results in a decreased hatch, but excessive moisture is seldom a problem in small incubators.) Too little moisture results in excessive evaporation, causing chicks to stick to the shell, remain in the pipped shells, and sometimes hatch crippled. Reducing the humidity during incubation lengthens the incubation period.

During the hatching period, the humidity in the incubator may be increased by using an atomizer to spray a small amount of water into the ventilating holes. (This is especially helpful when duck or goose eggs are hatching). Increased relative humidity during the last 3 d has the following functions:

1. To keep the beaks of pipping chicks wet thereby facilitating the process of chicks coming out of the shells
2. Some of the hair-like feathers (down) from the chicks will be floating in the hatcher due to air movement created by the inlet fans. These feathers absorb the additional moisture provided and settle down due to their own weight.

8.3. Turning

8.3.1. Characteristics of embryos in unturned eggs (Deeming, 1990)

1. Retardation of growth of the *area vasculosa*
2. Reduced amount of sub-embryonic fluid
3. Poor growth of chorio allantoic membrane
4. Failure to form albumen-sac
5. Poor orientation between the embryo, yolk-sac and albumen-sac
6. Reduced amount of allantoic fluid on 12th d
7. Retarded movement of albumen into amniotic fluid
8. Reduced amount of amniotic fluid on 14th and 16th d
9. Poor utilization of albumen

10. Large amounts of residual albumen in unhatched eggs
11. Low oxygen tension in blood
12. Retarded embryonic growth during second half of incubation
13. Low hatchability

8.4. Ventilation

The best hatching results are obtained with normal atmospheric air, which usually contains 20-21 % oxygen. It is difficult to provide too much oxygen, but a deficiency is possible. Make sure that the ventilation holes are adjusted to allow a normal exchange of air.

8.5. Others

1. The older the egg at setting, the longer the incubation period.
2. The larger the egg, the longer the incubation period.
3. The longer a female is in production, the longer the incubation period of her eggs.
4. Chick weight increases as breeders turn older; in case of broiler breeders, egg size increases from 50 to 70 g and therefore, size of the chick increases from 32 to almost 45 g. Further, as breeder age increases, chicks tend to have larger residual yolks in their abdomen which will act as a ready supply of feed and water thereby making them more sturdy during holding period (Leeson, 2003).
5. Certain stresses and diseases in the breeder flock will lengthen the period of incubation.
6. Embryos with certain malpositions require a longer time to emerge from the shell.
7. Some forms of egg dipping which is used to destroy bacteria on the shell, delay the hatching time.
8. The smaller the breed, the shorter the incubation period.
9. Do not help the chicks from the shell at hatching time. If it doesn't hatch, there is usually a good reason. Also, prematurely helping the chick hatch could cripple or infect the chick. Humidity is critical at hatching time. Don't allow your curiosity to damage your hatch.
10. As soon as the chicks are dry and fluffy or 6 to 12 hr after hatching, remove the chicks from the incubator.
11. It is good practice to remove all the chicks at once and destroy any late hatching eggs.
12. Hatching time can be hereditary and you can control the uniformity of hatching by culling late hatchers. If you keep every chick which hatches late, in a few years each hatch could last 4 d or longer.

13. No matter what type of incubation you use, it is important that you thoroughly clean and disinfect the incubator before and after you use it. It is just as important that the incubation room and egg storage area are kept equally clean. The lack of sanitation will decrease hatchability. Incubator (Setter) should be cleaned and disinfected before each use. Wash the unit with a warm detergent solution and rinse with a disinfectant solution. This reduces the chances of carrying disease-causing organisms from one batch of eggs to another.
14. Immediately after each hatch, thoroughly clean and disinfect all hatching trays, water pans and the floor of the hatcher. Scrape off all egg shells and adhering dirt. Wipe clean surfaces thoroughly with a cloth dampened in quaternary ammonium, Clorox or other disinfectant solution.

9. Life in 21 days

One of the greatest miracles of nature is the transformation of the egg into the chick. A chick emerges after a brief three weeks of incubation and the major events are summarized in Fig 24.4, Table 24.6 and Plate 24.1.

9.1. Before oviposition

Cell division begins soon after fertilization, even while the rest of the egg is being formed. Cell division will continue if the egg is kept warmer than 19.4°C. The first cell division is completed about the time the egg enters the isthmus. Additional cell divisions take place about every 20 min; so, by the time of lay, several thousand cells form two layers of cells called a "gastrula."

At this time the egg is laid, it cools, and embryonic development usually stops until proper environmental conditions are established for incubation. After incubation begins, the cellular growth resumes. At first, all the cells are alike, but as the embryo develops, cell differences are observed. Some cells may become vital organs; others become a wing or leg.

9.2. 1st d

Soon after incubation is begun, a pointed thickened layer of cells becomes visible in the caudal or tail end of the embryo. This pointed area is the primitive streak, and is the longitudinal axis of the embryo. Before the first day of incubation is through, many new organs are forming. The head of the embryo becomes distinguishable; a precursor of the digestive tract, the foregut, is formed; blood islands appear and will develop later into the vascular or blood system; the neural fold forms and will develop into the neural groove; and the eye begins.

9.3. 2nd d

On the 2nd d of incubation, the blood islands begin linking and form a vascular system, while the heart is being formed elsewhere. By the 44th hr of incubation, the heart and vascular systems join, and the heart begins beating. Two distinct circulatory systems are established; an embryonic system for the embryo and a vitelline system extending into the egg. The neural groove forms and the head portion develops into the parts of the brain. The embryo is developed enough that flexion and arching of the embryo begins, the ears begin development, and the lens in the eyes are forming.

9.4. 3rd d

At the end of the 3rd d of incubation, the beak begins developing and limb buds for the wings and legs are seen. Three visceral clefts (gills) have formed on each side of the head and neck. These formations are important in the development of the arterial system, eustachian tube (in the ear), face, jaw, and some ductless glands. The fluid-filled amnion has surrounded the embryo to protect it: it helps maintain proper embryonic development. The tail appears, and the allantois is seen. The allantoic vesicle is a respiratory and excretory organ. Nourishment from the albumen and calcium from the shell are transported to the embryo through the allantois.

9.5. 4th to 6th d

Torsion and flexion continue through the fourth day. The chick's entire body turns 90° and lies down with its left side on the yolk. The head and tail come close together so the embryo forms a "C" shape. The mouth, tongue, and nasal pits develop as parts of the digestive and respiratory systems. The heart continues to enlarge even though it has not been enclosed within the body. It is seen beating if the egg is opened carefully. The other internal organs continue to develop. By the end of the 4th d of incubation, the embryo has all organs needed to sustain life after hatching, and most of the embryo's parts can be identified. The chick embryo cannot, however, be distinguished from that of mammals.

Many complex physiological processes take place during transformation from the egg to the chick. They include: the use of highly nutritious food materials in the egg; the respiration of gases, or the taking in of oxygen and the removal of carbon dioxide; and the building of living energy within the chick.

9.6. 7th d and onwards

The embryo grows and develops rapidly. By the 7th d, digits appear on the wings and feet, the heart is completely enclosed in the thoracic cavity, and the embryo looks more like a bird.

After the 10th d of incubation, feathers and feather tracts are visible, and the beak hardens.

On the 14th d, the claws are forming and the embryo is moving into position for hatching. The supply of albumen is exhausted by the sixteenth day, so the yolk is the sole source of nutrients.

After 20 days, the chick is in the hatching position, the beak has pierced the air cell, and pulmonary respiration has begun. The yolk sac is contained completely within the body cavity in preparation for hatching.

9.7. Hatching

The normal position of the chick for hatching is with the head in the large end of the egg, under the right wing, with the legs drawn up toward the head. If the head is positioned in the small end of the egg, the chick's chances of survival are reduced by at least one-half. This is a serious malposition, or wrong position, for hatching.

After 21 d of incubation, the chick finally begins its escape from the shell. The chick begins by pushing its beak through the air cell. The allantois, which has served as its lungs, begins to dry up as the chick uses its own lungs. The chick continues to push its head outward. The sharp horny structure on the upper beak (egg tooth) and the muscle on the back of the neck help cut the shell. The chick rests, changes position, and keeps cutting until its head falls free of the opened shell. It then kicks free of the bottom portion of the shell. The chick is exhausted and rests while the navel openings heal and its down dries. Gradually, it regains strength and walks. The incubation and hatching is complete. The horny cap will fall off the beak within days after the chick hatches.

Weight of the embryo (White Leghorn) from 1st d to 20th d has been recorded as 0.0002, 0.003, 0.02, 0.05, 0.13, 0.29, 0.57, 1.15, 1.53, 2.26, 3.68, 5.07, 7.37, 9.74, 12.00, 15.98, 18.59, 21.83, 25.62, 30.21 g, respectively. The newly hatched chick usually weighs 60 to 65% of the weight of the egg from which it hatched.

Chicks in hatcher are usually fumigated with 1X concentration for 3 min which gives an attractive bright yellow color to the chicks at hatch (details of fumigation are given in a separate Chapter). This process has also been found to cause severe damage to the epithelium of the respiratory system of the chicks (Gerrits and Dijk, 1991); but, fumigation of chicks in the hatcher is still practiced due to non-availability of any other convenient method in its place.

Newly-hatched chicks can be shipped long distances (up to 72 hr travel time) without food. Chicks have to be provided with feed and water on the first day of life so they can learn to eat and drink immediately. The yolk is largely unused by the embryo and is deposited within the chick's body on the 19th d, just before it hatches. The yolk is highly nourishing and provides proteins, fats, vitamins, minerals, and water for several hr after hatching. The yolk is consumed gradually during the first 10 d of the chick's life.

9.7.1. Hatching maturity (Tazawa and Whittow, 2000)

Hatching maturity differs considerably in different Avians species and the hatchlings fall into the following four categories based on criteria such as mobility amount of down and ability to feed themselves: 1. Precocial (most mature) which can walk, swim or dive soon after hatching 2. Semi-precocial 3. Semi-altricial and 4. Altricial (least mature) which are naked with their eyes closed and incapable of locomotion.

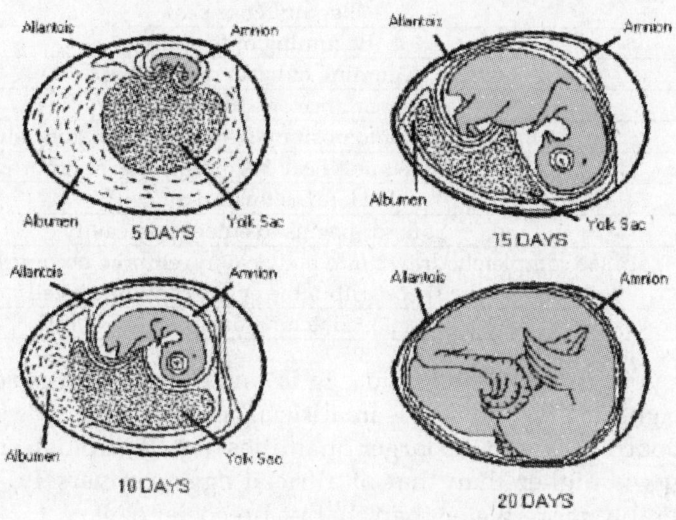

Fig 24.4 Stages of embryo development

Table 24.6 Summary of embryo development

Before Egg Laying	
Fertilization, Division and growth of living cells, Segregation of cells into groups of special function (tissues)	
Between Laying and Incubation	
No growth; stage of inactive embryonic life	
During Incubation	
Day	Development
1st d	
16 hr	First sign of resemblance to a chick embryo
18 hr	Appearance of alimentary tract
20 hr	Appearance of vertebral column
21 hr	Beginning of nervous system
22 hr	Beginning of head
24 hr	Beginning of eye
2nd d	
25 hr	Beginning of heart
35 hr	Beginning of ear
42 hr	Heart beats

3rd d	Table 24.6 contd...
60 hr	Beginning of nose
62 hr	Beginning of legs
64 hr	Beginning of wings
4th d	Beginning of tongue
5th d	Formation of reproductive organs and differentiation of sex
6th d	Beginning of beak
8th d	Beginning of feathers
10th d	Beginning of hardening of beak
13th d	Appearance of scales and claws
14th d	Embryo gets into position suitable for breaking shell
16th d	Scales, claws and beak becoming firm and horny
17th d	Beak turns toward air cell
19th d	Yolk sac begins to enter body cavity
20th d	Yolk sac completely drawn into body cavity; embryo occupies practically all the space within the egg except the air cell
21st d	Hatching of chick

The amount of yolk in the freshly laid egg is much greater in precocial species (69% of egg weight in Kiwi) than it is in altricial species (only 16% of egg weight in Red-footed Booby). Due to the larger quantities of yolk (lipids), energy content of precocial eggs is higher than that of altricial eggs; conversely, altricial eggs contain considerably more water than that of precocial eggs.

9.8. Extra-embryonic circulation system

In later stages of embryonic development, there are two distinct extra-embryonic blood systems. One system, the vitelline system, transports nutrients from the yolk to the growing embryo. Before the 4th d, it oxygenates blood. The other blood system, made of allantoic vessels, is concerned with respiration and the storage of waste products in the allantois. When the chick hatches, both circulatory systems cease to function.

10. Utilization of egg contents and shell by the embryo

10.1. Water utilization (Tazawa and Whittow, 2000)

Water loss from the egg inevitable due to the presence of shell pores. The rate of water loss designated M_{water} (mg/d) is determined by:

1. Water vapor conductance of the shell and shell membranes (mg/d/torr) designated G_{water} – is largely a function of the number of pores in the shell and the shell thickness both of which increase in larger eggs; the former increases while the latter decreases M_{water}. M_{water} is greater in larger eggs and therefore increase in number of pores with increasing size is the predominant factor.
2. The difference in water vapor pressure between the contents of the egg and egg's micro-environment (torr) designated ΔP_{water} – is largely a function of

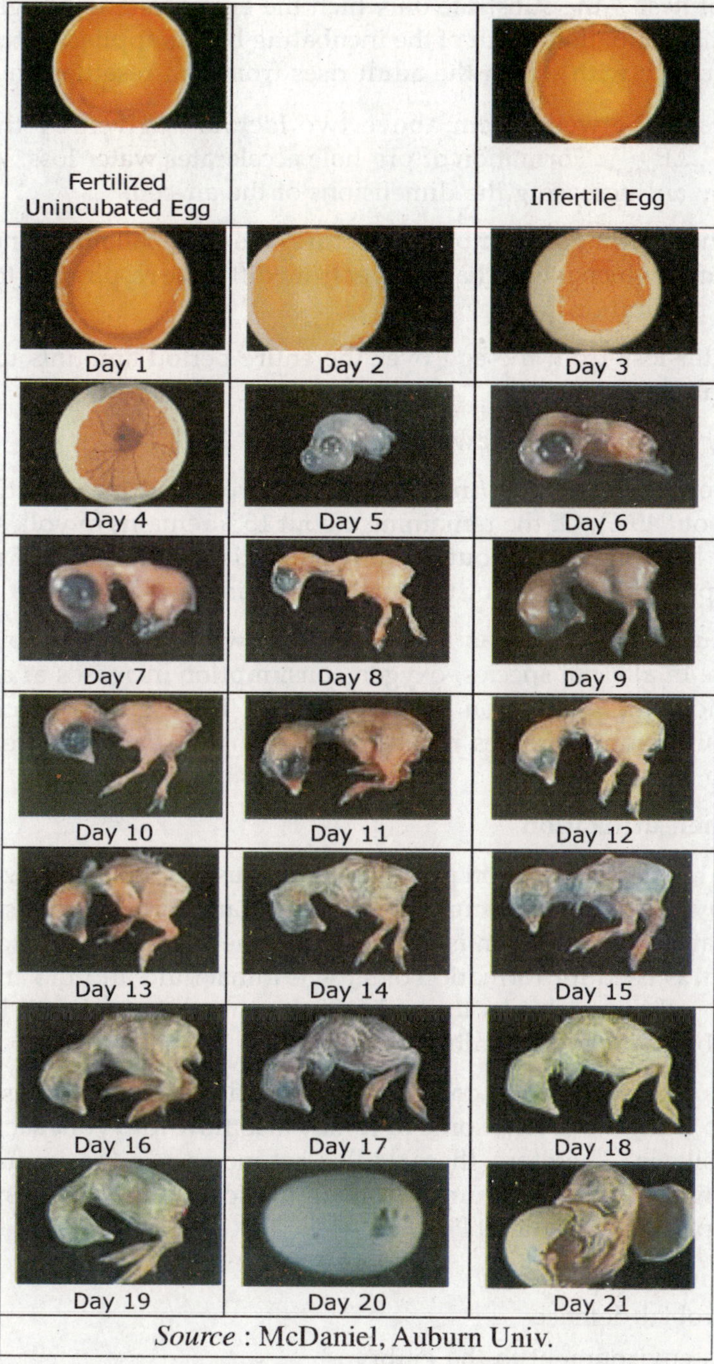

Source : McDaniel, Auburn Univ.

Plate 24.1 Chick embryo development

nature of nest / the substrate on which the egg is laid and "tightness" with which the incubation patch of the incubating bird is applied onto the egg and the frequency with which the adult rises from and resettles on the egg.

The relationship between them above two factors is given by the equation: $M_{water} = G_{water} \cdot \Delta P_{water}$. Formation of pip-hole accelerates water loss. Water so lost is replaced by air increasing the dimensions of the air-cell.

During internal pipping, the embryo punctures the chorioallantoic and inner shell membranes after which it will change over from diffusive respiration to pulmonary (convective) respiration.

The total water loss from the egg over the entire period amounts to 18% of the mass of freshly laid egg.

10.2. Energy utilization (Tazawa and Whittow, 2000)

Most of the energy contained in the egg is incorporated into the tissues of the hatchling (about 45%). Of the remaining, about 15% remains in yolk sac (which is utilized after hatching) and about 6% in residual tissues. The cost of utilization of energy is approximately 34% of the total energy in the fresh egg.

The energy usage increases as the embryo grows as indicated by its oxygen consumption. In altricial species, oxygen consumption increases at an increasing rate throughout the incubation period whereas, in case of precocial eggs, the oxygen consumption decreases just before pipping and again increases sharply after pipping.

10.3. Albumen utilization

Begins from 13th d of incubation presumably because of the small size of embryo. Albumen moves through sero-amniotic connection and through yolk-sac umbilicus. However, water from albumen is utilized from the very beginning of incubation and most of it is used for formation of sub-germinal fluid and the amniotic fluid; consequently albumen drinks towards the narrow pole and will be gradually surrounded by CAM to form albumen-sac.

Perforation of sero-amniotic connection around 13th d facilitates moment of albumen into amniotic cavity; some of it may also flow into yolk-sac, presumably through umbilicus. Therefore, there will be net increase in the amount of routine in yolk during this period. Albumen from amniotic fluid is actively swallowed by the embryo particularly after 14th d.

10.3.1. Functions of albumen

1. antimicrobial defence
2. aqueous environment to the embryo
3. reservoir of water and protein and

4. supply of specific proteins like Ovotransferrin (conalbumin) and Calcium binding proteins.

Table 24.7 Composition of albumen during incubation

Day of incubation	Water	Total solids	Proteins	Carbohydrates
0	29.5	4.2	3.5	0.310
5	15.0	3.5	3.2	0.250
10	7.0	3.3	3.2	0.175
13	5.0	3.3	3.2	0.100
15	4.0	2.3	2.8	0.025
17	3.0	0.8	0.8	0.010
18	1.5	0.2	0.2	Nil
				Source : Romanoff, 1972

10.4. Yolk utilization

In the beginning, at the formation of primitive streak, the endodermal cells digest yolk after phagocytosis; the end products are sent to embryo through blood vessels. Endoderm of the yolk-sac also secretes enzymes which can digest yolk material extracellularly from 2nd d of incubation itself.

However, significant utilization begins from 5th d of incubation. The products of digestion and undigested larger molecules are transported into the embryo via vitelline circulation.

10.4.1. Functions of yolk-sac membrane

1. digestion and absorption
2. synthesis of specific proteins, amino-acids and blood
3. source of precursor cells for thymus and bursa of Fabricius
4. glycogen storage and
5. temporary excretory organ.

It is interesting that although yolk-sac is an extension of gut, very little material, if any, from yolk-sac directly goes to gut. Endoderm first takes-up yolk, passes on to mesodermal blood vessels on the way to embryo in presence of trimetaphosphatase enzyme. This is particularly important because there is no lymphatic system in the embryo. Most of the proteins also follow the same circuitous route i.e., yolk to albumen to amniotic fluid to digestive tract and finally to circulation.

Proteolytic enzyme activity begins as early as 4th d of incubation and protein digestion is mainly extracellular although intracellular digestion is also possible. After 13th d, proteins from albumen enter yolk-sac umbilicus (?) and this is continuous till the end of incubation.

Lipids are presumably absorbed intact (without hydrolysis) especially after 13th d of incubation.

Table 24.8 Composition of yolk during incubation

Day of incubation	Water	Total solids	Lipids	Proteins	Carbohydrates
0	9.00	10.0	6.30	3.2	0.150
5	11.00	10.0	6.10	3.2	0.160
10	9.50	9.0	6.00	2.5	0.230
13	7.00	8.0	5.50	2.1	0.275
15	6.00	7.0	5.20	2.4	0.250
20	2.50	6.0	2.75	3.0	0.180
21	2.25	5.0	1.75	2.8	0.150
					Source : Romanoff, 1972

The small proportion of amino-acids absorbed is used for synthesis of glucose *de novo* because the glucose present in egg is exhausted by 10th d of incubation. Gluconeogenesis begins by 4th d and reaches a peak by 16th or 17th d of incubation and declines thereafter. Ala, Glu and possibly Ser are the amino-acids which might be utilized by the embryo for synthesis of glycogen in the liver which begins differentiating on 6th or 7th d. Most of the lipids are stored in liver was glycogen is stored more in yolk-sac membrane than in liver.

10.5. Calcium utilization and acid-base balance

Shell has considerable resistance for diffusion and hence CO_2 content increases in embryonic blood resulting in respiratory acidosis. This is buffered by plasma HCO_3 obtained from kidney and/or shell. Kidneys exchange H^+ for Na^+ (which is excreted as NH_3 or with chloride resulting in acidic urine) and reabsorb HCO_3. This is facilitated by pH of allantoic fluid which reduces from 8.0 to 6.0.

However shell is a major source of HCO_3. Embryos have to obtain Ca mainly from the shell and, in fact, calcium salts are deleterious and even teratogenic to early development possibly because of their calcium binding properties.

Egg shell Ca is thought to be resorbed in a reverse reaction of shell formation catalyzed by the same carbonic anhydrase enzyme through CAM from 10th d of incubation and onwards. CAM touches the inner shell membrane on 10th d and covers greater part of it by 14th d. About 100 mg of Ca is resorbed between 10th d till the end of incubation and hence of 150 mg of HCO_3 are available for buffering activity. Hence, level of HCO_3 increases after 10th d. Surprisingly, that much of HCO_3 also is not required and excess of HCO_3 must be excreted by kidneys (which start functioning by 11th d) lest pH increases to the detriment of the embryo itself.

10.6. Energy requirements, excretion and water balance

10.6.1. Energy requirements

Small quantity of glucose (about 500 mg) available in the egg is used during 3rd to 6th d through pentose phosphate shunt which becomes predominant in brain during 12th to 15th d. Lipids supply most of energy required and hence formation of ketone bodies is inevitable (120 μg/ml at 14th d, 177 μg/ml at 17th d, 100 μg/ml at hatch)

10.6.2. Sub-embryonic fluid

This is located just below the blastoderm/germinal disc formed by endoderm (blastodermal) from 2nd d reaching a maximum volume by 7th d and persisting till 15th d of incubation. Most of the water is contributed by albumen and partly by yolk. This fluid, rich in minerals, provides proper physiological medium and supports normal development of the embryo without itself being a source of nutrients. Hence, reduced sub-embryonic fluid, as in unturned eggs, results in the reduced hatchability.

10.6.3. Excretion and water balance

The major nitrogenous excretory product of chicken embryo is uric acid which can be formed from 5th day and onwards. Urine is formed from 4th or 5th d and is collected in allantoic sac; its volume increases to 7 ml by 11th d consisting mainly of NH_3, urea and uric acid. Allantoic endoderm can also reabsorb water from 13th d and onwards. The volume of hypotonic urine progressively reduces from 13th d and intimately becomes a semi solid precipitate of urates. Resorption of water accompanies active transport of Na^+ and Cl^- by the allantoic endoderm. By 18th d, degeneration of allantois begins, and in newly hatched chick, probably cloaca with hind gut performs extra-renal absorption of water and Na^+.

Oxidation of nutrients also produces water, referred to as metabolic water, which is available for the embryo. For instance, a standard egg weighing 56.7 g has 39 g water; when incubated, a chick weighing 40 g containing 32 g of water will hatch. During incubation, about 10 g of water is lost which indicates that the newly hatched chick has 29 g of water from the egg and 3 g from its metabolic activities.

10.7. Gaseous exchange

Oxygen is primarily required for respiration/production of energy. Therefore, as expected, oxygen uptake increased from about 1.3 ml/d to 537 ml/d while the production of carbon dioxide increased from 5.8 ml/d to 429 ml/d. The total uptake of oxygen, per egg, was about 4.6 l whereas total carbon dioxide produced was about 2.8 l in the setter. This enormous gaseous exchange is impossible without loss of water; the latter should be within limits to prevent dehydration and death of the embryo. Hence, there is loss of water, especially during the beginning of incubation which eventually results in increase in air-cell volume from 0.15 ml to 11 ml or with progressive increase in partial pressure carbon dioxide and decrease in that of oxygen.

Gaseous exchange, after 15 d of incubation, tends to become constant whereas growth of embryo continues at relatively constant even after 12th d. Therefore, oxygen intake per g of embryo (respiratory intensity) reduces. In other words, this arrival of the embryo itself is very delicately balanced after 15 d of incubation.

Therefore embryo compensates with the following adaptations to ensure sustained oxygen supply:

1. Early development of blood vascular system; Hb synthesis begins as early as 35 hr of incubation.
2. Increased blood volume – maximum by 15th d
3. Increased number of RBCs
4. Heart begins to function as early as 33 to 38 hr of incubation and heart rate is as high as 275/min by 10th d; increased cardiac output – 4.8 ml/min at 12th d to 6.3 ml/min at 17th d
5. Increased mean Hb concentration from about 4 g/100 ml at 8th d to 9 g/100 ml at hatching
6. Embryonic Hb has higher oxygen affinity than that of adult Hb

Table 24.9 Gross O_2 uptake and CO_2 production by chicken embryo

Age (d)	O_2 uptake Romanoff, 1941	O_2 uptake Romijn and Lokhorst, 1960	CO_2 output Romanoff, 1967	CO_2 output Romijn, 1954
1	1	1	3	3
2	2	2	3	4
3	4	3	3	4
4	6	7	5	10
5	9	13	8	11
6	14	20	13	13
7	24	23	20	20
8	28	40	25	25
9	35	53	34	34
10	53	80	48	46
11	66	132	71	69
12	101	193	102	114
13	125	278	133	155
14	193	343	167	209
15	226	398	204	221
16	264	417	232	239
17	289	395	260	265
18	300	420	270	261
19	3003	373	298	262
TOTAL (ml) *	2942	4609	2740	2830

* Note : µl/min x 1.44 = ml/d — *Source* : Freeman and Vince, 1974

Table 24.10 Moisture content of shell and shell membranes (%)

Days of incubation	Fertile egg	Infertile egg
1	70	70
2	64	70
5	55	70
8	45	70
10	44	70
14	42	68
17	41	62

Source : Freeman and Vince, 1974

Therefore, it follows that ventilation requirements for chicken embryos are particularly important during the 3rd week of incubation.

Oxygen concentration required during incubation and hatching is 21% and for every 1% reduction in oxygen level, there will be a concomitant reduction to an extent of 5% in hatchability. If the oxygen concentration is less than 18%, the decline becomes more rapid and at less than 13%, none of the eggs will hatch.

In forced-draft setters and hatchers, fresh air is forcibly brought inside the anti-clockwise rotation of the fans located at the back of the machines and in proximity to an opened window. Hence in such machines, requisite level of oxygen is met; availability of oxygen reduces at higher altitudes due to reduced partial pressure; under such conditions, additional oxygen may have to be pumped into the setters and hatchers.

10.7.1. Functions of CO_2

Carbon dioxide is essential for

1. Dissociation of calcium from shell and subsequent transport to the embryo for skeletal development
2. Giving muscle twitch/muscle tone when the chicks are pipping and coming out of shells
3. Initiating respiration in the embryo (when it pips) by stimulating respiratory reflex.

The optimum concentration of carbon dioxide is 0.04% although embryos can tolerate 0.4 to 0.6%. In forced-draft incubators, the problems of carbon dioxide concentration is rare.

Duck eggs hatch better in still-air incubators and hence, operational temperature must be increased by 1.2 to 1.8°C in both setter and hatcher. In case of goose eggs, sprinkling warm water or dipping in luke-warm water for 30 sec each day during last half of incubation has been found to improve hatchability.

11. Embryo mortality pattern

In case of chicken eggs (and at proportionate time in other species) it is generally agreed that embryo mortality during the 1st and 2nd week is due mainly to inadequate breeder nutrition and that during the last week is due to faulty incubation management (Table 24.11).

Mortality during the 1st week of incubation is normally due to the way the fertile eggs are handled from farm to the incubation facility and/or to the health of the frock in question (diseases and/or deficiencies). Eggs must be treated with great care at this stage. They must be properly transported with as little shaking/ jarring as possible. Contaminated eggs (floor eggs which are dirty, cracked eggs, double-yolk eggs etc.) should not be transported to the incubation facility. Other than nutritional deficiency of the breeders, the most common cause of early dead germs is improper egg storage.

Mid-term (2nd week) mortality are confined to either nutritional deficiencies are to a carry-over of earlier contamination problems. This period is a rapid growth period and major structures are formed. The most common lesions associated with nutrition are clubbed down (riboflavin deficiency), micromelia and parrot beak (deficiency of biotin, genetic defect).

Embryo mortality during the last week of incubation is mainly due to failure of the embryo to position itself within the egg in "hatching position" i.e., parallel with the long axis of the egg, the head the blunt end and the beak under the right wing pointing to the air-cell. Any position which differs from the above is likely to result in failure or a delay in hatching.

In addition, up to 19th/20th d, the embryo breathes through chorioallantois which grows around the inner shell membrane inside the egg; in other words, the chorioallantois acts as embryonic lung. After the penetration of air-cell, the embryo gradually starts to relate more and more upon the lung function for exchange of gases than on chorioallantois. The yolk must also be withdrawn into the body cavity to sustain the embryo during immediate post-hatch period.

Therefore, proper management of temperature, humidity, turning and other requirements during the entire period of incubation, in general, and during the last 3 d of incubation, in particular, is mandatory to obtain desirable results.

Table 24.11 Embryo mortality pattern

Peak	Days of incubation	Reason (s)
Major	2nd, 3rd, 4th d	a) Faulty heart formation – heart starts beating during 33 to 38 hr of incubation and if there is any failure, embryos die at this early stage and are referred to as early embryonic death b) Nutritional deficiency of the breeding flock, especially of Vitamin A and K c) Storage hygiene d) Incubation handling – turning e) Old eggs f) Respiratory infection and damage to oviduct
	18th, 19th, 20th	a)The embryos are changing over from poikelothermy to homoeothermy b) beginning of pulmonary respiration. If hatchery management is defective (high/low temperatures, vitamin deficiency/ies, damage at transfer, high 1st week temperatures), well-grown embryos but dead, referred to as, dead-in-shells are recorded c) Storage hygiene d) Disease(s)
Minor	4th, 5th, 6th	Embryos are changing over form carbohydrate protein metabolism; hence there will be accumulation of carbon dioxide, ammonia and lactic acid. If ventilation is improper, the embryos are likely to die
	14th, 15th	a) The organs of the embryo organize themselves (organogenesis) and liver and other organs will be functioning. Hence, deficiency of nutrients in the egg, especially of Vitamin B_2, results in failure of organogenesis and the embryos die with characteristic clubbed-down syndrome. b) Hygiene in the setter

12. Nutrition and metabolism of chicken embryo

Embryos require amino-acids (Table 24.12) and all other important nutrients (Table 24.13) required by a normal healthy chick for various functions. Physiologically, high K to Na ratio (K/Na = 5) appears to be important in osmotic relationships in the yolk-sac fluid.

Table 24.12 Amino-acid requirements of chicken embryo

	For growth	For survival
Required	Leu, Lys, Met, Phe	Leu, Lys, Met, Phe, Trp, Val
Not required	Arg, Pro	Asp, Pro
Doubtful	Arg, Trp, Val, Tyr	Arg, Tyr

13. Evaluating performance

Even under natural conditions, neither all the eggs laid will be fertile nor all eggs will hatch; hence, with artificial incubation also, there will be eggs which will be either infertile or fail to hatch for various reasons already discussed.

Therefore, it is necessary that performance of the breeding farms well as the artificial incubation facility be evaluated to take up suitable measures to obtain maximum profit out of hatchery business.

In view of the above, the following parameters are estimated to assess the efficiency of both breeding farm as well as incubation facility:

$$\text{Fertility, \%} = \frac{\text{Number of eggs fertile}}{\text{Total number of eggs set}} \times 100$$

$$\text{Hatchability, \% (on TES)} = \frac{\text{Number of chicks hatched}}{\text{Total number of eggs set}} \times 100$$

$$\text{Hatchability, \% (on FES)} = \frac{\text{Number of chicks hatched}}{\text{Number of fertile eggs}} \times 100$$

Among the above estimates, fertility is the prerequisite for incubating any egg and hence it can not account for efficiency of artificial incubation and hatching process; however, it definitely indicates health and management of the breeding stock.

Hatchability on total eggs set (on TES) is less accurate because the denominator includes all the eggs set, whether fertile or not and hence, will be adversely affected as fertility reduces as indicated by the following hypothetical example (Table 24.14):

Table 24.13 Functions of nutrients to the embryo

Nutrient	Function (s)
Amino-acids	Survival, growth, down-feather formation
Glucose	Growth and development; energy source
Linoleic acid	Essential fatty acid

Nutrient		Function (s)
Minerals	Ca	Skeletal development
	P	Skeletal development, energy transfer
	Mn	Enzyme cofactor, skeletal and feather development
	Zn	Maturation of hepatocytes, calcium transport from Shell to embryo, feather development
	Cu	Early development of chick, synthesis of connective tissue (blood vessels)
	I_2	Thyroid development and function
	Fe	Heam synthesis, feather pigmentation
	Na and K	Osmoregulation
	Mg	Energy metabolism, muscle contraction
Vitamins	A	establishment of circulatory system in a.v, bone mineralization
	D_3	Calcium metabolism
	E	development of proper circulation
	K	Blood coagulation
	B_1	Carbohydrate metabolism
	B_2	Myelination and development of nerves
	Niacin	normal muscle development
	B_6	Growth, protein utilization
	Biotin and Folic acid	Skeletal development, purine synthesis
	Pantothenic acid	Capillary health and acid-base balance
	B_{12}	Lipid metabolism, uptake of iodine by the thyroid

In both the situations, if 1000 eggs are incubated, 675 chicks are obtained. In situation A, embryos failed to hatch and in situation B, many eggs failed to contain an embryo.

Therefore, it is clear that a poor hatchability on TES alone is a poor index for evaluating incubation performance and it must be studied in conjunction with hatchability on FES.

Table 24.14 Hypothetical incubation results

	Fertility, %	Hatchability, %	
		On fertile eggs (FES)	On total eggs (TES)
Situation A	90	75	67.5
Situation B	75	90	67.5

1. If hatchability on TES and FES are low, problem(s) lie(s) both before and after the eggs are received at the hatchery.
2. If hatchability on TES is low and hatchability on FES is high, problem(s)

lie(s) before the eggs are received at the hatchery.

3. If fertility of the eggs is satisfactory (say above 95 %), and hatchability on FES is low, problem(s) lie(s) after the eggs are received at the hatchery.

Therefore, calculating fertility and hatchability (both on FES and TES) help identify the actual problem(s) so that suitable remedial measures can be initiated.

A good breeding farm cum hatchery must record a fertility of at least 95% and hatchability on TES and FES of 90% and 95%, respectively.

13.1. Performance losses in incubation : Critical management points (Hill, 2004 a,b)

Incubation determines the embryonic development of the chick, which includes the development of the heart, intestine, and immune system. This means that the incubation determines the value potential, or field performance and mortality.

Nature defines a physiologically mature chicken at hatch by

$$5.13 = \frac{\text{Incubation period X Conductance of the egg}}{\text{Egg weight}}$$

All the factors in the equation are controlled by the hen and excepting incubation period, the others are fixed. Therefore, it is incubation period which has to be optimized. But, incubation period is dependent on various factors like heat production by the embryo, age of the breeder flock, season, breed, airflow within the egg mass, type of incubation, altitude etc.

13.1.1. Heat produced by the embryos

The embryos do not produce significant heat until approximately 9 d of age; therefore, the embryo temperatures of less than the air temperature for the 1st 9 d. Once the embryo begins to produce heat, the embryo temperature rises above that of the surrounding air. Hence, it becomes mandatory that any changes that are made must be a suitable compromise for both critical stages in the incubator, especially in case of multistage incubators. Current research suggests that the optimal embryo temperature is 37.8°C.

The actual temperature of the embryo depends on the age of the embryo and where in the machine the embryo is located. In the 1st 9d, the embryo in the bottom of the machine, where the air is cooler, generally, have a lower temperature and are smaller at transfer than those in the warmer areas of the machine. As indicated earlier, unfortunately, there is no method to obtain the actual temperature of the embryo.

Actually embryo temperature is a function of the heat production of the embryo, the airflow over the embryo and the relative humidity of the air. The increased heat production by theembryo during the latter part ofincubation is an important factor in the actual embryo temperature. Therefore, variability in the egg mass becomes an extremely important variable dictating the actual embryo temperature. At the commercial conditions, inconsistent air movement among the setter trays will result in differences in cooling of eggs. It follows that the best way is to promote airflow through the egg mass for cooling, instead of excessive moisture or excessive cooling coil is usage.

13.1.2. Age of the breeder flock

Chicks from the youngest breeder flock (26-32 weeks) are smaller in winter and chicks from all other age flocks are smaller in summer. In the typical multistage hatchery, chick size is smaller in the flocks 50 weeks of age is or greater, than in the prime flocks. This explains why small dehydrated chicks are obtained in winter, especially in the youngest breeder flocks whereas, in case of oldest breeder flocks, chicks quality problems are common during the summer.

13.1.3. Season

The effect of season is primarily due to the temperature of the incoming air and its relative humidity; the latter indicating the ability of the air to transfer heat to the embryos and to remove heat from them. Therefore, humidity is a key factor during winter. Artificially humidified air is inferior to naturally humidified air because it is inconsistent, inadequate and cools the air.

If air is heated, it gets dried out; conversely, if air is humidified by traditional means, the air gets cooled and triggers heating system of the incubating machinery which, in turn, decreases the airflow to counteract the cold air. It is for this reason that hatcheries are installing steam humidification so that cooling quality of the air is improved without actually cooling the room.

During summer, the problems are the other way round; the heat is not adequately removed, especially during the last week of incubation, to maintain the optimum 37.8°C embryo temperature. Since the eggs from the oldest breeder flocks produce the most heat, overheating problem increases as flock age increases.

The other aspect of summer problem occurs when the machine must use moisture for cooling and the embryo must lose the moisture in the limitation of a 21-d incubation cycle.

Recent research also indicates that increasing incubation period to 22 d during summer improves feed conversion and breast meat yield. Incidentally, selection for breast meat yield has resulted in concomitant changes in conductance of the egg shell.

Therefore, to hatch a physiologically mature chicken with the yield breeds of today, the incubation period must change to balance the equation for an optimally incubated bird. To do this, the embryos must be incubated to 22 d for growth and yolk utilization provided that embryo temperatures are uniform and optimum.

13.1.3.1. Critical management points – Winter

1. Emphasis should be given on the setter operation – the room should be warm and humid.
2. The setter should be in the cooling not the heating mode because when the setter is heating, the embryos in the first 9 d of incubation are cooler than when the setter is cooling.

13.1.3.2. Critical management points – Summer

1. Emphasis is on the hatcher operation – it should cool with minimal spray because, if moisture enters egg mass, both hatch and chick quality are affected adversely.
2. To make spray as effective as possible, nozzle maintenance and pressure are critical.
3. Relative humidity requirements of the embryo and the machine must be properly balanced. For instance, with heavy breeds, conductance of the shell is reduced making it more difficult for them to lose moisture. In such cases, reducing relative humidity may be beneficial; but, this results in reduced heat loss by the embryos.
4. Under hot humid conditions, air conditioning the setter and hatcher rooms is an ideal solution to maintain optimum room condition for optimum machine function.

13.1.4. Breed

In single-stage incubation, different breeds have different preferences for incubation conditions. In multistage incubation, it is not possible to give consideration for breed and/or variety preferences, even if such differences do exist.

13.1.5. Altitude

With increasing altitude, that heat capacity of the air decreases. For example, at an altitude of 2000 m, the heat capacity of the air is only 78% of that at sea level. Reduced moisture levels at high altitude, further complicates because the machine sprays more and cooling becomes more diffuse. Therefore, spray management becomes crucial with increasing altitude.

13.1.6. Transfer day

As a general rule, the setter is a better environment than the hatcher. Within the normal transfer time for each machine, longer the eggs are in the setter, the better the result in hatchability and chick quality. Transfer time should be dictated by management, not embryo age. The embryo does not automatically need to be in a lower temperature environment at 17, 18 or 19 d. To optimize hatch and quality, the transfer should be adjusted to maintain the optimum embryo temperature within the incubation system.

At pip and hatch, the embryos are susceptible to chilling and therefore preferably kept at setter temperatures until hatch and then stepped down to maintain the comfort of the chick after hatch indicated by the rectal temperature of 40.0-40.6°C. Unfortunately, it is extremely difficult not only to monitor rectal temperatures but also to recommend a common program for all types of eggs, machines and other conditions. Further research is needed to make these findings useful under commercial set-up.

14. Factors affecting incubation performance

14.1. Fertility

Fertility of an egg cannot be determined when the egg is intact. Fertility is inherited, but its heritability value is very low (0.05). Individual males and females vary in their ability to produce viable embryos. This may depend both on the quantity and quality of semen, mating ratio, mating efficiency etc. Certain mutations are correlated with infertility. Homozygosity of the R gene (rose comb) is associated with poor fertility in males but not in females. Breed differences are recorded; fertility of Cornish birds is less than other breeds.

The following are the suggested mating ratios (Table 24.15); however, commonly 10 to 12 females per male in case of Single Comb White Leghorns and 6 to 8 females per male in case of meat-type birds is practiced.

Table 24.15 Suggested mating ratios

Cockerel	Pullet	Males/100 females	
		On litter	On slat and litter
Mini-Leghorn	Standard leghorn	8	9
Standard leghorn	Standard leghorn	8	9
Medium size	Medium size	9	10
Standard meat-type	Mini meat-type	9	10
Standard meat-type	Standard meat-type	10	11

14.2. Hatchability

14.2.1. Temperature

Embryos are poikelotherms during the first 18 d of incubation and hence are very sensitive for fluctuations in setter temperature. Maximum temperature causing little effect on hatchability seems to be 38°C above with detrimental effects depending directly on severity and duration of recorded. Embryo subjected to heat stress exhibited clubbed, wiry down and unsteady gait.

Embryos tolerate cold better than heat and during first 19 d of incubation, reducing the temperature to as low as 18.3°C will not seriously affect hatchability. However, since the embryos are poikelotherms, cooling during the 1st 2 weeks is more detrimental than during the next 5 d. In any case, cooling lengthens the incubation period and the effect is cumulative; cooling during the 1st 19 d also increases the embryo malpositions.

Reduction in temperature during the last 3 d is highly detrimental because the embryos will be changing over to homoeothermy and beginning their pulmonary respiration to accelerate the heat production. Development of hypothalamus is also affected due to cooling during the last 3 d.

14.2.2. Humidity

If the humidity is lower than the required 55% during the first 19 d, excessive evaporation of the contents occurs resulting in smaller and drier chicks. If the humidity is higher than required, there will be reduced moisture loss leading to wet and larger than normal chicks. In both cases, the embryo is weakened thereby the chick quality and survivability will be lowered. Higher humidity results in earlier hatching and *vice versa.* Humidity in the hatcher must be increased gradually and optimum humidity may vary between machines (65 to 75%). Lower humidity results in chicks smeared with egg or shell, stuck down and partial dehydration whereas high humidity causes unclosed navels and chicks smeared with it contents

Effects of both temperature and humidity, in general, and humidity in particular, all related to egg weight (size) and shell quality. Smaller eggs and eggs with porous and thinner shells dehydrate faster than do the larger eggs and eggs with normal shell quality. Smaller eggs have higher relative surface area and hence dehydrate faster. Shells with microscopic cracks or eggs cracked during setting must be removed from the setter.

14.2.3. Ventilation

On 18th d of incubation, 1000 eggs require 420 l of O_2 which can come from 2000 l of fresh air. An incubator of 14,000 eggs capacity would normally measure 1.8 m in length, breadth and height; which means a volume of 5.832 m^3 (5832 l) of which about 1000 l is occupied by eggs in the setter trays. Thus, actually about 4800 l of air volume is available. When the incubator is of full capacity, the 14,000 eggs require 28,000 l of fresh air; in other words, the air in the incubator may have to be changed (28,000/4800 = 5.83) about 5 to 6 times a day or once every 4 hr.

However, in forced-draft setters, the problem of O_2 availability is considerably rare. Similarly, on 18th d of incubation, 1000 eggs produce 270 l and 14,000 eggs 3780 l of carbon dioxide. Therefore, with the airflow ensured for O_2 availability, the expulsion of carbon dioxide will also be affected. Limit of CO_2 tolerance is 0.5% and hatchability reduces proportionately above that level and drastically when the concentrations exceed 1.5 to 2.5%.

The speed of airflow does not seem to influence hatchability. During the 1st 13 d of incubation, embryos require heat whereas after 13 d, there will be a necessity of the dissipation and hence air flowbecomes critical only after 13 d of incubation.

14.2.4. Air pressure

At higher or altitudes, barometric (or atmospheric) pressure is reduced which not only decreases the effective O_2 tension in the atmosphere but also increases the effective egg shell conductance or egg shell permeability, facilitating passive diffusion of gases through the shell pores. Therefore, reduced oxygen availability due to reduced partial pressure is further complicated by excessive loss of CO_2 and excessive loss of water (weight) by the incubating embryos. The atmospheric pressure sea level is 760 mm Hg and the partial pressure of O_2 is 159 mm Hg which corresponds to about 20.93% oxygen in the air. At an altitude of 3500 m, the barometric pressure goes down to 493 mm Hg and the partial pressure of O_2 is only 103 mm Hg; hence, O_2 is 35% less efficient at 3500 m than at sea level (Salazar, 2001).

In other words, at higher altitudes, atmospheric pressure reduces and therefore air expands resulting in reduced concentration of O_2 in a given volume – for every 300 m of elevation above sea level, there will be 2.5% reduction weight of air. The reduction in hatchability becomes particularly perceivable above an altitude of 1050 m. Hatchability reduces by about 5% for every 300 m elevation above 1050 m. Under such conditions, O_2 may have to the injected into the incubating cabinets so as to maintain an O_2 concentration of 23 to 23.5% but it may not be cost-effective. However, O_2 concentration should not exceed 26%.

Birds naturally living in higher altitude adapt by reducing

1. Total pore area
2. Number of pores
3. Water loss during incubation
4. Water vapor conductance of shell in relation egg mass.

Therefore, under natural conditions, birds have been found to reproduce normally even at altitudes of 3048 to 3568 m.

However, under artificial incubation of eggs (produced at a lower altitude) at higher altitude, effect of altitude is sure to precipitate. When O_2 is pumped into incubators, ventilation must also be restricted to conserve O_2. This definitely leads to build up of CO_2 which is detrimental to the embryo. Hence, CO_2 level has to be monitored, on most occasions, by a special instrument. Since CO_2 is a heavy gas, it accumulates at the bottom of the machines further complicating the ventilation problems especially in still-air incubators.

Improvements in hatchability have also been recorded when setter temperature was increased by 0.15°C and the relative humidity is maintained at a constant 60%. Opening the setter air inlets more and/or for longer periods to increase volume of air under circulation so as to expose embryos to more oxygen is also suggested. However, relative humidity must be maintained at 55-60% to limit weight loss by the embryo to less than the critical levels of 12-14% (Salazar, 2001).

14.2.5. Light during incubation

Light stimulates embryo growth in an unknown way. White light of wavelength 295 mμ used continuously by fixing a bulb close to the top of the broad-end of the egg has been found to

1. Advance hatching by 20 to 48 hr (depending on egg size)
2. Improve chick weight by 15% (especially in large eggs)
3. Reduce embryo mortality
4. Improve the body weight of broilers at market age and
5. Advance age at sexual maturity in Leghorn pullets.

However, light does not seem to influence either egg production or male fertility. Bulbs should not move during the turning of eggs.

14.2.6. Position and turning the eggs

The normal position of eggs is broad-end up to help development of head of the chick in the broad-end. If set small-end up, head develops in the small-end and fails to break the air-cell while hatching. The embryos orient themselves with head uppermost near the air-cell during the 2nd week of incubation. Eggs have to be turned back and forth along their long-axis. They must not be turned in a circle since such a rotation can result in rupture of allantoic sac. Most of the eggs are turned to a position of 45° from the vertical then reversed in the opposite direction to a similar position; that is, a total of 90°. Although chicks can come out when the eggs are placed broad-end up, the eggs are kept horizontal to help the process.

The most critical period for turning commercial broiler hatching eggs during incubation is the 1st 7d and the single-most critical period was days 0 to 2 (Elibol and Brake, 2004).

14.2.7. Diseases

Aflatoxicosis, Arizona disease, Aspergillosis, E. coli infection, Fowl typhoid, Infectious Bronchitis, Infectious laryngotracheitis, MG infection, MS infection, Newcastle disease, Omphalitis, Pullorum disease etc. have been found to adversely affect hatchability of eggs. These diseases are dealt in detail in separate chapters on poultry diseases.

14.2.8. Other factors

Eggs produced both during early or latter part of the laying cycle, do not hatch well. Eggs from birds producing at higher rate hatch well; eggs laid in longer clutches as well as near the end of clutch hatch better. Eggs laid during the extremes of weather as well as from older hens hatch poorer than during normal weather and young breeders.

15. Non-linear weight loss method of incubation

(Degraeve, 2004)

In order to replicate the optimum environment required by the embryo in the commercial incubator, the rate of fluid loss and gaseous exchange is controlled by dictating the partial pressures experienced by the egg.

Relatively high levels of water and CO_2 are generated during the initial endothermic stage of incubation in order to create the partial pressures required to stimulate the actions of the mother bird. Conversely, during the exothermic stage, levels of water and CO_2 generated are significantly lowered continuing the replication of the mother bird's actions.

The profiles generated are such that all parameters become non-linear and hence the name "non-linear weight loss method". Improvements in hatchability are obtained throughout incubation process, especially in control of late embryonic mortality. However, the concept is yet to be adopted on a commercial scale.

15.1. Scientific basis

The fluid and gaseous control has been shown to affect several areas of incubation process, particularly during the initial embryo development stage which requires the incubator to operate no air exchange, reducing the introduction of contaminates and improving the temperature stability and overall differential. The restricted loss of fluid and CO_2 improves the egg's efficiency in maintaining its embryonic fluid's pH balance.

It is also likely that specific levels of CO_2 at critical periods during the embryo's development can result in improvements in heart size, volume and in the efficiency of the oxygenation of the blood circularity system.

During the endothermic stage incubation, the energy source required to commence full embryonic development is heat whereas, subsequently, the rate of heat exchange has to be controlled. During exothermic stage of incubation, the optimum rate of embryo growth directly correlates, within the genetic limitations, to the rate of the exchange.

Air temperature, as a general guide, is a valid indicator as to the thermal conditions experienced by the embryo. The embryonic heat generated within the confines of an incubator will be the dominant influence; but the reading shown by the temperature sensor will be a combination of both embryonic heat and a) other heat sources, motors etc., b) ambient conditions (air intake temperature), c) the activity of the cooling system utilized and d) the activity of humidification system utilized.

Therefore, a method to offset the variable effects of the above factors and to identify the specific variant of embryonic heat is necessary to optimize both hatchability as well as post-hatch performance. With the available technology, shell temperature sensing, which has been recently introduced, offers an accurate, non-invasive indication of embryo temperature. Dynamic weight loss measuring equipment are being introduced as a first step to commercialize this concept in anticipation of improvements in both hatchability and post-hatch performance.

16. Nutrition and incubation performance

Table 24.16 Nutrition *Vs* incubation performance

Nutrient deficiency	Description of embryonic death
Vitamin A	Early embryonic death (2-3 d of incubation), faulty heart formation
Vitamin D_3	Stunted chicks, soft bones
Vitamin E	High mortality between 1 and 3 d, exudative diathesis (edema), bulging of eye(s)
Vitamin K	Hemorrhages and blood clots in the embryo and extra-embryonic blood vessels
Riboflavin	High mortality at 9 to 14 d, faulty organogenesis, edema, clubbed down, curled toes and dwarfing
Pantothenic acid	Abnormal feathering, subcutaneous hemorrhage
Biotin	High mortality between 1 and 7 and 18 and 21 d, shortening of long bones (micromelia), shortened and twisted bones of feet, wings and skull, webbing between 3rd and 4th toes, parrot-beak
Vitamin B_{12}	High mortality between 8 and 14 d, head between legs, edema, short beak, curled toe, poor muscle development
Folic acid	High mortality during 18 to 21 d, symptoms similar to biotin deficiency
Calcium	Short and thick legs, shortened wing and lower mandible, pliable beaks and legs, bulging of forehead, edema of neck and protruding abdomen
Phosphorus	High mortality between 14 and 18 d, soft beak and legs
Zinc	Skeletal abnormalities (absence of wings and leg may be seen), tufted downs
Manganese	High mortality between 18 and 21 d, short wings and legs, abnormal head, parrot-beak, retarded growth, edema and abnormal down
Se	Subcutaneous edema, exudative diathesis
Se toxicity	Crooked toes, high mortality
PCB toxicity	Severe reduction in hatchability but egg production and fertility unaffected (unlike in case of DDT toxicity)

17. Analyzing poor hatch

Table 24.17 Causes of poor hatchability

Observation	May be caused by
Clear eggs	Infertile, very early mortality
Blood-ring (embryonic death 2-4 d)	Diseased breeding flock, incubator temperature too high or too low, old eggs
Dead embryos (2nd week of incubation)	Temperature too high or too low, eggs not turned, inadequate breeder ration, too much CO_2 in air (not enough ventilation), eggs not cooled prior to incubation
Late hatch	Temperature and/or humidity too low 1-19 d, incorrect thermometer leading to improper temperature and/or humidity, large eggs, old eggs, temperature too low in hatcher, variable room temperature
Early hatch	Temperature and/or humidity too high 1-19 d, incorrect thermometer, small eggs, Leghorn eggs Vs meat-type eggs
Air-cell too small	Humidity too high 1-19 d, inadequate breeder ration, large eggs
Air-cell too large	Humidity too low 1-19 d, small eggs
Fully development embryo dead with beak not in air-cell	Temperature and/or humidity too high 19th d, inadequate breeder ration
Fully development embryo dead with beak in air-cell	Temperature and/or humidity too high 20-21 d, inadequate breeder ration, incubator air circulation poor
Chicks pipping early	Temperature too high 1:19 d
Chicks dead after pipping shell	Temperature too high 20: 21d, inadequate breeder ration, inadequate air-circulation 20: 21d, incorrect temperature 1-19 d, temperature too low immediately after eggs were transferred to hatcher
Chicks dead before pipping shell	CO_2 content of air too high 20: 21d, thin-shelled eggs, disease in breeding flock
Trays not uniform in hatcher with chick	Inadequate incubator air-circulation, eggs of different sizes/ of different breeds/ of different ages when set, disease or stress in some breeder flocks
Sticky cheeks (shell sticking to chick's down)	Humidity too low 20-21 d, temperature too high 20-21 d, eggs transferred too late

Observation	May be caused by
Sticky cheeks (albumen sticking to chick's down)	Temperature too low 20-21 d, inadequate air in hatcher, air speed to slow 20-21 d, humidity too high 20-21 d, old eggs
Chicks too large	Large eggs, humidity too high 1-19 d
Chicks too small	Small eggs, humidity too low 1-19 d, thin, porous shells, eggs produced during hot weather
Crippled chicks	Variation in temperature 1-21 d
Mushy chicks	Unsanitary incubator conditions
Unhealed naval, dry	Humidity too high 20-21 d, temperatures too low 20-21 d, inadequate breeder ration, humidity not lowered after completion of hatch
Chicks cannot stand	Improper temperature 1-21 d, humidity too high 1-19 d, inadequate breeders ration
Unhealed navel, wet and odorous (mushy chicks)	Omphalitis, Unsanitary hatcher and incubator
Soft chicks (abdomen)	Humidity too high 1-19 d, Temperature too low 1-19 d
Closed eyes	Loose down in hatcher, down collectors not adequate, temperature too high 20-21 d, humidity too low 20-21 d
Chicks dehydrated	Humidity too low 20-21 d, eggs set too early, chicks left in hatcher too long after hatch is complete
Malpositions	Continuous light in incubators, inadequate breeder ration

18. Operating a Still Air Model Incubator

It is recommended that the still-air incubator is operated with a small quantity of inexpensive eggs at the beginning to be assured of operating procedure and the performance of the incubator, before attempting to hatch large number of eggs or expensive eggs.

In still-air incubators, position the thermometer bulb about 2.5 cm above the screen floor. The thermometer must not touch the eggs or side of the incubator. Rarely is the humidity too high in a still-air incubator. The water pan should cover at least one-half of the floor area during the first 18 d. At hatching time, increase humidity by adding another pan of water or a wet sponge. The humidity is raised by increasing the water surface area.

Eggs are placed on their sides with the small end pointed slightly downward. This enables the embryo to turn itself into the proper position for hatching.

An excellent method for knowing whether the eggs have been turned is to mark an "X" on one side of the shell and an "O" on the opposite side. It is easy to determine if the eggs have been turned by observing whether all eggs have the correct mark turned upward. When turning, hands must be clean. During the first week of incubation, maximum care has to be exercised while turning. The developing embryo has delicate blood vessels that may rupture if jarred or shaken.

The ventilation openings must be opened both above and below the eggs for proper air exchange. Eggs should not be set in more than one layer and the incubator is kept closed to conserve the heat and humidity.

The relative humidity in the incubator can also be varied by changing the size of the water pan or by putting a sponge in the pan to increase the evaporative surface. The pan should be checked regularly while the incubator is in use to be sure that there is always an adequate amount of water. Adding additional water pans to small still-air incubators is also helpful to increase humidity. Whenever water is added to an incubator, it should be about the same temperature as the incubator to avoid stress on the eggs or the incubator. A good test is to add water just warm to the touch.

18.1. Location

The location of the machine is important to successful operation. A room temperature of 21.1-26.6°C is ideal, and fresh air without drafts is necessary. A well-ventilated basement room is often just right. No direct sunlight should strike the incubator.

An incubator is designed to bring normal room temperature to the desired temperature. Room temperature of 15.6°C or below will reduce the temperature in the incubator. Room temperature changes of 5.6°C or more will change the temperature in the incubator. The change is more pronounced below a temperature of 21.1°C.

18.2. Setting-up the incubator

The incubator has the windows, heater and thermostat.

18.2.1. Thermometer instructions

The thermometer is placed at the center of right side of the bottom, against rim and facing towards the center. This is the correct position when using an automatic turner and can be seen from the window. Thermometer stand is adjusted such that it can be seen from the window.

Thermometer is placed on the floor of the incubator where it can be read through the window. When using the automatic turner, the thermometer must be placed on the right side.

18.2.2. Thermostat Operation

Electric cord is plugged into electrical outlet and thermostat adjusting screw is turned clockwise until pilot light goes out.

Adjusting screw is turned four complete turns counterclockwise. The pilot light will come on. Any time the pilot light is on, the heater and fan (if there is one) will be on. Any time the pilot light is off, the heater will be off.

As incubator warms up, the thermostat wafer will expand, turning off pilot light and heater.

When pilot light goes out, temperature is checked. If it has not reached 37.8°C, adjusting screw is turned counterclockwise one or two full turns and incubator is allowed to heat up until light goes out.

If temperature has not reached 38°C, process is repeated. If it does not hold exactly on 38°C, it is regulated in such a way that it turns on and off the same above and below 38°C. Temperature setting can vary as much as ½°C above and ½°C below the desired temperature.

18.2.3. Temperature

Temperature requirement differs for different size eggs, when on their side on the floor and when they are in the automatic turner. When setting eggs of different sizes, an average half way between temperatures in the chart below can be used. Goose eggs should not be set in automatic turning machines.

Table 24.18 Operating Temperature (°C) for still-air incubators

Manually turning eggs set on wire-floor		
Quail eggs		38.0
Bantam and Pheasant size eggs		37.8
Chicken and other large eggs (including Goose)		38.0
Eggs in automatic turner		
	Summer	Winter*
Quail eggs	37.2	36.7
Bantam and Pheasant size eggs	36.7	37.2
Chicken and other large eggs	37.2	35.6
Duck eggs	35.0	34.4

* Winter operation requires lowering operating temperature to prevent overheating top of eggs. Hence, these temperatures must be used throughout the incubation period. Three days before eggs are to hatch, they are removed from the turner, kept side on the wire floor, and temperature increased by 2°C above the operating temperature for hatching.

18.2.3.1. Precautions

1. Temperature is regulated for desired setting and ensured that it holds for 2 to 3 hrs before setting eggs in incubator.
2. Check temperature daily.
3. When the turner is removed for hatching, adjusting screw is turned one full turn counterclockwise to correct for the heat produced by the motor on turner.
4. About half way through incubation process, an increase in temperature is expected and can be corrected by adjusting thermostat down nearly one full turn. This is normal and is caused by the embryos forming into chicks and generating heat.
5. When cold eggs are loaded into the incubator, it can take three hours or more for eggs to warm up and temperature to stabilize at the setting. Also, when incubator is opened, it can take up to two hours for temperature to stabilize.
6. If chicks hatch out a day early, it indicates temperature was a little too high, so on next setting lower temperature by ½°C for entire incubation period. Conversely, if chicks hatch a day late, raise temperature by ½°C for entire incubation period.

18.2.4. Setting and Turning Eggs

18.2.4.1. Manual Egg Turning

1. Place eggs on their side with small end pointed slightly down. Do not over-crowd the eggs.
2. The eggs should be turned three times a day. Turning the eggs is best done by removing about a dozen from the center and rolling the rest of them toward the center. Palms of the hand are placed on the eggs and are rolled around until it is sure that all have been turned, and then eggs taken from the center are placed around the outer edge.
3. During turning eggs shocks or jars must be avoided; otherwise, it may result in rupture to the blood vessels of the germ. Eggs should not be left standing on end. Eggs should be kept flat, pushing the pointed ends down a little with the hand. With a soft lead pencil, puta small "X" on one side of egg and "O" on the other side so as to help check turning all the eggs.

18.2.4.2. Automatic Egg Turning

The turner runs very slowly—only one revolution in 4 hours. The turner is placed on bottom of incubator with the motor side to the back of the incubator (rim of bottom with notches is the back) and slid as close as possible to the front rim of bottom of incubator. The turner must sit flat on the wire floor. Goose eggs should not be set in incubator with turner.

18.2.5. Relative humidity

Moisture in an incubator prevents excessive drying out of the natural moisture in the egg. It is impossible to give any set rule for supplying moisture. If the incubator is operated in a damp cellar or in a room with considerable natural moisture, then it may not be necessary to supply artificial moisture. If operating in a dry climate or in dry room, moisture will be needed. The important thing to watch is the air space in the egg. When testing eggs for fertility, note the size of the air space. If the air space is too large, provide moisture.

Moisture in the incubator is controlled by putting water in the small inner trough of the bottom. The small trough by itself will increase the humidity to take care of most climates. If climate is extremely dry larger outer trough instead of the small trough will be needed. During time of hatching, higher humidity is required to prevent excessive drying of hatched chicks. The water trough should be checked regularly and filled as and when needed, usually, twice a week. During winter, three days before time to hatch, put water in both troughs of bottom to compensate for extra dryness of air.

18.2.5.1. Spraying water during incubation

Duck and goose eggs are sprayed thoroughly with water twice each week, and at least three times a week during the last ten days.

18.2.6. Hatching

Three days before total incubation and hatching time, turning eggs is discontinued. The automatic turner must be removed from the incubator or the eggs must be moved to a separate incubator for hatching. Eggs should not be hatched while the turner is in the incubator, as the slow turning egg racks could crush the chicks. Eggs are kept on wire floor with small end pointed slightly down.

18.2.7. Plastic vent plugs

The front vent plug (just below label) is used to regulate humidity and the back vent plug (by electric cord) is used when there is excessive humidity, as follows:

1. When incubator is over 75 % of capacity, the front vent plug is removed one week before hatch date.
2. The day that chicks start to hatch, the back vent plug is removed.
3. If incubator is over 90 % of capacity and contains large chicks, it may be necessary to prop one side of the incubator top up about 0.3 cm to get chicks dry and its left side propped up just long enough for most of moisture to clear on windows, but no longer than one hour at a time.
4. When incubator is from 25 to 75 % of capacity, the front vent plug is removed the day chicks start to hatch.
5. When there is moisture condensed on the windows, the front vent plug is

removed.

6. The vent plugs have to be replaced before next setting of eggs; if vent plug is lost, the vent hole is closed with scotch tape.
7. Chicks may be removed 24 hr after they start to hatch. Extremely wet chicks should be left in incubator to dry.

Chicks are usually removed once a day, as every time incubator is opened, warm moist air escapes. Avoid chilling of wet chicks. Some chicks may be late in hatching, so remaining unhatched eggs are left up to two days longer.

18.2.8. Specials points to remember

1. Do not bother the regulator unless it is absolutely necessary. The working of the machine may be affected if the regulator is tampered with excessively.
2. If the machine does not heat, carefully investigate and see if you have all connections properly made.
3. Do not overcrowd the eggs.
4. Keep the eggs clean. Perspiration from the hands or any sort of grease stops up the pores of the shells.
5. Clean you incubator after each hatch with bleach water. Scrubbing of moisture troughs may cause leaks.

19. Incubation and hatching of ratite eggs

19.1. General

The incubation of ratite eggs is the first step to the beginning of the growing period. Care must be exercised throughout the incubation process to ensure the best possible hatch, considering the environment that the developing embryo is placed in (Jeffrey *et al.*, on internet).

19.1.1. Hatchery location and design

The hatchery room is much more than a shelter for the incubator. The room acts as a "plenum" chamber that helps pre-condition air prior to movement of the air into the incubator. Most incubators depend on proper room temperature and humidity to make the work of maintaining incubator conditions easier. Typically the temperature of the hatchery will be approximately 5.6-8.3°C cooler than the temperature set on the incubation cabinet). This temperature difference is needed for cooling the eggs as the embryo develops, and to ensure proper heating of the cabinet early in the incubation period (Jeffrey *et al.*, on internet).

As the embryo grows in the shell, the consumption of oxygen (as well as the release of CO_2) increases with egg age. This is most apparent with the eggs in the hatcher. The birds will "pip" through the inner shell membrane, and breathe air that is conducted through the shell. The air in the cabinet has to have a level of oxygen high enough to ensure proper levels of oxygen in the egg for growth and the strenuous job of hatching out of the egg (Jeffrey *et al.*, on internet).

Because of this high demand, the most frequently overlooked consideration in ratite hatchery design and construction is the placement of air handling devices and windows to help move air efficiently throughout the hatchery. A one-way flow of air is the optimal situation that will ensure that fresh air is flowing through the hatchery. A general rule is to have as much fresh air moving through the hatchery room as possible without changing the room temperature / humidity requirements. Drafty conditions can create cold spots in the room that could be detrimental to proper air mixing. Because cold air drops to the floor as it enters a room, baffles or circulation fans should be in place to ensure proper air mixing (Jeffrey *et al.*, on internet).

The quality of the air is an important consideration as well. The location of the hatchery should be as far away from the other pens as possible, to avoid dust and other particulates from being pulled into the hatchery. Air Filters should be monitored and changed frequently to maintain good air quality (Jeffrey *et al.*, on internet).

Humidity levels can be controlled to a certain degree by ventilation, if external conditions permit. By increasing air flow, humidity levels will drop. As the temperature drops the % relative humidity will increase. This is the basis of most of the room dehumidifiers that are available for ratite hatcheries that need to reduce % RH to assure proper water loss. Humidity is adjusted for the average eggs that are incubated, as some eggs may fall outside this range. Some producers will have several cabinets set for different humidity climates to ensure proper weight loss for all eggs being incubated (Jeffrey *et al.*, on internet).

The Hatchery needs to be designed for one-way traffic of personnel and eggs to reduce the possibility of contamination. Both air flow and personal traffic should move from the incubation (setter) end and exit at the hatcher end. All surfaces within the hatchery should be washable with cleaners and disinfectants commonly available (Jeffrey *et al.*, on internet).

19.1.2. The egg

Most ratite eggs are laid in nests that the male has dug in the soil of the pen. The eggs are exposed to sun, rain, temperature extremes, and the microbial inhabitants. At the time of lay, the egg is wet and an ideal environment exists for microorganisms to enter the moist pores as the egg cools and the contents contract. Because of these conditions, contamination is often a significant source of loss of ratite eggs. Decontamination by fumigation or disinfection is ineffective after the organisms are inside the shell. Therefore, preventative measures should be diligently practiced (Wilson, on internet).

The egg is the common unit that is thought of while operating a hatchery. Made mostly of water and nutrients, it is a self-contained unit that ensures that all the physiological needs of the developing embryo are met during and after the incubation period. Although ratite eggs of the same species are very similar in appearance, there are differences in shell porosity and size depending on the age, genetic background and nutritional plane of the breeder hen. This plays an important role in how the eggs need to be handled and incubated (Jeffrey *et al.*, and Wilson on internet). Under large commercial conditions this large variation would be intolerable and uneconomical. The most rapid and satisfactory progress in improved hatchability is likely to be achieved through selecting hens that lay eggs with good shell quality and adequate, uniform shell porosity (Wilson, on internet).

19.1.3. Egg selection

Eggs should be kept dry and as clean as possible. The shell of the egg is covered by the cuticle, a natural barrier to microbial invasion. In addition to the cuticle, the shell membranes and albumen of the egg are natural barriers to bacteria. Any egg that happens to have a large accumulation of soil or fecal material should not be set in the incubator (Jeffrey *et al.*, on internet).

Eggs that are seeping or giving "off" smells should be pulled from the incubator. Very dirty eggs could be washed, but care must be taken to prevent the invasion of the wash water into the egg. Eggs should be washed in water that is 5.6° C higher than the temperature of the egg (Jeffrey *et al.*, on internet).

Egg shells are very porous, and water loss can be as much as several grams of water per day, depending on how many pores that an egg has and how the egg is handled. Water has to be lost from an egg in order to make room for water created via the metabolism of materials found in the egg. Egg washing will change the rate of water loss from the eggs and incubator conditions may need to be changed in order to compensate for this loss (Jeffrey *et al.*, on internet).

19.1.4. Egg storage

Embryonic development before oviposition is similar to that in chicken and turkeys. The environmental conditions in the oviduct during egg formation and during the storage period from oviposition (egg laying) to placement in the incubator can have significant effects on subsequent hatchability. Time of exposure to the various conditions is also a major factor. However, the environmental factors that are most critical to the normal development of the embryo are those that occur during the incubation and hatching processes. These factors include incubation temperature and time, humidity, egg orientation, egg turning, ventilation, and sanitation (Wilson, on internet).

Eggs can be stored in a cooler prior to incubation, provided that the cooler is regulated to 13.3-15.6°C and 75% relative humidity to prevent desiccation of the eggs. Cooling eggs serve two major purposes:

1. To synchronize eggs into batches, to allow for budgeting of time and incubator space, and
2. To allow for the formation of an air cell in fresh eggs, which is very important to the proper orientation of ratite eggs (Jeffrey *et al.*, on internet).

19.1.5. Incubation period

Incubation time (i.e., time from set to hatch) often is an indicator of incubation temperature. Temperatures above optimum shorten incubation time, whereas temperatures below optimum lengthen incubation time. However, caution should be used in changing incubation temperatures based on incubation times, because incubation time is a very heritable trait. Therefore, obvious differences in incubation times can be seen in eggs from different hens.

The "average" incubation times for ratite eggs at normal incubator temperatures are 39 d for the rhea, 42 d for the ostrich, and 56 d for the Emu (Wilson, on internet).

19.2. Environmental factors for artificial incubation

The key environmental factors that are needed for artificial incubation are: temperature, humidity, airflow, position (turning), and sanitation. Each factor, either singularly or in combination can be the cause of many problems seen in poor hatches. Through proper hatchery design and management, good hatches can be realized (Jeffrey *et al.*, on internet).

19.2.1. Incubation temperature

Temperature is the most important factor in the incubation of any avian egg. While the other factors must be controlled during incubation, temperature is the most critical for maximum hatchability. There has been insufficient research to determine the optimum incubation temperature for ratite eggs. The optimum temperature is almost certain to change with increased age of the embryo. Therefore, temperature settings of single-stage incubators (all eggs set at the same time and all embryos one age) would be different, probably decreased, as the embryos become older. In multi-stage incubators (eggs set at different times and embryos of various ages) a single, "average" temperature would be used continuously.

Common incubator temperatures for ratites range from 35.9 to 36.5°C for multi-stage machines. Optimum temperatures will vary with individual incubators, manufacturer and model, geographical and within-building location, number of eggs in the machine, and many other factors. Incubators without fans (still-air) are normally run about 1°C higher than those with fans (forced-air incubators). Several influences are introduced due to breeder genetics, age, health, nutritional status, and other factors. Although the optimum incubation temperature is usually understood to mean the one which results in maximum hatchability,it also results in the most efficient embryonic growth and the highest quality chick. Furthermore, post-hatch growth also may be affected.

The hatcher temperature should be approximately 0.5°C lower than that of the incubator (setter) because of the large amount of heat generated by the late-stage embryos (Wilson, on internet).

An egg can, to a certain degree, compensate for various insults to its well-being except for temperature extremes. A 0.25° C change in incubator temperature can have a profound effect in overall performance of a group of ratite eggs because they are incubated for at least 40 d. Extremes in temperature (high or low) can cause problems with embryo growth, or in many cases death (Jeffrey *et al.*, on internet).

Incubator temperature is for the average sized egg of the flock. A survey of the ratite industry finds a variety of settings that are used for ratite eggs. Most range between 36.4–37.2°C, and there are a variety of explanations as to the choice of a particular setting. The key is to match the temperature setting with the performance of a particular incubator and eggs. The temperature of the incubator controls the rate at which the embryo develops and hatches. Eggs that were stored should be allowed to come to room temperature for a few hrs prior to setting in the incubator. This helps reduce temperature fluctuations, cold spots in a cabinet, or stress fractures in the eggshell. The temperature in the hatcher has traditionally been set 0.25-0.5°C lower than the incubator (setter) temperature in chickens and turkeys. Although this practice could be followed for use in ratite eggs, it has been shown that hatch will take place satisfactorily at incubation temperature itself (Jeffrey *et al.*, on internet).

Eggs that are hatching early or very small chicks at hatch (more than the norm for the farm) are an indication of incubator temperatures being too high. Eggs that take 2 or more days to hatch a live bird could be associated with too low of an incubator temperature. The first few fertile eggs in the season are a good indication of what steps need to be taken (if any) to adjust the temperature of the incubator to promote proper growth (Jeffrey *et al.*, on internet).

Problems do occur when the cabinet temperature is too high, as early embryonic death (within the first 2 weeks) can be detected. These deaths can be caused by other factors but are easily found through candling. Hence, all eggs are candled by the 14th d of incubation to determine that the embryos are progressing in growth (analogous to fertility testing at 6th-7th-8th d in case of chicken eggs). Eggs that appear clear are removed and broken to ascertain that they are infertile.

In an incubator that has had a series of late hatches, a 0.5°C increase in incubation temperature could produce favorable results by pushing the embryo towards accelerated growth and earlier hatch. Care must be taken to adjust for proper humidity and ventilation rates to compensate for the changes in incubator temperature (Jeffrey *et al.*, on internet).

The incubator room should be held at a constant temperature, to keep the incubation environment as stable as possible. Wide temperature variances in the incubator room could cause overheating of the eggs (Jeffrey *et al.*, on internet).

It is advisable not to set to the full capacity of the incubator at the same time, unless it is specifically designed for single-stage incubation. Care must be taken in single stage incubation to accommodate the high rise in egg temperature towards the end of the incubation cycle. Eggs have to be set uniformly throughout the cabinet to reduce hot and cold spots in the incubator as well (Jeffrey *et al.*, on internet).

Temperature of hatchery room and the incubator must be constantly monitored throughout the day and recorded to be sure that the optimal environment is maintained. A mercury thermometer is placed in the incubator to serve as a standard to check the other incubator sensors against and it is checked against a standard, such as an ice water slurry (0°C) to determine its accuracy periodically, or when the accuracy of any of the instruments is in question.

At the beginning of the incubation season, or when installing a new machine, temperature of the incubator is set higher than the safety level (dead-man) switch to test its accuracy and to ensure that the circuit is operating correctly. As the temperature exceeds the safety temperature the heater circuit should be interrupted and the cooling system (if so equipped) should be at its maximal exertion. At no time should a heating and cooling circuit be running on an incubator at the same time. On double-wafer safety units, the safety set point should be no more than 0.5-0.75°C higher than the incubator set temperature (Jeffrey *et al.*, on internet).

19.2.1.1. Heat Stress

In general, the detrimental effect of high temperature on hatch increases with increased temperature, with increased exposure time, and with younger embryos. However, some early stages are more resistant and embryos near hatching are very susceptible to high temperature stress. Some of the age effects may be associated with oxygen consumption, heat production, and the growth stage of the chorioallantois (the embryonic membrane that attaches to the shell). The circulatory and nervous systems seem to be the most susceptible to heat stress (Wilson, on internet).

19.2.1.2. Cold Stress

Sensitivity of the embryo to cold stress increases as embryonic age increases. The major effects of cold stress appear to be retarded and abnormal development of the circulatory system and heart, reduced growth and development of the embryonic membranes, and subsequently, reduced respiratory efficiency and utilization of yolk, albumen and shell. Duration of cold stress exposure and embryonic age are major factors contributing to the effects of the exposure. Short-term cold exposure of embryos at certain ages may not even be detrimental, although incubation time will be increased (Wilson, on internet).

19.2.2. Relative humidity

19.2.2.1. Water loss during incubation

Most ratite eggs will perform well when a moisture loss of 12- 16% is achieved. This loss, as measured as a percentage of the set weight over 42 d incubation, will be adjusted somewhat by the embryo as it creates water as the by-product of digestion of foods found in the egg. By weighing the egg at set time, and monitoring egg weight periodically through incubation, best humidity profile for optimal hatch can be ensured. Although the primary effect of humidity is on water loss from the incubating egg, it also affects utilization of minerals from the shell, gas exchange and other functions.

It is interesting to note that the water content of the newly hatched chick (~76%) is essentially the same as the water content of the combined yolk and albumen of a fresh egg. Egg weight at setting, debris weight after hatch, and weight loss during incubation are the primary determining factors for chick weight at hatching.

Insufficient water loss results in large, sluggish, edematous chicks which are often in a malposition in the egg-causing problems in pipping the shell and in hatching. Excessive water loss results in small, dehydrated, weak chicks that may not be strong enough to hatch. Chick weights of 60 to 65% of egg weight at setting appear to be in the normal range for ratites, as it is for other domestic species (Wilson, on internet).

Incubator manufacturers usually recommend optimum relative humidity settings for each model and each species of bird. Adjustments normally are needed in each machine to obtain the desired egg weight loss.

Present data indicate that normal hatchability should be expected in ostrich eggs that lose 12 to 17% of initial weight from setting to 38 d of incubation. Weight loss at 38 d can be projected from 7-d or other intermediary-loss calculations. The 38-d actual loss will usually be about 0.5 to 1.5 % less than that projected from 7-d loss.

The major factors that determine egg weight loss are shell porosity and relative humidity. Secondary factors are egg size, altitude, and incubation temperature.

Approximate relative humidity requirements are 15-20% for the ostrich, 25-40% for the Emu, and 35-65% for the rhea.

19.2.2.1.1. Measurement of water loss

Water loss can be calculated several ways. One method is to calculate the per cent water. loss relative to the egg's set weight for a given period of time by simple difference in weight. When choosing a scale or balance, that has a measuring weight range that places the average egg in the middle of that range should be chosen to obtain accurate weights (Jeffrey *et al.*, on internet).

19.2.2.1.2. Egg losing more water

If the eggs are losing too much water, increase the humidity in the incubator cabinet by adding water trays or increasing the misting frequency. If only one or more eggs are showing this rise in water loss, it can either be placed in another incubator set for a higher level of humidity, or taped near the equator (just below the air cell) with electrical tape. Taping the egg (up to 10% surface area) reduces the number of pores that are conducting water to the outside air (Jeffrey *et al.*, on internet).

19.2.2.1.3. Egg losing less water

Increasing the loss of water, especially in humid climates, can be a challenge. Room desiccators, dehumidifiers and air conditioners coupled with dehumidifiers have been used to drop humidity within the incubator room. Increasing air velocity/air exchange within the cabinet also aids in the removal of water vapor from the incubator environment (Jeffrey *et al.*, on internet).

19.2.2.2. Humidity in the hatcher

When eggs are transferred to the hatcher, the humidity should remain the same as in the incubator until the embryos begin to pip the shells. At that time the humidity should be increased to prevent membranes from drying too quickly which causes the embryo to stick in the shell. Humidity during the hatching process should be approximately 40% for ostrich, 60% for Emu, and 70% for rhea eggs. These are estimates, due to the lack of research data.

In eggs that lose 10% or less of their weight through 38 d, some advantage may be gained by drilling about four 2mm holes in the shell over the air cell at the time of transfer to the hatcher. This has little effect on egg weight loss but does allow the chick to survive by lung respiration after pipping into the air cell. Many of these embryos will require some assistance in hatching because they will be in some degree of malposition due to their edematous condition (Wilson, on internet).

19.2.3. Egg Turning and Orientation

19.2.3.1. Considerations

Egg turning during incubation involves turning frequency, axis of setting (egg orientation), axis of rotation, turning angle, plane(s) of rotation, and stage of embryonic development requiring turning. The absence of egg turning has been shown to result in adhesion of the embryo to the inner shell membrane, premature or abnormal adhesion of the embryonic membranes to the inner shell membrane or other structures, increased incidence of malpositions (abnormal embryo position), decreased albumen utilization, abnormal fluid distribution in the egg, decreased oxygen exchange surface of the chorioallantois (large embryonic membrane attached to the shell), and a poorly developed yolk sac (embryonic membrane enclosing the yolk).

The critical period for turning ostrich eggs is thought to be during the first two weeks of incubation, and turning after 24 to 28 d of incubation (or after closure of the chorioallantois) probably has little or no beneficial effect (Wilson, on internet).

19.2.3.2. Frequency and methods

An optimal turning frequency has not been determined for ratite eggs. However, 24 times per d is practical and appears to give satisfactory results. Neither tilting and returning to the original position nor shaking of the egg is a viable substitute for turning. Eggs from most species, when set with large end (air cell) up and turned around the small axis, hatch as well as, or better than, when set horizontally and turned around the long axis (the natural way). Eggs set air cell end up hatch best when turned 90° to rest at a 45° angle, whereas those set horizontally hatch best when turned approximately 180°. Turning horizontally set eggs only in one direction will cause rupturing of membranes and blood vessels resulting in mortality (Wilson, on internet).

Eggs should never be turned only in one direction,as this will cause malformation of the chorioallantoic membrane (CAM) and other structures within the egg. The number of turns by automatic incubators is typically 12 or 24 turns in any given 24 hr period. If turning eggs by hand, an odd number (minimum of 5) should be chosen, and distributed throughout the day. This eliminates the eggs from sitting on the same side overnight (Jeffrey *et al.*, on internet).

Egg positioning in the incubator and turning will ensure that the embryo is fully developed and in position to hatch. Common practice in the chicken is to set the eggs vertically with the air cell at the top of egg. Studies have shown a 1-2% increase in hatchability over that of eggs placed on their sides. Because of the translucence of the ostrich eggs, they can be set vertically as well (Wilson, on internet).

Ostrich eggs are candled to determine the position of the air cell. If the air cell is found on the side of the egg, position the egg so that the air cell is in line with the axis of the turn. This promotes the migration of the air cell towards the top of the egg. Eggs should be turned at least through a 90° angle (Wilson, on internet).

Emu eggs and other opaque shelled eggs should be placed on their sides, unless the air cell can be definitively found. A number of infrared viewers are showing some promise of detecting the air cell in opaque eggs. The accuracy of this evolving technology is still being evaluated (Jeffrey *et al.*, on internet).

Any eggs that are positioned on their sides should be turned on their long axis. This means that if the pointed ends of an egg were north and south the egg would turn from side to side (east to west) along the equator (Jeffrey *et al.*, on internet).

Eggs are normally transferred to a hatcher 1 or 2 d prior to expected pipping (breaking of the shell) of the first egg and are not turned thereafter. (Wilson, on internet).

Setting eggs with the air cell down causes most embryos to be malpositioned with the head up and away from the air cell. If the egg is set with the air cell down for the first week and then placed with the air cell up thereafter, the embryo will usually orient properly (Wilson, on internet).

19.2.4. Ventilation

Ventilation within the incubator and hatcher provides an adequate O_2 supply, removes excess carbon dioxide, and is a factor in maintaining the proper humidity level. Ventilation recommendations may vary considerably from one type or model of incubator to another. Furthermore, single-stage incubators require adjustment of ventilation in respect to the age of the embryo.

Oxygen consumption and CO_2 production increase rapidly in the last two-thirds of incubation, and will essentially double during the last week. Hatchability may be reduced at altitudes of more than 1000 m and may reach zero at altitudes of more than 3600 m. The reduced barometric pressure at high altitudes decreases O_2 tension and increases the effective conductance of the egg shell. Therefore, the embryo suffers from lack of O_2 and an excessive loss of water and CO_2. O_2 supplementation and decreased ventilation usually corrects the problem.

Recommended incubator levels are 21% O_2 and 0.05 to 0.10% CO_2 for ratite eggs incubated at less than 1000 m altitude. However, few supportive data are available (Wilson, on internet).

19.2.5. Assistance at hatching time

At times during the incubation season, producers feel the need to assist eggs that are not progressing in hatch. Care must be taken when assisting a hatch to prevent problems arising from the assistance procedure (Jeffrey *et al.*, on internet).

Typically in ostrich, a period of 12-18 hr should be allowed from the time the bird breaks through the internal shell membrane into the air cell until it breaks a hole in the egg shell before hatching assistance is offered. If the chick fails to break the shell after the twelve hours duration:

1. A series of 1 cm holes could be drilled into the upper and lower edges of the air cell of the egg to allow enough oxygen to flow to the chick.

2. Humidity should be increased and condition of membranes noticed after few hrs. The membrane should be somewhat dried and bleed very little if torn, before assistance can continue.
3. Slowly, larger holes are broken in the egg, starting with the two holes from the air cell end of the egg.
4. Condition of the chick and, if possible, the location of the yolk sac should be checked. If a small part of the yolk sac is visible, the chick may be left in the egg for a period of time to allow for the navel to close fully (Jeffrey *et al.*, on internet).

Timing is critical to this operation and assistance of hatch should not continue for more than 1½ d in length. Prolonged assistance usually results in weak chicks that may not be able to stand and eat. Further delays could lead to death. There isn't any guarantee that an assisted chick will survive, and historically assisted chicks have not performed as well as un-assisted chicks (Jeffrey *et al.*, on internet).

19.2.6. Unhatched eggs

After the hatching process, surveying the hatch waste is very important. It is necessary to look for signs of stress by noting how much fecal (meconium) material is left on the hatcher tray. A large amount of material signifies that the chick may have been in the hatcher longer than necessary (Jeffrey *et al.*, on internet).

All unhatched eggs that may have been pulled during the incubation period, or eggs that failed to hatch that were left in the hatcher are invariably broken open to observe color, odor and appearance of the albumen, yolk and membranes within the egg along with the size and position of the embryo (if found) within the egg. Small sized chicks with little feather covering denote early embryonic death. Larger sized chicks covered well with feathers denote late embryonic deaths. (Jeffrey *et al.*, on internet).

19.2.6.1. Common symptoms found in breakout

1. Smelly, brown-colored albumen contamination of eggs
2. Missing or small lower jaws high / low selenium, low calcium chicks with head away from air cell
3. Extremely small embryo or chick, no feathering seen (early death) indicating high incubator temperatures during beginning period of incubation
4. Little formation of air cell, failure to pip by embryo indicating high humidity in incubator
5. Low water loss within eggs, edema seen in chicks also indicating high humidity in incubator, Vitamin E /calcium / selenium deficiencies
6. Chicks with dry membranes stretched and surrounding chick indicative of low humidity in incubator and possibly within the hatching cabinet

7. Liquid seeping from eggs due to contamination of eggs
8. Chicks hatched with large amounts of yolk exposed which results due to late high temperatures in incubator and hatcher, low water loss during incubation
9. Spraddled legs, problems standing slick hatcher trays, giving improper footing recorded due to low calcium in some cases
10. No embryo seen within egg, undeveloped germinal disk indicating infertility and lack of breeding activity

19.2.7. Miscellaneous

Documentation is the key to successful incubation management.

1. Egg movement through the hatchery must be noted so that it will be easy and accurate to identify parents (if possible) for future reference.
2. Water losses and chick hatch weights must be routinely recorded.
3. Any unusual events that take place in the hatchery should also be noted like the purchase of eggs and power outages, visitors log etc. (Jeffrey *et al.*, on internet).
4. A maximum-minimum thermometer will also be useful for recording data about the room condition during the day (Jeffrey *et al.*, on internet).

Chapter **25**

Climate and Climatic Indices

1. Climate classification

(Smith, 1981)

After several modifications, Koeppen-Geiger system is the widely accepted one for climate classification which is summarized below:

1.1. I Level classification

This depends on average temperature (A and B) and precipitation to evaporation ratio (C).

Table 25.1 Koeppen-Geiger Classification: I Level

Type	Features
A	Tropical climate with each month having an air temperature > 18°C; with large annual rainfall . which exceeds annual evaporation.
B	Dry climate with mean temperature < 18°C (k subclass) or > 18°C (h subclass); evaporation > rainfall and therefore, no water surplus and no permanent streams.
C	Mesothermal climates with mean temperature of the coolest month between (-) 3°C and + 18°C.
	Source : Smith, 1981

1.2. II Level classification

This is primarily based on amount and distribution of rainfall and is summarized below:

Table 25.2 Koeppen-Geiger Classification: II Level

Type	Subdivision	Features
A	Af	Climate with no dry season
	Am	Climate with short dry season and monsoon-type rainfall in wet season
	Aw	Rainfall of driest month of cold season is less than 10% of rainfall during wettest month of hot season
B	BW	Arid: Rainfall < 250 mm/year
	BS	Steppe: Rainfall 375-750 mm/year

Type	Subdivision	Features
C	Cs	Dry summer with rainfail in warmest month < 1/3 rds the rainfall in the latest month of cold season; and is < 40 mm
	Cf	Similar to Af
	Cw	Similar to Aw
		Source : Smith, 1981

1.3. III Level Classification

This depends on temperature and is summarized as follows:

Table 25.3 Koeppen-Geiger Classification: III Level

Type	Features
Bh and Bk	Similar to BW and BS, respectively
Csa, Cfa, Cwa	Warmest summer month average temperature > 22°C
Csb, Cfb, Cwb	Warmest summer month average temperature < 22°C
Note: Arbitrarily, the boundary between Bh and Bk and between Csa, Cfa, Cwa and Csb, Cfb, Cwb can provide a boundary between subtropics and cooler climate.	
	Source : Smith, 1981

Table 25.4 Characteristics of Tropical climate

Type	Features
Wet equatorial (Af, Am)	Between ± 5° of the Equator, average temperature is close to 27°C the year round, with an annual range of 2-4K only which is less than the diurnal range of 8-11K, annual rainfall > 2000 mm. Such a climate is expectable in Asia between 10 and 25°N on West coasts of India and South-East Asia. Climate is warm and wet large annual rainfall with a dry month having < 60 mm rainfall. Dry season is characterized by low sun period when North-East monsoon prevails. Similarly, along East coasts of Central and South America, Madagaskar, South-East Asia, the Philippines and North-East Australia in the latitudes 10 to 25°S also have this climate. Easterly trade winds produce heavy rainfall when moist air rises up coastal slopes. During high sun solstice period tropical cyclone are common. At higher latitudes, annual average temperature can be 5-7K cooler than the zone 5° of the Equator.

Type	Features
Tropical desert and steppe (BW and BS)	Very dry climate mostly in areas between 15 to 35° which are truly arid zones with rainfall < 250 mm per annum (BWh) and steppe or semi arid zones with rainfall 375-750 mm per annum (BSh). In extremely arid regions relative humidity falls to 15-20% at 1300 hrs in the hottest months. Rainfall is sporadic and variable from year to year. Temperature in high sun months will have a daily mean of about 37.5°C with the mean daily maximum of 45°C. Annual range of monthly average temperature is 17-22K and average daily range is similar. Steppe are wetter; they border the tropical deserts to the north, South and East. Equatorward steppe climates merge with tropical wet-dry climate (Aw) and have higher annual temperature than desert climate; therefore annual range is smaller. Poleward steppe graduate into a Mediterranean climate (Cs). At the border of BS and Aw climates, during hot-wet season, it is difficult to manage livestock.
Tropical wet-dry (Aw)	Constitutes area between wet Af and Am and dry BW and BS, the intermediate zones with a tropical wet-dry climate with a wet season during the high sun period and dry season during low sun period. Central and South America, Africa and Australia fall into this category. Aw climate occurs in the latitudes 5 to 25° (pushed to 10°N and 30°N in Asia). Except in Asia, higher dry-bulb temperature occur in the period approaching high sun or before the season and the average daily temperature in this season will be 27-30°C. In the low sun seasons of wet-dry climates, temperature decreases by 4-6K. Asian Aw climate is controlled by the monsoon and in extreme contrast exists between wet and dry seasons. Poleward extension of Aw is the Cw which has a greater animal temperature range and cooler low sun temperature. Cwa climates are included in the subtropical groups because of warmest air temperature being ≥ 22°C.
Subtropical, humid Summer (Cf)	South-east of United States of America, East China, Japan, South-East Brazil, Uruguay, East Argentina and East Australia fall into this category. Rainfall is adequate throughout the year but is highest in warm months. Temperature ranges between 25-30°C during summer and is sometimes similar to wet equatorial zone. Annual range of temperature is 10-20K and therefore, winters are mild to cool and humid. In South-East Asia, this climate is modified by intense monsoon thereby winters become drier and summers wetter.

Type	Features
Subtropical, dry summer (Cs)	Recorded between latitudes 30-45° on western coasts. In contrast to Cf, wet winter and dry summer or characteristic. Located between a dry West coast tropical desert on the equatorial side and a wet West coast on the Poleward side. Average temperature during the warmest month will be 22-27°C. Mediterranean lands are known to have this climate and has referred to as "Mediterranean climate". Annual range of temperature will be 13-16K.
K = Kelvin	*Source* : Smith, 1981

1.4. Tropical zone

Tropical zone lies between Tropic of Cancer and Tropic of Capricorn in the latitudes 10-25° and subtropical zone lies between 25-35° latitudes on either side of the Equator towards Poles. Totally about 57% of the land area lies between the zones 35°S and 35°N. Three broad zones may be defined within the tropical zone namely,

1. Equatorial: a zone extending 10° north and South of Equator where insolation is intense throughout the year and day and night are roughly of same duration.
2. Tropical zones North and South: zones astride the tropics of Capricorn and Cancer in the latitudes 10° to 25°. In the zones, the sun's path lies close to the zenith at one solstice but is appreciably lower at the other solstice. There are marked seasonal cycles but the annual total insolation is large.
3. Subtropics: these are transitional areas between the tropics and the cooler mid-latitudes, lying between 25° and 35°; they may extend a few degrees Poleward or Equatorward of these parallels.

1.5. Problems of poultry production in tropics

Any particular geographical location, a combination of total daily solar radiation, solar altitude and azimuth, cloud cover, rainfall distribution and intensity, wind speed and direction, ground cover and shade, and thermal characteristics of housing determine the "effective temperature" for the birds in an enclosed space.

1.5.1. Design temperature

The main problem of tropics is of heat – manifested as high wet-and/or dry-bulb temperature. Summary of outdoor maximum and minimum temperature for a particular month/season can be utilized to design a building ("design temperature") which is based on the frequency distribution of hourly temperature and expressed as that exceeded during a certain percentage of hours during winter or summer. Commonly used are 1, 2½ or 5% design temperature.

For New Delhi 29°N BSh climate, 2½% design temperature (°C) are 5.0, 41.7, 27.8 and 14.4 for winter (dry bulb), summer (dry bulb), summer (wet bulb) and daily range of dry bulb, respectively. In other words, during summer at New Delhi, temperatures exceed 41.7°for 2½% of hours (36 min); corresponding value for winter is 5°C (for 36 min the temperature will be less than 5°C). Design wet bulb temperature is important during summer.

It is likely that in Af or Am climates where diurnal variation of dry bulb temperature is = annual range, poultry gets acclimatized to the constantly high and uniform temperature and produce as well as under cool climate.

If seasonal rainfall differences are marked, as in some Am and most Aw climates, the wet season differences productivity depending upon the age of the bird at the onset of wet season.

2. Assessment of climate

(Starr, 1981)

Effect of climate on animals depends on various factors such as age, breed, posture, nutritional status of the animal, degree of acclimatization, duration of stress and animal's response time etc.

Temperature of both air and surroundings is important along with wind speed and direction. Altitude influences temperature as well as atmospheric pressure. Humidity is important when loss through sensible means is reduced / minimized. Precipitation is not an important factor for poultry. Air movement affects connective heat loss. In tropics "advection" of hot air over the animal increases the heat stress and conversely, in temperate zone, air movement increases cold stress. Solar radiation is important when the animal is kept out-door. Shelter oriented East-West may intercept more solar radiation and therefore provide cooler environment. Effect of atmospheric pressure is not totally understood; but, housing cannot influence effect of a positive pressure, if any.

2.1. Climatic indices

2.1.1. Wind-chill factor (K, in W/m²)

$K = 1.163\ (10u^{1/2} + 10.45 - u)\ (33 - T_a)$ where T_a = air temperature (°C) and u = air speed (m/sec). This is useful as a rough measure of heat loss under forced convection. This factor is more applicable to human beings (1watt = J/sec).

However, chicken cooling effect corresponding to air speed of 6, 15, 30, 76 and 152 m/min has been estimated to be 0, 0.5, 1.7, 3.4 and 5.6°C, respectively (North and Bell 1990).

2.1.2. Wet Bulb Globe Temperature (WBGT,°C)

$WBGT = 0.7\ T_w + 0.2\ T_g + 0.1\ T_a$ where T_w = Wet bulb temperature and T_g = Globe (black) thermometer reading.

WBGT = 0.7 T_w + 0.3 T_a for shaded regions (inside the house). WGBT is more applicable to human beings.

2.1.3. Equivalent temperature (Teq, °C)

This is the temperature of a uniform enclosure in which under still air, a sizeable black body at 24°C would lose heat at the same rate in the given environment.

$T_{eq} = 0.522\ T_a + 0.478\ T_w - 0.207\ (37.8 - T_a)\ u^{1/2}$.

However, this is not appropriate when humidity effects are important particularly when T > 20°C

2.1.4. Operative temperature (T_{op}, °C)

T_{op} combines the effect of radiation, air movement and air temperature. T_{op} also ignores effects of humidity and skin temperature has tobe known for calculations.

$$T_{op} = \frac{h_r \bar{T}_R + h_c T_a}{h_r + h_c}, \text{ where } h_r = \text{radiative heat transfer, } h_c = \text{convective heat transfer}$$

and $\bar{T}_R$ = mean radiant temperature; since h_c is proportional to $\sqrt{u}$,

$$T_{op} = \frac{1}{1+k}\left[k(17.7+\bar{T}_R) + \frac{u}{u_0}(17.7+T_a) - \left(\frac{u}{u_0}\right)^{0.5}(17.7+t_{sk}) - 0.55\right] - 17.7$$

where $k = \frac{h_r}{h_c}$, u_0 = standard air velocity (0.2 m/sec), t_{sk} = skin temperature

2.1.5. Temperature-humidity index (THI)

THI = 0.72 (T_a + D) + 40.6 where D = dew point temperature.

Many other climatic indices have been proposed; but, most of them are either influenced by several factors or impracticable under commercial set-up. In general, wind chill factor and temperature-humidity index are popularly used indices along with dry- and wet- bulb temperature and relative humidity.

Chapter **26**

Thermoregulation

1. General

There are about one million animal species of which only 20,000 species are homoeothermic; majority of the species bred by man are homoeotherms (Poczopko, 1981).

1.1. Altricial animals

These homoeotherms are born/hatched blind with bare skin (Poczopko, 1981).

1.2. Precocial animals

These homoeotherms are born with open eyes and skin covered with pelage or down and are able to move about soon after birth; most of the domestic animals fall into this category (Poczopko, 1981).

Many intermediary forms also distinguished. However, all juveniles (both altricial and precocial) have a narrower zone of thermoneutrality and lower critical temperature more than that of adult animals; therefore, they are more sensitive to cold (Poczopko, 1981).

Birds are homoeotherms and hence are capable of balancing rates of thermolysis and thermogenesis so that they are able to regulate their body temperature within a narrow range over a wide span of thermal conditions. The general features of thermoregulation in Mammals and Avians is essentially same; but, there are some behavioral and functional differences (Table 26.1, Dawson and Whittow, 2000).

Table 26.1 Behavioral and functional differences between Mammals and Avians

	Avians	Mammals
Seasonal dormancy (hibernation)	Very rare	Recorded
Non-shivering thermogenesis	Absent or less prominent	Prominent
Plumage	Well developed, aerodynamic and insulative	Not well developed
Distribution of subcutaneous fat	More localized and hence tissue insulation is not uniform	More uniformly distributed
Sweat glands	Absent	Present
Panting mechanism	Present	Restricted

2. Body temperature

(Dawson and Whittow, 2000)

2.1. Deep body ("Core") temperature

Body temperature of Avians shows more variability than other mammals; body temperature of newly-hatched chick (39.7°C) is less than that of adults (40.6 to 41.7°C) which is achieved by 3 weeks of age. Smaller breeds, males within a breed and more active birds will have higher body temperature.

Therefore, for inter-specific comparisons, deep body temperature is generally used (Table 26.2) because it is maintained within fairly narrow limits whereas temperature of more superficial areas comprising the "shell" may vary substantially depending on the bird's activity states and external conditions as well as from one anatomical region to another. Appropriate thermal gradients between the core and the shell will be maintained by the bird in the process of controlling heat exchange with its environment.

2.1.1. Measurement of deep body temperature

Deep body temperature of birds is commonly measured with a thermistor or thermocouple inserted into the proventriculus or large intestine (rectum); under experimental conditions, temperature of brain or spinal cord and also frequently measured.

Deep body temperature of large flightless birds like Emu, Ostrich etc. and certain aquatic species like Penguins, Albatrosses etc. is lower than other Avian species.

Table 26.2 Deep body temperature of selected species of birds

Species	Body mass (kg)	Temperature (°C)
Domestic fowl	2.4	41.5
Domestic duck	1.9	42.1
Domestic turkey	3.7	41.2
Domestic goose	5.0	41.0
Emu, male	39.7-40.7	37.7
female	37.0-45.4	38.2-38.3
Ostrich	100.0	38.3

Source : Dawson and Whittow, 2000

2.2. Regulation of body temperature and energy exchange

(Etches *et al.*, 1995)

Excessive flow of energy from or into the body mass, both, can lead to death; as seen in very low ambient temperature or very high body temperature, respectively.

If exposure to heat results in increased body temperature by 1-2°C, birds cannot tolerate for a long time. Lower critical temperature ('a') is the minimum environmental temperature which, even if maintained over a period of days, is compatible with life (Fig. 26.1).

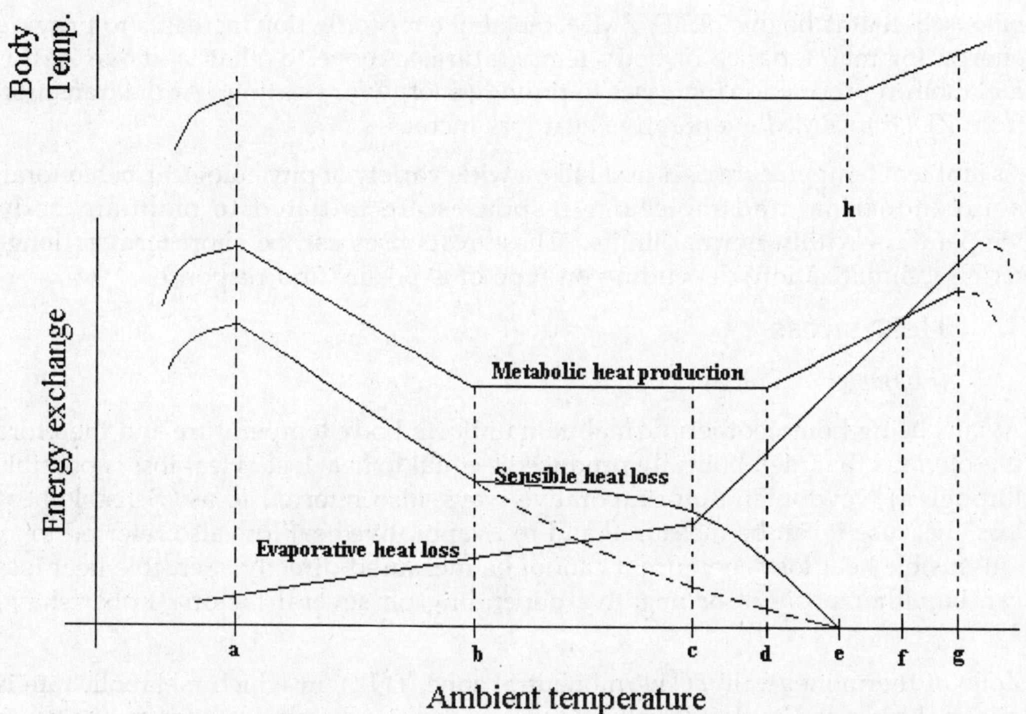

Fig 26.1 Ambient temperature *Vs* heat loss and body temperature

Below a → Zone of hypothermia
b → Lower critical temperature
a – h → Zone of normothermia
a – g → Thermoregulatory zone
b – c → Zone of least thermoregulatory effort
b → Critical temperature
b – d → Zone of minimum metabolism
d → Upper critical temperature
c → beginning of intense evaporative loss
e → Ambient temp = Body temp
f → Metabolic heat prodn = Evaporative heat loss or Sensible heat loss = 0
h= Point of incipient hyperthermia
Above h → Zone of hyperthermia
g → Critical thermal maximum

Source : Etches *et al.*, 1995

Below 'a', body temperature begins to decrease and the prolonged, leads to death within zone of hypothermia. Conversely, above 'h', body temperature increases till critical thermal maximum ('g') is reached; and if prolonged, terminates in death.

Within 'b' and 'c' referred to as zone of least thermoregulatory effort (ZLTE), metabolic heat production reduces the minimum with a constant sensible heat loss and limited evaporative heat loss due to normal respiration and exposure of non-insulated areas of the body.

Within 'b' and 'd' referred to as zone of minimum metabolism (ZMM) which extends from ZLTE Tumors higher ambient temperature which can be accommodated by increasing both evaporative and sensible heat loss.

Below 'b' that is below ZLTE/ZMM, metabolic reproduction increases to provide energy for maintenance of body temperature. Above 'd' that is above ZMM, metabolism production increases to provide energy for panting. At the transition from ZLTE to ZMM evaporative heat loss increases.

As ambient temperature rises and falls, a wide variety of physiological, behavioral, neuroendocrinal and molecular responses are initiated to maintain body temperature within normal limits. These responses can be short term or long-term (acclimatization) depending on type of exposure and response.

3. Heat stress

(Etches *et al.*, 1995)

Avians being homoieothermic maintain uniform body temperature and therefore it is obvious that metabolically producedis equal to heat lost. Heat loss is possible through a) Newtonian non-evaporative ways, also referred to as "Sensible heat loss" because it can be measured and b) Evaporative heat loss also referred to as "Insensible heat loss" because it cannot be measured directly. Sensible heat loss can be either positive or negative depending on several factors (Robertshaw, 1981).

Zone of thermoneutrality (Thermoneutral zone, TNZ), in which metabolic rate is minimal bounded by lower critical temperatures and upper critical temperature (hyperthermic point), and the values of its boundaries and its width depends on species, breed and the individual in question. The width is a consequence of the ability to regulate evaporative heat loss; the lower critical temperature is largely determined by insulation and the production.

It is generally agreed that optimum temperature for chicken during brooding is 35°C at day-old with a drop of about 0.5°C daily until house temperature is reached; for broilers, 16-25°C in house with a practical limits of 13-30°C and for layers 20-25°C (Sainsbury and Sainsbury, 1988)

Species adapted to high environmental heat loads will have low insulation, well-developed evaporative heat loss mechanisms and, in many cases, lower paces metabolism. Cold-adapted species, on the other hand, tend to have higher metabolic rate and greater insulation. Therefore, if temperature decreases below the lower critical temperature, maintenance requirement increases appreciably. Similarly, when temperature exceeds hyperthermic point, maintenance requirement again increases although the magnitude will be less (Robertshaw, 1981).

3.1. Behavioral responses

All the behavioral responses are preceded by molecule response to the stress which is mediated by heat-shock proteins.

As ambient temperature exceeds comfort zone, there will be decreased feed consumption with concomitant increase in water intake, reduced time on walking and standing and increased evaporation. Birds splash water on the combs and wattles and reduce time on social behavior and changing posture. In cages, they keep distance from each other, pant, often stand with wings drooped and lifted slightly to increase sensible heat loss. Layers prefer cooler cage temperature during night and during periods of low feeding activity. Back of the birds appears to be most thermally sensitive part.

3.2. Physiological responses

3.2.1. Consumption of feed and water

Feed consumption decreases after several hours whereas water consumption increases immediately. Fasting for 1-3 d increases survival time under heat stress. Due to reduced feed consumption, body weight decreases.

When heat gain is greater than evaporative heat loss, body temperature increases; but in reality, rectal temperature increases and heat dissipation during night increases. Both acute and chronic heat exposures stimulate release of cortisol which in turn promotes protein catabolism (Robertshaw, 1981).

Level of food intake is directly proportional to level of thermoneutral metabolism. Heat tolerant species will decrease the food intake less than those with their adapted to cooler environment. High producing animals have relatively high thermoneutral metabolic rate and therefore are poorly adapted to heat conditions. It follows that such animals decrease food intake a greater extent than low producing ones. Increased temperature also increases digestibility probably because of reduced thyroid secretion; but the magnitude is quite small (Robertshaw, 1981).

3.2.2. Sensible heat loss

3.2.2.1. Specialized heat exchange mechanisms

Arteriovenous heat exchange mechanism in legs and feet increases heat loss. At high ambient temperature, shunts in the vascular system which regulate volume of blood flow through arteriovenous network bring cool venous blood to close proximity of arterial blood for maximum heat dissipation to environment. Hens roost on pipes in which cold water is circulating.

Rete opthalmium, arteriovenous heat exchanger between optic cavity and brain help dissipation of heat through cornea, the eye, the buccal cavity, the beak and nasal passages.

3.2.2.2. Feather cover

Loss of feather increases sensible heat loss and hence birds may shed feathers. In case of birds with Na gene, better growth rate, FCR, viability, egg weight and egg production are recorded due to nakedness at the neck region facilitating heat loss.

3.2.2.3. Respiratory rate and blood pH

Panting begins at 29.4°C ambient temperature and increases water evaporation from 5 to 18 g/h when the ambient temperature increases from 29 to 35°C with a relative humidity of 50-60%. Increased carbon dioxide loss results and reduced partial pressure of carbon dioxide and reduced plasma bicarbonate and hydrogen ion concentration precipitating in increased plasma pH (alkalosis) and reduced shell thickness (More details given in the Chapter on Egg formation and structure). However, in case of turkeys, respiratory alkalosis is recorded during acute hyperthermia but not during chronic hyperthermia.

It is obvious therefore that dissipation of heat demands increased respiratory rate whereas loss of carbon dioxide demands the opposite. Therefore, as a compromise, birds resort to "Gular flutter", a rapid resonant vibration of upper throat passages driven by hyoid apparatus so that nasal cavity, nasopharynx and trachea which are not involved in exchange are actually utilized for heat loss. In case of pigeons, slowed deep breathing accomplishes the same result.

Anions in the form of ammonium chloride or carbonated water can be offered to birds under severe heat stress in anticipation of reduced detrimental effects on shell thickness.

3.3. Hormonal responses

3.3.1. Neurohypophyseal hormones

3.3.1.1. Arginine vasotocin

During heat stress, release of arginine vasotocin results and decreased temperature of shank, comb, cloaca and foot. At least in pigeons, it also controls BMR. The hormone mobilizes free fatty acids to meet energy requirement for respiratory muscles for panting. In case of pigeons, there will be reduced BMR due to reduced T_3.

3.3.1.2. Mesotocin

This hormone is the Avian analogue of oxytocin whose levels of reduced during heat stress due to the increased levels of arginine vasotocin. Mesotocin is likely to be a diuretic hormone in birds and hence, lowered levels of this hormone during heat stress help conserve body fluids.

3.3.1.3. Growth hormone

During the stress, levels of growth hormone increases in order to enhance fatty acid mobilization to support panting.

3.3.2. Hypothalamic-pituitary adrenal axis

3.3.2.1. Corticosterone

This is the principal steroid of the Avian adrenal cortex. During acute heat stress, there will be a surge of adrenal corticosterone and aldosterone. On the other hand, during prolonged stress, there will be an initial surge followed by decline, hypothermia and death.

Aldosterone along with arginine vasotocin increase renal absorption of water and prevent dehydration. Increased corticosteroids and ACTH reduce immune response.

3.3.2.2. Catecholamines

Epinephrine and nor-epinephrine released during acute heat stress (not during chronic heat stress) affect body temperature more directly through hypothalamus.

3.3.2.3. Melatonin

This hormone, the major pineal hormone (which is also produced by extra-pineal tissues like retina and Harderian gland), facilitates heat dissipation by peripheral tissues through vasodilation and increased blood flow particularly to feet. Melatonin may also lower the set point of the main "thermostat" which may be present in hypothalamus.

3.3.2.4. Reproductive hormones

During heat stress, hypothalamus appears to be the target organ; corticosterones produced during the heat stress period reduce production of leutinizing hormones as well as leutinizing hormone releasor hormone (LHRH). Brain monoamines may also reduce hypothalamic function culminating in reduced LHRH.

3.3.2.5. Thyroid hormones

During heat stress, there will be reduced levels of thyroid hormones which helps the tolerance; however, levels are not consistent in response to heat stress.

3.3.3. Heat-shock proteins (HSPs)

These are synthesized because of nutrient deprivation, oxygen-starvation or presence of heavy medals, oxygen radicals or alcohol. Depending on their molecular weight, they are referred to as HSP 70, HSP 90 etc. Some of them are required for performance of vital functions of normal cells whereas some others are required to withstand the toxic effects of fixed in temperature. Generally, HSPs are involved in the assembly or disassembly of proteins/protein-containing complexes during life and death of a normal cell. Under heat stress, there will be impairment of transcription, RNA processing, translation, post-translational processing, oxidative metabolism, membrane structure and function, cytoskeletal structure and function etc.

Response of cells/organism to heat-shock is extremely rapid, but transient; the response includes redistribution of pre-formed HSPs within the cell as well as immediate translation of pre-formed mRNA into HSPs, immediate transcription of genes encoding HSPs and cessation of transcription/translation of other genes/ mRNA.

Splicing of introns to form functional RNAs is likely to be inhibited under the stress contributing to thermotolerance by allowing production of functional mRNAs from a range of genes for protection of normal functions of a cell.

4. Heat balance

(Dawson and Whittow, 2000)

For an animal not performing external work, the familiar heat balance equation is as follows:

$$S = H - E \pm R \pm C \pm K, \text{ where}$$

S = the gain of loss of bodily heat

H = metabolic heat production

E = evaporative heat loss

R = radiative heat gain or loss

C = convective heat gain or loss

K = conductive heat gain or loss

It is obvious from the above equation that if S is positive, body temperature rises and *vice versa*; and body temperature is unchanged when S = 0 (Fig 25.1).

4.1. Units of measurement

The Systeme International (SI) units such as watts or joules are recommended in place of calories or kilocalories or British Thermal Units (Btu). However, the relationship between these units is as follows:

1 kcal = 4187 J = 4.187 kJ; 1 kcal/hr = 1.163 W; 1.055 kJ = 1 Btu

4.2. Heat production (H)

4.2.1. Measurement

Heat production is usually determined from measurements of oxygen consumption when the bird is at rest. The dry volume of oxygen consumed per unit time is connected to standard conditions of temperature (0°C) and pressure (760 mm Hg = 101.3 kPa) i.e. to STPD and multiplied by 20.1 kJ/l O_2 to convert into energy units.

The other methods of measurement include temperatures of skin and plumage surface, he transferred through plumage etc.

4.2.2. Definitions

4.2.2.1. Basal metabolic rate (BMR)

This is defined as the minimal heat production of a normothermic bird that is awake while resting and fasting (i.e. has reached a post-absorptive state) at an ambient temperature requiring neither regulatory thermogenesis nor expenditure of energy in active evaporative cooling. The range of the ambient temperature within which BMR occurs is the zone of thermoneutrality (TNZ).

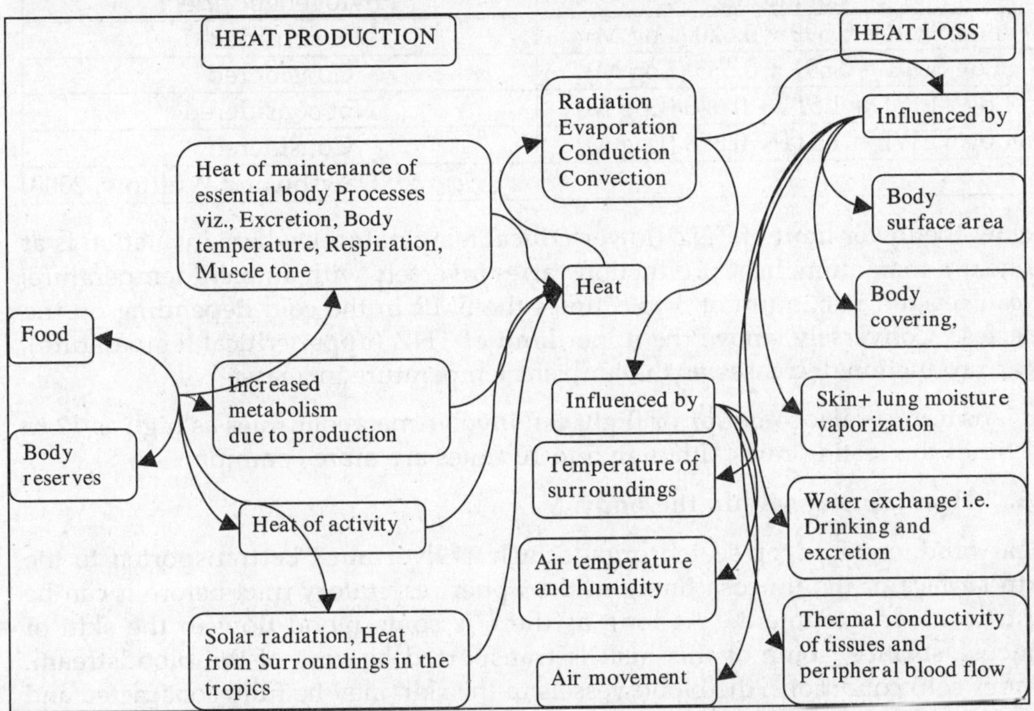

Source : Sainsbury and Sainsbury, 1988

Fig 26.2 Major routes of heat production and heat loss

4.2.2.2. Standard metabolic rate (SMR)

Results of metabolic measurements for small birds (which reach post-absorptive phase in just a few hr) resting in a post-absorptive state in the TNZ are referred to as standard metabolic rates (SMR) when the conditions are clearly specified.

4.2.2.3. Resting metabolic rates (RMR)

Measurements of birds resting in the TNZ, but not in a post-absorptive state, represent resting metabolic rates (RMR).

4.2.3. Influence of ambient temperature on metabolic rates

Over and intermediate range of ambient temperature, metabolism remains constant standard or resting rate, depending on the nutritional state of the bird. The RMR tends to be higher due to heat increment of feeding. TNZ represents the temperature range over which heat production is independent of ambient temperature.

Table 26.3 Equations for calculating SMR (in W) and Total evaporative water loss (TEWL) at 25°C (in ml /d) in relation to body mass (M, kg) (for all birds)

Equation	Phylogenetic effect
Log SMR = 0.599 + 0.670 (Log M)	Not considered
Log SMR = 0.591 + 0.635 (Log M)	Considered
Log TEWL = 1.511 + 0.678 (Log M)	Not considered
Log TEWL = 1.511 + 0.678 (Log M)	Considered
	Source : Dawson and Whittow, 2000

Below the lower limit of TNZ (lower critical temperature), where insulation is at or near a maximum, heat production varies inversely with ambient temperature; it can reach a maximum of 3 to 8 times the SMR in the cold depending on the species. Conversely, above the upper limit of TNZ (upper critical temperature), heat production decreases as the ambient temperature increases.

Avians terrestrial locomotion or flight can involve metabolic rates as high as 12 to 14 times the resting rates although 5 to 10 times are more common.

4.3. Heat transfer within the body

Heat produced in deep-seated organs such as liver must be transported to the skin surface or the mucosa lining in the upper respiratory tract before it can be lost to the environment. As long as there is some blood flow to the skin of mucosa surface, some of this heat is transported by way of the bloodstream. Under cold conditions, the blood vessels in the skin may be fully constricted and heat transfer then has to occur by conduction through the tissues.

4.3.1. Vascular heat exchange

Arteriovenous anastomoses in the skin are a more ubiquitous vascular arrangement to permit a larger volume of warm blood to flow to the skin and facilitate heat loss. This mode of heat loss occurs mostly in unfeathered skin in domestic fowl representing 17, 53 and 83% of total blood flow in cold, thermoneutral and hot conditions, respectively.

4.3.2. Cold vasodilatation

The possibilities of issues in the cold extremities might freeze is circumvented by periodic increases in blood flow to the extremities (cold vasodilatation) which occurs at the expense of increased heat loss from the bird.

4.3.3. Thermal conductance of the tissues

Tissue thermal conductance (W/°C or W/kg/°C) is the ratio of difference between metabolism production (W or W/kg, computed from oxygen consumption) and respiratory evaporative heat loss (W or W/kg) to that between core and mean skin temperatures (°C).

4.4. Heat loss

Heat transported to the surface of skin may be lost by evaporation or by non-evaporative means. As most of the bird's surface is covered with feathers, this heat and any water evaporated on the skin must traverse the plumage before being lost to the environment. The depth of the plumage is not fixed; it varies inversely with ambient temperature.

4.4.1. Non-evaporative heat loss – total thermal conductance

It is relatively difficult to estimate non-evaporative heat loss mainly because of multiple pathways through which it can occur i.e. conduction, convection and radiation. However, total thermal conductance provides an indication of the facility with which heat is transferred through the tissues and plumage and lost (when body temperature is greater than ambient temperature) to the surrounding medium at a given difference between body temperature and ambient temperature. This includes conductance of tissues after as a coefficient describing the heat loss from the bird to the surroundings.

4.4.2. Evaporative heat loss

4.4.2.1. Total evaporative heat loss (E)

Total evaporative water loss (TEWL) at 25°C in birds from arid environments is statistically lower than that among mesic birds. For every g of water evaporated, heat equivalent to latent heat of vaporization of water (2.4 kJ) is lost. Evaporation of water in birds involves both respiratory (including the buccopharyngeal cavity) and cutaneous (including corneal surfaces) components.

Table 26.4 Total evaporative heat loss (E) of birds as percentage of metabolic heat production (H) at high ambient temperatures

Species	Body mass (kg)	Ambient temperature (°C)	E ÷ H (%)
Japanese quail	0.100	43	144
Ostrich	100.0	44.5	100
Rock pigeon	0.315	44.5	118
			Source : Dawson and Whittow, 2000

4.4.2.2. Respiratory evaporative heat loss

This is the primary means of heat defence in most Avians almost without exception under sufficiently challenging circumstances.

4.4.2.3. Cutaneous evaporative heat loss

This is measured depending on the resistance of Avian skin to the diffusion of water. In most of the species, cutaneous evaporation represented half or more of the total evaporation at low humidities and moderate to warm ambient temperature (25 to 25°C or 37.5°C). Cutaneous water loss is apparently affected by the hydration state of the bird; water deprivation severely reduces cutaneous evaporation.

However, the importance of cutaneous evaporation for birds in very hot environments appears to be secondary to respiratory evaporation in most of the domestic birds like chicken, ducks, geese, pheasants, quail, etc.

4.4.2.3.1. Thermal tachypnea (polypnea)

Under heat challenges, birds resort to open-mouth respiration, or thermal panting. The increased ventilation is achieved by panting sufficiently to overcome the effect of a concurrent reduction in tidal volume (type I panting) so that increased ventilation is confined to the respiratory dead space so as to facilitate substantial increase in respiratory minute volume and respiratory evaporative cooling without substantially affecting blood carbon dioxide and acid-base balance.

Typically, in a hyperthermic, panting bird respiratory minute volume may increase six-fold. However, as the body temperature increases to very high levels, respiratory frequency reaches a maximum and subsequently declines. These are usually accompanied by an increased tidal volume so that respiratory minute volume actually increases further (type II panting) eventually declining as the limit of animal's thermal tolerance is approached.

However, significant hypocapnia and alkalosis (mean values for arterial carbon dioxide tension and pH, 14.2 torr/1.89 kPa and 7.70, respectively Vs 27.1 torr/ 3.61 kPa and 7.51 for birds at TNZ) have been observed in a number of species during heat challenges may be due to induction of type II panting.

4.4.2.3.2. Gular flutter

Many birds, including domestic ones, supplement panting with rapid fluttering of the gular area. Rhythmic inflation or pulsation of the well-vascularized esophagus contributes the effectiveness of this activity in pigeons and doves. This mechanism appears to have two advantages i.e.

1. The air movement only affects surfaces that do not participate in gas exchange thereby avoiding the problems of hypocapnea and alkalosis.
2. The energetic cost of fluttering is considerably less than that for moving thoraco-abdominal structures leading to greater cooling efficiency.

During heat challenges, an estimated 20% and up to 35% of total evaporative heat loss occurred through gular flutter in Japanese quail and domestic chicken, respectively. Frequencies of gular flutter ranged from 176 to 1000 cycles/min depending on the species and in most species gular flutter preceded thermal tachypnea.

4.4.3. Site of evaporative cooling

The principal sites of the evaporation produced by thermal tachypnea are the nasal, buccopharyngeal and upper tracheal regions with some possibility of participation of the walls of the air sacs in the ostrich. Respiratory surface temperatures (°C) of the domestic fowl during thermal panting are as follows: mouth – 40.7, pharynx – 40.7, cervical esophagus – 41.5, thoracic esophagus – 42.3, anterior thoracic air sac – 43.1, abdominal air sac – 43.2 and rectal temperature – 43.0.

4.4.4. Osmotic state and respiratory evaporative cooling

In addition to thermal factors, the onset and extent of thermal tachypnea are affected by the osmoregulatory state of the animal; dehydrated individuals tending to delay its appearance until body temperature increases during thermal stress.

4.4.5. Heat conservation

During inspiration, air is warmed nearly to body temperature and saturated with water vapor while still in the nasal passages. Heat is removed from the nasal mucosa in the conversion of water from a liquid to vapor state and the mucosal temperature falls, in some cases to a level cooler than the initial temperature of the inhaled air.

Expired air, essentially saturated with water vapor, leaves the lungs and air sacs at body temperature. Passage over the cool nasal surfaces during exhalation cools this air and a portion of the water content condenses on the mucosa.

This action of the nasal passages is particularly effective below the lower critical temperatures, where Avian evaporative water loss tends to decline with decreasing ambient temperature despite the opposite trend in metabolism. At higher ambient temperatures, respiratory evaporative cooling is required and therefore the birds shift to open-mouth respiration characterizing thermal panting.

4.4.6. Acclimatization and acclimation

Acclimatization refers to a modification in response to stressful changes (seasonal or geographical) imposed by one or more climatic factors in nature, whereas acclimation is reserved for modifications in physiological response induced in the laboratory by controlled changes in specific climatic factors (like ambient temperature).

Acclimation and acclimatization have a number of effects on birds which influence their responses to thermal stress. Secretory unit densities in the lateral nasal glands and arteriovenous anastomoses in the nasal mucosa was significantly greater in the rostrum nasal conchae of domestic fowl exposed to heat for 4 hr/d over a period of two months than in control birds indicating a possibility of increased evaporative cooling from the nasal mucosa. Birds acclimated to high temperature had an improved capacity of panting as well as to maintain almost normal acid-base balance.

Acclimation to temperature changes occurs in 3 phases namely, 1. Neuronal →seem immediately, increased BMR then changed from 35 to 5°C for 2h. 2. Hormonal →after 2d, a second rise in BMR 3. Morphological→long-term changes. Chicken exposed to high ambient temperature develop enlarged combs and wattles, contain less fat and reduce feather covering (Kampen, 1981).

Within a day of exposure to changed temperature, respiration rate, blood pressure, pulse and shell thickness change in either direction. Rectal temperature adjusts in about 2-5 d; heat production after 3-12 d; food intake, egg production and egg weight after 1-2 weeks (Kampen, 1981).

Acclimation to high temperature takes longer time than that for low temperature. For adjustments of body weight under high temperature, it takes about 6 weeks after which, generally, bird is well acclimatized (Kampen, 1981).

4.5. Partition of heat loss

Measurement of heat loss from domestic fowl using a gradient-layer calorimeter indicated that most of the heat loss at low ambient temperature occurred by non-evaporative means (primarily convection or radiation depending on the rate of air movement, sensible heat loss) whereas, evaporative heat loss (insensible heat loss) became important as ambient temperatures increased; especially at are above 40°C, the entire heat loss may be due to evaporation of water (Table 26.5; Note: 3.6 kJ/hr = 1 W).

Table 26.5 Effect of temperature on heat output

Temperature (°C)	Sensible heat (%)	Insensible heat (%)	Sensible heat output (W/kg)
4.4	90	10	20.9
15.6	80	20	18.4
26.7	60	40	14.1
37.8	40	60	10.0

Source : North and Bell, 1990

5. Behavioral thermoregulation

Behavioral changes allow birds to save energy, conserve water and generally reduce thermal stress. It typically involves some form of movement leading to a change in position, posture or orientation to wind and/or Sun, though in some instances remaining quiescent is an effective means of minimizing thermal stress.

Hypothalamus appears to have a prominent role in control of behavioral thermoregulation. However, major portion of the thermogenic response of the birds to cold appears to be driven by input from deep body for thermosensors located outside the central nervous system.

5.1. Heat loss reduction / Heat gain

In cold environments, postural adjustments reducing the surface area in the form of hunching/huddling contribute to heat conservation. Unfeathered feet and tarsi when enclosed within ventral contour feathers while squatting also reduces heat loss substantially (20-50%). Birds seek shelter and avoid ruffling their feathers by facing into the wind during cold in an attempt to reduce heat loss. Whenever possible, birds reduce requirements for regulatory thermogenesis by sunbathing.

5.2. Facilitating heat loss

In hot environments, birds frequently minimize their heat loads and requirements for evaporative cooling by seeking shade and remaining inactive under a shelter. Domestic fowl splash water over its comb and wattles so as to facilitate heat loss.

6. Development of thermoregulation

6.1. Embryo

(Tazawa and Whittow, 2000)

Chicken are essentially poikelothermic during embryonic stages, even at external pipping, and rapidly develop the capacity to maintain body temperature upon cold exposure soon after hatching. Being precocial, they exhibit a feeble, incipient metabolic response to cooling, indicating endothermic homoeothermy before hatching. This coincides with enhanced activity of the thyroid gland and increased concentrations of peripheral thyroid hormones during the last stages of incubation.

The transition for a precocial bird takes place in four stages:

1. An Arrhenius-limited stage in which the metabolic rate is directly related to the temperature with a Q_{10} value of approximately 2
2. An oxygen-conductance-limited stage in which the oxygen consumption is limited by the rate of diffusion of oxygen through the shell and the chorioallantoic membrane in relation to oxygen demand by the embryo
3. A power-limited stage in which the embryo has the limited capacity to generate heat in response to cooling, a function of the maturity of tissues and development of thyroid activity and
4. "full-blown" homoeothermy.

An altricial species, on the other hand, never passes through the oxygen-conducted-limited stage; it is subject to the Arrhenius-limitation until after it hatches.

6.2. Hatchling

Birds vary in their thermoregulatory capacities immediately after hatching and have been classified as follows:

1. Precocial species – the hatchlings are covered with down and they are able to respond effectively to heat and to cold
2. Altricial species – the hatchlings are naked when hatched and have little ability to regulate body temperature at the ambient temperatures below 35°C or above 40°C.
3. Semi-precocial, semi-altricial – are intermediate between precocial and altricial species in their thermoregulatory capacities.

6.3. Impact of environmental factors

(Tzschentke, 2005)

Peri- natal thermoregulation is characterized by peripheral as well as central nervous thermoregulatory mechanisms that are already developed during prenatal ontogeny. However, during early postnatal development, physiological mechanisms are not completely mature; for instance, heat production capacity.

Hatchlings have a large surface area relative to the body mass (high metabolic body weight) and their low insulation (due to absence of feathers) makes them susceptible for increased heat loss.

In spite of the above, poultry hatchlings have a highly effective thermoregulatory behavior to keep body temperature constant. For instance, poultry hatchlings are able to select their specific temperature range in a temperature gradient with high accuracy immediately after hatching.

To save energy hatchlings develop different mechanisms like a lower the body temperature caused by a lower thermoregulatory set-point during the first days of post-hatching in an attempt to reduce the thermal gradient between the animal and its environment.

Prenatal environmental influences may have a training effect on the postnatal efficiency of the thermoregulatory system. At the end of the embryonic, or during the early postnatal, period a qualitative change occurs in the reaction of adaptive body function, such as thermoregulation, due to environmental stimulation. Changes in the environmental conditions induce, as a rule, first un-coordinated and immediately non-adaptive reactions.

It appears that during the early development of body functions it is not important for the organism that adaptive reactions occur but rather the fact that the reaction occurs anyway is important for the adaptability during postnatal development.

The uncoordinated and/or immediatelynon-adaptive "training" reactions change into coordinated and/or adaptive reactions. During the course of perinatal period, most of the functional systems develop from an open loop system without feedback into a closed control system with feedback. During this process, during so-called "sensitive or critical" phases, environmental influences may have a strong and long-lasting influence on the set-point of a regulatory system via epigenetic adaptation mechanisms.

To a suboptimal prenatal environment, the embryo may adapt changes that are beneficial for survival before hatch, but these changes may be mal-adaptive in the later postnatal life (as in the case of induction of diseases, metabolic and behavioral disorders). On the other hand, epigenetic adaptation mechanisms might be specifically used to adapt the organism to the postnatal climatic conditions; for instance, induction of epigenetic temperature adaptation by changes in the incubation temperature. It is already been recorded that, compared to the control birds which were incubated at 37.5°C, a low incubation temperature induced postnatal cold adaptation and *vice versa.*

7. Thermoregulation in juveniles

(Poczopko, 1981)

7.1. Physical factors

7.1.1. Effect of size and body covering

Physiological mechanisms that control internal environment in juveniles are often different from those in adults although they are neither poorly developed nor ineffectual.

Reduced surface area to mass ratio helps thermoregulation but development of body cover more important. In precocial animals, species vary in their capacity to regulate body temperature presumably because of differences in rate of heat production. Subcutaneous fat is important only in adults as insulators but not in juveniles because the latter have only low levels of fat in their body.

Among domestic birds, turkeys are least cold-resistant and ducks and geese, the most; in either case, in all young birds, brooding is required.

7.1.2. Physical thermoregulation

Insulation in animals depends on 1. Vasomotor and 2. Pilo- or Ptero-motor mechanisms; the latter processes increase insulation 1.55 times in goslings when environmental temperatures decrease from 25 to 17°C. However, the effect is not uniform in all species.

The better the thermal insulation and its regulation, the lower is critical temperature for an animal and wider the thermoneutral zone.

7.2. Changes in standard metabolic rate (SMR) with age

SMR designated $M = aW^b$ where W is body weight in kg and 'a' and 'b' are constants obtained by plotting M against W in a log-log paper. In other words, Log M = Log a + b log W. Calculations that several authors has revealed that b = 0.75 and therefore, $W^{0.75}$ is referred to as 'Metabolic body size'. Daily heat production ÷ Metabolic body size = Metabolic level. Metabolic level of adults is approximately 70 kcal/kg$^{0.75}$/d = 293 kJ/kg$^{0.75}$/d = 3.4 W/kg$^{0.75}$. This can be used as a reference point, generally referred to as "Kleiber's level". (Note: 1 kcal/hr = 1.163 W or 3.6 kJ/hr = 1 W).

Table 26.6 Body weight, thermoneutral zone (TNZ) and oxygen consumption in Rhode Island Red birds

Age (wks)	Body weight (g)	TNZ (°C)	O_2 consumption at TNZ (mg/g/h)	Cold exposure		Metabolic quotient
				Temperature (°C)	O_2 consumption at TNZ (mg/g/h)	
< 1	36	34-36	1.25	21	2.85*	2.28
2	60	33-35	1.40	21	2.45*	1.75
5	260	30-34	1.35	10	2.45*	1.85
8	590	29-33	1.10	0	2.30*	2.09
12	1030	27-32	0.90	- 15	1.90*	2.11
18	1610	23-31	0.77	- 12	1.30	-
23	1960	20-29	0.65	- 12	1.05	-
52	2430	16-27	0.62	- 12	0.90	-

* Summit metabolism

Source : Poczopko, 1981

Changes in metabolic rate with size within a single species are more complex and are represented by a curve containing several straight sections representing different metabolic phases. Energetic cost of growth, transition from poikelothermy to homoeothermy and stabilization of endocrine system are involved; endocrine factors include primarily thyroid hormones.

Table 26.7 SMR (W/kg$^{0.75}$) in young *Vs* adults

Species	Age (d)	SMR	Age (d)	SMR	SMR in adults
J. quail	1	3.79	21	9.20	6.40
Dom. Pigeon	3	5.15	11	7.42	3.50
RIR – hen	1-6	4.26	30	7.02	4.60
Goose – White Italian	2-4	2.81	15	6.83	4.20
Turkey	1-3	3.25	60	6.31	3.34
Dom. Duck – Pekin	1-2	3.92	27	6.53	4.70

Source : Poczopko, 1981

7.3. Cold-induced increase in thermogenesis

Heat production increases both with activity and level of feeding; thermoneutral metabolism also increases depending on level of feeding as well as quality of feed. Poorer the quality higher the heat increment. Physical activity also increases heat production. Both feeding and exercise can substitute for cold thermogenesis (Robertshaw, 1981).

Fat is the main source of energy under cold exposure; surprisingly, in animals, it will not result in ketosis (Robertshaw, 1981).

At a constant food intake, due to increase metabolism during cold, ME retained for production decreases almost linearly at temperatures below 15°C. This is compounded by decreased digestibility, decreased retention time and increased gut motility which are independent of changes in food intake. There will be a concomitant increase in thyroxine production (Robertshaw, 1981).

Both shivering and non-shivering contribute to thermogenesis. In addition, thermoregulatory tonus due to an increase in muscular tone accompanied by increased heat production, voluntary muscular activity may also play roles.

7.3.1. Shivering thermogenesis

Altricial animals are generally unable to shiver soon after birth at least for some period depending on species. Precocial animals can shiver early; newly hatched, even wet goslings can shiver. Even invisible shivering may be working.

7.3.2. Non-shivering thermogenesis (NST)

This includes the production mechanisms liberating chemical energy due to processes which do not involve muscular contraction.

7.3.3. Obligatory or Basal NST

Refers to NST under conditions of basal metabolism or regulatory NST. This occurs at temperatures below TNZ due to the cold environment stimulating adrenergic system leading to release of nor-adrenaline at sympathetic nerve endings. Nor-adrenaline being lipolytic mobilizes free fatty acids for oxidation resulting in heat production. Brown adipose tissue which is the main target of nor-adrenaline is absent in birds and therefore thyroid hormones or glucagon are likely to mediate the process. In any case, NST appears to be absent in birds and therefore, free fatty acids mobilized in cold exposure are likely to be used for shivering thermogenesis.

7.4. Summit metabolism

Summit metabolism is defined as the maximal rate of heat production that is achieved in response to cold and can be sustained for sometime without hypothermia. In adults, summit metabolism is about 3-4 times the SMR. SMR of newborn animals is usually less than that in adults and also less than Kleiber's interspecific mean ($3.4W/kg^{0.75}$) because of large surface area and poor thermal insulation especially in case of altricial animals.

Domestic fowl is a precocial bird and its SMR or changes with age. In 1-6d old chicks, summit metabolism occurring at 27°C, is only 2.2 times the SMR with basal values of $4.3W/kg^{0.75}$.

7.5. Changes in deep body temperature with age

Generally, the body temperature of a newborn is regulated at a somewhat lower level than that of adults and it increases gradually. Difference between body temperature of one day-old animals and adults is greater in altricial than precocial animals. Time necessary to stabilize body temperature varies from several hours to one month.

Chapter **27**

Principles of Ventilation

1. General

Birds balance their heat production and heat loss to maintain homoieothermy. Birds lack sweat glands and have feathers which restrict heat loss from the body surface. Therefore, as temperature increases above the thermoneutral zone (12.8 to 23.8°C) insensible heat loss increases.

By not not having sweat glands the birds' dependence on insensible heat loss increases as the environmental temperature increases. At 29.4°C, body temperature increases by 0.1 to 0.4°C which results in stimulation of brain and subsequent panting - characterized by increased rate of respiration (polypnoea). This leads to the dramatic rise in water loss through evaporation from 5g to 30 g per hour. Heat lost by the bird to evaporate water from the respiratory tract is the main avenue of heat loss in birds subjected to a temperature above 29.4°C. At temperatures about 32°C, thermoregulation breaks down and respiration rate increases to 140 - 170 per min. Upper lethal temperature is 47°C at which the body temperature shoots up to 44°C initially and declines thereafter. Birds seek shade, reduce the food intake and increase water consumption to cope up with evaporative heat loss through panting. There will be corresponding increase in wet manure. Excess body temperature is indicated by rise in skin temperature of the extremities (combs and wattles), increased body temperature, vasodilation and panting.

Birds tolerate cold better, particularly, because they have feathers which are excellent insulators. However, birds will be very uncomfortable in dampness. Below the freezing point of water, birds have difficulty in regulating their body temperature. Hence, they hunch to reduce the surface area to conserve as much heat as possible. Birds fluff-out feathers and huddle to increase insulation, tuck their head under the wing and increase feed consumption when exposed to cold. Temperature of the extremities and skin decreases, food and water consumption increases at temperatures below 10°C; this leads to increased excretion of water in feces and consequent damp litter. The lower lethal temperature is (-) 30 to (-) 32°C.

In an under-ventilated building, there willbe stagnant air warmed-up and humid, with reduced O_2 which are beneficial to pathogenic organisms; hence, the birds drop production and show poor health. Such birds are more prone to respiratory diseases. Cold weather further compounds the problems because the excess humidity will condense on the bedding (litter) causing wetness and uneven distribution of birds (and hence the excreta and exhaled air also) which, in turn, accentuates the respiratory problems and predisposes to gastro-intestinal diseases.

Any system of ventilation, therefore, should primarily remove stale air and toxic gases and replace it with fresh air. However, during the winter, ventilation should be such that the animal heat is retained and draft is avoided.

2. Ventilation systems

2.1. Gravity convection

This is the basis of natural ventilation. Warm, moist air produced by the birds being lighter than normal rises and escapes through the opening in the top, all along the length of the building. This causes a partial vacuum at the floor allowing fresh air to enter from the sides of the building. Thus, ventilation can be controlled by adjusting the opening along both sides of the building which may be covered with adjustable curtains to regulate the area of the air inlet. Fans may be used in the house to circulate air although it may not help to move air into or out of the house.

2.2. Forced-air

This is the basis of mechanical ventilation. Fans are used to create either positive or negative pressure.

2.2.1. Positive-pressure ventilation system

A fan is mounted at the end of the house that pushers air into a duct along the ceiling spanning the inside length of the house. The duct has small holes along the length on either side, totally equaling the area of air inlet. The fresh air drawn in (forced-air) mixes with house air and pressure is built up in the house which pushes the stale warm air through an opening near the floor at the end of the house and up fruit and exhaust tower on the outer wall.

2.2.2. Negative-pressure ventilation system

Consists of exhaust fans mounted along one side of the building that pull air out of the house. Fresh air enters through openings under each eave, which can be controlled by adjustable openings. In the system, drafts can be expected; to avoid it, the exhaust fans should move 0.06 to 0.23 m^3 of air/min/bird which can be increased or reduced depending on the local conditions. In humid conditions, a minimum of 0.01m^3/min/bird is mandatory to control humidity

In cold climates, the outside air maybe at or below freezing temperatures; under such conditions, multiple speed fans with or without supplemental heat may be installed.

2.2.3. General requirements

Under temperate climates, minimum ventilation rate required to provide oxygen and remove carbon dioxide and ammonia in order to support maximum level of production is primarily controlled by levels of ammonia which, in turn, is influenced by manure handling and litter condition. In any case, for all weights and for all species of poultry, a minimum value of 2 m^3/sec/tonne of food consumed/d appears to be suitable. On the basis of body weight, 1.5 x $10^{-4} m^3$/$kg^{0.75}$/sec appears to be the minimum required and it is increased if ammonia concentration exceeds 25 ppm (Charles, 1981).

3. Heat balance equation

(Starr, 1981)

At steady-state, M – W = C + E + R where M = metabolic energy production, W = mechanical working, C, E and R are the heat loss by convection, evaporation and radiation, respectively, all measured in W/m^2.

$R = h_r(T_o - T_a) - R_{ni}$ where h_r = linearized radiative transfer coefficient with approximately given by $4\sigma T_a^{\ 3}$ (where σ = Stefan's constant), T_o is the interface temperature, R_{ni} is the additional radiation gained from surfaces not at T_a, also referred to as isothermal net radiation.

$\sigma (T_a^{\ 4} - T_a^{\ 4}) = h_r(T_o - T_a)$; σ is also called surface emmisivity which is assumed to be equal to unity.

$C = h_c (T_o - T_a)$ and $E = h_e (e_o - e_a)/\gamma$ where h_c and h_e are the heat transfer coefficient for convection and evaporation, respectively; e_o and e_a are mean vapor pressure (in mbar) of interface and surrounding air, respectively and γ is the psychrometric constant (66 kPa/K at sea level).

$$\therefore\ M - W = C + E + R = (h_c + h_r)(T_o - T_a) + h_e \frac{e_o - e_a}{\gamma} - R_{ni}$$

The equation can be simplified by amalgamating sensible and latent heat terms.

If apparent equivalent temperature, $\theta = T + \frac{e}{\gamma}$ where $\gamma = \frac{h_c + h_r}{h_e} = \frac{h_\alpha}{h_e}$

θ-T is equivalent of humidity increment proportional to vapor pressure and dependant on wind speed, weekly, when it is > 5 m/sec, and h_α is the combined heat transfer coefficient for sensible and latent heat terms.

$\therefore\ C + E + R = h_\alpha(\theta_o - \theta_a) - R_{ni}$ where θ_o and θ_a are apparent equivalent temperature of interface and air, respectively.

Thermal radiation increment is the increase in air temperature that will be needed to compensate for remaining external source of radiant energy.

$\theta_o - \theta_a = \dfrac{R_{ni}}{h_\alpha}$ where θ_e is the apparent equivalent temperature of the environment

$\therefore \quad C+E+R = h_\alpha(\theta_o - \theta_e)$; therefore, $M\text{-}W = h_\alpha(\theta_o - \theta_e)$

h_α and h_e both depend on size and shape of the animal and θ_o on wind speed.

The above equation can be modified depending on hair coat; several other factors can also be introduced depending on the requirement.

It is necessary to note that no single equation can incorporate all factors satisfactorily; but heat balance equation appears to be the best because it incorporates all environmental and physiological variables.

3.1. Heat loss from animals

(McArthur, 1981)

Metabolic heat production in a thermoneutral environment, M_{tn}, ranges between 50 and 200 W/m² depending on species and level of production. With strong sunshine outdoors, heat gained by solar radiation may exceed 3-4 times M_{tn} when the animals are outdoors.

3.1.1. Exposure to cold

If rate of heat loss is greater than M_{tn}, the latter is increased to prevent reductions in body temperature; it can reach up to 500 W/m² in response to cold at summit metabolism but, it cannot be maintained for a long time. Increased metabolic rate to cold is due to oxidation of fat or by increased feed intake; in either case, it reduces productivity. If metabolic rate exceeds the supply of ME (from feed), body mass decreases.

3.1.2. Exposure to heat

At high temperature, sensible heat loss is less than M_{tn} and the animal relies more on evaporative heat loss to maintain body core temperature;but, skin temperature increases along with respiration rate resulting in reduced appetite and decreased productivity. Increased respiratory rate raises the metabolic rate above M_{tn} because of increased muscular activity associated with panting.

3.1.3. Estimation of heat loss

Rate of sensible heat flow, G (W/m²), across unit area of an insulating layer with thermal resistance r (sec/m) and volumetric specific heat of air at an arbitrary temperature (usually 0°C) ρC_p (=1.29 x 10³ J/m³/K) is given by $G = \rho c_p \Delta T / r$ where ΔT is the temperature difference across the layer (in K).

The corresponding values for evaporative heat transfer, E(W/m²) from a wet surface is given by $\lambda\varepsilon = E = \rho c_p \Delta e / \gamma r_v$ where λ is latent heat of vaporization of water (≈ 2400 J/g), ε (g/m²/sec) is the rate of water loss by evaporation, Δe (kPa) is the difference in vapor pressure between the surface and surrounding air, r_v is the resistance to mass transfer (sec/m) and γ = 0.066 kPa/K, the psychrometric constant. Thermal resistance of 1 sec/m is equal to an insulation of 0.078 m^2 K/W (Note: 3.6 kJ/h = 1W).

3.1.3.1. Total thermal resistance

For animals kept indoors (like domestic poultry), total thermal resistance (r_t) is the summation of average resistance provided by body tissues (r_b), the coat (r_c) and the environment (r_e). ($r_t = r_e + r_c + r_b$)

Therefore, $G = \rho C_p (T_b - T_a)/r_t$ where T_b and T_a are the temperature of the body core and environment, respectively.

3.1.3.1.1. Environmental resistance (r_e)

Sensible heat transfer between the outer surface of an animal's coat and environment takes place by thermal radiation exchange with its surroundings, in a building primarily to the roof, floor, walls and other animals and a convection to the air.

Thermal radiation exchange, $L_n = \rho C_p (T_{cs} - T_r)/r_r$ where T_{cs} is the mean radiative temperature of the coat surface, r_r is the resistance to radiative heat transfer and convective heat exchange, $C = \rho C_p (T_{cs} - T_a)/r_a$ where r_a is the resistance to convective heat transfer by the boundary players. If r_e is the combined resistance to convection and thermal radiation transfer of the boundary layer given by $r_a r_r/(r_a + r_r)$, then $C = \rho C_p (T_{cs} - T_a)/r_e$ and the total flux density of sensible heat, $G = L_n + C$.

The resistance to radiative transfer, r_r, is approximately equal to $\rho C_p/(4\sigma T^3)$ where T (in K) is the average of T_{cs} and T_r. For values of T between 273 and 303 K, 2.8 > r_r > 2.0 sec/cm.

Boundary layer resistance depends on rate of air movement, geometry of the body and nature of interface. With decreasing air movement, buoyancy forces are important; and because of free convection, r_a decreases with increasing temperature difference ($T_{cs} - T_a$). At wind speeds more than 0.5 m/sec, buoyancy can be ignored and r_a decreases with increasing wind speed in forced convection. Rate of air movement depends on type of stock and environmental temperature. Within a poultry house, wind speed is usually about 0.1-0.3 m/sec, free/forced-convection regimes will have a wind speed of 0.2 m/sec and forced ventilation under hot climate will have a wind speed of 0.5 m/sec

For simplicity of estimation, birds are considered as spheres and boundary layer resistance under low rate of air movement with ($T_{cs} - T_a$) = 20K (constant) is given by $r_a = [248d/\{1+63(ud)^{0.5}\}]$ where d is the cylinder's diameter (0.165 m) usually measured as trunk diameter, u is the wind speed. For free convection, the equation can be modified as follows: $r_a = [248d/\{1+30d^{0.75}(T_{cs} - T_a)^{0.25}\}]$.

At high wind speeds, forced convection is predominant whereas at low wind speeds, free convection dominates. At u = 0.46 m/sec, boundary layer resistance for forced and free convection become equal (2.3 sec/m).

3.1.3.1.2. Combined radiation and convection

When $T_a = T_r$, $G = \rho C_p (T_{cs} - T_a)/ r_e + \rho C_p (T_a - T_r)/r_r$ will reduce to $G = \rho C_p (T_{cs} - T_a)/ r_e$. In animal houses, most of the times $T_a \neq T_r$; still, the equation is valid for estimation of sensible heat loss if measured by using the formula $T_e = T_a + \Delta T$ where $\Delta T = (T_r - T_a)/(1+ r_r/r_a)$; in other words, temperature difference between environmental temperature sensed by the animal and the air temperature.

Therefore, rate of air movement should be minimized in the vicinity of young stock, if $T_r > T_a$, T_e has to be maximized by using infrared heaters because as the wind speed increases, the boundary layer resistance is reduced.

3.1.3.1.3. Coat resistance

The coat resistance $Z_c = r_c/l$ where l = mean coat depth, in cm. Sensible heat transfer through coat occurs simultaneously by conduction, convection and radiation; but the relative magnitude depends on coat structure, depth, wind speed and mean temperature difference across the coat layer.

A. In still air: Coat resistance in still air depends on molecular diffusion, thermal radiation and free convection; the former two can be theoretically estimated. Conduction along fibers is too small and hence, neglected.

Z_d, resistivity for molecular diffusion, $= r_d/l$ and it depends on the air trapped within the coat; it is estimated to be about 5.0 s/cm² when temperature difference across the coat is 5 to 30 K); similarly Z_r, resistivity to radiant exchange, $= r_r/l$ and it depends on the size, density and orientation of fibers. On the same lines, resistivity to free convection, $Z_f = r_f/l$, it is obvious that molecular diffusion, thermal radiation and free convection are reciprocals of Z_d, Z_r and Z_f, respectively. In still air Z_f will be high and therefore its reciprocal becomes negligible.

Therefore, Overall resistivity of the coat, $Z_c = (Z_d^{-1} + Z_r^{-1} + Z_f^{-1})^{-1}$ and Z_c of White Leghorn chicken is estimated to be 2 to 3 s/cm²

B. In wind: Reciprocal of coat resistance, coat conductance that is 1/rc, increases linearly with wind speed, u as follows: $1/r_c = 1/r_{c(0)} + \alpha u$, where α is a constant depending on coat type and $1/r_{c(0)}$ is conductance in still air.

Wind destroys insulation of a layer of coat, thickness of which depends on velocity. Both forced and free convection occurs within the exposed fleece; close to the skin, heat transfer occurs through molecular diffusion, thermal radiation and free convection; but, between a wind penetration depth to and outer surface of the coat, the heat transfer is mainly by forced convection and coat resistance in this outer section is approximately 0.

Penetration depth, $t = l\alpha u/[\alpha u + 1/r_{c(o)}]$ and it increases with wind speed. At wind speed of 5 m/sec, about 50% of insulation by fleece is destroyed. In case of poultry, the orientation of wind is having a substantial effect on wind penetration and therefore on coat resistance.

3.1.3.1.4. Tissue resistance

This component forms a very small portion of total resistance coated animals. The mean thermal resistance between body core and skin surface is given by $G_b = \rho C_p (T_b - T_{sk})/ r_b$, where G_b is the total heat flow density through the body tissue and r_b the tissue resistance which will be maximum during vasoconstriction and minimum during vasodilation.

3.1.4. Sensible heat loss in relation to environmental temperature

Sensible heat loss from each region depends on environmental temperature and local thermal resistance of the body tissue, coat and environment. At 10°C, only a small percentage of total heat loss of birds occurs through legs whereas the converse is true at 35°C. The physical characteristics of the coat does not change but environmental temperature changes, there will be a vasomotor response bringing about apparent changes in coat resistance. In case of poultry, at 10°C, sensible heat loss is 65 W/m^2 the bird is well-feathered; if the bird is poorly feathered, the value will be 125 W/m^2.

3.1.4.1. Decreasing ambient temperature

In case of well-feathered birds, an increase in total thermal resistance from 3.8 at 35°C to 7.3 s/cm at 0°C cannot be explained from tissue resistance alone because significant contributions to overall thermal resistance is possible through increase in coat resistance due to fluffing of feathers, vasomotor action and postural adjustments through protection of legs beneath the feathers. Therefore, relationship between sensible heat loss and environmental temperature is non-linear for well-feathered birds.

On the contrary, increase in total thermal resistance with decreasing environmental temperature for poorly feathered birds is comparatively small and can be explained by the increased tissue resistance alone. Unlike well-feathered birds, poorly feathered birds shows little change in coat resistance because of behavioral responses or non-homogeneity of heat flow due to poor coat insulation. Consequently, sensible heat loss in case of poorly feathered birds increases almost linearly with decreasing temperature between 35 and 0°C mainly due to near constant total resistance in this range of temperature.

3.1.4.2. Increasing ambient temperature

At high environmental temperature, rate of sensible heat loss is almost independent of its feather status. For instance, at 35°C, sensible heat loss is about 23 W/m^2 in both types which is equivalent to a heat flux of about 3.5 W. The metabolic rate is about 60 W/m^2. In well-feathered birds, thermal resistance is available on the "trunk" (about 6 s/cm) and therefore much of the sensible heat loss must be occurring through extremities (legs and comb) rather than through plumage. Surface area of the extremities is at least 0.02 m^2, about 15% of the total surface area. Assuming that 30% of the total sensible heat loss occurs through the extremities at 35°C and environmental resistance (r_e) for the extremities is 1.0 s/cm, then, the local tissue resistance must have reduced to about 0.5 s/cm because of vasodilation. Therefore heat loss from extremities at high environmental temperature is highly dependent on the environmental resistance (r_e); the latter can be reduced by increasing air movement over the legs and head which will help the birds to lose more heat through extremities.

3.1.5. Panting

Faster wind speeds will relieve thermal strain at high environmental temperature by decreasing the rate of respiratory evaporation required to prevent increasing body core temperature.

When sensible heat loss is insufficient for balance, poultry begins to pant – heat dissipation by evaporation from respiratory system; the rate of panting, $E_r = \rho C_p (e_s - e_a)/\gamma r_{vr}$ W/m^2 where e_s is the saturation vapor pressure at the site evaporation, e_a is the vapor pressure of air inside the house, r_{vr} is resistance of water vapor loss from respiratory system. However, e_s can be considered a constant (8 kPa for poultry) determined by the body temperature and r_{vr} can be estimated as 600A/V where V is the respiratory flow = 1/min and A in the surface area of the body (m^2).

As respiratory rate increases from 12 to 350/min, resistance r_{vr} decreases from 100 to 25 s/cm. In other words, if the bird has to dissipate 40 W/m^2 by respiratory evaporation (for instance at 35°C), the respiratory rate must be about 250/min (r_{vr} about 35 s/cm) at e_a of 1 kPa and about 400/min (r_{vr} about 20 s/cm) at e_a of 4 kPa. Respiratory rate (RR) required at different r_{vr} can be estimated by $1/r_{vr} = (9.1 \times 10^{-5}\ RR) + 9.0 \times 10^{-3}$.

3.1.6. Air velocity

(Kampen, 1981)

Forced convection decreases boundary layer around the body. If the air temperature is less than skin temperature, heat loss increases and body temperature may increase if air temperature is greater than skin temperature; at this juncture, the only way the animal can lose heat is by evaporation.

Increasing air velocity when ambient temperature is greater than critical temperature at the temperatures lower than body temperature will decrease heat stress. If air temperature is less than critical temperature, there will be increased heat loss leading to increased heat production necessitating increased food intake.

Therefore, if ambient temperature is very low, body weight reduces.

Air flow at a rate of 1.4 m/s on either head/body of a sitting WL at 20°C, causes an immediate increase of the production by 15% in comparison to that in still air. Increasing air velocity even from 0.1 to 0.3 m/s at fluctuating temperatures of 26-35°C can bring about 9% increase in food intake and 5% increase in egg weight with concomitant retardation of body weight loss.

ME intake of domestic fowl is also linearly related to (air velocity)$^{1/2}$ and constant wind speed, ME intake increases by 8.5 times as ambient temperature decreases from 35 to 5°C. This relationship between wind speed and ME intake is particularly true for naked birds and for others, the curve is definitely inverse. In other words, increased air movement is equivalent to reduced ambient temperature.

3.1.7. Activity

(Kampen, 1981)

Movement of birds causes forced convection which, in turn, disturbs boundary layer of air and feather cover. Therefore, heat loss or gain is increased with increased heat production due to activity. In case of chicken, standing up increases heat loss by 20-40% and also contributes to heat production *per se*. At low temperature, forced movement increases heat loss whereas at high temperature, heat loss is less than heat production. Therefore, forced movement decreases upper lethal temperature.

It is obvious therefore that birds at low temperature huddle whereas at high temperature they are restless. Food-restricted chicks are more active than those fed *at libitum* ; but are more active than starving chicks.

Activity of the hens is lowest between 27-33°C. Activity was constant for a WL over a range of 10-30°C and for a group of WL at 25°C.

4. Thermal influences on poultry

(Kampen, 1981)

When a hen has ME (intake) = ME (output), then ME (intake) is utilized for heat production and as net energy for egg production; ME (intake) as well as heat production depends on climatic factors and therefore, net energy for egg production changes unless hen is gaining/losing body mass or changing body composition.

Body reserves that can be used are limited and if, as a change of temperature, ME intake was initially insufficient, it has been corrected within and acclimation period by a lower maintenance requirement concurrently lower body weight, by an increase in ME intake or decrease in egg production.

Changes in food intake and temperature may also cause a change in water intake especially higher temperature. The evaporative loss of water is considerable, although it depends on relative humidity. The water lost through evaporation has to be compensated by increased water intake of reduced water loss via excreta. In general, drinking water temperature follows ambient temperature and water is usually supplied *ad libitum*. Therefore, controlling amount and temperature drinking water may have several beneficial effects.

General response of laying birds to different temperature ranges is summarized in Table 27.1.

Table 27.1 Effect of temperature on laying birds

Temperature range (°C)	Remarks
37.8 – 42.0	Danger of heat prostration and death; 47°C is upper lethal temperature
29.4 – 37.8	Further stress on the bird; sharp increase in water intake and fecal excretion of water resulting in damp litter → higher humidity in the house accentuating the effects of high temperature
23.8 – 29.4	At 29.4°C, panting characterized by increased rate of respiration (polypnoea) begins → increased vaporization of water through which birds lose heat. Since the birds lack sweat glands, adverse effects of high temperature begins soon in the form of reduced egg size and production, increased thin-shelled eggs
12.8 – 23.8	Zone of thermoneutrality – basal heat production is minimal; body temperature is maintained within normal range through variation in body heat losses
0 - 12.8	Litter is likely to be damp which causes discomfort to the birds
0	Freezing point of water
0 – (-) 9.4	Hens will be uncomfortable and at or below (-) 6.7°C, hens fail to maintain both body temperature and egg production
Below (-) 9.4	Reduced activity, severe drop in egg production; combs begin to freeze
(-) 30 – (-) 32	Lower lethal temperature

Source : Nesheim *et al.*, 1988

Several mathematical models of the proposed for predicting ME requirement/ intake by birds.

1. $ME_m = 388\ W^{0.75}\ e^{0.027(22-T)} + 8.67E_g$ where ME_m is the minimum ME required (kJ/d), W is body weight in kg, T is the ambient temperature (°C) and E_g is the egg production (g/d).
2. $ME = (558 - 5.58\ T)^{0.75} + 3.35\ \Delta W + 2.3\ E_g$ where ME is ME/bird/d (kJ) and ΔW is weight gain (g/d).

3. For a bird of 1.5 kg body weight, ME = 1690 – 20.1 T.

Many data suggest that there is no clear zone of thermoneutrality in case of fowl; but only a narrow zone (in the range between 30 and 40°C) where metabolism is lowest has been suggested. It also depends on strain, acclimation, food intake, production level and physical activity.

Table 27.2 Adverse effects of temperature on birds

	Low temperature	High temperature
Chicks	Increased mortality due to huddling / suffocation (asphyxia) due to collapse of air-sacs, reduced feed efficiency	Reduced food intake and performance, feather coverage poorly developed and hence, pecking results
Adults	Reduced feed efficiency, reduced quality and number of eggs produced	Egg size, specific gravity and shell thickness reduced when temperature exceeds 26.6°C. If temperature exceeds 29°C, egg production is reduced and mortality is increased at temperatures above 35°C
Breeders		Fertility and hatchability, frequency of mating, quality and quantity of semen reduced; Feed intake reduced at 1.7% for every 1°C rise in temperature

4.1. High ambient temperature

4.1.1. Influence on food intake

Table 27.3 Effect of temperature on White Leghorn layers (1.816 kg)

Temperature (°C) →	4.4	10.0	15.6	21.1	26.7	32.2	37.8
Feed intake (g / bird/ d) (*x*)	118	116	110	100	87	70	48
Water intake (g / g of feed) (*y*)	1.3	1.4	1.6	2.0	2.9	4.8	8.5
Water intake (ml/d) (*x y*)	155	163	178	201	254	337	409
Feces produced (g/d) (*q*)	166	162	153	140	121	97	67
Fecal water (g/d) (*p*)	131	130	124	115	101	82	57
Fecal water, % *** ($z = 100\, p \div q$)	79	80	81	82	83.5	84.5	85
Respired water (g / d) $A = x\,(y - 0.014\, z)$	21	29	51	88	153	255	345

Temperature (°C) →	4.4	10.0	15.6	21.1	26.7	32.2	37.8
Fecal + Respired water (g / d) $B = x y$	152	160	176	203	254	337	402
A as % of B $= 100 - (1.4 z \div y)$	13.8	18.1	29.0	43.3	60.2	75.3	85.8
A as g / g of feed $= y - 0.014 z$	0.18	0.25	0.46	0.88	1.76	3.64	7.19
*** Droppings produced @ 1.4 g/g of feed				*Source* : North and Bell, 1990			

4.1.2. Influence on heat production

Heat production decreases with increasing ambient temperature up to about 35°C; a limited part of this can be linear; above the critical temperature, metabolic rate increases. In the temperature range which includes critical temperature, the relationship is quadratic or can be described by two linear equations which intersect at critical temperature. In short-term experiments non-acclimatized hens, the critical temperature is about 33°C.

Difference in feather cover also contributes to the difference in slope. The difference in heat production between poorly and well-feathered layers increases with decreasing temperature and reaches a factor of 2 at about 0°C. Comb and wattles are highly vascularized. In White Leghorn, surface area of head appendages will be as much as 7-8% of total body surface and about 25% sensible heat loss can occur through them. Therefore, White Leghorns have high maintenance ME requirement. The decrease in heat production for every K rise in temperature is greater for cockerels than for layers. It follows that egg production at lower ambient temperature will be very low because the regression

Table 27.4 Heat production of fowls (kJ/BW$^{0.75}$/d)

Ambient temperature (°C)	Cockerels, 2.5 to 2.8 kg @ 15 to 34°C *	Layers, 1.65 to 1.90 kg @ 7.2 to 35°C **	Layers, 1.50 to 1.90 kg @ -5 to 40°C ***
10	573	586	639
15	526	562	575
20	479	539	523
25	433	516	484
30	386	493	457
35	339	470	442
Compound % change / °C	2.12	0.89	1.49

* M = 667 – 9.37 T (O'Neill *et al.*, 1971) ** M = 632 – 4.64 T (Davis *et al.*, 1973)
*** M = 803 – 18.9 T + 0.245 T^2 (Kampen, 1974); the difference in slopes for layers is because of behavioral temperature regulation in birds housed singly and in groups, the latter having lower slope than the former Note: 3.6 kJ/h = 1W

lines for heat production against ambient temperature for layers and cockerels intersect in the low temperature range. Therefore, ME intake and heat production lines are not parallel in the lower and upper ambient temperature indicating that egg production cannot be maintained unless stored energy is used. Hence, birds subjected to low temperature lose weight. Relatively lower ME intake may also be due to consequence of filling capacity of the intestinal tract and the length of the light period.

4.1.3. Influence on production characteristics

4.1.3.1. Egg production

Decreases when temperature exceeds 26-30°C; but under diurnal variation, it is difficult to establish exact temperature from when egg production reduces. Heat stress is greater when highest temperature is associated with narrow diurnal variation. As temperature increases above a threshold point of 26-29°C, diurnally recycling temperature is needed to maintain "normal" egg production. As daily maximum temperature increases, the magnitude of diurnal variation must increase to avoid reduction in egg production; if daily maximum increases above the upper threshold (29°C), production is reduced regardless of magnitude of diurnal variation. An increasingly higher constant temperature, the decrease in egg production is progressively more severe. Higher temperatures act as though the birds are undergoing an increasing rate of senescence. Birds of smaller body weight can produce more than those with larger body weight under continued exposure to high temperature (Smith, 1981).

Between 30-35°C, heat production increases in general, and a temperature-time effect may influence the results. Linear decrease in heat production with increasing ambient temperature is unlikely to be related to changes in activity. Minimal activity is expected in the range of 25-32°C when body temperature begins to increase and metabolic rate becomes minimum (Charles, 1981).

Relation between ambient temperature, heat production and ME intake is not parallel; that means, NE for egg production is not constant without concurrent change in body mass. In a range of 10-30°C, energy available for egg production and egg production *per se* are almost constant while ME intake reduces with increase in ambient temperature. Food cost for an egg is minimum at 30°C but not for a kg of egg because of inverse relation between egg size and ambient temperature. At temperatures above 30°C, both egg production and egg weight reduce whereas at temperatures below 10°C, only egg production decreases; in either case, both food and body energy are used for egg production (Charles, 1981).

Egg production can be maintained with limited food intake though accompanied by substantial loss of body weight, and in hens subject to high ambient temperature. Therefore, if diet is properly formulated, production is relatively constant in the range of 10-30°C. Reduction in food consumption while maintaining egg production at high ambient temperature increases the number of eggs per unit of food. However, food/kg of egg is minimum at about 25°C because of changes in egg size with increasing ambient temperature. Egg size varies inversely with temperature and at temperatures about 30°C, both production and size reduce; the former due to reduced food intake, a direct effect of high temperature. Reduced energy intake results and decreased body weight and egg weight; therefore, smaller eggs are likely to be a reflection of smaller body mass.

It is possible that birds with smaller amount of body fat may synthesize less lipids and therefore, layers at high ambient temperature have lower mature body weight, contain less fat and produce smaller eggs. Loss of body weight in hens on restricted energy intake or at high ambient temperature is because of loss of body fat. In eggs, weight of yolk is reduced more than that of albumen in energy restricted hens at 38°C. But, due to high variation in body fat, correlation between body fat mass and egg weight is not significant.

Table 27.5 Effect of temperature on egg production

Mean temperature (°C) *	Relative production (%)	Relative egg size (%)	Relative food required/egg (%)	Relative food required per unit egg mass (%)
16	100	100	100	100
18	100	100	95	95
21	100	100	91	91
24	100	99	88	89
27	99-100	96	86	89
29	97-100	93	85	81
32	94-100	86	84	98

* Range in daily temperature = 4.7-7.1 K

Hens at 21°C with the same food intake as those of hens at 32°C produced heavier eggs; body weight of the latter was less than the former. Night heat stress causes are significant reduction in egg production. Reduced egg weight may be due to decrease blood flow to the ovarian follicles.

Time interval between ovipositions also increases and high ambient temperature with a concurrent reduction in clutch size and egg production. A change in temperature from 38 to 21°C, immediate change in egg shell thickness is recorded but egg size remains unchanged because decrease in egg size is not due to direct effect of temperature.

Egg production is also decreased by low temperature (especially if the drop is a sudden one) but low temperature has no effect on egg size.

4.1.3.2. Egg weight

Above 26°C, 80% of depression in egg production is due to heat stress and the rest due to inadequate energy in the diet. Heat stress reduces shell weight (metabolic alkalosis). Magnitude of loss of egg weight depends on 1. Energy balance set up by interaction between genotype, environmental and diet and 2. Energy expense of thermoregulation and general adaptation syndrome (Smith, 1981).

As temperature increases above 10°C, egg weight gradually decreases, rapidly between 26-35°C. In a diurnal cycle, when maximum temperature is above the 26-29°C threshold, egg production decreases and egg weight will be maintained at that expected for constant temperature regime (at average of the diurnally cycling regime) (Smith, 1981).

4.1.3.3. Mortality

Mortality increases with increasing temperature; and this is not related to relative humidity, winds speed especially under longer periods at high temperature. Birds are more susceptible at the onset of or just after attaining sexual maturity (Smith, 1981).

4.1.3.4. Fertility and hatchability

Seasonal variation is recorded in fertility and hatchability under Aw climates. Males are less affected than females by temperature up to 30°C; both semen production and fertilizing capacity of that semen are likely to be affected when temperature reaches 38°C. In any case, males are less sensitive to high environmental temperature as far as fertility is concerned (Smith, 1981).

4.1.3.5. Growth

4.1.3.5.1. Broilers

After brooding, maximum growth is achieved at 15-20°C; on either side of this range, growth depression is similar. Over a wide range of temperature, FCR improves at high temperature. If average daily temperature is kept nearer to "optimum" from hatch to market age, growth and FCR will be affected if the diurnal variation in temperature exceeds 11K (16K for post-brooding temperature) (Smith, 1981).

Maximum dry bulb temperature of 40°C can be tolerated provided the diurnal variation is = 16K. For daily maximum temperature > 30°C, food intake and growth increase as the diurnal variation widens (Smith, 1981).

If roof is not properly insulated, although dry bulb temperature may not be excessive, long-wave radiant heat load from roof can increase black-bulb temperature resulting in decreased growth and increased mortality, especially in males (Smith, 1981).

At temperatures > 29°C, with relative humidity constantly > 80% or < 40%, permanently depress growth. Abrupt increase in temperature with low (< 40%)

or high (> 80%) relative humidity severely depress growth and FCR; in other words, hot-wet conditions are more detrimental on growth (Smith, 1981).

4.1.3.5.2. Laying hens

When average temperature is > 26.7°C, heavier (older) birds show growth depression. In Cwa climate of Southeast United States of America, spring-hatched pullets will be too small at sexual maturity to sustain maximum egg production (Smith, 1981).

4.1.3.5.3. Turkeys

The growth is contained between 10-20°C and FCR reaches optimum somewhere between 21 and 27°C. Depression of growth rate is as marked as in meat chicken although range of temperature for maximum growth is twice that of meat chicken (Smith, 1981).

4.1.4. Constant Vs fluctuating the ambient temperature

If constant temperature = mean of the cycling temperature, no difference in food intake, egg production and egg weight is recorded. However, if the fluctuations are outside 10-30°C, ME intake and heat production are not parallel; the egg production will be worse than at equivalent constant temperature. While egg production is constant in the temperature range of 10-30°C, egg size and food intake vary; and at 30°C, both will be minimal and at 10°C, maximal.

5. Water intake

(Kampen, 1981)

Intermittent or restricted water programs may result in saving of water, reduced production of wet excreta, increased FCR and the reduction in food consumption. However, controlling temperature drinking water may also have beneficial effects during low/high ambient temperature.

5.1. Factors affecting water intake

Several factors including food intake, composition of diet, growth, egg production, housing and watering system, water temperature and climatic conditions have a direct bearing on water intake. Effects of ambient temperature and drinking water temperature are summarized below:

5.1.1. Effects of ambient temperature on water intake

Water, on most occasions, is considered a "forgotten nutrient" because it is cheap and, till it is not at all available, no significant effects are noticed. Poultry being a uricotelic species, does not have urinary bladder and thus, cloaca (the distal part of the gut) is essential for electrolyte and water balance. Water is absorbed along with salt in a non-enzymatic process; glucose and amino-acids facilitate absorption of water.

Fecal composition depends primarily on water intake. If there is shortage of water in water pools of the body, the thirst is evinced to stimulate birds to drink water. If surplus water is available, it is useful for including in eggs, evaporation to lose heat, tissue growth and excretion in urine and feces. Of the above, water intake and excretion show high degree of variability; the former being primarily a function of environmental temperature (Table 27.3).

Water consumption increases at the rate of 3 g/d/°C up to 27°C and at a rate of 11 g/d/°C above 29°C; breed/strain/individual differences have been reported.

A temporary water deficit is expectable decrease (from 35 to 5°C) in temperature, especially on the 1st day because the water intake is so severely reduced with concomitant increase in food intake bringing down water to food ratio to around 1.07. On the other hand, such a water deficit is not expected for the temperature suddenly increases from 5 to 35°C.

5.1.2. Effect of drinking water temperature on water intake

Effect of drinking water temperature on water consumption generally follows that of ambient temperature. WL at high ambient temperature of 35°C drink 10% less during 1st 8h and water temperature was 0°C than it was 35°C. It is well-known that water consumption has a positive correlation with food consumption and therefore, it follows that provision of cool drinking water can help maintain production during heat stress.

5.1.3. Factors other than temperature affecting water intake

Factors other than environmental temperature also affect water consumption (Table 27.6).

It is very clear from Table 27.3 that as temperatures increase above the panting temperature (29.4°C), there will be a steep increase in water consumption per g of food consumed with a concomitant reduction in the latter. At 37.8°C, 1000 layers consume only 48 kg of feed but drink over 400 l of water every day. Of the water consumed, 86% (about 345 l) is added back to the air inside the building resulting in acute humidity build-up further complicating the adverse effects of high temperature.

If the water temperature is not optimum, water consumption itself reduces leading to severe depression in food consumption and dehydration resulting in heavy mortality as well as losses in production. Maintaining optimum water intake as well as relative humidity inside the poultry house during thermal stress itself is a big challenge for poultry farmers.

Several factors affect water balance and they are shown in fig 27.1.

5.2. Water vapor production

At > 25°C, feather cover is disadvantageous and internal heat transport (convection via blood) will be directed more to head appendages and respiratory system. Respiratory evaporation accounts for most of the late in heat loss about 25°C and it depends on water vapor pressure of inspired and expired air, and on respiratory minute volume; in turn, on metabolic rate. Plane of nutrition and state of acclimation determine the metabolic rate. Evaporative heat loss decreases with increase in ambient water vapor pressure @ 0.7 mg of water/g of live weight/h/kPa increase in water vapor pressure. A hen produces 37 to 42 g/d of water vapor at 5°C, 59 to 62 g at 20°C and 133 to 204 g at 35°C.

Table 27.6 Factors affecting water intake (other than temperature)

Factor	Remarks on water intake
Feed composition and intake	Fiber, fat, protein, salt and K increase
Water quality	Salt and sodium sulfate increase; Magnesium sulfate and Zinc sulfate (as in some ground-waters) reduce
Genetic	Larger breeds consume more water
Age	Older birds need more water
Performance	At age of sexual maturity, sudden increase in water intake; after peak production, age can't influence water intake
Type of waterers	Trough and bell-drinkers - more water intake than with nipple drinkers
Water temperature	Negatively related

When humidity increased at 25°C, a new balance between heat loss and heat production can be achieved by 1. Decreasing insulation of feathers 2. Decreasing heat production (in other words, decreased food intake) 3. Increasing body temperature or 4. By increased ventilation of the respiratory tract. The latter three are most effective but culminate in reduced production; therefore, effects of increased relative humidity are similar to that of increased ambient temperature.

Water content of fresh excreta increases from 76% at 5°C to 86% at 35°C. Therefore, water evaporation from excreta increases at high ambient temperature depending on relative humidity. If evaporation of moisture from excreta is prevented, there is no reduction in excreta output at high ambient temperature although food intake reduces.

6. Moisture condensation during winter

One of the keys to understanding poultry house ventilation during cold weather is remembering one simple rule: In order for air to hold moisture, energy or heat is required. The warmer the air, the more moisture it can hold. For example, the air contained in a 120 m broiler house can hold up to 30 kg of water at 4.4°C. But, if the same amount of air were heated to 21.1°C, the amount of moisture the air can hold increases to 83.5 kg.

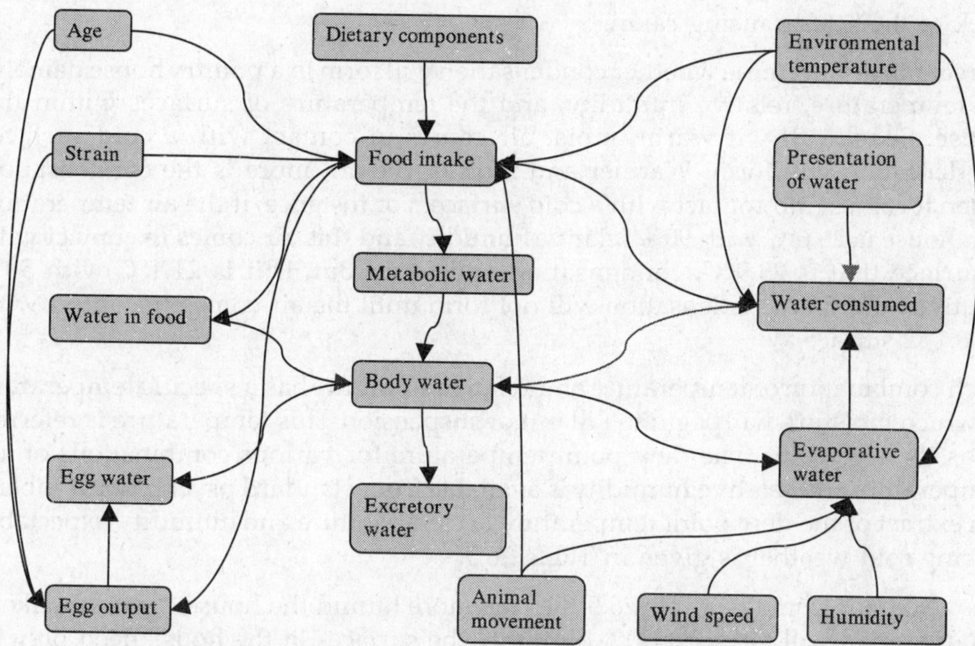

Fig 27.1 Variables affecting water balance in chicken (Belyavin, 1991)

This phenomenon is a very important consideration when trying to keep litter dry. During the winter, heat given off by the birds and furnaces/brooders collects near the ceiling. If cold outside air is drawn in through inlets by exhaust fans and is directed along the ceiling, it will mix with the warm air near the ceiling and heat up. As the temperature of the fresh air increases, its moisture holding ability increases. The warmer and dryer air then moves down toward the floor and across the litter, removing moisture and thereby preventing caking. The now moisture-laden air is removed from the house by exhaust fans.

This moisture removal process is an essential tool for the poultry grower. On the other hand, if warm, moisture-laden air comes in contact with a cold surface, the heat in the air will be transferred to the cold object. This of course decreases the temperature of the air and, in turn, the moisture-holding ability of the air. The now cooler air cannot hold the same amount of moisture so the moisture drops out of suspension on to the object which cooled the air. This process is commonly referred to as condensation.

Depending on where it occurs, condensation can cause a number of problems. If condensation forms in the attic space, the moisture can ruin insulation and decrease the structural integrity of the building. In open-ceiling houses, if moisture from condensation gets between the metal roof and board insulation, the life of the roof can be decreased and the moisture can drip onto the litter and equipment. Condensation forming on side walls and curtains returns the evaporated moisture back to the litter, causing caking.

Three things determine whether condensation will form in a poultry house namely, air temperature, relative humidity, and the temperature of surfaces within the house. Specifically, if warm, moist air comes in contact with a cold surface, condensation will form. Warmer and moister the air, more is the condensation when it comes into contact with a cold surface. For instance, if the air temperature in a house is 26.6°C with 90% relative humidity and this air comes in contact with a surface that is 23.9°C, condensation will form. But, if it is 21.1°C with 50% relative humidity, condensation will not form until the air comes in contact with a 16.1°C surface.

Each combination of temperature and relative humidity has a specific temperature at which moisture will begin to fall out of suspension. This temperature is referred to as its dew point. The dew point temperature for various combinations of air temperature and relative humidity is available from standard psychometric tables. An extract of the dew point temperatures at temperature and humidity expectable during cold weather is given in Table 26.5.

It is apparent from the Table 26.5 that the more humid the house, the more likely condensation will form. At 90% humidity the surfaces in the house need only to be approximately 1.7°C cooler than the air and condensation will begin to form, whereas at 40% humidity, house surfaces must be about 13.9°C cooler for condensation to form.

Most condensation problems can be remedied by simply bringing in an adequate amount of fresh air to keep the moisture from building up in the house. It is important to keep in mind that there is a tremendous amount of moisture being added to the litter and air in a poultry house. During its lifetime, a 2.3 kg broiler will consume about 8.3 L of water. Only 20% of this water (1.7 L) is retained by the bird. The rest is added to the litter and to the air in the form of water vapor from breathing.

Table 27.7 Dew-point temperatures (°C)

		Air Temperature (°C)				
		15.6	18.3	21.1	23.9	26.7
Relative humidity (%)	40	2.2	4.4	7.2	9.4	18.9
	50	5.0	7.8	10.6	18.3	17.8
	60	7.8	10.6	18.3	15.6	21.1
	70	10.0	18.3	15.6	18.3	23.3
	80	12.2	15.0	17.8	20.0	25.6
	90	13.9	16.7	19.4	22.2	27.8

The key to keeping this moisture from building up is to continuously remove excess moisture with timer fans. Ideally, a producer would bring in enough fresh air so that house relative humidity would be approximately 60%. At this level of humidity, condensation is much less likely to form. Ventilation charts like the designed to help keep house moisture down without causing the producer to use excessive amounts of fuel are available.

Even in a well-ventilated house, condensation can form if the air comes in contact with a cold surface. This is why it is important that a poultry house be well insulated. This is especially true of any exposed metal surface where condensation is most likely to form. In houses with open ceilings it is crucial that no air is allowed to seep between the sheets of insulation. Using tongue and grove insulation board can dramatically decrease the amount of condensation that forms. On dropped-ceiling houses, it is crucial that any hole in the ceiling surface be patched. Any warm air that travels through these holes will cause condensation to form on the metal roof which will then drip down onto the ceiling insulation, reducing insulating value as well as life. Metal end walls should always be insulated. The insulation value of most 2" wooden walls is usually high enough to minimize the amount of condensation forming as long as house humidity is not too high.

6.1. Adverse effects of humidity

With dry air (low Absolute Humidity, AH), evaporative loss increases at a rate of 0.03 mg/(g/h/°C) at temperatures between 0 and 22°C; as the temperature rises to about 23 to 40°C, the corresponding evaporative loss will be 0.17 mg/(g/h/°C). A hen weighing 2 kg produces about 43 kJ/h of heat at 0°C and 27 kJ/h at 35°C. At low AH, it produces 30 kJ/h at 0°C and 21 kJ/h at 35°C; the latter accounting for 78% of total heat loss. If AH increases towards saturation, evaporative loss is curtailed by about 50%.

At normal temperature, equal amount of moisture is lost through feces and respiration; and as the body weight increases, water lost through excreta reduces. At 21.1°C, 100 layers produce as much as 20 lit of water which has to be eliminated from the house.

Effects of humidity have been found to precipitate when RH is about 75% and temperature is above 27°C in the form of increased rectal temperature, depressed food intake and retarded body weight gain. At RH above 72%, wet-litter condition appears causing contact dermatitis which includes hock-burn, breast-blisters etc. especially between 21 and 35d in broilers. Coccidiosis becomes a problem to reckon with especially with wet-litter in broilers. Air-sacculitis also is expectable under high temperature and high humidity as well at under low temperature and low humidity.

Hence, high temperatures and high humidity cannot be tolerated by birds regardless of age because, if the atmosphere is saturated with moisture, even panting has to be accelerated (which may not be even effective) leading to prostration and death. (Note: 3.6 kJ/h = 1W)

6.2. Heat and moisture production

6.2.1. Calculation of heat production

Basal heat production by fasting hen	11.5 kJ/hr/kg
Activity increment @ ½ the basal heat	5.75 kJ/hr/kg (Litter)
@ 1/3 the basal heat	3.83 kJ/hr/kg (Cages)
Heat production due to feeding	5.80 kJ/hr/kg
Total heat produced: Deep litter system	23 kJ/hr/kg
Cage system	21 kJ/hr/kg

Hence, standard Leghorn layers weighing 1.8 kg, under ideal temperature, produce 41.4 and 37.8 kJ/hr of heat.

However, an allowance for latent heat of vaporization, which is not available for warming of air, especially at winter temperature, is necessary. This accounts for about 20% of total heat production. Hence, under winter conditions, standard Leghorn layers (1.8 kg) on deep litter and cages produce 33.12 and 30.24 kJ/hr of total heat. (Note: 3.6 kJ/h = 1W)

For every % increase in temperature, the heat production in the body reduces by 5%. As a rule of thumb, a standard Leghorn layer, a Brown-egg layer and a meat-type layer produce 42.2, 47.5 and 58.0 kJ/h of total heat. Broilers of same weight produce more heat than layers due to their higher growth rate (metabolism) and feed consumption. At ideal temperature of 21.1°C, about 75% of the total heat is lost by sensible means (North and Bell, 1990).

Table 27.8 Heat, moisture and fecal production by broilers at 21.1°C

Age (weeks)	Av. Body weight (kg)	Heat production (kJ/hr/kg)	Water output (g/bird/d)	Fecal output (g/bird/d)
2	0.20	60.4	59.0	31.8
3	0.34	53.4	90.7	49.9
4	0.59	41.8	127.0	77.1
5	0.79	32.5	158.7	97.5
6	1.09	30.2	181.4	117.9
7	1.36	23.2	231.2	142.9
8	1.70	23.2	226.8	167.8

Source : North, 1972

Table 27.9 Heat and moisture production by chicken at 21.1°C

Average body weight (kg)	Total heat production (kJ/hr/kg)	Moisture (g/bird/d)			Feces (g/d)
		Fecal	Respiratory	Total	
0.5	46.42	72	24	96	44
0.9	33.65	92	49	141	82
1.4	26.69	105	71	176	114
1.8	23.21	114	88	203	140
2.3	20.89	124	94	218	162
2.7	19.00	133	102	235	178

Source : North and Bell, 1990

Table 27.10 Heat and moisture production by layers (per bird/h)

Temperature (°C)	Total heat (kJ)	Moisture (g)	
		Respired	Fecal
3.9	46.23	2.86	6.58
1.7	45.23	3.76	6.58
7.2	38.89	3.81	5.85
12.8	38.89	4.72	5.81
15.6	38.89	5.17	5.76
26.7	38.69	6.49	6.53
35.0	24.62	9.07	4.67

Source : North, 1972

7. Temperature requirements

(Charles, 1981)

7.1.1. Broilers

7.1.1.1. Brooding

27-28°C decreased gradually @ 0.5-1.0 K/d. Brooding area – 21°C. Maximum – 30°C. Finishing – 21-23°C. Breeders – 20°C

Body weight and food consumption (up to 49d) decrease linearly with increasing temperature within a range of 19-26°C @ 16.2 and 52.9 g/K/bird, respectively; economic optimum temperature – 21°C.

(K is the wind-chill factor in W/m^2 estimated as $K = 1.163\,(10u^{1/2} + 10.4 - u)(33 - T)$ where u=air speed in m/s and T = air temperature in °C)

The following equations explain the relationship between body weight (kg/bird, W), cumulative feed intake (kg/bird, F), age (days, A) and mean dry bulb temperature (°C):

$$W = 0.141T - 1.373 - 0.0016\,(T - 21.75)^2 + 0.074A - 0.0013TA \text{ (Females)}$$

$$W = 0.141T - 1.499 - 0.0016\,(T - 21.75)^2 + 0.085A - 0.0013TA \text{ (Males)}$$

$$F = 0.111T - 6.739 + 0.211A + 0.0014\,(A - 49)^2 - 0.0034TA \text{ (Females)}$$

$$F = 0.111T - 7.202 + 0.229A + 0.0014\,(A - 49)^2 - 0.0034TA \text{ (Males)}$$

7.1.2. Layers

Over a range of 15-25°C, food consumption *ad libitum* decreases at a rate of 1.5g/d/K/bird/°C but up to 30°C, it is not associated with reduction in egg production if intake of nutrients is maintained; in fact, this leads to saving of feed or, in other words, an increase in egg production @ 1 egg/hen housed/°C. However, egg weight decreases with increasing temperature @ 0.3g/K.

Upper limit of temperature where egg production is affected depends on strain, feathering, acclimatization and number of birds/cage. It is generally accepted that the minimum house temperature under market conditions is 21°C.

7.1.3. Turkeys

7.1.3.1. Brooding

29°C at hatch decreased gradually @ 3 K/week; finishing – 17-20°C

Food intake decreases @ 3 g/d/K/bird, total weight gain by 30 g/K/bird above 18°C; but total weight gain was unaffected by temperature between 10-18°C up to 18 weeks of age. Reduction in cumulative food consumption will be about 150 g/bird of 14 weeks of age and body weight loss will be about 75 g/K

(K is the wind-chill factor in W/m^2 estimated as $K = 1.163\,(10u^{1/2} + 10.4 - u)\,(33 - T)$ where u=air speed in m/s and T = air temperature in °C).

Chapter **28**

Natural Ventilation

Most livestock producers are aware of the importance and benefits of a good ventilation system for removing excess moisture and heat and for improving the building environment in general. So the question is not whether there is merit in good ventilation, but how to obtain it (Jones *et al.*, 1980).

There are two methods of ventilating livestock housing: Natural (gravity or non-mechanical) and Mechanical (fan). Air exchange removes moisture and heat produced by the birds and controls contaminants generated from manure, feed, and the poultry themselves. When air exchange relies on buoyancy and wind forces, it is called natural ventilation. When air exchange is accomplished with fans, it is called mechanical ventilation (Janni and Jacobson, 2003).

Natural ventilation is air exchange caused by buoyancy and wind-induced forces through designed inlets and outlets in a building. The predominant buoyancy force is due to thermal differences and the chimney effect.

It is important to recognize that naturally and mechanically ventilated buildings operate under different principles. Mechanically ventilated buildings use fans to exchange air, which can be controlled to provide the desired air exchange rate. Thermal buoyancy and wind are both dependent on uncontrollable weather. This makes natural ventilation control different and difficult as well.

1. Principles of natural ventilation

(Janni and Jacobson, 2003)

1.1. Thermal buoyancy

Warm air rises because it is less dense than cool air. This principle is called thermal buoyancy or the chimney effect. In naturally ventilated houses, warm air rises and exits the building through openings in the ridge. The exiting air is replaced by cooler fresh air that enters through inlets in the sidewalls. Thermal buoyancy requires both a temperature difference between inside and outside the house and a height difference between the ridge opening and the sidewall openings.

Thermal buoyancy increases as the temperature difference between inside and outside increases. In cold weather, the temperature difference can be quite large making thermal buoyancy an important driving force for air exchange. In warm weather, the temperature difference between inside and outside can be quite small making thermal buoyancy less effective for air exchange.

Thermal buoyancy is greater if the height difference between the ridge outlet and the sidewall inlets is large. Therefore, roof slopes between ¼ and ½ are recommended for naturally ventilated buildings.

1.2. Wind

Natural ventilation due to wind increases as wind speed increases. Wind speed can easily vary by 100% within minutes. Wind direction also fluctuates widely within minutes and varies year-round. Wind blows air into the building through the upwind sidewall opening. Air can exhaust through the ridge opening and the downwind sidewall opening.

To take advantage of wind, it is important to avoid having obstructions near naturally ventilated houses. Naturally ventilated houses should be built on small rises or open flat terrain. Naturally ventilated buildings should be avoided at the bottom of hills or in valleys where the natural terrain blocks the wind.

Trees, feed bins, and other barns disturb airflow for a distance of 5 to 10 times their height downwind. A minimum of 15 m is recommended between trees, small buildings and a naturally ventilated house. Greater separation distances are recommended between larger buildings and complexes with several large buildings.

To take advantage of wind in warm weather when thermal buoyancy is minimal, naturally ventilated buildings should be oriented perpendicular to the prevailing summer winds. High wind speeds during cold weather can cause excessive air exchange if the openings cannot be adequately closed.

Building maintenance is important for providing year-round environmental control.

2. Advantages and disadvantages

2.1. Advantages

2.1.1. Energy

Natural ventilation does not require energy to operate fans or a furnace, which means fuel conservation and cost savings.

2.1.2. Animal safety

Because natural ventilation systems do not require fans, confinement animals would not be affected by electrical power failures and consequent suffocation etc. expected in environmentally controlled houses.

2.2. Disadvantages

2.2.1. Lack of control

Probably the most serious drawback of a natural system is the lack of precise control of air flow, which only fan ventilation can ensure. Natural ventilation depends somewhat on the difference between inside and outside temperature but mostly on the wind, which can change in both speed and direction every few minutes. This means a building runs the risk of under-ventilation on calm, hot days and over-ventilation on cold days. In general, natural ventilation is not efficient in very hot, air-less summer or calm "muggy" weather. A sudden change in weather, which is not uncommon, causes problems.

Light control also is difficult in naturally ventilated buildings.

2.2.2. Building location

Because natural ventilation depends largely on prevailing wind currents, a location where wind would be deflected or blocked is unacceptable for a natural system, although it might be ideal from the stand-point of feed and animal handling.

Natural ventilation is efficient only in narrow-span and less intensive type of houses.

2.2.3. Difficult-to-correct problems

Good natural ventilation in a livestock building is the result of the 'right' design, 'right' location and 'right' construction. Therefore, a naturally-ventilated building which does not function properly is often difficult and expensive to correct. Sometimes the only solution is to revert to mechanical ventilation.

3. Types of natural ventilation

(Janni and Jacobson, 2003)

Naturally ventilated buildings can be categorized by the thermal environment maintained in cold weather. Naturally-ventilated buildings have indoor temperatures within a few degrees of the outside temperature. Modified environment naturally-ventilated buildings keep the indoor temperature above freezing using only animal heat. Automatically controlled naturally-ventilated buildings have sophisticated controllers and heaters that maintain indoor temperatures at specified levels similar to mechanically ventilated barns.

For poultry to maintain feed efficiency in cold weather, automatically-controlled naturally-ventilated houses are used. These buildings use both thermal buoyancy and wind to provide the necessary air exchange.

Many mechanically-ventilated turkey grower houses use natural ventilation only in the summer to reduce costs for operating fans. Some use manually-adjusted doors on the north side and an automated curtain on the south side. These hybrid barns, mechanically-ventilated in the winter and naturally ventilated in the summer, rely on wind to provide the natural ventilation.

3.1. Naturally-ventilated systems without automatic controls

Natural ventilation methods date back to the first time animals were confined in shelters, when farmers left the barn dooropen to reduce moisture or heat buildup. Today's naturally-ventilated livestock shelters, although more sophisticated, operate on exactly the same principles.

Natural ventilation occurs primarily because of the difference in wind pressure across a building, and to a lesser extent because of a difference inside and outside temperature. A natural system works best in a building with ceiling but having small openings at the eaves and ridge (peak) of the roof, and having large sidewall openings (Fig. 28.1).

3.1.1. Winter ventilation

Occurs as wind blows across the open ridge of the gable-roof building. Suction is created which draws warm, moist air out of the building and fresh air in through the eave openings. If wind velocity is great enough, the downwind (leeward) eave openings can also act as air outlets.

On calm winter days, the hot, moist air still rises and eventually finds its way out of the ridge opening. This chimney or 'stack' effect accounts for only about 10 percent of the total ventilation, because there is not a great difference between inside and outside temperatures in most naturally-ventilated buildings, except on very cold days.

3.1.2. Summer ventilation

Provided by opening up large portions (typically 1/3 to ½) of each sidewall to allow a cross-flow of air. The ridge opening has little effect in summer.

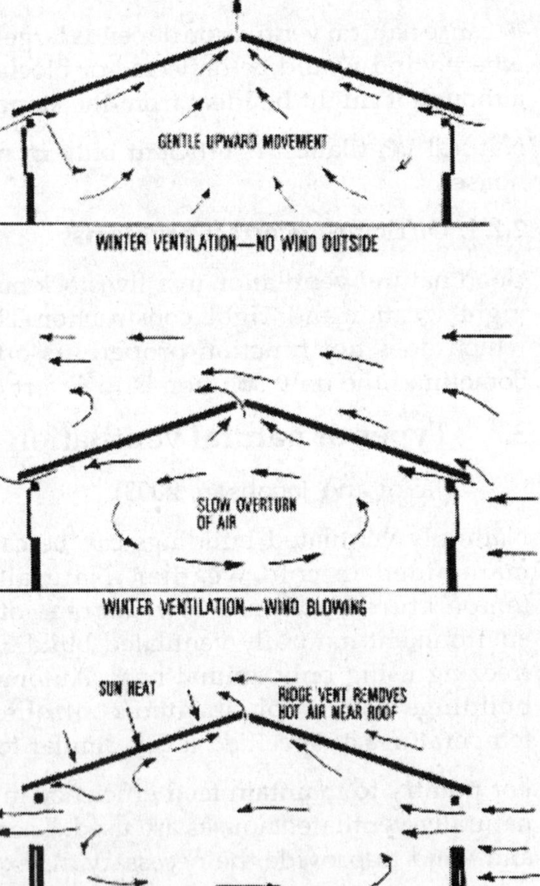

Fig 28.1 Natural ventilation in a gable-roof building

3.2. Natural ventilation with automatic controls

(Janni and Jacobson, 2003)

Automatically-controlled natural ventilation systems consist of three major components namely openings, heaters and controls. The openings serve as either inlets or outlets. Air exchange rates are adjusted by changing the opening sizes. The heaters provide supplemental heat needed to maintain the desired indoor temperature during cold weather when the birds do not produce enough heat to keep the house at the desired temperature. Controls adjust the opening sizes and the supplemental heating rate as weather, bird age and size change.

3.2.1. Openings

Air exits the building through outlets and enters through inlets. Ridge openings should always be an exhaust while sidewall openings can serve as either inlets or outlets depending on the wind. Wind blowing around the end of a building can produce unusual flow patterns which make uniform ventilation difficult.

Both inlets and outlets should be uniformly distributed along the length and in both sides of the building to provide good fresh air distribution and mixing. The best ventilation performance is achieved when the total inlet area is nearly equal or slightly larger than the total outlet area. Continuous ridge and sidewall openings along the entire length of the building work well.

3.2.1.1. Outlets

The most effective outlets are located at the highest point in a naturally ventilated building. In gabled or peaked roof buildings this is the ridge. Wind velocity greatly affects airflow through ridge outlets. Obstructions and constrictions that restrict air-flow though ridge openings greatly reduce ridge airflow.

Upstands, even 5 cm tall ones, greatly increase airflow. Upstands also reduce snow and rain penetration into the barn through ridge outlets. An alternative outlet, developed in Canada, is a series of well insulated chimneys 0.6 m x 0.6 m in dimension.

Ridge caps greatly reduce airflow and are not recommended. In very cold weather ridge openings can freeze shut or bridge over with frost. These situations will reduce air exchange.

Supplemental heat may be needed to prevent freezing or frosting over.

3.2.1.2. Inlets

Inlets in naturally-ventilated buildings are located in the sidewalls. Air enters through the sidewall inlets, flows through the building, and either exits through the ridge opening or through the sidewall opening in the opposite wall. Continuous sidewall openings are recommended for good air distribution. Various curtain systems and mechanical door systems are used to adjust sidewall opening size.

3.2.2. Heaters

Supplemental heat is needed in naturally-ventilated grower houses to maintain desired indoor temperatures during cold weather.

Different types of heaters are used for supplemental heating in poultry houses including, radiant, space, and make-up air heaters. Radiant heaters work well for improving bird comfort but do not heat the room air directly. Radiant heaters warm surfaces, which give up heat to warm the room air. Unit space heaters heat room air directly. Make-up air heaters heat in-coming ventilation air.

Unvented heaters add both heat and the products of combustion into the building. The products of combustion include gases that can create health and safety problems within the building if gas concentrations accumulate. For this reason unvented heaters are often not recommended. Proper maintenance and burner adjustment is critical for effective heater operation and minimizing the amount of undesirable combustion products like carbon monoxide from being produced. If unvented heaters are used, increase the cold weather ventilating rate by 0.07m^3/MJ/hr of heater capacity because of the moisture and products of combustion added to the building. For a 100 MJ/hr (typical) heater this would mean an increase of 7 m^3 of airflow by the continuous ventilating fan. (Note: 1 MJ/hr = 277.78 W).

3.2.3. Controls

Automatic controls are needed to maintain the indoor temperature and provide air exchange as weather changes hourly and seasonally. Natural ventilation system controllers are available to regulate air exchange, by adjusting inlet and outlet opening sizes. Controllers also regulate the supplemental heating rate. Solid state controllers and computer systems capable of controlling the inlet and outlet opening and supplemental heaters are available. They can use both time and temperature to provide the desired ventilation strategy. Thermostatic control is typically used to turn on and off supplemental heaters as needed.

The inlet and outlet openings are adjusted to control the air exchange rate. The inlets need to provide the minimum air exchange necessary for moisture control during cold weather when supplemental heat is needed. In mild and warm weather the inlets and outlets need to provide sufficient air exchange to maintain the desired inside temperature. Various devices can be used to adjust the opening size such as pneumatic systems, either manual or motorized cable and winch system and motorized mechanical arms. Typically the opening control units make small adjustments, either increasing or decreasing the sidewall and ridge opening size, on a frequent (every 10 minutes) basis to either increase or decrease the air exchange rate.

4. Building location considerations

Proper location of the building with respect to prevailing winds and surrounding trees, structures and land formations is essential to the success of a natural ventilation system. Locating on high ground is best, because it permits necessary drainage as well as exposing the building to the effects of wind.

4.1. Building Orientation

An open-front building is usually constructed with the long axis running east-west to maximize summer shade. Facing the open side to the south also minimizes the effect of the prevailing South-West and North-East monsoon winds in India. In general, for maximum air movement, houses must be oriented perpendicular to the prevailing wind direction with sufficient space between the buildings.

4.2. Separation Distance

The presence of upright trees and other farm buildings can greatly affect air flow around a livestock shelter. As a rule, naturally-ventilated units should be placed on the west or south side of the complex, with fan-ventilated units on the east or north.

When locating an individual naturally-ventilated building on the farmstead, it should be placed preferably upwind of other buildings or any obstructions that would block air flow. If it must be placed downwind, enough room should be allowed to overcome the swirling effect caused by wind passing around and over the other building or obstruction.

To determine minimum separation distances (a greater distance would always be better), the following equation* can be used:

$D_{min} = 0.73\ h\ (L)^{1/2}$, (* Original equation modified to SI units) where D_{min} = the minimum distance needed (m), h = building height at the ridge (m) and L = length of the obstructing building (m).

For instance, a poultry shed is located downwind from another poultry shed 27 m long 9 m wide and 3.3 m high. Then, minimum distance to maintain adequate natural ventilation will be $0.73 \times 3.30 \times (27)^{1/2} = 12.52$ m.

5. Sizing and design of ventilation opening

In India, not much is in practice as far as sizing and design of ventilation opening. However, with commercial poultry-rearing coming-up in all climatic zones of the country, the days are not far off when exact sizing and design of the ventilation becomes a necessity.

Since the wind can be toward either side of the building, the intake or vent opening on each side should be at least as large as the ridge vent opening on top. Size of the vent openings can be determined using the equation: $A = (0.18\ Q) \div V$, where A = area of vent inlet and/or outlet openings (m^2), Q = ventilation rate (m^3) and V = wind velocity (kmph).

For example, winter ventilation needed on the basis of $1m^3$ per 10 layers and a 10 kmph wind velocity. Then, area of the vent openings must be $(0.18 \times 1) \div 10 = 0.018\ m^2$ (i.e., 180 cm^2) for 10 layers.

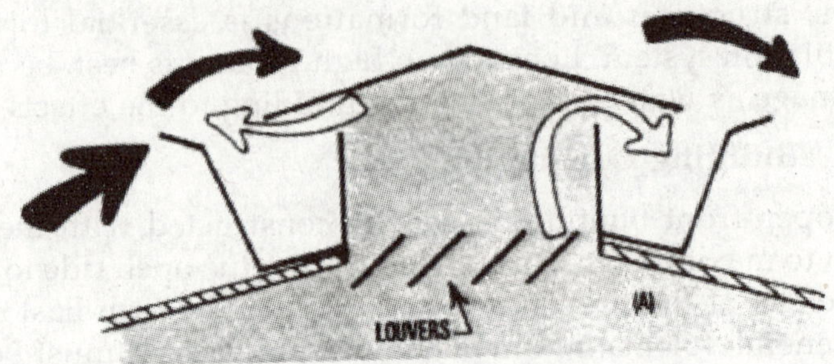

5.1. Ridge Vent Design

After calculating the total inlet/outlet area, length and width of the opening cal be obtained depending on the building dimensions. It is advisable to compensate for the reduced actual vent area by installing one that is at least 25 percent larger than needed.

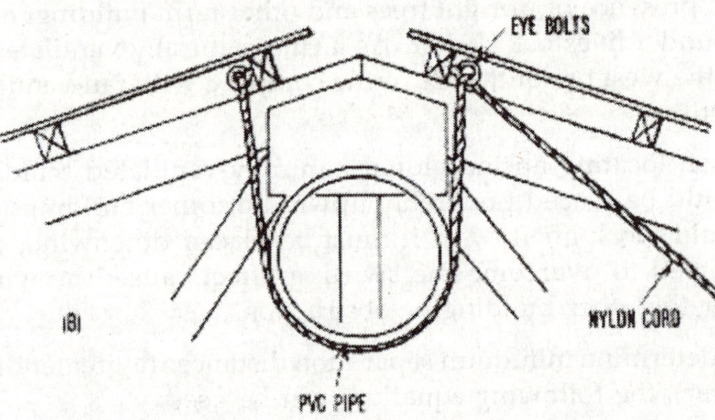

Figure 28.2 Ridge ventilators with adjustable openings

Some commercial ridge ventilators have adjustable openings which can be closed as needed during severe weather (Fig. 28.2). This is a useful option in buildings where freezing temperatures may harm animals and water supply lines.

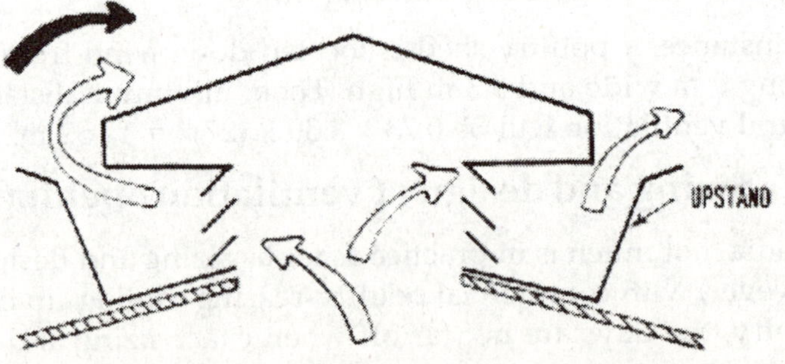

Fig. 28.3 Ridge ventilators with 'upstands'

'Upstands' that deflect wind and snow up over the cap are available for some commercial ridge vents (Fig. 28.3). However, regardless of the type of ridge vent used, snow blow-in will still occur at times when snow blows parallel to the ridge, since the downwind portion of the roof will be under pressure.

5.2. Eave Vent Design

Eave vents (winter air inlets) can simply be the open spaces between the rafters or trusses at the top of the wall. This area is usually large enough to meet the eave opening requirement. Buildings with insulated roof panels that obstruct the space between trusses must have other means for allowing adequate air to enter.

Eave vents should always be protected against direct wind, which can create unwanted drafts in the animal area. Forcing air up through a hollow-wall vent or through a rafter or truss cavity protected by a large facia board, helps break up wind gusts. Also, bringing in winter air high on the sidewall allows it to warm up some before reaching animal level and improves air circulation in the building.

5.3. Sidewall Vent Design

Several types of doors are being used satisfactorily for sidewall ventilation, including the pivot door (A), top- or bottom-opening door (B), curtain (C) and sliding door (D) (Fig. 28.4).

6. Other considerations

6.1. Roof slope

This is important for good ventilation. As a rule, steeper the slope, better the ventilation. If possible, slopes less than ¼ (i.e. 25 cm of rise per m of run) should be avoided. And generally, the wider the building, the steeper the roof slope should be.

6.2. Obstructions

Protrusions from the underside of the roof, such as deep purlins, can trap moist air and increase metal corrosion and wood deterioration. Therefore, wide truss spacings which require deep exposed purlins must be avoided. Purlins deeper than 15 cm (10 cm preferred) should not be used, if the building is to be ventilated naturally.

6.3. Sidewall height

Stud height can also affect natural ventilation. For instance, if walls are not high enough, mechanical bunks can disrupt proper air flow through the building in summer. Also, winter sun cannot penetrate open-front buildings adequately if wall height is insufficient. Stud height must be at least 2.1 m (2.4 to 3.0 m preferred); and can be raised at will.

Wall height becomes more important as building width increases, if enough sidewall vent area is to be available for summer air flow.

6.4. Insulation

Proper insulation prevents condensation and also helps conserve animal heat. All livestock produce water vapor through waste production and respiration. In a

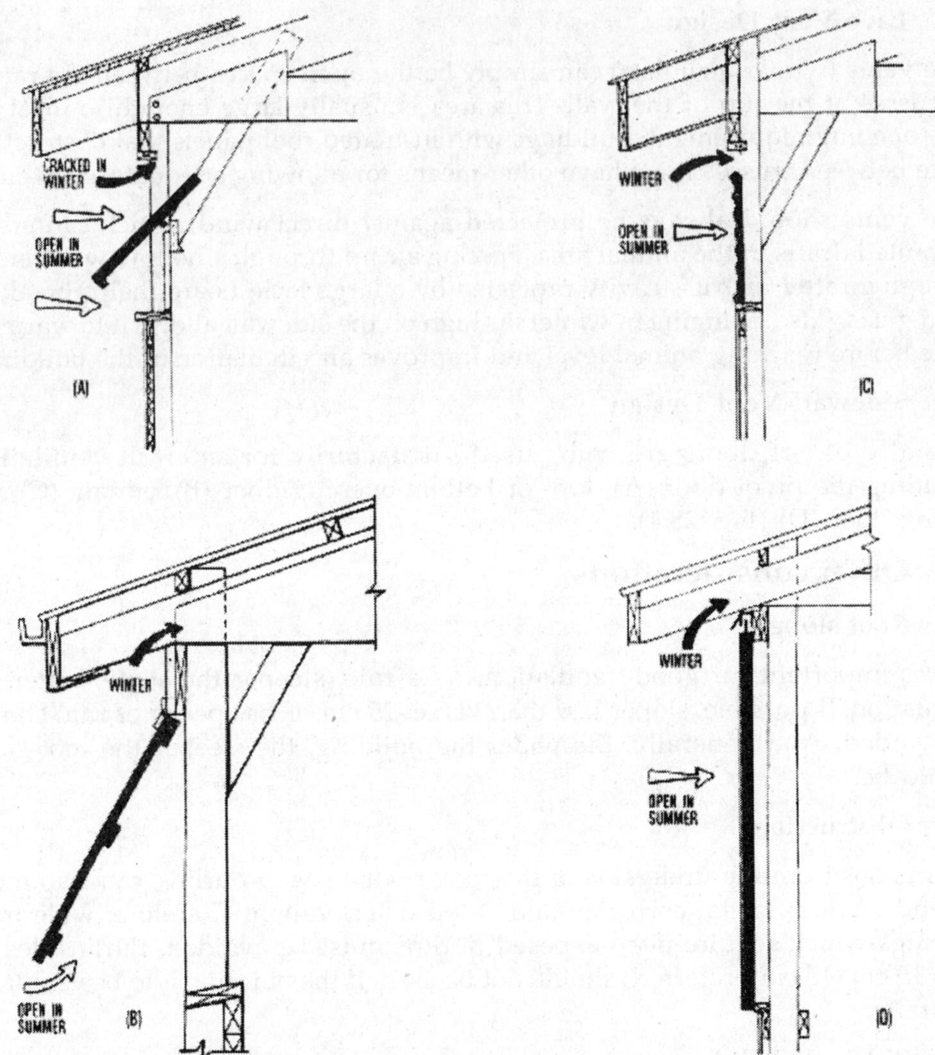

Fig. 28.4 Sidewall air inlets for summer ventilation

layer farm at ideal temperature of 21.1°C, about 200 ml water vapor/bird/d is added. This water vapor rises and can condense if the air carrying it cools off enough. Insulation helps keep the roof surface warm enough to prevent condensation when moist air contacts it.

If a livestock building is operated as a cold building, such as an open-front unit, the temperature difference between inside and out is only about 2.8°C; therefore, condensation problems are minimal. In this case, insulation is of marginal benefit unless needed to reduce the solar heat load in summer.

If a building is mostly closed in winter, the difference between inside and outside temperature is around 11.2 to 14.0°C, and condensation may occur. A minimum amount of insulation (R-value of 4 to 5) placed beneath the roof will prevent the problem.

7. Management and operation of natural systems

During periods of changing weather conditions, sidewall doors and ridge vent openings should probably be adjusted several times a day to avoid a sudden temperature change in the building. Eave vents are seldom closed except in the case of blowing snow. In some modified open-front buildings, sidewall doors are cracked open in winter instead of using eave openings. If this is done, flashing is needed at the bottom of the door to keep cold air from coming in there (Figure 28.4 D).

Winter ventilation control is accomplished by closing the sidewall opening, ridge opening, or both.

7.1. Stack effect

(Sainsbury and Sainsbury, 1988)

Stack-effect refers to the heat around the bird and is greater if there is larger difference between inside and outside temperatures or the distance between inlet and outlet is greater.

7.1.1. Stack-effect ventilation

This can be calculated by the formula $S = K_h A_i (h_d t_d)^{½}$ where S is rate of airflow (m^3/hr), K_h is a constant which is the function of ratio of inlet area to outlet area (r) as $K_h = 565.2r \div (1+r^2)$ where A_i is the area of inlets (m^2), h_d is the difference in height between inlets and outlets (m), and t_d is the difference between mean internal and external temperatures (°C). Hence, if r is = 2, stack-effect ventilation will function efficiently. Ventilation by wind can be calculated by $S = K_w A_i V$ where K_w is a constant calculated as $844R \div (1+r^2)$ and V is the mean wind speed (Kmph). On exposed sites, V can be as high as 16 and on sheltered sites as low as 1. Therefore, Poultry sheds must be preferably located in open sites for simple reason that wind can be restricted but can't be created.

7.1.1.1.1. An example

Assuming a ratio of outlet to inlet area (r) of 0.4, $K_h = (565.2 \times 0.4) \div 1.16 = 195$; $h_d = 1.5$ m (ridge ventilation), $t_d = 10°C$, S = 11.16 m^3/min/broiler (49 d, 2.68 kg body weight, maximum ventilation rate), for 1000 broilers, area of inlet required (A_i) = $S \div \{K_h(h_d t_d)^{½}\}$ = 15 m^2 (approximate) and hence, area of outlet required is 6 m^2. In other words, 1000 broilers require 90 m^2 of floor space and if they are housed in a building measuring 12 m x 7.5 m, with inlet opening of 1.25 m width on both the side walls and ridge ventilation of width 50 cm, on either side, is sufficient.

7.2. Summer ventilation

Maximum summer ventilation rate to limit the rise in temperature due to metabolic heat (T_p whose practical figure is about 3°C) is estimated by $T_p = H \div (B+V_h)$ where H is the heat production by the birds (kW / animal), B is heat loss through buildings and V_h is ventilation heat loss; the latter two being measured in terms of kW/bird/°C. The ventilation rate (S in m³/min/animal for T_p = 3°C) is given by $S = 50[(H/T_p) - B]$. Adequate housing must provide optimum temperature and air circulation to optimize performance of birds. (Note: 3.6 kJ/h = 1W)

7.2.1. Cooling open-sided poultry house

During summer, birds are uncomfortable especially when the outside temperature is =35°C. Under the conditions, the following can be practiced either singly or combination

a) Sprinkling the entire roof wet
b) Sprinkle surrounding ground area – it will increase RH
c) Use of foggers within the house
d) Use of fans - suspended from interior building structure and positions about 1.2 m above the litter and in the centre of the building to move air down the length of the building. Fans have to be spaced 6-15 m apart and tilted about 8° from vertical to direct air down the birds. Fans should have a protective grid to prevent injury to birds as well as to personnel. Fans are placed on the wind-ward side of the house to increase the velocity of air as it blows through the house. If outside temperature is high, the fans are better located inside the house to blow the air length-wise. Regardless of location of the fans, high-speed fans are preferred. Typical ventilating fans used are 0.9 m diameter on 0.746 W motor which delivers about 283 m³/min of air. Fans of 1.4 m diameter can deliver 850 m³/min of air. Airflow required is (0.024 + 0.00135*T) m³/min/kg where T is the ambient temperature (°C). Roof can be covered with thatches. (Note: 1 HP = 0.746 W)

Vertical cooling fans may also be used when temperature is above 37.8°C and are located preferably 3.7 m above the birds and 7.6-15.0 m apart.

The side walls can be covered with gunny clothes which are kept wet by spraying water regularly. It is also been found that wood-wool curtains, knitted and fitted in a within wire-net, hung over the inlets (side walls) with a provision of perforated plastic pipe of 0.6 cm diameter fitted on the top of the wood-wool curtain all along the shed to provide running water to wet the pads continuously can help reduce the inside temperature to the tune of 10 to 13°C. The cost of such pad system has been found to be economical on a long run. Reflective roof coatings (white) are also means of combating high temperature. These coatings have no R-value but still can reflect considerable solar radiation and reduction up to 1°C can be expected. During the warmest period, there can be 4% (3.34 kJ/hr ≈ 0.93

W) increasing radiant heat lost; this would be still greater if there is no ceiling insulation.

7.3. Removal of moisture

Airflow for removal of moisture (M, in m^3/hr/animal) can be calculated as follows: $M = W_a \div (g_i - g_o)$ where W_a is the total respiratory moisture + added moisture from evaporation from excreta (g /hr / animal) g_i and g_o are absolute humidity inside and outside, respectively (g/m^3). Fecal water and respired vapor are the major sources of moisture in poultry houses. Spillage of water from waterers also adds to the moisture in the house. Major route of elimination of moisture is increased by air movement and hence efficiency of moisture removal is inversely in proportion to the RH of the incoming air. Deep-litter can have as low as 5 to 10% moisture during dry weather and as high as 70 to 80% during wet weather; but optimum for growing birds is 20 to 40% and for older birds 10 to 30%. As the RH of the inside air approaches 50%, litter moisture approaches 25%. As temperature increases, water consumed per g of food consumed dramatically increases, especially above 26.6°C. Thus, fecal excretion of water also increases during hot weather; the water so excreted evaporates due to heat and causes humidity build-up in the house. However, this can be alleviated by simply increasing the airflow during hot weather as air outside will have low RH and can easily drive out excess humidity. In cold weather, real problems of humidity build-up arise because, birds consume more food and hence more water; thus, total fecal matter and water excreted increases. On the other hand, he has to be retained by curtailing the airflow which eventually leads to humidity build-up. Thus, RH increases, which, with evaporation reduced due to lower air temperature, can lead to moisture condensation and wet-litter. Therefore, insulation to retain heat to avoid condensation during cold weather is equally important.

7.4. Automatic or manual vent control

To eliminate much of the worry associated with sudden temperature changes, many producers use automatic door controllers. Several types are available commercially. Most work much like overhead garage door openers, except that they operate in response to a thermostat instead of radio waves.

Generally, controllers are used with only plastic fabric curtains or butterfly doors, because they require less power to operate than bottom- or top-hinged doors.

Plastic curtains, being relatively lightweight, lend themselves well to automatic control. They fasten at the bottom and are raised or lowered from the top, using a system of cables and pulleys.

Butterfly doors are usually hinged just above their center to minimize the force on the controller, so they tend to be self-closing when pressure on the cable is released. If butterfly doors are to open completely to the horizontal position, the individual cables must be connected to arms located on the outside of the doors

(Fig. 28.5). The cables run up over the top of the doors and are connected through a pulley to the main cable attached to the controller.

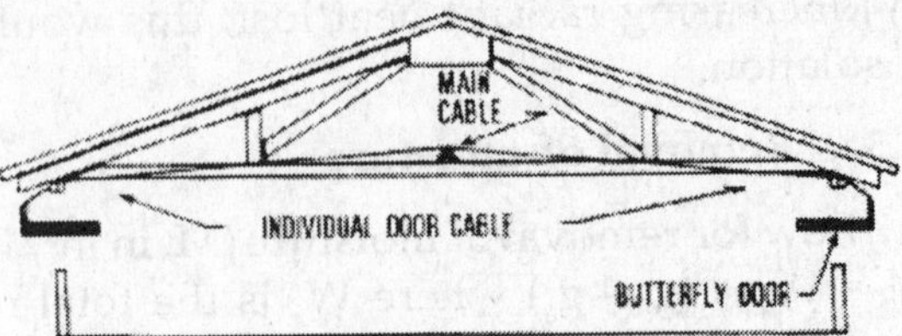

Fig. 28.5 Butterfly doors with bottom-mounted cables (open position)

The controller is located at the center of the building so that equal force is exerted from both directions. This also ensures less strain on the controller supports than when located at one end of the building and pulling all of the doors from only one direction.

If using automatic controllers, it is wise to have standby power units (invertors or generators) that will engage in case of power failure.

Manual control of vent intakes usually means a wider fluctuation in building temperature but, of course, is less expensive. Make adjustment of doors as easy as possible to encourage continued proper regulation of openings. One suggestion would be winches located just inside the building entrance and connected to a system of cables and pulleys so that several doors can be opened at one time; similar to an automatic system.

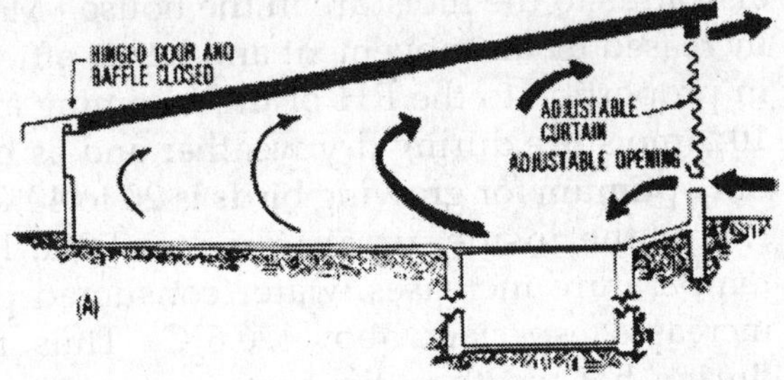

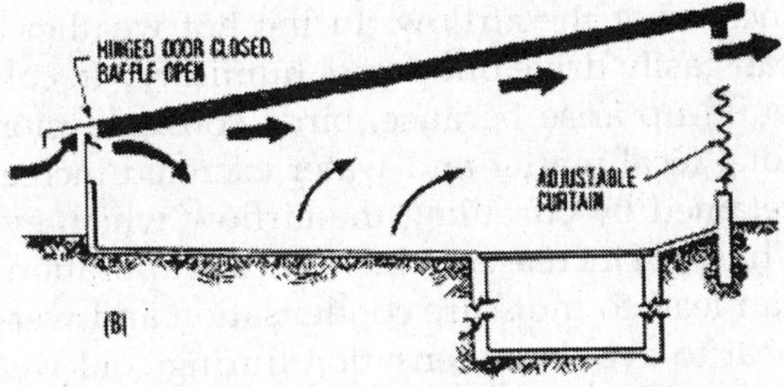

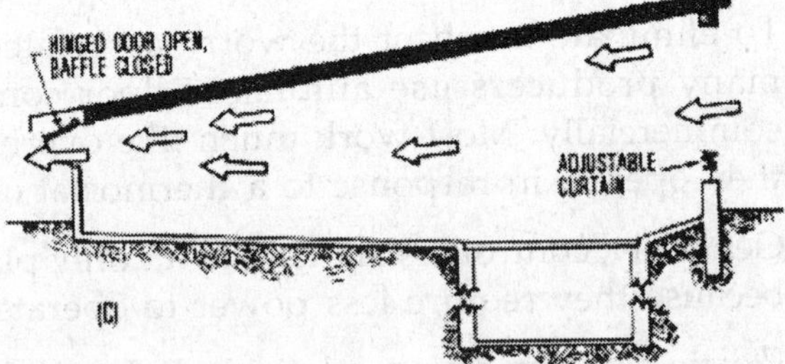

Fig. 28.6 Desired air flow patterns in a modified open-front mono-slope-roof building for (A) cold, (B) mild and (C) hot weather.

8. Mono-sloped building ventilation management

The main difference between a mono-slope (entire roof slopes in one direction) and gable-roof building is that the mono-slope has no center ridge opening to adjust. It is usually oriented with the high or open-front facing the south. Summer ventilation is accomplished using a cross-flow of air from one sidewall to the other, while winter ventilation is accomplished by air entry and discharge through openings in the highest wall (Fig. 28.6).

The building should be oriented so that the high wall opening catches the prevailing summer breeze, while blocking out the prevailing winter wind. Then, since building height decreases as air nears the low sidewall, the 'funneling' effect increases air velocity and thus aids in cooling.

8.1. Draft Control

In winter, as long as sufficient ventilation air moves through the building, poor air distribution is not generally a problem. But occasionally, draft conditions do occur, particularly in open-front buildings.

When the wind blows against an end-wall of a naturally-ventilated building, air tends to enter at the upwind end and leave at the downwind end, creating drafts throughout the length of the building (Fig. 28.7). In modified open-front buildings, these drafts can be broken up by making every other pen cross-partition (solid, about 7.5 m apart). In open-front buildings, cross-partitions should be as high as the sidewall opening and located at spacing 2½ times the building width (but no more than 24 m apart).

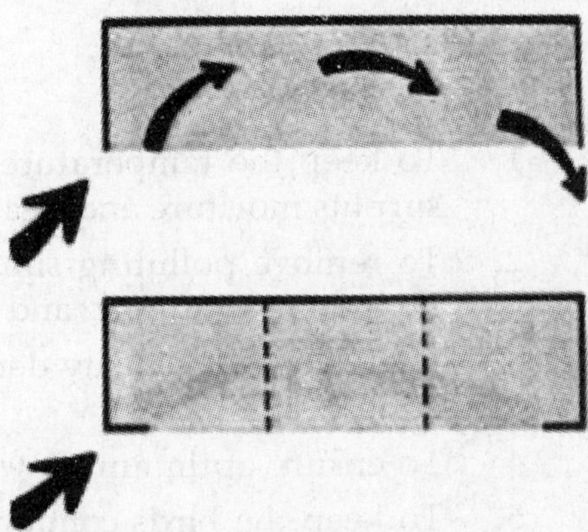

Fig. 28.7 Cross partitions to prevent draft.

Dead air zones or 'hot spots' can be a summer ventilation problem in some buildings. Pen partitions running parallel to the length of the building should be open or 'porous' to allow a cross-flow of air currents. Large feeders should not be placed along the end of a pen where they would block air flow across the building.

However, if porous partition as the lower part of the outside wall is used, it must be closed in winter.

Buildings with high eave heights are more subject to drafts than low-profile buildings because of their larger volume. The only solution is either to drop the ceiling to 2.1 to 2.4 m high and use mechanical means to ventilate, or to add plywood or polyethylene sheet hovers at the rear of the pens to provide sleeping areas where animals can move out of the draft.

Chapter 29

Mechanical Ventilation

1. Objectives of a ventilation system

(Alchalabi, 2001b)

1.1. General

1. To keep the temperature and air humidity in balance, removing the surplus moisture and heat produced by the animals
2. To remove polluting substances like ammonia, carbon dioxide, dust, hydrogen sulfide etc. and keep the level of oxygen always the optimum
3. High stocking density demands large air volume changes without creating draughts
4. To ensure optimum growth and reproduction
5. To keep the birds comfortable
6. To offer protection in times of power failure or other shortcomings.

1.2. During winter

(Oderkirk, 2001a on internet)

During winter, the chief function of a ventilation system is to remove the respired moisture (in the form of water vapor) that is produced by poultry. This is accomplished by drawing outside air through the building to absorb this moisture and exhausting the resultant moisture-laden air from the building. The problem is that the incoming outside air is already holding almost all the moisture it can. During the winter months the relative humidity (R.H.) of the outside air is quite high (80% to 90% and it often reaches 100%). Unless this incoming air is heated, little or no additional moisture can be added to it. As this incoming air mixes with inside air, it is warmed to room temperature and is capable of holding much more moisture. For every increase in temperature of 10 (°C) the moisture holding capacity of air is approximately doubled.

Good ventilation controls humidity by removing the water vapor as fast as it is produced. For example, one kilogram (kg) of cold outside air occupies 700 l and holds only 0.4 grams (g) of water even when saturated (100% R.H.). This cold air is drawn into the barn through the air inlets where it then mixes with warm air inside. Warming one kg of dry air up to 15°C causes it to expand (from 700 to 820 L) and in addition, greatly increases its capacity to hold water vapor. This air now warmed and dry, still carrying 0.4 g of water vapor would be at only 3% relative humidity. If necessary (at 100% relative humidity), it could carry over 30 times as much water vapor as the cold air outside. However, it is not desirable to allow the room air to go up to 100% relative humidity; 75% is a more practical maximum for a dry comfortable house and healthy birds. So moisture from broilers and litter evaporates into the room air until each kg of mixed room air is at 75% relative humidity, and holds 8 g of water. Exhaust fans remove this warm moist air at a controlled rate. Each kg of air comes through the inlets carrying 0.4 g of water and later goes through the exhaust fans carrying 8.0 g, thus removing 8.0 - 0.4 = 7.6 g water vapor.

Since research has shown that a broiler chicken on litter produces about 4.8 g of water vapor per hour, the ventilation rate required to control humidity can be calculated as follows:

Ventilation Rate = (4.8 g/h water vapor) ÷ 7.6 g water/kg air = 0.63 kg air/h. Since exhaust fans are rated in m^3/min not kg of air per hour (kg/h), this is changed as follows for each broiler:

Table 29.1 Poultry ventilation rates

Stock	Age (d)	Weight (kg)	Maximum		Minimum	
			m^3/min/bird	Fans *	m^3/min/bird	Fans *
Pullets and layers including breeders		1.8	0.138	0.9	0.0150	0.10
		2.0	0.150	1.0	0.0162	0.11
		2.2	0.162	1.1	0.0174	0.12
		2.5	0.180	1.2	0.0192	0.13
		3.0	0.204	1.4	0.0216	0.15
		3.5	0.228	1.5	0.0246	0.16
Broilers	7	0.17			0.0024	0.02
	14	0.43			0.0048	0.03
	21	0.80			0.0084	0.05
	28	1.25			0.0114	0.08
	35	1.74			0.0144	0.10
	42	2.23			0.0174	0.12
	49	2.68	0.186	1.3	0.0198	0.13
Turkeys		0.5	0.054	0.4	0.0060	0.04
		2.0	0.150	1.0	0.0162	0.11
		5.0	0.300	2.0	0.0318	0.21
		10.0	0.504	3.4	0.054	0.36

* 61 cm diameter fans / 1000 birds (900 rpm and 70 Pa static pressure)

Source : Charles, 1981

Ventilation Rate = (0.63 kg/h x 0.82 m^3/kg) ÷ 60 min/h = 0.0086 m^3/min for each broiler. Calculations similar to this are used to develop the winter moisture control rates for all classes of poultry. Ventilation rate for poultry is given in Table 29.1. (Note: 1 ft^3/min or 1 cfm = 0.0283 m^3/min)

During cold weather, in a poorly insulated broiler house, the total heat lost will be greater than the heat gained, therefore the ventilation rate must be reduced to maintain the building temperature. This will result in a damp or wet building. There are two possible solutions - either to add supplemental heat or to add additional insulation to reduce building heat loss.

Supplemental heat is required even in well insulated poultry buildings to maintain the desired building temperature and to allow enough ventilation to remove the moisture produced by the birds.

1.3. Interlocking of ventilation and heating systems

(Oderkirk, 2001a on internet)

Interlocking these two systems means that the switching of the heating system is connected, with a range, to the set ventilation temperature. The distance between these two temperatures is called the "neutral zone" and is expressed in °C, usually 2°C. The purpose of the neutral zone is to prevent the heating system being on while the fan is operating above the minimum required ventilation rate and thus ventilates more than is required for moisture removal and fresh air supply. This "competition" between heating and ventilation system results in energy wastage. Modern type regulators and computer ventilation systems have a built-in neutral zone.

1.4. Comfort Zone

(Oderkirk, 2001a on internet)

An animal has a certain temperature range in which it does not use any extra feed to compensate for a lower temperature and also it does not have to use extra energy to expel heat. This temperature zone is called the comfort zone. An animal feels comfortable in this zone and the feed required for maintenance is minimal. Only at the lower end of this comfort zone is it necessary that the heating system is on. Also, the fan needs only to increase in ventilation rate at the upper end of this zone. In most regulators, this heating neutral zone has been set at 1°C. In most computer ventilation systems this heating neutral zone is built in to allow the heating system to operate at 100% capacity at the low end of the heating neutral zone and decreases its operating capacity gradually to zero as the temperature reaches the upper end of the "neutral" heating zone.

1.5. Intake Control

(Oderkirk, 2001a on internet)

Intakes in cross ventilated houses can be controlled by temperature. The intake opening temperature is set 0.5 to 1°C above the set house temperature. As the temperature increases the fan will increase its exhaust capacity first. The intake will open up as the temperature further increases. The opening of the intake is always a step behind the increase in exhaust capacity which results in the desired negative pressure in a barn.

The above discussed ventilation techniques are difficult to practice with conventional thermostat operated ventilation systems. However, electronic regulators and/or computer-operated ventilation systems can easily apply and interconnect the various ventilation components.

1.6. During summer

During the summer the ventilation system must remove excess heat from the house. Typically large volumes of air are required for this purpose and can be as great as a complete air change every minute. A major misconception is the thought that fans will cool a house in summer. No number of fans can cool a house below outside temperature. This can only be done with a refrigeration (cooling) system.

All ventilation systems work on the principle of heat balance. That is, the heat produced by poultry must balance the heat lost through the ceiling, walls, floor and foundation plus the heat required to warm incoming ventilation air plus the heat loss through the ventilation fans. If a balance cannot be achieved, then the building temperature must fluctuate until a balance can be reached.

Extremely cold or hot weather can cause a heat balance problem. The value of sufficient insulation cannot be over emphasized as it plays such an important role in achieving a good heat balance year round.

Heat balance means that heat loss equals heat gain and temperature fluctuations allow this law to operate. That is, if the heat loss is greater than the heat provided, then the temperature in the space will go down until heat loss exactly equals heat gain. This is a dynamic situation, always changing, due to such changing things as outside temperature, solar energy, wind, bird's activity, growth of bird or ventilation adjustments.

In a poultry house, about 75% of the heat is lost through ventilation. The remainder is lost through the walls, ceiling, floor, doors and windows. Heat is gained by heat produced by the birds (thermal energy originally brought into the barn as feed), plus supplemental heat produced by miscellaneous sources such as electric lights and motors. Heat produced by livestock is substantial. For example, 2,500 hens produce about the same amount of heat as an average house furnace.

Broilers need 0.17 m^3/hr of ventilation in the beginning which increases to 7 m^3/hr for full-weight broilers during summer. Hence, air changes required varies between less than once to 30 to 40 changes per hr. This has to be achieved with natural and/or artificial ventilation; which may include fans with or without duct system.

2. Poultry house ventilation

Most of the discussion below pertains to environmentally-controlledhouse because in case of conventional/open-sided/windowed poultry houses, it is extremely difficult to accurately calculate the various parameters.

It must be noted that all the operations in an environmentally-controlled house are power-driven. The entire house is a closed one and power failure results in drastic reductions with concomitant build-up of heat.

Therefore, it is mandatory that an environmentally-controlledhouse is constructed with a standby electric plant preferably with an automatic changeover facility.

2.1. Movement of air

(North, 1972)

2.1.1. Negative pressure system

Usually, air is brought into the house at the front wall and exhausted by fans installed in the opposite back wall. The amount of air to be moved determines the size and number of fans necessary. The amount of air exhausted should be slightly more than the amount of air coming into the building so as to create a "negative pressure or partial vacuum" inside the building; sometimes, this is known as the "exhaust system".

The partial vacuum created inside the house offers the following advantages (Alchalabi, 2001b):

1. Air enters the house evenly through every opening
2. The pattern of air movement inside the house is not affected by external winds
3. Air enters the house with enough speed so as to mix with the air already present in the house
4. Exhaust fans will provide a uniform ventilation because in the absence of partial vacuum, they can provide ventilation only for those birds in the immediate vicinity and those further away are affected only by external winds

Therefore, to be effective, it is important that the inlet openings match the number of exhaust fans. If the inlet is too big, the static pressure will be too low resulting in poor air distribution, draughts, litter-caking and high fuel costs. On the other hand, insufficient inlets produce excess static pressure leading to reduced fan performance and undue strain on ceilings. However, range of acceptable static pressure is fairly wide and hence, under practical situations, high static pressure is not an important factor (Alchalabi, 2001b).

The amount of air to be forced out of the poultry house by the exhaust fans is to be determined by the type, age and size of the birds and the outside temperature and humidity. When the houses are exceptionally wide, air circulation within the house will not be adequate under the above system. Under such circumstances, air inlets are placed at the front and back of the house and the air is exhausted through a cupola of fans placed in the centre of the ceiling.

2.1.1.1. Heat removal from the building

At ideal temperature (21.1°C), birds produce about 4 kJ/hr/kg (0.067kJ/min/kg) body weight of heat; heavier birds producing slightly less heat than the lighter ones. (Note: 3.6 kJ/hr = 1 W)

As a rough approximation, for each 0.56°C (1°F) rise in temperature, 1.42 m^3 of air can remove 1.005 kJ of heat; in other words, if the house temperature is increased by 5°C, 1.42 m^3 of air can remove 9.045 kJ of heat. As a rule of thumb, an air flow of (0.024 + 0.00135*T) m^3/min/kg body weight, where T is the air temperature, is needed under practical conditions.

Table 29.2 Typical air flow requirement at 30-60% relative humidity

Temperature (°C)	Air flow (m3/min/kg body weight)
4.4	0.030
15.6	0.045
26.7	0.060
37.8	0.075
43.3	0.083

Source : North and Bell, 1990

Depending upon the airflow required to remove the heat produced within the building (Table 29.2) when the outside temperature in the highest, fan capacity must be calculated. This is for a simple reason that airflow produced by the fans can be reduced whenever necessary but it cannot be increased beyond the capacity of the fans installed.

Airflow can be lowered by fixing rheostats of the fans (either manually-operated or automatic) or by operating only a part of the fans (some fans may be stopped when the air requirement is less or two or three fans placed side by side, smaller ones running continuously and other cut-off one at a time as less air is needed).

2.1.1.2. Air intake

As indicated, slightly less air is admitted than exhausted to create a difference of about 10 Pa (0.1 cm of water) static pressure so that the fans will function at almost full capacity ensuring a proper air circulation.

2.1.1.2.1. Total air intake

A general rule is to allow 56.78 cm^2 of intake opening for each m^3/min of exhaust; when light traps are used, air intake is increased by 25% to 70.98 cm^2 for each m^3/min of exhaust.

2.1.1.2.2. Shape of air intake

For uniform distribution of air throughout the building, air must enter at a gushing rate and hence through a narrow entrance, usually referred to as "intake slot". The width of this slot will be usually about 5 cm when the outside temperature is 21.1°C usually running almost a length of the house to create adequate air movement.

To ensure a static pressure of 10 Pa (0.1 cm of water), it is necessary to have adjustable slot intake (either manual or by a winch and cable system) so that whenever required, air intake can be regulated.

2.1.1.2.3. Location of slot intake

The slot intake should be as high as possible on the wall opposite to the fans. Air may be brought directly into the building from outside or may enter from an attic where it is warmed during cold weather.

2.1.1.2.4. Baffle to direct incoming air

An adjustable baffle will be handy not only to direct the incoming air but also to control the amount of air coming through the slot. At normal temperatures, the baffle should be almost horizontal to mix the incoming air adequately with that in the house. When more incoming air is needed, the inside of the baffle edge is dropped so as to deflect the air towards the floor.

2.1.1.2.5. Velocity of incoming air through the slot

When all details of the ventilating system are operating at average, the velocity of the incoming air through the slot will be between 12.78 and 13.68 kmph.

2.1.2. Positive pressure/Pressurized system

This system involves forcing of air into the poultry house by fans and regulating the outlets in such a way as to provide a slight positive pressure inside the building. This necessitates use of large, low-speed fans to move a large volume of incoming air at a slow rate. Therefore, even the air distribution ducts must be large and located about 30.5 cm below the ceiling of the house to induce the best air circulation. In narrow span houses, the ducts may be located near the leeward wall and in wide span houses; they should be placed at the middle of the house, with dampered exits directing the air to both sides.

This system is not popular because its installation as well as maintenance is difficult and it requires intricate devices and thermostats which often call for services of a ventilating engineer. In addition, the system is expensive and in dry climates, it may be necessary to install dust collectors on the exhausts further adding to the costs.

2.1.3. Combined system (Alchalabi, 2001b)

A combined system is sometimes used during summer utilizing two types of fans viz. the usual exhaust fans to create the negative pressure in the house and second blower fans outside the house to pump cool air (evaporative coolers). Therefore, this system keeps the birds cool in summer and distributes air properly by drawing the incoming air through the evaporative coolers instead of inlets.

2.2. Cooling the poultry house

When the environmental temperature is above 32.2°C, it is necessary to cool the incoming air in order to provide comfortable environment to the birds.

Evaporative cooling is the most practical method wherein the incoming air evaporates water before it enters the house.

2.2.1. Volume of air required

The volume of air needed to be moved through the poultry house having pad and fan cooling depends upon age, number and weight of birds; distance from pad exhaust fans, type of building insulation and maximum outside temperature. Approximately, 0.09 to 0.12 m^3/min/kg body weight of air movement is required; as a general rule, it is advisable to provide 0.11 m^3/min of the exhausted air per kg live birds in poultry houses.

2.2.1.1. Temperature of exhaust air

The cool air entering into the poultry house from one end becomes warm at the other end from where it is exhausted out of the house; this rise in temperature should not be greater than 2.8°C. Otherwise, more air should be made to pass through the house by increasing the speed of airflow so that not more than 2.8°C is gained by the air exhausted from the building.

2.2.1.2. Systems of evaporative cooling

2.2.1.2.1. Pressurized system

The evaporation coolers are placed outside the house and air is sucked through the evaporative pads of the cooler; the cooled air is forced into the poultry house. Exhausts are provided to remove the air allowing for a slight pressure build-up of 10 to 25 Pa (0.1 to 0.25 cm water) in the building.

2.2.1.2.2. Pad and fan system

This system is used only in poultry houses. The evaporator-pad is placed in the wall at one end of the poultry house, and exhaust fans at the other so that air is sucked through the evaporator-pad.

2.2.1.3. Thermostatic control

Evaporative cooling is used only during hot hours. During cooler hours, as in case during night, water entering into the pads is shut off so that the ventilating fans continue to move air which makes the pads dry and ventilation system becomes a conventional one. Considering that the distance between air inlet and exhaust is large, a faster air movement becomes necessary when the pads are dry.

Thermostats, set to allow flow of water at the above 26.7°C, are necessary to control flow of water into the pads. When more than one exhaust fan is used, it is advisable to have separate thermostats for each of the fans.

2.2.2. Pad and fan specifications

In case of building not more than 60 m long, pad may be placed at one end of the building, the fans in the other. If the building is more than 60 m long, pads may be placed in both ends of the house and exhaust fans in the centre or *vice versa*. Pads must be placed in the windward wall of the house.

2.2.2.1. Pads

Usual height of the pads is 1.22 m and the length depends upon the amount of air required to enter into the building. They are made of excelsior or similar material in narrow trough with holes drilled at the bottom to allow uniform flow of water over the pad is constructed just above the pad. At the bottom of the pad, and other trough is placed (without holes) to collect excess water running off the pad which is conveyed into a tank for recirculation by use of a water pump. A water tank to govern the amount of water onto the pads with an incoming water line as well as the excess water from the water pump is constructed at a suitable location nearer to the pads. This water tank is provided with a float valve to govern the amount of water flowing into the tank and maintain uniform level of water.

Pads measuring 45 cm which may be placed horizontally in a hood outside the house, running the entire length. It may be kept wet by foggers placed just above it and the air drawn directly into the house through the wet pad and hood.

2.2.2.1.1. Pad dimensions

It is recommended that the rate of flow of air is 45.7 m/min through the pads. Therefore, to calculate the area of the pad, the amount of air moved per min by the exhaust fans at 25 Pa (0.25 cm) of static pressure is divided by 45.7 to obtain the total pad area required in m^2. Then, the length of the pad can be obtained by dividing the total pad area by the height of the pad.

For instance, 5000 Leghorn layers, each weighing 1.8 kg, require 5000 x 1.8 x 0.11 = 990 m^3/min of air, and dividing by 45.7, 21.66 m^2 of total pad area. Assuming that the pads are 1.22 m high, the length of the pads required will be about 18 m.

2.2.2.1.2. Cooling pad efficiency

Cooling pad efficiency should be such that the difference between dry bulb reading inside the pad and wet bulb reading outside should not exceed 1-1½ °C (Ernst, 1995).

2.2.2.2. Exhaust fans

These are located farthest from the evaporator-pad in the opposite or end wall. One or more fans will be required to move a specified amount of air depending on the number of birds and length of the house. Fans with slow revolutions and large air volume movement are preferred.

2.2.3. Two-stage cooling, a new innovation (Alchalabi, 2001a)

Evaporative cooling works best under hot-dry climates due to low relative humidities especially during the hottest part of the day. However, as the humidity remains high of the day, as is expectable at coastal regions, evaporative cooling becomes less effective and often counter-productive because it increases relative humidity which further accentuates the effects of high temperature.

Generally, soil below the surface will be at a more constant temperature because it heats and cools with the seasons and there is a time lag as it does so. The temperature variation increases with the depth beneath the surface because of storage capacity of the soil. For instance, the soil surface temperature reaches its maximum during the summer but at depth, it may not reach a peak until three months later.

This thermal lag helps both the heating and cooling performance of heat exchange systems located deep enough inside the soil. It is interesting that during winter, the temperature of the soil 3 m below the surface season the autumn season level; *vice versa* is true during summer months when the soil at a depth of 3 m is at its spring level as for the temperature is concerned. However, soil temperature is also affected by soil type, depth, moisture content, time of the year and geographical location. Sandy soils tend to have more temperature variations than clays and soil moisture also increases the range. Seasonal rains can affect soil temperature during winter.

2.2.3.1. The concept

The air is cooled in two steps to reach the final target economically and practically. The first stage uses sensible cooling to reduce the temperature with constant moisture content (subsoil cooling) and the second stage is to reduce the temperature by evaporative cooling by adding moisture of the air and increasing the relative humidity.

2.2.3.2. A practical two-stage air-cooling system

A metal tube, 30 m long and 30 cm diameter, was bent suitably and buried about 3 m deep inside the soil so that one end is open as air inlet and other end is connected to the evaporative cooling unit. Under experimental conditions, air flow rate was fixed at 20 m^3/min and the power of the motor to pump water into the evaporative cooler at 200 W. The results indicated that the outside temperature ranged between 23.5 and 43.0°C (a range of 19.5°C) whereas in the house the range was just 3.8°C (19.1 to 22.9°C). Relative humidity increased from 20% at the air inlet to 75% in the poultry house.

It is interesting that the two-stage air cooling not only reduced the range of temperatures which offered stable inside temperatures to the birds but also reduced the peak temperature by approximately 20°C which is in itself enough to provide alleviation from heat stress. Although relative humidity increased to 75%, it is likely to have little effects because temperature was below 24°C.

This system can be used continuously because it helps the soil to lose heat gained during the day and it heats the air at night so that the system will be ready to work efficiently the next day.

During winter, the system can be used without the evaporative cooling step. The relatively warm soil helps to heat the air and reduce fuel costs to heat the incoming air.

2.3. Typical calculations

(Ernst, 1995)

Temperature in evaporative-cooled houses depends on 1. Heat entering from walls and roof 2. Heat given-off by birds 3. Volume, temperature and relative humidity of the air which is being moved through the house to remove moisture, heat and gases.

2.3.1. Heat entering through walls and roof

This depends on area of walls and roof, insulation of the material and the difference between the ambient and inside temperatures.

$Q_b = ARt_d$, where Q_b = heat entering through walls and roof, A = area of the roof (about 1.25 times the floor area) and t_d, the average temperature difference = (design dry bulb reading - design wet bulb reading) - 3.9; the design dry and wet bulb readings can be obtained from the local weather station.

2.3.2. Heat produced by birds

Q_a = NwH, where Q_a = heat produced by birds in W, N = number of birds, w = Average body weight in kg and H = heat production/min/kg. Generally, H = 0.29 kJ for adults, 0.35 kJ for young birds weighing = 1.5 kg and 0.23 kJ for large birds weighing = 3 kg. (Note: 4.187 kJ = 1 kcal)

Total heat produced, $Q = Q_b + Q_a$

2.3.3. Cooler capacity

The main objective of using coolers is to keep the inside temperature about 3.9°C above wet bulb temperature. A cubic meter of air absorbs 0.7652 kcal of heat when warmed by 2.8°C.

Therefore, cooler capacity = total heat load/0.7652

2.3.4. Pad area required

Pad area, m² = cooler capacity/design air speed for the type of pad used; for aspen fiber pad (5 cm thick), design air speed is 46 m/min.

2.3.5. An example

House = 150 m x 12 m; N = 20,000 broilers; w = 2.5 kg each; R = 14 for roof and 8 for walls; heat conductivity = 0.0059 kcal/m²/min/°C; design dry bulb = 46°C; design wet bulb = 30°C.

Building heat, Q_b = [(150 x 12) x 1.25] [0.0059] [(46 - 30) - 3.9]
= 161 kcal/min (674 kJ/min)

Bird heat, Q_a = 20,000 x 2.5 x 0.0686 =3,430 kcal/min (14361 kJ/min)

Total heat produced = 161 + 3,430 = 3,591 kcal/min (15035 kJ/min)

Therefore, cooler capacity = 3591/0.7652 = 4693 m³/min and

Pad area = 4693/46 = 102 m² for an aspen fiber pad 5 cm thick.

(Note: 1 kcal/hr = 1.163 W)

3. Summer ventilation

(Barnwell, 2003)

The most important factors to be considered for former ventilation are

1. Air velocity across the birds
2. Air exchange in the house
3. Air distribution through the house
4. Control of relative humidity

3.1. Air velocity, exchange and distribution

Air speed in m/min created by a fan can be calculated as the ratio of air capacity of the fan (m³/min) at actual working pressure to cross section of the house (m²). For instance, a poultry house with a width of 9 m, stud height of 2.1 m and a rise of 1 m will have a cross sectional area of 23.4 m². A fan with 4 blades of diameter 91.4 cm mounted on a ½ Hp motor can produce an air speed of 339.6 m³/min / 23.4 m² = 14.38 m/min; air speed in excess of 15.27 m/min will be expensive to install. If air speed calculated is low, use of baffles may be considered.

Incorrect use of baffles can create problems as well. Baffles are sometimes installed where they are not needed or too low to the floor which creates an overload on the fans leading to higher utility costs. In addition, air volume through the fans is decreased and hence more fans will be needed to do the same job.

If baffles are necessary, their height above the litter should always be calculated so as to obtain the correct height and desired air speed. The distance at which baffles must be fixed from the floor is given by the ratio of required cross-section of the house by width of the house. Required cross-section of the house is the ratio of total m^3/min rating of all the fans at actual working pressure to the desired air velocity through the house.

Assuming that an air speed of 122.16 m/min is required in a house of width 12.2 m having a total rating of all fans at actual working pressure of 3635.2 m^3/min, required cross-section = 3635.2 m^3/m / 122.16 m/min = 29.76 m^2; therefore, baffles must be fixed at 29.76 m^2/12.2 m = 2.44 m above the litter to ensure an air velocity of 122.16 m/min.

The speed of air entering the house must be based on the width of the house or the distance the air should move to reach the centre of the house. The pad area for the evaporative systems and inlet area for all tunnel ventilated houses must match the fan capacity.

3.2. Control of relative humidity

The pumps on the cooling system should be operated based on both temperature and relative humidity with a solid-state control, humidistat or cycle timer. The pumps on the cooling system should never be allowed to run below 27.8°C or if the relative humidity is above 70%.

All the summer fans should be an operationin any house before the pumps on the cooling system are allowed to run because enough friction needs to be generated in the cool cell pads from air velocity for efficient evaporation of water of the surface area.

When foggers are used, moisture should never be added to the air directly in the inlet opening when the air velocity is more than 152.7 m/min to prevent wetting of the floor and of the birds. In addition, if the baffles are not properly spaced, it will add to humidity problems at the fan end of the house. As a rule of thumb, the distance between the baffles must be less than 10% of the average air speed.

The air velocity through a house should be higher when foggers are used than in a house using evaporative cooling because higher air velocity with evaporative cooling leads to increased relative humidity from the inlet end of the house to the fan end which ultimately precipitatesin bird weight and egg productionproblems throughout the house.

When foggers evaporative cooling are not used, the required air speed should be higher because no moisture is being added to the air and when foggers are used, the air speed is reduced and finallywith evaporative cooling, it is further reduced.

3.3. Guidelines

Table 29.3 Summer ventilation guidelines (Barnwell, 2003)

	Effective temperature (°C)*	Air exchange time (min)	Maximum air velocity (m/min)	
			Lower limit	Upper limit
Broilers	21.1	< 1.3	137.43	145.06
Heavy breeders	18.3	< 1.3	122.16	137.43
Blackout rearing	23.9	< 1.5	91.62	106.89
*Temperature the birds "feel"; depends on dry bulb temperature, RH and air velocity Desired relative humidity for all the above is 50%				

Table 29.4 Poultry ventilation guidelines

Type of poultry	Temperature (°C)		Rate of flow (m^3/min)		
	Range	Recommended	Minimum	Recommended	Maximum
Chicks	30-35	30-35	0.0014	0.0014	0.0014
Pullets	20-30	20-30	0.0014	0.0284	0.1136
Layers in cages	10-30	18-21	0.0142	0.1846	0.2130
Layers on litter	10-30	18-21	0.0142	0.1136	0.14
Breeders	10-30	18-21	0.0142	0.1420	0.1704
Broilers (litter)	10-30	20-22	0.0142	0.0284	0.1136
Turkeys	10-21	20-22	0.0142	0.0620/kg	0.0775/kg
Note : relative humidity should be within the range of 50 and 75%					

3.3.1. Heat stress index number

This is the total of dry bulb temperature (in °F) and relative humidity (in %). Hence, it can indicate the effect of heat stress (effective temperature).

Any time the heat stress number is too high, and the relative humidity is the major part of the problem, slight increase in dry bulb temperature will reduce the relative humidity and alleviate the problem.

If the dry bulb temperature is 90°F, and relative humidity is 75%, the index is a problem because increase in dry bulb temperature is not without risk on the comfort and production performance of the birds. However, if the dry bulb temperature is increased by 5°F, relative humidity decreases to 60% reducing the heat stress index to 155 where the birds are likely to show a better performance than at 90°F and 75% relative humidity.

3.3.2. Air leaks

Cold air entering into the poultry house mixes with hot air, it removes moisture from the cold air, and it makes the warm air up at the peak of the roof travel down to bird level. This movement of hot air helps keep the house warm, litter dry, oxygen levels up, birds comfortable and utility costs reduced. To exploit these advantages of movement of hot air, the house must be as air tight as possible and it should not have leaks especially at ridge cap or top of the house, around or close to the fans or anywhere down close to the floor where the cold air would travel along the floor surface and create severe chill factors on the birds. An air tight house also maintains a proper pressure for optimum operation of air inlet and exhausts.

Table 29.5 Heat stress index number *Vs* comfort of birds (Barnwell, 2003)

Heat stress index number	Remarks
≤ 150	No problem with heat stress or heat prostration
155	Borderline of beginning to lose performance
160	Reduced feed intake and performance; increased water intake
165	Beginning of mortality and permanent damage to lungs and cardiovascular system
170	High mortality

4. Ventilation – a model spreadsheet

(Alchalabi, 2004)

4.1. Inputs

Number of birds: 20,000 broilers, Body weight: 2 kg

Building: Length – 90 m, Width – 12 m, Stud height – 2.3 m, Height at the center – 3.5 m

Number of fans: 7, Fan size: 120 cm, Air movement: 623 m^3/min

Air speed: 119 m/min

4.2. Calculations

4.2.1. Building

Shed area: 90 x 12 = 1080 m^2, Rise = (3.5- 2.3) = 1.2 m

Cross-sectional area: (12 x 2.3) + ½ (12 x 1.2) = 34.8 m^2

Shed volume = (34.8 m^2 x 90 m) = 3132 m^3,

Bird density: 20,000/1080 m^2 = 18.5 birds/ m^2

4.2.2. Ventilation

4.2.2.1. Summer conditions

Total airflow required @ 0.1009 m^3/min/kg = 4036 m^3/min

Number of fans operated = 7

Total airflow supplied by the fans = 623 x 7 = 4361 m^3/min

Air exchange (per min) = (4036/3132) = 1.4/min

Estimated air speed in the middle = 2.1 m/s (from standard tables)

Estimated air speed @ 95% efficiency = 1.98 m/s

Actual airflow supplied by fans = (4361 x 0.95) = 4141 m^3/min

Total CO_2 produced = 26 m^3/h (@ 650 cc/h/kg body weight)

4.2.2.2. Winter conditions

Min. airflow required @ 0.00985 m^3/min/kg = 394 m^3/min

Number of fans operated = 1

Total airflow supplied by the fan = 623 m^3/min

Actual airflow supplied by the fan = (623 x 0.95) = 592 m^3/min

Air exchange (per min) = (394/3132) = 0.13/min

Total CO_2 produced = 1.18 m^3/h (@ 29.5 cc/h/kg body weight)

5. Fans – selection, efficiency and performance

5.1. Fan selection

(Arnold and Veenhuizen, 1994; ohioline.ag.ohio-state.edu)

Ventilation is an air exchange process that brings fresh air into a building, thoroughly mixes the air, and exhausts moist contaminated air from the building.

A properly designed and installed ventilation system is more than just a fan. It includes air inlets, outlets, and controls. The fan is the "heart" of a mechanical ventilation system that provides the driving force for the needed air exchange in livestock facilities.

Fans are used to force air into, pull air out of, or circulate air inside livestock buildings. Energy (electricity) is required to operate fans. To most efficiently use the energy inputs and achieve desirable results, all components of ventilation system must be carefully designed, selected, installed, and managed.

When selecting ventilation fans, consider: Fan type, required ventilation capacity, quantity of air delivered at different static pressures, system static pressure, energy efficiency, m^3/min/watt, service and support, durability, reliability, and expected life, suitability for intended use and cost.

Some of these factors, such as service and support, depend on location and availability of local suppliers. Others, such as cost, durability, and suitability are subjective evaluations and must be considered on a case by case basis. Factors such as fan type, ventilation capacity, quantity of air delivered, static pressure, and energy efficiency can be quantitatively measured and reported. All are important criteria for selecting a fan or group of fans.

5.1.1. Fan Type

Fans can be axial flow or centrifugal. Axial fans are most commonly used in livestock buildings. Centrifugal fans are often utilized where inflatable curtains or ridge-closing bladders are found.

Fans can be axial flow or centrifugal. Axial flow fans pass air straight through the fan parallel to the fan blade drive shaft. Centrifugal fans (squirrel cage fans) bring air in through a central inlet, the air makes a right angle turn and is discharged perpendicular to the fan axis.

Axial flow fans are designed to operate at low static pressures (typically less than 500 Pa). They are commonly used for livestock building ventilation. The initial cost is less and performance is influenced less by dirt buildup on the fan blades. (Note: 1 cm water pressure = 100 Pa)

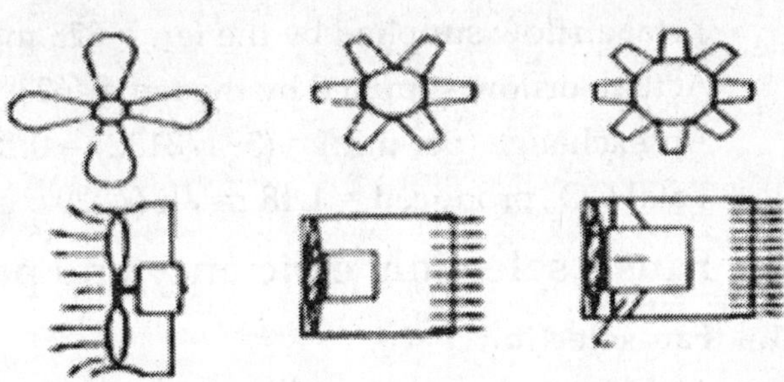

Fig 29.1 Propeller, Tube-axial and Vane-axial fans

Propeller fans have propeller-shaped blades mounted in a circular ring or orifice plate and a drive motor. Propeller fans scoop air and move it away from the fan blade. Blade tip clearance is an important factor in fan performance. A small, uniform clearance is preferred to prevent air from flowing back around the propeller. These fans move large volumes of air at low static pressure. Propeller fans are typically 15 to 180 cm in diameter. They are designed to operate at static pressures of 0 to 500 Pa of Static pressure.

Another type of propeller fan is the circulation fan used to help improve air distribution, reduce temperature stratification, increase air velocity in the animal zone, and eliminate dead air spots. Circulation fans usually operate at zero to low static pressure. Paddle fans mounted to the ceiling are an effective method to circulate large volumes of air against zero or low static pressure.

Tube-axial fans consist of a tube-shaped housing, a propeller-shaped blade, and a drive motor. Tube-axial fans can operate at higher static pressure because of a larger hub and reduced blade tip clearance, so less air flows back through the fan. Tube- axial fans are typically used in low and medium pressure duct air distribution systems. A rapidly growing use for this fan is manure pit ventilation. They are typically 30 to 120 cm in diameter and operate at static pressures as high as 3500 Pa of static pressure.

Vane-axial fans similar in design and application to tube axial fans. The major difference is that air straightening vanes are added either in front of or behind the blades to reduce the circular motion of the air. This results in a more efficient fan capable of operating at a slightly higher static pressure. Typical uses are grain drying, aeration, and sometimes pressure duct air distribution.

Centrifugal fans are quieter and can operate at much higher static pressures than axial flow fans. Centrifugal fans operate by throwing air away from the fan blade tips. These blades can be forward curved, straight, or backward curved. Centrifugal fans should be used for handling relatively clean air because dirt accumulation on centrifugal fan blades affects efficiency and airflow capacity. Centrifugal fans are commonly used in grain drying and forced air heating systems where high static pressures are required.

5.1.2. Ventilation Capacity

A ventilating system is used to control the environment. Ventilation capacity is based primarily on the amount of air exchange needed to remove moisture in the winter and excess heat in the summer. Airflow requirements vary with animal size and outside weather conditions. Ideally, ventilation air must vary from just enough air to maintain good air quality during very cold weather up to a maximum rate to eliminate heat stress during hot weather.

A properly sized ventilation system should provide at least three seasonal ventilation rates—cold, mild, and hot weather.

Cold weather ventilation is the minimum continuous rate that provides enough air to remove excess moisture. Cold weather fans run continuously with an emergency shut-off thermostat. Mild weather ventilation increases the air exchange rate to modify temperature and remove moisture. Mild weather fans are typically turned on and off with thermostats to maintain a desired air temperature. Hot weather ventilation reduces heat build-up and increases air movement past the animals. Table 29.6 gives recommended ventilating rates for several livestock types (Also see Tables 29.14 and 29.15).

Table 29.6 Recommended Ventilation Rates (m^3/min for 1000 birds)

Weather →	Cold	Mild	Hot
Animal		Poultry	
Broiler 0-7 d	1.14	14.20	28.40
Broiler over 7 d	2.84	14.20	28.40
Layers	2.84	14.20	28.40-42.60
		Turkeys	
Poults	5.68	19.88	28.40-113.60
Growers	2.27	9.94	22.72
Breeders 20-30 weeks	1.42	4.26	14.20

5.1.3. Fan Capacity

Fan capacity is measured as the volume of air and it is usually expressed as m^3/min.

The amount air a fan moves depends on the blade diameter, blade shape, fan speed (revolutions per minute, rpm), motor power, and housing. Two most common measurements used to describe the characteristics of a fan are blade diameter and motor power.

Where unvented heaters are used, provide at least an additional 0.070 m^3/min continuous ventilating capacity per 1.055 kJ/ hr of heater capacity (or 0.25 m^3/min/W).

5.1.4. Static Pressure

Static pressure is the pressure that a fan is capable of developing. Static pressure is also a measure of the resistance to airflow or the pressure developed as a given rate of airflow moves through system. Static pressure or air pressure difference is usually measured between the inside and outside of a building with a manometer and is expressed in Pa.

Static pressure is low in poultry houses; however, high static pressures reduce fan efficiency and low static pressures offset uniform air circulation. Hence, and air speed of 152 m/min through the inlets and the static pressure of 30 Pa (0.3 cm of water) in a negative pressure house is generally recommended.

Resistance to airflow occurs throughout the entire ventilation system. In a mechanically ventilated house, the exterior air inlets, baffled slot inlets, building space, fan safety guard, fan housing, louvered shutters, and exterior weather hood all contribute to the system's resistance to airflow.

Typically, the system resistance is about 100 to 125 Pa. For this reason, ventilation fans are typically selected to deliver the desired airflow at 125 Pa. For any ventilation system, the resistance to airflow or static pressure increases at higher airflow rates. (Note: 100 Pa = 1 cm of water)

When purchasing fans for livestock buildings a fan performance chart has to be obtained from the manufacturer. The performance chart is needed to determine the airflow capacity in m^3/min at various static pressures.

5.2. Fan efficiency

(Jacobson and Chastain, 1994 on internet)

5.2.1. Fan Capacity

The amount of air a fan moves depends on the diameter of the blades, shape of the blades, speed at which the blades turn (revolutions per minute or rpm), horsepower (hp) of the motor, design of the shroud, and other attachments such as louvers. These combined factors establish the air-moving capacity of a fan. Fan capacity is measured in terms volume of air moved per minute (m^3/min). (Note: 1 hp = 745.7 watts or 0.746 kW)

5.2.2. Static Pressure

When a fan operates, it creates a static pressure difference between the inside and outside of a building. This air pressure difference can be measured with a manometer in Pa.

Depending on the type of ventilation system, there may be a negative, positive, or neutral pressure inside the house compared with the outside. The negative pressure system, where air is exhausted with fans and enters the barn through inlets, is by far the most common system. A positive pressure system, used with certain types of furnaces, is used to supply both airflow and heat in a livestock unit. A neutral pressure system exhausts from and blows air into the house simultaneously, creating an approximate zero pressure difference. Most of the commercial heat exchangers marketed today for use in animal buildings use a neutral pressure system.

5.2.3. Fan Ratings

The air moving capacity of a fan (m^3/min) depends on the size of the operating static pressure difference. Table 29.7 shows the variation of m^3/min values with static pressure and the ratings for a group of commercial fans. The maximum m^3/min value occurs at zero static pressure, or "free air." As static pressure increases, m^3/min values are reduced until a point is reached where there is no airflow.

The common recommendation for poultry sheds is 100 – 120 Pa static pressure. This is because most systems operate in the range of 30 to 100 Pa static pressure. So, by choosing pressures slightly above the operating range, sufficient airflow is assured.

5.2.4. Factors Affecting Fan Capacity

Table 29.7 shows fan capacity differences among similar diameter blade fans due to such variations as motor horsepower and rpm. Louvers or shutters (used to close the fan opening when not in operation) can also restrict airflow. Some fan manufacturers rate fans both with and without shutters to indicate the effect on fan performance. Table 29.8 compares m^3/min values in both instances for a small capacity fan.

Fan ratings also depend on motor speed. Table 29.9 shows that as rpm values are reduced, so is the air delivery (m^3/min) of variable speed fans. At low speeds there is concern whether or not variable speed fans can maintain sufficient static pressures to provide reliable air exchange against any wind pressure present.

Table 29.7 Sample performance data for exhaust fans

Diameter (cm)	Speed (rpm)	Motor Size (W) 1hp = 0.746W	Airflow in m3/min at static pressure (Pa)			
			0	100	125	250
20	1,650	0.015	11.36	8.97	8.21	–
20	3,500	0.050	16.30	14.80	14.46	11.79
25	1,550	0.015	16.87	12.98	11.73	–
25	3,416	0.124	35.78	34.65	34.34	32.38
30	1,600	0.062	33.74	30.47	29.39	23.49
30	1,741	0.186	47.71	43.17	41.24	–
35	1,752	0.249	74.12	67.88	66.14	56.80
40	1,140	0.062	47.57	40.90	39.02	–
40	1,670	0.186	96.84	84.35	81.05	36.92
40	1,725	0.249	71.97	67.93	66.83	60.83
45	1,140	0.124	76.28	69.86	68.02	–
45	1,648	0.249	127.52	116.44	113.68	95.42
45	1,725	0.466	115.45	111.33	110.19	104.57
52.5	1,140	0.186	108.26	102.21	100.54	–
52.5	1,725	0.560	139.56	135.47	134.62	128.08
60	855	0.249	133.22	122.40	118.71	–
60	1,071	0.249	186.30	161.31	154.50	104.51
60	1,139	0.373	198.52	179.49	174.46	143.99
60	1,140	0.653	177.61	170.12	168.12	155.35
75	855	0.746	287.55	275.48	271.93	245.38
90	460	0.373	303.88	258.44	222.94	82.36
90	505	0.373	269.80	232.88	219.39	–
90	635	0.373	286.84	252.76	241.63	–
90	849	0.373	329.44	275.48	258.92	156.20
90	851	0.373	289.68	252.76	242.34	–
90	570	0.466	300.93	271.50	261.85	–
105	490	0.746	443.89	406.83	397.46	–
120	363	0.746	559.48	474.28	451.33	255.60
120	385	0.746	579.36	502.68	468.12	–
120	495	0.746	548.12	494.16	475.93	–

Table 29.8 Fan capacity with and without louvers for a 18.75 cm diameter blade fan, 50 W motor turning at 3400 rpm (1 hp = 0.746W)

Shutters	Air delivery in m3/min at indicated static pressures (Pa)						
	0	50	100	125	150	200	250
No	16.70	16.16	15.51	15.14	14.71	13.66	8.89
Yes	8.12	7.33	6.70	6.70	7.64	8.55	5.74

Table 29.9 Fan capacity of 35 cm diameter blade with 93 W motor turning at the four indicated rpm (1 hp = 746 W)

	Air delivery in m3/min at indicated static pressures (Pa)						
Rpm	0	50	100	125	150	200	250
1675	61.68	59.98	57.60	56.46	54.87	52.26	42.03
1355	50.41	46.06	41.83	39.33	36.86	28.54	16.61
855	31.52	23.17	7.75	—	—	—	—
586	13.69	2.36	—	—	—	—	—

More reliable airflow rates are provided by single speed fans since they have better pressure characteristics. This is important to remember when selecting continuous-running fans (providing the minimum ventilation rate). This is especially important if the house has manure storage below a slatted floor and the associated "pit" fan(s).

5.2.5. Energy Efficiencies

Another characteristic that is becoming important when selecting fans for an animal ventilation system is energy efficiency. This is expressed as airflow per unit of input energy, or m³/min /W. Tests on 90 cm diameter blade fans showed a wide variation in both fan performance (181.76 to 369.20 m³/min) and energy efficiencies (8.3 to 18.6 m³/min/W) at 100 Pa static pressure.

Choosing fans with high energy efficiencies is most important for the mild and warm-weather fans since more air is being moved when these fans are operating. Small continuous-running fans need to have good performance ratings, since their most important job is to guarantee that the air exchange will occur under almost all conditions to insure good air quality in the facility.

Energy efficient fans are usually more expensive. However, energy-efficient fans can save the extra cost of the fan in about 2 to 3 years. Good quality fans can last more than ten years, the money saved is more than the cost of the fan.

If fan efficiency ratings are not available, the following guidelines can be helpful:

1. For a given airflow and static pressure, a large diameter fan is more energy-efficient than a small diameter fan.
2. For a given ventilating rate, one large diameter fan is more efficient than several smaller diameter fans.

3. When two fans have the same diameter, the fan with the lower motor current input rating is usually more efficient.
4. If two fans deliver the same airflow at the same static pressure, the fan with the slower speed is usually quieter and more efficient.

As an example, let's take the most energy efficient 90 cm diameter blade fan of the fans and compare it to the least efficient of that group. At 100 Pa static pressure the most efficient fan has a 0.53 m^3/min/W energy efficiency value, while the least efficient fan only has an efficiency of 0.24 m^3/min/W (about 45% of the most efficient one). Using the m^3/min ratings and efficiencies at 100 Pa static pressure and assuming a rate of Rs 2.50 per kWh (electric cost) and 120 days (24 hrs/day) of operation for a warm-weather fan (utilizing 60 W/hr, 0.062 W), one would save roughly Rs 522 per year (Rs 954 *Vs.* Rs 432) by using the more energy efficient fan for producing the same air volume. Typically, the more energy-efficient fan would cost an additional Rs 400 to Rs 600, but the payback with the above assumptions would be only 1 or 2 years. If one assumes a 10-year fan-life, that would result in a Rs 5220 gross savings over the life of the fan!

5.2.6. Maintenance of Fan Systems

Fans operating in poultry houses are exposed to vast amounts of dust and moisture, and accumulate dirt on blades, louvers, and shrouds. Dirt on fan blades has little or no effect on fan performance, but dirt on louvers and guards can reduce airflow by as much as 40 %. Fan louvers and guards should be cleaned regularly and lubricated (using graphite to prevent dirt accumulation) to prevent large airflow reductions in fans. Louvers should be removed where fans are running continuously to avoid the potential for restricting airflow, but guards should remain in place to prevent personal or animal injury.

Table 29.10 Performance and efficiencies of most and least efficient of the 90 cm blade fans

Fan	40 Pa static pressure		100 Pa static pressure	
	m3/min	m3/min/W	m3/min	m3/min /W
Most efficient	314.10	0.62	276.90	0.53
Least efficient	312.40	0.27	281.16	0.24

Not only is the environment in a poultry house dirty, but there are corrosive gases, especially in facilities with manure pits inside the facility. Continuously running fans or fans that exhaust from pit areas need a non-corrosive housing, such as fiberglass, to prevent deterioration. Whether the fan is controlled by a simple switch or a microprocessor, those controls should also be protected from the environment by a watertight and dust-tight electrical box to eliminate electrical contacts and switches from being exposed to corrosive gases.

5.2.7. Size of Openings

Although fans are a necessary part of a ventilation system, they are not the only thing to consider. Also needed is some way to bring air into a negative pressure system (inlets), or to allow its escape in a positive pressure system (outlets). In both instances, a standard rule for sizing air openings is to provide 1 m^2 of area for every 245 m^3/min of fan capacity.

5.3. Fan performance

(Arnold and Veenhuizen, 1994; ohioline.ag.ohio-state.edu)

Wind pressure blowing against a building also affects fan performance.

Table 29.11 Wind effect on static pressure *

Wind speed, km/h	8	16	24	32	40
Static pressure, Pa	20	50	100	200	280
* Fan exhausts against the wind with no hood or wind shielding					

If the wind is blowing in the same direction as the fan airflow, then a negative pressure or suction effect can reduce the pressure across the fan increasing airflow capacity. However, wind blowing against the fan air increases the static pressure the fan has to operate against (Table 29.11). For example, when a 40 km/h wind is blowing into the fan, the pressure from the wind is 280 Pa. This is added to the operating pressure of the building of about 15 Pa static pressure to give a total static pressure of 340 Pa. Fan capacity will be greatly reduced by the wind pressure. To reduce the effect wind has on fan performance, fans have to be located on the leeward side or an adequate weather hood has to be provided to reduce wind effects.

A Pin wheel anemometer is moved uniformly over the fan opening for 1 min and the result is multiplied by area of the opening which gives the air volume delivered by the fan; this value should be similar to the manufacturer's ratings (Ernst, 1994).

5.3.1. Fan Maintenance

Fans operating in livestock housing environments are exposed to large amounts of dust and moisture. Maintenance practices can have a significant effect on fan performance. Dust on fan blades has only a slight effect on fan performance, but dust accumulations on shutters and guards could reduce air flow by as much as 40%.

Electric currents for fans are shut off before checking and servicing. Fuses are checked for a tight fit and wiring for deterioration and good connections.

Fan blades, shutters, and guards have to be regularly cleaned and fan shutters lubricated with graphite to reduce dust accumulations. A pressure washer can be used to clean the fan, housing and hood thoroughly. Be sure the fan motor has a totally enclosed housing so water and dirt cannot get into the motor windings. Otherwise, take the motor off and clean it separately. Most fan and motor bearings are sealed and don't require oil. Oil should be used sparingly, when necessary, as too much oil attracts dust and soaks into motor windings.

After cleaning and servicing the fan, place a sticker on the fan housing with the current service date. Locate the sticker and check regularly as a reminder to perform regular maintenance. If the date cannot be read or the sticker can't be seen, it is definitely time to clean the fan.

Livestock environments also include corrosive gases, especially in buildings with manure storage. The combination of high moisture, gas concentrations, and dust can create a corrosive environment. Fan blades and fan housings should be made of non- corrosive materials such as high density plastic and fiberglass. Protect controls from the effects of the environment with air-tight enclosures or by locating them in a separate room and utilizing remote sensors.

6. Triangle of ventilation

(Alchalabi, 2001b)

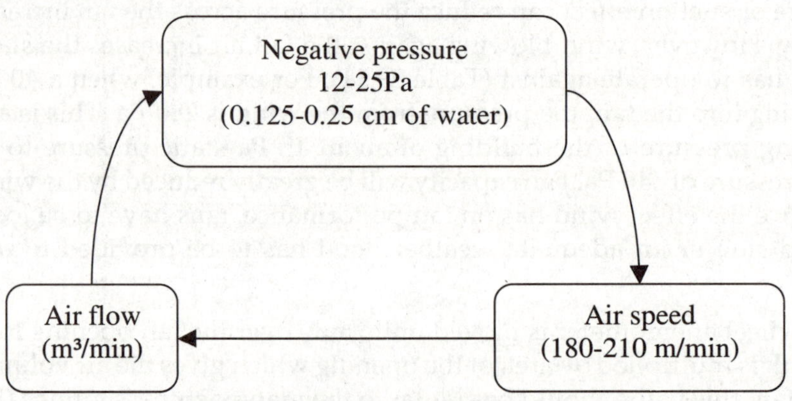

Fig 29.2 Triangle of ventilation

There are three steps to follow to reach an adequate ventilation scheme in poultry house and these steps are repeated so the equilibrium is obtained between the elements of the triangle namely, air flow, negative pressure and air velocity (speed). The number, age and weight of birds determine air flow requirement, area of inlets determine the static pressure and proper balance between the two is cardinal for proper distribution of incoming air.

The first step is to calculate the required airflow and activate the right number of fans to achieve it. Then, the static pressure has to be monitored with the help of a static pressure meter so as to read between 50 and 100 Pa and inlet flaps have to be adjusted suitably. Finally, on the same lines, air speed at the inlet should be checked to read 180 to 210 m/min.

If there are "dead" spots in the house, they can be overcome with internal circulating fans which can be added to reduce heating costs during winter.

6.1. Heat and moisture production - turkeys

During cold weather, the ventilation and heating system in chicken and turkey houses needs special attention. During winter months, the main function of a ventilation system is to provide a minimum (continuous) ventilation rate (Table 29.6) for removal of moisture and supply of fresh air. This minimum rate is required in order to lose as little heat as possible from the house and still maintain optimum litter and air quality conditions. During warm weather the ventilation system, in addition to moisture removal and fresh air supply, has to remove body heat in order to prevent excessive build up of temperature (Oderkirk, 2001a).

6.2. Indoor set point temperature

Table 29.12 Typical indoor set point temperatures and ventilating rates for turkeys

House	Bird age (weeks)	Tom weight (kg/bird)	Set point temperature (°C)	Ventilating rate m3/min/1000 birds
Brooder	1	0.14	26.7	1.14
	2	0.3 to 0.4	23.9	2.27 to 3.41
	3	0.5 to 0.8	21.1	4.54 to 7.10
	4 to 6	1.0 to 2.8	18.3	11.36 to 31.24
			21.1	19.88 to 59.64
			18.3	45.44 to 139.16
Grower	7 to 9	2.8 to 5.6	15.6	34.08 to 56.80
			18.3	56.80 to 122.12
			21.1	139.16 to 278.32
	10 to 12	5.0 to 9.0	15.6	51.12 to 73.84
			18.3	119.28 to 198.80
			21.1	NAT
	13 to 15	8.2 to 12.7	15.6	68.16 to 93.72
			18.3	187.44 to 278.32
			21.1	NAT
	16 to 18	11.8 to 16.3	15.6	79.52 to 99.40
			18.3	258.44 to 369.20
			21.1	NAT
	19 to 21	15.0 to 19.5	15.6	85.20 to 102.24
			18.3	312.40 to 426.00
			21.1	NAT
	22 to 24	17.7 to 22.2	15.6	90.88 to 110.76
			18.3	369.20 to 482.80
			21.1	NAT
Breeder		15.9	15.6	51.12
			18.3	150.52
			21.1	497.00

6.3. P-Band

P - Band is short for proportional band. This band is a range of temperatures at which a fan operates from minimum to maximum revolutions. This band is expressed in degrees Celsius (°C). In most modern type ventilation regulators such as electronic and/or computer-operated ventilation systems, a P - band of 4 °C is built in. During the winter period, this P - band may even be set higher than 4°C because it is a waste of energy to immediately exhaust extra heat in a barn during the winter period. The solution is a variable P - band which by high outside temperatures can be shortened and by low outside temperatures can be lengthened. This way, wide variations in house temperatures during warm days and cold nights will be minimized. Most modern regulators and computer ventilation systems have an adjustable P - band. In most ventilation computer systems, the adjustment of the P - band takes place automatically on the outside temperature (Oderkirk, 2001a).

Table 29.13 Heat and moisture production estimates for turkeys

Bird age (Week)	Tom weight (kg)	Total heat kJ/d•bird	Sensible heat kJ/d•bird	Moisture, kg / d • bird Respired	Fecal
1	0.14	125	99	0.011	0.006
2	0.41	338	270	0.029	0.019
3	0.77	577	461	0.050	0.035
4	1.32	890	712	0.077	0.112
5	2.00	1223	979	0.106	0.145
6	2.77	1534	1228	0.133	0.166
7	3.63	1821	1457	0.158	0.194
8	4.58	2079	1664	0.180	0.215
9	5.58	2291	1833	0.198	0.251
10	6.67	2484	1983	0.214	0.285
11	7.85	2629	2110	0.228	0.319
12	9.03	2750	2197	0.238	0.348
13	10.22	3112	2484	0.269	0.384
14	11.49	3497	2793	0.302	0.421
15	12.71	3859	3092	0.334	0.405
16	13.94	4245	3389	0.367	0.376
17	15.16	4607	3688	0.399	0.369
18	16.34	4969	3975	0.430	0.352
19	17.48	5306	4252	0.460	0.329
20	18.57	5644	4515	0.488	0.298
21	19.57	5958	4759	0.515	0.289
22	20.52	6247	4990	0.540	0.290
23	21.38	6512	5200	0.563	0.289
24	22.16	6730	5388	0.583	0.285

6.4. Minimum ventilation during cold weather

Table 29.14 Minimum ventilation requirement of straight-run chicken during cold weather

Age	Body weight (g)	m^3/min/1000 birds
Day-old	40	0.36
1 week	130	1.17
2 weeks	280	2.52
3 weeks	580	5.22
4 weeks	890	8.01
5 weeks	1,220	10.98
6 weeks	1,590	14.31
7 weeks	1,990	17.91
8 weeks	2,400	21.60
9 weeks	2,840	25.56
10 weeks	3,280	29.52

Source : Oderkirk, 2001a

6.5. Interlocking of Ventilation and Heating Systems

Interlocking these two systems means that the switching of the heating system is connected, with a range, to the set ventilation temperature.

The distance between these two temperatures is called the "neutral zone" and is expressed in degrees Celsius, usually 2 °C. The purpose of the neutral zone is to prevent the heating system being on while the fan is operating above the minimum required ventilation rate and thus ventilates more than is required for moisture removal and fresh air supply. This "competition" between heating and ventilation system results in energy wastage. Modern type regulators and computer ventilation systems have a built-in neutral zone (Oderkirk, 2001a).

6.6. Comfort Zone

An animal has a certain temperature range in which it does not use any extra feed to compensate for a lower temperature and also it does not have to use extra energy to expel heat. This temperature zone is called the comfort zone. An animal feels comfortable in this zone and the feed required for maintenance is minimal. Only at the lower end of this comfort zone is it necessary that the heating system is on. Also, the fan needs only to increase in ventilation rate at the upper end of this zone. In most regulators, this heating neutral zone has been set at 1°C. In most computer ventilation systems this heating neutral zone is built in with a P - band. This allows the heating system to operate at 100% capacity at the low end of the heating neutral zone and decreases its operating capacity gradually to zero as the temperature reaches the upper end of the "neutral" heating zone (Oderkirk, 2001a).

6.7. Intake Control

Intakes in cross ventilated barns can be controlled by temperature. The intake opening temperature is set 0.5 to 1°C above the set barn temperature. As the temperature increases the fan will increase its exhaust capacity first. The intake will open up as the temperature further increases. The opening of the intake is always a step behind the increase in exhaust capacity which results in the desired negative pressure in a house (Oderkirk, 2001a).

These ventilation techniques are difficult to practice with conventional thermostat-operated ventilation systems. However, electronic regulators and/or computer-operated ventilation systems can easily apply and interconnect the various ventilation components (Oderkirk, 2001a).

7. Air inlets

(Oderkirk, 2001b)

7.1. Functions of an Air Inlet

The air inlet is the most important part of every ventilation system. Without proper air inlets no ventilation system can function satisfactorily. The air inlet is responsible for providing good air movement throughout the house - not the fan. The fan is only the air pump.

The functions of an air inlet are

1. provide fresh air throughout the barn
2. to maintain a fast inlet air velocity so that good air mixing occurs
3. adequate air circulation within the house is provided

If these functions are not achieved, then poor air distribution, uneven temperatures and drafts are the result

7.2. Sizing

When air flows through any opening, the cross-sectional area of the issuing jet is reduced to 60%-80% of the total free-area of the opening. This phenomenon, known as the vena-contracta effect, increases the velocity of the air emerging from the opening and should be taken into account when sizing the air inlets. A common rule of thumb sizesair inlets at 0.2 m² of inlet openingper 1000 L/s of air exhausted. However, when the vena-contracta effect is considered these inlets should be sized on the basis of 0.33 m² per 1000 L/s. Expressed another way one can calculate the size of the air inlet opening by substituting a velocity of 3 m/s in the formula:

$A = Q/1000\ V$ where

A = cross-sectional area of inlet in square meters

Q = fan exhaust capacity in L/s

V = inlet velocity in metres per second

This design prevents the air intake opening to the outside of the house from developing much static pressure ahead of the inlet control baffle. Too little air intake opening can destroy all attempts to control the air flow within the barn space and of course can overload the fans due to the restriction.

7.3. Location

Since this fresh air must be distributed throughout the building it is necessary to have either a continuous slot-type inlet or point source inlets with an effective distribution mechanism such as a nozzle assembly on an intake fan or an air tube distribution system. Introducing air in blocks (such as windows) tends to cause drafts at that location and dead spots where there is no inlet.

Regardless of the type of inlet, location is critical and obstructions or restrictions to the air flow before and after it enters the intake must be avoided. Restrictions before can cause a pressure drop on the intake air even before it reaches the inlet baffle. Obstructions after the inlet, such as electrical conducts, a corrugated steel ceiling, beams and joists will deflect air down into the pen area and possibly cause drafts as well as destroy good air circulation within the barn.

7.4. Direction of Air Flow

It is desirable to inlet air across the ceiling to avoid cold drafts on the birds. In any barn containing sensitive livestock, this incoming air should be warmed up by the warm room air before it reaches the area where the birds are located. Also, the air velocity at bird level should be within acceptable limits to avoid drafty conditions.

7.5. Air Inlet Adjustment

The air inlet must be adjustable in order that a fast inlet air velocity is maintained. When cold air enters a warm barn, it is heavier than surrounding warm air and sinks rapidly to the floor causing cold drafts. To solve this problem, the cold air must be directed along the ceiling and jetted forcefully (4-5 m/s) into the room to promote rapid air mixing and prevent sinking. Adjustment is also necessary to provide a good air pattern within the building. Easy adjustment is a very important consideration. For typical cross-flow systems, the inlet must be adjusted each time the exhaust rate changes. Even the best operator will not perform this adjustment should it be difficult or cumbersome. A hand winch control is often used. However, there are a number of automatic static pressure controllers on the market which are designed to help the farmer better manage his ventilation air inlets. These controls help to ensure a constant environment and eliminate another chore.

7.6. Slot-type air inlet with baffle board

This is probably the most common type of air inlet used today. It is found in most cross-flow (negative pressure) ventilation systems either as a side or center air inlet. Firstly, be sure you have sufficient air inlet opening to handle the full range of fan capacity. Provide at least $0.33m^2$ of inlet opening per m^3/s of air (1000 L/s) exhausted. Note that $0.4m^2/m^3/s$ (1000 L/s) is even better.

An air inlet along one side wall is generally sufficient. Wide barns will require an air inlet along both side walls. Be sure to choose an inlet that is compatible with the pen layout. The air inlet should be as continuous as possible. Therefore, you must remember to keep wall fans at least 0.5 meters below the ceiling so that they will not interfere with a side air inlet running the entire side wall length. A 2.75 m or higher side wall is desirable for this arrangement. If two side air inlets are made, they should be of equal length to guarantee a balanced air flow.

In cross-flow systems an inlet air speed of 4-5m/s is desired to promote good mixing and prevent heavier, cooler air from dropping directly on the animal. This air speed is maintained by adjusting the baffle board such that the exhaust fans create a slight negative pressure within the barn. Since a constant incoming air speed is desired, a steady static pressure must be maintained regardless of the number of fans operating. This static pressure should be kept in the range of 10-25 Pa (0.10-0.25 cm of water) static pressure. During cold weather, keep the static pressure at the high end of this range and lower during warm weather. This is due to the fact that cold air is heavier and because of the small amount required we do not gain the mass flow effect as we do in the summer.

Each room must be set individually to obtain the correct air circulation pattern for that room. A smoke generator is beneficial in determining the correct pattern. A simple static pressure gauge will aid you in maintaining the correct air inlet opening or you can install an automatic static pressure controller which controls the baffle board automatically.

8. Litter moisture

(Czarick and Lacy, 1997)

As a general rule, for every kg of feed a bird eats it will drink almost 2 kg of water. For instance 24,000 four-day-old birds will eat approximately a 454 kg of feed a day but will drink close to one ton of water a day. As birds get older, feed consumption increases dramatically and so does the amount of water they drink. For instance, 24,000 seven-week-old birds will eat over 3½ tonnes of feed a day and drink over 8 tons of water a day. By the end of a seven-week grow-out, a house with 24,000 birds has consumed over 100 tons of feed and well over 200 tons of water!

Where does the water go? Actually, less than a third of this water is retained by the birds and the rest is either exhaled into the air in the form of water vapor or, to a larger extent, deposited into the litter. When a grower operates fans, the moisture-laden air in the house is replaced with drier outside air, thereby removing water from the house. The drier air that the fans bring into the house will not only pick up water that the birds are placing into the air but also remove water from the litter, keeping it drier. Of course, if a producer does not operate exhaust fans enough, water will build up in the air as well as litter leading to house sweating and litter caking.

When it comes to removing moisture from a house, it is important to know that the amount of water that air can hold varies dramatically depending on temperature. For instance, 28.4 m^3 of 4.4°C air can only hold about 225 g, but 2.84 m^3 of 29.4°C air can hold approximately 900 g of water. At 63.9°C, 2.84 m^3 of air can hold over 3.6 L! As a general rule, the moisture-holding ability of air doubles for every 11.2°C rise in temperature.

Monitoring relative humidity is a good way to make sure that the air in the house can remove moisture from the litter. Ideally, the relative humidity should be between 55 and 65%. If the relative humidity is below 55% the litter dries out too much. If relative humidity is above 65% for a prolonged period, litter caking and house sweating can occur.

That warmer air holds more moisture than colder air is important to keep in mind when while trying to keep the litter dry. For instance, let's say with older birds and house temperature was 21.1°C and the relative humidity of 65%, litter started to get a little damp and timer fan thermostats are turned down to 18.3°C to bring in more air. Since the temperature of the air is lower, the moisture-holding ability of the air would also decrease, resulting in the relative humidity increasing to approximately 78%. This will make it more difficult to remove water from the litter.

On the other hand, if the house temperature is increased by just 1.7°C, and left the timer fan settings the same, the relative humidity would decrease to approximately 58%, making it significantly easier to draw water out of the litter.

Another option would have been to leave the house temperature the same and to turn up the timer fan settings. This also will work provided that the house temperature does not fall. In some instances this would mean adding heat to the house in order to maintain the desired house temperature.

The key to keeping a house dry is quality of air not quantity of air. Bringing in a lot of cold air and letting the house temperature fall does not usually do as good of a job as bringing in a moderate amount of air and warming it up.

It is important to note that using warm air to dry out litter does not always mean increased heating costs. Take advantage of the fact that as outside temperatures increase during the day that the moisture-holding ability of the air increases. So during the day when the sun is out and the air temperature is warmer, timer fan settings are increased to take advantage of the relatively drier air.

9. Suggested ventilation rates for chicken and turkeys

Table 29.15 Suggested ventilation rate for chicken and turkeys

Animal	Weight (kg)	Unit	m3/min/unit		
			Cold weather	Mild weather	Hot weather
Turkeys					
0 to 1 wk		Bird	0.001	0.020	0.028
1 to 2 wk		Bird	0.002	0.020	0.028
2 to 3 wk		Bird	0.005	0.020	0.057
4 to 6 wk		Bird	0.011 to 0.018	0.020 to 0.034	0.099
7 to 9 wk	2.27 to 3.86	kg	0.013	0.022	0.051
10 to 12 wk	4.54 to 6.81	kg	0.009	0.022	0.051
13 to 15 wk	2.26 to 10.0	kg	0.009	0.022	0.051
16 to 18 wk	10.44 to 12.71	kg	0.007	0.022	0.051
19 to 24 wk	13.62 to 18.16	kg	0.004	0.022	0.051
Breeders	15.89	kg	0.002	0.009	0.031
Chickens					
Broilers					
0 to 7 days		Bird	0.001	0.006	0.011
over 7 days		kg	0.006	0.031	0.062
Layers		kg	0.006	0.031	0.062 to 0.094

Source : Jacobson and Janni, 2003

10. Hatchery ventilation

(North, 1972)

Uniformity of temperature and humidity throughout the hatchery is most important and therefore, forced air should be used to ventilate hatchery units. Air should not be allowed to circulate from one room to another; instead, each room must be ventilated as a separate unit. Incoming air should be filtered, pre-heated during winter and cooled during summer. A duct system for circulating the air uniformly in each room is the ideal with rheostats to control air flow on all ventilating fans.

Since forced air is used into the hatchery rooms, a slight pressure is built up within the hatchery. Therefore, there must be some type of exhaust to make the system work efficiently. Exhaust fans should move slightly more (approximately 10%) air than the intake fans, creating a negative pressure in the rooms to ensure uniform distribution of air in the hatchery and also to avoid condensation of moisture during cold climate.

10.1. Cooling the hatchery

During periods of hot weather, hatchery rooms have to be cooled; especially the chick room where heat build-up is expected followed by hatching room. Setters operate better if the room in which they are located has a temperature of 21.1 to 22.2°C.

The most economical method of reducing the temperature in hatcheries is with evaporative coolers in which moisture is provided by a moisture-laden pad and air is sucked through the pad to cause absorption/ evaporation of moisture resulting in cooling of air. The cooled air is then circulated into the hatchery through the ducts as discussed above.

Table 29.16 Airflow through hatchery rooms (m^3/min)

Outside temperature (°C)	Egg handling room		Incubating room		Hatching room		Chick-holding room	
	Per 100 m3		Per 1000 eggs		Per 1000 eggs		Per 1000 chicks	
	Min	Opt	Min	Opt	Min	Opt	Min	Opt
- 1.1	11.0	12.4	0.6	0.8	1.5	2.0	2.8	3.8
15.6	16.6	18.7	0.9	1.1	2.3	3.0	4.1	5.7
32.2	19.1	25.1	1.3	1.7	3.4	4.5	6.4	8.5
Optimum air change/ hr at 36.7°C	15		20		30		60	

Table 29.17 Capacities of exhaust fans

Motor		Fan blades		Air capacity at 0 static pressure, (m3/ min)
W	rpm	Number	Diameter, cm	
0.093	1725	4	30.4	46.7
0.186	1725	4	45.7	82.1
0.186	1140	4	45.7	50.9
0.249	1140	5	45.7	101.9
0.373	1140	5	60.1	150.0
0.249	630	4	60.1	175.5
0.249	473	4	76.2	178.3
0.373	412	4	91.4	339.6

Caution : Many fans do not move their designated or rated amount of air. Loose belts, dirty blades, improper pitch of the blades and high static pressure in the building contribute to inadequacies. Note: 1 hp = 0.746 W

However, the amount of cooling the evaporative cooler depends on the temperature and relative humidity of the incoming air; hence, its efficiency will be restricted if temperature is low or relative humidity is high. Standard tables are available to calculate the cooling effect when the outside temperature and percent humidity are known.

10.2. Gaseous environment for eggs under incubation

(Deeming, 2005)

Carbon dioxide is a useful gas in incubation. It is relatively cheap and easy to measure and it is the inverse of oxygen consumption, which is difficult and expensive to measure in a practical environment.

The oxygen consumption and carbon dioxide production of embryos change with developmental time and this means that determining carbon dioxide could be direct measure of the metabolism of the embryo inside the setter and hatcher. Such information could also be used to control the amount of fresh air entering the incubating machinery to optimize the oxygen supply for the embryos. Therefore, carbon dioxide control is now almost standard on control panels of modern incubators; carbon dioxide desired levels can be set so that the machine will open up the damper flap too let in more fresh air in this limit be reached or exceeded.

The industry standard for the amount of fresh air entering an incubator is 3.398 m^3/hr/1000 chicken eggs; in other words, 3.398 l/hr/chicken egg. Considering that fresh air has 21% oxygen, each egg will be getting about 714 cc of oxygen/hr whereas the actual requirement on 18th day of incubation is only about of 630 cc! That means, more than 20 times the requirement of oxygen is being provided.

In view of the above, there is a possibility of excessive amounts of cool air entering into the setter and it might affect the pattern of heating and cooling; the sensors may become numb to the actual changes in the temperature because too much cool air is entering into the setter. In addition, egg temperature might increase if the thermometer does not respond and/or respond slowly or if the thermometer responds too quickly, it leads to cooling of eggs; the former being most common. Hence, the ultimate result is increased ventilation leading to loss of humidity followed by more spraying of water, cooling of air and over-influence on the temperature sensor. In any case, hatchability and/or chick quality is affected adversely. This can be overcome by use of an infra-red egg thermometer.

Therefore, there is still scope for improving hatchery performance by use of more sophisticated equipment in order to have a sound understanding of what the incubation environment should be and how it is maintained.

11. Ventilation in slaughter house

(Heber *et al.*, 1995)

The shackling, killing, scalding, and picking areas of the poultry slaughtering and processing plants emit airborne microorganisms, moisture, and dust. These contaminants are unwanted in the processing and packing areas of the plant because they can affect product quality and safety. They also pose a potential threat to the health and well-being of the workers in the plant.

11.1. Typical Ventilation Systems

Poultry slaughtering plants are typically ventilated with negative-pressure systems. Most of the air-moving capacity is provided by large roof fans above the scalding and picking equipment. The balance of the air is moved with local exhaust fans in other locations. This ventilation strategy is based on the theory that fresh air will flow one-way from relatively clean meat cutting and packaging operations to locations with bioaerosol (bacterial pathogens and spoilage microorganisms) emissions and finally to the exhaust fans. However, they do not always achieve this desired airflow pattern. The following aspects of the ventilation system are critical for achieving the desired airflow pattern.

11.1.1. Fresh Air Intakes

The distribution of air in any ventilated room depends on the size and location of ventilation intakes. In a poultry plant, many rooms are ventilated with air entering and leaving wall openings. These wall openings consist of shackle line openings and doorways. Shackle line openings are for transferring product from one processing operation to another. Doorways are for the convenience of worker and vehicle traffic. Room ventilation is apparently not considered when these wall openings are located. However, their sizes and locations have a significant impact on room air distribution.

With a negative pressure system, air leaks into the room around doorways and other unplanned openings. Many doors in poultry plants consist of double doors, large garage doors, doors to walk-in coolers, and shipping/receiving docks. These large doors can have crack areas with up to 14.2 m^3/min of air flowing through them. Cold air leaking into a room due to negative pressure makes occupants uncomfortable and can cause considerable condensation and frost as warm moist air mixes with weak cold air jets and contacts cold surfaces.

The plant's largest exhaust fans are located in relatively small picking and scalding rooms and move about 852 to 1420 m^3/min of air. These fans create a large negative static pressure that draws in air from all openings to the room. Much of the air flows inward from the line openings to the outside shackling area, typically 30 to 90% of the exhaust airflow. The rest of the air comes in from the plant. The overall room air balance is influenced by the size of the openings to the room.

More air could be drawn through the plant if the openings into the shackling area were reduced and the openings to the evisceration room were enlarged.

11.1.1.1. Arbitrary Modifications

The original ventilation system for a poultry processing plant is sometimes modified as rooms are added for more processing capacity. Additionally, exhaust fans are sometimes added in attempts to eliminate condensation. These exhaust fans can move air opposing the main exhaust fans in the picking and scalding rooms. Air movement from clean areas to bioaerosol emission sources can be reversed if enough fans are added.

11.1.1.2. Hypothetical Example

A plant has three fans in the scalding room with a total capacity of 1420 m^3/min. However, 852 m^3/min is drawn through large line openings from the outdoor shackling area which is immediately adjacent to the scalding room. The remaining 568 m^3/min is drawn through line openings from the eviscerating room. Therefore, the original design causes 852 m^3/min to flow through the plant from the processing rooms to the eviscerating area and then to the scalding and picking room.

A year later, the plant adds three, 113.6 m^3/min fans in the packaging room to eliminate moisture, and prevent condensation occurring on the ceiling. Two years later, new vacuum sources requiring 71 m^3/min are added to the processing areas, a 99.4 m^3/min exhaust system is added to a cooler room to remove carbon dioxide.

After these modifications, 511.2 m^3/min are exhausted at other locations besides the picking and scalding rooms. A major portion of this air comes from the evisceration room thus producing some reverse flow in the plant and causing an even greater portion of the air to flow into the picking and scalding rooms from the shackling area.

11.1.2. Fan Maintenance

The fans and intakes in poultry meat plants are above the roof line and are easily ignored because most of the attention is given to the operation on the floor ("out of sight, out of mind"). However, fan capacity can be reduced by up to 50% when the guards and back-draft shutters are laden with dust and feather material. Intakes can become completely plugged by particulate emissions from nearby exhaust fans. Ventilation equipment maintenance should be a routine task of the plant maintenance crew.

11.1.3. Multi-Purpose Rooms

Air distribution should be taken into consideration when laying out processes within the room. Sometimes, "clean operations" for final product such as cutup and packaging are conducted in the same room as "dirty operations" such as evisceration. The "dirty operations" closer to the raw receiving area emit bioaerosols that can move to and deposit on to clean product. Thus, room-air distribution is very important in preventing unnecessary food contamination.

11.1.4. Fan Mismatches

A food manufacturing plant may have several types and sizes of fans exhausting from the same room or adjoining rooms. High-pressure power ventilators, tube-axial fans, centrifugal fans, and vacuum systems sometimes operate in parallel with low-pressure propeller fans. As a consequence, weaker exhaust fans operate at reduced efficiency and air flow. Also, relatively moist air condenses on the inside of fan housings as air flow decreases. Additionally, air flow from areas with contamination sources to clean areas such as meat packaging may be created due to reverse flow. This increases product contamination and decreases food safety.

In one case, about 142 m^3/min of air actually reversed through a large, belt-driven, exhaust fan running at slow speed in a feather washroom. This occurred because, in addition to strong exhaust fans in adjoining rooms, a furnace with a high speed centrifugal fan exhausted air from the same room and blew hot air into the adjoining rooms via ductwork. The negative static pressure created by the furnace fan was felt by the slow-moving fan as excessive back pressure. The airflow through the weak exhaust fan became positive again after the furnace was shut off. To correct this situation, weak fans should be replaced with stronger or higher pressure fans. Or the speed of the weak fan should be increased. In this particular case, the supply air for the furnace could have been taken from another room.

11.1.5. Effects of Wind Pressure

Wind blowing into a large opening such as open doors or windows can pressurize a room causing undesired flow of air. In one case, air flowed through door cracks and other openings from a compressor and refrigeration room into a meat processing room when an overhead door was opened to "cool the compressors,". The door was opened to an oncoming 21.6 kmph wind even though the compressor room had its own exhaust ventilation system. Upon shutting the overhead door, the pressure in the compressor room became negative compared to that of the processing room and air flowed from the processing room to the compressor room, as desired, to keep contaminants away from clean product.

In another case, an overhead door in the picking room facing an 8-mph wind was opened on a hot summer day. The resulting 511.2 m^3/min of air flowing inward through this door created high humidity and airborne microbial concentrations in the adjoining evisceration and processing room. The picking room was equipped with large exhaust fans, but the wind pressurized the "dirty" room in such as a way as to induce airflow into the "clean" room through some of the line openings between the two rooms. Shutting the overhead garage door allowed the exhaust fans to create the negative pressure required to draw air through all openings from the evisceration and packing room.

Chapter **30**

Insulation

1. General

The main aim of insulation of a poultry house is to reduce structural heat flow which in consequence offers the following advantages (Wathes, 1981):

1. Desired house temperature can be readily achieved
2. Condensation on the interior surfaces eliminated
3. Concentrations of airborne pollutants can be lowered by exceeding minimum ventilation rates and still maintaining a satisfactory temperature lift.

While providing insulation to a livestock building, the following must be considered (Wathes, 1981):

1. Physiological needs of the animal
2. Cost of insulation
3. Thermal behavior of the building
4. Losses expected if insulation is not provided

Insulation of animal houses depends on physical properties of insulation material, energy and moisture balance of a wall or roof structure, the measurements of and design values for the thermal transmittances of a building structure and finally, the economic level of insulation (Wathes, 1981).

A well-insulated building shell is needed to successfully naturally ventilate a poultry house. Insulation helps prevent condensation on the building's inside surfaces, reduce heat loss in cold weather, and reduce solar heat gain in warm weather.

Thermal buoyancy is enhanced by reducing building heat loss through the building shell. Condensation occurs when the building's inside surface temperature dips below the indoor air's dew-point temperature. It can be prevented by providing sufficient insulation to maintain inside surface temperatures above the dew-point temperature. Insulation also reduces building heat loss, however, only 20% of the total heat (i.e., building and air exchange) is lost through the walls, ceiling, and perimeter in most poultry facilities in cold weather (Janni and Jacobson, 2003).

The majority of total heat is lost through the cold weather air exchange needed to control moisture and maintain acceptable air quality. The insulated building shell also reduces solar heat gain in the summer; especially insulation located on the underside of the roof (Janni and Jacobson, 2003).

The insulation of the vascular and fatty tissues can be modified. Under tropical climate, animals tend to have little subcutaneous fat. Further, breeding of farm animals for low fat content also has resulted in reduced fatty tissues. Fat as an insulator has an overall thermal conductivity about half that of muscular tissue. Vascular insulation arises from vasomotor control. Extremity such as the nose, ears and feet of livestock have strong vasoconstrictor regulation of blood flow; of these, feet is only relevant in case of poultry (Macfarlane, 1981).

2. Physical properties of insulation material

The physical properties of insulation material are thermal conductivity (k), bulk density, water vapor resistivity and loading characteristics; of these, the former two are dependent upon the moisture content of the material.

2.1. Thermal properties

The thermal conductivity of the material reflects its resistance to heat flow and is the flow per unit area per unit temperature gradient between the two faces of the material. Similarly, water vapor resistivity, describes a material's resistance to water vapor movement and hence, it is the reciprocal of vapor diffusivity which, in turn, is determined by vapor pressure gradient measured in Pa. Therefore, vapor resistivity is measured in Ns/kg/m (Wathes, 1981).

Heat transfer within a material occurs by conduction, convection and radiation and hence thermal conductivity has a positive relationship with all these three factors.

Bulk density influences thermal conductivity either by limiting (convection at high densities) or by facilitating (conduction) heat transfer processes (Table 30.1). For instance, convection in glass fiber ceases to exist when bulk density exceeds 12 kg/m^3. Conduction along the solid fibers or cells accounts for little of the actual conductivity whereas in gas, density has a prominent role to play; higher densities impede heat transfer. Radiative heat transfer also accounts for increase in thermal conductivity with thickness and the magnitude of increase depends on pore size (Wathes, 1981).

Interstitial condensation within insulants increases the thermal conductivity in addition to causing physical deterioration. Therefore, the rate at which materials absorb surface water also determines their suitability as an insulating material (Table 30.2). It is important to note that thermal conductivity of straw, wood shavings and dry and wet sawdust at compressed bulk densities of 43, 370, 150 and 1100 kg/m^3 was 0.12, 0.04, 0.11 and 0.72 W/m/K, respectively indicating that maintaining dry litter is extremely important as a routine. Further, it is extremely desirable that structural insulants are neither exposed to wetting nor high vapor pressure gradients.

Table 30.1 Bulk density, thermal conductivity and vapor resistivity of some common building material

Material	Condition	Bulk density kg/m³	Thermal conductivity W/m/K	Vapor resistivity GNs/kg/m
Asbestos cement sheet	Conditioned	1600	0.400	1.60-3.50
Asbestos insulating board	Conditioned	750	0.120	-- --
Asphalt roofing	Dry	1600-2325	0.430-1.150	4.35-100.0
Brickwork	Function of density and moisture content (See Table 30.2)			25-167
Concrete hollow clockwork				32.5
				Source : Wathes, 1981

Table 30.2 Thermal conductivity (k) of brickwork and concrete as a function of moisture content and bulk density

Density kg/m³	Thermal conductivity, W/m/K		
	Brickwork protected from rain (moisture 1%)	Concrete protected from rain (moisture 3%)	Brickwork or concrete exposed to rain (moisture 5%)
200	0.09	0.11	0.12
400	0.12	0.15	0.16
600	0.15	0.19	0.20
800	0.19	0.23	0.26
1000	0.24	0.30	0.33
1200	0.31	0.38	0.42
1400	0.42	0.51	0.57
1600	0.54	0.66	0.73
1800	0.71	0.87	0.96
2000	0.92	1.13	1.24
2400	1.49	1.83	2.00
			Source : Wathes, 1981

Solar absorptivity and long-wave emissivity over a temperature range of 0-100°C of common building material is given in Table 30.3; the absorptivity of most material ranges between 0.60 and 0.90 whereas long-wave emissivity of 1.0 represents an acceptable value for normal use (Wathes, 1981).

2.1.1. Distance between animals (Housing density)

Radiant energy is exchanged between animals, and between animals and their surroundings, in the thermal infra-red, which peaks at about 10μm for surface temperatures of 30-37°C. Therefore, space between animals is important. Removal of heat by convection is also affected by the distance between animals; it is for this reason that birds huddle to decrease dissipation of heat when exposed to cold conditions. Height of the building also allows hot, humid air to rise above the animals, thereby reducing radiant load from the roof. When there is cloud, high roofs without walls are not an advantage because clouds reflect solar energy into the shelters (Macfarlane, 1981).

2.2. Degradation of insulation material

Rats, mice and some insects like mealworm beetle (*Alphitobius diaperinus*) and hide beetle (*Dermestes maculatus*) may cause damage to the insulation of the buildings. The only way to avoid such degradation is to obviate environmental conditions that promote their breeding. Laminated material like extruded polystyrene bonded to steel panels can be a great value in eradication; but cost of such installation can be an impediment (Wathes, 1981).

In general, if electric cables are installed in cavity insulants, the current rating must be reduced, because there is a danger of overheating of cables.

Insulants may also be degraded during routine operations such as cleaning between batches. Wood and wood products and material with smooth hard surfaces such as glazed tiles and laminated plastics are the most difficult to clean whereas those coated with resins and bitumen paints are the easiest; therefore, the latter group shows least degradation over extended periods of time (Wathes, 1981).

2.2.1. Moisture retarders

It is critical to protect insulation from the moisture produced in a poultry house with some type of vapor retarder (formerly called vapor barrier). Generally, the vapor retarder is a 4 or 6 mm thick polyethylene film that is placed on the warm side of the insulation. This prevents water vapor inside the house from moving into the insulation and condensing in the insulation inside the cold wall. Polyethylene film or sheets should always be used even if the insulation has an attached vapor retarder (i.e., aluminum foil backing on fiberglass blankets). The large moisture load in a poultry facility can cause significant moisture problems with even very small breaks or cracks in the vapor retarder along studs, ceiling joints, and electrical outlets (Janni and Jacobson, 2003).

Protecting the insulation from rodents (mice and rats) is also very important. Rodent control is difficult in poultry housing facilities, but is necessary to safeguard the insulation. Crushed rock around the perimeter of a building to prevent rodents from burrowing under walls and maintaining a bait and trap system throughout the farm to hold down rodent populations, are highly recommended for preventing insulation deterioration in walls and ceilings (Janni and Jacobson, 2003).

3. Insulating animal sheds

In case of animal shed, it is the roof through which most heat is lost and most

Table 30.3 Solar absorptivity, long-wave emissivity over the temperature range of 0-100°C for some common building material

Material	Solar absorptivity	Long-wave emissivity
Aluminium – metal	0.15-0.26	0.08-0.14
foil	-- --	0.03-0.09
paint	0.54	0.29-0.55
Asbestos – cement	0.61-0.75	0.95
Insulating board	-- --	0.93-0.96
Asphalt – new	0.91	-- --
weathered	0.85	0.96
Bitumen felt	0.88	0.91
Bricks – based on color	0.40-0.89	0.94
Concrete (rough)	0.65	0.94
Glass (smooth)	0.83	0.92-0.95
Galvanized iron – new	0.64-0.66	0.22-0.28
aged	0.90	-- --
Slates	0.86-0.93	-- --
Tiles – based on color	0.43-0.91	-- --
		Source : Wathes, 1981

sunrays strike. Therefore, insulation of the roof is, by far, the most important. The efficiency of any insulating material or combination of material or type of construction is rated by its ability to resist the transfer of heat through it. k-value refers to the thermal conductivity and is the amount of heat, in Watts, which passes through 1 m^2 of the material when a temperature difference of 1°C is maintained between opposite surfaces of 1 m thickness. Thus, lower the k-value better the insulating ability of the material. Since most of the livestock buildings are composite nature, amount of heat, in Watts, that is transmitted through 1 m^2 of the construction from the air inside to the air outside where there is a 1°C difference between inside and outside, referred to as U-values, are more useful.

Table 30.4 R-values recommended for different climates

Climate	Roof and ceiling	Walls
Hot	4	2
Medium	8	2.5
Cold	12 to 14	8 to 10
		Source: North and Bell, 1990

U-values are similar to k-values in indicating insulating ability of a material. In environmental controlled houses, it is economical to have U-values = 0.4 for the roof and floor and 1.0 to 1.5 for the walls. Reciprocal of k-values are the thermal resistance or R-values which are directly proportional to the insulating ability of a material (North and Bell, 1990).

Table 30.5 R-values of common building material

Item	Thickness (cm)	R-value*
Asbestos cement	0.3	0.03
Concrete	20.3	0.61
Concrete block	20.3	1.11
Hard-board	0.6	0.18
Plywood	1.2	0.63
Surface, inside	-	0.61
Surface, outside	-	0.17
Shingles, Asbestos	-	0.18
Vapor barrier	-	0.15
* Sum of R-values of the material used		*Source* : North and Bell, 1990

Higher "R" values are sometimes used in facilities in cold climates like the northern U.S. and Canada. Since the primary function of insulation in poultry facilities is to prevent condensation, excessively large insulation values (R values greater then 25 in the walls and 40 in the ceiling) have limited benefit (Janni and Jacobson, 2003).

3.1. Design R-values for tropical climate

Table 30.6 Suggested design R-values for tropical housing

Expected high temperature (°C)	Recommended R value* Roof	Walls	Heat conductivity (kcal/m/min/°C)
32.2	5	2	0.01611
37.8	8	4	0.01025
43.3	11	6	0.00732
48.9	14**	8	0.00586
54.4	17**	10	0.00439
* Under mechanical ventilation ** With reflective material for open housing			
			Source : Ernst, 1995

3.2. Foundation insulation

Building foundations are another important place to insulate. Perimeter insulation will keep concrete floors and inside wall surface temperatures warmer, making it more comfortable for animals during cold weather. Perimeter insulation also eliminates condensation and frost in these areas. Rigid board insulation is recommended with an R value between 6 and 8, extending 0.6 to 0.9 m below ground level.

4. Heat production by birds

4.1. Basal Heat Production

2.75 g kcals/hr/kg of live weight. If average live weight is M kg,

Basal Heat Production = 2.75 M kcal/hr or 66 M kcal/d

4.2. Activity increment =

50% of basal heat in birds on deep-litter, 1/3 rds the basal heat for birds in cages; cages being more popular for housing layers. Therefore, activity increment = (2.75 M)÷3 = 0.917 M kcal/hr or 22 M kcal/d

4.3. Heat production due to feeding

68 kcal/100 g of dry matter intake. If feed consumption = F g/hen/d and the feed has dry matter of D%, then,

Heat production due to feeding = 0.0068DF kcal/hen/d or 0.000283DF kcal/hen/hr

Total heat produced = (2.75M + 0.917M + 0.000283DF) kcal/hen/hr = (3.367M + 0.000283DF) kcal/hen/hr or (88M + 0.0068DF) kcal/hen/d

4.4. Heat lost as latent heat of vaporization

20% of total heat produced which does not heat surroundings.

4.5. Net heat produced

0.8 (3.667M + 0.000283DF) kcal/hen/hr

or 0.8 (88M + 0.0068DF) kcal/hen/d

Considering F = 100 g/hen/d and D = 91%, and M = 1.8 kg net heat produced is 7.34 kcal/hen/hr or about 176 kcal/d (about 205 W) or 30.72 kJ/hen/hr or 8.54 W.

By calculating the surface area and total volume of the building where the birds are housed, heat produced per unit surface area and per unit volume can be estimated.

4.6. Example

N = 3600 in cages, L = 46 m, B = 9 m, H = 2.4 m

Total volume (ignoring inclination of the roof) = LBH = 993.6 m^3

Volume/bird = 0.276 m^3

Total surface area = LB + 2H(L + B) = 678 m^2 and 0.19 m^2/bird

Heat production/m^3 of volume of the building

= 7.34/0.276 = 26.59 kcal/hr/m^3 or 2672 kJ/d/m^3

Heat produced/m^2 of surface of the building

= 7.34/0.19 = 38.63 kcal/hr/m^2 or 3882 kJ/d/m^2

(Note: 3.6 kJ/hr = 1 W)

5. Moisture production by birds

5.1. Fecal moisture

Amount of feces produced = 1.4 g/g of feed; for "F" g of feed/hen/d, total feces produced = 1.4F g/d; 20% of feces will be free water and therefore, Total fecal moisture = 0.28F g/d = 0.00028F kg/day/bird.

5.2. Respiratory moisture

5 g/hen/hr; therefore, 0.120 kg/day/bird

5.3. Total moisture produced

(0.00028F + 0.120) kg/d/bird

As in case of heat produced, moisture production per unit surface area and volume of the building where the birds are housed can be computed.

5.4. Example

N = 3600 in cages, L = 46 m, B = 9 m, H = 2.4 m, F = 100 g/d

Total volume (ignoring inclination of the roof) = LBH = 993.6 m^3

Volume/bird = 0.276 m^3

Total moisture produced = 532.8 kg

Total moisture produced/m^3 of the building volume

= 532.8/993.6 = 0.536 kg/d

6. Moisture removal

Fresh poultry excreta has 75 to 80% moisture and hence, it is the primary source of moisture in a poultry house at ideal temperature of 21.1°C. Respired moisture and spillage of water from waterers also add moisture into poultry house. Major route for elimination of moisture is by increased air movement and therefore, efficiency of moisture removal is inversely proportional to the relative humidity of the incoming air.

If relative humidity (RH) of outside air = M_o, outside (ambient) temperature = T_a, inside temperature = T_i then, 0.01M_o grains of moisture is present per cubic foot of air; since 15.44 grains = 1 g and 1 m^3=35.314724 cubic feet, the outside air has 0.02287M_o g/m^3 of moisture.

Assuming that $T_i > T_a$, air entering in gets heated up and hence its moisture-holding capacity increases. Keeping in view that the RH desired inside the house is M_i, maximum moisture-holding capacity of air at temperature T_i and RH of M_i can be found by using standard charts available or by equation given in Table 30.8. The chart gives the value in grains per cubic foot which can be converted to g/m^3 by multiplying by a factor of 2.287; let it be designated Y g/m^3 of moisture.

Therefore, the extra moisture the outside air can remove to keep the inside RH at M_i is given by (Y - 0.02287 M_o) g/m^3. Assuming M_o = 70%, T_a = - 9.4°C, T_i =15.6°C, M_i = 75%, maximum moisture-holding capacity at RH of 75% (M_i) is 4.4 grains/cubic foot or Y = 10.064 g/m^3 (that is, 4.4 X 2.287). Hence, the outside air can remove (10.064 - 0.02287x70) = 8.463 or 8.5 g/m^3 of moisture; let this be designated *x*.

Further, to remove total moisture of y g/m³/d, $y \div x$ air changes/d or $y \div (24x)$ air changes/hr is required. Calculation of y is shown under "moisture production". In the example given above, y = 536 g/m³ and x = 8.5 g/m³ of moisture, 536/8.5 = 63 air changes per day or about 3 (63/24 = 2.6) air changes per hr are needed.

Assuming that N number of birds, each consuming F g/d of feed, are housed in a building measuring (in m) L, B and H in length, breadth and height, respectively, air changed required/hr can be calculated as :

$$\frac{N(0.00028F + 0.120)}{24LBH(Y - 0.0229M_o)}$$

Air changes or air circulation to remove moisture depends on T_a and T_i and their difference ($T_d = T_i - T_a$) as well as M_o and M_i. If $M_o > M_i$, as can be expected in severe winter (not common in our country), either T_a should increase or M_o should decrease; otherwise, dampness within the house increases. This can be alleviated by providing built-up litter to absorb moisture or by increasing T_i by artificial heat of the increasing insulation too retained heat released by hens. Inadequate removal of moisture during cold weather can lead to condensation of moisture on in the surface of poultry houses, which when prolonged, can damage walls and roofing.

It is in the cold weather the real problems of humidity arise because birds consume more feed and hence, more water; consequently, total moisture built-up increases accentuated by reduced evaporation of water from the litter. In addition, air flow has to be reduced to contain heat. Therefore, the excess moisture so accumulated starts condensing on the litter at the prevailing low temperature leading to wet-litter.

Table 30.7 U-values required to overcome condensation

Relative humidity (%)	Maximum temperature difference (internal - external), °C				
	15	20	25	30	35
60	4.6	3.0	2.6	2.3	1.9
65	3.6	2.8	2.3	1.9	1.6
70	3.0	2.4	1.9	1.5	1.4
75	2.6	1.9	1.5	1.3	1.1
80	2.0	1.5	1.2	1.0	0.9
85	1.5	1.1	0.9	0.7	0.8
90	1.1	0.8	0.6	0.5	0.4

Therefore, insulation to retain heat simultaneously avoiding condensation is important. U values to overcome condensation are shown in Table 30.7; however, under Indian conditions, condensation of moisture on to the litter is seldom recorded.

7. Calculation of R-value for a house

Heat produced per hen per unit surface area as calculated in the example above is 38.63 kcal/hr/m². If the R value is 0.40 (or 2 in terms of °F or Btu) with a T_d of only 0.4°C and with no direct ventilation (tightly closed house), a population of 3,600 hens in a house measuring 46 m X 9 m x 2.4 m can maintain the T_i (38.63 x 0.4) = 15.4°C above that of outside air (T_a). If T_a should be maintained at 2.8°C when T_a drops to –23.3°C, and insulation value of (T_i – T_a) ÷ heat produced per unit surface area is required; in the given example, it will be 38.7 ÷ 38.63 which is approximately equal to 1 on metric scale (5 in terms of °F or Btu). It also follows that, with lower densities, heat production is reduced and hence R value required increases to ensure a particular T_d within the house.

Let "V" be the volume of air change in m³/hen/hr. It is known that 1 Btu is required to warm 50 cubic feet of air by 1°F; when converted to metric units, it becomes 1 kcal/°C/3.125 m³ of air. Therefore, total heat produced to warm "V" volume of air through T°C = [(T_dV) ÷ 3.125] kcal/hen/hr. Heat produced (P) by the birds is also lost through the surface area available in the house. This loss is directly proportional to T_d and surface area per hen (A), and inversely proportional to Thermal resistance (R) of the walls and roofings. Hence, heat lost through this means is [(T_dA) ÷ R] kcal/hen.

It follows that total heat produced by a hen should be more than or equal to the sum of the above two heat losses. In other words,

$$P = 0.8(3.667M + 0.000283DF) \geq \frac{T_d V}{3.125} + \frac{T_d A}{R} = T_d \left\{ \frac{V}{3.125} + \frac{A}{R} \right\}$$

In the given example, assuming Td = 14°C, R = 18 in terms of Btu and R = 3.6 in metric units, we get V = 1.47 m³/hen/hr or 5.3 air changes/hr. This amount of air change can also take care of moisture removal since the latter requires only 2.6 air changes/hr.

Insulation value, R, is expressed as the temperature difference in °C that will permit a heat loss of 1 kcal/hr through 1 m² of expose surface. In terms of Btu, R value represents the temperature difference (Td) in °F to allow a heat loss of 1 Btu/hr/ft² of surface area. In other words, R value of 2 means that a T_d of 2°F will allow heat loss of 1 Btu/hr/ft². In metric units, it will be 10.763915 Btu/hr/m² = 10.818 kJ/hr/m² or 2.584 kcal/hr/m². As T_d of 2°F (i.e., R = 2) will be allowing 2.584 kcal/hr/m² of heat loss, 1 kcal/hr/m² can be lost when T_d = 0.774°F = 0.43°C (i.e., R = 0.43 on metric system) (1 W = 3.6 kJ/hr).

Ultimately, R value of 2 in terms of °F or Btu = R value of 0.43 in terms of °C. In other words, R value in metric units is about 21½ % or about 1/5th of that in terms of °F or Btu. On most occasions, T_d and A will be known whereas P can be computed; and hence, the above equation can be used to calculate V or R when one of these is known; particularly for the calculation of R by the rearranging of the above equation as:

$$R = \frac{3.125T_dA}{(3.125P - VT_d)}, \text{and similarly, } v = 3.125\left\{\frac{P}{T_d} - \frac{A}{R}\right\} m^3/\text{hen/hr}$$

It must be emphasized that the above calculations are valid for window-less/ environment-controlled houses under cold climate wherein heat has to be conserved within the house. In tropical and subtropical climate, most often it'll be necessary to drive out heat from inside the building and hence, most of the houses are open-sided; more particularly, cage-layer houses whose side-walls are made almost completely of expanded metal. In open-sighted houses, it is necessary to estimate the requirements for removing the hot, humid and stale air from the house. These can be assessed the calculation of stack effect, summer ventilation rate and ventilation heat loss, if the temperature on various parameters must be considered before any conclusion is drawn.

8. Effect of temperature on various parameters

Environmental temperature has a profound influence on feed and water intake on one hand and respiratory and fecal moisture output on the other (Table 30.8) and accordingly suitable adjustments have to be made.

Table 30.8 Effect of temperature (X, in °C) on various parameters

Parameters	Equation	Standard error	Most suited range (°C)
Feed intake, g/hen	$Y = 116.82+0.53X-0.06X^2$	0.3347	4.4 to 37.8
Respired moisture g/hen/hr	$Y = 3.15 + 0.04X+0.003X^2$	0.1834	- 1.1 to 32.2
Respired moisture g/hen/d	$Y = 34.92-4.28X+0.33X^2$	7.7607	4.4 to 37.8
Respired+Fecal moisture g/hen/hr	$Y = 10.31+0.07X+0.002X^2$	0.1294	- 1.1 to 32.2
Respired+Fecal moisture g/hen/d	$Y = 163.03-3.27X+0.26X^2$	8.0502	4.4 to 37.8
Water intake, cc/hen/hr	$Y = 13.43- 0.74X+0.03X^2$	0.4963	- 1.1 to 32.2
Water intake, g/g of feed	$Y = 2.59-0.26X+0.01X^2$	0.5177	4.4 to 37.8
Moisture holding capacity of saturated air, g/m³	$\text{Log10}Y=0.68+0.03X-0.0002X^2$	0.0120	- 15 to 37.8

Chapter **31**

Shelter Engineering

1. Building material

Livestock buildings consist of several materials so as to provide suitable shelter in anticipation of optimum returns. Some of the material used in buildings to house poultry is enumerated below:

1.1 Bricks

For poultry house construction, bricks must be of 1. Uniform size, shape, texture and color 2. Well-burnt and regular 3. Free from cracks and flaws and 4. Should give ringing sound when struck.

Porous bricks absorb water and weather badly; hence they should never be used in foundation or external walls. Moderately rough or irregular, or chipped bricks may be used for partitions or walls which will be plastered subsequently.

Brick-work done in cement and lime-mortar is more durable, water-resisting, strong and of low depreciation than lime-mortar alone.

A standard modular brick with frog (a depression) measures 19 cm long, 9 cm wide and 9 cm high with a frog of 10 cm long, 4 cm wide and 1 or 2 cm deep on one surface (Fig. 31.1). However, the nominal size for standard modular brick is 20 cm x 10 cm x 10 cm. Actual thickness of wall with 1 cm mortar joint is 11½ cm for ½ brick, 23 cm for 1 brick, 34 cm for 1½ bricks, 45 cm for 2 bricks and so on. For calculations, they can be taken as 11, 23, 34, 45 cm, respectively.

A brick-tile will have an actual size of 19 cm x 9 cm x 4 cm and a nominal size of 20 cm x 10 cm x 5 cm.

Traditional bricks, on the other hand, will be of 22.9 cm x 11.2 cm x 7.0 cm actual size and 22.9 cm x 11.4 cm x 7.6 cm nominal size. Thickness of walls is measured as multiples of ½ brick thickness i.e. 11.4 cm.

Other types of bricks are facing bricks, enameled bricks, salt-glazed bricks, blue bricks, Dutch-clinkers, perforated bricks and air-or ventilating-bricks.

1.2 Tiles

Tiles are used mainly for roofing. They are red in color and made of good plastic clay. Tiles should never be laid flatter than a pitch of 45° to avoid seepage of water. If the pitch of the house is less than 45°, pan tiles which are rough and heavy can be used. Glass pan tiles can be inserted to obtain more light into the house.

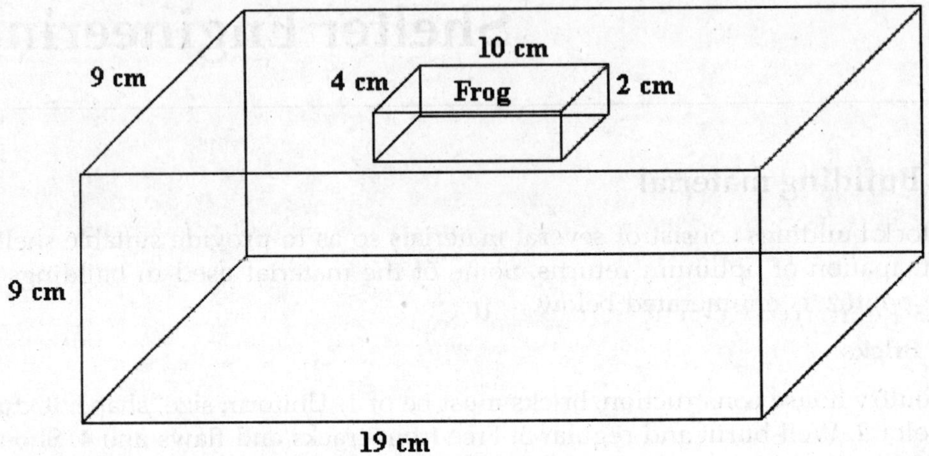

Fig 31.1. Modular brick with frog

1.3 Limes, mortars and cement

1.3.1 Lime

Lime for building purposes (plastering, mortar) is obtained by heating limestone/ shells/chalk-stone in a kiln to remove carbonic acid and moisture.

1.3.2 Mortars

Ordinary building mortar is prepared by placing one part of lime on a cement floor surrounded by three parts of sand; lime is made wet, thoroughly allowed to cool following effervescence, mixed with the peripheral sand and left for 4-6 weeks for weathering before use. Ox or goat hair or mantilla fiber at 7.5 kg/m³ of coarse stuff is added and mixed with plaster lime to bind the material into position in the building.

1.3.3 Cement

1.3.3.1 Portland Cement (PC)

This is the most important of all building material and is used in foundations, floors, paving, walls etc. It is similar in appearance to Portland stone; hence, the name.

1.3.3.2 Cement mortar

Consists of three parts of clean sharp pit/river sand and one part of PC mixed with water which is free from organic matter and other impurities. This must be used fresh and left-over mortar at the end of the day should be discarded.

1.3.3.3 Cement Concrete (CC)

This is used for foundations, drainage etc. Ordinary foundations normally consist of CC made of stones or broken bricks four parts, clean sand two parts, PC one part and water quantum sufficient (q.s). The material are mixed on a platform of stone or cement pavement but not on soil; turned over twice dry and once after being wet; turned into foundation trenches immediately thereafter.

1.3.3.4 Cement plaster

This is made up of clean sharp sand 3-4 parts and PC one part in water q.s. This is used to plaster inner and outer surface of walls, partitions etc. On outer face of walls, rough casting may be done on the cement plaster with pebble or crush-stone dashing.

1.3.3.5 Reinforced Cement Concrete (RCC)

RCC is the fine concrete reinforced with mild or tempered steel rods or expanded steel sheets which are so placed that they take-up tensile stress set-up in the concrete. RCC is used in tanks, poles, beams, load-bearing walls, stanchions etc.

1.4 Timber

Timber is becoming increasingly costly due mainly to reduced availability. A good timber should have the following characteristics: 1. straight in grain. 2. Free from large/dead/lose knots and waney edges. 3. Thoroughly seasoned and uniform in color. 4. Free from sap-wood. 5. Give fresh small, clear and firm cut-surface when cut and cut freely. 6. Free from dull or spongy appearance. 7. Clear ringing sound when struck. 8. Have closely-set, dark-colored annual rings indicating strength.

Perfect timber is difficult to get. The most serious defect of timber is sap-wood which is spongy in grain, with large annual rings (usually bluish in color), lacks strength and liable to decay especially by "dry-rot". Sap-wood indicates an immature tree but for timber.

1.4.1 Dry-rot

This is caused by fungi, especially *Merulius lacrymans* which prefers unventilated, warm and damp atmosphere. Its spores spread by wind and tools of the carpenter. The timber will be reduced to powder by this rotting; sap-wood is particularly susceptible. The condition is detected by a characteristic musty odor emitted. The fungus may spread to joints, brick-work and plaster; but, derives nourishment only from timber.

There is no cure for dry rot after the attack has set-in; affected portion and adjoining parts be carefully cut (without allowing the dust or spores to be blown-out), the material is burnt and the portion so removed is replaced with sound wood.

Other defects in timber include shakes (cracks or faults due to uneven contraction during seasoning or drying of timber), knots and waney edges.

1.5 Roofing material

Different roofing material will be discussed along with construction of roof separately.

2. Construction

2.1 Wall

Walls have the following functions:

1. offer shelter from cold, wind and rain
2. Support roof and upper floor, if any and
3. Support stall-divisions or partitions, if any.

Walls should not absorb moisture either from rain or from capillary action from the ground.

2.1.1 Foundation

Walls are generally constructed on foundation, preferably of concrete.

For concrete foundations, a trench equal to the thickness and width of the foundation is dug to remove top-soil and humus so that firm clay, gravel or rock is available. Width of the foundation will be more than that of the wall, generally 25 cm and preferably 30 cm, to offer necessary bearing area. Monolithic, concrete foundation is preferred since it will be strong and water-resistant. Otherwise, brick-work foundation with stepped increase in thickness of wall (greatest at the base; projection called as footings) can be practiced with or without concrete foundation (Fig. 31.2)

2.1.2 Plinth-wall

Poultry sheds are normally constructed over a plinth wall made of brick in cement mortar, 45-60 cm high.

2.1.3 Damp-proofing Course (DPC)

A 2.5 cm thick rich CC, 1:1½:3 or 2 cm thick rich cement mortar 1:2 mixed with standard water-proofing material (Cemseal or Imperm) 1 kg/bag of PC provided at the level of plinth-wall to its full width. Hence, floor of the house will be at least 47-62 cm above the ground level.

2.1.4 Wall

Usually made of brick-work, 25 cm thick, in PC, lime and sand. The wall will be strengthened by piers (pillars) which also help anchor the roof structures. Wherever feasible, concrete block-work, pre-cast or *in-situ* concrete walls or hard-plastic etc. can also be used for walls. Height of the side walls (stud) depends on type of the house; however, on the East and Western ends, full walls are constructed with opening for doors.

Internal finishing of the walls must be smooth, unbroken and white-washed.

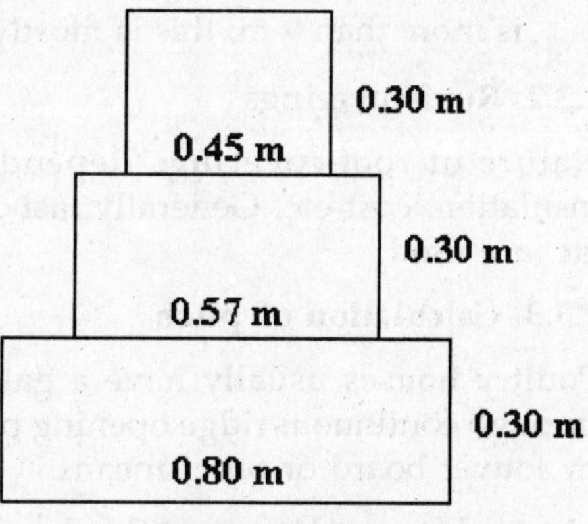

Fig. 31.2 Foundation

2.2 Sidewalk

Sidewalk on all sides of the building (as projection of the floor outside the side-walls and front-walls) is usually provided for poultry buildings to help observation of the birds without entering into the building. It is made of CC (1:5:10) to 45-50 cm width, 7.5-9.0 cm thick and smoothened on all surfaces so that it can act as rat- and snake-proofing structure as well.

2.3 Roof

Normally roof is hoisted on truss-work made of wood or steel; the latter is durable and common. The roof protects the internal structures and also offers protection to birds from extremes of weather conditions.

Roof-trusses are framed structures used to fix roof-coverings. They should not cause obstruction to air and light and present smallest surface for dust. Parts of a steel-truss are shown in Fig. 31.3.

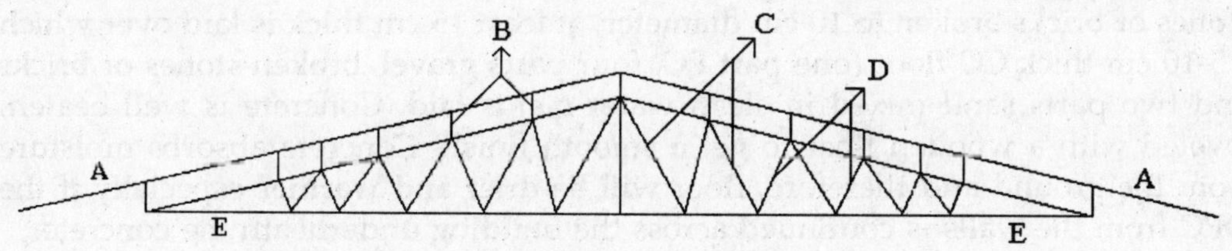

A Roofing material B Purlins C Struts D Ties E Main rafters
Fig. 31.3 Steel truss with roofing material

2.3.1. Types of trusses

1. Collar-beam truss having two rafters is used when span is 4-6 m
2. King-truss having two rafters, one tie, one hanger and two struts is used when span is up to 6 m
3. Queen-truss having two rafters, two ties, two hangers and two struts is used when span is up to 9 m

4. Truss with multiple struts having two rafters and one tie is used when span is more than 9 m; this is mostly of steel and is highly durable.

2.3.2 Roof-coverings

Nature of roof-coverings depends on various considerations like climate, insulation, cost etc. Generally, asbestos, aluminium, galvanized iron (GI), tiles etc. are used.

2.3.3 Calculation of pitch

Poultry houses usually have a gable-roof having two slopes, roof-ventilation through continuous ridge opening protected against seepage of rainwater suitably by louver board or other means.

The angle formed between the slope of the roof and the imaginary line connecting the side-walls at the eaves is called "Pitch" of the roof. Vertical height between eaves and ridge of the roof is called "Rise" and the distance between opposite eaves is called "Span"; one-half of the span is called as "Run" (Fig. 31.4).

It is evident from the Fig 31.4 that Pitch = $\tan^{-1}$ (Rise ÷ Run). However, as a rule of thumb, it can also be approximated as the ratio of Rise to Span times 100.

2.4 Floor

Floor of a poultry house must be hygienic, non-porous, easily cleanable, durable, quick-drying, non-slippery, hard-wearing, non-absorbent of moisture and least expensive. However, solid floor is constructed in case of deep-litter and all-slat systems. Before construction of the floor, the trench formed by the construction of the plinth wall is suitably filled. Concrete flooring is the best because it is durable, hygienic and most impervious and cheap. Layer of hard-core, a layer of stones or bricks broken to 10 cm diameter, at least 15 cm thick is laid over which 7.5-10 cm thick CC floor (one part PC, four parts gravel, broken stones or bricks and two parts sand mixed in clean water q.s) is laid. Concrete is well beaten, leveled with a wooden float to get a smooth finish. Concrete absorbs moisture from the ground and therefore, floor will be drier and warmer especially if the DPC from the walls is continued across the building underneath the concrete.

A slope of 1 cm for every 40-60 cm of the solid floor will help in cleaning when the birds are removed.

In cage-system (California-type), just below the cages, there will be no flooring so as to allow absorption of fecal moisture by the soil. In case of high-rise houses, no flooring is required since the catwalk and the cages are raised 2 m from the ground level.

3. Estimates of cost of construction

Estimate is a computation for calculation of quantities required and expenditure

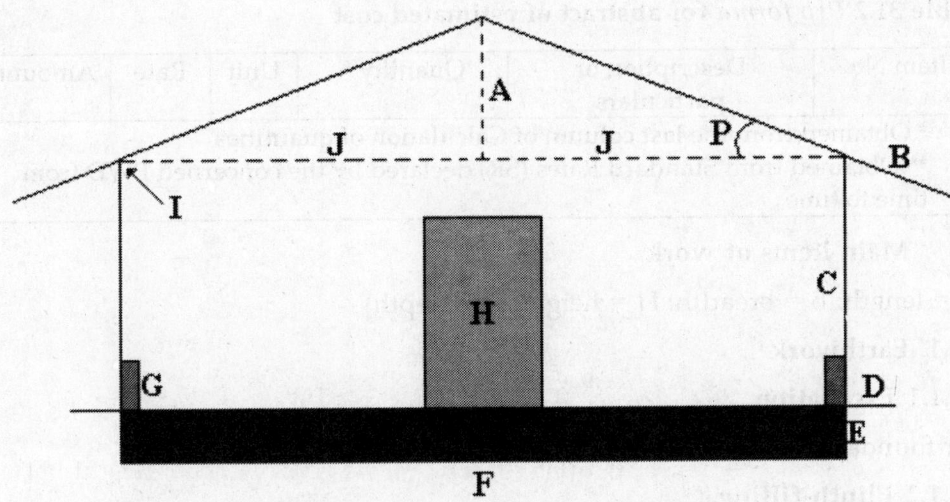

A Rise B Overhang C Stud height D Side projection E Plinth F Width G Side-walk H Doo I Eaves J Run P Pitch, angle depending on A and J or F (normally < 30°)

Fig. 31.4. Calculation of pitch

likely to be incurred in construction of a work. This would help knowing cost of a work beforehand. However, actual cost of a work can be known only after the completion and it should not differ much from the estimated cost.

Working-out a detailed estimate involves the following:

1. Calculation of quantities of different items of work and
2. Developing the abstract of estimated cost based on quantities

3.1 Details of measurements and calculation of quantities

The whole work is divided into different items like Earth-work, foundation, brick-work etc. and each work is measured in detail in the following *pro forma*:

Table 31.1 *Pro forma* for calculation of quantities

Item No	Description or particulars	No.	Length	Breadth	Height or Depth	Content or Quantity	Total quantity or content

3.2 Abstract of estimated cost

Cost under each item of work is calculated from the total quantity worked-out. Further, ½% is added for contingencies, is 7 ½% for internal electrification, 5% for external electrification, 4% for sanitary and water-supply works and 2 ½% for work-charged establishments. The *pro forma* for abstract of estimated cost is as follows:

Table 31.2 ***Pro forma*** **for abstract of estimated cost**

Item No	Description or particulars	Quantity *	Unit	Rate	Amount **
* Obtained from the last column of Calculation of quantities ** Obtained from Standard Rates (SR) declared by the concerned PWD from time to time					

3.3 Main items of work

(L = length; B = breadth; H = height; D = depth)

3.3.1 Earthwork

3.3.1.1 Excavation

For foundation; calculated in m^3 as LBD

3.3.1.2 Plinth-filling

Calculated in m^3 for interior dimensions discounting the thickness, usually about 20%, of earthwork. Excavated earth can be used and the extra earth after filling may be used for leveling or dressing of site or carted away.

3.3.2 Concrete in foundation

Calculated as LBD where D = 20-45 cm (usually 30 cm); consists of lime concrete @ 1:4:8 or 1:5:10.

3.3.3 Soling

A layer of dry-brick or stone applied below the foundation when the soil is soft or bad is called soling. It is calculated in m^2 by specifying thickness.

3.3.4 DPC

Calculated in m^2 and is usually 2.5 cm thick.

3.3.5 Masonry

Computed in m^3 by LBH in two groups: 1. Foundation and plinth and 2. Superstructure. In case of storeyed buildings, calculated separately for each floor. Walls are measured as solids with splayed/rounded edges considered rectangular and taking extreme dimensions. The partition walls are measured in m^2 specifying thickness. Stone-masonry is considered similar to brick-masonry

3.3.5.1 Deductions in masonry

Opening up to 0.1 m^2 – nil; ends of beams, rafters, posts etc. – nil; rectangular opening (doors, mesh work in side-walls) – LBH

3.3.6 Flooring

Calculated in m^2 by LB taking interior dimensions and specifying thickness.

3.3.7 Roofing

Roof covering is measured in m^2 by LB; measured flat including overlaps. Fittings and supporting trusses are considered as separate items.

3.3.7.1 Roof-fixing

Estimated as 5% of cost of roofing material.

3.3.8 Plastering

Usually 12 mm thick; calculated in m^2 by LH for walls, whole face of the wall is measured for both sides as solids and deductions allowed as follows: Ends of beams etc. – nil; up to 0.5 m^2 – nil; more than 0.5m^2 but less than 3.0 m^2 – one face; more than 3.0 m^2 – both faces.

3.3.9 Pillars

Volume calculated by their geometric shape; usually square pillars are constructed in which case calculated in m^3 by LBH. Plastering of pillars is calculated in m^2 and taken under the item "Plastering".

3.3.10 Doors and windows

Frames in m^3 and doors in m^2 specifying thickness.

3.3.11 Wood-work

Beams, trusses etc., if used, in m^3 by LBH

3.3.12. Iron-work

In weigh (kg or quintal) or running m times weight per m.

3.3.13. Whitewashing

In m^2 similar to plastering; hence, the same values are taken

3.3.14. Painting

Calculated in m^2

3.4 Methods of estimation

3.4.1. Individual wall method

This method also referred to as Separate method/Long-wall and Short-wall method/General method is as follows:

1. External length of walls running in one direction are measured; long-walls out-to-out and short-walls in-to-in (Fig. 31.5).
2. Quantities calculated by LBH.
3. Same rules are applied for excavation in foundation, concrete in foundation and masonry, changing breadth as footing to be noted carefully.

This method is easier and hence is widely practiced.

3.4.2. Centre-line method

The procedure in this method is as follows:

1. Some-total length of all centre lines of all walls, long and short, of same type foundation and footing or calculated.
2. Quantities estimated by total LBH.
3. For cross-or partition-walls and footings, for every junction, ½ breadth of respective item to be deducted (Fig. 31.6).
4. For buildings with different types of walls, each set of walls built separately; as indicated above, ½ breadth of the concerned wall is subtracted at the junction.
5. No subtraction necessary at corners where the walls meet.

When number of footings is involved, L of the first footing is determined by deducting ½ breadth of footing per junction from the total centre line and length of the subsequent footing by adding one offset of footing.

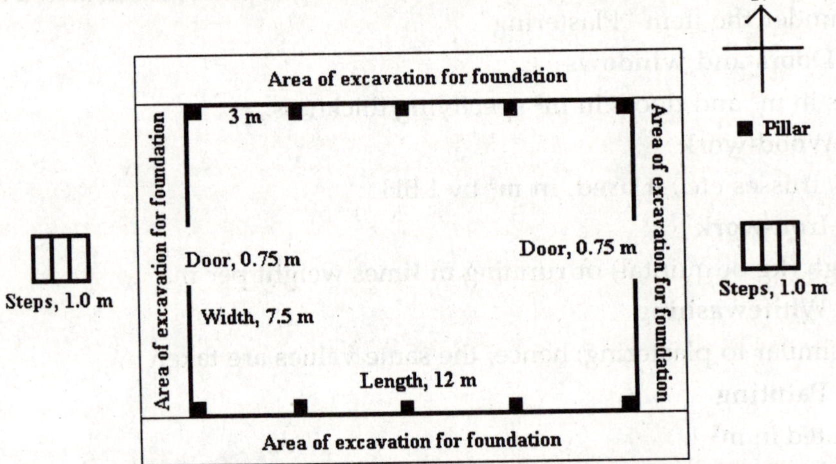

Fig. 31.5 Individual-wall method – Floor diagram

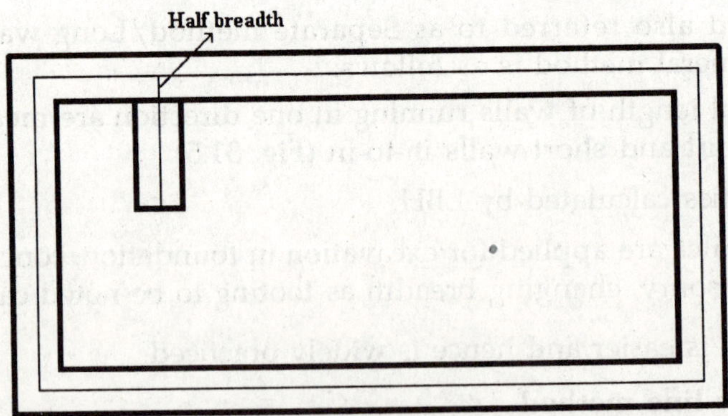

Fig. 31.6. Center-line method

4. Standards for some selected items

Table 31.3 Standards for selected items

Sl. No	Item	Description	
1	Aluminium sheets	2.0-3.6 m x 0.75 m	
2	Asbestos cement sheets	Width 1.05 m; Length 1.50, 1.75, 2.00, 2.25, 2.75 and 3.00 m; corrugations 7/sheet; lengthwise overlap 15 cm; side overlap for cm	
3	Brick – with frog	Nominal 20 cm x 10 cm x 10 cm; 10 cm width/ ½ brick	
	Traditional	Nominal 22.9 cm x 11.4 cm x 7.6 cm; 11.4 cm width/½ brick	
4	Door – collapsible	2.4 m x 2.4 m	
	ordinary	Single-door	Double-door
		0.75 m x 2.10 m	1.20 m x 2.10 m
		0.75 m x 1.80 m	1.05 m x 2.10 m
		0.60 m x 1.80 m	0.90 m x 2.10 m
5	GI sheets, corrugated	Width 0.8 m, length 1.2, 2.2, 2.5, 2.8 and 3.2 m; corrugations 10/sheet; lengthwise overlap 15 cm; side overlap 15 cm	
6	Gate	30 kg weight	
7	Iron pipes	To be inserted at least 0.6 m inside CC	
8	Maximum distance	Between item between iron pipes 2.0 m, iron truss 3.0-4.5 m, wooden poles 1.5 m	
9	Maximum span	Wooden poles or iron pipes 6.0 m	
10	Poles	Inserted at least 0.6 m deep inside the soil	
11	Purlins	With wooden poles, 0.9 m apart, maximum	
12	Rise of a thatched roof	1/3 to ½ of the span	
13	Windows	1.2 m x 0.9 m	
14	Wire-mesh	Width 0.9, 1.2 m	

5. Current rates of certain works and material

Table 31.4 Current rates (Finished) of certain works and material

Sl. No.	Work/Material	Unit	Rate (Rs)
1	Angle iron	kg	43.04
2	Asbestos roofing	m^2	160.93
3	Bricks	m3	1433.40*
4	Concrete foundation 1:4:8	m3	1643.13
	1:5:10	m3	1496.47
5	Doors, 4 cm thick	m2	694.85
6	Door frames, ordinary wood	m3	29720.44
7	DPC 1:2:4, 2.5 cm thick	m2	88.34

Sl. No.	Work/Material	Unit	Rate (Rs)
8	Excavation (Earth-work)	m3	80.88
9	Flooring, concrete, 4 cm thick	m2	121.67
	mud	m2	
	stone-slabs	m2	609.22
10	Gate, 30 kg	kg	1288.00
11	Masonry, foundation to plinth	m3	1433.40
	superstructure	m3	1620.00
12	Painting	m2	28.54
13	Pillars, brick	m3	1532.40
14	Plastering, 12 mm thick	m2	42.12
15	Poles, Eucalyptus	kg	4.00
	Iron pipe	Running m	42.43
16	Steel truss, ≈ 225 kg weight, 9 m span	kg	42.43
17	Thatches, with poles	m2	75.00
	without poles	m2	35.00
18	Tubular truss, 5 m span	kg	34.83
19	Trench-filling (plinth-filling)	m3	87.88
20	White-washing	m2	5.24
21	Windows – with door, 1.2 m x 0.9 m	m2	2551.00
22	Wire-mesh, chicken mesh	m2	62.50
	welded mesh	m2	215.64
23	Wooden truss	m3	29720.44
24	Roof fixing	Roof cost	5%
25	Electrification – internal	Total cost	7.5%
	external	Total cost	5.0%
26	Sanitary and water supply	Total cost	4.0%
27	Contingencies	Total cost	0.5%
28	Work-charged establishment	Total cost	2.5%

Source : Public Works Department, Pondicherry, India; 2004-05

6. Worked-out example

Preparation of an estimate for housing 1000 broilers on deep-litter till market in the conventional house assuming the following:

1. Foundation – Step-foundation 0.9 m deep (Fig. 31.2) made of CC at the base and brick-work for the rest.
2. Plinth – 0.6 m.
3. Side-wall – 2.4 m of which 0.6 m is solid wall, the rest expanded metal
4. East and West – full walls with 4 cm thick doors measuring 0.75 m x 1.8 m on both ends.

5. Pillars – 10 Nos supporting steel trusses.
6. Roof – asbestos on steel-truss, trusses 3 m apart, overhang 0.9 m, rise 1.2 m.
7. Floor – CC.
8. Steps – on either side (Fig. 31.7)
9. Doors – Single door, 0.75 m x 1.8 m on either side

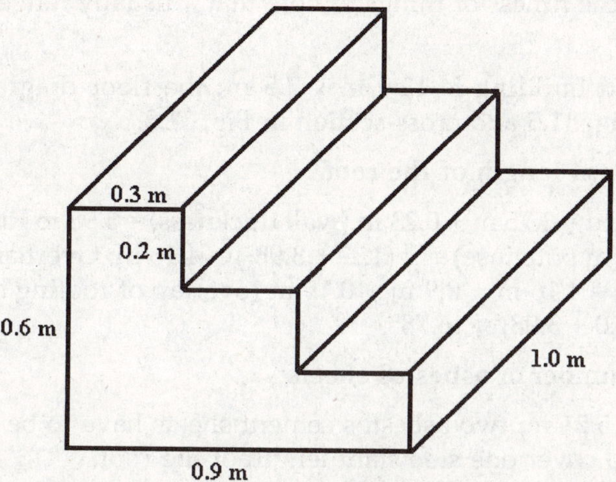

Fig. 31.7. Steps

Thickness of walls and pillars 0.3 m, DPC and white-washing to be included.

A cross-section of the house is shown in Fig. 31.8

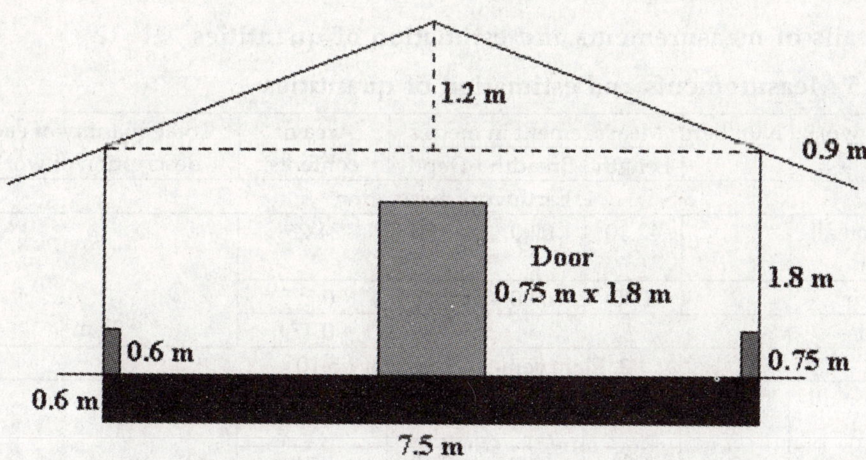

Fig. 31.8. Cross-section of a house for estimation of quantities

6.1 Calculation of dimensions of the building

Floor space required for is 900 cm^2/bird (0.09 m^2/bird, A) and therefore, total floor space required for 1000 broilers (N) is 90 m^2 (NA). Assuming that length of the house (L) is k times its width (W), it is obvious that $LW=kW^2=NA$. Keeping in view that width of the conventional house cannot exceed 9 m, it is evident that k = (NA÷81); therefore, keeping k=1.6 or L=1.6W, W = v (NA÷k) = 7.5 m and L = 1.6W = 12.0 m. It is in the fitness of things to note that k usually ranges between 1.5 and 4.0.

Therefore, the required building is 12.0 m x 7.5 m; the floor diagram of this building is shown in Fig. 31.5 and cross-section in Fig. 31.8.

6.2 Calculation of slant length of the roof

W = 7.5 m, therefore, Run = 3.75 m + 0.23 m (wall thickness) = 3.98 m Rise = 1.2 m, therefore, slant height (hypotenuse) = v ($1.2^2 + 3.98^2$) = 4.16 m; Overhang = 0.9 m and hence, slant width = 4.16 m + 0.9 m + 0.15 m (overlap of roofing material) = 5.21 m. Pitch = $\tan^{-1}(1.20 \div 3.98) = 16.78°$.

6.3 Calculation of number of asbestos sheets

For the slant length of 5.21 m, two asbestos cement sheets have to be fixed with an overlap of 0.15 m to cover one side slant length of the roof.

Total length of the building (including wall thickness of 0.23 m + additional projection of 0.37 m on either end) = 13.2 m, width of the asbestos sheet = 1.05 m less overlap of 0.08 m (on either sides) = 0.97 m. Therefore, number of asbestos sheets for one side is = (13.2 ÷ 0.97) x 2 = 27.21 (i.e. 28). Hence, for covering the roof of the entire building, 56 asbestos sheets are required.

6.4 Details of measurements and estimation of quantities

Table 31.5 Measurements and estimation of quantities

Details of work	Number	Measurement in metres			Area or contents	Total quantity of each description of work
		Length	Breadth	Depth		
1. Earthwork excavation						
Foundation all-round	1	42.30	0.90	0.90	34.26	
For steps	2	1.00	0.90	0.15	0.27	
Sundries					0.47	35 m^3
2. Plain cement concrete, 1:5:10						
Foundation all-round	1	42.30	0.90	0.10	3.81	
For steps	2	1.00	0.90	0.10	0.18	
Sundries					0.01	4 m^3

Details of work	Number	Measurement in metres			Area or contents	Total quantity of each description of work
		Length	Breadth	Depth		
3. Brick-work in cement mortar, 1:6						
I footing	1	41.68	0.67	0.15	4.19	
II footing	1	41.24	0.56	0.15	3.46	
III footing	1	40.80	0.45	0.15	2.75	
IV footing	1	40.36	0.34	0.15	2.06	
Up to plinth	1	39.92	0.23	0.80	7.34	
Above plinth	Sub-total, up to plinth ≈ 20 m³					
Long wall	2	12.00	0.23	0.60	3.31	
Short wall	2	7.96	0.23	1.80	6.59	
Doors, dedn.	2	0.75	0.23	1.80	(-) 0.62	
Δ portion	2	½x7.96	0.23	1.20	2.20	
Pillars	10	0.22*	0.45	0.35	0.35	
Pillars	10	0.45	0.22	1.80	1.78	
Side-walk						
Long wall	2	12.46	0.90	0.15	3.36	
Short wall	2	7.96	0.90	0.15	2.15	
Sundries	Sub-total, superstructure ≈ 19 m³				0.08	39 m³
4. Damp-proofing course (DPC), 2.5 cm thick						
All round	1	39.92	0.23		9.18	
Sundries					0.82	10 m²
5. Cement plastering in cement mortar, 1:6						
Long wall						
Inside	2	12.00		0.60	14.40	
Outside	2	12.46		0.75**	18.69	
Short wall						
Inside	2	7.50		1.80	27.00	
Inside, Δ	2	½x7.50		1.20	9.00	
Outside	2	7.96		1.80	28.66	
Outside, Δ	2	½x7.96		1.20	9.55	
Doors, dedn.	2	0.75		1.80	(-) 2.70	
Sub-total (5 A): 107.30					≈ 110 m²	
Sidewalk						
Long wall	2	12.46	0.90		22.43	
Long wall, sides	4	13.36	0.15		8.00	
Short wall	2	7.96	0.90		14.33	
Table 31.5 contd.						
Short wall, sides	4	8.86		0.15	5.32	
Pillars	10	0.89		1.80	16.02	

Details of work	Number	Measurement in metres			Area or contents	Total quantity of each description of work
		Length	Breadth	Depth		
Steps						
Top	6	1.00	0.30		1.80	
Vertical face	6	1.00		0.20	1.20	
Sides – 1	4	0.30		0.60	0.72	
Sides – 2	4	0.30		0.40	0.48	
Sides – 3	4	0.30		0.20	0.24	
Rear side	2	1.00		0.60	1.20	
Sundries					0.94	
Total quantity of cement plastering						180 m²
6. Whitewashing						
Quantity as per sub-total 5 A						110 m²
7. Plinth-filling						
Inside area	1	12.00	7.50	0.60	54.00	54.00 m³
8. Flooring, cement concrete 1:2:4						
Flooring	1	12.00	7.50		90.00	90 m²
9. Wire-mesh						
Long wall	2	12.00	1.80		21.6	44 m²
10. Roofing, Asbestos cement sheets						
Roofing	1	12.30	7.80		95.94	96 m²
11. Roofing, trusses, 50 mm δ MS pipes (Main rafters), 25 mm δ MS pipes (Struts and Ties)						
MS pipe, 50 mm	4	18.00			72.00	
MS pipe, 25 mm	4	12.00			48.00	≈ 450 kg
MS pipe, 50 mm δ 72 x 5.10 kg/m					367.20	
MS pipe, 25 mm δ 48 x 1.68 kg/m					80.64	
12. Door frame						
Door	2	4.35	0.08	0.10	0.069	
Sundries					0.031	0.10 m³
13. Door shutter						
Door shutters	2	0.75		1.80	2.70	
Sundries					0.30	3.00 m²

* Pillars make a part of the wall itself and therefore, wall thickness of 0.23 m subtracted.

** Includes 15 cm plastering in foundation

1. Long wall and short wall differ only in superstructure. Therefore, earthwork, concrete and brickwork in foundation as well as DPC is calculated by center-line method; masonry, brickwork, plastering and whitewashing are calculated by individual wall method.
2. The building described above has no internal partitions; estimates of quantities change according to the number, type and structure of partitions.
3. The pillars are made of brick; estimates change if concrete pillars are needed.
4. 4 Trusses (Fig. 31.9) fixed at 4 m intervals; two on short wall and the others over the span.

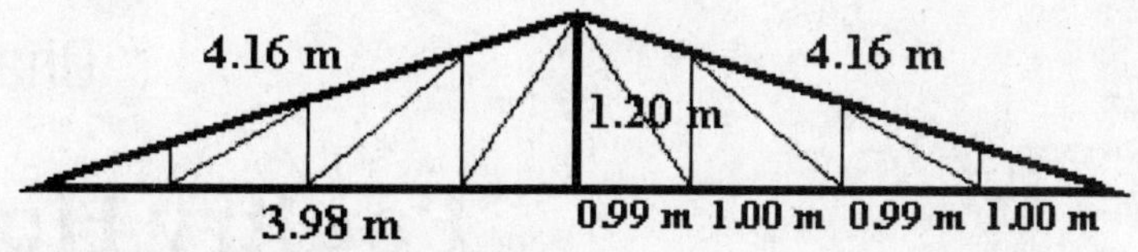

— 50 mm dia MS pipes — 25 mm dia MS pipes

Main rafters: 50 mm dia MS pipes (Total - 17.48 m; say 18 m)

Horizontal - 3.98 x 2 = 7.96 m Slanting - 4.16 x 2 = 8.32 m Vertical - 1.20 m

Struts and Ties: 25 mm dia MS pipes (Total - 5.85 x 2 = 11.70 m; say 12 m)

Strut lengths: 1.56, 1.34 and 1.16 m - Total: 4.06 m

Tie lengths: 0.89, 0.61 and 0.29 m - Total: 1.79 m

Fig. 31.9. Roofing truss details (Pitch = 16.78°)

Table 31.6 Abstract of estimated cost

Sl No.	Description or particulars	Qty	Unit	Rate (Rs)	Amount (Rs)
1	Earthwork and excavation	35	m³	80.88	2830.80
2	Plain cement concrete	4	m³	1496.47	5985.88
3	Brick-work up to plinth	20	m³	1433.40	28668.00
	Brick-work above plinth	19	m³	1620.00	30780.00
4	DPC, 2.5 cm thick	10	m²	88.34	883.40
5	Cement plastering	180	m²	42.12	7581.60
6	White-washing	110	m²	5.24	576.40
7	Plinth-filling	54	m³	87.88	4745.52
8	Flooring, cement concrete	90	m²	121.67	10950.30
9	Wire-mesh (welded)	44	m²	62.50	2750.00
10	Roofing, Asbestos cement	96	m²	160.93	15449.28
11	Roofing, trusses	450	kg	34.83	15673.50
12	Doors, frame	0.10	m³	29720.44	2972.04
13	Door shutters	3	m²	694.85	2084.55
TOTAL 131931.27					(≈ 132000)
Roof-fixing charges @ 5% of cost of roofing material					6600.00
Electrification @ 7½% of total cost					9900.00
Water supply @ 4% of total cost					5280.00
Contingencies @ ½% of total cost					660.00
Work-charged establishment @ 2½% of total cost					3300.00
Electricity, establishment etc.					25740.00
GRAND TOTAL: 157740.00					≈158000.00
Cost of construction/m² of building = Rs 1755.56 (≈ Rs 1760/m²)					

Chapter **32**

Poultry Housing

Domestication has necessitated construction of poultry houses to suit the physiological needs of the bird. Otherwise, the birds in the wild used to scavenge for food and the mother accompanying, especially during the chick stage taking care of the brooding needs as well as training the chicks for scavenging.

Basically, a poultry house should ensure

1. Confinement
2. Protection from predators and environmental extremes
3. Facilitate light control
4. Facilitate bird management
5. Warmth during cold
6. Cooling during hot weather
7. Proper relative humidity
8. Adequate air movement and
9. Reduce ammonia concentration.

Vast information is available on poultry houses in terms of design, construction, materials etc. The following gives the basic guidelines for poultry house design.

1. Site selection

(Berry, on internet)

A number of factors must be considered before building a new building or modifying an existing one. If environmental and production concerns are to be addressed, proper site planning is a must.

1.1. Utilities

Availability of utilities like water, electricity etc. is high on the list of factors to consider during site selection. In some areas, high mineral content of groundwater makes it undesirable for animal use.

Another important factor in site selection is roads. All-weather roads are a must to allow feed trucks, chick delivery trucks, and live haul trucks access to all buildings during any time of the year. On-farm roads should be properly crowned and ditched to insure usability in all types of weather. Although access is important, visitors should not be allowed on to the poultry farm without proper precautions. If precautions are not taken, diseases can easily spread from one farm to another.

1.2. Topography

Topography is also an important factor to consider. It is very important for the building site to avoid low-lying areas with flooding potential. The site must have adequate drainage to easily control storm water. This is particularly important if several buildings are being considered because of the large amount of roof area from which runoff must be controlled. Storm water cannot be allowed to stand near buildings because there is danger of flooding the broiler houses during an extremely heavy rain.

The area around buildings may need to be graded after construction is complete and again on an as needed basis to prevent water from standing or running toward the houses. Storm water should also not be allowed to flow rapidly away from buildings in unplanned ditches because of the potential danger of erosion and damage to building foundations and footings. Perhaps thebest way to control roof water is to use gutters or grass covered ditches to carry water where it should go during run-off periods.

It is also very important that storm water not become contaminated with litter that may have been spilled during the clean out of a poultry house. Careful removal of any spilled litter around the poultry houses should help avoid this problem.

In addition to drainage considerations, topographic concerns should include elevation differences which could hinder the natural ventilation. It is not usually desirable to build a poultry house on the hillside opposite prevailing winds.

1.3. Prevailing winds

The prevailing wind direction is also important from another standpoint. In order to maximize benefit from natural ventilation the poultry houses should be oriented with the long side exposed to the wind. This directional orientation has another benefit, it minimizes the amount of sunlight which enters the house in the summer months. In contrast, during the winter months when the sun is lower on the horizon, sunlight through the windows can help warm the house.

1.4. Existing buildings

1.4.1. Farm buildings

Existing buildings are an important consideration in site selection. Planning for the future is always wise when building plans are on the drawing board. Another aspect to be considered is the effect of existing structures on the natural ventilation. Allowing a good distance on the downwind side of existing structures is always a good idea.

1.4.2. Residence

The residence, if one exists on the site, is also a very important consideration. It is always suggested that poultry houses be constructed on the downwind side of the residence and at a distance such that odors, feathers, and noise from trucks are not a problem. There are times when constructing a poultry house upwind from existing houses and buildings are the best choice. When upwind site selection is necessary the appropriate distance may vary with different locations but a suggested distance is at least 450 m upwind from existing residences on the farm and 900 m from public buildings (i.e., schools, churches), residences off the farm, and commercial fruit or vegetable farms. The downwind distance from other buildings may need to be determined by the wind shed. Wind shed is a term applied to the pattern of the wind on the downwind side of an existing building. The distance between buildings should be such that air flow is very close to returning to normal after passing the existing building.

The location of neighbors' homes must be taken into consideration as well. The neighbors' house and its location in relationship to the crop land area where litter will be spread at clean out time should also not be overlooked as an important consideration. Even though house-cleaning may only occur on an annual basis it may significantly impact neighbor relationships. When considering building sites close to property lines, growers should remember that there is no guarantee the adjoining property will always be used for agricultural purposes.

1.5. Building permits

Building permits, which include waste disposal plans, wherever applicable, must be obtained in advance of construction. Even if existing laws do not require approval of a waste disposal plan, it is advisable to develop a plan for use when construction is complete. Land application is the best use for poultry house litter. The suggested application rate is based on the annual fertilizer need of the crop being grown. Therefore if sufficient land is not available for proper disposal of litter an alternate site should be selected.

1.6. Disposal of dead birds

Proper disposal of dead birds should also be a consideration when considering a building site. Former practices such as disposal-pit burials are not recommended and in some areas not permitted. A composter may be a good alternative, but they too must be included in site considerations.

When new buildings or changes in existing buildings are being considered, a comparison of sites, based on the factors discussed above, should be made. The site which best satisfies all factors considered would likely be the best choice.

2. Structural constraints

(Charles *et. al.*, 1994)

Intensive rearing of poultry with high genetic potential mandates provision of enough ventilation, precise temperature control etc. In cold and temperate climates, insulation is also compulsory. Building dimensions and configurations like length, width and height depend not only on number of birds but also on several other factors like manure pit, ammonia accumulation, humidity build-up etc. Broiler and turkey houses may require a clear span design to facilitate cleaning and catching.

Internal surfaces usually need to be smooth for cleaning and the external surfaces should have proper color, appearance and design for aesthetic reasons.

Choice of structural strengths and design of structural members depends on weight of the birds, food, water and equipment to be carried. However, readymade equipment are available specifying the number of birds and/or weight of birds equipment can handle which greatly reduces the need of calculation of the structural matters by a farmer.

3. Orientation

Poultry houses, conventional/open-sided/windowed or environment-controlled/window-less, are constructed with their length facing east-west. This is to obviate direct sunlight, draft and rainfall into the building. In case of open-sided houses, direct sunlight complicates the control of temperature within the house, direct draft causes problems of dustiness due to agitation of litter and direct rainfall results in damp litter and the consequent disease problems. The roof of the poultry house also projects 0.60 to 0.90 m outside the side-wall in order to drive away the rainwater; this projection is referred to as overhang. Buildings for younger birds should be constructed upwind from those for adult birds.

4. Open-sided houses

These houses are common because they are cheaper for both construction and maintenance. Further, detailed calculation of airflow requirement is not always required because, airflow, in itself, is not under the control of the poultry farmer.

4.1. Plinth

Poultry houses are constructed in an elevated area with a plinth of at least 0.60 to 0.90 m in order to keep the floor well above the ground to obviate seepage of moisture from the surroundings.

4.2. Width

Poultry excreta has a digested food material, uric acid etc. which when acted upon by enzymes elaborated by the bacteria present in it (deaminases, uricase etc) liberate ammonia; hence, there is the accumulation of ammonia in the poultry house which is accentuated by the reduced draft due to orientation of the building itself. Therefore, to keep the ammonia concentration within the acceptable limits of 15 ppm, the width of a conventional poultry house (which are windowed) should not exceed 12.20 m; preferably, 9.0 m.

4.3. Length

The length of the house is fixed depending on the type and age of birds as well at system of rearing.

4.4. Side walls and roof

Stud or height of the side wall of a poultry house will be 2.10 to 2.40 m and the height at the centre will be 3.00 to 3.30 m. In areas where temperatures are likely to be high, the stud height can be 3.00 m. Hence, there will be a gradient with a pitch of 25 to 33° in the roof on both the sides. The roofing material is preferably asbestos; otherwise, tiles, thatches etc. can also be used as roofing material. An overhang should be provided as protection from seepage of rain-water into the poultry house.

A brooder house (where young chicks will be grown up to eight weeks of age) will have one-half to two-thirds the side wall from the floor level made of brick and the remaining covered by the expanded metal. This is to conserve heat during the brooding period. If the climatic conditions are expected to be warm throughout the year, the entire side wall, excepting the brickwork sufficient to fix expanded metal, can be covered with it (expanded metal).

In case of grower house (8 to 18 weeks) and layer houses (after 18 weeks) under deep-litter system, the side walls will be of one-third brick with the remaining portion covered by expanded metal. In case of caged layer house (both conventional and high-rise types), the entire side wall is open and covered only by expanded metal. This is , because, housing density is higher in cage system (nearly twice as much as in deep-litter system) which results in a) excess heat b) excess carbon dioxide and c) excess excreta being produced; the latter, also results in production of excess ammonia due to microbial action and excess moisture due to evaporation; consequent on these, higher ventilation becomes mandatory to drive out excess ammonia , carbon dioxide, heat and moisture from the building so as to keep the layers comfortable.

4.5. Floor

The flooring, in case of deep-litter system, must be preferably of reinforced cement concrete to contain the rats making a way through the flooring and also to minimize seepage of moisture. In case of conventional caged layer house, dropping pits are constructed, amidst the concrete flooring, just below the cages to suit the fall of droppings from the birds so that part of the moisture in the excreta will be absorbed into the soil. In case of high-rise houses, no flooring is required.

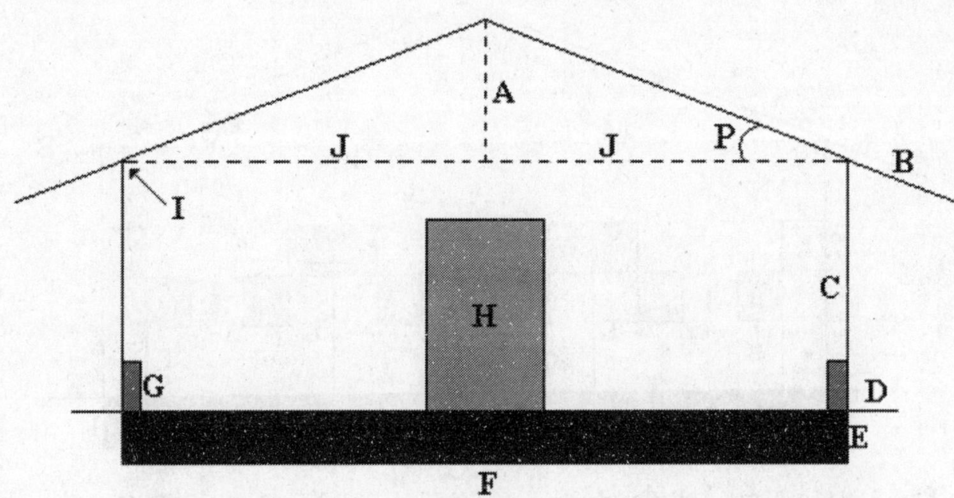

A Rise, 0.60 to 1.20 m B Overhang C Stud height, 2.10 to 2.40 m
D Sidewalk, 0.45 to 0.60 m E Plinth, 0.30 m F Width, 9 m (Max)
G Side-wall, 0.30 to 1.20 m depending on climate H Door, 1.80 x 1.20 m I Eaves J Run, 4.5 m (Max) P Pitch, angle depending on A and J or F
(Note: Pitch normally does not exceed 30°)
Fig. 32.1. Cross-section of a poultry house – Deep-litter system

4.6. Sidewalk

It is a good practice to allow a 0.45 to 0.60 m projection of the floor outside the side walls with all the surfaces being smoothened; this helps control of rats and snakes. Similarly, a gap of 0.90 to 1.20 m between the steps and the poultry house itself also helps in control of rats.

A cross-section of deep-litter and conventional cage-system is shown in fig. 32.1 and 32.2, respectively; a floor-diagram of conventional cage-system indicating dropping-pits is given in Fig. 32.3.

In case of conventional caged layer houses, the problem of elimination of excess moisture, carbon dioxide, ammonia and heat is accentuated by the fact that a) the height of the house is only 3.30 m and b) only 0.60 to 0.75 m space is available between lower-most cage and the floor.

Hence, the new innovation in housing players in cages is by raising the passages to 2.1 m above the floor over concrete pillars and the cages being fixed above the catwalk so formed. Consequent on this, a clear 2.10 m space is available below the cages for optimum ventilation and quick drying of the excreta and the stud or height of the side wall increases to 4.10 m; thereby facilitating ventilation and manure-handling. Such houses are referred to as high-rise or improved caged layers houses. Inside such houses, the cages can be arranged as per the needs of the farmer. Cross-section of two popular high-rise house designs is shown in Fig. 32.4 and Fig. 32.5.

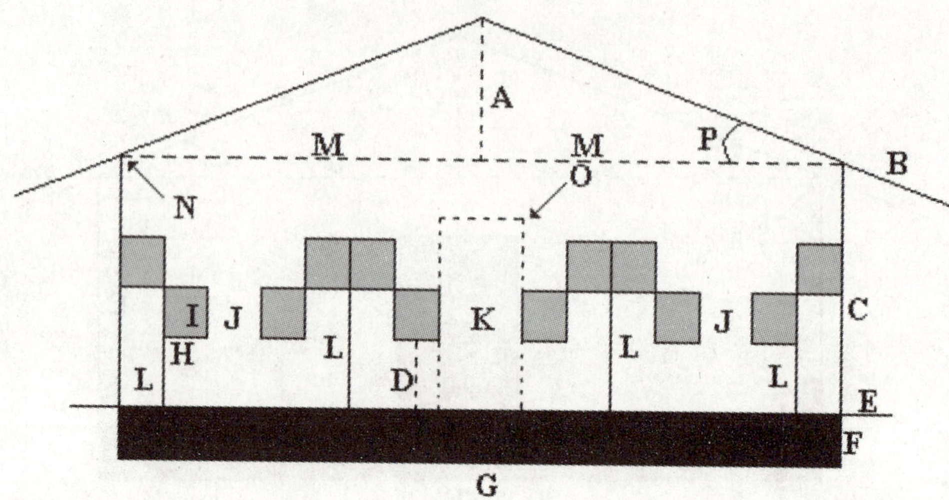

A Rise, 0.60 to 1.20 m B Overhang C Stud height, 2.10 to 2.40 m D Height up to cages, 0.60 to 0.90 m E Sidewalk, 0.45 to 0.60 m F Plinth, 0.30 m G Width, 7.8 m H Depth of the cage, 0.375 m
I Height of the cage, 0.45 m J Side passages, 0.90 m
K Central passage, 1.20 m L Cage support, 0.08 to 0.10 m M Run, 3.90 m
N Eaves O Door, 0.75 m x 1.80 m
P Pitch, angle depending on A and G or M (usually does not exceed 30°)

Fig. 32.2. Cross-section of a poultry house – Conventional cage system

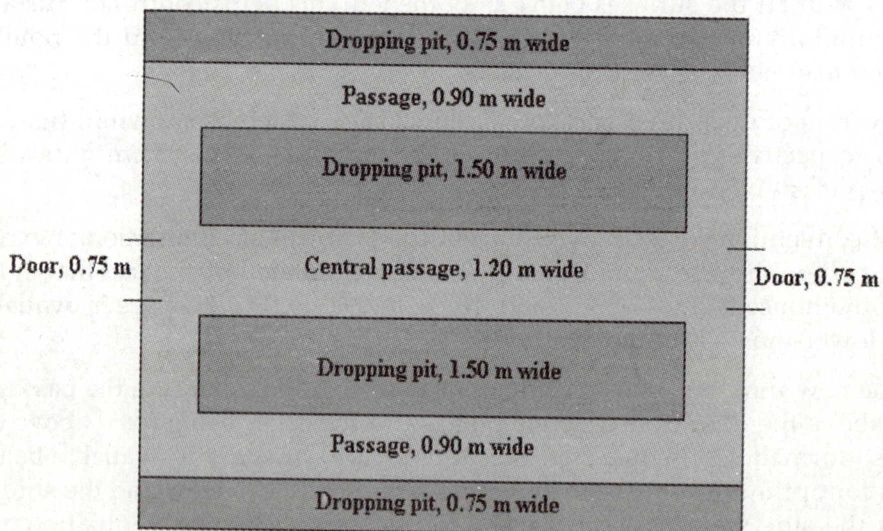

Fig. 32.3. Floor diagram – Conventional cage-system

5. Environment-controlled poultry houses

(North and Bell, 1990)

In the poultry houses, the optimum requirements of the bird is provided inside a completely enclosed and insulated house without windows. Hence, fresh air is brought-in and stale air is removed mechanically. Artificial light is provided to illuminate the interior. On most occasions, animal heat is used and additional heat is used mainly during brooding period.

Structural requirements of environmentally-controlled houses are much similar to that of open-sided houses with the following specifications:

5.1. Structure

1. They should have good foundation and a gable roof
2. Insulation is compulsory
3. Overhang need not be as extensive as in case of open-sided houses because, sides are completely covered
4. Both sides and the top of the building should be given protection

5.2. Ventilation

1. Details of exact ventilation requirements have to be worked-out and even distribution of air both during hot and cold weather must be ensured.
2. Width of the poultry house can be greater than that of open-sided houses because, mechanical ventilation takes care of the removal of stale air; therefore, width can be 12.2-16.5 m.
3. Size and number of fans must be properly calculated and positioned; their blades and shutters must be cleaned regularly. As a rule of thumb, 0.9 m diameter fans working on 0.373 W can deliver 283 m^3/min of air whereas, 1.4 m diameter fans can deliver 850 m^3/min.
4. Air flow calculated as described under chapter on "Mechanical ventilation" has to be provided; however, a general rule is to provide an airflow of (0.024 + 0.00135*T) m^3/min/kg where T is the ambient temperature. In the range of 30-60% relative humidity, when the air temperature is 4.4, 15.6, 26.7, 37.8 and 43.3°C, airflow required is 0.030, 0.045, 0.060, 0.074 and 0.082 m^3/min/kg.
5. It is advisable to operate fans with rheostats so as to alter the speed of motors driving the fans as and when required.
6. Depending on the air movement required, some of the fans may be cut-off at intervals and some others operated continuously
7. The volume of air move by a fan to its electrical consumption is referred to as ventilating efficiency ratio; higher the ratio of the fan installed, the more economical it will be.

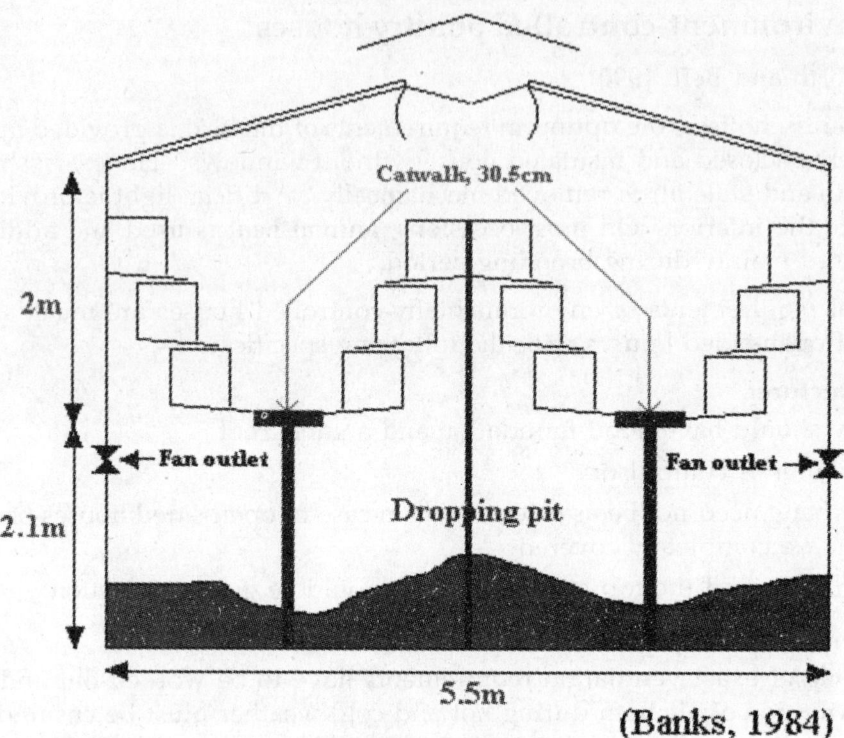

Fig. 32.4. High-rise house (2L, 1M)

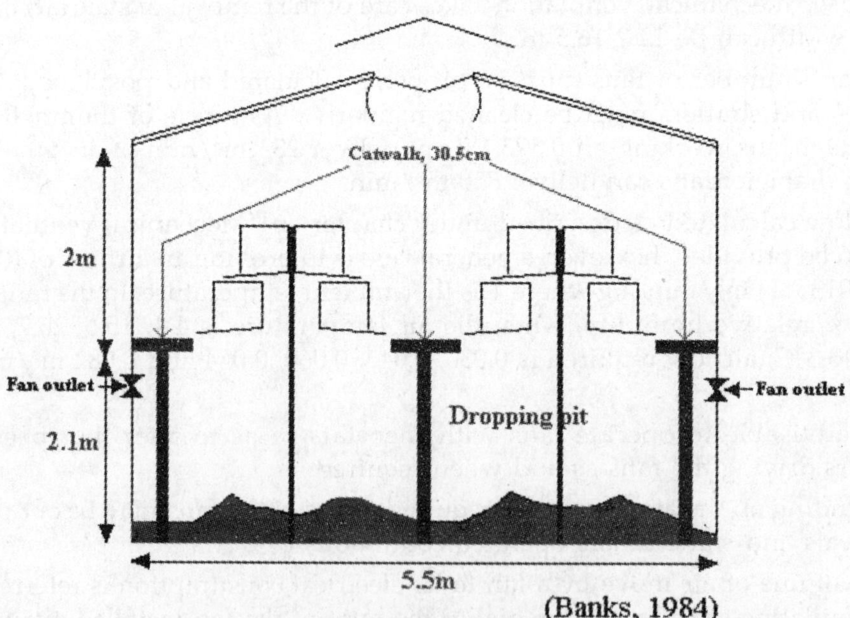

Fig. 32.5. High-rise house (2Ms)

8. As discussed under the chapter "Mechanical ventilation", it is advisable to have a static pressure of 10 Pa (although it reduces the fan efficiency to 85% of full capacity) in order to have optimum air circulation.
9. Shape of air inlets, their locations, use of baffles, velocity of incoming air etc. has been detailed under "Mechanical ventilation". However, as the rule of thumb, 57 cm^2 of intake opening for every m^3/min of air exhaust; with light-traps over the fan openings, intake opening may be increased by 25% to 71 cm^2.
10. When ductwork is used in a ventilating system with positive pressure, static pressure is likely to drop through the ducts resulting in reduced movement of air. Under most air velocities (183 to 366 m/min) corresponding to static pressures of 5 to 225 Pa, the fan velocity can be approximated as V(1 - 0.0625*vh), where V is velocity of air in m/min and h is static pressure in Pa (Note: Original equation in FPS units is converted to metric units). For instance, at 25, 50, 75, 100, 125, 150, 175, 200 and 225 Pa, air speed of 275 m/min reduces to 189, 153, 126, 103, 83, 65, 48, 32 and 17 m/min, respectively.

5.3. Cooling

When outside temperature exceeds 29.4°, it is mandatory that poultry houses must be cooled by any of the following methods:

5.3.1. Low-pressure fogging system

Fogging nozzles are installed at regular intervals throughout the building or over the birds in cages so that, as and when necessary, fogging can be practiced.

5.3.2. Pad-and-fan system

The details of calculation of pad and fan requirements are given under "Mechanical ventilation". However, the two types of this system namely,

5.3.2.1. Pressurized system

Evaporative coolers are placed outside the house and air is sucked through the evaporative pads and forced into the poultry house. A slight positive pressure is built within the house to facilitate air exhaust through the opening provided.

5.3.2.2. Vacuum system

Evaporative pad is placed in the wall at one end of the house and exhaust fans at the other; the operation of the latter creates enough pressure to cause air to be sucked through the evaporative pad, thus reducing the temperature of the incoming air. Other considerations are similar to that for pressurized system.

In either of the above two systems, exhaust fans should be located in the opposite wall or end from the evaporative pads. Fans that are large in diametre and that revolve slowly, yet move large volumes of air, should be preferred. The amount of air movement required has already been discussed under "Mechanical ventilation"; however, as a rule of thumb, airflow required is 0.11 m^3/min/kg of live weight of birds in the house.

It is necessary to remember that the cool air entering the house gets warmed by the time it reaches the other end of the house. Therefore, care has to be exercised to limit the heat build-up to less than 4.5°C.

5.3.3. Fog-and-fan system

This is similar to pad-and-fan system except that incoming air is drawn through a hood in which high-pressure foggers have been installed to cool the air as and when it is drawn through the fog.

5.3.4. High-pressure fogging system

This system consists of a hood over air inlets and high-pressure fogger nozzles placed in a pipe just below the lower limit of the hood so as to emit a water mist into the hood; the water mist is then sucked into the house by incoming air created by the exhaust fans. The nozzles operate under about 35 kg/cm^2 pressure.

A slot at the top of the wall and that the top of the hood is used for better mixing of the fogged air as it enter the building. Exhaust fans are located on the opposite walls, and covered with a hood.

Since very high pressure is involved, it is highly advisable to use filtered water to minimize sedimentation and disruption of water flow through the nozzles. Similarly, thermostatic control of the fogging system is also desirable.

5.3.5. Additional cooling

Whitewashing of the roof of an uninsulated house significantly reduce the interior temperature of the house by reflecting the sun's rays. The following combinations can be used for whitewashing:

1. 9.1 kg hydrated lime + 18.9 l water
2. 9.1 kg hydrated lime + 9.1 kg white cement + 22.7 l water + 118 cc bluing
3. 9.1 kg hydrated lime + 18.9 l water + 946 cc polyvinyl acetate

5.4. Light control

Environment-controlled houses are light-proof; where fans are installed in the side, light-traps must be installed to prevent light seeps into the building by providing a hood on the outside of the building over an opening extending down far enough to prevent light from entering. However, light-traps are likely to obstruct air movement and hence a floor requirement increases.

6. Types of houses

Most commonly employed housing types (Deep-litter, cage system and high-rise houses) have been already discussed under "open-sided houses". These types can be environment-controlled also provided they are totally enclosed (windowless) and the internal environment is mechanically controlled. In addition, slat-and-litter and all-litter houses can also be constructed, with or without windows.

6.1. Deep-litter house

In this system, birds are reared on floor covered with a litter material. Cross-section of a deep-litter house is given in Fig. 32.1.

6.2. Cage system

In this system, the birds are confined inside cages. Hence, the cages should have sufficient structural strength to support the weight of birds, feed, eggs, water and droppings without failing and/or sagging.

6.2.1. Conventional cage system

Birds are reared in cages of varying dimensions, depending on the number of birds it can hold, fixed at about 45 cm above the floor level. The floor is left uncovered just underneath the cages for absorption of moisture in the fecal matter by the soil (Fig. 32.2 and 32.3).

6.2.2. High-rise houses

Floor of the conventional cage house along with cages is raised by 2.1 m and the passages are converted into catwalks. Therefore, droppings fall into a 2.1 m deep pit making it easy for in-house drying, sheltered accumulation and handling (Fig. 32.4 and 32.5). High-rise houses are becoming popular due to the obvious advantages.

It is obvious that a high-rise house is a two-storey house; the top floor of the birds in cages (sometimes on an all-slat floor) and the bottom floor, with no ceiling, directly underneath the birds used for accumulation of manure; each about 2.1 m high.

During cold weather, when manure drying becomes the problem, supplementary fans may have to be operated. The fans should blow air down the front of the pit and back on the opposite side, creating a circular movement over the dropping at about 120-180 m/min. A fan blowing at 283 m^3/min is required for every 186 $m^2$of floor space. The fans should be hung from the top of the pit, and be reversible.

The manure pit can be cleaned annually; in fact, it is deep enough to accommodate droppings of seven years. After cleaning the manure pit, it is advisable to leave a pad of manure of about 15-30 cm to help bacterial action on subsequent droppings.

6.3. Slat-and-litter house

In this type of house, part of the floor is covered with slats (60% of the floor area) leaving the rest for deep-litter either at the centre or adjacent to the side walls (down the edges of the house); the former is preferred because on deep-litter, even if there is seepage of water, it will not cause problems associated with damp/ wet litter. On the contrary, slats at the centre of the house has an advantage in environment-controlled houses because the feeders and sulfur are placed on the slats will be close together. Slats measuring 2.5-5 cm in width and spaced 2.5 cm apart are fixed running lengthwise of the building. The top of the slats should be at least 70 cm above the floor level to allow space below the slats for a year's droppings to accumulate; care has to be exercised while fixing the slats at a too far higher level from the floor since sagging will be a problem.

Usually breeders are reared in slat-and-litter house; birds aged less than 20 weeks need to be trained to use the slats. Welded-wire is poor substitute for slats. However, if wire has to be used, it should be heavy enough not to sag; therefore, supports at about every 30 cm may be required. Mesh size desirable is 2.5 by 5.0 cm with the long part of the mesh running crosswise of the building.

Slat or wire floors to be constructed in sections to facilitate their easy removal, cleaning and disinfection; it will also help removal of droppings and cleaning of the entire house.

6.4. All-slat house

The main advantage of the house is that the birds require less floor space than deep-litter. The details regarding slat size, spacing in construction are similar to that for slat-litter system

7. Poultry housing under hot climate

7.1. Factors affecting poultry house design

(Ernst, 1994)

7.1.1. Climate

Under typical housing densities (2.5 kg/m^2), birds produce about 15 times more heat than that enters through roof and walls of an insulated poultry house. This heat can be removed by increasing the airflow through the building or the evaporation of moisture into the air. When the ambient temperature approaches 41°C, air movement cannot help heat removal and therefore, the building must be closed to allow minimum ventilation required to remove moisture. Evaporation inside the building may be increased byfogging water or direct wetting of poultry and by running fans inside the house.

7.1.2. Roof type

Gable/shed roof is used for poultry with ridge ventilation to allow hot air escape at roof peak. Shed-type roof is useful with weight less than 10 m and ridge ventilation is beneficial when hot weather is accompanied with extended periods with little wind. Hence, ridge ventilation does not work well with open-sided houses.

Low-conductance material like loose fill, batts or blankets, boards, spray-on foams or blown-in material which dry to form a rigid fill can be used to decrease heat absorption from roof depending on cost and availability. In regions where temperature may go below 15.6°C, condensation will be a problem which necessitates use of vapor barriers so as to maintain effectiveness of insulation. Insulation material must be installed without gaps, tears or cracks.

7.2. Location

(Smith, 1981)

A choice of good site may take advantage of lower daily maximum temperature and prevailing winds. A site suitable for green cover and shading trees would be especially advantageous in a B/C climate. In As, Am, Aw, Bwh and Bsh climates, cooling may be a daily routine whereas in C climate, cooling may be needed only during hottest months and are used only on hottest days.

7.3. Heat transfer into the building

(Smith, 1981)

In tropics, vertical surface receives least energy and a horizontal surface the most. In addition, vertical surface facing the Equator receives less direct solar radiation than one facing the east or west. Hence, building-oriented east-west receives less solar radiation.

Diffuse radiation contributes 15-20% of total radiation and is available even when the sky is overcast. Even smaller amount of cloud can increase radiation sometimes although many times, clouds decrease solar radiation. Cloud cover, simultaneously, obstruct loss of heat from the Earth leading to narrower diurnal range. In the temperature range of 0-20°C, the sky temperature is cooler by 20K; at 38°C it is 16K cooler and at 43°C it is 12K cooler with complete cloud cover. Cloud surface is usually 2K cooler than ambient temperature.

Ambient temperature may exceed 40°C on a clear day and if cloud cover becomes complete in the late afternoon, the low long-wave radiation losses will result in only a small nocturnal fall in temperature; then, it may be necessary to cool even during the night.

Atmosphere against most of its energy in directly – via connection and thermal radiation from the earth's surface. In tropics, surface of concrete can go up to 60°C. Temperature gradient over 2-3 m is likely to be in the order of 30K. However, temperature of air close to the turf is essentially the same as ambient temperature.

7.3.1. Radiant heat *Vs* shade

Direct irradiance of the wall can be decreased by shades on the building. Roof overhang will shade to a vertical wall most of the time of the day. But reflected short-wave and long-wave radiation emitted by the ground and surroundings too contribute to heat load. Amount of overhang needed depends on solar altitude angles for a specific time, data and latitude; for instance, at the latitudes 20-35°, for 90d of highest noon Sun (45 d on either side of summer solstice), the Sun's zenith angle is about 45° from 0900 to 1500 hrs.

A shade with an angle of 45° allows the use of minimum material to shade the whole wall. With the sun above 45° zenith angle, the shade length needed is 70% of the wall height. Therefore, before and after 0900 to 1500 hrs, up to ½ of the wall would be unshaded. From 0700 to 0900 hrs, when radiation intensity on east-and west-facing walls is greatest, only the bottom quarter of the wall would be exposed.

By intercepting the sun with a shade, the radiant heat load may be reduced by 30%. Within the boundaries of the shade, the radiant heat load on chicken comes from 1. Atmosphere 2. Roof 3. Shaded ground and 4. Sunny ground. The amount of radiation received by the bird depends on fraction of total energy from various sources intercepted by the bird which, in turn, depends on bird's height above the ground; the higher it is, the more is the total radiation. Long-wave radiation forms about 80% of the total radiation heat load. The sunny ground component is a maximum when the ground is bare; a cover of turf will decrease the load by about 25%.

Heat load is minimum under shade when the distance from the ceiling is maximum. Hence, birds on floor are cooler than those in elevated cages; in other words, litter is preferred under hotter Aw climate prevalent in India.

Radiant heat load may further be reduced by shading the building with trees and shrubs. Trees capable of shading the walls and roof are highly advantageous for buildings with a north-south orientation particularly if placed to take advantage of prevailing winds. The trees also contribute a microclimate – when the ground is covered with vegetation, the heat absorbing surface (previously the soil) is transferred to top of the plants. Solar radiation absorbed by plants is largely carried away by forced convection with a small fraction being re-radiated to cooler bodies.

High foliage density of some shrubs curtail wind flow which is a disadvantage; they can be good wind-breakers. Loosely foliated shrubs can shade the walls of a building while allowing passage of air. Layer of air beneath trees and moving through shrubs receive the water of transpiration which has a substantial cooling effect decreasing the temperature to less than that on open turf.

7.3.2. Shelter

If shrubs at the perimeter are impractical, radiation may be reduced by intercepting reflected short-wave and emitted long-wave radiation by the walls. Location of such solid parts depends on location of stock – floor or cages; wooden slats decrease short-wave radiation considerably but slightly of long-wave. Complete enclosure of a shelter with walls could decrease radiant heat by 10%. Different construction maferial also can help decrease radiant heat; for instance, reflective and low-emissive paints to coat interior and exterior building surfaces and use of water to cool building surfaces.

7.3.3. Building material

Flow of heat through a building is a function of thermal conductivity, k (W/m/K), the density, ρ (kg/m^3) and specific heat, c (J/kg/K) of the material. Thermal diffusivity, k/(ρc) (m^2/s) is the property of the material.

Walls made of low thermal diffusivity produced large time lags between outdoor and indoor maximum temperature which results in reduced diurnal variation; this is helpful under A and B climates which have maximum temperature to the extent of causing even fatal heat stress. It is also suitable for smaller units. However, with large flocks, higher ventilationis more advantageous and practical under commercial conditions.

When solar radiation is intense, high outdoor dry bulb temperature and radiation decide heat flow. The calculated dry bulb temperature that would produce the same heat flow if the radiation exchanges were to have been absent is called "sol-air" temperature. Difference between sol-air temperatureand indoor temperature together with materials of construction decide the heat flow. It is preferable to use material of lower than a higher thermal diffusivity; thermal diffusivity of wood < asbestos cement < dense concrete < steel < aluminium.

7.3.3.1. Surfaces and surface coatings

All radiation (long-/short-wave) when strikes a material is either reflected from/ transmitted through/absorbed by the material; and the summation of all these three is unity for all materials although each of them may be different for different materials. The differences depend upon absolute temperature of the material, is chemical and physical characteristics and wavelength of the incident radiation. A song energy is transformed to heat and is conducted: parts of the material or emitted as long-wave radiation. Each material has a different capacity to emit long-wave radiation and it is referred to as its emissivity, ε.

Materials for and coatings on the outside of buildings would absorb less heat if they had a lower absorptivity for short-wave radiation and also had a high ε. Material and coatings on the inside which reduce radiation load on the animals if they had high short-wave absorbance and low ε. Therefore, it is obvious that ratio of absorptivity to ε is important in deciding the material for surface coating. Material with low ratio are suitable for outside of the buildings (for instance white paint of aluminium and whitewashing on GI with ratios of 0.22 and 0.24, respectively). On the contrary, material with high ratio are suitable for inside of the buildings (new and old GI and aluminium foil with ratios 5.0,2.9 and 3.0, respectively). Blue nickel oxide on aluminium has a ratio of 15.

GI is universally available with higher ratio; therefore, it is used wherever others are either not available and/or expensive. Aluminium foil bounded to plywood sheets, reinforced bituminous paper sandwiched between two aluminium foil layers etc. can be used to decrease the summer thermal conductivity value of a wall/roof with one or more air spaces.

7.4. Methods of cooling

(Smith, 1981)

7.4.1. Evaporative cooling

By use of roof-top sprinklers where water is available, fogging can be practiced where buildings are not totally closed and it is more effective than roof-top cooling. However, with frequent use of fogging, feed and litter become wet. Foggers can be operated for 15 min/hr when temperature is above 32°C.

In completely closed buildings, evaporative cooling is necessary when outside temperature is above 30°C, foggers, pad and fan cooling, refrigerated air cooling etc. can also be practiced wherever practicable. However, pad and fan appears better than evaporating cooling

Evaporating cooling are effective in B climates and in certain A and C climates in the dry season; but of little use in humid tropics. As a rule, relative humidity should not be more than 65% and 75% on litter or in cages, respectively for use of evaporative cooling. Broilers appear to tolerate higher relative humidity if temperature is reduced by evaporative cooling.

7.4.2. Air movement

In all climates, increase in air velocity increases heat losses. Velocity up to 2.5 m/s will decrease physiological response to dry bulb temperature up to 40°C. But, at > 40°C, increased velocity exacerbates the physiological response. Weight gain benefits in broilers at 2.5 m/s velocity and 18°C dew point, over a range of dry bulb temperature from 21-35°C was nearly directly proportional to (air velocity)$^{1/2}$.

In humid tropics, increasing air velocity up to 2.5 m/s is probably the only option to increase heat loss. Evaporative cooling may further benefit as long as indoor relative humidity is less than 70-75% beyond which, the only alternative left is air-conditioning.

Radiation heat loss/unit of surface is independent of size of the animal but convective heat loss is dependent on geometry of the body and air velocity. For young birds, increasing air velocity is more effective than reducing wall temperature and for large turkeys, both are equally effective.

In Am and Af climates, growth and FCR of young broilers can be improved by increasing rate of air movement alone rather than evaporative cooling up to the practical limits of relative humidity.

Chapter **33**

Rearing Systems

Before rearing birds exclusively for commercial purposes became popular, they used to be reared in the backyard with no sophisticated housing, equipment, feeding and, for that matter, disease prophylaxis. In other words, rearing birds was more a hobby with whatever useful product that comes in the way as a bonus.

The above system, popularly referred to as "Extensive system" has serious limitations as far as rearing high-producing modern poultry; some of the salient shortcomings are as follows:

1. As the genetic potential of the birds improved, requirements from the environment to express that potential to the fullest extent also became more and more complicated.
2. Nutrient requirements of high producing poultry cannot be met with backyard grazing.
3. Similarly, high producing birds are more susceptible to changes in the environment like temperature, humidity, ventilation etc.
4. High-producing birds are also more susceptible to diseases and parasites and hence, growing them in the backyard was surely exposing them to various infective agents in addition to nutritional inadequacies.
5. Protecting the birds from predators is another problem because high producing birds are normally non-aggressive and slow to react to predators.

Therefore, gradually a temporary shelter was introduced at the backyard itself where the birds were given protection against inclement weather and also additional nourishment for improved production. This system is referred to as "Semi-intensive or Semi-extensive system" depending on the proportion to which the birds are subjected to indoor environment.

Research regarding space requirements (feeding and drinking) and nutrient requirements helped develop farming systems wherein birds are totally confined inside a "poultry shed or barn" and feed with requisite nutrients, clean drinking water and other requirements are provided by the farmer. Rearing of birds within a confinement is referred to as "Intensive system". Simultaneous developments in control of diseases and development of strains and varieties with specific characters in view, augmented poultry industry development.

The intensive system of rearing is under continuous refinements starting from all-litter, cages, all-slats, slats-and-litter to the recent "enriched or furnished cages" and "alternate systems". Each of the systems has its own advantages and disadvantages; therefore, the farmer has to select the one depending on his needs.

1. All-litter (Deep-litter) system

In this system, the birds are reared on floor covered with six to eight cm of litter material which is made of paddy husk (rice hulls), wood shavings, peanut shells (peanut hulls) or any other suitable material. The birds are free to move and the feeders and waterers are arranged at convenient height so as to be available to the birds at all times on the litter floor.

1.1. Qualities of a good litter material

(North and Bell, 1990)

A good litter material should

1. be light in weight
2. have a medium particle size or fairly coarse to prevent caking/packing
3. be highly absorbent
4. dry rapidly by releasing absorbed moisture into the incoming air
5. be soft and compressible
6. show low thermal conductivity
7. absorb a minimum of atmospheric moisture
8. be inexpensive
9. be compatible when sold as fertilizer
10. be easily available during all times of the year

1.2. Commonly used litter material

Commonly used litter material along with their moisture-absorption capacity is given in Table 33.1.

1.3. Advantages

1. Deep-litter keeps cool during summer and warm during winter. Hence, the birds will be comfortable during all seasons; that they can move freely also adds to their comfort.
2. Birds derive certain un-identified growth factors (presumably, vitamin B_{12}, animal protein factor etc.) from the litter. This has been found to be of specific advantage for breeding flock in terms of fertility.
3. There will be no incidence of breast blisters in case of broilers.
4. Usually, there'll be no problem of ammonia accumulation and house-flies in a well-managed deep-litter flock.

5. Incidence of broken eggs is very minimal.
6. There wi'' be no problem of caged layer fatigue.
7. There wiıı be uniform distribution of light in the layer house.
8. Initial investment is less when the land cost is low.
9. Provides for all the welfare requirements of the birds

Table 33.1. Moisture-holding capacity of litter material (g/100 g)

Litter material	Moisture held
Pine straw	207
Peanut hulls	203
Pine shavings	190
Chopped pine straw	186
Rice hulls	171
Pine stump chips	165
Pine bark and chips	160
Pine bark	149
Corncobs	123
Pine sawdust	102
Clay	69

Source : North and Bell, 1990

1.4. Disadvantages

1. Housing density is lower than in cage system.
2. There will be more feed wastage due to spilling etc.
3. Birds consume more feed since they move about more freely wasting some energy; hence, feed efficiency is inferior to that of birds in cages.
4. Litter-borne diseases can occur, especially coccidiosis, costing severe economic losses particularly in broiler industry.
5. Diseases spread faster due mainly to free movements.
6. Cannibalism, if starts, will be severe when compared to cage system. Similarly, feather-pulling, picking and other vices are possible.
7. Incidence of unclean or soiled eggs is higher.
8. Birds consume more feed per dozen eggs.
9. Maintenance of individual records will not be as accurate as in case of cage system and hence, culling of birds requires careful observation and even handing of birds individually.
10. Birds are likely to produce slightly fewer and lighter eggs than in cage system; but this is still debatable.
11. Broodiness can be a problem.

12. Nests have to be provided and the eggs have to be collected regularly.
13. There is a chance of egg-eating vice, especially when the eggs are left in nest boxes or on the floor for a long time.
14. Fighting among breeding cockerels/cocks is noticed especially in pens with many cocks.
15. Birds of different color patterns cannot be mixed and reared. However, if chicks of different plumage are reared together from the brooding stage itself, no problems will be encountered; otherwise, serious fighting among the birds, sometimes leading to considerable mortality among cockerels, can result.

2. Cage system

In this system, the birds are reared in enclosures made of, usually, wire mesh or expanded metal. Therefore, the birds do not have access to their own fecal matter. The enclosures, referred to as "cages" can be in the form of "tiers" depending on the convenience and the type of birds. Hence, more number of birds can be reared per unit floor space than in deep-litter system or any other system for that matter.

Housing design of both conventional and high-rise cage houses are discussed in "Poultry housing".

2.1. Conventional cage-system

2.1.1. Special features of cage-layer houses

2.1.1.1. Side-walls

One of the main advantages of cage-system is higher density possible; approximately, twice the number of birds as on all-litter can be reared in the same space. Increased density has its own drawbacks: a) higher sensible and insensible heat produced b) more fecal matter per unit space and therefore more ammonia production as well as more evaporated water c) increased ventilation requirements to meet oxygen demand and eliminate both ammonia and carbon dioxide produced d) fecal matter, particularly with delayed drying, is an excellent breeding ground for flies and hence, fly nuisance is a common problem in most of the cage-layer houses.

In order to minimize the above problems, side-walls of a cage-layer house will be covered entirely with expanded metal to facilitate easy air moment and ventilation.

2.1.1.2. Feeding and watering channels

Regardless of the type of cage, feeding channel is fixed at the level of the back of the birds and the watering channel about 10 cm above it. This arrangement serves two purposes; a) it is convenient for the birds to drink because, they cannot suck water from lower levels due to absence of lips and b) if watering channel leaks, water falls into the feeding channel but not on to the droppings.

Due to high density of the birds, there are already multiple problems in the form of removal of heat and moisture produced by the birds and ammonia produced in the droppings. Further, fly problem is common in most of the cage-rearing facilities due to excess moisture content in droppings which acts as an excellent breeding ground for flies. Therefore, it is beneficial if, in case of water leaks, additional water is not added into the droppings.

2.1.2. Advantages

1. Higher housing density possible.
2. Movement of birds is restricted and hence they consume less feed than on other systems.
3. There will be less wastage of feed because movement is restricted.
4. No problem of litter-borne diseases, especially coccidiosis.
5. Spread of disease is slower than in other systems.
6. Incidence of cannibalism is minimal.
7. Birds lay more and heavier eggs; this is still controversial.
8. Consume left feed per dozen eggs.
9. Eggs will be cleaner than in other systems.
10. Maintenance of individual records is easy and accurate; hence, culling is also easy and accurate.
11. Broodiness is avoided.
12. Labor requirement minimum

2.1.3. Disadvantages

1. Birds are uncomfortable because they are not able to move freely.
2. Birds suffer from boredom. It is common to fix a colored plastic wheel in all cages so that the birds can peck at them and play; but it cannot substitute for free movement and association between birds.
3. Under tropical climate, birds are most uncomfortable due to high temperature accentuating the effects of high density and humidity.
4. Feed must be accurately balanced since the birds do not get unidentified nutrients from the litter.
5. Under humid conditions, there will be problems of house flies and ammonia concentration within the house.
6. More number of cracked eggs is expectable since the eggs roll on a wire/ metal floor.
7. Caged layer fatigue is unique to birds reared in cages; when such birds are shifted to deep-litter, they recover but, when put back to cages, they once

again show the symptoms. The actual cause of the disease is still incompletely understood.

8. Distribution of light will not be uniform; birds located in the lowermost tiers receive lower intensity than those above them.
9. Broilers show higher incidence of breast blisters leading to loss of carcass quality and grade. This is because, when the birds rest, breast region comes in contact with the cage material.
10. Layers show higher incidence of bumble foot, breast blisters and other injuries due to cages.
11. Initial investment is high on equipment.
12. Natural mating is not possible unless special type of cage is installed for the purpose. Hence, if fertile eggs are required, artificial insemination has to be resorted to.
13. Handling the birds is generally difficult and stressful to the birds.
14. Manure handling is difficult because of limited space available below the cages.
15. The system is seriously objected to by Animal Welfare groups and rightly so. Details of Poultry Welfare are given in a separate Chapter.

2.2. Raised platform/High-rise houses

Most of the shortcomings under conventional cage-houses were due to a) high temperature and moisture build-up b) fly problems c) difficulty in handy manure and d) ventilation problems due to limited air volume. In order to minimize many of the above problems, the passages (platforms) in the conventional cage-houses were raised by 2 m above the ground and hence, such houses are referred to as "raised platform" houses.

It is obvious that in a raised platform house, manure pit is 2 m deep and hence it is easy for drying process as well as for handling. Mechanization of manure removal is also possible. In addition, side-walls will be of more than 4 m height which increases the total volume of the building, thereby facilitating ventilation to a considerable extent.

It is for the above reasons that raised platform houses are becoming more popular for housing layers.

3. Slat-and-litter system

In this system, part of the floor area is covered with slats. This system is used normally for meat-type breeders (see Chapter "Management of breeders" for figure and further details); when used for others, especially younger birds, they must be trained to use the slats. Slats must be 2.50 to 5.00 cm wide and 2.50 cm apart running lengthwise of the building. They must be fixed at 68 cm above the floor to allow collection of manure over a period of one year for which the birds are grown. The wire (making the slats) should be strong enough not to sag due to weight of the birds and their own. Slats can be located against the side-walls with deep-litter area at the centre or vice versa; in open-sided house, the former is better because, it avoids wetting of litter if rainwater, for some reason, enters the house.

3.1. Advantages

1. More eggs per unit area.
2. Floor space requirement is lesser than on deep-litter.
3. Fewer floor eggs than on all-slat system.
4. Less labor per bird because of increased density.
5. Fertility higher than on all-slat system.
6. Coccidiosis is minimized.
7. Manure handling is easier.

3.2. Disadvantages

1. Costlier than deep-litter system.
2. Fewer eggs than on all-slat or all-litter systems.
3. Crowding may increase incidence of dirty eggs.

Note: Other advantages and disadvantages of deep-litter and cage systems are expected; for instance, availability of unidentified growth factors, uniform distribution of light, incidence of coccidiosis, chances of soiling the eggs etc.

4. All-slat system

In this system, the birds are reared on floor completely covered with slats. Therefore, they do not get access to their own fecal matter and hence do not get unidentified nutrients; at the same time, they are free to move as in case of deep-litter system. Slats must be 2.50 to 5.00 cm wide and 2.50 cm apart running lengthwise of the building. They must be fixed at 68 cm above the floor to allow collection of manure over a period of one year for which the birds are grown. The wire (making the slats) should be strong enough not to sag due to weight of the birds and their own. In general, this system is a combination of most of the advantages of both deep-litter system and cage system.

4.1. Advantages

1. Lesser floor space than on deep-litter; floor space required is approximately midway between deep-litter and cage systems.
2. Coccidiosis can be avoided.
3. There will be no incidence of breast blisters in case of broilers.

4. Usually, there'll be no problem of ammonia accumulation and house-flies in a well-managed all-slat flock.
5. Incidence of broken eggs is very minimal.
6. There will be no problem of caged layer fatigue.
7. There will be uniform distribution of light in the layer house.

4.2. Disadvantages

1. Costlier than deep-litter system.
2. Carcass condemnations due to breast blisters may increase especially when slats are made of wire.
3. Feed must be accurately balanced since the birds do not get unidentified nutrients from the litter.
4. Initial investment is higher than in deep-litter system.
5. Alternate systems

Due to continuous pressure from Animal Welfare activists, several alternate systems are being innovated and work is still in progress to develop a system satisfying all concerned. Member countries of the European Union responded favorably for the demands of the welfare of animals and passed a directive which will certainly have a wide-ranging impact on the poultry productionin the World, in general, and in European Union, in particular.

A brief description of the European Directive and the systems currently under use/trial/consideration is given below:

5.1. The European Directive

The European directive (1999) provided for the following (http://europa.eu.int/eur-lex/en/index.html):

5.1.1. Article 4

1. .... from 1 January 2002 all newly-built or rebuilt systems of production referred to in this chapter and all such systems of production brought into use for the first time comply at least with the following requirements: All systems must be equipped in such a way that all laying hens have: (a) either linear feeders to provide at least 10 cm per bird or circular feeders providing at least 4 cm per bird; (b) either continuous drinking troughs providing 2.5 cm per hen or circular drinking troughs providing 1 cm per hen. In addition, where nipple drinkers or cups are used, there shall be at least one nipple drinker or cup for every 10 hens. Where drinking points are plumbed in, at least two cups or two nipple drinkers shall be within reach of each hen; (c) at least one nest for every seven hens. If group nests are used, there must be at least 1 m^2 of nest space for a maximum of 120 hens; (d) adequate perches, without sharp edges and providing at least 15 cm per hen. Perches must not

be mounted above the litter and the horizontal distance between perches must be at least 30 cm and the horizontal distance between the perch and the wall must be at least 20 cm; (e) at least 250 cm² of littered area per hen, the litter occupying at least one-third of the ground surface.

2. The floors of installations must be constructed so as to support adequately each of the forward-facing claws of each foot.
3. In addition to the provisions laid down in points 1 and 2, (a) if systems of rearing are used where the laying hens can move freely between different levels, (i) there shall be no more than four levels; (ii) the headroom between the levels must be at least 45 cm; (iii) the drinking and feeding facilities must be distributed in such a way as to provide equal access for all hens; (iv) the levels must be so arranged as to prevent droppings falling on the levels below. (b) If laying hens have access to open runs: (i) there must be several potholes giving direct access to the outer area, at least 35 cm high and 40 cm wide and extending along the entire length of the building; in any case, a total opening of 2 m must be available per group of 1000 hens; (ii) open runs must be: of an area appropriate to the stocking density and to the nature of the ground, in order to prevent any contamination; equipped with shelter from inclement weather and predators and, if necessary, appropriate drinking troughs.
4. The stocking density must not exceed nine laying hens per m2 usable area.
5. However, where the usable area corresponds to the available ground surface, Member States may, until 31 December 2011, authorize a stocking density of 12 hens per m² of available area for those establishments applying this system on 3 August 1999. Member States shall ensure that the minimum requirements laid down in paragraph 1 apply to all alternative systems from 1 January 2007.

5.1.2. Article 5 (rearing in unenriched cages)

1. .... from 1 January 2003 all cage systems referred to in this chapter comply at least with the following requirements: a) at least 550 cm² per hen of cage area, measured in a horizontal plane, which may be used without restriction, in particular not including non-waste deflection plates liable to restrict the area available, must be provided for each laying hen; b) a feed trough which may be used without restriction must be provided. Its length must be at least 10 cm multiplied by the number of hens in the cage; c) unless nipple drinkers or drinking cups are provided, each cage must have a continuous drinking channel of the same length as the feed trough mentioned in point (b). Where drinking points are plumbed in, at least two nipple drinkers or two cups must be within reach of each cage; d) cages must be at least 40 cm high over at least 65 % of the cage area and not less than 35 cm at any point; e) floors of cages must be constructed so as to support adequately each of

the forward-facing claws of each foot. Floor slope must not exceed 14 % or 8 %. In the case of floors using other than rectangular wire mesh, Member States may permit steeper slopes; f) cages shall be fitted with suitable claw-shortening devices.

2. .... rearing in the cages referred to in this chapter is prohibited with effect from 1 January 2012. In addition, with effect from 1 January 2003 no cages such as referred to in this chapter may be built or brought into service for the first time.

5.1.3. Article 6 (rearing in enriched cages)

.... after 1 January 2002 all the cages referred to in this chapter comply at least with the following requirements:

1. laying hens must have: (a) at least 750 cm^2 of cage area per hen, 600 cm^2 of which shall be usable; the height of the cage other than that above the usable area shall be at least 20 cm at every point and no cage shall have a total area that is less than 2000 cm^2; (b) a nest; (c) litter such that pecking and scratching are possible; (d) appropriate perches allowing at least 15 cm per hen;
2. A feed trough which may be used without restriction must be provided. Its length must be at least 12 cm multiplied by the number of hens in the cage;
3. each cage must have a drinking system appropriate to the size of the group; where nipple drinkers are provided, at least two nipple drinkers or two cups must be within the reach of each hen;
4. to facilitate inspection, installation and depopulation of hens there must be a minimum aisle width of 90 cm between tiers of cages and a space of at least 35 cm must be allowed between the floor of the building and the bottom tier of cages;
5. Cages must be fitted with suitable claw-shortening devices.

Table 33.2. The European directive – a summary

Variable	2003 (existing)	2003 (new cages)	2005	2013
Space allowance (cm^2/hen)	450	550	750*	750*
Feeding space/hen (cm)	10	12	12	12
Cage height (cm)	40	40	45	45
Perch (cm/hen)	na	na	15	15
Abrasive strip	na	na	present	present
Nest box	na	na	present	present
Dust bath	na	na	present	Present

* includes 600 cm^2/hen-free space and an additional 150 cm^2/hen of nest and dust bath space na = Not available

In relation to the above recommendations for cages, and assuming they will not be changed during 2005, some possible options are a) to use current cages until 2013 and b) to use an existing style of cage that has a 45 cm cage height as a normal cage for 13 years by which time, part of every other cage division could be removed and a nest, dust bath, perch and abrasive strip installed.

5.2. Alternative systems

Of the 'five freedoms', 'freedom to express normal behaviors' is seriously infringed for birds reared in cages. Thus, there was an acceptance that any change closer to the 'natural environment' inherently improves welfare. On these lines, a number of alternate systems were proposed and investigated.

The two key issues in the alternate laying systems are bird density and the ability to move around; the former being a key issue in profitability and the latter being the major constraint for the former.

A number of different variations of alternative systems are in vogue. They fall into 3 categories viz. barn, aviary and free range (comprising a barn or aviary as the housing component plus access to a prescribed outdoor area) in commercial environments and a category (furnished cages) in a research environment.

Within each of the above categories, many variations are possible depending on the local preferences and constraints. The ensuing information is, therefore, only of general nature and needs to be modified as per the resource availability, market premiums and several related factors.

Eggs produced on floor systems such as barn and free range as well as from aviaries are usually referred to as "non-cage" eggs; eggs produced from enriched cages, however, are still referred to as "cage" eggs.

5.2.1. Barn

5.2.1.1. Shed and equipment

All sheds have full concrete bases and walls of timber, tin sheet or brick. They all are solidly built and well insulated. Sheds are either fully slatted (timber or recycled plastic battens, wire mesh, or plastic panels with slots) with an additional undercover scratching area, or partially slatted with a scratching area at the center constituting about 1/3 of the floor space. Feeders and drinkers are located on the slats.

Nest boxes are usually down the centre and are either single or double tier, generally with a bird excluder, and have either a front or rear roll-out to the nest belt. Feed and water are provided on the slatted area, either bell feeders or chain feeders, with the latter being more popular, and water is provided via bell drinkers, nipple drinkers or shallow open drinkers. Rollers on the top of the nest boxes were generally used to discourage perching. If distribution of eggs along the nests is a problem, baffles can be used about every 8 nests to prevent birds walking to the ends of the shed.

Estimated artificial light level of at least 100 lux is provided.

5.2.1.2. Birds and management

Birds are housed at a density of 7-12/m² (≈ 850 to 1425 cm²/bird), based on shed floor space within the shed. Pullets are introduced at 16 weeks of age and, to discourage floor eggs, they are generally locked on to the slatted area for about 5 weeks until lay commences. If birds have to be beak-trimmed, it is generally done as a single trim at about 6 weeks of age.

5.2.2. Aviary

5.2.2.1. Shed and equipment

These are a logical progression from the barn system in that, instead of having nests on the outside walls (as occurred during early development of the aviaries) the nests were incorporated onto the platforms. Thus, as in barn systems, there is a logical movement of birds from the scratching area to the slats, to the feeders, drinkers and nest boxes.

The main difference from the barn system is the 3-dimensional use of space (commonly there are three tiers of platforms for the birds) and hence a higher density of birds at 16 birds/m² of floor space (range = 10-25) and would have 3 tiers of birds (range = 3-5). Aviaries typically consist of three-tired slatted areas separated by litter space; feeders, drinkers, nests and perches are located on the slats. Tiers next to each other are stair-stepped so that the birds can easily jump from one level to the area. At night, the birds tend to move to the highest level to sleep. Perches are extensively used by the birds especially at night.

Nests are arranged similar to barn systems. Aviaries generally use a manure belt collection system for each of the tiers. The construction and ventilation systems are similar to the barn systems. Dust levels will be generally greater than in barn systems.

Whenever it becomes necessary to treat birds in a shed (e.g. vaccinations during pullet rearing) and at de-stocking, the process is complicated by the additional freedom the birds have. False walls, if installed, it is possible to collect birds easily and if necessary treat them and pass them through a hole in the false wall into an emptied sub-compartment. However, there is no denying the increased labor requirement and the difficulties in sheds where birds have access to the entire shed.

5.2.2.2. Birds and management

Success of aviary systems requires specialist pullet rearing facilities with components (such as platforms and perches) similar to the system used for laying hens (except for nest boxes).

A typical rearing involved introduction of one-day old chicks on to the middle platform of a 3-tier system and locking them in by lifting up the front 'platform'. After 3 weeks, the platform is opened and ladders are used so that birds can have access to the ground and the lower tier. Later on, as the birds grow, they are given access to the top tier and perches on the side wall are also made available. Most birds are beak-trimmed at 6 weeks of age.

An alternative was to rear birds for the first 8 weeks on the floor with access to frame and side-wall perches and transfer them to a pullet aviary rearing facility at 8 weeks of age. In this system of rearing it was possible to control light levels and by providing very low light levels for the first 8 weeks it was unnecessary to beak-trim the birds.

To encourage birds to roost at night at the top of the aviary, lights are generally progressively turned off at night starting at the floor and progressing up the tiers. Eggs are collected either once or twice daily to maintain quality. If more eggs appeared to be laid at the ends of the cage rows, 'baffles' can be used to minimize the distance traveled by birds.

5.2.3. Free range

5.2.3.1. General design

Free range systems generally use either barn or aviaries as the housing component with access to an outdoor area via pop-holes; access either directly to outdoors or via an enclosed verandah area. From either the shed or the verandah area, pop-holes are made which are manually opened daily, to the free range area.

The free range component can have several configurations depending on the sheds' location. If sheds are close to other sheds, the free range area can be provided directly on one side and on the other side and back of the shed can be utilized as grass area.

Water and limited shade are made available outdoors. No feed is supplied outdoors to prevent wild birds and also to encourage return to the sheds at night. The birds usually have access to outdoors during the entire daytime. To minimize the amount of dirt carried back into the sheds, wire mesh grates are provided in front of the pop-holes.

If required, beak-trimmed at < 10 d of age and routine vaccination, parasite control (including coccidiosis) programs are followed.

Small rocks, gravel, wood chips, wood shavings are spread along the length 5-10 m away from the shed to prevent the area becoming muddy from excess bird activity. However, it must be removable, to improve parasite control.

5.2.3.2. Ventilation

As a guide, maximum ventilation rate may typically be achieved using the equivalent of one 610 mm fan or by allowing 2.6–4.3 m^2 of inlet and outlet per 1000 layers.

5.2.3.3. Layout of equipment and perches

Inadequate, poorly designed or poorly laid-out feeders, drinkers, nest-boxes and perches will create stress in the flock and will compromise both bird welfare and performance. There are a number of key issues to consider: numbers of feeders, drinkers and nest-boxes must meet the minimum requirements set out by the concerned country. Perches should be arranged so that birds can move easily between them and the other equipment, thereby reducing the risk of collisions and subsequent bruising and/or other damage. Consideration should be given to minimizing bird stress and downgrading during catching at the end of the laying period; if the equipment can easily be removed from the pens or winched up out of the way, it is desirable.

Because of the importance of thorough cleaning between flocks, equipment, fixtures and fittings should be selected and installed for ease of cleaning. The ability to remove equipment and fittings in order to achieve a really thorough clean pen is usually the best option, although this is often not practicable in the case of nest-boxes, perches and slatted floors.

5.2.3.4. Pasture management

In free range-laying systems, good pasture management is essential if the ground is to remain in good condition and the problems of poaching and the build-up of parasitic intestinal worms and coccidial oocysts are to be avoided. Worm infestation and coccidiosis can seriously compromise bird health and welfare.

The signs of a worm infestation include a general loss of condition, diarrhea, and loss of egg size and numbers and possibly of yolk color too. Gapeworm (*Syngamous trachea*) causes dyspnoea or 'gaping' and head shaking. Left untreated, heavy worm infestations will result in mortality. Many of the external symptoms and effects of coccidiosis are similar to those of worm infestation, although blood may also be present in the droppings. Low level infestations of worms and coccidiosis can increase stress in the flock and precipitate pecking.

Both worm eggs and coccidial oocysts will survive for some time in the soil. This means that the threat of re-infestation is always present and will inevitably carry over to subsequent flocks.

Although effective paddock management is the key to avoiding problems during lay, birds should be wormed before delivery and rearing programs should enable the birds to develop resistance to coccidiosis through the use of coccidiostats in the feed and/or by vaccination.

5.2.3.5. Management of paddocks

The land surrounding the laying house should be divided into a series of paddocks which the birds are allowed to use for periods of up to 6 - 8 weeks each. The number of paddocks used will depend on the specific arrangement of the house in relation to the available area, but normally six paddocks is a practical compromise. Similarly, the length of time that the birds are allowed to use individual paddocks will vary depending on soil type, drainage, grass cover and weather conditions.

Paddocks should be maintained in good condition by the judicious use of chain harrows. This breaks up the soil surface and promotes good drainage and helps reduce the parasite burden by exposing the worm eggs to sunlight.

The use of short, resilient grass varieties and clover are generally considered to be most appropriate for free range, and the grass should be kept short to reduce the risk of crop impaction.

The area immediately outside the poultry house tends to suffer the greatest amount of damage, so the ground adjacent to the pop holes should be covered with large, rounded stones/pebbles or slats. As well as providing health and welfare benefits the birds' feet will be cleaned as they enter the building, which will help to reduce the number of dirty eggs produced. The incorporation of a covered verandah along the side(s) of the laying house is now a common practice. It can be very beneficial as an aid to effective range management, and to welfare and production.

Providing shelters on the range area, perhaps with some trees, can encourage the birds out to range, and spread the wear on the pasture. Shelters also provide protection from the sun in hot weather and from rain and wind during inclement weather. Birds find open space a threat, as it increases the risk of being spotted by predators, so shelters and trees can provide the cover they need to feel safe.

5.2.3.6. Protection from predators

Free range layers are attractive to predators like foxes, mink, dogs, badgers etc. which often kill or maim large numbers of birds - far more than they are able to consume. In addition to direct attacks on the birds, the presence of these predators can cause panic and hysteria in the flock. This can cause losses through smothering and trigger outbreaks of feather pecking.

Location of the unit can influence the likelihood of such problems. Thick cover, such as conifer plantations, close to the range area will encourage predators, and in some cases hens will be taken in broad daylight.

Permanent fencing substantial enough to exclude predators can be extremely expensive and is not normally warranted. Wherever possible and permissible, flexible electrified fencing powered by a mains transformer (rather than a battery powered unit) may be installed.

5.2.3.7. Advantages

(Savory, 2003)

1. Birds are quiet and often easier to handle
2. Freedom from movement and more exercise
3. Enriched environment
4. Access outdoors and ability to range extensively and eat fresh grass
5. Opportunity to dust-bath in soil
6. Provision of perches and nest boxes at different heights allowing greater use of space
7. Most behavioral needs satisfied
8. Perching and increased exercise can increase bone strength
9. Choice of nest boxes
10. More space for birds to avoid aggression and cannibalism

5.2.3.8. Disadvantages

(Savory, 2003)

1. Stocking density can be too high in places, increasing risk of smothering
2. Increased mortality
3. Increased risk of feather pecking and cannibalism
4. Greater need for low light intensity and/or beak trimming to control pecking damage
5. Risk of some birds being denied access to food and water due to aggression
6. Increased risk of disease, due to contact with droppings and wild birds
7. Increased incidence of internal parasites, due to contact droppings and/or consumption of worms containing eggs and/or larvae
8. Increased incidence of external parasites, especially red mite
9. Increased risk of collision and bone injury
10. Floor eggs; increased risks of egg breakage and egg eating
11. Reduced control of environment, especially near open pop holes
12. Risk of crop impaction due to consumption of uncut grass
13. Exposure to predators and bad weather
14. Less efficient conversion of food to eggs, due to egg breakage, increased energy costs and consumption of food by wild birds
15. House layout and equipment makes it hard to inspect and de-populate birds

5.2.4. Furnished cages

5.2.4.1. Welfare assessment

5.2.4.1.1. The Homeostasis approach

The definition of animal welfare that underpins this approach is "The welfare of an individual is its state as regards its attempts to cope with its environment". In this definition, the "state as regards attempts to cope" refers to both how much has to be done by the animal in order to cope with the environment and the extent to which the animal's coping attempts are succeeding. Attempts to cope include the functioning of body repair systems, immunological defenses, physiological stress response and a variety of behavioral responses. Therefore, using such a definition, the risks to the welfare of an animal by an environmental challenge can be assessed at two levels: firstly the magnitude of the behavioral and physiological responses and secondly the biological cost of these responses.

These behavioral and physiological responses include the stress response while the biological cost includes adverse effects on the animal's ability to grow, reproduce and remain healthy.

The "homeostasis" approach appears to offer science the best assessment of the welfare of animals. As a research tool, this approach involves comparing housing or husbandry systems and risks to welfare are assessed on the basis of relative changes in biological (behavioral and physiological) responses and corresponding decreases in fitness. Assessing motivation using preference testing has the potential to measure the animal's important underlying needs, and thus provides a valuable addition to the homeostasis approach in studying animal welfare.

5.2.4.1.2. Perches

The literature suggests that perches can overcome some welfare problems associated with the very high calcium turnover and mobilization in laying hens without seriously affecting egg quality. Notwithstanding some potential problems with perches if they are incorrectly designed and placed, such as an increase in cracked and dirty eggs and keel bone deformation, they are inexpensive and appear to contribute to 'fitness' by reducing the potential for injuries. They should be included in cages.

Current recommendations for perches are for elliptical wooden perches with flattened tops and bottoms (vertical cross section of 3.1 cm and horizontal cross section of 3.6 cm) installed 17 cm from the back of a 48 cm deep cage and 7-7.5 cm above the floor and with sufficient perch space (15 to 18 cm per hen), so that all birds can perch simultaneously. This shape of perch reduces the incidence of bumble foot compared to rectangular perches. Plastic perches increase the incidence of bumble foot. Based on the physiological benefits to fitness of perches, and notwithstanding some potential production problems, they should be incorporated into cages.

5.2.4.1.3. Nest-boxes

Many reports suggest that birds are highly motivated to use a nest box (particularly if they have experience of laying eggs in a nest box) and some data on the behavior of birds without nest boxes is used as evidence of frustration. However, an argument that their use is not warranted is based on the very high reproductive rate (a sensitive indicator of welfare) recorded in cages without nest boxes. Thus, while nest boxes may result in an improvement in some subtle aspects of welfare, there is no evidence that fitness is improved; i.e. the magnitude and cost/benefit of any improvement has not been demonstrated and there are no data available on the magnitude of any adverse consequences for welfare of not having a nest.

5.2.4.1.4. Dust bath

The literature on dust baths is far less convincing than for nest boxes. It is likely that any welfare benefits are relatively small and there are no data on the magnitude, on the basis of fitness variables, of any adverse consequences of not having a nest box or any improvement in welfare from the presence of nest boxes. Hence, it difficult to agree with the European recommendation that dust baths is a requirement in cages.

5.2.4.2. General design

Commercially available furnished colony cages consist of 2 tiers of 12 cages, totalling 24 furnished cages. Each of the cages is 120 cm wide and 110 cm deep, housing 26 birds and providing 450 cm^2 of floor space/bird.

A metal dust bath, measuring 60 cm wide by 20 cm deep, is provided in 12 randomly selected cages (6 per tier). To deter hens from nesting in the dust bath, the facility has to be opened daily at 1300h, and closed one hour before lights are turned off; it is filled with peat moss.

Nest boxes, constructed of metal, measuring 60 cm wide and 50 cm deep, providing an extra 115 cm^2 per bird are also added. A single doorway measuring 20 cm wide is provided to allow entry into the artificial turf-lined nest box.

5.2.4.3. Advantages

(Savory, 2003)

1. Less labor for attendants
2. Total control of environment (lighting, temperature, ventilation)
3. Small group size
4. Reduced risk of birds being denied access to food and water by other birds
5. Birds separated from their droppings
6. Reduced risk of disease
7. Easier control of disease
8. Easier control of external parasites
9. Reduced risk of damage due to aggression, feather pecking and cannibalism
10. Beak trimming not always necessary
11. Low mortality compared to other systems
12. Reduced risk of smothering
13. No risk of predation

14. Provision of suitable claw shortening devices and resources intended to allow expression of wing flapping, wing stretching, group pecking, ground scratching, nesting and perching behaviors
15. Perching and increased exercise can increase bone strength

5.2.4.4. Disadvantages

(Savory, 2003)

1. Environment more barren than in non-cage systems
2. Depending on cage resource design, prevention of full expression of foraging, dust bathing and perching behaviors
3. Variation between cage tiers (feather pecking and cannibalism most common in top tier)
4. Reduced exercise can reduce bone strength
5. Cage structure may cause damage to feathers (abrasion) and feet (bumble foot)
6. Risk of entrapment
7. Inability to escape aggression, feather pecking and cannibalism when it occurs
8. Inspection of birds by attendants can be difficult, especially in top and bottom tiers

5.2.4.5. Conclusion

While many of the practical problems of nest boxes and dust baths have been overcome, furnished cages appear to require some development prior to their introduction into the commercial industry. It has been estimated that these modifications will increase egg production costs by 10-20 % over conventional cages. Also, surveys of public opinion suggest that modified cages are only slightly more acceptable than conventional ones and they need to be taken into account in a cost benefit analysis and more focus should be put on bird welfare rather than public perceptions.

Notwithstanding the above statements regarding the incorporation of nest boxes, perches and dust baths into cages, reports indicate better physical condition (and either no differences in mortality or a lower mortality) of birds in furnished cages that incorporate all 3 items of furniture. Thus, there may well be an interaction between the items of furniture that improves fitness and this aspect requires further research.

5.2.5. Organic poultry production

(Lampkin, 1997)

The term "Organic farming" is being used recently; organic farming can be defined as an approach to agriculture where the aim is to create integrated, humane, environmentally and economically sustainable agricultural production systems producing acceptable levels of crop, livestock and human nutrition, protection from pests and diseases, and an appropriate return to the human and other resources employed. Maximum reliance is placed on locally or farm-derived, renewable resources and the management of self-regulating ecological and biological processes and interactions. Reliance on external inputs, whether chemical or organic, is reduced as far as possible. In many European countries, organic agriculture is known as ecological agriculture, reflecting this reliance on ecosystem management rather than external inputs.

In order to achieve the animal welfare, environmental, resource-use sustainability and other objectives, certain key principles are adhered to. Those relevant to poultry production include:

1. Management of livestock as land-based systems (i.e. excluding feedlots and intensively-housed pig and poultry units) so that stock numbers are related to the carrying capacity of the land and not inflated by reliance on 'purchased' hectares from outside the farm system, thus avoiding the potential for nutrient concentration, excess manure production and pollution
2. Reliance on farm- or locally-derived renewable resources, such as biologically-fixed atmospheric nitrogen and home-grown livestock feeds, thereby reducing the need for non-renewable resources as direct inputs or for transport
3. Reliance on feed sources produced organically, which are suited to the animal's evolutionary adaptations (including restrictions on use of animal proteins) and which minimize competition for food suitable for human consumption
4. Maintenance of health through preventive management and good husbandry in preference to preventive treatment, thereby reducing the potential for the development of resistance to therapeutic medicines as well as contamination of workers, food products and the environment
5. Use of housing systems which allow natural behavior patterns to be followed and which give high priority to animal welfare considerations, with the emphasis on free-range systems for poultry
6. Use of breeds and rearing systems suited to the production systems employed, in terms of disease resistance, productivity, hardiness, and suitability for ranging.

5.3. Minimum standards for laying hens and broilers

Minimum standards for laying hens as per the European Council directive 1999/74/EC are given in table 33.3; usable area is defined as area at least 30 cm wide; floor slope not exceeding 14%, headroom of at least 45 cm; nesting areas shall not be regarded as usable area.

European directive for the protection welfare of broilers is also likely to be released soon; it is expected that broiler stocking density is likely to be restricted to a maximum of 30 kg/m^2 and up to 38 kg/m^2 by a "Competent authority" for establishments that comply with certain requirements like records (for each house) of ambient temperature, humidity, ammonia, mortality and conditions of the birds at slaughter (Savory, 2003).

Table 33.3 Minimum standards for laying hens

Facility	Standards required
Barn (Wright, 2005)	
General	Maximum 9 birds/m^2 usable area; maximum 4 levels; headroom between the levels must be at least 45 cm; prevent droppings falling on to the levels below
Feed	Linear feeders – at least 10 cm/hen; circular feeders – at least 4 cm/hen
Water	Continuous drinking troughs – at least 2.5 cm/hen; circular drinking troughs – at least 1 cm/hen; nipple drinkers – at least 1 nipple/10 hens
Nests	Single nest – at least 7 hens/nest; group nests – maximum 120 hens/m^2 nest space
Perches	At least 15 cm/hen; not to be mounted above the litter; horizontal distance between perches at least 30 cm; distance between perch and wall at least 20 cm
Litter	Friable material; litter area at least 250 cm^2/hen and 1/3 of the ground surface
Enriched cages (Wright, 2005)	
General	Space of at least 35 cm between the wall of the building and the bottom tier of cages; minimum aisle width of 90 cm between tiers of cages
Area/height	At least 750 cm^2/hen, 600 cm^2 of which is usable; total area at least 2000 cm^2; height other than that above the usable area at least 20 cm
Equipment	Nest; litter such that pecking and scratching are possible; at least 15 cm/hen perch-space; claw-shortening devices
Feed	At least 12 cm/hen trough space
Water	At least two nipple drinkers within the reach of each hen
Deep-litter aviary *	
Day length	Minimum 8 hr with an average indoor light intensity of 10 lux
Darkness	Minimum 6 hr are natural dark phase
Feeder	10 cm trough space/bird
Water	100 birds/circular drinking trough; 10 birds/nipple
Distances	Maximum 8 m from feeder/drinker
Nest boxes	5 hens/individual nest; 120 hens/m^2 in community nests
Litter	At least 1/3 of floor area
Dropping box	Deep enough to hold the droppings from one batch
Perches	15 cm/hen; perches 4 cm wide and 30 cm apart
Pop holes to exterior	One pop hole/600 hens; minimum size 45 cm high x 2 m wide
Duration of access	Pop holes open for at least 8 hr to outside run

* Int. Poultry Prodn. 10(5): 11

5.4. Prospects of alternate systems

5.4.1. Areas of concern

Aviaries still result in some human discomfort. The dust levels and the difficulties of inspection of birds, bird pick-up and handling for vaccinations in aviaries are of concern. The low level of real information provided to consumers (e.g. eggs may be marketed on the basis that birds are fed a combination of grains, organic food or are free range) and the apparent lack of interest in welfare. The potential compromises to welfare from purely market-driven production systems (i.e. those organic-free range systems associated with a high mortality).

5.4.1.1. Comparative production

The poorer production in the aviary was partly due to periods of broodiness in the hens. Labor requirements will be 33-50% higher for the aviary system (Table 33.4).

Table 33.4. Comparative performance – Cage *Vs* Aviary system

Parameter	Cage	Aviary
Egg number, 17-80 weeks of age	320	310
Egg weight, g/egg	63	62
Feed consumed, g/bird/d	110	125
Mortality, %	5	5-8*
* 7-10 if not beak-trimmed		

5.4.2. Areas of conflict

Future price premiums, if any, are still unclear. Based on current market shares, approximately 80 % of all consumers bought eggs on the basis of price (i.e. currently purchase cage eggs). However, assuming the cost of production in alternative systems is higher, it is unknown how price increases will be handled i.e. the prices could be absorbed by the farmer by increasing efficiency or by the supermarkets cutting margins or by consumer paying a higher price.

Some supermarket chains not selling cage eggs could be for both marketing reasons and higher margins per unit of shelf space rather than for any concern over welfare.

There appeared to be a conflict between food safety and perceptions of welfare. It is likely that biosecurity in free range and barn system is at a greater risk of compromise than in cages and this type of information is not provided to the consumer.

Whatever may be the advantages of furnished cages with regards to welfare of the birds, they are, after all, nevertheless cages and eggs produced in these are still referred to as "cage-eggs" anyway. Therefore, there are unlikely to fetch premium prices in comparison to those produced on floor systems (barns, aviaries or free-ranges). Furnished cages are undergoing considerable evaluation and it is likely that some design modifications will be recommended as a result of this work. Therefore, they cannot be considered a currently viable alternative production system. The major problem with furnished cages is that the birds are, after all, still in cages. The major objection to conventional cages is the lack of space and while space allocations are considerably larger, when the cages contain birds there is certainly no 'feeling' of the birds having a freedom to move around, i.e. the space allocations do not meet their goal.

5.4.3. Homeostasis approach

Research needs to be undertaken to incorporate the feelings based approach to assess welfare into the widely-accepted homeostasis approach. This involves the need to conduct research to establish the consequences of emotions experienced by birds on the magnitude of behavioral and physiological responses and the cost of these responses, for example, on growth, reproduction and health. On current scientific evidence, there is no demonstrable welfare benefit, on the basis of improved fitness, from incorporating nests into cages as recommended in the EU Directive.

On the basis of fitness there appears to be no scientific evidence that incorporating dust baths into cages, as recommended in the EU Directive, will improve bird welfare. However, there may be welfare benefits, on the basis of increased fitness from reduced mortalities, from incorporating certain enrichment devices into cages. It is recommended that the welfare benefits of enrichment devices be further researched.

Based on the physiological benefits to fitness of perches, and notwithstanding some potential production problems, it is recommended that perches should be incorporated into cages. Notwithstanding the above statements regarding the incorporation of nest boxes, perches and dust baths into cages, reports indicate better physical condition and either no differences in mortality or a lower mortality of birds in furnished cages that incorporate all 3 items of furniture. Thus, there may well be an interaction between the items of furniture that improves fitness. It is recommended that research be conducted to determine if furnished cages *per se* improve fitness.

Chapter 34

Brooding

1. Need for brooding

Newly-hatched chicks cannot regulate the body temperature efficiently because:

1. They have a higher metabolic body size (surface area per unit body weight) than that of adults; therefore, they lose heat quickly.
2. Their body temperature is higher (41.7°C) than that of adults and hence, tends to lose heat.
3. They lack feathers which are excellent insulating material which further facilitates heat loss
4. They have changed over from poikelothermy to homoieothermy just three days prior to hatching and hence the hypothalamus, the thermoregulatory centre, will not be completely functional. Poults are known to be least equipped at hatch as far as thermoregulation is concerned.

Therefore, it is mandatory that newly-hatched chicks have to be provided with artificial heat during the initial stages - duration of which depends on the environmental temperature.

2. Brooding temperature

The temperature required, for chicken, at the beginning is 35°C which has been reduced at a rate of 2.8°C per week until the brooding temperature equals environmental temperature; this duration varies between two weeks to six weeks, and usually four weeks; cooler the environment temperature the longer the brooding period and vice versa. However, specifications are available regarding brooding temperature requirements of different species of poultry (Table 34.1). Brooding temperature has to be reduced every week to facilitate development of homoieothermy and also to stimulate normal feather growth.

To conserve heat within the brooder house, windows may be closed by curtains during the night and/or daytime depending on the season and environment temperature. As the birds age, the curtains are lowered to open the upper portion of the windows to facilitate easy exit of hot air which, being lighter, rises up.

2.1. Brooding temperature Vs body temperature

Brooding temperature is lower than body temperature (41.7°C) because, birds, being homoieothermic, used only about 50% of the actual energy released by NADH+H$^+$ and FADH$_2$ in the respiratory chain (in mitochondria) is converted into the chemical energy (ATP); the remaining 50% of energy is released into the cell medium for thermoregulation. The latter is useful only when the environment temperature is lower than the body temperature. Therefore, all homoieotherms will be uncomfortable at temperatures at or above their body temperature. They may have to use heat to drive the extra heat to the environment against temperature gradient (heat to keep the body cool). This is especially true in case of poultry because they lack sweat glands and their metabolic body size is very high. Therefore, at once the environmental temperature exceeds the zone of thermoneutrality, the birds pant and depend more and more on insensible heat loss.

Table 34.1. Brooding temperature (°C)

Species	1-2 d	3-7d	2nd week	3rd week	4th week
Chicken, Quail, G. fowl	35.0 to 37.7*	35	32.2	29.4	23.9 to 26.6
Poults	37.7	36	32.2	28.3	24.4
Ducklings	29.0 to 32.0	26.0 to 29.0	23.0 to 26.0		
Goslings	32.0	29.0	22.0		

* Whole house start at 33°C; then 29°C after 12 hr

3. Brooder house

The house in which the chicks are brooded is referred to as "brooder house" and it should not be located near houses accommodating older birds to minimize the chances of disease transmission. A minimum distance of 90 m is generally advised (North and Bell, 1990). Prevailing air movement (winds) must be from brooder area to other poultry areas and not *vice versa*. It is also desirable that the house is properly fenced and secured.

Brooder house should be isolated not only by distance but also by management; workers must be exclusively allotted to work in brooder house and not shifted to work in houses for older birds. All personnel entering the brooder house area have to preferably shower and change into clean clothing; visitors are generally not permitted and contact with other poultry is strictly prohibited.

3.1. All-in, all-out system

It is highly preferable that a brooder house has only one batch of chicks and the house is cleaned and disinfected before the next batch of chicks arrive. This system of housing in which all birds entered the house at once and all move out of the house at the same time is referred to as "all-in all-out system". This system is highly convenient for vaccination, change of feed and other managemental procedures.

However, under practical conditions, it may not always possible to follow all-in all-out system especially by small and medium farmers.

4. Brooding methods

4.1. Heating systems

The desired temperature of the brooding area is maintained by supply of heat which can be by any of the following methods:

4.1.1. Spot-heating

Floor Brooders (canopy/hover brooders or infra-red bulbs or propane-fuelled systems) are use to keep a specific area(s) within a building at the desired temperature. Even battery brooders, which provide the warmth at specific part of each of its tiers, fall into this category.

Spot-brooding is common in conventional/windowless housing systems with manually operated or automatic curtains fit to the side-walls. Whereas propane-fuelled brooders are frequent in developed countries, canopy/hover brooders or infra-red bulbs are popular among farmers in other countries.

When floor brooding is practiced, it is necessary to use brooder surrounds/ brooder guards to contain chicks within brooder area. Brooder guards are usually 30 cm high and made of cardboard; under hot-humid conditions, even brooder guards made of wire-mesh can be used to facilitate air movement.

Caution needs to be exercised in observing for comfort of the birds and adjusting the brooder temperature accordingly.

4.1.2. House heating

This method is used in environmentally controlled houses wherein entire brooder house or part of it is heated by using propane or electricity; the former being common.

4.1.2.1. Whole house heating

In this system, it is very convenient to alter the temperature precisely as per the requirements of the chicks. Chicks, on the other hand, can move freely in the entire area because, there are no brooder guards restricting their movement. This also helps easy observation of the chicks.

4.1.2.2. Partial house heating

This is accomplished by means of temporary plastic walls partitioning off about 1/3 of the brooder house. Generally, 20-25 chicks/m^2 are brooded; no need of canopy/hover or infra-red bulb or a battery brooder and hence observation of the chicks is easy. The chicks move freely within the area delineated by the plastic walls. As in case of whole house heating, requisite brooding temperature can be easily provided.

4.2. Spot brooding

Spot-brooding of chicks can be done either on floor or in battery cages; the former can be either by use of canopy or infra- red bulbs.

4.2.1. Floor brooding

4.2.1.1. Canopy brooding

In this method, an umbrella-like canopy with two to three incandescent bulbs (40 to 100 W each, depending on the season) fixed at the centre, is inverted and hung in such a way that the birds can move freely in and out of it. The bulbs, when put on, heat the air and the hot air is trapped by the canopy preventing the escape of hot air thereby providing warmth to the chicks.

4.2.1.1.1. Brooder guard

The brooder area is delineated by the brooder guard which is 30 cm high and 60 to 75 cm away from the edge of the canopy. The brooder guard is arranged in a circular fashion so as to avoid corners and dark areas. The brooder guard contains chicks nearer to the brooder area and prevents the chicks from straying away from the heat source. A canopy measuring 90 cm in diameter can accommodate about 250 chicks. As the chicks grow, brooder guard is moved away from the canopy to avoid requisite floor space.

4.2.1.1.2. Paper on litter

For the first seven to ten days, paper is spread on the litter material and feed or maize grit is sprinkled on it. This is primarily to avoid the chicks eating the litter material which might result in choke and possible death. In addition, birds when comfortable move freely throughout the brooder area; and hence, if the droppings are uniformly distributed on the paper, it indicates the correctness of the brooder temperature. On the contrary, if the droppings are distributed only within the canopy area, it suggests that the temperature was insufficient and the chicks were mostly inside the canopy. If the droppings are noticed all-round but only near the brooder guard, it is indicative of a high temperature inside the canopy forcing the chicks to move to the periphery. If the droppings are noticed as a triangle with the apex towards the centre of the canopy, it is suggestive of draft blowing from the direction corresponding to the base of the triangle. All the above have been schematically represented in Fig. 34.1.

4.2.1.1.3. Feeders and waterers

The feeders are arranged like spokes in a wheel in the brooder area with half the length of the feeders inside the canopy area (Fig. 34.2). The waterers are arranged at the age of the canopy interspersed between the features. This is done to ensure both feed and water within 30 cm of distance from any part of the brooding area, especially at the beginning of brooding. Later on, both feeders and waterers

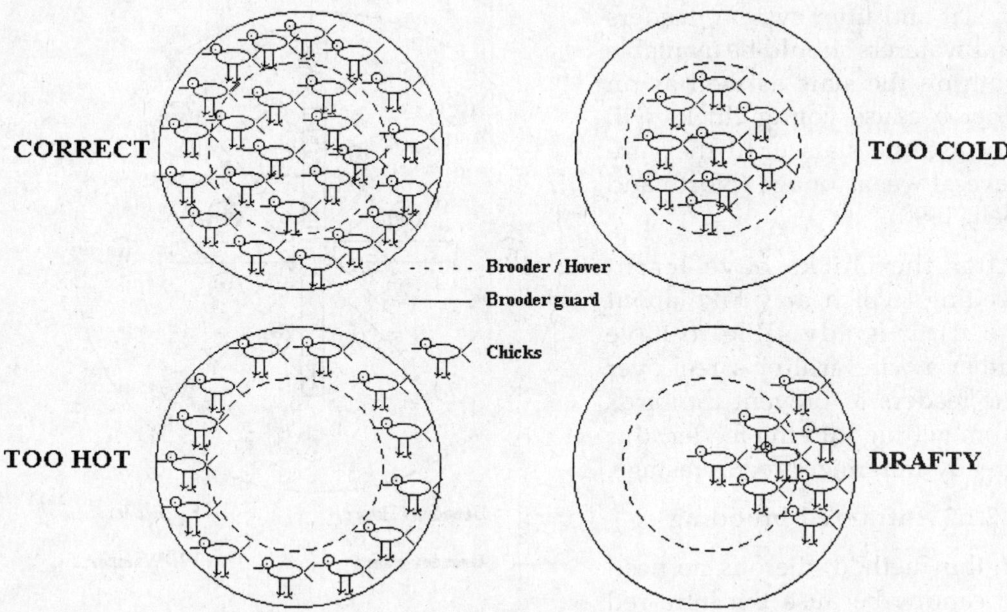

Fig. 34.1. Brooding temperature *Vs* comfort of chicks

should be uniformly distributed so that no bird need move more than 3 m to have an access of either feed or water.

The person looking at the brooder house should have the "chicken sense "and must be able to judge the comfort of the birds without the aid of a thermometer. The brooder temperatures required for chicks is warm for human beings and hence there will be slight sweating in properly-managed brooder house at least during first two weeks. Should the temperature be recorded, it is done at a height of 10 cm above the litter level at the edge of the canopy, since it is the height at which the most portion of the body of the chick is exposed to the brooding temperature. On the same lines, as the birds grow, the height at which the temperature has recorded also increases.

The height at which the canopy is fixed depends on the height of chicks; in other words, age of birds. In the beginning, edge of the canopy should be about 8 to 10 cm above the litter level; as the birds age, the canopy is raised till it is about 0.9 m above the birds and after the brooding is over, the canopy can be removed, cleaned and disinfected for further use.

Waterers are not placed inside the canopy area because the water gets heated and the chicks do not drink hot water; in addition, there is a risk of drowning especially during the initial period because, the chicks will be learning to drink and eat.

In slat and litter system, feeders and waterers should be available both on the slats as well as on litter because young chicks will not use the slats until they are several weeks of age (North and Bell, 1990).

After the chicks have learnt feeding (which may take about 2-3 d), it is advisable to have either a wire grill or a reel over the feeders to prevent the birds from getting into the feeder; this greatly minimizes feed wastage.

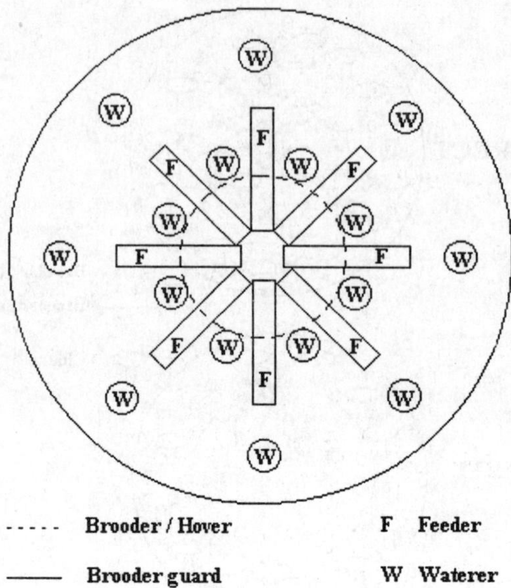

Fig. 34.2. Placement of equipment during brooding

4.2.1.2. Infra- red brooding

In this method, there is no need of canopy because the infra-red light heats any object that comes in contact with it, by radiation, but not the air. Infra-red red and infra-red white bulbs of 150 and 250W are available which can be suffice for 75 to 90 and 125 to 150 chicks, respectively. The bulbs must be hung at least 25 to 30 cm above the litter floor; otherwise, the litter material itself may catch fire (due to radiation heat), especially during summer.

Spread of paper on litter, placement of feeders and waterers and rest of the managemental procedures are similar to that under canopy brooding.

4.2.1.2.1. Advantages

1. Since there is no canopy, observation of the chicks is easier.
2. Accidental mortality due to improper handling of canopy is avoided.
3. Infra-red light has been found to have some germicidal effect and hence survivability of the chicks will be higher in this method.
4. Infra-red light has been found to reduce cannibalism; infra-red red bulbs are popularly used for improving broiler chicks.
5. Brooding cost, on a long run, has been found to be lower.
6. Although debatable, some authors feel that infra-red enhances Vitamin D synthesis.

4.2.2. Battery brooding

In this method, birds are reared in a battery brooder which consists of 4 to 5 tiered batteries each of which has the heating space comprising one thirds of the total area and the remaining portion as the "run space". The heating unit consists of an electric heater with a thermostatic control; the heating space is also covered with a false-roofing made of GI to trap the hot air. Paper is spread of the mesh flooring and after 10 days it is removed so that the droppings fall directly into the fecal trays.

Usually, each of the tiers measures 150 to 180 cm long, 75 to 90 cm wide and 30 to 40 cm high; each of these tiers can accommodate 25 to 35 broiler chicks until market or 75 to 100 chicks up to three weeks of brooding. Therefore, it is evident that in an area where only 18 broilers can be grown till market age, nearly 100 broilers can be reared in a battery brooder (Fig. 34.3).

4.2.2.1. Advantages

1. Housing density is high.
2. There will be no litter-borne diseases.
3. Since movement is restricted, birds spend less energy and hence feed efficiency will be improved.
4. Survivability of the chicks will be higher.
5. Labor requirement is minimal.
6. Control of temperature is more accurate.

4.2.2.2. Disadvantages

1. Initial investment is high.
2. Birds are uncomfortable.
3. Broken legs, breast blisters and other carcass defects may appear, especially in case of broilers.
4. Handling of fecal matter is cumbersome.

5. Preparation for arrival of chicks

1. Cleaning and disinfection of brooder house.
2. Cleaning and disinfection of feeders and waterers followed by sun-drying.
3. House must be thoroughly examined for leaking roof and gutters, drafty doors, functioning of fans, switch-boards, shutters etc. The floor may be given a coat of lime solution.
4. The house and equipment may be fumigated using 3X concentration of formaldehyde especially in case of window-less houses; in case of conventional houses, the side walls may be closed with curtains and then fumigated.
5. Floor of the building has to be cleaned with a suitable detergent and disinfected with a commercial product suitable for floor-cleaning.
6. All weeds and debris, if any, surrounding the building must be cleared and the area sprayed with a commercial disinfectant.

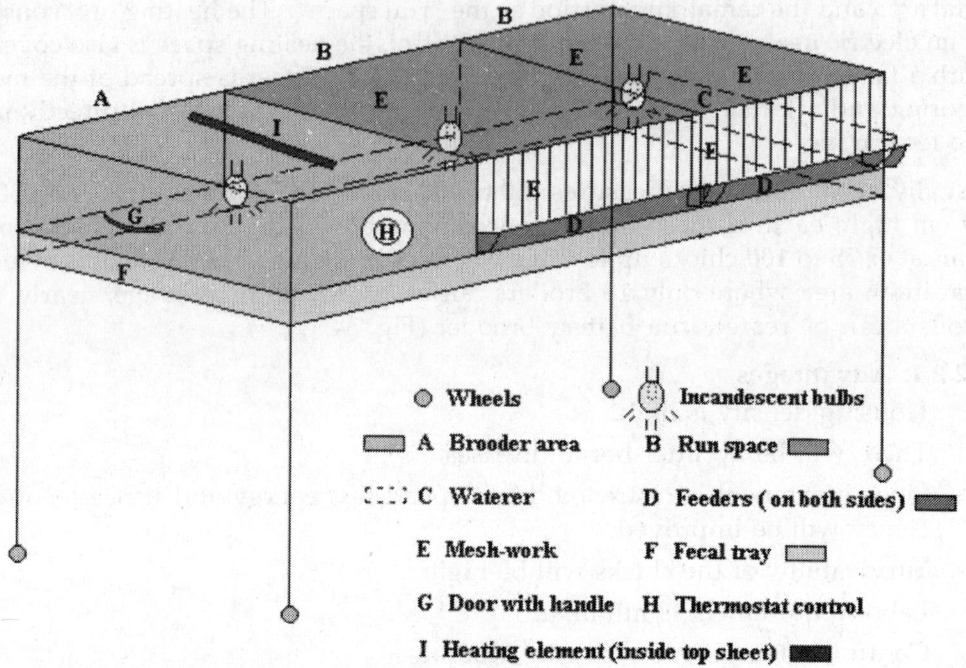

Fig. 34.3. Battery brooder

7. Checking of brooders, feeders, waterers and other equipment required during the brooding period.
8. New and clean, dry, mould-free absorbent litter has to be spread evenly on the floor making a thickness of about 6 to 8 cm; care has to be taken not to close all the windows when the litter is being spread so that dust can settle down easily (not applicable for battery brooding).
9. Paper is spread on the litter (applicable only for floor-brooding).
10. Brooder, feeders and waterers are arranged at least 6 to 8 hours before the arrival of chicks; brooder is put on so that requisite brooder temperature is attained before the arrival of chicks; if required, windows may have to be covered with gunny sacks to conserve heat. This also helps to maintain drinking water temperature at = 18°C.
11. Anti-stress factors (B-complex vitamins, vitamin C etc.) may be added in water.
12. Feed or maize grit is sprinkled on paper to help chicks identify the feed. During the first few days, it is mandatory that the chicks should have feed easily available; for this purpose, feed can be kept in inverted chick box lids, being egg flats or any such large flat containers.
13. It is desirable to have a standby generator; otherwise, charcoal/coal, kero-

sene or LPG must be available to make exigencies in case of power failure.

14. All data sheets required for recording various aspects like medication, vaccination, feed consumption, mortality etc. should be kept ready.

6. Practices after arrival of chicks

1. It is advisable to examine a sample of chicks from each of the chick boxes to ascertain the quality and thriftiness of the chicks; it also helps to note mortality, if any, during transit.
2. It is a good practice to dip the beaks of the newly-hatched chicks in water at waterers and leave them into the brooder area. Otherwise, the chick box can be inverted quickly to dump all the chicks nearby the brooding area.
3. They have observed carefully to identify weak ones and also to help those which are not able to reach feed and water.
4. It is equally necessary to observe and ensure that all the chicks are actually eating and watch out for "starve-outs" by examining for well distended crop.
5. Waterers have to be cleaned and freshwater to be given at least twice a day; similarly, feed is also offered at least twice a day.
6. Care has to be exercised by placing the waterers – they have to be kept at horizontal place and it must be ensured that water is not flowing out onto the litter.
7. In case of Japanese quails, which have a tendency to get drowned, it is a good practice to put colored beads so as to avoid drowning as well as to attract them to waterers.
8. Standard balanced ration (chick starter in case of laying-type of chicks and broiler starter for broiler chicks) has been offered *ad libitum.*
9. As the birds advance in age, automatic/nipple drinkers and automatic/linear/hanging feeders can be introduced; replacing part of the existing feeders and waters at a time.
10. Floor space and waterer space must be adequate to avoid unnecessary competition followed by cannibalism, poor growth and wastage of feed (See Chapter "Management requirements and specifications").
11. Litter has to be raked regularly to avoid wet spots and caking of litter; this can be done simultaneously while feeding and watering.
12. Standard vaccination schedule have to be practiced to contain diseases like Newcastle and Infectious Bursal Disease etc. in laying as well as broiler chicks.
13. Proper records to be maintained with regards to feeding, vaccination, medication, mortality, post-mortem reports, income and expenditure involved during the brooding period.

7. Space requirements

7.1. Rule of thumb

As the rule of thumb, waterer space required is about the width of the head and feeder space required is about the width at shoulders. However, floor space required depends on the system of rearing; on all-litter system, floor space required is approximately equal to a square whose side is equal to 1½ times the length of the bird from base of the neck to the tip of pubic bones. In case of cage system, floor space required, including the area for passages, is about ½ of that on all-litter system. Floor space requirement under all-slat system is intermediate between all-litter and cage systems. With regards to slat-litter and/or slat-wire systems, floor space required is intermediate between all-litter and all-slat systems.

Floor, feeder and waterer space requirement of chicks under all-litter and cage system of rearing during brooding period is given in Table 34.2. (Space requirements under other systems of rearing have been tabulated in chapter "Management requirements and specifications").

Table 34.2. Space requirements during brooding (per bird)

	Leghorn			Medium-size			Meat-type	
	Egg type	Breeders		Egg type	Breeders		Breeders	
	♀	♂	♀	♀	♂	♀	♂	♀
	Floor space, m²							
All-litter	0.070	0.093	0.079	0.079	0.093	0.093	0.116	0.093
Cages	0.0155			0.0181				
	Feeder space, cm							
All-litter	5.0							
Cages	5.1			5.6				
	Waterer space, cm							
All-litter	1.5						2.5	1.9
Cages	1.9			2.0				

8. Other important considerations

8.1. Routine feeding

After the first couple of days when the chicks learn to eat and drink, feed is offered in linear/automatic feeders. Whenever linear feeders are used, care has to be taken to ensure that they are fixed at the level of the back of the birds to prevent the chicks entering the feeders. In addition, the linear feeders are usually equipped with grills/reels and are filled only to the level of ½ or 1/3 to minimize wastage of feed.

Feed is generally offered at least twice a day during the brooding period; if need be, more frequently.

8.2. Routine watering

On the first day of brooding, addition of sugar (sucrose) or glucose in drinking water is being practiced by many commercial poultry farmers. It may be very useful especially when the chicks were under transport for more than 6-8 hours. Generally, an 8% sugar solution along with water-soluble vitamins and electrolytes is provided during the first 15 hr after the chicks are placed in the brooders to minimize the effect of stress on chicks.

Chick founts (fountain waterers) and other watering equipment have to be cleaned thoroughly everyday. Water left-over should be discarded, the equipment cleaned, filled with freshwater and offered. It's also a common practice to add sanitizers in drinking water, especially during first 3-4 weeks of brooding to minimize losses due to diseases. However, water sanitizers should not be used while administering vaccines through water.

Waterers are initially kept on the floor and are raised as the chicks become older and fixed in such a way that the water level is even with the tops of the backs of the birds. This is particularly important because, birds cannot suck water kept at a lower level due to lack of lips. In addition, such an arrangement will prevent regurgitation of feed into the water.

8.3. Light

8.3.1. Duration

During brooding 23hr light + 1 hr darkness is provided. Under canopy brooding, light serves to purpose of both illumination for visibility and also heating the air by convection to provide the required brooding temperature. Under infra-red brooding (light emitted by the infra-red bulbs) and battery brooding, light primarily serves the purpose of visibility.

In all the methods of brooding, 1-hr darkness at a specified time of the day is provided mainly to acclimatize chicks for power failure; this would help minimize panic huddling of chicks and subsequent mortality consequent on sudden power failure for whatever reasons.

8.3.2. Intensity

The recommended intensity of light is 37.66 lux for the first 48 hr. This can be effected by use of 10 W of light bulb per m^2 of floor space; the bulb being fixed at about 2.4 m above the floor. This intensity is sufficient enough to brightly illuminate feeders and reflect off from the drinking water to attract the chicks.

After 48 hr of brooding, intensity at the floor level can be reduced to 10.76 lux which can be supplied by use of 2.70 W of light bulb per m² of floor space; the bulb being fixed at about 2.4 m above the floor.

8.4. Beak trimming

There is growing controversy on beak trimming and hence, it is dealt in detail in a separate chapter. However, general considerations are outlined below and they have to be implemented depending on the existing laws.

8.4.1. Objectives

The main objective of beak trimming is to prevent cannibalism, a vice, prevalent among chicken. Causes for cannibalism are many; one among them is the sharp beaks with which the birds peck at each other either while fighting to establish peck order, feather pulling, vent nipping and/or due to natural instinct of attacking an injured bird. It is also recorded that birds peck at themselves and aggravate their injury. Cannibalism can spread within a flock and if not prevented, the entire flock may be affected.

The age at which it is most ideal to trim the beaks is not only variable but also controversial; in any case, it's always done prior to the onset of egg production. It is therefore obvious that beak trimming should create as little stress as possible and the same time should ensure that beaks do not grow again. Hence, this is a very precision operation demanding utmost care lest it affects future performance and/or the beaks regain their original size.

8.4.2. Advantages

(North and Bell, 1990)

1. Toe picking is reduced
2. There is less stress in the flock
3. It helps prevent feather picking and cannibalism
4. Free deficiency is improved as a result of less feed wastage
5. Livability is better, with fewer culls
6. There is more uniformity of the birds in the flock

8.4.3. Disadvantages

(North and Bell, 1990)

1. Birds lose weight for 1-2 weeks after beak trimming
2. Growth rate is reduced for a long period; it may take from 10-20 weeks before beak-trimmed birds compensate
3. Sexual maturity is delayed along with body weight at sexual maturity, egg production and, sometimes, egg weight.

In most of the countries, beak trimming is still practiced and, in a decade or so, most of the developed countries are likely to ban is procedure.

8.5. Vaccinations

Details of vaccinations are dealt in a separate chapter.

8.6. Dubbing

(North and Bell, 1990)

Removal of comb is referred to as "dubbing". It prevents comb from injury due to contact with equipment and/or during fighting, provides better vision (especially in those breeds known to have large combs which droop to one side obstructing vision) and lowers the damage due to frost bite under cold climate. However, dubbing is not advisable on birds reared under tropical (warm) climate because, combs have an important role in thermoregulation (heat loss).

Whenever dubbing is practiced, it is best done at day-old or at least within the first few weeks of life to minimize hemorrhage. At day-old, a pair of manicuring scissors is used to cut the comb close to the head, running the shears from the front to the back of the comb with the concave side of the scissors facing upwards.

8.7. Toe-clipping

(North and Bell, 1990)

The inside and back toes of all breeding males are clipped to prevent tearing the backs of the females during mating. It is best done at the hatchery; otherwise, in conjunction with 6-8 d beak trimming. Special equipment is available with which the toes at the outer joint, just above the toenail, are trimmed.

Breeding and/or commercial pullets are not generally subjected to toe clipping although removal of front, middle and back toes at the outer joint at the day-old has resulted in improved egg production in some commercial strains.

8.8. Dewinging

(North and Bell, 1990)

This is a common practice with turkeys to prevent flight; but it induces severe stress and seldom practiced in chicken.

8.9. Coccidiosis control

Details of coccidiosis are dealt under "Protozoan diseases".

8.10. Unabsorbed yolk

The normal practice of feeding chicks soon after they are left under the brooder slows down the absorption of yolk left over in the abdominal cavity. Therefore, presence of unabsorbed yolk during the first 14d is not a serious condition unless it is associated with bacterial infection and/or severe stress. Higher brooder temperature given the first 2d also contributes to unabsorbed yolk. The unabsorbed yolk may assume various colors like yellow, green and orange even in absolutely healthy chicks.

8.11. Adverse weather conditions

8.11.1. Hot weather

Hot weather precipitates in reduced feed and increased water intake which, in turn, leads to wet-litter condition. Increased ventilation does help alleviate the condition. However, under severe conditions, about super phosphate at the rate of 250 g/m^2 floor space may be added to the litter and mixed thoroughly.

8.11.2. Cold weather

Supplementary heat may be required under the brooder depending on the severity of cold weather condition. Sufficient heat should also be provided to maintain the temperature of the rest of the house at about 18°C.

9. Feeding and disease control

Nutrient requirements, feeding and vaccination programs are discussed separately in Chapters "Nutrient requirements and specifications" and "Poultry diseases - vaccines and vaccination", respectively.

Chapter **35**

Beak-trimming

1. Beak-trimming

Most of the information presented below is from the report for the Rural Industries Research and Development Corporation, South Africa presented by Glatz (2000).

1.1. Present status

Beak-trimming is commonly performed in the commercial egg industry in order to prevent injury and mortality in pullets from aggressive pecking and cannibalism. If birds are not trimmed mortality of up to 25-30% will occur. However welfare groups have expressed concern that beak-trimming may lead to unnecessary pain and loss of sensory function. It is possible to rear pullets without beak-trimming if there is effective light control. However, the majority of commercial poultry sheds in many countries are open to natural light and once an outbreak of cannibalism has occurred it is difficult to control. Under some conditions it might be possible to manage laying hens without beak-trimming, but for most producers, the risk of such a policy is too high.

There have been substantial genetic studies, which have shown there is potential for commercial breeders to develop strains of birds, which are not aggressive and hence do not need to be beak-trimmed. In the interim, however, as the beak-trimming procedure faces increased welfare scrutiny, there is a need to develop quality assurance guidelines to ensure beak-trimming is conducted consistently and effectively with minimal welfare effects on birds.

There has been little change to the method of beak-trimming since it was first developed fifty years ago. New technology methods of beak-trimming have been introduced to the Industry, but with limited success.

To resolve the key issues associated with beak-trimming a research and development plan has been recommended to

1. Ensure consistent and effective beak-trimming
2. establish high technology beak-trimming methods
3. improve technology transfer and
4. Develop a beak-trimming accreditation system for operators.

A review of beak-trimming methods available or under development indicated that beak-trimming of commercial layer replacement pullets is a common yet critical management tool that can affect the performance for the life of the flock. The most obvious advantages of beak-trimming are:

1. Reduction in cannibalism although the extent of the reduction in cannibalism depends on the strain, season, and type of housing, flock health and other factors.
2. Beak-trimming also improves feed conversion by reducing food wastage.
3. A further advantage of beak-trimming is a reduction in the chronic stress associated with dominance interactions in the flock.

Beak-trimming of birds at 7-10 days is favored by Industry but research over the last 10 years has shown that beak-trimming at day-old causes the least stress on birds and efforts are needed to encourage Industry to adopt the practice of beak-trimming birds at day-old.

Consultation with experienced beak-trim operators in Australia revealed that the main problem facing the practice is achieving appropriate beak length and shape of beak to minimize further pecking. Operators report variations in beak hardness, bleeding of beaks following trimming, and a lack of experienced well trained personnel to conduct the operation.

Proper beak-trimming can result in greatly improved layer performance but improper beak-trimming can ruin an otherwise good flock of hens. Re-trimming is practiced in most flocks, although there are some flocks that need only one trimming. Re-trimming may not be required in many birds, but to avoid the potential problem of cannibalism as a result of beak re-growth most birds are re-trimmed at 10-14 weeks. Given the continuing welfare scrutiny of using a hot blade to cut the beak, attempts have been made to develop more welfare-friendly methods of beak-trimming.

1.2. The debate

A concern expressed by welfare groups is that beak-trimming is "a discredited mutilation and farmers who still practice it should be brought into line by law. It is a last-ditch measure to avoid the consequences of bad management. If stocking densities are too high, the diet not balanced, or if lighting is at fault, stress-aggression occurs. To overcome it the beak is burned off cutting through a bed of highly sensitive nerve tissue, similar to the quick of the human finger nail".

On the other hand the prevention of cannibalism is a positive contribution to animal welfare as the pain and suffering resulting from cannibalism is much greater, resulting in the death of the bird. Many producers believe that on balance the practice of beak-trimming is to be favored, provided that the operation is performed properly.

Decision about beak-trimming includes the age to trim, amount of beak to remove, temperature of the blade and length of time taken to cauterize the beak. These factors coupled with differences in beak growth characteristics have the potential to create an endless number of combinations, many of which may be harmful to the individual bird.

The beak-trimming method needs to be tailored to the strain. One method may work with one strain but maybe quite inadequate for another strain. In general, however, if the beak is cut near the tip of the bone, at the correct angle and blade temperature, trimming will be successful.

The problem facing producers is that they are sometimes unsure of the age to trim chickens and the method to use. Because of the continuing welfare scrutiny of the beak-trimming procedure, attempts have been made by researchers to develop more welfare friendly methods. Beak-trimming is one of the most critical programs in use and extreme care must be taken to assure minimum harm to the flock, while at the same time necessary control of cannibalism.

1.2.1. Welfare comments

1. Purpose of beak-trimming is to limit the damage the hens can inflict on one another living in abnormally cramped quarters. In addition to causing pain to the birds beak-trimming has drawbacks from a production point of view. "A small portion of the chicks cannot simply survive the shock of the operation.
2. The minimum-waged employees who burn tiny beaks off chicks all day are not known for their care and precision work. Inevitably, some of the chicks have their tongues inadvertently burned during beak-trimming or they may suffer other facial injuries, which lead to certain death.
3. All beak-trimmed birds suffer a setback in growth and development as they recover from the trauma.
4. The accompanying beak tenderness adds to the problem by making it difficult for them to eat. When birds are moved to the layer house each bird is given a check to see if her upper beak has grown back or if it was not cut enough originally".
5. Beak-trimming is a discredited mutilation. It is a last ditch measure to avoid the consequences of bad management. If stocking densities are too high, the diet not balanced, or if lighting is at fault, stress-aggression occurs. To overcome it the beak is burned off cutting through a bed of highly sensitive nerve tissue.
6. Another form of aggression is the competitive aggression. Birds brought together in any system will have a go at each other in order to establish the peck order. On free range or in straw yard systems the more timid hens retreat and dominance is established without bloodshed. But in a battery

cage system the timid hen can neither take flight nor show submission and the confrontation often becomes a blood business leading to cannibalism and death.

1.2.2. Beak-trimming welfare research

The beak is essential to the bird for feed particle prehension, exploration of the environment, preening and social defence. It is an efficient tool. The beak epidermis contains dermal papillae, which plays an important role in precise tactile discrimination. The prevention of cannibalism is seen as a positive contribution to animal welfare as the pain and suffering resulting from cannibalism, may be much greater resulting in the death of the bird. Many producers believe that on balance the practice of beak-trimming is to be favored, provided that the operation is performed properly.

Using cannibalism and mortality as criteria, beak-trimming could be interpreted as having a stress alleviating effect, but when other criteria are used, beak-trimming has a stressful effect on the bird. Objections to the use of beak-trimming include its removal of sensory receptors, with a subsequent reduction in feed intake, pecking efficiency, pecking preferences drinking ability, permanent loss of temperature and touch responses and behavioral evidence (hyperalgesia and guarding behavior) for persistent pain. Severe beak-trimming (two-thirds of beak at 65°C) produces a long-term passivity in hens which has long-lasting consequences. In birds, which have short blunted beaks, dry mash tends to block their nostrils, and in cages they are unable to clean up their food trough.

1.2.2.1. Acute pain

It is generally considered that beak-trimming results in acute pain to the birds which passes quickly with the birds behaving normally a few minutes after the amputation. The results from the research suggest that the acute pain from beak-trimming is indeed short lived and some of the birds may not be in pain for 24 hours after the amputation.

1.2.2.2. Chronic pain

Following the pain free period, the birds may experience chronic pain for long periods of time. It seems likely therefore that not all birds subjected to partial beak amputations will suffer phantom and stump pain, but for those, which do suffer, it will be a welfare problem. Traumatic-neuromas in the beak stump after trimming have been implicated as a cause of chronic pain in commercial hens

1.2.2.3. Neuromas

If birds are beak-trimmed earlier than 10 days old both neuroma formation and pain is reduced. The tip of the incisive bone including the zone of ossification is normally removed with trimming. Ossification of the bone continues after beak-trimming but the order of ossification is disturbed. Traumatic-neuromas consist of swollen tangled masses of regenerating axon sprouts. These may form as either large masses or may develop as small scattered multiple fascicles of axons to form microneuromas. After several weeks, the nerve fibers re-grow, the excess axon sprouts degenerate and the neuroma regresses.

Occasionally, the neuroma mass persists and can discharge spontaneous action potentials that are perceived as chronic pain. Neuromas were present in all beaks at 10 weeks, but neuromas were not found at 70 weeks after moderate trimming at hatch. As neuromas were not observed in adult hens that had been moderately trimmed at hatch, the results indicate that they develop and persist for at least 10 weeks, before resolving.

The presence of neuromas is not in itself evidence of chronic pain as neural activity may cease following resolving of neuromas and normal feeding and pecking behavior could be restored.

Recent research with 6 and 21-day-old turkeys and in 1 and 10-day-old chickens suggest that neuromas do not develop after beak-trimming. Generally, however, neuromas will form irrespective of age of beak-trimming but have a greater potential to resolve if chicks are trimmed conservatively early in life.

Birds beak-trimmed at day-old and re-trimmed at 12 weeks do not exhibit feeding and pecking behaviors in the long term that indicate they are suffering severe chronic pain. There was some evidence that beak-trimmed birds used less force when pecking and might be more sensitive to pain when drinking hot water.

It is now currently accepted in the scientific community that beak-trimming correctly at day-old (relative to trimming at other ages) allows birds to return to apparently normal feeding and pecking behaviors with evidence of some beak sensitivity as neuromas resolve.

1.3. Need for beak-trimming

Beak-trimming is performed early in the life of commercial hens to decrease injuries caused by the behavioral vices of cannibalism, bullying as well as feather and vent pecking and to avoid feed wastage. Beak-trimming is known to help flocks with a hysteria problem. For the majority of birds beak-trimmed in the world today, it involves the partial removal of the upper and lower beak using an electrically-heated blade. Without a correct beak-trimming program, the egg producer risks heavy losses of chickens and pullets from cannibalism and in the laying stage from protrusion and vent pick outs. In many cases these losses represent the major part of mortality not caused by disease. If birds are not trimmed, mortality of up to 25-30% of the flock will occur and can be financially as disastrous as a disease outbreak.

1.3.1. Cannibalism

1.3.1.1. Etiology

Outbreaks of cannibalism are clearly recognizable. The evidence of blood-stained birds, broken skin, raw wounds and injured vents which are occasionally mistaken for prolapse are all clear indicators of cannibalism. Many causes of cannibalism have been suggested. Often outbreaks occur in one pen, while similar environmental conditions or feeding practices in other pens on the same farm do not cause any difficulty. Cannibalism develops either as a result of misdirected ground pecking or is associated with the dust bathing behavior.

Of the several factors thought to initiate cannibalism, the following are considered important among innumerable causes reported:

Strain, feeding only pellets, restriction of feed or water, or insufficient feeder or drinker space, periods of high humidity and high temperature, light leakage from inlets and outlets resulting in "torch-like" beams of light on birds' feathering, attraction to reproductive tract exposed during egg laying, injuries attracting pecking at damaged tissue, boredom, overcrowding, insufficient nesting space, lame and/or dead birds left in flock, keeping different ages or birds with different feather colors together, aggressive individual birds, sick, weak, small, or odd colored birds which are susceptible for attack by other normal birds as a survival instinct. feather change from fluff to feathers and poor ventilation.

1.3.1.2. Symptoms

1.3.1.2.1. Vent pecking

If a bird has died due to a pick-out in a cage, other birds in the cage may be next, this is usually caused by an individual bird, engaging in vent pecking. In this situation, all birds should be taken out of the cage and oviduct everted to examine the mucosa inside the cloaca and the lower part of the reproductive tract. It is likely that there will be damage observed, caused by the pecking bird. One or more wounds, produced by the pecking will be seen while the other hens were laying eggs on the previous day. The damage affects the muscular activity of these tissues and produces a vent prolapse a few days later.

1.3.1.2.2. Tissue-picking in bare areas

Forceful pecking is directed at bare skin leading to hemorrhage (and usually death follows) attracting other birds to join in the pecking.

1.3.1.2.3. Vent-pecking

Vent-pecking appears unrelated to the tissue pecking although it shares some of its features. It may be related to hormonal changes. Picking of the vent region or region of the abdomen several cm below the vent is the severest form of cannibalism. Predisposing conditions are prolapse or tearing of the tissues by passage of an abnormally large egg. Alternatively, pecking may be directed at the small downy feathers below the cloaca. After birds have tasted blood they will continue their cannibalistic habits without provocation.

Cannibalistic pecking is responsible for at least 80% of all vent prolapse cases and often results from poor beak-trimming with the offender usually being a cage mate or a bird in an adjacent cage that has been improperly beak-trimmed.

The true prolapse is a condition that occurs during the first stages of lay when the pullet has suffered a rupture of the tissues of the lower part of the reproductive tract. Usually prolapse occurs because of poor muscular elasticity or tone and may be the result of a hen's laying too large an egg for its age. Prolapse occurs where obese pullets have been put into production, where pullets have been indirectly light-stimulated or where pullet's flocks have been reared non-uniformly and has under- developed members.

1.3.1.2.4. Feather-pecking

Feather pecking is the precursor of cannibalism. Naked areas appear on the bird's body especially back, tail and the vent. This is sometimes believed to be caused by feather mites or by the hen molting. Slow feathering strains are more susceptible to feather-pecking. In its mildest form it has been observed as barb pecking and ignored by the recipient. Mould growth and mycotoxins in feed can cause abnormal feathering, contributing to feather-pecking. Feather abnormalities have also been attributed to deficiencies of zinc, tryptophan, lysine, glycine, leucine, arginine, valine, isoleucine, phenylalanine and tyrosine. Dietary deficiencies of pantothenic acid, folic acid, Vitamin B12, Vitamin E, pyridoxine, and biotin also contribute to poor feathering.

1.3.1.2.5. Feather-pulling

Feather-pulling is most frequently seen in flocks in close confinement with a lack of sufficient exercise. Nutritional and mineral deficiencies may be contributing factors. Sometimes the feather is removed and eaten. Irritation caused by lice and mites may induce this vice.

1.3.1.2.6. Toe-picking

Toe-picking is most commonly seen in domestic chicks and is often initiated by hunger, excessive warmth and toe-trimming. It is a particularly serious vice among young chicks reared on dark colored litter and can lead to an increase in mortality and a reduction in growth. Strong light illuminates the blood in the quick of the toes attracting the attention of the other chicks.

1.3.1.2.7. Head picking

Head-picking is directed by dominant birds at birds low in the pecking order causing the recipient to vocalize. In severe cases the areas above the eyes are black and blue with subcutaneous hemorrhage, wattles are dark and swollen with extravasated blood, and ear lobes are black and necrotic. Even though birds have trimmed beaks and are kept in separate cages, they will reach through the wire and peck at a neighbor or grasp its ear lobe or wattles and shake their heads.

1.3.1.3. Remedies to stop cannibalism

Beak-trimming is by far the primary remedial measure for preventing cannibalism. The other measures reported include hanging cabbages or sugar beets in the pen, Putting pine boughs on the floor, painting windows red, applying Stockholm tar to picked birds, using no pick salves, using repellent sprays, adding salt to the feed and water, feeding oats, feeding vitamin preparations or including vitamin B complex in drinking water, feeding DL methionine (in feed/water), manganese sulfate and horn meal, applying Vicks Vapor rub on wound, feeding whole grain diets.

1.3.2. Feed wastage

1.3.2.1. Quantitative feed wastage

Birds have a natural tendency to scratch the feed and search for grains especially when feed is in the mash form. In the process, there will be spillage of feed out of the feeders.

1.3.2.2. Qualitative feed wastage

Birds do establish a peck order within the pen. The stronger birds access feed first and preferentially pick and eat the gains (also a natural instinct) if beaks are not trimmed. It is well known that the grains are energy-rich and poor in all other nutrients ; hence, the stronger birds become weaker. When the weaker birds reach the feeders after the stronger ones have left, they will be left with only powdery feed which they cannot eat because of sharp beaks. Therefore, they also are subjected to nutrient deficiency and become weaker.

Ultimately, the entire flock shows a poor feed conversion and the farmer is at loss. If the beaks are trimmed, the birds cannot search for grains; instead, they have to scoop the feed and eat thereby making available all components of the feed to all the birds ensuring uniform growth, production and reproduction.

1.3.3. Egg-eating vice

When layers are reared on floor, there is possibility that sometimes eggs are laid on floor which attract the birds. They peck at it and if they get to the taste of egg, they start breaking eggs just as a habit. This vice is picked up by other birds also and quickly spreads within a flock. If beaks are trimmed, the chances of egg breakage are greatly reduced. Therefore, the vice is not a common occurrence among beak-trimmed layers.

1.3.4. Handling of breeding males

Beak-trimming facilitates easy handling of breeding birds, especially aggressive males expectable among Cornish and New Hampshire breeds.

1.4. Beak-trimming machines

1.4.1. Hot blade

Despite the developments in design of hot blade beak-trimmers the process has remained largely unchanged. That is, a red-hot blade cuts and cauterizes the beak. The variables in the process are blade temperature, cauterization time, operator ability, severity of trimming, age of trimming, strain of bird and beak length. This method of beak-trimming is still overwhelmingly favored in industry and there appears to be no alternative procedures that are more effective.

1.4.1.1. Gas beak-trimming

This machine consists of a hot plate and cutting bar operated by means of a foot lever. The efficiency of the machine varies with gas pressure and wind conditions. Generally it is slow to use and it is a useful portable machine for beak-trimming small numbers of birds. Producers can currently purchase a pocket style machine for trimming pullets (not available in India) which uses gas from a cigarette lighter as its heat source.

1.4.1.2. Electric Soldering Iron

This is a simple inexpensive device consisting of an ordinary electric soldering iron. A disk or coin made of brass or copper is welded to the tip of the soldering iron and the projecting edge of the circumference of the disc was sharpened like that of a blade. When the soldering iron is connected to the wall plug the temperature of the sharpened disc at the tip attains the maximum temperature within a few minutes which is quite sufficient for cauterizing the beak.

1.4.1.3. Hot blade machines

Following the development of the "debeaker" in 1943 there have been refinements to the machine including some control of cutting and cauterization and control of blade temperature. However control of blade temperature is still assessed mostly by the color of the blade, although thermocouples are available for measuring blade temperature. The most commonly used is the dark (dull) red heat with an approximate temperature of 650-675°C. Cherry red color (850-895°C) is used for toe clipping.

Precision beak-trimming of 6-10 day-old chicks is one of the most accurate methods available by using hot-blade machines which have a timed cauterization of 2 sec; this beak-trimming will suffice for the productive life of the bird.

Older bird, beak-trimming is performed with heavier blades on birds that are up to 12 weeks of age; two-thirds of the upper beak is removed but no closer than 0.3 cm to the nostril. If the lower beak is trimmed, it should protrude beyond the upper beak by 0.3 cm.

Beak-trimming birds over 12 weeks are generally accomplished by removing two-thirds to three-quarters of the upper beak again determined by the bird's age and maintaining a distance of 0.3 cm from the nostril.

1.4.2. Cold blade

Sharp secateurs have been used to trim the upper beak of both layers and turkeys. Bleeding from the upper mandible ceases shortly after the operation, and despite the regrowth of the beak, a reduction of cannibalism has been reported. Very few differences have been noted between behavior and production of the hot blade and cold blade cut chickens. This method has not been used on a large scale in industry. There are many reports of cannibalism outbreaks in birds with regrown beaks.

1.4.2.1. Temporary trimming

By using a sharp jackknife, a nick is made in the beak about 0.6 cm from the tip, with the thumb holding the cut portion of the beak against the blade. The knife is rolled around the tip of the beak tearing off the horny portion and exposing the quick. If properly done, there is little bleeding. It is not recommended to cut into the quick without cauterization.

1.4.2.2. Cutting with secateurs

By using a pair of secateurs at 1, 6 or 21 days, to trim the upper beak of turkeys. There will be bleeding from the upper mandible, which ceases shortly after the operation. Despite the beak growing back, a reduction of cannibalism was noted.

1.4.3. Robotic beak-trimming

A robotic beak-trimming machine was developed in France, which permitted simultaneous, automated beak-trimming and vaccination of day-old chicks of up to 4500 chickens per hour.

However, if the birds are not loaded correctly they could drop off the line, receive excessive beak-trimming or very light trimming because they were not positioned correctly on the holding cups. In addition it was observed that the machine could not beak-trim chickens effectively if there was a variation in the weight or size of chickens. Hence, the instrument has not been successful commercially.

1.4.4. Chemical beak-trimming

By the use of capsaicin applied at the time of conventional hot blade beak-trimming to retard beak growth. Capsaicin is a cheap non-toxic substance extracted from hot peppers. Applied topically or orally to mammals it induces a short-term burning sensation. In contrast to this effect in mammals, capsaicin is reported to induce only mild behavioral responses when applied topically to birds.

Although capsaicin's long-term effect in birds is not known it can cause degeneration of sensory nerves in mammals. It is well known that if the nerve supply is removed or prevented from re-innervating a particular tissue, then the tissue will degenerate. Hence, capsaicin decreases the rate of beak re-growth, and hence the need for re-trimming by its action on the sensory nerves; but, operators must avoid contact with the substance during its application to the beak.

The feeding ability of birds improved with capsaicin administration in the feed and, therefore, has the potential to reduce the percentage of starve-outs.

1.4.5. Laser beak-trimming

A laser method, which cuts the beaks of day-old chickens with a laser beam has been reported. By 16 weeks, the beaks of laser-trimmed birds looked similar to untrimmed beaks. Unfortunately, feather pecking and cannibalism during the laying period were highest among the laser-trimmed hens. These results suggest that the severity of beak-trimming by laser was insufficient, enabling regrowth of the beaks.

It might be expected that the use of laser beak-trimming, would enable greater uniformity in beak-trimmingand improved welfare as the beak would not require cauterization. Laser beak-trimming may represent a welfare advance but further work is required with this technology before it can be applied in industry.

1.4.6. Bio beak-trimming

The most innovative of more recent developments in the last 10-15 years has been the bio beaker, which uses a high voltage electrical current to burn a small hole in the upper beak of chickens. In the 1980's the bio-beaker was developed which used a high voltage arc (1500 Volt AC electric current) across two electrodes to burn a small hole in the upper beak of chickens. Up to 2000 day-old chicks can be beak-trimmed in an hour using this process.

The chicks being bio-beaked struggle as the beaks are inserted into the mask of the instrument and also when the current is passed. It takes 0.25 seconds to burn a hole in the beak. The primary advantage of the bio-beaker is that an adequate beak-trim is achieved during the first day of life, making the unit ideal for use in the hatchery. This allows treated chickens to eat and drink normally for the first few days with their beaks intact.

It was originally hoped that after a period of 3-7 days, the portion in front of the hole (tip of the beak) would die and slough off leaving a rounded stump. The aim was to burn a hole in the upper beak at a point just beyond the horny projection. In about 4 days the chick should begin to lose that portion of the upper beak from the hole to the tip and by 10-14 days-of-age, this portion of the beak should be completely lost from all the chickens.

Unfortunately in many chicks the tip of the beak did not slough off and birds had to be re-trimmed using conventional equipment. In turkeys, however, the bio-beaker was more successful with the beak tip falling off in 5-7 days and the wound healed by 3 weeks. This method is used for trimming the upper beak of turkeys, but operator errors and inconsistencies have caused welfare problems for the turkeys.

1.4.7. Freeze-drying method

In this method, liquid nitrogen is used to de-claw emus but, the conventional hot blade method was more effective. Further, the method was costly, time-consuming and re-growth of the claws occurred. Development of equipment thatcould freeze and cut the beaks, however, may be worth investigating.

1.5. The best age to beak-trim

Unless the beak-trimming process is done very carefully and accurately, the effects from it are frequently worse than the problem it is supposed to save. Decision about beak-trimming includes the age to trim, amount of beak to remove, temperature of the blade and length of time to cauterize the beak. These factors, coupled with differences in beak growth characteristics have the potential to create an endless number of combinations, many of which are harmful to the individual bird.

Beak-trimming will set a flock back by 2-3 weeks through a reduction in feed consumption and body weight. Feeders should be kept filled, adequate drinking water provided and procedures, which might stress the bird, should be minimized. Extra care is required to ensure that the birds using nipple drinkers are able to consume sufficient water after beak-trimming.

The beak-trimming method should be tailored to strain. One method may work with one strain but maybe quite inadequate for another aggressive strain.

Many producers are able to beak-trim their flocks at seven days with the precision method. However, this method sometimes fails to hold the flock for its entire life.

Chickens can be beak-trimmed in the hatchery before delivery, at 5 to 9 days (precision or block beak-trimming) or at 6, 10 to 12 weeks, during the growing period. Many hatcheries are reluctant to carry out beak-trimming prior to delivery. Hatchery operations point to the combination of stresses to which baby chicks are exposed - sexing, vaccination, and transportation, to which they say beak-trimming seems a further addition. This is despite the knowledge that day-old trimming results in fewer birds needing to be re-trimmed, that age of trimming does not influence mortality and performance is enhanced by earlier trimming.

Thus, it appears that if beak-trimming has to be performed it should be done at hatch. One of the problems, however, is that many hatcheries dealing only in pullet chicks do not have the skilled operators to perform long-lasting beak-trimming.

Probably the most popular time for beak-trimming in the USA is over 7-10 days-of-age as the chick has grown and operators feel that it is easier to perform, with the required precision. This situation also applies in Australia, although 9-10 days of age is more popular. At the earlier age (7-10 days), birds are often much easier to catch and to handle than 6 week-old chickens and the overall operation can be conducted with much less stress on both birds and beak-trimming personnel. Beak-trimmed birds need access to open water troughs.

At 10-12 weeks, beak-trimming is more costly to perform and imposes greater stress on the birds although all birds should be checked at this age to decide if re-trimming is required. Whenever beak-trimming is contemplated, the timing should be such that it does not coincide with other stresses.

1.5.1. Day-old beak-trimming

Beak-trimming at the hatchery before delivery of the chicks may be convenient and relatively cheap, but causes higher mortality and the beaks are likely to re-grow. Birds beak-trimmed at day-old have a permanent reduction in body weight but eat less food without affecting egg weight and production.

Many hatcheries are reluctant to carry out beak-trimming prior to delivery because the birds are already under enough stress from sexing, vaccination and transportation but recommend beak-trimming between 5-10 days or 10-12 weeks as the best time for performing a permanent job which can last throughout the laying period.

On the other hand, recent welfare research has shown conclusively that the best age to beak-trim chickens is at day-old or soon after hatch. Therefore, field trials are needed to convince the industry one way or the other that beak-trimming at day-old is effective.

1.5.2. Early precision beak-trimming (5-10 days)

E.L. Bramhall developed trimming at 5-10 days in the early 1960s at the University of California. 7-9 days is the best age to trim as the beak is more horny and if the beak-trimming job is done correctly there is no reason why any bird should be re-trimmed. However, the precision method at 5-10 days sometimes fails to hold the flock for its entire life, and hence, precision method is best performed at 6-10 days, with a follow-up trim at 10-14 weeks. The moderate method is performed at 10-14 weeks involving removal of one-third of the lower beak and two-thirds of the upper beak.

The severe method is also done at 10-14 weeks. It involves removing two-thirds of both the upper and lower beak and is often performed in the USA as a follow up to the earlier precision trim method.

High quality beak-trimming at 7-10 days probably has a minimum effect on weight gain, in contrast with the effects of beak-trimming at 10 to 12 weeks.

To beak-trim a 5-10 day-old and a day-old chicken it is held with the thumb on the back of its head and forefinger under the throat. Alternatively, the thumb can be wrapped around the upper body of the chicken during beak-trimming, or be placed on one side of the head, with the forefinger curved around the neck on the other side (to prevent operators pushing too hard on top of the head causing internal bleeding in the brain with no outward signs of why the bird died).

The blade of the beak-trimmer should be heated to a cherry red color and should be sharp and straight edge. Overheating the blade will cause it to warp under pressure and cause blisters in the mouth. A cold or blunt blade will fracture the beak and cause sensitive bulb-like growths on the cut end.

The closed beak is inserted into the 0.43 cm hole in the gauge plate of precision trimmers or operators judge the correct amount of beak to remove. Light pressure is exerted on the throat to pull back the tongue and prevent it from being cut or burnt. This also withdraws the lower beak slightly so that is a little longer than the upper beak when beak-trimming is completed. Care should be taken not to sever or burn the tongue.

Treadle-type beak-trimmers are used as well as power units with a motor-controlled cam that pushes the hot blade through the beak, automatically timing the operation. The blade remains in the cutting position for 2 + seconds during which time cauterization of the remaining tissue is accomplished.

If precision debeaking is done correctly it eliminates the need for a second beak-trimming.

1.5.3. Beak-trimming at 4-6 weeks

This is the time when birds are transferred from brooder cages to rearing cages and hence, the beak-trimming at this age would last throughout the laying year.

With older pullets the birds mouth is opened with the index finger of the right hand holding the tongue down and back to prevent it from being burnt or severed. The upper beak is rested on the cutting bar so that it rests evenly and a cut is made by slowly moving the hot blade down on the beak by means of a foot pedal. Sufficient pressure is exerted to cut the beak slowly, cauterizing it at the same time.

The lower beak is then cut in the same fashion. Care should be taken to cauterize the beaks thoroughly by rubbing up against the hot blade momentarily after cutting, and also to round off the corners. Some producers beak-trim their flocks at 4 weeks-of-age if the pullets are being grown on nipple-type drinkers. The idea behind this is to minimize any weight loss.

1.5.4. Beak-trimming at 6-8 weeks

One advantage of beak-trimming at 6-8 weeks is that pullets have more time to regain lost body weight compared to those beak-trimmed closer to maturity. If the crew is experienced, beak-trimming at this age generally results in more accurate results compared with beak-trimming at an early age. In light controlled housing it is recommended to beak-trim at 7-8 weeks.

1.5.5. Beak-trimming at 8-16 weeks

Late non-precision beak-trimming is usually performed with a heavy-duty cutting blade. Two-thirds of the upper beak and one-third of the lower beak are removed in 12 week-old birds followed by a brief untimed cauterization. Amount of beak removed is determined by visual inspection.

1.6. Severity of beak-trimming

Beak-trimming one half or three quarters of the beak did not result in any loss of egg production, but reduced food consumption and body weight gain. In severely trimmed birds there were changes in the consistency of the outer covering of the re-grown portion of the beak.

It is generally accepted that removal of half the upper beak and one-third of the lower beak at the tip of bone at 5-10 days results in reduced mortality compared to more moderate levels of beak-trimming

Birds with one-quarter of the beak removed showed considerable re-growth, 50% of the birds with one-half of the beak removed showed re-growth whereas those birds with two-thirds of the beak removed showed virtually no re-growth. As the temperature of the blade was increased from cold, medium to hot the proportion of birds showing re-growth decreased emphasizing the point that cauterization is required with a hot blade to reduce re-growth. Beak re-growth is dependent on cauterization time and amount of beak removed.

The general recommendation is that beaks be trimmed leaving 2 mm beyond the nostril at 6-10 days, and to avoid re-trimming, leave 1 mm beyond the nostril. For birds 10-12 weeks-of-age, beaks should betrimmed 6-7 mm beyond the nostril with 2 sec of cauterization.

In case of breeding males, both upper and lower beaks to be cut equally so as to facilitate them to catch the females while mating.

Adequate feed depth must be maintained in flocks with long lower beaks. Water must also be maintained at a level that will enable chickens with long lower beaks to drink without difficulty. During periods of high-water consumption, it is urgent to increase the water flow in trough systems so that all birds can drink regardless of their location on the water line.

1.6.1. Blade temperature and cauterization time

Blade temperature is commonly evaluated by looking at the blade color, which can vary with the light conditions in the area where beak-trimming is being performed. A bright room will cause the blade to look duller while a dark room

Fig 35.1 Severity of beak-trimming (Gleaves, 1997)

will make the blade look hotter. Standard light conditions or a device to measure blade temperature is critical, if beak-trimming is to be consistent. The use of a small light that illuminates the blade while the rest of the shed is at normal lighting is an effective practical method for standardizing lighting.

Cauterizing helps stop bleeding, promotes healing and prevents the possible infection especially of *E. coli* and *staphylococcus.* The use of a rolling motion to cauterize the beak is recommended as a means of inhibiting regrowth and to prevent the formation of sharp edges on the outer edges of the beak.

The relationship between color of blade, temperature, age of birds and cauterization is summarized in Table 35.1.

Table 35.1. Temperature and duration of cauterization

Blade color and cauterization time for 6-10 d beak-trimming		
Blade color	Temperature (°C)	Cauterization time (s)
Light red	500-550	3.0
Dull red	650-750	1.0 * or 2.5 **
Bright red	850-950	1.0 * or 2.0 **
Yellowish red	1050-1150	Not used
* Non-permanent, only end of beak singed, red in color		
** Permanent, no further growth		
	Cauterization time (s)	
Age	690°C	860°C
1 d	2.5	2.0
6-10 d, Permanent	2.5	1.5
6-10 d, Non-permanent	1.0	0.75
7-8 weeks	1.0	0.5 ***
10-12 weeks	1.0	0.5 ***
*** Two cuts, time per cut		

1.6.2. Angle of the cut

The angle of the cut should be given close attention. Generally a straight cut is made at 0-10 days-of-age although it is possible to get an angled cut at 10 days by holding the head slightly lower than horizontal and cut both beaks at one time (block trimming). Slanting the cut of the upper beak is more possible after 6 weeks-of-age, but more difficult to do on the lower beak. The angled cut of about 5-10° is more important to achieve when the birds are re-trimmed. If the cut is made at an improper angle, the beak can become deformed or impacted and the bird rendered worthless.

1.6.3. The need for hot blade beak re-trimming

There can not be a single recommendation regarding beak re-trimming; however, beak-trimming at 6-10 days, then touch up at 7-8 weeks or 10-12 weeks appears satisfactory. Such an early schedule helps to avoid early pecking. It also allows more permanent beak-trimming to be completed well before the flock comes into production.

Hot weather, open housing, high bird density and large colony sizes generally necessitate more extreme beak-trimming and re-trimming, but these effects can be reduced with less aggressive strains.

1.7. Care before and after beak-trimming

1.7.1. Birds

1. Administering vitamin K through drinking water 2-3 days prior to trimming can be considered as an option.
2. Mechanical feeders should be operated frequently to stimulate feed consumption. If cart fed, birds should receive feed at least twice daily.
3. Birds should not be subjected to stress from housing, vaccination or worming during the week prior to or the week after trimming.
4. Give no medicines which will give a bad taste to the feed or water.
5. Sick birds should not be beak-trimmed
6. Never beak-trim in combination with vaccinations except for fowl pox, when moving birds or birds on medication

1.7.2. During beak-trimming

The following care has to be exercised while handling birds aged 6 weeks or more while debeaking:

1. The operator should hold the bird in such a way that it neither shakes it head nor suffocates.
2. The beaks are opened with the help of index finger and the tongue is held back.

3. The upper beak is cut first to the recommended level. The beak is held against the blade and circular motion is given for at least 2 sec while holding to effect proper cauterization.
4. Lower beak is then cut as per the recommendation.
5. Proper cauterization is once again ensured before the bird is left into the pen.

Fig. 35.2. Beak-trimming procedure

6. In case of breeding males, both upper and lower beaks to be cut equally so as to facilitate them to catch the females while mating.

1.7.3. After beak-trimming

1.7.3.1. Birds

1. Immediately after trimming, birds will suffer acute pain. A mixture of bupivicaine (local anesthetic) and phenylbutazone (anti-inflammatory) can be swabbed on to beaks immediately after beak-trimming, or beaks dipped into ice-cold water.
2. Feeders must be kept full with feed to help birds eat easily; probably this is the only occasion when feeders are full with feed.
3. Vitamins (B-complex and C) and vitamin K can be given through water to help alleviate stress.
4. All the birds must be observed carefully for any bleeding, especially in the upper beak. If any bird shows bleeding it must be separated at once, suitably treated/cauterized; otherwise, there is a likelihood of cannibalism.

1.7.3.2. Blades

1. When using the precision method, the plate should be taken off to scrape away the dried blood and tissue to enable a proper cut.
2. The plate should be adjusted such that the blade is snug against the plate. If there is a space between the blade and the plate a poor cut will result. If the

blade is too snug against the plate, odd shaped beaks will result from trimming. To make sure that the blade and plate are against one another the unit can be triggered to bring the blade edge down to the half-way point on the plate. Push the plate against the blade to make a snug fit and then tighten.

3. The blade should be sharp and scraped every 15 minutes to remove the fine grey residue, which accumulates from the birds during beak-trimming.

1.7.4. Common errors in beak-trimming

1. The upper and lower beaks are cut too short.
2. Lack of cauterization.
3. Nostrils of the chicks are burnt.
4. The beaks are cut on the side, leading to crooked beaks on account of rushing and holding too many chicks in one hand while trimming birds using the other hand.
5. The beaks are cut at the wrong angle.

1.8. Alternatives to beak-trimming

Although many alternatives have been proposed to beak-trimming, none seems to really fit for commercial use; some of them are enumerated below:

1.8.1. Implanting hormones

Progesterone and estrogen increased but testosterone reduced pecking; but the hormonal implants adversely affected egg production; therefore, hormonal implant is not practicable.

1.8.2. Blind chickens

Genetically blind chickens remain fully feathered, produce more eggs and eat less feed, and do not exhibit feather pecking; but, this too is not a practicable proposition.

1.8.3. Spectacles (restricting vision of birds)

The use of spectacles (anti-pecking devices made of a colored flexible polyethylene material), when fitted on the nares of the hens allows birds to look to the side or down but not directly ahead. The use of spectacles was effective in controlling feather pecking. Pecking damage was higher for control hens compared to hens with spectacles, which were better feathered after 11 months of lay. The use of spectacles has the advantage in reducing social stress by limiting visual contact and breaking down the social hierarchy.

Mechanical devices also have disadvantages since they can only be put on birds of pullet size or larger, they are relatively expensive and they take considerable time to fit to the bird. They cannot be used in cages because they interfere with eating and drinking, can be easily dislodged and these devices are held in place by metal clips which pierce the nasal septum. Most important of all, the spectacles are not commercially suitable.

1.8.4. Contact lenses

Red contact lenses were first used in the 1960s for layers as an alternative to beak-trimming. They failed to gain popularity because they caused considerable eye irritation, eye infections, abnormal behavior, and they were not retained well. The lenses were redesigned recently in attempt to eliminate or reduce these problems. Birds can be fitted with the lenses at the rate of approximately 10 birds/minute. Egg production was lower and mortality higher due mainly to the inability of birds to find the feed. More recently colored contact lenses for laying hens have been introduced. Hens fitted with red lenses (rosy glasses) appeared to be the least stressed. Cost and practical considerations definitely obviate use of contact lenses under commercial set-up.

1.8.5. Environmental enrichment devices

Environmental enrichment reduces aggressive behavior and mortality and improves feather condition and egg production. Hens rapidly habituate to simple objects hung in their cage. Studies have been undertaken on commercial farms using environmental enrichment devices like small rotating wheel, biting/pecking devices, tin pants etc. The devices failed to show significant effects within experiments but when all the data was combined a significant improvement in economic returns from using the devices was observed Enrichment devices may need to be introduced to chickens at a very early age for them to be effective.

1.8.6. Changing the light intensity

The intensity of light can be manipulated in poultry houses to reduce social interactions among the flock mates. Under low intensity light it is possible that the birds cannot see each other well which may help reducing antagonistic encounters and aggressive behavior among them. It is important to eliminate areas where bright sunlight strikes the floor. The use of very low light during the first 3 weeks of rearing may discourage any pecking vices developing in poultry avoiding the need for beak-trimming.

1.8.7. Rearing under red light

For many years it was practice to brood and rear chickens under red light to prevent cannibalism, but, there is little direct evidence that this practice is effective. Low intensity white light is satisfactory.

1.8.8. Provision of straw, grain and whey blocks

A number of authors have suggested that feather-pecking and subsequently cannibalism in poultry may be considered as redirected ground pecking, based on strong similarities in the performance of both behaviors. Adequate substrate like scratch grain during the rearing period affect the development of feather-pecking. Similarly, locating semi solid milk or whey blocks around the house, hanging green leafy vegetables and spreading grass clippings can help prevent feather pecking.

1.8.9. Use of anti-pick compounds

Applying anti-pick compounds (commercial anti-pick, pine tar or axle grease to wounded areas reduces pecking. Commercially available sprays can be applied to wounds and bare areas on the body to improve feather cover and minimize pecking.

1.8.10. Genetic strategies

Primary breeders need to select strains with a low tendency to pick. Breeders have been reluctant to do this in the past because they would have to relax selection pressure on production traits and thus put themselves at a commercial disadvantage to their competitors. A behavioral trait in a commercial breeding program is generally resisted by breeding companies although there is potential for changing the propensity for feather-pecking and cannibalism if the underlying low fearfulness of a strain can be selected.

Current gene mapping efforts will provide information at the molecular level of gene function and could provide alternative strategies for improving welfare than by conventional selection. Biotechnology has the potential for suggesting novel methods for improving welfare such as the manipulation of a receptor for a neurotransmitter involved in the control of pecking behavior. The existence of genetic markers for welfare traits will permit markers-assisted selection to take place, possibly obviating the need to measure complex behavioral traits in commercial selection programs.

1.9. Implications of not beak-trimming

It is possible to rear pullets without beak-trimming if there is effective light control. However, once an outbreak has occurred, it is difficult to control cannibalism. Poultry producers should leave beaks intact only if they are satisfied that there is little if any risk involved. The more flighty aggressive strains need to be reared at lower light intensities than others are. Control of pecking is quite different and more problematical in the laying stage when a higher light intensity up to 5 lux needs to be provided. Under some conditions it might be possible to manage laying hens without beak-trimming, but for most producers, the risk of such a policy is too high.

1.10. Alternative production systems and beak-trimming

It is not possible to ban beak-trimming due to the risk of damage caused by pecking activity especially in alternative systems or in open-sided houses where light intensity is high. Feather pecking and cannibalism amongst birds kept in virtually all of the alternative systemsremains unpredictable and a major problem yet to be solved. Increased pecking and cannibalism is considered to be caused by the larger group sizes. In welfare terms current alternative systems, housing large flocks of laying hens have a chance of an outbreak of cannibalism providing a major argument against the system. No strategy guarantees that feather-pecking will not develop in practical poultry-keeping and beak-trimming may be required in specific cases to prevent the greater risk of welfare problems caused by cannibalism.

2. Dubbing

Dubbing refers to removal of all or part of the comb. It can also mean removal of wattles and possibly, earlobes.

2.1. Functions of comb

1. Attracting females
2. One of the means of recognition in chicken
3. Thermoregulatory function - during summer, birds dip their combs and wattles in water. When the water evaporates, it takes away heat from the body thereby cooling the blood that is being supplied to the combs and wattles. The cool blood enters the body and the warm blood that comes again will be cooled. In this way, combs and wattles are very helpful in thermoregulation, especially in tropical climate.

However, when the combs are too big and drooping, they might cover the eyes and be an interference to feeding and stimulation of egg production. They are also exposed to physical injuries by sharp objects such as feeders, waterers, wire cages etc. which complicates further into infection. In very cold climates (temperate countries) big comb is likely to be frost-bitten.

2.2. Method of dubbing

Dubbing is mostly practiced when the climate is very cold. The comb of the day-old chick will be removed with a small manicuring scissors running the shears from front to back. The concave side of the scissors should face upwards. Due to dubbing, the hens will be on a more equal social basis and therefore reduces the peck-order or bossing in the flock and hence, the flock will be quieter and tamer. The dubbing also reduces the chances of injury during fighting among breeding males.

Chapter **36**

Broilers and Roasters

Tender chicken, usually less than 8 weeks of age, weighing about 1.8 to 2.0 kg are popularly known as "Broilers"; the term "fryer" can also be used to mean a broiler. Broilers if maintained till about 3 kg body weight, they are referred to as "Roasters". If tender meat, probably the most tender, is obtained by removal of testes, the bird is referred to as a "Capon". However, capon production is no more a common practice due mainly to the procedural difficulties in case of surgical caponization and public health concerns in case of use of hormones or chemicals for caponization.

Managemental practices, therefore, are essentially same for both broilers and roasters; only the latter are held for few more weeks to attain the desired weight and hence proportionate increments are allowed in space, feed, water and other requirements. In addition, after 8 weeks of age, they are fed on diets specially developed, keeping in view their nutrient requirements.

1. Housing

Broilers are usually reared on floor, in general, and all-litter system, in particular. Typical cross section of an open-sided all-litter house is described in Chapter "Poultry housing". In case of environmentally-controlled houses, the descriptions hold except that the side-walls are completely closed but for openings made for ventilating and insulating equipment as per requirement.

1.1. Space requirements

Table 36.1 Space requirements of broilers and roasters

Type	Floor space, m^2/bird	Feeder space, cm*	Waterer space, cm*
Broilers			
1.36 kg	0.05		
1.82 kg	0.06	5.0 – 7.6 **	2.0 ***
2.27 kg	0.08		
2.72 kg	0.09		
Roasters	0.11	7.6	2.8

* Linear cm ** 5.0 up to 5 weeks ***For all body weights

Source : North and Bell, 1990

Under winter conditions, floor space allowance can be reduced to a tune of 10% in order to conserve animal heat within the building. However, floor space requirement is very critical for broiler/roaster performance. Reductions in floor space can have adverse effects on almost all broiler parameters like growth, feed consumption/FCR, survivability and feathering while increasing ventilation requirement, mortality, incidence of breast blisters and carcass condemnations. However, broiler weight produced per m^2 area of housing is increased albeit uneconomically.

1.2. Ventilation

Ventilation is required to maintain good air quality for poultry and appropriate litter moisture for a healthy environment. Air exchange is necessary to remove carbon dioxide and ammonia from poultry houses and to bring in oxygen; however, removal of heat and moisture from litter houses usually requires greater air exchange than required for carbon dioxide and ammonia removal. (Stull, 1998).

Ideal moisture levels for litter are 25 to 35 percent. Lower levels result in excessive dust which is detrimental to the respiratory system. Higher moisture levels result in excessive caking of litter which can contribute to breast blisters, disease problems and lameness (Stull, 1998).

In open-sided houses (natural ventilation), it is extremely difficult to regulate air flow to the exact requirements. However, it is possible to ensure comfort of birds by judicious handling of curtains and other simple inlet controls (if installed). In addition, ceiling fans are also used by some farmers under tropical climate.

1.2.1. Windowless houses

In windowless houses, ventilation requirements should be regulated to the following requirements as per the season.

1.2.1.1. Winter

During winter, the cold outside air brought into the house is warmed by the birds and it simultaneously reduces the relative humidity as well; the latter helps in absorption of litter moisture thereby keeping the litter dry.

However, airflow should be kept in such a way that the birds do not feel "chilly"; for this reason, airflow is restricted to = 0.73 km/h. Under winter conditions, width of windowless houses will be normally = 12.2 m and hence, if air can be replenished across the width every minute, the requisite airflow is automatically ensured.

Table 36.2 Ventilation requirements of broilers (m^3/min/1000 birds)

	Average house temperature (°C)			
Age (weeks)	10.0	15.6	21.1	26.7
0 – 2	8.49	11.32	14.15	16.98
3 – 4	33.96	36.79	42.45	48.11
5 – Market	65.09	79.24	90.56	99.05

Source : North and Bell, 1990

1.2.1.2. Summer

Unlike during winter, air is brought through evaporative cooler pads located at one end of the house and it is drawn by the exhaust fans operated at the other end. If the length of the building is x times its width (12.2 m), to replace all the air in one minute, the airflow should be x times that calculated for winter conditions.

Assuming that $x = 10$, that means, length = 122 m, wind speed should be at least 7.3 Km/h in order to completely replenish the air inside the house once every minute.

At the evaporative cooler, the temperature of air will be 6.7°C lower than the ambient temperature which while moving across the birds can give a "wind-chill" effect of additional 5.6°C to them. The air speed of 7.3 km/h or more can also remove the heat produced by the birds quite efficiently.

1.3. Light

As a general rule, once a lighting program is put on operation, it should never changed to another program for the same batch of birds. It is also advisable to measure light intensity at floor level occasionally and also to clean both bulbs and reflectors at least once a fortnight.

For catching the birds for market, in order to reduce bruising, night-time with a low intensity red/blue light has been found to be advantageous.

1.3.1. First 3-5 d of brooding

A continuous light with an intensity of 35 lux at floor level is provided all through the day except for 1 hr darkness at midnight to train the birds not to panic in the event of sudden power failure due to any reason. This is true for both open-sided as well as windowless houses.

1.3.2. After 3-5 d of brooding

In either system of housing, under continuous system of lighting, 23hr light and 1 hr darkness is preferred for the reason mentioned above. However, in windowless houses, intermittent dim light can also be provided.

1.3.2.1. Open-sided houses

In case of open-sided houses, light control is extremely difficult and hence, broilers are normally exposed to higher intensity of light which might, sometimes, induce higher activity, cannibalism etc. Notwithstanding these facts, bulbs are provided similar to that in windowless houses.

1.3.2.2. Windowless houses

In case of windowless houses, light control to limit intensity to 3.5 to 5.0 lux at the bird level is recommended. Light intensity required is only to facilitate location of feed and water while minimizing unnecessary movement and activity of the birds.

It is also advisable to use red colored light which is found to reduce cannibalism. However, it must be remembered that the birds should never be changed back to white light after being exposed to red light because such an action will trigger cannibalism almost occasions.

To provide a light intensity of 5.0 lux, 2 bulb-Watts/m² are required. Incandescent bulbs with square reflectors measuring 25 to 31 cm on each side fixed at the height of 2.1-2.4 m with distance between two bulbs at 3.15-3.60 m will meet the requirements. In the above example, for the building measuring 122 m x 12.2 m, 76 numbers of 40 W bulbs can be fixed at a height of 2.15 m in two rows with 3.25 m space between two bulbs.

If intermittent dim light has to be provided, 1L : 3D cycles, providing a total of 6 hr light for feeding, is advisable. However, during hot weather, 1½L : 3D is referred in order to increase the total feeding time. It is also necessary to increase feeder and waterer space by 50% to help all the birds eat and drink at one time. Obviously, investment on equipment increases under the system of lighting while there will be saving on power and power costs. In addition, intermittent light has been found to improve growth rate by about 6% over and above that on continuous light in open-sided houses.

2. Brooding and growing

Broilers are normally reared straight-run. In the event that the sexes have to be reared separately, sexing can be done at the hatchery by vent method, feather sexing or by color sexing (details in Chapter "Qualitative genetics").

2.1. Temperature

Preparations required before and after arrival of chicks are already been discussed in Chapter "Brooding". Brooding temperature for broilers about 5 cm above the surface of the litter under the brooder (hover/canopy) should be 30.0°C during the 1st 2 weeks, followed by 27.2 and 23.9°C during 3rd and 4th week, respectively (North and Bell, 1990). This brooding with temperature under the canopy/hover warmer than the room itself, is sometimes referred to as "cool room brooding" (Stull, 1998).

Under winter conditions, fuel required to maintain brooder temperature increases significantly; for instance, when the ambient temperature is 4.4°C, 2.2 times the fuel required at 21.1°C will be required.

2.1.1. Limited area brooding

During the extreme winter conditions, a limited area within the brooder/broiler house can be delineated with plastic curtains to conserve heat, preferably in the middle of the house so as to provide 325 cm² floor space per chick. In this limited area, brooding is carried on up to 3 weeks of age and this process is referred to as "limited area brooding"; this procedure is likely to reduce fuel costs and improve survivability (North and Bell, 1990).

2.2. Feed and water

By the time the chicks arrive, the water temperature should be about 24°C; this temperature is maintained for the first few days after which, it is lowered and cold water is made available. Water is kept in small fountain waterers at the beginning; later on, nipples or automatic waterers or watering channels, as the case may be, may be used to provide water for the birds.

The chicks are dumped near the brooder and allowed to drink water for about 3 hr after which broiler mash/crumbles are offered on the feeder lids or sprinkled on the paper.

2.3. Feeding and drinking equipmet

Regular feeders (linear or tube-and-pan or any other type) are introduced as soon as possible, even on the 2nd day itself. With linear feeders, feed level is always kept low; if the feeders are 2/3 full, 10% of the feed will be wasted; if ½ full, 3% and when < 1/3 full, about 1%. Therefore, feeders are never filled more than 1/3 full. It is absolutely necessary to note that birds can be fed as frequently as feasible, say 3-5 times a day, but at specific time of the day. Frequent feeding, especially with linear feeders and mash type of feed, stimulates feed consumption and therefore, the growth rate.

Feeding and watering equipment used to start chicks must be properly sized so that chicks can eat and drink comfortably. The edge of the feeder or waterer should be located at the average level of back of chicks. Feed is often placed on egg flats or in plastic trays during the first few days to help chicks locate feed. Changes in placement of feeders and waterers should be made gradually so that the chicks find the new feed and water locations. Chickens should always be grown on diets which meet their nutrient requirements (Stull, 1998).

2.4. Beak-trimming

The present day broilers grow at the phenomenal rate of nearly 2 g/hr from hatch to market (about 6 weeks). Therefore, there are already under severe physiological stress and hence, beak-trimming is generally not advisable.

Beak-trimming is primarily performed to prevent cannibalism which is not that common among broilers because they are basically non-aggressive. However, the open-sided house where light control is not possible, during warm summer months, high light intensity sometimes stimulates hyperactivity leading to cannibalism. Under such circumstances, beak-trimming may be necessary.

Although trimming at 5-10 d of age can produce excellent results, handling of birds not only requires labor but also inflicts strain on the fast-growing birds. Therefore, beak-trimming at hatchery itself is preferable. A hot blade trimmer with a blunt blade should be used to notch the upper beak. By 10 days of age the tip of the upper beak will separate. An electric spark trimmer can also be used. The spark arcs between two electrodes and leaves a small hole in the beak. After a few days the tip of the upper beak will separate leaving a trimmed upper beak. These methods leave the tip of the beak intact until the chick learns to eat and drink which reduces early stress (Stull 1998; see Chapter "Beak-trimming").

In case of windowless houses, controlled light intensity and/or color of light (red color) should, on most occasions, prevent cannibalism; and hence, beak-trimming is not generally required.

3. Rearing broilers sexes separate

Notwithstanding that broilers are normally grown straight-run, it is very important to note that unlike layer chicks which are sexed and only female chicks are reared. Therefore, it is very important to note that the rate of growth of males is definitely higher than that of females and hence, their nutrient requirements are also conspicuously different. In other words, in a straight-run flock, the sex forming 50% of the population is not offered a balanced diet. In addition, it is well-known that females attain their mature body weight earlier to males and hence it is imperative to infer that in a straight-run flock, the females are continued to be fed even after they are ready for market thereby reducing the returns.

Contrary to one of the most common misconceptions that yield of males is better than that of females, the yield from females is generally higher due to the larger breast portion and better conformation than males. Further, females yield more meat and skin and less bones than males. Thus, the gross quantitative yield of males is higher mainly because of their higher body weight.

In view of the above, it is possible to produce uniform weight-spread not only in terms of weight but also in terms of maturity and in "desired/targeted" weight if the sexes could be grown separately. For example, in a flock wherein sexes are grown separately, the male and female mean body weights were 2.0 and 1.6 kg, respectively with a coefficient of variation (CV) of 5% in both. Therefore, standard deviation (sd) of males and females will be 100 and 80 g, respectively.

Assuming a normal distribution of the body weights, 96% of the birds will be within 2sd limits on either side of the mean. That means, 1.80 to 2.20 kg and 1.44 to 1.76 kg among males and females, respectively. If the same flock was reared straight-run, the spread of weights would have been 1.44 to 2.20 kg with the mean body weight of 1.80 kg; which means, relatively fewer birds will be near to the mean body weight.

Recent reports further indicate that male broilers require more protein whereas the females require less energy and less protein. Feeding method and/or type of feed may have to be modified to ensure that the sexes get the nutrients they require. Feeders fixed at a level suitable to only males may be one of the ways of offer additional nutrients to males. Females have to be de-populated earlier and processed before their feed conversion ratio deteriorates. This would also help males to grow better because they get more floor space due to removal of females.

3.1. Advantages

(North and Bell, 1990)

1. It is possible for the processor to meet the market demands by processing males when they are younger than females.
2. Sex-separated flocks, being more uniform, are easier to handle in the processing plant and in the longer run, there can be significant reductions in condemnations.
3. Different feeds for different sexes, probably at the different periods, are possible and the same may help reduce feeding cost.

3.2. Disadvantages

(North and Bell, 1990)

1. Increased costs of sexing. Feather sexing is costlier than color sexing
2. Larger breeding flocks are necessary to meet the demand for male and female broilers chicks. In turn, hatcheries may require that their customers always purchase equal number of male and female chicks.
3. Integrated broiler growing contract becomes more complicated.
4. If for any reason some processes in demand small and large birds, it would be difficult with sex-separated flocks which are marketed at the same average weight.

4. Growth and feed consumption

4.1. Growth characteristics

1. Shows a slower juvenile growth during the first 2 weeks followed by an explosive growth till market. Male broilers tend to grow faster than females; the effect progressively increasing with advancing age. Males, about 1% heavier than females at hatch, grow to about 17% more than females at market. Since gonads are not yet functional during this period, the differential growth must be a consequence of other physiological mechanisms presumably affecting growth hormone. Male roasters are 25% heavier than their female counterparts indicating a further divergence in weight spread.
2. Higher the average weight in a straight-run flock, the larger will be the

difference between the sexes.

3. Males grow faster than females; males attaining same weight of females approximately 4 d earlier to females.
4. Weight gain reaches a peak at about 7 weeks of age and declines thereafter.
5. Feed required per unit weight gain progressively increases with age; initial weight gains being cheaper than the subsequent ones.
6. Males require less feed (about 9 to 10%) than females to put on a kg of body weight.

4.1.1. Uniformity

Although it is ideal if plotting of percentage of birds under each weight group against the weights results in a bell-shaped Normal distribution curve, more often than not it does not occur under commercial conditions. At the same time, producers prefer to sell broilers weighing within a certain range of weights and utilize those which fall away on either sides of the range for preparation of by-products or selling at a discount.

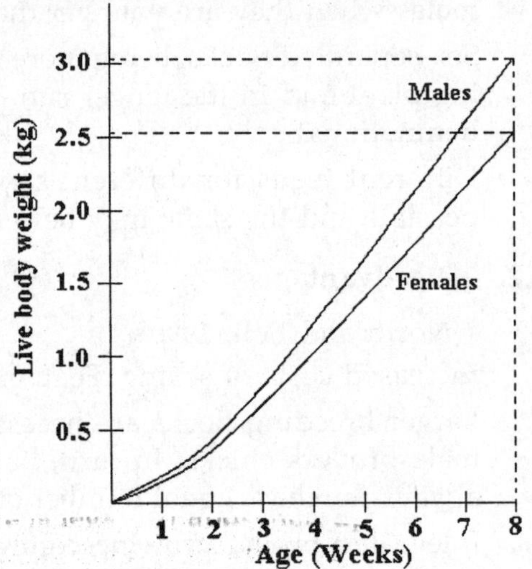

Fig. 36.1. Growth rate of broilers

Approximately 75% of the males and 78% of females should be within 10% of the average weight of each sex for the flock to have satisfactorily "uniformity" at maturity (North and Bell, 1990). A thorough examination of brooding, growing, feeding and other aspects of broiler production may have to be looked into if a flock falls into "poor" or "very poor" classification (Table 36.3).

Table 36.3. Uniformity of flock

Classification	Percentage within 10% of mean sex weight *	
	Male	Female
Excellent	≥ 86	≥ 89
Good	79-85	82-88
Average	72-78	75-81
Fair	65-71	68-74
Poor	58-64	61-67
Very poor	≤ 57	≤ 60
* Sensitivity of weighing balance ≤ 30 g		*Source* : North and Bell, 1990

4.2. Measuring growth efficiency

4.2.1. Live weight at market

This is the simplest and most easily obtainable measures because each bird is invariably weighed before selling/processing. Records have to be maintained of age and weight at market using which average live weight can be easily calculated. The live weight at market is compared with the standard value specified by the chick supplier or the value specified for the strain and suitable conclusions drawn for future course of action.

4.2.2. Feed conversion ratio (FCR)

FCR is the kg of feed required per kg weight gain; hence, lower the value of FCR, better is the performance of the birds. This value can be easily calculated utilizing the farm records on body weight, feed consumption, mortality and age at market giving due weightage to each of these. Actual calculation is dealt in a separate Chapter.

4.2.3. Age to reach the desired weight

In most developed countries, it is necessary that broilers of a specific weight range are demanded by the customers and processors. Therefore, producers have to calculate the age at which their flock attained that desired weight; earlier it is attained, the better.

4.3. Production efficiency indices

(North and Bell, 1990)

4.3.1. Point spread

Point spread is the difference between live weight (in lb) and FCR times 100; with Metric units, modification is required value of FCR exceeds live weight even up to 6 weeks of age and sometimes even afterwards. Further, point spread increases with advancing age and hence, it can be used to compare flocks only when the live weights are similar.

4.3.2. Performance index

This index also is based on FPS system; it is the ratio of live weight (in lb) to FCR times 100. Since it is a ratio, it can be converted to metric units straight away and used. In any case, this index also suffers from the same shortcoming as that of point spread in that it increases with advancing age and hence it is a poor criterion for comparing different flocks.

5. Feeding and disease control

Nutrient requirements, feeding and vaccination programs are discussed separately in Chapters "Nutrient requirements and specifications" and "Poultry diseases –

vaccines and vaccination", respectively.

6. Broiler problems

6.1. Leg and bone problems

Improved growth rate in broilers induced by genetic maneuvers has resulted in greater strain on legs (tibia), therefore, in the beginning, due to very high specific-growth rate in broilers, there will be some sort of bending of tibia which, later on, gets corrected by itself when the specific-growth rate falls. Many diseases and/or disorders manifest with impaired gait or reluctance to walk. Etiology of each of the leg disorders is not very easy to be defined. However, some of them are outlined below :

6.1.1. Chondrodystrophy, enlarged hock disorders, turkey-syndrome '65

Most rapidly growing bones are most affected and the bones become shortened, thickened and more frequently, secondary bowed. Hock joints become enlarged and knobby with the birds becoming cow-hocked or bow-legged with or without slipping of gastrocnemius tendon. Deficiency of manganese, choline, niacin, folic acid, zinc and pyridoxine also can cause similar condition. However, with modern-day balanced diets, it is rare to expect deficiency conditions. In any case, in turkeys, deficiency of niacin can cause these leg disorders of a considerable degree. Wire floors and high mineral content in rations increase its incidence.

6.1.2. Rickets and osteomalacia

Bone maturation by mineralization occurs within first fortnight of age. Thus, defective mineralization can cause rubbery bones which, and growing birds, is called rickets and in adults, osteomalacia. Thus, the condition is most frequent during 2 to 3 weeks of age and detected early can be treated with water-soluble Vitamin D_3 preparations. Calcium and/or phosphorus deficiency also causes osteomalacia and rickets. In calcium deficiency, severe standing will always be noticed before skeletal changes and most of the problems of rickets and osteomalacia is due to deficiency of Vitamin D_3 and/or available phosphorus. Mycotoxicosis (aflatoxins and ochratoxins) too can predispose for skeletal deformities similar to rickets by interfering with mineral absorption. Malabsorption of nutrients due to an infective agent is also suspected.

6.1.3. Twisted, bent and bowed legs and rotated tibias

Seen in broilers with an incidence not exceeding 5%, normally in those flocks which are grown to older ages. The syndrome is higher in caged birds than on floor. It could be due to autosomal recessive gene. High levels of thyroid-blocking agents (like thiouracil), or 5% propylene glycol, high levels of rape-seed meal, certain types of sorghum, fungal toxins, manganese deficiency etc. There is no reduction in skeletal strength.

Lack of exercise which weakens the leg muscles, low feed consumption during hot weather and new litter are also possible causes.

6.1.4. Tibial dyschondroplasia (focal osteodystrophy)

This is quite common in all meat-type birds although it does not cause leg weakness. It is characterized by uncalcified plugs of avascular, hypertrophic cartilage in the proximal metaphyses of the tibiotarsi or, sometimes, tarsometatarsi. The etiology is not yet clear but high chloride diet, excessive pressure due to primary skeletal or postural abnormality resulting from very rapid early growth has been suspected. If the plugs bulge outwards into bone cortex, locomotor problems can appear following bowing and/or fracture of proximal end of tibia. The condition could have genetic implications. Lack of exercise or low calcium with high phosphorus in diet also causes the condition. Addition of 2.5% dried Brewer's yeast to the breeder diet completely eliminates the condition in broilers.

6.1.5. Spondylolisthesis (Kinky Back)

Seen only in broilers at the level of 6th thoracic vertebra during 3 to 6 weeks of age. In severe cases, leg weakness or paralysis may follow. This condition is probably genetic and severe restriction of early growth can prevent this condition.

6.1.6. Aseptic necrosis of femoral head

Head of the long bones becomes porous and discolored to yellow or brown; the condition may extend to tibia and metatarsus. This condition is seen usually in turkeys; response to dietary supplementation of 0.5 to 2.5 ppm of molybdenum (as sodium molybdate).

Arthritis caused by reovirus also is a probable cause; a vaccine is available for the condition in developed countries.

6.1.7. Staphylococcal arthritis

Usually a secondary infection following viral arthritis seen in 2-4 week old chicks.

6.1.8. Plantar pododermatitis

Characterized by lesions on ventral foot-pads in meat-type chicken and turkeys. It may be due to some corrosive substance in the litter, possibly arising from droppings. The initial change is necrotic degeneration of the epithelial cells followed by ulceration and inflammation. This is more common in overweight birds and the lesions are seen on pressure points such as keel and backs of hocks.

6.1.9. Synovitis

Infection of the synovial joint of the leg and adjacent tendon sheaths by *Staphylococcus* or a virus or by *Mycoplasma synoviae*.

6.1.10. Fragile bones

Osteoporosis condition due to withdrawal of calcium phosphate from the bones expectable in the deficiency of vit D_3, calcium or phosphorus in the diet.

6.1.11. Osteomyelitis

A bacterial infection causing leg weakness without bone deformities.

6.1.12. Redhocks

Newly hatched chicks will have pink or red hocks; some showing recovery and others worsening. Vitamin deficiencies in breeder diet have been suspected to be the cause of this condition which is not fully understood.

6.2. Metabolic disorders

6.2.1. Fatty liver kidney syndrome

Fatty liver syndrome is, by far, the most important metabolic disorder in poultry. Two conditions are described under this syndrome of which Fatty liver kidney syndrome is reported in broilers and broiler breeders.

Etiology : Diet, sex and age of birds, diet of breeding flock, environmental temperature and other stress factors. Of these, diet is most consistently a predisposing factor. Wheat-based diets, low in fat or protein, produce high mortality due to FLKS.

Nutrient of crucial importance is biotin; its availability being the prime factor. However, birds with FLKS may not show biotin deficiency symptoms and the birds fed biotin-deficient diets (otherwise adequate) may not show any growth depression. But, increased fat and protein reduces FLKS but increases biotin deficiency symptoms; in contrast, reduced fat & protein reduced the appearance of biotin deficiency symptoms.

Incidence of this condition may vary from 1% in broiler flocks to 10% in broiler breeding flocks. Symptoms : Lethargy is the most predominant symptom, severely of which may vary from slight or moderate dropping of head or inability to even stand with head sinking progressively lower. Symptoms may develop rapidly with birds moribund and dead within few hours. Necropsy findings : Enlargement of liver and keep, characteristic massive accumulation of lipid in the liver and kidney.

Treatment : Supplementation of diets with biotin, or increase in fat or protein content can cause significant reductions in incidence of FLKS.

6.2.2. Oily bird syndrome

Excessive amounts of subcutaneous fat which is released while on processing line making it a problem carcass. It is sometimes recorded in birds which are not obese but having higher hydroxyproline and hydroxylysine with associated low acid insoluble collagen in the skin. Certain weed seeds have been incriminated to produce this condition; but, it remains still an incompletely understood disorder.

6.3. Others

6.3.1. Breast-blisters

(North and Bell, 1990)

Breast-blisters constitute one of the greatest difficulties, especially in roasters. Heavier roasters and males, in particular, have a higher tendency to sit than lighter birds and females. Breast-blisters are responsible for one of the most important causes for downgrading of carcasses; sometimes, even the entire carcass may be condemned if breast-blisters are very extensive.

It is extremely difficult to totally eliminate the problems of breast-blisters. But, maintaining litter in good condition and providing proper thickness of litter (10 cm) are very important. The birds must be fed or stirred (mild disturbance) more often so that they are stimulating to move rather than sitting on the litter. Any object or equipment on which the birds can roost should be removed. Feed should be formulated especially for roaster production and although growing period may be extended, only female roasters can be produced.

6.3.2. Bare-backs

Probable causes: 1. Picking among litter-mates and 2. Certain anticoccidials which increase flightiness and induce the condition.

6.3.3. Scabby-hip syndrome

This is a kind of dermatitis between feathers or feather follicles appearing at about 3 weeks of age. Reducing density or any other means to reduce feather pulling alleviates the condition. In feather-sexed broilers, males are slow-feathering and hence, the condition is seen more in males

6.3.4. Hock-burn and related problems

Hock-burn is a pressure-induced necrosis of the skin; related problems are breast-blisters, pododermatitis and scabby-hip syndrome; and are collectively referred to as "contact dermatitis". Cause is mainly poor litter condition. Initially they appear as brown-black erosions, later on, the skin is damaged followed by microbial invasion. The incidence is variable between 15 and 20%. The economic importance is due to carcass downgrading.

Higher moisture content of litter is the primary cause and birds recover if wet-litter is avoided/replaced. Causes of wet-litter are many but primarily due to diarrhea which results in increased moisture elimination through feces. Hence, all nutritional, infective and managemental causes for diarrhea should be attended to.

7. Marketing

(Stull 1998)

7.1. Bird-handling and transportation

The manner in which birds are handled will affect their reaction to the restraint.

Birds should always be held in a manner that provides as much support as possible and reduces the likelihood of the bird struggling which may injure its wings or legs. An appropriate manner in which to hold an adult chicken is to support the bird's breast in the palm of the hand while restraining the bird's legs between the handler's fingers or by holding both legs. When carrying an adult chicken from one location to another, the bird can be carried by both legs or both wings. The period of time a bird is held in a vertical position, with its head down, should be minimized. After handling, birds should be placed directly onto the floor or released from a low height that allows them to land feet down, without flying.

7.2. Catching

Feed is withdrawn 8-12 hours before slaughter to reduce carcass contamination. Water should not be withdrawn. Elimination of grit feeding at least 2 weeks prior to marketing time also helps prevent bruising. Starving the birds does cause shrinkage to a tune of 2-6%, but it will be more than compensated during chilling of the carcass.

All floor equipment must be removed soon after dark just before catching the birds. The catching process should be as efficient as possible, minimize the stress on the birds, and be done in such a way as to prevent piling of the birds. Dimming of the house lights (with red or blue lights) or catching at night has long been employed to create a less stressful environment for catching.

The manner in which a market age bird is caught can have significant effects on carcass quality. All personnel involved in bird-catching and transport operations should be given training in appropriate bird-catching and handling methods.

7.3. Loading

The crates, cages or bins used for live-haul should be properly constructed to allow loading, transportation, and removal without injury to the birds. All containers must be kept clean and in good repair. Stocking density will depend on the size of the crate, the size of the birds, and transportation conditions (transit time and temperature). When the crate is full, there should be sufficient floor space so that all birds can be resting simultaneously on the floor and each bird should have free head movement.

During hot weather, birds should be moved during the cooler part of the day whenever possible. Additional space in transport containers will minimize heat stress. Trucks waiting to unload poultry at processing plants should be provided shade, fan ventilation and during very hot weather, evaporative cooling. To provide for bird comfort, ventilation and bird density in transport containers should be adjusted with changes in weather. The time birds are kept on trucks should be minimized.

7.3.1. Vehicles

Vehicles must be kept clean and in good repair. The truck driver is responsible for assuring that all crates, cage doors, and bins are secured before leaving the live production site. The vehicle should not be brought close to the house before dark.

7.3.2. Transportation Conditions and Route

The driver should be prepared to cover the vehicle and protect the birds from possible severe winds and/or rain. Care must be taken that the loaded vehicle is not left standing for any extended period of time. In the case of required stops (inspection stations, truck maintenance, etc.), the birds must be protected from environmental extremes and provided adequate air circulation.

While the shortest route between the live production facility and the processing plant may seem most logical, the driver should avoid passing other sensitive poultry establishments, agricultural operations or residential areas.

7.4. Processing

Procedures at processing plants should be designed to prevent unnecessary pain, ensure maintenance of optimal meat quality and microbial safety, and preserve the product's visual appeal.

7.4.1. Unloading

Care taken in unloading at the plant, will prevent injury of birds. While speed is of importance, it should not be achieved by inappropriate handling of birds, crates, or bins. The crates and bins should never be tossed or dropped.

If a mechanized system is used for unloading the crates from the truck, care should be taken so that the crates are never at an angle which would cause piling of birds.

7.4.2. Shackling

Birds must be hung carefully on the shackles to avoid injury. The shackles should be size-appropriate to the age and species of birds being slaughtered.

7.4.3. Stunning and slaughtering

State or Federal Guidelines must be followed depending on the type of inspection being done at the processing plant. When stunning is employed, the electrical current is applied in such a manner that the animal be rendered insensible to pain. Stunning immobilizes the bird and prevents wing breakage, dislocation of joints and bruising. The instrument used for stunning should be safe for plant personnel and should deliver the required electrical shock (appropriate amperage may vary with equipment, age of bird, etc.). Appropriate State and Federal regulations may be consulted for any exemptions based on the nature and/or size of the operation and those related to ritual slaughter (religious) practices.

Other welfare aspects of Poultry slaughter are discussed in a separate Chapter.

8. Yield

All the edible portion after processing is collectively referred to as "yield" and it is very important for a producer/processor to maximize the yield because, 1% additional yield in the processing plant can make substantial differences in the overall returns for the processor as well as the producer. It is it estimated (North and Bell, 1990) that 1% additional yield is equivalent, in trade-offs, to:

1. 25-28 additional hatching eggs per breeder hen
2. 13-15% better hatchability per breeder hen
3. 9.1-11.4 kg less feed per breeder bird
4. 5 points better broiler feed conversion

There appears to be no effect of sex on yield of cut-up parts of chilled, ready-to-cook broilers; the average yield in broilers is around 70-75%.

9. Rearing broilers in cages

Table 36.4. Rearing broilers in cages

Advantages	Disadvantages
Higher density of rearing possible	Higher incidence of breast-blisters which increases carcass condemnations
Easy to catch the birds at market time and hence reduces bruising	Higher incidence of crooked-keel
No expenditure on litter	Wing bones will be more brittle which will be a disadvantage for the processor also
No incidence of coccidiosis	Higher feather-follicle infection
Less labor requirement	Birds are not having access to the unidentified growth factors
Reduced cannibalism	Cleaning fecal-trays is not labor-friendly
Cleaning and disinfection easier	Highly initial investment on cages
Better growth and feed efficiency	Birds will be uncomfortable especially during summer

However, due to advantages of higher density and absence of coccidiosis, rearing broilers in cages is becoming increasingly popular. The floor space required. 0.02 m^2/kg mature body weight. Assuming that the mature weight will be 1.50 to 2.00 kg, a floor space of 0.03 to 0.04 m^2/bird is essential. However, under commercial conditions, the floor space of 0.06 to 0.07m^2/bird is normally allowed. Several types of cages have been designed; but, none has been able to eliminate breast-blisters. The cage material tried includes wire, wood, plastic, wire fabric etc.; usually, softer the material better the results.

10. Squab broilers

Very small broilers often sold as "Cornish game hens" or "Rock-Cornish cross" are broilers weighing, on average, 0.9 kg (1.02-1.14 kg). The present-day broilers will attain this body weight as early as 28-30 d of age itself. The Squab broilers, usually raised straight-run, are processed and sold as fresh or frozen whole-body birds, completely eviscerated.

11. Capon production

(North and Bell, 1990)

Capons are castrated male chickens used for meat purposes. Capon meat is considered most tender, juicy and flavored than any other meat. Capons are grown to weights more than that of roasters, and they are processed at 20-24 weeks of age when their body weight ranges between 5.4 and 6.4 kg. They are also marketed, similar to Squab chicken and roasters, as whole-body, eviscerated.

Chemical castration by subcutaneous implantation of a paste of *estradiol* or any analogue of estrogens in the nape of the neck to the birds when they are at least 4 weeks of age is currently banned from use for capon production. A withdrawal period of 6 weeks was also necessary following chemical castration.

11.1. Age and weight

It is most ideal to caponize cockerels when they are 2½ to 4 weeks of age and weighing approximately ½ to 1 kg.

11.2. Surgical procedure

(Ensminger, 1993; Nesheim *et al*., 1979)

The cockerels are off-fed 12 to 24 hr before caponization and depending on season, water can also be withdrawn 6 to 12 hr before surgery. This helps in keeping the intestine empty and enable easy access to testes. Antibiotic is administered for a week either through feed or water prior to or injected just before surgery.

11.2.1. Step I

Feathers on right side between thigh and 3rd rib from the point of the rib down to the rib joints plucked. Feathers are pulled in an upward motion towards the back.

11.2.2. Step II

Wings and legs are usually are fastened and the bird is kept on lateral recumbence with right side upwards.

11.2.3. Step III

With the left hand fingers, last two ribs are located and the skin is slid upward and backward toward the thigh. Keeping the underlying thigh muscles away, knife is forced through the skin and flesh between the last two ribs and an incision of about 3 cm is made extending from height of the hip joint down to the rib joints. Care has to be exercised not to cut either the rib joints or the thigh muscles. If incision is proper, no major blood vessel will be involved and very little, if any, bleeding will be noticed.

The incision is made in the last intercostal space to avoid injury to the pleural membrane and lungs.

11.2.4. Step IV

A spreader is inserted soon after to keep the ribs 1-2 cm apart to expose the peritoneum. If the membrane is not pierced, the incision is extended upwards to cut it at the place where it is most adhering to the ribs.

11.2.5. Step V

By using the caponizing hook peritoneum is opened and with a probe or any blunt instrument the intestines are moved aside to expose the testes.

The testes will be creamy yellow in color varying in size from a size of a wheat kernel to navy beans, located in front of kidneys and back of lungs.

11.2.6. Step VI

Left testis (present at lower aspect) is removed first; even if it is not visible, it must be lifted into view by the forceps by moving the surrounding intestines away. It is properly grasped and the entire organ and the connecting portion of the spermatic duct must be removed. The entire organ so exposed is carefully secured by the remover (instrument) and then drawn out with a slight twisting motion. Followed by the removal of the left testis, the right testis is removed by the same procedure. Extreme care is essential to ensure that no major blood vessel is torn in the process.

Both testes have to be removed; otherwise, the left over testis grows in compensation and produces the same amount of testosterone as that from both testes making the entire procedure futile.

11.2.7. Step VII

Spreader is removed and the bird is let free; the skin and thigh muscles slip back over the incision and afford natural protection. This is augmented by relatively high normal body temperature of the birds. Therefore, on most occasions, neither suturing nor antiseptics are necessary; as there is no danger of infection, no special care is needed.

11.3. Care and management

Capons are fed and managed on the same lines of that for broilers up to 8 weeks of age after which, they are provided with the following space allowances (Table 36.5). Capons, similar to roasters, are fed on diets specially developed keeping in view their nutrient requirements.

Table 36.5. Space allowances – Capons (per bird)

Age (weeks)	Floor space, m²	Feeder space, cm*	Waterer space, cm*
6-11	0.19	7.6	2.8
12-15	0.28	12.7	3.8
? 16	0.37	17.8	6.3

* Linear cm

Source : North and Bell, 1990

11.4. Problems of capon production

Capon production also has certain problems which are, more or less similar to broiler and/or roaster production. Some of them are as follows:

1. Slips – the term is used on improperly castrated cockerels wherein some testicular material is left over which can produce sufficient testosterone making the bird a cockerel/cock. If both testes are removed entirely, slips do not result.
2. Wind-puffs – a swelling at the site of operation. This is due to escape of air abdominal air sacs (which are punctured due to the procedure) before the incision between the ribs is healed. Capons must be examined at intervals and a puncture is made, sometimes 4 to 5 times, on the swellings to allow the air to escape. Normally, birds recover without further complications in about 2 weeks tie.
3. Leg problems – being castrated, capons behave like females but are very heavy in body weight. Therefore, they tend to sit for extended periods and develop leg problems. Hence, they have to be frequently stirred/disturbed so that they move around and exercise their legs. Sprinkling whole oats on to the litter floor will also help in inducing scratching by the birds to find the oat grains thereby exercising their legs.
4. Breast-blisters – similar to broilers/roasters; the same precautions listed under broilers can be practiced.

Chapter **37**

Restricted Feeding

Birds during growing period (8-20 weeks) are generally subjected to feed restriction. This is particularly true for breeding birds; even commercial laying pullets are subjected to feed restriction. Broiler breeding birds are feed restricted from as early as 3-4 weeks of age itself.

However, several welfare concerns are being expressed against this practice and hence, alternate methodology may have to be developed to replace feed restriction.

1. Advantages

1.1. Broiler breeding flocks

(www.animallaw.info, 2005)

1. As a result of the genetic selection of broilers for increased growth rate and lower (more efficient) feed conversion ratio it has become necessary to severely restrict the feed intake of the broilers intended for breeding to enable them to survive into adulthood and reproduce successfully.
2. If the breeding flocks were fed *ad libitum* they would become obese and suffer thermal discomfort, a high incidence of lameness, and high mortality due to skeletal disorders and heart failure.
3. Excessive body weight is also associated with reduced disease resistance and hence feed-restricted birds are likely to survive better.
4. Increased incidence of multiple ovulations in females, resulting in lowered production of hatching eggs; poor egg shell quality and reduced fertility in males are recorded in breeding birds fed *ad libitum.*
5. Minimizes multiple ovulations and disrupted egg production

1.2. General advantages

In addition to the advantages listed under broiler breeder pullets, the following general advantages go in favour of feed restriction which is practiced even on commercial layers during growing period:

1. A considerable saving on feed cost because, only 80% of the calculated feed requirement will be offered.
2. They are likely to consume less feed per dozen eggs even during laying period when they are offered *ad libitum* feed.

3. The pullets accumulate less fat and therefore produce more eggs.
4. It is easier to identify weaker birds at an early age during feed restriction. Culling of such birds helps not only saving feed but also promoting layer house survivability because, healthier birds will be moving to layer house
5. Layers feed-restricted during growing period have been found to produce heavier eggs in longer clutches than those fed *ad libitum*.

2. Disadvantages

1. Feed restricted birds mature late; but this is more than compensated by sustained production of heavier eggs.
2. There will be reduction in grower house survivability because, weaker birds will be culled; but this is reflected in the form of higher layer house survivability. In fact, it is possible to save feed which would have been offered to weaker birds had they entered the layer house.
3. Feed-restriction requires technical supervision.
4. Does not conform to welfare requirements - is discussed separately.

3. Methods

Feed restriction can be effected by several ways like

1. Quantitative feed restriction - offering 92-93% (commercial egg-type pullets) 75-85% or less (meat-type/egg-type breeder pullets) of the calculated feed requirement / skip-a-day program in which the birds are fed on alternate days from 9 weeks to sexual maturity.
2. Increasing fiber content which increases bulk in the intestine thereby reducing feed intake
3. Reducing protein content
4. Reducing lysine content and
5. Reduction in energy content (similar to increasing fiber).

However, under commercial conditions, quantitative feed restriction is commonly practiced on Leghorn-type laying pullets mainly because it is easier for execution and the chick-supplier will be providing a readymade chart of the quantity of feed to be offered during the growing period.

4. Egg-type pullets

(North and Bell, 1990)

4.1. Commercial

It is desirable that egg-type growing pullets attain correct body weight at the correct age with respect to the strain in question. Several factors, including date/season the hatch influence sexual maturity and/or egg production; mediated mainly through variations in environmental temperature and/or light.

It is theoret :ally estimated that egg-type pullets require 15% protein during 6-14 weeks and 12% thereafter; but, under commercial conditions, the requirements are slightly higher than the estimated. In any case, the protein requirement is not constant and it definitely reduces between 14-20 weeks of age. Therefore, phase-feeding growing pullets (egg-type) with a step-up/step-down program has been tried. However, phase-feeding is not yet a commercial practice among many poultry operators.

Most breeders also feel that growing pullets (egg-type) should be full fed; a low level of feed in the troughs is allowed, both in case of automatic or non-automatic feeders; but the feed is offered at least 2-3 times a day (non-automatic feeders) or intermittently running approximately 20 min with an idle time of 20 min in-between (automatic feeders).

Table 37.1. Standard body weights – egg-type growing pullets

Age (wks)	Standard Leghorn	Medium-size *
1	0.065	0.13
2	0.121	0.18
3	0.186	0.27
4	0.262	0.36
5	0.335	0.46
6	0.427	0.59
7	0.513	0.68
8	0.593	0.77
9	0.671	0.86
10	0.754	0.95
11	0.828	1.04
12	0.904	1.14
13	0.968	1.23
14	1.030	1.32
15	1.092	1.36
16	1.157	1.45
17	1.211	1.50
18	1.259	1.54
19	1.311	1.64
20	1.362	1.68

* brown-egg producers — *Source* : North and Bell, 1990

It is recommended that the body weights of the pullets (Table 37.1) should be maintained on the weekly schedule beginning from 7/8 weeks of age. It must be remembered that egg-type strains, being smaller, are more sensitive to feed restriction; therefore, they do not respond positively for any feed restriction in excess of 7-8%. A sample of birds must be weighed regularly during the growing period to ensure desired growth rate.

4.2. Breeders

Same as that of commercial egg-type pullets; although breeder males eat 25% more feed than the females during the growing period, they will consume the same amount of feed per day as a female during the laying period. Therefore, when reared along with females on the same feed during the growing period, the males rarely become overweight. Hence, feed-restriction program must be started early in chick's life to be effective; it would be difficult to control body weight of meat-type pullets after 12 weeks of age than earlier.

5. Meat-type or broiler breeders

(North and Bell, 1990)

5.1. Females

Breeder pullets should attain sexual maturity at a specific age and they also showed attain a recommended body weight (Table 36.2). For this reason, they are subjected to feed restriction, often severely, because they have a genetic potential of a rapid growth rate which makes them susceptible to become overweight when fed *ad libitum*. Therefore, the main aim is to restrict calorie intake to produce pullets that the smaller and older when they lay their first eggs.

5.1.1. Severity

(www.animallaw.info, 2005)

A female broiler breeder receives 52g of feed daily at 7 weeks of age, while a commercial broiler (intended for slaughter at around 6-7 weeks) will consume 182g of feed daily. The impact of this restriction can be seen in the fact that at 7 weeks of age a female breeder weighs 780g, while a female commercial broiler weighs around 2440g. Similarly, a male breeder receives 78g of feed daily at 7 weeks of age and weighs just 1100g, compared with a male commercial broiler of the same age that will consume 205g of feed and weigh around 2897g.

In the UK, broiler breeders are generally fed on restricted rations from the age of 15 days. Feed allowances during the rearing period are typically 60-80% less than the birds would consume *ad libitum*, and may be 25-50% less during the laying period. This results in a reduction in adult body weight to approximately 45-50% that of *ad libitum*-fed birds.

The feed is usually supplied in a single daily feed, whichis generally consumed in less than 10 minutes.

The pedigree flock are subjected to particularly extreme feed restriction. In order to be able to identify those birds that have the most desirable traits for increased

Table 37.2. Body weight recommendations for breeders

Age (wks)	Standard Leghorn		Medium-size brown egg-type		Standard size, meat-type		
	♂	♀	♂	♀	♂	♀	♂ ÷ ♀
1	0.14	0.09	0.18	0.13	0.15	0.14	
2	0.18	0.14	0.22	0.18	0.25	0.22	
3	0.27	0.22	0.32	0.27	0.46	0.41	
4	0.36	0.27	0.45	0.36	0.58	0.50	1.15
5	0.46	0.36	0.59	0.46	0.68	0.59	
6	0.55	0.41	0.73	0.59	0.82	0.64	
7	0.68	0.50	0.86	0.68	0.91	0.77	
8	0.77	0.59	1.00	0.77	1.04	0.86	1.19
9	0.91	0.68	1.09	0.86	1.13	0.96	
10	1.00	0.73	1.22	0.95	1.29	1.05	
11	1.04	0.82	1.32	1.04	1.41	1.14	
12	1.14	0.91	1.45	1.14	1.50	1.23	1.23
13	1.23	0.96	1.54	1.23	1.63	1.32	
14	1.32	1.04	1.63	1.32	1.77	1.41	
15	1.36	1.09	1.73	1.36	1.91	1.50	
16	1.46	1.14	1.82	1.45	2.04	1.59	1.27
17	1.50	1.19	1.91	1.50	2.13	1.68	
18	1.55	1.23	1.96	1.54	2.27	1.77	
19	1.64	1.27	2.09	1.64	2.40	1.86	
20	1.68	1.32	2.13	1.68	2.54	1.96	1.31
21	1.73	1.36	2.18	1.73	2.68	2.05	
22	1.77	1.41	2.27	1.77	2.90	2.18	
23	1.86	1.45	2.32	1.82	3.09	2.32	
24	1.90	1.50	2.36	1.86	3.36	2.50	1.35
25	1.96	1.55	2.45	1.96	2.58	2.63	
30	2.00	1.59	2.54	2.00	3.86	2.73	1.42
40	2.09	1.64	2.59	2.05	4.14	2.96	1.40
50	2.13	1.68	2.64	2.09	4.32	3.09	1.40
60	2.18	1.73	2.72	2.18	4.41	3.18	1.39
70	2.27	1.77	2.82	2.23	4.46	3.27	1.36
80	2.32	1.82	2.94	2.27			

Source : North and Bell, 1990

production in future generations, they are reared to their maximum potential growth rate up to at least 6 weeks of age, at which point selection takes place. This creates birds that are very heavy and they are then severely feed-restricted to ensure they are not overweight for the laying period.

5.2. Males

Meat-type cockerels also should attain recommended body weight (Table 37.2). They are grown separately wherever possible so that feed can be controlled and at the same time the pullets need not have to compete with cockerels for feed.

Poor fertility in overweight males is related to the inability to mate and in case of underweight males (< 3.5 kg), poor fertility is due to inability to compete with larger males; hence, uniform and acceptable weights are extremely important for optimum fertility. Therefore, males should be fed gradually increasing the amount of feed so that they gain at least 1.5 kg from 30 to 60 weeks of age (Woolford, 2003).

5.3. Effects of feed restriction

(North and Bell, 1990)

1. Delays onset of sexual maturity from a few days to 3-4 weeks depending on severity of restriction
2. Reduces the body weight of the bird at sexual maturity, usually by reduction in body fat
3. Mortality during growing period is not normally affected unless the restriction approaches starvation
4. May lead to certain nutritional deficiencies
5. Although cost of raising a pullet is reduced, the additional time required to reach sexual maturity would nullify such benefits
6. Generally, livability during egg production improves
7. Egg production is not generally affected
8. Egg weight normally improves because of delayed sexual maturity

5.4. Welfare concerns

(www.animallaw.info, 2005)

Fowls would naturally spend a considerable portion of their day engaged in activities associated with foraging, and when given the choice prefer to work for at least part of their daily feed intake rather than eating it all from a free supply. Feed-restricted broiler breeders, however, consume their feed in a very short space of time and are chronically hungry. This is demonstrated by the fact that they are strongly motivated to consume feed at all times. Indeed, their level of feeding motivation is 3.6 times greater than that of *ad libitum*-fed birds subjected to 72 hours of feed deprivation, and is just as high one hour after their daily meal as it is one hour beforehand.

Feed-restricted birds are hyperactive, and they show increased pacing before the expected feeding time and increased drinking and pecking at non-food objects afterwards, compared with *ad libitum* fed birds. The expression of these activities is often stereotyped in nature, is characteristic of frustrationof feeding motivation, and is positively correlated with the level of feed restriction imposed. Also, feed-restricted males are more aggressive than *ad libitum*-fed males. It is concluded that restricted fed broiler breeders are "chronically hungry, frustrated and stressed" and that the first of the 'Five Freedoms' is being contravened. The Five Freedoms are a widely recognised approach to assessing animal welfare; the first freedom is freedom from hunger and thirst.

There is also evidence that physiological indices of stress, such as heterophil/ lymphocyte ratio, basophil and monocyte frequencies, and plasma corticosterone concentration, are higher in feed-restricted than in *ad libitum* fed birds, and are positively correlated with the level of feed restriction imposed. According to a recent review "broiler breeders show evidence of physiological stress as well as an increased incidence of abnormal behaviours, and are also chronically hungry".

It has been suggested that qualitative feed restriction, such as dietary dilution to allow the birds to consume large amounts of feed without increasing their energy intake, could alleviate hunger. However, recent research demonstrates that broiler breeder welfare is not improved by using qualitative rather than quantitative feed-restriction methods.

Compassion in World Farming (CIWF) believes that the proper approach to avoid the health problems that affect broiler breeders is not to restrict their feed, but to end the use of fast-growing genotypes. Instead, slower-growing genotypes should be used as these would not be vulnerable to a high incidence of leg and heart problems arising from fast growth rates, and so feed-restriction would not be needed.

CIWF believes that the severe feed-restriction necessary to maintain health and fertility of broiler breeders results in unacceptable health and welfare problems, and that the need for feed-restriction should be removed by ending the use of fast-growing genotypes. This view is supported by the conclusions of the Farm Animal Welfare Council that "the problem of hunger in broiler breeders is not easy to solve with present strains of birds and is likely to get worse if selection for fast growth continues. A long-term solution is to change the genetic strains but, in any case, breeders must avoid exacerbating the problem and reduce their demand for ever-increasing growth rates".

The March 2000 report by the European Commission's Scientific Committee on Animal Health and Animal Welfare (SCAHAW) states that the "chronic quantitative food restriction" to which broiler breeders are routinely subjected leads to them being "very hungry". The SCAHAW condemned this practice stating that "the severe feed restriction…resultsin unacceptable welfare problems" and they insisted that "the welfare of breeding birds must be improved".

Notwithstanding the above, feed restriction remains a commercial practice in both egg-type and breeder-production facilities.

Chapter **38**

Lighting Management

Light is a very important factor stimulating egg production. Although extra-retinal receptors of light have been documented, eye (optic nerve) appears to be primarily involved as far as stimulation of egg production is concerned.

Sun provides the natural light and intensity of sunlight depends on various factors like position of the Sun, cloudiness, dust, moisture in the air etc. Similarly, length of the sunlight also varies depending on the related position of Earth to the Sun and also on whether the location is in Northern or Southern hemisphere. In the former, the longest day is June 21st and shortest day is December 21st; whereas, the *vice versa* is true in case of the latter.

The time-phase between sunrise and sunset is considered as "light-day (Photo-period)" although, due to curved surface of the Earth, day-light is noticed 15 to 30 minutes before sunrise and after sunset.

1. Light intensity

Birds, similar to other animals, can perceive light represented by wavelengths between 400 and 700mμ. All birds have color vision, especially the latter part of the visible spectrum (red, orange and yellow; possibly, blue). Light required to see and eat is quite low (even less than 2.69 lux) but nearly 5.38 to 10.76 lux (minimum) is required to stimulate a production.

In feed storage/processing area, egg-handling room, egg-processing room, office and shop, the intensity recommended is 100-200, 500, 700-1000, 400-500 and 500-1000 lux, respectively.

It is also highly desirable to make the lighting dimmable by use of rheostats; especially in broiler houses where the lights may be required to be dimmed during night time just to allow visibility.

1.1. Measurement of light intensity

1.1.1. Candela

is the unit of luminous intensity, defined as 1/600000th of the luminous intensity of 1m^2 of a black body maintained at the freezing point of platinum (1773°C). One candela is actually equal to 0.982 times the luminous intensity of original candle of sperm-wax (weight 75.7 g, diameter 2.22 cm, burning at a rate of 8 g/hr; height of the wick 4.5 cm).

1.1.2. Lumen

is defined as the flow of light energy per second through $1m^2$ of a surface of 1 m radius, when a source of one international candle power is placed at the centre of curvature.

1.1.3. Lux

is defined as the amount of light falling on a $1m^2$ spherical surface of radius 1 m, when a source of one candle power is kept at the centre of curvature. 1 lux = 1 lumen/m^2; hence, lux is also referred to as "meter-candle". A bright sunny day will have an intensity of about 80,000 lux.

1.1.4. Watts

The amount of electricity necessary to light a bulb is measured in Watts; one Watt is defined as the rate of doing 1J/s of work i.e., Watt=1 J/s. The number of lumens of light per Watt of bulb is an indicator of the efficiency of the light source. With incandescent bulbs, 1 Watt = 12.56 lumens. However, lumen per Watt is inversely related to the size of the bulbs. Fluorescent bulbs produce more lumens per Watt.

1.1.5. Foot-candle

is defined as the intensity of light striking each and every point on a segment of an imaginary 1 foot radius sphere 1 candle power source at the centre. 1 foot-candle = 1 lumen / sq ft = 10.76 lumen / m^2 = 10.76 lux.

All light generated by a bulb is not available to chicken. About 30% is absorbed by the walls, ceiling, equipment etc.; another 33% may be lost due to cloudiness or dirtiness of the bulb. Therefore, for practical calculations, only 49% of the total lumens produced are considered available. Therefore, 6.15 lumens per Watt is the available light intensity from incandescent bulbs. In addition, height at which the bulbs are fixed also affects light intensity.

2. Types of light

2.1. Incandescent light

Cheapest, low light efficiency, reflectors are essential and has a short bulb-life of 750 to 1000 hr. Produces 12.56 lumens per Watt.

2.2. Fluorescent light

Light efficiency is four times than that of the incandescent light, costly and has a longer bulb-life (7000 to 9000 hr). Produces 67 lumens per Watt. Fluorescent bulbs have peak efficiency when the surrounding temperature is between 21 and 27°C and efficiency reduces at higher or lower temperatures.

2.3. Mercury vapor

Efficiency is similar to fluorescent bulbs, can work efficiently with temperature fluctuations and has a very long bulb-life (2400 hr). The bulb requires several minutes to warm-up and they must be fixed at a height of at least 3.3 m which precludes their use in low-ceiled houses. Besides, they cast too many shadows from that height. They produce 50 lumens per Watt.

Metal halide lamps producing 50-110 lumens per Watt and high-pressure sodium lamps producing 50-140 lumens per Watt are also available; but, they are seldom used in poultry houses due to their cost, maintenance, minimum height for fixing them and other practical problems.

However, under most circumstances, fluorescent bulbs are quite efficient provided that recommended light intensity with incandescent bulbs is maintained.

Table 38.1. Effect of color of light on poultry

Item	Color of light				
	Red	Orange	Yellow	Green	Blue
Growth stimulant				*	*
Depresses FCR			*	*	
Advances sexual maturity				*	*
Delays sexual maturity	*	*	*		
Eye enlargement					*
Reduces nervousness	*				
Reduces cannibalism	*				*
Increases egg production	*	*			
Decreases egg production			*		
Increases egg size			*		
Improves male fertility				*	*
Lowers male fertility	*				

Source : North and Bell, 1990

3. Color of light

The oil droplets in the retina of chicken filter out some of the shorter rays such as green, blue and violet. However, white light is commonly provided under commercial conditions. Red and orange-colored lights (wavelength 664 to 740 mμ) seem to have beneficial effects on egg production characteristic (Table 38.1). Since the incandescent bulbs emit more of red and orange colored light, it is probable that they stimulate a production better than fluorescent bulbs.

4. Reflectors

It is preferable to use flat-type reflectors rather than cone-type, since the latter confines light to a small area and corners are left without illumination. Size of the reflector is a square of side 25 to 31 cm. Cleanliness of the bulb and reflector is also important to produce calculated quantity of light (Table 38.2). Clean reflectors, on most occasions, improve the light efficiency by about 50% since they reflect the light that would have been absorbed by the ceiling.

Table 38.2. Effect of cleanliness of bulbs and reflectors

(All Values in Watts)		BULB	
		Clean	Dirty
REFLECTOR	Clean	100	80
	Dirty	80	60
	Absent	60	40

5. Arrangement of bulbs

When the birds are reared on floor, the rule of thumb is to fix the bulbs in such a way that the distance between two bulbs is 1½ times the distance from the bulb to the bird level and the bulb. The reflector, a square of side 25 to 31 cm, with the reflecting surface facing the bulb, is fixed in such a way that it will not sway due to wind/draft etc. If the bulbs move, they cast moving shadows, especially in cage system, which frighten the birds and may lead to reduced egg production.

Table 38.3. Height for fixing bulbs in a poultry house (m)

Lamp (W)	To obtain 5 lux intensity		To obtain 10 lux intensity	
	With reflector	Without reflector	With reflector	Without reflector
25	2.0	1.4	1.4	0.9
40	2.7	2.0	2.0	1.4
60	4.3	3.1	3.1	2.1
100	5.8	4.1	4.1	2.9
Source : North and Bell, 1990				

Height at which bulbs have to be fixed to obtain 5 or 10 lux intensity at the bird-level is given in Table 37.3. It is evident that to provide 5 lux intensity, when the bulbs are to be fixed at a convenient height of around 2 m, a 25 W incandescent bulb with reflector or a 40 W incandescent bulb without reflector is necessary; the corresponding requirement for providing 10 lux intensity is 40 W and 60 W bulbs, respectively.

In case of cage system, the bulbs are arranged over the passages (Fig. 38.2) on the same rule of thumb so that light always falls on feed and water.

Incandescent bulbs supply 12.56 lumens/W; assuming only 49% of it is available, 6.15 lumens/W of which, under normal farming conditions, only about 40% of light intensity will be available to the birds. That means, about 2.5 lumens/W is finally available. In general, it is recommended to supply about 11 lumens/m^2 (1 foot-candle = 1 lumen / sq ft = 10.76 lumens/ m^2) at the bird-level and this can be

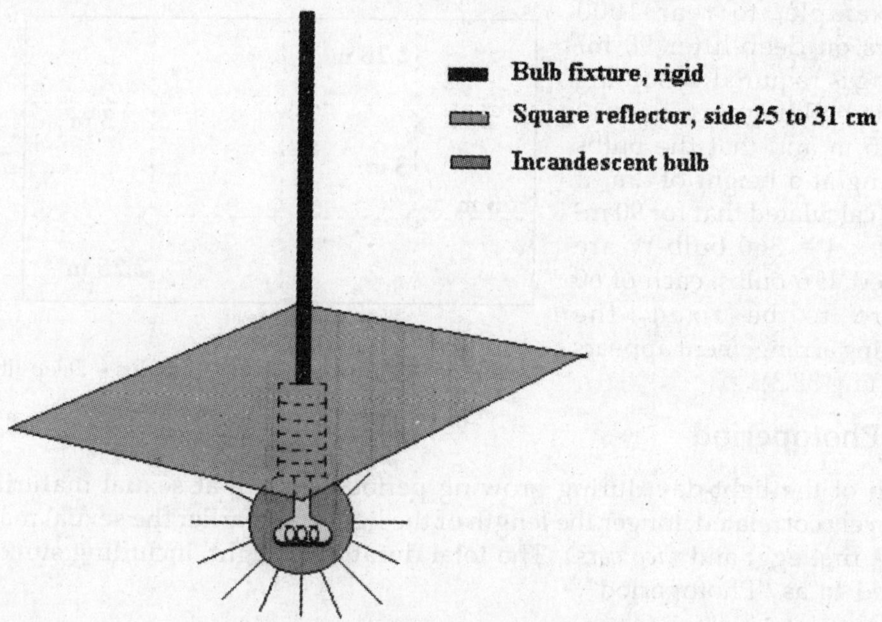

Fig. 38.1. Bulb-reflector assembly

achieved by supplying 4 bulb-Watt for each m² of floor space. The bulbs are fixed at a height of 2.1 to 2.4 m above the floor and 3.15 to 3.6 m apart. It is preferable to use many smaller bulbs than fewer larger bulbs to effect proper distribution of light.

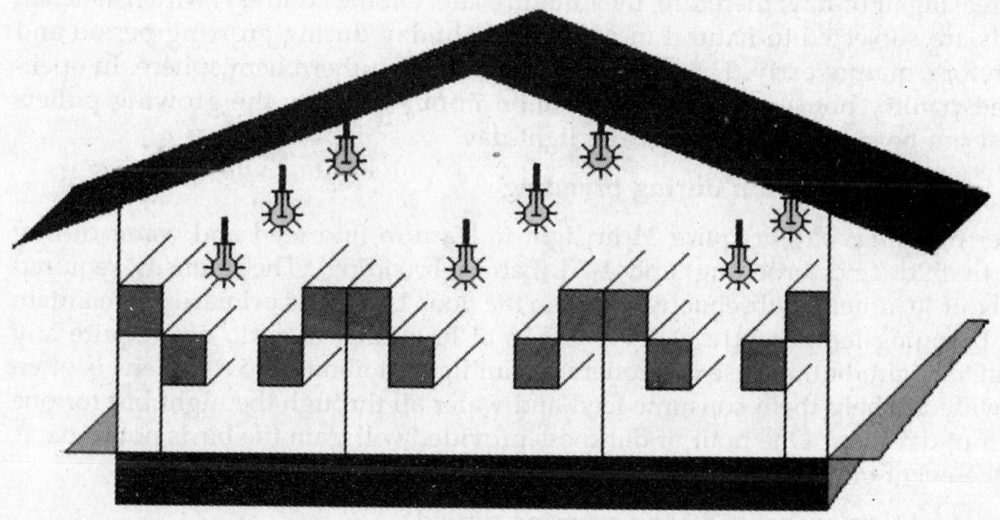

Fig. 38.2. Arrangement of bulbs – Cage system

For example, to rear 1000 broilers on deep-litter 90 m^2 building is required; assuming that the building measures 12 m x 7.5 m and that the bulbs are hung at a height of 2m, it can be calculated that for 90 m^2 are 90 x 4 = 360 bulb-W are required. If 6 bulbs, each of 60 W, are to be fixed, the following arrangement appears ideal (Fig. 38.3):

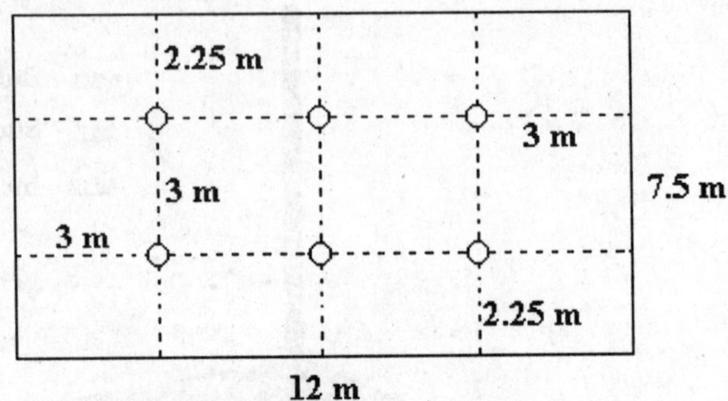

Fig. 38.3. Distribution of 60W bulbs - Deep-litter house

6. Photoperiod

Length of the light-day during growing period and age at sexual maturity are negatively correlated; longer the length of the light-day, earlier the sexual maturity (age at first egg) and *vice versa*. The total duration of light, including sunlight is referred to as "Photoperiod".

6.1. General

Under natural sunlight, in the Northern hemisphere, light-day increases gradually and reaches maximum by June 21st and thereafter starts reducing gradually and reaches the lowest by December 21st. Hence, spring-hatched birds complete half of their growing period under increasing light-day and the remaining half during decreasing light-day; therefore, they mature late. On the contrary, winter-hatched birds are subjected to natural increases in light-day during growing period and therefore, mature early. The reverse is true in the Southern hemisphere. In open-sided poultry houses, which are common in our country, the growing pullets must not be subjected to increasing light-day.

6.2. Lighting program during brooding

After hatching, birds require 24 hr light to learn to find feed and water during the first 2d (floor-brooding) and 4-7d (battery-brooding). The intensity required is about 40 lumens. Subsequently, light in the floor-brooder is primarily to maintain the brooding temperature. After 4 weeks of brooding, they do not require any artificial light; but, in case of broilers, a dim light of about 4 to 6 lumens is often provided to help them consume feed and water all through the night but for one hour of darkness. One hour of darkness provided will train the birds not to panic in the event of power failure.

6.3. Lighting program during growing period

Lighting programs are among the most important environmental-management tools affecting flock performance. As there is an interaction between lighting programs used during growing and laying period, it is important to plan a coordinated program for the life of each flock (Ernst, 2002a).

The age at which pullets mature (Age at sexual maturity or age at 50% egg production) has a direct influence on their performance. It is shown that each genetic stock has an optimum age at sexual maturity (± 5 d) to produce the maximum possible egg mass (egg number x egg weight). Managers should continually evaluate the optimum age to light stimulate pullets even if they are using the same strain because breeders are continually selecting for earlier sexual maturity (Ernst, 2002a).

Producers usually want a lighting program whichwill maximize the value of eggs produced by the flock during the subsequent production cycle. This optimum program may differ with genetic stock, cycle length, housing date and possibly other factors (Ernst, 2002a).

The age at sexual maturity is affected to some extent by changes in photoperiod which occur between hatch and subsequent photostimulation. Relatively consistent effects on age of sexual maturity have been demonstrated using 24 hour photo-schedules (Ernst, 2002a).

In any case, it must be remembered that the light intensity during growing period should be maintained below the threshold value of 5 lux and length of the light day should never increase for growing pullets.

6.3.1. Environmentally-controlled/Windowless houses

These houses are light-proof and even the fan and other openings are equipped with light traps to prevent light entry. Therefore, any of the programs described under conventional houses preferably constant light program (8L:16D) can be practiced (Program 1 or 2 in Fig. 38.4).

6.3.1.1. Abrupt changes in day-length

(Ernst, 2002a)

Abrupt changes in day-length are sometimes made during the growing period with the assumption that they will have little affect on sexual maturity; however, research has shown that decreases in day-length as early as 3 weeks of age will delay sexual maturity. Abrupt increases or decreases in day-length have a much greater affect on age at sexual maturity if they occur at older ages but before egg production begin (Program 3 in Fig. 38.4).

6.3.1.2. Intermittent lighting

In light-controlled houses, growing Leghorn pullets with alternating hourly dark-light cycles (15 min dark : 45 min light) i.e. 6L:18D, has been as effective as any other lighting program; in addition, this method helps save on energy costs and also helps save on feed costs to a tune of 5%.

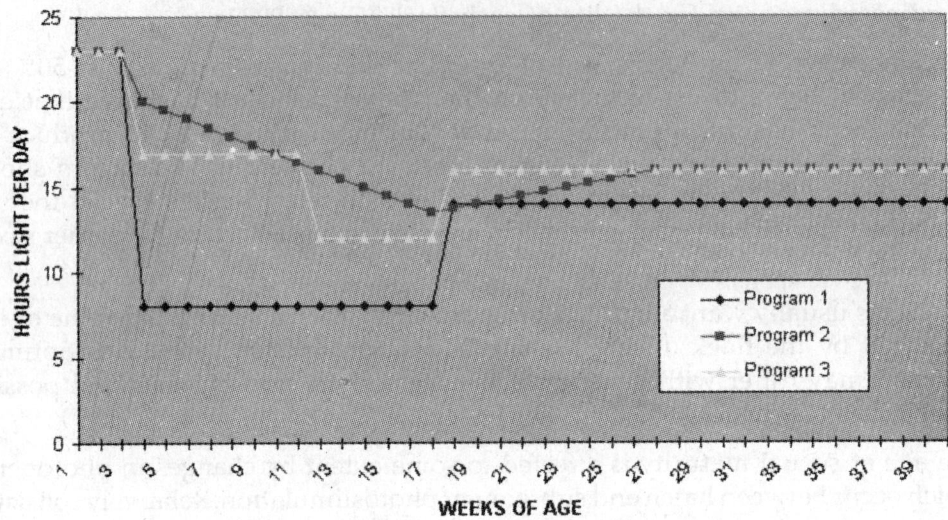

Fig. 38.4. Lighting programs for growing pullets (Ernst, 2002a)

6.3.2. Conventional/Windowed/Open-sided houses

6.3.2.1. Growing during a period of decreasing light-day

To this category included are the birds hatched from March 1st to August 31st, since during their latter part of the growing period they will be under decreasing natural light-day. These are called "in-season" flocks. These require no supplementary light during the growing period.

6.3.2.2. Growing during increasing light-day

Birds hatched during September 1st to February 28th fall into this category. They will have increasing light-day during latter part of their growing period. These are called "out-of-season" flocks. These flocks are usually the problem flocks and lighting program in case of these birds can be one of the following:

6.3.2.2.1. Constant light-day program

From a local meteorological table, the number of hours of day-light during the longest day until the flock reaches 20 weeks (egg-lines) or 22 weeks (meat-lines) is found out and this amount of light (natural + artificial) is given when the chicks are three days of age till the pullets begin to lay.

6.3.2.2.2. Decreasing light-day program

From a local meteorological table, the number of hours of day-light during the longest day until the flock reaches 20 weeks (egg-lines) or 22 weeks (meat-lines) is found out and 7 hrs are added to this figure. That much amount of light is given to the chicks from the first week (starting from three days) and reduced at a rate of 20 min per week till the pullets reach sexual maturity. However, decreasing light-day program is not very effective because the birds will be receiving light well above the threshold value to stimulate a production in conventional open-sided poultry houses (Program 2 in Fig. 38.4).

6.4. Lighting program during laying period

A sudden change in duration of light and/or feed should not be done and only a gradual change must be made in length of the light-day (photoperiod). When the first eggs are laid, is advisable to consider the change of lighting program (it may be 20 weeks for egg-type pullets and 23 weeks for meat-type pullets). Egg-lines require a light-day of 16 hr whereas meat-lines require a light-day of 15 to 16 hr. The additional light other than the natural day-light may be given either before the sunrise or after sunset or a combination of both; the latter is more convenient as it coincides with the normal working schedule.

Layers must never be given continuous (24 hr) light; excess light results in excess feed consumption, fat accumulation, hyperexcitability, cannibalism, prolapse, reduced egg production and others.

Light intensity threshold of layers appears to be about 10 lux at the bird level; excess light doesn't produce additional benefits. Under natural sunlight, maximum egg production is stimulated when light is provided 11 to 13 hr after dawn ("Open period" – See Chapter 7 for details).

Length of the light day should never be reduced for laying pullets.

6.5. Stimulatory lighting program

(Ernst, 2002b)

Laying pullets are usually exposed to a stimulatory lighting program at 16 to 23 weeks of age. A stimulatory lighting program is defined as one which provides light of adequate intensity in the photo-sensitive period. Chickens perceive dawn as the first light following the longest dark period in the 24 hour cycle. Therefore, photo-schedules often contain one longer dark period to establish the photo-sensitive phase and time of oviposition. Continuous light or short repeated cycles such as 1 hour light followed by 2 hour darkness (1L:2D) are photo-stimulatory because there is always adequate light within the photo-sensitive period regardless of the light period perceived as dawn. There are an infinite number of photo-schedules which would be stimulatory for chickens of an appropriate age. The challenge is to find a schedule which will optimize the profitability of the flock.

6.5.1. Age to light pullets

Research has shown that there is an optimum age for pullets to reach sexual maturity (50% lay) for each genetic stock and each set of economic conditions. In general, very early sexual maturity for the strain (produced by photo-stimulation at an early age) results in more total eggs but large numbers of small and peewee eggs. In contrast, if sexual maturity is delayed too long very few small eggs will be produced but the total mass of eggs produced will be reduced.

6.5.2. Decreasing or increasing day lengths

Decreasing day-lengths inhibit secretion of reproductive hormones from the pituitary gland and should never be allowed to occur when egg production is desired. Increasing day lengths stimulate a reproductive response even when the light does not extend into the photo-sensitive period (e.g. increasing from an 8 hr day). Decreases in daylight are often used as part of a molting program to inhibit egg production.

6.5.3. Light intensity

The minimum recommended light intensity to photo stimulate hens is 5 lux measured at the feed trough in the darkest part of the house. The lighting system should be designed with a slightly higher light intensity because lamps decrease in light output as they age or if they are not clean. However, under practical situations, 10 lux is considered as the light intensity required to be provided for the layers.

6.5.4. Skeleton photo-periods

In photo-stimulatory photo-periods the light phase of a 24-hour cycle can be interrupted by any number of dark periods so long as none of the dark periods exceeds the length of the major dark period which determines dawn. If the latter occurs, the chicken will perceive this dark period as night and first light following it will become dawn, resulting in a new photosensitive period (often called phase-shifting). The short repeating cycle of 1L:2D is a skeleton photo period for a 24-hr. light cycle (it affects the neuro-endocrine system of the chicken like continuous light).

It is well documented that a light-day > 11 to 12 hr stimulates egg production, it must be at least 14 hr for maximum production. Most of the programs recommend 15 to 16 hr of total photoperiod. In any of the programs, it is mandatory that the photoperiod should never be reduced.

6.5.4.1. Step-up program

This program starts from a constant short (usually 8 hour) or decreasing day length. At the age when stimulation will begin the day length is usually increased by increments of 15 or 30 minutes per week. Often a larger increase in day length is given during the first week to stimulate earlier sexual maturity. The combination of growing period day-length, age step-up is started, and presence or absence of an initial abrupt increase, are all important in determining how pullets will respond. A step-up program from a constant 8-hour-day length will result in later sexual maturity than either a program where a large increase is given before the step increases are started, or a program which increases day length abruptly to a photo-stimulatory photo-period. In general pullets respond more slowly to step-up programs and show a less pronounced peak but lay more eggs later in the cycle so that production to 65 weeks of age is the same. Step-up programs which start from a 12-hour-day-length result in earlier sexual maturity than those which start from an 8 hour day length. Step increases are usually discontinued when day-length reaches 14 to 17 hours.

6.5.4.2. Intermittent lighting program

6.5.4.2.1. Environmentally-controlled/Windowless houses

This reduces energy costs. It is started by abruptly changing pullets from a constant non-stimulatory day length (e.g. 8L:16D) to a 14-hour-day-length with a photoschedule of 2L:4D:8L:10D. The hens respond to this program as if the 2L:4D:8L were all light. The longer dark periodis essential if the bird is to respond to the first light period (2L) as dawn. A widely-used modification of this program (0.5L:5.5D:8L:10D) is very similar but uses slightly less energy.

Short periods of darkness within each hour of light can also be used to reduce electrical cost and improve feed conversion. For example, at 36 weeks of age, the changes are made using the following schedule:

Table 38.4. Intermittent lighting program for layers

Age (weeks)	Lighting program
36	45 min. light/15 min. dark*
37	30 min. light/30 min. dark
38	15 min. light/45 min. dark during half of the periods
39	15 min. light/45 min. dark in-all time periods

*the last hour of the light period must end with 15 minutes light

Changes are made gradually to allow the birds to adjust their eating patterns to the light periods. Attempts to use the program during early lay have depressed egg production, probably due to the reduced eating time. This program assumes moderate temperatures in the house. If temperatures become extremely cold it may be necessary to temporarily substitute 30-minute-light periods in all or part of the cycles.

6.5.4.2.2. Conventional/Windowed houses

The program utilizes one hour of light from 3:30 to 4:30 a.m. and one hour from 7:30 to 8:30 p.m. plus natural daylight as available. This program has resulted in performance comparable to traditional programs (e.g. morning and evening lights to maintain a constant stimulatory day-length).

6.5.5. General program

The most commonly practiced lighting program (in conventional houses) for layers grown on any of the lighting programs during their growing period is as follows: (Meat-type and egg-type strains should receive 15½ and 16 hr of total light, respectively).

6.5.5.1. Light-proof houses

When egg-type birds are 20 weeks of age and meat-type birds are 22 weeks of age, the light-day is abruptly increased to 13 hr; then, light-day is increased at the rate of 1hr per week (meat-type) or 15 min per week (egg-type) till the required light-day is achieved.

6.5.5.2. Conventional houses

If they were exposed to < 13 hr day length, total light is increased to 13 hr and then light-day is increased at the rate of 1hr per week (meat-type) or 15 min per week (egg-type) till the required light-day is achieved. If they were already exposed to > 13 hr day length, then also light-day is increased at the rate of 1hr per week (meat-type) or 15 min per week (egg-type) till the required light-day is achieved.

6.6. Ahemeral Lighting

Ahemeral lighting means creating a day lasting for less or more than the normal 24 hr. Obviously, this is possible only in light-proof houses.

6.6.1. Longer day

(Ernst, 2002b)

Research on ahemeral lighting (other than 24-hour-day-length) has demonstrated that longer light cycles increase egg size and egg specific gravity but may reduce egg numbers. Twenty-eight hour cycles have been successfully used, on commercial farms to increase early egg size. The schedule used was as follows:

Table. 38.5. Ahemeral lighting

Age (Weeks)	Light Schedule	Cycle length
17 to 26	11L:17D	28 hours
27	13L:13D	26 hours
28 to 70	15L:9D	24 hours
		Source : Ernst, 2002b

Under this schedule pullets produced larger eggs with higher specific gravity without loss of egg numbers. The improved egg size continued after the birds were returned to a 24-hour-cycle. If this lighting schedule is used commercially it must be followed exactly.

The main requirement of ahemeral lighting is that it is possible only in light-proof houses and not in conventional ones. Further, the scheme also suffers with disadvantages of special working schedule for the entire farm and reduced egg numbers; however, shell quality and albumen content of the eggs increase (North and Bell, 1990).

6.6.2. Shorter day

(North and Bell 1990)

Of many day-lengths tested, the 22-hr day seems to be most practical; the scheme has yielded 2% additional eggs. But, this scheme also requires light-proof housing and special working hours for the entire farm.

Table 38.6. Lighting scheme – Commercial laying-type birds

Age / Stage	Intensity options (lux)		Duration options (hr)	
	I	II	I	II
Brooding, 1-3 d	20	20	23	23
Growing				
4 d–2 weeks	5	5	9-11	23
2-3 weeks	5	5	9-11	21
3-4 weeks	5	5	9-11	19
4-5 weeks	5	5	9-11	17
5-6 weeks	5	5	9-11	15
6-8 weeks	5	5	9-11	13
8-9 weeks	5	5	9-11	11
9-20 weeks	5	5	9-11	11
Laying, 20-72 weeks	10-30		Increase light-day by ½ hr per week till a maximum of 16-17 hr	

7. Practical lighting programs

It is evident from the above discussion that many lighting programs which can satisfactorily stimulate egg production performance are possible during brooding, growing and laying period for commercial layers as well as breeders. The exact schedule depends on 1. Type of housing and 2. Local preferences and experiences. Some of the lighting programs (other than ahemeral) for chicken and turkeys (Winchell, 2001) are given in Tables 38.6 to 38.9 to help develop a suitable schedule depending on local conditions keeping in view the physiological requirements of the birds:

Table 38.7. Lighting scheme – Broiler and Commercial laying-type breeders

Age / Stage	Intensity (lux)	Duration (hr)
Brooding, 1-3 d	20	23
Growing		
4 d-3 weeks	5	15
3-19 weeks	5	11
Laying, 20-60 weeks	50-60	Increase light-day by ½ hr per week till a maximum of 16-17 hr

Table 38.8. Lighting scheme – Broilers and Roasters

Age /Stage	Intensity options (lux)				Duration options (hr)			
	I	II	III	IV	I	II	III	IV
Brooding, 1-3 d	20				23			
Growing, 3-10 d	5	5	5	20	23	23	8	18
10-15 d	5	5	5	5	23	16	12	8
15-21 d	5	5	5	5	23	16	16	12
21-28 d	5	5	5	5	23	16	18	16
28-35 d	5	5	5	5	23	16	18	18
35-42 d	5	5	5	5	23	16	23	18

Table 38.9. Lighting scheme – Turkeys

Age /Stage	Intensity options (lux)		Duration options (hr)	
	I	II	I	II
Brooding, 1-3 d	50		23	
Growing, 3 d-8 weeks	5-10		23	2.5L : 3.5D
Growing, 8-12 weeks	5-10		23	2.5L : 2.5D

Chapter **39**

Judging Pullets

Judging/culling, in the broad sense, as practiced by poultry-men, refers to the sorting of the desirable and undesirable hatching eggs, chicks, pullets, cockerels, hens, or breeding males. The greatest emphasis, however, has been placed on the sorting of hens, not only to eliminate the non-layers but also to determine when and how long the remainder have been laying.

Culling hens refers to the identification and removal of the non-laying or low producing hens from a laying flock. Unless the birds are diseased, they are suitable for marketing or home cooking.

It is obvious that it would be more profitable to eliminate to be poor- and/or non-layers early in life, if that were possible, than to wait until later. But, a successful method of selecting high and low producing pullets before they start laying, beyond eliminating the obviously weak, has not been definitely worked out. Birds of low vigor which are crow-headed or have long rangy bodies, or slow-maturing, off-type individuals, should be culled during the growing period itself before they are shifted to the layer house.

Removing the inferior birds reduces the cost of producing eggs, reduces the incidence of disease, and increases the available space for more productive hens. Hens eat feed whether or not they are laying. Removing the cull birds will make more feed and space for more productive birds.

It is a commercial practice that when the birds are shifted to the layer house they are visually examined and birds which are weak and unthrifty are culled; this is popularly referred to as "sight culling".

1.1. Sight Culling

Sight culling is removal of obviously undersized, underdeveloped, weak, crippled, or diseased birds which have very little chance of becoming good laying hens when being placed in the laying house. It is a general practice of all commercial operators. However, one should not be too critical when evaluating the pullet's size and development, since some good laying hens mature late.

Any bird which has a permanent genetic or injury-produced deformity such as crossed beak, slipped wing, one or both eyes blind, or any leg deformity that can interfere with the bird's ability to mate or to reach feed, water, or the laying nest is removed. It is most economical to remove these birds from the flock as soon as they are noticed.

Similarly, sick or unthrifty birds will often have short, narrow, emaciated bodies and appear listless or droopy; small, pale combs and wattles generally indicate chronic poor health. These birds have to be removed from the flock as soon as possible to avoid disease problems that may spread to the flock.

Culling at night is recommended, since the birds are less likely to be frightened and reduce egg production. A flashlight with the lens covered with blue cellophane will make it easier to detect poor layers without disturbing the flock. Birds should not be subjected to unnecessary handling and culling is delayed if a significant portion of the flock is suffering or recovering from a minor disease or molt. Culling a diseased or molting flock often removes some of the better laying birds.

However, even after the pullets are housed in layer houses, some of them may be uneconomical due to poor production abilities. Such birds are identified by judging them at or after 26 to 28 weeks of age. Pullets can be judged for both present production as well as persistency of production; the latter popularly referred to as "past production".

2. Present production

Present production may be determined by examining the vent, pubic bones, comb, wattles, and earlobes (Table 38.1). Non-layers are the birds old enough to have produced eggs but have not yet started laying and poor layers are the pullets which have started laying but are producing fewer than expected number of eggs.

2.1. Comb and wattle

A good layer will have more active and well developed ovary and hence more sex steroids are secreted which includes testosterone which is responsible for good development of combs and wattles.

2.2. Eyes

Good layers tend to open their eyelids much more than poor-and non-layers in order to receive more light stimulus required for egg production.

2.3. Vent

When the egg is being laid, there will be relaxation of the pubic bones to help passage of egg and oviposition. When it occurs frequently, as in case of good layers, the pubic bones remain relaxed. This relaxation causes the otherwise round vent (cloaca) to become oblong. Added to this, during oviposition, there will be vaginal secretions to help passage and laying of egg which makes the vent moist. Pink color is due to loss of xanthophyll pigments from round the vent.

Therefore, vent of a laying hen is large, moist, and dilated, and tends to become oblong in shape. The lower edge appears flat and extends almost straight across, and the upper edge blends evenly into the surrounding tissue which has a smooth, loose, pliable appearance. In case of non-layers, it will be small, contracted, and dry; the corners are drawn in giving the vent a round appearance with thin, prominent edges. The region around the vent is puckered, rough, and hard.

2.4. Distance between pubic bones

In case of good layers, due to frequent relaxation of the pubic bones, they become more flexible, thinner and remain separated by a larger distance. In non-layers, the bone will be stout and very hard.

Table 39.1. Judging present production

	Good layer	Poor layer	Non-layer
Comb	Large, red, warm	Small, less warm, shrunken	Underdeveloped
Eyes	Big, bright and active	Comparatively looks smaller and less active	Appears dull and inactive
Vent	Oblong, moist and pink	Less oblong, maybe moist and pink	Round, dry and has a yellow rim
Distance between two pubic bones	At least three fingers	Less than three fingers	Maximum one finger
Distance between tip of the breastbone and pubic bones	At least four fingers, the region being soft and pliable	Less than four fingers, not very soft	Hardly two fingers, very hard and rubbery
			One finger is approximately 1¼ to 1½ cm

2.5. Distance between tip of the breast bone and pubic bones

This is a measure of abdominal capacity. Good layers consume more feed than poor- or non-layers. They will also have well developed ovary and oviduct (about 20 times as large as the same organs of a nonproductive hen). Hence, there will be some structural change in the skeletal system to accommodate these changes. In addition, due to constant pressure of the viscera and the weight of egg in the oviduct, the abdominal muscles become flabby in case of good layers.

Capacity to produce eggs is shown by the depth or distance from the front of the keel to the center of the back, the space between the end of the keel and the pubic bones, the width and length of the back, and by the width and length of the keel. These dimensions are more or less fixed in the adult bird except that the end of the keel moves up or down. The downward tendency is somewhat regulated by the demand for more internal space which usually indicates a greater intensity in laying. A depth of 4 to 5 fingers from the end of the keel to the pubic bones is associated with good rate of production, while a depth of 2 or 3 fingers indicates fair to poor production.

The position of the keel relative to the back can be determined by facing the bird and running one hand down the top line and the other along the bottom line. The finger tips usually point downward in laying hens of high intensity. This test is applicable to laying hens only.

3. Persistency of production

3.1. Molting

Persistence of production is measured also by the condition of the plumage; as long as the hen lays regularly she usually retains her old feathers, but, if for any reason other than sickness or broodiness she stops laying, the feathers begin to drop, she is then said to be molting.

Each year chickens molt, or lose the older feathers, and grow new ones. Most hens stop producing eggs until after the molt is completed. The rate of lay for some hens may not be affected, but their molting time is longer. Hens referred to as "late molters" will lay for 12 to 14 months before molting, while others, referred to as "early molters," may begin to molt after only a few months in production. Late molters are generally the better laying hens and will have a more ragged and tattered covering of feathers. The early molters are generally poorer layers and have a smoother, better-groomed appearance. In addition, early molters drop only a few feathers at a time and may take as long as four to six months to complete the molt. The advantage of late molters is that the loss of feathers and their replacement takes place at the same time. This enables the hen to return to full production sooner.

3.1.1. Order of molting

The order in which birds lose their feathers is fairly definite. The feathers are lost from the head first, followed in order by those on the neck, breast, body, wings, and tail. A definite order of molting is also seen within each molting section, such as the loss of primary flight feathers before secondary flight feathers on the wings.

The primary wing feathers determine whether a hen is an early or late molter. These large, stiff flight feathers are observed on the outer part of each wing when the wing is spread. Usually 10 primary feathers on each wing are separated from the smaller secondary feathers by a short axial feather.

Molting birds lose the primary feathers in regular order, beginning with the feather nearest the axial feather and progressing to the outer wing-tip feathers. Late molting hens will lose primary feathers in groups of two or more feathers, whereas early molters lose feathers individually. Replacement feathers begin to grow shortly after the old feathers are shed. Late molting birds can be distinguished by groups of replacement feathers showing similar stages of growth.

The neck molt is rather common at any season of the year even in good layers, but if the molt progresses to the back, the primary feathers of the wing generally molt also. This stage is seldom reached unless the hens have stopped laying. In other words, the cessation of laying is likely to bring on a general molt. Soon after the old feathers are dropped, new ones grow in to take their place.

3.1.2. New feathers

A molting condition can be determined easily by examining the base of the feather. The web appears new and glossy, the quill is large, full, soft, and pink, and the accessory plume is not visible. It is obvious that the later a hen molts the longer has been her laying season, and hence the greater is likely to be her production. This means that the early molting hens have taken vacation periods and are probably low producers.

The new plumage of the early molting hen is easily recognized by its clean, bright color, with every feather perfect in shape, while the plumage of the heavy laying hen that has not molted is rough, soiled, dry, and worn threadbare. The tail and wing feathers, as well as those about the back, head, and neck, are frequently ragged, worn, and broken.

3.1.3. Wing Molt

When a general molt, as opposed to a partial molt, starts, that fact is registered in the primaries of the wing and the length of time a hen has been molting can be calculated by the number and length of the new feathers in the primary. The primary feather next to the axial feather drops when the body molt starts and 6 weeks on the average is required for it to grow to full length. Feather 2 is dropped about 2 weeks later and so on with each subsequent feather until the 10 primaries have been renewed.

Hence, in calculating the time elapsed since a molt started, 6 weeks should be allowed for the first full-grown feather and 2 weeks more for each additional mature new feather; in other words, the time a bird has been molting can be determined by examination of the large primary wing feathers. Length of molt can be estimated by allowing six weeks for the first mature group of primaries and two weeks for each additional feather or group of feathers. If the primary feathers are not fully grown, the time of molt can be estimated based on the present stage of growth of feathers.

Often pullets undergo a partial molt, involving the neck and tail feathers. This condition can usually be eliminated by purchasing pullets hatched in April or later in each year and by following proper management practices. The length and incidence of a molt are influenced considerably by the bird's body weight, physical condition and environmental conditions such as nutrition and management.

The new primary feathers make about ½ of their growth in 2 weeks, 2/3 in 3 weeks and 1/3 in last 3 weeks. That is, if the feathers when mature are 15 cm in length, they will grow 7 ½ cm in 2 weeks, 10 cm in 3 weeks and 5 cm the last 3 weeks. The application of this growth rate is made in determining the beginning of a molt before any of the new primaries attain full length.

The new mature primary feathers may be identified by their clear, bright appearance, with the web slightly wider and less pointed. The base of the shaft on the underside of the wing shows less of the clear horny quill than is seen in the old feathers. The molt is usually at the same stage in both wings.

The above description holds very well for the early, slow-molting, poor-producing hen, but is not so reliable for the high-producing late molters. The principal difference comes in the rapidity with which good hens molt. Instead of dropping the primary feathers separately, they frequently drop, two, three, four, or even five at a time. (Fig. 39.1C)

When two or more new feathers grow in together they are counted as one in calculating time. This works very well until the feathers reach full length when it is impossible to judge whether they have come in one at a time or all together. Such a wing with four new feathers might indicate 6, 8, or 12 weeks molt, that is, the feathers may have come in individually, in pairs, or all together. In such cases, the length of new feathers over the body, hardness of quill, and the number and position of the additional new wing feathers may throw some light on the situation.

Heavy producers frequently shed nearly all of their feathers at once. These grow in again quite rapidly, requiring usually 5 to 8 weeks to become full grown, after which time the hens are again ready to lay. Such quick molters do not always renew all of the primary feathers but retain some of them a second year. The low producers usually molt very slowly, requiring several months to renew their coat of feathers.

3.2. Pigmentation

Xanthophyll, the yellow pigment responsible for yolk color will be present in various parts of the body viz., round the vent, earlobes, eyelids, beaks, shanks etc. from where it is continuously oxidized. Before the birds begin to lay (non-layers), the oxidized pigment is replenished regularly by the feed the birds eat (principally from yellow corn). But, when the birds begin to produce eggs, the xanthophyll will be taken for yolk coloration and hence, the tissues from where the xanthophyll is oxidized do not get the pigment; and therefore, they get discolored or bleached or de-pigmented.

It can also be noted that non-layers are characterized by presence of yellowish color round the vent whereas both poor-and good layers will definitely have bleached vent.

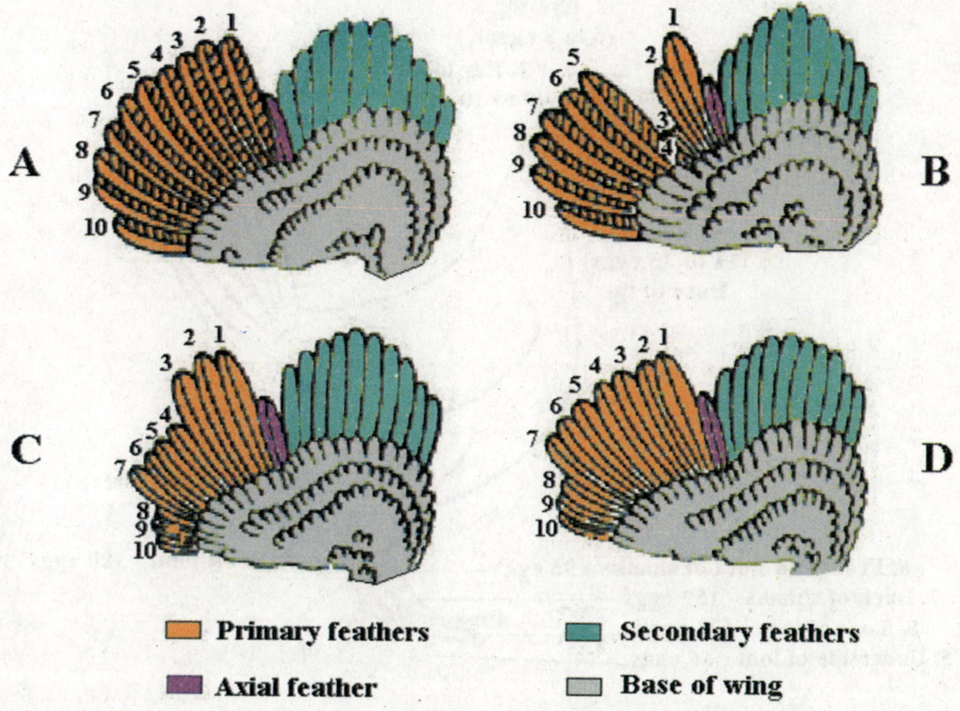

A. Normal wing

B. Slow molter: At 6 weeks. 1st Primary feather fully grown. 2nd 4 weeks old, 3rd 2 weeks old, and 4th just appeared. Others not yet dropped.

C. Fast molter: At 6 weeks. All primaries dropped. 1st to 3rd primaries fully grown; 4th to 7th (dropped in a group) 4 weeks old; 8th to 10th 2 weeks old.

D. Fast molter: At 8 weeks. 1st to 7th primaries fully grown; 8th to 10th 2/3 grown; molt completed by 10 weeks.

Fig. 39.1. Molting *Vs* persistency of production

This bleaching occurs in a specific order (Table 39.2, Fig. 39.2) and hence, can act as a guide to assess persistency of production. In non-layers, the pigment will not be bleached in any of the tissues. When the birds are in fag-end of production (last 20 weeks), the pigments reappear, in the same order, but at twice the rate as they disappeared.

Therefore, any one of the three variables, namely - egg production, age of the layer/age at sexual maturity/number of days after first egg and tissue depigmented, can be found out by knowing the other two.

3.2.1. Examples

3.2.1.1. To obtain persistency (% Egg production)

Given: Age at sexual maturity(ASM): 154 d and age at judging: 280 d Area

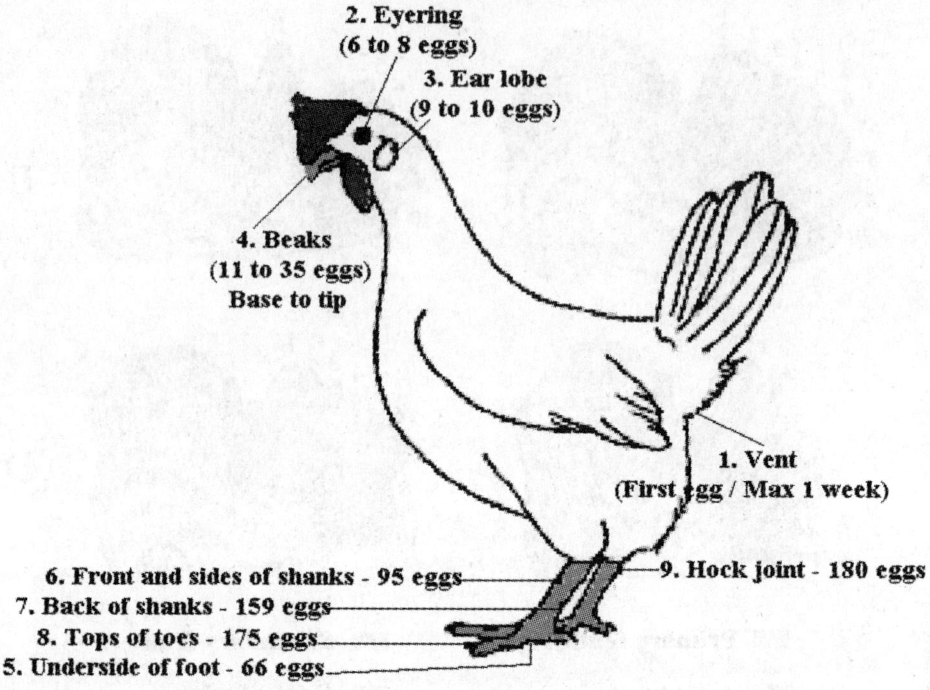

Fig. 39.2. Order of bleaching of pigments

Table 39.2. Order of bleaching of pigments

Tissue bleached	Number of eggs
Vent	When first egg is laid
Eyelids	6-8
Earlobes	9-10
Beaks	11-35
Underside of foot	66
Front of shanks	95
Back of shanks	159
Tops of toes	175
Hock joint	180

Source : North, 1972

bleached: Front of shanks. Therefore, number of eggs produced = 95 and number of days in production = (Age – ASM) = 126. That means, in 126 d, 95 eggs are produced or (95÷126)*100 = 75.40% Egg production.

3.2.1.2. To obtain age of pullets

Given: ASM: 154 d; Egg production: 44% and Tissue bleached: underside of foot.

That means, 66 eggs have been produced at a rate of 44%. In other words, pullets required 100 days to produce 44 eggs and therefore, should have taken (66÷44)*100 = 150 d to produce 66 eggs; obviously, age of the pullets should be 150 d over and above ASM; i.e. 150+154 = 304 d.

3.2.1.3. To obtain ASM

Given: Age of pullets: 400 d and Number of eggs produced: 175 and rate of egg production: 77.78%. Since 175 eggs are produced, tops of toes are likely to be bleached. Since the birds have produced 175 eggs at a rate of 77.78, they have taken (175÷77.78)*100 = 225 d to produce them. Further, 400 days' old pullets have been in production for 225 days and hence, their sexual maturity is (Age - Number of days in production) = 175 d.

4. Judging breeding males

4.1. Vigor

The first requirement in a good breeding cock or cockerel is vigor. It is identified by an alert, active, and commanding disposition. Upon close examination it will be seen that a vigorous male has a short, well-curved beak, broad, deep head, full face, with a prominent clear, bright eye and a deep red color in the comb, face, and wattles. The neck is short, arched, and blends well into the shoulders. The legs, of medium length, are set squarely under the body and the knees are straight and wide apart. Pronounced masculine characters should dominate all sections of the bird, even to the bright sheen on the plumage.

4.2. Type

In addition to the general discussion of types, other points emphasized for the male birds include a prominent breast, breadth across the shoulders, and great depth from center of back to center of keel. Well-bred cocks, unlike hens, are usually narrow between the pubic bones and show but little capacity in the abdomen because the keel of the male is inclined upward at the tip since the reproductive organs of the male needs far less space than those of an actively laying hen.

4.3. Other factors

In selecting a breeding male one should not be too quick to choose the precocious cockerel, that is, the first to show the red comb and to crow and take command of the flock. Such a bird will often show high fertility in hatching eggs and other good breeding characteristics, but is likely to be fine-boned and undersized for the breed. The continuous use of early-maturing males may have a tendency to reduce the size of the offspring and eventually affect the size of the egg. The small size of the male cannot be readily overcome by the size of the females with which he is mated, for it is well known that the highest-producing hens are often a 250 g or more below the standard weight for the breed.

Chapter **40**

Management of Breeders and Commercial Layers

Production of high producing day-old chicks, egg type as well as broilers, is the primary requirement for the poultry industry. Both these, require production of hatching eggs from breeding birds selected for the purpose. In certain countries, pullets that lay brown eggs form a separate category.

Therefore, breeding birds falling to three categories namely, Standard Leghorns to produce white eggs, Medium-sized pullets to produce brown eggs and Meat-type broiler breeder parents to produce straight-run broiler chicks.

1. Management prior to lay

While obtaining chicks, 10 to 12 cockerel chicks for each 100 pullet chicks in case of Standard Leghorns and Medium-sized breeding stock have to be ordered whereas, in case of Meat-type breeding stock, 12-15 cockerel chicks for each 100 pullet chicks are required.

Most of the breeding stock is grown in brood-grow houses on all-litter floor or slat-and-litter (or plastic-and-litter) floor. Floor, feeder and waterer space requirements are given in chapter "Management requirements and specifications". Roosts are not required for growing birds, but if provided, can help reduce first floor eggs.

1.1. Brooding

Brooding principles and practices are essentially as described in chapter "Brooding"; however, floor, feeder and waterer space requirements should be provided depending on the type of birds under brooding. The details of space requirements are also given in chapters "Brooding" and "Management specifications and recommendations".

Breeding birds are generally grown straight-run up to the start of laying period. However, sexes can be separated at hatch (vent sexing or color sexing) or at 8-10 d of age (feather sexing). Normally, Leghorns are vent-sexed, medium-sized chicks color-sexed (male chicks having silver color background and females gold colored) and broilers feather-sexed.

Even when cockerels are raised separately, about 5% of the cockerels are mixed with pullets when they are about 12 weeks of age and it about 18 weeks of age, by which time inferior cockerels must have been culled, remaining cockerels are mixed with pullets; care has to be exercised to perform this mixing during night to minimize fighting among the males.

1.1.1. Toe- and Comb-trimming

Toes of Meat-type cockerel chicks are trimmed at hatchery to prevent injury to the backs of the females during mating. Outer first joint of the back toe and inside toe of each foot is trimmed by using an electric beak-trimmer or toe-clipper.

In addition, under temperate climate, male chicks are usually dubbed to reduce comb injury and problems related to peck order.

Toe-and Comb-trimming also helps identify sexing errors so that such birds can be removed from growing flock before they enter into breeding pens (Table 40.1); it is also likely that chicks whose sex was not properly identified, if allowed to reproduce, identification of sex in their progeny also may be erroneous under commercial conditions.

Table 40.1. Identifying sexing errors

Condition	Cockerel chicks	Pullet chicks
Combs trimmed	Keep	Sexing error
Toes trimmed	Keep	Sexing error
Combs untrimmed	Sexing error	Keep
Toes untrimmed	Sexing error	Keep

Source : North and Bell, 1990

1.2. Growing

Most of the poultry breeders use the same house for brooding and growing the breeding stock; some even up to the end of laying period; the former is referred to as "brood-grow house" and the latter "brood-grow-lay house". Brood-grow house is advantageous in terms of utilization of space as well as ease of vaccination programs.

Further, raising sexes separate is also a general recommendation for meat-type breeders till 21 weeks of age after which 12 cockerels for every 100 pullets are mixed and at sexual maturity (21-22 weeks of age), number of cockerels per 100 pullets is reduced to 9-11. Even when they are grown straight-run, sexes are raised separately for the first fortnight during which beak-trimming is performed for cockerels and by 2 weeks of age, equal number of male and female chicks is allowed under the brooder.

Whenever separate male feeding system is introduced, care has to be exercised to fulfill the following criteria: 1. All males should have access to uniform feed levels 2. Females and dominant males should be prevented from stealing feed 3.

Males must be allowed to feed quickly and easily 4. Feeders must be removable whenever necessary, especially when males are culled 5. Feeder space per bird should be maintained at a constant level 6. Female feeders should be run first to encourage females to be away from the male feeders 7. Feed must be uniformly delivered throughout the house (Nicholson, 2003).

1.2.1. Feed restriction

Broiler breeding stock (Meat-type chicks) must be subjected to restricted feeding in order to control excessive growth of both males and females; at 24 weeks of age, it is ideal if cockerels attain a body weight of 3.36 kg and pullets 2.50 kg.

Standard Leghorns as well as Medium-sized chicks meant for breeding purposes are also subjected to feed restriction during growing period (For details see Chapter "Feed restriction").

During feed restriction, birds consume more water to have a satiated feeling and therefore, wet-droppings are expectable. However, feed restriction being mandatory to attain optimum body weight, water restriction has to be followed only when temperature does not exceed 26.7°C. In other words, the amount of water made available on days with feed and without feed is different.

1.2.1.1. Daily feeding, but quantity restricted

During feed days, when calculating quantity of feed is given, water is supplied an hour before the first feed is given and continued till one hour after all the feed is consumed. The birds consumed feed rapidly because the amount offered is restricted. The birds are given access to water again only in the evening.

1.2.1.2. Alternate day feeding

If ambient temperature is below 21.1°C, they are subjected to program similar to birds which are fed daily on restricted quantities.

If the ambient temperature is between 21.1 and 26.7°C, water supplied 1½ hr before the first being; the rest of the program is similar to birds which are fed daily on restricted quantities.

Body weight recommendations, feed and water consumption data pertaining to growing periods are provided in Chapter "Management requirements specifications".

1.2.2. Importance of correct body weight

The body weight specifications have to be strictly adhered to especially in case of meat-type breeders. Differences between strains and varieties are also recorded; therefore, it is advisable to follow the recommendations of the breeder/hatchery supplying the day-old chicks.

1.2.2.1. Checking for body weight

In case of meat-type breeders, in particular, body weights are monitored from as early as 3 weeks of age to ensure that accelerating phase of growth is extended so that pullets, in particular, and cockerels attain sexual maturity with good body fleshing but without excess fat accumulation.

Many factors influence body weight/growth rate; important ones being chick weight, feed quality/quantity and season/temperature. Weekly body weights have to be maintained as per the stipulations for the breed/strain/variety given by the hatchery. Alterations may have to be made in quantity of feed offered to correct over-or under-weight problems; as the rule of thumb, for every 1% over-are under-body weight, feed allocation is decreased or increased by 1%, respectively.

1.2.2.2. Sample weighing

For assessing uniformity of the flock, it is not necessary to weigh the entire flock. Instead, a small proportion sufficient to represent the flock can be weighed. If the flock is housed at different locations, additional samples have to be taken from each of the locations. It is definitely true that higher the sample size, greater the accuracy and weighing of all the birds is the most ideal, though commercially neither possible nor necessary.

Prior to lay, birds are weighed every week on a sufficiently sensitive balance (sensitivity = 30 g) starting from 3 weeks of age itself. It is desirable that the birds are weighed at the same time of the day, preferably, in the afternoon; in case of skip-a-day program of restricted feeding, weighing is advisable on no-feed day.

When the pullets are floor-reared, birds are selected from several locations segregated by use of a catching fence and weighed individually. In case of cage-reared pullets, birds from various locations in the room/line are weighed; the cages can be marked so that the same birds can be weighed at each weighing.

A general guideline as a number of birds to be stamped is given in Table 40.2:

Table 40.2. Sample size *Vs* Flock size

Flock size	Sample size	Flock size	Sample size
< 500	60	4000-6000	150
500-1500	80	6000-8000	175
1500-3000	100	8000-10000	200
3000-4000	125		

Source : North and Bell, 1990

1.2.2.3. Body weight uniformity

Uniformity in body weight of growing pullets is a very important indicator of not only good management but also of homozygosity of the strain in question.

Standard statistical parameters can be estimated, (assuming normal distribution curve), like standard deviation (sd), coefficient of variation (CV), skewness, kurtosis etc. to assess the spread of body weights around the arithmetic mean.

However, under commercial conditions, uniformity is measured as a percentage of birds within 10% of the mean weight of all the birds. Generally, it is acceptable if at least 80% of the flock is within mean ± 10% of mean ("Average" score); however 85% (mean ± 7½% of mean) or even 90% (mean ± 5% of mean) is achievable even under commercial conditions. It has been suggested that for every % improvement in uniformity, the overall advantage works out to be 1 egg/breeder pullet.

The individual body weights of the sample of birds are fit into a frequency distribution (with sensitivity of the balance as class interval) and percentage of birds on either sides of the mean value is computed. Utilizing the value so obtained and based on the score card (Table 40.3) for degree of uniformity, the flock is assessed.

Table 40.3. Score card for degree of uniformity

Score	Percentage of pullets within Mean ± 10% of Mean *
Superior	≥ 91
Excellent	84-90
Good	77-83
Average	70-76
Fair	63-69
Poor	56-62
Very poor	≤ 55
* Sensitivity of weighing balance ≤ 30 g	*Source* : North and Bell, 1990

Advent of computers and their use becoming common, the body weight data can be subjected to statistical analysis to obtain mean, sd and CV; the latter two being the measures of dispersion of values around the mean. A CV = 8% (lower the better) is generally accepted as representing "uniform" flocks (Leeson and Summers, 2000)

It is very important to note that sensitivity of the weighing balance significantly affects percentage weight spread around the mean; higher the sensitivity (smaller weight divisions), lower the percentage of pullets within mean ± 10% of mean in the same flock (Table 40.4).

Table 40.4. Sensitivity of balance *Vs* uniformity

Sensitivity (g)	Percentage of pullets within Mean ± 10% of Mean
5	68
30	73
45	78
	Source : North and Bell, 1990

However, under commercial conditions, weighing balances of sensitivity as low as 5 g cannot be used to weigh larger body weights expectable among growing birds; in addition, to weigh birds on very sensitive balances is highly time-consuming and hence impracticable. Therefore, it is generally accepted to use balances of sensitivity 20-30 g for weighing growing flock.

1.2.2.4. Frame measurements

Although body weight is most commonly used parameter to assess uniformity, nowadays breeders recommend measuring the shank of a representative sample of pullets every alternate week beginning from 4 weeks of age till sexual maturity. The basis of this criterion is that skeleton is practically fully developed by 10 weeks of age and hence, pullet quality can be assessed at an early age. In any case, shank measurements are yet to replace body weight uniformity assessment.

2. Management during lay

Floor, feed and waterer space along with light (photoperiod) requirements are given in Chapter "Management requirements and specifications". Both egg-type and meat-type cockerels do not mate well on slat/wire-floors; a drop in fertility by 2-3% and 5-7% has been recorded, respectively. Therefore, it is common that the breeding birds of either type are reared on all-litter or litter-and-slat floors.

2.1. Nests

Breeding birds, of either type, require nests at the rate of 25-30% of the number of birds; nests for meat-type layers being slightly larger than those for egg-type birds. A perch should be in front of all nests and should be constructed in such a way that it may be used to close the nests at night. Roof of the nests must have a slope so that birds cannot perch on them (Fig. 40.1). Nesting material should be maintained clean, dry and resilient in order to reduce microbial load on hatching eggs.

2.2. Floor management

2.2.1. All-litter

Litter moisture recommended is 25-30% which is ideal for maintenance of good health, cleanliness of eggs, and for minimizing worm infestation and coccidiosis. It also helps prevent litter material adhering to feet of pullets which, in turn, will be carried into the nests. Hatching egg cleanliness is of paramount importance to optimize returns at the hatchery.

2.2.1.1. Floor eggs

Floor eggs are the general problem when laying birds are reared on floor. Elimination of floor eggs depends on efficient management. Pullets before lay make a peculiar sound which should be identified by the attendant and the bird should be shifted to nest manually to train it for using the nest. Adequate number

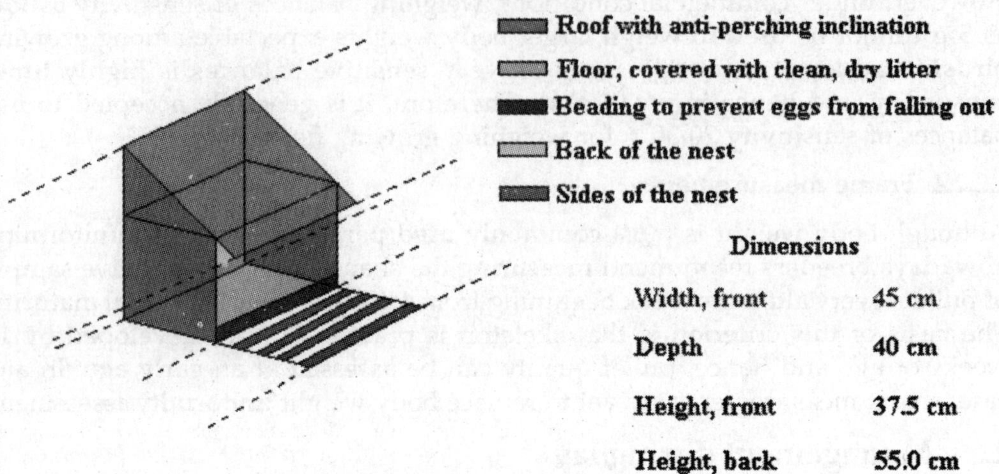

Fig. 40.1. Nests for layers

of nests properly arranged also helps reducing floor eggs. Pullets prefer nest sections placed crosswise of the house. Frequent collection of eggs, at least hourly once, during the peak production period of the day (10 a.m. to 3 p.m. in case of chicken), also helps reduce floor eggs. Slat-and-litter system of rearing greatly reduces frequency of floor eggs.

2.2.2. Slat-and-litter

This system of rearing is particularly popular for meat-type breeding birds. Although there are some drawbacks like a) difficulty in ventilating litter area located at the centre b) reduced egg numbers than on all-litter c) higher layer-house mortality than on all-litter and d) analogous to floor eggs, some birds lay eggs on slats in preference to nests; floor eggs on litter area which also be higher in the system of rearing., advantages in terms of fertility will clearly outweigh all the disadvantages. Besides, cost of maintaining breeder hen as well as producing dozen hatching eggs is same as that on all-slat.

2.2.3. All-slat

This system of rearing is generally not advised for breeding birds mainly because of lower facility and most of the floor eggs being broken. Broken eggs have to be removed as quickly as possible to prevent a possibility of development of egg-eating vice among layers. If such houses are used, the height at which the slats are fixed will be usually more than in case of all-slat houses for brooding chicks; this provision is mainly to help removal and handling of manure.

2.2.3.1. Drawbacks.

1. Egg-type pullets tend to be more flighty, especially if the house is not partitioned into smaller cubicles but allowing room for the birds to walk under

the partitions in the entire house. This requires special design and fixing adding to the rearing cost.

2. Birds tend to get heavier on slats than on litter system because they exercise less and higher density of birds further contributes to that effect. Therefore, feed restriction program needs to be extremely stringent and accurate.
3. Bottom of the nests should be closer to the floor than in case of litter floors so that the pullet can nearly walk into the nests. Although floor eggs are reduced, routine collection of eggs manually is more difficult. However, with automatic egg collection, this problem can be overcome.
4. Foot and leg problems as well as breast blisters are more common than on litter floors

2.3. Managing optimum fertility

It is generally known that although males constitute about 10% of the total breeding population, they contribute the same proportion of genes as that of females. More attention is required on males because, defect in male is exaggerated through its mates; it could be in the form of severe reduction in number of fertile eggs or abnormal increase in embryo abnormalities or abnormally low hatchability or substandard performance of its progenies. Therefore, much care and attention are needed in managing males for high fertility.

2.3.1. Facts about mating in chicken

1. Males attempt to mate 10-30 times a day depending on competition, number of female available, social order, temperature, light etc.
2. In the early part of the day, volume and concentration of spermatozoa in semen will be higher; the former ranging between 0.1 and 1.0 cc.
3. Males may mate several times a day with the same hen.
4. Pullets in the middle of social order are likely to be mated in preference to more precocious or timid ones.
5. Only about 2/3 of the attempted matings will be completed
6. More matings occur early in the day; but, fertility rate is not related to time of mating.
7. If new males are introduced in place of existing ones, fertile eggs produced after 3d will be the result of matings by the new males.

2.3.2. Mating ratio

Mating ratio refers to number of females allowed per male; for commercial purposes, it is generally indicated as number of males per 100 females. The current ratio of males to females depends on type and size of the breeding individuals involved. More males are usually allowed in slat-and-litter and all-slat systems than on all-litter system. In case of conventional cages/raised platform cages, artificial insemination is the only option for obtaining hatching eggs.

Under all conditions, it is rather mandatory to grow 10-20% more males are actually needed and also to allow few more males than the suggested making ratio to provide for culling at a later date without affecting fertility. Suggested mating ratios are given in Table 40.5.

2.3.3. Care of cockerels

Whenever cockerels have to be caught, it must be done by catching both the legs and, if possible, with a wing; otherwise, there is all the chance of a permanent injury to the legs making them unfit for further mating. Inferior cockerels must be removed and soon be replaced with another one; because, preferential mating (a male mating with specific female/s) is not uncommon.

Table 40.5. Suggested mating ratios (Males per 100 Females)

Parents	Progeny	All-litter	Slat-and-litter
Standard Leghorn	Commercial layers	8	9
Medium-size	Commercial medium-size layers	9	10
Standard meat-type	Commercial broilers	10	11
			Source : North and Bell, 1990

Exercise is also equally important. Spreading some grain on the litter in the afternoon induces scratching thereby exercising the legs; similarly, separate male feeders fixed at a height to force the males jump to get feed is also a way of keeping the males exercised.

If timid males are noticed, sufficient feed was made available for them and sometimes adding few more cockerel feeders facilitates them to eat feed and maintain body weight.

When the breeders are grown on slat-and-litter, the males should have access to litter for at least 8 hr/d to maintain optimum fertility; they should not be fenced off from the litter area during the day. In addition, wherever slat and/or wire make part or the entire floor, foot pad infection can occur in some of the males. Such males have to be identified and culled.

On slat-and-litter system there is yet another problem: males prefer to remain on slats because the slats are at a raised level from litter portion dealing with a feeling of "roosting". However, pullets prefer to be mated on litter. Therefore, overall fertility in the flock declines. Spreading some grain on the litter area in the afternoon will definitely stimulate cockerels to come to the litter area for eating grains by scratching; this helps exercise their legs as well as promote mating with the pullets.

2.3.4. Spiking the flock

As the breeders age, there is a steady fall in fertility which is natural especially in case of meat-type of breeders than in egg-type. To rectify, new younger males are introduced into the flock after about 2/3 of the egg production period is over which is referred to as "spiking". However, cost of maintaining these extra males, on many occasions, will be more than the benefits in the form of fertility and incubation performance.

3. Hatching eggs

3.1. Optimum age at sexual maturity

It is extremely difficult to define a particular age as optimum for obtaining hatching eggs. However, pullets brought into production at 25-27 weeks have shown the following advantages:

1. Longer egg production period
2. Higher % egg production and therefore more number of eggs
3. More number of settable eggs
4. Eggs of higher quality, especially shell quality
5. Consume less feed per dozen eggs

3.2. Minimum weight for hatching eggs

A general recommendations regarding minimum weight of settable eggs is given in Table 40.6. It should be noted that minimum weights for hatching eggs differ on either sides depending on breed/strain/variety.

Table 40.6. Minimum egg weight for settable eggs (g/egg)

Parents	First 12 weeks of egg production	After 12 weeks of egg production
Standard Leghorn	52.0	54.3
Medium-size, egg-type	54.3	54.3
Standard meat-type	49.6	52.0
		Source : North and Bell,1990

3.3. Hatching egg storage

Details pertaining to hatching egg storage are given in Chapter "Incubation and hatching". However, the size of the egg cooler room as well as cooling required depends on the number of eggs that is likely to enter into the cooler room which, in turn, depends on the flock size. Table 40.7 summarizes cooler room requirements:

Table 40.7. Flock size *Vs* egg cooler requirements

Flock size	Size of cooler (min)	Cooling required (kJ)
5,000	1.8 m X 2.4 m	4,750
10,000	2.7 m X 2.7 m	8,000
20,000	3.1 m X 4.3 m	12,500
30,000	4.3 m X 4.9 m	21,000
		Source : North and Bell, 1990

4. Broiler breeders

(Vest, 1997)

In addition to the management details given above, the following are relevant to the broiler breeders, in particular:

4.1. Uniformity and Body Weight

Average body weight and body weight uniformity are inseparable topics. A flock that is uniformly off the targeted body weight will probably perform better than a non-uniform flock with the correct average weight. A uniform flock with the proper weight should out-perform both. Uniform flocks have several advantages: (a) are more efficient, (b) have higher peak production, and (c) come closest to expressing their full genetic potential.

A weighing program is essential in orderto know whether pullets are on schedule or behind standard weight values. This is especially important during the first 10 to 12 weeks of age. When the pullet is grown improperly during the early growing period, it will be difficult to improve the pullet during the remaining growing period.

Starting at four weeks of age, a random sample of one percent of the pullets is weighed, individually every week. Pullets from different areas of the house are taken. The birds should be weighed on the "no feed" day during the growing period or late afternoon in case of every day feeding program (after 24 weeks). Every bird in the sample should be weighed. The extra labor involved in weighing and recording body weight and uniformity will result in the greatest profit potential. Otherwise, the flock may over-consume feed, get heavy, mature too early, lay fewer eggs or have reduced fertility. The use of electronic in-house weighing equipment has reduced labor costs and in many cases improved accuracy.

In order to have a better idea of uniformity, it would be beneficial to determine what percent of the pullet weights are within a range of 10 percent above and below the average for the flock. Flock uniformity guidelines are given in Table 40.8.

Table 40.8. Flock Uniformity Guideline

Age (Weeks)	Birds within flock average ± 10% of flock average
4-6	85-95%
7-11	80-85%
12-15	75-80%
? 20	80-85%

4.2. Factors associated with uniformity

4.2.1. Baby Chick

The primary breeders should send chicks from grandparent flocks of approximately the same age. Chicks from young parent stock will be smaller than those from old flocks. If there are differences, the grower should be notified and the boxes identified so the chicks can be grown separately, if possible.

4.2.2. Feed Restriction

Féed restriction should begin early (4 weeks of age) to prevent the flock from becoming overweight and developing a large skeletal frame. It is difficult to reduce body weight after overfeeding and resultant overweight condition; it makes very difficult to maintain flock uniformity as well. The art of growing breeder pullets centers upon supplying the essential level of nutrients without wastage.

4.2.2.1. Weighing Feed

An accurate feed scale for weighing the feed is essential. Regardless of the type of scale used, it should be checked periodically for accuracy. Volumetric measurement is not recommended because it can be misleading, especially with formulation changes.

4.2.2.2. Feed and Water Space

Feed and water space requirements should be as recommended by the chick-supplier or primary breeder. Aggressive birds often take more than their fair share of feed and water space. Insufficient space becomes detrimental, especially during periods of peak requirements.

4.2.2.3. Feed Distribution

Feed distribution is another critical area. The feed should make a complete cycle in 10 minutes or less. If the feed is moved slowly it may be necessary to use supplemental feed hoppers.

The feed trough should never be empty until all the feed allotted for the day has been fed. Feeding several times allows the aggressive birds an opportunity to make several trips to feed, which further restricts the birds on the lower end of the social scale.

During the growing period, every other day of feeding can be followed. Care must be taken to control growth rate or more harm than good may result. Having ample feeder space and good feed distribution are critical.

4.2.3. Nutrient Density of Feed

Whatever the nutrient density of the ration, all birds should have an opportunity to consume equal quantities of feed. If the feed supplied does not last over three hours during the lay period, it is less likely that all birds have an opportunity to consume the same amount of feed. Thus, the nutrient density of the feed should be reduced to facilitate longer feeding times.

4.2.4. Beak-Trimming

Beak-trimming exerts considerable influence on uniformity. Excessively long bottom or top mandibles and uneven cuts restrict feed intake, resulting in reduced growth.

4.2.5. Stress

Management factors may place undue stresses on the birds.

4.2.5.1. Environmental Conditions

Extremes in environmental temperature cause the birds to use more energy to maintain their normal body temperature. High humidity and fluctuating temperature may also influence nutrient demands. Birds kept at uniform temperatures of around 21.1°C during the winter and not over 26.7°C during the summer exhibit maximum feed efficiency, assuming good air quality. It is known that noxious gases such as ammonia are detrimental to growth. Thus, it is advantageous to maintain good air quality and a comfortable temperature.

4.2.5.2. Litter Condition

Wet, caked litter is a hindrance to the development of a uniform flock. Litter problems are greatest during the cold months of the year. Often these problems are created by improper ventilation, by lack of insulation and/or by increasing bird density.

The more severe and prolonged the stress, the more pronounced effect it has on the health and eventual uniformity of the flock. Many times these stresses will reduce performance for the life of the flocks. For instance, males become sterile because of excess heat.

4.2.6. Intestinal Health

A bird's ability to fully absorb available nutrients is impaired when the intestinal tract is not healthy. Coccidiosis, enteritis, intestinal parasites or diseases of the intestinal tract will interfere with nutrient absorption, resulting in poor growth and non-uniform flocks. Often it is necessary to add drugs, antibiotics, vitamins, etc., to the feed in order to maintain the birds' health and promote the efficient utilization of nutrients.

4.2.7. Lighting Program

Artificial lighting is needed to compensate for seasonal day length changes. Light intensity and length of daily light periods change the age at which pullets reach sexual maturity. Hold pullets on a constant day-length until you have reached your target body weight and are ready to stimulate. Response to light stimulation is expected two or three weeks after the stimulus. Hens in production, in open-sided houses, should receive a minimum of 16 hours of light (natural and artificial) daily. They are seldom given over 18 hours of total light daily.

4.3. Male Management

In hatching egg production, the males are important for ensuring adequate fertility. Since males are very important in producing good hatching eggs, they need extra attention. The following are recommended:

1. If a male becomes weakened or crippled, remove from the breeding pen immediately.
2. If male birds appear to be less than normal in appearance or health, submit a bird with the typical symptoms and a healthy bird to the nearest poultry diagnostic laboratory for a potential disease diagnosis.
3. Care should be taken that males are not gaining excess weight. Also, make every effort to see that all males have a chance to consume the same amount of feed.
4. It may be necessary to spike a flock with additional males if fertility is low or you have lost an abnormal number of males.

4.4. Hatching Egg Care

Hatching eggs have to be properly handled and stored; the guidelines are:

1. Be sure the nests are available two to three weeks before the first eggs are laid. Train hens to use nests. Floor eggs are a liability.
2. Provide ample nesting area for all hens. Consult the breeder guide or integrator representative.
3. To aid in keeping nests clean, close the nests at night to keep the hens from roosting in them. Be sure the nests are open when the lights come on in the morning.
4. Eggs must be gathered frequently to maintain high hatchability and to prevent dirty eggs. The sooner an egg can be placed in proper storage conditions the healthier the embryo will be.
5. The incidence of dirties, floor eggs, checks and cracks can be substantially reduced by frequent gathering of hatching eggs.

6. Remove droppings and soiled litter from nests promptly.
7. Separate clean eggs from dirties as they are gathered.
8. Remove all eggs not suitable for hatching (unless otherwise instructed by hatchery personnel). Eggs not suitable for hatching are eggs with thin, rough or porous shells, eggs with rings or ridges around them, round or long pointed eggs, eggs with checks and cracks, and eggs above or below the weight specified by the hatchery.

Hold eggs at 65 degrees to 70 degrees F temperature and 75 percent to 80 percent relative humidity. Do not allow eggs to sweat.

5. Feeding and disease control

Nutrient requirements, feeding and vaccination programs are discussed separately in Chapters "Nutrient requirements and specifications" and "Poultry diseases – vaccines and vaccination", respectively.

6. Broiler breeders in cages

Meat-type birds are generally reared on deep-litter up to laying and subsequently either transferred to deep-litter or slat-and-litter laying houses or sometimes, cage-layer houses where artificial insemination is practiced.

However, an alternative method would be to brood, up to five weeks, in battery system and grow them up to laying stage in cages (120 cm with x 30 cm depth x 35 cm height for four female birds) to be subsequently transferred to floor-housing. Males can continue to grow on deep-litter. Advantages of this method are:

1. Saving of feed cast at about 800 g/bird
2. Saving on medication (coccidiostats) and litter costs
3. Early sexual maturity and peak production
4. Uniform body weight
5. Number of culls reduced and
6. More number of eggs produced.
7. Commercial layers

Brooding, growing, beak-trimming, restricted feeding, lighting and judging are as discussed in relevant Chapters. Space requirements, record-keeping, cage management, force-molting, nutrition, disease diagnosis and control and other related details will be provided in the forthcoming Chapters. Management practices during the laying period are essentially the same as those for breeding birds.

8. Depopulation

Many factors determine when the breeder flock has to be liquidated, the main ones being rate of egg production, number of settable eggs, mortality among breeding flock and hatchability of eggs set. It is based on these factors that meat-type pullets are generally liquidated after 38-40 weeks (rarely 44 weeks) of production after which maintenance of the breeding flock becomes economically unviable. In case of Standard Leghorn as well as Medium-size breeders, breeding flock is generally maintained for 60 weeks or more in egg production.

Commercial egg-producing hens are usually retained up to 90 weeks of age or more.

Several factors also prompt the population of individual birds/part of the flock/the entire flock depending on local circumstances; they include, out-of-season flocks, overweight pullets, poor layers, layers that do not peak in production, non-uniform birds, partially molted flock, poor shell quality, insufficient mating ratio, wet-litter due to abnormally high water consumption etc.

Chapter 41

Litter Management

Litter material provides a media for absorbing moisture from fecal matter and also protects the feet of the birds from direct exposure to the floor and thereby minimizing the incidence of bumble foot, foot-sores etc. Commonly used litter materials are paddy-husk, groundnut kernels, ground maize-cobs, wood-shavings, sawdust, sugarcane bagasse, wheat-straw etc... Of these, paddy-husk is cheapest, easily available and hence, commonly used in our country. Initially, 5 to 8 cm thick litter material is spread which increases to 12 to 15 cm by 20 weeks of age of birds. Bacteria decompose and convert the litter into crumbly and powdery form referred to as "compost litter " or "build-up litter " in about 6 to 8 weeks of time. The pH of normal compost litter will be alkaline (at least 8) and hence only a few yeasts, molds and bacteria can grow. The fermentation also produces heat which helps drying of litter. When outside temperature is less than 10°C, moisture content of litter exceeds 30%.

1. Testing of condition of litter

In a well-managed litter, moisture will be between 25 and 30%; such manure when pressed into a ball in hand and fist is opened, breaks into about three pieces whereas a wet litter forms a solid ball and the dry litter falls out like powder.

2. Reuse of old litter

Proper management of litter in the poultry house will reduce the need to remove litter between flocks and will aid in developing a cleanout schedule that allows direct application of manure to cropland without intermediate storage (Collins, 1996).

Reuse of litter is not a normal practice. If old litter has to be used, then a) it must be from a healthy flock b) at the centre of the heap, temperature should be about 51 to 52°C which ensures safety from worm eggs and coccidial oocysts. The heap should be headed this temperature for about 3 to 4 days c) if litter is dry, water may be added d) litter has to be raked frequently e) hydrated lime to be added that the rate of 1 kg per sq m of floor area.

The parameter having the greatest influence on temperature build-up of the stacked litter is water activity; litter with higher water activity will be more hotter than those with low water activity. Bacterial survival in stacked litter has been found to be poor; for instance, following inoculation, the maximum time for which *Campylobacter*, *Escherichia coli* and *Salmonella* could be recovered was 2, 32 and 28 hr, respectively with an approximate 4-6 log reduction in numbers.

If the litter is to be temporarily in open storage and/or stockpiled, it should be covered with plastic sheeting (6 mm minimum thickness) held in place with old tires, by burying the edges of the sheeting, or by other anchoring systems. If this practice is used often, a reinforced, ultraviolet resistant cover will last longer and may be a good investment. Sites should be selected carefully, as described earlier; location near windbreaks will help protect the plastic covering. Compacting of litter is not necessary, but more manure can be stored in a smaller area and with less plastic sheeting if compaction is provided. Sheeting must be applied with care to prevent tearing. Sheeting is anchored by laying the edges across a small trench approximately 30 cm deep and backfilling with soil. Lay used tires over the sheeting, similar to methods used on bunker silos, to prevent loosening and damage in the wind. It is preferable to leave the pile sealed until all litter can be spread or otherwise utilized (Collins, 1996).

3. Keeping litter in good condition

Two primary factors relate to good litter conditions: proper heating and ventilation and selection and operation of bird-watering systems to minimize spillage on the litter.

Investment on proper selection and use of water systems and attention to good management provide economic and environmental returns to all phases of bird and litter management (Collins, 1996). The following are precautions for maintaining good litter condition:

1. Proper floor space for birds to ensure normal moisture content in litter
2. Proper ventilation to remove excess moisture
3. The floor should be raised at least 0.3 to 0.5 m above the ground level to avoid seepage of water
4. Concrete flooring is preferred; otherwise, thickness of litter has to be increased
5. Overhangs to the roof must be and equate to protect the interior of the house from rainwater
6. Leaky and/or overflowing waterers should never be used; waterers must be kept on a horizontal plane; watering channels should never overflow. Any common type of watering system can be used effectively if maintained properly. Careful adjustment of height, water depth, and other operating factors will help assure minimum spillage onto the litter. Reducing water spillage will: save water, improve bird quality, improve production environment, reduce ammonia release from litter, reduce volume of wet manure cake, and extend time between litter cleanout.

7. Wet and/or moldy litter should be replaced at once with good dry litter .
8. Litter should be raked several times daily to help drying as well as to obviate caking
9. For wet-litter, hydrated lime at a rate of 1 kg or super phosphate of lime at a rate of 0.75 kg per sq m can be used
10. Nutritional causes, if any, for wet-litter should be checked for, and suitable corrective measures to be taken up

4. Litter quality and in-house environment

Of the many gases and odors inside a poultry house, ammonia is, by far, the most predominant. Litter pH has an important influence on ammonia emission. Effective way of reducing ammonia is to acidify the litter. The pH is usually higher at the sides than in the middle of the house. Reuse of litter also contributes to increase ammonia production. High brooding temperatures and low ventilation rate are also important factors leading to ammonia build-up. Reducing litter moisture is an efficient way to reduce ammonia. As better moisture is in equilibrium with relative humidity, increasing ventilation rate will help reduce ammonia concentration due to dilution effect by the incoming air. Proper arrangement of drinkers not only helps maintain litter quality but also minimizes ammonia concentration as well. Similarly, nutritional management of fecal moisture helps control ammonia concentration (Alchalabi, 2002).

5. Poor litter quality

Many factors may give rise to poor litter quality and it is necessary to understand in order to plan suitable measures to avoid them.

5.1. Factors leading to poor litter quality

5.1.1. Litter Moisture

Litter moisture is the key to the burnt hock problem. It is unusual for there to be burnt hocks when the litter condition is friable and dry. Litter moisture is affected by drinker design; air change rate; litter material and depth; stocking density; diet and flock health. Wet litter causes degeneration of the outer scales on the hocks and feet (i.e. on the pressure points). As live-weight increases and mobility decreases the pressure becomes greater and contact with the litter more prolonged. Males are therefore more affected by wet litter than females.

5.1.2. Greasy Capped Litter

When there is too much fat in the feed or it is of poor quality, the fat content of the feces increases. Consequently, the litter also has a higher fat content which causes it to lose its friability more quickly. A cap forms and the pressure points on the bird's legs and breast are then vulnerable to damage.

5.1.3. Nitrogen in the litter

Experimental evidence suggests that the worst burning tends to occur when the nitrogen content of the litter exceeds 5.5%. The quality and amount of protein in the feed should be examined if litter nitrogen levels are high. At these times, the moisture content of the litter is also often found to be high.

5.1.4. Drinker Design and Management

Of all the factors that affect litter moisture, probably the most important is the design and management of the drinkers. This is not surprising when one considers that a poor drinker design will waste as much as 750 ml of water per broiler over a 49 day-growing period.

It is essential for drinkers to be at the optimum height for the birds. This too reduces wastage. A nipple or cup-drinking system at the correct height can lead to up to a 7% reduction in litter moisture. A ring of capped litter is often found under bell drinkers. Nipple drinkers can reduce litter moisture by up to 7 per cent.

5.1.5. Air-Change Rate and House Environment

In poultry houses, three factors have to be considered together, because their control is interdependent. They are the environmental temperature, ventilation rate and humidity. Ideally their control should be interlinked.

The humidity of the poultry house environment is affected by the number and size of the birds and therefore by their respiratory output and also, of course, by the relative humidity of the air being drawn into the house by the ventilation system. When the relative humidity in the house exceeds 70%, the moisture content of the litter tends to increase, leading to poorer conditions. The aim should be to maintain a relative humidity level in the house of between 50 and 70% by supplying sufficient air and added heat when necessary.

The ventilation rate must always be maintained at a level sufficient to ensure that ammonia does not approach the threshold level of 25 ppm. In cold weather this may necessitate increasing the heating levels within the house.

It is important to prevent cool moist air from falling to the litter which is expectable especially in temperate countries. With the correct controls and inlet design this can be achieved.

On the same lines, wherever extremes of winter are routine, roofs and walls of poultry houses must be adequately insulated to prevent condensation. Insulation with a U value of $0.4 W/m^2$ °C or better is necessary. To prevent deterioration of the insulation, a vapor seal between the birds and the insulant is essential unless self-sealed insulation is used. Likewise cool water pipes and tanks should be lagged and the dwarf walls should be insulated. Ideally the concrete floor of the house should have a waterproof membrane to prevent rising damp.

5.1.6. Litter Material and Depth

The most effective litter material for poultry is wood shavings. The most commonly used alternative is chopped straw. The action of chopping the straw makes it more absorbent. Recently introduced pelleted, dried and treated products are now available and these are used at a lesser depth than the usual 5 cm of wood-shavings or 5 - 10 cm of chopped straw. In general, neither the choice of material nor its depth has a consistent effect in preventing the problems associated with poor quality litter. With friable litter, increasing the depth can obviously improve the ratio of litter to faeces. Poultry litter that is friable has a temperature that increases to about 27°C. This warmth is due to bacterial breakdown of the faeces. When it is capped and cold the litter material may need 'topping up' with fresh litter so that direct contact with the wet cap can be avoided and the welfare of the birds enhanced. An extreme example of this occurs with ducks where frequent topping up is essential because of their very wet faeces.

5.1.7. Stocking Density

Floor space recommendations for different species of poultry are given in Chapter "Management requirements and specifications". For table chickens, maximum stocking density recommended is 34 kg/ m² and for adult-laying birds not more than 7 birds / m².

Most of the water, fat and nitrogen found in the litter have been excreted from the birds as faeces. Therefore the higher the density the more of these factors the litter has to absorb. The rate of evaporation of moisture from the litter falls as stocking density increases, so particular attention must be paid to litter quality as the birds approach killing weight.

5.1.8. Nutrition

Any dietary factor that makes the birds increase their water consumption is likely to lead to wetter litter and therefore higher risks to the birds' welfare. The following factors should not exceed their optimum levels:

1. Sodium and chloride: use sodium bicarbonate as well as salt
2. Potassium: beware of molasses, manioc, excessive Soya bean meal
3. Crude protein and amino acids: as well as increasing the moisture content of the faeces, excess nitrogen is excreted. This increases the severity of burnt hocks.
4. Poorly digested feed ingredients can lead to wetter litter, higher litter nitrogen and therefore an increased welfare risk. For this reason protein quality and the amino acid balance must be optimal. Enzymes such as â-glutamate can aid the digestibility of cereals in the feed.

5. Added fat is an essential ingredient in achieving adequate dietary energy levels for poultry meat production. The quality, composition and quantity of the fat have a direct effect on the level of the fat in the faeces. High fecal levels can lead to the litter capping at lower moisture levels. The fat blend must have optimal ratios of saturated and unsaturated fats. Saturated fats are less well-digested especially early in life.

It is vital that those responsible for formulations realize that the effects of each of the possible nutritional anomalies are additive.

5.1.9. Flock Health

There are several infectious and non-infectious diseases and conditions that may increase the severity of hock burn, pododermatitis and breast blisters.

Any disease or skeletal abnormality that reduces the birds' mobility is likely to affect their welfare adversely, as they will have increased contact with the litter.

Enteritis and disorders such as the so-called malabsorption syndrome can result in an increased excretion of water in the faeces. Likewise, infectious bursal disease (Gumboro Disease) is well known to result in extremely wet and foul-smelling litter almost overnight.

Further, not all hock burn or breast blisters are simply a result of poor litter quality. If birds spend excessive amounts of time squatting down due to leg problems or other diseases they will be more likely to suffer from these lesions regardless of litter condition.

5.2. Conditions associated with poor litter quality

A poultry flock kept on well-maintained litter is healthier and more profitable than one kept on poor quality litter. Diseases such as pododermatitis, hock burn and breast blisters are all a consequence of poor litter (Fig. 41.1A to 41.1D). In turkeys the so-called "shaky leg syndrome" sometimes follows foot ulceration associated with poor litter quality. These disorders cause unnecessary suffering to the birds and can also result in downgrading of the end-product at the slaughterhouse.

  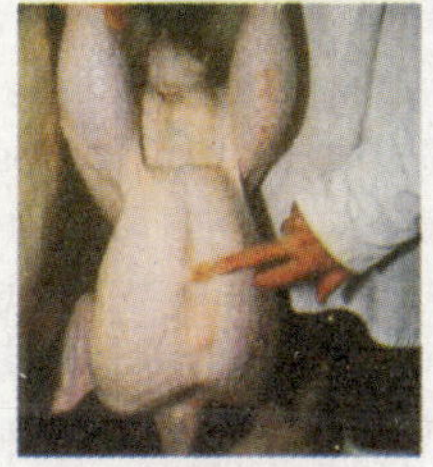 

A. Pododermatitis B. Pododermatitis C. Breast blisters D. Burnt hocks

Fig 41.1 Conditions related to poor litter management

(*Source* : http://www.defra.gov.uk/animalh/welfare/farmed/meatchks)

Chapter **42**

Cage Management

Notwithstanding new welfare-promoting rearing systems becoming popular in European countries, USA, Australia and other advanced countries, cage rearing is till popular in many other countries which are contributing significantly to the World's poultry production. Therefore, for some more time to come, cage system will definitely be utilized by several producers. It is for this reason, some salient features of cage-rearing poultry is presented in the forthcoming pages. It is necessary to note that general principles of rearing, feeding, lighting and other management are essentially the same as in case of other systems of rearing already discussed. However, in this Chapter, rearing chicken entirely in cages from day-old till the end of production period will be discussed.

1. House

Dimensions of the house (brooder house or grower house or brood-grow house) depends on the number and dimensions of the cages to be housed in it. As many decked cages are common, stocking density will be very high compared to any of the floor systems of rearing; hence, extreme care has to be exercised in providing for ventilation requirements. This aspect may be troublesome in case of open-sided houses under hot-humid climates. In environmentally controlled houses, special care has to be taken while calculating inlet and exhaust details for the house.

2. Brooding cages

Brooding is done in a battery-brooding cage (See Chapter "Brooding") made of 4-5 tiers, each 31 to 41 cm high. The size of the tier can be 55.9 cm X 61.0 cm or 61.0 cm x 61.0/68.6/91.4 cm. The floor is usually made of welded mesh 1.3 cm x 5.1 cm or 2.5 cm x 2.5 cm, 14 gauge; in the latter case, the floor has to be covered with rough paper (corrugated paper in case of Japanese quails) during the first 2 weeks of brooding to prevent legs of the chicks getting entangled. Either plastic meshwork or plastic-covered wire mesh are being used for durability. There is no need to provide slope to the floor. Other details of heating, feeding and watering arrangements are as described under "Brooding".

Paper on the floor during brooding serves for the following:

1. Solid floor when the chicks are young

2. Mesh work with larger openings can be used so that the same equipment is useful for growing birds also
3. Provides area for sprinkling first feed to the chicks
4. Keeps the brooding area warmer
5. Facilitates free movement of chicks on the mesh floor

3. Growing cages

Brooding cage can be used for growing Leghorn-type pullets as well. However, separate growing cage specifications are also available. They will be of the same size as that of the brooding cage but, the floor is covered with wire mesh of larger size to allow droppings to fall through easily into the dropping tray below.

4. Brood-grow cages

The chicks are reared till they are to be moved to laying cages in these cages which are larger than brooding cages.

4.1. Advantages

1. The birds are moved only once into the layer cages and hence, minimum stress on them.
2. Downtime (time for which the equipment is kept empty)is shorter and hence more profitable.
3. Less labor required and even beak-trimming can be combined with transfer of birds into laying cages.

4.2. Disadvantages

1. When ambient temperature is low, the cage and the house where it is located are cooler during brooding due to the large size of the cage unit.
2. Heating units are kept idle during the growing period.

The brood-grow cage units are available as single-, two- or three-deck units. Single deck units are not very popular mainly due to greater housing cost. In both two- and three-deck units, heating element is installed in one of the decks and the chicks are uniformly thinned-out after 4-6 weeks of age.

Droppings fall into the dropping board below which are suitably inclined so that droppings fall on to the floor below; the floor need not be made of concrete although concrete flooring helps move the cage unit and scrape the floor efficiently.

5. Space and heat requirements

See Chapters "Brooding" for brooding temperature requirements and "Management requirements and specifications" for space requirements.

6. Light

Provision of uniform distribution and intensity of light is difficult in cage system, in general, and multi-deck cages, in particular. Shadows produced by the cage frame complicate the problem further. In open-sided houses, natural light easily supports feeding and drinking but in environmentally-controlled houses, light control is a major challenge to the producers. Lighting program generally prescribed for brood-grow period is as follows:

1. First 4 d – continuous light, intensity 35 lx at bird level to help chicks learn to eat and drink
2. After 4 d – lighting program depending on predetermined schedule (constant-day-length or decreasing day-length or any other depending on the producer) till sexual maturity. Intensity of light suggested is 5 lux for single-deck cages or 3 lux at the location of lowermost deck in multi-deck cages. Rheostats are advisable to facilitate adjusting intensity of light as and when required.

All other management procedures like beak-trimming, watering, feeding, feed restriction etc. are similar to those for rearing birds on any other floor-rearing systems.

Lights are generally arranged on the passages at about 2.1-2.4 m height and 3.15-3.60 m apart. Incandescent bulbs are preferred because they are easy to install, clean and are most efficient in stimulating egg production.

7. Laying cages

Advantages and disadvantages of cage system for rearing chicken are already given in Chapter "Rearing systems"; discussion on alternate systems of rearing is also provided. In ay case, laying cages are still very popular in many countries and they are likely to continue to be so for some more time.

Performance of layers is influenced by a) Cage area/bird and b) colony size.

7.1. Cage area/bird

When the cage area per bird is reduced that is, when overcrowding occurs, they will be increased mortality and reduction in egg production (both hen-housed and hen-day), shell quality and net profit. It is recommended to calculate the cage floor space requirement as 255 cm^2 per kg body weight of the pullet. However, commercially, some amount of overcrowding may not reduce the profits, although the exact degree of overcrowding is not determined. Few believe that the amount of feeder space available to the bird is a better criterion than the cage area/bird.

7.2. Colony size

Keeping the cage area allowed per bird as constant, size of the colony (number of birds/cage) seemed to affect the performance of birds. When the colony size was increased, the birds were heavier at the end of laying period and there was a lower incidence of soft-shelled eggs. However, the number of broken eggs and incidence of flightiness increased. Therefore, it seems reasonable to give more weightage to cage area/bird rather than colony size especially when the number of birds per cage is less than five to ten.

Conventional laying cages measured 40.6 cm high at the rear side with variable floor dimensions depending on number of birds it can accommodate. Some of them even had watering channel at the back and feeding channel in front; this was adding problems for lighting both feed and water especially in multi-deck cages.

7.3. Reverse (Shallow) cage

When cages for 2, 3 or 4 birds are reversed by making the longer sections of the cage across the front rather than of the side (Fig. 42.1), the layers have been found to perform better. More birds can be accommodated and about 50% more feeder space is available for the birds and hence reduces competition for feed.

It seems possible that egg production improves by 2 to 3% without alterations in egg size, mortality and incidence of cracked eggs; the latter due to reduced depth of the cage thereby reducing the roll-out distance. There will be reduced incidence of cannibalism because the birds cannot turn completely but have to stand side by side. However, reverse cage involves higher housing and cage cost and feed consumption per dozen eggs is also increased. In any case, the advantages outweigh the disadvantages and hence, reverse cages have become extremely popular.

Some cage arrangements are shown in Fig. 42.2.

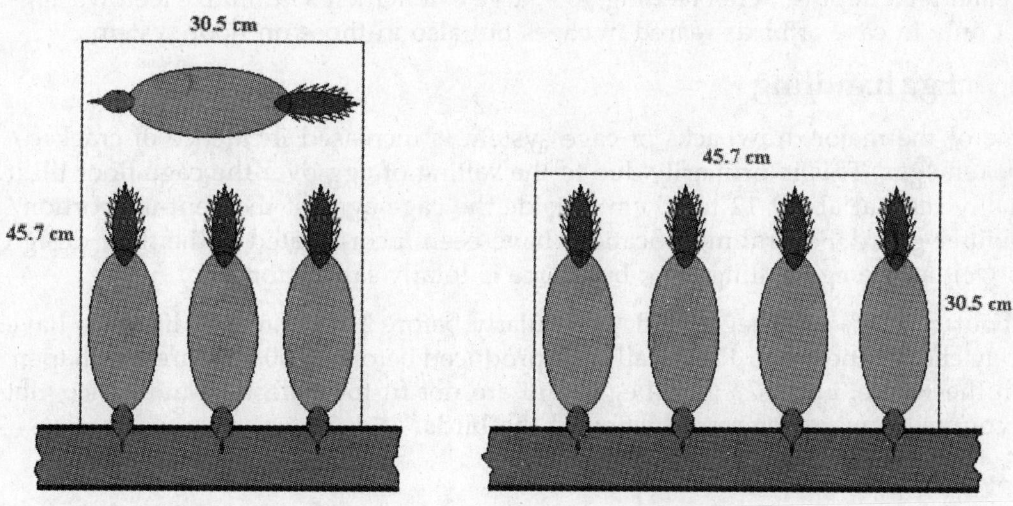

A. Conventional cage B. Reverse cage
Fig 42.1 Conventional and reverse cages

(Moreng and Avens, 1985)

7.4. Feed and water troughs

Laying cages conventionally have feeding and watering troughs; the latter above the former to facilitate drinking by use of gravity. It is obvious that a slope is required in water troughs fed by running water; the suggested slope is 7.6 cm per 30.5 m length of the watering channel. This works out to be 1.43° from the horizontal. No slope is needed for feeding channel.

With nipple drinkers becoming popular, problems of leaking watering channels as well as maintaining proper water levels throughout the length of the house have been minimized.

It is necessary to note that caged layers drink ¼ of their daily water consumption during the 2 hr immediately preceding the time the lights go off or the Sun sets regardless of the house temperature. During the remainder of the day, hourly water consumption is fairly constant. Further, caged layers consume more water than those on floor systems and hence, wet-litter is always a problem. Intermittent watering does help reduce wet-litter but, it is not advisable to restrict water during and prior to peak production. The problems further complicate during hot weather accentuating by high humidity conditions.

Feeding troughs should never be filled to the brim; increasing the frequency of feeding not only reduces feed wastage but also stimulates higher feed intake. Excessive feed depth stimulates the birds for picking and scratching in an attempt to search for grains; this is particularly applicable with mash type of feed and birds with fully grown beaks. However, poorly beak-trimmed birds may require greater feed depths. Pellet feeding, to a large extent, helps minimize feed wastage not only in case of birds reared in cages but also in those on floor systems.

8. Egg handling

One of the major drawbacks of cage system is increased incidence of cracked/ broken eggs. This is primarily due to the rolling of egg over the cage floor till it finally rests at about 12 to 15 cm outside the cage against the bent-up portion/ bumper guard. Several modifications have been incorporated to the floor design as well as the egg rolling area; but, none is totally satisfactory.

About ½ to 3% of all eggs laid, particularly before 10.00 a.m. are likely to have body checks and up to 15% of all eggs produced before 10.00 a.m. are misshapen. On the whole, up to 7% of all eggs laid are not fit for collection and hence, not accounted towards egg production of the birds.

8.1. Cracked eggs

Eggs can be gathered either manually or by automatic egg collection systems. Manual collection may be difficult when the number of layers is more than 30,000 unless as attendant is specially assigned the work; the latter increases labor costs. Cracking of eggs does occur during the collection process also; it may go up to

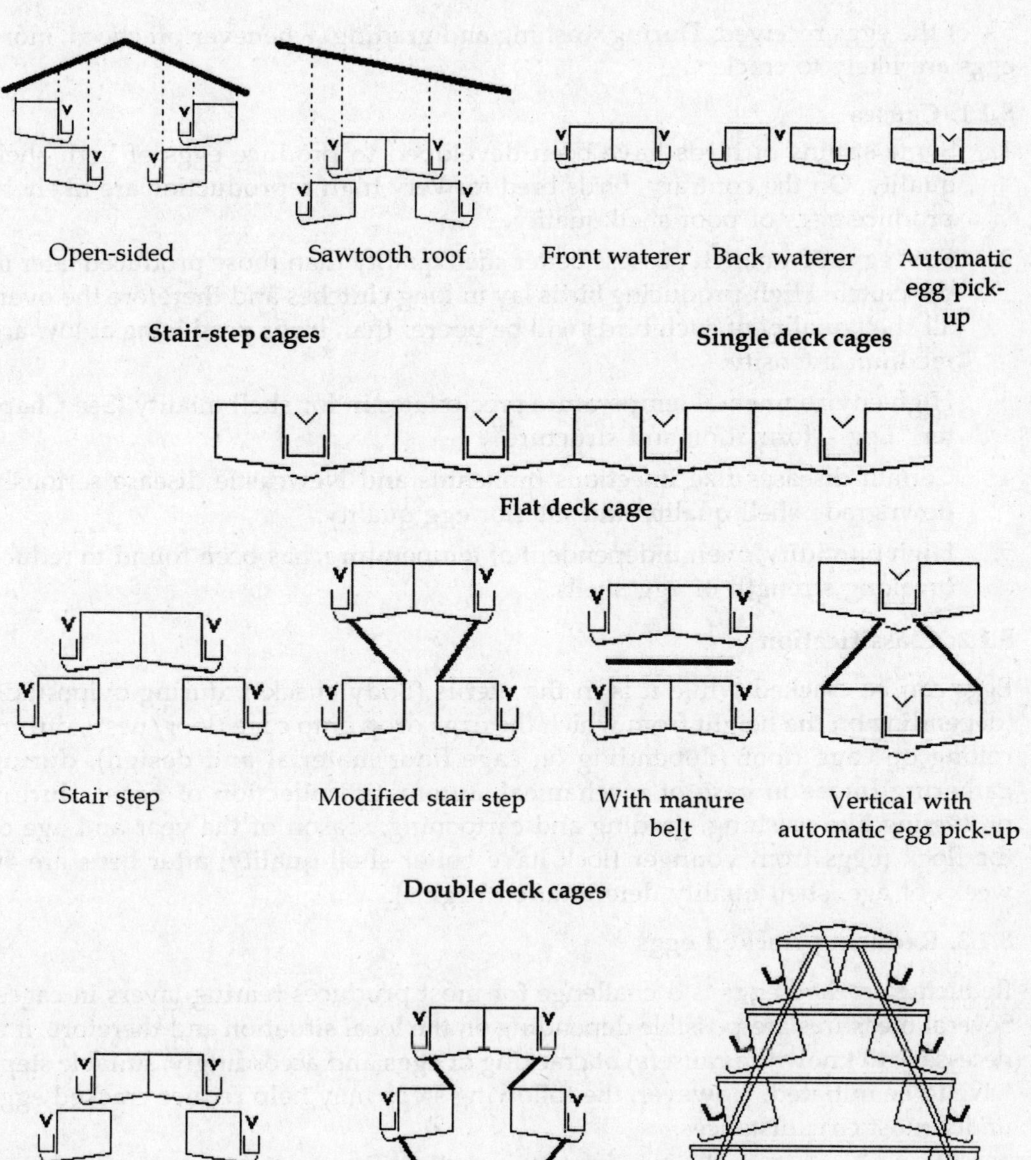

Fig 42.2. Cage arrangements (North and Bell, 1990)

5% of the eggs received. During washing and grading, whenever practiced, more eggs are likely to crack.

8.1.1. Causes

1. Some strains of birds have been developed to produce eggs of high shell quality. On the contrary, birds bred for very high reproduction are likely to produce eggs of poor shell quality.
2. First eggs of a clutch possess better shell quality than those produced later in the clutch. High-producing birds lay in long clutches and therefore the overall shell quality of such birds will be poorer than birds producing at low are medium intensity.
3. High environmental temperature precipitates in for shell quality (See Chapter "Egg - formation and structure")
4. Certain diseases like infectious bronchitis and Newcastle disease seriously downgrade shell quality and interior egg quality.
5. High humidity, even independent of temperature, has been found to reduce breaking strength of egg shells.

8.1.2. Classification

Eggs can be cracked while it is in the uterus (Body checks), during oviposition (depending on the height from which the eggs drop onto cage floor/nest), during rolling on cage floor (depending on cage floor material and design), during gathering (more in case of mechanical/automatic collection of eggs), during processing like washing, grading and cartooning, season of the year and age of the flock (eggs from younger flock have better shell quality; after hens are 40 weeks of age, shell quality deterioration begins).

8.1.3. Reducing cracked eggs

Reducing cracked eggs is a challenge for most produces rearing layers in cages. Several measures are possible depending on the local situation and therefore, it is necessary to know the cause(s) of cracking of eggs and accordingly, suitable steps have to be initiated. However, the following steps may help reduce cracked eggs under most circumstances:

1. Purchase of strain known for better shell quality.
2. Careful handling of eggs particularly during latter part of egg production.
3. Attendants must be properly trained and their efficiency regularly monitored by candling the eggs collected by them separately.
4. Cushion bumper must be provided at the front of the egg collection area of the cages.
5. Reducing cage density does help on some occasions.
6. Frequent collection of eggs particularly during summer months and from older flocks top

7. Eggs have to be collected on filler-flats and not more than 6 flats must be stacked one above the other. If baskets are used for collection of eggs, they should not be more than half full.
8. Most of the birds are producing eggs of poor shell quality, feed quality should necessarily be checked.
9. Any type of stress, cannibalism, fright can lead to cracking of eggs; hence, they have to be avoided.
10. If automatic egg collection is practiced, it must be regularly checked for belt material, speed, angling and corner devices.
11. During egg-laying period all house activities must be avoided.
12. If egg washing is practiced, the machinery should be regularly checked and personnel involved properly supervised.

9. Problems in caged layers

Problems in caged layers include cage layer fatigue, fatty liver syndrome, prolapse etc. The former two, being metabolic disorders, will be discussed in Chapter "Poultry diseases – metabolic"; prolapse, on the other hand, is more a management-related problem.

Prolapse is evagination of oviduct (vagina) and rectal region through vent; that is invariably followed by pecking leading to irreversible damage. Salient features of prolapse are:

1. More prevalent in chicks hast in spring coming to lay during June to August – maybe because of increasing day lengths there are subjected to.
2. Early-maturing and heavier birds are more susceptible.
3. High-producing birds subjected to improper management.
4. The duration of light is increased during growing period and hence, birds matured earlier without proper physical development predisposing such birds to prolapse; particularly, the lower strength of abdominal muscle expectable in early-maturing birds.
5. If sufficient nests are not provided, birds lay eggs on floor, get infected on to the oviduct during oviposition.
6. If very big egg has to be laid, prolapse may occur.
7. Overcrowding of birds leads to competition for space, feed and water.
8. Birds which have enteritis are alsoproned to prolapse due to excessive bowel movements.
9. Nutritional deficiency of calcium can also cause prolapse.

The following precautions can be taken avoid prolapse:

1. Debeaking of all birds.

2. Smaller birds may be removed and reared separately.
3. Proper floor space depending on age of birds.
4. Birds have to be weighed at least once a week or fortnight to calculate food requirements.
5. Balanced poultry feed has been offered to birds.

Chapter **43**

Molting : Natural and Induced

1. Natural molting

(Ellis, M.R. on internet)

Molting is the process of shedding and renewing feathers. During the molt the reproductive system of the bird is allowed a complete rest from laying and the bird builds up its body reserves of nutrients. Normally, under natural conditions, molting in adult birds will occur once a year, though it may occur in certain individuals twice in one year, and more rarely only once in a period of two years.

1.1. Before sexual maturity

The chick goes through one complete and three partial molts during its growth to point of lay, after which the mature bird normally undergoes one complete molt a year, usually in autumn although this depends on the time of the year at which the bird commenced laying. Generally, complete molting occurs from 1-6 weeks and partial moulting at 7-9 weeks, 12-16 weeks and 20-22 weeks; during latter molt the stiff tail feathers are grown.

1.2. During laying period

Natural molting usually begins sometime during March-April and should be completed by July when egg production recommences. The three main factors which bring about molting are:

1. Physical exhaustion and fatigue
2. Completion of the laying cycle.
3. Reduction of day length, resulting in reduced feeding time, and consequent loss of bodyweight.

Eleven months continuous production is expected from pullets hatched in-season (March 1 to August 31), so that if a flock of pullets commenced laying in March at six months of age, they should continue laying until the following February, although the odd bird may molt after laying for a few weeks. These few birds, however, should begin laying again after June 22 (the shortest day of the year) and continue in production until the following autumn. Pullets coming into lay in June should lay until the following April thereby giving eleven months continuous egg production without the aid of artificial light. Pullets coming into lay in spring

(August) should lay well into April (9 months) but unless artificial lighting is provided, most of them will molt during May and June.

1.3. Good layers and moulting

The time at which a laying hen ceases production and goes into her molt is a reliable guide as to whether or not it is a good egg producer. Poor producing hens molt early (November-December) and take a long time to complete the process and resume laying from six to seven months; they seldom cast more than a few feathers at a time and rarely show bare patches.

High producing hens molt late, molt for a short period (no more than 12 weeks) and come back into production very quickly. Rapid moulting is not only seen in the wing feathers of good producers, but also in the loss of body feathers generally. Because of this, it is common to see a late and rapid moulting hen practically devoid of feathers and showing many bare patches over the body.

1.4. The molting process

Molting takes place in a fairly definite order. Feathers are confined to definite tracts or areas of the body surface, with bare patches of skin between. The first plumage is lost from the head and neck, then from the saddle, breast and abdomen (body), then the wings and then from the tail.

While the first feathers are being dropped from the neck and body, good layers will often keep laying, but when the wing feathers begin to drop, laying usually ceases.

The main wing feathers consist of four tiny finger feathers on the extreme tip of the wing, then 10 large primary or "flight" feathers, the small axial feather, and the 14 secondary feathers, which are smaller and softer than the primaries (Fig.43.1).

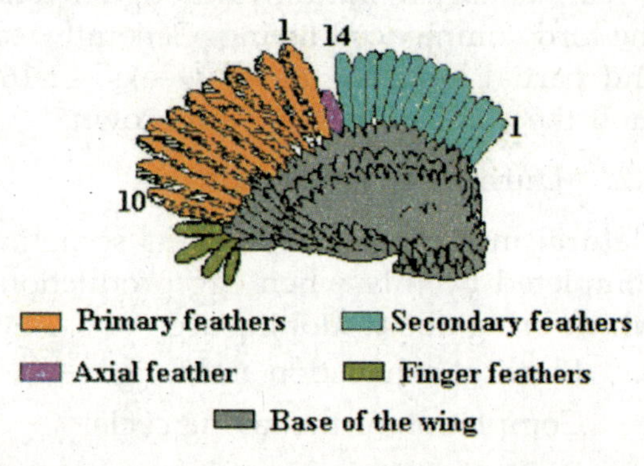

Fig. 43.1. Feathers on the wing

When the wing molts, primary feathers are shed first, from the axial outwards to the end of the wing, and then the secondaries, which are not shed in such a set order as the primaries. The axial feather is dropped at the same time as the secondary next to it. The new quill starts to grow as soon as the old feather is out and takes approximately 6-7 weeks to grow.

The molt is complete when all primary flight feathers on the wing are replaced. The feathers of the molted bird are large and full, softer, cleaner, brighter and glossy in contrast to the feathers before molting which were small and hard, dry, frayed and tattered.

1.4.1. Rapid and slow molters

The difference between a rapid and slow molter is not due to a difference in growth rate of the individual feather, but because the rapid molter renews a large number of feathers at the same time. With this knowledge, the rate of molting can be ascertained by examining the number of flight feathers on the wing being replaced simultaneously. If a hen is found to have grown some of her primaries before starting to molt her secondaries, it may be assumed that she laid well into the moult and was therefore a good layer (See Chapter "Judging pullets").

Sometimes, high producing hens do not molt all their primary feathers but carry them on for another year. Generally a layer molts when production ceases although if the bird has an inherited tendency for high production, molting will probably precede cessation of production, and *vice versa* if it is a poor producer. Modern laying breeds should molt in late autumn because they have been bred specifically for egg production i.e. to lay at a higher rate and for a longer period of time.

1.4.2. Vacation molts, neck molts or partial molts

Old feathers are usually retained by a laying bird which lays regularly. If it ceases production for any other reason than for mild sickness or broodiness, it will molt.

If a hen ceases production during spring or summer, it may molt one or two primaries, then stop molting and come into lay again. This is known as a vacation molt. When it starts its full molt later in the autumn, it will drop the next feather in sequence and molt in order of the remaining primaries. A neck or partial molt is sometimes experienced by a bird without any loss of production, but if the molting extends beyond the neck molt stage (to wings) the hen ceases production.

The presence of "pin" feathers (new emerging feathers) usually indicate a short or partial molt.

Some birds molt continuously and can be easily detected in the flock by the spotless condition of their new feathers. These birds are poor producers and should be culled.

1.5. Stress factors and molting

Natural molts can occur any time of the year due to birds being subjected to stress. A bird is stressed when the environment or management present a challenge to which the bird cannot respond without suffering a harmful effect. A hen subjected to a mild stress condition in late spring when in full production will suffer a drop in egg production whereas the same stress condition applied to a bird in the autumn will lead to cessation of lay and molt.

The following are common stress factors which can induce molting: lighting (decreasing daylight, decreasing artificial light), loss of bodyweight, disease, internal parasites, climate (excessive cold, heat waves), feed, feeding and feedstuffs (deficiencies of essential ingredients, irregular feeding, insufficient feed), predators, fright, peck order, prolonged broodiness, mismanagement (overcrowding, movement to another house, water deprivation, insufficient feed and water space, faulty ventilation, wet litter, debeaking, vaccinations, exposed housing, etc.)

2. Induced molting - concept and effects

Because of increasing economic pressures, the commercial egg industry must make maximum use of its resources. High interest costs and the need to lower production costs have led many enterprises to use induced-molting programs.

An induced molt causes all of the hens in a flock to go out of production for a period of time. During this time regression and rejuvenation of the reproductive tract occurs, accompanied by the loss and replacement of feathers. After a molt, the hen's production rate usually peaks slightly below the previous peak rate, and egg quality is improved.

Induced molting can extend the productive life of a flock to an age of 105 weeks and its timing has to be adjusted as part of a total profit plan that maximizes egg production over the life span of the hens and matches periods of highest egg production to periods of highest egg prices.

The purpose of an induced-molting program is to rejuvenate the reproductive system of the hen. For complete rejuvenation and optimum post-molt performance, the reproductive tract must experience complete regression; that is, egg production must completely stop. Complete regression results in the flock being totally out of production for 14 to 17 days. Because weight loss is closely associated with reproductive tract regression, body weight is closely monitored throughout a molt.

2.1. Advantages

The advantages of keeping hens during the molt and the following year are as follows:

1. It is cheaper to carry a bird through a moult than to buy replacement pullets
2. Fewer replacement pullets may be needed, and buying can often be deferred, which can mean a saving of money, time and transport
3. Moulted birds are hardier, and not as prone to disease
4. If strict culling is carried out during the first year, only high producing, efficient birds will be retained.
5. To tide over low price periods and obtain eggs when high prices are likely to exist.
6. Economic problems in maintaining egg production at low prices of eggs.

7. Larger eggs produced when the hens resume production.
8. If replacement stock is not available, the same flock can be recycled.
9. Egg production throughout the year can be achieved by purchasing pullets at point-of-lay in the autumn to provide sufficient eggs while the older birds are molting.
10. To the egg producer, a major advantage of induced molting is the reduction in time that the house is not producing income. Because flocks are replaced less frequently, the laying house will be empty less often; in other words, down time is reduced significantly.

2.2. Disadvantages

The main disadvantages of molting are:

1. Although molted birds eat less feed than pullets, they also lay less eggs. Overall, their conversion of feed into eggs, and feed cost per dozen eggs is higher.
2. During the moult the birds continue to eat but remain unproductive
3. If the birds are to be slaughtered for the table after two years of laying they will not be as tender to eat
4. Too few birds may be retained to provide sufficient eggs the following year.
5. Cost of feeding during molting may sometimes be more than the extra profits expected in second cycle.
6. Lower egg production. After molting, the second year of egg production will be between 10 and 30% less than that achieved by the birds in their first year of lay. Birds which have molted twice and are laying for their third year will lay only 70 to 80% of their second year's eggs i.e. about 60% of their first year's production.
7. Poorer interior quality of the egg than in first cycle.

2.3. Effects

2.3.1. Egg size

For the egg marketer, induced molting offers a means of matching the egg supply with market conditions. Molted flocks produce a higher proportion of larger eggs than do first-cycle flocks.

Periods when high prices or premiums are paid for the larger eggs (usually during the summer) favor molted flocks. Periods when the pricespread between medium and large eggs is small (usually in the winter and spring) favor production from pullet flocks. Regardless of the time of year, if flock placements and molting schedules can be adjusted to take advantage of anticipated market conditions, molted flocks can produce greater returns per hen than single-cycle flocks.

2.3.2. Hen depreciation

An even greater benefit of induced molting is that it reduces hen cost per dozen eggs because it lengthens the productive life of the hen. On the average, the cost of a pullet is spread over 240 eggs, and a replacement flock is purchased annually. A molting program allows the pullet cost to be spread over an average of about 370 eggs, and replacement flocks are purchased less often; typically, three times in five years. This difference may reduce egg production cost substantially if pullet costs are high. Even when pullet prices are low, the savings are possible although to a lesser magnitude.

2.3.3. Feed efficiency

In the case of induced molting, a major disadvantage is that it leads to poorer feed efficiency.

The amount that feed efficiency will drop depends on several factors, including strain, season, equipment and housing type, nutrition, and whether the molting technique is applied properly. If feed efficiency for a first-cycle flock is calculated by including the feed needed to grow a pullet and for the molted flock by including the molt feeds, then the molted flock will have a better overall feed efficiency. If, however, feed efficiency is calculated from 50 percent production for first-cycle and for molted flocks, molted flocks have a poorer feed efficiency. Obviously, the latter is logical and acceptable.

2.3.4. Egg production and quality

The most noticeable effect will be a reduction in the total number of eggs sold (about 7%). This difference will obviously result in lower egg income unless specific arrangements are made to offset it. Interior egg quality also is lowered due to induced molting.

3. Induced-molting methods

There are many methods of inducing a molt in laying hens. Any method employed should induce sufficient stress. It can be achieved by feed and/or light and/or water withdrawal. All programs involve feed withdrawal and restriction. Normally, feed is withdrawn in the beginning and later on, only grains are offered to create nutritional stress. Most of the programs also include withdrawal of water for 1 or 2 days with care not to withdraw water during summer season when temperature is above 35°C. It is very difficult to reduce light to 9 to 10 hrs/d in open-sided poultry houses with the common in our country. However, artificial light can be discontinued.

Feed and light restriction/withdrawal is effective in summer, whereas, water along with feed or light withdrawal/restriction is ideal during winter.

Induced molt by feed restriction is, by far, most popular due mainly to its ease in application. Hence, this program is outlined below. However, many other programs are in vogue involving water with or without feed restriction, with or without light restriction. They are more practicable in temperate climate (water restriction) and in environmentally controlled houses (light restriction).

3.1. Types

1. Two-cycle: One molt and two cycles; force-molting done after 10 to 12 months of lay and brought back to lay for another 6 months.
2. Three-cycle: Two molts and three cycles; first molting after 9 months of lay, second and third lay are successively of short duration. Hence, laying life of the flock is increased by several months. Interior egg quality as well as shell thickness improve in eggs produced in the third cycle; but, factors such as egg number, feed per dozen eggs etc. make this schedule less profitable. Hence, commercially, two-cycle programs are popular.

3.2. North Carolina molting program

(North Carolina University Website)

3.2.1. Factors to be considered

The following factors are essential to a successful induced molt: age of the flock, nutrition, lighting, flock history, house and equipment design, season of the year and variations among strains. Of these, the age of the flock has a profound influence on the success of an induced molt. Attempts to molt a flock less than 57 weeks old will be hampered by the hen's resistance to ceasing production and hence, some of the hens will not experience an adequate regression and rejuvenation of the reproductive tract. On the other hand, if the flock is more than 67 weeks old, the potential for restoring shell quality is greatly diminished, and the overall economic advantage of an induced molt is considerably reduced. The second-laying cycle of the flock should end at 100 to 105 weeks of age.

3.2.2. Phases

An induced-molting program consists of three phases: a pre-molt period, a period of fasting and weight loss, and a return to production after the fast.

3.2.2.1. Pre-molt Phase

3.2.2.1.1. Body weight sampling

The success of a molt depends on accurate body weight sampling. The pre-molt weight of the hens is one of the most important information in theentire program. One week before withdrawing feed, all the birds in a cage are weighed at several locations throughout the house. Sample cages should be selected from all decks, rows, and areas in the house. The cages so selected can be marked so that the same birds can be weighed subsequently.

3.2.2.1.2. Targeting body weight loss

The amount of weight loss necessary for complete regression of the reproductive tract depends on the pre-molt weight of the hens. Table 43.1 indicates target weight loss for various pre-molt weight ranges.

3.2.2.1.3. Lighting program

To induce the birds to stop lay abruptly, they should be "conditioned" by exposing them to constant light (24 hr per day) for seven days before withdrawing feed. The hens will then experience the maximum decrease in day length at the time of feed withdrawal.

Table 43.1. Weight-loss targets (%)

Pre-molt weight (kg)	Target
≤ 1.63	30
1.63 to 1.73	33
> 1.73	35

3.2.2.1.4. Pre-molt calcium

The addition of supplemental calcium to the feed during the final two days before the feed is removed improves the shell quality of the final eggs laid before production ceases. Adding 4½ to 9% oyster shell in addition to normal ingredients produces the best results. Alternately, oyster shell can be top-dressed in the house at the rate of 20 g/bird.

3.2.2.2. Fasting and body weight-loss phase

3.2.2.2.1. Monitoring weight loss

All the hens in the same cages as were sampled during the pre-molt period are weighed on the 7th and 9th day after feed withdrawal, the average weight loss per day calculated. From this estimated rate of loss of weight, it is possible to predict as to when the birds will achieve the target weight. Hens in the sample cages are weighed every other day until 2 d before they are predicted to reach their target weight loss. Then, they are weighed every day.

Many factors influence weight loss. For instance, a cool environment causes birds to lose weight more quickly and *vice versa*. Many factors such as temperature differences and other ventilation problems, feed equipment problems, or localized problems with parasites or diseases may cause poor uniformity in body weight. Rate of weight loss during a fast varies considerably with strain of the bird.

3.2.2.2.2. Livability

Livability should be more than 98% through the fasting period. There will be a notable decrease in livability as the flock approaches the target weight loss.

If the flock has experienced some sort of challenge (such as disease, exposure to mycotoxins, or environmental stress) that has significantly affected egg production or livability in the first cycle, livability during the fasting period may decrease below 98 percent. If this challenge was severe and very recent, it might be wise not to molt the flock.

3.2.2.2.3. Lighting program

Appropriate management of the lighting program for the flock is critical during the fasting and weight loss phase. The fundamental requirement is to provide constant or decreasing day length for 21 days after feed withdrawal. The best way to accomplish this depends on the house type and season.

3.2.2.3. Return to production

When the target weight loss has been achieved, the flock must be closely managed as feeding resumes and production begins.

3.2.2.3.1. Resuming full feed

Hens that have been fasted must never be returned immediately to full feed. For the first two days, only about 45 g feed per hen is offered to prevent severe crop impaction. After this adjustment period, the hens should be given full feed.

3.2.2.3.2. Nutrition during the recovery period

Before the onset of production, the hens must be fed diets that promote rejuvenation of the reproductive tract and maximize feather growth. Special diets have to be formulated to meet these requirements depending on availability of feedstuffs and feed prices.

3.2.3. Second-cycle performance

The success of an induced molt is measured by the performance of the flock during the second laying cycle. No exact standards exist for second-cycle production. It is most accurate to express second-cycle performance in reference to the first-cycle production of the flock. Generally, molted flocks will lay 60 fewer eggs per hen housed, have a hen-day egg production rate 15 percent lower, and peak 10 percent below first-cycle performance.

3.3. Conventional molting program

This method, also referred to as "On-again, Off-again program" involves a combination of feed, water and light restriction (Table 43.2).

3.4. California molting program

This method is suitable during periods of hot weather because there is no water restriction (Table 43.3). Feeding additional calcium during feed withdrawal is desirable. Production is likely to cease by 5th or 6th d and body weight drops by about 25% after 10 d. Body weight will be regained in about 7 weeks after resumption of feeding. Mortality has to be restricted within 1.25%.

Table 43.2 Conventional molting program for layers

Day	Feed	Water	Light
1-2	None	None	8 hr
3	Egg-type – 45 g/hen Meat-type – 68 g/hen	Given	
4	None	None	
5, 7, 9	Same as 3rd d	Given	
6, 8	None	None	
10 to 55-60	75% of full-feed intake	Given	
61	Full feed layer/breeder ration	Given	14-16 hr

Source : North and Bell, 1990

Table 43.3 California molting program (with water *ad libitum*)

Period (d)	Feed	Light
1-10	None	Discontinued or 8 hr
11-28	Full-feed cracked grains	
≥ 29	Full-feed layer mash	16 hr

Source : North and Bell, 1990

3.5. Other methods

In all the programs described above (The North Carolina program, conventional and California programs), feed and/or water and/or water withdrawal is an essential part which is in violation of Animal Welfare codes which are being strictly enforced in European countries, USA, Australia and other advanced countries. Therefore, soon it may be necessary that alternate method has to be developed or induced molting should be done away with.

Some alternate methods have been tried by manipulation of composition of diet with variable consistency and success. However, commercially, none of the methods has been convincing for inducing molt as are the methods involving use of chemicals/hormones.

However, certain dietary manipulations/drugs in feed have been tried; for instance,

1. High levels of zinc in the diet: Incorporating 2.75% of ZnO (73% Zn) in the diet (Zn content in the diet 20,000 ppm) which also has 3.5% Ca; the diet is fed for 5 d along with light restriction and then the birds are fed normal diet (0.005% Zn or 5 ppm) and provided regular lighting regimen. Feed intake normally falls by 80% and body weight reduces by 350 to 450 g. Cessation of egg production is recorded by 5th d of Zn feeding. Birds return to normal after 7 d of normal feeding and lighting regimen. Some of the sources of Zn are contaminated with Pb; care has to be exercised to avoid such sources (North and Bell, 1990)

2. Reducing Na levels in the diet to 0.04% and discontinuing artificial light in open-sided houses (8 hr light in light-controlled houses) will induce molt and the molt will be complete by 6 weeks. Subsequently, normal layer diet is provided for the birds. Most of the low-sodium diets are grain-rich and it is advisable to supplement high fiber and reduce Ca and P to that of grower ration. There is no feed withdrawal/restriction and hence, it is not violating the animal welfare codes.
3. Enheptin at 1000 ppm (0.1%) in feed causes cessation of the production in 10 days and return to production in 4 weeks.
4. Methallibure in feed at 10 ppm (0.001%) for two weeks will stop egg production in 1 week and returned to egg production in 2 months.

In the beginning of any of the above programs, birds stop laying eggs and mortality should be limited to as minimum as possible. Body weight may reduce by up to 20%. After molting, birds get rest of 3 to 5 weeks during which body weight is regained. After rest period, birds recover and reach peak production in 6 to 10 weeks of the new cycle.

Use of hormones has been banned in most of the countries.

4. Induced molting – breeders

(North and Bell, 1990)

4.1. Purpose

1. To supplement normal hatching egg production; in egg-type breeders, seasonal demand for day-old chicks calls for recycling flocks for a second cycle.
2. To compensate for high grower house mortality leading to reduction in supply of hatching eggs.
3. Unexpected demand for day-old chicks

4.1.1. Cockerels

Like hens, cockerels also molt, and while in this condition are nearly always infertile due to loss of bodyweight and because their reproductive physiology is undergoing a resting phase. Care must be taken to ensure that cockerels do not lose more than 25% of their bodyweight while molting as this can lead to sterility. Same program as that for females can be used.

4.1.2. Medium-size breeder hens

Any of the programs commercially used for egg-type hens can be used. Medium-size layers may take 1-2 weeks longer to return to egg production than a commercial hen.

4.1.3. Broiler breeder hens

Feed withdrawal for 10-14 d is needed for 25% weight loss and molt. Mortality should be restricted to < 3%. Weight reduction is very important to ensure adequate fat reduction in the reproductive system.

Feather renewal is mandatory, particularly for meat-type hens, to resume egg production and hence, a properly balanced rationhas to be given post-molt (Table 43.4).

Table 43.4. Minimum nutrient levels of recovery diets

Nutrient	Diet – I *	Diet – II **
Crude protein, %	16	17.5
Metabolizable energy, MJ/kg	11.75	12.00
Total sulfur amino acids, %	0.65	0.70
Lysine, %	0.80	0.95
Calcium, %	2.0	3.75
Available phosphorus, %	0.4	0.4
* Till 5% egg production ** From 5 to 50% egg production		
Source : North Carolina University Website		

Chapter **44**

Poultry Manure

Poultry manure production occurs as a result of the normal everyday processes of the poultry industry. It is a valuable by-product of this industry and has valuable potential uses beyond the traditional one of fertilizer.

It is interesting that the major product of any animal feeding system is manure, not animal protein. India produced 160 m layers and 1200 m broilers during 2004. Assuming that 20 layers or 300 broilers produce 1 ton manure, about 12 m tons of manure was produced. During 2004, egg production was 1.89 m tons and chicken meat output was 1.65 m tons; total of only 3.54 m tons (about 1/3 of total manure produced)!

Even when protein excretion is calculated, protein excreted through manure is more than in the form of eggs and meat. Assuming 11 and 18% protein content in eggs and chicken meat, respectively, total protein output during 2004 was about 0.5 m tons whereas, on the assumption that protein content (on as such basis) in poultry manure is 9%, total protein rejected through excreta is 1.08 m tons ! Hence, poultry droppings fit best as an example for the phrase, "a waste is simply a resource in the wrong place."

In other words, 60 to 70% of ingested nitrogen is excreted in droppings (Table 44.1) and it is worthwhile that systems have to be developed to utilize the same for other useful purposes. In addition, in today's mass production poultry farms is that alternative uses for poultry manure must be employed as most of these farms do not own enough land to simply use the poultry manure as fertilizer.

Table 44.1. Nutrient excretion (% of intake)

Type of chicken	N	P	K
Chicken broiler	61	69	80
Laying hen	70	68	87
Source : Agriculture and Agri-Food Canada, Poultry Section, 1990			

These large quantities of poultry manure production pose serious socio-economic problems, the most prominent of which is the protection of our environment and environmental resources.

Majority of poultry manure is produced in broiler and layer operations and there is a basic difference in the set-up of layer and broiler operations which leads to a difference in the type of litter produced.

Caged systems are used mostly for laying operations but their use is growing in the broiler industry as well. In the laying systems, the depth of the droppings pit will vary with the form in which the manure is handled. A shallow pit usually means a liquid type of flushing is used every few days while a deep pit means the manure is handled in solid form and need only be cleaned out once or several times a year.

Litter or floor systems are used mostly for broiler production. An absorbent litter material is usually laid down on the floor. The removal of this litter is handled in solid form and can be done after each brood or yearly or can be left for longer periods to add more litter and produce a "deep littering" system.

Table 44.2. Systems used in high-density, large volume operations

Dry Systems	Liquid Systems
High rise houses	Aerobic: Oxidation ditch, Surface aeration
In-house drying	Anaerobic: Soil injection, Anaerobic digestion
Dehydration	
Source : Agriculture and Agri-Food Canada, Poultry Section, 1990	

It is important to note that all of these systems have merit. A certain system may work for a particular operation but not for another operation because of certain circumstances, such as location, climate, size, land availability, crops, and markets.

A brief description of each of the systems is given below; for further details see A Review of Poultry Manure Management: Directions for the Future, Agriculture and Agri-Food Canada, Poultry Section, 1990.

1. Chemical Composition

The chemical composition of poultry manure isimportant in that it has very direct bearings on the environmental effects of poultry manure as well as the specific applications of poultry manure. Because it is so high in certain macronutrients, excessive land applications can lead to water pollution and soil toxicity.

Table 44.3. Gross composition of poultry manure (%)

	Water	Nitrogen	Phosphorus	Potassium
Fresh manure	77-80	1.0	0.9	0.5
Manure - Dry matter	--	5.0	3.9	2.4
Source : Agriculture and Agri-Food Canada, Poultry Section, 1990				

Chemical composition of poultry manure will vary because of several factors such as source of manure, age, feed and condition of animals, manner of storage and handling and litter used. However, average values are presented in Tables 44.3 to 44.6. It is clear from the Tables that poultry manure is very rich in terms of nitrogen, phosphorus, and potassium, as well as some other elements. Hence, poultry manure should not be considered a waste but should be considered a product which is nutrient-rich resource.

Before poultry manure can be applied in these beneficial ways, it must be processed in one of many various ways. These methods of processing dictate the kind of use the poultry manure will be subject to later on.

Table 44.4. Nutrient composition of manure (Average)

Composition, dry matter basis (%)	Dried poultry waste	Broiler litter
Crude protein	28.0	26.8
True protein	14.6	15.8
Digestible protein (ruminant)	12.6	22.6
NPN x 6.25	9.7	7.6
Ammonia x 6.25	-	5.1
Ether extract	2.2	2.4
Crude fiber	13.0	21.2
Nitrogen free extract	33.4	27.5
Total digestible nitrogen	52.3	58.9
Acid digestible fiber	24.7	30.4
Non-digestible fiber	52.4	47.4
Lignin	1.4	9.7
Gross Energy *	12.75	15.29
Digestible energy (Ruminant) *	10.28	10.21
Metabolizable energy (Poultry) *	5.48	-
Metabolizable Energy (Ruminant) *	7.95	6.81
Dry Matter (%)	81.7	80.6
* MJ/kg	*Source* : Agriculture and Agri-Food Canada, Poultry Section, 1990	

2. Processing Poultry Manure

There are several ways in which poultry manure can be collected and processed. Several factor such as operation size, climate, animal type etc. will determine what type of system is used in what kind of circumstances. It should be noted that in many instances, the strongest influence on which system is used is the economics (costs) of the system. Each system has its own merits and costs, but careful consideration must be used in order to select a system which will make the most efficient use of the factors in which it will be operated.

Table 44.5. Mineral composition of manure (Average)

Composition, dry matter basis	Dried poultry waste	Broiler litter
Ash (%)	27.6	18.6
Aluminum (%)	0.11	0.05
Calcium (%)	8.07	2.60
Chlorides (%)	0.87	0.35
Iron (%)	0.54	0.07
Magnesium (%)	0.50	0.39
Phosphorus (%)	2.29	1.81
Potassium (%)	2.24	1.78
Sodium (%)	0.60	0.38
Sulphur (%)	-	0.24
Arsenic (ppm)	1.5	4.1
Cadmium (ppm)	0.94	0.86
Chromium (ppm)	4.9	6
Copper (ppm)	66.0	50
Lead (ppm)	4.6	2.3
Manganese (ppm)	320	211
Mercury (ppm)	< 0.04	0.06
Selenium (ppm)	0.68	0.44
Zinc (ppm)	376	187

Source : Agriculture and Agri-Food Canada, Poultry Section, 1990

2.1. Anaerobic Processing

Anaerobic processing of manure occurs in almost all storage piles, pits and ponds. The idea behind anaerobic processes is that they occur in the absence of oxygen and they rely on the degradation of the manure by anaerobic bacteria. There are two basic types of bacteria involved in the process.

The first type convert fats, carbohydrates and proteins in the manure into simpler compounds and are rapidly reproducing bacteria that are not sensitive to environmental changes. They produce the highly odorous gases and volatile substances associated with ordinary manure storage units.

The second type is methane-producing bacteria which control odors and produce energy. They are small in number, reproduce slowly, and are generally sensitive to their environment, especially oxygen.

If the anaerobic process is functioning properly, the end products are methane, carbon dioxide, water, new bacterial cells, inert solids, traces of hydrogen and hydrogen sulfide, ammonia, water vapor and other gases. The two most widely used anaerobic systems for manure management are anaerobic lagoons and anaerobic digesters.

Table 44.6. Amino acid composition of manure (Average)

Amino Acid, dry matter basis (%)	Dry poultry waste	Broiler Litter
Arginine *	0.39	0.84
Cystine	0.06	0.22
Glycine	1.65	2.12
Histidine *	0.20	0.29
Leucine *	0.64	1.11
Isoleucine *	0.40	0.64
Lysine *	0.41	0.69
Methionine *	0.16	0.30
Phenylalanine *	0.38	0.64
Tyrosine	0.31	0.48
Valine *	0.52	0.88
Alanine	0.67	0.94
Proline	0.58	1.34
Glutamic Acid	1.33	2.66
Serine	0.52	0.76
Threonine *	0.45	0.67
Aspartic Acid	1.03	1.27
Tryptophan *	0.53	-
Total amino acids	10.23	15.85
Essential amino acids	3.63	5.39
Essential amino acids (% of total)	35.5	34.0

* Essential amino acid

Source : Agriculture and Agri-Food Canada, Poultry Section, 1990

2.1.1. Anaerobic Lagoons

In anaerobic lagoons, the bacterial activity reduces solids but often results in the production of odorous gases which makes a lagoon unsuitable except in isolated areas. Anaerobic lagoons liquefy and break down manure solids, but not all wastes are totally degraded. The unprocessed solids settle at the bottom of the lagoon and accumulate as sludge.

A lagoon has low initial cost and is easy to operate. Further labor can be saved by using irrigation to dispose of liquids. Long storage times permit pumping flexibility while bacteria break down solids, which results in a high degree of stabilization, resulting in a reduction of odors during spreading. The process also leads to a great reduction in the amount of nitrogen present, which is an advantage if the liquid is to be spread over a small area. This avoids large nitrogen leaching.

However, under low temperatures, the decomposition rate will be very low, which results in the filling of the lagoon with unstable solids which produce obnoxious odors. Often, because of poor design and management, these lagoons simply become holding basins. Another disadvantage to the anaerobic lagoon system is that the nutrient value of the liquefied manure is greatly reduced. Up to 80% of the nitrogen is lost in an anaerobic lagoon, while most of the phosphorus precipitates to the bottom and can only be recovered when the bottom sludge is removed.

With respect to poultry, an anaerobic lagoonis used anytime a water wash system is used in a pit system. If a liquid pitting system is to be used, the manure is flushed from the poultry house every 1-3 days. With a medium-depth pit, manure can also be diluted with water to form an "indoor lagoon" within the poultry house which is drained once or several times a year. In this case, special precautions must be adopted as the production of noxious gases and fumes can greatly affect the laying chickens.

In any case, if an anaerobic lagoon is being considered, there are special requirements. It should not be used close to living areas and should be located where space for expansion is available. Surface drainage should be prevented from entering and the lagoon contents should not be able to escape. A properly designed and managed anaerobic poultry lagoon presents an effective and cost-efficient way of treating poultry manure. But, on the other hand, when an anaerobic lagoon is properly functioning, it is wasteful of the manure nutrient content. In cases where high-density large volume poultry operations functionin areas where little agricultural land for manure spreading exists, the use of anaerobic lagoons is an economically viable and environmentally-adequate alternative.

2.1.2. Anaerobic Digesters

The second anaerobic processing system which is widely used for manure management is the anaerobic digester. The digester itself is usually a circular, airtight structure which varies in height and volume. A digester is equipped with various types of mixing and heating devices to keep the manure at 35°C. The primary motivation for constructing a digester and keeping the temperature around 35°C is to utilize poultry manure to produce biogas. In this case, the anaerobic digestion produces a biogas which is a combination of methane and carbon dioxide. There are three realistic options for using poultry biogas:

1. Use it directly for cooking, lighting, space heating, water heating, grain drying or gas-fired refrigerating and air-conditioning
2. Transform it into electricity by burning it in an engine that turns a generator; or
3. Vent it into the atmosphere.

In most digesters, a generator is used to produce electricity and the heat from the generator engine is used to maintain the 35°C temperature of the digester. Poultry manure also produces more biogas per unit live-weight than any other common manure.

There are several advantages to operating an anaerobic poultry manure digester. It is a stable and reliable process as long as the digester is loaded daily with a uniform quantity of waste; the digester temperature is kept constant, and antibiotics in the waste do not slow down biological activity. The process converts the biodegradable organic portion of poultry manure into biogas. The remaining semi-solid is relatively odor-free and retains all the nitrogen, phosphorus, and potassium of the original poultry manure which can be spread on agricultural fields.

There are several disadvantages to operating an anaerobic poultry manure digester, of which the cost of such a system is the most prohibitive. To be economical, digester capital cost must be offset by energy savings, fertilizer use, and bedding replacement.

Another disadvantage is that fresh poultry manure can be hauled as solid manure, but if digested, the dilution water would increase the original volume by about four times. In addition, the biogas produced has low energy per unit volume and can only be used for on-site operations such as running an electrical generator. Other problems associated with digesters include manure handling-pumping, grinding, mixing, and screening of miscellaneous debris.

A major problem encountered with poultry manure digesters is manure grit which must be removed by a settling tank or the digester will have to be cleaned on several occasions. Other problems are gas leakage (methane is explosive at 5-15% in air) and pipe and valve corrosion.

Although economically prohibitive, poultry manure digesters are environmentally friendly as well as possibly being a good on-site energy source.

2.2. Aerobic Processing

Aerobic processing of poultry manure requires the presence of bacteria that need oxygen in order to decompose organic matter. The decomposition occurs when a mixture of diluted organic wastes is supplied with oxygen. When these conditions occur, the aerobic bacteria use the diluted poultry manure as a food source in various biochemical and oxidation reactions to reproduce themselves. When the aerobic processing of poultry manure is functioning properly, the end products are: new bacterial cells, carbon dioxide, and, primarily, water. In reality, not all of the poultry manure will be digested aerobically and a certain accumulation of these stabilized solids along with fixed solids will occur.

2.2.1. Aerobic Lagoons

The aerobic lagoon works very much on the same principle as does an anaerobic lagoon except that the aerobic lagoon is aerated in some way. There are basically two types of aerobic lagoon. They are – the naturally-aerated lagoon (sometimes called an oxidation pond), and the mechanically-aerated lagoon.

The naturally-aerated lagoon is very similar in construction to an anaerobic lagoon except in depth. They are usually shallow (up to 1 m in depth) and bacteria and algae are expected to process the organic matter. The advantages of a naturally aerated lagoon are its flexibility, its capability to minimize odors, and its low initial costs. A disadvantage to the naturally-aerated lagoon is that, if winter is severe, it is subject to freezing, which completely stops the decomposition process.

The mechanically-aerated lagoon is also very similar in construction to the anaerobic lagoon. In this case, a mechanical aerator which is a pump or blower is designed to float in place in the lagoon.

There are several advantages to mechanically-aerated lagoons. There is certain flexibility with regard to existing building as well as fairly low initial costs. There is also a very large reduction in the total organic content and nitrogen content when compared to the original poultry manure. Along with these reductions, another benefit of the mechanically-aerated lagoon is the ability to control decomposition odors.

Mechanically-aerated lagoons are often used in poultry operations where odor control and land for application are at a premium. It has been found that in poultry operations over 90% of the organic content can be removed along with over 80% nitrogen removal. The large decrease in organic content is due to aerobic bacteria action while the high level of nitrogen loss is due to ammonia volatizing by the surface aerator.

In the case of both the naturally and mechanically-aerated lagoons, properly designed and managed systems are an effective and cost-effective way of treating poultry manure. The aerobic action in both types of lagoon removes most of the constituents out of the poultry manure which are useful in normal manure application. In regions where high density, large volume poultry operations function and a premium is placed on odor control and terminal land application, the use of aerobic lagoons is economically-inviting and is an adequate method of treating poultry manure.

2.2.2. Oxidation Ditches

The oxidation ditch is very similar in function to the mechanically aerated lagoon. It is an open channel pit shaped like a racetrack in which a paddle, brush type rotor or an air pump supplies oxygen to the liquid manure and keeps the liquid contents of the ditch in circulation. They are often used under caged floor systems for poultry.

2.2.2.1. Advantages

1. A properly designed, installed and operated-oxidation ditch can reduce odor production dramatically.
2. Similar to aerobic lagoon, the organic content and the nutrient content of the

poultry manure can be significantly reduced by the aeration of the ditch by a rotor.

3. The ditches generally require less space than a lagoon and the costs of collection and transfer are fairly low.
4. The system works well in cold climates provided that a majority of the ditch is within the confinement facility.

2.2.2.2. Disadvantages

1. Not only can the construction and installation costs be high, but maintenance can be a problem.
2. Foaming and scum formation of the ditch contents often occur.
3. A ditch that is not properly designed, installed, and maintained can become a hazard to livestock and an eyesore.

A properly designed and managed oxidation ditch can be an effective method for treating poultry manure. Although costly in construction and maintenance, it has reduced manure collection and transfer costs. The aerobic action of the bacteria reduces organic and nitrogen levels. Again, in regions where high density, large volume poultry operations function and premiums are placed on odor control and land application, oxidation ditches are a feasible way of treating poultry manure. The cost effectiveness of this system can be enhanced if treated poultry manure can be applied to a pasture.

2.2.3. Composting of Poultry Manure

One of the most promising ways in which to aerobically process poultry manure is by composting it. Composting is a relatively fast aerobic process in which organic matter is degraded by bacteria and fungi to produce a relatively stable humus-like material. The aeration of the poultry manure can be achieved in several ways. It can be done by mechanical scraper or windrower. High rate composting can be accomplished by using large rotating drums into which forced air is supplied. The composting process is self-heating to about 60°C and can produce compost in about 10 d if favorable conditions are maintained. Poultry manure can be effectively composted if the moisture level is kept between 50-60% and a good mixing ingredient such as bedding or paper wastes is mixed with it.

2.2.3.1. Advantages

1. Stabilizing justifiable organic matter.
2. Killing pathogens and weed seeds.
3. Producing a uniform, sterile, relatively dry end product, free from odors.
4. Conserving the nutrient content and organic matter found in the raw poultry manure.
5. Conducting the process free from insects, rodents, and odors, as inexpensively and dependably as possible.

6. Producing a valuable fertilizer and soil conditioner.

2.2.3.2. Disadvantages

1. The process can become expensive if high rate composting is desired. It requires special equipment for the aerating and mixing of the compost.
2. Composting is both labour-intensive and time consuming. Good composting usually requires daily mixing.
3. The compost is not rich in fertilizer value, containing 0.5% nitrogen, 0.4% phosphorous, and 0.2% potassium.

Compost addition to soils improves moisture retention of light soils and pore volume of heavy soils while providing a soil structure which is relatively stable and resists erosion. The extent and speed of the composting cycle are affected by moisture content, particle size, aeration temperature and initial carbon-nitrogen ratio. Composting is gaining wide popularity as a means of recycling biodegradable wastes. It only makes sense that composting poultry manure would be an environmentally and economically desirable way of processing poultry manure. It may not be cost-effective to compost poultry manure in high density, large volume poultry operations.

2.2.4. Dehydration of Poultry Manure

Poultry manure has a higher dry matter content than any other manure. It is costly to add water to poultry manure because,

1. Water is expensive, adds weight and volume.
2. Liquid poultry manure requires special collection transfer, processing, and disposal equipment, which can be costly.
3. In several circumstances, water addition allows for anaerobic processes to start in the manure, leading to high nutrient loss and decomposition odors.

It is obvious that poultry manure would have a great potential in its drier form. Poultry manure is highest in nutrients, the driest to begin with, and can be further dried in the poultry house by several methods of ventilation and heating. There are several ways in which poultry manure can be dehydrated. These include: the deep pit system over high-rise poultry buildings, the in-house manure drying system on slats, the in-house manure drying-system on belts, and the dehydration of poultry manure by mechanical dryer systems.

2.2.4.1. Deep-pit drying system

This system involves having the collection of poultry manure under the cages in deep pits. Air-drying occurs when circulating fans in the pit dry the poultry manure to a moisture content of 50% or less. With sufficient circulation and ventilation, the dried poultry manure can be handled as a solid and the problem of ammonia and odors of wet pits are avoided. As well, a better overall environment for staff and stock is provided when poultry manure is dried.

2.2.4.1.1. Advantages

a) low collection, transfer, and processing costs, b) low air pollution and easy stockpiling of dried manure and c) dried manure can be fed to animals (if dried to 10% moisture) or applied on horticultural land (if dried to 35% moisture).

2.2.4.1.2. Disadvantages

a) high nutrient losses b) possibility of presence of pathogens c) non-uniform in texture d) slow process e) feasible in arid and semi-arid climates and f) only solid fraction is utilized.

2.2.4.2. In-house drying system

Air circulation by fans is necessary to dry the poultry manure and stirring the poultry manure by mechanical means enhances the drying effect.

2.2.4.2.1. On slats

The in-house manure drying system on slats is an efficient and economical method of in-house manure drying. It is a more efficient and economical way than drying poultry manure in the deep pits system. It requires less mechanical drying and uses the existing ventilation for drying on wooden slats. The system uses slats of wood to collect the poultry manure. The wooden slats are set up on a two-level system. On the top level, slats and gaps of 10 cm are used to allow half the manure to pass on to the lower level and provide unrestricted air circulation. The lower level has 12.5 cm slats with 7.5 cm gaps. The manure collects in tall columns and is continuously dried until the manure is removed.

The efficiency of the system derives from the facts that: a) fresh manure adheres continuously, producing tall columns with a large surface area b) the warm ventilation air passes over these columns before being exhausted below the slats through windproof outlets in the pit walls c) heat is provided by stock as they metabolize the energy of the food, and air movement by the existing ventilation system.

When the birds are removed from the house, the slats are removed and the manure is deposited on the pit floor. The manure can then be handled in solid form by tractor and loader. A certain amount of poultry manure is deposited on the pit floor originally.

Normal poultry manure has enough moisture for it to adhere to the manure columns on the slats. The poultry manure which does not adhere to the columns is much drier and falls directly into the pit. Narrow slats with narrow gaps result in faster drying, but because of the narrow gaps, "bridging" of the poultry manure occurs and the drying process slows down or the slats have to be emptied earlier.

The slat drying method overcomes the problem of ammonia encountered in deep pit high-rise houses. The manure is in a much drier form and consequently less ammonia is produced. Since the levels of ammonia are lowered, a very significant decrease of in- house odor will result.

2.2.4.2.1.1. Advantages

a) manure moisture content can be reduced to 12-15 percent at no fuel cost b) the dried material produced is convenient to handle and can be ground to yield a pleasant, odor-free product c) this product has an enhanced value over wet manure and can be utilized as a fertilizer or feed ingredient d) an amenable pollution-free house environment is provided for staff and stock e) the drying system does not produce offensive odors as in wet pit houses.

2.2.4.2.1.2. Disadvantages

a) although the system is not costly or labor intensive, the wooden slats and forms must be purchased and constructed b) usually, the slats must be emptied every six months or so to avoid "bridging" of the poultry manure c) several of the disadvantages listed in the natural air drying in the deep pit system are also applicable here.

The slat drying system is a cost-effective and environmentally friendly way of simply drying poultry manure. This dried product can then be used as a fertilizer or feed ingredient. It appears that this system is a very cheap, effective way of processing poultry manure.

2.2.4.2.2. On belts

This system is based on a two-phase drying system. The first phase involves the pre-drying of the poultry manure on belts by means of air. The second phase involves the post-drying of the manure in covered storage by means of internal heating. This method requires heat exchangers to speed up the drying process. The poultry manure is allowed to accumulate and dry on the belts for one week. After drying on the belts for one week, the poultry manure has a dry matter content of about 45% and hence, is unsuitable for transport and storage in the open air. Anaerobic processes readily take place in such manure, the result being sticky, malodorous manure which is difficult to process. The manure is further dried by storing in sealed containers and allowing spontaneous internal heating to occur. After a heating period of six weeks, poultry manure is obtained with a dry-matter content of 55-60%. This manure can now be easily transported and stored as a solid without objectionable odors.

2.2.4.2.2.1. Advantages

a) the overall process is speeded up when compared to the deep pit method and the slat method b) it is a method that is applicable anywhere in the world, including humid areas c) because of frequent removal, the environment for staff and flock in the house is good d) odor emissions are limited e) this product can only be used as fertilizer and must be further dried if it is to be used as a feedstuff f) like composting, the internal heating process can destroy some weed seeds and pathogens g) the manure belt batter drying system is less expensive than even the deep pit drying system.

2.2.4.2.2.2. Disadvantages

a) the cost of purchase and construction of the drying system including the belts and covered store for internal heating b) it is more labor-intensive because of frequent manure removal from the belts c) this method requires energy in order to more rapidly dry the poultry manure d) as with the other air dry systems, the nitrogen and nutrient losses can be high.

The losses of nutrients can be minimized if frequent manure removal from the belts occurs. Experience with in-house manure drying systems has demonstrated that 50% of organic matter and nitrogen are lost during prolonged storage in-house. The spontaneous heating process used in this system stimulates the drying process without promoting excessive nutrient loss. The belt drying system for dehydrating poultry manure is an economical and environmentally safe way of treating poultry manure. The obtained product can be used as a substitute feedstuff. It requires a greater energy input than do the previous two dehydrating systems but is cost-efficient because of the rapid drying of the poultry manure.

2.2.4.3. By mechanical dryers

Characteristics which are desirable in a mechanical dryer are

1. Sterility - for animal feeds, by maintaining a high enough temperature for a sufficient time.
2. No odor problems - the drying process should not emit odors or should be amenable to easy treatment for odor removal.
3. Low labor requirement and simple-to-operate (especially for on-farm units).
4. Flexibility - dryer should be capable of handling a variable composition feed and giving an adjustable product moisture content.
5. Dryer should be capable of handling "foreign bodies"
6. Materials of construction should be compatible with product specification and have good corrosion resistance.
7. Farm units should have simple, quick start-up/shut-down procedures to maintain efficiency.
8. Economy - in terms of capital cost, installation and running costs.

Mechanical dryer system is a much more energy intensive system. Because of this heated air drying principle, there are special advantages and disadvantages to using such a mechanized system.

2.2.4.3.1. Advantages

1. Dry material is easy to incorporate as fertilizer or feed. It is also easy to stockpile.
2. High temperature kills pathogens.
3. Dry material is deodorized.

2.2.4.3.2. Disadvantages

1. Air pollution may occur during processing, requiring odor-control equipment.
2. Processing (dehydration) plant may be affected by zoning and/or regulatory restrictions.
3. Drying energy costs are high (these may be prohibitive in the future).
4. Equipment is fairly costly.
5. Time and energy requirements are high for collecting and for transporting to and from dehydrators.

There are several ways of mechanically drying manure. They include: Tray dryers, continuous band dryers, batch agitated dryers, direct-heated rotary drum dryers and pneumatic (flash) dryers.

2.2.4.3.3. Tray dryers

Tray dryers are of relatively low capacity being applicable to the drying of fine chemicals. They are inappropriate to the drying of poultry manure because of costs, high labour, utility requirements and odor problems. Continuous hand dryers present odor problems due to low gas velocity. Afterburners, stacks, filters and scrubbers require high gas velocities. Drying rates are low and labour requirements are high making this type of dryer unattractive. The batch agitated dryer seems to be best suited for farm use because of its flexibility of operation.

2.2.4.3.3.1. Advantages

1. Operates satisfactorily at intermittent rates, with variable feed moisture content, and with foreign matter other than stones.
2. Low labour, running capital costs.
3. Almost certain product sterility.

2.2.4.3.3.2. Disadvantages

1. Maximum capacity is limited, thus losing economy of scale.
2. Low thermal efficiency.
3. High depreciation rate.
4. Nutrient losses can be high.

It is believed that the batch agitated dryers are the most attractive means of drying poultry manure for small-scale farmers (with less than 50,000 head).

2.2.4.3.4. Rotary drum dryers

Another system of drying poultry manure is the direct heated rotary drum dryer. It uses very high heat to dry the poultry manure. Because of high heating, product cooling is necessary as the material is discharged at 100°C.

2.2.4.3.4.1. Advantages

1. Product quality is very high and the system can handle variations in feed moisture.
2. Labour requirements are low.
3. Foreign matter is readily handled.
4. Dryers can be constructed from various materials.

2.2.4.3.4.2. Disadvantages

1. Capital costs are high in comparison to batch agitated and pneumatic dryers.
2. Thermal efficiency is poor during intermittent operation.
3. Nutrient losses can be high.

From a technical point of view, rotary drum dryers are the most acceptable for poultry manure drying. The low labour requirements for the small scale farmer and the capital costs to the large scale farmer make this drying system economically non-competitive.

2.2.4.3.5. Pneumatic (flash) dryers

2.2.4.3.5.1. Advantages

1. They can approach 90% thermal efficiency when run continuously.
2. Low gas flow rate and high velocity help in odor control.
3. The space requirements are low.
4. It is a gentle drying process and reduces nutrient loss.

2.2.4.3.5.2. Disadvantages

1. Problems of sterility due to short residence times and relatively low temperatures.
2. Tendency to form balls.
3. Necessity to macerate foreign matter in the feed.

It is believed that with further development, the pneumatic dryers' problem of product sterility can be solved. The dryer system represents the best drying system for large-scale farms (over 1 million head).

Mechanical drying of poultry manure is becoming more feasible with the rising prices of animal feeds and fertilizers. The trend towards recycling and protecting the environment makes the drying of poultry manure suitable as well as economically advantageous.

Depending on the size and requirements of the poultry farm, there is a dryer system which could provide an appropriate means of processing the poultry manure produced by such a facility.

2.2.5. Incineration

The burning of poultry manure is a very wasteful and ineffective way of processing the manure. The incineration of the manure allows for the escape of all the beneficial nutrients into the atmosphere. It produces air pollution due to odors and the release of particulate matter. Since poultry manure has a high organic content, incineration still yields a product which is very high in ash content. The result is that 10-30% of the initial dry matter still remains as ash. As well, collection and transportation of the manure to the incineration site make burning an expensive way to process poultry manure. Hence, the burning of poultry manure as a means of processing and disposal appears to be very unsuitable as a proper manure management alternative.

3. Handling poultry manure

3.1. Facts about poultry manure

(Bird and Munroe, 1996)

1. Freshly-produced poultry manure does not give off offensive odor.
2. The odor produced by stored manure increases as the moisture content increases. The least odor results from drying the manure soon after it is produced.
3. As produced, droppings are 70 to 80% moisture. The moisture content depends on such factors as salt in the diet, bacteria in the water supply, health of the birds, pen temperature and strain of bird.
4. As produced, poultry manure without bedding weighs about 1,000 kg/m^3 and contain about 250 kg/m^3 of dry matter.
5. If poultry manure is to be handled and stored as liquid manure, then an amount of water equal in volume to the manure must be added.
6. Poultry litter, which is poultry manure and bedding mixed, usually ranges from 15 to 30% moisture. In this range it is crumbly and doesn't stick together. Below 15% moisture it is dusty, and above 30% moisture it becomes sticky and slippery.
7. In the 15 to 30% range, it weighs about 320 kg/m^3. The volume of poultry litter ranges from 1 to 2 times the volume of manure without bedding depending on the amount of bedding material used. In addition, the volume may be reduced somewhat due to handling and compaction during removal.
8. Poultry manure is often stored in situ for part or all of the storage period. This may be in the form of litter, as with broilers, droppings under cages, or a combination of droppings and litter as with breeder flocks. Droppings

stored in situ without air circulation for drying purposes usually adjust to 65% moisture. With circulated air drying, moisture levels as low as 40 to 50% can be achieved. If drying conditions are poor, such as in hot, humid weather, or if there is water leakage, or if the birds are producing very wet droppings, then the droppings stored in situ may be 80 to 85% moisture.

9. Shavings or chopped straw is spread to a depth of 5 cm. This is approximately 3.5 kg/m^2. Litter is managed to maintain the moisture content in the 15 to 30% range. To do this, adequate ventilation and air circulation are required.

3.2. Quantity of manure produced

Typical manure production in layer and broiler systems is given in Table 46.7. Assuming that fresh droppings have 85% moisture and manure 30-35%, weight of manure is calculated as 50% of total feces.

Table 44.7. Manure production by chicken

	Layers in cages	Broilers		
		On litter	In cages	Breeders on litter
Duration reared	1 year	6 weeks	6 weeks	20 weeks
Feed/bird	100 g/d	4 kg	4 kg	200 gm/d
Total feed (kg)	36.5	4	4	28
Total feces (kg) *	≈ 50	5.6	5.6	≈ 40
Manure (kg) **	25	2.8	2.8	20
Litter (kg)	#	0.5	--	#
Total/bird (kg)	25	3.3	2.8	20
No of birds/tonne	40	300	≈ 350	50
* @ 1.4 g/g of feed intake		** 50% of total feces		# Negligible

3.3. Volume of poultry manure

One of the best is to measure the rate of production from the operation in questions. This may not be possible as the storage design may be needed before production begins. A second method is to estimate the manure volume based on the amount of feed consumed as m^3 dry bedded manure = tonnes of feed fed x 1.14. If a liquid manure system is used, these manure storage volume requirements should be doubled to account for dilution water.

Typical poultry manure storage volume requirements for various production systems and classes of poultry are given in Table 46.8. The volumes are given for the housing periods listed which is the usual duration of a flock. If storage is to be constructed for a longer or shorter period, required storage volume from the table has to be multiplied by the number of flocks raised during the planned storage period.

Manure handling is generally done in the following sequence: collection, transfer, storage, removal, transport and incorporation. The principles behind the selection of a poul :ry manure handling system are:

1. Most systems use the poultry manure as a soil fertilizer. Only a few systems currently have other end uses such as re-feeding.
2. All of the systems are compromises between investment, labour, convenience, aesthetics, and regulations.

No system is best. Each has advantages and disadvantages. The ideal system for a specific operation depends on capital and labour, waste sources, soil type, cropping practices, personal preferences, and a number of other factors.

Table 44.8. Manure storage volumes required

Class of poultry	System of housing	Period (d)	m^3/bird
Egg strain caged-laying hens	Liquid manure system	365	0.085
Egg strain caged-laying hens	Dry(solid) manure	365	0.043
Egg strain-laying hens	Partial or total litter	365	0.048
Meat strai- laying hens	Partial or total litter	365	0.063
Caged egg strain pullets	Liquid manure system	140	0.013
Caged egg strain pullets	Dry (solid) manure	140	0.006
Egg strain pullets	On litter	140	0.009
Meat strain pullets	On litter	140	0.011
Chicken broilers and roasters	On litter	42	0.003
Ducks	Dry(solid) manure	56	0.007
Ducks	Liquid manure	84	0.012
Turkey breeder hens	On litter	180	0.074
Turkey breeder toms	On litter	180	0.131
			Source : Bird and Munroe, 1996

3.4. Collection

The following are the methods of collecting poultry manure:

Confinement Housing: Caged

Shallow pit - flush
Shallow pit - scrape
Medium pit - collector water - drain
Medium pit - scrape
Medium pit - dropping boards - scrape
Deep pit - scrape
Deep pit - collection water - drain

Confinement Housing: Litter or Floor Systems

Litter - scrape
Deep Litter - scrape

3.5. Transfer and Storage

The collected manure is transferred and stored in various methods depending on whether it is in liquid or solid form.

In the litter system, the transfer is accomplished by using a tractor and bucket loader if the storage area is close. If the storage area is located at a distance, then the manure must be loaded on to a dump truck or manure spreader to be hauled to storage.

The high-rise house uses "in-situ" storage for the poultry manure. In most cases, the manure is deposited on the floor of pits located under the cages. The manure is then air dried in this storage area. This system requires a circulation and exhaust system to further dry the manure and to remove airborne odors and chemicals. Poor drying conditions occur in hot, humid weather. As well, excessive water consumption and spillage can result in water collection within the pits. This gives rise to odor production due to anaerobic conditions as well as the production of certain noxious chemicals such as ammonia. In the more "conventional" type houses, the manure is transferred to storage by a cross auger. This completes the manure collection by receiving the manure and conveying to the side of the house where it is elevated into a liquid manure storage tank. If a below-ground storage tank is used for storage, no cross auger is necessary.

The type of storage facility used for "conventional" type houses is a liquid manure tank. The location of such a tank should be convenient to the house but at a site where expansion of the facilities and storage areas can occur. It should be easily accessible to allow easy transport to andfrom the storage area. The storage facility should be located far enough away from local residences that it will not be a nuisance. In below-ground storages, areas with a high-water table and surface runoff should be avoided. In above-ground storages, the soilshould be compacted enough to prevent settlement of the storage structure. In the liquid storages, agitation or aeration may be required to control odor emissions. In addition, it may be necessary to cover these storage areas to avoid odor emissions and rain water collection. If an earthen storage is used, it must have manure-tight soil conditions or a water-proof lining is required.

The type of storage facility used for broiler litter can range from an open pile on a well-drained site to a covered storage facility. An existing concrete slab or a horizontal silo also makes good storage areas. Storage next to a bucking wall is recommended so that loading of the manure by a tractor-mounted bucket is easier. It is also highly recommended that if the litter is already fairly dry, it should be stored in a covered area in order to keep the litter dry and relatively odor-free.

The transfer and storage of poultry manure is dependent on the liquid and the solid consistency of the manure. In a solid system, the poultry manure is usually stored in a dry location for further processing or for application as fertilizer. In a liquid system, the manure is stored in a tank or lagoon where it is left until it can be applied a a liquid fertilizer.

3.6. Removal, transport, and incorporation

The removal, transport and incorporation of poultry manure are fully dependent on the form in which the manure has been stored.

The "in situ" storage of dried poultry manure usually presents special removal problems. These problems are related to the very large volume of manure to be removed by loader bucket and the long travel distance required to retrieve each load in large "high-rise" poultry barns. This removal time can be shortened if more than one removal entrance to the poultry house is provided. Since most "high-rise" houses allow their poultry manure to collect for several months or years, the removal of the poultry manure can be expected to take a fairly lengthy amount of time.

The dry poultry manure can be transported and spread with any conventional type of box spreader. The poultry manure should be spread as evenly as possible. The application of the solid poultry manure is usually by broadcast. To avoid excessive nutrient loss, the poultry manure should be spread at a time when the land can be immediately plowed or tilled. The poultry litter which is obtained from broiler operations can be removed, transported, and incorporated in the same way as caged layer solid waste. Compared to other dry manures, broiler litter is likely to be spread more thinly. In order to control odors and to preserve plant nutrients, quick incorporation after spreading should occur.

The liquid poultry manure can be handled as a liquid if the wastes are up to 4% solids. From 4-15%, the waste is semi-solid but can be handled as a liquid provided special equipment such as chopper pumps for cutting fibrousmaterials and piston pumps for handling waste with bedding is used. If large quantities of liquid are to be handled, it may be preferable to use a pipeline instead of tank transports.

Irrigation equipment disposes of liquid poultry manure while adding fertilizer and water to crops. This system can help pay for itself by improved crop production resulting from the irrigation. There are several ways of irrigating agricultural land with poultry manure.

These methods include: a) Surface irrigation b) Sprinkler irrigation and c) Soil injection.

3.6.1. Surface irrigation

Done through a portable or stationary pipe and spread by a gated irrigation pipe or an open ditch with siphon tubes.

3.6.1.1. Advantages

1. Low cost.

2. Low power requirements.
3. Few mechanical parts.

3.6.1.2. Disadvantages

1. They require a high degree of management skill to avoid runoff and to get uniform distribution.
2. They are inflexible with respect to land area.
3. They require a moderate amount of labour.
4. They cannot be used on lands with greater than a 2% slope.

Unless quickly absorbed by the soil, there can be a high degree of nutrient loss with respect to irrigation methods.

3.6.2. Sprinkler irrigation

This allows for the distribution of liquid poultry manure on rolling and irregular land.

3.6.2.1. Advantages

1. Labour requirements are low.
2. Some systems can be automated.
3. Application can be more uniform.

3.6.2.2. Disadvantages

1. Initial and operating costs are higher than for surface irrigation.
2. Odors from sprinkled manure can be a nuisance.
3. Nutrient losses from a sprinkler system can be very high.

Sprinkler systems are a very efficient way of delivering liquid poultry waste to farm acreages. If such a sprinkler system is used that is well planned and managed, then it can also be an efficient way of irrigating the field with liquid poultry manure.

3.6.3. Soil injection (or knifing) of liquid poultry

This method involves the use of soil injectors which push the liquid poultry manure, while under pressure, from the tanker through tubes located behind deep cultivator teeth.

3.6.3.1. Advantages

1. This holds the greatest potential for odor control.
2. It can lengthen the time manure can be applied in the spring.
3. It can incorporate liquid manure into hay and pasture crop without completely destroying the crop.
4. It also achieves an acceptable rate of application.

3.6.3.2. Disadvantages

1. Refinements are still needed to avoid buildup ahead of the injector unit.
2. Work is needed to ensure that there is adequate coverage behind the injector unit.
3. Injectors should be made suitable for row crop application under a wide range of crop types, crop sizes, and soil conditions.

Considering that most of the nutrients contained in poultry feed are processed into manure, it only makes sense that using this product to fertilize agricultural lands or to feed livestock is a very good, sustainable environmental approach.

4. Specific applications of poultry manure

4.1. As feedstuff

4.1.1. Dried poultry manure (DPM)

Poultry manure in its dry form can be used as a fertilizer or alternatively, it can be used as a feedstuff ingredient for animals. The use of animal manures as feedstuffs is conceptually attractive because it has the potential to reduce feed costs and to provide a partial solution to manure management and environmental problems. The value of manures as feedstuffs can be determined by the assessment of:

1. Nutrient composition of animal manures to determine sound utilization strategies and to estimate their value.
2. Monetary benefits that could result from reduced feed costs and increased revenue from meat, milk or eggs as indicated by feeding trial results.
3. The pollution control benefits that might result from a re-feeding strategy.

There have been several re-feeding experiments using DPM as a component of feedstuff. The dried manure has been considered comparable to protein feeds such as soyabean and cottonseed meals because of similarities in crude protein content. But, in reality, dried poultry waste appears to be nutritionally most comparable with forages such as corn silage and hays rather than with energy or protein feeds. DPM has a low digestible protein level and fairly low-energy content and, as a feed material, it equals or exceeds its value as a fertilizer.

It is important to realize that DPM used as a feedstuff can only be tolerated by animals up to certain concentration levels. It has been found that the maximum and optimum levels of incorporating DPM into laying hen and ruminant rations, on the basis of animal performance, varied from 5% to 20%. Typically, when the level of DPM exceeds these levels, the consumption of feed will increase while the production will decrease. This results in DPM not being as economically viable as when the percentage of DPM in the feeds is kept below these threshold levels.

4.1.1.1. Economic analyses

Analyses have indicated that when DPM was used at its optimum level, its monetary value as a feed exceeded its monetary value as a fertilizer. This benefit is quickly reduced as the amount of DPM in the feed is increased above optimum levels.

DPM can also be used as a re-feeding component of poultry diets. Results concluded that DPM can be safely included up to 10% in broiler rations without any detrimental effect on their growth rate, feed efficiency and performance; but, at levels above 10%, there will be a very significant depression of growth rate.

In addition, it should be noted that DPM can be a carrier of disease if it has not been processed properly. It is obvious that great care and planning must be used in order to successfully supplement the diets of animals with dried poultry manure. Use of dried poultry manure is both economically viable as well as being an environmentally attractive way of disposing of poultry manure.

4.1.2. Ensiled poultry manure

Another way that poultry manure can be used as a feedstuff is by ensiling it. Ensiling consists of adding 20% poultry litter and 30% cage layer manure to 50% ground corn while adding some molasses and bacteria to ferment the mixture After mixing and storing in sealed containers, the mixture is left for 3 weeks to ferment .

4.1.2.1. Advantages

1. Nutrient losses are low.
2. Ensiling improves palatability and decreases pathogens.
3. Ensiling is economical and uses less fuel than dehydration.
4. It provides a procedure other than dehydration that is applicable for use in animal feeds.
5. Ensiling allows for a use of a greater proportion of poultry manure to be used.
6. Material can be stockpiled.
7. Odors are controlled.
8. Liquid and solid fractions are utilized.

4.1.2.2. Disadvantages

1. Diluting material is often used at ensiling time.
2. Proper ensiling takes time. It can be reduced from 6-8 weeks to less than 3 weeks by adding molasses and the proper bacteria culture.
3. It requires a suitable storage container which is fairly air-tight to prevent contamination.

4. Handling or labour is required for harvesting, transport to storage, ensiling material, and transport from storage to feed bunk.
5. Forage diluting materials are not always available because of seasonality.

The process of ensiling can actually produce a more complete feed. The digestibility of ensiled materials is very good and the nitrogen available in the poultry manure is efficiently utilized. Ensiling of poultry manure is another way in which the manure can be used as a component of feedstuff. It is a process which is very economical and is an environmentally-friendly way of disposing of poultry manure.

4.1.3. Processed poultry manure

Other than dehydrating and ensiling, there are other ways of converting poultry manure into animal feedstuff. Although dehydration occupies a vast majority of the waste re-feeding market and ensiling is a distant second, processing of poultry manure does occur in small quantities. These methods include: cooking, autoclaving and chemical processing.

4.1.3.1. Cooking

The process is to cook 2 parts manure in 1 part water. After it has been cooked for 24 hours, it is dried by oven.

4.1.3.1.1. Advantages

1. Nutrient losses are fairly low.
2. Dry material is easy to incorporate into diet and stockpile.
3. Dry material is deodorized.
4. Boiling temperature of water kills pathogens.
5. Animal acceptance is good.

4.1.3.1.2. Disadvantages

1. Cooking and drying energy costs are high.
2. Equipment for cooking and drying is costly.
3. Time and energy requirements are high for collecting and transporting.

Although not as prevalent as the methods of dehydrating and ensiling, the cooking of poultry manure may in some cases be an acceptable way of treating poultry manure in order to re-feed it to animals.

4.1.3.2. Autoclaving

This is not commonly used. It is a method of quickly drying poultry manure. The advantages and disadvantages of this method of treating poultry manure are much the same as they are in the use of dryers. It is a very expensive process which appears to only be useful on a small scale. It is only mentioned in passing because some scientific experiments use this process in order to obtain dried poultry manure to be used in re-feeding studies.

4.1.3.3. Chemical treatment

In this method, the poultry manure is collected and is chemically treated with substances such as Grazon. Other ingredients such as corn are mixed in to form the desired ration.

4.1.3.3.1. Advantages

1. Process increases animal acceptability.
2. Immediate harvesting and re-feeding reduce losses.
3. No storage is required.
4. Odor is controlled.
5. Energy and labour requirements are low.
6. Liquid and solid fractions are utilized.

4.1.3.3.2. Disadvantages

1. Daily harvesting and processing are required.
2. Short shelf life does not permit extensive stockpiling.
3. Mixing equipment is required.
4. Chemicals are costly.

In this instance as well, the treatment of poultry manure by chemicals to produce an animal feedstuff is practiced on a very small scale when compared to dehydration and ensiling. In some cases, it can be a very acceptable way of treating poultry manure in order to re-feed it to animals.

The poultry production systems throughout the world produce a large quantity of poultry manure. Since many of the production facilities have a very limited land base, it is very important that we develop alternative uses for poultry manure that its usual application as a fertilizer. Poultry manure has great potential as an animal feedstuff.

It is believed that its economic value as a feed material outweighs its economic value as a fertilizer in the markets of today. Although this may be true from an economic standpoint, it should be noted that there is still very strong opposition to using poultry manure in animal feed. It is still primarily used as a fertilizer. When the application of poultry manure to available land is properly planned and managed, it can be a very positive influence on crop growth and yield. On the other hand, as feed prices keep increasing in cost, poultry manure as a substitute feed becomes increasingly attractive. The use of poultry manure as a means of re-feeding in the poultry production cycle presents a very sound, sustainable, environmentally-friendly way of maintaining a specific branch of agriculture which is the poultry industry.

4.2. As fertilizer

The most convenient and practical way of disposing of poultry manure is to spread it on agricultural land as a means of fertilizing the soil. The constituents of poultry manure that make it a suitable fertilizer are its levels of nitrogen, phosphorus and potassium.

4.2.1. Crop needs

Crops generally have a higher requirement for nitrogen than for the other major elements. To better understand the use of poultry manure as a fertilizer, it is necessary to understand a few basic principles about soil nutrient chemistry.

4.2.1.1. Nitrogen

The nitrogen in freshly-excreted manure is in organic form which is converted to ammonium-nitrogen. Ammonium is firmly adsorbed to soil particles but can volatize and vast quantities can be lost from the soil in this way. Bacteria in the soil convert the ammonium-nitrogen to a nitrate form which is readily used by plants. This form of nitrate does not adsorb to soil particles and is easily subject to leaching by water movement

4.2.1.2. Phosphorus

The phosphorus in the soil is usually taken up by plants in the form of mineral phosphates which usually arise from the weathering of bedrock. Phosphates are virtually immobile in soils, so that at any time, a small proportion of the soil phosphorus will be available to plants.

4.2.1.3. Potassium

The potassium stores of the soil are generally on the order of 10 times greater than are the nitrogen or phosphorus stores. A majority of the potassium added to the soil in poultry manure is readily available and will either be used by the crop or will be adsorbed on the soil to be used on another occasion. Potassium is not subject to leaching like nitrogen.

4.2.2. Non-treated

When managed properly, poultry manure is an effective means of fertilizing agricultural soils. It has been recognized for a long time that poultry manure is a very good source of nutrients for the production of corn, small grains, fruits and vegetables. The most important nutrient in poultry manure for crops is nitrogen. The problem with applying poultry manure to a crop occurs when the applied manure exceeds the crop requirements. This results in the leaching of soil nitrates which can lead to the contamination of groundwater.

The application rates of poultry manure have often been based on the amount required to maximize crop yields. Unfortunately, this can result in an excessive amount of nitrogen being applied to the soil. It has been found that only small amounts of poultry manure are needed to optimally fertilize most crop lands.

The over-application of manure to any soil over a long term will build up the nutrient content (the phosphorus level is also of concern) far beyond crop requirements. This may cause water pollution resulting from excessive phosphorus as well.

4.2.3. Composted

One of the problems with using poultry manure as a fertilizer is that high levels of nutrients can be lost with normal methods of collection, storage and spreading on the land. One way to try and limit the nutrient losses is by composting poultry manure and the details are already given earlier in this Chapter.

Composting allows for a lag-period between removal and spreading because the composted material can be stored. Its advantages make it more suitable for a system which requires longer storage and ease of application. The only problem associated with composting is that it requires additional storage areas and can require special equipment.

4.2.4. Digested

Both anaerobically and aerobically digested poultry manure are used as land fertilizer after they have been treated. The nutrient value of the obtained treated liquid depends on the treatment method used. In both anaerobic and aerobic digestion, there can be a large amount of nutrient loss from the system.

However, it must be stressed that proper planning and management of the whole system is needed in order to prevent waste and pollution. When a liquid obtained from anaerobic or aerobic digestion is properly applied to a crop, it is a very effective and efficient way of maintaining a sustainable method of poultry waste disposal.

4.2.5. Dehydrated

Dehydrating poultry manure is another way one can obtain soil fertilizer. Although dehydrating poultry manure can remove a large amount of the nutrients from the original manure, the dried form is very easy to use as a solid fertilizer.

Chapter **45**

Pest Management

Extreme competitiveness in the poultry industry has resulted in narrow cost-profit margins. Poultry operations can neither allow ectoparasites and nuisance insects to interfere with production nor can they afford high pest control expenses. For production to be profitable, poultry production must be managed to reduce the incidence of insect pests.

Poultry are infested with a variety of insects and mites that live on the skin and feed on skin debris, feathers, and blood. This activity can lower growth rates, reduce egg production, and, if the infestation is heavy, cause debilitation and death of the birds. In addition, poultry houses can be a breeding source for a variety of flies that may be vectors of poultry diseases or, at the least, be a nuisance which may bring about litigation from neighbors. The ecto- and endo-parasites will be dealt with in a separate Chapter and in this Chapter, flies and rodents which cause nuisance and economic loss are discussed. For further information, please refer to an article "Poultry pests and their control" by Hoelscher at The Texas A&M Univ. Website.

1. Flies

One of the largest management problems facing the poultry producer of today is filth fly control. The shift from many small farm flocks to fewer large poultry operations has greatly increased fly problems by creating concentrated breeding areas and large volumes of waste that cannot be removed frequently. As urbanization and rural non-farm residences increase, poultry producers face increasing pressures to reduce fly population. Fly populations (manure-breeding flies) may cause a public health nuisance, resulting in poor community relations and threats of litigation. A dedicated effort is necessary to achieve an acceptable level of fly control.

Several kinds of flies are common in and around poultry houses, in general and caged layer houses, in particular. Probably the most common flies are the house fly and the little house fly. About 95% of problems involve the house fly. Both of these flies can move up to 32 km from the site of development, but normally no more than 2 to 3 km from the initial source.

1.1. Types of flies

1.1.1. House flies

Musca domestica (L), About 1.5 cm long, breed in moist, decaying-plant material, including refuse, spilled grains, spilled feed, and in all kinds of manure. For this reason, house flies are more likely to be a problem around poultry houses where sanitation is poor. These flies prefer sunlight and are very active, crawling over filth, people, and food products.

This fly is the most important species because it can carry and spread human and poultry diseases and cause flyspecking problems of the eggs. For example, house flies are the intermediate host for the common tapeworm in chickens, and they carry millions of bacteria.

1.1.2. The little house fly

Fannia canicularis (L), about 0.5cm long, is somewhat smaller than the house fly. This fly prefers a less-moist medium than the house fly for breeding and reproduction. The little house fly will choose poultry manure over most other media. This fly also prefers shade and cooler temperatures and is often seen circling aimlessly beneath hanging objects in the poultry house, egg room, and feed room. It is less likely to crawl about on people and food. However, it does cause people living near poultry establishments to complain about fly problems. The little house fly may hover in large numbers in nearby garages, breezeways, and homes because it prefers shade.

1.1.3. The black garbage fly

Ophyra aenescens, is slightly smaller than the house fly and shiny bronze-black in color. The wings are held straight back. This fly tends to stay on the food source at night rather than resting on the ceiling or on outdoor vegetation, as does the house fly. The female fly doesn't seem to fly great distances, but has been found about 8 km from its breeding area. Although black garbage fly larvae have been known to exterminate house fly populations, they should not be considered entirely beneficial because these flies can build large populations on the farm and disperse as adults to nearby communities. All stages are found throughout the year under suitable conditions, and they show rather good tolerance to cold weather. The life cycle is similar to that of the house fly.

1.1.4. Blow flies

Sometimes known as green or bluebottle flies, these are slightly larger than house flies and sometimes live in poultry houses. They prefer to breed and reproduce in decaying animal and bird carcasses, dog manure, broken eggs, and wet garbage. Generally, a good sanitation program will hold these flies in check.

Other flies found on the poultry establishment include soldier flies, small dung flies, fruit flies, and rat-tailed maggots.

1.2. Fly Biology

All flies develop through four life stages: egg, larva, pupa, and adult. Adult flies lay small, white, oval eggs on the breeding medium, and creamy white larvae (maggots) develop in this moist (wet) material. Mature maggots crawl out of this material and move to a drier place for the pupal stage. The brown, seed-like pupae finally yield adult flies. Development from egg to adult fly may take just 7 to 10 days under ideal conditions.

Adult house flies live about 3 to 4 weeks, and females lay two to 20 batches of 75 to 200 eggs at 3- to 4-day intervals. At this rate, a pair of flies beginning operation in April, if all offspring were to live, would result in 191 quintillion, 10 quadrillion flies by August. Allowing 2 cc to a fly, this number would cover the Earth 47 feet deep. Of course, this does not happen because beneficial predators and parasites keep the populations under control. Flies can be present in poultry houses year-round if there are warm temperatures and no true diapauses.

1.3. Surveillance of flies

It is important to monitor fly population to make wise-control decisions. Visual observations alone can be misleading. One needs to know the fly's behavior patterns and history. Documentation is very helpful in legal defense if needed.

1.3.1. Moving tape count

This is the best surveillance method, taking about five minutes each day walking on a 300 m-walk to catch 25 to 75 flies. Walking down and back in each house, in the same pattern and at the same time of the day, while carrying the sticky fly tape is cheap and easy.

1.3.2. Sticky fly tapes

Tapes that hang often tell nothing. Tapes fill up fast during summer months within a chicken house. However, one can determine fly species. Some hang sticky fly ribbons along aisles. Captured flies are counted weekly and ribbons replaced. A weekly count of 100 flies per ribbon may indicate fly control is required. Ribbons may become ineffective after 2 to 3 days because of dust and fly covering. Tapes are messy to use and location is important.

1.3.3. Speck fly count

A 2.5 x 7.5 cm white file card fastened flush against feed troughs, ceilings, braces or other fly resting areas, left for a period of several days to a week, will provide documented evidence as to the number of "fly specks" counted on a given date, over a period of time within a given house. The cards are placed on head rafters (three cards per house) and fly specks are counted on one side. Cards are changed once each day or week, depending on populations present. Fifty or more spots per card per week may indicate fly-control measures are required. Cards are placed in the same position of each renewal. Fly species cannot be determined from the spots. The spot card method is very economical.

1.3.4. Baited jug trap

This is more expensive than other sampling methods, but offers greater sensitivity to fly population changes. A plastic milk jug, with four access holes (5 cm in diameter) around the upper part of the jug with a wire attached for hanging about 0.9 m above the floor around the pit periphery, may indicate need for control. The jug is baited with a commercial fly bait (about 30 ml) placed inside the jug bottom. Fly pheromone can be used for effectiveness.

1.3.5. Larval sampling

It is most important to walk the pits to determine "hot spots" where the manure appears flattened and wet and contains heavy population of fly eggs and maggots. A hoe or trowel is carried along to sample the larvae present. Pits have to be kept walkable, clean, and water-free. "Hot spots" usually appear where water was standing in the manure. Some producers may carry a knapsack sprayer to treat only the "hot spots" to halt excessive fly larval breeding.

Treating manure widely and excessively will kill beneficial agents. It is best not to treat with chemicals in the manure pits. Pit manure should appear tall, narrow, capped, and dry, perhaps with beetles to assist in aerating the manure, making it drier. (Manure will cone with proper fly management.) Beneficial arthropods should be monitored and establishment encouraged for suppressing fly population.

1.4. Control methods

House fly, Lesser House fly, and Soldier Fly control in poultry operations, including layer and breeder chickens, should include appropriate sanitary and management practices to reduce the number and size of fly breeding sites. A successful sanitary and management program may allow less than constant use of insecticides. This, in turn, should prolong the effective life of such control agents.

1.4.1. Cultural Control

Manure management is the most effective way to control flies. As many as 2,000 house flies can complete development in 1 kg of breeding material. Fresh poultry manure contains 75 to 80% moisture, which makes it ideal for fly breeding. Flies can be practically eliminated in this material by reducing the moisture content to 30% or less or by adding moisture to liquefy it. Drying manure is preferred because the product occupies less space and, usually will have, less odor.

1.4.1.1. Dry Manure Management

Frequent removal of manure (at least once a week) prevents fly breeding because it breaks the breeding life cycle. It is important to scatter the manure lightly outdoors to kill the eggs and larvae by drying. Piles or clumps of manure must be avoided. If enough land is available, manure can be spread thinly; this keeps excessive amounts of nutrients from building up in the soil. Quantity of manure spread should suit the local agronomic conditions.

In-house storage of manure requires drying it to a 30% moisture level and maintaining this level where sufficient storage space is available. Dry manure can be held for several years. Any practice that limits moisture in the droppings or aids in rapid drying is important for fly control.

1.4.1.2. Water Management

Managing the water content of manure is important in controlling flies. The following steps can help minimize water content:

1. Prevent leaks in water troughs or cups.
2. Regulating water flow to an on/off cycle may help eliminate moisture problems.
3. Provide abundant cross-ventilation beneath cages, especially during hot weather. Using 0.90 m pit fans blowing across the manure can be very effective.
4. Placing a curtain above the manure every 30 m helps keep air moving over the manure.
5. Adequate house ventilation is important at all times.

If the water table in the area is high, or if there is a danger of water running into the house from the outside, the floor-grade relationship should be adjusted so that the house floor is higher than the surrounding ground outside. Surface water from the building must run away smoothly. Low areas around the houses must be properly drained and filled.

1.4.2. Sanitation

Sanitation is the most important aid in successful fly control. Often, certain conditions in and around the poultry operation will encourage fly outbreaks. These must be eliminated. The following steps help improve sanitation:

1. Quickly remove and dispose of dead birds and broken eggs. Dispose of them far from the poultry premises by burning in an incinerator or other approved management method.
2. Clean up and dispose of feed spills and manure spills, especially if wet, immediately.
3. Clean out weed-choked water drainage ditches.
4. Install proper eave troughs and downspouts on poultry houses to carry rain water far from buildings. Provide proper drainage in poultry yards.
5. Minimize the migration of flies from other fly-infested animal operations close to the poultry house.

1.4.3. Biological Control (Beneficials)

1.4.3.1. Fly-parasites

Entomologists encourage the use of biological control in poultry houses. In biological control, only beneficial insects (also called "beneficials") adapted to the climate in the area should be introduced. These fly parasites, actually very tiny wasps, are the naturally-occurring enemies of manure-breeding flies. They destroy flies in the pupal stage. These wasps, *Spalangia nigroaenea, Spalangia cameroni, Spalangia endius* and *Muscidifurax raptor*, are about the size of the head of a house fly (0.2-0.4 cm) and live in the manure, depositing eggs in fly pupae. Adult female wasps lay an egg on the fly pupa within the puparium (the hard case containing the pupa). Then the developing wasp larva consumes the pupa and emerges as an adult. These fly parasites are specific to flies and attack nothing else. They are bite-less and sting-less to people and usually go unnoticed by those living near poultry operations. They self-propagate in the process of controlling pest flies.

However, mass releases are needed. Also, the wasp lays fewer eggs than the fly over the same period, making it necessary to start with an initial wasp release and follow up with weekly supplemental releases; these releases have to be made before and during the fly season.

Whenever beneficial insects are used, one must be very careful with insecticides. Chemical sprays must be discontinued in areas of the poultry house where these wasps are used. And the entire manure surface should never be treated with insecticides, with the exception of cyromazine; otherwise, beneficial insects as well as the pest flies will be killed.

To improve the chances of successful biological control with these wasps, a strict sanitation program is also essential, involving manure management, water management, weed mowing, etc. Manure has to be kept dry, since wet manure promotes fly breeding and inhibits beneficial insect-breeding. Also, when the poultry house is cleaned, areas of old dry manure are left to provide a reservoir (seed) of beneficials to repopulate the house as new flies occur.

1.4.3.2. Fly-predators

Other beneficials in poultry manure include mites and beetles. Both are major predators in caged-layer operations. The macrochelid mite, *Macrochelis muscaedomesticae*, is reddish brown and less than 0.2 cm long. It feeds on house fly eggs and first-instar larvae. These mites, found on the outside layer of manure, can consume up to 20 house fly eggs per day. Another mite is the uropodid mite, *Fuscuropoda vegetans*, which feeds only on first-instar house fly larvae deeper in the manure.

A hister beetle, *Carcinops pumilio*, is black and about 0.3 cm long and feeds on house fly eggs and first-instar larvae. This effective beetle predator, common in both broiler and layer houses, can consume 13 to 24 house fly eggs per day. Both adult and immature hister beetles live in the surface layers of manure. Another hister beetle, *Gnathoncus nanus*, is present at lower numbers on poultry farms. Insect members from *Staphylinidae* which can function as predators have also been identified.

Using fly parasites and predators for biological control would reduce chemical residues to people, birds, eggs, and the environment. However, to date, claims that wasps will provide long-term fly control have not always been backed by scientific research results. When using biological control methods, habitat management to keep the manure dry is cardinal for biological control. Accumulations of poultry manure left undisturbed over long periods of time will support large populations of native fly parasites (wasps and mites) and fly predators (beetles). Native strains of beneficials already present in the dry manure should be encouraged to populate.

Manure is removed only during the fly-free time of the year and insecticide sprays are avoided in manure pits.

1.4.4. Mechanical Control

Many types and styles of fly traps appear in the market each year. These traps are usually electrical, employing a black light with an electrically-charged grid to kill the insects. Some traps are baited with a fly attractant material.

Traps do appear to be helpful in tight, enclosed areas such as egg rooms—where there is a breeding fly population, if good sanitation practices are followed. However, in areas of heavy fly population, traps are not effective in reducing fly numbers to satisfactory levels. Traps are used in the middle of the night away from doors and windows. Most entomologists feel that fly traps used alone are not effective in controlling flies, especially in and around livestock and poultry operations.

Flies will not move against the wind into the egg room or other work area and therefore, ensure that air moves through a screened doorway from the egg room or other work area into the main poultry house. There are commercial electric-powered air curtain fans. However, certain state health departments may require solid doors between the egg room and other main work area into the main poultry house. Wherever appropriate, sticky fly strips can be used.

1.4.5. Chemical Control

Insecticides should be considered supplemental to sanitation, and management measures must be directed to prevent fly breeding. Accurate records should be kept on insecticides and dosage rates used. Resistance to insecticides has developed at different levels in various poultry house locations, depending somewhat on prior exposure.

The use of a variety of different classes or families of insecticides can minimize the development of resistance. Rotate the use of organophosphate, carbamate, pyrethroid, and other classes of insecticides, when necessary.

1.4.5.1. Residual Sprays

Residual sprays usually are the most effective and economical method for controlling potentially heavy population of adult flies of any species present. These sprays should be applied in spring at the beginning of the fly season. Application after manure removal will reduce fly build-up that usually follows house clean-out. A second application should be made 5 to 6 weeks later. (Two sprays are required.)

Surfaces on which flies locate, such as poultry house framework, the ceiling, walls, trusses, wires supporting cages, electric light cords, and other areas marked by fly specking should be targeted. Outside the poultry house around openings and on shrubs and other plants where flies rest must also be treated (Table 45.1).

Table 45.1. Residual sprays for fly control

Insecticide	Dilution level in water	***
Dichlorvos	0.1% solution of 40.2% EC	1
Malathion	2.5% solution of 57% EC	6-7
Permethrin	½ to 1% solution of permethrin 25% WP	5-6
Permethrin	2.5% solution of 5.7% EC in water or 0.5% solution of 25% WP	5-6
Permethrin	1.25% solution of permethrin 11% EC	5-6
Permethrin	0.25% solution of 40% permethrin	4
Tetrachlorvinphos and dichlorvos	4% solution of 28.7% EC	4
Tetrachlorvinphos	2-4% solution of 50% WP	4-8
λ- cyhalothrin 10%	1.5% solution	5-6
*** lit of diluted mixture per 100 m² floor area for coarse, wet spray		
	Source : Hoelscher on internet	

Coarse, low-pressure sprays are applied to the point of runoff at pressures of 5-6 kg/cm², using a power sprayer or good proportioned-type sprayer. Depending on the insecticide used and the type of surface sprayed, treated areas may remain toxic for 2 to 15 weeks.

Contamination of feed, water, and eggs has to be avoided during spraying. Birds should never be sprayed upon.

1.4.5.1.1. Portable Mechanical Foggers and Misters

It is often impractical to treat large poultry houses with residual sprays. Portable, light-weight, mechanical fogging machines are convenient, efficient, and labor-saving in caged bird operations to quickly reduce adult fly population, providing quick fly knockdown with poor residual action. Spraying micron-particle-size spray droplets (≤ 30μ), is a very effective contact application with little or no residual effect. Space applications should fill the room with fog or mist.

For indoor space application to kill flies, windows and doors are closed. Natural pyrethrins, used inside for adult fly control, are easy to use at 1% pyrethrin + 5% piperonylbutoxide; 1 ml per m^2 of space is required. Spray is directed toward upper areas of the room. The room is closed for at least 1 hour and ventilated before re-entry. Application is repeated as required.

To kill flies in open areas near buildings, an outdoor ground application is needed, preferably when the temperature is cool (= 23.8°C) and wind velocity is approximately = 8 km/h. A rate of 3 ml per 100 m^2 area is sufficient. Spray drift is allowed to penetrate dense foliage. The application may be repeated, if required.

Other treatments (per 100 m^2 area) include using 200 cc of dichlorvos 1% oil base (ready-to-use) or 125 cc permethrin 5.7% undiluted or 200 cc permethrin 10% EC. Spray equipment must be rinsed after application.

Generally, fly kill will be good and spray can be repeated, if needed, only after 2 weeks and label directions and safety precautions should be strictly adhered to.

1.4.5.1.2. Stationary Building Atomizers

Treatments are especially useful in closed egg rooms or other work areas where there is little or no air movement. Pyrethrum oil-base space spray (0.06% to 0.1% pyrethrins) plus piperonyl butoxide can be used as a mist or fog in the air throughout the poultry house at the rate of ½ lit per 1000 m^3 on an "as-needed" basis for best fly control.

1.4.5.2. Baits

Baits are a supplement to residual and aerosol sprays. Baits are placed outside cages upstairs in the high-rise house. They can be effectively applied on clean walkways by using a simple fertilizer spreader. (Baits falling into the pit may destroy beneficial parasites.) These selective adulticides suppress low fly population, maintaining them at a low level.

Baits should never be accidentally eaten by the birds or mixed into feed. Dry sugar baits of methomyl are effective. To reduce potential resistance, baits are rotated. Methomyl is a carbamate insecticide, whereas other baits, wet or dry, using dichlorvos, trichlorfon and tetra-chlorvinphos, mixed with sugar are organophosphate insecticides.

1.4.5.2.1. Resin Strips and Fly Belts

Ready-to-use dichlorvos 20% resin strips can be used at the rate of 35 strips per 1,000 m^3 of enclosed area. Strips will need to be replaced as they lose their effectiveness in about every 3 months.

Methomyl fly belts can be attached to surfaces out of the reach of food-producing animals. The belt may be cut to any desired length and attached to surfaces such as walls and ceilings.

Both resin strips and fly belts may become dusty and dirty if used for long periods.

1.4.5.2.2. Feed Additive

An insect-growth regulator known as a cyromazine, when blended into a poultry ration, will control manure breeding flies in and around caged or slatted flooring in layer and breeder chicken operations not to be fed to broilers).

Cyromazine 1% Premix kills fly larvae before adulthood and does not adversely affect natural predators and parasites. It is blended into feed at the rate 0.05%(W/W). However, potential fly resistance must be monitored. Resistant flies have developed in large poultry operations where label directions have not been followed. Cyromazine should never be fed continuously throughout the year.

Adult flies are first monitored in and near the poultry house. When the population reaches a level to cause concern, an adulticide, such as pyrethrins, is sprayed or fogged to reduce the breeding potential. Adults are sprayed for as long as possible. Then, the manure is checked first at "hot spots" in the pits for maggot activity. If maggots are active, Cyromazine is introduced in the ration. Feeding could begin and continue as directed for 4 to 6 weeks (minimum of 4 weeks) and, if little or no maggot activity is observed in the manure, cyromazine is discontinued. This is usually enough time to break the fly population life cycle. Manure pits are continuously monitored and if maggots become active again, the procedure is repeated.

Baits, sprays or fogs are used, as needed, during and between Cyromazine feeding periods to control flies. Manure pits are not to be sprayed. During winter months or periods of low fly pressure, Cyromazine is discontinued for at least 4 consecutive months per year.

Cyromazine use in poultry is limited as a feed-through in chickens only and may not be fed to any other poultry species. To avoid illegal residues, Cyromazine – treated feed must be removed from layers at least 3 days (72 hours) before slaughter.

Manure from animals fed Cyromazine maybe used as a soil fertilizer supplement. Manure is not to be applied more than 7½ tons per ha per year. It is not applied to small grain crops that will be harvested or grazed as well because, illegal residues may result.

1.4.5.2.3. Liquid Spray

Cyromazine is available as a soluble concentrate, which when diluted with water suitably acts as a larvicide to control fly species developing in poultry manure and refuse by breaking the life cycle at the maggot stage.

2. Rodents

It is unusual to find a poultry farm that does not have at least a few rats or mice, and, more often than not, the population is much larger than suspected. In addition to eating and contaminating a great deal of food, rodents do considerable damage to buildings. They undermine foundations, destroy curtains and insulation, damage equipment, and cause fires by gnawing electrical wiring. In rare cases, rats have been known to kill poults and young chickens. Finally, since they are able to carry a variety of diseases and ectoparasites, rodents can affect flock health and performance.

2.1. Identification

Two species of rodents, the Norway rat and the house mouse, are common pests on most poultry farms. The following descriptions provide information on their biology and behavior that is useful in their managing these pests:

2.1.1. Norway Rat (*Rattus norvegicus*)

Adults are up to 46 cm long from head to tail. The tail is hairless and shorter than the body; the fur is reddish, grayish brown, or black with the underside gray or yellowish white. (Varicolored forms may occur). Common names are brown rat, house rat, barn rat, sewer rat, and wharf rat.

Rats burrow under and along foundations and feed bins, and in secluded spots near poultry houses (fields, trash piles, and banks of ditches and lagoons). They may also burrow into manure under slats or cages. Their burrows are large, often with conspicuous piles of dirt nearby.

They are active at night. If they are seen above ground during the daytime, it indicates a large population. Capsule-shaped droppings about 13 mm long may be seen along walls and areas where rats move or congregate.

Rats prefer fresh food when available. They are cautious. They may not take baits immediately unless placed directly in their path. They may pass up baits even when correctly placed if a better food source is readily available. They are excellent climbers, able to enter buildings by a variety of routes.

2.1.2. House Mouse (*Mus musculus*)

Adults are between 13 and 18 cm long from nose to tail. Their hairless tail is as long as their body. They have light brown to black fur with a white underside.

Mouse burrows are small, ranging from 6 to 13 mm in diameter. They are found along foundations, under boards, near feed bins, in manure under slats and cages, and in other similar areas. They also nest in walls, ceilings, and curtains left down for extended periods.

Mice feed throughout the day, with greatest activity at dawn and sunset.

Droppings are smooth and about 6 mm long. They may be seen along interior walls, on sills, and in secluded areas where mice move and congregate.

Mice are very curious and will investigate bait stations and bait placed in their path. They are excellent climbers, able to enter buildings by a variety of routes.

2.2. Rodent Biology

In general, rodents have three basic requirements: food, water, and harborage (places to hide and nest). If one or more of these items is missing from the area, rodent population will remain low.

Unfortunately, all three are usually abundant in and around poultry houses. An adult rat eats about 30 to 60 g of food each day, whereas a mouse will eat far less, about 3 g per day. Individually, this is not a lot of feed, but a large population can account for several tons of food each year.

Although both rats and mice need water to survive, mice are often able to get what little moisture they need from the food they eat. This ability allows them to nest and feed in locations where water is not abundant. Rats are not so adaptable. They cannot extract enough moisture from their food and must be relatively close to a source of water.

Rats and mice are both burrowing animals, but mice also build nests above ground in hidden, secluded areas such as walls and ceilings. Rats, on the other hand, generally nest almost exclusively underground and come out only to find food or water.

The reproductive capacity of rats and mice is quite high. Both breed throughout the year, producing 4 to 8 litters annually. Rats are sexually mature at 3 to 5 months of age and have 6 to 12 young per litter. Mice reproduce when younger (1 to 2 months) and deliver 5 to 6 young per litter. Based on reproductive potential alone, a pair of rats could produce 1,500 offspring in a year. Fortunately, other factors such as predation, food availability, and population density limit reproduction and survival in nature. Even so, rat and mouse numbers can rise quickly if ignored.

2.3. Control

Three elements make up a good rodent management program: sanitation, rodent proofing, and rodent killing.

Sanitation and rodent proofing, the first lines of defense, include a number of cultural practices easily incorporated into the overall management of the poultry farm. Successful management of rodent infestations also involves some method of killing rats and mice. Rodenticides (baits, concentrates, tracking powders, and fumigants) are the most efficient method of rodent control but should be selected and used carefully to be effective.

2.3.1. Sanitation

Sanitation is nothing more than a combination of cultural practices that deny rodents harborage and food. Rats and mice are most troublesome where there is a good cover to hide their movements and provide sites in which to nest. A well-maintained poultry facility not only exposes them to predators but also makes it easier to spot any burrowing activity. This latter aspect is a big help when using rodenticide.

The following items are important to the success of a good sanitation program:

1. The area around poultry houses should be kept mowed.
2. A clear zone of not less than 15 m from field borders and fence or tree lines needs to be maintained.
3. Old equipment, lumber, and trash from around buildings should be removed. Lumber may be stacked near buildings if it is elevated 0.3 to 0.6 m above the ground.
4. If side curtains are dropped for the summer, they are raised and lowered twice a week to stop mice from nesting in the folds.
5. Feed spills inside poultry houses and at feed bins are cleaned at once.

2.3.2. Rodent-Proofing

Completely rodent-proofing a poultry house is impractical, if not impossible. However, sealing-up obvious entry points makes it a little more difficult for rodents. Even small cracks and holes in walls, foundations, and screens should be patched. Mice need only a 0.6 cm hole to gain entry into a building.

Seal openings around water pipes, drain spouts, and vents with concrete or heavy mesh. Cover openings and floor drains with mesh. In new structures with corrugated siding, use flashing to seal both top and bottom of the siding. Be sure that corner seams are tight.

Sidewalk on all sides of the building (as projection of the floor outside the side-walls and front-walls) is usually provided for poultry buildings to help ease the observation of the birds without entering the building. It is made of CC (1:5:10) to 45-50 cm width, 7.5-9.0 cm thick and smoothened on all surfaces so that it can act as rat- and snake-proofing structure as well.

2.3.3. Killing Rodents

Glue-boards and traps are devices that can be used to control rodent population. In small areas, these devices can be efficient; in a house of size > 1000 m^2, however, baits are often more practical.

2.3.3.1. Rodenticides

A wide variety of rodenticidal compounds and formulations are available. The selection of the right material for a specific situation is important.

An understanding of the basics is necessary to know how to use a particular compound and formulation most effectively. Rodenticides are formulated as pellets, bar-baits, tracking powders, and concentrates.

Pellets are formulations of poison mixed with grain products and a binder. They may be packaged loose or in individual pitch packs. Bar-baits are formulated with rodenticide, grain products, and a binder with a high wax content to withstand moisture for long periods. Tracking powders are compounds formulated with talc or other inert material. They are intended for use along rodent runways. Rodents pick up the poison on their fur, tails, and feet and ingest it during grooming. Concentrates are designed to be mixed with feed or water.

The most important thing to know about a rodenticide is that the type of active ingredient determines how the material is to be used. Failure to use a particular rodenticide correctly will result in poor control and may present a hazard to non-target animals.

In general, rodenticides can be classified as either multiple- or single-dose poisons. All of the multiple-dose and some of the single-dose products affect the rodent's nervous system or other bodily functions.

Multiple-dose poison must be eaten every day for 7 to 21 days in order for the rodent to accumulate a lethal dose. Any interruption of exposure breaks the cycle, and, although rodents may become ill, they will not die. In such cases rodents may learn to avoid the bait.

Active Ingredients in Multiple-Dose rodenticide are Warfarin, Fumarin, Chlorophacinone and Diphacinone.

Single-dose poison have a decided advantage over multiple-dose rodenticides in that rats and mice receive a lethal dose after only one or two feedings. Active Ingredients in Single-Dose rodenticide are Brodifacoum, Bromadiolone, Bromethalin, Cholecalciferol and Zinc phosphide.

Once a rodenticide has been selected, it must be used properly to be effective. Random placement of bait or tracking powder around a poultry facility is rarely successful. It must be remembered that rodents will not go out of their way to eat poison bait if they have other food readily available. Similarly, tracking powders will not work as intended if rodents do not run through the material on their way to and from their feeding or nesting areas.

2.4. Baiting methods

2.4.1. Rat Baiting

Rats are much easier to bait than mice. Their burrows are conspicuous and, once located, can be baited by placing the rodenticide directly in the burrow. The following method is called pulse baiting and is an effective way to kill rats.

1. All burrows located and marked.
2. All burrows sealed with newspaper or soil.
3. Burrows inspected the next day and a packet of bait placed well inside each open burrow. It is not necessary to open the packets.
4. All open burrows baited for 2 consecutive days if using a single-dose and 10 to 14 days if using a multiple-dose rodenticide.
5. All burrows closed and waited for 1 week.
6. Step 4 repeated.
7. All burrows closed and monitored for activity.
8. All new burrows baited when they first appear.

Where rat burrows are located in inaccessible areas such as the manure under slats and along steep banks, bait stations should be used. Place stations against walls or on rodent run-ways with the first station as close to the burrows as practical. Orient the entry holes along the wall or path. Well-used runways are easy to spot (by the large amounts of droppings, rodent tracks, greasy looking rubbing marks, etc.) and are the best place to locate bait stations. Position stations at 3 to 5 m intervals to cover a large area and give rats ample opportunity to find the bait before reaching their normal feeding sites. One or two stations are located around feed bins as well. The stations are inspected every few days and 30 g of fresh, loose bait is added, as needed. If the bait has not been taken within several days, the station is shifted to a new location.

2.4.2. Mouse Baiting

Large numbers of mice will often nest in the walls and ceilings of poultry houses. Consequently, mouse baiting is a matter of quantity and persistence. Bait bars or stations baited with 30 g of loose bait should be placed at 1½ to 3 m intervals throughout the house. Sill plates and horizontal wall braces are often good locations to place bait in breeder, turkey, and other open-floor houses. Bait stations may also be placed along alleyways in facilities where birds are caged or penned. Egg rooms, offices, and attic spaces should also be baited.

As with rat baiting, placement is important. Bait stations and bar baits should be placed next to walls or on the horizontal surface of sills and braces. Corners are good locations, as are cool cells and housings for exhaust fans. In all cases, care should be taken to attach stations firmly and in such a way that birds cannot reach the bait.

Once bait has been placed, inspect the locations frequently and replenish the bait as needed. Inspection intervals depend on the severity of the mouse infestation and type of bait used. Where multiple-dose baits are used or the infestation is heavy, check locations daily for the first week. Try increasing the amount of bait placed in each station if mice have eaten it all. This will allow for less-frequent inspections. In time, it will be necessary to inspect stations only once a week.

Bar-baits are partially useful in locations such as sill plates that are near the ceiling. Most bars have a hole in the center that allows the bait to be nailed in place. Bar-baits generally last much longer than loose baits because mice cannot carry them to their nests. It is not uncommon for bar baits to last several months. Evidence of feeding is also easier to gauge by noting the condition of the bar bait.

Bait stations are easily made from a variety of materials. One of the simplest consists of a 45 to 60 cm section of 2.5 cm polyvinyl chloride (PVC) pipe (5.0 or 7.5 cm for a rat station) or plastic drain tile. More elaborate stations that fit nicely into corners or behind materials in storage can be made with PVC pipe and a variety of T-U joints. Plastic pails with lids also make good bait stations. 7.5 to 10.0 cm-diameter holes are made (5 cm diameter holes for rats) at the base of the pail.

Chapter **46**

Carcass Disposal

A practical and sanitary system for disposing of dead poultry will help prevent the spread of disease, prevent odors and fly breeding, and meet local regulations for water and air pollution. Regardless of the method of disposal chosen, access to the carcasses by scavenger animals such as dogs, birds, cats etc. must be prevented. Further, a disposal method that best fits the management system and location of the farm can be selected. The criteria for selecting a most suitable disposal method needs to include the following (www.cps.gov.ca):

1. Compliance with existing laws
2. Economics of each method
3. Rate of mortality
4. Capital costs
5. Equipment availability
6. Cost of labor
7. Reliability of each method
8. Degree of biosecurity

The pattern of mortality is also important. Carcass mass is fairly consistent in a breeder or layer operation but a grow-out operation will have increasing volumes as body size increases with age. Catastrophic losses can create havoc with any disposal method and alternative procedures should be in place in case of a severe disease outbreak or a management problem, such as ventilation failure (in windowless houses) which may cause high losses (www.cps.gov.ca).

It is necessary to note that, India ranks 6th and 5th in World egg and poultry meat production (Year 2004), respectively (www.wattpoultry.com). Annually about 2000 m birds are being reared. Even assuming a mortality of 1%, 20 m carcasses (approximately 20,000 tonnes) are being disposed. Hence, it is not far off that serious issues like environment, public relations etc. are going to crop up. It is in the fitness of things that proper carcass disposal methods are adopted by the producers depending on the local agro-climatic conditions as well as economic and practical feasibilities.

1. Pre-processing

Pre-processing of carcasses on-site increases biosecurity and will increase the number of process options available to utilize mortalities. It isa good pre-requisite for most of the carcass disposal methods. Pre-processing methods are: Freezing, Grinding, Fermentation and grinding and sterilization.

1.1. Freezing

This is most suitable to normal day-to-day mortalities, in which the carcasses are stored in a freezer until they can be taken to a central processing site. Freezing is currently being used by some large poultry producers.

The temperature required depends on the length of storage time required. For extended periods (weeks instead of days)or situations where the mortalities must be transported for several hours before reaching the central processing unit, the carcasses must be frozen.

Freezing should have no direct negative effects on the environment.

1.2. Grinding

A large portable grinder can be taken to an affected farm to grind up to 15 tons of animal carcasses per hour. The processed material could be preserved with chemicals or heat and placed in heavy, sealed, plastic-lined roll-off containers. The containers could then be taken off-site to a central processing facility.

Grinding of carcasses will make most mortality-processing systems more effective and more rapid by increasing the surface area where chemical and biological processes can occur. It is required as a component for fermentation of carcasses, which is a method of storing carcasses on the farm for several months, until they can be transported to a central site for disposal.

Grinding will speed biological decomposition of a carcass, so ground material should be used rapidly for additional processing/disposal, or a preservative may be warranted if the material will be stored. Grinding will also increase the potential for odor.

1.3. Fermentation

Carcasses can be stored for at least 25 weeks. Fermentation is an anaerobic process that proceeds when ground carcasses are mixed with a fermentable carbohydrate source and culture inoculants and then added to a watertight fermentation vessel. It produces a "silage" end project that is nearly pathogen-free. Fermentation of carcasses typically proceeds at near ambient temperatures in sealed containers that are vented for carbon dioxide.

Fermentation should not pose a threat to the environment, as long as the fermentation container is watertight and as long as no spills occur while preparing materials for the fermentor or while removing material from it.

2. Carcass-disposal methods

Many methods have been developed and they are comprehensively reviewed by Nutsch and Spire (2004). The following is an excerpt from their report. Readers are advised to look into the original review for exhaustive information on each of the following carcass-disposal methods: 1. Burial 2. Incineration 3. Composting 4. Rendering 5. Lactic acid fermentation 6. Alkali hydrolysis 7. Anaerobic digestion and 8. Miscellaneous methods.

2.1. Burial

In spite of potential logistical and economic advantages, concerns about possible effects on the environment and subsequently public health have resulted in a less favorable standing for this method. Landfills represent a significant means of waste disposal throughout the world, and have been used as a means of carcass disposal in several major disease eradication efforts, including the 1984 and 2002 avian influenza (AI) outbreaks. The term "mass burial site " is used to refer to a burial site in which large numbers of animal carcasses from multiple locations are disposed, and which may incorporate systems and controls to collect, treat and/ or dispose of leachate and gas.

2.1.1. Techniques

2.1.1.1. Trench burial

Disposal by trench burial involves excavating a trough into the earth, placing carcasses in the trench, and covering with the excavated material (backfill). Relatively little expertise is required to perform trench burial, and the required equipment is commonly used for other purposes. The primary resources required for trench burial include excavation equipment and a source of cover material. Cover material is often obtained from the excavation process itself and reused as backfill.

Important characteristics in determining the suitability of a site for burial include soil properties; slope or topography; hydrological properties; proximity to water bodies, wells, public areas, roadways, dwellings, residences, municipalities, and property lines; accessibility; and the subsequent intended use of the site. Although many sources concur that these characteristics are important, the criteria for each that would render a site suitable or unsuitable vary considerably.

2.1.1.1.1. Advantages

1. Economical
2. Convenient, logistically simple and relatively quick
3. Done on farm and no transport is required
4. More discrete than other methods and hence does not attract public attention

2.1.1.1.2. Disadvantages

1. Potential for detrimental environmental effects, specifically water quality issues, as well as the risk of disease agents persisting in the environment.

2. Trench burial may be limited by constraints or exclusions due to lack of sites with suitable geological and/or hydrological properties, and the fact that burial may be prohibitively difficult when the ground is wet or frozen.
3. In some cases, the presence of an animal carcass burial site may negatively impact land value or options for future use.
4. Burial of carcasses does not generate a useable by-product of any value.

2.1.1.2. Landfill

Modern landfills (under Subtitle D regulations of the USA) are highly regulated operations, engineered and built with technically-complex systems specifically designed to protect the environment.

2.1.1.2.1. Advantages

1. Infrastructure availability especially for large quantity of carcass.
2. Since the site will be usually pre-approved, poses little risk to the environment
3. Wide geographic dispersion

2.1.1.2.2. Disadvantages

1. Landfill sites that do accept animal carcasses may not be open for access when needed or when convenient.
2. Potential spread of disease agents during transport of infected material to the landfill
3. Expensive because the site does not usually belong to the producer.

2.1.1.3. Mass burial

The total amount of space required for a mass burial site would depend on the volume of carcass material to be disposed and the amount of space needed for operational activities. In general, the resources and inputs required for a mass burial site would be similar in many respects, although likely not as complex, as those required for a landfill. However, whereas the infrastructure at an established landfill would be pre-existing, the resources for a mass burial site likely would not.

2.1.1.3.1. Advantages

1. High volumes of carcass can be disposed of
2. Poses little risk to the environment

2.1.1.3.2. Disadvantages

1. Expensive
2. Long-term environmental consequences are not yet known
3. No usable by-product produced

2.1.2. Implications to the environment

Animal carcass decomposition from the point at which an animal succumbs to death, degradation of bodily tissues commences, the rate of which is strongly influenced by various endogenous and environmental factors. Soft tissue is degraded by the postmortem processes of putrefaction (anaerobic degradation) and decay (aerobic degradation). Putrefaction results in the gradual dissolution of tissues into gases, liquids, and salts as a result of the actions of bacteria and enzymes. A corpse or carcass is degraded by microorganisms both from within (within the gastrointestinal tract) and from without (from the surrounding atmosphere or soil).Generally body fluids and soft tissues other than fat (i.e., brain, liver, kidney, muscle and muscular organs) degrade first, followed by fats, then skin, cartilage, and hair orfeathers, with bones, horns, and hooves degrading most slowly.

Relative to the quantity of leachate that may be expected, it has been estimated that about 50%of the total available fluid volume would "leak out"in the first week following death, and that nearly all of the immediately available fluid would have drained from the carcass within the first two months.

Regarding the gaseous by-products that may be observed from the decomposition of animal carcasses, the composition would be approximately 45%carbon dioxide, 35% methane, 10%nitrogen, with the remainder comprised traces of other gases such as hydrogen sulfide; gas production, including methane, increases over time, rather than decreases.

The time required for buried animal carcasses to decompose depends most importantly on temperature, moisture, and burial depth, but also on soil type and drainability, species and size of carcass, humidity/aridity, rainfall, and other factors. In addition to actual carcass material in a burial site, leachates or other pollutants may also persist for an extended period. Although much of the pollutant load would likely be released during the earlier stages of decomposition, mass burial sites could continue to produce both leachate and gas for as long as 20 years.

2.1.3. Environmental impacts

The primary environmental risk associated with burial is the potential contamination of groundwater or surface waters with chemical products of carcass decay. Since precipitation, amount and soil permeability are key to the rate at which contaminants are "flushed out" of burial sites, the natural attenuation properties of the surrounding soils are a primary factor determining the potential for these products of decomposition to reach groundwater sources. The most useful soil type for maximizing natural attenuation properties was reported to be a clay-sand mix of low porosity and small to fine grain texture.

High levels of ammonia, total dissolved solids (TDS), biochemical oxygen demand

(BOD),and chloride in the monitoring well closest to the burial site of turkey carcasses (within 2 ft) were observed, and average ammonia and BOD concentrations were observed to be very high for 15 months. However, little evidence of contaminant migration was observed more than a few feet from the burial site.

2.2. Incineration

Incineration has historically played an important role in carcass disposal. Advances in science and technology, increased awareness of public health, growing concerns about the environment, and evolving economic circumstances have all affected the application of incineration to carcass disposal.

Today there are three broad categories of incineration techniques: open-air burning, fixed-facility incineration, and air-curtain incineration.

2.2.1. Open-air burning

Open-air burning should be conducted as far away as possible from the public. Material requirements for open-air burning include straw or hay, untreated timbers, kindling wood, coal, and diesel fuel. Tires, rubber, and plastic should not be burned as they generate dark smoke. To promote clean combustion, it is advisable to dig a shallow pit with shallow trenches to provide a good supply of air for open-air burning. Kindling wood should be dry, have low moisture content, and not come from green vegetation. Open-air burning, particularly in windy areas, can pose a fire hazard.

Open-air burning of carcasses yields a relatively benign waste (ash) that does not attract pests. However, the volume of ash generated by open-air burning can be significant. Open-air burning poses additional clean-up challenges vis-à-vis groundwater and soil contamination caused by hydrocarbons used as fuel.

2.2.2. Fixed-facility incineration

Small animal carcass incinerators have been used to dispose of on-farm mortalities for years. Fixed-facility incinerators include (a) small on-farm incinerators,(b)small and large incineration facilities, (c)crematoria, and (d)power plant incinerators.

Fixed-facility incineration is wholly contained and, usually, highly controlled. They are typically fueled by diesel, natural gas, or propane. Newer designs of fixed-facility incinerators are fitted with afterburner chambers designed to completely burn hydrocarbon gases and particulate matter (PM) exiting from the main combustion chamber.

Compared to open-air burning, clean-up of ash is less problematic with fixed-facility incineration; ash is typically considered safe and may be disposed of in landfills. Although more controlled than open-air burning, fixed-facility incineration also poses a fire hazard.

2.2.3. Air-curtain incineration

Air-curtain incineration involves a machine that fan-forces a mass of air through a manifold, thereby creating a turbulent environment in which incineration is greatly accelerated — up to six times faster than open-air burning. This is a relatively new technology for carcass disposal in the wake of natural disasters. In air-curtain incineration, large-capacity fans driven by diesel engines deliver high-velocity air down into either a metal refractory box or burn pit (trench).

Air-curtain incineration also poses a fire hazard and the requisite precautions should always be taken. Air-curtain incineration, like other combustion processes, yields ash. From an ash-disposal standpoint, air-curtain incineration in pits is advantageous if the ash may be left and buried in the pits.

2.2.4. Advantages and disadvantages

Open-air burning can be relatively inexpensive, but it is labor-and fuel-intensive, and depends on favorable weather conditions, poses environmental problems, and has poor public perception.

Fixed-facility incineration is highly biosecure. However, fixed-facility incinerators are expensive and difficult to operate and manage from a regulatory perspective. Most on-farm and veterinary-college incinerators are incapable of handling large volumes of carcasses that typify most carcass disposal emergencies. Meanwhile, larger industrial facility incinerators are difficult to access and may not be configured to handle carcasses.

Air-curtain incineration is mobile, usually environmentally-sound, and suitable for combination with debris removal. However, air-curtain incinerators are fuel-intensive and logistically-challenging.

2.2.5. Environmental implications

2.2.5.1. Air pollution

It is generally accepted that open-air burning pollutes. The nature of open-air emissions hinges on many factors, including fuel type. Studies focused on dioxins, furans, polyaromatic hydrocarbons (PAHs), polychlorinated biphenyls (PCBs), metals, nitrogen oxides, sulphur dioxide, carbon monoxide, carbon dioxide, organic gases, and particulate matter (PM) — PM especially less than 10 micrometers in diameter that can be drawn into the lungs. The fear of dioxins and smoke inhalation, along with the generally poor public perception of pyres, eventually compelled the discontinuation of the use of mass burn sites. In contrast to open-air burning, properly operated fixed-facility and air-curtain incineration pose fewer pollution concerns. When compared to open-burning, air-curtain incineration is superior, with higher combustion efficiencies and less carbon monoxide and PM emissions.

2.2.5.2. Groundwater and soil pollution

Incineration is a legitimate alternative when factors related to hydrology (e.g., a high water table) or geology (e.g., soils of high permeability) rule out burial. Unfortunately, however, open-air burning itself poses problems for groundwater contamination, primarily in the form of the hydrocarbons used as fuel. Dioxins and PCBs, both of which are known to emanate from pyres, are also of soil-and food-pollution concern. Nevertheless, the general risks that incineration, particularly open-air burning, pose to groundwater and the soil are real and should always be minimized.

2.3. Composting

Composting is becoming an increasingly preferred alternative for disposing of mortalities at animal feeding operations. Carcass composting offers several benefits, including reduced environmental pollution, generation of a valuable by-product (soil amendment), and destruction of many pathogens. Because finished compost is different than the original materials from which it was derived, it is free of unpleasant odor, easy to handle, and can be stored for long periods.

2.3.1. The process

2.3.1.1. I Phase (Developing or heating phase)

This is characterized by high oxygen-uptake rates, thermophilic temperatures, and high reductions in bio-degradable volatile solids (BVS).This phase, which may last 3 weeks to 3 months, is also characterized by a potential for significant odor because the organic materials of mortalities break down into relatively small compounds, soft tissue decomposes, and bones soften partially..

2.3.1.2. II Phase (Maturation or curing phase)

The remaining materials (mainly bones)break down fully and the compost turns to a consistent dark brown to black soil or "humus" with a musty odor containing primarily non-pathogenic bacteria and plant nutrients, may require one month or longer for completion. In this phase, aeration is not a determining factor for proper composting, and, therefore, it is possible to use a low-oxygen composting system. A series of retarding reactions, such as the breakdown of lignins, occurs during this maturation or curing stage and requires a relatively long time, could be as long as five months at temperatures below 40°C.

Carcass composting systems require a variety of ingredients or co-composting materials, including carbon sources, bulking agents, and biofilter layers.

2.3.2. Composting material

2.3.2.1. Carbon sources

Various materials can be used as a carbon source, including materials such as sawdust, straw, corn stover (mature cured stalks of corn with the ears removed and used as feed for livestock), poultry litter, ground corn cobs, baled corn stalks, wheat straw, semi-dried screened manure, hay, shavings, paper, silage, leaves, peat, rice hulls, cotton gin trash, yard wastes, vermiculite, and a variety of waste materials like matured compost. A 50:50 (w/w)mixture of separated solids from manure and a carbon source can be used as a base material for carcass composting. Finished compost retains nearly 50%of the original carbon sources. Use of finished compost for recycling heat and bacteria in the compost process minimizes the needed amount of fresh raw materials, and reduces the amount of finished compost to be handled.

A carbon-to-nitrogen (C:N)ratio in the range of 25:1 to 40:1 generates enough energy and produces little odor during the composting process. Depending on the availability of carbon sources, this ratio can sometimes be economically extended to 50:1.As a general rule, the weight ratio of carbon source materials to mortalities is approximately 1:1 for high C:N materials such as sawdust, 2:1 for medium C:N materials such as litter, and 4:1 for low C:N materials such as straw.

2.3.2.2. Bulking agents

Bulking agents or amendments also provide some nutrients for composting. They usually have bigger particle sizes than carbon sources and thus maintain adequate air spaces (around 25-35%porosity)within the compost pile by preventing packing of materials.

They should have a three-dimensional matrix of solid particles capable of self-support by particle-to-particle contact. Bulking agents typically include materials such as sludge cake, spent horse bedding (a mixture of horse manure and pinewood shavings), wood chips, refused pellets, rotting hay bales, peanut shells, and tree trimmings.

The ratio of bulking agent to carcasses should result in a bulk density of final compost mixture that does not exceed 600 kg/m^3. As a general rule, the weight of compost mixture in a 19 l bucket should not be more than 11.4 kg; otherwise, the compost mixture will be too compact and lack adequate airspace.

2.3.2.3. Biofilters

A biofilter is a layer of carbon source and/or bulking agent material that 1)enhances microbial activity by maintaining proper conditions of moisture, pH, nutrients, and temperature, 2)deodorizes the gases released at ground level from the compost piles, and 3) prevents access by insects and birds and thus minimizes transmission of disease agents from mortalities to livestock or humans.

2.3.3. Site selection

A compost site should be located in a well-drained area that is at least 90 cm above the high-water table level, at least 90 m from sensitive water resources (such as streams, ponds, wells etc.), and that has adequate slope (1-3%) to allow proper drainage and prevent pooling of water. Runoff from the composting facility should be collected and directed away from production facilities and treated through a filter strip or infiltration area. Composting facilities should be located downwind of nearby residences to minimize potential odors or dust being carried to neighboring residences by prevailing winds. The location should have all-weather access to the compost site and to storage for co-composting materials, and should also have minimal interference with other operations and traffic. The site should also allow clearance from underground or overhead utilities.

2.3.4. Preparation and management of compost piles

2.3.4.1. Staging mortalities

Mortalities should be quickly removed from pens, or houses and transferred directly to the composting area. In the event of a catastrophic mortality loss or the unavailability of adequate composting amendments, carcasses should be held in an area of temporary storage located in a dry area downwind of other operations and away from property lines (ideally should not be visible from off-site). Storage time should be minimized.

2.3.4.2. Preparation and monitoring of compost piles

Co-composting materials should be ground to 2.5-5 cm and mixed. Compost materials should be lifted and dropped, rather than pushed into place (unless carcasses have been ground and mixed with the co-composting materials prior to the composting process).Compost piles should be covered by a biofilter layer during both phases of composting. If warranted, fencing should be installed to prevent access by livestock and scavenging animals.

The moisture content of the carcass compost pile should be 40-60%(wet basis),and can be tested accurately using analytical equipment or approximated using a hand-squeeze method. In the hand-squeeze method, a handful of compost material is squeezed firmly several times to form a ball. If the ball crumbles or breaks into fragments, the moisture content is much less than 50%.If it remains intact after being gently bounced 3-4 times, the moisture content is nearly 50%.If the ball texture is slimy with a musty soil-like odor, the moisture content is much higher than 50%.

A temperature probe should be inserted carefully and straight down into each quadrant of the pile to allow daily and weekly monitoring of internal temperatures at depths of 25, 50, 75,and 100 cm after stabilization during the first and second phases of composting. During the first phase, the temperature at the core of the pile should rise to at least 55-60°C within 10 d and remain there for several weeks. A temperature of 65°C at the core of the pile maintained for 1-2 d will reduce pathogenic bacterial activity and weed seed germination.

Proper aeration is important in maintaining uniform temperature and moisture contents throughout the pile during the first and second phases of the composting process. Uniform air flow and temperature throughout a composting pile are important to avoid clumping of solids and to minimize the survival of microorganisms such as coliforms, *Salmonella*, and fecal *Streptococcus*. During composting, *actinomycetes* and fungi produce a variety of antibiotics which destroy some pathogens; however, spore-formers will survive.

After the first phase of composting, the volume and weight of piles may be reduced by 50-75%. After the first phase the entire compost pile should be mixed, displaced, and reconstituted for the secondary phase. In the second phase, if needed, moisture should be added to the materials to reheat the composting materials until an acceptable product is achieved. The end of the second phase is marked by an internal temperature of 25-30°C, a reduction in bulk density of approximately 25%, a finished product color of dark brown to black, and the lack of an unpleasant odor upon turning of the pile.

Odor can be evaluated by placing two handfuls of compost material into a re-sealable plastic bag, closing the bag, and allowing it to remain undisturbed for approximately one hour (5-10 min is adequate if the sealed bag is placed in the sun).If, immediately after opening the bag, the compost has a musty soil odor (dirt cellar odor), the compost has matured. If the compost has a sweetish odor (such as slightly burned cookies),the process is almost complete but requires a couple more weeks for adequate maturation. If the compost odor is similar to rotting meat/flesh, is overpowering, is reminiscent of manure, or has a strong ammonia smell, the compost process is not complete and may require adjustments.

After the primary and secondary phases of composting are complete, the finished product can be recycled, temporarily stored, or, if appropriate, added to the land as a soil amendment.

2.3.5. Changes in pile properties during composting

The most important changes that occur in a carcass compost pile are weight and volume loss, pH changes, and production of gases and odors.

2.3.5.1. Weight and volume loss

The biochemical reactions of the composting process transform large organic molecules into smaller ones, and produce different gases and odors. As a result, the weight of the end-product becomes much less than that of the parent materials. Due to their different natures, carcasses and co-composting materials have different rates of shrinkage during the compost process.

2.3.5.2. pH

A high-alkali or low-acid environment is not well-suited to the composting process. Since the bio-degradation process releases CO_2, a weak acid and NH_3, a weak base, the compost process has the ability to buffer both high and low pH back to the neutral range as composting proceeds. Based on this fact, the right amount of carbon and nitrogen sources (for production of these two essential gases)is very important and a proper carbon-to-nitrogen (C:N)ratio keeps pH in the range of 6.5 to 7.2, which is optimum for composting. If the pH approaches 8, ammonia and other odors may become a problem. They suggested that the pH could be reduced by adding an inorganic compound, such as granular ferrous sulfate.

2.3.5.3. Gases and odors

Fermentation and oxidation of carcasses during composting produces unpleasant gases (CO_2, NH_3, H_2S etc.) and odors associated with the liquid or solid biomass. A properly covered compost pile that is biodegrading carcasses under aerobic conditions should generate little or no odor.

2.3.6. Pre-processing (grinding) of carcasses

One factor being evaluated is preprocessing (e.g., grinding) of carcasses; this pre-processing step can be used in combination with almost any composting configuration. Any process that minimizes composting time will result in a more efficient operation that is easier to manage. This is particularly true for large animal carcasses. Grinding has the following advantages:

1. Diminishes the composting time due to increased surface area and thus management cost.
2. Reduces the co-composting materials up to one-fourth of the conventional system.
3. Decreases the risk of odor production and risk of scavengers.
4. Allows better control over key composting parameters such as temperature pattern, pH, particle size, and color.
5. Produces a more uniform product.

2.3.7. Design parameters

Under standard conditions, for every 4.5 kg of mortality, there is a need for about 4.25 l space. For poultry, Daily composting capacity =Theoretical farm live weight / 400 where, theoretical farm live weight = Farm capacity x market weight. A value of 10 days was used as a minimum for poultry composting work.

2.3.8. Biosecurity

In terms of biosecurity, composting facilities should not be located directly adjacent to livestock production units, and the vehicles associated with operation should be sanitized with appropriate cleaning and disinfecting agents for each trip. The site should be downwind from residential areas, provide a limited or appealing view for neighbors or passing motorists, and possibly have a pleasing appearance and landscape. In addition to conserving energy and moisture content and minimizing odors, a biofilter also excludes insects and birds (as the most important carriers of disease microorganisms)from the compost pile, thus minimizing or preventing transmission of microorganisms from mortalities to livestock or humans.

2.3.9. Site selection *Vis a Vis* environmental factors

Disposal of animal carcasses may generate different environmental and health hazards. Improper carcass disposal processes might cause serious environmental and public health problems, including:

2.3.9.1. Odor and air quality

Odor nuisance, resulting from the anaerobic breakdown of proteins by bacteria, reduces the quality of life and decreases property values. Pathogens which may be present in decomposed material are capable of spreading diseases in soil, plants, and in animals and humans. Leaching of harmful nitrogen and sulfur compounds from carcasses to groundwater. Attraction of insects and pests as potential vectors of harmful diseases for public health.

Location of a compost facility has an important role in meeting environmental interests. Choosing an appropriate site will help to protect water and soil quality, increase biosecurity, prevent complaints and negative reactions of neighbors, decrease nuisance problems, and minimize the challenges in operating and managing the composting operation. A composting operation should:

1. Protect surface and groundwater from pollution.
2. Reduce the risk of the spread of disease.
3. Prevent nuisances such as flies, vermin, and scavenging animals.
4. Maintain air quality.

2.3.9.2. Water

The location of the composting pile should be easily accessible, require minimal travel, be convenient for material handling, and maintain an adequate distance from live production animals. Sites near neighbors and water sources or streams should be avoided. Additionally, surface runoff and other pollution controls should be employed at the site.

2.3.9.3. Soil

Compost piles should be underlain with a water barrier in order to prevent compost leachate from penetrating and contaminating the soil or base underneath. A plastic barrier may complicate turning of the pile or windrow; hence, a concrete or asphalt base (pad)is recommended instead of plastic materials.

2.4. Rendering

Rendering has historically been defined as separation of fat from animal tissues by the application of heat. However, the definition can be refined thus: rendering is a process of using high temperature and pressure to convert whole animal and poultry carcasses or their by-products with no or very low value to safe, nutritional, and economically-valuable products.

Rendering of animal mortalities involves conversion of carcasses into three end products —namely, carcass meal (proteinaceous solids),melted fat, and water — using mechanical processes (e.g., grinding, mixing, pressing, decanting and separating), thermal processes (e.g., cooking, evaporating, and drying), and sometimes chemical processes (e.g., solvent extraction).The main carcass rendering processes include size reduction followed by cooking and separation of fat, water, and protein materials using techniques such as screening, pressing, sequential centrifugation, solvent extraction, and drying. Resulting carcass meal can sometimes be used as an animal feed ingredient. If prohibited for animal feed use, or if produced from keratin materials of carcasses, the product will be classified as inedible and can be used as a fertilizer. Fat can be used in livestock feed, production of fatty acids, or can be manufactured into soaps.

Livestock mortality is a tremendous source of organic matter. A typical fresh carcass contains approximately 32% dry matter, of which 52% is protein, 41%is fat, and 6% is ash. Rendering offers several benefits to food animal and poultry production operations, including providing a source of protein for use in animal feed, and providing a hygienic means of disposing of fallen and condemned animals. The end products of rendering have economic value and can be stored for long periods of time. Using proper processing conditions, final products will be free of pathogenic bacteria and unpleasant odors.

2.4.1. Handling and storage

Animal mortalities should be collected and transferred in a hygienically safe manner because raw materials in an advanced stage of decay result in poor-quality end-products. Carcasses should be processed as soon as possible; if storage prior to rendering is necessary, carcasses should be refrigerated or otherwise preserved to retard decay.

The cooking step of the rendering process kills most bacteria, but does not eliminate endotoxins produced by some bacteria during the decay of carcass tissue. These toxins can cause disease, and pet food manufacturers do not test their products for endotoxins.

2.4.2. Rendering process

Generally rendering process is accomplished by receiving raw materials followed by removing undesirable parts, cutting, mixing, sometimes preheating, cooking, and separating fat and protein materials. The concentrated protein is then dried and ground. Additionally, refining of gases, odors, and wastewater (generated by cooking process)is necessary. Rendering processes may be categorized as either "edible" or "inedible."

2.4.2.1. Edible rendering

Carcass by-products such as fat trimmings are ground into small pieces, melted and disintegrated by cooking processes to release moisture and "edible" fat. The three end product portions (proteinaceous solids, melted fat, and water)are separated from each other by screening and sequential centrifugations. The proteinaceous solids are dried and may subsequently be used as an animal feed, water is discharged as sludge, and the edible fat is pumped to storage for refining.

2.4.2.2. Inedible rendering

This process is to convert the protein, fat, and keratin materials found in carcasses into fat, carcass meal (used in livestock feed, soap, production of fatty acids, etc.), and fertilizer, respectively. Similar to the edible process, raw materials in the first stage of an inedible process are dehydrated and cooked, and then the fat and protein substances are separated. The pre-cooking processes mainly include removal of skin and viscera and thorough washing of the entire carcass. The carcasses are crushed and transported to a weighing bin and then passed through metal and non-metal detectors. These devices in turn sort out nearly all of the magnetic and non-magnetic metal materials (tags, hardware, etc.).Metals that may be associated with the carcasses are removed by strong magnets attached to conveyors.

The use of carcasses in advanced stages of decomposition is undesirable because skin removal and carcass cleaning is very difficult, and the fat and protein resulting from such carcasses is generally of low quality. In the event of a disaster situation, decayed carcasses without entrails along with dumped viscera should be segregated and processed separately.

2.4.3. Environmental impacts

2.4.3.1. Odor

Because carcasses are typically not refrigerated for preservation prior to rendering, they begin to putrefy and give rise to a number of odorants. Due to this, rendering is often perceived by the public as an unpleasant or "smelly" industry. A significant environmental issue for the rendering industry is controlling various odors generated during pre-rendering, rendering, and post-rendering processes. Only certain chemical compounds are responsible for odor constituents. The threshold levels at which humans can detect (smell)various odorants are shown in Table 46.1.

Amines, mercaptans, and sulfides are generally expected to be present in gases from rendering plants. All emitted odors should be treated in condensing units followed by either chemical scrubbers or incinerators (afterburners)and/or biofilters for non-condensable odors. For chemical deodorization of rendering units, use of hypochlorite, multi-stage acid and alkali scrubbing followed by chlorination, and incineration of the final gases in boilers is recommended.

2.4.3.2. Wastewater

The main criteria for determining the acceptability of wastewater discharged from rendering facilities have been levels of BOD, suspended solids, and organic substances. However, available nutrients (nitrogen, phosphorus, and perhaps potassium)within wastewater may play increasingly important roles.

Microorganisms require ratios of carbon, nitrogen, and phosphorus (C:N:P)of approximately 100:6:1 to grow. Bacteria in pond systems are unable to use high loadings of nitrogen and phosphorus that may be present in rendering wastewater. Treatment of wastewater to address these constituents, specifically phosphorus, is very important. Continued use of wastewater for irrigation tends to accumulate nitrogen and phosphorus in the soil.

Table 46.1. Odor threshold concentrations of selected compounds from a rendering plant

Compound	Odor threshold (ppm by volume)
Acrolein	0.21
Butyric acid	0.001
Ammonia	46.8
Pyridine	0.021
Skatole	0.220
Methyl amine	0.021
Dimethyl amine	0.047
Trimethyl amine	0.00021
Allyl amine	28
Ethyl mercaptan	0.001
Allyl mercaptan	0.016
Hydrogen sulphide	0.0047
Dimethyl sulphide	0.0025
Dimethyl disulphide	0.0076
Dibutyl sulphide	0.180

Mechanical aeration and oxidation of wastewater can reduce nitrogen, and to some extent phosphorus, contents. Addition of appropriate chemical flocculants, such as aluminum sulfate, to wastewater converts available phosphorus to insoluble phosphorus, which can be removed by settling processes. These chemical procedures will make rendering wastewater treatment more complex and more expensive.

2.5. Lactic acid fermentation

This process provides a way to store carcasses for at least 25 weeks and produce an end product that may be both pathogen-free and nutrient-rich. Lactic acid fermentation should be viewed as a means to preserve carcasses until they can be rendered. The low pH prevents undesirable degradation processes.

The process of lactic acid fermentation is simple and requires little equipment. Indeed, the process needs only a tank and a grinder. Fermentation is an anaerobic process that can proceed in any sized non-corrosive container provided it is sealed and vented for carbon dioxide release. During this process, carcasses can be decontaminated and there is a possibility of recycling the final products into feedstuff. Fermentation products can be stored until they are transported to a disposal site.

Carcasses are ground to fine particles, mixed with a fermentable carbohydrate source and culture innoculant, and then added to a fermentation container. Grinding aids in homogenizing the ingredients. For lactic acid fermentation, lactose, glucose, sucrose, whey, whey permeates, and molasses are all suitable carbohydrate sources. The carbohydrate source is fermented to lactic acid by Lactobacillus acidophilus.

Under optimal conditions, including a fermentation temperature of about 35°C,the pH of fresh carcasses is reduced to less than 4.5 within 2 days. Fermentation with *L.acidophilus* destroys many bacteria including Salmonella spp. There may be some microorganisms that can survive lactic acid fermentation, but these can be destroyed by heat treatment through rendering.

Biogenic amines produced during putrefaction are present in broiler carcasses. A presence of a single amine (tyramine) at a concentration above 550 ppm indicates a real risk of toxicity to animals being fed. This concentration is higher in the final product after rendering because the rendered product has less moisture than the fermentation broth. Thus, efforts should be made to reduce putrefaction. Properly prepared products will remain biologically stable until they are accepted for other processes such as rendering.

Lactic acid fermentation is commonly referred to as pickling because microorganisms are inactivated and the decomposition process ceases when the pH is reduced to approximately 4.5.

Given its capacity to inactivate microorganisms and decompose biological material, lactic acid fermentation is used for decontamination and storage of carcasses in poultry production. For poultry producers, the utilization of lactic fermentation to store carcasses reduces the cost of transportation to rendering facilities by 90.

2.5.1. Risk of contamination

There is a risk of toxic products that may be present following lactic acid fermentation and rendering. If the rendered product is used as an animal feed, it is important to realize that certain toxic agents can survive this treatment. If carcasses are sterilized after particle size reduction and prior to inoculation, the risk of contamination is reduced significantly.

During the fermentation of broiler carcasses, certain amino acids have been shown to undergo decarboxylation and become biogenic amines (Table 46.2). Necrotic cellular debris in the intestines of carcasses has been associated with biogenic amines in animal protein products. The level of biogenic amines depends on the state of decomposition of carcasses that are used in lactic acid fermentation.

Table 46.2. Common biogenic amines and their precursors

Biogenic amine	Amino acid precursor	Putrefaction	Fermentation
Cadaverine	Lysine	Produced	Produced
Histamine	Histidine	Produced	Produced
Phenylethylamine	Phenylalanine	Produced	Produced
Putrescine	Arginine, Methionine	Produced	Produced
Spermine, Spermidine	Arginine, Methionine	Produced	Reduced
Tryptamine	Tryptophan	Produced	Produced
Tyramine	Tryosine	Produced	Produced

2.5.2. Implications to the environment

Lactic acid fermentation does not have any significant environmental effects if the products of fermentation are rendered and/or processed into marketable products. The process allows the carcasses to be stored until they can be processed.

2.5.3. Advantages and disadvantages

The advantages and disadvantages associated with lactic acid fermentation are presented in Table 46.3.

Table 46.3. Advantages and disadvantages of fermentation process

Advantages	Disadvantages
Decontamination of carcasses	All pathogens are not destroyed
Possibility of recycling into a feedstuff	Risk of contamination
Possibility of storage	Problem of corrosion
Potentially mobile process	Need carbohydrate source and culture of *Lactobacillus acidophilus*

2.6. Alkaline hydrolysis

Alkaline hydrolysis represents a relatively new carcass disposal technology. It has been adapted for biological tissue disposal (e.g., in medical research institutions)as well as carcass disposal (e.g., in small and large-managed culls of diseased animals).

2.6.1. The process

Alkaline hydrolysis uses sodium hydroxide or potassium hydroxide to catalyze the hydrolysis of biological material (protein, nucleic acids, carbohydrates, lipids, etc.)into a sterile aqueous solution consisting of small peptides, amino acids, sugars, and soaps. Heat is also applied (150 ^{o}C)to significantly accelerate the process. The only solid byproducts of alkaline hydrolysis are the mineral constituents of the bones and beaks. This undigested residue, which typically constitutes approximately two percent of the original weight and volume of carcass material, is sterile and easily crushed into a powder that may be used as a soil additive.

Proteins — the major solid constituent of all animal cells and tissues — are degraded into salts of free amino acids. Some amino acids (e.g., arginine, asparagine, glutamine, and serine)are completely destroyed while others are racemized (i.e., structurally modified from a left-handed configuration to a mixture of left-handed and right-handed molecules).The temperature conditions and alkali concentrations of this process destroy the protein coats of viruses and the peptide bonds of prions. During alkaline hydrolysis, both lipids and nucleic acids are degraded. Carbohydrates represent the cell and tissue constituents most slowly affected by alkaline hydrolysis. Both glycogen (in animals)and starch (in plants)are immediately solubilized; however, the actual breakdown of these polymers requires much longer treatment than is required for other polymers.

Once broken down, the constituent monosaccharides (e.g., glucose, galactose, and mannose)are rapidly destroyed by the hot aqueous alkaline solution. Significantly, large carbohydrate molecules such as cellulose are resistant to alkaline hydrolysis digestion. Items such as paper, string, undigested plant fibers, and wood shavings, although sterilized by the process, are not digestible by alkaline hydrolysis.

Alkaline hydrolysis is carried out in a tissue digester that consists of an insulated, steam-jacketed, stainless-steel pressure vessel with a lid that is manually or automatically clamped. The vessel contains a retainer basket for bone remnants and other materials (e.g., indigestible cellulose-based materials, latex, metal, etc.).The vessel is operated at up to 70 psig to achieve a processing temperature of 150°C. An individual can load and operate an alkaline hydrolysis unit. In addition to loading and operation, personnel resources must also be devoted to testing and monitoring of effluent (e.g., for temperature and pH) prior to release into the sanitary sewer system .Once loaded with carcasses, the system is activated by the push of a button and is thereafter computer-controlled. The weight of tissue in the vessel is determined by built-in load cells, a proportional amount of alkali and water is automatically added, and the vessel is sealed pressure-tight by way of an automatic valve. The contents are heated and continuously circulated by a fluid-circulating system.

The process releases no emissions into the atmosphere and results in only minor odor production. The end-product is a sterile, coffee-colored, alkaline solution with a soap-like odor that can be released into a sanitary sewer in accordance with local and federal guidelines regarding pH and temperature. This can require careful monitoring of temperature (to ensure release of the effluent at or above 190°C, a temperature below which the effluent solidifies), pH, and biochemical oxygen demand (BOD).

The total process time required for alkaline hydrolysis digestion of carcass material is three to eight hours, largely depending on the disease agent(s)of concern. For conventional (e.g., bacterial and viral)contaminated waste, fourhours is sufficient.

2.6.1.1. Advantages

1. Combination of sterilization and digestion into one operation
2. Reduction of waste volume and weight by as much as 97 percent
3. Complete destruction of pathogens, including prions
4. Production of limited odor or public nuisances, and
5. Elimination of radioactively contaminated tissues.

2.6.1.2. Disadvantages

1. At present, limited capacity for destruction of large volumes of carcasses and
2. Potential issues regarding disposal of effluent.

2.7. Anaerobic digestion

Anaerobic digestion, sometimes referred to as biomethanization and biodigestion is suited for large-scale operations, reduces odor, and reduces pollution by greenhouse gases due to combustion of methane. Anaerobic digestion has been used for many years for processing a variety of wastes. Carcasses have higher nitrogen content than most wastes, and the resulting high ammonia concentration can inhibit anaerobic digestion. This limits the loading rate for anaerobic digesters that are treating carcass wastes.

Anaerobic digestion involves a transformation of organic matter by a mixed culture bacterial ecosystem without oxygen. It is a natural process that produces a gas principally composed of methane and carbon dioxide.

2.7.1. End-products

2.7.1.1. Biosolids

After several appropriate treatments, sludge becomes a biosolid that may be used for land disposal .The first treatment of sludge involves thickening of the watery mass. During thickening, biodegradation continues and converts some of the sludge into biogas, which reduces a part of the solid concentration. Thickening process reduces disposal costs by reducing the quantity of biosolid. Furthermore, the water-eliminated liquor may be treated before discharge and used for agricultural purposes. With the addition of lime, chlorine, or heat, biosolids may be stabilized to reduce both odor and the number of pathogens present.

2.7.1.2. Liquor

Liquor produced from anaerobic digestion can be used as a liquid fertilizer because of its wide range of nutrients; however, if it has a low level of nutrients and a high level of water, it could simply be used in irrigation. Liquor can be used for "fertigation"(a combination of fertilizer and irrigation)on agricultural lands, but it cannot be used in greenhouses as it contains some particles that block feeder pipes.

2.7.1.3. Methane

Methane from anaerobic digestion can be used for the production of heat, which could be used to maintain a proper temperature in the digester or to heat a local farm or factory. Methane can also be used for the production of electricity. Such use of biogases reduces the consumption of fossil fuels and may reduce the cost of electricity at the particular digestion facility. If produced in sufficient quantities, methane-produced electricity can be sold to local energy distribution networks.

2.7.2. Implications to the environment

There are several environmental advantages to anaerobic digestion. The process reduces greenhouse gas problems, decreases the consumption of fuel, and transforms waste into fertilizer. The consumption of water is a problem in dry areas; however, water from the process can be used for irrigation.

From a public relations perspective, people generally accept biodigesters. However, they should still be constructed far from residential areas for reasons of biosecurity and to minimize odor problems.

2.7.3. Advantages and disadvantages

Table 46.4. Advantages and disadvantages of anaerobic digestion.

Advantages	Disadvantages
Couples the treatment of waste and production of energy	Cost of construction is expensive
Reduction of odors	Sludge disposal is a problem in some locations
Suited for large-scale operations	Larger than other installations such as lactic acid fermentation
Methane is used in place of fossil fuels	Difficulty of storage of gas (corrosive)

Advantages	Disadvantages
Reduces chemical and biological oxygen demand, total solids and volatile solids of the carcass	Problem of management of the sludge
Destroys, or reduces to acceptable levels, coliform bacteria, pathogens, insect eggs and internal parasites	Does not destroy all pathogens: Thermo resistant bacteria (e.g., *Bacillus cereus*)

2.8. Miscellaneous methods

There are several unconventional options for disposing of animal mortalities. These miscellaneous methods include the following: Thermal depolymerization, Plasma arc process, Re-feeding, Napalm, Ocean disposal, Non-traditional rendering (including flash dehydration, fluidized-bed drying, and extrusion/ expeller press) and Novel pyrolysis technology.

2.8.1. Non-traditional and novel-disposal methods

A variety of traditional carcass-disposal methods are used to address daily mortalities in animal production operations. These non-traditional or novel methods can be adapted for daily mortalities, catastrophic losses, or both.

The selection and implementation of mass carcass disposal strategies should be considered within a decision-making process that is part of a national and/or state-level contingency plan. The goals and objectives of a contingency plan should endeavor to:

1. Minimize the number of animals to be slaughtered in the case of disease or injury
2. Minimize disruption of farming activities and food production
3. Minimize potential damage to the environment
4. Minimize damage to the economy
5. Contain and eradicate infectious disease outbreaks
6. Provide a safe, rapid attainment of disease-free status
7. Maximize use of existing infrastructure
8. Minimize cost to the taxpayer
9. Protect public health and safety and
10. Retain the confidence and support of the public.

2.8.1.1. Thermal depolymerization

This is an intriguing possibility for processing large-scale mortality events. This is a relatively new process that uses high heat and pressure to convert organic feedstock (e.g., pre-processed carcasses)into a type of fuel oil. The plasma arc process relies on extremely hot plasma-arc torches to vitrify and gasify hazardous wastes, contaminated soils, or the contents of landfills. It can reduce material volume by up to 90 %. The process also generates fuel gases that can be collected and sold to help defray operational costs. But, this method is not yet common for animal carcass disposal.

Thermal depolymerization is to reduce complex organic materials to light crude oil, is a promising method of processing waste organic materials, including animal mortalities. The process has been described as follows: It mimics the natural geological processes thought to be involved in the production of fossil fuels. Under pressure and heat, long chain polymers of H, O, and C decompose into short-chain petroleum hydrocarbons. Many methods to create hydrocarbons use a lot of energy to remove water from the materials. This method instead requires water, as the water both improves the heating process and supplies H and Ofor the chemical reactions.

Environmental implications should be minimal; site residues from the thermal depolymerization and pyrolysis processes are inert. The materials are held in sealed containers before and duringprocessing, and emissions to the environment should be contained.

2.8.2. Re-feeding (primarily to alligators)

Re-feeding of animal carcasses is already important in the poultry industry. There are currently a number of poultry producers using predators, particularly alligators, to consume mortalities. There is typically very little processing involved in the re-feeding process, with most carcasses being fed whole. However, this may not be practical everywhere.

There is potential for pathogens present in carcasses fed to alligators to be transmitted via the excrement of the alligators. Rodents and birds could also transport pathogens present in carcasses awaiting consumption by predator species, particularly if the predators are raised in uncovered enclosures. The re-feeding of diseased carcasses should be avoided.

2.8.3. Napalm

Napalm is a mixture of gasoline, benzene, and a thickening agent. Hence, it is very expensive.

If handled improperly, napalm fuel could contaminate soil or groundwater, just as other petroleum-derived products can. Smoke and particulates resulting from carcass burning could affect air quality, and ash remaining after burning could be a potential groundwater contaminant. Substituting kerosene for diesel fuel in the mixture may reduce the black smoke. Transportation of napalm on public roads may be of concern to public safety.

2.8.4. Ocean Disposal

Carcasses would be loaded on to open barges or into containers, floated beyond territorial limits, and emptied overboard. Eliminating all floating debris may require some sort of packaging or pre-processing.

The mortalities should provide a protein and energy source for aquatic life in the area. Specific environmental impacts are not fully defined, but will likely be minimized if steps are taken to ensure there is no floating debris resulting from disposal. In addition, nutrient loading limits at the ocean disposal sites need to be defined to reduce potentially-negative environmental impacts.

2.8.5. Non-Traditional Rendering

Instead of using conventional rendering procedures, ground non-disease related mortalities can be converted into a feedstuff by fluidized-bed drying, flash dehydration, or extrusion.

In fluidized-bed drying or flash dehydration, the material flows along a channel of super-heated air. Flash dehydration can be used to dry many types of wet wastes, but it is most applicable for drying animal by-products and offal.

In extrusion and expeller press processing, finely-ground, high-moisture material is mixed with an organic carrier to a moisture content of about 30% and then subjected to processing by friction heat, shear, and pressure within the dry extruder barrel.

If a disease agent can be neutralized by this process, the resulting material could be used in a composting system, deposited in a landfill or utilized an energy recovery system such as in a fixed hearth plasma arc furnace or a thermal depolymerization oil recovery system. The recovery of plant nutrients or energy would reduce costs and the effects on the environment.

2.8.6. Novel Pyrolysis

In this method electromagnetic waves are concentrated and directed at solid, liquid, or gaseous targets. Hence, this technology directly couples electromagnetic energy with a target material to produce heat. The target absorbs energy, generating temperatures that exceed the melting or vaporization points of the target materials. Advantages of the technology reportedly include instantaneous, controllable heating, a lack of hydrocarbon pollutants or harmful emissions, and reduced energy requirements. Additionally, the technology is reportedly scalable and lends itself to continuous operation and automation. However, the technology has yet to be tested on actual intact carcasses.

3. Decontamination of sites and carcasses

Disinfectants for decontamination of sites and carcasses along with their properties and toxicity are given in Tables 46.5 and 46.6.

Table 46.5. Use and toxicity of various types of disinfectants

Disinfectant type	Bacteria	Mycobacteria	Viruses	Bacterial spores	Yeasts	Molds	Toxicity
Chlorine	++	++	++	+	++	++	Medium
Formaldehyde	+	+	+	+	+	+	High
Phenolics	+	+	±	-	+	+	High
Peracetic acid	++	++	++	++	++	++	Low
QACs	±	-	-	-	++	+	Low
Hydrogen peroxide	++	+	+	±	+	+	Low
Iodophors	++	++	++	+	+	+	Medium
Sodium hydroxide	+	+	+	+	+	+	High

++ kills rapidly, + kills most, ± kills some, - negligible kill,
QACs = Quaternary ammonium compounds

Table 46.6. Background information on six major disinfectant groups

Disinfectant Group	Form	Contact Time	Applications	Precautions
		Soaps and detergents		
Quaternary Ammonium Compounds (QACs)	Solid or liquid	10 min.	Use for thorough cleaning before decontamination and for Cat. A viruses	N/A
		Oxidizing Agents		
Sodium hypochlorite	Concentrated liquid	10-30 min.	Use for Cat. A, B, and C viruses except in the presence of organic material	N/A
Calcium hypochlorite	Solid	10-30 min.	Use for Cat. A, B, and C viruses except in the presence of organic material	N/A
Sodium hydroxide	Pellets	10 min.	Cat. A, B, and C if no aluminum	Caustic to eyes and skin
Sodium carbonate	Powder/crystals	10-30 min.	Use with high concentrations of organic material	Mildly caustic

Disinfectant Group	Form	Contact Time	Applications	Precautions
Acids				
Hydrochloric acid Concentrated	Liquid	10 min.	Corrosive, use only if nothing better is available	Toxic to eyes, skin, and respiratory passages
Citric acid	Powder	30 min.	Use on clothes and person	N/A
Aldehydes				
Gluteraldehyde Concentrated	Liquid	10-30 min.	Cat. A, B, and C viruses	Avoid eye and skin contact
Formalin	40% Formaldehyde	10-30 min.	Cat. A, B, and C viruses	Releases toxic gas
Formaldehyde gas	Gas	15-24 hours	Cat. A, B, and C viruses	Releases toxic gas

Chapter **47**

Environmental Implications of Poultry Farming

Environmental implications of any animal farming are manifold and poultry farming is no exception. The primary concern is through manure followed by carcasses, gases and dust from poultry facilities.

Details regarding manure and its disposal are given in a separate Chapter "Poultry manure" and carcass disposal also is dealt as a separate Chapter. However, a brief account of manure as far as environmental issues are concerned is included in this Chapter.

1. Manure

The use of poultry manure as a feedstuff and as a fertilizer is a strong indication that the environment is becoming a very serious concern among the public at large. Soon other aspects like water and air pollution, odor and dust will be on agenda for serious consideration in developing countries as well.

1.1. Water Pollution

The application of excessive amounts of poultry manure can result in the leaching of nutrients through the soil and into groundwater. Efficient manure management ensures that manure or its constituents cannot gain access to rivers, streams, lakes, or water supplies.

Pollution of water by poultry manure application can occur in several ways including:

1. Direct dumping into surface water.
2. Runoff from feedlots or stockpiles.
3. Overflow from manure storages of inadequate capacity.

There probably exist several ways in which water pollution can occur. The addition of poultry manure constituents such as nitrates and phosphates may cause or contribute to eutrophication and the resultant unsightly growth of algae. Excess nutrients in surface water can cause algae blooms, impaired fisheries, fish kills, odors and increased turbidity.

Water pollution by poultry manure can result in several consequences. The oxygen level in the water is depleted because bacteria decomposing the manure constituents utilize oxygen and hence, dissolved oxygen concentrations are seriously depleted making it unsuitable aquatic life such as fish; consequently, water becomes septic and unpleasant.

The pollution of water resulting from poultry manure can also present a health hazard to both humans and livestock. In addition, the pollution of water by poultry manure may be responsible for nitrate poisoning in both animals and humans, particularly infants. Finally, the pollution of water by poultry manure can greatly affect drinking water quality. Manure can impart taints and odors to drinking water.

1.2. Manure Gases

The aerobic decomposition of poultry manure is basically an odorless process which produces stabilized organic matter, some carbon dioxide and water. Anaerobic decomposition is typical of liquid manure-handling systems and characteristic of collection pits, holding tanks and storage lagoons. The anaerobic decomposition process is characterized by obnoxious odors and the production of considerable amounts of gases which are hazardous to both man and livestock.

Gases in poultry house reach higher concentrations when the manure is agitated usually when it is being removed. If ventilation is insufficient and/or mechanical system of ventilation, if employed, fails, the concentration of gases can reach toxic levels. The house temperatures that are optimum for broilers and layers are also optimum for fermentation process in the litter by microorganisms. Humans can be overcome by the high concentrations of these gases because all of these gases are colorless.

The following gases are particularly important: a) CO_2 b) NH_3 c) H_2S d) CH_4, e) CO and f) HCHO.

1.2.1. CO_2

This is a colorless, odorless gas considerably heavier than air and highly soluble in water. Normal air has 300 ppm and more is released through respiration of animals and by manure decomposition. Being heavier, it settles down and in ventilation is not proper, it causes O_2 deficiency. Threshold Limit Value (TLV) for man is 5000 ppm (0.5%). 100,000 ppm (10%) causes violent panting and at 250,000 ppm (25%) for a few hours can cause death. Death due to CO_2 asphyxiation occurs very rarely; any death will probably be the result of heat stress and O_2 deficiency.

1.2.2. NH_3

This is a colorless gas with a characteristic pungent odor. It is lighter than air and soluble in water. Air mixed with more than 16% NH_3 is considered an explosive.

It is released from fresh manure and during anaerobic decomposition of organic matter. TLV for-man is 50 ppm to protect against irritation to the eyes and 5000 ppm causes suffocation. In chicken, 50 to 100 ppm retards growth and feed consumption, and causes kerato-conjunctivitis. However only in under-ventilated poultry houses its concentration is likely to exceed 50 ppm.

Because of its high water solubility, NH_3 can be more easily controlled in liquid systems than in solid systems. Low concentrations can greatly affect animal health and performance. There are several ways in which the concentration of NH_3 can be controlled. These methods include:

1. Rapid manure removal from the facility.
2. Adequate floor slopes to ensure good drainage.
3. Liberal use of bedding and increased ventilation rates.
4. Use of a liquid system to dissolve NH_3.

1.2.3. H_2S

This is a colorless gas with a rotten egg smell. Soluble in water and produced during anaerobic fermentation in a normally ventilated house, a concentration of 0.09 ppm is expectable; but may rise to above 0.3 ppm if poorly-ventilated. Its presence is indicated by formation of black accumulations of copper sulfide on copper thermostats and electrical wiring; white deposits of zinc sulfate on galvanized metals; black discoloration where lead pigmented paints are used due to lead sulfide.

Of all of the manure gases, H_2S is the most toxic and is potentially the most dangerous. TLV for this gas is set at 10 ppm which is not commonly expectable in poultry houses.

It is flammable and can be explosive in an O_2 mixture. The greatest risks from this gas occur when emptying tanks and pits, above which concentration levels can reach lethal levels in as little as a few minutes. There are several ways to minimize the hazards of H_2S. These include:

1. Do not agitate when emptying pits located within a building. At the same time, provide maximum ventilation.
2. Remove stock from facility if agitation is necessary. If stock cannot be removed, use a windy day to ensure maximum air exchange through open doors and windows.
3. Use gas traps in sewage channels between covered outside pits and barns to prevent gas back-up.
4. Empty pits as frequently as possible, especially in warm weather.

1.2.4. CH_4

This is a colorless and odorless gas much lighter than air, not very soluble in water and highly inflammable; it burns with a blue flame but at higher concentrations there is a real danger of explosions. This is one of the gases incriminated to cause thinning of ozone layer surrounding the Earth and subsequent greenhouse effect. However, in poultry houses this gas does not pose serious problems.

Since CH_4 is lighter than air, it has a tendency to rise and pool at the top of stagnant corners or in tightly enclosed manure storage pits. There are many ways in which methane-related risks can be controlled, which include:

1. Prohibit smoking or use of naked flames in and around manure building and storage facilities.
2. Use explosion-proof electric motors on fans and equipment.
3. Adequately vent-covered storages to outside air.
4. Use U-bends or gas traps to prevent re-entry of gases from outside tanks to the facility.
5. Ensure that a facility with inside manure storage is continuously ventilated.

1.2.5. CO

Carbon monoxide is a colorless gas of about the same density as air and is a rare gas expectable in animal houses. However, in such houses where brooding is done by burners and/or burning of charcoal/coal and ventilation system has suddenly failed which is not been noticed for a sufficiently long time, CO may be produced in place of CO_2. It is exhausted from gas engines, gas, and oil and coal heaters. The best way of preventing a toxic buildup of CO is to ensure that all engines are vented to the outside of the facility. If this is not possible, then sufficient ventilation to prevent gas buildup should be provided. This gas is highly poisonous but, the probability of its occurrence in poultry houses is very low.

1.2.6. HCHO

This is the most common fumigant employed in poultry operations. With modern practice of control fumigation, chances of toxic levels of HCHO are very remote.

Entering poultry manure storage areas and pits at any time should be done with caution, especially during or following emptying of the storage. As all gases produced are colorless, extreme caution should be taken when entering these facilities. There are several precautionary steps which are recommended. These include:

1. Always enter such a facility with an air-breathing apparatus and harness.
2. Omit the breathing apparatus only when positive ventilation is provided to ensure that manure gases are purged.
3. Ensure that there is a constant supply of fresh air to the facility.

4. Do not follow an animal that has fallen into a manure collection pit.
5. Do not enter the facility alone. Ensure that someone is there to pull you out if you become dizzy.
6. Never introduce a flame or spark near a manure collection area unless it has been ventilated first.

The production of manure gases is inevitablebecause of the aerobic and anaerobic decomposition processes. The solution to the potential hazards caused by these gases is to recognize that they exist and to exercise some common sense to avoid their effects.

The use of solid poultry manure management principles is the best defence against the potential hazards and risks associated with manure gases.

Table 47.1. Gases in poultry houses (%)

Gas	Lethal limit	Practical limit
CO_2	Above 30	Below 1
CH_4	Above 5	Below 5
H_2S	Above 0.05	Below 0.04
NH_3*	Above 0.05	Below 0.0025
O_2	Below 6	Optimum 21

1% = 10,000 ppm; * For every 10 ppm increase, body weight is depressed by 1%; if slight odor is detectable, 1 to 2% reduction in body weight and if eye irritation is noticed, 25% reduction in body weight expected.

2. Odor

The greatest public complaint about most poultry production facilities is about the smell that surrounds such an establishment. The odors are the result of the biological breakdown of the poultry manure within storages, whether these are piles, lagoons, or indoor pits. This microbiological process occurs in the absence of dissolved oxygen in stored manure (West, 1996).

At or near the litter surface, the presence of oxygen from the air creates aerobic conditions under which uric acid, proteins and animal fats are biodegraded. The aerobic processes produce nitrogen-containing odorants such as ammonia, amines, indole, skatole and volatile fatty acids. In the presence of oxygen, sulfide containing compounds such as methionine are oxidized by microbes into sulfur containing odorants such as hydrogen sulfide, dimethyl disulfide, and dimethyl trisulfide (Jiang and Sands, 2000).

The supply of oxygen is limited within the litter and on poorly managed farms, in pads of caked manure. Where oxygen supply is limited, anaerobic conditions may occur. Under anaerobic conditions, sulfur-containing compounds are biodegraded into thiols, volatile organic sulfides and mercaptans (Jiang and Sands, 2000).

Odor generation processes tend to accelerate with increasing bird age. As the birds grow, the faeces accumulate in on and in the litter. Consequently it may be expected that the odor concentration will increase as a batch ages (Jiang and Sands, 2000).

Odorous compounds, produced by aerobic biodegradation at the litter surface or anaerobic biodegradation inside the litter, occur as gases. The large amount of litter and bird body acts as a filter, trapping the odorous compounds that have been generated. The capacity of the filter to act as a temporary sink to the odors generated in broiler sheds is probably substantial. However movement of air across the litter and bird surfaces results in a transport process of odorants from the litter to the air in the shed (Jiang and Sands, 2000).

Over one hundred different odor compounds have been identified as a result of this naturally-occurring process (West, 1996). Most of the public complaint results when the manure is being spread as a fertilizer on an agricultural field. Components of odor are: odor quality - comparing with a known odor; odor strength - the amount of fresh air needed to dilute the odorous air to the threshold odor level; and odor occurrence - the frequency and total length of time the odor persists.

Most odors in poultry manure are the result of gases or vapors. However, particles and aerosols can also affect the sense of smell. Hence, odor is a subjective response by people to an airborne compound or particulate. Some substances are considered pleasant or inoffensive, but some may be extremely offensive. A substance's nuisance level may depend upon its intensity and concentration as well as the subjective response by the person experiencing the odor. Adaptation to an odor can occur and is the adjustment of the observer to the odor with the odor response diminishing with time even though the stimulus is applied at a steady rate. Livestock operators themselves have little ability to make unbiased odor judgments because of their constant exposure to the odors in the livestock environment around them. Fatigue or complete exhaustion to the sensitivity of an odor may occur at very high levels of odor intensity. This is particularly relevant when toxic gases such as hydrogen sulfide (H_2S) are involved (West, 1996).

Concentration may significantly change the odor quality. Mixtures of odors have been studied and it has been found that the odor level of the combinations was between the sum and the average of the individual odors.

Other factors affecting an individual's response to odor are age, sex, and habits such as smoking. Perhaps the greatest factors of all influencing a person's perception of odor as normal or offensive are their background and experience (West, 1996).

Odors from livestock facilities are well below occupational health minimums. They are not considered to be hazardous to human health but rather nuisance pollutants. Distress associated with the nuisance may cause real or perceived illness (West, 1996).

Nuisance is a function of odor frequency, intensity, duration, and offensiveness. This has been referred to as the FIDO relationship (West, 1996).

2.1. Measurement of odor

(West, 1996)

A means of accurately assessing odor intensity and quality has not been successfully developed. Measurement is complicated since the subjective perception of a nuisance odor is not only dependant upon the intensity but also the frequency and duration to which the receptor is subjected to the odor. Idiosyncrasies further complicate the issue.

Two approaches have been taken; organoleptic (sensory) and physical/chemical. Odor measurement can focus on its quality (offensiveness, intensity) or on the measurement of specific odor compounds as in physical/chemical analysis (ppm, %). Each approach presents special problems.

2.1.1. Organoleptic measurement

2.1.1.1. Threshold Determination Level (TDL method)

A sample of odorous air is diluted with clean air until the odor can no longer be detected. The number of volumes of air required to dilute the odorous air to the threshold level is called the threshold odor number or odor units (ou) per unit volume.

Some researchers recommend beginning with non-odorous air and gradually adding odorous air until the threshold level is reached. This avoids the desensitizing of the olfactory sense that occurs when starting with odorous air and then adding non-odorous air until the threshold point is reached. The device for determining the TDL is called an olfactometer.

2.1.1.2. Pyridine Equivalent

Odor measurements are made by a panel in an odor laboratory by comparing emission air to a comparison gas (pyridine).

2.1.1.3. Fabric Swatch

Cotton fabric swatches are suspended in odorous air for a specified time. Then they are tested for odor quantity and quality.

2.1.2. Physical and chemical measurement

2.1.2.1. Gas Chromatography

This provides the most dependable albeit expensive procedure for identifying odorous manure compounds. While a very useful measurement technique, the gas chromatograph cannot measure the quality of the odor. Often sensory-measurements are combined with gas to measure the quality of the odor.

2.1.2.2. Gas Detectors

Detects levels of individual gases such as NH_3, H_2S, CO_2, CH_4, and CO. Knowing the levels of these gases is useful from a safety standpoint since high levels of H_2S and CO are lethally toxic. CO_2 is an asphyxiant and CH_4, an explosive. A high NH_3, level is suspected to adversely affect livestock health and production.

Measurements from gas detectors are not useful for odor determinations since odor threshold levels are much below the measurable range of most gas detectors. Several compounds which are suspected ascontributing to odor are NH_3, *p*-cresol, skatole, and various volatile organic acids.

2.2. Sources of odor

2.2.1. Manure

There are several sources of odor coming form the manure resulting from the everyday poultry operation. These sources include:

1. Feed odors associated with feeding of waste materials or fermented products.
2. Odorous compounds from manure-covered surfaces or treatment facilities.
3. Roofed facilities often have high odor production because of high animal density, large manure inventory, and limited air exchange.
4. Other structures such as manure storage tanks, anaerobic lagoons and manure-covered floors produce obnoxious odors.
5. Any wet area supports widespread anaerobic decomposition and can be a source of odors.

2.2.2. Animal Odors

Animals and their fresh feces have characteristic odors. Unsanitary housing conditions where animals are dirty may increase odor. Body heat and moisture promote bacterial growth and odor production. Odor is also a function of animal type, feed, and housing system.

2.2.3. Open Lots

Feedlots, paved or unpaved, consist of large surface areas of exposed manure which are subject to odor production. Odor emission depends on amounts of manure accumulated and the moisture content. Weather conditions have a great impact on surface conditions and odor production. In arid and windy regions, dust may contribute to the nuisance problem.

2.2.4. Anaerobic Storages

Manure allowed to undergo anaerobic decomposition will produce malodors.

Open storages are particularly significant in this regard. Under pit storage and collection pits may contribute to high yard odors. Any action which disturbs the storage volume, such as top filling a lagoon, or wind disturbance, will result in odor release.

2.2.5. Ventilation Air

Exhaust air from livestock buildings is a lesser source of odor, dust, and feathers. Increased attention to in-barn sanitation and air quality reduces odors in the discharged air.

Therefore, any handling and/or treatment which inhibit this biological anaerobic action will be effective in decreasing the production of some of these gases.

2.3. The nature of odors

(Jiang and Sands, 2000)

When odorous molecules are inhaled, they bind to specialized proteins, known as receptor proteins that extend from the cilia on sensory neurons in the olfactory epithelium in the upper region of the nose. The binding process triggers an electrical signal to the brain producing the sensation of odor. Poultry farm odor is made up of a mixture of odorous molecules, the actual composition ofwhich is too complex to determine by chemical analysis.

Odor annoyance is considered to occur at the point in time at which exposure to an odor is perceived by a person to be unwanted. Factors often considered to be the most relevant to annoyance are:

1. Concentration (odor or chemical concentration level)
2. Duration of exposure to the odor
3. Frequency of odor occurrence
4. Intensity of perceived odor (a mixture of offensiveness, odor character and hedonic tone)
5. Tolerance degree and expectation of the receptor

2.4. Odor control

2.4.1. Odors outside the facility

Since every poultry facility is unique, the most critical and effective means of reducing odor complaints is by proper site selection. In terms of odors, facilities should be located downwind of residential developments.

Most odor complaints occur because of poultry manure application to crop land which is offensive to residents in the neighboring regions. Morning application of manure is more desirable than late afternoon application,which limits potential drying time. Humans are generally the most sensitive to odor problems in the early evening when utilizing outdoor recreational facilities.

In addition, spreading of manure on land adjacent to residences or main roads should be avoided unless it is directly injected into the soil or it is immediately cultivated. These methods not only limit odor production, but also minimize nutrient losses.

2.4.2. Odors within the facility

Odors can also become a problem within the poultry production facility. This can best be controlled by a high standard of hygiene within the facility. The thorough cleaning and frequent manure removal to storages coupled with good drainage and the liberal use of bedding or litter where appropriate all contributes to maximizing the hygienic effects within a facility .

The potential sources of odors in the manure of poultry operations can be controlled by the following control principles:

1. Some volatile compounds can be converted to a less volatile form by pH control, chemical conversion or biological conversion to a less odorous or less volatile compound.
2. Another method is to inhibit the anaerobic decomposition of manure.
3. A third potential for odor control is the physical confinement of the odor sources. Covered manure storage tanks and anaerobic manure treatment devices are effective in controlling the escape of odorants.
4. Anaerobic processes may be limited by reducing the ingress of water into the litter, by increasing the exposure of the litter to air (O_2) by providing greater space for movement by birds. It may also be limited by feeding balanced complete rations with sulfur compounds of high biological availability particularly early in the growth cycle of broilers (Jiang and Sands, 2000).

In recent years, manure odors have been the subject of intense research around the world. Widespread attention has been put on adding odor control chemicals to manure storage tanks or animal-feeding areas. Materials preventing the release of odorous compounds, inhibiting their formation, or masking their odor are most effective in odor control.

1. Oxidizing agents such as potassium permanganate have potential as odor control chemicals. Some of these agents have been found to significantly suppress the release of odorous gases.
2. Enzymes and other digestive aids have also been proposed for controlling livestock production odors. Unfortunately, product secrecy prevents manufacturers from disclosing the composition of these agents, which makes these products hard to evaluate. In most cases, these products do not significantly reduce manure odors and most of these agents are expensive.

The solution for most of these sources of odor is good, "common sense" management (Chastain, 2000).

1. Provide adequate bedding for each flock of birds.
2. Remove manure from the building as often as possible.
3. Repair all leaky waterers or pipes.
4. Clean feeding equipment regularly.
5. Remove spoiled feed regularly.
6. Remove dead animals and dispose of them promptly.
7. Cover litter stockpiles completely with a tarp weighed down with rocks, limbs, or old tires as soon as a house is cleaned out.
8. Avoid excess moisture in stacking sheds since excess moisture increases the amount of odor generated due to anaerobic decomposition.
9. Make sure ventilation fans are cleaned regularly and that airflow rates are adequate for the season and stage of growth.
10. Avoid orienting buildings so that ventilation fans blow exhaust air towards neighbors or adjacent road ways.
11. In high-rise layer houses, the airflow should be equipped with lower level, or basement, exhaust fans that pull air in through well-designed inlets, past the birds, and out past the manure storage area to promote manure drying and to improve indoor air quality.
12. Manure in high-rise buildings must be dry enough to maintain a conical pile to control odor and flies (less than 45% moisture).

2.4.2.1. Ventilation system in odor control

(Chastain, 2000)

The major carriers of odors are: gases from manure, dust, and water vapor. A well-designed and managed ventilation system will control the levels of all three and is an important factor in controlling odors from poultry buildings.

2.4.2.1.1. Evaluating ventilation system

A practical and objective way of evaluating the effectiveness of a ventilation system for odor control and indoor air quality is to take gas measurements during cold weather. That is, when ventilation system works at minimum airflow conditions.

Carbon dioxide and NH_3 are the two most important gases to measure. CO_2 concentrations are set by the ventilation rate and the number of animals in the building. NH_3 concentrations are influenced by both the ventilation system and the waste system. If the concentration of these gases is at or below recommended levels, then other gases are typically within recommended levels. Recommended concentrations of some of the most important gases in animal buildings are shown in Table 47.2.

Table 47.2. Maximum gas concentration (ppm) for air quality and odor control in poultry buildings.

Gas	Odor	Concentration
CO_2	None	3,000
NH_3	Sharp, pungent	15
H_2S	Rotten egg smell	3
CO	None	50
1% = 10,000 ppm		*Source* : Chastain, 2000

2.4.2.1.1.1. Mechanical Ventilation

Exhaust mechanical ventilation is the most common type of system used in modern poultry facilities. It takes three basic components: properly-sized fans, properly-sized and distributed fresh air inlets, and controls. The fans and inlets must be designed to provide at least three stages of ventilation:

1. A minimum, continuous ventilation rate for winter
2. A mild weather rate for temperature control during the fall and spring
3. A maximum rate to control the temperature rise of the building in summer.

For details see Chapter "Mechanical ventilation"

2.4.2.1.1.2. Natural Ventilation

Natural ventilation uses local wind and thermal buoyancy, often called the stack effect, to move air through the structure. Fans are not used. Instead, the quantity of airflow is determined by the size and placement of openings, roof slope, and orientation of the building with respect to prevailing wind direction. Thermostatically-controlled side wall-curtains are used in some poultry buildings to control inside temperature to some extent. However, the air temperature of heated spaces, such as during brooding, can be controlled better with mechanical ventilation. For details see Chapter "Natural ventilation"

3. Dust

Dust, to a degree, is a feature of animal production units. It can be a severe problem in poultry operations, particularly in an operation where a dry manure-end product is produced. Good air circulation and adequate ventilation will limit the effects of dust on both the poultry flock and the human operators. Proper management with a high standard of hygiene can greatly reduce the problems of dust associated with poultry manure.

Dust particles can carry gases and odors. In fact, a large portion of the odor associated with exhaust air from mechanically-ventilated poultry buildings are dust particles that have absorbed odors from within the building. Therefore, dust control in the buildings can reduce the amount of odor carried outside by the fans. High dust concentrations can also be a health risk for workers in poultry facilities as well as the animals. Control of dust improves the working conditions for the producer and helps significantly in odor reduction (Chastain, 2000).

Dust is generated from feed, manure, and the animals themselves. Factors determining the amount of dust include cleanliness of the buildings, animal activity, temperature, relative humidity, ventilation rate, stocking density and feeding method.

The exposure to dust and other harmful gases in confined livestock buildings is at a high level among livestock producers. Air pollutants have been implicated in causing respiratory diseases for livestock producers and in making them worse. Poultry workers have a high prevalence of wheezing and chronic bronchitis (Atia, 2004).

3.1. Dust measurement

Dust is mostly measured as mg (airborne dust) per m^3 of airspace (mg/m^3). The size of the dust particles varies from less than 0.1 micron (μm) to over 100 μm. The important fractions are inhalable dust (less than 10 μm) and respirable dust (less than 5.0 μm). About 80 to 90 per cent of the dust inside swine and poultry buildings is smaller than 5 μm and can be inhaled deeply into the lungs.

3.2. Factors affecting dust concentration

The concentration of dust in the air is affected by temperature, relative humidity, ventilation systems, feeding practices, stocking density, cleanliness of the buildings, bedding materials and animal activities.

3.3. Dust is an occupational hazard

The most harmful dust is aerosol dust (size range of 2.5 micrometers and smaller). These particles are small enough to reach the lungs when inhaled by birds or humans. Barn workers that inhale poultry dust and gases have an increased risk of pulmonary disease, the primary symptom of which is pulmonary inflammation due to either hypersensitivity or allergic reaction (Paulson, 1999).

The majority of agricultural safety groups consider working in livestock facilities as a high-risk occupation. A worker may experience high risk because of work area (location) or work procedure (task). The following are important considerations when assigning tasks for workers in poultry units:

1. Proximity to contaminant sources
2. Frequency of proximity to contaminant
3. Worker complaints and illness

3.3.1. Respiratory diseases

3.3.1.1. Bronchitis

Symptoms include cough, phlegm, tightness of chest, shortness of breath, wheezing.

3.3.1.2. Chronic farmer lung disease

This condition may occur with repeated dust exposures although it is possible to develop it after only one attack. Symptoms include chronic coughing, increasing and severe shortness of breath with slight exertion, weakness and body aches, and occasional fever.

3.3.1.3. Occupational asthma

Symptoms include tightness of chest, shortness of breath and wheezing.

3.3.1.4. Organic dust toxic syndrome (ODTS)

This condition is more common in swine producers characterized by one or more episodes. Symptoms include fever episodes, headaches, muscle aches, flu-like illness and shortness of breath.

3.4. Aerosols and broiler dust (Paulson, 1999)

Aerosols are solid particles in the atmosphere, either formed in the air by reactions among gases or injected into the air by processes on the ground. They consist of a variety of materials and vary in size from < 1 μm to the size of a sand grain. In poultry units, concentrations of dust aerosols may be up to 25 times greater than those found in an urban atmosphere.

Broiler dust comprises feather fragments, fecal material, skin debris or dander, feed particles, mold spores, bacteria, fungus fragments and litter fragments. Broiler dust is approximately 92% dry matter; chemically it is 60% crude protein, 9% fat and 4% fiber; calcium, derived from the feed, is the most common inorganic element, followed by magnesium and small amounts of copper, iron and zinc. Various gases which result from microbial decomposition of feces, are also common in poultry house air.

High dust levels enhance the awareness of odors. Odors are carried by airborne particles. Odorants such as acids, carbonyl compounds, and phenols attach to dust particles. The concentration of odorants on aerosol particles is approximately 40 million times greater than that found in an equal volume of air.

3.5. Factors affecting dust levels (Paulson, 1999)

Dust levels in the barn vary with age of flock, the number of birds and the season of the year. Dust and gas concentrations increase with flock size and with flock age. Inside dust levels are higher in the winter because ventilation is decreased to conserve heat. Dust levels rise in the barns during clean-out and when birds are moved. Litter tilling can result in high levels of dust and odor in the barns.

Computerized ventilation has helped to reduce the extreme levels of dust in both the poultry house and in the air discharged from the buildings throughout the production cycle.

Dust levels can vary due to bird activity, temperature, relative humidity, ventilation rate, ceiling height, feeding method and type of bedding. The quantity of dust produced in a broiler barn depends on the type of litter used and increases with litter age. Temperatures of 15.5 to 21°C result in higher dust production than lower or higher temperatures.

Particle dust less than 0.5 μm in diameter is highest at humidities between 45% and 60%, and declines at humidities of less than 45%. The lowest levels of aerosol dust occur at humidities of greater than 60%. More dust is produced during periods of illumination than during darkness due to higher bird activity.

Dust levels outside the barn may be an irritation to some neighbors particularly during clean-out. During clean-out there are two areas of particular concern:

1. When the litter is loaded into the truck via a conveyor, litter falling into the truck creates large amounts of dust. Feathers also blow from the conveyor into the air.
2. When air is used during "blow down" the dust levels exiting the fans can be very high. Also when the fan hoods are blown out clouds of dust are emitted which can drift for several hundred m.
3. Impact of disinfectants: Broiler barns are thoroughly cleaned between flocks. Cleaning includes more than just the removal of litter. The barns are washed thoroughly with water and detergent. Dust is blown down from feeders, waterers and fans. The barns are then disinfected with phenols, quaternary ammonium compounds or other types of disinfectants. These disinfectants represent little risk to the environment or neighborhood health or safety when applied according to specifications (Paulson, 1999).

3.6. Dust control

(Chastain, 2000 and Paulson, 1999)

Management practices that can greatly reduce the amount of dust in poultry buildings are described below:

1. Clean interior building surfaces regularly. Modern poultry production facilities are designed around an "all-in, all-out" style of management is very convenient for this purpose. This helps reduce occurrence of fine particles.
2. Reduce dust from feed. Addition of oil to dry rations significantly reduces the amount of dust in a building.
3. Dust can be reduced in poultry houses by ensuring that pelleted feed is of high quality.

4. Manage the relative humidity (RH) in poultry houses. If the air in a broiler or turkey house is too dry (low RH), then the amount of dust in the exhaust ventilation air will be excessive. This will lead to excess odor and results in poor indoor air quality for the animals and the people who work in the facility. This would prevent formation of dust-clouds. The recommended range for humidity is 60 to 75%. The minimum humidity is 40% and maximum 80%. Note that the generally recommended humidities for broilers of 50 to 75% are within the range where maximum dust is observed (45 to 60%). Humidity levels may have to be adjusted when dust levels rise.
5. Remove airborne dust using air-cleaning devices and workers provided with dust masks.
6. Properly managed mechanical ventilation
7. Maximizing the distance between barn and residences.
8. Established a row of trees between your poultry house and residences. This will assist in reducing larger-sized dust particle drift but is less effective in reducing light aerosol drift. A tree or vegetation buffer is recommended where vented dust becomes a problem to neighbors.
9. The buildings should be set up so that the vent fans are directed away from neighboring properties.
10. 60 and 90 cm fans should be hooded to ensure that dust and feathers leaving with exhausted air are directed towards the ground.
11. Choose a time for clean-out that minimizes inconvenience to residents.
12. To minimize the potential for dust complaints, wash out fan blades, screening and hoods with water rather than blow them out with air
13. Use modern up-to-date ventilation equipment to minimize exiting dust levels. Implementing state of the art ventilation is one of the best methods of avoiding excessive dust accumulation and emissions and ensuring efficient movement of air.
14. Strive to maintain litter moisture conditions which are not excessively dry or damp by establishing appropriate ventilation rates.
15. Molds and their contribution of spores and toxins to dust can be controlled by changing litter frequently and/or monitoring and controlling humidity.
16. Reduce the temperature during light periods. This will increase the humidity and reduce dust during the time when birds are active and feeding. Then, increase the temperature during dark periods when the birds are less active. This will minimize dust as well as improve feed efficiency.
17. Keep bird density levels within the recommended level.
18. A reduction in dust can be attained by adding a negative electric charge to moist fog.

19. Control of darkling beetles will reduce the contribution of insect fragments to the dust load.

4. Other environmental considerations

4.1. Infective agents

4.1.1. Viruses

Residents should not be overly concerned about viruses found in poultry dust. There are very few poultry viruses that can affect humans. The Avian Influenza strains of viruses that caused the Hong Kong flu is still a matter of concern. However, if chicken is grown in high biosecurity operations which exclude all wild birds, even Avian Influenza may not be of public health concern.

Newcastle virus can cause conjunctivitis in the eyes of vaccinators if eye protection is not used. This requires large exposure. Bronchitis, Bursal and Adenovirus are not known to affect humans; there is no avian tuberculosis and mycoplasma has no effect on people.

4.1.2. Bacteria

Generally, dust levels are highly diluted by air as the vented dust from the barns enters the environment. There is almost no risk of humans beings infected by bacteria associated with dust particles from poultry operations. A normally healthy individual has an immune system that is capable of resisting bacterial infection. If a person has reduced immunity, they may be more susceptible to infection. Very young or old people may have less ability to resist infection from bacteria. However, the chances of a person being infected with a bacteria from drifting poultry dust are very low.

4.1.3. Parasites

There are but a very few parasites (protozoa, internal and external parasites) that are communicable from poultry to humans; hence, measures to control other infective agents will take care of parasites as well.

4.2. Pests

Manure makes an ideal substrate for various living things to breed and feed upon. It is an ideal breeding ground for various insects, including flies, and tends to attract birds and even rodents . All of these living things tend to be a nuisance, but more importantly, they can become carriers of disease The solution to this potential problem lies again in the proper planning and management of poultry manure.

Although these concerns might not be as obvious as gases, odors and water pollution, they can become very serious. These can be controlled by proper planning and managing of poultry manure. After a system has been developed to properly take care of these concerns, it is a simple matter of keeping the system in working order to avoid the development of possible problems (See Chapter "Pest management" for details).

5. Management principles regarding environmental concerns

The operation of poultry facilities will always have some environmental impact. The extent to which this impact is detrimental will depend on the way the poultry operation is managed . Although most poultry operations will have some common features, each operation will be unique in some way. The best way to minimize the environmental hazard from poultry manure is to examine the management options in each case. These options involve the following:

1. When possible, use cropland for manure application, having regardfor water pollution and making for the most efficient use of the plant nutrients in the manure.
2. If the land base is insufficient, investigate alternative means of manure disposal to arrive at an environmentally satisfactory solution.
3. Ensure sufficient manure storage capacity to eliminate uncontrolled release of manure into the environment. Storage mustbe of a type that will eliminate seepage into ground water.
4. Avoid land application of manure during winter and permit most effective crop use of nutrients.
5. Locate poultry production facilities to avoid nuisance complaints about odors, dust, flies, noise and aesthetics, by providing adequate separation and suitable screening.
6. Use weather conditions to best advantage to minimize odor nuisance during field spreading of manure.
7. Use soil injection or rapid soil cover of manure, and gain the benefits of greatly reduced nitrogen losses and control of possible pollution arising from surface runoff.
8. Control manure gases, particularly from stored liquid manure, to ensure the safety and health of both humans and animals, either by frequent transfer of manure from animal facilities to separate storages or by exhausting enough air from the headspace in indoor storages to prevent gas build-up.

In any specific poultry operation, some or all of these management principles may require closer scrutiny to achieve both an economically and environmentally-acceptable poultry manure management system. The problems associated with poultry manure management are greatly reduced in a properly planned and managed operation. The management of poultry manure can provide substantial physical and economic benefits. Since the emphasis has now rightly shifted on to sustainable agricultural development, it is quite obvious that environmental poultry manure management plays an integral part in minimizing the impact of poultry manure operations on the environment.

Chapter **48**

Farm Data – Recording and Processing

1. Farm records

In any commercial farm operation, records of all transactions have to be meticulously maintained in order to arrive at the economic returns of the enterprise. The records must be such that they are accountable, auditable, easy to maintain, easy to understand and convenient to arrive at the economic returns of the enterprise. The records to be maintained are as follows:

1.1. Registers to be maintained

1.1.1. For accounting and audit

1. Journal 2. Ledger-feed 3. Ledger-chicks 4. Ledger-medicines 5. Ledger-equipment 6. Ledger-electricity, water and miscellaneous 7. Ledger-sales (livestock) 8. Ledger-sales (eggs) 9. Ledger-sales (manure) 10. Ledger-sales (G.bags) 11. Cash book 12. Miscellaneous register-to maintain records of land, buildings etc.

1.1.2. For farm operations (auditable)

1. Livestock register 2. Feed procurement and issue register 3. Feed ingredient procurement and compounding register 4. Medicines and biologicals register 5. Building maintenance register 6. Equipment purchase and maintenance register 7. Miscellaneous expenditure register 8. Sales register (eggs) 9. Sales register (birds) 10. Sales register (manure) 11. Sales register (G.bags) 12. Bank transactions

The registers mentioned above are neither exhaustive nor compulsory; additions/deletions can be made depending on the requirements of the poultry facility.

1.2. Pro forma for various farm registers

1.2.1. Livestock register

Month and year:				Breed/Strain:		Date of hatch:	
Date	Receipt	Issue	Sold	Died/Killed/Culled	Total debit	Closing balance	Initials

1.2.2. Feed procurement and issue register

Month and year:				Breed/Strain:		Date of hatch:
Manufacturer's name and address:				Feed:		Batch No.
Date	Receipt	Quantity	Amount	Issue	Balance	Remarks/Initials

1.2.3. Feed ingredient procurement and compounding register
(Separate registers/pages to be allotted to each ingredient)

Month and year:				Ingredient		
Supplier's name and address:				Batch No.		
Date	Receipt	Quantity	Amount	Issue/Discard	Balance	Remarks/Initials

1.2.4. Medicines and biologicals register
(Separate registers/pages to be allotted to each medicine/biological)

Month and year:					Ingredient				
Supplier's name and address:									
Date	Receipt	Quantity	Amount	Issue/Discard	Batch No.	Date of manufacture	Expiry date	Closing balance	Initials

1.2.5. Building maintenance register

Date	Building constructed	Cost of construction	Name of construction firm	Remarks

1.2.6. Equipment purchase/maintenance register
(Separate sheets for each of the equipments)

Date	Receipt	Quantity	Amount	Supplier details	Issued/condemned/sold	Closing balance	Remarks

1.2.7. Miscellaneous expenditure register

Date	Item	Quantity	Supplier details	Amount	Issued	Closing balance	Remarks

1.2.8. Egg production register

Month and year:				Breed/Strain			Date of hatch	
Total hens housed					Age at sexual maturity			
Date	Opening balance	Live hens	Eggs produced	Broken	Saleable eggs	Closing balance	cumulative	Initials

1.2.9. Egg sales register

Month and year:				Breed/Strain			Date of hatch	
Date	Opening balance	Receipt	Sold	Broken or discarded	Total debit	Closing balance	Selling price	Initials

1.2.10. Livestock sales register

Month and year:				Breed/Strain		Date of hatch
Date	Opening balance	Receipt	Sold	Selling price	Purchaser's details	Initials

1.2.11. Manure sales register

Month and year:				Breed/Strain		Date of hatch
Date	Opening balance	Receipt	Sold	Selling price	Purchaser's details	Initials

1.2.12. G.bags sales register

Month and year:				Breed/Strain		Date of hatch
Date	Opening balance	Receipt	Sold	Selling price	Purchaser's details	Initials

1.3. Pro forma for registers – Bank transactions

1.3.1. Journal

Date	Particulars	Ledger folio Debit Credit	Debit (amount)	Credit (amount)

1.3.2. Ledger

Separate ledger or separate sheets within a ledger have to be maintained for each of the items like livestock, feed, medicine, sales etc.

Debit				Credit			
Date	Particulars	Journal folio	Amount	Date	Particulars	Journal folio	Amount

1.3.3. Cash-book

Date	Receipt		Credit	
Opening balance				
	Particulars	Amount	Particulars	Amount

2. Data processing

2.1. Production indices

2.1.1. Layers/Breeders

1. Hen-day egg production

$$\text{A) For one day} = \frac{\text{Number of eggs produced}}{\text{Number of live hens}} \times 100$$

$$\text{B) For a long period} = \frac{\text{Total number of eggs produced over the period}}{\text{Total number of hen - days}} \times 100$$

2. Hen-housed egg production

$$\text{A) For one day} = \frac{\text{Number of eggs produced}}{\text{Number of hens housed at the beginning of the laying period}} \times 100$$

For long period: Similar to hen-housed egg production for one day excepting that the numerator will be the average number of eggs produced per day during the concerned period.

For a given flock, hen-day and hen-housed egg productions will be same only when all birds housed survive throughout the period and whenever there is mortality, the former criteria will be greater than the latter due to reductions in the magnitude of the denominator.

3. Egg-mass: calculated as a product of egg number and egg weight
4. Feed consumed per dozen eggs
5. Number of settable eggs produced per dam (for breeders only)
6. Number of saleable chicks produced/dam (for breeders only)
7. Cracked eggs (%)
8. Blood-spots
9. Performance efficiency index (PEI)

PEI = (KWP) ÷ F where K = x W ÷ BW, W = Av. Egg weight (g), BW = Body weight (g), P = Egg production (%), F = Feed consumed/bird (g/bird/d); x is developed on the basis that egg weight, in case of chicken, is 1/30th of the body weight (x = 30) and hence, in hens producing eggs of average weight, the K value will the unity (x = 30). Birds producing smaller eggs will have K < 1 and those producing bigger eggs > 1. On the same basis, the value of x can be modified to 12.5 in Japanese quails (Sreenivasaiah, 1980) and to 100 in turkeys.

2.1.2. Broilers

1. Average body weight at market (5, 6, 7 or 8 weeks of age)
2. Carcass yield

The following are the standards for carcass yields of broilers:

Table 48.1. Broiler carcass yield

	Males	Females
Live weight (kg)	1.84	1.50
	Yield (%)	
Breast	26.6	28.3
Thigh	17.5	17.2
Drumsticks	16.3	15.5
Wings	11.7	12.1
Neck (without skin)	3.8	3.5
Tail back	9.6	9.4
Rib back	8.9	8.4
Heart	0.6	0.6
Liver	2.6	2.4
Gizzard	2.6	2.7

3. Feed Conversion Ratio (FCR)

   FCR is defined as the kg of food of required per kg live-weight gain; weightage for mortality, if any, may be given to get accurate values.

4. Feed cost per kg broiler produced

   This is calculated as the product of FCR and cost per kg of feed.

5. Performance efficiency factor (PEF)

   PEF is 100 times the ratio of Average body weight in kg to FCR.

6. Gross margin per unit floor space

   This is the ratio of gross income over and above the feed cost to the total floor space in m^2.

7. European efficiency factor (EEF) or Protein efficiency factor (PEF):

$$EEF = \frac{\text{Livability}(\%) \times \text{Live weight (kg)}}{\text{Age(d)} \times \text{FCR}} \times 100$$

8. Livability (%)

   This is 100 times the ratio of number of birds marketed to the number of birds started.

9. European broiler index (EBI)

   EBI is 10 times the product of growth rate (g/bird/day) and survivability (%) divided by FCR to obtain this index.

10. Uniformity in weights

    Gives spread of weight; calculated as coefficient of variation (CV) which is 100 times the ratio of range of weight (In kg) to the product of average weight (kg) and F value (Table 48.2).

Table 48.2. F values for calculation of uniformity in weights

Sample size	F	Sample size	F
10	3.08	60	4.64
15	3.54	65	4.70
20	3.73	70	4.76
25	3.94	75	4.81
30	4.09	80	4.87
35	4.20	85	4.90
40	4.30	90	4.94
45	4.40	95	4.98
50	4.50	100	5.02
55	4.57	150	5.03
			Source : Poultry International, 1991

Good uniformity is indicated by CV < 8%, Average uniformity by CV between 8 and 12%, and poor uniformity by CV of about 16%.

3. Data processing

The data recorded during various farm operations have to be processed to make them comparable with similar data from other poultry facilities and also to know whether the farm operations are proceeding at optimum levels or not. Hence, the following hypothetical examples are provided as a guideline for processing of larger volume of data.

3.1. Layers/Breeders

3.1.1. Hen-housed (HH) and Hen-day (HD) egg production

The data in Table 48.3 pertains to a 20 days' record of a 1000-bird layer farm (N=1000). The data is to be processed for calculation of HH and HD a production both on daily basis as well as for the entire period. The procedure can be extended as per the requirements.

HH egg production for the entire period of 20 days can be calculated utilizing the totals as 100 times the ratio of total egg produced to the product of total hens housed at the beginning and the period; that is {16925÷(1000*20)}*100 = 84.625% which is same as the average of daily HH egg production values. It can also be calculated as 100 times the ratio of average number of eggs produced to total number of hens housed at the beginning; that is (846.25÷1000)*100 = 84.625%.

Similarly, HD egg production for the entire period of 20 days can be calculated for the entire period as 100 times the ratio of total eggs produced to total hen days; that is (16925÷19860)*100 = 85.23% which is same as the average of daily HD production values. It can also be calculated as 100 times the ratio of average number of eggs produced to average hen-days; that is (846.25÷993)*100 = 85.22%.

The commercial layers are expected to produce at the rate of 280-300 eggs HH. In a commercial farm it is highly desirable to calculate HH and HD egg production values so that it would be convenient to monitor the performance of the laying birds on a day-to-day basis.

The HH and HD calculations reveal the following:

1. HH and HD egg production of a particular day are same only when there was no mortality.
2. HD egg production values can be higher than HH values; but not *vice versa*
3. HD production values are more accurate than HH egg production values. Although, from the farmer's point of view, the latter is more important.

Table 48.3. Calculation of hen-housed and hen-day egg production

Day	Birds live (Hen-days) A	Eggs produced (Number) B	Feed consumed (kg)	Egg production of the day	
				Hen-housed (B÷N)*100	Hen-day (A÷B)*100
1	1000	820	106	82.0	82.0
2	1000	828	108	82.8	82.8
3	998	830	110	83.0	83.2
4	998	823	105	82.3	82.5
5	996	840	112	84.0	84.3
6	995	840	110	84.0	84.4
7	995	842	113	84.2	84.6
8	995	845	110	84.5	84.9
9	995	843	115	84.3	84.7
10	995	850	118	85.0	85.4
11	993	847	112	84.7	85.3
12	992	850	115	85.0	85.7
13	990	850	117	85.0	85.9
14	990	858	110	85.8	86.7
15	990	855	113	85.5	86.4
16	990	857	115	85.7	86.6
17	990	849	117	84.9	85.8
18	987	863	116	86.3	87.4
19	986	865	118	86.5	87.7
20	985	870	118	87.0	88.3
TOTALS	19860	16925	2260	1692.5	1704.6
AVERAGE	993.0	846.25	113.0	84.625	85.23

3.1.2. Feed efficiency

Feed efficiency during laying period is generally calculated as feed consumed per dozen eggs. In the above example, fed consumed per dozen eggs = (2260*12)÷16925 or (113*12)÷846.25 = 1.60 kg.

However, unlike egg production, feed consumed per dozen eggs does not change on giving weightage to mortality because it leads to proportionate increments in both numerator and denominator of the ratio. It can be verified; HD egg production was 85.22% which means a bird required 14.08 d to produce dozen eggs. On bird-day basis, feed consumed/bird/d = 2260÷19860 or 113.0÷993=0.1138 kg. Therefore, after giving weightage to mortality, feed consumed per dozen eggs is 0.1138*14.08 = 1.60 kg which is same as that calculated without giving weightage to mortality.

3.1.3. Persistency of egg production

See Chapter "Judging pullets" for details.

3.2. Broilers

3.2.1. Feed conversion ratio (FCR) – weekly

Feed conversion ratio (FCR) is the amount of feed consumed (kg) per kg live weight gain. Giving weightage to mortality is essential to get accurate values. But, commercially, ratio of total feed consumed to total weight of birds sold is commonly calculated.

In the following example, hypothetical data from a broiler farm with 1000 birds (N) the shown; it is assumed that there was an outbreak of IBD during the 3rd week and cecal coccidiosis during the 5th week. However, higher body weights, survivability and better FCR are expected in healthy flocks.

Table 48.4. Weight gain, feed consumption and mortality in a broiler farm

Week	Number of birds alive							Feed intake (kg)*	Weight gain (g)**
	1	2	3	4	5	6	7		
I	1000	995	992	992	991	990	990	100	60
II	990	990	990	986	986	986	986	245	150
III	980	976	960	940	930	925	922	490	250
IV	920	920	920	920	918	918	918	575	300
V	910	906	900	900	896	896	895	570	300
VI	895	895	895	895	895	895	895	800	300
* Weekly	** Average gain during the week							N = 1000	

It can be noticed that weekly weighted FCR values are higher than the FCR calculated ignoring mortality. The difference narrows if mortality tends to 0 and *vice versa*. It is also worth while noting that birds housed and bird days are similar to hen-housed and hen days in case of laying hens.

Table 48.5. Calculation of FCR (Birds housed, BH = 1000)

	Bird days (BD)		Feed intake	Feed intake (g/bird)*		Weight gain	FCR	
	Total/week	Average/d	(kg)*	BH	BD	(g/bird)**	BH	BD
Week	A	B=A÷7	C	D=C÷N	E=C÷B	F	D÷F	E÷F
I	6950	992.86	100	100	100.72	60	1.67	1.68
II	6914	987.71	245	245	248.05	150	1.63	1.65
III	6633	947.57	490	490	517.11	250	1.96	2.07
IV	6434	919.14	575	575	625.58	300	1.92	2.09
V	6303	900.43	570	570	633.03	300	1.90	2.11
VI	6265	895.00	800	800	893.85	300	2.67	2.98
Totals	39499	940.45	2780	2780		1360	2.04	2.17
* Weekly	** Average weight gain during the week							

3.2.2. Calculation of FCR up to specified periods

Under commercial conditions, it may not be necessary to calculate weight gain, feed consumption and FCR on weekly basis; it is sufficient if these are calculated soon after the birds are marketed. Hence, calculation of FCR up to a specified period is shown below:

3.2.2.1. Calculation of FCR up to the end of x weeks of age

3.2.2.1.1. On BH basis

If F is the cumulative feed consumption (kg) and W is the total body weight produced (kg), FCR up to x weeks of age = F÷W; alternatively, average values (per bird or per bird/d over the entire period) can also be computed and utilized.

In the above example, considering x = 6 weeks, F = 2780 kg or 2.78 kg/bird or 0.0662 kg/bird/d and W = 1360 kg or 1.36 kg/bird or 0.032 kg/bird/d. Therefore, FCR = 2780/1360 or 2.78/1.36 or 0.0662/0.0324 = 2.043.

3.2.2.2. On BD basis

Calculation of bird days facilitates giving weightage to mortality. In the above example, considering x = 6 weeks, F = 2780 kg, out of a possible 42,000 bird days (42 days x 1000 birds housed at the beginning), only 39,499 bird days were effective and the rest lost due to mortality. Therefore, for each bird day, (2780÷39499) kg of feed, and for x weeks (42 d), (2780÷39499)*42 = 2.956 kg of feed was consumed by each bird. This value can also be obtained directly by dividing total feed consumed by average number of birds per d; i.e. (2780÷940.45) = 2.956 kg. With cumulative weight gain per bird being 1.36 kg, weighted FCR = (2.956÷1.360) = 2.174.

It can be noted that FCR on BD basis will be usually greater than that on BH basis and they will be same if and only if mortality = 0%; analogous to HH and HD egg production.

3.2.3. Simplified method of calculating bird days

Instead of developing a table of number of birds living (Table 48.4), it will be convenient if a table of mortalities is prepared as follows:

Table 48.6. Simplified method of calculating bird days

	Number of birds alive								
Week	1	2	3	4	5	6	7	Feed intake (kg)*	Weight gain (g)**
I		5	3		1	1		100	60
II				4				245	150
III	6	4	16	20	10	5	3	490	250
IV	2				2			575	300
V	8	4	6		4		1	570	300
VI								800	300
* Weekly	** Average gain during the week								N = 1000

Considering mortality to have occurred in the morning itself and not considering the birds from that very day, it is obvious that 5 x 41=205 bird days are lost due to the death of 5 birds on day 2 of week I. Similarly 3 x 40, 38, 37, 32 x 4, 28 x 6 ...... 8 bird days; that means, totally 2501days are lost due to mortality. Subtracting from the possible 1000 x 42 = 42,000 bird days, actual bird days spent (39,499) is obtained which is same as in Table 48.4.

With the bird days calculated, further computation of FCR on BH and BD becomes easy. The same method can be adopted for calculating hen-days while calculating HD egg production.

3.2.4. Mortality/Livability

Survivability can be calculated either on weekly basis or for a specified period by utilizing data in a form either as in Table 48.4 or Table 48.6; the latter being more convenient. Mortality is generally calculated as % of the number of birds housed in the beginning.

Chapter **49**

Ducks, Turkeys, Geese, Quails and Other Poultry

Exhaustive information is available on rearing and management of each of the species; but, management procedures like brooding, growing, laying, lighting, litter management, manure handling, etc. are generally similar to that of chicken; some differences specific to the species concerned are indicated below. Floor, feeder and drinker space specifications are given in Chapter "Management requirements and specifications". Further, details regarding nutrient requirements and diseases will be dealt separately in the relevant forthcoming Chapters.

1. Ducks

1.1. Floor and swimming water

Slabs of welded wire (1.25 cm x 1.25 cm of 8 gauge) 10 cm above the concrete are preferred since it is easy to wash the manure and avoid dampness. Brooding is similar to chicken and ducklings do not require swimming water. On the contrary, stagnant water can act as a source of infection. After 4 weeks of age, welded wire (2.5 cm x 2.5 cm of 8 gauge) is required up to the end of fattening period (8 weeks).

Ducks cannot tolerate shortage of water and cold water is fatal to overheated ducklings. They are also intolerant to sun soon after eating.

If swimming water has to be provided, concrete troughs 0.9 m wide, 20 to 30 cm deep opposite the breeder house at the end of the yard is preferred. Water is a very important requirement for ducks because they are one of the Waterfowls. At 20°C, water required for 1000 ducks aged 1, 4 and 8 weeks is 28, 120 and 330 l/d, respectively; duck breeders require about 250 ml/bird/d.

1.2. Debilling

Overcrowding soon precipitates in feather-pulling which usually appears at around 4 weeks of age. The vice can be minimized by "debilling" (similar to debeaking) to remove the horn at the front of the top bill.

1.3. Egg production

It is advisable that ducks are at least seven months old when they start laying eggs to avoid small eggs. For this purpose, a photoperiod of 14h per day is provided 3 weeks prior to the expected date of laying. Drakes are provided with light 4 to 5 weeks before their mates are likely to start production. Egg-type ducks reach more than 90% production within 5 weeks. Most of the eggs are laid before 7 am and hence are collected at around 7am. Eggs are preferably washed soon after laying, fumigated and stored. Incubation of eggs is similar to chicken excepting that a relative humidity of 75% is required throughout. If the eggs are held for more than one week before setting, they have to be turned daily.

1.4. Hatching eggs

Drakes are identified by their typical belching voice and Drake feather and females by their hak voice at 6-7 weeks of age. A mating ratio of 6 to 8 ducks per drake is recommended and hatching eggs are collected one month after the drakes are allowed with their mates. Nests must be clean to ensure eggs free from Salmonella with the most common with duck eggs.

1.5. Aflatoxin in feed

Ducklings are most susceptible to Aflatoxicosis and hence it is extremely important to make sure that the feed is devoid of aflatoxins. It is for this reason that groundnut cake is generally not used in duck rations. For details regarding aflatoxins see Chapter "Toxins in poultry feeds". Unlike chicken, ducks prefer pellets to mash since it is easier to feed upon and feed efficiency is superior; pellet size of 0.3 cm for starter and 0.5 cm for others. Breeder ration is provided one month prior to the expected onset of lay. As in case of chicken, oyster shells are offered free-choice in separate hoppers. Feed restriction is practiced similar to chicken. The conversion ratio is around 3.0.

2. Muscovy ducks

(Gourley on internet)

2.1. Incubation

Muscovy ducks require an incubation period of 35 days. In forced-draft incubators, turning is as per the manufacturer's recommendations, usually once every hour. If the eggs must be turned manually, it should be done 3 or more (an odd number) times per day. A dry bulb temperature of 37.5°C and a wet bulb temperature of 31.5°C (65% relative humidity) are recommended.

For still-air incubators the temperature within the incubator should be maintained at approximately 102°F (39°C) with 60 to 65% relative humidity. Water pans should be filled just prior to use and water replaced every 3 to 4 days throughout the incubation period. The eggs should beplaced horizontally and turned 180 degrees on the long axis, 3 or more times per day (an odd number).

Eggs should be transferred to hatching machines approximately 3 d prior to hatch (this can be done anytime after 24 days if desired). Recommended settings during hatch are 37°C temperature and a wet-bulb temperature of 31°C (66% relative humidity). The duckling holding room should be maintained at 23.9°C and 75% relative humidity.

Any unhatched eggs or cull ducklings must be humanely-euthanized immediately after the hatch is pulled. Acceptable methods include cervical dislocation and maceration.

2.2. Range Rearing

Ducks can be successfully reared on range; the traditional system involves rearing ducks completely on pasture. Feed should be provided as a supplement to the pasture in order to ensure proper nutrition. Shelter from wind, rain and sun should be provided. Ducks should be confined to the shelters at night as protection from predators. When first placed on range young ducklings should be carefully monitored to assure they are not chilled. Another range system allows ducks access to outside runs or pens that are attached to the building. Doors to the run should be opened only during the daytime when weather permits. Each of these systems can be an effective type of enrichment without excessive cost.

Beak and Claw Trimming

2.3. Debilling

Muscovy ducks have a problem with feather-pecking. Pecking generally begins around three weeks of age when the adult plumage begins to grow. Trimming the bill is stressful and probably causes some pain. Unfortunately, with Muscovy ducks, no other viable alternatives exist at this time. Trimming results in a slight shortening of the upper bill relative to the lower. This keeps the duck from being able to grasp feathers or flesh during pecking.

Trimming should take place before the birds are three weeks of age. Trimming can be done at the hatchery by searing the nail of the upper bill with an electric beak trimmer. This method, however, may not be ideal since re-growth can occur and the process may need to be repeated. Accuracy is also more difficult since the beak of a day-old duckling is small. Trimming can also be done at 7 to 21 days of age. The upper bill is cut at the mid-point of the nail; about ¼ inch is removed. This procedure can be done with an electric debeaker or very sharp straight scissors.

Muscovy ducks are very strong and their nails are extremely sharp. The nails can be trimmed to reduce scratching of pen mates or risk of injury to workers. It can be performed as early as 10 days of age. However, it may be more economical and less stressful to the birds if it is done at the same time as bill trimming. This procedure can be done with little discomfort to the duck if performed properly. The claw should be trimmed close to the base but extreme caution should be taken to ensure it is not trimmed too closely.

3. Turkeys

3.1. Rearing

Turkeys are the most susceptible to cold among domestic birds. They are reared intensively at least up to 16 weeks. All in all-out system is recommended. They are weaned off heat at 8-10 week of age and most of the early mortality is due to inanition or failure to drink.

3.2. Shooting the red

After 6-8 weeks of age, head parts become bright red referred to as shooting the red; stags can be easily separated at this stage. However, vent-sexing (similar to chicken) can also be practiced.

3.3. Debeaking, wing-clipping/wing-notching

Debeaking is practiced at 3-5 weeks of age (similar to chicken). In addition, desnooding (removal of snood or dew bill) can be practiced to prevent head injuries and spreading of Erysipelas. It can be done by thumb-nail or thumb pressure at day-old or by cutting it off close to head by use of sharp, pointed scissors at about 3 weeks of age.

Similarly, to prevent flight, wing-clipping or wing-notching can be practiced; the former by clipping wing feathers at 1-14 d of age and the latter by severing the tendon crossing the centre of outer most wing point by use of a vertical red-hot steel bar (in an electrical debeaker) at 1-3 weeks of age. However, for market and breeding turkeys, none of the above flight-preventing procedures is practiced.

3.4. Hatching eggs

The breeding females are preferably saddled to avoid injury while mating; if AI is practiced, saddling is not required. For single male matings, number of hens/ tom of small, medium and large breeds is 20, 18 and 16, respectively; for flock mating, the corresponding values are 14, 12 and 10, respectively. AI, if practiced, is done 10-15 d after the start of additional light. Second and third inseminations are done 5d apart and subsequently once every 10d till 10 weeks followed once weekly to achieve optimum fertility. Lighting is similar to chicken. During laying period, nests at a rate of 20 to 25% of the number of hens have to be provided and most of the eggs are laid in the afternoon. Hatching eggs are fumigated with 3X concentration and cooled slowly. If the eggs are held for more than 15 days, they have to be turned daily in the cold storage.

4. Geese

4.1. Rearing

Goslings must be let out on the short grass as soon after hatching as possible.

Litter management is very important because the fecal matter contains more moisture than in chicken.

4.2. Sexing

Except in case of Pilgrim, sex determination is by the examination of reproductive organs which is done as follows: Goose is kept tail-end away over the edge of the table/knee so that it can be readily bent downwards. After doing so, vaseline-smeared index finger is inserted 1-2 cm into cloaca and moved in a circular manner till the sphincter relaxes; subsequently, pressure is applied directly below and on side of the vent (similar to eversion of vagina in chicken while inseminating) to expose the accessory sex organ of the gander.

4.3. Hatching eggs

While breeding, a mating ratio of 5 geese/gander is allowed for at least some weeks in separate pens for establishment of sets. Afterwards, ganders can identify their mates even when all are run together. Specific observation has to be made to avoid preferential mating. Number of nests required is 30% of the number of females.

If eggs are held for more than 2 days in the cold storage, they have to be turned. Incubation period is 29 to 31d except in case of Egyptian geese which take 35d. Artificial incubation is difficult as they require higher humidity and therefore, natural incubation under chicken, turkeys or ducks is practiced. If chicken are used, they cannot turn goose eggs and hence turning has to be done manually. Sprinkling water or dipping hatching eggs in water for 30 sec daily during last half of incubation increases hatchability. Hatchet gosling should be quickly removed to a warm place otherwise, the brooding goose leaves the unhatched eggs to take care of newly-hatched one.

4.4. Geese as weeders

Geese can be employed as good weeders on crops like cotton, hops, onions, garlic, strawberries and other hoe crops (not cabbage or lettuce on which the geese themselves feed). For the purpose, 6 week-old goslings are fed only a light feed of grain the night before to keep them hungry and let out on the field the next day morning. Sufficient shade and water must be provided all through the crop area (Ernst and Coates, 1977).

5. Japanese quails

Species or subspecies of the genus *Coturnix* are native to all continents except the Americas. One of them, *Coturnix coturnix japonica*, was introduced into the United States by bird fanciers around 1870. This subspecies is called Japanese quail but is also known by other names: Common quail, Stubble quail, Pharaoh's quail, Eastern quail, Asiatic quail, Japanese Grey quail, Red-throat quail, Japanese migratory quail, King quail, and Japanese King quail (Woodard *et al.* 1973).

Certain properties of coturnix, such as its ability to produce 3 to 4 generations per year, make it an interesting laboratory animal. Depending on the day length, some females start laying at 35 days of age (average 40 days) and are in full production by 50 days of age. Under favorable environments they produce for long periods, averaging 250 eggs per year. Coturnix is relatively inexpensive to maintain and some 8 to 10 quail can occupy the same space as one chicken (Woodard *et al.* 1973).

5.1. Rearing

The day-old chicks are only about 6 g in weight and hence require extreme care during brooding. A 250 W infrared bulb can take care of 275 to 300 chicks. On deep litter system, there may be higher incidence of accidental mortality which will be compensated by better survival during growing and laying periods.

5.2. Growth

Growth rate in quails has been found to be the fastest in reaching mature body weight; earliest than any of the gallinaceous species reported so far. Females grow relatively faster than males and both reach maximum body weight by 8 weeks of age. During egg production, the feed consumed is 25 to 28 g/bird/d with an average feed efficiency on egg mass basis is around 3.8 (Sreenivasaiah, 1977). However, during maximum production period, Feed conversion ratio can be 3.3. Performance efficiency during maximumproduction period willbe around 28. Total photoperiod required during laying period appears to be similar to that for chicken.

5.3. Sex dimorphism

The two sexes can be distinguished outwardly at about 3 weeks of age. The adult male is identified readily by the cinnamon-colored feathers on the upper throat and lower breast region. The voice of the male is described as a loud, castanet-like crow, describing sound as "pick-per' wick" or "ko-turro-neex". Young birds begin to crow at 5 to 6 weeks old. During the height of the normal breeding season, coturnix males will crow throughout the night (Woodard *et al.* 1973).

The female is similar to the male in coloration except that the feathers on the throat and upper breast are long, pointed, and much lighter cinnamon. Also, the tan breast feathers are characteristically black-stippled. In addition, feathers on the above regions in the females will be long and characteristically pointedagainst blunt and shorter feathers in males (Woodard *et al.* 1973).

Males weigh less than the females – at 8 weeks, females weigh 200 g as against 160 g in case of males (meat line); 150 g and 120 g, respectively in case of egg line. Age at sexual maturity – 8 weeks.

5.4. Egg production

Peak egg production is attained by 15 weeks of age; Average egg weight is 10 g (about 8% of the hen's body weight); clutch size ranges from 2-3 to 40 (average 6.65); average pause duration is 1.68 d. Most of the eggs are laid between 3 p.m. to 8 p.m. The eggs are highly colored although a separate white-egg line has been developed. Mating ratio for obtaining good fertility is 1 : 2 to 1 : 3.

It has been that an individual hen lays eggs of a shape, size, and color pattern characteristic of that hen and hence, it is possible to distinguish the eggs of individuals in mixed clutches of coturnix eggs with a high degree of accuracy (Woodard *et al.* 1973).

5.5. Egg quality and meat yield

The egg quality parameters are – share index 73-80, albumen index 0.09 to 0.10, yolk index 0.4 to 0.5, Shell thickness 0.17 to 0.18 mm, Haugh Unit Score 87 to 90. Average dressing percentage ranges between 67 and 77 (6 to 8 weeks). The meat is predominantly red and contains 1.5% fat and 20% crude protein.

Dressing percentage in case of quails is around 75%. Meat from Japanese quails has around 20% protein, 5% fat and 1.5% ash. The most preferred product is fried meat on a low fire with fat and spices. Quail meat is essentially red meat. Cooked quail meat can be pickled in a medium containing (w/v) 8% salt, 2% ginger, 2% spices mixture and 2% garlic in a medium of vinegar and water in equal quantities. 600 to 700 g of quail meat can be pickled in one lit of the above recipe.

6. Pheasants

Origin : Orient. Mainly are game birds; they are bred, released and shot for sport. Chinese Ring neck pheasant is most common game breed. Pheasants lay in clutches of 10-12 eggs and the eggs hatch in 23-24 d. They are debeaked at 7-9 d of age. The common diseases encountered are coccidiosis, botulism, Pullorum and gapes.

7. Bobwhite quail

Origin : Canadian border to Mexico and Cuba. It is similar to Japanese quails excepting that the incubation period is 23-24 d.

8. Guinea fowl

Origin : West Africa; Varieties – Pearl, White and Lavender. They are highly sensitive to light and produce 90-170 eggs per annum which are smaller but with thicker shells than chicken eggs. Keets (young ones) grow slower than chicken. Males are differentiated by their cry, larger helmets and wattles making their head looked coarser. A sex ratio of 4-5 hens per male is satisfactory. The eggs hatch in 26-28 d.

9. Pigeons

Origin : Europe, Asia and near East. Pigeons are reared for fancy, meat, flyers and performers and for sport of racing. Of over 200 breeds, Homer, White King and Swiss Mondaine are important. Sex differentiation is difficult. They mate in pairs and remain faithful throughout. The house for pigeons is called loft. They usually lay eggs on alternate days. Males incubated eggs during midday and females during the rest of the time; the eggs hatch by 17th day and young one is referred to as squab. Both the parents take care of the young ones by regurgitating the crop contents (called as crop milk or pigeon milk). Pigeons, like geese, are very fast growing.

10. Swans

Origin : Europe. These are mainly ornamental. Common breed is Mute Swan which has no voice. They mate in pairs and, like pigeons, remained faithful throughout. Females breed as long as 30 years and males live up to 60 years. They lay only 6-8 large, greenish-white eggs per annum which hatch in 35-40 d (six weeks). The young ones are called cygnets and they can be looked after similar to geese. They are susceptible to Mark's disease, Salmonellosis, Pasteurellosis, Staphylococcosis, botulism, Tuberculosis, fungal diseases, coccidiosis and internal and external parasites.

11. Partridges

Origin : Europe (Hungarian Partridge or Hungarian Grey Cannon balls) and Asia (Chukar Partridge). Sex is determined by wing coverts. They are monogamous and mate in single pairs. They fight, even while young, when put in groups. Hence even brooding is preferably done with less than 30 chicks in one group. On average, 30 eggs are normally laid in spring and summer which, stored up to 10d, can hatch in 24-25 d.

11.1. Chukar partridge (Red-legged partridge)

(Woodard, 1982)

Chukar partridges (*Alectoris chukar chukar*) can be easily identified by a black band across forehead through the eyes, down the neck and meeting as a gorget between white throat feathers and upper breast. Lower throat and back are ash-gray and feathers at flanks have a gray base with 2 black bands at the tip giving a uniform stipled appearance to the bird. In both the sexes, bill, legs and feet are orange-red.

Day-old chicks weigh about 13-14g hatching after an incubation of 23-25 d. Sexes can be identified when they are adults by looking for a cone-like protuberance at the center of the cloacal fold in case of males.

Birds mature by 35-40 weeks and produce more eggs in the second cycle than the first. Eggs are oblong (4.2 cmx 3.1 cm) weighing 16-25 g (average 21 g) yellowish white in color studded with speckles of brown of different shapes and sizes all over with a shell of thickness 0.228 mm and shell membranes 0.047 mm thick. Proportion of shell : yolk : albumen is approximately 3 : 7 : 10.

12. Peafowl

Origin : India, Myanmar, Java, Sri Lanka, Malaya, Congo. Indian blue is the common breed with Java green and Congo being the other breeds. A male can mate with 5 females and laying begins after 2 years of age. Usually, 10-12 eggs are laid in one year. Black head and coccidiosis are the common diseases.

13. Ostrich

Ostriches are being farmed in South Africa for over 100 years mainly for feathers and also for skin (leather industry), eggs and meat. The two species identified are *Srtuthio camels* (North African ostrich) and *Struthio australis* (South African ostrich). They stand at 2.5 m high, weigh 80-120 kg and produce up to 60 eggs a season. Hatchability of eggs is good but chick mortality is high. They also are good foragers like geese and feed efficiency during the first 8 weeks is estimated to be 3.0. Chicks, similar to broilers, suffer from leg problems due primarily to large body size; can be overcome by calcium supplementation.

14. Emu

Origin : Australia; Zoological name - *Dromaius novaehollandiae*. Life-span about 25-30 years. Average height of the adult is 1.5-1.8 m each weighing about 40-50 kg. Age at sexual maturity is 18-24 months and eggs are produced during winter/ spring (October to March). Each season 20-30 eggs are produced each weighing, on average, 600-750 g, dark green in color (appears black). Incubation period is 47-53 d (Average 52 d). An adult bird consumes about 1.4-1.5 kg feed/d and each pair requires a space of 18 m x 9 m with 2.4 m x 2.4 m shelter space. Skin is useful in the industry, fat/oil is used in cosmetics and medicine, meat is high in protein and lowers in cholesterol, eggs are mainly for hatching although they are edible and feathers are used for manufacturing fancy garments. They are resistant to most of the diseases and hence no vaccination is needed.

Under Indian conditions, the growth parameters estimated at 1, 2 and 3 months of age were as follows: body weight (g) –1065, 2970 and 6120; growth rate (g/d) – 23, 61 and 105; feed consumption g/bird/d – 50, 149 and 232; and FCR – 2.08, 2.34 and 2.23, respectively. In addition, fertility, hatchability (TES) and hatchability (FES) were 70.4, 61.5 and 87.3%, respectively. the day-old chick weight was 327 g which was 66.8% of egg weight (Menezes *et al.* 2001)

Chapter **50**

Management Requirements and Specifications

1. Floor, feeder and waterer space

1.1. Chicken, Turkeys, Ducks and Geese

Table 50.1. Space requirements – Major species

Age / System	Chicken		Turkey		
	Egg-type	Meat type	Turkey (large)	Ducks	Geese
Floor space (m²/bird)					
Hover space	0.001 to 0.002	0.0015 to 0.0025	0.003	0.003	0.0035
0-4 wk	0.065	0.090	0.135	0.072	0.135
4-8 wk	0.120	0.1125	0.180	0.135	0.180
8-12 wk	0.090	0.135	0.270	0.180	0.270
>12 wk	0.135	0.225	0.450	0.270	0.450
Adult	0.135 to 0.270	0.225 to 0.360	0.720	0.450-0.540	0.720
Cage	0.0387-0.0645	0.125	0.250	----	----
Feeder space (cm/bird, linear)					
0-1 wk	1.5	1.5	3.0	5.0	5.0
1-2 wk	2.5	5.0	6.25	5.0	6.25
2-4 wk	2.5	5.0	7.5	6.25	7.5
4-8 wk	3.75	7.5 to 15.0*	10.0	6.25	10.0
>8 wk	6.25	8.75 to 15.0*	12.5	7.5	12.5
Adult	10.0	12.5 to 15.0*	15.0	12.5	15.0
Waterer space (cm/bird, linear)					
0-1 wk	1.5	1.75	2.5	1.75	1.75
1-4 wk	1.5	1.75	2.5	1.75	2.5
4-8 wk	1.5	2.0	2.5	1.75	2.5
>8 wk	2.0	2.5	3.0	2.0	3.0
Adult	2.5	3.75	3.5	2.5	3.5

* Usually under feed restriction — *Source* : Wilson *et. al.*, 1997

1.2. Pheasant, Bobwhite, Guinea, Ostrich and Emu

Table 50.2. Space requirements – Other species

Requirement	Pheasant	Bobwhite	Guinea	Ostrich	Emu
Floor space	cm²/bird			m²/bird	
Hover space	25-50	25.0-37.5	25-50	0.18	0.135
0-4 wk	900	225	450	0.90	0.72
4-8 wk	1800	270	720	1.80	1.35
8-12 wk	2700	450	900	3.60	2.70
>12 wk	2700	450	1350	7.20	5.40
Adult	2700	900-1350	1350-2700	0.1-1.2*	0.04-0.12*
Cage	387-645	322-644	387-644	----	----
Feeder space	(cm/bird)				
0-1 wk	1.5	1.5	1.5	10	10
1-2 wk	2.5	1.5	2.5	15	15
2-4 wk	3.75	1.5	2.5	30	22.5
4-8 wk	5.0	2.5	3.75	30	22.5
>8 wk	7.5	2.5	6.25	45	22.5
Adult	7.5	4.0	10.0	45	22.5
Waterer space	(cm/bird)				
0-1 wk	0.75	0.50	1.25	25	15
1-4 wk	1.25	0.60	1.25	25	15
4-8 wk	1.50	0.75	1.25	25	15
>8 wk	2.50	0.75	2.00	25	15
Adult	2.50	0.75	2.50	25	15

* in hectare (ha) for 2 to 4 birds — *Source* : Wilson *et. al.*, 1997

1.3. Chicken on all-litter system

Table 50.3. Floor, feeder and waterer space requirements of chicken (per bird) (All litter system)

Type	Floor space, m²			Feeder space, cm		Waterer space, cm		
	Brood	Grow	Lay	Grow	Lay	Brood	Grow	Lay
Standard Leghorn								
Egg type	0.070	0.09 (0.14*)	0.14	6.4	8.75	1.50	1.90	2.50
Breeder pullets	0.079	0.16	0.19	6.4**	9.40			
Breeder cockerels	0.093	0.16	0.19	7.6	9.40			
Medium size								
Egg type	0.079	0.11 (0.16*)	0.16	7.6	10.5		2.20	
Breeder pullets	0.093	0.18	0.21	7.6**	10.6			
Breeder cockerels	0.093	0.20	0.21	8.9	10.6			

Type	Floor space, m²			Feeder space, cm		Waterer space, cm		
	Brood	Grow	Lay	Grow	Lay	Brood	Grow	Lay
Meat type								
Breeder pullets	0.093	0.23	0.28	15.0**	15.0	1.9	2.50*	3.10
Breeder cockerels	0.116	0.28	0.28	20.0	15.0	2.5	3.20*	
Broilers***								
1.36 kg	0.05							
1.82 kg	0.06					2.0****		
2.27 kg	0.08							
2.72 kg	0.09							
3.18 kg	0.11							

* up to 22 weeks ** for strait-run flocks *** decrease floor space by 10% during winter **** up to 8 weeks, 2.8 cm for roasters For broilers, feeder space required is 5.0 cm up to 5 weeks, 7.6 cm up to market (7 weeks); for all others, 5.0 cm All feeder and waterer space are in linear cm; in case of circular feeders and waterers, 20 to 30% extra birds can be fed and watered, respectively.

Source : North and Bell, 1990

1.4. Chicken in cages

Table 50.4. Floor, feeder and waterer space – Cage system

Type	Floor space (cm²/bird)			Feeder space (cm/bird)			Waterer space (cm/bird)			Pullets/nipple			Pullets/cup		
	B	G	L	B	G	L	B	G	L	B	G	L	B	G	L
Std. Leghorn	155	290	389	5.1	4.1	7.6	1.9	2.5	3.8	15	10	8	25	15	12
Medium size	181	348	452	5.6	6.9	8.4	2.0	3.1	4.3	12	8	6	19	13	10

B – Brooding (0 to 5 wks), G – Growing (6 to 18 wks), L – Laying (> 18 wks)

Source : North, 1984

1.5. Chicken on different systems of rearing

Table 50.5. Floor space (m²/bird) for chicken under diffcrent systems of rearing

Type	Standard Leghorn	Medium size	Standard Meat-type
Layers			
Slat and litter*	0.12	0.14	
Wire and litter*	0.12	0.14	
All slat	0.09	0.12	
All wire	0.09	0.12	
Breeders			
Slat/Wire and litter*	0.16	0.19	0.23
All slat	0.12	0.14	0.19
All wire	0.12	0.14	

* 40% litter and 60% slats

Source : North and Bell, 1990

1.6. Cage dimensions

Table 50.6. Cage dimensions (cm)

	Width	Depth	Birds	Height	Welded wire mesh size	Slope
Brood – cum – grow cages	55.9	61.0		35.6 to 40.6	1.3 x 5.1 or 2.5 x 2.5, 14 gauge; if more than 1.25 cm, cover with paper during first 2 weeks	NIL
	61.0	61.0				
	61.0	68.6				
	61.0	91.4				
Laying*	25	41	2	40.6	2.5 x 5.0; uppermost wire at right angles to length of the house or from back to front of the cage	4.1 for each 30.5 cm of depth; 7.66°
	31	41	3			
	31	46	4			
	31	51	4			
	36	41	4			
	36	46	5			
	41	46	6			
	41	51	6			
	61	46	7			
	61	91	15			
	91	122	30			

* Feeder/Waterer space: 12.7 to 15.3 cm/bird *Source* : North and Bell, 1990

2. Feed and water consumption

2.1. Laying type chicken

Table 50.7. Feed and water intake by Standard Leghorn pullets (21.1°C)

Brood-grow period					Laying period (Cages)		
Age (wks)	Av. Weight (kg)	Feed g/bird/d	Water intake % of Av. Weight	Water intake g/g of feed	Week of egg prod.	Feed g/bird/d	Water intake ml/bird/d
1	0.068	13.2	18.3	0.94	1	82	155
2	0.118	16.3	25.5	1.84			
3	0.177	24.5	31.9	2.31	2	84	193
4	0.241	31.8	32.2	2.42			
5	0.309	36.3	29.3	2.47	3	85	208
6	0.377	38.1	27.0	2.65			
7	0.472	39.0	23.4	2.82	4	87	220
8	0.531	40.9	22.1	2.86			
9	0.617	43.1	19.8	2.87	5	88	227
10	0.699	45.4	18.4	2.84			
11	0.763	47.7	17.1	2.74	6-7	91	239
12	0.849	49.9	15.8	2.68			
13	0.922	52.2	14.8	2.61	8-12	95	227
14	0.990	54.5	14.0	2.54			

Brood-grow period					Laying period (Cages)		
Age (wks)	Av. Weight (kg)	Feed g/bird/d	Water intake		Week of egg prod.	Feed g/bird/d	Water intake ml/bird/d
			% of Av. Weight	g/g of feed			
15	1.053	56.8	13.4	2.48	13-18	105	220
16	1.108	59.0	12.9	2.43			
17	1.158	61.3	12.5	2.37	19-38	105	208
18	1.203	63.6	12.2	2.31			
19	1.244	65.8	11.9	2.26	39-49	105	201
20	1.280	68.1	11.7	2.20			
21	1.312	70.4	11.6	2.15	50-60	105	193
22	1.335	72.6	11.5	2.10			
Source : North and Bell, 1990							

2.2. Broilers and Roasters

Table 50.8. Growth and feed consumption of broilers and roasters (21.1°C)

Age (wk)	Males			Females			Straight-run		
	Av. Weight (kg)	Cumulative		Av. Weight (kg)	Cumulative		Av. Weight (kg)	Cumulative	
		Feed intake (kg)	FCR		Feed intake (kg)	FCR		Feed intake (kg)	FCR
1	0.15	0.12	0.80	0.15	0.12	0.80	0.15	0.12	0.80
2	0.41	0.42	1.05	0.38	0.40	1.05	0.39	0.41	1.05
3	0.72	0.89	1.23	0.67	0.84	1.25	0.69	0.86	1.24
4	1.12	1.56	1.40	1.00	1.43	1.42	1.06	1.49	1.41
5	1.54	2.40	1.56	1.37	2.20	1.60	1.45	2.30	1.58
6	2.01	3.48	1.73	1.75	3.10	1.77	1.88	3.29	1.75
7	2.52	4.79	1.90	2.15	4.94	1.94	2.34	4.49	1.92
8	3.03	6.26	2.07	2.53	5.34	2.11	2.78	5.80	2.09
9	3.51	7.87	2.24	2.88	6.57	2.28	3.19	7.22	2.26
10	3.97	9.58	2.41	3.18	7.82	2.45	3.58	8.70	2.43
11	4.40	11.34	2.58	3.45	9.05	2.62	3.93	10.20	2.60
12	4.47	13.11	2.75	3.67	10.24	2.79	4.22	11.69	2.77
Source : North and Bell, 1990									

2.3. Straight-run broilers

Table 50.9. Feed and water intake by Straight-run broilers (21.1°C)

Age (wks)	Av. Weight (kg)	Feed intake g/bird/d	FCR at end of week (Cumulative)	Water intake (ml/bird/d)
1	0.15	16.8	0.80	30
2	0.39	41.4	1.05	61
3	0.70	65.0	1.24	95
4	1.06	90.5	1.41	133
5	1.46	115.0	1.58	174
6	1.89	143.7	1.75	216
7	2.34	170.9	1.92	254
8	2.78	188.2	2.09	288

FCR = kg feed per kg weight gain *Source* : North and Bell, 1990

3. Body weight recommendations

3.1. Egg-type growing pullets

Table 50.10. Standard body weights – egg-type growing pullets

Age (wks)	Standard Leghorn	Medium-size *
1	0.065	0.13
2	0.121	0.18
3	0.186	0.27
4	0.262	0.36
5	0.335	0.46
6	0.427	0.59
7	0.513	0.68
8	0.593	0.77
9	0.671	0.86
10	0.754	0.95
11	0.828	1.04
12	0.904	1.14
13	0.968	1.23
14	1.030	1.32
15	1.092	1.36
16	1.157	1.45
17	1.211	1.50
18	1.259	1.54
19	1.311	1.64
20	1.362	1.68

* brown-egg producers *Source* : North and Bell, 1990

3.2. Breeders

Table 50.11. Body weight recommendations for breeders

Age (wks)	Standard Leghorn		Medium-size brown egg-type		Standard size, meat-type		
	♂	♀	♂	♀	♂	♀	♂ ÷ ♀
1	0.14	0.09	0.18	0.13	0.15	0.14	
2	0.18	0.14	0.22	0.18	0.25	0.22	
3	0.27	0.22	0.32	0.27	0.46	0.41	
4	0.36	0.27	0.45	0.36	0.58	0.50	1.15
5	0.46	0.36	0.59	0.46	0.68	0.59	
6	0.55	0.41	0.73	0.59	0.82	0.64	
7	0.68	0.50	0.86	0.68	0.91	0.77	
8	0.77	0.59	1.00	0.77	1.04	0.86	1.19
9	0.91	0.68	1.09	0.86	1.13	0.96	
10	1.00	0.73	1.22	0.95	1.29	1.05	
11	1.04	0.82	1.32	1.04	1.41	1.14	
12	1.14	0.91	1.45	1.14	1.50	1.23	1.23
13	1.23	0.96	1.54	1.23	1.63	1.32	
14	1.32	1.04	1.63	1.32	1.77	1.41	
15	1.36	1.09	1.73	1.36	1.91	1.50	
16	1.46	1.14	1.82	1.45	2.04	1.59	1.27
17	1.50	1.19	1.91	1.50	2.13	1.68	
18	1.55	1.23	1.96	1.54	2.27	1.77	
19	1.64	1.27	2.09	1.64	2.40	1.86	
20	1.68	1.32	2.13	1.68	2.54	1.96	1.31
21	1.73	1.36	2.18	1.73	2.68	2.05	
22	1.77	1.41	2.27	1.77	2.90	2.18	
23	1.86	1.45	2.32	1.82	3.09	2.32	
24	1.90	1.50	2.36	1.86	3.36	2.50	1.35
25	1.96	1.55	2.45	1.96	2.58	2.63	
30	2.00	1.59	2.54	2.00	3.86	2.73	1.42
40	2.09	1.64	2.59	2.05	4.14	2.96	1.40
50	2.13	1.68	2.64	2.09	4.32	3.09	1.40
60	2.18	1.73	2.72	2.18	4.41	3.18	1.39
70	2.27	1.77	2.82	2.23	4.46	3.27	1.36
80	2.32	1.82	2.94	2.27			

Source : North and Bell, 1990

4. Incubation, hatching and brooding specifications

4.1. Chicken, Turkeys, Ducks and Geese

Table 50.12 Incubation, hatching and brooding specifications – Major species

Requirement	Chicken (egg-type)	Chicken (meat-type)	Turkey (large)	Ducks	Geese
		Storage of hatching eggs			
Time	≤ 8d	≤ 7d	≤ 8 d	≤ 7d	≤ 7d
Temp.(°C)	12.8-18.3	12.8-18.3	12.8-18.3	12.8-18.3	12.8-18.3
RH (%)	70-85	70-85	70-85	75-85	75-85
		Incubation and hatching			
Length (d)	21	21	28	28 *	30-35
Setter (°C)					
Dry bulb	37.5	37.5	37.5	37.3	37.2
Wet bulb	30.0-30.6	30.0-30.6	30.6-31.1	34.4	31.7-32.2
RH (%)	56 – 59	56 – 59	59 – 62	82	67 – 70
Hatcher (°C)					
Dry bulb	36.9	36.9	36.7	36.9	37.2
Wet bulb	32.2-33.3	32.2-33.3	32.2-33.9	35.6	31.7-32.2
RH (%)	71 – 77	71 – 77	72 – 82	92	67 – 71
Still air (°C)	38.9-39.4	38.9-39.4	38.9-39.4	38.9-39.4	38.9-39.4
		Brooding (°C)			
0-1 wk	32.2-35.0	29.4-35.0	32.2-35.0	29.4-32.2	26.7-32.2
Decrease/wk	2.8	2.8	2.8	2.8	2.8
Brooder room	15.6-26.7	15.6-26.7	15.6-23.9	15.6-23.9	15.6-23.9

* 35 d for Muscovy ducks — RH = Relative humidity

Source : Wilson *et. al.*, 1997

4.2. Pheasant, Bobwhite, Guinea, Ostrich and Emu

Table 50.13. Incubation, hatching and brooding specifications – Other species

Requirement	Pheasant	Bobwhite	Guinea	Ostrich	Emu
		Storage of hatching eggs			
Time	≤ 8d	≤ 14 d	≤ 8 d	≤ 7d	≤ 7d
Temp.(°C)	12.8-18.3	12.8-18.3	12.8-18.3	12.8-18.3	12.8-18.3
RH (%)	70-85	70-85	70-85	60-75	60-75
		Incubation and hatching			
Length (d)	24	24	28	42	52-56
Setter (°C)					
Dry bulb	37.5	37.8	37.5	36.3	36.4-36.7
Wet bulb	31.7-32.2	29.4-30.0	29.4-30.0	20.0	23.9
RH (%)	65 – 68	52 – 55	53 – 56	20	34

Requirement	Pheasant	Bobwhite	Guinea	Ostrich	Emu
Hatcher (°C)					
Dry bulb	37.2	37.2	36.7	35.8	36.1
Wet bulb	27.8-29.4	31.1-33.3	29.4-30.0	25.0	27.2
RH (%)	47 – 55	63 – 76	57 – 60	40	50
Still air (°C)	38.9-39.4	38.9-39.4	38.9-39.4	37.5	37.5
Brooding (°C)					
0-1 wk	35	35.0-37.8	35.0-37.8	29.4-32.2	32.2-35.0
Decrease/wk	2.8	2.8	2.8	2.8-4.4	1.7-2.8
Brooder room	15.6-29.4	15.6-29.4	15.6-26.7	15.6-23.9	15.6-23.9

RH = Relative humidity   *Source* : Wilson *et. al.*, 1997

5. Light

5.1. Chicken, Turkeys, Ducks and Geese

Table 50.14. Specifications for light and natural mating – Major species

Requirement	Chicken		Turkey (large)	Ducks	Geese
	Egg-type	Meat-type			
0-2d	24h, 2.8w/m2	24h, 2.8w/m2	24h, 2.8w/m2	24h, 10w	24h, 10w
2d-2wk	23h, 2.8w/m2	23h, 2.8w/m2	23h, 2.8w/m2	23h, 10w	23h, 10w
2-10wk	Natural or <12h	Natural or <12h	Natural or <12h	Natural	Natural
10-20wk	Decreasing or 8h	Decreasing or 8h	Decreasing or 8h	Natural	Natural
Layer or breeder	15h, 10.76 lx, or add 15min/wk	16h, 32.28 lx, or add 15min/wk	16h, 53.8 lx, or add 15min/wk	15h, 10.76 lx, or add 15min/wk	15h, 10.76 lx, or add 15min/wk
Females/male	12-15	10-12	7-10	5-6	2-5
Females/nest	4	4	5	3	5
Sex. maturity (months)	5-6	5-7	7-8	7-8	6-9

w = J/s   lux (lx) = 0.0929 foot candle (fc)   *Source* : Wilson *et. al.*, 1997

5.2. Pheasant, Bobwhite, Guinea, Ostrich and Emu

Table 50.15. Specifications for light and natural mating – Other species

Requirement	Pheasant	Bobwhite	Guinea	Ostrich	Emu
0-2d	24h 1.1w/m2	24h 1.1w/m2	24h, 1.1w/m2	24h, 1.1w/m2	24h, 1.1w/m2
2d-2wk	Natural or <12h	Natural or <12h	Natural or <12h	Natural	Natural

Requirement	Pheasant	Bobwhite	Guinea	Ostrich	Emu
2-10wk	Natural or decreasing	Natural or decreasing	Natural or decreasing	Natural	Natural
10-20wk	Natural or decreasing	Natural or decreasing	Natural or decreasing	Natural	Natural
Layer or breeder	17h, 53.8 lx, or add 30 min/wk	17h, 53.8 lx, or add 30 min/wk	16h, 53.8 lx, or add 15 min/wk	Natural or 16h, 53.8 lx	Natural
Females/male	6-8	1-4	4-8	1-4	1-2
Females/nest	5-6	3-5	3-5	1-4	1-2
Sex. maturity (months)	4-5	5-6	5-6	20-36	20-36
w = J/s lux (lx) = 0.0929 foot candle (fc)					*Source* : Wilson *et. al.*, 1997

5.3. Broilers

Table 50.16. Lighting scheme for broilers

Age (d)	0	1	2	3	4	5	6	7	8	9	? 10
Max. intensity*	40	40	30	20	20	20	10	10	3.5	3.5	2.5
* in lux											*Source* : Charles *et.al.*, 1994

5.4. Broiler breeder parents

Table 50.17. Lighting program for broiler breeders in a black-out house

Age	Males		Females	
	Light (hr/d)	Remarks	Light (hr/d)	Remarks
1 d	24	37.67 lux, full feeding	24	37.67 lux, full feeding
6 d	8	10.76 lux, controlled feeding	8	10.76 lux, controlled feeding
19 wk	10	Males moved to breeder house	14	
20 wk				
21 wk	14			Females moved to breeder house
22 wk				First eggs produced; breeder feed offered
23 wk	15		15	
24 wk				5% HD egg prod.
25 wk				Saving of hatching eggs
26 wk				
27 to 60-70 wk	16	Lights off no later than 20.00 hrs	16	50% HD egg prod. by 27th week
lux (lx) = 0.0929 foot candle (fc)				*So urce* : North and Bell, 1990

5.5. According to housing type

Table 50.18. Lighting program according to housing type

Age (d)	Black-out growing Black-out breeding	Black-out growing Open-sided breeding	Open-sided growing Black-out breeding	Open-sided growing Open-sided breeding
0-3	24 hr, 30-40 lux	24 hr, 30-40 lux	24 hr	24 hr
3-21	18 hr, 30-40 lux	18 hr, 30-40 lux	18 hr	18 hr
21-140	8-10 hr, 10-30 lux	8 hr, 10-30 lux	Corresponding to natural maximum in this period	Corresponding to natural maximum in this period
140-147	12 hr, > 30 lux	14 hr minimum or natural daylength	1-2 hr increment depending on season	1-2 hr increment depending on season
> 147	Increase @ 1 hr/week, 14-15 hr maximum	Increase @ ½ hr/week, 17 hr maximum	Increase @ ½ hr/week, 17 hr maximum	If natural daylength is ≤ 14 hr, Increase @ 1 hr/week, 17 hr maximum. If natural daylength is ≥ 15 hr, Increase @ ½ hr/week, 17 hr maximum.

Source : Leeson and Summers, 2000

6. Mating ratio

Table 50.19. Suggested mating ratio (Males/100 Females)

Parents	Result of mating	On litter	On slats-and-litter
Standard Leghorn	Commercial Standard Leghorn pullet	8	9
Medium - size	Commercial medium-size pullet	9	10
Standard meat – type	Commercial broiler	10	11

Source : North and Bell, 1990

7. Ventilation

7.1. Heat and moisture production

Table 50.20. Heat and moisture production by broilers at 21.1°C

Broilers *				Layers **		
Age (weeks)	Av. Body weight (kg)	Heat production (kJ/hr/kg)	Water output (g/bird/d)	Av. Body weight (kg)	Heat production (kJ/hr/kg)	Water output (g/bird/d)
2	0.20	60.4	59.0	0.5	46.42	96
3	0.34	53.4	90.7	0.9	33.65	141
4	0.59	41.8	127.0	1.4	26.69	176
5	0.79	32.5	158.7	1.8	23.21	203
6	1.09	30.2	181.4	2.3	20.89	218
7	1.36	23.2	231.2	2.7	19.00	235
8	1.70	23.2	226.8			
Note: 3.6 kJ/hr = 1 W			*Source* : * North, 1972; ** North and Bell, 1990			

7.2. Mechanical ventilation

1.1.1. Ventilation rates required

Table 50.21. Poultry ventilation rates

Stock	Age (d)	Weight (kg)	Maximum		Minimum	
			m³/min/bird	Fans *	m³/min/bird	Fans *
Pullets and layers including breeders		1.8	0.138	0.9	0.0150	0.10
		2.0	0.150	1.0	0.0162	0.11
		2.2	0.162	1.1	0.0174	0.12
		2.5	0.180	1.2	0.0192	0.13
		3.0	0.204	1.4	0.0216	0.15
		3.5	0.228	1.5	0.0246	0.16
Broilers	7	0.17			0.0024	0.02
	14	0.43			0.0048	0.03
	21	0.80			0.0084	0.05
	28	1.25			0.0114	0.08
	35	1.74			0.0144	0.10
	42	2.23			0.0174	0.12
	49	2.68	0.186	1.3	0.0198	0.13
Turkeys		0.5	0.054	0.4	0.0060	0.04
		2.0	0.150	1.0	0.0162	0.11
		5.0	0.300	2.0	0.0318	0.21
		10.0	0.504	3.4	0.054	0.36

* 61 cm diameter fans / 1000 birds (900 rpm and 70 Pa static pressure)

Source : Charles, 1981

7.3. Exhaust fan efficiency

Table 50.22. Sample performance data for exhaust fans

Diameter (cm)	Speed (rpm)	Motor Size (W) 1hp = 0.746W	Airflow in m3/min at static pressure (Pa)			
			0	100	125	250
20	1,650	0.015	11.36	8.97	8.21	—
20	3,500	0.050	16.30	14.80	14.46	11.79
25	1,550	0.015	16.87	12.98	11.73	—
25	3,416	0.124	35.78	34.65	34.34	32.38
30	1,600	0.062	33.74	30.47	29.39	23.49
30	1,741	0.186	47.71	43.17	41.24	—
35	1,752	0.249	74.12	67.88	66.14	56.80
40	1,140	0.062	47.57	40.90	39.02	—
40	1,670	0.186	96.84	84.35	81.05	36.92
40	1,725	0.249	71.97	67.93	66.83	60.83
45	1,140	0.124	76.28	69.86	68.02	—
45	1,648	0.249	127.52	116.44	113.68	95.42
45	1,725	0.466	115.45	111.33	110.19	104.57
52.5	1,140	0.186	108.26	102.21	100.54	—
52.5	1,725	0.560	139.56	135.47	134.62	128.08
60	855	0.249	133.22	122.40	118.71	—
60	1,071	0.249	186.30	161.31	154.50	104.51
60	1,139	0.373	198.52	179.49	174.46	143.99
60	1,140	0.653	177.61	170.12	168.12	155.35
75	855	0.746	287.55	275.48	271.93	245.38
90	460	0.373	303.88	258.44	222.94	82.36
90	505	0.373	269.80	232.88	219.39	—
90	635	0.373	286.84	252.76	241.63	—
90	849	0.373	329.44	275.48	258.92	156.20
90	851	0.373	289.68	252.76	242.34	—

7.4. Power consumption by motors

Table 50.23. Running current requirements of electric motors

Power (W)	230 V, Single phase, AC		230 V, 3-phase, AC	
	Current load, Amp	Wire size, Gauge	Current load, Amp	Wire size, Gauge
0.124	2.2	14		
0.186	2.9	14		
0.249	3.6	14		
0.373	4.9	14	2.0	14
0.560	6.9	14	2.8	14

Power (W)	230 V, Single phase, AC		230 V, 3-phase, AC	
	Current load, Amp	Wire size, Gauge	Current load, Amp	Wire size, Gauge
0.746	8.0	14	3.5	14
1.119	10.0	14	5.0	14
1.492	12.0	14	6.5	14
2.238	17.0	10	9.0	12
3.730	28.0	8	15.0	10
At 240 V through 14, 10, 12 and 8 Gauge wire requires 15, 20, 30 and 40 Amp fuse (circuit breaker), respectively.				
Source : North and Bell, 1990				

8. Calculation of feeders and drinkers required

8.1. Feeders

8.1.1. Linear feeders

Linear feeders can offer feeding space on both sides. Therefore, if length of the feeder is L cm, actually 2L cm of feeder space is available for the birds.

Keeping the above in view, number of birds a linear feeder of length L cm can support be calculated by dividing 2L by the feeder space requirement (F) of the birds in question. Conversely, for N number of birds, number of feeders, each L cm long, required = (NF ÷ 2L).

8.1.2. Circular feeders/Hanging feeders/Automatic feeders

Total feeder space a circular feeder of diameter D cm can provide is given by its circumference, πD. When the birds eat from circular feeders, their body will be away from their head and hence, more space is available for the birds to stand round the feeder. In other words, feeder space available increases radially in case of a circular feeder whereas no matter how far away the body of the bird from its head, no extra space is available in case of linear feeders. Therefore, circular feeders can accommodate 25% more birds than the linear feeder when $2L = \pi D$ (Fig. 50.1)

Obviously, number of birds, each requiring F linear cm feeder space, a circular feeder of diameter D cm can support $=1.2\pi D \div F$ or $6\pi D \div 5F$.

8.2. Drinkers

8.2.1. Linear drinkers

Normally, linear waterers running at the center of the pen (in floor systems) so as to provide drinking space on both sides are not common; in most facilities, watering channel will be on either side of the passage and hence, only one side of the drinker is accessible by the birds. In case of cage system, only one side of the

Linear feeder/drinker

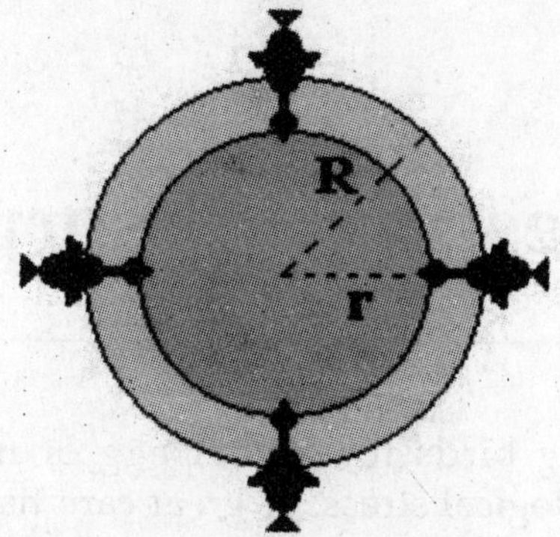

Circular feeder/drinker (R > r)

Fig. 50.1. Comparison of linear and circular feeders/drinkers

watering channel is available for the birds.

Therefore, number of birds a drinking channel of length L cm can support be calculated by dividing L by the waterer space requirement (W) of the birds in question. Conversely, for N number of birds, number of watering channels, each L cm long, required = (NF ÷ L).

8.2.2. Circular drinkers/Water basins

Calculations are similar to those for circular feeders; feeder space requirement is replaced by waterer space requirement.

Chapter **51**

Digestion, Absorption and Excretion

Breeding birds for higher egg or meat yield has imposed a tremendous physiological stress. A great care has been exercised in providing a suitable environment for proper/optimum expression of the genetic worth. Poultry-feeding is one of the most important environmental factors deciding the yield of the bird. Confinement of the birds following domestication has further accentuated the importance of balanced diet to poultry.

Notwithstanding the above, cost on feeding birds constitutes nearly 70 to 80% of the recurring expenditure in poultry farming. Hence, a rigorous control of quality of feed is mandatory in order to maximize the returns of the enterprise.

1. Nutrients

Nutrients comprise the external chemical environment needed for optimum functioning of the many metabolic chemical reactions in the cells inside the animal involved in growth, maintenance, work, production and reproduction. The process of furnishing the nutrients is referred to as nutrition and it includes procurement, ingestion, digestion, absorption and transportation of nutrients so that the cells can assimilate and use them.

The nutrients ingested by the animal undergo series of changes in the course of metabolism comprising both anabolism and catabolism. Hence, these organic and inorganic nutrients are also referred to as metabolites; meaning, those chemicals participating or undergoing metabolism. Some of these metabolites can be synthesized in the body of the animal by using certain precursors while the other metabolites have to be obtained/supplemented from the environment; the latter group is referred to as essential nutrients. The chemical nature, quantity and number of essential nutrients vary both between and within species. For chicken, it is known that at least 40 essential nutrients in an available (ready to use) form has to be supplied in the diet in adequate amounts and in proper ratio to one another to ensure maximum absorption.

2. Essential nutrients

2.1. Water

Water is the most abundant liquid on earth, covering about ¾ of the earth's surface and hence, on most occasions, most neglected of the nutrients. Pure water is colorless, odorless and tasteless. It is essential for survival, growth and development of all animals. A day-old chick has 74% body water which decreases with age (replaced by fat) to 55% in adult-laying hen.

Water intake varies with age, rate of egg production, dietary composition and ambient temperature. The latter factor contributes most to variations in water consumption since respiration is the major route of heat loss from the body at high temperatures, which greatly increase water requirements (See Chapter "Thermoregulation" and "Principles of ventilation").

2.2. Carbohydrates (Sugars)

Hexoses, in general, and glucose in particular, are preferentially absorbed. Pentoses are absorbed only to a limited extent by chickens (maximum 10% of the diet). Polysaccharides like starch and oligosaccharides like sucrose and maltose and monosaccharides (D-forms) like glucose, fructose, galactose and mannose are used chiefly for energy purposes by chicken. Pentoses are utilized during synthesis of nucleic acids.

Most of the energy in the body is produced by various reactions involving metabolism of glucose (Embden-Myerhoff-Parnas scheme and Citric acid cycle) and stored in the form of high energy phosphate i.e., Adenosine phosphates and Creatine phosphate. Whenever the body requires energy, the high energy phosphates are broken down.

2.3. Proteins (Amino acids)

Biological value of protein depends on the number, ratio and availability of the essential amino-acids it contains. The essential amino-acids for poultry are arginine, histidine, isoleucine, leucine, lysine, methionine, phenylalanine, threonine, tryptophan and valine. However, glycine (to support uric acid synthesis) and serine are essential for very young chicks and proline is essential in diets with crystalline amino-acids. Cystine, hydroxy lysine and tyrosine are synthesized from limited sources like methionine, lysine and phenylalanine, respectively; hence, in diets adequate in latter three amino-acids, the former three are not essential. The other amino acids viz., alanine, aspartic acid, asparagine, glutamic acid, glutamine and hydroxy proline are not essential since they are synthesized from simpler sources.

2.4. Fats (Fatty acids)

All the essential fatty-acids are unsaturated and they are linoleic, linolenic and arachidonic acids. Animals cannot synthesize a double bond between 6th and 7th carbon counting from terminal methyl group and linoleic acid having this specific double bond is the truly essential fatty-acid and other essential fatty-acids can be biosynthesized from linoleic acid. Chicken do not require linolenic acid.

2.5. Vitamins

Vitamin refers to inorganic compound distinct from other components (like carbohydrates, fats, proteins, minerals and water), presenting minute quantities in foods, essential for normal tissue and for health, growth and maintenance, when absent from the diet (or not properly absorbed or utilized) results in a specific deficiency disease or syndrome and cannot be synthesized by the animal and hence is an essential nutrient to be supplied in the diet.

2.5.1. Classification

2.5.1.1. Fat-soluble

The vitamins in this group are Vitamin A, D, E and K

2.5.1.2. Water-soluble

The vitamins in this group are Vitamin B_1, B_2, B_6, B_{12}, Niacin, Pantothenic acid, Folic acid, Biotin, and Choline, collectively referred to as "B-complex vitamins", and Vitamin C.

2.5.2. Essential vitamins

Most of the vitamins are of plant origin; however, microorganisms can synthesize water-soluble vitamins, provitamin A (β-carotene) and menaquinines (Vit K_1). Vitamin B_{12} cannot be synthesized either by plants or by animals. Since vit B_{12} is absolutely essential, all animals (including man) have to depend on microbial synthesis. Since vitamins synthesized by microbes are of little use to the birds, they are highly susceptible for vitamin deficiencies.

The fat-soluble vitamins and vitamin B_{12} are stored in animals whereas other water-soluble vitamins are excreted in urine when given in excess. Therefore, water-soluble vitamins (except B_{12}) have to be supplied daily to avoid deficiencies. Chicken require all the fat-and water-soluble vitamins excepting vitamin C. Chicken biosynthesize vitamin C utilizing glucose in kidneys. Since the embryos also can produce their own vitamin C, hen does not include vitamin C into the egg.

Usually vitamins are supplied more than their minimum requirements to obviate losses during storage, to ensure safety against non-uniform mixing of small quantities of vitamins, especially in case of mash-type of diets, and to safeguard against fluctuations in food consumption (due to environmental temperature, energy content in feed etc.) and vitamin requirement of birds.

2.6. Inorganic elements

2.6.1. Distribution of inorganic elements

The inorganic elements (ash constituents) are distributed in the body rather in a non-uniform fashion. For example, most of the iron is concentrated in blood, calcium in bones, potassium in muscles, iodine in thyroid gland etc... Besides, the content of important inorganic elements in various species is relatively uniform. These elements have varied functions; for example, calcium and phosphorus from the muscle structures, sodium, potassium and chlorine are concerned with homeostasis (osmotic pressure and pH), iron as a part and parcel of hemoglobin, copper, zinc etc. as parts of hormones or enzymes or as activators of enzymes.

2.6.2. Classification of inorganic elements

Some of the inorganic elements are required in fairly large quantities and these are the "Major elements" – grouped in this category are calcium, phosphorus, sodium, potassium and chlorine; other elements are also required, but in small quantities and hence referred to as "Trace elements" - they are magnesium, manganese, zinc, iron, copper, molybdenum, selenium, iodine and cobalt. Few others like fluorine and chromium also can be beneficial though not nutritionally essential. There are elements which produce toxicity in birds referred to as "Toxic elements" – cadmium, vanadium, lead, arsenic, beryllium,mercury and tungsten.

Chicken require the following inorganic elements / groups:

Table 51.1. Essential inorganic elements for poultry

	Cations	Anions / anionic groupings
Monovalent	Na, K	Cl, I
Divalent	Ca, Mg, Mn, Cu, Zn, Co	MoO_4, SeO_3
Trivalent	Fe	PO_4

3. Ingestion

Notwithstanding period of feed restriction during growing period of commercial layers and breeders, feed is generally offered *at libitum* to all species and classes of poultry. Quality of feed is controlled by the producer and the bird itself regulates feed intake. Several factors like dietary energy, protein, body weight, volume etc. have been found to influence feed intake. However, the exact physiological mechanism controlling voluntary food intake by birds is not totally understood.

3.1. Theories on regulation of feed intake

Several theories have been put forward explaining the mechanism of feed intake by poultry; each of these is based on a variable monitored by the central nervous system.

3.1.1. Glucostatic theory

This theory assumes that blood glucose levels determine appetite. However, this appears to be more applicable to mammals than to poultry because blood glucose or a change in blood glucose does not seem to control appetite in case of domestic birds.

3.1.2. Thermostatic theory

The basic assumption for this theory is that the output (loss) of body heat or energy drives a bird to conserve body heat and/or consume food. Obviously, this requires the existence of sensors to detect temperature changes peripherally and a central-processing mechanism (brain?). Such heat detecting centers (thermoreceptors) have been identified in bill, skin and spinal cord; however, much evidence is still required to demonstrate that thermostatic mechanism is a primary controller of feed-intake.

3.1.3. Lipostatic theory

This theory is based on the feedback from fat depots to the brain for long-term regulation of feed intake. In any case, the evidences are not yet totally conclusive to consider this mechanism to be the primary one controlling feed-intake in poultry.

3.1.4. Aminostatic theory

It has been found that addition of single amino-acid or groups of amino-acids depresses growth rate; force-feeding prevents the typical growth depression due to amino-acid imbalance in chicks. Therefore, amino-acid imbalance suppresses feed intake. Conversely, proper balance of amino-acids is important to sustain appetite. Histidine is one of the amino-acids presumed to have some effect on appetite; however, conclusive evidences are still lacking.

3.1.5. Ionostatic theory

The extracellular role of sodium and calcium ions within the brain (hypothalamus) has marked effects on both body temperature and appetite. The ratio of sodium to calcium appears to regulate set point for body temperature whereas, excess calcium precipitates in hypothermia. Further studies are still required to confirm this theory.

3.1.6. Role of brain

A few brain loci connected with feed intake or body composition have been identified. Obesity without hyperphagia due to lesions within medial basal hypothalamic region has been recorded. Ventromedial hypothalamic nucleus has been thought to be critical in controlling feed intake. Bilateral lesions in and around natural hypothalamic-thalamic and mid-brain areas resulted in transient aphagia or hypophagia.

Therefore, it is evident that mechanism of regulation of feed and water intake in case of poultry is not totally understood. However, it is generally agreed that feed-intake is primarily the function of energy concentration of diet; of the environmental factors, temperature is probably having the maximum influence on feed-intake.

3.2. Estimating feed-intake

3.2.1. Laying hens

During the laying period, there is a continuous change in daily feed consumption up to the attainment of peak production after which, feed consumption remains consistent. In addition to level of egg production, body weight of the hen, weight of egg produced and environmental temperature are the other important factors that determine feed intake. Several equations have been proposed to predict feed-intake of laying hens and the following equation developed and modified by National Research Council (NRC) is popularly used for predicting apparent ME:

$ME = W^{0.75}(173-1.95T) + 5.5\Delta W + 2.07EM$ where, W = body weight (kg), T = ambient temperature (°C), ΔW = change in body weight (g/d) and EM = daily egg mass (g).

3.2.2. Broilers

Broilers are always fed *ad libitum* and hence, estimate of energy requirements for a) maintenance of body weight b) per gram of tissue synthesis and c) modifications due to environmental temperature have to be calculated. An equation consisting of all these three components seems to be not available. Maintenance needs of ME of a broiler chick from 8 to 22d and 28 to 42d post-hatch can be estimated as 153 and 175 kcal/kg$^{0.75}$, respectively.

3.3. Dry matter intake

3.3.1. Laying hens

Table 51.2. Estimate of dry matter-intake by laying hens

Age (Weeks)	Egg production (%)	Egg weight (g)	Body weight (g)	Daily gain (g)	Dry matter intake (g/d) at 25°C			
					*	**	***	****
20	5	47.7	1317	7	60.2	56.0	59.7	61.9
24	62	50.7	1513	6	82.2	78.2	81.5	83.9
28	91	55.0	1653	6	98.0	94.1	96.7	99.3
32	89	57.6	1737	3	93.2	89.4	94.6	97.2
36	87	59.3	1821	2	92.6	88.8	95.1	97.9
40	85	60.4	1877	0	88.5	84.9	93.0	95.8

* $F = (0.534 - 0.004T)\,W^{0.75} + 2.76\Delta W + 0.80EM$

** $F = (0.259 - 0.00259)\,W^{0.75} + 2.76\Delta W + 0.80EM$

*** $ME = 130\,W^{0.75} \pm 1.015\Delta T + 5.50\Delta W + 2.07\,EM$ **** $ME = W^{0.75}(173-1.95T) + 5.5\Delta W + 2.07EM$ where, F = feed/hen/d (g), W = body weight (kg), T = ambient temperature (°C), ΔW = change in body weight (g/d), ΔT = (25 – ambient temperature) and EM = daily egg mass (g).

Note: Diet having 2.89 kcal/g ME and 12% moisture

Source : National Research Council, 1994

3.3.2. Broilers

Table 51.3. Estimate of dry matter-intake by broilers*

Age (d)	Body weight (g)	Daily gain (g)	Estimated ME (kcal/d)			Daily feed intake (g)**	Daily dry matter intake (g)***
			Maintenance	Gain	Total		
7	130	27	47.4	55.3	102.7	32.2	28.3
14	320	34	85.9	69.7	155.6	48.6	42.8
21	560	43	124.4	88.1	212.5	66.4	58.4
28	860	56	165.1	114.8	279.9	87.5	77.0
35	1250	63	211.3	129.2	340.5	106.4	93.6
42	1690	59	257.8	120.9	378.7	118.4	104.2
49	2100	60	297.6	123.0	420.6	131.4	115.6

* ME = $1.91\ W^{0.66} + 2.05\Delta W$ where, W = body weight (g), ΔW = daily weight gain (g)

** Feed having 3200 kcal/kg ME   *** F eed having 12% moisture

Source : National Research Council, 1994

4. Digestion

The birds obtain various types of chemical compounds (metabolites) from the food they eat. The main ones are carbohydrates, proteins and fats which are to be broken down into smaller units so that they can be absorbed and subsequently utilized for useful purposes. This process is referred to as digestion.

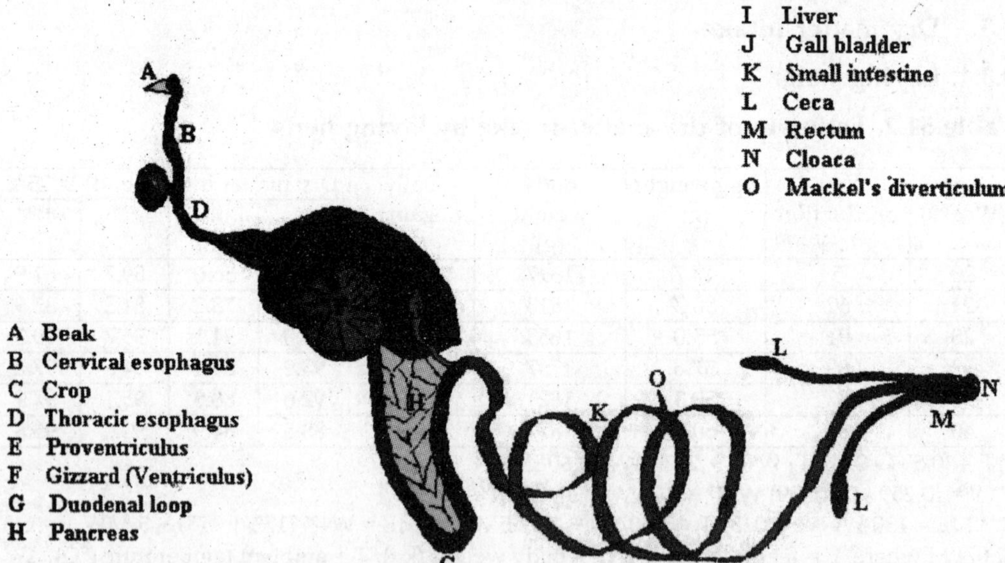

Fig 51.1. Digestive system of chicken

4.1. Carbohydrates

Carbohydrates are easily digested since they are easily soluble. Certain polysaccharides like cellulose are not digested by poultry since they lack the enzyme cellulase. However, cecal bacteria do digest cellulose but minimum absorption takes place as the food stays for a short period in ceca and mainly water reabsorption takes place. Similarly, pectins and pentosans are also not fairly digested.

Birds do not retain food in the mouth for a long time and quickly pass it on to the crop. Hence, action of salivary juice (ptyalin enzyme) is very limited. Similarly, if the gizzard is empty, the ingested food will be quickly propelled into proventriculus and hence, chemical changes in the food during its stay in the crop is very limited. However, considerable literature is available regarding the role of microflora in the crop as well as in intestine and ceca on digestion in poultry. Action of ptyalin, if any, occurs in the crop.

In the proventriculus (true stomach), whose secretions are controlled neurohumourally amylase (starch-splitting enzyme) or other carbohydrate-splitting enzymes have not been demonstrated.

In the gizzard, the food material will be ground which is greatly assisted by the presence of grit and violent muscular contractions. This grinding increases the surface area available for various digestive enzymes to act by many folds.

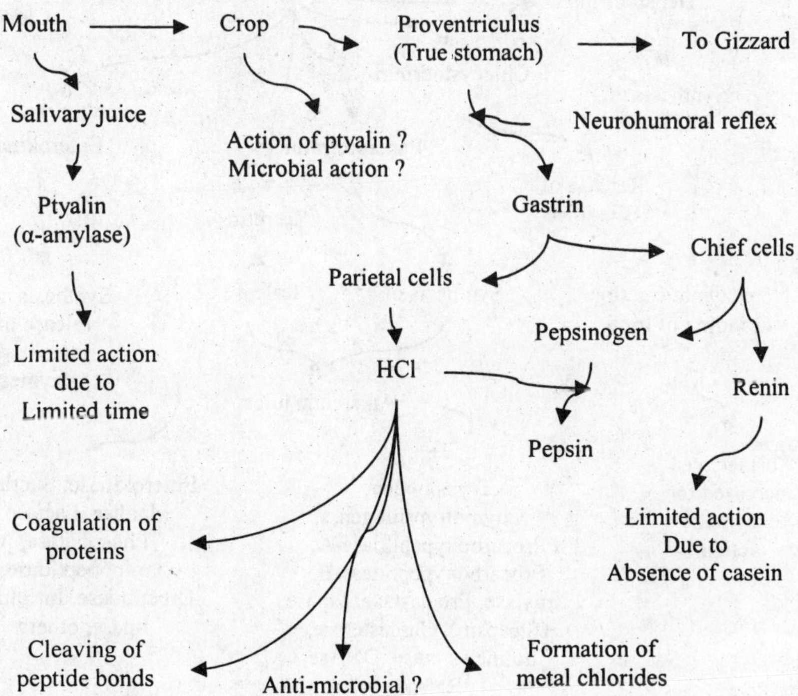

Fig. 51.2. Process of digestion – Mouth, crop and proventriculus

Subsequently, the food enters the duodenum and stimulates the glands of Brunner and glands of Liberkuhn (through a hormone enterocrinin) to secrete saccus entericus which has both proteolytic and amylolytic enzymes. The amylolytic enzymes include specific disaccharidases like α-amylase, sucrase, maltase lactase and phosphatase; the phosphatase attacks hexose phosphates (which are commonly present in the feed the birds eat) and removes the PO_4 and makes the hexose available. α-amylase is similar to ptyalin in action i.e., it converts starch into maltose. α 1-4 linkages in starch (amylose) is acted upon by the amylase. The α 1-6 linkages are either attacked by another specific enzyme (debranching enzyme) or there may be an isomutase to attack these linkages. Maltose units so formed will be attacked by maltase and converted into glucose units. Similarly, sucrose and lactose will be broken down into their monomeric units by sucrase and lactase, respectively.

Hence, the end-products of digestion of carbohydrates include glucose, fructose, galactose, mannose etc. and they are ready for absorption.

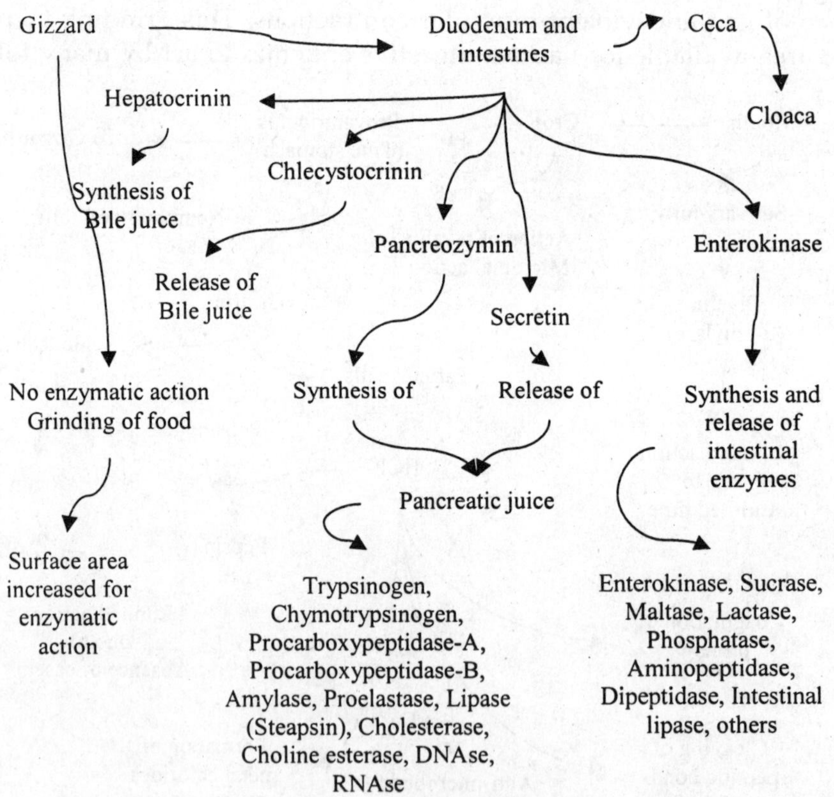

Fig. 51.3. Process of digestion – Gizzard, duodenum and intestines

4.2. Proteins

Digestion of proteins begins in the proventriculus. The parietal cells secrete HCl which causes

1. Denaturation of proteins – creating more area and sites for action of proteolytic enzymes
2. Activates pepsinogen to pepsin. The activity of pepsin is optimum between pH 2 and 3.

The entry of food into the proventriculus stimulates vagus which, in turn, results in release of hormonal called gastrin by the gastric pylorus. Gastrin stimulates the chief and parietal cells to elaborate pepsinogen and HCl. Renin and lipase are also secreted. Pepsin produced initially (by the action of HCl on pepsinogen) later on autocatalyzes the conversion of pepsinogen to pepsin. Pepsin acts on the coagulated protein and converts them into peptides and peptones (smaller units of protein)

Food then enters the intestines which stimulates the production of two hormones namely secretin and pancreozymin which, in turn, result in secretion of pancreatic juice. Pancreatic juice contains the inactive forms of enzymes viz., Trypsinogen, Chymotrypsinogen, Procarboxypeptidases A and B and proelastase. Food in the intestine also stimulates the release of yet another enzyme, from the intestinal mucosa, called enterokinase which converts the inactive trypsinogen to trypsin. Trypsin subsequently not only autocatalyzes its production but also converts the other inactive enzymes into active forms i.e., chymotrypsin, carboxypeptidase A and B and elastase.

Thus, trypsin is central in the development of full proteolysis activity. The consequence of trypsin inhibitors found in certain legumes and cereals is poor digestibility of dietary proteins.

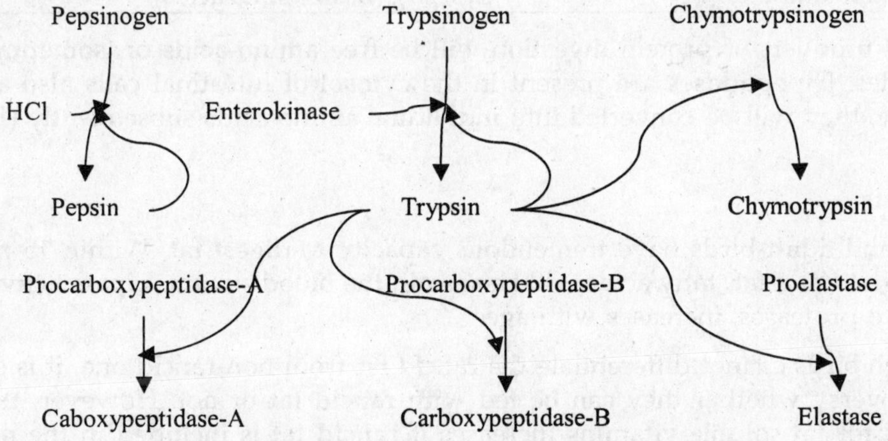

Fig. 51.4. Formation of active proteolytic enzymes

There seems to be an increase in enzyme activity as the bird gets older. Trypsin activity in the intestinal contents to increase around 10-fold and Chymotrypsin activity by 8-10 fold by 20-30 d of age.

Other enzymes of the pancreatic juice include lipase, amylase, choline-esterase, cholesterase, DNAses and RNAases. Intestine produces aminopeptidase and dipeptidase.

Elastases are relatively non-specific with regard to the type of peptide bonds. Carboxypeptidase A is inhibited in sequence hydrolysis by the proximity of charged side-chains and proline, while carboxypeptidase B is also inhibited by proline but requires cationic side-chains for maxima1 activity. The pancreatic secretion also contains certain collagenases that catalyze the hydrolysis of collagen to small peptides. The jejunal wall secretes the proteolytic enzymes peptidase (erepsin) and polynucleotidase which catalyze the hydrolysis of small peptides into amino acids and dipeptides, and also convert nucleic acids to mononucleotides, respectively, which can then be absorbed.

Table 51.4. Summary of actions of proteolytic enzymes

Enzyme	Amino acids
Pepsin	C=O of peptide bond contribuṭed by Phe, Trp, Tyr (Aromatic amino acids)
Trypsin	C=O of peptide bond contributed by Lys, Arg (Basic amino acids)
Chymotrypsin	C=O of peptide bond contributed by Phe, Trp, Tyr (Aromatic amino acids)
Carboxypeptidase-A	Aromatic amino acids at C-terminal end
Carboxypeptidase-B	Basic amino acids at C-terminal end
Aminopeptidase	One amino acid at a time from N-terminal end
Enterokinase	Lys, Arg (Basic amino acids)

The end-products of protein digestion will be free amino-acids or, sometimes, dipeptides. Dipeptidases are present in the cytosol of intestinal cells also and therefore, they will be converted into individual amino-acids subsequently (Fig. 51.6).

4.3. Fats

Chicks and adult birds have tremendous capacity to digest fat. Within 15 min after ingestion of fat, fatty acids will be seen in the blood stream. Lipase activity, similar to proteases, increases with age.

Although birds cannot differentiate the rancid fat from non-rancid one, it is still a controversy whether they can be fed with rancid fat or not. However, their demand for fat-soluble vitamins increases if rancid fat is included in the diet. Facts also help in the absorption of calcium, carbohydrates, vitamins and fat itself.

Although lipase-activity or lipase has been detected in gizzard and proventriculus, the digestion of fats is extremely limited due to lack of emulsification and improper pH, respectively.

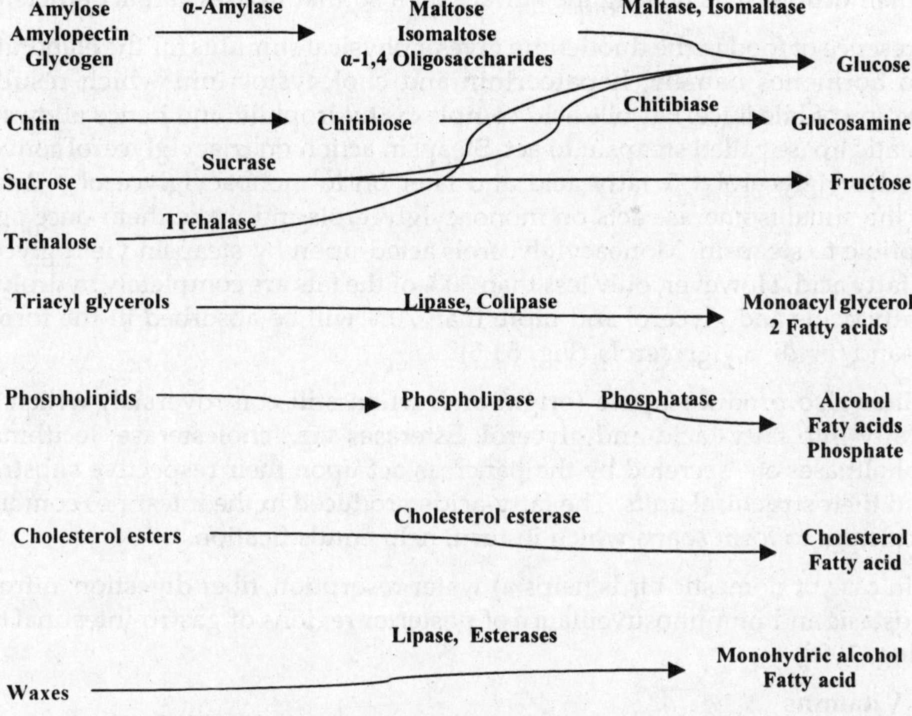

Fig 51.5. Autoenzymatic digestion in birds – Carbohydrates and lipds (Klasing, 1998)

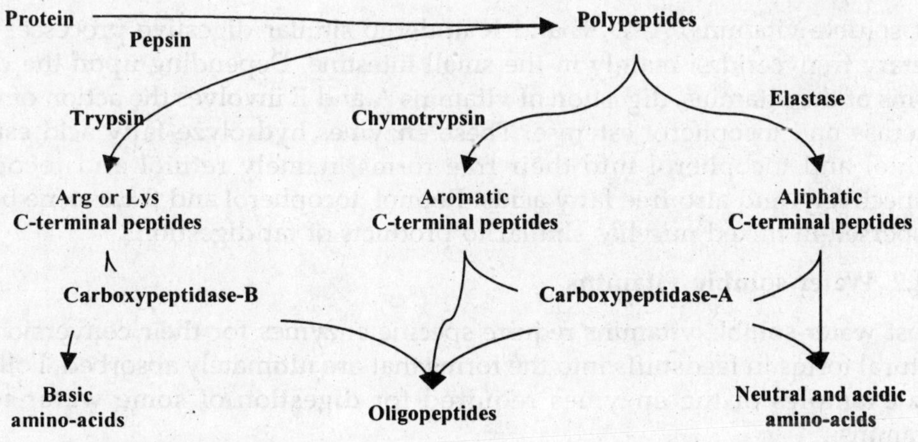

Fig 51.6. Autoenzymatic digestion in birds – Proteins (Klasing, 1998)

As lipases are water-soluble, the fats also must be made miscible in water to effect digestion. This is assisted in the intestines(duodenum) by the bile secreted by liver which contains salt derived from cholic acids (cholesterol origin) reduce the surface tension resulting in formation of an emulsion. The fat is broken down into small droplets, increasing the surface area so that lipases can act efficiently.

The presence of food in the duodenum gives a physical stimulus for the elaboration of two hormones namely, hepatocrinin and cholecystocrinin which result in production of bile juice. Fat-bile acid complex is hydrophilic and hence allows the pancreatic lipase called steapsin to act. Steapsin action on triacylglycerol converts it into diacylglycerol + 1 fatty acid and later on to monoacylglycerol + 2 fatty acids. Intestinal isomerase acts on monoacylglycerols and make them once again susceptible to steapsin. Monoacylglycerols acted upon by steapsin yield glycerol and a fatty acid. However, only less than 30% of the fats are completely hydrolysed into fatty-acids and glycerol and more than 70% will be absorbed in the form of mono-and/or di-acylglycerols (Fig. 51.5).

Intestines also produce lipase (origin and action still controversial) which can split fatty into fatty- acid and glycerol. Esterases viz., cholesterase, lecithinase, phospholipases etc. secreted by the pancreas act upon their respective substrates to yield their structural units. The fatty-acids produced in the intestines combined with calcium to form soaps which in turn, help emulsification.

Ceca in case of domestic birds helps a) water resorption, fiber digestion, nitrogen homeostasis and immunosuveillance of posterior regions of gastro-intestinal tract (Klasing, 1998)

4.4. Vitamins

4.4.1. Fat-soluble vitamins

Fat soluble vitamins, A, D, E and K undergo similar digestive processes as do dietary triglycerides, mainly in the small intestine. Depending upon the dietary forms of the vitamins, digestion of vitamins A and E involves the action of retinyl esterase and tocopherol esterase. These enzymes hydrolyze fatty acid esters of retinol and tocopherol into their free forms, namely retinol and tocopherol respectively, and also free fatty acids. Retinol, tocopherol and β-carotene become dispersed in mixed micelles similar to products of fat digestion.

4.4.2. Water-soluble vitamins

Most water-soluble vitamins require specific enzymes for their conversion from natural forms in feedstuffs into the forms that are ultimately absorbed. Following are examples of the enzymes required for digestion of some water-soluble vitamins:

1. Biotin is often covalently linked to lysine as biocytin, and that requires biotinidase that is secreted from the pancreas and mucosal cell for proteo-

lytic release of the free biotin.

2. Pyridoxine and thiamine in feedstuffs exist as active phosphorylated forms that need to be dephosphorylated before absorption. They, therefore, require non-specific alkaline phosphatases from intestinal cells for their hydrolysis into free absorbable forms. These vitamins exist as dephosphorylated forms in plant feedstuffs that require no digestion.
3. Most folic acid exists in plant feedstuffs naturally attached to a polymer of glutamic acid. This is hydrolysed into the absorbable monoglutamate form at the brush border membrane by the enzyme folate conjugase or pteroryl polyglutamate hydrolase.

5. Digestibility of nutrients

(Leeson and Zubair, 2004)

5.1. Carbohydrates

Digestibility of complex carbohydrates depends upon the type of carbohydrate and on whether animals possess the enzymes or microbial system necessary to degrade it to monosaccharide units ready for absorption. Cellulose cannot be digested by poultry. About 65% of ingested starch is digested by the time it reaches the terminal duodenum, while about 97% disappears by the time digesta reaches the terminal ileum. The disaccharide lactose (milk sugar) is only partly utilized by chickens because they lack sufficient lactase for hydrolysis. In chicks, about 97% of the glucose that is ingested or released from digested starch is absorbed, mainly in the duodenum, with the rate of absorption decreasing in the small intestine.

5.1.1. Factors affecting carbohydrate digestibility

5.1.1.1. Type of carbohydrate

Cellulose complex in plant feedstuffs is not digested by poultry simply because they do not posses cellulose. Adult turkeys can utilize about 10% of cellulose by microbial hydrolysis in ceca.

Other complex carbohydrates that occur in poultry feedstuffs are hemicelluloses, pentosans and oligosaccharides such as stachyose and raffnose. Hemicelluloses refer to the components of plant feedstuffs that are insoluble in water, soluble in dilute alkali and readily degraded by dilute acid hydrolysis. The polysaccharides found in the hemicellulose fractions of plant feedstuffs that are of importance to poultry are the pentosans. Xylans yield the pentose sugar xylose upon hydrolyses. Although chickens do not possess the enzymes required to hydrolyze these pentosans, they may derive some energy due to the xylans susceptibility to acid hydrolysis in the proventriculus and the gizzard.

Pentosans or β-glucans depress performance and wheat pentosans comprise principally arabinoxylans, which are linked to other cell wall components thereby further reducing their digestibility. The pentosans and β-glucans found in barley, rye, oats and wheat increase the viscosity of digesta and, thereby, interfere with digestion and absorption in poultry. Barley has glucan which causes growth depression accompanied by sticky droppings. It is possible that the limited ability of chicks for the absorption of pentoses, which are products of xylan digestion, is a contributing factor to the watery and sticky droppings. In practice, the rate of inclusion of such cereals as wheat, barley, rye and oats in poultry diets is often limited due to these antinutritional effects.

It has been demonstrated that the addition of exogenous enzyme sources to diets that contain any of these cereals may improve overall diet digestibility. Highly viscous digesta from wheat, barley and oat-based diets inhibits the access of digested nutrients to the gut epithelial cells. Pectins in vegetable proteins also cause a more viscous digesta, the extent of which is affected by their concentration and degree of esterification. Such inhibition of nutrient absorption has been attributed to:

1. Intestinal contractions mix the luminal contents and bring materials from the center of the lumen closer to the epithelium.
2. Nutrients have to diffuse across the thin, relatively unstirred layer of fluid lying adjacent to the epithelium.

Both these processes are greatly hindered by the presence of highly viscous materials in the lumen. The added/exogenous enzymes hydrolyze the xylan backbone of the arabinoxylans to smaller fragments that are less viscous in solution. Enzymes in poultry diets are dealt in a separate Chapter.

Soyabean meal contains substantial levels of the α-galactosaccharide family of oligosaccharides such as sucrose, stachyose and raffinose, which make up to 12% of the carbohydrate fraction. Such soluble carbohydrates have been reported to reduce the digestibility and lower the ME of soyabean meal-based diets. Poultry cannot metabolize the olligosaccharides due to the absence of α-galactosidase enzyme activity in the intestinal mucosa resulting in reduced ME, and wet and sticky excreta; the litter condition due to wet excreta predispose birds grown on litter to foot/leg disorders and breast blisters.

Possible nutritional strategies to improve the digestibility of diets with high levels of soluble carbohydrates include the use of higher dietary fiber to delay transit time of digesta and an enhanced cecal environment for microbial hydrolysation of polysaccharides. Due to lesser quantity of water soluble carbohydrates acted upon by microbes, lower level of VFAs is produced, and hence, higher pH of cecal contents results which supports digestibility of carbohydrates.

5.1.2. Effects of processing

Diets formulated with raw potato starch are poorly digested by poultry due to presence of phosphate groups in such starch interfering with α-and β- amylase degradation so that two or three residues are left unhydrolyzed on either side of the phosphorylated glucose unit. Cooking potato starch increases its digestibility and bird performance, probably due to cleavage of the phosphate bonds.

Similarly, roasting of corn grains at 160°C improved digestibility. Increase in digestibility related to using the higher cooking temperature was attributed to lower density and greater solubility and water absorption capacity of the grains, thereby permitting greater accessibility by digestive enzymes. This is consistent with the idea that precooking of feedstuffs may increase their metabolizable energy values in poultry, as may the steam and pressure used in pelleting of feed.

The effect of heat treatment on carbohydrate digestibility of feedstuffs is also influenced by moisture content during processing. Under the high temperature and moist conditions of pelleting, combined with physical stresses, starch granules are gelatinized and resistance to amylase activity is reduced. The improvement in starch digestibility associated with steam pelleting is therefore partly a result of the gelatinization process which gives better solubilization of the starch granules and greater accessibility for digestive enzymes. Heat treatment might be expected to improve the digestibility by decreasing the intermolecular bonds of starch. It is also possible that the greater improvement in the ME values of heat-treated legume compared to cereals is due to the partial inactivation of trypsin inhibitors, if any.

5.2. Proteins

Feedstuffs commonly used in poultry diets and the same feedstuffs under different conditions vary widely in content and digestibility of protein. Different factors, such as strain of plant, and source and processing conditions in case of animal feedstuffs, are likely the causes of variation. Digestibilities of synthetic amino acids and their analogues are much higher than for comparable amino acids fed as intact proteins.

5.2.1. Factors affecting protein digestibility

5.2.1.1. Naturally Occurring Anti-Nutritional Compounds

The feeding value of grain sorghum (Milo) is markedly influenced by its tannin content that causes a binding and precipitation of dietary proteins and digestive enzymes; consequently, depress growth. Tannins also reduce amino acid digestibility in chicks. The effects are more pronounced in younger birds probably due to the more mature digestive processes thus accommodating the anti-nutritional effects of tannins.

Feeding raw soyabeans is well known to cause growth depression, poor feed efficiency and pancreatic enlargement in young chickens, and also small egg-size in laying hens. These effects are due to the Trypsin inhibitors in soyabeans that reduce digestibility of proteins; this is compounded by reduced activities of other proteolytic enzymes that require trypsin for activation.

Heat treatment is known to be effective in deactivating the anti-nutritional compounds and trypsin inhibitor activity of conventional soyabean meal is substantially decreased during commercial heat treatment and solvent extraction processing. Adding enzymes to soyabean meal has also been shown to improve its digestion and utilization.

Cottonseed meal contains gossypol which, when heated during processing, forms indigestible complexes with lysine. Processing methods such as prepress solvent extraction significantly lowers the free gossypol content, thereby improving lysine digestibility.

The nutritive value of alfalfa meal is sometimes influenced by its saponin content. The mechanism of action of saponins is not entirely known, but they are thought to inhibit the functions of digestive enzymes, and so lower the absorption of amino acids, sugars and other nutrients.

Proteins found in animal hide, scales, hair, feather, and bone are not easily digested and contain high concentrations of keratin and collagen. The major component of feather meal, for example, is keratin, which necessitates that it be pre-hydrolysed to make it digestible by poultry. Heating feather meal might help in cleavage of disulfide bonds of cystine, allowing better digestion by protease in growing chicks. Due to differences in processing conditions, commercial feather meals vary widely in their protein digestibilities.

Table 51.5. Crude protein (CP) and digestibility of common feedstuffs

Feedstuff	CP (%)	Digestibility (%)			
		CP	Lys	Met	Cys
Vegetable sources (cereals)					
Yellow maize	8	82 - 86	81	91	85
Wheat	12	78 - 82	81	87	87
Barley	10	70 -82	78	79	81
Sorghum	10	67 - 72	78	89	83
Vegetable sources (oil seed meals)					
Peanut meals	49	88 - 91	83	88	78
Soyabean meals	46	83 - 87	91	92	82
Cottonseed meal	43	61 - 76	67	73	73
Animal sources					
Blood meal *	88	82-92	86	91	76
Fish meal *	66	86 - 90	88	92	73
Meat meal *	60	75 - 80	79	85	58
Feather meal	87	36 - 77	66	76	59
* Defatted			Leeson and Zubair, 2004		

5.2.1.2. Effects of Processing

Excess heat applied during processing may either destroy or render unavailable certain heat- sensitive amino acids thereby greatly reducing the nutritive value of such feedstuffs. This can happen due to either very high processing temperature, prolonged processing time or both. Many of the adverse effects of overheating result from reduced protein quality and decreased amino acid digestibility. The largest decreases in digestibility were observed for lysine and histidine; the former largely due to formation of Maillard reaction products created during heat treatment.

Many of the Maillard products result from the non-enzymatic browning reactions between lysine and oligosaccharides such as fructose, stachyose, sucrose and raffinose. Soyabean meal contains substantial quantities of these soluble sugars, making it very susceptible to Maillard reactions during heat treatment. Chemical treatments, such as ethanol extraction of soyabean meal, have been reported to remove the oligosaccharides, thereby decreasing the potential for lysine destruction occurring during heat treatment.

Table 51.6. True digestibility coefficients (%, average) of selected amino-acids in poultry feedstuffs

Ingredient	CP	Lys	Met	Cys	Arg	Thr	Val	Ile	Leu	His	phe
Alfalfa, dehydrated	17	59	73	40	82	71	75	77	80	74	78
Barley, grain	10	78	79	81	85	77	81	82	86	87	88
Blood meal	81-89	86	91	76	87	87	87	78	89	84	88
Corn gluten meal	60	88	97	86	96	92	95	95	98	94	97
Corn grain	8.8	81	91	85	89	84	88	88	93	94	91
Fish meal	60-63	88	92	73	92	89	91	92	92	89	91
Meat meal	50-54	79	85	58	85	79	82	83	84	80	84
Oats, grain	11	87	86	84	94	85	88	89	92	93	94
Peanut meal	46	83	88	78	84	82	88	91	92	83	94
Rice bran	13	75	78	68	87	70	77	77	75	82	77
Soyabean meal, dchulled	48	91	92	82	92	88	91	93	92	88	92
Sesame meal	44	88	94	82	92	87	91	92	91	89	93
Sorghum, grain	8.8	78	89	83	74	82	87	88	94	87	91
Sunflower meal, dehulled	45	84	93	78	93	85	86	90	91	87	93
Wheat bran	16	72	82	72	79	72	76	79	79	80	84
Wheat, grain	11-17	81	87	87	88	83	86	88	91	91	92

Source : NRC, 1994

5.3. Fats

The digestibility of fats by poultry depends upon many factors including type of fat, level of saturation, age of bird, level of fat inclusion in the diet and presence of other dietary components. The energy value of fats depends mainly upon the digestion and absorption of fatty acids from the intestinal tract. Because fatty acids are not excreted in the urine, their metabolizable energy values are directly related to their absorbability.

5.3.1. Factors affecting digestibility

5.3.1.1. Impurities and anti-nutritional factors

Impurities are most often referred to collectively as M.I.U. (moisture, impurities and unsaponifiables). Most of these compounds will have little nutritional value, and hence, digestibility values must be adjusted corresponding to total fat content. During oxidation at both high and low temperatures, a vast range of unusual polymers can be produced culminating loss in digestibility as low as 30%. A number of naturally occurring fatty acids can also adversely affect overall fat utilization, for instance, erucic acid present in rapeseed oils and some other *Brassica sp.,* and the cyclopropenoid fatty acids in cottonseed. However, their mode of action is most likely via general well-being of the animal rather than through any specific mode of action related to digestion or absorption, etc.

5.3.1.2. Fatty acid composition

Fat composition can influence overall fat digestion because different components can be digested and/or absorbed with varying efficiency. Efficiency of fat utilization is dependent upon there being an adequate supply of bile salts and an adequate balance of unsaturates : saturates; polar unsaturated fatty acids and monoacylglycerols forming micelles which are easily absorbed.

Table 51.7. Digestibility (%) and metabolizable energy (MJ/kg) of feed-grade fats

Fat Type	Age		Age	
	0-21d	>21d	0-21d	>21d
	Digestibility		Metabolizable energy	
Tallow	80	86	30.976	33.488
Poultry Fat	88	97	34.325	37.674
Fish oil	92	97	36.000	37.674
Vegetable oil	95	99	36.837	38.511
Coconut oil	70	84	27.209	32.651
Palm oil	77	86	30.139	33.488
Vegetable soapstock	84	87	32.651	33.907
Restaurant grease	87	96	33.907	37.255

Leeson and Zubair, 2004

While unsaturated fatty acid content of a diet has a marked effect on overall fat digestion, there is often concern over a fat's content of free-fatty acids (FEA).

Although a large proportion of fatty acids are released in the lumen after hydrolysis, the presence of monoacylglycerols plays an important role in solubilizing non-polar long-chain saturates. This type of research data suggests feed manufacturers should be wary of fats containing high levels of FFA.

Another important factor associated with micelle formation is availability of bile salts. The age-related effect on digestion of fats is partly accounted for by inefficient bile-salt recycling in very young birds.

While it is possible to define ratios and levels of fatty acids needed to optimize digestion, but, it may not be possible to incorporate them while formulating poultry feeds.

5.3.1.3. Intestinal factors, rate of passage and other ingredients

The status of the intestinal lumen will obviously have an effect on the digestion and/or absorption of any nutrient. Digesta pH can influence fat digestion in that acidic conditions reduce micelle solubilization. Hence, use of organic acid mold inhibitors as feed additives can modify gut pH and fat digestion. Age-related availability of fatty acid binding protein (FABP) also can be a factor.

Birds infected with coccidiosis often exhibit inferior fat digestibility. Steatorrhoea occurring with intestinal coccidiosis may, therefore, result from the loss of reconstituted fat globules following the rupture of parasitized epithelial cells. Fat, *per se*, and linoleic acid in particular may also affect the microbial population in the intestine. Increased levels of cellulose apparently result in reduced fat digestion, possibly through complexing of fiber with bile salts.

Diet fat, *per se*, can affect rate of passage of digesta, and this can influence overall diet digestibility.

5.3.1.4. Age

Young birds cannot digest fats as well as older birds (Table 51.4. The effect of age on ability to digest fats is most pronounced for the saturates. The reason for this phenomenon, and particularly saturated fats, is not clear but it may be attributed to the fact that young birds recycle bile salts less efficiently and fatty acid binding protein is not produced in adequate quantities by young birds.

5.3.1.5. Soap Formation

Once fats have been digested, free-fatty acids have the opportunity to react with other nutrients within the digesta. One such possible association is with minerals to form soaps that may or may not be soluble. If insoluble soaps are formed, there is the possibility that both the fatty acid and the mineral will be unavailable to the bird. Soap formation in the digesta of broiler chicks is substantial and most pronounced with saturated fatty acids and with high levels of diet minerals. Soap production seems to be less of a problem with older birds. This is of importance to laying hens that are fed high levels of calcium. In addition to calcium, other minerals, such as magnesium, can form soaps with saturated fatty acids.

In older birds and some other animals, there is an indication that while soaps form in the upper digestive tract, they are subsequently dissolved in the lower tract due to changes in pH. Under these conditions both the fatty acid and mineral are available to the bird. Control over digesta pH may, therefore, be an important parameter for control over soap formation.

5.3.2. Extra metabolic (calorigenic) effect of fat

Fats and oils are likely to inhibit stomach emptying and intestinal digesta movement. Decreased rate of passage suggests that digesta spends more time in contact with digestive enzymes, carriers or co-factors and absorptive sites, etc. Addition of fats to the diet may, therefore, lead to increased digestion of non-fat components of the diet.

5.4. Vitamins

5.4.1. Factors affecting digestibility of vitamins

5.4.1.1. Source of Vitamin

Table 51.8. Content and availability of vitamins

Source	Vitamin	Content (μg/kg)	Availability (%)
Corn	Biotin	45	≈ 100
Sorghum		300	20
Oats		210	30
Barley		150	20
Wheat		110	≈ 0
Groundnut meal		1630	53
Sunflower meal		1000	40
Cereal grains	Niacin		Low
Wheat	Vitamin E		17
Corn			14
Barley			11
			Leeson and Zubair, 2004

Stability of vitamins is affected by other compounds that are present in the premixes or complete diets and by the conditions occurring during storage. Factors such as length of storage, temperature, pH, humidity, presence of enzymes and exposure to ultraviolet light decrease vitamin stability. The presence of polyunsaturated fatty acids (PUFA) can cause oxidative rancidity which will lead to destruction of vitamins A, D and E. In diets that are formulated with PUFA, an effective antioxidant such as ethoxyquin is required to prevent oxidative damage. The presence of minerals in the premix or diets can enhance oxidative destruction of such vitamins, and this can be further enhanced by the presence of highly hygroscopic compounds, such as choline chloride.

5.4.1.2. Effects of processing

While pelleting can improve the nutritive value of feeds, it can also decrease the stability and bioavailability of heat sensitive vitamins. The stability of vitamins A, D, K, C and thiamin, for example, is substantially reduced in pelleted feeds. This effect is attributed to the high temperature, moisture and pressure that the feed ingredients are subjected to during pelleting. The bioavailability of certain vitamins like biotin and niacin that are often present in bound forms can, however, be improved by pelleting.

5.5. Minerals

5.5.1. Factors affecting availability

Between 50 and 70% of organic phosphorus in poultry diets is present in the form of phytate which is not available to birds, because of absence of endogenous phytase enzymes. The addition of microbial phytase to cereal-based diets improves phosphorus utilization and reduces phosphate excretion. Reducing the phosphorus intake and addition of microbial phytase to broiler diets improved phosphorus digestibility. Synthetic phytase may also improve digestion of other nutrients in the diet, such as nitrogen and amino acids.

Most synthetic phosphate sources are 95- 100% bioavailable, while natural bone meals are sometimes as low as 85% available. Other chemical forms of calcium with poor bioavailability are oxalates and oxides, because of poor solubility in the intestine. It is likely that utilization of phosphorus may improve with steam pelleting.

6. Absorption of nutrients

6.1. Carbohydrates

Sodium ion concentration within the intestinal content is critical to this active transport system. A high Na+ concentration facilitates rapid absorption, while a low Na+ concentration reduces the rate of sugar absorption. Some hexoses and pentoses are absorbed through diffusion, but this process is quite slow and not quantitatively important to the bird.

Hexoses are preferentially absorbed over pentoses. The absorption has been found to be active. Results on glucose have shown that glucose is phosphorylated in the intestinal cells and principally transported to hepatic circulation where it is once again dephosphorylated. Other monosaccharides are also likely to undergo similar changes but are transported to general circulation via lymphatic circulation. Natural (D-form) monosaccharides are preferentially absorbed.

Dietary fiber composed predominantly of non-starch polysaccharides and lignin present in cell walls of plant sources is not digestible and hence excreted in feces. Legumes store considerable non-starch polysaccharides within the cells of their seeds. The structure of plant cell wall consists mainly of cellulose fibers in the matrix of hemicellulose and pectin encrusted mainly by lignin. The ratio of these components vary between feedstuffs.

Functionally, dietary fiber is subdivided into soluble fiber (pectins, gums, β-glucans and some hemicelluloses) and insoluble fiber (lignin, cellulose and some hemicelluloses).

Domestic birds can ferment soluble fiber in their ceca and obtain useful amount of energy in the form of volatile fatty acids; in some species, even insoluble cellulose can also be fermented. However, the advantage of energy obtained through fermentation is negated by anti-nutritive aspects of soluble fiber.

Fiber increases viscosity of intestinal contents and soluble fibers, especially pectins, gums and β-glucans, which impairs the diffusion of substances and digestive enzymes in the intestines thereby hindering the digestive and absorption processes. Enzymes of microbial origin are being used to reduce viscosity in grain-based diets; for details, see Chapter "Performance enhancers in poultry feed".

In addition to the anti-nutritive properties, pectins which have a high charge density ionically chelate dietary cations like manganese, iron, zinc, copper etc. decreasing their absorption. High viscosity also predisposes for change in the type of microorganisms due to be increased residence time of digesta thereby reducing oxygen tension and stimulating more anaerobic bacterial population.

6.2. Proteins

Peptide uptake is most rapid in the jejunum whereas amino acid uptake is most rapid in the ileum. Peptides are absorbed into the mucosal cells by pinocytosis. The amino acids enter the blood stream from the mucosal cell by facilitated diffusion as free amino acids, with only trace amounts of peptides found in plasma.

The natural isomers (L-amino acids) are generally absorbed more rapidly than the D-forms. A common but L-preferring site exists for the transport of both isomers of methionine. Other neutral L-amino acids have a high affinity for this site, whereas the D-isomers, except for D-methionine, have a very low affinity.

Dipeptides, when absorbed, are also converted into individual free amino acids by dipeptidases present in the cytosol of intestinal cells. Transport of amino-acids is also active and is dependent on sodium and Vitamin B_6. The energy for the transport is actually spent on pumping out sodium which enters the cell along with a carrier molecule (carrying amino acid). There are separate carrier molecules for basic, acidic, neutral, aromatic, branched chain amino acids.

Therefore, there will be competition between the amino-acids under each of the above categories; of these, the one with maximum practical importanceis between lysine and arginine. Lysine is a highly essential amino-acid and if there are more than 3 molecules of arginine for every molecule of lysine, lysine absorption is likely to be hindered. Therefore, poultry feed should have a lysine : arginine ratio of less than 1:3. The amino-acids absorbed enter the portal circulation and later on, general circulation. Natural (L-form) amid acids are preferentially absorbed.

6.3. Fats

Fatty acids cannot pass intestinal barrier and therefore, are esterified by bile salts in lumen to form water-soluble compounds which enter the intestinal wall. Monoacylglycerols and glycerol, being water-soluble, can directly pass through the intestinal wall. Absorbed glycerol is utilized for energy production. Fatty acids and glycerol (from glycolysis) are esterified to form triacylglycerols in the intestinal cell. Monoacylglycerols enter the lymphatic stream whereas triacylglycerols enter hepatic circulation.

Cholesterol esterase, also secreted from the pancreas, hydrolyses cholesterol-fatty acid esters into cholesterol and free fatty acids. Short-chain fatty acids and free glycerol are then absorbed directly into the mucosa of the small intestine and transported to the portal circulation.Other free fatty acids, monoacylglycerols and cholesterol molecules are emulsified by bile salts, forming micelles which are essential for normal fat absorption. Water-insoluble compounds such as polar unsaturated fatty acids and monoacylglycerols that cannot form micelles alone, readily form stable mixed micelles with conjugated bile salts. Saturated fatty acids such as palmitic and stearic acids on the other hand, which are non-polar and that have high melting points are only slightly soluble in emulsions with bile salts. They are, however, markedly solubilized in the presence of a mixed micelle. In this form, the fatty acids and other lipid-like materials are solubilized in the aqueous phase of the lumen and are transported to the mucosal cell membrane. Consequently, the levels of dietary saturated : unsaturated fatty acids and bile salt secretion are important factors in fat absorption.

In poultry, unlike mammals, lipoproteins rich in resynthesized triglycerides are absorbed directly into the portal blood system and transported to the liver. These types of lipoproteins have been referred to as portomicrons as opposed to chylomicrons in mammals.

6.4. Vitamins

6.4.1. Fat-soluble vitamins

The mixed micelles comprised bile salts, monoacylglycerols and long-chain fatty acids, together with vitamins A, D, E and K, and β-carotene, facilitates their absorption into the mucosal cells. Absorption takes place mostly at the proximal site of the small intestine with little uptake for vitamins D and K at the more distal sites. Under normal conditions, where there is adequate bile and monoacylglycerols, both D-and L-α-tocopherol are shown to be well absorbed, but equal dietary levels produce a lower blood concentration of L-rather than D-α-tocopherol.

Competition among the fat-soluble vitamins for absorption has been demonstrated. Thus, in diets containing barely adequate levels of vitamins A, D, E and K, a marked increase in the dietary level of vitamin A may cause decrease in growth or egg production. This has been attributed to the induced deficiency of one or more of the other fat-soluble vitamins rather than to any toxic effect of vitamin A.

Inside the intestinal cells, much of the β-carotene is converted into vitamin A by the enzyme β-15,15′ dioxygenase. Most of the retinol and tocopherol are then converted to mixed esters, the nature of which depends to a great extent upon the type of fatty acids being absorbed with these vitamins. Palmitic acid appears to be used preferentially with vitamin A. Vitamins A and E are transported in plasma both as a free alcohol and in the esterified form.

6.4.2. Water-soluble vitamins

Unlike fat-soluble vitamins that are absorbed mostly by passive diffusion, absorption of water-soluble vitamins involves different active carrier systems. The mechanism of uptake of vitamin B_{12} is not adequately understood, but it is believed to involve pinocytosis. Similar to fat-soluble vitamins, water- soluble vitamins are mostly absorbed in the proximal small intestine.

6.5. Minerals

Regardless of the form in which the inorganic cations occur in the feedstuffs and/or diets, they are principally converted into their respective chloride (by HCl in the proventriculus), thereby making them highly soluble and therefore, absorbable. Excess of certain elements can cause deficiency of others; for example, excess of calcium and phosphorus form insoluble $Ca_3(PO_4)_2$ in the alkaline conditions of the intestines, which being the macro-molecule, adsorbs many other elements especially Mn, and later on is excreted through feces.

Absorption of minerals occurs throughout the jejunum and ileum. The rate of absorption depends on factors such as pH, carriers, type of the minerals and presence of other minerals that will influence competition for absorption. Numerous mechanisms of mineral absorption have been elucidated. Many minerals, such as iron, sodium and zinc, require active transport systems and absorption is regulated. Absorption of non-metal minerals such as sulfur, selenium and chlorine are very efficient, unregulated and depends mainly on the source and presence of other minerals in the gut. Calcium absorption involves both carrier proteins and simple diffusion mechanisms. Absorption rate is regulated to maintain blood calcium level by factors such as parathyroid hormone and vitamin D_3. Such regulation means that calcium absorption is usually proportionally lower compared to phosphorus which is unregulated and where absorption efficiency can be as high as 95%. Minerals that are complexed to simple carbohydrates, amino acids and other chelators like EDTA are more bioavailable.

7. Excretion

(Klasing, 1998)

7.1. Carbohydrates

All carbohydrates that is absorbed is metabolized; excess is stored as glycogen and, when these pools are replete, as triacylglycerols. Monosaccharides are almost completely reabsorbed in the kidney.

Several types of disorders in regulation of blood glucose levels have been reported in chicken; diabetes, malabsorption, sucrose intolerance etc.

7.1.1. Dietary fiber and intestinal viscosity

Dietary fiber composed predominantly of non-starch polysaccharides and lignin present in cell walls of plant sources is not digestible and hence excreted in feces.

7.2. Proteins (Amino acids)

Birds are uricotelic and hence excrete most of the waste nitrogen as uric acid rather than urea or ammonia. This unique excretion strategy is important adaptation to help the growing embryo to precipitate the waste ammonia and maintain osmotic balance during incubation. In addition, excretion of uric acid requires minimum water which is beneficial for the birds in the sense that they need not have to drink and store large quantities of water during their flight and normal nesting activities. In other words, in adult birds, osmotic advantage helps water conservation.

Uric acid is a purine and is synthesized mainly in the liver and to some extent in kidneys utilizing the same metabolic pathway as the synthesis of other purines namely, adenine and guanine. The origin of the carbon and nitrogen atoms of uric acid (Fig.51.7) indicates that disposal of excess amino-acid nitrogen by birds imposes greater requirements for arginine, methionine and glycine; the latter is therefore semi-essential in very young chicks because the biosynthesis may not suffice the metabolic requirements as one mole of glycine is consumed in the synthesis of each mole of uric acid. Formyltetrahydrofolate contributes methyl groups which in turn originate from methionine; therefore, requirement of the latter increases.

Excretion of uric acid is more energy consuming than that of urea (3.75 *Vs* 2.00 ATP per mole).

Although birds are uricotelic, when fed high-protein diets, excrete about ¼ of their nitrogen as ammonia and this proportion increases with dietary protein.

7.3. Lipids

Triacylglycerols and cholesterol circulating as lipoprotein complexes and free fatty

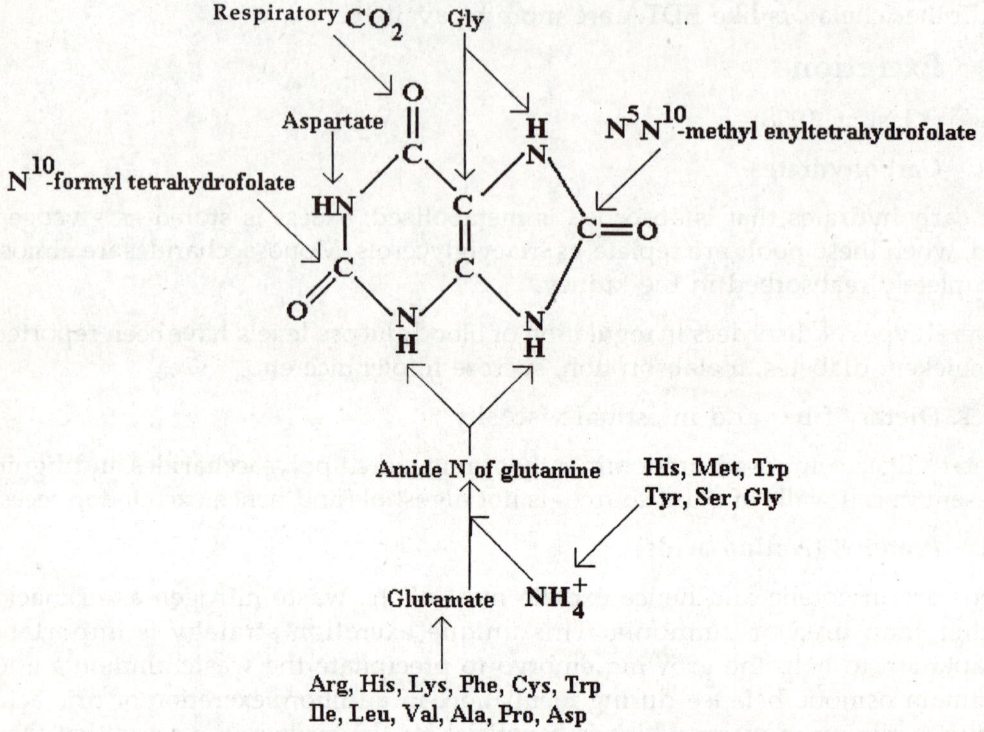

Respiratory CO_2 Gly
Aspartate
$N^{5}N^{10}$-methyl enyltetrahydrofolate
N^{10}-formyl tetrahydrofolate
Amide N of glutamine
His, Met, Trp
Tyr, Ser, Gly
Glutamate NH_4^+
Arg, His, Lys, Phe, Cys, Trp
Ile, Leu, Val, Ala, Pro, Asp

Fig. 51.7. Metabolic synthesis of uric acid (Klasing, 1998)

acids bound to albumin are both too large to escape glomerulus. Bile secretion into the small intestine is the only appreciable form of lipid excretion. Excess cholesterol, cholesterol esters and other non-polar lipids from peripheral tissues are transferred to liver as components of HDL and LDL; they are also ultimately excreted through bile.

Oily secretion of the uropygial gland is a local secretion and does not represent in excretory route for lipids.

7.3.1. Rancidity

Dietary fats and oils undergo rancidity either by enzymatic hydrolysis (by endogenous or microbial lipases to produce free fatty acids, monoacylglycerols, diacylglycerols and glycerol) or by oxidation (peroxidation) of unsaturated fatty acids; the former reduces palatability without markedly affecting nutritive value whereas, the latter, reduces ME, destroys essential fatty acids and produces potentially toxic end-products.

Oxidative rancidity can be minimized/prevented by antioxidants. Xanthophylls and vitamin E are naturally occurring antioxidants. Synthetic antioxidants like ethoxyquin and butylated hydroxytoluene (BHT) are commonly used in poultry feeds.

Chapter **52**

Nutrients – Functions, Deficiency and Excess

Each of the nutrients has specific metabolic functions and hence, its deficiency/excess will precipitate in one or the other form of disease resulting in poor performance by the bird. Hence, each of the nutrients is discussed with respect to its metabolic functions *Vis-à-vis* symptoms due to its deficiency/excess.

1. Water

1.1. Metabolic functions

The importance of water to the chicken is related to its chemical and physical properties which are as follows:

1. Its polarity and use as a solvent in the digestion and absorption of food, the transport of nutrients in the body and the elimination of waste products via the urine.
2. Its low viscosity contributes to its ability to move freely through the small blood capillaries throughout the body, thereby distributing nutrients to, and removing waste products from tissues. Hence, it is a major constituent of blood, intra-and extra-cellular fluids
3. It also has a high specific heat (4.186 J/g) and latent heat of vaporization (2.26 kJ/g) which means that large amounts of heat can be transported in the blood and dissipated through respiration and *vice versa.*
4. Homeostasis – by participating in reactions involved in maintenance of pH, osmotic pressure and electrolyte balance.
5. Most of the reactions of the body occur in aqueous medium and many of them involves either in addition (hydration) or removal of water.
6. A satisfactory intake of water is necessary for the maintenance of normal food intake, the growth of tissues and egg production.
7. Metabolic water: produced during normal metabolic processes; has an important role to play in normal physiological processes; for Ex. A standard egg weighs about 57 g and contains 39 g water. The chick hatched-out from this egg weighs about 40 g and has about 32 g of water. The egg during incubation loses about 10 gm water which means that the newly-hatched

chick has about 3 g of excess water which is produced by the normal oxidation of nutrients. For every 100 g, proteins, carbohydrates and fats yield 42, 60 and 100 g of metabolic water, respectively.

The following figure summarizes the various factors involved in water balance in chicken

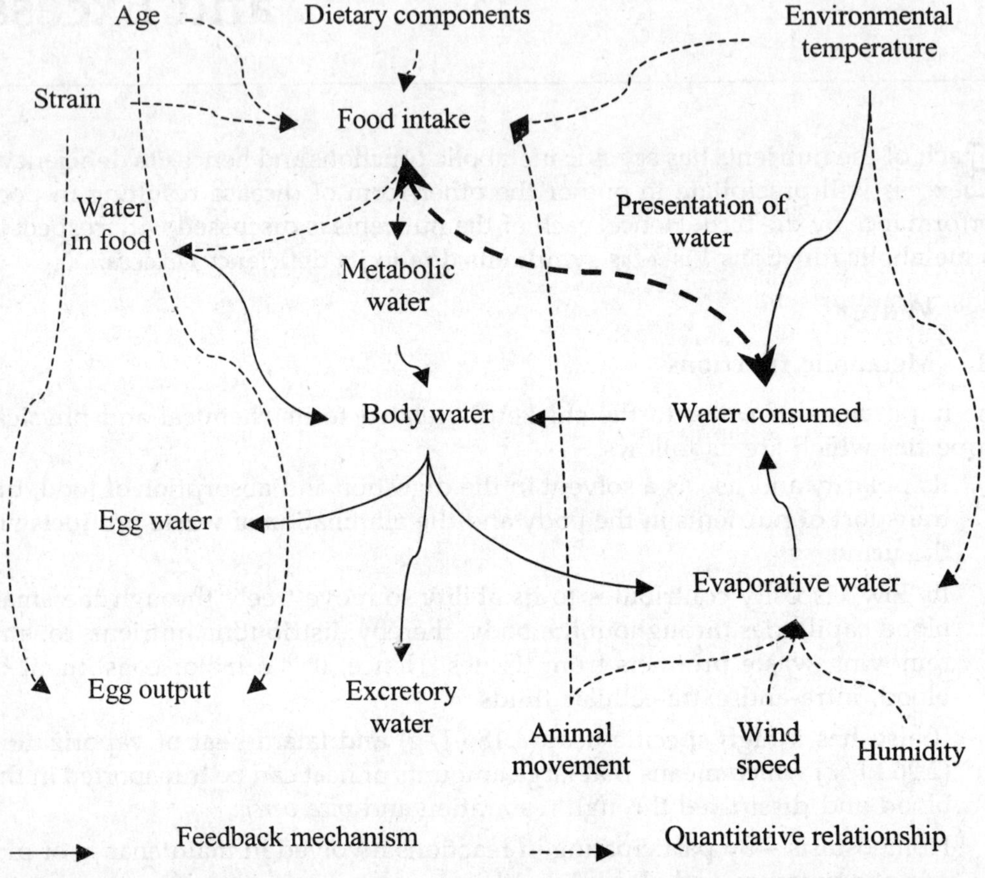

Fig. 52.1. Inter-relationship between variables affecting water balance (Belyavin, 1991)

1.2. Water deficiency

Birds tolerate starvation better than water deprivation. Feed efficiency is severely affected presumably because movement of food from the crop is delayed. Circulation of water from body fluids to crop and back seems to regulate the movement of ingesta from crop to gizzard. Stunted growth and, when prolonged, necrosis, polycythemia and signs of dehydration precipitate. In adults, necrosis of ovaries, inflammation of proventriculus, nephrosis, reduced egg-and shell-weights, followed by cessation of egg production.

1.3. Excess water

Very rare under practical conditions. The signs, however, will be reduced food intake and subsequent retarded growth. However, Waterfowls deprived of water for 48 hrs when given access to water, drink so much water that they even die of water intoxication (drunken syndrome). Otherwise, increased water intake occurs at high temperature, or when Na, K, lactose or others is (are) high in the diet.

2. Carbohydrates (Energy)

The main function of carbohydrates being the source of energy, the following discussion is centered on energy aspects in poultry nutrition.

It is generally agreed that all animals eat to meet their energy requirements. Animals derive energy and various other nutrients from the food they eat. Of course, there are exceptions for this; for example: Laying hens have a special hunger for calcium and they do search for ingredients rich in calcium. Hence, it is up to the nutritionist to balance the ration in such a way that the animal gets all the other nutrients required within the amount of feed it consumes to satisfy its energy needs. This concept was originally proposed with respect to proteins by Donaldson (1955); he referred it as calorie-protein (CP) ratio. After more knowledge accumulated on amino-acids and other nutrients, their requirements are usually given with a standard energy base.

Very fast-growing animals such as broilers (which grow between 1.5 and 4 g/hr) need more protein and hence the CP ratio is narrower for them than those which have completed their growth. Hence, when the energy consumption/requirement reduces (as in the case of summer), related proportion of other nutrients must be increased; in other words, CP ratio must be narrowed and *vice versa*.

Energy is defined as capacity to do work and is stored in molecules of the nutrients which are converted into kinetic energy of the chemical reactions of metabolism in the animal system. The energy so liberated is utilized for maintenance (thermoregulation) and production (growth, production, reproduction etc.)

2.1. Unit of measurement

Energy is measured in terms of calories and calorie defined as the heat required to raise the temperature of 1g of water from 16.5 to 17.5°C. 1000 calories = 1 kcal and 1000 kcal = 1Mcal. The recent Le Systéme International d'Unites (SI; International System of Units) unit of measurement is a Joule (J) which is equal to 10^7 ergs and 1 erg is the amount of energy expended to accelerate a mass of 1 g by 1 cm/sec; 4.186 J = 1 cal.

Birds, similar to any other animal, require energy for maintenance of body temperature, activity, growth, production of eggs etc. They derive energy from carbohydrates, fats and proteins and over and above required for maintenance, growth and production, if available, cannot be excreted and hence is stored as fat.

Food consumption is closely related to the energy demand created by the metabolic reactions and still regulation of appetite in poultry is incompletely understood. Several factors like blood glucose level, appetite regulating mechanism in hypothalamus, taste (which is a minor importance in chicken), energy content in food, environmental temperature, physiological status etc. have been documented to control for intake.

Since energy content in food is of major importance, it is essential to know the exact energy requirement of birds at different stages. It is very well known that in diets containing adequate amounts of all the required nutrients, the efficiency of feed utilization depends upon the metabolizable energy content of the diet.

2.2. Partition of energy

Any molecule containing C and H which can be oxidized to carbon dioxide and water by animal cells, represents potential energy in the molecule. Total energy present in the food is referred to as gross energy (GE) and it can be estimated by burning a known quantity of the sample in a Bomb Calorimeter in the presence of oxygen. The heat produced will be transferred to known quantity of water and the rise in temperature will indicate the GE.

The GE content of the feed does not indicate the actual energy available to the living system. Part of the GE is lost in the form of feces/excreta which contains partially digested/un-digested portion. Thus, GE less the fecal energy (FE) is referred to as digestible energy (DE). There will be further loss from the DE primarily in the urine and hence, DE less the urinary energy (UE) is referred to as metabolizable energy (ME) which indicates the actual amount of energy available at the metabolic level. In case of poultry, since urine and feces are passed simultaneously, ME = GE – FE.

However, there will be further losses in the body from ME; the cause for this is the fact that the body can trap only about 50% of the free energy released during the oxidation of nutrients and the remaining part is lost as heat; it will be useful to the animal only when it is placed at the temperature lower than the body temperature; otherwise, a waste. Similarly, energy is lost while masticating food, for propulsion of food in the alimentary tract etc. These losses are collectively referred to as heat increment (HI) or specific dynamic effect of food.

The ME less the HI is the net energy (NE) and it is the energy that is available to the animal for useful purposes. NE is principally partitioned into:

1. That it is used for maintenance (NE_M) – used to perform work within the body and leave the animal as heat and
2. That used for growth and fattening and production (eggs, meat etc.) which

is either stored in the body or leaves it as chemical energy – referred to as net energy (production) (NE_P) or as animal's energy retention.

2.2.1. Metabolizable energy

The ME value of food and feedstuffs are easily evaluated and fairly consistent, whereas the NE is influenced by various factors both intrinsic to the animal and the extrinsic ones. Hence, the most popular system followed is the ME system. In case of poultry, NEP, on average, will be 73% of ME.

Table 52.1. Average GE and ME of nutrients

Nutrient	Gross energy (GE)		Metabolizable energy (ME)	
	kJ/g	Kcal/g	kJ/g	Kcal/g
Carbohydrates	17.35	4.15	16.72	4.00
Proteins	23.62	5.65	16.72	4.00
Fats	39.29	9.40	37.62	9.00
ME as the proportion of GE is low in case of proteins because there will be more loss in the form of UE (urea or uric acid)				

2.3. Sources of energy for poultry

Not all the sources of potential energy are useful to birds. For example, cellulose (fiber), lignin, pectins, inulin, chitin, agar and hemicellulose are not digested by chickens. Pentoses are absorbed only to a limited extent by chickens (maximum 10% of the diet). Polysaccharides like starch and oligosaccharides like sucrose and maltose and monosaccharides (D-forms) like glucose, fructose, galactose and mannose are used chiefly for energy purposes by chicken.

Most of the energy in the body is produced by various reactions involving metabolism of glucose (Embden-Myerhoff-Parnas scheme and Citric acid cycle) and stored in the form of high energy phosphate i.e., Adenosine phosphates and Creatine phosphate. Whenever the body requires energy, the high energy phosphates are broken down.

2.4. Energy content of nutrients

Glucose, whose molecule of formula is $C_6H_{12}O_6$, has lesser number of carbon and hydrogen in relation to oxygen in comparison to fat; a typical fat (tripalmitate) has a molecular formula $C_{51}H_{98}O_6$. Therefore, fats yield more energy than carbohydrates following oxidation in the body. Since fatty acids contain still higher number of carbon and hydrogen, they yield slightly more energy than fats (Table 52.2).

Nutritive value of fats is not affected by hydrolytic rancidity whereas, oxidative rancidity, resulting in formation of free radicals, causes a severe Reduction energy value of fat. Since vegetable oils are better absorbed (due to higher oleic acid and other unsaturated fatty acids) their metabolizable energy values are also higher than animal fats. Proteins are not normally catabolized for energy purposes; however, the potential energy in proteins is 5.65 kcal/g.

Table 52.2. Gross energy of some compounds

Compound	C : O	H : O	Gross energy	
			Kcal/g	kJ/g
Glucose	1 : 1	2 : 1	4.3	17.97
Tripalmitate	8.5 : 1	16.3 : 1	9.4	39.29
Stearic acid	9 : 1	18 : 1	9.53	39.84

2.5. Energy requirement

Although it is theoretically most ideal to know the energy requirement per d for maintenance, growth and production, it is extremely difficult, especially for young chicks since the requirement changes (increases) day by day. Thus, the general practice is to give a range of ME requirement per kg of diet which the bird consumes to meet its energy requirement

However, it must be borne in mind that food consumption is inversely related to an ambient temperature (a variation to tune of ± 20 to 30% is possible) and they cannot adjust their energy intake exactly. They consume slightly more energy as energy content in the diet is increased and subsequently tend to put on more fat. Conversely, with lower energy diets, they fail to consume sufficient food to support normal growth and hence, will not deposit normal amount of fat in their body tissues. Hence, broiler finisher diets are formulated to contain slightly lower protein than required for maximum growth and more energy so that the desired body finish results by deposition of body fat. Similarly, during growing period (8 to 22 weeks) of pullets and cockerels, the dietary energy must be adjusted so that the pullets must not put on excess adipose tissue and cause reduced egg production. Under moderate environmental temperatures, during laying period, White Leghorns require about 300 to 320 kcal (1254 to 1338 kJ) whereas broiler breeders require around 400 to 450 kcal (1672 to 1881 kJ) and medium to heavy egg-producing breeds require around 360 kcal (1505 kJ) per bird per d.

2.6. Energy deficiency

Chicken consume more of low energy feed to meet their needs. Hence, to precipitate energy deficiency symptoms, the bulk (fiber) content must be increased so that the food ingested must fail to supply the required energy. It is logical therefore, that low energy diets are bulky to the birds. This lower limit of energy is about 10.868 and 10.032 MJ (2600 and 2400 kcal) per kg under cool or moderate and warm environment, respectively. Energy efficiency is characterized by reduced growth and reduced fat deposition in the carcass. If the energy level is reduced even below the maintenance levels, birds progressively lose weight (since proteins are burnt for energy purposes) and if prolonged, vital functions cease, resulting in death.

2.7. Energy excess

Slight excess energy in diets does not result in detectable symptoms although slight reduction in growth and extra deposition of fat may be observed. When energy content is too high, the feed consumption reduces so severely that deficiency of many or all nutrients results leading to growth depression/cessation of growth with excess fat accumulation accompanied by protein and vitamin deficiency.

3. Proteins and amino acids

The importance of proteins, in general, and amino acids, in particular, has been clearly documented. Of several amino acids nature, about 20 are involved in various combinations to form different tissue proteins; of these, 12 are essential nutrients for poultry. Food protein, regardless of origin, containing amino acid proportions, especially essential amino acids, similar to that of the animal tissue protein is most suitable to the animal in question.

3.1. Metabolic functions

The term protein means first or primary substance. The proteins have several functions like:

1. They are structural parts of soft tissues like muscle, collagen etc. and also of hard parts like feather, beaks etc.
2. Blood proteins are essential for regulation of osmotic pressure, homeostasis and supply of amino acids.
3. Carrying oxygen to the cells and lipoproteins supply fat-soluble vitamins.
4. As nucleoprotein, glycoprotein and enzymes.
5. In blood clotting – fibrinogen, thromboplastin etc.
6. Defense functions – antibodies (γ-globulins)

3.2. Amino-acid deficiency

Mild or partial deficiency results only in decreased growth which will be directly proportional to the severity of deficiency. Since all nutrients are expressed in terms of energy, protein deficiency is otherwise a case of excess energy. Hence, excess deposition of fat is expectable since berries insufficiency of amino acids leading to reduce growth and production and subsequently, more energy is available for deposition as fat. Severe deficiency precipitates in immediate cessation of growth and growth loss at the rate of 6 to 7% of the body weight per day

3.3. Amino-acid excess

Slight decreasing growth, reduction in body fat and increase in uric acid levels in blood results even when the essential amino acids are properly balanced. The excess uric acid excretion, birds might consume more water thereby the moisture content in litter is likely to increase. Swelling of adrenal glands and production of adrenocorticosteroids are also noticed which indicate that excess of proteins amino acids induce a stress to the birds.

4. Fats (Fatty acids)

4.1. Functions

Fats have several functions like:

1. Increase palatability and taste of feed.
2. Facilitate easy movement of ingesta in the alimentary canal.
3. Carriers of fat-soluble vitamins.
4. Extra-calorigenic effect – increase the metabolism energy from the other foodstuffs also.
5. Decrease dustiness of, especially mash type of, feeds.
6. Increase the energy content of the feed.
7. Help in binding the feed ingredients during pelleting.
8. Linoleic and arachidonic acids are structural components of the cell particularly of phospholipids and many membranes. They may be essential for transport of lipids from the liver.

Essential fatty acids were first recognized by Evans and Burn in 1926. All the essential fatty-acids are unsaturated and they are linoleic, linolenic and arachidonic acids. Animals cannot synthesize a double bond between 6th and 7th carbon counting from terminal methyl group and linoleic acid having this specific double bond is the truly essential fatty-acid and other essential fatty-acids can be biosynthesized from linoleic acid. Chicken do not require linolenic acid.

Major source of linoleic acid is vegetable oils. Maize contains about 2% oil which contains 50% linoleic acid. Since 40 to 60% of most of the poultry diets is made of maize, linoleic acid deficiency is not common. Corn distillers dried solubles are also good sources. Arachidonic acid is found only in Animal Fats.

4.2. Fatty acid deficiency

Deficiency results in disruption of membrane structure which when occurs in mitochondria leads to deranged oxidative phosphorylation and basal metabolism. Hence, in chicks, it results in reduced growth, enlarged and fatty livers. Chicks become more susceptible to respiratory infections. In adults, egg size and hatchability are reduced.

5. Vitamins

5.1. Vitamin A

Discovered in 1913 by McCollum and Davis

5.1.1. Unit of measurement

1 International Unit (IU) = activity of 0.3 µg of retinol or 0.344 µg of retinyl acetate. Of various stereo-isomers of Vitamin A, biological activity of alcoholic form, *trans*-retinol ($C_{15}H_{30}O$) is considered 100; and of several precursors of Vitamin A, β-carotene has a biological activity of 10 (in comparison to *trans*-retinol). Other compounds having some vitamin A activity for chicken are retinal (aldehyde form) and retinoic acid (acid form).

Table 52.3. Vitamins in foods and their bioavailability

Vitamin	Form of vitamin	Amount	Activity
A	Retinol	1µg	3.33 IU
	Retinol acetate	1µg	2.91 IU
	Retinol palmitate	1µg	1.82 IU
	β-Carotene	1µg	1.67 IU
D	Cholecalciferol (D_3)	1µg	40 IU
	Ergocailciferol (D_2)	1µg	? 40 IU
E	DL-α-Tocopheryl acetate	1 mg	1.00 IU
	D-α-Tocopheryl acetate	1 mg	1.36 IU
	D-α-Tocopherol	1 mg	1.49 IU
	DL-α-Tocopherol	1 mg	1.10 IU
	D-γ-Tocopherol	1 mg	0.07 IU
	Tocotrienols	1 mg	1 – 30 IU
K	Phylloquinone (K_1)		100%
	Menaquinone (K_2)		100%
	Menadione (K_3)		60 – 140%
B_1	Thiamin		100%
B_2	Riboflavin		100%
Niacin	Nicotinic acid		100%
	Nicotinamide		125%
B_6	Pyridoxal		100%
	Pyridoxol		100%
	Pyridoxamine		100%
Pantothenic acid	D-Pantothenate		100%
	DL-Pantothenate		50%
Biotin	D-Biotin		100%
	DL-Biotin		50%
Folic acid	Folic acid		100%
	Polyglutamyl folacins		?
B_{12}	Cyanocobalamin	1µg	1 USP
Choline	Choline		100%
C	Ascorbic acid		100%
	Dehydroascorbic acid		100%

Source : Klasing, 1998

5.1.2. Sources

Fish liver oils for the principal sources. Green leafy vegetables, carrots and corn gluten meal are also fair sources of the pro-Vitamins. β-carotene is converted to various vitamin A forms in the intestinal mucosa, more efficiently by adults than chicks under four weeks of age. A molecule of β-carotene normally can yield two molecules of retinal.

5.1.3. Absorption, transport and metabolism

Retinyl esters (RE) in the diet or hydrolyzed in gastro-intestinal tract into retinol and absorbed into mucosal cell where it is re-esterified usually with palmitic or stearic acid. The RE forms a part of chylomicrons and travels through lymphatic blood strain to be ultimately stored in liver.

RE (liver) are hydrolyzed by enzymes and free retinol is transported by retinol-binding protein (RBP) to tissues wherever there is need for vitamin A. RBP is a small protein synthesized liver which complexes with both pre-albumin (PA) (which transports thyroxine) and retinol thereby protecting retinal from oxidation and catabolism. In protein deficiency, RBP and PA levels reduce; and thus levels of these two in blood indicate nutritional status of the animal. Liver contains as much as 95% of vitamin A in the body.

5.1.4. Functions

Except for vision and reproduction, retinoic acid can function as vitamin A

5.1.4.1. Vision

The energy trapped in 11-*cis* retinal helps its combination with scotopsin spontaneously in dark two produce rhodopsin enabling night vision for the bird. Therefore vitamin A deficiency results in reduced levels of 11-*cis*-retinal and thereby precipitating in night blindness (Fig. 52.2).

5.1.4.2. Reproduction

Although retinoic acid can support egg production, it is not deposited in the egg resulting in faulty heart formation and death of embryo at 3d of age (early embryonic death). Supplementation in the diet or injection of retinol into the egg restores the hatchability.

5.1.4.3. Cell permeability

It is hypothesized that vitamin A plays an important role in stabilizing the lipoprotein membranes.

5.1.4.4. Mucous membranes

Vitamin A is necessary for proper differentiation of the mucous membrane. Low levels of vitamin A result in formation of stratified squamous epithelium, often keratinized, instead of mucus secreting columnar epithelium. This can block the

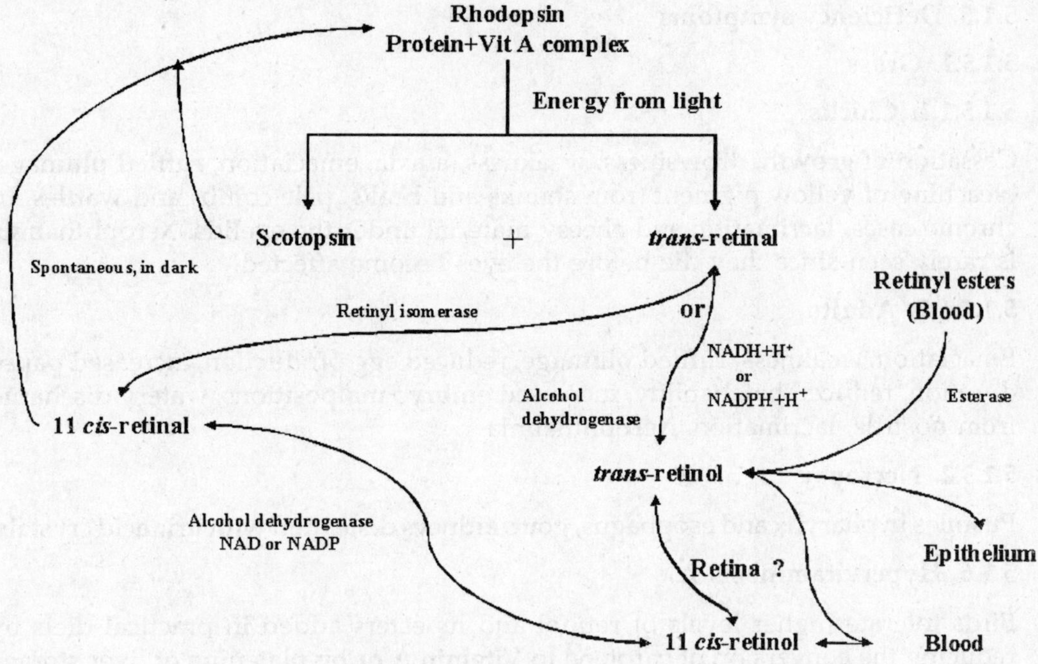

Fig. 52.2. Vitamin A and optic cycle

ducts of various glands in digestive, respiratory and genito-urinary system as well as in lacrimal gland. Hence, pustules in esophagus and pharynx, reduced absorption and growth, keratinization of air sacs resulting in nutritional roup, blockage of urinary tubules resulting in impaired uric acid excretion and subsequent visceral and/or articular gout, reduced spermatogenesis with concomitant reduction in hatchability, blockage of lacrimal glands resulting in dryness and cracking of eyeball followed by microbial invasion and Xerophthalmia.

It also hypothesized that vitamin A is required for transport of calcium ions from outside to the inside of the cell; thus, keratinization results due to calcium accumulation outside the glandular epithelium, causing blockage over a period of time.

5.1.4.5. Bone development

Vitamin A is associated with endochondrial bone development since it is concerned with synthesis of mucopolysaccharides.

5.1.4.6. Cerebral spinal fluid (CSF) pressure

Vitamin A is required for maintenance of proper CSF pressure. Deficiency results in increased CSF pressure and ataxia in chicks.

5.1.5. Deficiency symptoms

5.1.5.1. Gross

5.1.5.1.1. Chicks

Cessation of growth, drowsiness, weakness, ataxia, emaciation, ruffled plumage, bleaching of yellow pigment from shanks and beaks, pale combs and wattles. In chronic cases, lacrimation and cheesy material under the eyelids. Xerophthalmia is rarely seen since they die before the eyes become affected.

5.1.5.1.2. Adults

Emaciation, weakness, ruffled plumage, reduced egg production, increased pause duration, reduced hatchability, increased embryo malpositions, watery discharge from nostrils, lacrimation, Xerophthalmia

5.1.5.2. Necropsy

Pustules in pharynx and esophagus, gout, kidneys distended with uric acid crystals.

5.1.6. Hypervitaminosis A

Birds tolerate higher levels of retinol and its esters added in practical diets by reducing the conversion of carotene to Vitamin A or by plateuing of liver storage and lowering conversion of retinoic acid. The toxic level of retinol and retinoic acid is about 500 and 50 to 100 times the minimum requirement, respectively. Hypervitaminosis A results in weight loss, reduced appetite, swelling and crusting of eyelids, inflammation of nares, mouth, skin adjacent to mouth and skin of the feet, reduced bone strength and mortality.

5.2. Vitamin D

Named by McCollum in 1925; also referred to as antirachitic Vitamin, sunshine Vitamin

5.2.1. Unit of measurement

1 International Unit (IU) or USP Unit = activity of 0.025 μg of Vitamin D_3 contained in the USP Vitamin D reference standard. 75% of 1 chick unit (AOAC) = 1 IU.

5.2.2. Sources

Fish liver oils are the richest sources of Vitamin D followed by eggs.

5.2.3. Absorption, transport and metabolism

Vitamin D is absorbed in presence of bile and rapidly reaches liver by circulation.

5.2.4. Functions

1. Absorption of calcium from intestinal tract. The active form of Vitamin D_3 is formed in the kidneys under influence of parathyroid hormone (PTH) during reduced serum calcium levels. The $1\alpha\ 25\ (OH)_2D_3$ results in the forma-

tion of calcium binding protein (CaBP) in the intestines thereby facilitating absorption of calcium

2. Vitamin D is also required for the release of calcium from kidney mitochondria under influence of PTH. It is likely that Vitamin D might help uptake of calcium by the subcellular particles.

5.2.5. Deficiency symptoms

5.2.5.1. Gross

5.2.5.1.1. Chicks

Retarded growth, rickets characterized by leg weakness, softness of beaks and claws, unsteady gait, poor feathering, abnormal pigmentation of feathers in breeds with colored feathers. Chronic deficiency – skeletal distortions, bending of spinal column at sacro-coccygeal region and denting of sternum – reduced thoracic and abdominal space resulting in crowding of visceral organs.

5.2.5.1.2. Adults

Increased thin-and soft-shelled eggs, reduced egg production and hatchability, leg weakness leading to Penguin-like squat, beak, claws and keel becomes soft and pliable, denting of sternum and inward curving of ribs.

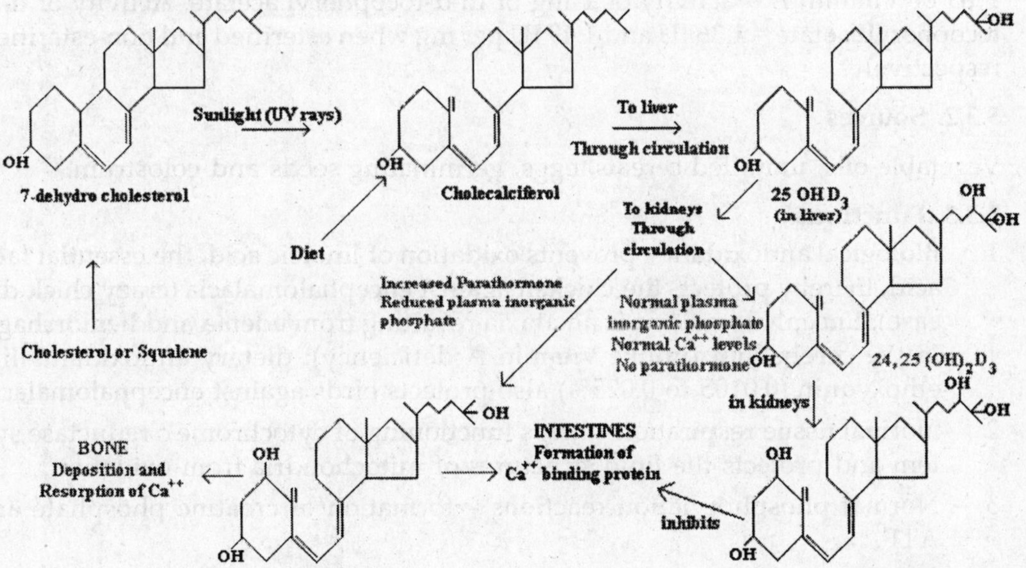

7-dehydro cholesterol
Sunlight (UV rays)
Cholecalciferol
To liver
Through circulation
25 OH D_3 (in liver)
Cholesterol or Squalene
Diet
To kidneys Through circulation
Increased Parathormone Reduced plasma inorganic phosphate
Normal plasma inorganic phosphate Normal Ca^{++} levels No parathormone
24, 25 $(OH)_2 D_3$
in kidneys
BONE Deposition and Resorption of Ca^{++}
INTESTINES Formation of Ca^{++} binding protein
inhibits
1 α 25 $(OH)_2$ D_3
1, 24, 25 $(OH)_3 D_3$

Fig. 52.3. Hormone-like properties of Vitamin D_3

5.2.5.2. Necropsy

5.2.5.2.1. Chicks

Beading of ribs at the junction with spinal column with downward and posterior bending, poor calcification of epiphyses of tibia and/or femur.

5.2.5.2.2. Adults

Bone of the soft and fragile, knobs on the inner surfaces of the ribs at sterno-vertebral junctions, spontaneous fractures in ribs, skeletal changes in vertebral column, pelvis and sternum, enlargement of parathyroid glands.

5.2.6. Hypervitaminosis D

Birds tolerate sufficiently high levels of Vitamin D than their requirement. Too high levels, which is very rare when practical diets are fed, can cause calcification of kidney, aorta and other arteries. Renal damage is the main consequence.

5.3. Vitamin E

Discovered by Evans and Bishop in 1922; referred to as anti-sterility vitamin. Biologically active form is d-α-tocopherol.

5.3.1. Unit of measurement

1 IU of Vitamin E = activity of 1 mg of dl-α-tocopheryl acetate. Activity of d-α-tocopheryl acetate = 1.36 IU and 1.49 IU per mg when esterified and non-esterified, respectively.

5.3.2. Sources

Vegetable oils, unmilled cereals, eggs, germinating seeds and colostrum.

5.3.3. Functions

1. Biological antioxidant – prevents oxidation of linoleic acid, the essential fatty acid, thereby protects the chicken against encephalomalacia (crazy chick disease). Encephalomalacia is an ataxia resulting from edema and hemorrhages in the cerebellum (unlike vitamin A deficiency); dietary antioxidants like ethoxyquin (0.0105 to 0.025%) also protects birds against encephalomalacia.
2. Normal tissue respiration – helps functioning of cytochrome c reductase system and protects the lipid structures of mitochondria from oxidation.
3. Normal phosphorylation reactions – formation of creatine phosphate and ATP.
4. In metabolism of nucleic acids.
5. Synthesis of ascorbic acid and ubiquinone.
6. Protects the capillary walls from glutathione peroxidase and thereby preventing exhibition of fluids from the capillaries – exudative diathesis. Dietary Se also is effective in preventing exudative diathesis.

7. Associated with sulphur amino acid metabolism – Vitamin E deficiency be-long with that of sulphur amino acids results in severe myopathy (especially of breast muscles at about four weeks of age) characterized by Zenker's degeneration with perivascular infiltration of eosinophils, lymphocytes and histiocytes. This is referred to as nutritional muscular dystrophy. High lev-els Se can reduce but cannot prevent completely the nutritional muscular dystrophy due to deficiency of Vitamin E.

5.3.4. Deficiency symptoms

5.3.4.1. Chicks

Encephalomalacia, exudative diathesis and nutritional muscular dystrophy.

5.3.4.2. Adults

Marked reduction in hatchability – embryosdie at 4th day, testicular degeneration in males.

5.4. Vitamin K

Discovered by Henrik Dam in 1929; unit of measurement: mg

5.4.1. Sources

Biologically active forms possess 2-methyl-1, 4-naphthoquinone nucleus. Example: Phylloquinone-4 in the natural vitamin K_1; Menaquinone-4, 5 and 6 (Vitamin K_2) and Menadione dimethyl pyrimidinol bisulfite – the synthetic menadiones are not as effective as natural phylloquinones in counteracting the action of Sulfaquinoxaline. Vitamin K_1 is present in plant sources especially green leafy vegetables whereas Vitamin K_2 are produced by the bacterial flora in animals. However, chicks are not benefited much by the Vitamin K_2 synthesized by the intestinal microflora.

5.4.2. Functions

1. Synthesis of prothrombin (factor II) in liver.
2. Synthesis of proconvertin (factor VII), plasma thromboplastin components (PTC, factor IX) and Stuart factor (factor X).

All the above factors are intimately concerned with intrinsic and/or extrinsic systems of blood clotting. Requirement of vitamin K by the birds is influenced by several factors including stress, Sulfaquinoxaline therapy, coccidiosis and others.

5.4.3. Deficiency symptoms

Prolonged clotting time, hemorrhages on breast, legs, wings, abdominal cavity and surface of intestines. Anemia and prolonged prothrombin time, hemorrhagic syndrome

5.4.4. Hypervitaminosis K

Vitamin K_1 and K_2 are non-toxic at very high levels also and hence is unlikely that practical diets can cause toxicity in chicken.

5.5. Thiamin (Vitamin B_1)

Discovered by Eijkman in 1897 – found that polyneuritis in chicken can be produced by feeding them on polished rice. Also referred to as Vitamin B_1. Metabolically active form: Thiamin pyrophosphate (Thpp); unit of measurement: mg

5.5.1. Sources

Cereal grains and their by-products, Soyabean meal, cotton seed meal, groundnut meal alfalfa meal

5.5.2. Functions

Thiamin is concerned with activation and transfer of "active aldehydes" viz., acetaldehyde succinic semi-aldehyde and glucaldehyde; the former two in TCA cycle and the latter in HMP shunt pathway

5.5.3. Deficiency symptoms

Conversion of pyruvic acid to acetyl SCoA forms a very important link between anaerobic glycolysis and aerobic TCA cycle. This also facilitates entry of 2-carbon units into the mitochondria for further aerobic oxidation. Hence, this reaction is very crucial and is catalyzed by the pyruvate dehydrogenase complex, of which, Thpp initiates the conversion being the first coenzyme in the complex. Therefore, deficiency of thiamin results in accumulation of pyruvic acid (especially in brain which utilizes glucose preferentially) resulting in polyneuritis, and start gazing posture (crazy chick disease) due to paralysis of neck muscles.

There will be severe depression in appetite (to the extent not seen in any other deficiency), followed by loss in weight, ruffled feathers, leg weakness and unsteady gait. Paralysis of leg muscles can follow resulting in start gazing posture. Hypothermia, progressive reduction in respiratory rate, hypertrophy of adrenals resulting in edema of tissues results. In chronic cases, atrophy of genital organs, especially testes occurs. Atrophy of heart, stomach and intestinal walls can also occur. All these are likely to be due to failure of energy production system i.e., TCA cycle.

5.6. Riboflavin (Vitamin B_2)

Discovered by Warburg and Christian in 1932; also referred to as Vitamin B_2. Metabolically active form: FAD, $FADH_2$; unit of measurement: mg

5.6.1. Sources

Riboflavin is not synthesized by any animal but the intestinal microflora, Green plants, Yeast and fungi can synthesize. Since birds are not much benefited by the microbial synthesis, this Vitamin is extremely critical and hence has to be supplied in diets. Besides, riboflavin-rich ingredients like yeast, liver, milk and eggs are not used in poultry diets normally which further make this Vitamin highly critical for birds.

5.6.2. Functions

Riboflavin is a prosthetic group of several enzymes (over a dozen) in the form of flavin mononucleotide (FMN) or flavin adenine dinucleotide (FAD) which is very important in oxidation-reduction reactions involved in cell respiration (electron transport system).

Few of the enzymes riboflavin is associated are cytochrome reductase, lipoamide dehydrogenase, xanthine oxidase etc.

5.6.3. Deficiency symptoms

Similar to that of thiamin deficiency, there will be a demolition of pyruvic acid in myelin sheath of major nerves like sciatic nerve. This produces a continuous flexor reflex resulting in curled-toe paralysis. Egg production may or may not be affected but all riboflavin-deficient embryos die due to impaired organogenesis (due to failure of both electron transport system and energy production).

Embryos show dwarfing, edema, degeneration of Wolffian bodies, and a characteristic clubbed down. Appetite will not be usually affected but growth depression is severe, birds become weak and emaciated, walk on hocks and atrophy of leg muscles is noticed. Epithelium also is very sensitive to riboflavin deficiency – skin becomes dry and harsh. Necropsy examination reveals marked enlargement of sciatic and brachial nerve sheaths (more specifically sciatic nerve) to read the diameter of 4 to 6 times the normal.

5.7. Niacin

Isolated by Warburg and Christian in 1936; also referred to as Niacin. Metabolically active form : Nicotinamide as seen in two coenzymes NAD, NADP; unit of measurement : mg

5.7.1. Sources

Widely distributed in grains and their by-products, and in protein supplements

5.7.2. Functions

NAD and NADP are involved in several reactions of carbohydrate (aerobic and anaerobic cycles which includes the pyruvate and α keto glutarate-dehydrogenase complexes), fat and protein metabolism; in rhodopsin synthesis.

5.7.3. Biosynthesis of niacin

Tryptophan in presence of pyridoxine can be converted into nicotinic acid. Gastro-intestinal microbes also synthesize niacin and much of it may not be available to the birds. Similarly, much of niacin in many sources is in a bound form and not available to birds. Hence, requirements of niacin is highly variable depending upon the above factors.

5.7.4. Deficiency symptoms

5.7.4.1. Chicks

Enlargement of hock joint, bowing of legs (similar to perosis, but without the slipping of tendoachilis from the condyles), inflammation of mouth, diarrhea and poor feathering.

5.7.4.2. Ducks and Turkeys

They are more sensitive to niacin deficiency than chicken.

5.7.5. Hypervitaminosis

In overdoses, it can cause pruritis, nausea and abdominal cramps.

5.8. Pyridoxine (Vitamin B_6)

Deficiency first described by Goldberger and Lillee in 1926; also referred to as Vitamin B_6. Metabolically active form are pyridoxal and pyridoxamine-phosphate; in plants, it is present as pyridoxol. Unit of measurement : mg

5.8.1. Sources

Widely distributed in most foods (as protein complexes) viz., muscle meats, liver, green leafy materials and whole grains

5.8.2. Functions

It is an important coenzyme in transamination, decarboxylation and desulfhydration reactions in amino acid (protein) metabolism. Conversion of DL-methionine hydroxy analogue to the essential amino acid (l-methionine) is an important reaction catalyzed by B_6. Oxidation of amines (histamine to imidazole, acetaldehyde and ammonia), Glycogen phosphorylase activity of the muscle and transport amino acids (neutral and basic amino acids, histidine, proline and hydroxy proline)

5.8.3. Deficiency symptoms

5.8.3.1. Chicks

Anorexia, poor growth and characteristic nervous symptoms – chicks run aimlessly flapping their wings, falling and rolling; and frequently ending in death due to complete exhaustion.

5.8.3.2. Adults

Reduced egg production and hatchability; anorexia, loss of weight and mortality.

5.9. Pantothenic acid

Deficiency described by Norris and Ringrose in 1930. Metabolically active form : Coenzyme A (CoASH). Unit of measurement : mg

5.9.1. Sources

Distributed in all living cells. Liver, yeast, eggs and green leafy plants

5.9.2. Functions

1. CoASH is important coenzyme in many reversible acetylation reactions in carbohydrate, fat and amino acid metabolism; the most important of the metabolites involving CoASH is acetyl SCoA which is the central metabolite plainly vital role in the intermediary metabolism.
2. CoASH may act as an acyl acceptor and later on pass it on to other acceptors.
3. It is also a part of the all-important pyruvate-and a keto glutarate-dehydrogenase complexes, and coenzyme for formation of citrate by condensation of acetyl SCoA and oxaloacetate.
4. Acetyl SCoA is intimately related to the synthesis of many important metabolites such as acetyl choline, acetyl glucosamine and the synthesis of steroids.
5. CoASH is also involved in synthesis of triacylglycerol and phospholipids.
6. Interrelationship of pantothenic acid with Vitamin B_{12}, B_6, folic acid and biotin have been suggested and it is also suspected to the necessary for synthesis of Vitamin C; its relationship with Vitamin B_{12} and folic acid is thought to be responsible for prevention of early chick mortality.

5.9.3. Deficiency symptoms

5.9.3.1. Chicks

Emaciation, crusty scab-like lesions in the corners of the mouth, eyelids stuck together with viscous exudate, skin of the feet also affected-it may peel off or cornify with wart-like growths.

5.9.3.2. Adults

Reduction in egg production and hatchability; embryos show subcutaneous hemorrhage and severe edema.

5.10. Biotin

Discovered by Allison, Hoover and Burk in 1933. Metabolically active form : Biocytin. Unit of measurement : mg

5.10.1. Sources

The richest sources of biotin are liver, yeast, black strap, molasses, groundnut and eggs. Green leafy plants are also good sources. Biotin occurs both in bound and free forms; the former making about 50% in feeds, is not available to animals

5.10.2. Functions

The main function is the transfer of CO_2 (activated CO_2, Biocytin) from one compound another in carbohydrate metabolism and hence bring about carboxylation and decarboxylation; for example, conversion of pyruvic acid to oxaloacetic acid, malic acid to pyruvic acid, succinic acid to propionic acid, oxalosuccinic acid to α keto glutaric acid and other dehydrogenation reactions also require biotin. Biotin also is related to lipid and protein metabolism. For example, conversion of carbon skeleton from leucine two oxaloacetate.

5.10.3. Deficiency symptoms

Typical dermatitis of the feet and skin round the beak and eyes (similar to that in pantothenic acid deficiency). Perosis is also a characteristic deficiency symptom. Embryos show congenital perosis, ataxia and characteristic skeletal deformities – shortened tibiotarsus, tarsometatarsus and bones of wing and skull, and bending of anterior end of scapula. Embryos also develop syndactylia (webbing between toes, especially 3rd and 4th), chondrodystrophy, parrot beak, crooked tibia and tarsometatarsus.

5.11. Folic acid

Deficiency symptoms first recorded by Wills in 1951. Unit of measurement : mg

5.11.1. Sources

Although folic acid is widely distributed in nature (animals, plants and micro-organisms), most of it is bound to two or more glutamic acid residues.

5.11.2. Functions

Intimately connected with single carbon metabolism viz., interconversion of serine and glycine, synthesis of many compounds involving carbon of glycine, histidine degradation, purine synthesis, and synthesis of methyl groups necessary for methionine, choline and thymine.

5.11.3. Deficiency symptoms

Being intimately involved in the synthesis of purines and pyramidines, the deficiency results in reduced nucleic acid production thereby leading to improper maturation of primordial RBCs which, in turn, leads to macrocytic anemia. Production of WBCs also is affected leading to thrombopenia, leukopenia and multilobed neutrophils. Chicks : Anemia is also associated with poor growth, poor feathering and perosis. In case of breeds colored feathers, loss of feather pigmentation results. Adults : Embryonic mortality seen after pipping with deformed beaks and bending of tibiotarsus. Oviducal growth is also inhibited.

5.12. Cyanocobalamin (Vitamin B_{12})

Identified as unknown factor by Minor and Murphy in 1926; referred to as Vitamin

B_{12}. Unit of measurement : μg

5.12.1. Sources

Vitamin B_{12} is primary of microbial origin, especially many bacteria and actinomycetes but not by yeasts and most fungi. Vitamin B_{12} is widely distributed in foods of animal origin viz., meat, milk, eggs, fish etc. fermentation products, kidney and liver. Planned sources are practically devoid of Vitamin B_{12}.

5.12.2. Functions

Involved in many interrelationships with folic acid, pantothenic acid, choline, methionine etc. It's a cofactor for methyl malonyl SCoA isomerase (mutase) and homocysteine transmethylase.

5.12.3. Deficiency symptoms

Slow growth, poor feed efficiency, mortality and reduced hatchability. Perosis may also occur.

5.13. Choline

Discovered by Strecker in 1862; unit of measurement : mg

5.13.1. Sources

Liver, fish meal, fish solubles, yeast, distiller's solubles and soyabean meal

5.13.2. Functions

It is apart and parcel of acetyl choline – the transmitter of para-sympathetic nerve impulses – therefore concerned with nerves like vagus and others. This also associated with contractions of oviduct and emptying of crop. Choline is a structural part of phospholipids – lecithin and sphingomyelin. Choline is a labile methyl donor helping formation of methionine from homocysteine and creatine from guanidino acetic acid. This methylation is highly essential for proper transport of depot fat into the cells and hence choline is one of the lipotropic factors.

5.13.3. Deficiency symptoms

Being a lipotropic factor, its deficiency leads to fatty liver, hemorrhagic kidneys and perosis. Perosis is characterized by enlargement and puffing around the hock joint followed by peticheal hemorrhages. Slowly, the tendoachilis slips off the condyles resulting in perosis. Adults can biosynthesize sufficient quantity of choline and hence, they do not have the requirement for choline. In chicks deficiency leads to poor growth and perosis.

5.14. Ascorbic acid (Vitamin C)

This Vitamin is synthesized in the kidneys by utilizing glucose; therefore, this is not an essential Vitamin for birds. However, this Vitamin has arole to play during stress like high temperature, disease outbreak, soon after debeaking, transport over a long-distance etc.

5.15. Other factors sometimes called vitamins

Myoinositol, *p*-amino benzoic acid, ubiquinones, lipoic acid, carnitine and pyrroloquinoline quinone have been found to be required for maximal growth under some circumstances. For example, under high levels of antibiotics or germ-free (xenobiotic) conditions, ubiquinones, lipoic acid and pyrroloquinoline become essential nutrients. However, normal poultry diets are adequate with respect to these vitamin-like compounds (Klasing, 1998).

6. Minerals

6.1. Calcium (Ca)

6.1.1. Sources

Bone meal, meat scrap, shell grit, limestone, fish meal etc.

6.1.2. Functions

Plays a major role in formation of bone, especially in growing birds. In laying birds, the main function is egg-shell formation. It is also necessary for normal neuromuscular transmission of impulses (muscular activity), normal heart beat, clotting of blood and acid-base balance.

6.1.3. Deficiency symptoms

Inanition, reduced growth, increased BMR, osteoporosis (low-calcium rickets), internal hemorrhage, polyurea, thin-shelled eggs, reduced egg production, low-calcium tetany and abnormal gait. Bones in layers fed on low-calcium diets becomes thin and highly susceptible to fractures. In case of caged layers, one of the causes for the cage-layer fatigue has been thought to be calcium deficiency.

6.1.4. Excess calcium

Growing birds show nephrosis, visceral gout and mortality; pullets show atrophy of parathyroid glands, hypercalcaemia and concomitant hypophosphatemia, inanition, delayed sexual maturity and reduced growth. Diets to pullets up to 18 to 20 weeks of age must not contain calcium levels above 1.2%.

6.2. Phosphorus (P)

6.2.1. Sources

Several inorganic phosphates can be used the sources of phosphorus. For example, dicalcium phosphate, bone meal etc. In plants, phosphorus is considered only two-thirds available since the other portion is bound to phytic acid of making it non-available to simple-stomached animals because they lack the enzyme phytase.

Phosphorus from animal sources is considered completely available. Ratio of calcium to available phosphorus is very critical determining not only the availability of calcium and phosphorus but also affect the absorption of elements like manganese, zinc and others. Ratio of calcium available phosphorus for growing chicken (up to 20 weeks) is 1.5 to 2.0 : 1 whereas for layers it can be 4.5 to 6.0 : 1.

6.2.2. Functions

Major constituent of bone. Besides, being the elements chosen by nature to store and dispense energy, plays a very important role in metabolism of carbohydrates, proteins and fats in the form of ATP, ADP and Creatine phosphate. It is an integral part of nucleic acids.

6.2.3. Deficiency symptoms

Inanition, weakness, stunted growth, rickets and, if prolonged, mortality.

6.3. Sodium (Na)

6.3.1. Sources

Salt, all feed ingredients have sodium

6.3.2. Functions

Chief cation in the extracellular fluid concerned with regulation of fluid volume, osmotic pressure and acid-based balance. It is intimately concerned with absorption of amino acids and transmission of nerve impulses. Firmly bound to inorganic phase of the bone.

6.3.3. Deficiency symptoms

Reduced growth due to reduced utilization of protein and energy, softening of bones, corneal keratinization, reduced activity of gonads, reduced cardiac output and thereby reduced arterial blood pressure, impaired adrenal activity (hypertrophy), reduced fluid volume and therefore increased hematocrit, increased blood urea (uric acid), decreased egg production and also cannibalism. Prolonged deficiency terminates in death.

6.3.4. Excess sodium

Moderate excesses result in no serious problems; but too high levels of salt exceeding the ability of chicks to consume excess water can result in toxicity. Increased water consumption results in excess water in feces and symptoms of water intoxication/dehydration depending on the amount of sodium loss.

6.4. Potassium (K)

6.4.1. Functions

Principal intracellular cation. It is involved in maintenance of intracellular acid-base balance, osmotic pressure and stimulation of many enzymes. It also favors relaxation of cardiac muscle and is likely to increase the membrane permeability facilitating absorption of neutral amino-acids.

6.4.2. Deficiency symptoms

Weakness of respiratory, cardiac and skeletal muscles; poor intestinal tone (thus resulting in distension). Hypokalaemia is also suggestive of adaptation to stress since along with the restored glycogen levels, potassium enters back to the liver and skeletal muscles.

6.5. Chlorine (Cl)

6.5.1. Functions

Principal anion in maintaining ionic strength of the extracellular fluid. It is essential for HCl production by the gastric mucosa.

6.5.2. Deficiency symptoms

Under practical feeding regimen, deficiencies very rare. However, with purified diets, deficiency has been found to cause poor growth, hemoconcentration, dehydration, hyperexcitability (chicks fall forward with legs extended backwards; recover quickly and spontaneously) and death.

6.6. Magnesium (Mg)

6.6.1. Functions

ATP in the body exist as Mg-ATP and hence, wherever ATP is involved, magnesium is required; therefore, magnesium is required for almost all functions of the body. Most of the enzymes catalyzed the magnesium or metal enzyme-complexes (metal not bound to enzyme). It activates alkaline phosphatase enzyme.

6.6.2. Deficiency symptoms

Stunted growth, inactive, panting, gasping, reduced egg production, egg weight and shell weight. Magnesium levels in blood, bone and egg yolk are reduced.

6.6.3. Excess magnesium

Wet droppings and reduced shell thickness. Hence, limestone containing high levels of magnesium (dolomite limestone) should not be used in poultry diets.

6.7. Manganese (Mn)

6.7.1. Functions

Formation of organic matrix of the bone (mucopolysaccharides), oxidative phosphorylation in mitochondria, fatty-acid and cholesterol synthesis and many other enzymes involved in amino-acid, nucleic acid and fatty-acid metabolism. It activates phosphatase enzyme.

6.7.2. Deficiency symptoms

The most characteristic symptom is perosis associated with enlargement and malformation of tibiometatarsal joint, twisting and bending of tibia, thickening and shortening of leg bones, followed by slipping of tendoachilis from its condyles. In layers, the symptoms include reductions in egg production and hatchability, increased incidence of thin-shelled and shell-less eggs. Manganese deficient embryos exhibit chondrodystrophy (shortened legs, wings and lower mandible, parrot beak, bulging of skull, edema, unabsorbed yolk and star-gazing posture.

6.8. Zinc (Zn)

6.8.1. Functions

Carbonic anhydrase is the metallo-enzyme with zinc as the prosthetic group. The enzyme is concerned with acid –base balance, release of CO_2 in the lungs, hydration of CO_2 in the gastric mucosa, calcification of bone and formation of egg shell. Zinc is also a constituent of glutamic acid-and alcohol-dehydrogenases, carboxypeptidase and various other dehydrogenases and peptidases.

6.8.2. Deficiency symptoms

Stunted growth, enlargement of hock joint, scaling of the skin, poor feathering and poor feed efficiency, inappetance and, in severe cases, mortality. In layers, reduced egg production and hatchability. Chicks from deficient embryos show polypnoea and dyspnoea and will be weak and ruffled or frizzled feathers, with characteristic impaired skeletal development (micromelia, curvature of spine, fused thoracic and lumbar vertebrae), sometimes without limbs.

6.9. Iron (Fe)

6.9.1. Functions

Prosthetic group in myoglobin and hemoglobin (transport O_2), cytochromes (electron-transport), peroxidases and catalase (decomposition of H_2O_2). NADH-cytochrome reductase and xanthine oxidase are the two other enzymes containing iron, the latter is involved in uric-acid synthesis. Hemosiderin and ferritin (ferric state) or the storage forms of iron.

6.9.2. Deficiency symptoms

Microcytic, hypochromic anemia, reduction in PCV and depigmentation of feathers (achromotrichia) in case of birds having colored feathers.

6.10. Copper (Cu)

6.10.1. Functions

Several oxidases contained copper; they include tyrosinase, amine oxidase and ascorbic acid oxidase. Copper is required for hematopoiesis and it is the prosthetic group of butyryl CoA dehydrogenase. Copper plays an important role in production of melanin from L-tyrosine by the copper-containing enzyme polyphenol oxidase. Amine oxidase is thought to be required for incorporation of lysine into the elastins (desmosines) of the aorta and other mucous membranes.

6.10.2. Deficiency symptoms

Anemia, achromotrichia, demyelination of spinal cord (causing ataxia and spastic paralysis), rupture of aorta and fragility of bones.

6.11. Molybdenum (Mo)

6.11.1. Functions

Xanthine oxidase enzyme which catalyzes the oxidation of purines (uric-acid synthesis), aldehydes pterins and NADH contains iron, molybdenum and FAD in a complex prosthetic group. It is likely to the absorption and deposition of fluorine.

6.11.2. Deficiency symptoms

Stunted growth and reduced xanthine dehydrogenase levels in liver

6.12. Selenium (Se)

6.12.1. Functions

Primary function is to prevent exudative diathesis in chicks. It also aids in absorption and retention of d α tocopherol (Vitamin E) and hence indirectly concerned with alleviating nutritional muscular dystrophy as well as encephalomalacia. However, Se alone can prevent encephalomalacia. Se inhibits enzymes like succinic dehydrogenase, choline oxidase, tyraminase and d-proline oxidase and inactivates l-proline oxidase.

6.12.2. Deficiency symptoms

In conjunction with Vitamin E deficiency, causes encephalomalacia, exudative diathesis and nutritional muscular dystrophy.

6.12.3. Excess selenium

When fed 100 times higher than required, it can precipitate in reduced growth, egg production and hatchability as well as increased embryo abnormalities.

6.13. Iodine (I)

6.13.1. Functions

Being the important component of thyroid hormone, iodine is intimately related with all the functions controlled by thyroxine viz., energy metabolism (BMR), differentiation and functioning of various tissues and organs, growth of integument and its outgrowths, affect other endocrine glands (pituitary and gonads), circulation etc.

6.13.2. Deficiency symptoms

Diminished thyroxine output triggers anterior pituitary to secrete excess TSH thereby ending in hypertrophy and hyperplasia of thyroid gland popularly referred to as goiter. Cessation of egg production and accumulation of fat in laying hens, feathers become abnormally long and lacy and deterioration of egg shell quality. Reduced hatchability, increased hatching time, unabsorbed yolk and enlargement of embryonic thyroid glands are noticed in iodine-deficient embryos.

6.14. Cobalt (Co)

Cobalt is part and parcel of Vitamin B_{12} and hence, all functions and deficiency symptoms of cobalt are similar to that of Vitamin B_{12}.

7. Signs of deficiency in embryos (NRC, 1994)

Table 52.4. Vitamin and mineral deficiency in chicken embryo

Nutrient	Deficiency signs
	Vitamins
Vitamin A	Death at about 48 hr of incubation from failure to develop the circulatory system; abnormalities of kidneys, eyes and skeleton
Vitamin D	Death at about 18 of 19 d of incubation, with malpositions, soft bones, and with a defective upper mandible prominent
Vitamin E	Early death at about 84 to 96 hr of incubation, with hemorrhaging and circulatory failure (implicated with Se)
Vitamin K	No physical deformities from a simple deficiency, nor can they be provoked by anti-vitamins, but mortality occurs between 18d and hatching with variable hemorrhaging
Thiamin	High embryonic mortality during emergence and no obvious symptoms other than polyneuritis in those that survive
Riboflavin	Mortality peaks at 60 hr, 14d and 20d of incubation, with peaks prominent early as deficiency becomes severe. Altered limb and mandible development, dwarfism, and clubbing of down
Niacin	Embryo readily synthesizes sufficient niacin from tryptophan. Various bone and beak malformations occur when certain antagonists are administered during incubation
Biotin	High death rate at 19 to 21d of incubation, and embryos have parrot beak, chondrodystrophy, several skeletal deformities, and webbing between the toes
Pantothenic acid	Deaths appear around 14d of incubation, although marginal levels may delay problems until emergence. Variable subcutaneous hemorrhaging and edema; wiry down in poults
Pyridoxine	Early embryonic mortality based on anti-vitamin use
Folic acid	Mortality at about 20d of incubation. The dead generally appear normal, but many have bent tibiotarsus, syndactyly, and mandible malformations. In poults, mortality at 26 to 28d of incubation with abnormalities of extremities and circulatory system

Nutrient	Deficiency signs
Cyanocobalamin	Mortality at about 20d of incubation, with atrophy of legs, edema, hemorrhaging, fatty organs, and head between thighs malpositions
Inorganic elements	
Manganese	Peak deaths prior to emergence. Chondrodystrophy, dwarfism, long bone shortening, head malformations, edema and abnormal feathering are prominent
Zinc	Deaths prior to the emergence, and the appearance of rumplessness, depletion of vertebral column, eyes under-developed, and missing limbs
Copper	Deaths at early blood stage with no malformations
Iodine	Prolongation of hatching time, reduced thyroid size, and incomplete abdominal closure
Iron	Low hematocrit; low blood hemoglobin; poor extra embryonic circulation in candled eggs
Selenium	High incidence of deadly embryos early in incubation

8. Deficiency symptoms *Vs* nutrients involved (NRC, 1994)

Table 52.5. Symptoms associated with nutrient deficiencies

Symptoms	Description	Species	Nutrients
Skin lesions	Crusting and scab formation around eyes and beak; bottoms of feet rough and calloused with hemorrhagic cracks	Chick, poult	Biotin, Pantothenic acid
	Scaliness on feet	Chick	Zinc, Niacin
	Lesions around eyes, eyelids stuck together.	Chick, poult	Vitamin A
	Mouth, inflammation of oral mucosa (chicken black tongue)	Poult, chick	Niacin
Feather abnormalities	Uneven feather growth, abnormally long primary feathers, feathers not lying smoothly	Chick, poult	Protein, amino-acid
	Frizzled and rough	Chick, poult	Zinc, Niacin, Pantothenic acid, Folic acid, Lysine
	Black pigmentation in breeds with red and brown feathers	Chick	Vitamin D
	Depigmentation	Chick, poult	Copper, Iron, Folic acid

Nutrients – Functions, Deficiency and Excess

Symptoms	Description	Species	Nutrients
Nervous disorders	Convulsions with head retraction	Chick, poult	Thiamin
	Convulsions with hyperexcitability	Chick, poult, duckling	Pyridoxine
	Hyperirritability	Chick, poult, duckling	Magnesium, NaCl
	Characteristic fright reaction which tetanic spasms	Chick	Chloride
	Spastic cervical paralysis, neck extended with birds appearing to look down	Poult	Folic acid
	Curled-toe paralysis, gross enlargement of sciatic and brachial nerves with myelin degeneration	Chick	Riboflavin
	Encephalomalacia, tetanic spasms with head retraction, hemorrhages lesions in cerebellum	Chick	Vitamin E
Blood and vascular system	Anemia: Macrocytic Macrocytic, hypochromic Microcytic, hypochromic Microcytic	All poultry	Cyanocobalamin. Folic acid. Iron, Copper. Pyridoxine
	Intramuscular or subcutaneous hemorrhage, internal hemorrhage from aortic rupture	Chick, poult	Vitamin K, Copper
	Exudative diathesis	Chick, poult	Selenium, Vitamin E
	Enlarged heart	Chick, poult	Copper
Muscle	Muscular dystrophy, white areas of degeneration in skeletal muscle	Chick, duck, poult	Vitamin E, Selenium
	Cardiac and gizzard myopathy	Poult	Vitamin E, Selenium
Bone disorders	Soft, easily bent bones and beak (rickets)	All poultry	Vitamin D, Calcium or Phosphorus deficiency or imbalance
	Hock enlargement	Poult, chick, gosling, duckling	Niacin, Zinc
	Perosis	Chick, poult	Biotin, Choline, Cyanocobalamin, Manganese, Zinc, Folacin
	Bowed legs	Duck	Niacin
	Shortening and thickening of leg bones	Chick	Zinc, Manganese
	Curled toes	Chick	Riboflavin
Diarrhea		Chick, duck, poult	Niacin, Riboflavin, Biotin

Chapter **53**

Nutrient Requirements and Specifications

1. Nutrient requirements

1.1. BIS (1992)

1.1.1. Broilers

Table 53.1. Nutrient requirements of broilers

Nutrient	Broiler starter (0-6 weeks)	Broiler finisher (6-9 weeks)
Moisture, % (max)	11	
Metabolizable energy, kcal/kg (min)	2800	2900
Crude protein, % (min)	23	20
Crude fiber, % (max)	6	
Lysine, % (min)	1.2	1.0
Methionine, % (min)	0.6	0.35
Total S-amino-acids, % (min)	0.9	0.7
Crude fiber, % (max)	6	
Acid insoluble ash, % (max)	3	
Calcium, % (min)	1.2	
Available phosphorus, % (min)	0.5	
Manganese, ppm	90	
Zinc, ppm	60	
Iodine, ppm	1	
Iron, ppm	120	
Copper, ppm	12	
Salt (as NaCl), % (max)	0.6	
Vitamin A, IU/kg	6000	
Vitamin D_3, IU/kg	600	
α-tocopherol, ppm	15	
Vitamin K, ppm	1	
Thiamin, ppm	5	
Riboflavin, ppm	6	
Pyridoxine, ppm	5	

Nutrient	Broiler starter (0-6 weeks)	Broiler finisher (6-9 weeks)
Pantothenic acid, ppm	15	
Nicotinic acid, ppm	40	
Biotin, ppm	0.2	
Folic acid, ppm	1	
Cyanocobalamin, µg/kg	15	
Choline chloride, ppm	1400	
Linoleic acid, %	1	

1.1.2. Laying-type

Table 53.2. Nutrient requirements of laying-type chicken

Nutrient	Chicks (0-8 weeks)	Growers (8-20 weeks)	Layers (20-80 weeks)	Breeders
Moisture, % (max)	11			
Metabolizable energy, kcal/kg (min)	2600	2500	2600	
Crude protein, % (min)	20	16	18	
Crude fiber, % (max)	7	8		
Lysine, % (min)	0.9	0.6	0.65	0.65
Methionine, % (min)	0.3	0.25	0.3	0.3
Total S-amino-acids, % (min)	0.6	0.5	0.55	0.55
Crude fiber, % (max)	7	8		
Acid insoluble ash, % (max)	4			
Calcium, % (min)	1		3	
Available phosphorus, % (min)	0.5			
Manganese, ppm	90	50	55	90
Zinc, ppm	50	60	75	100
Iodine, ppm	1			
Iron, ppm	120	90	75	90
Copper, ppm	12	9	9	12
Salt (as NaCl), % (max)	0.6			
Vitamin A, IU/kg	6000		8000	
Vitamin D_3, IU/kg	600		1200	
α-tocopherol, ppm	15	10		15
Vitamin K, ppm	1			
Thiamin, ppm	5	3		
Riboflavin, ppm	6	5	5	8

Nutrient	Chicks (0-8 weeks)	Growers (8-20 weeks)	Layers (20-80 weeks)	Breeders
Pyridoxine, ppm	5			8
Pantothenic acid, ppm	15			
Nicotinic acid, ppm	40	15		
Biotin, ppm	0.2	0.15	0.2	
Folic acid, ppm	1	0.5		
Cyanocobalamin, μg/kg	15	10		
Choline chloride, ppm	1300	900	800	
Linoleic acid, %	1	1	1	1

1.2. NRC (1994)

1.2.1. Chicken

1.2.1.1. Broilers

Table 53.3. Nutrient requirements of broilers

Nutrient	0-3 weeks	3-6 weeks	6-8 weeks
Metabolizable energy base, kcal/kg	3200		
Linoleic acid, %	1.00		
Crude protein and amino acids, %			
Crude protein	23	20	18
Arginine	1.25	1.10	1.00
Glycine + Serine	1.25	1.14	0.97
Histidine	0.35	0.32	0.27
Isoleucine	0.80	0.73	0.62
Leucine	1.20	1.09	0.93
Lysine	1.1	1.0	0.85
Methionine + Cystine	0.90	0.72	0.60
Methionine	0.5	0.38	0.32
Phenylalanine + Tyrosine	1.34	1.22	1.04
Phenylalanine	0.72	0.65	0.56
Threonine	0.80	0.74	0.68
Tryptophan	0.20	0.18	0.16
Valine	0.90	0.82	0.70
Inorganic elements (Major), %			
Calcium	1.00	0.90	0.80
Phosphorus, non-phytin	0.45	0.35	0.30
Potassium	0.30	0.30	0.30
Sodium	0.20	0.15	0.12
Chlorine	0.20	0.15	0.12
Magnesium	0.06		

Nutrient	0-3 weeks	3-6 weeks	6-8 weeks
Inorganic elements (Trace), ppm			
Manganese	60		
Zinc	40		
Iron	80		
Copper	8		
Iodine	0.35		
Selenium	0.15		
Vitamins, fat-soluble (per kg)			
A, IU	1500		
D_3, ICU	200		
E, mg	10		
K, mg	0.50		
Vitamins, water-soluble, ppm			
Riboflavin	3.6		3.00
Pantothenic acid	10		
Niacin	35	30	25
Cyanocobalamin	0.010		0.007
Choline	1300	1000	750
Biotin	0.15		0.12
Folic acid	0.55		0.50
Thiamin	1.80		
Pyridoxine	3.50		3.00

1.2.1.2. Laying-type

Table 53.4. Nutrient requirements of laying-type chicken*

Nutrient	Age (weeks) / Stage					
	0-6	6-12	12-18	18-1st egg	Laying	Breeders
ME, kcal/kg	2850			2900		
Linoleic acid, %	1.00					
Crude protein and amino acids, %						
CP	18	16	15	17	15	
Arginine	1.00	0.83	0.67	0.75	0.70	
Glycine + Serine	0.70	0.58	0.47	0.53	--	
Histidine	0.26	0.22	0.17	0.20	0.17	
Isoleucine	0.60	0.50	0.40	0.45	0.65	
leucine	1.10	0.85	0.70	0.80	0.82	
Lysine	0.85	0.60	0.45	0.52	0.69	
Methionine + Cystine	0.62	0.52	0.42	0.47	0.58	
Methionine	0.30	0.25	0.20	0.22	0.30	
Phenylalanine+Tyrosine	1.00	0.83	0.67	0.75	0.83	

Nutrient	Age (weeks) / Stage					
	0-6	**6-12**	**12-18**	**18-1st egg**	**Laying**	**Breeders**
Phenylalanine	0.54	0.45	0.36	0.40	0.47	
Threonine	0.68	0.57	0.37	0.47	0.47	
Tryptophan	0.17	0.14	0.11	0.12	0.16	
Valine	0.62	0.52	0.41	0.46	0.70	
Inorganic elements (major), %						
Calcium	0.9	0.80		2.00	3.25	
Phosphorus, non-phytin	0.4	0.35	0.30	0.32	0.25	
Potassium	0.25				0.15	
Sodium	0.15					
Chlorine	0.15	0.12		0.15	0.13	
Magnesium	600	500	400		500	
Inorganic elements (trace), ppm						
Manganese	60	30			20	
Zinc	40	35				45
Iron	80	60			45	60
Copper	5	4			?	?
Iodine	0.35					0.10
Selenium	0.15	0.10			0.06	
Vitamins, fat-soluble (per kg)						
A, IU	1500				3000	
D_3, ICU	200			300		
E, mg	10	5				10
K, mg	0.50					1.00
Vitamins, water-soluble, ppm						
Riboflavin	3.60	1.80		2.20	2.50	3.60
Pantothenic acid	10					7
Niacin	27	11			10	
Cyanocobalamin	0.009	0.003		0.004		0.080
Choline	1300	900	500		1050	
Biotin	0.15	0.10				
Folic acid	0.55	0.25				0.35
Thiamin	1.00		0.80		0.70	
Pyridoxine	3.00				2.50	4.50
* Daily requirement, mg						

1.2.1.3. Laying-type chicken (Brown eggs)

Table 53.5. Nutrient requirements of brown-egg laying chicken

Nutrient	Age (weeks) / Stage				
	0-6	6-12	12-18	18-1st egg	Laying *
Body weight, kg	0.50	1.10	1.50	1.60	
ME, kcal/kg	2850			2850	
Linoleic acid, %			1.00		1100
Crude protein and amino acids, %					
CP	17	15	14	16	16500
Arginine	0.94	0.78	0.62	0.72	770
Glycine + Serine	0.66	0.54	0.44	0.50	--
Histidine	0.25	0.21	0.16	0.18	190
Isoleucine	0.57	0.47	0.37	0.42	715
leucine	1.00	0.80	0.65	0.75	900
Lysine	0.80	0.56	0.42	0.49	760
Methionine + Cystine	0.28	0.23	0.19	0.21	330
Methionine	0.59	0.49	0.39	0.44	645
Phenylalanine+Tyrosine	0.51	0.42	0.34	0.38	520
Phenylalanine	0.94	0.78	0.63	0.70	910
Threonine	0.64	0.53	0.35	0.44	520
Tryptophan	0.16	0.13	0.10	0.11	175
Valine	0.59	0.49	0.38	0.43	770
Inorganic elements (major), %					
Calcium	0.90		0.80		3600
Phosphorus, non-phytin	0.40	0.35	0.30	0.35	275
Potassium			0.25		165
Sodium			0.15		165
Chlorine	0.12		0.11		145
Inorganic elements (trace), ppm					
Magnesium	570	470		370	55
Manganese	56		28		2.20
Zinc	38		33		3.90
Iron	75		56		5.00
Copper	5		4		?
Iodine			0.33		0.004
Selenium	0.14		0.10		0.006
Vitamins, fat-soluble (per kg)					
A, IU			1420		330
D_3, ICU		190		280	33
E, mg	9.5		4.7		0.55
K, mg			0.40		0.055

Nutrient	Age (weeks) / Stage				
	0-6	**6-12**	**12-18**	**18-1st egg**	**Laying ***
Vitamins, water-soluble, ppm					
Riboflavin	3.40		1.70		0.28
Pantothenic acid			9.40		0.22
Niacin	26		10.30		1.10
Cyanocobalamin	0.009	0.003			0.0004
Choline	1225	850		470	115
Biotin	0.14		0.09		0.011
Folic acid	0.52		0.23		0.028
Thiamin			1.00		0.08
Pyridoxine			2.80		0.28

* Amount required/d

1.2.1.4. Meat-type breeding hens

Table 53.6. Nutrient requirements of meat-type breeding hens

Nutrient	In diet	Daily intake, mg
ME	2850 kcal/kg	(on 90% dry matter basis)
CP	14.5 %	19500
Arginine	0.74 %	1110
Histidine	0.14 %	205
Isoleucine	0.57 %	850
Leucine	0.83 %	1250
Lysine	0.51 %	765
Methionine + Cystine	0.55 %	700
Methionine	0.35 %	450
Phenylalanine+Tyrosine	0.75 %	1112
Phenylalanine	0.41 %	610
Threonine	0.48 %	720
Tryptophan	0.13 %	190
Valine	0.63 %	750
Calcium	2.75 %	4000
Phosphorus, non-phytin	0.25 %	350
Sodium	0.10 %	150
Biotin	0.25 ppm	0.016

1.2.1.5. Meat-type breeding males

Table 53.7. Nutrient requirements of meat-type breeding males

Nutrient	0-4 weeks	4-20 weeks	20-60 weeks
	In diet		Per bird/d
ME, kcal/kg			350-400
Protein	15 %	12 %	12.0 g
Lysine	0.79 %	0.64 %	475 mg
Methionine	0.36 %	0.31 %	340 mg
Methionine + Cystine	0.61 %	0.49 %	490 mg
Arginine			680 mg
Calcium	0.90 %	0.90 %	200 mg
Phosphorus, non-phytin	0.45 %	0.45 %	110 mg

1.2.1.6. ME requirement of layers

ME (kcal/hen/d) is estimated by the following formula:

$ME = W^{0.75}(173\text{-}1.95T) + 5.5\Delta W + 2.07EE$ where, W = body weight (kg), T = ambient temperature (°C), ΔW = change in body weight (g/d) and EE = daily egg mass (g). The results at T=20°C, egg weight = 60 g and $\Delta W = 0$ (W ranging from 1.0 to 3.0 kg) and egg production range of 0 to 90% is shown in Table 53.8.

Table 53.8. ME requirement (kcal/hen/d) of layers *Vs* egg production and body weight

Body weight (kg)	Rate of egg production, %					
	0	50	60	70	80	90
1.0	130	192	205	217	229	242
1.6	177	239	251	264	276	289
2.0	218	280	292	305	317	350
2.5	259	321	333	346	358	371
3.0	286	358	370	383	395	408

1.2.2. Turkeys

1.2.2.1. Pre-breeding / Pre-lay

Table 53.9. Nutrient requirements of immature turkeys

	Age (weeks) of growing turkeys, males and females					
Males	0-4	4-8	8-12	12-16	16-20	20-24
Females	0-4	4-8	8-11	11-14	14-17	17-20
Nutrient	2800*	2900*	3000*	3100*	3200*	3300*
Linoleic acid, %	1.0		0.8			
	Protein and amino acids, %					
CP	28	26	22	19	16.5	14

	Age (weeks) of growing turkeys, males and females					
Arg	1.6	1.4	1.1	0.9	0.75	0.6
Gly+Ser	1.0	0.9	0.8	0.7	0.6	0.5
His	0.58	0.50	0.40	0.30	0.25	0.20
Ile	1.1	1.0	0.8	0.6	0.5	0.45
Leu	1.90	1.75	1.50	1.25	1.0	0.8
Lys	1.6	1.5	1.3	1.0	0.8	0.65
Met	0.55	0.45	0.40	0.35	0.25	0.25
Met + Cys	1.05	0.95	0.80	0.65	0.55	0.45
Phe	1.0	0.9	0.8	0.7	0.6	0.5
Phe + Tyr	1.8	1.6	1.2	1.0	0.9	
Thr	1.00	0.95	0.80	0.75	0.60	0.50
Trp	0.26	0.24	0.20	0.18	0.15	0.13
Val	1.2	1.2	0.9	0.8	0.7	0.6
	Inorganic elements (Major), %					
Ca	1.20	1.00	0.85	0.75	0.65	0.55
P, non-phytin	0.60	0.50	0.42	0.38	0.32	0.28
K	0.7	0.6	0.5		0.4	
Na	0.17	0.15	0.12	0.12	0.12	0.12
Cl	0.15	0.14		0.12		
Mg	0.05					
	Inorganic elements (Trace), ppm					
Mn	60					
Zn	70	65	50	40		
Fe	80	60			50	
Cu	8		6			
I	0.4					
Se	0.2					
	Vitamins, fat-soluble (per kg)					
A, IU	5000					
D_3, ICU	1100					
E, mg	12		10			
K, mg	1.75	1.5	1.0	0.75		0.50
	Vitamins, water-soluble, ppm					
Cyanocobalamin	0.003					
Biotin	0.250	0.200	0.125		0.100	
Choline	1600	1400	1100		950	800
Folic acid	1.0		0.8		0.7	
Niacin	60		50	40		
Pantothenic acid	10	9				
Pyridoxine	4.5		3.5		3.0	
Riboflavin	4.0	3.6	3.0		2.5	
Thiamin	2.0					
* ME values (kcal/kg)						

1.2.2.2. Breeders

Table 53.10. Nutrient requirements of breeding turkeys

Nutrient	Holding	Laying hens
ME, kcal/kg	2900	
Linoleic acid, %	0.8	1.1
Protein and amino acids, %		
CP	12	14
Arg	0.5	0.6
Gly+Ser	0.4	0.5
His	0.2	0.3
Ile	0.4	0.5
Leu	0.5	
Lys	0.5	0.6
Met	0.2	
Met + Cys	0.4	
Phe	0.4	0.55
Phe + Tyr	0.8	1.0
Thr	0.4	0.45
Trp	0.1	0.13
Val	0.5	0.58
Inorganic elements (Major), %		
Ca	0.5	2.25
P, non-phytin	0.25	0.35
K	0.4	0.6
Na	0.12	
Cl	0.12	
Mg	0.05	
Inorganic elements (Trace), ppm		
Mn	60	
Zn	40	65
Fe	50	60
Cu	6	8
I	0.4	
Se	0.2	
Vitamins, fat-soluble (per kg)		
A, IU	5000	
D_3, ICU	1100	
E, mg	10	25
K, mg	0.5	1.0

Table 53.10 contd.

Nutrient	Holding	Laying hens
Vitamins, water-soluble, ppm		
Cyanocobalamin	0.003	
Biotin	0.10	0.2
Choline	800	1000
Folic acid	0.7	1.0
Niacin	40	
Pantothenic acid	9	16
Pyridoxine	3	4
Riboflavin	2.5	4
Thiamin	2	

1.2.3. Japanese quails

Table 53.11. Nutrient requirements of Japanese quails

Nutrient	Starting and growing (up to 6 weeks)	Breeding (after 6 weeks)
ME, kcal/kg	2900	
Linoleic acid, %	0.8	1.1
Protein and amino acids, %		
CP	24	20
Arg	1.25	1.26
Gly+Ser	1.15	1.17
His	0.36	0.42
Ile	0.98	0.90
Leu	1.69	1.42
Lys	1.30	1.00
Met	0.50	0.45
Met + Cys	0.75	0.70
Phe	0.96	0.78
Phe + Tyr	1.80	1.40
Thr	1.02	0.74
Trp	0.22	0.19
Val	0.95	0.92
Inorganic elements (Major), %		
Ca	0.8	2.5
P, non-phytin	0.30	0.35
K	0.4	
Na	0.15	
Cl	0.14	
Mg	0.03	0.05

Nutrient	Starting and growing (up to 6 weeks)	Breeding (after 6 weeks)
Inorganic elements (Trace), ppm		
Mn	60	
Zn	25	50
Fe	120	60
Cu	5	
I	0.3	
Se	0.2	
Vitamins, fat-soluble (per kg)		
A, IU	1650	3300
D_3, ICU	750	900
E, mg	12	25
K, mg	1	
Vitamins, water-soluble, ppm		
Cyanocobalamin	0.003	
Biotin	0.3	0.15
Choline	2000	1500
Folic acid	1	
Niacin	40	20
Pantothenic acid	10	15
Pyridoxine	3	3
Riboflavin	4	
Thiamin	2	

1.2.4. Bobwhite quail

Table 53.12. Nutrient requirements of Bobwhite quails

Nutrient	0-6 weeks	After 6 weeks	Breeding
ME, kcal/kg	2800		
Linoleic acid, %	1.0		
Protein, %	26	20	24
Met + Cys, %	1.0	0.75	0.90
Calcium, %	0.65		2.40
Phosphorus, non-phyin, %	0.45	0.30	0.70
Sodium, %	0.15		
Chlorine, %	0.11		
Iodine, ppm	0.30		
Choline, ppm	1500		1000
Niacin, ppm	30		20
Pantothenic acid, ppm	12	9	15
Riboflavin, ppm	3.8	3.0	4.0

1.2.5. Ducks

Table 53.13. Nutrient requirements of ducks

Nutrient	0-2 weeks	3-7 weeks	Breeding
Metabolizable energy base, kcal/kg	2900	3000	2900
Crude protein and amino acids, %			
Crude protein	22	16	15
Arginine	1.1	1.0	
Isoleucine	0.63	0.46	0.38
Leucine	1.26	0.91	0.76
Lysine	0.90	0.65	0.60
Methionine + Cystine	0.70	0.55	0.50
Methionine	0.40	0.30	0.27
Tryptophan	0.23	0.17	0.14
Valine	0.78	0.56	0.47
Inorganic elements (Major), %			
Calcium	0.65	0.60	2.75
Phosphorus, non-phytin	0.40	0.30	
Sodium	0.15		
Chlorine	0.12		
Magnesium	0.05		
Inorganic elements (Trace), ppm			
Manganese	50	*	*
Zinc	0.20	*	*
Selenium	60	*	*
Vitamins, fat-soluble (per kg)			
A, IU	2500		4000
D_3, ICU	400		900
E, mg	10		
K, mg	0.50		
Vitamins, water-soluble, ppm			
Riboflavin	4		
Pantothenic acid	11		
Niacin	55		
Pyridoxine	2.50		3.00

* Not available

1.2.6. Geese

Table 53.14. Nutrient requirements of Geese

Nutrient	0-2 weeks	3-7 weeks	Breeding
Metabolizable energy base, kcal/kg	2900	3000	2900
Crude protein and amino acids, %			
Crude protein	20	15	
Lysine	1.0	0.85	0.60
Methionine + Cystine	0.60	0.50	
Inorganic elements (Major), %			
Calcium	0.65	0.60	2.25
Phosphorus, non-phytin	0.30		
Vitamins, fat-soluble (per kg)			
A, IU	1500		4000
D_3, IU	200		
Vitamins, water-soluble, ppm			
Riboflavin	3.8	2.5	4.0
Pantothenic acid	15.0	10.0	
Niacin	65	35	20
Choline	1500	1000	*

* Not available

1.2.7. Ring-necked Pheasants

Table 53.15. Nutrient requirements of Ring-necked Pheasants

Nutrient	0-4 weeks	4-8 weeks	9-17 weeks	Breeding
ME, kcal/kg	2800		2700	2800
Linoleic acid, %	1.0			
Crude protein and amino acids, %				
Crude protein	28	24	18	15
Glycine + Serine	1.8	1.55	1.0	0.50
Lysine	1.50	1.40	0.80	0.68
Methionine + Cystine	1.0	0.93	0.60	
Methionine	0.50	0.47	0.30	
Inorganic elements (Major), %				
Calcium	1.00	0.85	0.53	2.50
Phosphorus, non-phytin	0.55	0.50	0.45	0.40
Sodium	0.15			
Chlorine	0.11			

Nutrient	0-4 weeks	4-8 weeks	9-17 weeks	Breeding
Inorganic elements (Trace), ppm				
Manganese	70		60	
Zinc	60			
Vitamins, water-soluble, ppm				
Riboflavin	3.4		3.0	4.0
Pantothenic acid	10			16
Niacin	70		40	30
Choline				

2. Water quality and consumption (NRC, 1994)

2.1. Total dissolved solids (TDS) in drinking water

Table 53.16. Guidelines for suitability of drinking water

TDS, ppm	Remarks
< 1000	No serious burden on any class of poultry
1000 - 2999	Satisfactory for all classes of poultry; may cause watery droppings, especially at higher levels; but, should not affect health
3000 - 4999	Poor water for poultry, often causes watery diarrhea, increased mortality and decreased growth (especially in turkeys)
5000 - 6999	Not acceptable for poultry, and almost always causes some problem, especially at upper limits where decreased growth and production or increased mortality probably will occur
7000 - 10000	Unfit for poultry; but, may be suitable for other livestock
> 10000	Unfit for all livestock
	Source : NRC, 1994

2.2. Contaminants in water

Table 53.17. Maximum acceptable levels of contaminants in water

Contaminant	Average level	Maximum level	Remarks
Total bacteria	0/ml	100 cfu/ml	0/ml desirable
Coliforms	0/ml	50 cfu/ml	0/ml desirable
Nitrate	2 ppm	20 ppm	At levels > 50 ppm, performance affected
Nitrite	0.4 ppm	4 ppm	10 times more toxic than nitrate
pH	6.8 – 7.5	-	High or low pH degrade medications and precipitate minerals in water
Total hardness	60 – 189 ppm		< 60 water is "soft"; > 180 water is "hard"
Calcium	60 ppm	100 ppm	Binds with tetracyclines, precipitates in water systems leading to scale build-up, bacterial build-up; high levels alter nutrition

Contaminant	Average level	Maximum level	Remarks
Magnesium	14 ppm	125 ppm	Wet droppings, interfere with nutrient absorption especially in presence of high sulfates
Chloride	14 ppm	250 ppm	If Na is high, low chloride is detrimental; high NaCl may reduce performance and reduce shell quality
Sodium	32 ppm	50 pm	If sulfate or chloride is high, performance affected
Sulfate	125 ppm	250 ppm	Laxative effect with magnesium, fast bleeding and edema
Iron	0.2 ppm	0.3 ppm	Precipitation clogs water systems
Lead	---	0.02 ppm	Kidney failure, nervous disorder
Cadmium	---	0.01 ppm	Kidney failure
Cfu = colony forming units			*Source* : Poultry year book, 2003-04

2.3. Water consumption

Table 53.18 Water intake by chicken and turkeys (ml/bird/week, 20-25°C) *

Age (weeks)	Broilers	Layers		Large white turkeys	
		White egg	Brown egg	Males	Females
1	225	200	200	385	385
2	480	300	400	750	690
3	725	na		1135	930
4	1000	500	700	1650	1274
5	1250	na		2240	1750
6	1500	700	800	2870	2150
7	1750	na		3460	2640
8	2000	800	900	4020	3180
9		na		4670	3900
10		900	1000	5345	4400
11		na		5850	4620
12		1000	1100	6220	4660
13		na		6480	4680
14		1100	1100	6680	4700
15		na		6800	4720
16		1200	1200	6920	4740
17				6960	4760
18		1300	1300	7000	na
19				7020	na
20		1600	1500	7040	na

* Varies considerably depending on temperature and diet composition

na = data not available *Source* : NRC, 1994

3. Growth and feed consumption (NRC, 1994)

3.1. Chicken

3.1.1. Broilers

Table 53.19. Growth and feed consumption by broilers *

Age (weeks)	Body weight (g)		Weekly feed intake (g)		Cumulative feed intake (g)	
	Male	Female	Male	Female	Male	Female
1	152	144	135	131		
2	376	344	290	273	425	404
3	686	617	487	444	916	848
4	1085	965	704	642	1616	1490
5	1576	1344	960	738	2576	2228
6	2088	1741	1141	1001	3717	3229
7	2590	2134	1281	1081	4998	4310
8	3077	2506	1432	1165	6430	5475
9	3551	2842	1577	1246	8007	6721

* Diet containing 3200 kcal/kg ME

3.1.2. Layers

Table 53.20. Body weight (g) and feed intake (g/week) of immature layers

Age (weeks)	White egg-laying strains		Brown egg-laying strains	
	Body weight	Feed intake	Body weight	Feed intake
0	35	50	37	70
2	100	140	120	160
4	260	260	325	280
6	450	340	500	350
8	660	360	750	380
10	750	380	900	400
12	980	400	1100	420
14	1100	420	1240	450
16	1220	430	1380	470
18	1375	450	1500	500
20	1475	500	1600	550

3.2. Turkeys

3.2.1. Pre-breeding / Pre-lay

Table 53.21. Growth and feed consumption by turkeys *

Age (weeks)	Body weight (kg)		Weekly feed intake (kg)		Cumulative feed intake (kg)	
	Male	Female	Male	Female	Male	Female
1	0.12		0.10			
2	0.25	0.24	0.19	0.18	0.29	0.28
3	0.50	0.46	0.37	0.34	0.66	0.62
4	1.0	0.9	0.70	0.59	1.36	1.21
5	1.6	1.4	0.85	0.64	2.21	1.85
6	2.2	1.8	1.10	0.80	3.31	2.65
7	3.1	2.3	1.40	0.98	4.71	3.63
8	4.0	3.0	1.73	1.21	6.44	4.84
9	5.0	3.7	2.0	1.42	8.44	6.26
10	6.0	4.4	2.34	1.70	10.78	7.96
11	7.1	5.2	2.67	1.98	13.45	9.94
12	8.2	6.0	2.99	2.18	16.44	12.12
13	9.3	6.8	3.20	2.44	19.64	14.56
14	10.5	7.5	3.47	2.69	23.11	17.25
15	11.5	8.3	3.73	2.81	26.84	20.06
16	12.6	8.9	3.97	3.00	30.81	23.06
17	13.5	9.6	4.08	3.14	34.89	26.20
18	14.4	10.2	4.30	3.18	39.19	29.38
19	15.2	10.9	4.52	3.31	43.71	32.69
20	16.1	11.5	4.74	3.40	48.45	36.09
21	17.0	*	4.81	*	53.26	*
22	17.9	*	5.00	*	58.26	*
23	18.6	*	5.15	*	63.41	*
24	19.4	*	5.28	*	68.69	*

* Females not marketed after 20 weeks of age

3.2.2. Breeding turkeys

Table 53.22. Body weight and feed intake of large-type turkeys

Age (weeks)	Females			Males	
	Weight (kg)	Egg production, (%)	Feed intake (g/d)	Weight (kg)	Feed intake (g/d)
20	8.4	0	260	14.3	500
25	9.8	0	320	16.4	570
30	11.1	0 *	310	19.1	630
35	11.1	68	280	20.7	620
40	10.8	64	280	21.8	570
45	10.5	58	280	22.5	550
50	10.5	52	290	23.2	560
55	10.5	45	290	23.9	570
60	10.6	38	290	24.5	580

* Light stimulation given at this stage

3.3. Ducks

Table 53.23. Growth and feed consumption by Pekin ducks

Age (weeks)	Body weight (kg)		Weekly feed intake (kg)		Cumulative feed intake (kg)	
	Male	Female	Male	Female	Male	Female
0	0.06					
1	0.27		0.22		0.22	
2	0.78	0.74	0.77	0.73	0.99	0.95
3	1.38	1.28	1.12	1.11	2.11	2.5
4	1.96	1.82	1.28	1.28	3.40	3.33
5	2.49	2.30	1.48	1.43	4.87	4.76
6	2.96	2.73	1.63	1.59	6.50	6.35
7	3.34	3.06	1.68	1.63	8.18	7.98
8	3.61	3.29	1.68	1.63	9.86	9.61

3.4. Geese

Table 53.24. Body weight and feed intake of Geese

Age (weeks)	Body weight (kg)	Feed intake (two-week period)	Cumulative feed intake (kg)
0	0.11		
2	0.82	0.96	
4	2.05	2.93	3.89
6	3.05	3.20	7.09
8	4.05	4.34	11.43
10	4.85	4.68	16.11

4. Toxicity of inorganic elements (NRC, 1994)

Table 53.25. Toxicity of inorganic elements and compounds

Element or Group	Species	Age	Compounds	Level (ppm)	Adverse effects
Al	Chicken	Immature	$AlCl_2$	500	Reduced growth
			$Al_2(SO_4)_3$	1000-2200	Reduced growth, rickets
		Mature	$Al_2(SO_4)_3$	3000	Reduced egg production
As	Chicken	Laying hen	As_2O_5	100	Reduced growth and egg production
Ba	Chicken	Immature	$BaCO_3$, $BaCl_2$	200-2000	Reduced growth, mortality
Br	Chicken	Immature	NaBr	5000	Reduced growth
Cd	Chicken	Immature	$CdSO_4$	25-40	Reduced growth
		Adults	$CdSO_4.H_2O$	12	Reduced egg production
	Turkeys	Immature	$CdCl_2$	20	Reduced growth
Cl_2	Chicken	Immature	Arginine HCl, NaCl, KCl	15000	Reduced growth
Cr	Chicken	Immature	$K_2Cr_2O_7$, $Cr_2(SO_4)_3$	300	Reduced growth
		Adult	$CrCl_3.6H_2O$	10	Egg quality reduced
Co	Chicken	Immature	$CoCl_2.6H_2O$	100-200	Reduced growth
Cu	Chicken	Immature	$CuSO_4.5H_2O$	250-800	Reduced growth, gizzard erosion, muscular dystrophy
	Turkeys	Immature		50-800	Reduced growth
Fl	Chicken	Immature	NaF	500-1300	Reduced growth, reproductive performance
I_2	Chicken	Laying hen	KI	625	Reduced egg production, size and hatchability
Fe	Chicken	Immature	$Fe_2(SO4)_3$	4500	Rickets
Pb	Chicken	Immature	Lead acetate	320-1000	Reduced growth, and activity; mortality
		Mature		200	Reduced egg production
	J. quail	Mature		10	Reduced egg production
Mg	Chicken	Immature	MgO $MgCO_3$	5700 6000-6400	Reduced growth, skeleton Reduced growth, mortality
		Adults	$MgSO_4$ $MgCO_3$	19600 11200	Reduced egg production
Mn	Chicken	Immature	$MnCl_2.4H_2O$	4000	Reduced growth
	Turkeys	Immature	$MnSO_4.H_2O$	4800	Reduced growth
Hg	Chicken	Immature	$HgSO_4$ $HgCl_2$ CH_3Hgdicyanamide CH_3HgCl	400 250-400 33 5	Reduced growth Reduced growth, mortality Reduced growth, mortality 50% mortality

Element or Group	Species	Age	Compounds	Level (ppm)	Adverse effects
Mo	Chicken	Immature	Na_2MoO_4 $Na_2MoO_4.2H_2O$	500 350	Reduced growth, mortality Reduced growth
		Laying hen	$Na_2MoO_4.2H_2O$	500	Reduced egg production and hatchability
	Turkeys	Immature	$NaMoO_4$	300	Reduced growth
Ni	Chicken	Immature	$NiSO_4$ or Ni acetate NiCl	500 400	Reduced growth Reduced growth
NO_3^-	Turkeys	Immature	$NaNO_3$	900	Reduced growth, mortality
NO_2^-	Chicken	Immature	KNO_2	658	Reduced vitamin A in liver, thyroid enlargement
Se	Chicken	Immature	Na_2SeO_3 + Se in wheat	10-20	Reduced growth
		Laying hen		5-10	Reduced hatchability
Ag	Chicken	Immature	$AgSO_4$ $AgNO_3$	200 900	Reduced growth Exudative diathesis
Na	Chicken	Immature	Na glutamate	8900	Reduced growth
		Laying hen	Na_2SO_4	12000	Reduced egg production
NaCl	Chicken	Immature		7000	Reduced growth, mortality
		Laying hen	Water	10000	Reduced egg production
			Diet with low Cl^-	40000-60000	Reduced egg production
	Turkeys	Immature	Water / Diet	4000-40000	Reduced growth, lung congestion, enlarged kidneys, pendulous crop, mortality
		Mature		60000	Reduced growth
	Ducks	Immature	Water	4000	Reduced growth
Sr	Chicken	Immature	$SrCO_3$	6000	Reduced growth
SO_4^{--}	Chicken	Immature	K_2SO_4, Na_2SO_4, $CaSO_4$	14000	Reduced growth
		Laying hen	Na_2SO_4	8100	Reduced egg production
W	Chicken	Immature	Sodium tungstate	500	Reduced growth
V	Chicken	Immature	NH_4VO_3, $Ca_3(VO_4)_2$, $VOSO_4$, $NaVO_3$	8-25, 30-200, 25 and 5, respectively	Reduced growth, mortality
		Laying hen	In $Ca_23(PO_4)_2$ NH_4VO_3	6 15-50	Reduced albumen quality Reduced albumen quality, growth, egg production and hatchability
Zn	Chicken	Immature	$ZnSO_4$	1500-3000	Reduced growth, Exudative diathesis, muscular dystrophy
			$ZnCO_3$ ZnO	1500 800-4000	Reduced growth Reduced growth, bone ash

5. Stability of vitamins

Table 53.26. Stability of vitamins and losses during storage

Vitamin	Stability factor						Losses (%/month)
	Moisture	Reduction	Oxidation	Minerals	Heat	light	
A	++	-	++	++	+	+	10
D	++	-	++	++	+	+	8
E	-	-	+	++	-	-	40 (Alcohol form)
K	+++	+	-	+++	+	++	17
Thiamin HCl	++	++	++	+	++	-	11
Riboflavin	-	+	-	-	-	+	3
Pyridoxine	-	-	-	+	-	++	4
B12	++-	++	+	+	+	++	1
Pantothenate	++	-	-	-	+	-	2
Folate	++	+	+	++	+	+	5
Biotin	-	-	++	-	-	-	4
Niacin	-	-	-	-	-	-	5
Ascorbic acid	++	-	+++	+++	-	+	30
Choline	+++	-	-	-	-	-	1

+ Moderately sensitive, ++ Sensitive, +++ Very sensitive, - Resistant   *Source* : Klasing, 1998

Chapter **54**

Feedstuffs for Poultry

Birds must get the nutrients they require from the food they eat daily. No single ingredient can supply of the requirements; but a proper combination of ingredients and other by-products and additives do. Hence, poultry diets are composed primarily of a mixture of several feedstuffs such as cereal grains, soyabean meal, animal by-product meals, fats, and vitamin and mineral premixes. This compounded ration can be offered in the form of mash (powder), crumbles or pellets. Such a balanced ration, together with water, provide the energy and nutrients that are essential for the bird's growth, reproduction, and health, namely proteins and amino acids, carbohydrates, fats, minerals, and vitamins.

Poultry diets also can include certain constituents not classified as nutrients, such as xanthophylls (the pigment that can impart desired color to poultry products), the "unidentified growth factors" claimed to be in some natural ingredients, and antimicrobial agents (benefits of which may include improvement of growth and efficiency of feed utilization).

Various ingredients used to formulate/balance poultry rations are classified as follows:

1. Energy sources

1.1. Cereal grains

These are the members of the family *Gramineae* whose seeds are used as food. These grains are rich in carbohydrates and hence, in energy. Most of the carbohydrates are concentrated in the endosperm and the dry matter content of cereal grains ranges between 80 and 90%. Cereals also contain proteins ranging between 9 and 14%, most of which is present in the embryo and aleuronic layer. Lipid content of cereals, on average, 2.5% and is highly unsaturated (rich in linoleic and oleic acids).

1.1.1. Maize

Zea mays is the most commonly used energy source in poultry feeds as it is easily available and highly digestible. It contains 8 to 11% (Average 9%) low quality crude protein (CP) and has 13.81 MJ (3300 kcal)/kg of ME. It has a very low quantity of fiber (2%). It can be used up to 60% levels in the feed. Yellow maize has abundant quantity of carotenoid pigments called xanthophylls which are also responsible for yellow pigmentation of yolk. It is also a fair source of Vitamin A activity, but 30% of it may be lost during storage. White maize is similar to yellow maize except in that it does not contain xanthophylls and vitamin A activity. High-lysine maize also has been developed and used in Western countries.

1.1.2. Barley

Energy and fiber content in barley (*hordeum sativum*) is three fourths and three times as much as in maize, respectively. Hence, its use in rations for young birds is very limited. Besides, it does not contain xanthophylls and vitamin A activity. Barley is a popular cereal available for poultry feeding in hot dry regions. It has an ME value of 12.5 to 13.2 MJ/kg dry matter with a fiber content of 5-6%, CP of 6-16% (average 9%) and a low fat content (2.5%). It also has an anti-nutritional compound, β-glucan, whose levels increase as the grain matures. Hence, addition of enzyme β-glucanase increases its nutritive value both for broilers and layers. It can be included at a maximum level of 20% above which both palatability and consumption of the feed are reduced. Barley can be fed to post-peak layers because, its inclusion is likely to increase feed required per dozen eggs.

1.1.3. Oats

Oats (*Avena sativa*) is richer in fiber than barley (has twice the amount of fiber as in barley) and has energy as much as in barley. This can therefore be used only in rations meant for adults. This also lacks xanthophylls and vitamin A of activity. Protein content ranges between 7 and 15% and it is of poor quality being very rich in glutamic acid. It is also rich in oil (6%).

1.1.4. Millets

Millets include several cereals of varying sizes and their feeding value is comparable to barley. CP content of millets is between 10 and 12%, ether extract 2-5% and crude fiber 2-9%. It has more linoleic acid and hence supports egg weight better than wheat. Pearl millet (Bajra) can be used up to 20% levels in rations for growing pullets and up to 30-32% levels in layer diets.

1.1.5. Rice

Although rice (*Oryza sativa*) can be used in poultry rations, it is a very important human food and hence its use in animal diets is not practicable in most countries.

1.1.6. Millets

The term "millet" applies to several species of cereals which produce small grains. They are popular in tropics. Popular millets are pearl (bulrush), proso (broomcorn), foxtail (Italian), finger (birdsfoot), kodo (ditch) and Japanese (barnard) varieties. The crude protein content ranges between 10 and 12%, fat between 2 and 5% and crude fiber between 2 and 9%. Nutritive value of millets is comparable to that of oats which is mainly due to indigestible fiber content.

1.1.7. Sorghums

Of several sorghum (*Sorghum bicolor*) grains, kafir and milo are usually incorporated in poultry rations. Sorghums contain more protein but lesser amounts of fat than maize. They also do not contain xanthophylls and vitamin A activity. They also contain tannins which has a growth-depressing effect. However, addition of methionine or Choline or fat along with adequate grinding of grains has been found to alleviate growth depression.

1.1.8. Wheat

Wheat (*Triticum aestivum*) has energetic value equal to 90% of that in maize but is one of the very important human foods. It contains 8 to 14% CP. Besides being gelatinous, if ground wheat is used in high proportions, it has a tendency to paste on the beaks which may sometimes lead to beak necrosis. In addition, a doughy mass may accumulate in the crop. Wheat also lacks xanthophylls and vitamin A activity.

The ME level in wheat is generally more variable than in corn. Wheat-based diets require less protein supplementation than corn-based diets. Broiler mortality may be higher with wheat-based diets. Wheat generally gives a firmer pellet during feed manufacture. Wheat should be coarse-ground for use in mash diets. Wheat contains a low level of pigmenting agents for yolk and skin color The body fat in broilers fed wheat-based diets is firmer.

1.1.9. Triticale

This cereal is a hybrid between wheat and rye; the latter being poorly digestible by poultry. The nutritive value of triticale is comparable with that of wheat and hence, it can replace wheat in poultry diets. This crop is usually grown in temperate climates.

1.2. Cereal by-products

1.2.1. Maize

During manufacture of starch and glucose from maize, germ, bran and gluten are produced; the former is very valuable as human food. Maize gluten (prairie) meal is a very rich protein (70%) and pigment source in broiler diets wherever it is available. Maize gluten also has about 8% fiber and 9-12.5 MJ/kg ME.

Sometimes, all the three by-products are mixed and sold as maize gluten feed; it contains 20-25% protein.

During dry corn milling process, hominy feed, corn bran, corn germ cake and corn germ meal are produced; of these, hominy feed, a mixture of corn bran, corn germ and starchy portion of kernels, is available in some places for poultry feeding. It contains fat 4.3 to 7.8% and fiber 3.4 to 5.1%. Hominy feed has an ME content of 2628 to 3366 kcal/kg depending on fat content. It is rich in linoleic acid; high fat hominy feed can be used up to 15% levels in poultry diets.

1.2.2. Barley

Many by-products of barley like malt culms, brewer's grains, spend hops and brewer's yeast are obtained during brewing. However, these by-products are not commonly available in all the countries; of these products, dried brewer's yeast which is rich and protein and B-complex vitamins is used in poultry diets wherever available.

By-products produced during distilling barley in water are not popular in poultry diets.

1.2.3. Oats

During preparation of oat meal for human consumption, several by-products are obtained, principally hulls, dust and meal seeds. None of these by-products is a popular ingredient in poultry diets.

1.2.4. Rice

Rice bran and rice polish are the two important by-products. Rice bran contains high amounts of fiber and therefore can be used in rations for adults. Rice polish has 12-14.5% crude protein and 11-18% highly unsaturated oil; this is a very valuable product used in poultry feeding. Non-deoiled rice polish is rich in energy containing up to 12.56 MJ (3000 kcal)/kg of ME, 13 to 18% fat and about 12% CP. Rice oil is highly unsaturated and hence, rice polish is usually deoiled to prevent rancidity. When deoiled, its energy value reduces but protein value increases to 14 to 18%. Rice polish is an excellent source of B-complex vitamins.

1.2.5. Wheat

The main by-products include middlings, thirds and bran; the latter is used in poultry diets. Wheat bran is the outer layer of wheat kernel. It has about 14% CP and 4.19 MJ (1000 kcal)/kg of ME. It is rich in both fiber and phosphorus and therefore not used in diets for young birds. Even for adults, it is not used for more than 10 to 15% levels.

1.3. Molasses

(Curtin, 1983)

Initially the term molasses referred specifically to the final effluent obtained in the preparation of sucrose by repeated evaporation, crystallization and centrifugation of juices from sugar cane and from sugar beets. Today, several types of molasses are recognized and in general, any liquid feed ingredient that contains in excess of 43% sugars is termed molasses. The extent to which molasses has been used in animal feeds varies from a small amount used to eliminate dust and feed wastage to serving as the major source of dietary energy.

Molasses is the only inexpensive and available energy source that can be used in livestock and poultry production. In addition to energy, molasses products also provide other advantages in rations like:

1. Increases the palatability of many types of rations
2. Energy in form of simple sugars is easily digestible
3. Molasses at times appears to exert a tonic effect
4. In many feeds it eliminates dust.

1.3.1. Types of molasses

1.3.1.1. Cane Molasses

Cane Molasses is a by-product of the manufacture or refining of sucrose from sugar cane. It must not contain less than 46% total sugars expressed as invert.

1.3.1.2. Beet Molasses

Beet Molasses is a by-product of the manufacture of sucrose from sugar beets. It must contain not less than 48% total sugars expressed as invert.

1.3.1.3. Citrus Molasses

This is the partially dehydrated juices obtained from the manufacture of dried citrus pulp. It must contain not less than 45% total sugars expressed as invert.

1.3.1.4. Hemicellulose extract

This is a by-product of the manufacture of pressed wood. It is the concentrated soluble material obtained from the treatment of wood at elevated temperature and pressure without use of acids, alkalis, or salts. It contains pentose and hexose sugars, and has a total carbohydrate content of not less than 55%.

1.3.1.5. Starch Molasses

It is a by-product of dextrose manufacture from starch derived from corn or grain sorghums where the starch is hydrolyzed by enzymes and/or acid. It must contain not less than 43% reducing sugars expressed as dextrose and not less than 50% total sugars expressed as dextrose. It shall contain not less than 73% total solids.

1.3.2. Composition

The composition and selected nutrient content of the various types of molasses is highly variable and average values are given below:

Table 54.1. Composition and nutrient content of molasses products

Item	Cane	Beet	Citrus	Extract	Starch
Total Solids (%)	75.0	77.0	65.0	65.0	73.0
Specific Gravity	1.41	0.41	1.36	1.32	1.40
Total Sugars (%)	46.0	48.0	45.0	55.0	50.0
Crude Protein (%)	3.0	6.0	4.0	0.5	0.4
Nitrogen Free Extract (%)	63.0	62.0	55.0	55.0	65.0
Ash (%)	8.1	8.7	6.0	5.0	6.0
Calcium, (%)	0.8	0.2	1.3	0.8	0.1
Phosphorus, (%)	0.08	0.03	0.15	0.05	0.2
Potassium, (%)	2.4	4.7	0.1	0.04	0.02
Sodium, (%)	0.2	1.0	0.3	---	2.5
Chlorine, (%)	1.4	0.9	0.07	---	3.0
Sulfur, (%)	0.5	0.5	0.17	---	0.05
Energy (kcal/kg)	1962	1962	---	---	---

In general, in comparison to the commonly-used sources of dietary energy, mainly cereal grains, the calcium content of cane and citrus molasses is high, whereas the phosphorus content is low. Cane and beet molasses are comparatively high in potassium, magnesium, sodium, chlorine and sulfur.

The trace mineral and vitamin contents of cane, beet and citrus molasses is presented in Table 54.2.

Cane and beet molasses are commonly used in poultry rations; they are rich in potassium and therefore, when used at more than 5 to 6% level in poultry feeds, it can cause diarrhea.

1.4. Fats and oils

These are excellent source of energy (29.3 MJ or 7000 kcal/kg of ME). Vegetable oils and animal fats can be used up to 8% levels in practical diets although birds can tolerate up to 16% fat in their feed. Birds utilize vegetable oils better than animal fats.

Table 54.2. Trace minerals and vitamins in molasses

	Cane	Beet	Citrus
Minerals, mg/kg			
Copper	36	13	30
Iron	249	117	400
Manganese	35	10	20
Zinc	13	40	---
Vitamins, mg/kg			
Biotin	0.36	0.46	---
Choline	745.0	716.0	---
Pantothenic Acid	21.0	7.0	10.0
Riboflavin	1.8	1.4	11.0
Thiamine	0.9	---	---

2. Protein sources

2.1. Animal protein controversy

(Pearl, 2002)

The world today is nothing more than a "global village" in which a food safety incidence anywhere is an incident everywhere. However, there are no scientific reasons that animal protein ingredients are not safe feed ingredients for poultry rations. During the 16-year-period following the advent of bovine spongiform encephalopathy (BSE) commonly referred to as "mad cow disease" drastically, perceptions of food safety have changed altogether. In any case, animal protein feedstuffs in poultry rations are being replaced more due to cost and quality rather than food safety considerations.

The advent of marketing programs that promote "No animal protein", "No animal byproducts used" and other natural/organic food claims, though opportunistic have promoted marketing concepts that infers a negative connotation to the use of animal-derived products. These have been evident in nearly every food animal species and especially so in poultry products both meat and eggs.

The feathers, blood, bone, viscera, skin, fat, other trim tissues forming about 30-40% of the live weight also must be utilized or discarded. The traditional utilization has been via rendering these tissues into valuable protein and fat ingredients. Other alternatives to rendering exist such as burial, burning, incineration, landfilling, composting or extruding. But when compared to rendering all are very unacceptable either due to human and animal health, environmental, ecological or economic challenges.

The future of animal protein in poultry diets in reality lies within the decisions made by the industry itself and the consumers of its products. There are no scientifically-based reasons for the exclusion of animal protein ingredients from poultry diets.

2.2. Animal protein

2.2.1. Fish meal

(Source: Fishmeal information network on internet)

Fishmeal is the brown flour obtained after cooking, pressing, drying and milling fresh raw fish. It is produced almost exclusively from fish for which there is no human food demand – the so-called industrial species. It is a primary product – not a by-product – and is manufactured in purpose-built plants that meet stringent safety and quality criteria. There are extensive controls and checks throughout the supply chain to ensure the quality, safety and integrity of fishmeal. However, diets containing fishmeal and fish oil must be stabilized with antioxidant and supplemented with vitamin E.

Benefits of incorporating fish meal in poultry diets are as follows:

2.2.1.1. Nutritional benefits

1. Fishmeal is a natural balanced feed ingredient that is high in protein, energy, minerals (calcium and phosphorus), a natural source of vitamins (including choline, biotin and vitamin B_{12}, A and E) and the micronutrients – selenium and iodine. It is the compositional quality of the nutrients in fishmeal that make it distinctive, in particular its content of essential amino acids and the very long chain polyunsaturated ω-3 fatty acids. Fishmeal is a rich source of the essential amino acids required for animal growth and maintenance, especially lysine, methionine, threonine and tryptophan. These are present in a readily digestible form. Lysine is often the first limiting amino acid when poultry diets are formulated from cereals and vegetable proteins.
2. More important than the amount of ω-3 fatty acids, is its ratio to omega-6 fatty acids. A desirable ratio would be about 4:1 of ω-6 to ω-3. The ratio of ω-6: ω-3 in fishmeal varies from 8:1 to 11:1; i.e. they contain about 10 times as much ω-3 as ω-6. This means the inclusion of fishmeal into the diet can produce the overall desired ratio of 4:1.
3. It has long been recognized that the antigenicity of fish protein is low and coupled with the anti-inflammatory properties of fishmeal, its inclusion in chick diets has been shown to improve disease-resistance in poultry.
4. Chicks challenged with salmonella, coccidiosis etc. showed a lower adverse effect on growth and reduced gut lesions when fish oil was included in the diet. Incidence of ascites in broilers was also reduced with fish oil in their diet.
5. The bone development of chicks has also been reported to have been improved following the inclusion of fish oil to improve the ratio of ω-6 to ω-3 fatty acids in the diet.
6. An increase in fertility of cockerels has been reported when salmon oil was fed; fertility increased from 91% to 96%.
7. Fishmeal is also a natural source of retinol (vitamin A) and tocopherols (vitamin E). Both of these vitamins act as antioxidants, as does ubiquinone which has a similar structure to vitamin E and is also found in fish. All these substances can help to maintain and protect cell membranes and increased levels of the protective antioxidant enzymes have been reported in the liver of animals-fed fish oil.

2.2.1.2. Welfare benefits

1. There was a positive effect on mortality and weight gain
2. condemnation of carcasses from blisters, inflammatory processes and cellulitis were significantly lower in the fishmeal-fed birds.

2.2.1.3. Public health benefits

1. Feeding fishmeal or fish oil has been shown to increase the PUFA content (ω-3 fatty acids, in particular) of poultry tissue without adversely affecting the eating quality.
2. Increasing the ω-3 content of eggs has been of particular interest, as the increase reported can make a significant contribution to the additional ω-3 fatty acids required in the human diet

2.2.1.4. Environmental benefits

Being highly digestible, protein excreted in excreta is reduced and hence, less pollution of air and water.

Two grades of fish meal are available – Grade A with 60% protein and Grade B with 50% protein. Fish meal contains 7.53 MJ (1800 kcal)/kg ME. It is also a good source of Vitamin B_{12}, calcium and phosphorus. It can be used up to a maximum of 10 to 12% and 5 to 6% in chick and layer rations, respectively; otherwise, meat and eggs will have fishy odor. Fish meal is likely to be adulterated with urea and quite often, the salt content will be too high causing problems of salt toxicity. Fish meal should not contain more than 3% salt.

Table 54.3. Rates of inclusion of fishmeal for optimum benefit

Poultry	% inclusion
Chick rearing	up to 3
Broiler	2–5
Layer	2
Breeder	1–5
Turkey	3–10
Pheasant/game	3–7
Bulk density: 550–650kg/m³	

2.2.2. Blood meal

Contains ground dried blood. Has 80% protein and is an excellent source of lysine of which about two-thirds is available to birds. But it is deficient in quality protein and only small amounts can be added in poultry diets.

2.2.3. Meat scrap

This has 50 to 55% protein, can be used between 5 and 10% levels since its amino-acid value is limited.

2.2.4. Meat-cum-bone meal

This has more calcium and phosphorus than meat scrap and this also can be used up to 5 to 10% levels and rations.

2.2.5. Poultry by-product meal

It consists of ground, dry-rendered poultry offal (excluding feathers). It can completely replace fish meal but its availability is limited.

2.2.6. Feather meal

This contains over 80% protein (keratin). Although it is rich in cystine, it is poor in methionine, tryptophan and lysine. Further, it must be prepared from hydrolyzed poultry feathers. It can be used up to 1 to 2% levels in poultry nations.

2.3. Proteins of vegetable origin

Most of the vegetable protein sources contain reducing carbohydrates which reacts with amino group of free amino acids or those hanging on a protein chain; especially, ε-amino group of lysine rendering the amino acid non-available for birds. This reaction popularly called "Maillard" or "non-enzymatic browning" reaction severely reduces biological value of vegetable proteins. Hence, extreme precautions have to be taken to minimize such a loss and biological value during the processing of vegetable proteins. Alternatively, supplementation of diets with synthetic lysine almost always produces a favorable response in the form of improved growth, production and reproduction. Addition of synthetic methionine, tryptophan, cystine etc. will further help improve the performance of birds.

In addition, many of the vegetable protein sources do contain performance-limiting (toxic) factors. However, most of those factors are inactivated during the processing of the raw material. Most of the so called "toxic factors" are Nature's way of protecting the plant seeds and hence, it is almost always not possible to eliminate them from such feedstuffs. It is, therefore, not advisable to feed any of the vegetable protein sources in raw state to poultry.

2.3.1. Peanut / Ground nut cake (GNC)

This is the most important vegetable protein source in our country. It has 45 to 52% protein and 9.63 MJ (2300 kcal)/kg ME. It can be used up to 30 to 40% levels in rations. The protein is limiting with respect to lysine, cystine and methionine. It is available in two forms - non-deoiled and deoiled; the former will have higher amounts of fat and energy whereas the latter contains higher protein. It is susceptible for aflatoxins.

2.3.2. Soyabean meal/cake

It has higher nutritive value than GNC but it is not available abundantly in our country. It can completely replace GNC. It contains 43 to 50% protein. Raw soyaeans should not be fed to poultry since they contain trypsin, chymotrypsin and other protease inhibitors. Most of these anti-nutritional factors are lectins. It is a poor source of B-complex vitamins and its oil has laxative effect; the latter leading to soft body fat. Soyabean meal contains about 0.1% genestein, a compound having estrogenic properties; however, its residues in poultry products appear to pose no risk for human beings.

2.3.3. Mustard oil cake

This is satisfactory source of both methionine and lysine, but, it can be used more than 10% levels in layer rations; otherwise, it causes mottling of yolks. It contains 35 to 38% protein and can be used to replace 50% of GNC. It also contains Senigrin, a sulfur-containing glucoside which when acted upon by endogenous myosinase releases a goitrogen called allyl isothiocyanate (a volatile compound). Senigrin is also present in Rapeseeds, and other seeds belonging to *Brassicaceae* family. It can be detoxified by $FeSO_4.7H_20$.

2.3.4. Cotton seed cake

Contains 40 to 42% protein. It contains a polyphenolic aldehyde (toxic factor) called gossypol (0.03 to 2.00%). If the gossypol content is less than 0.036%, it can completely replace GNC. Lysine is the limiting amino-acid followed by methionine and cystine. If used in poultry diets, it must be ensured that gossypol content is less than 0.01%. Toxicity of gossypol is characterized by edema of body cavities, degenerative changes in liver and spleen, hemorrhages in liver, hypoprothrombinemia, anemia (because gossypol binds Fe) and diarrhea.

Cyclopropene fatty acids (malvalic and sterculic) in cotton seed cake cause pink discoloration of egg white and enhance deposition of Stearic and Palmitic acids in depot fats and egg fats (yolk discoloration).

2.3.5. Sunflower seed meal

It contains around 42% protein. It can be used at a maximum of 15% since the meal is sticky and may cause necrosis of the beak at higher levels; besides, is also rich in fiber (22%) which limits its use in diets of young birds. It yields 8.1 MJ/kg ME and can be used up to 10% levels in grower and layer diets. Deficient in lysine and can replace 50% of soyabean meal. In eggs, yolk color score is reduced and cholesterol content increased when sunflower meal is incorporated in layer diets.

2.3.6. Guar meal

This drought-resistant legume has Trypsin inhibitor, mucilage and gums. It contains 45% CP, 4% fat, 6% fiber and 4.5% ash. Its protein is rich in lysine and methionine; but limited by the gums present in the legume. Pectinase and cellulose can digest the gums. The gums cause fecal stickiness and growth depression. Normally, it is not used more than 10 to 15% levels in poultry diets.

2.3.7. Sesame meal (Gingili, Til)

It has 47% protein. It is inadequate in lysine; otherwise, a good vegetable protein supplement. It can form 50% of the vegetable protein in poultry rations. It is rich in phytic acid limiting phosphorus availability, in particular, and those of Calcium and Zinc, in general.

2.3.8. Coconut meal

Coconut meal contains 2.5-6.5% oil which is highly unsaturated. But, the protein is limiting with respect of lysine and histidine; it is also rich in fiber (12%) and hence, it is not a popular ingredient in poultry diets. Rancidity of its fat is also of concern. With suitable amino-acid supplementation, it can be used to a maximum of 10% in grower diets.

2.3.9. Safflower meal

This protein source is limited by its high fiber content (31%) and low contents of lysine and methionine. However, wherever available, it can be used up to 15% levels in diets for growing and laying poultry.

2.3.10. Tamarind seeds, spent coffee/tea powder, mango seed kernel

All these contains tannin and hence producing poultry rations is very limited.

2.3.11. Neem seed cake

This contains toxic factor called nimbin and hence not much used in poultry rations.

2.3.12. Castor seed cake

It has a toxic factor called ricin and hence not generally used in poultry diets.

Table 54.4. Limiting amino-acids in selected protein sources

Ingredient	Limiting amino-acid			Suggested levels (%)
	# 1	# 2	# 3	
Soyabean meal	SAA	Threonine	Lysine	Maximum
Mustard seed meal	Lysine	SAA	Arginine	2 – 4
Peanut meal	Lysine	SAA	Threonine	8 - 12
Sunflower meal	Lysinc	SAA	Threonine	10 – 15
Sesame meal	Lysine	Tryptophan	Threonine	5 - 10
Coconut meal	Lysine	SAA	Threonine	5 - 7
Fish meal	SAA	Arginine	Threonine	5 - 7

SAA = Sulfur amino-acids *Source* : Indian Poultry Year Book 2003-04

3. Green leafy products

These are excellent sources of xanthophyll, carotene and unknown growth factors; few are highly Vitamin K.

3.1. Alfalfa meal

This contains 13 to 20% protein and 20 to 30%fiber. Usually used as pigment source at 3 to 4% levels.

3.2. Subabul meal (Ipil ipil leaf meal)

It contains a toxic amino-acid called mimosine (levels varying from 2 to 5%). Can be used to 10% levels. It has 24% CP, 3.25% fat, 14% fiber and 530 ppm β-carotene activity.

4. Other feedstuffs

4.1. Single-cell protein

Availability of this protein source is extremely limited and confined to few countries. Yeasts and bacteria grown on nuclear substrates constitute single-cell protein. Several microorganisms have been grown and used in livestock feeding; 0.2-0.5% in broiler diets and 1% in layer diets have been recommended. Even algae have been harvested and tried as a poultry feedstuff with variable success.

Single cell proteins are critical with respect to methionine and arginine, in that order. Further, palatability is reduced by use of this feedstuff; hence, maximum level of inclusion is 5 and 10% in broiler and layer diets, respectively.

4.2. Cassava root meal (Tapioca)

Tapioca, a popular crop in tropics and subtropics, is rich in digestible carbohydrates; but, it is limited by its linamarin content. However sweet cassava after drying is fit for poultry feeding and if drying process is properly regulated, it can replace ¾ of corn. Dried cassava has 1.8% CP, 1.8% fat, 85% nitrogen free extract, 1.8% fiber, 0.3% calcium, 0.35% phosphorus, 16 ppm manganese.

4.3. Palm kernel meal

Generally unsuitable for poultry feeding because of low palatability and high fiber content although its protein quality is good; however, with animal protein (fish meal) or corn oil in the diet, it can be fed to poultry.

4.4. Sal seed meal

Sal seed is available in certain tropical regions; it has 9.8% CP, 2.2% fat and 45% available carbohydrates; but also has tannins (11.7%) limiting its nutritional value. It can be incorporated to a limited extent in layer diets.

4.5. Bambara groundnut meal

This is an African leguminous plant containing trypsin inhibitors which can be destroyed by boiling for 30 min. Heat-treated meal has ME value of 3880 kcal/kg, 0.11% phosphorus, 36 ppm zinc. Raw beans are not fit for poultry feeding.

4.6. Jojoba meal

This is a by-product of jojoba oil production in Africa. It has several toxins like glycosides, polyphenols, phytic acid, trypsin inhibitors and Simondsin; the latter related to growth depression in poultry. This ingredient can best be a distress feedstuff.

4.7. Distiller's dried grains with solubles (DDGS)

In the dry mill production of ethanol two products are produced – liquid solubles and grain residue. Each could be dried separately but are mixed together to form DDGS as a dry ingredient. DDGS as a feed ingredient has a moderate protein content and energy level similar to soyabean meal but, found to be limiting in tryptophan and arginine after lysine. In addition, its digestibility also is low.

An early use of DDGS in poultry diets was primarily as a source of unidentified factors that promote growth and hatchability. Distillers dried solubles (DDS) or DDGS, if available, can be used in diets at low levels of inclusion usually less than 10%.

Table 54.5. Anti-nutritional factors in poultry feedstuffs

Source	Anti-nutritional factor
Canola meal	Glucosinolates, Sinapine, pectins, oligosaccharides
Castor seed cake	Ricin
Coconut oil meal	Unidentified toxic factor?
Copra meal	Mannans, high fiber
Corn gluten meal	Mycotoxins, high xanthophylls
Cotton seed meal	1. Gossypol, Growth depression 2. Cyclopropene fatty acids, increased deposition of Stearic and palmitic acids in egg and depot fat 3. Tannins
Guar meal	Trypsin inhibitor, gums and mucilage
Jojoba meal	Glycosides, polyphenols, phytic acid, trypsin inhibitors and Simondsin
Linseed meal	1. Anti-pyridoxine factor and cyanogenic glycoside (Linamarin). 2. Rich in mucilage and hence, is almost indigestible by poultry; generally not used in poultry diets
Lupin meal	Quinolizidine alkaloids, pectins, saponin, oligosaccharides, high Mn
Mustard oil cake	Senigrin (glucoside)
Neem seed cake	Nimbin
Palm kernel meal	Galactomannans, high fiber
Peanut meal	Mycotoxins, tannins, oligosaccharides, lectins
Peas	Trypsin inhibitor activity, tannins, lipoxygenase
Rapeseed meal	1. Glucoside (Senigrin): on hydrolysis by myrosinase yield isothiocyanates and 5-vinyloxazolidine-2-3 thione, a goitrogen. 2. Tannic acid (3%): reduced protein digestion and absorption. 3. Erucic acid, sinapine, pectins, oligosaccharides

Source	Anti-nutritional factor
Raw Soyabeans	1. Trypsin and chymotrypsin inhibitors. 2. Lectins: produce pancreatic hypertrophy, growth depression and reduce ME of the diet 3. Hemaggutinatinin activity
Sal seed meal	Tannins
Sesame meal	Phytate, oxalate
Soyabean meal	Oligosaccharides, phytin, saponin
Subabul meal	Mimosine (amino-acid)
Sunflower meal	Chlorogenic acid, high fiber, Trypsin inhibitor activity, tannins, lipoxygenase
Tamarind seeds, spent coffee powder, mango seed kernels	Tannin
Tapioca (Cassava) meal	Linamarin
Wheat and barley germ	Trypsin inhibitor activity

5. Mineral sources

Table 54.6. Mineral sources in poultry feeds

Mineral source	% calcium	% phosphorus
Dicalcium phosphate (DCP)	23	18
Steamed bone meal	26	13
Limestone	38	--
Oyster shell (Shell grit)	36	--
Gypsum	22	--

5.1. Salt

Highly essential for normal absorption of nutrients (especially amino-acids) and normal electrolyte balance. Feed ingredients also supply some quantities of sodium and chlorine, but free salt is added between 0.25 and 0.50% in poultry rations.

6. Other feed constituents

Included in this category are Antibiotics, Arsenicals, Antioxidants (to prevent rancidity of fats and destruction of fat-soluble Vitamins; usually ethoxyquin and butylated hydroxy toluene (BHT) are used at 0.0125% level), Coccidiostats (to prevent coccidiosis disease. Several Coccidiostats are being used including Amprol (0.5 g/kg), Coyden, Bifuran etc., Amino-acids, Tranquillisers, Vitamin supplements (several commercial preparations are available; most commonly used is Vitablend AB_2D_3E at a rate of 250 ppm).

7. Rates of inclusion

Table 54.7. Rates of inclusion of common feed ingredients

Ingredients	Chicks	Growers and Layers
Energy sources		
Maize	0-50	0-50
Molasses	0-5	0-5
Rice polish (deoiled)	0-30	0-30
Rice polish (fresh)	0-50	0 50
Tapioca meal	0-10	0-10
Vegetable protein sources		
Groundnut cake (expeller/deoiled)	0-50	0-40
Cotton seed cake	0-5	0-5
maize gluten meal	0-10	0-10
Mustard oil cake	0-5	0-10
Safflower cake	0-15	0-15
Sesame seed meal	0-10	0-10
Soyabean meal	0-30	0-30
Sunflower cake	0-10	0-10
Guar meal	0-6	0-6
Animal protein sources		
Fish meal	3-15	3-10
Blood meal	0-3	0-3
Feather meal (hydrolyzed)	0-5	0-5
Meat meal	0-7	0-7
Miscellaneous ingredients		
Alfalfa meal	0-3	0-5
Bone meal	0 2	0-2
Limestone	0-2	2-6
Salt	0-0.5	0-0.5

8. Composition of feedstuffs

Composition of various feed ingredients is available in literature and significant variation in composition is all but expected because several factors like soil, soil maintenance, crop program etc. influence it. It is further compounded by the fact that the analytical procedures themselves do contribute differences in the final composition of an ingredient.

Notwithstanding the above, composition of commonly-used feed ingredients for poultry, especially in India, is given in the tables 54.8 to 54.11; the values are as per the report of National Research Council (1994):

9. GMO crops in poultry feeds

(Thomison and Loux, 2001)

As market restrictions for various transgenic (genetically-modified organism or GMO) crops (e.g., Bt-corn, Roundup Ready soyabeans and corn) continue, there is increasing interest among growers in determining the presence of GMOs in crops. Growers producing non-GMO grains for specialtiy markets need to verify that there is no GMO contamination, or that contamination levels meet tolerance levels established by an end-user. The default standard for certification as GMO free has been taken to be zero in many cases, although experience shows that meeting such a standard will be difficult. There have been proposals for setting maximum allowable levels in the range of 1 to 3%, and it is likely that some tolerance level above zero will be accepted in the future.

Japan recently established new legislation that sets a zero tolerance for seed and food imports containing unapproved biotech material. The European Union (EU) has recently proposed rules on the labeling and traceability of foods containing GMOs. According to these new rules, accidental traces of GMOs that have been cleared by the EU's scientific advisers, even if they have not received final official approval, will be allowed in food and feed up to a maximum of 1%, without being subject to labeling requirements. Tolerances similar to those of Japan and the EU are being considered by other countries importing U.S. grains.

Table 54.8. Composition of feed ingredients – Gross, on as-fed basis

Ingredient	Number*	Dry matter (%)	ME (kcal/kg)	TME (kcal/kg)	CP (%)	EE (%)	Linoleic acid (%)	CF (%)
Alfalfa meal, dehydrated	1-00-023	92	1200	1011	17.5	2.5	0.47	24.1
Barley, grain	4-00-549	89	2640	2900	11.0	1.8	0.83	5.5
Blood meal, vat dried	5-00-380	94	2830	--	81.1	1.6	--	0.5
Corn gluten meal	5-28-242	90	3720	3811	62.0	2.5	--	1.3
Corn, yellow, grain	4-02-935	89	3350	3470	8.5	3.8	2.20	2.2
Fish meal, mech. extract.	5-01-985	92	2580	--	64.2	5.0	0.20	1.0
Meat meal, rendered	5-00-385	92	2195	--	54.4	7.1	0.28	2.7
Meat meal, with bone	5-00-388	93	2150	2495	50.4	10.0	0.36	2.8
Millet, pearl, grain	4-03-118	90	2675	3367	14.0	4.3	0.84	3.0
Millet, proso, grain	4-03-120	90	2898	--	11.6	3.5	--	6.1
Oats, grain	4-03-309	89	2550	2625	11.4	4.2	1.47	10.8
Peanut kernels, expeller	5-03-649	90	2500	--	42.0	7.3	1.43	12.0
Peanut meal, sol. extract.	5-03-650	92	2200	2462	50.7	1.2	0.24	10.0
Rice bran	4-03-928	91	2980	3085	12.9	13.0	3.57	11.4
Rice grain, brewer's	4-03-932	89	2990	3536	8.7	0.7	--	9.8
Rice polishing	4-03-943	90	3090	--	12.2	11.0	3.58	4.1

Ingredient	Number*	Dry matter (%)	ME (kcal/kg)	TME (kcal/kg)	CP (%)	EE (%)	Linoleic acid (%)	CF (%)
Rye, grain	4-04-047	88	2626	2931	12.1	1.5	--	2.2
Safflower seed meal	5-07-959	92	1921	--	43.0	1.3	--	13.5
Sesame seed meal	5-04-220	93	2210	1978	43.8	6.5	1.90	7.0
Sorghum, grain	4-20-983	87	3288	3376	8.8	2.9	1.13	2.3
Soyabean meal, sol. extr.	5-04-612	90	2440	2485	48.5	1.0	0.40	3.9
Sunflower meal, sol. extr.	5-04-739	93	2320	2060	45.4	2.9	1.59	12.2
Triticale, grain	4-20-362	90	3163	3144	14.0	1.5	--	4.0
Wheat, bran	4-05-190	89	1300	1725	15.7	3.0	1.70	11.0
Wheat, grain, hard red	4-05-268	87	2900	3167	14.1	2.5	0.59	3.0

* International feed number CP = Crude protein EE = Ether extract CF = Crude fiber
-- Data not available

Table 54.9. Composition of feed ingredients - Amino-acids (%), on as-fed basis

Number*	CP	Arg	Gly	Ser	His	Ile	Leu	Lys	Met	Cys	Phe	Tyr	Thr	Trp	Val
1-00-023	17.0	0.69	0.82	0.72	0.57	0.67	1.19	0.73	0.24	0.19	0.81	0.81	0.69	0.23	0.84
4-00-549	11.0	0.52	0.44	0.46	0.27	0.37	0.76	0.40	0.18	0.24	0.56	0.35	0.37	0.14	0.52
5-00-380	81.1	3.63	4.59	3.14	3.52	0.95	10.53	7.05	0.55	0.52	5.66	2.07	3.15	1.29	7.28
5-28-242	60.2	1.82	1.67	2.96	1.20	2.45	10.04	1.03	1.49	1.10	3.56	3.07	2.00	0.36	2.78
4-02-935	8.5	0.38	0.33	0.37	0.23	0.29	1.00	0.26	0.18	0.18	0.38	0.30	0.29	0.06	0.40
5-01-985	65.0	3.81	3.68	2.51	1.59	3.06	4.98	5.07	1.95	0.65	2.75	2.22	2.82	0.78	3.46
5-00-385	54.4	3.73	6.30	1.60	1.30	1.60	3.32	3.00	0.75	0.66	1.70	0.84	1.74	0.36	2.30
5-00-388	51.6	3.28	6.65	2.20	0.96	1.54	3.28	2.61	0.69	0.69	1.81	1.20	1.74	0.27	2.36
4-03-118	15.7	0.74	0.47	0.74	0.31	0.37	1.14	0.45	0.25	0.24	0.56	0.35	0.48	0.08	0.49
4-03-120	9.1	0.35	0.31	0.40	0.22	0.35	1.14	0.21	0.16	0.17	0.47	0.34	0.29	0.08	0.44
4-03-309	11.4	0.79	0.50	0.40	0.24	0.52	0.89	0.50	0.18	0.22	0.59	0.53	0.43	0.16	0.68
5-03-649	40.0	4.35	2.18	1.83	0.87	1.27	2.42	1.26	0.45	0.52	1.97	1.47	1.01	0.37	1.53
5-03-650	49.0	5.33	2.67	2.25	1.07	1.55	2.97	1.54	0.54	0.64	2.41	1.80	1.24	0.48	1.87
4-03-928	13.7	0.96	0.70	0.59	0.35	0.45	0.91	0.59	0.26	0.27	0.60	0.42	0.48	0.12	0.68
4-03-932	10.0	0.74	0.50	0.44	0.26	0.37	0.74	0.43	0.22	0.21	0.48	0.33	0.36	0.10	0.54
4-03-943	12.2	0.78	0.71	1.36	0.24	0.41	0.80	0.57	0.22	0.10	0.46	0.63	0.40	0.13	0.76
4-04-047	12.1	0.53	0.49	0.52	0.26	0.47	0.70	0.42	0.17	0.19	0.56	0.26	0.36	0.11	0.56
5-07-959	43.0	3.65	2.32	--	1.07	1.56	2.46	1.27	0.68	0.70	1.75	1.07	1.30	0.59	2.33
5-04-220	41.0	4.68	2.04	1.72	0.99	1.51	2.68	0.91	1.22	0.72	1.93	1.48	1.40	0.62	1.91
4-20-983	9.1	0.35	0.31	0.40	0.22	0.35	1.14	0.21	0.16	0.17	0.47	0.34	0.29	0.80	0.44
5-04-612	47.5	3.48	2.05	2.48	1.28	2.12	3.74	2.96	0.67	0.72	2.34	1.95	1.87	0.74	2.22
5-04-739	36.8	2.85	2.03	1.49	0.87	1.43	2.22	1.24	0.80	0.64	1.66	0.91	1.29	0.41	1.74
4-20-362	11.8	0.57	0.48	0.52	0.26	0.39	0.76	0.39	0.26	0.26	0.49	0.32	0.36	0.14	0.51
4-05-190	15.4	1.02	0.81	0.67	0.46	0.47	0.96	0.61	0.23	0.32	0.61	0.46	0.50	0.23	0.70
4-05-268	13.3	0.60	0.59	0.59	0.31	0.44	0.89	0.37	0.21	0.30	0.60	0.43	0.39	0.16	0.57

* International number; refer Table 54.7 for ingredient name -- Data not available

Table 54.10. Composition of feed ingredients – Inorganic elements, on as-fed basis

Number*	Ca	NpP	P	K	Cl	Fe	Mg	Mn	Na	S	Cu	Se	Zn
	%	%	%	%	%	ppm	%	ppm	%	%	ppm	ppm	ppm
1-00-023	1.44	0.22	0.22	2.15	0.47	480	0.36	30	0.09	0.17	10	0.34	24
4-00-549	0.03	0.36	0.17	0.48	0.15	78	0.14	18	0.04	0.15	10	0.10	30
5-00-380	0.55	0.42	--	0.18	0.27	2020	0.16	5	0.32	0.32	10	0.01	4
5-28-242	--	0.50	0.14	0.35	0.05	400	0.15	4	0.02	0.43	26	1.00	33
4-02-935	0.02	0.28	0.08	0.30	0.04	45	0.12	7	0.02	0.08	3	0.03	18
5-01-985	3.73	2.43	--	0.69	0.60	220	0.24	10	0.65	0.54	9	1.36	103
5-00-385	8.27	4.10	--	0.60	0.91	440	0.58	10	1.15	0.49	10	0.42	103
5-00-388	10.30	5.10	--	1.45	0.69	490	1.12	14	0.70	0.50	2	0.25	93
4-03-118	0.05	0.32	0.12	0.43	0.14	25	0.16	31	0.04	0.13	22	--	13
4-03-120	0.03	0.30	0.14	0.43	--	71	0.16	--	--	--	--	--	--
4-03-309	0.06	0.27	0.05	0.45	0.11	85	0.16	43	0.08	0.21	8	0.30	38
5-03-649	0.16	0.56	--	1.15	0.03	156	0.33	25	0.06	0.29	15	0.28	30
5-03-650	0.20	0.63	0.13	1.15	0.03	142	0.04	29	0.07	0.30	15	--	20
4-03-928	0.07	1.50	0.22	1.73	0.07	190	0.95	250	0.07	0.18	13	0.40	30
4-03-932	0.08	0.08	0.03	0.13	0.08	--	0.11	18	0.07	0.06	--	0.27	17
4-03-943	0.05	1.31	0.14	1.06	0.11	160	0.65	12	0.10	0.17	3	--	26
4-04-047	0.06	0.32	0.06	0.46	0.03	60	0.12	58	0.02	0.15	7	0.38	31
5-07-959	0.35	1.29	0.39	1.10	0.16	484	1.02	39	0.04	0.20	9	--	33
5-04-220	1.99	1.37	0.34	1.20	0.06	93	0.77	48	0.04	0.43	--	--	100
4-20-983	0.04	0.30	--	0.35	0.09	45	0.15	15	0.01	0.08	10	0.20	15
5-04-612	0.27	0.62	0.22	1.98	0.05	170	0.30	43	0.02	0.44	15	0.10	55
5-04-739	0.37	1.00	0.16	1.00	0.10	30	0.75	23	0.20	--	4	--	98
4-20-362	0.05	0.30	0.10	0.36	--	44	--	43	--	0.15	8	--	32
4-05-190	0.14	1.15	0.20	1.19	0.06	170	0.52	113	0.05	0.22	14	0.85	100
4-05-268	0.05	0.37	0.13	0.45	0.05	60	0.17	32	0.04	0.12	6	0.20	34

* International number; refer Table 54.7 for ingredient name NpP = Non-phytate P -- Data not available

Table 54.11. Composition of feed ingredients – Vitamins (ppm), on as-fed basis

Number*	Biotin	Choline	Folacin	Niacin	Pantothenic acid	B_6	B_2	B_1	B_{12}	E
1-00-023	0.30	1401	4.20	38	25.0	6.5	13.6	3.4	0.004	125
4-00-549	0.15	990	0.07	55	8.0	3.0	1.8	1.9	--	20
5-00-380	0.08	695	0.10	29	3.0	4.4	2.6	0.4	0.044	--
5-28-242	0.15	330	0.2	55	3.0	6.2	2.2	0.3	--	24
4-02-935	0.06	620	0.40	24	4.0	7.0	1.0	3.5	--	22
5-01-985	0.23	4408	0.20	100	15.0	4.0	7.1	0.1	0.352	4
5-00-385	0.17	2077	0.30	57	5.0	3.0	5.5	0.2	0.068	1
5-00-388	0.14	1996	0.30	46	4.1	12.8	4.4	0.8	0.070	1
4-03-118	--	793	--	53	7.8	--	1.6	6.7	--	--

Number*	Biotin	Choline	Folacin	Niacin	Pantothenic acid	B_6	B_2	B_1	B_{12}	E
4-03-120	--	440	--	23	11.0	--	3.8	7.3	--	--
4-03-309	0.27	946	0.30	12	7.8	1.0	1.1	6.0	--	20
5-03-649	0.33	1655	0.40	166	47.0	10.0	5.2	7.1	--	3
5-03-650	0.39	2396	0.40	170	53.0	10.0	11.0	5.7	--	3
4-03-928	0.42	1135	2.2	293	23.0	14.0	2.5	22.5	--	60
4-03-932	0.08	800	0.20	30	8.0	28.0	0.7	1.4	--	14
4-03-943	0.61	1237	0.20	520	47.0	--	1.8	19.8	--	90
4-04-047	0.06	419	0.60	19	8.0	2.6	1.6	3.6	--	15
5-07-959	1.67	3248	1.6	22	39.1	11.3	2.4	4.5	--	1
5-04-220	0.34	1536	--	30	6.0	12.5	3.6	2.8	--	--
4-20-983	0.26	668	0.20	41	12.4	5.2	1.3	3.0	--	7
5-04-612	0.32	2731	1.30	22	15.0	5.0	2.9	3.2	--	3
5-04-739	1.45	2894	--	220	24.0	16.0	4.7	3.1	--	--
4-20-362	--	462	--	--	--	--	0.4	--	--	--
4-05-190	0.48	1232	1.20	186	31.0	7.0	4.6	8.0	--	14
4-05-268	0.11	1090	0.40	48	9.9	3.4	1.4	4.5	--	13

* International number; refer Table 54.7 for ingredient name -- Data not available

There are several commonly-used GMO testing protocols, including biological tests, as well as ELISA and PCR tests, for herbicide and insect tolerance. Growers and end-users should consider the advantages and disadvantages of the various testing methods before harvest. Exporters should probably resign themselves to the most rigorous testing protocol, to anticipate the additional scrutiny their products will receive overseas. Some major end-users, i.e., large food processors, are currently using a combination of tests for identity-preserved (IP) grains.

9.1. Herbicide bioassay

Herbicide bioassays are used to detect GMO herbicide resistant traits in Roundup Ready and Liberty Link soyabeans and corn. The tests involve placing seeds in a germination media, moistened with a diluted solution containing the herbicide, or spraying the herbicide on seedlings. Seeds that test positive for the presence of the herbicide-tolerant GMO trait will germinate and develop normally, whereas those that die or do not develop normally will be GMO-free. This procedure is widely used by seed and grain companies exporting soyabeans.

9.1.1. Advantages

It is relatively inexpensive, user friendly, and produces straightforward results.

9.1.2. Disadvantages

It takes up to a week to complete, its use is limited to the Roundup Ready and Liberty Link herbicide-resistant GMO crops, and the seeds need to germinate for the test to work. Herbicide bioassays can also be used to detect herbicide-resistant traits in non-GMO corns, such as hybrids which are tolerant to imidazolinone herbicides.

9.2. ELISA (enzyme-linked immunosorbent assay)

ELISA tests for the presence of the specific protein that the genetically modified DNA produces in the plant. ELISA procedures use antibodies that react with specific proteins produced by the GMO. There are different versions of the ELISA method used for GMO detection. One version uses lateral flow strips and delivers results in two to five minutes.

9.2.1. Strip test

Strip tests are commonly used at grain elevators, where a rapid assessment to determine the presence or absence of GMOs is needed. These tests are referred to as the "dipstick" procedure by some companies marketing this ELISA technology for seed testing.

As the number of GMO traits increases, it will become more costly to monitor the presence of GMOs in crops, since each different gene requires a separate test. However, if the demand for non-GMO crops increases, it is possible that tests for different genes may be combined on the same ELISA test strip.

9.2.1.1. Advantages

It is quick, easy to use, and low cost.

9.2.1.2. Disadvantage

It cannot quantify how much GMO is present.

9.2.2. Plate test

This provides some indication of the quantity (percentage) of the tested sample that is the GMO in question. Intensity of color indicates the amount of the protein present. The plate test can take two to four hours and is more laborious, and costly than the strip test.

ELISA tests have limited application for testing GMOs in processed foods because heat processes denature the proteins, thereby making detection of proteins difficult.

9.3. The PCR (polymerase chain reaction) method

This is more sensitive than the ELISA method and tests for the presence of the specific DNA sequence of the gene itself. (See Chapter "Biotechnology").

9.3.1. Advantage

Sensitivity, i.e., detection of GMOs at very low levels. PCR is the only one of these methods that can effectively detect GMOs in processed foods.

9.3.2. Disadvantages

Length of time needed (two to three days), and cost. PCR tests also require more sophisticated equipment and greater expertise. While more sensitive to GMOs, PCRs in some cases tend to show false positives. Given the expense, time, and expertise required, PCR testing has limited potential in the field or at grain elevators.

Chapter **55**

Feeds and Feeding

1. Poultry feeds

Poultry can be fed on different types of feed (like Mash or Pellets) or allowed to choose and eat at their choice (Choice feeding). Each of these has its own advantages and disadvantages which are outlined below:

1.1. Mash feed

Mash quality is assessed by the size and uniformity of its particles. A positive correlation between the increase in feed particle size and broiler growth has been demonstrated.

Good uniformity of particle size is essential because birds prefer bigger particles. Thus the dominant birds will quickly eat those bigger cereal particles, while the rest of the birds will eat the finer particles. The improvement in performance with feed particle size and uniformity is explained by the lower energy output birds make when they ingest bigger particles. The number of pecks to eat the given amount of feed is reduced when particle size increases.

Being grain eaters, birds have a digestive tract designed to quickly ingest large amounts of feed, that are stored in the crop to be 'hydrated" and 'acidified' by lactic acid secretion before going through the proventriculus. In the proventriculus, hydrochloric acid and pepsin and mucus secretions are increased when feed particle size increases. The gizzard carries out feed grinding, feed impregnation and pre-digestion of the feed by the secretions from the proventriculus, as well as the regulation of feed in-flow and out-flow. This will have an effect upon three digestive flows: from gizzard to proventriculus; from jejunum to duodenum; from rectum to ceca. The intestinal peristaltic motility slows down the feed flow, allows better absorption of the nutrients by the intestinal viol, and helps to stabilize the intestinal flora (Trevidy, 2005).

In hot climates, fine mash is not recommended because of its adverse effect on ingestion. Mash should be coarse. The concentrate part of the feed (premix, minerals and proteins) may be presented as a crumble to reduce the amount of fine particles. When this is possible, it is of real interest to give a crumbled starter feed, made from an initially coarse mash (Trevidy, 2005).

1.1.1. Particle size

1.1.1.1. Layer diets

Feed that has been ground too fine can greatly reduce consumption. In hot climates a correct feed particle size will help to reduce the underfeeding observed in the summer months. That is why it is recommended that at least 80% of the particles should be between 0.5 and 3.2 mm. For laying hens it would appear that coarsely ground maize is probably the best manner in which they should be fed.

1.1.1.2. Broiler Diets

The ideal particle size that should be used in broiler diets is possibly more complex than the situation that exists in laying hens.

In the case of broilers the issue is more complex. It is true that more coarsely ground maize leads to improved nutrient utilization by the birds. This remains the case even after pelleting. If young birds, during the so-called starter phase, are being fed a mash diet, it would seem that the advantages of a larger particle size are outweighed by the disadvantages of selective eating. Hence, broilers prefer varying particle size depending on age.

1.2. Pelleted feed

Most feed for meat birds worldwide is either in the form of crumbled or whole pellets. The advantages of pelleting include (Dozier, 2005):

1. Reduced selective feeding
2. Increased nutrient availability
3. Decreased energy required for feed consumption
4. Reduced feed pathogen load
5. Increased bulk density (which lowers trucking cost)
6. Reduced shrinkage (because less material is lost as dust), and
7. Improved handling in automatic feeding equipment.
8. The above benefits can dramatically lower live production cost.

The positive effects of pelleting are well documented: higher feed density, no feed ingredient separation, better bacteriological quality, easier ingestion, improved growth and FCR. However, these may vary according to the quality of the raw materials, and that of the grinding and pelleting processes.

1.2.1. Pellet quality

If pellet quality is less than favorable, certain management techniques may rectify the problem without incurring large increases in feed manufacturing cost via capital expenditure and energy usage. However, these techniques vary in both effectiveness and cost.

Pellet quality is not easy to assess. Pellet binding must have good adhesion to reduce the production of fine particles during transportation, storage and distribution, yet they must not be too hard in order to avoid possible drops in consumption. Maize-based feed pelleting is more difficult than that of wheat based feed.

The two main physical indicators of pellet quality are:

1.2.1.1. Hardness

Hardness is measured by pellet resistance to breaking when submitted to external pressure. This indicates the physical integrity of the finished feed pellet in handling and transport with minimum generation of fines and broken pellets.

1.2.1.2. Durability

This is measured by the level of "fines" produced during transportation from the feed factory to the farm, and distribution in the feeding system at farm. Typically this is measured as percentage pellets or fines in the feed as fed, or by pellet durability index (PDI), which represents the percentage of pellets by weight that survive a standardized durability test.

1.2.2. Pellet quality and performance

1.2.2.1. Broilers

Pellet quality varies among the birds used for meat production. Pellet quality is especially important with ducks, where the mill may seek PDI of 96% to achieve optimum bird performance. The PDI target for turkeys is often 90% and the target for broilers is typically 80% (Dozier, 2005).

Broiler reaction to the above two quality criteria is not easy to assess. Pellets always produce better results when compared to the same fine mash as used to make a good pellet, even more so when the energy level is low. Hence, the major effect of pelleting appears to be through improvement of ingestion. However, a high-energy feed presented either as a coarse mash containing whole grain, or a medium-quality pellet because of its fat content, will give very similar results in growth, FCR and fat deposit. Unlike high-energy feed, low or medium energy feed is easy to pellet. When high-energy feed is presented as a coarse mash, or as free-choice grain and concentrate, the difference between the results is not as significant as when compared to pellet. Therefore, free-choice feeding may be the wiser choice to reduce the risk of metabolic diseases. (Trevidy, 2005).

Broilers may also show high sensitivity to pellet hardness. A hard pellet may be eaten less readily than a softer one, the latter being more likely to yield more fine particles. Still, surprisingly, cereal whole grain ingestion remains easy for the birds (Trevidy, 2005).

1.2.2.2. Turkeys

Compared with broilers, turkeys are more adversely affected as the proportion of fines increases in the feed. Pellet quality is critical with turkeys because the birds spend more time on feed, thus sub-optimum pellet quality causes more feed wastage.

1.2.2.3. Ducks

Duck performance is more sensitive to pellet quality than either turkeys or broilers. When a duck consumes mash feed, a sticky paste forms on its bill. The caking discourages optimum feed consumption and, when the duck washes its bill, feed wastage increases. Duck feed conversion could be improved by 2.8% by decreasing the amount of fines in the feed to 0%.

In general, Pellet quality influences performance to a larger degree in ducks than either turkeys or broilers. It is more important in turkeys than broilers because of the longer grow-out period and greater opportunity for waste due to poor quality pellets.

1.2.3. Pellet-quality control

Ingredients have a very significant effect on pellet quality. Certain types of starches and proteins have natural, inherent, binding capabilities. Ingredients' fiber, mineral and fat content also affect pellet quality. Wheat, barley and canola meal, for example, contain inherent binders that form physiochemical bonds during processing, resulting in higher pellet quality. However, corn and soyabean meal are low in inherent binders.

In any case, most diets for meat birds are formulated by least cost or optimal cost methods that generally do not consider ingredients' "pelletability". But other management techniques offer cost-effective means to improve pellet quality. For feeds with high coarse grain content, therefore, the focus should be on proper management of ingredient particle size, mash conditioning, pellet-die performance and fat.

1.2.3.1. Ingredients

Wheat and wheat co-products have inherent binding capacity and can improve pellet quality. However, manipulating a diet formula to improve pellet quality can reduce flexibility in least cost formulation and may prove costly on a long-term basis. To optimize pellet quality cost-effectively, a poultry feed manufacturer should be sure to put a priority on managing steam, fat and pellet dies.

1.2.3.2. Particle size

Grinding grains to a fine particle size enhances pellet quality. A smaller particle has a larger surface area that allows heat and moisture to penetrate more quickly to the core of the particle during conditioning. This larger area increases starch cell rupture and gelatinization. As a practical matter, however, optimum particle size for pellet durability in corn-soya poultry diets is probably in the range of 650-700 microns.

Table 55.1 American Feed Manufacturers' Association sieves and pore sizes

USDA sieve number	4	6	8	10	14	20	28	35	48	65	100	150	200	270
Pore size (μ)	4760	3360	2380	1680	1190	841	595	420	297	210	149	105	74	53
Relative pore size can be calculated by keeping pore size of screen 4 as 100														
Adopted from : Narahari, 2004														

1.2.3.3. Cold mash moisture

Moisture content of the pre-conditioned or cold feed mash has a linear effect on pellet quality. New moisture monitoring and control technology allow the moisture to be added in the mixer, which can be particularly beneficial when using low-moisture grain. Disadvantages of adding moisture in the mixer include increased weight per volume of feed, which would hurt transportation efficiency, poorer feed conversion on an as-fed basis.

1.2.3.4. Fat application

Adding more than 2% fat in the mixer causes a very rapid, perhaps exponential, decline in pellet quality. High concentrations of supplemental fat in the mash tend to reduce the friction between the feed, roll and die. This prevents the roll from pushing the feed through the die.

However, modern post-pelleting application system can add fat at the die at a high rate of accuracy (0.5%) without compromising pelletquality. Currently, there is a preference to add fat just before load-out, using modern pressurized spray and non-pressurized, rotating disk-coating systems. In such downstream liquid application systems there is less mess. Fat application may also be combined with the application of enzymes and other heat-sensitive micro-ingredients. Fat-at-the-die systems also have the problem of heat and moisture simultaneously escaping from the pellet.

1.2.3.5. Die maintenance

Maintaining optimum die performance is vital for manufacturing high quality pellets. Some of the common problems that affect pellet quality include die face wear, rollover, pitting and corrosion. These problems compromise pellet quality by decreasing die effective thickness and die-hole compression ratio.

1.3. Free-choice (Cafeteria) system

(Pousga *et al.* 2005)

Under free-choice feeding, birds are usually offered a choice between three types of feedstuffs: an energy source, a protein source, and in the case of laying hens, a calcium source. The realization that it is necessary for animals to be able to differentiate between foods with different nutrient compositions by color, taste and/or position, and that they need to be taught to associate the sensory properties of foods with their yields of nutrients, has made it possible to envisage a learned appetite for each of the essential nutrients. Birds can improve the balance between their nutrient requirements and their nutrient intake if they can be taught to select an appropriate diet. Protein is an expensive dietary constituent and there is more interest in optimizing its dietary concentration in commercial practice. Choice feeding of layers has financial advantages for rural small-scale poultry production.

Free-choice feeding is an infinitely more natural and delicate system of feeding. Each bird can accurately select its balance of nutrient to meet its particular physiological requirement. As long as the hens are given the opportunity to make clear, easy nutritional choices, they will be healthy and productive. They can indulge their natural appetites, can develop a normal digestive tract and a degree of natural resistance to coccidiosis. Changes in requirements, as birds grow, are also met by appropriate changes in selection.

1.3.1. Choice feeding of chickens

1.3.1.1. Factors in the feed determining feed selection by birds

The birds are selecting foods with a nutritional purpose. This is strengthened if their food intake over long periods is compared with the amount which is selected against when choice is given. Flavors can initially influence intake and preference but soon lose this ability as the birds learn that there is no nutritional implication in the different flavors.

It has been observed that when a new diet formulation is introduced, or a new type of food is presented, animals will often refuse to eat the food for a period of time or intake is reduced. Furthermore, it has been noted that this may also occur when a new batch of the same diet formulation is presented to broiler chickens or turkeys. If vitamin and mineral premixes are placed in separate feeders, some birds might not eat them because they will not like the taste, while other individuals may over-consume them and suffer toxic side effects.

1.3.1.2. Comparative choice-feeding studies on laying hens

Up to about 14 weeks of age growing pullets require protein only for feather development and a relatively slow rate of muscle deposition. However, around fifteen weeks of age, there is rapid development of the ovaries and oviduct, which would be expected to increase the protein demand, and thus protein intake, in a choice-feeding situation. A marked increase in the protein content of their diet about 2 weeks before the onset of lay in choice-fed pullets has been recorded. Once in full lay, the birds are capable of choosing a balanced diet when given the opportunity.

Laying hens are normally fed a mash diet. The use of whole grain would not only save the energy cost of grinding and mixing but is also accompanied by increased efficiency of food utilization. Further, grinding or pelleting wheat, barley, oats or maize may not result in a consistent increase in the metabolizable energy of poultry feed. Moreover, intake of whole grain is accompanied by increased efficiency of feed utilization. It has also been found that hens increase their calcium intake when ovulation was about to occur during the laying period and calcium appetite appeared to be a learned response; the laying hen has the ability to selectively consume calcium to meet maintenance and production requirements.

Hens offered the choice-fed diet are reported to lay heavier eggs and eggs with thicker shells and better Haugh unit scores than those fed the complete diet. The heavier egg size of the choice-fed hens accounted for the better food conversion ratio obtained by these hens compared to those fed the complete diet.

Hens are in many cases able to adapt their feed intake to their requirement in a quantitative (amount of feed ingested) and also qualitative (level of protein, energy, minerals etc.) way. However, both the scavenging and the confined birds on choice-feeding failed to eat sufficient amounts of the supplement to meet their protein requirements. On the other hand, the energy intake of both scavenging and confined hens under choice feeding was in excess of requirement (by 22-29%). Under choice-feeding, scavenging hens laid smaller eggs compared to the confined groups.

1.3.1.3. Choice-feeding of growers

During the rearing period (7-16 weeks) the pullets offered the choice-fed diets gained more weight than those fed the commercial grower diet. These pullets consumed approximately 7 g maize per d more than the control pullets, but had a lower total food intake. This suggests that the pullets given whole maize were able to utilize dietary energy more efficiently than those given ground maize. In addition, pullets fed the free-choice diets consumed more than twice the amount of limestone than those offered the complete diet

1.3.1.4. Choice feeding of broiler chickens

From the evidence available, laying hens seem less able than broilers to balance their protein intake when offered high-protein and low protein foods (mature birds learn more slowly than rapidly-growing ones). Broiler chickens when provided with 2 or 3 foodstuffs containing just one protein source, on a free-choice basis, which in some proportion will meet their requirements, effectively select a combination which maximizes their biological performance.

Under conditions of heat stress, with day-time temperatures rising to 33°C, broiler chickens will reduce their grain (energy) intake by 34% but their protein intake by only 7%, compared to similar birds in a cooler (20°C) environment.

Remarkably, choice-fed birds have been found to have a "protein memory" and consume early the next day, before the temperature rises, the protein they did not eat during the previous hot day. Thus, the performance of choice-fed broilers was very significantly better than that of the same birds fed the most sophisticated complete diets in hot environments.

1.3.1.4.1. Regulation of amino acid intake

Regulation of protein intake in choice-feeding situations appears to be only possible when the diet offered has adequate sulphur amino acids. However, growing birds appear to have the ability to compensate for short periods on amino acid imbalanced foods.

1.3.1.4.2. Lysine

Given a choice between L- and D-lysine, birds ate more of the L form. There is, therefore, some evidence of nutritional wisdom, but not sufficient to give a properly balanced diet.

1.3.1.4.3. Methionine

Selection for methionine also is not enough to prevent a decline in egg production under deficiency conditions.

1.3.1.5. Biological control of coccidiosis by choice-feeding

Evidence produced in a laboratory suggests that an active functioning gizzard may well play a role in the chickens' resistance against coccidiosis and that chickens free-choice fed on a high (42% CP) protein concentrate and whole wheat were even more resistant to coccidiosis than those on complete high-fiber diets. In male broiler chickens, occyst output has been shown to be negatively correlated to relative gizzard size, whether fed on conventional complete diets or free-choice fed. Finally, the feeding of insoluble grit has been shown to reduce oocyst output from chickens fed either complete diets or free-choice fed whole grains. Further work under field conditions is still required to draw practical conclusions.

1.3.1.6. Importance of behavior in choice-feeding

Broiler chickens take about ten days to learn to accurately balance their protein concentrate and whole grain intakes. They need to be in groups of at least eight birds and be offered the protein concentrate (in mash or crumble form) and whole grain in identical, adjacent troughs, or even better, in the same feed trough. Similarly, both growing and laying birds also need a period of learning before becoming proficient in the selection of feedstuffs.

Egg-type stock adapt quicker than broilers. Brown egg layers seem to adapt more readily than white or tinted egg layers. However, most strains of commercial layers and broilers can learn in ten to fourteen days to very accurately balance their energy and protein intakes to maximize production and optimize economic returns.

1.3.2. Choice-feeding of turkeys

Choice-fed turkeys consumed 10% less food, 44% less protein and the same amount of energy yet laid a similar number of eggs as those fed conventionally. Broodiness tended to be reduced by choice-feeding but fertility and hatchability were lower. There was no significant difference in food intake but choice-fed birds had a higher energy intake as they selected a higher proportion of the low protein food.

1.3.3. Choice-feeding of ducks

Ducklings prefer high protein to low protein feeds, resulting in excess protein intake, and higher protein conversion ratios. Therefore, choice feeding does not appear to be an economically viable system for growing meat ducks.

1.3.4. Feeding small and backyard stocks

Choice-feeding can be an easy way for small flock owners to feed their laying hens because, when hens are choice-fed, grinding the grain or mixing it accurately with a supplement is not necessary. Instead, separate feeders for the grain and supplement should be provided and the birds allowed choosing how much of each they want to eat. A third feeder full of one of calcium sources can be provided. Laying hens can make good nutritional choices and pick out the amounts of grain, supplement and limestone or oyster shell that they need to be healthy and productive. The following basic rules can be followed:

1. Do not give the hens too many choices. Hens can handle up to three choices quite well (grain, supplement and limestone or oyster shell). If more than one grain is used, such as wheat and barley, they should be mixed together in the same feeder.
2. Give the hens choices that are nutritionally distinct. For example, grain is high in starch and energy, the supplement is high in protein and vitamins and limestone is high in calcium. When provided such clear choices, the hens learn which feeders to go to and how much to eat to meet their basic nutritional needs.
3. Introduce the whole grain and choice-feeding a month before the onset of lay (at about 15 weeks of age for commercial layers and 24 weeks for village hens). This adjustment period will allow the birds time to learn how to choice-feed themselves before they are exposed to the nutritional demands of egg production. It will also allow the pullets the opportunity to increase their calcium consumption and build up the calcium reserves in their bones before they start to lay eggs. Finally, it takes the gizzard three weeks to build muscle mass to enable the hen to be able to efficiently grind the grain once egg production begins.

4. Vitamins or micro-minerals (e.g. copper, zinc etc.) should not be provided in a separate feeder, rather mixed with the supplement. If vitamins or micro-minerals are placed in a separate feeder, some birds may not eat them because they do not like the taste while other birds may over-consume them and suffer toxic side-effects.
5. Birds should be given adequate feeder space. For large backyard flocks, several feeders for each ingredient may be needed. For a one hundred-hen flock, two hanging feeders each of grain, supplement and limestone are suggested.
6. A supplement designed to be mixed with grains or grains and limestone to provide a complete laying hen diet should be used. A supplement formulated in this manner will contain a range of 25% to 40% CP. A grower supplement may be used prior to the start of egg production but a laying-hen supplement used once the birds begin laying eggs.
7. It is not necessary to grind grains when choice-feeding hens. The birds will readily eat whole wheat, whole oats or whole barley (but they can have difficulty eating whole maize). After about three weeks of eating whole grains, the hens' gizzards will increase in muscle mass and will grind the grain as efficiently as a hammer mill. Hens can successfully consume70% of their diet as whole grain when it is choice-fed. It is important to note that if the grains, supplements and limestone are provided in different feeders, these separation problems are avoided.

1.4. Mash *Vs* pellet

1.4.1. Viscosity of gut contents

Pellets made using fine grinding and high dye temperature increase the viscosity of the intestinal content. This effect is enhanced when saturated fat (palm oil) is added to the formula. In this case, the osmotic pressure of the intestinal content is increased so much that the resorption of liquids normally occurring in the intestine is blocked. Such effects are recorded in all classes of birds viz., broilers, growers, layers and breeders. Consequent on impaired Resorption of fluids, nutrient assimilation in the duodenum-jejunum loop is impaired, favoring bacteria proliferationin the lower digestive tract. This results in a disruptionof the microbial balance leading to development of coliforms, salmonella and clostridia. The intestinal microbiological balance can be restored only by using antibiotics; but, use of antibiotics is not an acceptable practice in most countries now.

1.4.2. Technical supervision

Pelleting may be a good method to conform to broiler requirements and to obtain the expected performance. When the pelleting process is achieved under good hygienic conditions, it contributes to lower the bacterial load of the feed (*enterobacteria, salmonella, E. coli*, etc.). However, it may become more risky where the farm technical level is low and the use of antibiotics is strictly limited if not forbidden.

Conversely, a coarse and uniform mash produced from well-controlled raw materials may yield very competitive results. This will be more obvious with higher energy levels. Mash presentation helps steady the digestive process, especially in farms where the technical or health level is not satisfactory.

In lieu of the above, changes are coming up in the feed manufacturing process: grinding speed reduction to 55m/s, mesh size increase (from 4 to 6 mm), use of wire-type grills with a higher % of holes and the addition of whole grain (wheat) before or after grinding.

Practically, the ideal balance between an expensive good quality pellet and a cheaper good coarse mash is not so easy to find. There are many different economical solutions in between these two options, in response to the complex equation of raw materials, power costs and housing conditions, etc. In one given environment, the best economical balance may even not be that of top-class broiler growth.

1.4.3. Feed type *Vs* broiler growth

(Trevidy, 2005)

Genetic selection has yielded tremendous progress in weight gain in broilers. As a consequence of faster growth, problems such as skeleton disorders (e.g. lameness) and metabolic disorders (e.g. ascites, sudden death) are expected. Lately, the tightening of restrictions regarding the use of antibiotics and coccidiostats has given rise to increased incidence of necrotic enteritis.

Practically, the above problems have mainly been addressed through the implementation of lighting programs. These aim at slowing down the growth while the birds are young, in order to achieve better skeletal and cardiovascular development, that allows them to sustain later compensatory growth and to decrease the incidence of late mortality. However, the gain in growth potential and therefore in broiler appetite has been such that the lighting programs were eventually unable to give satisfactorily control. This has been even truer in conditions where the minimum light duration was 16 hours and light intensity was high; especially in open-sided houses in tropics. Growth control in this case was obtained by quantitative restriction, with controlled feeding schedules. The use of such techniques showed an improvement in feed efficiency because feed was better digested compared to *ad libitum* feeding on one hand, and reduced late mortality on the other. Obviously proper implementation of the above solutions requires an extremely good level of management and housing conditions.

Under field conditions, pelleted wheat-based feed presented as a mash following pellet grinding, has been found to reduce feed consumption and increase starch digestibility. Hence, the pelleting effect was essentially explained by a higher feed intake leading to impaired feed efficiency because of an alteration in intestinal feed-flow regulation. Consequently, the gizzard may not be able to properly play its regulatory role through its grinding activity and satiety; thus, its hormonal or nervous communication with the intestine may be impaired.

The practical consequences of this are :

1. Quantitative restriction of broilers improves feed efficiency because of better regulation of intestinal feed flow and
2. The gizzard plays a major role in the above feed flow regulation, as long as feed particle size remains coarse.

1.5. Feed factory technology

(Trevidy, 2005)

A coarse and uniform mash feed is certainly a good solution for broiler production. Coarse mash that may even be used to produce a pellet is an important factor to regulate feed digestion. Feed particle size depends mainly upon grinding. Two types of grinder are available (cylinder grinder and hammer grinder).

Grinders with fluted cylinders are not designed for heavy production. They are more sensitive to deterioration by foreign bodies, but they are less power-consuming and the feed produced is more uniform in particle size.

Grinders with hammers are more often used. Grinding is achieved both by contact between feed particles and the hammers and the abrasive effect of the grills. Thus grinding control depends upon two main factors: hammer-peripheral speed, and grill-mesh size and the percentage of holes.

1.5.1. Hammer peripheral speed

Hammer peripheral speed is a combination of grinder diameter and rotation speed. For one given raw material, the higher the hammer speed, the wider is the distribution range of the feed particles. Peripheral hammer speed (m/s) can be calculated as follows:

If grinder rotation speed is S rpm, grinder diameter is D m, then peripheral hammer speed = $(\pi*D*S)/60$ m/s; for example, if the grinder of diameter 0.7 m runs at 3,000 rpm, the speed is 110 m/s.

For poultry feed, 55 m/s is the most frequently used speed. Grinders with variable speed allow adapting speed to the raw materials and to the targets for feed particle size.

1.5.2. Grinder grills

The two important criteria are mesh diameter (from 2 to 10 mm), and percentage of holes in the grill (from 27 to 52 %). The higher these two values, the higher the average feed particle size and the feed particle size distribution range is.

1.5.3. Hammer and grill wear

The feed particle size and the average distribution range of the feed particle size must be regularly monitored. Excessive variation is a sign of hammer or grill wear.

When blades are worn, the distance between the blade and grill (normally 8mm) is increased. The peripheral feed particle layer therefore becomes thicker and particle ejection is slowed down. The abrasive effect at grill level is increased. Grinder yield diminishes and more fine particles are produced.

In the same manner, worn grills will tend to reject particles back to the grinder instead of letting them out.

For most poultry species, the relevant range for feed particle size is -0.5 to 2 mm. Under 0.5mm, particles are less readily ingested, but this size is essentially composed of vitamins and minerals. Above 2 mm, comprises mostly of the cereals, which may give rise to feed particle selection by the birds.

Grinders with variable speed improve the uniformity of particle size and diminish the amount of particles outside the desired range. Grinders with 55 m/s speed, together with post-grinding sifting to exclude those particles above 3 mm, give good results when working with larger diameter mesh grills to reduce the production of fine particles.

1.5.4. Manufacturing costs

Power consumption for grinding and pelleting processes, which represents nearly 80% of the total power requirements, is taken into account.

1.5.4.1. Grinding

Power consumption varies with raw materials and particle size produced; power consumption and material loss increase with moisture at a rate of 10% per % moisture.

1.5.4.2. Pelleting

This is the highest power-consuming operation. Yet, comparing mash and pellet on the basis of the same feed formula is often a biased exercise, as the mash should be finer to obtain a good pellet. To compare pellet with a coarser mash, taking into account the additional grinding and pelleting costs, is more relevant.

1.6. Feed sampling procedures

(Harrman, 2001)

A systematic sample involves random selection of one unit, and then repeated collection of sampling units at equal intervals thereafter. Systematic samples are easier to perform than a simple random sample and often provide greater information per unit cost than does simple random sampling.

The bulk container may be stratified and sampled if there is an inferior portion of grain in the carrier. Each sample is a minimum of 2 kg in weight and is obtained in a balanced manner.

The following are the guidelines for equipment, collection, handling and storage of feed samples:

1.6.1. Collection of samples

Bag shipments of basemix, premix, and medicated-feed articles should be sampled with a bag probe.

Drums or barrels of liquid ingredients such as fat or molasses can be sampled using a sufficiently long tube of glass or stainless steel, 1 to 1½ cm in diameter, referred to as a drum thief. Sample at least 10 percent of the containers and collect a minimum of 1 pint. Bulkshipments of liquid ingredients may be sampled using a bomb sampler or core sampler. In all cases, liquid ingredients should be subject to some stirring action (e.g., rolling drums) prior to sampling to ensure ingredient distribution.

Finished feed can be sampled as it is transferred to the delivery vehicle if feed is in bulk form.

1.6.2. Sampling Equipment

1.6.2.1. Slotted grain probe

Slotted grain probes may be used to collect a representative sample from grain, soyabean meal, or finished feed. The grain probe should be long enough to penetrate at least ¾ of the depth of the feedstuff. Official grain samples are collected using a 2 cm diameter probe that consists of two tubes, one inside the other. The inner tube is divided into compartments that enable the individual collecting the sample to detect inconsistencies in grain quality across the profile (depth) of the carrier. This procedure is more labor-intensive since the contents of the probe must be emptied on to a tarp or trough and inspected before the grain is transferred into a container.

1.6.2.2. Open-handles grain probe

Open-handled grain probes, in which the inner tube is not divided into compartments, may be used for sampling feed ingredients including grain. The probe's contents are emptied from the handle end and mixing will occur, making it difficult to perform a visual inspection for load inconsistencies by depth.

1.6.2.3. Open-handled spiral probe

An open-handled spiral probe is designed such that openings on the inside tube rotate around so it opens first at the bottom and then in gradual steps to the top. This assures a fair portion of the sample is collected across the profile (bottom to top) of the material. However, incorrect use of this probe can result in the opposite effect if the inside tube is rotated in the opposite direction, resulting in a disproportionate amount of sample collected from the top.

Probes come in standard lengths of 100, 130, 170, and 180 cm, 2.4, 3.0, and 3.6 m. The probe should be inserted into the grain or feed ingredient at a 10° angle from the vertical, with the slots facing upward and completely closed. A 10° angle is used to obtain a cross section of material, while placing the end of the probe as close to the bottom of the carrier as possible. The slots must be kept closed until the probe is inserted as far as it will go. If the probe's slots are open as it enters the grain, a disproportionate amount of material from the top will fill the probe.

After the probe is fully inserted, open the slots and move the probe up and down quickly in two short motions. Close the slots completely, grasp the probe by the outer tube, and withdraw it from the grain.

1.6.2.4. Pelican grain sampler

The Pelican grain sampler is used for on-line grain sampling. The Pelican is a leather pouch, approximately 18 cm deep and 45 cm long, with a band of iron inserted along the edge to hold the pouch open. The pouch is attached to a long pole. Pelicans are designed to catch grain as the pouch is swung or pulled through a falling stream of grain. The Pelican grain sampler would be useful for collecting grain, soyabean meal, or complete feed samples while a truck is unloading.

1.6.2.5. Tapered bag-triers

Tapered bag-triers are constructed of stainless steel and are characterized by a sharp point, a tapered body, and an open throat. These triers are available in lengths from 15 to 30 cm. Tapered bag-triers are used to sample closed bags of powdered and granular commodities.

1.6.2.6. Double-tube bag-triers

Double-tube bag-triers are constructed of stainless steel or chrome-plated brass. These triers are available in various lengths and diameters, in both close-ended and open-ended models. These triers are used to sample closed and opened bags of powdered and granular ingredients.

1.6.2.7. Single-tube, open-ended bag-triers

Single-tube, open-ended bag-triers are constructed of stainless steel tubing and are used to sample opened bags of dry, powdery commodities when removal of a core of material is desired.

1.6.2.8. Bomb or zone samplers

Bomb or zone samplers are used to collect liquid ingredients from bulk carriers. These samplers consist of a closed cylinder ranging in size from 30 cm long by 3 cm in diameter to 40 cm long by 7½ cm in diameter; with capacities of 120 and 960 ml, respectively.

1.6.3. Sample reduction

The contents of each probe location should be mixed together prior to reducing the sample. Sample reduction may be performed using a riffler, Boerner Divider, or by quartering the sample. The end result of this process should produce a working sample of ½ to 1 kg and a retained sample that should be kept for a predetermined time (usually until the meat animal is marketed and processed).

Complete feed and feed ingredients may be partitioned into uniform sub-samples using a riffler. The sample is poured into the hopper, which is divided into equal portions by two series of chutes that discharge alternately in opposite directions into separate pans.

The Boerner Divider is the grain industry's standard for splitting samples. A sample of grain is placed in the hopper and then released down a cone, where grain is cut into 38 separate streams, which rejoin into two streams and then empties into the pans.

Quartering is a method for reducing the sample size of high-roughage feed (e.g., cattle feed) to a convenient amount for analysis. Spread the mixed composite sample on clean plastic or paper to form an even layer. Mark into quarters. Take two opposite quarters, mix, and repeat until the two quarters selected give the desired sample size.

1.6.4. Storage of samples

Heavy plastic bags, zip-lock bags, or plastic containers with lids make excellent sample containers for dry ingredients or finished feed. Label samples as they are taken, identifying the date, sample number, and the contents (or ingredient to assay). Preservation of samples is highly important. Immediately freeze high-moisture feedstuffs, silage, or green forage. Store other materials in cool, dry locations.

1.6.5. Sampling frequency and retention

With few exceptions, all ingredients should be sampled upon arrival and inspected for identity, physical purity, and compared with a reference sample. These samples, at a minimum, should be retained until the complete feed has been consumed by the animal and performed according to the label.

Commercial feed mills should collect and retain a sample of complete feed for each run of a given product. The sample should be retained as long as potential liability exists (e.g., until the meat animal is marketed and processed).

Medicated feed sampling and evaluation must conform to regulatory requirements. Three representative samples of medicated feed containing each drug or drug combination used shall be collected and assayed by approved official methods every year.

1.6.6. Receiving procedures

All feed processors should develop and follow a set of procedures for receiving feed ingredients. This should include inspection of the carrier's paperwork to ensure the correct material is on the carrier, a sensory inspection of ingredients collected from the sampling process, and documentation of receipt of those ingredients.

When receiving bulk material, inspect the transport documents for ingredient identification, mill and supplier, and name of the individual hauling the cargo. Inspect the ingredient label and compare to the previous labels. Usually feed ingredients (with the exception of grain), are not unloaded until a label can be supplied. Check the label for the correct ingredient and analyses guarantees.

Ingredients should be examined for sensory characteristics and ingredients that do not pass this initial inspection are not unloaded; particularly with bulk ingredients.

A receiving report that documents receipt of ingredients will augment a sampling program. This report should include the date, ingredient name, supplier, carrier name, license, bill of loading, purchase order or invoice number, time received, weight, bin number where the ingredient was placed, sensory or physical qualities, and signature of the individual who unloads the material.

1.6.7. Sampling for livestock health problems

Proper treatment of a livestock problem depends on correct identification of the causative agent. A carefully compiled history will usually provide an adequate background for immediate and practical emergency treatment. A correct diagnosis of the causative agent usually rests upon sample assays of feed, water, surrounding environment, or animal tissue and body fluids.

2. Poultry feeding systems

Poultry can be managed under different feeding systems, depending on the husbandry skills and the feed available. The systems of feeding are:

1. *ad libitum* feeding
2. Restricted feeding
3. Skip-a-day feeding
4. Sex separate feeding
5. Choice feeding
6. Phase feeding

Of the above, *ad libitum* feeding means that feed is always available and the birds eat feed whenever they want depending on their appetite.

Restricted-feeding and skip-a-day feeding are outlined in Chapters "Restricted feeding" and "Molting: natural and induced", respectively. Sex separate feeding is dealt in Chapter "Broilers and Roasters". Choice-feeding has been described in the earlier part of this Chapter.

An outline of the Phase feeding is give below:

2.1. Phase feeding

A careful observation of egg production cycle of chicken (Fig. 55.1) indicates the following:

1. Egg production steadily increases after sexual maturity to reach a peak, and then starts reducing
2. Egg weight increases along with egg production and continues to increase even when the egg production is receding
3. Consequent on the above, egg mass (product of egg number and egg weight) increases during the first 20 weeks of egg production and later on decreases.
4. Body weight increases from about 1.3 kg to 1.8 kg by the end of laying period.
5. Consequent on increased body weight, birds increase their food consumption as they grow older, in general, and as they increase their egg mass output initially, in particular.

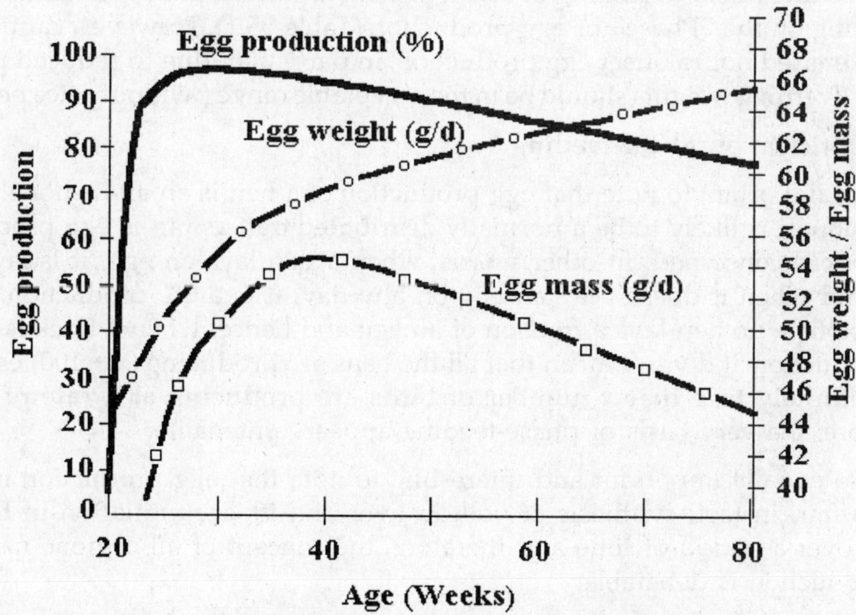

Fig. 55.1. Egg production parameters during a laying cycle

(Adopted from Leeson, 2001)

Table 55.2. Protein, amino acid and mineral requirements of hens in different phases of egg production

Age (Weeks)	(%)				
	Egg production	Protein	Met	Calcium	Available phosphorus
< 35	90	17.0	0.35	3.8	0.42
45	85	16.0	0.32	4.0	0.40
55	80	15.5	0.31	4.1	0.38
70	75	15.0	0.30	4.2	0.36
80	70	14.5	0.29	4.3	0.34
					Source : Leeson, 2001

Therefore, it is logical to predict that protein requirement of layers can be reduced after peak production is achieved so that feed cost on egg production can be reduced; in addition, it is necessary to note that, in most countries, there is no economic advantage of producing bigger than normal eggs because they are not likely to attract premium price (Leeson, 2001). It is generally agreed that Standard Leghorns require 17 g protein per day. The protein requirement can be reduced to 16 and 15 g/d when egg production comes down to 80 and 70% respectively. Assuming that birds under ideal conditions consume 100 g feed/d, it means a protein content of 17, 16 and 15%, respectively in the layer diets.

Phase-feeding refers to feeding of laying hens with diets differing in composition depending on the "Phase" of egg production (Table 55.1). However, caution has to be exercised not to offset egg production, in particular, due to reduced protein in feed. Even egg weights should be in the acceptable range (without price penalty).

2.2. Criticism on phase-feeding

It is very important to note that egg production of a hen is an all-or-none trait in expression. It is likely to be a normally distributed trait as far as egg production of a flock is concerned. In other words, when a hen lays an egg, it is laying at 100% and when it doesn't lay an egg on any day, it is a 0% production. To be more explicit, no hen lays a fraction of an egg and hence, when a flock is at x % egg production, it doesn't mean that all the hens are producing ($x \div 100$) eggs but it is definitely true that x number of birds are producing at a rate of 100%. Therefore, the very basis of phase-feeding appears untenable.

It is also equally important and interesting to note that egg production is not a day's affair; in fact, synthesis of yolk in liver and its accumulation in follicles occurs over a period of time and therefore, the concept of all-or-none nature of egg production is debatable.

Therefore, phase-feeding is not totally unscientific and wherever it is practiced, it has yielded benefits to the producers in terms of saving on feed (protein); the latter is likely to contribute to reductions in environmental pollution through reduced nitrogen excretion in excreta.

Chapter **56**

Performance Enhancers in Poultry Feed

Commercialization of poultry production has led to several approaches to maximize/optimize returns of theenterprise. The development of high-producing birds was only the beginning. For full expression of the genetic potential, optimum environment is mandatory; after all, phenotypic expression is simply the total of genotypic potential and environmental influences. Under the environment, the entire management consisting of rearing, feeding, disease control etc. is included.

Probably the use of anabolic hormones either as implants or in the form of its analogues in feed was the earliest method of improving performance, especially growth. Production of capons also comes under this category. But, the hormones, being banned universally, have little significance in the present-day poultry production.

The next stage was to use antibiotics infeed at levels sufficient to enhance growth. This particularly was successful where the general hygiene was not to the desired level. But, identification of development of antibiotic resistance not only in poultry but also transmitted to humans due to presence of antibiotic residues in poultry products has virtually restricted the use of antibiotics as growth promoters. However, judicious use of antibiotics is still acceptable even in developed countries.

In the post-antibiotic feeding era, enzymes, competitive exclusion (CE) products (probiotics) and others are under thorough scrutiny for use in poultry diets to enhance production.

In this chapter, a brief discussion on use of antibiotics, enzymes, probiotics and others is given; the reader is advised to look into exhaustive reviews on each of these topics on internet for further details.

1. Antibiotics

1.1. Growth-promoting effects of antibiotics (Doyle, 2001)

The three primary effects of the antibiotic growth promoters are:

1. Increased growth

2. Improved feed efficiency, and
3. A lower incidence of certain diseases

The exact mechanisms by which this occurs are not completely understood. Since feed antibiotics provide a relatively greater improvement in farms with poor hygiene, their effectiveness is at least partially due to suppression of some pathogenic bacteria in livestock.

The antibiotics alter the normal, non-pathogenic flora of the gut and these changes have a beneficial effect on digestive processes and the utilization of nutrients in feed. Antibiotics added to feed alter activities of microbial population and may thereby prevent some of this loss to microbial fermentation.

Intestinal bacteria also inactivate pancreatic digestive enzymes and metabolize dietary protein with the production of ammonia and biogenic amines, such as cadaverine. Antibiotics inhibit these activities and increase the digestibility of dietary protein.

Antibiotics also appear to prevent irritation of the intestinal lining and may enhance uptake of nutrients from the intestine by thinning of the mucosal layer. The effects of subtherapeutic antibiotics may extend beyond digestion in the intestine and stimulate metabolic processes throughout the animal. However, as producers improved hygiene, sanitary and other husbandry practices, incidence of disease decreased and antibiotic use declined steadily.

In evaluating the possible alternatives to subtherapeutic use of antibiotics, one must consider not only their relative short- and long-term costs but also their ability to produce the same positive effects as the antibiotics. It is also possible that some of these alternatives may have unintended negative effects that are not immediately apparent.

1.2. Development of antibiotic resistance

(Doyle, 2001)

Antibiotics have been used in animal agriculture since shortly after their discovery. Concerns of antimicrobial resistance have existed for nearly as long, but recent concerns regarding the prevalence of antibiotic-resistant infections in humans have raised the controversy to new heights.

Bacteria are very adaptable organisms because of their very short generation time and their propensity for sharing genetic information even among different species of bacteria. The presence of an antibiotic may kill most of the bacteria in an environment but the resistant survivors can eventually re-establish themselves and pass their resistance genes on to their offspring and, often, to other species of bacteria.

Both medical and veterinary uses of antibiotics have resulted in the appearance of resistant strains of bacteria. Resistant bacteria which are human pathogens may cause diseases that are difficult to treat; even if the resistant bacteria are not human pathogens, they may still be dangerous because they can transfer their antibiotic resistance genes to other bacteria that are pathogenic.

Antibiotic-resistant strains of bacteria, including *Salmonella* spp., *E. coli*, and *Campylobacter* spp., have been isolated from farm animals in many countries. Fluoroquinoline-resistant *C. jejuni* have recently been detected in chicken following the introduction of fluoroquinolines into feed for chickens.

Resistance to antimicrobials existed even before antimicrobials were used. However, this intrinsic form of resistance is not a major source of concern for human and animal health. The vast majority of drug-resistant organisms have instead emerged as a result of genetic changes, acquired through mutation or transfer of genetic material during the life of the microorganisms, and subsequent selection processes. Mutational resistance develops as a result of spontaneous mutation in a locus on the microbial chromosome that controls susceptibility to a given antimicrobial. The presence of the drug serves as a selecting mechanism to suppress susceptible microorganisms and promote the growth of resistant mutants.

Spontaneous mutations are transmissible vertically. Resistance can also develop as a result of transfer of genetic material between bacteria. Plasmids, which are small extra-chromosal DNA molecules, transposons and integrons, which are short DNA sequences, can be transmitted both vertically and horizontally and can code for multi-resistance. It is believed that the major part of acquired resistance is plasmid-mediated, although the method of resistance transfer varies for specific drug/bacteria combinations.

In addition, *de novo* resistance arises as a result of single or multiple genetic mutations. This is a phenomenon to which all cells are susceptible but which increases in the face of a selection pressure. Further, resistance may be acquired through the horizontal flow of genetic material. Conjugative plasmids may carry resistance elements between different genera of bacteria. However, it is also likely that multiple resistance elements can be, and are, transferred between completely different species of micro-organisms. Integrons, distinct families of DNA elements, are involved in the capture and recombination of DNA "cassettes", which normally encode for a single resistance mechanism. Multiple cassette insertions can, and do, occur and this confers the property of multi-drug resistance. *S. typhimurium* DT104 is a well-know example of this phenomenon (Revington, 2002).

1.3. Judicious use of antibiotics

(The American Veterinary Poultry Association website)

There are fifteen general principles which emphasize preventive actions to avoid disease, consideration of other options before choosing to use antimicrobials, and consideration of use of less important drugs before using the drugs of last resort, especially those that are very important to human or animal medicine. The principles are:

1. Preventive strategies, such as appropriate husbandry and hygiene, routine the ealth monitoring, and immunizations, should be emphasized. The objective is to prevent disease to the greatest extent possible so that antimicrobial treatment is not required. In food animals, antimicrobial use should always be part of, and not a replacement for, integrated disease control programs.
2. Other therapeutic options should be considered prior to antimicrobial therapy. Altering the temperature in the poultry house may aid the birds in recovering from the disease.
3. Judicious use of antimicrobials, when under the direction of a veterinarian, should meet all the requirements of a valid veterinarian-client-patient relationship.
4. Prescription, Veterinary Feed Directive, and extra-label use of antimicrobials must meet all the requirements of a valid veterinarian.
5. Extra-label antimicrobial therapy must be prescribed only in accordance with the existing laws of the concerned country. No drug can be marketed unless its quality, safety, and efficacy have been demonstrated.
6. Regimens for therapeutic antimicrobial use should be optimized using current pharmacological information and principles.
7. The choice of the right antimicrobial needs to take into account pharmacokinetic parameters, such as bioavailability, tissue distribution, apparent elimination half-life, and tissue kinetics to ensure the selected therapeutic agent reaches the site of infection. Duration of withdrawal times may be a factor in choosing suitable products. Consideration must also be given to the available pharmaceutical forms and to the route of administration. Prolonged oral use should be avoided, as most of the concerns with regard to resistance are associated with the selection and transfer of resistant, zoonotic bacteria that inhabit the gut.
8. Antimicrobials considered important in treating refractory infections in human or veterinary medicine should be used in animals only after careful review and reasonable justification.
9. Use narrow spectrum antimicrobials whenever appropriate. Generally, antimicrobials with a broad spectrum of activity lead to development of resistance in non-target microorganisms more rapidly than those with a narrow spectrum because they exert a selection pressure on a greater number of microorganisms. The theory is that narrow spectrum antimicrobials will have lessened effect on non-target species of bacteria and therefore will lessen the chances of resistance development in commensal bacteria.

10. Utilize culture and susceptibility results to aid in the selection of antimicrobials when clinically relevant.
11. Therapeutic antimicrobial use should be confined to appropriate clinical indications. Inappropriate uses such as for uncomplicated viral infections should be avoided.
12. Therapeutic exposure to antimicrobials shouldbe minimized by treating only for as long as needed for the desired clinical response. Theoretically, infections should be treated with antimicrobials only until the host's defense system is adequate to resolve the infection, but that period is difficult to judge.
13. Limit therapeutic antimicrobial treatment to ill or at risk animals, treating the fewest animals indicated.
14. Minimize environmental contamination with antimicrobials whenever possible. Unused antimicrobials should be properly disposed. Also some antimicrobials may be environmentally stable in manure. If the antimicrobials are not bound in an inactive form, environmental exposure could contribute to resistance development. Consideration may need to be given to disposal methods that will not recycle resistant organisms to humans or animals.
15. Accurate records of treatment and outcome should be used to evaluate therapeutic regimens. Outcome records can greatly assist with design of future empiric treatment regimens.

In any case, many countries have banned use of antibiotics in animal production for human food. This loss of antibiotic growth promoters (AGPs) can be compensated by:

1. Pathogen reduction
2. Augmentation of the immune response and
3. Nutritional strategies and/or additives that either improve performance in their own right, or help to directly modify the gut microbial flora.

However, alternative feed additives or supplements may have different mechanisms of action and other positive or negative effects which must be considered. It may be necessary to combine two or more alternative feed ingredients or to combine a new feed supplement with a change in husbandry practices to achieve the best effects (Doyle, 2001).

2. Pathogen reduction – CE products

(Jeffrey, 1999)

Competitive exclusion (CE) is a term that has been used to describe the protective effect of the natural or native bacterial flora of the intestine in limiting the colonization of some bacterial pathogens. Competitive exclusion products are also called probiotics, direct-fed microbials or CE cultures and they may provide a significant tool for the poultry industry in combating the occurrence of intestinal disease and reduction of food-borne pathogens.

2.1. Physiologic maturation of the intestine

The newly-hatched chicken or turkey gut is devoid of bacteria. In the first few hours to days of life, the normal gut bacteria (microflora) that inhabit the intestine become established. Intestinal microflora function to break down ingested food, produce some vitamins and most importantly provide a naturalbarrier to harmful bacteria that enter the host.

Under natural brooding under a hen, the bacteria shed in the feces of the healthy adult hen provided the inoculums for the establishment of a similar microflora in the chicks. With the advent of modern incubation, the first bacteria the chick or poult is exposed to are those in the incubator, chick box, and litter of the poultry house.

2.2. Factors influencing the microflora of the gut

The normal microflora of the intestinal tract is made up of a diverse population of bacteria. Some of these bacteria are anaerobic, some are aerobic and some are facultative anaerobes or microaerophilic; all on competition for survival. They compete for attachment sites and nutrients from the ingesta passing through the intestine.

Each species of bacteria has specific requirements for growth. They are affected by relative acidity or alkalinity (pH) of their environment and by-products produced by neighboring bacteria. The environmentof the intestine can be altered by the composition of the diet of the bird or by disease (e.g. coccidiosis, viral enteritis).

2.3. Mechanisms of protection

The exact mechanism by which CE products infer resistance to pathogens is still under investigation. Three mechanisms have been proposed to explain how CE products work:

1. Physical obstruction of attachment sites for salmonella by the native flora lining the intestine
2. Competition for essential nutrients by the native flora limits the ability of salmonellae to grow
3. Protective flora may produce volatile fatty acids (especially in the ceca) that limit the growth of salmonellae.

2.4. Defining and reproducing protective microflora in the laboratory

The first CE products were simply fecal contents from healthy adult chicken suspended in an aqueous solution and placed in the crop of the newly-hatched chicks. Later, undefined mixtures of intestinal bacteria were serially cultured under anaerobic conditions. These preparations were shown to be highly effective. A concern about undefined cultures arises from their undefined nature. Serial subculture has been used for the dilution to extinction of harmful parasites or viruses that may be present in the original cultures. The known pathogens of poultry that are shed in the feces are not capable of withstanding the culture techniques for CE products. Screening tests for known pathogens are conducted but doubt has been raised that all pathogens can be identified by current screening methods.

Some research groups have developed defined mixtures of bacteria for use as CE products. The number of bacterial strains and species is critical to the effectiveness of the product. In general, products that contained a single or only a few bacterial strains have not been protective. There appears to be difficulty in maintaining a stable culture of numerous strains of bacteria over long periods of time.

2.5. Administration of CE products

In laboratory studies, the protective microflora is usually applied directly into the crop of the chick. For commercial application this technique is not feasible, therefore, CE products have been produced in liquid and lyophilized (dried) forms for practical use. In field studies, CE products have been administered in the drinking water, by spraying hatching eggs or chicks in the hatching trays or shipping boxes, within feed slurries, or sprayed on agar plates for the chicks to eat. All methods have had some success but no method has been 100% effective.

2.6. Effect of feed or water additives

The effect of feed additives such as antibiotics and coccidiostats regularly fed in commercial poultry diets and the effect of antibiotics used for disease treatment on the protection inferred by CE products must also be considered because antibiotics can alter intestinal flora. The presence of high levels of chlorine or other disinfectants in water available to poultry flocks may also limit the usefulness of CE treatment. Medicated feed containing about 200 ppm of bacitracin, furazolidone, gallimycin, penicillin, streptomycin, chlortetracycline and tylosin, or 10 ppm of nitrivin did not adversely affect the efficacy of CE treatment. Common antibiotic feed additives used as growth promoters (5-50 ppm) had mixed effects on CE treatment. For example, bacitracin or virginiamycin improved the performance of CE cultures, flavomycin had no effect and avoparcin reduced the level of protection.

The positive benefits from commercial preparations of CE cultures are well accepted in Europe after more than 10 years of commercial use. CE products have been efficacious against colonization by numerous salmonellae serovars, including Salmonella enteritidis PT4, but not against campylobacter. The microflora of the intestine are in a dynamic state, undergoing continuous change, and subject to alteration by numerous forces. Therefore, the use of CE products is not a panacea against salmonellae infection throughout the life of the bird, however, in conjunction with good husbandry, CE products can make a valuable contribution to flock health and the safety of poultry products as food.

3. Augmentation of immune response

Increasing reliance on vaccination and development of improved vaccine delivery may provide opportunities to minimize feed-borne medications. It seems to be the systemic, acute phase response of the animal to disease challenge that confers significant nutrient requirement and therefore detracts from productive efficiency. Conjugated linoleic acid is one compound that has been investigated for its apparent abilities to alleviate the immune-associated anorexic response.

Immunologically-active compounds affectthe working of the immune system and may enhance resistance to disease. These substances include antibodies, cytokines, spray-dried plasma, and other compounds. Some or all of the growth-promoting effects of sub-therapeutic antibiotics in feeds may result from their action against sub-clinical infections or competitive intestinal bacteria. Therefore, it has been suggested that addition of antibodies or other immunoactive compounds to feed may accomplish the same purpose.

3.1. Vaccines

Vaccines may be important in preventing some diseases which may arise when antibiotics are withdrawn. But vaccines may not replace antibiotics as growth promoters because they specifically target pathogens while antibiotics affect general bacterial populations in the gut.

3.2. Cytokines

Some cytokines (which are normal regulators of the immune response) can act as growth promoters perhaps by stimulating the immune system to ward off pathogens. As such, it has been proposed that these compounds may be a substitute for sub-therapeutic antibiotics in feed. Avian cytokine genes have been cloned and can be delivered to chickens in a viral vector.

3.3. Conjugated linoleic acid

Conjugated linoleic acid (CLA) has also been shown to affect immune function in laboratory animals by increasing production of T-cells and interleukin-2. CLA refers to a mixture of positional and geometric isomers of linoleic acid with conjugated double bonds in the region of carbon atoms 8–13. It has been demonstrated to have anticarcinogenic effects and, in laboratory animals, reduces the proportion of body fat and increases lean tissue (Doyle, 2001).

3.4. Mannan oligisaccharides

While AGPs act primarily by inhibiting the growth of gram-positive bacterial populations, mannan oligosaccharides (MOS) and other oligosaccharides have been shown to act as "decoy" attachment sites for gram-negative pathogens. For pathogenesis to occur, bacterial cells must colonize the surface of the host enterocyte through specific fimbrionic attachments. Oligosaccharide preparations of yeast cell wall have been shown to block fimbrionic receptors and prevent colonization. In addition, MOS may also improve the structural integrity of the gastrointestinal tract, through mechanisms not yet clearly understood (Revington, 2002).

3.5. Organic acids

Organic acids are also known to have strong antibacterial effects. Use of acidifiers in baby pig feeds has proven beneficial, and organic acids have been used as *salmonella*-control agents in feed and in water supplies for livestock and poultry. Typically, blends of organic acids representing an array of pKa optima are more effective than single acids alone (Revington, 2002).

3.6. Specific antibodies

One promising alternative to antibiotics involves the feeding of specific antibodies to neutralize pathogenic organisms. In simple terms, hens are exposed to particular antigens and their systems stimulated to produce immunoglobulins. The immune proteins are then harvested from eggs and processed for inclusion into feed. This approach has met with some success in the swine industry. There may be problems with mode of delivery since heat treatment is obviously detrimental to the functionality of protein, as is the digestive process of the animal (Revington, 2002).

3.7. Bacteriophages

These are particles which infect bacterial cells and may destroy them by lysis. They can be very specific to certain pathogens. This idea is not new, but has been stalled perhaps because of the prevalence and success of antibiotics in fighting bacterial disease. Recent work with poultry has suggested that the bacteriophage may be a useful replacement for antibiotics (Revington, 2002).

4. Nutritional strategies and additives

Notwithstanding the concerns about antibiotic resistance, there are few alternatives available today that can meet the benefits of the AGPs that they purport to replace.

The link between diet and the incidence of enteric disease in birds is well known. Diets based on wheat, barley or rye, for example, seem to confer greater susceptibility to necrotic enteritis. The provision of substrate to the lower gastrointestinal tract increases the risk of bacterial overgrowth. Thus, any approach that improves digestibility is typically of benefit to the bird.

Some of the alternatives include enzymes, CE products and others; it appears that, when applied in tandem, they may match the efficacy of growth promotion by use of AGPs.

4.1. CE products

CE products have been described above in the light of pathogen reduction. However, they can be added in feed to stimulate positive reactions in the intestines.

Probiotics are live cultures of microbes – often lactic acid bacteria but also some other species – which are fed to animals to improve health and growth by altering intestinal microbial balance. Some bacterial cultures are used specifically for competitive exclusion (CE): They are fed in one or a few doses to newborn or newly-hatched animals in order to quickly establish an intestinal flora that will prevent colonization by pathogenic bacteria. Competitive exclusion preparations are not always pure cultures of bacteria, and their microbial composition may not be completely known. Some CE cultures have proven effective in protecting chicks from *Salmonella* infections (Doyle, 2001).

Probiotic microorganisms added to feed may offer protection from intestinal pathogens by several possible mechanisms, sometimes referred to as competitive exclusion:

1. Adherence to intestinal mucosa thereby preventing attachment of pathogens
2. Production of antimicrobial compounds such as bacteriocins and organic acids
3. Competition with pathogens for nutrients
4. Stimulation of intestinal immune responses
5. Probiotics may affect the permeability of the gut and increase uptake of nutrients.

However, difficulties arise in defining the specific culture to be used and in application, since the heat-treatment conferred by pelleting is obviously deleterious to live-cell products. Much work remains to be done to refine the commercial application of probiotics (Revington, 2002).

4.2. Enzymes

Addition of enzymes to feed may be a useful strategy to increase its digestibility. Dietary enzymes may supplement the animal's own digestive enzyme activity or enable it to utilize the energy in complex carbohydrates which normally pass unchanged through the gastrointestinal tract.

Some of the enzymes that have been used over the past several years or have potential for use in the feed industry include cellulase (β-glucanases), xylanases and associated enzymes, phytases, proteases, lipases, and galactosidases (Table 56.1). Enzymes in the feed industry have mostly been used for poultry to neutralize the effects of the viscous, non-starch polysaccharides in cereals such as barley, wheat, rye, and triticale. These anti-nutritive carbohydrates are undesirable, as they reduce digestion and absorption of all nutrients in the diet, especially fat and protein (Marquardt, 2005).

Table 56.1. Some enzymes that are or can be used in the feed industry

Enzymes	Substrate	Function	Benefits or use
β-glucanases	Barley Oats	Viscosity reduction	Enhanced digestion and utilization of nutrients
Xylanases	Wheat, rye Triticale Rice bran(?)	Viscosity reduction (other effects?)	Enhanced digestion and utilization of nutrients
β-galactosidases	Grain legumes Lupins	Viscosity reduction (other effects?)	Enhanced digestion and utilization of nutrients
Phytases	Plant feedstuffs	Release of phosphate from phytate-P	Enhanced phosphate absorption
Proteases	Proteins	Hydrolysis of protein	Increased digestion of proteins
Lipases	Lipids	Hydrolysis of fats	Use in young animals
Amylases	Starch	Hydrolysis of starch	Supplemental amylase for young animals

Source : Marquardt, 2005

Recently, considerable interest has been shown in the use of phytase as a feed additive, as it not only increases the availability of phosphate in plants but also reduces environmental pollution. The enzyme phytase can decrease the antinutritional effects of phytate which binds 50–75% of the phosphorus in vegetable matter. Phytate also appears to interfere with the digestion and absorption of other minerals such as calcium.

Several other enzyme products are currently being evaluated in the feed industry, including protease to enhance protein digestion, lipases to enhance lipid digestion, β-galactosidases to neutralize certain antinutritive factors in non-cereal feedstuffs, and amylase to assist in the digestion of starch (Marquardt, 2005).

Addition of carbohydrate-degradingenzymes (amylase, glucanase, glucoamylase) to a barley- or wheat-based diet and improved feed conversion (but not average growth). The efficacy of enzyme additives appears to depend on several factors including the other components of the diet, and the source of the enzymes. Carbohydrate-degrading enzymes, isolated from different bacteria and molds, usually differ somewhat in their activity against specific compounds (Doyle, 2001).

By removing fermentable substrates from the ileum, ileal populations of microflora are reduced. The relative contribution of an enzyme preparation is greater when the quality of the cereal source is lower. While exogenous enzyme addition appears to limit microbial growth in the ileum, the products of enzymatic breakdown may provide fermentable substrates to the cecal flora. Increases in volatile fatty acid (VFA) productionand changes in the VFA profile serve to favor the beneficial organisms *(Bifidobacteria,* for example) and suppress population of deleterious organisms (*Campylobacter, Salmonella, Clostridium*). The combination of xylanase, amylase and protease enzymes has been shown to improve protein, amino acid and energy utilization, improve performance/uniformity and impact microbial population in a beneficial manner in the upper and lower intestine (Revington, 2002).

4.3. Water-soluble non-starch polysaccharides (WSNSPS) for chicken

4.3.1. β-glucans

β-glucans constitute the most abundant class of naturally occurring polysaccharides because of the wide occurrence of the 1,4-β-glucan, cellulose. However, many other β-glucans are produced by both microbial and non-microbial (plant) sources.

The production of β-glucan-degrading enzymes is a characteristic attributable to a wide variety of organisms, although the fungi are the most common producers of this enzyme. The many β-glucan-hydrolyzing enzymes are classified according to the type of β-glucosidic linkage(s) they cleave and their mechanism of substrate attack (Table 56.2).

Cellulases are the most widely found -glucanases in fungi hydrolyzing the 1,4-β - glucan (cellulose). The 1,3-β- and 1,2-β-glucanases appear to be widely distributed in both fungi and yeast.

Partial hydrolysis of the β-glucans, particularly by the endo form of the enzyme, has been shown to not only reduce the viscosity of the -glucans but also to improve the nutritional value of cereals. The exo-β-glucanases, although capable of releasing glucose from β-glucan, have relatively little effect on the viscosity of β-glucans and therefore are of little benefit as an agent for reducing their viscosity. The complete degradation of β-glucans to glucose is accomplished by the synergistic interaction of both endo- and exo-β-glucanases. However, dietary enzymes are, at best, only able to partially hydrolyze β-glucans. Therefore, they probably do not directly increase the availability of glucose from the β-glucans.

Table 56.2. Nomenclature and action of β-glucan-degrading enzymes

Common name	Systematic name	Action
Cellulase	1,4-(1,3;1,4)- β-D-glucan 4-glucanohydrolase	Endohydrolysis of 1,4 linkages in cellulose and β-D-glucans containing 1,3 and 1,4 linkages
Laminarinase	1,4-(1,3;1,4)- β-D-glucan 3(4)-glucanohydrolase	Endohydrolysis of 1,3 or 1,4 linkages in β-D-glucans when the glucose residue whose reducing group is involved in the linkage to be hydrolyzed is itself substituted at C-3

Common name	Systematic name	Action
β-glucosidase	β-D-glucoside glucohydrolase	Hydrolysis of terminal non-reducing β-D-glucosyl residues, with the release of β-D-glucose
Endo-1,3-β-glucanase	1,3-β-D-glucan glucanohydrolase	Endohydrolysis of 1,3 linkages in 1,3-β-D-glucans
Exo-1,3-β-glucanase	1,3-β-D-glucan glucohydrolase	Exohydrolysis of 1,3 linkages in 1,3-β-D-glucans, with the release of α- glucose
Endo-1,2-β-glucanase	1,2-β-D-glucan glucanohydrolase	Endohydrolysis of 1,2 linkages in 1,2-β-D-glucans
Lichenase	1,3-1,4-β-D-glucan 4-glucanohydrolase	Endohydrolysis of 1,4 linkages in β-D-glucans containing 1,3 and 1,4 linkages
Exo-1,4-β-glucanase	1,4-β-D-glucan glucohydrolase	Exohydrolysis of 1,4 linkages in 1,4-β-glucans
Endo-1,6-β-glucanase	1,6-β-D-glucan glucanohydrolase	Endohydrolysis of 1,6 linkages in 1,6-β-glucans
		Source : Marquardt, 2005

4.3.2. Arabinoxylans

The arabinoxylans extracted from wheat flour consist predominantly of the pentoses (arabinose and xylose) and hence are often referred to as pentosans. However, in many instances, hexoses and hexuronic acids are also minor but important constituents. For this reason they are more correctly and descriptively referred to as heteroxylans. In addition, they may contain different phenolic acids, such as ferulic acid and acetyl esters. The hemicellulose fraction is composed mainly of xylans that are linked to cellulose fibrils by hydrogen bonds.

4.4. Improvements obtained with different enzyme preparations

Enzymes added to poultry diets, especially diets containing cereals such as wheat, barley, and rye, not only enhance the nutrient availability of these diets but also produce many other benefits (Table 56.3).

Table 56.3. Benefits obtained from the addition of enzymes to poultry feeds

Benefits
Reduced viscosity in the diet and digesta
Enhanced digestion and absorption of nutrients especially fat and protein
Improved AME value of the diet
Increased feed intake, weight gain, and feed-gain ratio
Reduced beak impaction and vent plugging
Decreased size of gastrointestinal tract
Altered population of microorganisms in gastrointestinal tract
Reduced water intake
Reduced water content of excreta
Reduced production of ammonia from excreta
Reduced output of excreta, including reduced N and P
Reduced output of bile salts in digesta
Source : Marquardt, 2005

The degree of improvement obtained by adding enzymes to the diet depends on many factors, including the type and amount of cereal in the diet; the level of antinutritive factor in the cereal, which can vary within a given cereal (for example, low- versus high-β-glucan barley); the spectrum and concentration of enzymes used; the type of animal (poultry tend to be more responsive to enzyme treatment than pigs); and the age of the animal (young animals tend to respond better to enzymes than older animals).

In view of the above, it is obvious that nutritional value of cereals that contain a high level of viscous WSNSPS can usually be improved to a marked degree by the addition of the appropriate enzyme to the diet. In general, barley and oats have a high content of soluble β-glucans and therefore respond to β-glucanase, whereas wheat and rye tend to have higher contents of arabinoxylans and therefore respond to xylanases and possibly other associated enzymes. Corn, in contrast, contains very low levels of WSNSPS; therefore, nutrient utilization is not depressed, and as a result, corn does not show a response to enzyme treatment. Enzyme addition to a diet high in WSNSPS has been shown to improve weight gain, feed–gain ratio, apparent protein digestibility, apparent lipid digestibility, and the AME of the diet. In addition, enzymes reduce the size of the gastrointestinal tract, the water content of excreta, and the incidence of vent pasting. When properly used, enzymes can produce many beneficial effects in the chicken and other kinds of poultry and monogastric animals, such as the young pig.

4.5. Alternative feed ingredients

4.5.1. Organic acids (Acidifers)

Organic acids contain one to seven carbon atoms. They are widely distributed in plants and animals and are also produced during microbial fermentation. These acids and their salts are often used as food preservatives and, since they are easy to handle, can be used to acidify feed.

In fact, although these compounds lower feed and gut pH, their effects are not simply a result of acidification because neither dietary HCl nor phosphoric acid improved growth or feed conversion. It is likely that the antimicrobial effects of the organic acid ions, which act by controlling bacterial population in the upper intestinal tract, are responsible for the beneficial effects of these acids.

Two problems may occur at higher organic acid levels:

1. Palatability may be decreased, leading to feed refusal and
2. Acidic feed is corrosive to cement and galvanized steel in poultry housing. In order to minimize these effects, the natural buffering capacity of feeds (related to mineral and protein content) should be evaluated to determine the minimum effective amount of acid to use.

Salts of organic acids, such as formates and diformates, are not as corrosive and can be used to significantly improve growth rate and feed conversion. Another strategy to extend the effectiveness of acid supplements and reduce corrosion damage to housing materials is the use of a slow-release form of acid. It consists of organic acids with fatty acids and mono- and diacylglycerols mixed to form microgranules. Tests show that use of these granules, as compared to use of free acids, results in greater feed intake and growth.

4.5.2. Other feed supplements

4.5.2.1. Minerals

Zinc or copper added to diets containing antibiotics significantly improved average daily weight gain, feed intake and feed efficiency.

4.5.2.2. Amino acids

Simply adding more protein to the diet may increase the absolute amounts of limiting amino acids but will not provide an optimum mix of amino acids and may increase intestinal disorders. Much of the additional protein may be "wasted" because it contains excess amounts of more common amino acids. Therefore, supplements of the limiting amino acids have been added to diets to improve performance.

4.5.3. Betaine

This osmolyte, has been used to improve the moisture characteristics of manure and to alleviate the effects of heat stress in birds. One concern over the loss of AGP availability is an expected increase in necrotic enteritis (NE) caused by *Clostridium perfringens*. NE is often associated with coccidiosis challenge. When fed with ionophore anticoccidials, betaine appears to enhance the ability of the ionophore to lower lesion scores (Revington, 2002).

4.5.4. Herbs, spices, plant extracts etc.

(Revington, 2002)

Herbs, spices and various plant extracts have received increasing attention as possible AGP replacements. There is evidence to suggest that some of these components have appetite-stimulating properties (menthol from peppermint), anti-bacterial effects (carvacrol from oregano) or may provide anti-oxidant functions (cinnamaldehyde from cinnamon).

To be effective on a practical scale, they may have to be provided in more concentrated form. These materials, although are "all natural", a safety point of view is mandatory. Many antibiotics are also "all natural" compounds in concentrated form; it is likely that those herbal extracts that prove most effective in modifying GIT microbial ecology will also face increasing pressure for regulatory approval.

4.5.5. Alternative Husbandry Practices

Although alternative feed supplements may compensate to some extent for the reduction or elimination of antibiotics in feeds, some changes in husbandry practices may also be important. Good hygiene is, of course, very important for preventing the spread of disease among livestock. If the routine use of antibiotics in feeds is discontinued, it may be necessary to be even more rigorous in maintaining a clean environment for livestock. Improvements in the environment which have demonstrated effectiveness include:

1. Attention to efficient-cleaning methods and effective sanitizer use to minimize spread of disease
2. Maintenance of an appropriate ventilation rate since pathogens may be spread through the air
3. Appropriate environmental temperatures
4. Stocking rates appropriate for size of the farm
5. All-in-all-out system
6. Careful record-keeping to identify problem areas

Chapter **57**

Ration Formulation

For formulating any ration, the following information is necessary:

1. Type of birds for which the ration is being formulated and their nutrient requirements.
2. Feed ingredients available, their composition, availability, quality and cost per unit.
3. Upper and lower limits of each of the ingredients in different rations.
4. Feed additives like minerals, vitamins, anticoccidials, antibiotics and other growth promoters.
5. Type of feed –mash-type or pellets; most popular being mash-type.

With the above information, ration formulation can be done by the following methods:

1. Iteration method

In this method, a tentative formula is made based on experience and various nutrients are calculated and compared with the requirements. If any alteration has to be made, once again the calculations are repeated after suitable correction. Iterative method can be greatly simplified by calculating the nutrients supplied by certain fixed ingredients like fish meal, vegetable oils etc. and then manipulating the other ingredients to supply the remaining requirements. The latter part of the computation can be done by grouping the ingredients into grain and grain by-products and protein supplements. The amounts of these two groups to be added to put in the desired protein level can be found by using the Pearson's square method. Average protein content in grains and grain products as well as in protein sources are used in the Pearson's square. A tentative formula is obtained in which the other nutrients are calculated. The final formula is obtained by incorporation of any other changes, if needed. However, this procedure is highly cumbersome and time-consuming. The following steps outline the iteration method:

1. Nutrient requirements are listed in a row.
2. The ingredients whose levels are fixed at the beginning itself are considered. The amount of nutrients contributed by the fixed ingredients are calculated and subtracted from responding values of requirements.
3. Variable ingredients are altered within the maximum levels of inclusion permissible till the actual requirements are met by a combination of ingredients.

One nutrient is taken at a time.

4. At once the first nutrient is satisfied, the same formula is used to calculate the next nutrient. If the next nutrient is not satisfied, slight modifications are made in the formula so that the levels required for both the nutrients are met by a common formula. The procedure is extended till all the nutrient requirements are met by a common formula.
5. The common formula along with the fixed ingredients and feed additives will be the final ration formula. Care has to be taken to see that the quantity of all ingredients and additives in the common formula add to 100. Therefore, this method is very tedious and time-consuming; and on most occasions, it is not possible to meet the exact requirements with a common formula.

1.1. Slack space

Slack space, also called as reserved space is generally used during ration formulation. This is particularly essential to adjust some of the nutrients like minerals, amino acids etc. without affecting the other nutrient components. For example, a ration is formulated meeting all the requirements excepting lysine and methionine. This deficiency can easily be met by adding synthetic amino acids which will not alter the levels of any other nutrient. However, such additions must be within the total of 100 kg or whatever the total weight of the feed may be. Therefore, it is advisable to keep a buffer space depending on the ration that is being formulated so that minor adjustments can be made easily. In case of poultry diets, vitamins and mineral premixes can be given a slack space of about 2% in diets for young birds and growers; in case of layers, a buffer space of up to 10% can be given to make adjustments in calcium and phosphorus levels. Fixing levels of certain ingredients also helps easy and accurate ration formulation; examples in this category are fish meal, fat, salt etc. in poultry diets.

1.2. Iteration process

1.2.1. Factors to be considered

It is obvious that ingredients, their composition, quality, availability, permitted levels of inclusion and cost, nutrient requirements of the type of poultry in question, level of fixed ingredients have to be carefully considered before including any feedstuffs in poultry ration. These information can be tabulated as follows:

Table 57.1. Variables for iteration

Ingredients	ME kcal/kg	Protein	Remarks
Maize	3300	9	Variable ingredient
Rice polish (non-deoiled)	3000	12	Variable ingredient

Ingredients	ME kcal/kg	Protein	Remarks
Groundnut cake	2800	40	Variable ingredient
Fat	7700	--	Maximum 6-8%
Fish meal	1800	60	Fixed at 10-12%
Requirements (broilers 3-6 weeks)	3200	20	NRC (1994)
Contribution by the fixed ingredient	216	7.2	Fish meal 12%
Requirement to be met by variable ingredients	2984	12.8	By difference

1.2.2. Tentative formula

By utilizing the available ingredients, a tentative formula is generated based on experience keeping in view the admissible levels of inclusion for each of the ingredients. In the tentative formula, oneof the nutrients in question is considered and its level is calculated based on composition of the ingredients with respect to the nutrient and % inclusion in the diet.

Table 57.2. Tentative formula - I

Ingredients	Level, %	ME contributed/kg of ration
Maize	50	(50 x 3300)/100 = 1650
Rice polish	10	(10 x 3000)/100 = 300
Groundnut cake	23	(23 x 2800)/100 = 644
Fat	5	(5 x 7700)/100 = 385
Total	88	2979

It can be seen that the tentative formula - I has only 5 kcal/kg less than the requirement. It can also be seen by the computation table that if one unit of GNC is replaced by maize, ME will increase by 5 kcal/kg. Therefore, the tentative formula to meet the ME requirements will be: Maize 51%, Rice polish 10%, GNC 22% and Fat 5%. Now, the next nutrient, the crude protein, will be calculated in this tentative formula to find out whether it would meet the requirements of crude protein also.

Table 57.3. Tentative formula - II

Ingredients	Level, %	Crude protein,% /kg of ration
Maize	51	(50 x 9)/100 = 1650
Rice polish	10	(10 x 12)/100 = 300
Groundnut cake	22	(23 x 40)/100 = 644
Fat	5	Nil
Total	88	14.59

But, the requirement is only 12.8%; hence, the formula has 1.79% more protein. This can be corrected by replacing five parts of GNC with maize and one part of GNC with rice polish. But, with these changes, ME with increased by (5 x 5 + 2) = 27 kcal/kg which is allowable under practical conditions. In any case, even this can be corrected by replacing ½% fat with maize which will result in reduction of ME by 22 kcal/kg and increment in protein content by a mere 0.045%. Now, a final formula is ready to meet the requirements of both ME and crude protein.

1.2.3. Final formula

A final ration composition is obtained which includes the last tentative formula with the fixed ingredients. Nutrient composition of the final formula can be verified to confirm that it is as per the desired requirements.

Table 57.4. Final formula

Ingredients	Level,%	ME, kcal/kg	Crude protein, %
Maize	56.5	1864.5	5.085
Rice polish	11.0	330.0	1.320
Groundnut cake	16.0	448.5	6.400
Fat	4.5	346.5	----
Fish meal	12.0	216.0	7.200
Total	100	3205.0	20.005

This procedure is extended for as many nutrients as desired. It is obvious from the above description that iteration method becomes more and more complicated as and when number of ingredients and/or number of nutrients increase. The method of Pearson's Square can be used to assist the process of iteration.

1.2.4. Pearson's square method

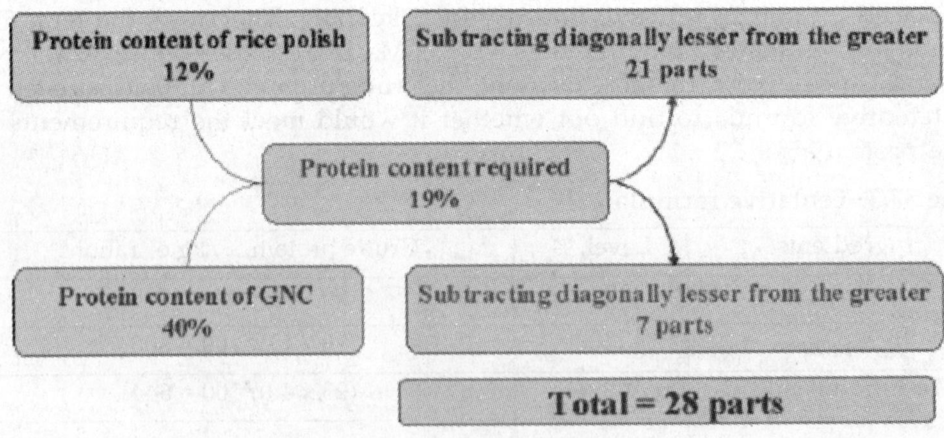

Fig. 57.1 Pearson's Square method

That means to say that, out of 28 parts of the mixture 21 parts (75%) must be rice polish and 7 parts (25%) must be GNC.

In other words, if X and Y are the ingredients whose composition with respect to the nutrient in question is C_X and C_Y, respectively and the composition of their mixture is C_{X+Y} then, the ratio of X and Y to be mixed to obtain the desired composition can be obtained by solving

$$\frac{C_X X + C_Y Y}{X+Y} = C_{X+Y}$$

In the present example, if X refers to rice polish and Y to GNC, {(12X + 40Y)/(X + Y)} = 19; on simplification, it condenses to 21Y = 7X; or X : Y = 3 : 1; or for every part of X there must be 3 parts of Y.

Alternatively, considering X + Y = 100 and therefore, Y = 100 – X, it is obvious that $[C_X X + C_Y Y] \div (X + Y) = C_{X+Y}$ or $[C_X X + C_Y (100 - X)] \div 100 = C_{X+Y}$; by simplification,

$$X = \frac{100[C_{X+Y} - C_Y]}{C_X - C_Y}\%$$

In the present example, X = 100 (19 – 40) ÷ (12 – 40) = 75 % and therefore, Y = 100 – 75 = 25 % or the ratio of X and Y is 3 : 1.

1.2.5. Double Pearson's Square

This method is used to develop a combination of 3 or 4 ingredients to exact requirements of 2 nutrients. Using a simple Pearson's square, one of the nutrients is balanced through a mixture of 2 ingredients (Mix I). The same nutrient is again balanced by using the third ingredient in combination with one of the ingredients in Mix I or 2 different ingredients (Mix II).

In the Mix I and II, the second ingredient is calculated and the values are used to calculate the combination of these two mixes to obtain the desired level of the second nutrient.

Depending on the proportions of Mix I and II required, the quantities of 3 or 4 ingredients can be computed by reflecting back on the ingredient composition of the mixes.

Therefore, 3 Pearson's Squares are required to solve for the required composition and ingredient profiles. However, if a 3rd nutrient has to be solved for, it requires 9 Pearson's Squares.

2. Matrix/Simultaneous equation method

In the above ration formula, if QM, QRP, QGNC, QF and QFM are the quantities; PM, PRP, PGNC, PF and PFM are the protein content (%); and MEM, MERP,

MEGNC, MEF and MEFM are ME values (kcal/kg) of maize, rice polish, ground nut cake, fats and fish meal, respectively, then

[{(QM*PM)÷100} + {(QRP*PRP)÷100} + {(QGNC*PGNC)÷100} + {(QF*PF)÷100} + {(QFM*PFM)÷100}] = % Protein requirement and

[{(QM*MEM)÷100} + {(QRP*MERP)÷100} + {(QGNC*MEGNC)÷100} + {(QF*MEF)÷100} + {(QFM*MEFM)÷100}] = ME requirement.

However, protein and ME content of the ingredients as well as their requirements are known; only quantity of each of the ingredients is the only unknown. Hence, it can be solved as simultaneous equations or by Matrix algebra provided that there are as many unknowns as are number of equations. In the above case, there are two equations and 5 unknowns; hence, either number of unknowns must be reduced or number of equations must be increased.

A composition matrix can be derived as 1/100*[Composition of the ingredients] which forms the Left Hand Side (LHS). The requirements in the form of one column forms the column vector and the Right Hand Side (RHS). Since only a square matrix can be inversed and then multiplied with an RHS, number of rows should be equal to number of columns on the LHS and RHS should have as many rows as that of LHS.

2.1.1. Characteristics of Composition matrix

LHS consists of composition and hence referred to as "Composition matrix of size (n x n) whereas, RHS is a column vector of size (n x 1). LHS has the following features:

1. It is mandatory that each of the element of the composition matrix must correspond to the (composition ÷ 100).
2. Number of columns = number of rows; in other words, number of ingredients = number of nutrients to be balanced.
3. Rows correspond to nutrients and columns to the ingredients.

Since RHS consists of requirements, it is also referred to as "Requirement vector"; the requirements must be in the same order as in the rows of composition matrix.

Solution for 'n' simultaneous equations can be obtained thus:

(Composition matrix)$^{-1}$ X (Requirement vector) = (Ration formula column vector); or (n x n)$^{-1}$ X (n x 1) = (1 x n)

For ease of demonstration of matrix/simultaneous equation method, the number of unknowns is reduced by fixing the levels of rice polish, fats and fish meal as 11, 12 and 4.5%, respectively. Now, maize and GNC have to be added to add 2307.5 kcal/kg of ME and 11.48% protein into the diet.

That means, the two simultaneous equations to be solved are:

$$33\ QM + 28\ QGNC = 2307.5$$

$$0.09\ QM + 0.40\ QGNC = 11.48$$

2.1.1.1. Doolittle method

The above equations can be solved by standard algebraic methods. For quick solution, Doolittle or Cross-multiplication method can also be employed thus:

Converting the above equations into a standard form,

$a_1x + b_1y = c_1$ and $a_2x + b_2y = c_2$ where, x and y are the quantities a_i and b_i are coefficients (composition) of maize and GNC, respectively; c_i are the requirements. The coefficients and requirements are arranged as follows:

b_1	c_1	a_1	b_1
b_2	c_2	a_2	b_2

Then, x and y can be obtained as follows :

$$x/(b_1c_2 - b_2c_1) = y/(c_1a_2 - c_2a_1) = -1/(a_1b_2 - a_2b_1)$$

2.1.1.2. Matrix method

The composition matrix in the above example will be:

a_1 b_1

a_2 b_2 whose inverse can be calculated as follows:

Determinant = $a_1b_2 - a_2b_1$ = **Det**

Elements of Cofactor matrix = b_2 $-a_2$

$-b_1$ a_1

Elements of inverse of the matrix = $b2 \div$ Det $-a2 \div$ Det

$-b_1 \div$ Det $a_1 \div$ Det

Requirement vector is c_1

c_2

Solution to the simultaneous equations is the column vector given by the product of inverse matrix and requirement vector.

In the given example, composition matrix is 33.00 28.00

0.09 0.40

Determinant (Det) = (33.00 x 0.40) – (0.09 x 28.00) = 10.68

Cofactor matrix is 0.40 - 28.00

- 0.09 33.00

Inverse of the matrix is 0.0374531 - 2.6217228

- 0.0084269 3.0898876

The requirement vector is 2307.50

11.48

Hence, quantity of maize in the ration is given by

0.0374531 x 2307.50 – 2.6217228 x 11.48 = 56.325651

and quantity of GNC in the ration is given by

- 0.0084269 x 2307.50 + 3.0898876 x 11.48 = 16.026838

In other words, 56.33% maize and 16.03% GNC will supply 2307.5 kcal/kg ME and 11.48% protein. It can be clearly noticed that the values of these ingredients obtained by simultaneous equations/matrix method are more accurate than the final formula by iteration method.

However, the matrix method has the following limitations :

1. The total of all the ingredients need not add up to 100; in such cases, neutral ingredients like sawdust, sand etc. can be used wherever the total of all the ingredients is less than 100; if the total exceeds 100, it is very difficult to make accurate corrections. Alternatively, a row can be incorporated in the matrix with all the elements as 1 on the LHS and 100 on the RHS to indicate that sum of all ingredients is 100.
2. It cannot consider the cost of ingredients; even if introduced, a certain cost can be fixed but least cost formulation is not possible.
3. Constraints such as maximum levels of inclusion of the ingredients and availability cannot be included.
4. Sometimes, non-realistic values (negative values) are obtained; no correction is possible for the situations.
5. The whole process has to be repeated even if a small alteration and/or correction are required; this process is time-consuming.

3. Linear programming (Operations Research)

Linear programming is a scientific approach to help a decision-maker in various fields including animal production. It is being particularly used in ration formulation to determine the amount of different ingredients which have to be included so as to minimize a cost of the diet and at the same time to meet all the nutrient requirements and other constraints.

Linear programming has the following advantages:

1. Number of nutrients need not be equal to number of ingredients
2. Any number of nutrients and / or ingredients can be incorporated

3. Constraints such as maximum level of inclusion can be easily included
4. Cost of each of the ingredients can be incorporated
5. Constraint on the total cost of the feed possible
6. Least cost ration can also be formulated
7. None of the ingredients will have abnormal (negative) values unlike Matrix method
8. Total of all the ingredients can be fixed and will not deviate from the given value
9. Constraints regarding maximum / minimum values of the nutrient requirements can also be added for computation
10. The method is very fast, accurate and highly reliable
11. Any number of alterations and/or modifications can be easily carried out with ease, accuracy and speed

3.1. Conditions to be satisfied

The conditions for a mathematical model to be a linear program are:

1. All variables continuous (i.e. can take fractional values)
2. A single objective (minimize or maximize)
3. The objective and constraints are linear i.e. any term is either a constant or a constant multiplied by an unknown.

3.2. Least-cost formulation

Under practical circumstances, ration formulation has to be done for multiple nutrient requirements utilizing multiple feed sources so that the final product costs least for the producer. For many of the nutrients, although requirements are exact, it appears ideal to have maximum and minimum values; for instance, protein, lysine and the like as minimum values and crude fiber, acid insoluble ash and the like as maximum value. Therefore, it is necessary that diets are formulated for nutrients required in terms of greater than or equal to (cost per kg of ration, proteins, amino-acid etc.) or less than or equal to (crude fiber, acid insoluble ash etc.) or maximum (cost per kg of the ration) in addition to simpler constraints employed in other methods of ration formulation.

In other words, the concept of inequalities has to be adopted by formulating cost-effective/least-cost diets. It is in this aspect of mathematical programming the modern computers are coming into play. It is beyond the scope of this book to completely elucidate "Linear programming" or "Least-cost formulation". However, basic steps regarding least-cost formulation are given below:

1. Nutrient requirements have to be determined; they are available from various sources like NRC, BIS etc.

2. Other constraints desired have to be specified. These primarily include slack space required for vitamin-mineral premix, fixed ingredients, synthetic compounds etc.
3. Several linear equations for each of the nutrients required are developed. This is similar to the composition matrix (LHS) and requirement vector (RHS) in matrix method of ration formulation. The only difference is LHS and RHS need not be related with the symbol "="; they can be related by any of the symbols like =, =, <, >, #, minimum, maximum and, of course, =.
4. Restrictions can be set up for even individual feedstuffs; for instance, fish meal should not exceed 5-6% in layer diets.
5. Objective function has to be set up; this function deals with price of the diet. Price in the same unit is the coefficient for corresponding quantity of each of the ingredients and their summation is linked to a restriction of "minimum price".
6. These linear equations are processed through appropriate software (Linear programming software, Least-cost formulation software etc.) to obtain the desired ration formula.
7. The formula so obtained is verified to confirm that the formula so developed is as per the constraints imposed and is appropriate.

3.3. Basic concepts

The term "Linear" indicates that all the relationships which are likely to be considered while solving is linear and the term "Programming" means a process of determining a particular program.

Mathematically, the general linear programming problem (LPP) is as follows:

Maximize or minimize Z = $c_1x_1 + c_2x_2 + \ldots. + c_nx_n$ subject to the following **set of equations/inequations/constraints:**

$$a_{11}x_1 + a_{12}x_2 + \ldots.\ a_{1n}x_n\ (=, =, =)\ b_1$$

$$a_{21}x_1 + a_{22}x_2 + \ldots.\ a_{2n}x_n\ (=, =, =)\ b_2$$

....

....

$a_{m1}x_1 + a_{m2}x_2 + \ldots.\ a_{mn}x_n\ (=, =, =)\ b_m$ and the set of non-negative restrictions of the general LPP stated as follows:

$$x_1, x_2, \ldots \ldots.\ x_n = 0$$

where,

1. The linear function **Z** which has to be maximized or minimized is the **objective function** of LPP.

2. $x_1, x_2, \dots \dots x_n$ are the **Decision variables.**
3. In the set of constraints, the expression (= , = , =) indicates that each constraint can assume any of the three signs.
4. c_j $(j = 1,2, \dots, n)$ represents per unit profit are cost of the jth variable.
5. b_i $(i = 1,2, \dots, m)$ is the requirement or availability of the ith constraint and
6. a_{ij} $(i = 1,2, \dots, m$ and $j = 1,2, \dots, n)$ is referred to as **Technological coefficient.**

3.4. Definitions

3.4.1. Basic Solution

x of $(Ax=b)$ is a *basic solution* if the n components of x can be partitioned into m "basic" and n-m "non-basic" variables in such a way that: the m columns of A corresponding to the basic variables form a nonsingular basis and the value of each "non-basic" variable is 0. The constraint matrix A has m rows (constraints) and n columns (variables).

3.4.2. Basic

This refers to a set of basic variables.

3.4.3. Basic Variables

A variable in the basic solution (value is not 0).

3.4.4. Nonbasic Variables

A variable not in the basic solution (value = 0).

3.4.5. Slack Variable

A variable added to the problem to eliminate less-than constraints.

3.4.6. Surplus Variable

A variable added to the problem to eliminate greater-than constraints.

3.4.7. Artificial Variable

A variable added to a linear program in phase 1 to aid finding a feasible solution.

3.4.8. Solution

Solution is a set of values of decision variables satisfying all the constraints of an LPP.

3.4.9. Feasible solution

A feasible solution is any solution; it also satisfies the non-negativity restrictions of the LPP.

3.4.10. Optimal feasible solution

Any feasible solution which maximizes or minimizes the objective function is called optimal feasible solution.

3.4.11. Unbounded Solution

For some linear programs it is possible to make the objective arbitrarily small (without bound). Such an LP is said to have an unbounded solution.

3.4.12. Optimization techniques

Optimization techniques are processes of obtaining the optimal values for a given LPP.

3.4.13. Feasible region

Feasible region is the common region determined by all the constraints and non-negativity restriction of an LPP.

3.4.14. Convex region

A region is said to be convex if the line segment joining any two arbitrary points of the region lies entirely within the region. Feasible region of an LPP is always a convex region.

3.5. Formulation of an LPP

The following are the main four steps in the mathematical formulation of an LPP as a mathematical model:

1. Identification of decision variables: these of the quantities whose values have to be determined; in other words, these are unknown variables. They are usually assigned symbols x and y.
2. Identification of set of constraints and expressing them as linear equations/ inequations in terms of decision variables. These constraints are the given conditions.
3. Identification of objective function and expressing it as a function of decision variables; in ration formulation, which is usually in the form of least-cost diet.
4. Imposition of non-negativity restrictions on the decision variables because negative values have no practical significance.

3.6. Examples

3.6.1. Example 1 (Two variables)

A feed manufacturer wants to mix two feed ingredients (Maize and GNC; Maize has 9% protein and 3300 kcal/kg of ME and GNC has 40% protein and 2800 kcal/ kg of ME) in such a way that the resultant mixture should contain at least 18% protein and at least 2900 kcal/kg of ME. F_1 costs Rs 8/kg and F_2 costs Rs 12/kg.

3.6.2. Formulation of an LPP

Let the mixture contain *x* and *y* kg of F_1 and F_2, respectively; the given data can be tabulated as follows:

Table 57.5. Hypothetical data for LPP

Nutrients	Ingredients		Requirements
	Maize (*x*)	GNC (*y*)	
Protein, %	9	40	18
ME, kcal/kg	3300	2800	2900
Cost, Rs/kg	8	12	

The minimum requirement of protein is 18%; therefore,

$$0.09\ x + 0.\ 40\ y = 18$$

Similarly, since the requirement of ME is 2900 kcal/kg,

$$33\ x + 28\ y = 2900$$

The non-negativity constraint is $x = 0, y = 0$

The total cost of purchasing *x* kg of maize and *y* kg of GNC is

$Z = 8\ x + 12\ y$ and this is the Objective function which has to be minimized subject to the above constraints.

3.6.3. Example 2 (more than two variables)

(Beasley on internet)

Assuming that a feed mix must contain a minimum quantity of the following nutrients:

Nutrient	A	B	C	D
Requirement (%)	9.0	5.0	2.0	0.2

The ingredients have the following nutrient composition and cost (Rs)

Nutrients	A	B	C	D	Cost/kg
Ingredient 1 (%)	10.0	8.0	4.0	1.0	40
Ingredient 2 (%)	20.0	15.0	2.0	-	60

3.6.3.1. Variables

In order to solve this problem it is best to think in terms of 100 kg of feed mix and it is made up of three parts - ingredient 1, ingredient 2 and filler; therefore, let:

x_1 = % of ingredient 1 in the ration, x_2 = % of ingredient 2 in the ration and x_3 = % of filler in the ration, where $x_1 >= 0$, $x_2 >= 0$ and $x_3 >= 0$.

Essentially these variables (x_1, x_2 and x_3) can be thought of as the ingredients making up 100 kg of ration.

3.6.3.2. Constraints

3.6.3.2.1. Nutrient constraint

$10.0x_1 + 20.0x_2 >= 9.0$ (nutrient A)

$8.0x_1 + 15.0x_2 >= 5.0$ (nutrient B)

$4.0x_1 + 2.0x_2 >= 2.0$ (nutrient C)

$1.0x_1 >= 2$ (nutrient D)

3.6.3.2.2. Total constraint

$x_1 + x_2 + x_3 = 100$

3.6.3.2.3. Cost constraint

To minimize cost, i.e. minimize $40x_1 + 60x_2$

3.6.3.3. Solution

The optimal solution to the above is:

x_1= 36.67%, x_2=26.67% and x_3=36.67 %

The linear programming model can be extended by

1. Increasing the number of nutrients considered
2. Increasing the number of possible ingredients considered - more ingredients can never increase the overall cost (other things being unchanged), and may lead to a decrease in overall cost
3. Placing both upper and lower limits on nutrients
4. Dealing with cost changes
5. Dealing with supply difficulties
6. Filler cost

3.7. Methods of solving LPP

3.7.1. Graphical method

This method is used when only two variables are involved. There are two techniques of solving LPP by graphical method; they are:

3.7.1.1. Corner-point method

Corner-point method assumes that the optimal solution to a LPP, if it exists, occurs at the extreme point (corner) of the feasible region. The steps involved in this method are:

1. Feasible region of the LPP is found.
2. Co-ordinates of each vertex of the feasible region is found either by solving the two simultaneous equations of the lines intersecting at that point or by inspection.

3. The objective function is evaluated at each corner point identified in Step 2. Depending on the type of objective function, maximum/minimum solution is the optimum value of Z; in case of feed formulation, minimum (least-cost) is the solution of Z.

However, this method is tedious especially when the number of constraints is increased.

3.7.1.2. Iso- profit or iso- cost method

In this method any suitable constant value, say **Z1** is given to the objective function and a line is drawn corresponding to this value; this line is called iso- profit or iso- cost line because, every point on this line will yield the same profit or cost **Z1**. Another line by assigning another value to object to function, say **Z2**, is drawn; obviously the line corresponding to **Z1** will be parallel to that corresponding to **Z2**. Likewise, lines are drawn as far as possible until farthest point (in case of maximization) or nearest point (in case of minimization) in the feasible region is touched by the line; any further displacement of the line will take it out of the feasible region. The co-ordinates of the point so obtained yield the maximum/minimum value of the objective function.

The following are the steps of this method:

1. Feasible region of the LPP is found.
2. A constant value, **Z1**, is assigned to **Z** and the line is drawn corresponding to the objective function.
3. Another constant value, **Z2**, is assigned to **Z** and another line is drawn corresponding to the object to function.
4. If **Z1 > Z2** (minimization) or **Z1 < Z2** (maximization), the line is moved by assigning another constant value to **Z** till nearest (minimization) or farthest (maximization) point within the feasible region is touched by the line.
5. The co-ordinates of the point gives the minimum/maximum value of the objective function.

Unlike corner-point method, by this method more equations/inequations can be easily handled.

3.7.2. Simplex method

This method can consider any number of variables and constraints; the solution is found through computers and hence, details of the procedure involved is beyond the scope of this book. However, a brief outline of the concept is given below:

3.7.2.1. Steps in Simplex method

(Waner and Costenoble, 2001 on internet)

1. Introduction of slack variables and conversion to a system of equations to turn the constraints into equations, and rewriting the objective function in

standard form.

2. Formation of the initial tableau.
3. Selection of the pivot column: The negative number with the largest magnitude is selected in the bottom row (excluding the rightmost entry). Its column is the pivot column. (If there are two candidates, either one can be chosen.) If all the numbers in the bottom row are zero or positive (excluding the rightmost entry), the basic solution maximizes the objective function.
4. Pivot in the pivot column is selected: The pivot must always be a positive number. For each positive entry ***b*** in the pivot column, the ratio ***a/b*** is computed, where ***a*** is the number in the answer column in that row. Of these test ratios, the smallest one is selected. Then, the corresponding number ***b*** is the pivot.
5. The pivot is used to clear the column in the normal manner and then the pivot row is re-labeled with the label from the pivot column. The variable originally labeling the pivot row is the departing or exiting variable and the variable labeling the column is the entering variable.
6. Repeated from Step 3.

3.7.2.1.1. Basic Solution

To get the basic solution corresponding to any tableau in the simplex method, all variables that do not appear as row labels are set to zero (these are the inactive variables).

The value of a variable that does appear as a row label (an active variable) is the number in the rightmost column in that row divided by the number in that row in the column labeled by the same variable.

The simplex method generates a sequence of feasible iterates by repeatedly moving from one vertex of the feasible set to an adjacent vertex with a lower value of the objective function Z. When it is not possible to find an adjoining vertex with a lower value of Z, the current vertex must be optimal, and termination occurs.

Algebraically speaking, the simplex method is based on the observation that at least ***n-m*** of the components of x is zero if x is a vertex of the feasible set. Accordingly, the components of x can be partitioned at each vertex into a set of ***m*** basic variables (all nonnegative) and a set of ***n-m*** nonbasic variables (all zero). If the basic variables are formed into a subvector, x_M ^{a} R''', and the nonbasics into another subvector, x_N ^{a} R'', the columns of variable matrix can be partitioned as B/N, where ***B*** contains the ***m*** columns that correspond to x_M. (Note that ***B*** is a square matrix)

At each iteration of the simplex method, a basic variable (a component of x_M) is reclassified as nonbasic, and vice versa. In other words, x_M and x_N exchange a component. Geometrically, this swapping process corresponds to a move from one vertex of the feasible set to an adjacent vertex. Therefore which component of x_N should enter x_M (that is, be allowed to move off its zero bound) and which component of x_M should enter x_N (that is, be driven to zero) need to be chosen. In selecting the entering component, it is clear that Z can be expressed as a function of x_N alone.

Further details of simplex method are available at http://www-fp.mcs.anl.gov/otc/Guide/OptWeb/continuous/constrained/linearprog/section2_1_1.html or http://www.ece.nwu.edu/OTC/

3.8. Exceptional cases of LPP

3.8.1. LPP having infeasible solution

When constraints are inconsistent, no point exists it satisfies all the constraints and such LPP are said to have infeasible solution.

3.8.2. LPP having un-bounded solution

In some LPP, the common feasible region may not be bounded and variables can take any value in the un-bounded region without violating any of the constraints; such LPP are said to have un-bounded solutions. However, feed formulation LPP seldom falls into this category.

3.9. Limitations of linear programming

1. Applicable only when the objective function and all other constraints can be expressed in terms of linear equations/inequations; some times, this may not be possible.
2. Solutions are possible only when all elements are quantified and hence, unsuitable for qualitative characters like behavior, relations etc.
3. Coefficients of the objective function must be known with certainty and should remain unchanged during the period the solution/s is/are applicable; practically, this may not be true.
4. Some solutions may be fractional which may not be suitable under all circumstances; however, as far as ration formulation is concerned, this is not a limitation.

4. Vitamin-mineral premix

A premix refers to a small amount of a total mixture. Commercial premixes are made of antibiotics, vitamins, trace minerals and various drugs and medicines which are used in animal feeding. Under almost all circumstances, the amount of material to be added will be very small and hence, it will be difficult to mix uniformly into large quantity of final product. Further, weighing small quantities every time that batch of feed has to be mixed is an avoidable botheration. Therefore, most feed manufacturers prefer to have these micro-ingredients mixed with some type of carrier, which may be soyabean meal, ground grain etc. which are often fortified with 2% stabilized fat to minimize loss of micro-ingredients as dust during mill operations (Valera-Alvarez, 1995).

For example, it is usual to ignore most of the vitamins and trace minerals present in feedstuffs while formulating poultry diets. A vitamin-mineral premix is usually added to meet the nutrient requirements. An example of a premix for broilers is given in Table 57.6.

Table 57.6. Vitamin-mineral premix for broilers

Vitamins				
Vitamin	Amount/kg	Content (/g)		Amount in premix (g)
A	1.323 x 10^6 IU	6.5 x 10^5 IU		4.62
D	0.440 x 10^6 ICU	2.0 x 10^5 ICU		5.00
E	440 IU	275 IU		3.64
K	220 mg	Pure		0.50
Riboflavin	1.323 g	0.50 g		6.00
d-Pantothenic acid	2.205 g	0.41 g		12.20
Niacin	8.818 g	Pure		20.00
Choline	76.545 g	0.434 g		400.00
B12	2.205 mg	1.32 mg		3.80
Folacin	88.185 mg	0.45 g		0.45
TOTAL				456.21
Minerals				
Element	Amount, g/kg	Source	Content (%)	Amount in premix (g)
Mn	23.986	$MnSO_4$	28	194.6
Fe	8.007	$FeSO_4$	21	86.5
Cu	0.802	$CuSO_4$	25	7.3
I	0.481	KI	69	1.6
Zn	1.102	$ZnSO_4$	36	6.9
TOTAL				296.9

Source : Varela-Alvarez, 1995

Chapter **58**

Toxins in Poultry Feed

Natural toxicants are reckoned to be chemicals with potentially toxic effects as a result of their natural occurrence in food; these toxicants are produced by living organisms like plants, bacteria algae and fungi.

Table 58.1. Routes of contamination by natural toxicants

Toxicant	Route
Bacterial	Generated by bacteria in food or in gastro-intestinal tract
Algae	Through the food chain upwards
Fungi	Contamination at all stages of food production
	Source : Watson, 1998

1. Classification

(www.fao.org/DOCREP/ARTICLE/AGRIPPA/)

Feedstuffs incorporated in compounded feeds are known to contain a wide variety of toxicants that are endogenous (Tables 58.2 and 58.3) or exogenous in nature and adversely affect the production of egg and meat. Amelioration of these toxicants is absolutely necessary to warrant the health and profitable poultry production. The feed-based toxins concerned in poultry feeding are divided into the following groups:

Table 58.2 Toxicants in feedstuffs

Group	Toxins
	Endogenous toxicants
Proteins	Haemagglutinns (lectins) and Protease inhibitors
Glucosides	Cyanogens, Estrogens, Goitrogens and Saponins
Phenols	Gossypol and Tannins
Others	Aliphatic nitro compounds, Anti-vitamins, Erucic acid and other fatty acids, Non-starch polysaccharides, Oxalates and Phytates
	Exogenous toxins
	Mycotoxins, Argemone, Insecticides and Pesticides

2. Endogenous toxins

(www.fao.org/DOCREP/ARTICLE/AGRIPPA/)

Toxins in feedstuffs for poultry are also discussed in Chapter "Feedstuffs for poultry"

2.1. Proteins

2.1.1. Protease inhibitors

These are widely distributed in plant kingdom, particularly, the legumes. Raw soyabean seeds contain trypsin inhibitors, that cause growth inhibition, hypertrophy of pancreas followed by a stimulation of its secretory activity. These also result in an endogenous loss of the pancreatic enzymes, trypsin and chymotrypsin which are rich in the sulphur-containing amino acids, and thus accentuating the deficiency of methionine, being the first limiting amino acid in soyabean.

The soyabean has received the most attention with respect to the effect of heat treatment on its trypsin inhibitory activity. Trypsin inhibitor activity of solvent-extracted soyabean meal has been reported to be destroyed by exposure to flowing steam for 60 min. or by autoclaving under0.35 kg/cm^2 for 45 min, 0.70 kg/cm^2 for 30 min, 1.05 kg/cm^2 for 20 min. or 1.40 kg/cm^2 for 10 min. Regardless of moisture content (5 and 19%), atmospheric steaming (100°C) destroys 95% of trypsin inhibitor content of dehulled and defatted raw soyabean flakes within 15 min and protein efficiency is higher at 19% than 5% moisture level.

2.1.2. Phytohaemagglutinins (lectins)

Phytohaemagglutinins, present in legume seeds, exhibit the unique phenomenon of being able to combine with glycoprotein component of cell membrane and in case of Red Blood Cells, this is accompanied by haemagglutination, occurs only if the lectin molecule has at least two active groups. Another effect of lectins is a non-specific interference with the absorption of nutrients like protein. This is attributable to the attachment of the lectins to the cells lining of the intestinal wall resulting in disorganization of the main absorption cells. Lectins are among toxic principles responsible for poor growth. Dry heat is not effective in elimination of this toxic component but soaking and cooking is helpful.

Some of phytohaemgglutinins are extremely toxic, such as ricin from castor bean; others, such as kidney beans. Reports of studies on castor meal suggest that autoclaving (1.055 kg/cm^2 at 122.2°C for 30 min) or water-cooking (addition of

Table 58.3. Endogenous toxic factors in feeds of plant origin

Feedstuffs	Toxic factors
Raw soyabean and its meal	Trypsin inhibitor, phytohaemagglutinin, antigens, lipoxygenase, goiterogen, saponin, estrogen, phytic-acid and oligosaccharides (NSPs – 30.3%)
Groundnut and its meal	Trypsin inhibitor, goiterogen, tannins, oligosaccharides and lectins
Mustard or rapeseed and its meal	Goiterogens (thioglucosides or glucosinolates), tannic acid, erucic acid, sinapine (cholinester), pectins and oligosaccharides (NSPs - 46.1 %)

Feedstuffs	Toxic factors
Safflower seed and its meal	Estrogenic factor, Two phenolic glucosides (Bitter flavour) and Fibre
Sunflower seed and its meal	Chlorogenic, quinic-acid and Fibre (Tannin like compounds)
Sesame seed and its cake	Phytate (5g/100g) and Oxalates (35 mg/100g)
Linseed and its cake	Linamarin (cyanogenic glucoside), antipyridoxine (Linatin) factor and mucilage (HCN Level → 10-300 mg)
Kapok seed meal (seeds of silk cotton tree)	Tannins, tyrosine and fatty acids with cyclopropene rings
Copra meal (coconut meal)	Fibre (mannans) and Estrogenic factor
Palm kernel meal	Fibre (half of the fibre–NDF high levels of galactomannans β-(1,4)-D mannans) and Sharp shells
Cotton seed and its meals	Gossypol (phenol like compound), cyclopropenoid fatty acids, tannins
Cotton seed and its meals	Gossypol (phenol like compounds) cyclopropenoid fatty acids, tannins
Guar meal (*Cyamopsis tetragonoloba*)	Guar gum (18-20%), antitrypsin factor and antivitamin E factor
Castor seed and its meal	Ricin (toxalbumin)-Phytohaemagglutinin Ricinine (Toxic-alkaloid) Ricinus allergen(Protein Polysaccharide)
Mahua cake (*Madhuca latifolia*)	Mowrin (Saponin) and Tannins
Maize	Selenoamino acids (seleniferous) Estrogen (mouldy) Trypsin inhibitor
Neem seed and its meal (*Azadirachta indica*)	Bitterness → Limonoids → Triterpenoids Bitter principles: Protomeliacins, Limonoids, Azadirone, Gedunin, Vilasinin and Secomeliacins Nonisoprenoid polypenolics-Flavanoids, Tannins and coumarin viz. Nimbin, Salannin and Azadirachtin Dried seeds – Limonoids – 0.001 to 0.1%, Azadirone, 0.45% and Epoxy Azadirone, 0.72%, Azadiradione, 0.7% and Salanin, 0.95%
Karanja cake (*Pongamia glabra*)	Fat bound toxic factors – Karajnjin and Pongamol (Flavanoids) (NSPs – 38%)
Lupin meal	Quinolizidine alkaloids, pectins, oligosaccharides, high manganese, saponin.
Peas	Protease inhibitors, tannins, Lipoxygenase and lectins

Feedstuffs	Toxic factors
Rubber seed meal	Hydrocyanic-acid (20-40mg/kg)
Wheat	Tyramine, Trypsin-inhibitor, NSPs (11.4%)
Rice	Estrogen and haemagglutinins
Rye	Amylase and protease inhibitor and NSPs (13.2%)
Oats	Amylase inhibitor and estrogens
Triticale	Trypsin and chymotrypsin inhibitor (NSPs)
Some varieties of barley	β- glucan (NSPs - 16.7%)
Some varieties of sorghum	Tannins
Rice-polish and rice bran	Trypsin inhibitor and antithiamine factor
Chunies	Antitryptic factor
Sal seed and its meal	Tannic - acid (tannins)
Tapioca meal (Cassava)	Cyanogenic glucoside (HCN - 1000 - 3000 mg/kg DM)
Fish meal and meat meal (prepared from spoiled or putrefied material)	Gizzerosine and histamine (Biogenic amines)

castor meal in boiling water 1: 5 w/v and heating for 30 min) reduces the heamagglutinating activity considerably in meal (from 160 to 20 HA unit) when compared with other processing methods viz. water-soaking (submerged completely in water 1: 10 w/v and left overnight with stirring, 40 HA unit), ammoniation (addition of 6 M ammonium hydroxide to raw castor 1: 4 w/v and hot air oven drying at 80°C for 45 min, 40 HA unit) and hexane extraction (hexaneextraction for 6 hrs, 80 HA unit). Castor meal processed either by autoclaving or cooking with water can be included as protein source in broiler diets up to 2.5% level without adversely affecting the performance and general health status of birds.

2.2. Glucosides

2.2.1. Glucosinolates

(Verkerk *et al.* 1998)

These are secondary plant metabolites found exclusively in cruciferous plants. They are sulfur containing glycosides a cutting the highest concentrations in the families *Resedaceae*, *Capparaceae* and *Brassicaceae*. Of these, plants belonging to family *Brassicaceae* produce majority of the glucosinolates.

2.2.1.1. General structure

Glucosinolates are (Z)-*cis*-N-hydroximinosulfate esters possessing a side-chain R and as sulfur linked D-glucopyranose moiety. They are named after the side-chain R.

2.2.1.2. Hydrolysis

Upon hydrolysis by the plant enzyme myrosinase, glucosinolates yield isothiocyanate, nitrile and thiocyanate; these products are responsible for the characteristic flavor of brassica vegetables. Isothiocyanates have anticarcinogenic properties.

2.2.1.3. Toxicity

Reduced growth rate, goitrogenicity, enlarged livers, kidneys, thyroid and adrenal glands. Rapeseed meal is rich in glucosinolates.

2.2.2. Goitrogens

Goitrogenic agents are found in rapeseeds and mustard. The use of these meals has been restricted due to the presence of thioglucosides or glucosinolates, which, upon hydrolysis, release products (2-OH-3 butenyl isothiocynate and 5 - Vinyloxazolidinine-2-thione- goitroin), which are goiterogenic and growth depressant. The predominant glucosinolate of rape and mustard is progoitrinin, which is a deleterious factor. Such goitrogens at dietary levels of 0.03 to 0.42% do not cause any apparent affect on performance or goitrogenicity in chicks. However, glucosinolates may be removed from meal by extraction with hot water, dilute alkali or acetone or decomposed with iron salts or soda ash. The goitrogenic products may be removed by extraction with acetone or water or by steam stripping of volatile isothiocynates.

2.2.3. Cyanogens

Intact glucosides are not toxic, but become toxic when they are hydrolyzed to release free HCN. The enzymes, the glucosidases, when active in plant tissue, may release HCN by autolysis, which is enhanced by moisture. However, in general spontaneous release of HCN from plant depends on the presence of specific glucosidase and water. Auto hydrolysis is enhanced, if plant is soaked in water after crushing. Crushing without soaking, however, will lead to slow release of HCN and it is well recognized that bruised cassava root is not suitable for consumption.

2.2.4. Saponins

Most of the biological activities of saponins arise from the surface activity and their ability to form complexes with sterols and proteins. In view of the ability to form stable complexes with cholesterol, it has been found possible to reduce liver and serum cholesterol by using dietary saponins.

Approximately 20% alfalfa in chick ration(equivalent to 0.37% saponin) depresses the growth rate. However, the growth of quails remains unaffected by 2% alfalfa top saponins but by 5% alfalfa root saponins depress growth.

2.2.5. Estrogens

Some plants to contain chemicals with steroid skeleton and they are referred to as phytoestrogens (Table 58.4). Zearalenone produced by fungi also have estrogen-like properties.

Zearalenone is a non-steroidal fungal toxin produced by the fungus *Fusarioum* and in particular *F.graminaerum* and *F.culmorum*. Chicken turkeys are highly tolerant to zearalenone.

A number of isoflavones having estrogenic activity have been reported to be present in soyabeans. One of these genistein (4, 5, 7-trihydro-xyisoflavone), in addition to estrogenic activity can cause growth inhibition, elevated levels of zinc in the liver and bones and increased deposition of calcium, phosphorus, manganese in bones. Solvent extracted soyabeans do not contain sufficient amount of estrogen to cause any adverse effect in chickens. Dry or moist heat treatment or solvent extraction can achieve inactivation of estrogens.

Table 58.4. Some phytoestrogen-containing plants

Food/plant	Phytoestrogen
Soya beans	Genestein, Daidzein, Genistin, Daidzin, Glycitin and Glycitin + acetyl and malonyl esters of the glucosides
Soya sprouts	Daidzein, Genestein and their glucosides
Subterranean clover (*Rifolium subterraneum*)	Coumesterol and Formontein
Alfalfa	Coumesterol and Formontein
Clover sprouts	Coumesterol, Formontein and biochanin A
Split peas	Daidzein
Broad beans	Genistein and Formontein
Flax seed	Seicoisolaricresinol
Red clover (*Trifolium pretense*)	Formonetin, biochanin A, a small amount of Genestein and trace amounts of Daidzein
White clover (*Triflium repens*)	Coumesterol plus traces of isoflavones
	Source : Aldridge and Tahourdin, 1998

2.3. Phenols

2.3.1. Gossypol

Gossypol is phenol-like compound (polyphenolic binaphthalene derivative) indentified in cottonseeds, its meal and oil. More than 0.06% free gossypol in the diet depresses growth in chicks. The depression is severe when the level increases to 1.2%. In laying hens, 0.024% reduces egg production and hatchability. A comparatively low level of gossypol (more than 0.005%) in the diet causes an olive green discoloration of the yolk. Dietary levels of 0.015% or less free gossypol is believed to be safe when cotton seed meal is used as protein supplement in balanced diet of poultry.

The cake after hydraulic pressing, screw pressing and solvent extraction, contains 0.04-0.1%, 0.02-0.05% and 0.02-0.03% free gossypol, respectively. The heat generated during the commercial production of cotton seed meal helps to bind 80 to 90% gossypol with protein rendering it non-toxic. Higher levels of gossypol can be tolerated, if iron salts are added to diet @ 1 to 2 ppm iron for every 1 ppm of free gossypol. Solid substrate fermentation involving certain fungi is capable of reducing 90% of free gossypol of cottonseed and eliminating its toxicity in chicks.

2.3.2. Tannins

Another group of phenolic compounds is termed as tannins and is widely distributed in nature. Rapeseed or mustard residue appears to contain 2.85 to 3.7% tannins. The tannin content of sorghum grain may range from 0.25 to 2.5%. Tannins at 0.5% level and above in diet cause reduction in growth and available energy value of feed, decreased availability of protein and severe mortality at higher levels (4% and above). They also inhibit enzyme activities (trypsin, amylase and lipase) or enzyme systems.

2.4. Other endogenous toxicants

2.4.1. Erucic acid and other fatty acids

Erucic acid appears to be a major factor affecting the utilization of mustard oil cake or rapeseed meal by poultry. Solvent-extracted mustard oil cake might be useful substitute to groundnut oil cake, if diets are otherwise comparable. Residual oil present in these meals contains about 40-50% erucic acid.

A diet constraining 0.6% erucic acid (as contained in 10% mustard oil cake) does not cause any adverse effect on performance of birds. However, erucic acid at levels beyond 0.605% in diet is known to cause growth depression, reduction in feed intake and efficiency in growing chicks.

Polyenoic fatty acids present in the fish oil impart fishy flavour to meat and eggs. A level of 4% of these marine polyenoic fatty acids in carcass fat is sufficient for detection of off flavours.

2.4.2. Non-starch polysaccharides (NSP)

Cereal grains are very rich in carbohydrates including sugars, starch and cell wall polysaccharides. The use of many cereals (oats, barley, bajra, ragi etc.) is limited in poultry because of the presence of non-starch polysaccharides. Poultry birds lack enzymes needed to digest β-linkages of β-glucans present in barley and oats and pentosans of rye, wheat and triticale which limit their utilization. Efforts have been made to render such substances to advantage through processing which include water-soaking to mobilize native enzymes to action, application of heat and milling to expose cell wall contents and enzyme supplementation to hydrolyze the substrate or development of non-waxy cultivars to contain less or negligible soluble β-glucans or pentosans. Water treatment also removes water soluble NSP of the cereal. In enzyme supplementation, hydrolysis of pentosans and β-glucans into simpler polymers, alter the ability of these polysac- charides to form highly viscous solutions, which inhibit nutrient diffusion and transport and thus improve their utilization. Other methods like gamma irradiation, acid treatment and milling are also reported to increase utilization of cereals (See Chapter "Performance enhancers in poultry feeds").

2.4.3. Phytates

Phytates (salts of phytic acid), are formed due to combination of six phosphate molecule with inositol, a cyclic alcohol with six hydroxy radicals similar to hexose sugar. The availability of phytin P is influenced by alimentary tract pH and the level of vitamin D_3, calcium, Ca to P ratio and Zn. Supplementation with adequate minerals, which are affected by phytates, is usually practiced. At present, dietary supplementation of phytase enzyme (250-500 units/kg) is practiced to enhance the utilization of phytate P in poultry. Autoclaving soyabean meal can reverse the increased requirement of metals. Cottonseed meal can be treated with phytase enzyme prepared from (*Aspergillus ficcum* or *A.niger*). In both, vegetable and animal kingdom oxalic acid is found as free and in salt form.

2.4.4. Antivitamins

Raw soyabean contains an enzyme iypoxygenase which catalyses oxidation of carotene, the precursor of vitamin A and can be destroyed by heating soybaeans for 15 min at atmospheric pressure. Autoclaving of soyabean protein or supplementation with vitamin D_3 for about 8-10 times can eliminate the rachitogenic activity. Raw kidney beans contain anti-vitamin E activity that is eliminated by autoclaving. The dicumarol in sweet clover produces severe hemorrhages due to reduced prothrombin levels in the blood interfering with the blood clotting mechanism. The effect is due to the reduction of vitamin K utilization needed for the production of prothrombin in liver. An antagonist of pyridoxine from linseed has been identified as 1-amino-D proline and occurs naturally in combination with glutamic acid as a peptide, which is known as linatine. The nutritive value of linseed meal for chicks can be considerably improved after extracting the meal with water and autoclaving and supplementing with pyridoxine hydrochloride.

3. Exogenous toxicants

3.1. Mycotoxins

Mycotoxins are secondary metabolites produced by certain species of fungi (Table 58.5).

Table 58.5. Important mycotoxins

Mycotoxin	Main fungi	Foods infected
Aflatoxins B_1, B_2, G_1 and G_2	*A.flavus, A.parasiticus* *	Rice bran, maize
Aflatoxins F_1 and F_2	*Metabolic products of aflatoxin B1 and B2*	
Sterigmatocystin	*A.versicolor, A.nidulans*	Cereals
Cyclopiazonic acid	*A.flavus, P.commune*	Cereals, pulses
Ochratoxin A	*P.verrucosum, A.ochraceus*	Cereals, field beans
Citrinin	*P.verrucosum*	Cereals
Deoxynivalenol, nivalenol	*F.graminaerum, F.culmorum, F.crookwellense*	Cereals
T-2 toxin	*F.poae, F.sporotrichioides*	Cereals
Fumonisins, B_1, B_2 and B_3	*F.moniliforme, F.prliferatum*	Maize, maize products
Zearalenone	*F.graminaerum, F.culmorum, F.crookwellense*	Cereals
Moniliformin	*Fusarium species*	Cereals
Alernariol, alternariol monomethyl ether, tenuazonic acid	*Alternaria alternate, A.tenuis*	Cereals

* Aflatoxins B_1 and B_2 only — *Source* : Scudamore, 1998

The determination as to whether or not a given concentration of mycotoxin is safe will mainly depend on the factors (Jones *et al.*, 1994):

1. Chemical class and chemical structure of the mycotoxin in question (Tables 58.5 and 58.6). For example, aflatoxin B_1 is reported to be the most potent naturally-occurring carcinogenic substance known, but if just one chemical bond is changed in the structure of the molecule, its toxicity can be reduced dramatically.
2. Presence of other mycotoxins. A number of studies have demonstrated that mycotoxins occur simultaneously in field situations. This simultaneous occurrence can profoundly affect the toxicity of the mycotoxins present.
3. Species and strain of the animals involved. Ducklings are 5 to 15 times more sensitive to the effects of aflatoxin than are laying hens, but when laying hen strains are compared, certain strains of hens may be as much as 3 times more sensitive than other strains. This fact, along with the fact that there is continuous genetic improvement of farm animals, can mean that the exact sensitivity of a given animal to one or more mycotoxins is unknown.
4. Health status of the animals involved. Stress, physiological state, nutritional standing, and disease status will independently and collectively determine the response of a given animal to a specific mycotoxin level or complex of

mycotoxins.

5. Criteria by which effects are determined. At a given dose, aflatoxin reduces weight gain in growing animals, but disease resistance in the same animal may be reduced by about half that dose.
6. Number of animals involved in judging the no effects level. It has been estimated that 400 groups of 10 broilers would be required to detect a 1% difference in growth rate. Yet, in integrated-poultry operations a 1% difference in growth rate would have a significant economic impact.
7. Sampling and assay procedures. It is imperative that sampling and assay procedures are accurate, since the results are the basis for deciding whether or not to use a given lot of feed or feed ingredient.
8. Length of time animals are exposed to the mycotoxin(s). The exact mycotoxin tolerance levels available are for a limited period exposure. Obviously, the risks of harm to animals from mycotoxins increase as exposure time increases.

Although between 300 and 400 mycotoxins are known, those mycotoxins of most concern, based on their toxicity and occurrence, are aflatoxin, deoxynivalenol (DON or vomitoxin), zearalenone, fumonisin, T-2 toxin, T-2-like toxins (trichothecenes, Ochratoxin A, T-2 toxin, rubratoxin B and citrin. Deoxynivalenol (DON), zearalenone, T-2 toxin, and fumonisin are all produced by molds of the genus *Fusarium.* Molds in this genus are very common contaminants of corn and collectively are capable of producing 70 different mycotoxins. Some strains of *Fusarium* may produce as many as 17 mycotoxins simultaneously. Thus *Fusarium* mycotoxins are the most frequently identified group of mycotoxins in grains and feeds.

Table 58.6. Partial list of known mycotoxins

Aflatoxins	Alternariol	Citreoviridin	Citrinin
Cyclopiazonic Acid	Deoxynivalenol	Diacetoxyscirpenol	Dicoumarol
Ergotamine	Ergo Toxins	Fumitremorgen	Fumonisins
Fusaric Acid	Fusariocin	Fusarins	Islanditoxin
Luteoskyrin	Moniliformin	Monoacetoxyscirpenol	Neosolaniol
Ochratoxins	Oosporein	Paspalitrems	Patulin
Penicillic Acid	Penitrem	Phomopsin	Roridins
Rubratoxin	Slaframine	Sporidesmin	Stachbotryotoxins
Sterigmatocystin	T-2 Toxin	Tremorgens	Zearalenone
			Source : Jones *et al.* 1994

Mycotoxin levels found in most field situations tend to be low. Yet the combination of low levels of mycotoxins with the stresses associated with commercial production situations and/or exposure to disease organisms can produce effects in poultry which are subtle, indirect, and sometimes ill-defined. Since the effects of mycotoxins on poultry are dependant upon the age, physiological state, and nutritional status of the animals at the time of exposure, and since mold growth at various points within the feed production and distribution system can magnify mycotoxin problems, mycotoxicoses can be difficult to diagnose in field situations (Jones *et al.* 1994).

Mycotoxins produce a wide range of harmful effects in animals. The economic impact of reduced animal productivity, increased incidence of disease due to immunosuppression, damage to vital organs, and interferences with reproductive capacity is many times greater than the impact caused by death due to mycotoxin poisoning.

In comparison to other animals, poultry species tend to be resistant to the effects of fumonisin, deoxynivalenol, and zearalenone. However, the presence of these mycotoxins within poultry rations is an indication that mold activity has occurred in the ration or in the ingredients within the ration. Since mold activity can generate numerous other mycotoxins as well as reduce the nutritive value and palatability of feeds, the presence of fumonisin, deoxynivalenol, or zearalenone in poultry feeds is cause for concern.

Tolerance limits of different mycotoxins along with general effects toxicity on consumption of contaminated feed is given in Table 58.7.

Table 58.7. Tolerance limits (ppm) for mycotoxin contamination of mixed feed and general symptoms of toxicity

Mycotoxin	Limit	Symptoms of toxicity
Aflatoxin	0.25	Growth depression, deterioration of egg-shell quality, decreased egg production, immunosuppression, subcut and intramuscular hemorrhage
Ochratoxins A	0.20	Growth depression, immunosuppression, impaired mineralization, vitamin-deficiency symptoms, kidney and liver damage, decreased egg production, impaired hatchability
T2 – toxin	0.50	Decreased egg production, impaired hatchability, deterioration of egg-shall quality, growth depression
Vomitoxin	1.00	Subcutaneous and intramuscular hemorrhage, diarrhoea, bone changes
Zearalenone	0.50	Swelling of combs and cloaca, enlargement of ovaries.

Source : Rao, 1982

3.1.1. Aflatoxins

3.1.1.1. Production

Aflatoxins are produced by the fungi *Aspergillus flavus* and *A. parasiticus*. These fungi are present in soil and decaying plant material, cause heating and the decay of stored grain, and may invade corn in the field. Crops and feed ingredients most susceptible to fungi and aflatoxin development include corn, peanuts, peanut meal, cottonseed, and cottonseed meal. Conditions that favor the invasion of corn by *Aspergillus flavus* in the field include drought stress or damage to the corn ear by ear worms or other insects, birds, hail, or early frost. High temperatures, high relative humidity around the kernels, and kernel moisture below 30 percent (wet basis) are ideal conditions for fungal invasion of the kernel. The optimum temperature for aflatoxin production in storage is between 25°C and 32°C. Kernels with a moisture content below 15 percent are at less risk of mold growth and aflatoxin production while an optimum kernel moisture is around 18 percent and an optimum relative humidity in the bin is 85 percent or higher (Herrnam, 2002).

3.1.1.2. Tolerance

Aflatoxins are both acutely and chronically toxic; aflatoxin B_1 is one of the most potent hepatocarcinogens known and hence, the long-term chronic exposure of the extremely low levels in human diet has a tremendous public health importance (Scudamore, 1998).

Young animals are particularly susceptible and males more than females. For most species, LD_{50} is between 0.5 and 10 mg/kg body weight. The liver is the principal target organ although, the site of hepatic effect varies with species. Effects on lungs, myocardium and kidneys have also been observed and aflatoxin can accumulate in the brain (Scudamore, 1998).

While young animals are most susceptible to the effects of aflatoxin, all ages are affected; and clinical signs include gastrointestinal dysfunction, reduced productivity, decreased feed utilization and efficiency, anemia, and jaundice. Aflatoxin causes a variety of symptoms depending on the animal species. However, in all animals, aflatoxin can cause liver damage, decreased reproductive performance, egg production, embryonic death, teratogenicity (birth defects), tumors, and suppressed immune system function, even when low levels are consumed (Jones *et al.* 1994).

Aflatoxin affects all poultry species. Although it generally takes relatively high levels to cause mortality, low levels can be detrimental if continually fed. Young poultry, especially ducks and turkeys, are very susceptible. As a general rule, growing poultry should not receive more than 20 ppb aflatoxin in the diet. However, feeding levels lower than 20 ppb may still reduce their resistance to disease, decrease their ability to withstand stress and bruising, and generally make them unthrifty (Jones *et al.* 1994).

Laying hens generally can tolerate higher levels than young birds, but levels should still be less than 50 ppb. Aflatoxin contamination can reduce the birds' ability to withstand stress by inhibiting the immune system. This malfunction can reduce egg size and possibly lower egg production. In addition, one must pay special attention to the use of contaminated corn in layer rations because eggs are promptly used as human food and aflatoxin metabolites have been found in egg yolks (Jones *et al.* 1994).

Food and Drug Administration (FDA) guideline has fixes maximum tolerance level of 20 ppb in feed for all species of animal with an exception of Cottonseed meal (300 ppb) and corn for mature poultry (100 ppb) (Herrnam, 2002).

3.1.1.3. Clinical signs

Aflatoxin B_1 is most pathogenic to poultry as compared to other aflatoxins viz. B_2, G_1 and G_2. Aflatoxin toxicity (aflatoxicosis) has been reported to cause serious health hazard to poultry and other avian species. Aflatoxicosis in chicken is characterized by listlessness, anorexia, poor pigmentation, jaundice and dehydration of combs and shanks. Sensitivity or resistance to aflatoxins is inherited as distinctive characteristics of breed and strain. The adverse effect of aflatoxins on performance of chicken is also dose and time-related. This toxin is primarily a hepatotoxin in young broiler chicken. One effect that is used as a diagnosis of aflatoxicosis in poultry is an enlarged, fatty, yellow and friable liver that occurs in broilers when they consume aflatoxins contaminated feed. This mycotoxin is also a nephrotoxin and some kidney pathology does results during aflatoxicosis.

Aflatoxins affect the commercial poultry production in many ways:

1. Reduce body weight gain
2. Make birds more susceptible to brushing
3. Decrease egg production, egg weight and hatchability
4. Suppress the immune system of poultry making them more susceptible to diseases
5. Disrupt bone development causing a rachitic type problem and
6. Can cause nutritional problems through its effect on nutrient absorption and metabolism.

Aflatoxin inhibits fat digestion with a consequent streatorrhoea by decreasing enzymes and bile acids resulting into high faecal fat and liver fat content. Aflatoxin causes reduction in nutrient retention, changes in haematological values and bio-chemical parameters. This toxin causes low or heavy mortality depending upon level and duration of intakc of contaminated feed. Reduction in size of bursa of Fabricious is observed in growing chicks during aflatoxicosis.

The tolerance level of aflatoxin B_1 in feeds have been depicted in Table 58.8.

Table 58.8. Tolerance levels of dietary aflatoxin

Type of birds	Tolerance (ppb)
Cross-bred broilers	400
Pure bred broiler chicks	200
White Leghorn chicks	150
Quail chicks	300
Quail layers	300
Guinea fowl keets	1500
Layers	600

3.1.1.4. Detection

Analytical techniques for the detection of mycotoxins continue to improve. Several commercial laboratories now test for a variety of mycotoxins. Although analytical costs can be a constraint, these costs may be insignificant compared with the economic consequences of production and health losses associated with mycotoxin contamination (Jones *et al.*, 1994).

Aflatoxin and other mycotoxins can be measured using relatively simple and inexpensive technology. The black light method only works for aflatoxin and is a presumptive (not a definitive) test. This technology is based on the use of ultraviolet light to produce a bright greenish-yellow fluorescence from kojic acid. This acid is produced by the same fungi that produces aflatoxin, thus, it indirectly indicates the potential presence of aflatoxins. The black light method is no longer used because of the high probability of false positives and negatives. Serological (immunodiagnostic) methods for detecting aflatoxin and other mycotoxins are available through a number of detection kits that are fairly accurate and reasonably priced.

These kits use antibodies that selectively capture a specific mycotoxin that has been extracted from the grain or feed ingredient. After the extract has been applied to the test kit and the mycotoxin captured by the antibody, a color indicator is added. Depending on the kit, the presence or absence of the mycotoxin is indicated by the intensity of color. Some of these kits may be referred to as an ELISA (enzyme-linked immunosorbent assay) test.

Commercial antibody test kits for screening or quantitation are currently available for aflatoxins, zearalenone, deoxynivalenol (DON), T-2 toxin, ochratoxin A, and fumonisins. These antibody methods, while they are still being improved, are good if used properly.

The use of chromatography, either thin layer, gas, or high performance liquid chromatography, is a more reliable method of mycotoxin detection used by commercial laboratories.

3.1.1.5. Sample Collection

The manner in which samples are obtained and processed is an important consideration when testing for aflatoxin and other mycotoxins. Samples of grain must be representative of the lot and of sufficient size to compensate for the uneven distribution of the contaminant as well as the ultra-low levels (parts per billion) that must be detected. Variability in grain samples used for mycotoxin analysis occurs because individual kernels do not contain the same amount of the contaminant, not all kernels are contaminated, and their distribution throughout the load may not be uniform.

3.1.1.6. Inactivation of preformed aflatoxins

The physico-chemical treatments that can be employed for inactivation of preformed aflatoxins in contaminated maize and groundnut cake are:

1. Raising the moisture level up to 20%. Autoclaving at 0.35 kg/cm² for one hour followed by drying in an oven at 80°C.

2. Adding sodium hydroxide (15 g/kg) and mixing. Raising the moisture content upto 20%, autoclaving at 0.35 kg/cm² for one hour and drying in an oven.

3. Agitation of one kilogram of feedstuff with 20 g Ca(OH)2 followed by addition and mixing of formaldehyde to raise the moisture content up to 15%. Autoclaving at 1.05 kg/cm² for an hour and drying.

4. Addition of liquor ammonia to yield 6% concentration. Raising of moisture content up to 20%. Storing airtight for 20 days. Heating at 35°C and drying in an oven.

Physico-chemical treatments for inactivation of preformed aflatoxin B_1 in feeds are given in Table 58.8.

3.1.1.7. Chemical detoxification of aflatoxin-contaminated feed (Coker, 1998)

Several chemicals including oxidizing agents, aldehydes, acids and bases (organic and inorganic) have been tried to detoxify aflatoxin-contaminated feed and feed ingredients (Tables 58.9 and 58.10). It is in the fitness of things to note that none of the methods till now has been successful incompletely eliminating/removing/detoxifying preformed aflatoxins in poultry feeds and treatment with ammonia appears to be most commercially suitable method.

3.1.1.7.1. Ammonia

Ammonia both as anhydrous vapor as well as a solution has been used with varied success. Large-scale ammoniation is being employed in some countries to treat peanut products. Anhydrous ammonia (1 to 4%), a moisture level of 12 to 17%, a pressure of the approximately 2.1 kg/cm² and a temperature of at least 80°C for a period of ½ to 1 hr is the general recommendation; detoxification to the extent of 95 to 98% has been achieved with this method.

Ammoniation also resulted in some degree of thermal/chemical denaturation of protein and increased residual ammonia. However, ammonia treatment successfully prevented development of hepatic lesions. In addition, meat and eggs from birds fed on ammoniated feed did not show toxic or carcinogenic effects.

Table 58.9. Additives for protection against aflatoxins in broiler diets

Detoxifying agents	%
Activated charcoal	0.10 – 0.20
Hydrated sodium calcium aluminosilicate (HSCAS)	0.10 – 0.20
Esterified glucomannan (EGM)	0.05 – 0.10
Herbal mixture (Acacia catechu. 25%, Phyllanthus niruri, 400%, Andrographis paniculata, 25%, base 10%)	0.50 – 0.075
Butylated hydroxyanisole	0.50 – 0.10
Dl-methionine*	0.10 – 0.20
Selenium**	2000 – 3000*
Butylated hydroxy toluene*	0.05 – 0.15
L-lysine HCI	0.15
Water soluble vitamins*	Double of the requirements
Increase the dietary protein level	Upto 26 to 28%
* ppm	

Reduction in weight gain and FCR was observed in birds fed on ammonia-treated feed; however, methionine supplementation helped overcome the shortcoming. Egg production, egg weight, fertility or hatchability were found to be unaffected by the ammonia treatment of the feed. Ammonia treatment of the feed was not found to interfere with the effectiveness of vaccinations.

3.1.2. Ochratoxin A

These are produced by various species of genera *Aspergillus* (*A. ochraceus, A. alliaceus, A. melleus, A. osteanus, A. pertakii, A. sclerotiorum and A. sulphureus*) and *Penicillium* (*P. viridicatum, P. commune, P. cyclopium, P. variable, P. purpurescens and P. palitons*) of which *A. ochraceus* is the principal and *P. viridicatum*, the frequent sources of ochratoxins. The toxins frequently contaminate cereal grains and feedstuffs and are not sure to be eliminated by cooking or processing.

Ochratoxin A (OA) is nephrotoxic and its de-chlorinated derivative is Ochratoxin B (OB) and ethyl-ester derivative is Ochratoxin C (OC). Methyl-esters of OA and OB are referred to as OA-M and OB-M, respectively; and 4-hydroxy Ochratoxin is Ochratoxin D (OD).

OA is most toxic and is chemically 7 carboxyl 5 chloro 8 hydroxyl, 3,4 di- hydro 3R methyl iso- coumarin linked to L-Phenylalanine ($C_{20}H18C_{l}NO_{6}$). It is photosensitive and thermostable. It binds to bovine serum albumin and therefore exploited to produce, in rabbits, antisera to OA for detection of OA by ELISA.

Table 58. 10. Methods of detoxificaton of aflatoxin

Process	Procedure	Moisture in feed (%)	Detoxification GNC (%)	Detoxification Maize (%)
Exposure to sunlight	Groundnut feed spread in 1 g layers – exposed to sunlight for 6h for 2d	7-10	50	Nil
Heat treatment	Feeds made to contain 10,15, 20% moisture – Autoclaved at 0.07 kg/cm² for 2½ h – Air dried	10-20	79	50
Alkali treatment	Feeds mixed with NaOH at 20 g/kg in water to yield 15, 20 and 30% moisture. Autoclaved at 0.07 kg/cm² for 1½ to 2 h. Dried at room temperature.	20	92	88
$Ca(OH)_2$ + HCHO treatment	Feeds agitated with $Ca(OH)_2$ powder at 2% - Formaldehyde (37% in water) added to above to yield 15, 20 and 25% moisture. Autoclaved at 0.7 kg/cm² for 1 h - Dried	15	94	92
Liquor ammonia	Feed mixed with liquor ammonia to yield 3,4 and 5% concentration in feed – Feed moisture 10, 15 and 20%; stored air-tight for 10d; heated at 40°C for 1 h and dried.	3% - 15-20 4% - 15-20 5% - 30 5% - 15	69 69 79 84	91 96 97 94

Ochratoxin was first reported in the US by Shotwell et al 1969 in corn. In UK, a report on poultry meat suggests a level of ochratoxins to a tune of 4.3 to 29.2 ppb. First reported poultry was made by Hamilton et al 1982.

LD_{50} of Ochratoxin is 2.14 ppm for day-old chicks and 3.6 ppm for 3-week old birds; 5.9 ppm/kg body weight in case of turkeys, 16.5 ppm/kg body weight for Japanese quails.

This is a potent toxin affecting mainly kidneys producing both acute and chronic lesions. Its dechloro derivative, Ochratoxin B, is non-toxic. For chicken, Ochratoxin A appears to have geneotoxic carcinogenic effect also.

Although, ochratoxin A is primarily a nephrotoxin in poultry, it does have some secondary hepatotoxicity. The major diagnostic lesion of this mycotoxin is pale and enlarged kidneys. Ochratoxin, like aflatoxin, can affect poultry in number of ways:

1. Depresses growth of poultry

2. Acts as an immuno-suppressant making birds more susceptible to diseases and brushing and
3. Can affect nutrient absorption and metabolism.

It is important to note that ochratoxin is three times more toxic than aflatoxins to broiler chicken. This toxin is produced by certain Pencillium (*P. Verrucosum*) and Aspergillus (*A.Ochraceus* and *A.alutaceus*) species. It has been implicated in total human disease Balkan Endemic Nephropathy.

Ochratoxins are produced by several species of *Aspergillus* and *Penicillium*. They can occur in cereal grains, dry beans, and moldy peanuts. Ochratoxins occur most readily in storage of high-moisture (> 22 %) grains. Therefore, the only recommended control is to keep grains cool and dry in storage (Herrman, 2002).

3.1.2.1. Clinical Signs of Ochratoxin A

In poultry, symptoms include retarded growth, decreased feed conversion, impaired kidney function, and mortality. Ochratoxin A can cause a decrease in egg shell quality and egg production (Herrman, 2002).

Moisture content and temperature are the most important factors that affect growth of toxin-producing molds. Generally, fungi do not produce ochratoxin on cereal grains stored at moisture contents lower than 15 percent. Antioxidants such as ascorbic acid have been shown to reduce the toxic effects of ochratoxin A in laying hens (Herrman, 2002).

3.1.3. T-2 toxin

This toxin, produced by fungus Fusarium tricincutum, is another mycotoxin that can produce severe adverse effects on broiler chicken. T-2 toxin is a radiomimetic toxin, which means that the toxicity of T-2 toxin is very similar to the effects of radiation. The principle presumptive lesion caused by T-2 toxin is a crusty lesion in the mouth of poultry on lower and upper mandible. T-2 toxin can limit poultry production by a number of methods:

1. Decreasing body weight
2. Acting as immuno-suppressant by depressing the regenerative process in the bone marrow and spleen and
3. Altering nutrient absorption and causing nutritional disease due to irritation, hemorrhage, and necrosis throughout the digestive tract. Affected animals show signs of weight loss, poor feed utilization, lack of appetite, vomiting, bloody diarrhea, and (in severe cases) death.

T-2 toxin and trichothecenes can cause mouth and intestinal lesions as well as impair the birds' immune response, causing egg production declines, decreased feed consumption, weight loss, and altered feather patterns. While much is yet to be learned, T-2 toxin and related compounds are currently thought to be the most potent Fusarium mycotoxin for poultry (Jones *et al.* 1994).

3.1.4. Vomitoxin (deoxynivalenol, DON)

Deoxynivalenol (DON), commonly referred to as vomitoxin, is produced by *Fusarium* species. *F. graminearum*, a *Fusarium* species which causes root, stalk, and ear rots of corn and sorghum also causes scab in wheat and produces DON in all of these grains. Field conditions that favor the invasion of these grain crops include warm, moist weather. Blight symptoms can develop within 3 days after infection when temperatures range between 25ºC and 30ºC and moisture is continuous. Plants appear most susceptible when they are infected at the flowering stage of development (Herrman, 2002).

DON is not known to increase in stored shelled corn or in small grains that come contaminated from the field since *Fusarium* growth requires a minimum moisture content of 19 to 25%.

3.1.4.1. Clinical signs of trichothecenes

DON belongs to a class of mycotoxins referred to as trichothecenes, which also include the mycotoxin identified by the name, T-2. General signs of trichothecene toxicity in poultry include weight loss, decreased feed conversion, feed refusal, vomiting, bloody diarrhea, severe dermatitis, hemorrhaging, decreased egg production, and death (Herrman, 2002).

Feeds containing more than 1 part per million (ppm) of DON may result in a reduction of feed intake and lower weight gain. Vomiting may occur in some cases (Herrman, 2002).

T-2 toxins in poultry may produce lesions at the edges of the beaks, abnormal feathering, reduced egg production, eggs with thin shells, reduced body-weight gain, and mortality (Herrman, 2002).

DON alone has few effects in poultry. However, in field situations the DON level is sometimes associated with reduced feed consumption in layers and broiler breeders. This means that DON may be an indicator that T-2 or other unknown *Fusarium* mycotoxins are present (Jones *et al.* 1994).

3.1.5. Rubratoxin B

This causes reduction in growth, atrophy of bursa hypertrophy of liver.

3.1.6. Tremortirn A

Tremortirn A affects central nervous system and causes tremors.

3.1.7. Citrinin

One of the symptoms of citrinin toxicity in poultry is a dramatic increase in the amount of water consumed and excreted which helps in diagnosing its toxicosis.

Citrinin often occurs with Ochratoxin A and hence is often missed during feed analysis. This is mainly found in rice and other cereals. This toxin also is toxic to kidneys.

3.1.8. Patulin

This toxin causes irritation of the gastro-intestinal tract; but does not seem to be of importance in poultry diets.

3.1.9. Zearalenone

Zearalenone is typically produced by the fungus *Fusarium graminearum*; however, other Fusarium species may produce some zearalenone. Corn is the major source of this mycotoxin. The field conditions described for the production of DON also apply for zearalenone. Exposure to conditions that retain the kernel moisture content between 22 and 25%, such as a delay in harvest for several weeks, will favor the growth of the fungus and production of zearalenone. Production of the toxin in storage is unlikely, unless moisture exceeds 22% (Herrman, 2002).

Zearalenone mimics the effect of the female hormone estrogen and, at low doses, increases the size or early maturity of reproductive organs.

3.1.10. Fumonisin

Fumonisin is a mycotoxin which has only recently been discovered. Thus it has not been extensively studied. Nonetheless, it is known that in most animals fumonisin impairs immune function, causes liver and kidney damage, decreases weight gains, and increases mortality rates.

Fumonisins are produced by *Fusarium verticillioides, F. proliferatum* and other Fusarium species that typically occurs in corn. These fungi infect corn roots, leaves, stalks, and kernels. *F.verticillioides* survives on crop residue in or on the soil surface and, under favorable conditions, can infect corn stalks directlyor through wounds caused by hail or insects. The fungus is commonly seedborne (Herrman, 2002).

Fumonisins are reported to occur on visibly healthy grains. Of the currently identified fumonisins, B_1, B_2, and B_3 are the most abundant in naturally contaminated foods and feeds and fumonisin B_1 generally comprises 75% of the total content. 50 ppm for poultry feed has been considered maximum safe (Herrman, 2002).

Toxicity of fumonisins is thought to be due to their effects on sphingolipid synthesis. However, fumonisin toxicity is rare in poultry species.

3.1.11. Sterigmatocystin and Cyclopiazonic acid

Both are of rare occurrence in poultry diets.

3.1.12. Moniliformin

Contaminated maize may contain a mixture of toxins produced by *Fusarium* species which includes moniliformin among others. Otherwise, moniliformin toxicity in poultry does not seem to be reported.

3.1.13. Alternaria toxins

The principal mode of action of Alternaria toxins appears to be the inhibition of protein synthesis by suppressing the release of newly-formed proteins from ribosomes into supernatant fluid. Hence, they are cytotoxic and exhibit antitumor, antiviral and antibacterial activity. They also have mutagenic activity.

3.1.14. Miscellaneous mycotoxins

There are innumerable mycotoxins occurring in animal feeds and forage; but many of them are of rare occurrence. They include toxins from *Aspergillus fumigatus* and *Aspergillus clavatus*, citreoviridin, ergot alkaloids, fusarin C, gliotoxin, lolitrem B, mycophenolic acid, 3-nitropropionic acid, penicillic acid, PR-toxin, roquefortine C, rubratoxin A, satratoxins G and H, sporidesmin, viomellei, vioxanthin and xanthomegnin.

3.1.15. Multiple mycotoxins

There are a number of ways in which feeds can become multiple mycotoxin contaminated. Several analytical laboratories have found that most mycotoxin contaminated samples contain more than just a single mycotoxin. When aflatoxin and Ochratoxin are co-contaminants of poultry feed, these interact in a synergistic manner. During dual exposure of these toxins, ochratoxin prevents the major effect of aflatoxin (i.e. fatty, yellow, enlarged and friable liver). This confuses the ability to diagnose aflatoxicosis in the field. The target organ in this interaction appears to be the kidney. Citrinin is a nephrotoxin similar to ochratoxin and is produced by some of the fungi that also produce ochratoxin. The interaction is characterized as antagonistic. During co-toxicity ochratoxin prevents the very important diagnostic index of citrinin (increase in the amount of water consumption and excretion) which would make it very difficult to diagnose citrinin in the field based upon symptoms, if ochratoxin was also a contaminant. Aflatoxin and T_2 toxin combination is like interaction between aflatoxin and ochratoxin and exhibit synergistic toxicity.

3.2. Minimizing mycotoxin production

3.2.1. Before manufacture

3.2.1.1. Mold control

(Jones *et al.* 1994)

Moisture is the single most important factor in determining if and how rapidly molds will grow in feeds. Moisture in feeds comes from three sources: (1) feed ingredients, (2) feed manufacturing processes, and (3) the environment in which the feed is held or stored. To control the moisture content of feeds successfully, moisture from all three sources must be controlled.

3.2.1.2. Moisture in Feed Ingredients

Since corn and other grains are a primary source of the moisture and molds found in feed, the first important step in controlling moisture in feed is to control it in the grains from which the feed is prepared. Since all feed ingredients contain moisture, they should be monitored and their moisture content controlled.

It is commonly believed that the amount of moisture in grain is too small to permit mold growth except in rare and unusual circumstances. However, moisture is not evenly distributed in grain kernels. A batch of grain containing an average of 15.5% moisture may, for example, contain some kernels with 10% moisture and others with 20% moisture. The moisture content of individual grain kernels is directly related to the amount of mold growth that occurs: that is, kernels with higher moisture contents were more susceptible to mold growth. In addition to moisture, the amount of mold growth is about five times greater for broken kernels than for intact kernels. Thus the fraction of commercial grain, known as broken kernels and foreign matter, can be expected to have a higher mold and mycotoxin content than the portion composed of whole kernels.

3.2.1.3. Moisture in Feed Manufacturing Processes

Grains are commonly ground with a hammer mill to aid in mixing and handling, to improve digestibility, and to improve the pelleting process. This grinding process creates friction, which causes heat to build up. If unchecked, temperature increases greater than 5-6°C will cause significant migration of grain moisture encouraging mold growth. This is particularly true in cold weather when temperature differences cause moisture to condense on the inside walls of bins. Air-assisted hammer-mill systems reduce heat buildup in the product and, in turn, reduce moisture problems.

The pelleting process involves mixing steam with the feed, pressing the mixture through a die, and then cooling the pellets to remove heat and moisture. Generally, heat and 3 to 5 percent moisture are added to the feed during the pelleting process in the form of steam. If the pelleting process is done correctly, this excess moisture is removed from the feed before shipment. If, however, this excess moisture is not removed when the pellets are cooled, mold growth will be encouraged. Since feeds containing moisture are warmer than normal, storing hot or warm pellets in a cool bin will cause moisture to condense on the inside of the bin.

Although pelleting of feed has been shown to reduce mold counts by a factor of 100 to 10,000, many mold spores remain in the feed after it has been pelleted. After pelleting, the remaining spores can grow if conditions are right. Thus the pelleting process can only delay, not prevent, the onset of mold growth and plays only a minor role in efforts to control molds. In addition, pelleted feeds may be more easily attacked by molds than nonpelleted feeds.

3.2.1.4. Moisture and Feed Storage Environment

To control mold growth, obvious sources of moisture in the feed handling and storage equipment must be eliminated. These sources may include leaks in feed storage tanks, augers, roofs (either at the barn or at the feed mill), and compartments in feed trucks.

A fact about feed moisture often overlooked is that it changes in relation to the feed's environment. Since animals kept in confinement housing add moisture to their environment by respiration and defecation, the air in these houses can be very humid. Feed that was initially very low in moisture content will gain moisture when placed in a humid environment. The humidity in confinement housing should therefore be controlled by providing adequate ventilation.

3.2.2. Manufactured feed

(Jones *et al.* 1994)

Time is required for both mold growth and mycotoxin production to occur. It is therefore important to have feeds delivered often so that they will be fresh when used. Feeds should generally be consumed within 10 days of delivery.

It is equally important to manage the feed delivery system to ensure that feeds are uniform in freshness. If the feeder system is allowed to keep the feed pans full at all times, the feed in the pans will be significantly older than that in the storage tank. The animals will tend to eat primarily the feed in the top layer, and the feed at the bottom of the pans will age, providing greater opportunities for molds to grow.

3.2.2.1. Equipment Cleanliness

When feed is manufactured and delivered to farms, it may come in contact with old feed that has lodged or caked in various areas of the feed storage and delivery systems. This old feed is often very moldy and may "seed" the fresher feed it contacts, increasing the chances of mold growth and mycotoxin formation. To prevent this problem, caked, moldy feed should be removed from all feed manufacturing and handling equipment.

3.2.2.2. Use of Mold Inhibitors

3.2.2.2.1. Types of Mold Inhibitors

The use of chemical mold inhibitors is a well-established practice in the feed industry. However, mold inhibitors are only one of several tools useful in the complex process of controlling the growth of molds, and they should not be relied upon exclusively.

The main types of mold inhibitors are (1) individual or combinations of organic acids (for example, propionic, sorbic, benzoic, andacetic acids), (2) salts of organic acids (for example, calcium propionate and potassium sorbate), and (3) copper sulfate. Solid or liquid forms work equally well if the inhibitor is evenly dispersed through the feed. Generally, the acid form of a mold inhibitor is more active than its corresponding salt.

3.2.2.2.2. Dispersion

Mold inhibitors cannot be effective unless they are completely and thoroughly distributed throughout the feed. Ideally, this means that the entire surface of each feed particle should come in contact with the inhibitor and that the inhibitor should also penetrate feed particles so that interior molds will be inhibited.

The particle size of the carriers for mold-inhibiting chemicals should be small so that as many particles of feed as possible are contacted. In general, smaller the inhibitor particles, greater the effectiveness. Some propionic acid inhibitors rely on the liberation of the chemical in the form of a gas or vapor from fairly large particle carriers. Presumably, the inhibitor then penetrates the air spaces between particles of feed to achieve even dispersion.

3.2.2.2.3. Effect of Feed Ingredients

Certain feed ingredients may also affect mold inhibitor performance. Protein or mineral supplements (for example, soyabean meal, fish meal, poultry by-product meal, and limestone) tend to reduce the effectiveness of propionic acid. These materials can neutralize free acids and convert them to their corresponding salts, which are less active as inhibitors. Dietary fat tends to enhance the activity of organic acids, probably by increasing their penetration into feed particles. Certain unknown factors in corn also alter the effectiveness of organic acid inhibitors.

3.2.2.2.4. Time Dependence

When mold inhibitors are used at the concentrations typically recommended, they in essence produce a period of freedom from mold activity. If a longer mold-free period is desired, a higher concentration of inhibitor should be used. The concentration of the inhibitor begins to decrease almost immediately after it is applied as a result of chemical binding, mold activity, or both. When the concentration of the inhibitor is reduced until it is incapable of inhibiting mold growth, the mold begins to use the inhibitor as a food source and grows. In addition, feeds that are heavily contaminated with molds will require additional amounts of inhibitor to achieve the desired level of protection.

3.2.2.2.5. Influence of pelleting

The widespread use of pelleted feeds in the feed industry is beneficial to the use of mold inhibitors. The heat that the feed undergoes during pelleting enhances the effectiveness of organic acids. Generally, the higher the pelleting temperature, the more effective the inhibitor. Once mold activity commences in pellets, however, it proceeds at a faster rate than in nonpelleted feed because the pelleting process that makes feed more readily digestible by animals also makes it more easily digested by molds.

3.2.2.2.6. Copper sulfate

Although copper sulfate in the diet has been shown to improve body weight and feed conversion efficiency in broilers, excessive levels of copper may be toxic to young animals and will accumulate in the environment. In addition, feeding copper sulfate to poultry itself can cause formation of mouth lesions similar to those formed by some mycotoxins.

3.2.2.2.7. Animal Management

If unacceptable mycotoxin levels occur, removal of the contaminated feed is preferable. Acidic diets may intensify the effects of mycotoxins and should be avoided in these situations. Increasing nutrients such as protein, energy (fats and carbohydrates), and vitamins in the diet may also be advisable. The addition of antioxidants to the animal assists in dealing with the effects of mycotoxins.

The possible use of inorganic binders (mineral clays which include zeolites, bentonite, bleaching clays from refining of canola oil, and hydrated sodium calcium aluminosilicates [HSCAS]) to bind mycotoxins, and prevent them from being absorbed by the animal's gut, has received a lot of research attention recently. However, it should be clearly understood that binding of some mycotoxins may be weak or nonexistent and that clay products differ in their ability to bind mycotoxins. Nonetheless, many clay products are GRAS (Generally Recognized As Safe) and are used as anti-caking or free-flow additives for feeds.

4. Argemone contamination

Argemona maxicana is a yellow flowered poppy belonging to the family *Papaveraceae* and genera, *Argemona.* Due to its high oil content and cheap availability seeds are used as adulterant with mustard seed and its cake in oil extraction. It caused glaucoma, cancer, swelling of leg, diarrhoea, increased intraoccular tension and atrophy of optic nerve in animals including chickens. However, it has been observed that heating of this oil to 240°C for 15 min makes it innocuous.

5. Pesticides and other industrial chemicals

Pesticide/insecticide residues also pose a health hazard to livestock including poultry. These compounds of diverse chemical nature may be categorized into (a) the chlorinated hydrocarbon i.e. organochlorine (OC), such as DDT, BHC, aldrin, endrin, dieldrin, methoxychlor, chlordane, texaphene, mirex, etc., which are of persistent nature but harmless, if used correctly; (b) the organophosphorus compounds (OP), such as parathion, malathion, sumithion, dimelthoate, diazinon, etc.; (c) the carbonate compounds, such as carbaryl, pyrolan, etc. and (d) the synthetic pyrethroids, such as permehrin, cypermethrin, deltamehrin, allethrin, fenvalerate, etc.

The organochlorine insecticides are much less toxic and yet are dangerous because of their persistence and cumulative character in the body tissues. The poultry and other livestock products, viz. milk, meat, eggs obtained from the animals and birds, which are being exposed to the residues of such persistent insecticides, could be also harmful for human consumers.

Acute oral LD_{50} of malathion has been reported to be 524.80 mg/kg body weight for desi poultry birds. Exposure to some organophosphorus esters causes organophosphorus -induced delayed neurotoxicity (OPIDN) in adult poultry characterized by degeneration of axons and followed by myelin in the peripheral and central nervous systems.

DDT, sprayed on litters, feeders, roofs or nests in poultry houses, could be detected in the fatty tissues of birds and eggs within the same week. Feeding of the measured doses of DDT at 15 ppm level causes no detrimental effect on egg production although residual DDT persisted in egg yolk and abdominal fat as long as 17 and 13 weeks. No significant level of the residue could be detected in eggs at a feeding level of 0.1 - 0.15 ppm. A build up of 0.05 ppm residue in 7 days and 0.14 ppm in 8 weeks in eggs could be detected on a diet containing less than 0.1 ppm DDT. Fertility, hatchability of eggs or progressive growth was not significantly affected even at 5, 15 and 50 ppm levels in the diet.

Exposure of laying hens to technical DDT at a level of 100 ppm for 40-50 days was found to cause disorders in eggshell formation reducing its thickness and strength. Sperm production could be greatly reduced by incorporating 0.1 - 0.3% DDT in diet of domestic fowl. Broilers may retain more residues since they are forming new tissues rapidly and their feed consumption is also higher. In broilers, dietary fenvalerate (FEN) and methyl parathion (MPA) adversely affected the performance beyond dietary levels of 100 ppm and 25 ppm, respectively. Zeolite + activated charcoal (13.75:1) at rate of 2 kg/ton had a beneficial role in protecting birds from lower levels of insecticidal toxicity.

Chapter **59**

Immune System

Among the avian species, the immune system of the chicken has been studied most extensively. The immune system of chicken comprises bursa of Fabricius, thymus, bone-marrow, spleen, Harderian glands, cecal tonsils and primitive lymph nodes. The main immune functions are carried out by specialized cells, the lymphocytes, assisted by the reticuloendothelial system. Birds respond to antigenic stimulation by generating antibodies as well as cellular immunity.

The immune system begins with lymphocyte stem cells, which originate in yolk sac of the embryo during the first week of incubation. These stem cells migrate to the thymus and bursa of Fabricius; in the former, they mature as T cells and in the latter as B cells. Both thymus and bursa of Fabricius are collectively referred to as "Central lymphoid tissues" and contributes B and T cells to the other lymphoid structures.

The T cells begin to leave the thymus 2 or 3 d before hatching and they are continuously released by thymus till it regresses by sexual maturity. The B cells begin their movement from about 15th d of incubation and most of the migration is completed before 9 weeks of age; bursal function disappears before sexual maturity by about the same time (Bains, 1994).

A portion of the T and B cells so released are seeded in other lymphoid organs and the rest are found in blood circulation; blood generally has 17 and 30% of circulating T and B lymphocytes, respectively.

Avian lymphatic system is responsible for bestowing normal immunity and is distinct from that in mammals in presence of bursa of Fabricius (normally referred to as bursa), a lympho-epithelial organ unique to birds, and in absence of mammalian-type lymph nodes performing filtering role. However, there are smaller and more simple lymphoid nodules associated with walls of lymph vessels, but with no filtering function. Functionally, avian lymphatic system is divided into bursa-dependent (B-system) and thymus-dependent (T-system) components (Firth, 1977).

B-system comprises bursa with secondary germinal centers and plasma cells spread throughout the body conferring humoral immunity. T-system, on the other hand, comprises thymus and scattered collection of lymphocytes throughout the body which is responsible for cellular immunity. Other lymphoid/lymphatic organs contain mixed amounts of cells derived from both bursa and thymus (Firth, 1977).

1. Central component

(Firth, 1977)

1.1. Bursa

Bursa is an ovoid hollow organ that is a duct leading to the dorsal part of proctodeum. The shape varies from tubular to pyriform. In chicken, the inner surface has 11 to 13 primary and 6 to 7 secondary plicae. Below the epithelium lining of the cavity, is the tunica propria where the lymphocytes mature.

1.1.1. Functions of bursa

1. Hormonal – exerts an influence on glycogenolysis in liver, iodine uptake by thyroids and Vitamins C repletion mechanism in adrenals
2. Immunological – Bursa is the central immunoglobulin (Ig)-synthesizing organ from where Ig-producing cells may originate. These cells, the B-cells, occur in peripheral tissues outside the bursa comprising the plasma cell system and germinal centers.
3. Source of immunologically-competent cells which are seeded out to other tissues.
4. May secrete a hormone necessary for development of immunocompetence by the Ig-synthesizing system of cells of non-bursal origin.
5. Local defense of the gut. The bursal plical mucosa was found to take up sample of intestinal antigens by pinocytosis and produce bursal-antibodies.
6. Role in erythropoiesis by means of cells and/or humoral factors. However, bursa does not seem to be concerned with antigen phagocytosis in blood but antigenic stimulation in young birds may result in early B-cell proliferation.

1.1.2. Classification of B-cells

The B-cells are classified into:

1. B-stem cells which are needed for restoring normal bursal morphology in association with bursal epithelial cells. B-cells help restoring germinal centers in spleen which is a B-cell recipient organ. The B-stem cells migrate to bone-marrow at the time of bursal involution. Subsequently, a similar cell appears in spleen and to some extent in thymus. Ig-bearing cells appear in bursa at day 14 and increase in number by day 18 to 20 (interestingly, embryonic thymus lacks Ig-bearing cells). B-cells with self-replication and memory functions were found to be seeded after 4d of age. Intra-bursal switch mechanism is thought to operate in differentiating Ig M, Ig G and Ig A lymphocytes and, at once these lymphocytes leave, they are independent of bursa but immunological in function.

2. Post-bursal cells: these do not form a part of structure-restoring system and they mature independent of bursa itself. Post-bursal cells are classified further into early-and late-types. The early post bursal cells can multiply in treated bursa and form germinal centers, whereas, late post-bursal cells have no effect on germinal centers but are the principal cells in the bursa and bone-marrow after 10 weeks of life.
3. The B cells emerging from bursa are coated with non-specific antibodies; they respond to an infection by producing 1. Memory cells and 2. Antibody-producing cells (plasma cells) (Bains, 1994).

1.2. Thymus

1.2.1. Morphology and involution

Thymus appears like a series of yellowish-pink or cream lobes along both sides of the neck. There are 5 pairs of bodies in young chicken superficially to the jugular vein and held in position by fascia. The margins may have fat deposits around them. Thymus is a feature of only young chicken. It reaches its maximum size by four months (chicken) or six months (turkeys) before sexual maturity. It regresses gradually and is absent in adults. Unusual regression indicates infection.

1.2.2. Functions of thymus

Thymus is responsible for cellular immunity (T-cell immunity) i.e., lymphocytes. The lymphocytes of the medullary origin only are immunocompetent. Involution of thymus involves only cortex but medulla remains throughout life. Cells originating from involuted thymus are, in fact, more immunocompetant than from non-involuted thymus and development of immunocompetance is earliest in cells originating from medulla of thymus. Thymus has little, if any, effect on antibody (Ab) production. Thyroxin stimulates and glucocorticoids inhibit growth of thymus.

Ig A lymphocytes are relatively independent of both thymus and antigen whereas their maturation into IgA plasma cells depends on antigenic stimulation and T-cell help. Further, lymphocyte population of bursa and thymus are not independent but mixed. More than half of lymphocytes in bursa fail to react with either T-cell or B-cell specific Ab. These "null cells (n cells)" are thought to be involved in cell-mediated Ab-dependent cytotoxicity.

T-cells are the main effector cells of cellular immunity. Functionally, T-cells either amplify or suppress the production of agglutinating Ab. The cells present in 16d embryo and disappear in 6 to 8 months of age. Significant migration of lymphocytes of thymic origin moves to bone-marrow 3d after vaccination of 6 weeks old birds. Cells from bursa are transported to thymus, spleen, cecal tonsils but such migration is not demonstrated in 14-week birds. The migrated cells localize in medulla of thymus and red pulp of spleen but not to bone-marrow. Both T-and B-cells can produce lymphocyte mitogenic factors when stimulated by specific Ag, but the B-cells require the presence of some T-cells in the population.

The mature T-cells develop an "antigen recognition system" and produce several substances which have important roles in cellular immunity. In addition, T-cells are also required for activation of B cells. Following infection, T-cells produce 1. Memory cells 2. Specific effector cells (lymphocytes) and 3. Suppressor cells (Bains, 1994).

2. Peripheral component

(Firth, 1977)

This consists of all other normal lymphocyte aggregation or proliferation outside the central component; the primary peripheral component is spleen and the other important ones are lymphocyte tissue of the alimentary tract occurring as diffused masses irregularly in the lamina propria and sub-mucosa, especially in the areas of roof of pharynx, around posterior nares, esophagus, proventriculus, gizzard, small intestine, cecal tonsils, colon and cloaca. Some of the aggregations are visible to naked eye like those in the terminal ileum. Accessory bursae lie on each side of the posterior part of bursal stalk and are embedded in cloacal wall. However, its functional relationship with true bursa is doubtful. Para-ocular and para-nasal organs - include lacrimal ducts (small lymphocytes and germinal centers), Harderian glands (small lymphocytic nodules), and ducts of lateral nasal glands (plasma cells). Mural lymphoid nodules - are closely located on one side of a lymphatic vessel. Others – include numerous small foci of lymphoid tissue scattered throughout many organs and tissue (connective tissue, bone-marrow, skin, liver, lung, kidney, pancreas, endocrine glands, peripheral nerves, larynx and trachea)

B-cells reach spleen along with bursa but cannot proliferate till bursa is intact. Cell migration from bursa to spleen and thymus (but not to intestines, bone-marrow and cecal tonsils) occurs between 18 and 20 d of incubation. Even after hatching, no significant moment of cells into the bone-marrow has been recorded. Bursa seeds out immunocompetant cells only after the lymphoid cells are mature and bursa itself has started involuting.

Spleen appears to contain small lymphocytes at hatch and reaches serological maturity by 5 to 6 weeks of age. Primordia for cecal tonsils appear at about 10d of incubation and on 18th d contain small numbers of lymphocytes. Maximum number of germinal centers is recorded at 8 weeks of age.

2.1. Functions of peripheral component

The peripheral lymphoid tract contains both bursa-derived and thymus-derived lymphocytes. They can be differentiated using specific antisera. Proportion (%) of bursa-derived cells in spleen, peripheral blood lymphocytes, cecal tonsils and Harderian glands is 20 to 30, 30, 50 to 70 (depending on age), and more than 80, respectively; the rest are thymus-derived ones.

2.2. Spleen

Avian species is not erythropoietic; on the contrary, soon after hatching, its main functions are:

1. Phagocytosis of RBCs by macrophages of red pulp
2. Ag uptake and Ab production by lymphoid cells in both red and white pulps and
3. Lymphocytopoiesis in white pulp.

Spleen has been shown to contain two groups of Ab-producing cells viz., plasma cells appearing 1d post-infection, reaching peak by 8 to 9d post-infection and germinal centers where the Ab-producing cells appearing 13d post-infection and remaining up to 42d. The early plasma cells which are not derived from germinal centers probably help localize the Ag in the white pulp as Ag-Ab complex. Three types of Ag-induced B-cell differentiation has been demonstrated in spleen viz., i) dependent on T-cells and macrophages, ii) independent of T-cells but dependent on macrophages and iii) native Ag.

2.3. Alimentary tract

Lymphocytes of alimentary tract are also of mixed origin – bursa and thymus. The distribution of amount of lymphoid tissue in small intestine and ceca are influenced by normal bacterial flora and infections such as coccidiosis; there is more lymphoid tissue in conventionally-reared birds than reared under germ-free conditions. Intestinal lymphoid foci appeared to play both the general role via Ab-production and local role relative to bacterial and other antigenic substances

2.4. Para-ocular and para-nasal tissue

Associated with local immune response at the site of entry into and shedding from upper respiratory tract.

2.5. Mural lymphoid nodules

Developed germinal centers and plasma cells after Ag stimulation but have no filtering function.

2.6. Other organs

Have little, if any, immune function

3. Immunoglobulins of domestic fowl

(Firth, 1977)

Domestic fowl has at least three classes of antibodies (immunoglobulins, Ig), sufficiently similar but not identical to that of mammals, designated Ig M, Ig G (also called Ig Y) and Ig A.

Following infection, antibody Ig M is produced first followed by Ig G; if the infection is repeated, the major antibody produced will be Ig G. The plasma cells in the sub-endothelial locations produce another antibody, Ig A, which is effective in the prevention of surface invasion (Bains, 1994).

3.1. Ig M

Ig M of domestic chicken is markedly similar to that of mammals. This is the initial Ig produced by birds in response to most Ag. Further, it is generally agreed that particulate Ag like viruses, bacteria or erythrocytes will trigger Ig M whereas soluble protein cause production of Ig G. In other words, Ig M forms major portion of agglutinating Ab to heterologous erythrocytes and is the most active Ig in indirect hemaggutination (HA) test to plasma Ag. Polymeric structure and multiple valence of Ig M confers it the ability of HA. Ig M Ab to protein Ag can be detected 3d post-incubation, reaching a peak by 4 to 8d and declining rapidly thereafter. Hence, Ig M is only a transitory immune body. Ig M is found, other than serum, in seminal plasma, bile and intestinal contents but not in yolk or in serum of newly-hatched chick.

3.2. Ig G

This forms the precipitating Ab of the fowl and is more similar to mammalian Ig A than mammalian Ig G although functionally similar to the latter. Ig G is found, other than serum, in many secretions also; for instance, respiratory tract secretions giving the possibility of antiviral immunity. Thus, it is speculated that Ig G constitutes secretory immune system of fowl. Ig G, being concerned with soluble proteins, is the vehicle of maternal immunity which passes on all important protection to the day-old chick through yolk-sac. Therefore, serum and yolk Ig G are nearly indistinguishable. Hence, in newly-hatched chick, Ig G level can be as high as 5 mg/ml, which decreases rapidly to less than 1 mg/ml by 2 to 3 weeks of age.

3.3. Ig A

This forms the secretory immune system in fowl found, other than serum, in bile, tracheal washings, seminal plasma and yolk but not in serum of newly-hatched chick. It is possible that Ig G- and Ig A-producing cells can arise independently of the prior production of Ig M by the bursal stem cell.

3.4. Ig E

Ig E has not been demonstrated in fowl.

Table 59.1. Concentration of immunoglobulins in serum of adult fowl

	Concentration		Half-life (d)		Synthesis	
Immunoglobulin	Mean ± SD	mg/ml	Mean	Range	mg/kg/d	Remarks
Ig M	0.71 ± 0.18	0.50 to 0.93	1.7	1.6 to 1.8	16	Transient Ig
Ig G	5.29 ± 1.35	4.10 to 7.30	4.1	3.5 to 4.8	58	Maternal-immunity
Ig A / Ig G ratio is 0.12 in serum, 1.3 in bile, 1.6 in tracheal washings and 0.69 in seminal plasma. *Source* : Firth, 1977						

4. Immunity development

Invasion by a foreign antigen (virus, bacteria etc.) triggers immune mechanism; the speed of such response depends on frequency of challenge by an antigen and the degree of response depends on immunocompetence of the host. The immunocompetence of the host, on the other hand, depends on a) Development of immune system b) Presence of immunosuppressive agents c) Degree of infection d) Nutritional status e) Mixed infection f) Superimposed infections (vaccinations) g) Others (Bains, 1994).

There are three major types of response in a bird's immune system; the specific, consisting of the humoral and cell mediated response, and the nonspecific.

The immune system in poultry, like that of humans, has developed several levels of defense strategies to cope with a wide spectrum of pathogens. Included are aspects of innate immunity such as physical and chemical barriers that prevent entry of the pathogen, and cellular and soluble components that are designed to eliminate the pathogen once it has gained entry. Although very effective, innate immunity is often not able to fight off the pathogen and prevent disease. At this point adaptive immunity is required to specifically focus defense mechanisms on that particular pathogen resulting not only in the elimination of the pathogen but also as protection in case of a repeat encounter with the same pathogen. It is the ability of adaptive immunity to recognize molecular features of the pathogen using highly specific antigen receptor-antigen interactions that conveys specificity to adaptive immunity and allows it to specifically focus immune activities on the invading pathogen. Additionally, immune mechanisms mediated by adaptive immunity are not only specific and very powerful but are also differentially tailored to effectively deal with pathogens that are in the body fluids and tissue spaces compared with such pathogens that are located inside cells. These broad categories of specialization within adaptive immunity are inherent in antibody-mediated (humoral) immunity and cell-mediated immunity, respectively (Erf, 2004).

The humoral and cell mediated responses need a processed antigen to stimulate response, and their response is to create a specific antibody for each particular antigen. The B (bursal produced) lymphocytes are associated with humoral response and the T (thymus produced) lymphocytes are associated with the cell-mediated response.

4.1. Antibody-mediated responses (Humoral immunity)

Antibodies are most effective in eliminating extracellular antigens, as their interaction with the antigen will activate/enhance effector mechanisms that lead to the removal or destruction of the antigen. Antibody-mediated effector mechanisms include (Erf, 2004):

1. Activation of the classical complement pathway directly on the antigen

2. Opsonization of the antigen so it can be phagocytosed more effectively by receptor-mediated endocytosis
3. Agglutination or precipitation of the antigen, which also facilitates removal of the antigen by phagocytosis
4. Neutralization of the antigen by binding to the antigen (e.g., poison) and thus preventing it from binding to soluble or cellular receptors used for invasion or to exert harmful effects on the host and
5. As long as antigens are expressed on the surface of cells, antibodies can also help in the elimination of cells that may be infected or neoplastic. In this process known as antibody-dependent cellular-cytotoxicity, the antibody forms a bridge between killer cells and the target cell. This close connection between the killer and the target cell mediated by antibodies then results in the activation of the killer cell and the destruction of the target cell.

4.2. Cell-mediated immune (CMI) response

When antigens have entered cells (e.g., by endocytic mechanisms; exogenous antigens) or are generated within the cell (e.g., viral or neoplastic proteins; endogenous antigens), humoral immune mechanisms involving direct antibody-antigen contact are no longer effective in eliminating the antigen. In this situation, CMI mechanisms that lead to the intracellular destruction of the antigen or to the elimination of the host cell are the most promising approaches to antigen elimination (Erf, 2004).

CMI responses, like most humoral immune responses, are tightly regulated and require "help" from T helper cells, specifically the type 1 T helper cells (Th1, hence, the name Th1-responses). Th1 cells are characterized by their production of cytokines such as γ-interferon (γ –IFN), α-tumor-necrosis factor (α-TNF), and interleukin-2 that drive CMI responses. The functional effectors of CMI responses are various immune cells including cytotoxic lymphocytes (cytotoxic T cells and natural killer cells) and macrophages. Cytotoxic lymphocytes and macrophages are specialized in the elimination of endogenous and exogenous antigens, respectively (Erf, 2004).

In poultry, as in humans and other mammals, the antigen-specific component of CMI is the T cell. Cells within this population express a repertoire of T-cell receptors (TCR) collectively capable of recognizing a great number of different antigens. There are different types of T cells that are distinguishable based on their functional capabilities and cell surface markers. T cells always express Cluster Designation Surface (CDS) complexes (together with the TCR molecules), independent of the type of TCR expressed (γ/δ, α/β, or α/β_2) or the antigen-specificity of the TCR. Thus, CDS is a pan-T cell marker; its presence on a cell indicates that the cell is a T cell. T cells that play primarily a regulatory role in adaptive immunity, whether cell-mediated or humoral, are referred to as T helper (Th) cells and typically express CD4 molecules on their surface. Both mammalian and avian adaptive immunity are critically dependent on CD4+ Th cells. Upon specific recognition of the antigen with their TCR, Th cells secrete soluble factors such as cytokines and express cell surface molecules that provide important activation signals to cells of innate and adaptive immunity (Er, 2004).

4.2.1. Memory cells

Certain lymphocytes in the B-system specialize as "memory cells" which have higher longevity and cause T-cells to "remember" former immune response and accelerate repeated responses. Hence, there will be greater response to a second vaccination against a specific disease (booster effect). This improved immune response exhibited by an individual with former exposure to specific antigen is called anamnestic response which will be, usually, quicker as well as stronger than the earlier response. Therefore, globe bursa and thymus regress, the T-and B-systems, especially the latter, can carry out the antibody production in adults. Hence, any disease that affects the thymus or bursa in the very young chick disrupts the early development of both B-and T-systems thereby rendering the bird highly susceptible to infections.

4.2.2. Suppression of T-system

Infectious bursal disease, Marek's disease, heat and cold, genetic, incomplete vaccine reaction, aflatoxins etc. are the known causes of suppressing T-system.

4.2.3. Suppression of B-system

Heat and cold, Lymphoid leucosis, nutritional deficiencies, toxins, low antibody production, Inclusion body hepatitis, aflatoxins etc. have been found to suppress B-system.

The antibodies produced a specific to the protein (antigen, pathogen) and the length of time of antibody production depends on number and virility of the organism, condition of the bird and type of organism. In addition, antibodies may not be able to kill all the organisms, and especially those with lodge in areas of minimum or no blood supply; for example, *Salmonella pullorum* in the atrophied follicles.

4.3. Non-specific response

The non-specific immune system responds to all antigens. Macrophages, heterophils and thrombocytes are the main cells associated with the non-specific immune system.

5. Types of immunity

Immunity in animals can be classified as follows:

1. The innate (Native) immunity – which the animal possesses without being triggered by extraneous factors.

2. Acquired immunity – which the animal obtains from its mother directly through yolk (maternal or "passive immunity") or by exposure to live antigen (recovery from a disease) or vaccinal antigen; latter two collectively referred to as "active immunity".

5.1. The innate (Native) defense system

(Richey, 2003)

This is the first line of defense functional in the healthy animal and does not require vaccination for initiation. The components are not disease specific but immediately respond when an infectious agent enters the body. The major components of the innate defense system are the natural defense system, the complement system, the phagocytic cells, and the interferon system.

5.1.1. The natural defense system

Each species of livestock has natural immunity to certain diseases. In other words, some diseases are species-specific.

5.1.2. The phagocytic cells

There are two primary phagocytic cell types; the granulocytes, which include the neutrophils and eosinophils, and the mononuclear phagocytes, which include monocytes circulating in the blood and macrophages located in tissue.

5.1.2.1. Granulocytes

Phagocytic cells are responsible for engulfing, killing and digesting invading bacteria; they also have an important role in controlling viral infections. Phagocytic cells are attracted to infected or inflamed sites by chemotactic substances which may be produced by certain microorganisms, be generated by cleavage of certain complement components, or be released by sensitized lymphocytes. As the phagocytes arrive at the infected area, they begin to engulf the infectious agent if it is susceptible to phagocytic activity. Most disease- causing microorganisms must be opsonized before they can be engulfed by phagocytes; bacteria are opsonized by the attachment of a specific antibody and/or complement to their surface. This, in turn, allows for easier destruction of the bacteria by phagocytic cells. Enzymes within the phagocytic cell will attempt to destroy the bacterium after they have been engulfed.

Phagocytes may also play an important part in controlling certain viral infections. Since it is necessary that viruses be within a cell (intracellular) to replicate and thus cause disease, the phagocyte will attempt to destroy the infected target cell if it has been properly tagged by a specific antibody produced by the acquired immune system. Even though the mechanism of destruction is not completely understood, the antibody presumably creates a bridge between the phagocyte and the infected target cell; the phagocyte will then attempt to destroy the infected target cell and thus destroy the virus. All phagocytic cell types are capable of the activities described above; however, the granulocyte and most specifically the neutrophil is the most proficient of all the phagocytes in these activities.

5.1.2.2. Mononuclear phagocytes

These originate in the bone marrow as monocytes and are released into the blood stream where they circulate before migrating into the tissues to become macrophages. Even though macrophages are capable of all the activities of the neutrophil, macrophages are described as the second line of defense. They are not as aggressive as the neutrophil and are much slower to arrive at the sites of infection. Macrophages, however, are capable of much more sustained activity than neutrophils and are able to kill certain types of bacteria that are resistant to killing by the neutrophils. In addition, macrophages process engulfed microorganisms and present fragments (antigens) of them to specialized cells for initiating the antibody-mediated (humoral) immune response and the cell-mediated immune response to infectious diseases agents.

5.1.3. The complement system

The complement system consists of at least 20 serum proteins which, when stimulated, respond with a series of chain reactions similar to the "domino effect," i.e., something happens to activate the first component which in turn activates the next component which in turn activates the next component, etc., until the reaction is complete. The primary effect of this chain reaction is to increase blood flow to an injured area of the body or an area that has been invaded by an infectious agent; the increased blood flow to the area is accompanied by fluid loss from the blood vessels. This "rush" of blood and the accumulation of leaked fluids in the affected area results in redness and swelling of the tissues and is recognized in/on the animal as a "hot and painful" area. This complement system reaction, which at times appears to be detrimental, hastens the accumulation of disease fighting cells (phagocytes) to the area by providing a fluid "pathway" and by producing chemotactic substances which attract phagocytes. Components of the complement system may also be able to lyse (rupture) the microorganism or opsonize (tag) the infectious agent which renders the infectious agent more vulnerable to a phagocytic cell.

5.1.4. The interferon system

Interferons are small proteins that cells immediately secrete when they are invaded by infectious agents, especially viruses. Interferon controls replication of certain viruses by inhibiting production of required viral protein in the infected cells and signals other body cells to initiate defenses which prevent replication of viruses in cells if they should be attacked. Interferon further enhances the immune system by increasing the activity of phagocytes in the destruction of engulfed microorganisms and engulfed infected cells. Interferons are not viral- specific, that is, interferon will cross-protect against a variety of viruses. In addition, interferon levels secreted by invaded cells can reach protective levels very quickly, thus preventing replication of viruses in other cells. Interferons also function in the active immune system.

5.2. Acquired immunity

(Richey, 2003)

Acquired immunity is the protection a previously susceptible animal

1. Actively develops because of a "triggering" of the immune system by foreign agents such as viruses, bacteria, extracts or metabolites of infectious agents, or vaccines or
2. Passively receives by transfer of protective substances such as antibodies and/or other cellular factors from a previously immune animal.

5.2.1. Active immunity

While the innate defense system, which becomes immediately functional when an infectious agent enters the body, is capable of killing some bacteria and viruses it is not capable of protecting the animal from high levels of diseases challenges. In addition, passive immunity (of particular importance for protection of the newborn) is also not useful in controlling high levels of disease challenges. Active immunity is usually required to stimulate protection against high levels of disease challenges. Active immunization involves the administration of a disease antigen derived from an infectious agent to stimulate specific immune responses to achieve resistance to the disease. It also became apparent that different diseases required different types of active immunity to stimulate protection in the animals. Antibody-mediated immunity (also known as Humoral immunity) functions primarily to control "extra-cellular" infectious agents; cell-mediated immunity is generally required to kill intra-cellular infectious agents; and mucosal immunity is necessary for mucosal membranes to resist invasion by infectious agents.

Specialized macrophage cells called Antigen Presenting Cells roam the animal's body, seeking out, ingesting, and partially digesting foreign (disease) material. The antigen presenting cells in turn display fragments of the disease antigen, combined with special molecules produced in the cell, on their surface. This combination (antigen fragment & cellular molecules) is then "screened" by a specialized group of white blood cells called Helper T-Cells.

When a helper T-cell encounters the one combination it is programmed to recognize, it secretes messenger molecules (Lymphokine) that in turn excite other components of the immune system. Among the excited components are B-lymphocytes that carry receptors that recognize one particular disease antigen. When a B-lymphocyte recognizes an antigen in the presence of lymphokine, it begins to divide; some daughter cells become Plasma Cells that produce antibodies against the antigen, others become Memory Cells. The antibodies are very specific; each B-lymphocyte produces an antibody that binds to a specific antigen. When initially exposed to disease antigens the immune system responds with low levels of Ig M antibodies. As the antigen-specific Ig M antibody concentration begins to increase in the blood, the helper T-cells signal some of the B-lymphocytes to switch from producing Ig M antibodies to production of Ig G or Ig A antibodies. Which type of antibody the switched B-lymphocytes produce depends largely upon the origin of the stimulating helper T-cell. If the stimulating helper T-cell originated in a lymph node or the spleen (through which tissue fluids and/or blood flows) then Ig G would probably be produced. If the stimulating helper T-cell originated under a mucous- producing surface then Ig A would tend to be produced.

The antigen-stimulated B-lymphocyte cells retain a "memory" of each specific antigen and, following a subsequent exposure to the same antigen, respond rapidly and produce a large quantity of antibody. The source of the antigens can either be naturally occurring, as in a disease challenge, or by vaccination. Antibodies alone are not capable of killing infectious agents; however, the antigen-specific antibodies help to control disease by binding the infectious agents for destruction by the complement or phagocytic immune systems. They can also attach to and neutralize toxins produced by infectious agents.

5.2.1.1. Humoral Immunity

Humoral immunity (antibody-mediated immunity) involves the production of antigen-specific Ig M and Ig G antibodies. The Ig M antibody is the first antibody type to be produced following the initial exposure to a disease antigen; however, it is the largest of the antibody types produced by the humoral immune system, which makes it difficult to move out of the bloodstream. Ig M antibodies provide some low levels of early protection; however, because of their size, they are of limited value in providing any degree of protection outside the bloodstream. Ig G antibodies, on the other hand, are smaller and are produced in large amounts following a second exposure to a disease antigen. Because of the size and the amounts produced, the Ig G antibodies can readily leave the bloodstream in sufficient amounts to attack disease antigens in the tissues of the body. Most killed vaccines use this principle to build immunity by requiring a second dose. The memory cells provide the immune system with a long-lived mechanism to remember previous antigen encounters and respond more quickly to subsequent invasions by the same disease antigen or vaccine antigen.

This "memory" mechanism can persist for months to years and usually requires a booster vaccination to keep the memory system primed. The frequency of the booster vaccinations depends upon the disease antigen and the level of the disease challenge to the animal.

5.2.1.2. Mucosal Immunity

Mucosal surfaces are those that line the internal passages of the body; they include the duct system of the mammary glands, the lining of the mouth, the salivary duct system, the esophagus, the stomach and the intestinal tract and the duct system of the gland of the eye. Mucosal surfaces are moist and warm and thus provide an excellent environment for bacterial growth. The antibodies responsible for humoral immunity (mainly Ig G) are found in the blood stream and tissue fluids, including those tissues under the mucosal surfaces; even though they are helpful in controlling disease agents that have penetrated the mucosal membranes, they are not very effective at controlling infection on the mucosal surfaces. If a disease organism penetrates the mucosal membrane, the disease agent is engulfed by a macrophage and presented to B-lymphocytes.

The B-lymphocytes, in turn, will begin producing Ig M antibodies against the disease organism, some of which will be switched by helper T-cells originating from the blood system to produce Ig G antibodies; however, because of the close proximity to the mucosal surface, helper T-cells that originated in the submucosal tissues stimulate the Ig M producing B-lymphocytes to switch to Ig A production. Ig A is then transported across the cells lining the mucosal surface and extruded onto the mucosal surface. As the Ig A is being transported across the cellular lining it combines with a cellular secretion and forms a substance called "secretory Ig A". Secretory Ig A is highly resistant to destruction by mucous secretions and will remain on the mucosal surface to interfere with the ability of disease agents to attach to and penetrate the mucosal lining.

Some of the activated Ig A-producing B-lymphocytes are able to enter the blood system and be transported to other mucosal areas of the body where they will also colonize and produce more Ig A in the new location. This mechanism allows more than just the invaded mucosal surface to receive protection from the disease agent; other mucosal surfaces also receive the needed protection.

5.2.1.3. Cell-Mediated Immunity

All disease-causing viruses and some disease- causing bacteria invade, live and replicate within selected cells of the body. Eventually the infected cells may rupture and the released disease organisms will invade other cells. In some diseases, the infected cell may "bridge" with susceptible cells thus allowing the disease organism to enter other cells without the infected cell being ruptured. In the first scenario, the disease organisms may briefly become extra-cellular and can be exposed to the humoral (antibody) immune system; however, when the disease organism invades a susceptible cell through the "bridge" mechanism, the disease organism will not become extra-cellular and thus is never exposed to the humoral immune system. The cell-mediated immune system is stimulated to react against intra-cellular viruses and intra-cellular disease causing bacteria.

When an animal is exposed to a disease or vaccine that stimulates cell-mediated immunity, B-lymphocytes and T-lymphocytes are activated by specialized helper T-cells. The activated B-lymphocytes respond as described under humoral immunity; however, the T-lymphocytes respond by dividing to produce daughter cells, by destroying the disease-infected tissue cells, or by secreting chemical messenger molecules that signal the invasion of additional susceptible cells by the disease organism or direct and encourage macrophages todestroy the disease-infected cell. The daughter cells will mature to function in the same manner as the activated parent T-lymphocyte and retain a memory of the specific disease antigen. Once the infected cell is destroyed, the replication of the disease organism within the cell is halted and the organism is released into the surrounding tissue fluid. Upon their release into the extra-cellular environment, the disease organisms become susceptible to the activated humoral (antibody) immune system. Like the humoral immune system, cell-mediated immunity is much stronger when subsequent exposures to a particular antigen occur.

When attempting to stimulate the active immune system by vaccination, the selection of the vaccine should depend upon the type of active (humoral or cell-mediated) immunity needed to protect against a specific disease. In general, killed vaccines are noted for stimulating a humoral immune response and modified live vaccines that are capable of replicating in the animal's body are needed to stimulate a cell-mediated immune response; however, recent research has indicated that certain adjuvant killed vaccines can stimulate a cell-mediated response (adjuvants are materials added to vaccines that help introduce vaccine antigens to the immune system in such a way that stimulates stronger and longer-lasting immunity). It should be noted that adjuvants which have historically been used in animal vaccines have not been very effective at inducing cell-mediated responses; however, new adjuvants have been and continue to be developed that have the ability to induce cell-mediated immunity. In addition, killed vaccines stimulate very little production of IgA to protect the mucosal surfaces; hopefully, new adjuvants and/or technology which will stimulate the mucosal immune system will soon be developed.

5.2.1.4. Vaccination program

The antigen has to be suitably modified and employed artificially to immunize the birds against the common diseases prevalent in the environment and antigen in question is termed vaccine and the immunity so developed is referred to as active immunity.

Live virus vaccine initiates a mild infection followed by virus specific antibody production. However, live virus vaccine should not be administered at the face of outbreak; otherwise, vaccination itself triggers severe infection in the flock already incubating the virus.

Success of vaccination program depends upon the level of maternal antibodies, immunocompetence, intercurrent infections and stress factors.

However, for protozoan diseases, immunity is by means other than antibodies whereas for fungal diseases there is no antibody production since the fungal protein (antigen) does not usually enter circulation but the toxins cause the disease.

Young birds are immunologically less competent than older birds in respect of many diseases (excepting few diseases like lymphoid leucosis and others). Parental immunity may not be effective with few diseases. Antibodies present in the egg yolk are absorbed into intestines just before hatching and they protect young chicks against most bacterial and viral infections. However, organisms that infect through respiratory tract (Example – Newcastle disease) can bypass the circulating antibodies and cause the disease. Certain other diseases like Infections laryngotracheitis, the disease occurs only after 4 or 5 weeks of age and hence, parental immunity is of no use and vaccination can be delayed till the birds are 4 weeks of age.

Under commercial conditions, suboptimal immune response is frequently observed in flocks subjected to clinical or sub-clinical infectious bursal disease, malabsorption syndrome, reovirus infections, adenovirus infections, reticuloendothelial virus infection, mycotoxicosis, coccidiosis etc. In addition, stress in the form of heat, severe cold, overcrowding, high ammonia concentration or extreme weather fluctuations also contribute to suboptimal immune response. In any case, the suboptimal immune response is due to: a) Partial or complete destruction of immune system b) Poor addition of nutrients c) Poor absorption of nutrients d) Improper vaccination programs e) Mixed or superimposed infections e) Reduced immunocompetence f) Chronic infection g) Certain therapeutics.

It is also important to note that, the commercial conditions, many of the common diseases are endemic and hence are transmitted horizontally from one of another; in some diseases, even vertically from one generation to another through eggs and hatchery.

5.2.1.5. Hypersensitive Reactions

Under certain circumstances, harmful reactions referred to as "allergic" or "hypersensitivities" manifesting as simple swelling of the eye lids, runny eyes/ nose and vaccine injection sites reactions such as granulomas or sometimes even life-threatening episodes such as anaphylactic shock can occur. However, such complications are extremely rare in poultry.

5.2.2. Passive immunity

Day-old chicks can derive antibodies from their mother, depending on the antibodies status of the mother. This is termed as maternal or passive immunity since the antibodies are not produced, but only transferred, to the host. Hence, passive immunity is short-lived and it wanes off to 50% of that at hatch by 3d, by the end of 2 weeks, very little portion of the maternal immunity is remaining and by the end of 3rd week, the maternal antibodies disappear.

6. Nutrition and immunity

(Butcher and Miles, 2002)

The immune system of the bird can be influenced by nutrition in the following ways:

1. Anatomical development of lymphoid tissues
2. Mucus production
3. Synthesis of immunologically active substances
4. Cellular proliferation
5. Cellular activation and movement
6. Intracellular killing of pathogens
7. Modulation and regulation of the immune process.

Therefore, a well-nourished bird is more immunologically competent and better able to cope with disease challenges than a poorly nourished bird. A change in feed intake leads to a change in nutrient and energy intake. Dietary concentrations of most nutrients for optimal growth are also necessary for optimal immunocompetence and hence, development and maturation of the immune system of young poultry fed nutritionally deficient diets will be compromised. However, young poultry in today's commercial poultry industry are usually not subjected to diets which are severely limited in nutrients and energy.

It is well known that immunological challenge in the bird is accompanied by a decrease in feed intake. Carbohydrate calories are the best choice to increase the energy density since immunological stress is known to impair triglyceride clearance from the blood, thus decreasing fat use.

A reduced humoral immune response in poultry has been associated with low mineral levels in the feed. For example, sodium and chloride have been associated with the humoral immune response in broilers. Broilers fed diets deficient in sodium or chloride have lower antibody concentrations in their bodies. The balance between sodium and chloride in the diet is important because a high chloride concentration in relation to that of sodium may result in a reduced immune response.

Even though most nutrients required by the immune system are present in the diet in sufficient concentrations there is evidence that increased dietary supplementation of certain nutrients, above that needed for maximum growth and feed efficiency, is of benefit to the immune response. Vitamins A, E and C as well as the amino acids methionine and valine have been shown to benefit immune function when added to the diet of poultry at higher concentrations than are required to maximize growth and feed efficiency.

However, too much of a good thing may also be detrimental to the immune system. For instance, too much dietary methionine will suppress the immune system. Usually, the immune suppression caused by excess dietary methionine is evident before changes in growth and feed consumption are detected. It is essential to ensure that poultry diets contain enough of a specific nutrient to ensure the desired growth, feed conversion and immune response, but not too much so as to impair the immune function either directly or indirectly.

The ω-3 and ω-6 fatty acids are also of benefit to the immune system at higher than normal concentrations in the diet. Several nutrients have been shown to have a negative effect on the immunological response in the bird when they are deficient in the diet. Sodium, chloride, zinc, methionine, valine, threonine, vitamin A, riboflavin, pantothenic acid, pyridoxine and selenium will have a negative impact on the immune system if they are deficient in the diet. Thus, these are considered essential to proper immune function.

One vitamin that has received a lot of attention with respect to its importance to the immune response in poultry has been vitamin E. This vitamin appears to be an immune system "booster." Vitamin E seems to exert a complementary effect on the immune system by inhibiting the synthesis of prostaglandins. These prostaglandins are produced in the cells following the oxidation of cellular membranes and are responsible for inhibiting the inflammation and immune response. Vitamin E prevents oxidation and thus, the production of prostaglandins.

Zinc is another element that is important in assisting the bird's immune system to overcome a challenge. Zinc is especially important in wound healing, thymic function and proliferation of lymphocytes.

7. Stress - its effect on nutrition and immunity

(Klasing *et al.* 1991 and Butcher and Miles, 2002)

A three-way interrelationship exists among stress, nutrition and immunity. Stress is the nonspecific response that the body of an animal has to any demand made upon it. An animal is under stress when it has to make extreme functional, structural, behavioral, or immunological adjustments to cope with adverse aspects of its environment.

Poor management is one of the greatest causes of stress in all types of poultry. The exposure to disease agents, poor nutrition and exposure to immunosuppressive agents also contribute to reduced performance. Successful identification and correction of problems in the poultry house and in the diet will benefit the bird.

7.1. Energy needs during physiological stress

Whenever poultry are confronted with physiological stress they have to adapt to the situation in order to survive. This process of adaptation is essential and requires energy. The energy for adaptation comes from the three energy-yielding nutrients: carbohydrates, lipids and proteins. These nutrients are only available from the feed and the nutrient reserves in the animal. During the first stages of stress, poultry will eat less initially, and then increase their feed intake. During stress, nutrients in the feed are not digested and absorbed efficiently, and the animal must rely on the nutrient reserves of the body. These reserves help sustain the animal during stress. The muscle and liver carbohydrate stores (glycogen) are immediately called upon to furnish energy. Protein is broken down to yield the glucogenic and ketogenic amino acids which, following deamination, will supply the bird with energy. This energy from carbohydrate and protein allows the bird to maintain its health and survival.

During stress the vital functions of the brain, liver, heart, lungs, kidney, etc. cannot be compromised. Therefore, the less important functions such as egg production, reproduction, growth and immunity are set aside to promote the vital functions of the body in stressful situations. Obviously, full genetic potential of the bird for growth and egg production is not expressed during stress.

The shift in metabolism during stress favors fat deposition. During stress the consumption of water increases as a result of the necessity to clear the additional uric acid excretion arising from protein breakdown. The increased water consumption is also probably necessary to maintain osmolality in the body fluids due to the increased sodium retention concurrent with the effects of corticosterone.

7.2. Adrenal gland, stress and nutrition

Corticosterone is the main stress hormone that is produced and released from the cortex of the adrenal gland when an animal is confronted with which is responsible for increased use of glucose and amino acids for energy. Muscle protein synthesis declines since the carbon skeletons of the amino acids are used for energy. Hence, when poultry are exposed to various stressors, their dietary requirement of several nutrients increases.

In addition to its effects on all growth, production and reproduction parameters, environmental temperature has an influence on the immune response of poultry mediated through increased levels of corticosteroids. However, the influence of environmental temperature variations on the immune response depends on the degree of adaptation of the animal and the time of immunization.

7.3. Vitamin C and stress

All poultry are capable of synthesizing vitamin C in the kidney tissue, but, the synthesis and use of vitamin C are not constant. The ability of the kidney to synthesize vitamin C in the amount needed changes with age, management, environment, disease, nutrition and stress. Some of the highest concentrations of vitamin C are found in the testicles, ovaries, and adrenal gland. In the adrenal gland, vitamin C functions metabolically to help control production of corticosterone.

Stressors in the environment have a direct influence on plasma and tissue levels of vitamin C. A controlled rate of corticosterone release from the adrenal cortex is preferred in coping with stress. Vitamin C plays a central role in the continued synthesis of corticosterone. The proposed mechanism for this effect is through inhibition of the 21-hydroxylase and 11 beta-hydroxylase enzymes in the steroid biosynthetic pathway in the adrenal cortex. Vitamin C supplementation to the diet and water during periods of stress causes reduced synthesis of corticosterone.

For best results, the use of vitamin C in the diet or water should begin at least 24 or 48 hours before the onset of stress and should continue throughout the stressful period. The recommended amount of vitamin C to use in the diet is usually between 100 and 150 ppm. Higher levels can be used, but are often cost prohibitive.

Vitamin C is not the only vitamin that should be supplemented during periods of stress. Research has shown that all vitamins should be increased in the diet if the stress level in poultry is high. Without these supplemental vitamins, acceptable performance and carcass characteristics will not occur. Vitamins also enhance disease resistance under conditions of stress. These vitamins exert their mechanism of action during stress by protecting immune tissue via a reduction of certain hormonal effects and oxidative damage, and enhancement of cell differentiation and production.

Various stimuli, including stress hormones, activate phospholipase A_2 which frees the 20-carbon fatty acid, arachidonic acid, from membrane phospholipids. As a by-product of this breakdown of cellular membranes and the production of prostaglandins from arachidonic acid, reactive oxygen molecules (H_2O_2, OH and O_2) are formed and the free radicals can cause damage to cell membranes. These free radicals can also result in the disruption of protein function, DNA structure, and energy production within the cell. Therefore, antioxidant compounds (vitamin E and β-carotene) are also beneficial during stress.

Stress also causes the animal to increase its metabolic rate, which through normal oxidative metabolic pathways produces increased free radicals. Increased mineral losses have been reported in stressful conditions, which reduce metallo-enzymes necessary for free radical and peroxide removal from the body. This is another reason that the need for antioxidants in the diet will usually increase during stress.

Vitamins E and C are very important during disease stress, such as in infection. In infection, phagocytes destroy ingested particles by generating free radicals which may prove self-destructive if antioxidant systems are inadequate. Phagocytic function and surrounding tissues can be protected by feeding vitamins E and C. Vitamin C helps regenerate vitamin E by reducing vitamin E radicals formed when vitamin E scavenges oxygen radicals.

Disease resistance is also a function of cell differentiation and one of the primary functions of vitamin A is to maintain proper epithelial tissue differentiation and prevent epithelial keratinization which occurs in a deficiency. Vitamin A is required for maintaining cellularity of lymphoid organs, which are essential in combating disease stress. Vitamin A is also required for enhancing both cellular and humoral immunity and enhances phagocyte activity. Vitamin A has reduced mortality in chicks infected with coccidial oocysts from *E. tenella* and *E. acervulina*. Since vitamins A, E, C and carotenoids are able to protect cells from free radical oxidation, reduce the detrimental effects of certain eicosanoids, and enhance humoral and cellular immune responses in disease stress, the levels of these vitamins may have to be altered during stress and disease.

7.4. Trace Mineral Requirements

Stress decreases the circulating levels of trace minerals in animals, causes greater endogenous loss, decreases the efficiency of trace minerals and increases the metabolic need to fight the stress. During stress, increased serum copper and decreased serum iron, zinc and manganese have been reported. One of the most serious stressors that can confront poultry is disease. Trace mineral metabolism is radically changed during disease challenge because of the demand by a family of acute phase proteins that are central to the early phases of the immune response.

Zinc concentrations in the serum are decreased as a result from the redistribution of zinc from the plasma pool to a newly synthesized metallo-thionein pool in the liver and other tissues. Metallothionein induction supplies zinc to be used as a cofactor for metallo-proteinases and other acute phase proteins.

Copper-containing ceruloplasmin serves as a protective antioxidant and is a necessary component of the acute phase response. The quantity of copper required to sustain this role is large relative to other functions of copper. A copper deficiency reduces the amount of ceruloplasmin which is released during an inflammatory stress.

One of the most interesting mechanisms of defense that animals have to fight infection is to remove the circulating iron from circulation. During infection, iron is removed from the circulation and sequestered into compartments which are nutritionally unavailable to bacteria and parasites; referred to as "nutritional immunity". The absorption of iron from the digestive tract is markedly reduced within hours after a disease challenge.

Manganese needs by the gastrointestinal tract, lungs and several other tissues in the bird are increased during an immune response because, manganese serves as a cofactor of superoxide dismutase and ameliorates damage induced by the immune response itself. However, there is very little change in the circulating manganese levels in the plasma during stress.

The goal of trace mineral supplementation during stress should be to provide a readily available form that can be absorbed and give replete storage pools prior to the time that stress is encountered. These storage pools of the trace minerals will buffer the low levels due to stress and facilitate the greater tissue accretion that follows an encounter with stress.

Chapter **60**

Profiling A Disease

Disease is in impairment of the normal functions of any body organ or part of the animal. It can result on account of various reasons like nutritional inadequacy, toxicity, external injury, stress or by infectious agents or parasites. The diseases due to nutritional inadequacies have already been enumerated in Chapter "Nutrients - functions, deficiency and excess". In the forthcoming Chapters, diseases caused by infectious agents (antigens) will be discussed.

The cause of the disease is termed etiology. The onset, duration and severity of disease caused by infectious agents depend on:

1. Number, type and virulence of the infectious agent
2. Route of entry of the causal agent and
3. Defense status of the host which, in turn, depends on immune and nutritional status, genetic ability, stress and type of disease-prevention program to which the host is subjected to.

1. Antigen

Most of the infectious agents are unicellular and they multiply within the host producing certain reactions that are either useful (Symbiotic organisms) or detrimental (pathogenic) to the host. The infectious (pathogenic) agents belong to the following categories:

1.1. Bacteria

Many bacteria have been found to be pathogenic poultry. They grow exponentially till the competition for the available food occurs and in the process, produce the disease and quite frequently both morbidity (sickness) and mortality (death). Bacteria may exist in several forms (variants, types) few of which may be due to mutations are due to inconducive environment in which they are forced to live. Bacteria can cause disease principally by production of toxins during the normal metabolism (multiplication) which is incompatible to the host. These toxins can be exotoxins (elaborated by the intact cell), endotoxins (released upon lysis of the bacterial cell), hemotoxins (which cause lysis of RBCs) or those which destroy the barriers to prevent the invasion of bacteria.

1.2. Viruses

These are smaller than bacteria and the obligatory commensals which can survive and multiply only inside the host cell. Since they are intracellular, the chemicals and antibiotics fail to act of them. Viruses produce diseases which vary in characteristics viz., respiratory symptoms, tumors, skin lesions etc.

1.3. Protozoa

These are unicellular and they usually invade a cell (intracellular), multiply in it and eventually destroy the host cell. They have a complex life-cycle consisting usually of sexual and asexual cycles. There are protozoa which are extracellular also, like Trichomonas sp., Hexamita sp etc.

1.4. Fungi

Fungi (molds and yeasts) usually are extracellular and toxin producers. Their toxins produce diseases in poultry.

2. Epidemiology

Epidemiology is the study of the determinants, occurrence, and distribution of health and disease in a defined population. Infection is the replication of organisms in host tissue, which may cause disease. A carrier is an individual with no overt disease who harbors infectious organisms. Dissemination is the spread of the organism in the environment.

Epidemiology is a descriptive science and includes the determination of rates, that is, the quantification of disease occurrence within a specific population. The most commonly-studied rate is the attack rate - a ratio of the number of cases of the disease to the size of population among whom the cases have occurred. Epidemiology can accurately describe a disease and many factors concerning its occurrence before its cause is identified. One goal of epidemiologic studies is to define the parameters of a disease, including risk factors, in order to develop the most effective measures for control.

Epidemiologic studies may be (1) descriptive, organizing data by time, place, and animal; (2) analytic, incorporating a case-control or cohort study; or (3) experimental. Epidemiology utilizes an organized approach to problem solving by: (1) confirming the existence of an epidemic and verifying the diagnosis; (2) developing a case definition and collating data on cases; (3) analyzing data by time, place, and animal; (4) developing a hypothesis; (5) conducting further studies if necessary; (6) developing and implementing control and prevention measures; (7) preparing and distributing a public report; and (8) evaluating control and preventive measures.

Proper interpretation of disease-specific epidemiologic data requires information concerning past as well as present occurrence of the disease. An increase in the number of reported cases of a disease that is normal and expected, representing a seasonal pattern of change in host susceptibility, does not constitute an epidemic. Therefore, the regular collection, collation, analysis, and reporting of data concerning the occurrence of a disease is important to properly interpret short-term changes in occurrence.

A sensitive and specific surveillance program is important for the proper interpretation of disease occurrence data. Almost every country has a national disease surveillance program that regularly collects data on selected diseases. The quality of these programs varies, but, generally, useful data are collected that are important in developing control and prevention measures.

2.1. Report of a disease

The methods of case reporting vary within each state. Passive reporting is one of the main methods. In such a case, physicians or personnel in clinics or hospitals report occurrences of relevant diseases by telephone, postcard, or a reporting form, usually at weekly intervals. In some instances, the report may be initiated by the public health or clinical laboratory where the etiologic agent is identified.

In an active surveillance program, the health authority regularly initiates the request for reporting. The local health department may call all or some health care providers at regular intervals to inquire about the occurrence of a disease or diseases. The active system may be used during an epidemic or if accurate data concerning all cases of a disease are desired.

The health care provider usually makes the initial passive report to a local authority, such as a city or county health department. This unit collates its data and sends a report to the next highest health department level, usually the state health department.

2.2. Infection and its spread

Infection is the replication of organisms in the tissue of a host; when defined in terms of infection, disease is overt clinical manifestation. In an inapparent (subclinical) infection, an immune response can occur without overt clinical disease. A carrier (colonized individual) is a person in whom organisms are present and may be multiplying, but who shows no clinical response to their presence. The carrier state may be permanent, with the organism always present; intermittent, with the organism present for various periods; or temporary, with carriage for only a brief period. Dissemination is the movement of an infectious agent from a source directly into the environment; when infection results from dissemination, the source, if an individual, is referred to as a dangerous disseminator.

2.2.1. Chain of infection

Infectiousness is the transmission of organisms from a source, or reservoir (see below), to a susceptible individual. An animal may be infective during the preclinical, clinical, post-clinical, or recovery phase of an illness. The incubation period is the interval in the preclinical period between the time at which the causative agent first infects the host and the onset of clinical symptoms; during this time the agent is replicating. An animal may transmit infection (infective) during incubation period and/or clinical period and/or convalescent phase, or may become an asymptomatic carrier and remain infective for a prolonged period.

The spectrum of occurrence of disease in a defined population includes sporadic (occasional occurrence); endemic (regular, continuing occurrence); epidemic (significantly increased occurrence); and pandemic (epidemic occurrence in multiple countries).

The chain of infection includes the three factors that lead to infection: the etiologic agent, the method of transmission, and the host. These links should be characterized before control and prevention measures are proposed. Environmental factors that may influence disease occurrence must be evaluated.

2.2.1.1. Etiologic Agent

The etiologic agent may be any microorganism that can cause infection. The pathogenecity of an agent is its ability to cause disease; pathogenecity is further characterized by describing the organism's virulence and invasiveness. Virulence refers to the severity of infection, which can be expressed by describing the morbidity (incidence of disease) and mortality (death rate) of the infection. An example of a highly virulent organism is velogenic strain of Newcastle disease which almost always causes severe disease in the susceptible host.

The invasiveness of an organism refers to its abilityto invade tissue. *Campylobacter* organisms are noninvasive, causing symptoms by releasing into the intestinal canal an exotoxin that acts on the tissues. In contrast, *E.coli* organisms in the intestinal canal are invasive and migrate into the tissue.

No microorganism is assuredly avirulent. An organism may have very low virulence, but if the host is highly susceptible, as when therapeutically immunosuppressed, infection with that organism may cause disease.

Other factors should be considered in describing the agent. The infecting dose (the number of organisms necessary to cause disease) varies according to the organism, method of transmission, site of entrance of the organism into the host, host defenses, and host species. Another agent factor is specificity; some agents (for example, *Salmonella*) can infect a broad range of hosts; others like Coccidia are host-specific.

The reservoir of an organism is the site where it resides, metabolizes, and multiplies. The source of the organism is the site from which it is transmitted to a susceptible host, either directly or indirectly through an intermediary object. The reservoir and source can be same or sometimes different. The distinction can be important when considering where to apply control measures.

2.2.1.2. Method of Transmission

Disease that spreads from one bird to another is called as contagious disease. The capacity to cause a disease (virulence) depends on various factors intrinsic to both host and invading organism and the environment. The infectious agents can reach the host through many sources (Table 60.1).

Hence, it is mandatory to exercise suitable control measures to minimize or obviate the disease outbreak or spread. These include isolation and proper construction of buildings, personnel control, provision of adequate sanitary environment (including proper disposal of manure and carcass, use of disinfectants, cleanliness of buildings and equipment), and hatchery management (including the management of egg-borne diseases).

The method of transmission is the means by which the agent goes from the source to the host. The four major methods of transmission are by contact, by common vehicle, by air or via a vector.

2.2.1.2.1. Contact

In contact transmission the agent is spread directly, indirectly, or by airborne droplets. Direct contact transmission takes place when organisms are transmitted directly from the source to the susceptible host without involving an intermediate object. Indirect transmission occurs when the organisms are transmitted from a source, either animate or inanimate, to a host by means of an inanimate object. Droplet spread refers to organisms that travel through the air very short distances, that is, less than 1 m from a source to a host. Therefore, the organisms are not airborne in the true sense. An example of a disease that may be spread by droplets is infectious coryza.

Table 60.1 Principal routes of transmission of diseases

Route	Remarks
	Horizontal transmission
Hatchery (type of airborne)	Large numbers of susceptible young chicks in environment of high humidity, high air flow and 37.2°C temperature (Aspergillosis, Salmonella, AE)
Airborne	Among pen mates House-to-house Farm-to-farm (Newcastle, Bronchitis)
Food-Borne	On-farm contamination of feed. e.g. Candidiasis; Contaminated prior to farm delivery (Salmonella) .
Vector-Borne	Man, equipment, free-flying birds, predators and rodents, insects, artificial insemination, litter and houses
Environment	Contaminated uncleaned poultry house and litter, e.g. Marek's Disease.
Recovered carrier birds	e.g. MG, ILT, Infectious coryza
	Vertical transmission
True Vertical	Ovarian or uterine infection, e.g. Pullorum, AE, MG.
Apparent Vertical	Fecal contamination of egg with subsequent penetration, e.g. most Paratyphoid salmonellae
	Source : http://caltest.vet.upenn.edu/poultry/syllabus

2.2.1.2.2. Common-vehicle

Common-vehicle transmission refers to agents transmitted by a common inanimate vehicle, with multiple cases resulting from such exposure. This category includes diseases in which food or water as well as drugs and parenteral fluids are the vehicles of infection. This mode of transmission includes food-borne and water-borne diseases, and bacteremia resulting from use of contaminated intravenous fluids.

2.2.1.2.3. Air-borne

The third method of transmission, airborne transmission, refers to infection spread by droplet nuclei or dust. To be truly airborne, the particles should travel more than 1 m through the air from the source to the host. Droplet nuclei are the residue from the evaporation of fluid from droplets, are light enough to be transmitted more than 1 m from the source, and may remain airborne for prolonged periods. Newcastle disease is an airborne disease; the source may be sneezing / coughing which creates aerosols of droplet nuclei that contain the virus particles. Infectious agents may be contained in dust particles, which may become re-suspended and transmitted to hosts. *E.coli* organisms can be transmitted through dust particles.

2.2.1.2.4. Vector-borne

Usually, arthropods are the vectors and the transmission may be external or internal. External, or mechanical, transmission occurs when organisms are carried mechanically on the vector. Internal transmission occurs when the organisms are carried within the vector. If the pathogen is not changed by its carriage within the vector, the carriage is called harborage; otherwise, if the organism is changed biologically during its passage through the vector, it is called biologic.

An infectious agent may be transmitted by more than one route. For example, *Salmonella* may be transmitted by a common vehicle (food) or by contact spread (human carrier); *E.coli* may be transmitted by any of the four routes.

2.2.1.3. Host

The final link in the chain of infection is the host. The organism may enter the host through the skin, mucous membranes, lungs, gastrointestinal tract, or genitourinary tract, and it may enter embryos through the hatching eggs. The resulting disease often reflects the point of entrance, but not always: Lymphoid leucosis virus causes disease at a different locations from the site/route of entry; it is mainly a vertically-transmitted disease entering through egg albumen but produces neoplastic lesions involving liver, spleen, bursa of Fabricius etc. Therefore, development of disease in a host reflects agent characteristics and is influenced by host defense mechanisms, which may be nonspecific or specific.

2.2.1.3.1. Non-specific defense mechanisms

Non-specific defense mechanisms include the skin, mucous membranes, secretions, excretions, enzymes, the inflammatory response, genetic factors, hormones, nutrition, behavioral patterns, and the presence of other diseases.

2.2.1.3.2. Specific defense mechanisms

Specific defense mechanisms or immunity may be natural, resulting from exposure to the infectious agent, or artificial, resulting from active or passive immunization.

In chicken, the heterophil appears to lack proteolytic enzymes. Therefore, the chicken does not form liquid pus but rather a caseous exudate. Following acute inflammation, within 12 hr, heterophils predominate along with mononuclear cells and basophils. This is followed by lymphoid hyperplasia within 36 hours when lymphocytes predominate. (see Chapter "Immune system").

2.3. Factors influencing occurrence of a disease

The environment can affect any link in the chain of infection. Temperature can assist or inhibit multiplication of organisms at their reservoir; air velocity can assist the airborne movement of droplet nuclei; low humidity can damage mucous membranes; and ultraviolet radiation can kill the microorganisms. In any investigation of disease, it is important to evaluate the effect of environmental factors. At times, environmental control measures are instituted more on emotional grounds than on the basis of epidemiologic fact. It should be apparent that the occurrence of disease results from the interaction of many factors (Table 60.2).

Table 60.2. Factors influencing occurrence of an infectious disease

Factors	Remarks
Pathogenic agent	Growth characteristics, stability, ability to form spores, possession of antibiotic resistant plasmids, expression of antigens, production of enzymes, pathogenecity (virulence and invasiveness), dose, reservoir, source, mode of dissemination, host specificity etc.
Host	Incubation period, non-specific defense mechanisms which include age, sex, skin, secretions, ciliary function, peristalsis, inflammation, nutrition, genetic factors, hormones, behavior, chronic disease etc., specific defense mechanisms like active and passive immunity as well as vaccination program
Disease transmission	By contact (direct, indirect, droplets), common vehicle (food, water, medicines, parenteral solutions), air-borne (droplet nuclei, dust etc.) and vector-borne (external or internal)
Environment	Temperature, humidity, rainfall, radiation, air currents etc.
	Source : Brachman on internet

2.4. Epidemiologic Methods

The three major epidemiologic techniques are descriptive, analytic, and experimental. Although all three can be used in investigating the occurrence of disease, the method used most is descriptive epidemiology. Once the basic epidemiology of a disease has been described, specific analytic methods can be used to study the disease further, and a specific experimental approach can be developed to test a hypothesis.

2.4.1. Descriptive Epidemiology

Data that describe the occurrence of the disease are collected by various methods from all relevant sources. All pertinent characteristics of the animal(s) should be noted: source, age, sex, nutrition, immunization history, presence of underlying disease, and other data. The data are then collated by time, place, and details of the animal(s) involved. The following time trends are considered in describing the epidemiologic data:

2.4.1.1. Secular trend

The secular trend describes the occurrence of disease over a prolonged period, usually years; it is influenced by the degree of immunity in the population and possibly nonspecific measures such as improved nutritional levels among the population.

2.4.1.2. Periodic trend

The periodic trend may indicate a change in the antigenic characteristics of the disease agent. A classical example is the change in antigenic structure of the prevalent avian influenza virus.

Additionally, a lowering of the overall immunity of a population or a segment thereof (known as herd immunity) can result in an increase in the occurrence of the disease. This can be seen with some diseases which can be immunized when periodic decreases occur in the level of immunization in a defined population. This may then result in an increase in the number of cases, with a subsequent rise in the overall level of herd immunity. The number of new cases then decreases until the herd's immunity is low enough to allow transmission to occur again and new cases then appear.

2.4.1.3. Seasonal trend

This trend reflects seasonal changes in disease occurrence following changes in environmental conditions that enhance the ability of the agent to replicate or be transmitted. For example, occurrence of coccidiosis during summer or rainy season (high humidity) leading to wet litter conditions.

2.4.1.4. Epidemic occurrence

The fourth time trend is the epidemic occurrence of disease. An epidemic is a sudden increase in occurrence due to prevalent factors that support transmission.

Once the descriptive epidemiologic data have been analyzed, the features of the epidemic should be clear enough that additional areas for investigation are apparent.

2.4.2. Analytic Epidemiology

Analytic epidemiology analyzes disease determinants for possible causal relations. The two main analytic methods are the case-control (or case-comparison) method and the cohort method.

2.4.2.1. Case-control study

The case-control method starts with the effect (disease) and retrospectively investigates the cause that led to the effect. The case group consists of animals with the disease; a comparison group (control group) has animals similar to those of the case group except for absence of the disease. These two groups are then compared to determine differences that would explain the occurrence of the disease. This method is relatively easy to conduct, can be completed in a shorter period than the cohort approach, and is inexpensive and reproducible; however, bias may be introduced in selecting the two groups and it may be difficult to exclude sub-clinical cases from the comparison group.

2.4.2.2. Cohort study

In this method, two populations are studied prospectively: one that has had contact with the suspected causal factor under study and a similar group that has had no contact with the factor. When both groups are observed, the effect of the factor should become apparent.

The advantages of a cohort study are the accuracy of collected data and the ability to make a direct estimate of the disease risk resulting from factor contact; however, cohort studies take longer and are more expensive to conduct.

2.4.2.3. Cross-sectional study

This method involves a survey of the population in question over a limited period to determine the relationship between a disease and variables present at the same time that may influence the occurrence of a disease.

2.4.3. Experimental Epidemiology

In the experimental approach, a hypothesis is developed and an experimental model is constructed in which one or more selected factors are manipulated. The effect of the manipulation will either confirm or disprove the hypothesis. An example is the evaluation of the effect of a new drug on a disease. A group of animals with the disease is identified, and some of them are randomly selected to receive the drug. Another group (similar in all respects to the selected group except that these are healthy) is concurrently raised without administering the drug. If the only difference between the two is use of the drug, the clinical differences between the groups should reflect the effectiveness of the drug.

2.5. Epidemic Investigation

An epidemic investigation describes the factors relevant to an outbreak of disease; once the circumstances related to the occurrence of disease are defined, appropriate control and prevention measures can be identified. In an epidemic investigation, data are collected; collated according to time, place, and animals in question, and analyzed and inferences are drawn.

2.5.1.1. Data collection

In the investigation, the first action should be to confirm the existence of the epidemic by noting from past surveillance data the number of cases suspected and comparing this with the number of cases initially reported. Additionally, the investigator should discuss the occurrence of the disease with veterinarians or others who have seen or reported cases after examining sick animals and reviewing laboratory and hospital records. These diagnoses should then be verified. A case definition should be developed to differentiate animals which represent actual cases, others which represent suspected or presumptive cases, and also those which should be omitted from further study. Additional cases may be sought or additional data on sick animals obtained, and a rough case count made.

2.5.1.2. Analysis of data

The data so collected must be organized according to time, place, and animal details. The population at risk should be identified and a hypothesis developed concerning the occurrence of the disease. If appropriate, specimens should be collected and transported to the laboratory. More specific studies may be indicated. Additional data from these studies should be analyzed and the hypothesis confirmed or altered. After analysis, control and prevention measures should be developed and, as far as possible, implemented. A report containing this information should be prepared and distributed to those involved in investigating the outbreak and in implementing control and/or prevention measures. Continued surveillance activities may be appropriate to evaluate the effectiveness of the control and prevention measures.

Necropsy examination of ailing and/or dead birds is an important tool in diagnosing a disease; the procedure is detailed in the next Chapter.

Laboratory diagnosis of a disease involves several tests on material collected during profiling a disease. The reader is referred to standard Veterinary Microbiology/Medicine/Biotechnology/Clinical diagnosis books for further details on each of the diagnostic tests.

2.6. Risk analysis/modeling

Risk can be defined as the probability to get harmed by a hazard which, in livestock production, includes various factors like disease prevalence, impaired welfare parameters, depressed productivity etc. Risk analysis/modeling consists of risk identification and risk quantification (Noordhuizen *et al.*, 2003).

Table 60.3. Risk analysis for salmonellosis in broiler breeder flocks

Factor	Odds ratio (OR)*
Univariate	
Feed mill (small *Vs* large)	6.63
Geographic region	3.46
Ventilation	1.46
Other poultry (> 1 km *Vs* < 1 km)	1.13
Other species on farm (no *Vs* yes)	1.14
Flock size (> 15000 *Vs* < 15000)	4.13
Number of house (> 1 *Vs* 1)	7.86
Frequency of egg collection (once *Vs* > once)	1.21
Hygiene barriers (poor *Vs* good)	2.84
Order and tidiness (poor *Vs* good)	1.89
Disinfection tub present (no *Vs* yes)	1.87
Age of houses	1.02
Farm type (litter *Vs* battery)	4.05
Multivariate	
Disinfection tub present (yes *Vs* no)	0.50
Hygiene barriers (good *Vs* poor)	0.70
Interaction between the above two (poor)	25.00
Feed mill (small *Vs* large)	5.30

* crude for univariate and adjusted for multivariate; 95% Confidence levels

Source : Noordhuizen *et al.*, 2003

Risk analysis of a disease begins with clear definition of the disease, collection of data from field surveys involving properly sampled poultry farms through a standard questionnaire and study of extensive literature available on the disease.

The data so collected is analyzed (multivariate preferably) as per standard statistical (epidemiological) method(s) to arrive at contribution (probability) of each of the factors related to the disease. The probability so estimated is normally an odds ratio (OR) which will indicate the riskof a farm in question to a particular disease.

An example univariate and multivariate analysis of factors related to salmonellosis in broiler breeder flocks is given in Table 60.3.

An OR of 6.63 for feed mill (univariate) indicates that a farm with a small fed mill has 6.63 times higher odds to be among the salmonella positive group. Similarly, an OR of 0.50 for disinfection tub present (multivariate) means that a farm with disinfectant tub in use has only half as much risk as that without the tub to be infected with salmonella.

The OR values also give the relative importance of different risk factors studied which is extremely useful in developing strategies for disease control. In simple terms, OR values quantitate the disease risk associated with each of the factors under investigation and helps ranking the risk factors.

The major setback for the risk analysis/modeling is the cost, men and material requirements; however, in the larger interests of the public, it should not be a constraint, especially regarding diseases like AI, salmonellosis, campylobacter infection etc.

2.7. Quality control

Animal health is a reflection of quality of livestock production and it includes both the process of production as well as the product produced. Several methods of quality control have been developed; for instance, NEN ISO 9000 series, Good manufacturing practice codes (GXP process), Hazard analysis of critical control points (HACCP) etc. The latter (HACCP) requires minimum data collection, monitoring and documentation and is inexpensive in addition to several other advantages. Therefore, HACCP is the most widely-practiced quality control method. See Chapter "Biosecurity" for details.

Chapter **61**

Necropsy Examination

A necropsy (postmortem) examination is performed to determine the cause of disease by gross and microscopic examination of tissues and by conducting appropriate serologic and microbiologic examinations. A postmortem examination is indicated whenever there is a decrease in production, there are overt signs of illness, or there is an increase in mortality (Butcher and Miles, 2003). In other words, the necropsy (post-mortem dissection) of poultry is a procedure to find reasons for the bird's sickness and/or death.

Diagnosis of disease is very important to effect proper control and eradication. Several techniques and procedures are available for diagnosis which also depends on factors like type of disease, nature of pathogen, site of predilection of pathogen etc. Collection of information and material for diagnosis must be done in a most orderly fashion so that none of the possible clues will be overlooked.

Necropsy will not reveal all causes of disease because a high percentage of disease problems are related to management, including poor nutrition, feed and/or water deprivation, improper ventilation, poor sanitation, chilling or overheating of birds, and overcrowding. Such conditions often require an on-site investigation to determine the cause of the problem. Necropsy is most likely to identify infectious disease processes, nutritional deficiencies, toxicities, parasitic disease, and tumors (Butcher and Miles, 2003). Whether essential are not, it is preferable to collect blood and tissue samples which can be discarded later, if found superfluous.

1. History

Collection of information about the history of the flock and the environment and also of any signs observed in the bird(s) before death is very useful in conjunction with gross and histopathological changes which will be available after the detailed examination of bird (s). The flock history should include: primary clinical signs, age, strain (or species), number affected, vaccinations, medications, recurrent aviary diseases, feeding problems, feed consumption, production, body weights, and mortality pattern. If the disease problem is in chicks, the history on hatching and brooding procedures is needed (Butcher and Miles, 2003). These details regarding managemental practices (feeding, watering, ventilation, floor space, lighting and beak-trimming) are of vital importance either singly or in combination for diagnosis making a list of possible causes of the problem (differential diagnosis) can be developed following study of the complete history.

A proforma for collection of flock history and recording necropsy findings is given at the end of this Chapter (Smith, 2000); but, suitable changes can be made depending on the local requirements.

2. External examination

Thorough external examination of the sick birds for respiratory, nervous or other visible symptoms has to be made before they are killed for necropsy examination. Few ailing birds may be retained to find out whether the observed symptoms are temporary or sustained. The sick birds must also be physically examined for presence of tumors, abscesses, skin changes, beak condition, cannibalism, injuries, diarrhoea, nasal and/or respiratory discharge, conjunctival exudate, feather-and comb-conditions, dehydration and body-fleshing condition etc. The external examination assumes an extra significance if the condition is a flock problem.

General condition and fleshing (presence of meat on the bone) of the bird is noted. Condition of the skin, and all natural body openings (nasal openings, mouth, ears, and vent) observed. Head, eyes, comb, and wattles are observed for evidence of swelling, canker lesions, unusual discharge or coloration. Signs of lameness, paralysis, or general weakness looked for. Any affected areas are examined for abnormalities or swelling that can give a clue to the cause. If a partial or complete paralysis is noticed, position the bird assumes is recorded; it is often an indicator of the cause of illness. The bird is inspected for external parasites such as mites, lice, ticks, and fleas (Smith, 2000).

3. Blood sample

Blood sample has to be collected from the diseased bird just before or immediately after killing the bird. Paired samples are preferred. Blood can be collected from wing (brachial) vein or jugular vein or by heart-puncture.

3.1. Venipuncture

Venipuncture of the brachial vein can be made by exposing the vein by plucking a few feathers from the ventral surface of the humeral region of the wing. The vein will be seen easier if the region is first dampened with 70% alcohol or any other colorless disinfectant. The needle should be inserted opposite to the direction of blood flow.

3.2. Heart-puncture

Heart-puncture can be made antero-medially or antero-posteriorly through the thoracic inlet. A general rule for the lateral puncture is to form an imaginary vertical line at the anterior end at the right angle with the keel, then palpate along that line. The heartbeat can be felt and the needle is inserted about 2.5 cm above and posterior to the point of keel at about 45° angle to the proper depth to obtain the blood sample. To make a heart-puncture through thoracic inlet, the birdies kept on its back with the keel up. The crop and contents are pressed out of the way and the needle guided along the vertical angle of the inlet. After penetrating the inlet, the needle is directed horizontally and posteriorly along the midline until the heart is pierced.

The site and length of the needle required for blood collection depends on the size of the bird; for young chicks and poults a 1.9 cm 20 gauge needle, for mature chicks 5 cm 20 gauge needle and large-sized needle for mature turkeys are usually recommended.

The blood should be removed aseptically and placed in a clean vial and labeled properly. Foremost serological tests, serum from 2 ml blood sample is sufficient. If plasma sample is required, blood is drawn into sodium citrate solution (2%) at a rate of 1.5 ml/100 ml fresh blood or deposited in a vial containing sodium citrate powder at a rate of 3 mg/ml cold blood and mixed quickly after the blood is drawn. For collecting sterile citrated blood sample, the collecting tubes having proper amounts of 2% sodium citrate solution may be sterilized and the moisture evaporated in an oven. Blood-collecting vials containing heparin as an anticoagulant can also be used.

4. Killing of birds

The most humane methods of killing the bird are injecting sodium pentobarbitol, electrocution, and dislocation of the head from cervical vertebrae. The first two methods are usually too expensive or dangerous for common usage. Cervical dislocation is the most practiced method of killing birds for examination (Smith, 2000).

To dislocate the head from the vertebra, the bird's head is directed toward the operator and grasped with a handshake grip. Thumb is kept behind the head at the base of the skull, all the remaining fingers allowed to extend under the throat. The bird's feet are held with the other hand and the bird is stretched until it is felt that head is separating from the neck vertebrae. It may be necessary to bend the head back slightly while stretching the bird (Smith, 2000).

Care should be taken to stop pulling when the spine separates or the head may be pulled off. The bird dies immediately when the spine separates.

The killing of small birds such as chicks, poults, or parakeets is often difficult because their heads are small and hard to grasp. Therefore, they can be killed by breaking the neck. The vertebrae may be separated by applying pressure with scissor handles at a joint between two vertebrae. It is best to apply pressure on each side of the neck rather than at the throat and back of the neck. This avoids unnecessary damage to the gullet and windpipe. The wings are held over the back with one hand and head with the other in such a way as to bend the head sharply vertically; at the same time, it is pulled firmly and quickly forward in a stretching manner the effect instant breaking of neck and spinal cord. The neck is held firmly in the final position to avoid regurgitation and aspiration of crop contents into the respiratory passages. Sudden decapitation with sharp scissors or breaking of neck by pressing it firmly against a sharp table edge can also be practiced in case of very young birds.

Larger chicken and turkeys can be killed using a Burdizzo-forceps. Electrocution by using an alternate current of 100 V is also a satisfactory method, though it may not be in much practice.

5. Precautions

Most of the poultry diseases are not communicable human beings but still there are certain diseases that can be transmitted from birds to human beings (Influenza, Ornithosis, Erysipelas etc.). Here must be taken to disinfect the carcass and necropsy table to use proper rubber gloves, not to inhale dust or aerosols from tissues or feces and not to puncture the skin during the necropsy examination. Equipment used for necropsy examination must be properly cleaned. The equipment which are normally required are shears to cut bones, enterotome scissors to incise the gut, knife to cut skin and muscle, scalpel and forceps. Sterile syringes, needles, vials and petri-dishes are essential for collection of samples and specimens depending on the situation.

6. Necropsy equipment

Necropsies can be performed with a limited amount of equipment. Required are a knife (4- to 6-inch), bone shears, tissue scissors (preferably with sharp/blunt blades), forceps with teeth, disposable gloves, disposable syringes (3 cc and 5 cc), needles (20 gauge, 1 inch for wing vein blood collection, and 1 1/2 inch for heart blood collection), sanitizer for cleaning instruments and table, tissue bottles with 10 percent neutral buffered formalin, black marker, and labeling tape. If serum samples are collected, blood collection tubes and serum vial are also needed (Butcher and Miles, 2003).

7. Necropsy procedure

Necropsy examination of sick and/or dead birds is an important pre-requisite for efficient diagnosis; it is rightly said, "Dead will speak by itself". Necropsy examination of birds can possibly be done either from the head downwards or from legs and abdomen upwards. In either case, the observations, recordings and collection of material are essentially the same. Given below is the procedure from head downwards as described by Butcher and Miles (2003).

1. Feathers are moistened with water containing detergent. If psittacosis (or other human pathogen) is suspected, the bird should be soaked in a 5 percent Lysol solution, and a laminar flow hood should be used while performing the necropsy.

2. With scissors, through one lateral commissure, the mouth is cut and the oral cavity examined.
3. The cut is extended as a longitudinal incision through the skin of the neck to the thoracic inlet. Skin is reflected laterally.
4. A longitudinal incision is made in the esophagus and crop; the content and odor recorded.
5. A longitudinal incision made and the larynx and trachea examined.
6. With bone shears, the upper beak removed with a transverse cut near the eyes to facilitate inspection of the nasal cavity; it will also expose the open anterior end of the infraorbital sinuses.
7. One blade of a pair of sterile scissors is inserted into the infraorbital sinus. A longitudinal lateral incision made through the wall of each sinus and examined. Sinuses are cultured if indicated.
8. The loose skin between the medial surface of each thigh and the abdomen incised. Legs reflected laterally, and hip joints disarticulated. Skin on the medial aspect of each leg incised, and reflected to expose the muscles and stifle joint.
9. Lateral skin incisions connected with a transverse skin incision across the middle of the abdomen. Skin of the breast reflected anteriorly, and of the abdomen, posteriorly.
10. A longitudinal incision is made through the pectoral muscles on each side of the keel and over the costochondral junctions. The anterior end of each incision should intersect the thoracic inlet and the dorso-ventral midpoint. With bone shears or scissors, cut is made through the coracoid and clavicle bones.
11. With sterile scissors, a transverse incision is made through the posterior part of the abdominal muscles. On each side, the incision is continued anteriorly through the costochondral junctions. The ventral abdominal wall removed and breast as one piece, observing the air sacs as they are torn during removal.
12. Without touching, the viscera and air sacs are examined *in situ*.
13. Using sterile instruments, any organs required is removed and any swabs needed for culturing is taken. The spleen can be exposed aseptically by freeing the left margin of the gizzard and reflecting that organ to the bird's right side. All unnecessary manipulations and delays prior to culture increase the probability of contamination.
14. The pancreas is examined. The esophagus is transected at the anterior border of the proventriculus. The entire gastrointestinal tract is reflected posteriorly by cutting the mesenteric attachments and then removed after transecting the rectum.

15. The liver and spleen are removed and examined.
16. The genitalia are examined. In the female, the ovary and oviduct are removed, and the oviduct opened longitudinally.
17. The ureters and kidneys are examined *in situ*. If indicated, they can be removed for closer examination.
18. Heart is removed and examined.
19. The lungs are examined by reflecting them medially from between the ribs.
20. A longitudinal incision through the proventriculus, ventriculus, small intestine, ceca, colon, and cloaca with the help of a scissors and examined for lesions and parasites.
21. Both brachial plexuses and sciatic nerves should be examined. The brachial plexus is most easily observed anterior to the first rib. The extrapelvic sciatic nerve is exposed by careful separation of the adductor muscles. The intrapelvic portion of the sciatic nerve is exposed by blunt dissection of the middle lobe of the kidney.
22. With bone shears or scissors, one femur is split longitudinally and the bone marrow examined.
23. To examine the brain, the head is disarticulated and skin removed. The calvarium is removed with strong scissors, using the same technique as for mammals.

As indicated above, the alternate procedure is legs and abdomen upwards which is as follows:

1. The bird is laid on its back and each leg is drawn outwards away from the body with simultaneous incision of the skin between leg and abdomen.
2. The leg is spent forward, downward and outward to free the head of the femur from the acetabulum so that the legs lie flat on the table.
3. The skin is cut midway between keel and vent and the cut edge is forcibly deflected forward, cutting as necessary, to expose the entire ventral aspect of the body including neck. Hemorrhages of the musculature, if present, can be detected.
4. With the necropsy knife, the abdominal wall is cut transversely midway between keel and vent and then through breast muscles in each side. The coracoid and clavicle are cut on either side with the help of the bone shears without severing any large blood vessel. The sternum and attached structures can be removed from the body. This can also be achieved in a reverse order i.e. cutting the coracoid and clavicle followed by cutting of breast muscles and abdominal wall. Alternatively, the abdominal muscle may be

incised and the ribs are cut through on the sides of the keel bone. The keel is grasped near the abdomen and pulled upwards to expose the internal organs and chest cavity.

5. Heart-puncture can be done at the stage also to collect the blood samples.
6. Impression smears with sterile slides, sterile tissues for bacterial culturing (if the tissues are suspected to be contaminated, they can be seared with a hot iron before inserting a sterile culture loop), bile samples, intact section of lower trachea, the bronchi, upper portions of lungs and air-sacs with or without exudate for respiratory virus isolation, and similarly, tissues from viscera for isolation of other pathogens may be collected aseptically depending on the situation and the disease suspected.
7. All other visceral organs must be examined for lesions and wherever felt essential, cultures have to be made and tissue specimens are taken before the gut is opened.
8. The lungs, which are attached to the ribs, can be gently teased out of the ribcage for further examination.
9. The intestine is laid out on a newspaper or on table-top for examination for inflammation, exudates, parasites, foreign bodies, malfunctions, tumors and abscesses.
10. The various nerves, bone structure, marrow condition and joints observed. Sciatic nerve is exposed by the setting the muscles medial to the thigh and observed. Brachial plexus on either side near the thoracic inlet are examined. Vagus nerve is examined entirely.
11. Thorough examination of the bones for the strength, appearance and abnormalities is also done.
12. With a sharp knife, the stifle and hock joints are cut through, looking for yellow or white pus-like material, blood, or excess fluid. Joints should appear shiny and white with just a small amount of clear, sticky fluid inside. Joint exudate, if present, is collected aseptically after plucking the feathers and searing the overlying skin.
13. To find the bursa of Fabricius, the cloaca a cut is made and a grape-like structure towards the rear of the bird is looked for. The older the`bird the smaller the bursa. The bursa diminishes in size as the bird reaches sexual maturity. The bursa is cut in half. It should have wrinkles running parallel to each other on the surface and be cream colored in appearance. Any discoloration or swelling recorded.
14. For obtaining brain specimens, the head is removed at atlanto-occipital joint and the lower mandible is removed. The skin is reflected forward over skull and upper mandible. A heavy jawed bone shear is used to cut both sides of

head starting from occipital foramen and proceeding towards and laterally to the midpoint of the anterior edge of the cranial cavity. The portion is removed to expose the brain.

In either procedure, the carcass must be disposed properly and surfaces and tools disinfected for reuse.

7.1. Specific examinations/observations

While the carcass is being examined both externally and after opening the viscera, many other observations and maneuvers are generally made to obtain information and/or specimen for further diagnostic procedures; they are listed in Table 61.1.

8. Care of specimens

When performing the necropsy, the following should be done to guarantee the quality of the specimens for histopathological examination (Butcher and Miles, 2003).

1. Formalin-fixed tissues to be submitted only from freshly killed (or deceased) birds.
2. At least 10 times the volume of 10 percent neutral buffered formalin to the volume of tissue taken to be used for histopathology.
3. Containers with wide openings only are to be used.
4. Tissues should not be frozen prior to or after fixation in formalin.
5. Tissues should be only a 1/4 inch (0.5 cm) thick.
6. Excess extraneous tissue should be trimmed from the specimen.
7. All hollow organs (including trachea, GI tract, bursa of Fabricius, cloaca and uterus) have to be opened prior to fixation. Segments are best if they are 1½ to 2½ cm long.
8. Inside surface of hollow organs should not be touched if submitting the tissue. Excess ingesta or blood can be rinsed in another container of formalin (or saline).
9. Tissues should neither be squeezed nor distorted with forceps.
10. Spleen, heart, and brain are cut in half to allow contact with fixative.
11. Bone needs to be stripped of skin and muscle and incised in order for fixative to enter the core.
12. If tissue floats, it can be covered by gauze or a paper towel.
13. Tissue cassettes are helpful when submitting small pieces of tissue, such as air sac or peripheral nerve.
14. Make sure containers and/or tissues are labeled and properly sealed to prevent leakage during transport.

Table 61.1. General observations during necropsy examination of birds

Organ / Area / Procedure	Observation
Squeezing the turbinate area	Excessive matter oozing from the area.
Eyes	Inflammation (unusual reddening), mucus, or discoloration.
The mouth and larynx	Abnormalities that indicate pox, mycosis, or other disease
The gullet	Presence of tiny nodules (bumps) or signs of injury by foreign materials.
Interior of larynx and trachea	Excessive mucus, blood, or cheesy material.
Breast muscles	Conditioning and the presence of hemorrhages.
Legs (skin removed)	Small pin-point hemorrhages.
Air sacs	Cloudiness and covering with mucus.
Liver	Unusual swelling, lesions, hemorrhages, or abnormal coloration.
Liver incised	Scar tissue and necrotic (dead) tissue.
Spleen	Hemorrhages, lesions, and swelling.
Pericardium	Cloudy, fluid-filled.
Digestive system	Abnormal nodules, tumors, or hemorrhages.
Crop contents	Sour smell.
Crop lining	Thickening, patch-like areas or necrotic ulcers. Capillary worms: by making a small cut and slowly tearing the crop wall as if it were a piece of paper; capillary worms appear as small, hair-like fibers extending across the base of the tear.
The proventriculus	Hemorrhages or a white coating on the lining.
Gizzard (opened)	Lining examined for unusual roughness or lesions. It is determined whether the lining is separating from the underlying muscles or not.
The intestine (slit lengthwise)	Contents examined for the presence of worms, free blood, and excess mucus; the lining for inflammation, ulcers, or hemorrhagic areas. If unusual conditions exist, in which one-third portion of the intestine the conditions are located must be recorded.
The ceca (opened)	Presence for cheesy cores and small, cecal worms are checked. If blood is found, it is washed and the lining examined for scarring and cecal worms.
The reproductive organs (ovary and oviduct in females, testes and *ductus deferens* in males)	Abnormalities before removing them from the body.
The kidneys and ureters	Unusual swelling or the presence of whitish salt deposits.
The sciatic nerve and brachial nerve	Swelling.
The lungs and bronchial tubes	Lesions and unusual accumulation of mucus.
Bursa of Fabricius	Discoloration, swelling, tumors, regression

8.1. Brain and spinal cord

For culturing and histological examination (HPE), the brain is cut anterior to posterior along the midline with a sterile scalpel and one portion is transferred to a jar containing formalin. The other half is aseptically transferred to paper petri-dish. The specimens should be properly put in a fixative. For this purpose, the specimens must be small enough to allow proper penetration of the fixative. The volume of the fixative will be usually 8 to 10 times the volume of the tissue. Zenker's fixative may be used for all tissues excepting brain and spinal cord.

8.2. Lung tissue

Lung tissue floats on the fixing solution because of trapped air and hence, absorbent cotton may be placed over the tissue.

8.3. Bone

In case of bone, after fixing, it is decalcified in a 10% solution of nitric acid in 70% alcohol before sectioning in a microtome. Decalcification requires at least a week's time.

8.4. Eyes

In case of eye, a window is cut through the scleral wall to allow fixative too quickly to penetrate the vitreous body.

8.5. General precaution

Any tissue should not be held for more than 48h in formalin; but transferred to 70% alcohol solution to avoid excessive hardening of the tissue.

9. Tentative diagnosis

Gross lesions and microscopic examination at times to strongly suggest the particular disease; for instance, wet-mount smear of mucosal scrapping from intestine and microscopic examination of cecal contents can indicate coccidiosis. Similarly blood parasites and fungi can be demonstrated microscopically using suitable samples.

10. Embryo inoculation

For virus isolation, centrifuged and/or filtered fine-ground suspension of suspect tissues or body-fluids and exudates may be inoculated into the chorio-allantoic cavity, yolk-sac or onto the chorio-allantoic membrane (CAM) of the embryo at various stages of incubation depending on the suspected virus to be isolated. Eggs must be obtained from specific pathogen-free flocks.

11. Specimen disposal

It is necessary to autoclave, incinerate or to render the carcass incapable of causing infection to laboratory and/or other personnel or other birds. Afterwards, the necropsy area, instruments and gloves should be washed with the disinfectant solution.

12. Material for diagnostic laboratory

Material collected during necropsy examination should be carefully handled and sent for further processing to a recognized diagnostic laboratory (Table 61.2).

Table 61.2. Material to be sent to diagnostic laboratory

Disease	Material in 10% formalin for HPE	Material for isolation and identification
Pullorum	Ovary and testes (adults), heart, liver, dead chick, blood smear	Long-bone packed in charcoal; ovary, liver, gallbladder (in ice), whole blood
Fowl cholera	Liver, heart, gizzard, long bone	Blood smear, long-bone packed in charcoal
Coryza	Basal sinuses and passages	Throat swab moistened with nutrient broth
Chronic respiratory disease	Nasal passages, trachea and bronchi	Throat swab kept moist, whole blood
Spirochetosis	Spleen, liver	Blood smear, spleen, liver, heart in ice
Aspergillosis	Lungs	Lungs and ice
Infectious bronchitis	Trachea, oviduct	Lungs, trachea/bronchi in 50% glycerol saline
Infectious laryngotracheitis	Trachea, larynx	Trachea in 50% glycerol saline, whole blood
Disease	Material in 10% formalin for HPE	Material for isolation and identification
Newcastle disease	Proventriculus, spleen, liver, gallbladder, heart and brain	Brain in skull bones, spleen, lungs in 50% glycerine, whole blood
Pox	Skin lesions, buccal mucous membrane and pharynx	Skin lesions, tissue scrapings of mucous membranes underneath canker in 50% glycerol saline
Lymphoid leukosis	Bursa of Fabricius and organs showing lesions	Bursa of Fabricius in 50% glycerol saline, whole blood
Marek's disease	Gonads and other visceral organs with lesions	Skin, gonads 50% glycerol saline, whole blood
Avian encephalomyelitis	Brain, spinal cord, proventriculus, heart, pancreas	Brain, spinal cord 50% glycerol saline, whole blood
Coccidiosis	Intestines, ceca	Intestines and feces
Gout	Kidney, joints, heart	----

Pro forma for necropsy examination

Owner ____________________

Address ____________________ Phone No. ____________________

Number in Flock ____________ Breed ____________________

Age ____________ Hatchery source ____________________

Type of operation (floor, cage, range, etc.) ____________________

Feeding program ____________________

Vaccination History ____________________

Date Illness First Seen ____________________

No. Affected by Illness ____________ No. Dead ____________________

Medication ____________________

____

____

Symptoms and Remarks ____________________

____

External Examination

Condition of Bird ____________________

Comb and Wattles ____________________

Eyes, Ears, Mouth ____________________

Vent Opening ____________________

External Parasites ____________________

Necropsy Results

Female ____________ Male ____________________

Head

Eyes ____________ Nasal Cavities ____________________

Mouth ____________________

Respiratory and Circulatory Systems

Larynx and Trachea (Windpipe) ____________________

Lungs and Bronchial Tubes ____________________

Pro forma for necropsy examination (contd.)

Air Sacs ____

Heart ____

Digestive System and Accessory Organs

Gullet (Esophagus) ______________________________

Crop ____

Proventriculus and Gizzard ________________________

Small Intestine __________________________________

____

Ceca ____

Cloaca ____

Liver ____

Spleen ____

Excretory and Reproductive Systems

Kidneys and Ureters ______________________________

Ovary and Oviduct ______________________________

Testes or *Ductus Deferens*

Muscles

Breast ____

Legs ____

Nervous System

Brachial Nerve

Sciatic Nerve __________________________________

Diagnosis

____

____

____

Treatment

____

____

Chapter **62**

Bacterial Diseases

1. Avian salmonellosis

This is a group of diseases of poultry caused by one or more members of bacterial genus *Salmonella*. The organisms are classified, based on motility, into motile and non-motile. Of nearly 1800 serotypes, only two are non-motile; they are *Salmonella pullorum* (causing pullorum disease) and *Salmonella gallinarum* (causing fowl typhoid); the rest of the species are motile *Salmonellae* which are designated as paratyphoid organisms (causing paratyphoid infections).

1.1. Pullorum disease (Bacillary white diarrhea)

1.1.1. Etiology

S. pullorum; ubiquitous, slender rod, Gram-negative, non-motile, non-sporeforming, facultative anaerobe, can decarboxylate ornithine (unlike *S.gallinarum*).

1.1.2. Hosts

Natural host is principally chicken and a lesser degree turkeys. Other birds which can be naturally infected are ducks, guinea fowl, pheasant, quail etc. In human beings, Salmonellosis caused by *S. pullorum* (when 1.3 to 4.0 billion organisms were ingested) is characterized by rapid onset, high fever, prostration and prompt recovery.

1.1.3. Transmission

Primarily through hatching eggs; cannibalism of infected birds, egg-eating, wound infection and contaminated feed can also transmit the disease.

1.1.4. Symptoms

1.1.4.1. Chicks and poults

1.1.4.1.1. Frequent

Newly-hatched chicks from infected eggs will be moribund and few may be dead. They exhibit somnolence, weakness, inappetance, shrill cry while defecating and chalky white diarrhoea; peak mortality occurs during 2nd or 3rd week.

1.1.4.1.2. Occasional

Lassitude, huddling, drooping of wings and greenish brown stain in and around vent. If respiratory tract is infected (aerosol), respiratory distress can be noticed.

1.1.4.1.3. General

Retarded growth, poor feathering and recovered birds or carriers. Rare - blindness and joint infections.

1.1.4.2. Adults

No distinct signs on most occasions, reductions in egg production, fertility and hatchability of also highly variable. In rare cases, acute infection results in depression, anorexia, diarrhoea and dehydration.

1.1.5. Morbidity and mortality

Highly variable; morbidity is always greater than mortality and the latter may go up to 100%.

1.1.6. Gross lesions

1.1.6.1. Chicks and poults

Enlargement and congestion of liver with streaks of hemorrhages, hyperemia of internal organs due to septicemia, distension of ureters with urates, unabsorbed yolk, necrotic foci and nodules in heart, liver, lungs, ceca, large intestine and gizzard, punctiform or focal necrosis in liver, spleen enlargement, congested or anemic kidneys, peritonitis, thickening of intestinal mucosa, yellowish-grey nodules and hepatization of lungs can also occur.

1.1.6.2. Adults

Chronic carrier hens exhibit characteristic misshapen, discolored and cystic ova, peritonitis and acute or chronic pericarditis. Frequently, the follicles are pedunculated with thickened capsule and contain oily and cheesy material. They may get detached and embedded abdominal fat. Abdominal ovulation, oviduct impaction, peritonitis and subsequent adhesions of viscera can occur. Ascites may occur in turkeys. Pericarditis is also frequently recorded with increased volume and turbidity of pericardial fluid; thickening of pericardium follows, in males, the infection is frequently found in reproductive organs.

1.1.7. Diagnosis

A tentative diagnosis can be made on gross symptoms and lesions but has to be confirmed by isolation and identification of the organism.

1.1.8. Treatment

No drug or combination of drugs is capable of eliminating infection from treated birds. In chicken, several drugs have been tried and found effective in reducing mortality; they include, Sulphonamides (Sulfadiazine, Sulfamerazine, Sulfaquinoxaline and Sulfamethazine at a maximum drug level of 0.75% in starter mash for the first 5 to 10d of age), Nitrofurans (Furazolidone in mash at 0.04% levels for 10 to 14d from day-old, Furaltadone - in water), Antibiotics (Chloramphenicol 0.5% in feed for 10d, Chlortetracycline 200 mg/kg feed). *In-vitro* studies have indicated that the organism can develop drug resistance for nitrofurans and antibiotics.

1.1.9. Prevention and control

Identification and elimination of carriers is most important. Stained antigen whole blood test developed by Schaffer *et al* (1931) is commonly used to identify the carriers in chicken and the test is not satisfactory in case of turkeys. However, serological evidence of infection should be confirmed by bacteriological examination of one or more carriers. Non-Pullorum reactors also pose problems in interpretation - for instance, coliforms, micrococci, streptococci (Lancefield group D) and paratyphoid (group D) organisms produce cross-reactions.

1.1.9.1. Whole-blood agglutination test

A measured drop (0.5 ml) of colored antigen is taken on a testing plate. A drop of blood (0.2 ml) is taken from the bird (wing vein) and mixed with the antigen. The plate is rotated in a circular motion to facilitate mixing and clumping within 15 to 20 sec. The carrier will have antibodies in the blood and hence, clear clumping can be seen during the test. Such birds are referred to as carriers which are eliminated from the flock.

A more reliable tube agglutination test developed by Jones (1913) is available; but, not commonly practiced since it requires a specialized technician and is costly.

It is highly recommended that the breeder flock is blood-tested and hatching eggs obtained from Salmonella-free flock. This is the best method for eradication of Pullorum disease.

1.2. Fowl typhoid

1.2.1. Etiology

S.gallinarum; ubiquitous, short, plump rod, Gram-negative, non-sporeforming, non-capsular, non-motile, facultative anaerobe, produced endotoxins and doesn't decarboxylate ornithine (unlike *S.pullorum*).

1.2.2. Hosts

Natural hosts are chicken, turkeys, guinea fowl, peafowl, ducklings, quail, grouse and pheasants.

1.2.3. Transmission

Egg-borne is the predominant route; men, material, rats and flies can also transmit the disease.

1.2.4. Symptoms

1.2.4.1. Chicks and poults

Similar to Pullorum disease i.e. moribund or dead chicks in the hatching trays, somnolence, poor growth, weakness, inappetance, adherence of whitish material to the vent and when lungs are involved, dyspnoea.

1.2.4.2. Adults

Acute outbreak is characterized by sudden drop in food consumption, droopy and ruffled birds with pale heads and shrunken combs. Fever and death after 4 to 10d of infection.

Turkeys exhibit increased thirst, inappetance, listlessness, green or greenish-yellow diarrhoea and fever.

1.2.5. Morbidity and mortality

Depends on various factors like source and severity of infection and immune status of the bird etc. Mortality in chicken may vary between 10 and 40%.

1.2.6. Gross lesions

1.2.6.1. Chicken

Peracute cases show little or no gross lesions; prolonged cases – swelling and redness of liver, spleen and kidneys, especially in young birds. Sub-acute and chronic cases exhibit greenish-brown or bronze and swollen livers. Other lesions include grayish brown miliary foci in liver and myocardium, pericarditis, peritonitis, catarrhal inflammation of intestines.

1.2.6.2. Turkeys

Similar to chicken but the duration is short and therefore less extensive. Breast muscle appears as though partially cooked. Heart swollen with necrotic foci and/or peticluae. Liver enlarged, friable and bronze to mahogany colored with pinpoint necrotic foci. When liver is cut, blood flows readily. Spleen enlarged, friable and mottled. Caseated abscesses in lungs, kidney enlargement with or without petichiae may also occur. Paralysis of digestive tract (indicated by food in crop and gizzard), anemic intestines with ulceration is of the mucous membranes (especially in the duodenal region) are also likely. In carriers (both males and females), reproductive system is affected.

1.2.7. Diagnosis

History of the flock, signs and lesions help derive a tentative diagnosis to be confirmed by isolation and identification of the organism. Polyvalent rapid whole-blood plate antigens can be used to detect carriers of *S.pullorum* and *S.gallinarum.*

1.2.8. Treatment

Sulphonamides – Sulfaquinoxaline 0.1% (feed) or 0.04% (Water) from 2 to 3d or additional 2d, is needed. It has to be withdrawn at least 10d before slaughter. Other sulphonamides can also be used. Nitrofurans - Furazolidone (0.011% for two weeks followed by 0.0055% in feed till marketing of the flock). The withdrawal period is 5d. Antibiotics - Streptomycin, Chloromycetin and Chlortetracycline are effective but not approved (in the USA) for use in poultry raised for food.

1.2.9. Prevention and control

Testing the breeding flock by rapid whole-blood agglutination test and elimination of carriers. Chicks and poults must be obtained from Salmonella-free flocks. Few countries permit use of live-modified vaccines against fowl typhoid.

1.3. Paratyphoid infections

1.3.1. Etiology

Salmonella organisms other than *S.pullorum* and *S.gallinarum*; Gram-negative, non-sporeforming, non-capsular, motile (peritrichous flagella), facultative anaerobe producing endotoxins. The important species are *S.typhimurium, S.montevideo, S.derby, S.meleagridis, S. Newport, S.bredeney* and others.

1.3.2. Hosts

Can occur in most species of warm and cold-blooded animals and among poultry, the most common hosts are turkeys and chicken. The infections or recorded in geese, ducks, pigeon and other birds.

1.3.3. Transmission

Direct ovarian transmission, although recorded by few workers in turkeys, do not seem to be the common route of transmission. However, shells of eggs were found to be frequently contaminated especially by the fecal matter; therefore, this is the most important mode of transmission. Poultry feed can be a very important source of infection. The organisms can migrate into the eggs, especially during the air-cell formation and subsequently multiply inside the eggs. In this way, incubators, hatchers and newly-hatched chicks may be contaminated. Bird-to-bird transmission can occur by inhalation, fecal contamination of water and feed and direct consumption of feces. Rats, mice and other species of birds also act as potential source of infection. Naval infection has also been suspected.

1.3.4. Symptoms

1.3.4.1. Young birds

Most common; in acute form, no signs may be exhibited excepting high mortality of embryos in hatcher/at a few days of hatching (brooding stage). In less severe form, the symptoms include somnolence, birds standing with lowered head, droopy wings and closed eyes, ruffled feathers, anorexia, increased water intake accompanied with watery diarrhoea and vent-pasting, huddling near the heat source and rarely, respiratory symptoms. In some cases, blindness may result due to vacuolation of corneal membrane and opacity of cornea

1.3.4.2. Adults

Usually no gross symptoms are seen and they act as potential carriers. Morbidity and mortality: The mortality seldom exceeds 10%. Gross lesions: Young birds – In very severe outbreaks, there may not be any lesion appreciable. In less severe cases, the common lesions are emaciation, dehydration, coagulated yolks, congested liver and spleen with hemorrhagic streaks or pinpoint necrotic foci, congested kidneys and pericarditis with adhesions. Heart and lung lesions are not common (unlike Pullorum disease).

1.3.4.3. In poults

Hemorrhagic enteritis, especially in duodenum is commonly noticed and sometimes cecal cores are seen. Adults – In acute infection, the lesions that may be seen are congested and swollen liver, spleen and kidneys and hemorrhagic or necrotic enteritis, pericarditis and peritonitis. In some cases, ovary and oviduct may also be involved. Arthritis (turkeys) and leg weakness may also be noticed in few cases.

1.3.5. Diagnosis

Gross symptoms and lesions with a supportive history may help drawing a tentative diagnosis. Isolation and identification of the organism confirms the disease.

1.3.6. Treatment

Sulphonamides (Sulfamethazine, Sulfadiazine and Sulfaquinoxaline) in feed can be used with suitable precaution to avoid toxicity, especially in turkeys. Antibiotic (Combistrep or Polymyxin) injection to day-old poult is a common practice to avoid paratyphoid infections. Of the nitrofurans, Furazolidone can be administered at a concentration of 0.011% and 0.0055% in feed continuously for birds up to 2 weeks and over 2 weeks of age, respectively, as a preventive measure. During the outbreak, a dosage of 0.011 to 0.022% in feed for 2 weeks is recommended.

1.3.7. Prevention and control

Infected and recovered flocks or carrier birds should not be used for breeding. Sanitation of poultry premises, hatching eggs and feed are also extremely important. Serological testing by macroscopic tube-agglutination test or by rapid serum plate-test or by rapid whole-blood test or by other tests may be practiced lighted by the reactors and eliminating them. Immunization procedures are not practicable.

2. E.coli infections

2.1. Etiology

Gram-negative, non-acid fast, non-sporeforming bacilli, most of which are motile (peritrichous flagella). The organism occurs in various serotypes and they produce endotoxins and hemolysins although both are not responsible for pathogenecity in birds. Only three serotypes have been isolated in chicken.

2.2. Transmission

The organisms are natural inhabitants of gastro-intestinal tract. Hence, the most important route is by fecal contamination of feed, water and litter as well as dust. Another important route of transmission is through hatching eggs.

2.3. Symptoms

2.3.1. Mushy chick disease

When the *E. coli* infect the yolk sacs; it is characterized by edema and infected yolk. Fecal contamination of eggs is considered the most important source of infection although salphingitis due to *E. Coli* can also be a route of entry into the chick embryo. Late-embryo mortality is very common. Omphalitis also occurs frequently in hatched chicks dying after 4d of age showing pericarditis and infected unabsorbed yolks. Low brooding temperature or starvation accentuates omphalitis.

2.3.2. Air-sac infection

Characterized by air-sac infection along with pericarditis and perihepatitis. This causes severe economic losses both due to mortality as well as due to condemnations of carcass. Inhalation of infected dust is the main route of entry for this condition.

2.3.3. Panophthalmitis

This is not a very common manifestation but is usually a sequel of *E. coli* septicemia. There will be hypopyon and blindness (usually of one eye) and most birds die shortly after the onset of disease.

2.3.4. Pericarditis

A common sequel of septicemia usually associated with myocarditis and perihepatitis.

2.3.5. Salphingitis

Due to infection of left thoracic air-sac or duty entry of *E. coli* through vagina. Chronic salphingitis is characterized by a large caseous mass in a dilated thin-walled oviduct.

2.3.6. Acute septicemia

Similar to Fowl typhoid and Fowl cholera; the most characteristic lesions are green liver and congested pectoral muscles. Pericarditis and peritonitis follow septicemia. In ducks, septicemia is characterized by moist, granular to curd-like exudate on abdominal and thoracic viscera and surfaces of air-sacs. The exudate has a peculiar odor. Liver is frequently swollen, dark and bile-stained and spleen will be swollen and dark.

2.3.7. Sinovitis

A sequel of septicemia localizing in joints.

2.3.8. Coligranuloma (Hjare's disease)

Occurs in chicken and turkeys. Characterized by granuloma of liver, ceca, duodenum and mesentery but not spleen. This form is uncommon but mortality may go upto 75% in affected flocks. Lesions of granuloma or similar to that in tuberculosis. Occasionally, Leucosis-like lesions are also produced.

2.3.9. Enteritis

This condition is quite rare; there may be watery and yellowish droppings.

2.4. Diagnosis

Isolation and identification of the organism.

2.5. Treatment

Since the organism can develop drug resistance, it is better to test for drug sensitivity before administering any drug. Many drugs have been found to be useful; they include Ampicillin, Chloramphenicol, Chlortetracycline, Nitrofurans, and Ciprofloxacin etc.

2.6. Prevention and control

Fecal contamination of hatching eggs being the most important route of transmission, fumigation of hatching eggs 1½ to 2h after the eggs are laid is mandatory. Contamination of egg handling equipment, breakage of infected eggs in the incubating machinery and fecal contamination must be avoided.

3. Avian Pasteurellosis

This is a group of diseases of poultry caused by bacterial genus *Pasteurella*; viz., Fowl cholera, Pseudotuberculosis, Goose influenza etc. Among many diseases caused by *Pasteurellae*, Fowl cholera is very important.

3.1. Fowl cholera

3.1.1. Etiology

P. multocida; ubiquitous, rod, Gram-negative, non-motile, non-sporeforming organism, characteristically stains bipolar, produces characteristic odor and organism is penicillin-sensitive (unlike other Gram-negative organisms). The organism produces endotoxin.

3.1.2. Hosts

Natural host is principally chicken and a lesser degree turkeys. Other birds which can be naturally infected are, most commonly, chicks, turkey, geese and ducks.

3.1.3. Transmission

Usually the disease appears when a newly-purchased flock is added to the existing flock indicating that carriers are the important source of infection. Contaminated material like crates, feed-bags, carcasses of infected birds etc. also are potential sources of infection. Other birds like sparrows and pigeons also act as transmitters. Within a flock, the main route of transmission is by the contamination of feed and water by the nasal and oral secretions.

3.1.4. Symptoms

3.1.4.1. Acute infection

Characterized by fever, anorexia, ruffled feathers, and mucous discharge from the mouth, greenish diarrhoea and polypnea. Most often the symptoms go unnoticed since the mortality occurs very rapidly. Cyanosis of un-feathered regions, especially combs and wattles, often occurs.

3.1.4.2. Chronic infection

Localized infection. Swelling of wattles, sinuses, leg-or wing-joints, foot-pads and sternal bursae are often noticed. Sometimes, lesions in conjunctiva and pharynx, torticollis, tracheal rales and dyspnoea may be observed.

3.1.5. Morbidity and mortality

Highly variable; morbidity is always greater than mortality and the latter may go up to 100%.

3.1.6. Immunity

Subcutaneous injection of killed *P.multocida* organisms (in an adjuvant) was practiced earlier. Although, newer live for cholera vaccine is available to be administered through water, several serotypes of the organism and several species of bacteria involved have made immunization against fowl cholera not practicable.

3.1.7. Diagnosis

By clinical signs and lesions, confirmed by isolation and identification. Serological diagnosis is a limited value in case of chronic cases and of no value in case of acute form.

3.1.8. Treatment

Sulphonamides (Sulfamethazine 0.5 to 1% in feed or 0.2% in water, Sulfamethazine 0.4% in feed or 0.2% in water, and Sulfaquinoxaline 0.01 to 0.05% in water) and others. Antibiotics like streptomycin, penicillin, oxytetracycline, chlortetracycline, novobiocin, chloramphenicol and others can be used as prophylactic drugs.

3.1.9. Prevention and control

Elimination of reservoirs/carriers is most important by stringent managemental and sanitation practices. Vaccination can be practiced in farms located in endemic areas using bacterins or live-vaccines.

4. Infectious coryza

4.1. Etiology

Hemophilus paragallinarum (H.gallinarum); ubiquitous, short rods or coccobacilli, Gram-positive, non-motile, non-sporeforming, stains polar, virulent forms are capsulated, requires factor V (NAD) in the culture medium for growth. The organism is inactivated rapidly within 5 to 6h outside the host.

4.2. Hosts

Most common in chicken of all ages; also recorded in pheasants, guinea fowl and Japanese quail. Turkeys, pigeons, ducks are refractory among the domestic birds.

4.3. Transmission

Carrier birds are the main reservoirs. Farms with multiple age groups will have higher incidence of this disease. Incubation period is very short and the disease develops within 1 to 3d.

4.4. Symptoms

Involvement of nasal passages and sinuses with serous to mucoid nasal discharge, facial edema and conjunctivitis. Swelling of wattles may be seen in males. Respiratory rales are also recorded. Diarrhoea, inanition, reduced egg production and a foul odor in diseased flocks also noted.

4.5. Morbidity and mortality

Usually very low mortality but very high morbidity. Secondary infection by various other organisms might complicate the course of the disease.

4.6. Gross lesions

Acute catarrhal inflammation of mucous membranes of nasal passages and sinuses, catarrhal conjunctivitis and subcutaneous edema of face and wattles, are commonly noticed whereas pneumonia and air-sacculitis may be seen, occasionally.

4.7. Diagnosis

Symptoms, isolation and identification of the organism and serological techniques. Inoculation of nasal exudate from infected birds into nostrils or eyes of healthy birds will produce the disease in 2d.

4.8. Treatment

Erythromycin, Oxytetracycline, Sulfachloropyrazine, Sulfadimidine and certain combination of drugs have been used successfully. However, carrier-state is not eliminated by therapy.

4.9. Prevention and control

Purchase of adult stock from elsewhere must be avoided to reduce the carriers. Birds of different age groups must be reared separately. Commercial bacterins (Monovalent or polyvalent) may be used or controlled exposure of birds (vaccinated at 15 to 18 weeks of age) to the disease at 20 weeks of age has been suggested. Flock that suffered the outbreak of coryza must be de-populated to avoid carriers.

5. Spirochaetosis

5.1. Etiology

***Borrelia anserina*, a loosely spiralled (5 to 8 spirals), motile and aerobic organism**

5.2. Hosts

Natural hosts are geese, turkeys, chicken, ducks and pheasants

5.3. Transmission

Infected birds can transmit the disease directly through droppings during the course of infection. However, the most important route of transmission is by vectors. Ticks and mosquitoes have been found to serve as true vectors whereas mites also can serve as mechanical carriers. The tick that is incriminated as vector is *Argas persicus* which gets infected within 6 to 7d after biting the host. The infected tick remains infective up to 430 d and transmit the infection throughout this period. Infection from ticks is either by biting of the host or by the ingestion of ticks by the host. The infected ticks lay infected eggs which will pass the infection to the next generation. In chicken, the incubation period is 3 to 8d.

5.4. Symptoms

Droopy and cyanotic heads with ruffled feathers. High fever, anorexia, greenish diarrhoea, increased thirst and finally paralysis. Anemia may also be seen.

5.5. Morbidity and mortality

Morbidity may vary between 10 and 100% and mortality may go up to 100% depending on severity of infection.

5.6. Gross lesions

Splenomegaly with mottling (ecchymotic hemorrhages) is a characteristic lesion.

Liver may be enlarged, hemorrhagic and necrotic. Occasionally, liver infarcts may be seen. Kidneys will be enlarged and pale. Greenish mucoid enteritis may also be seen.

5.7. Diagnosis

Demonstration of spirochetes in blood smears (Giemsa stain) particularly during early part of the disease. Isolation and identification of the organism and serological techniques.

5.8. Treatment

The organism is sensitive to penicillin (1000 units). Many other antibiotics can also be used.

5.9. Prevention and control

Tick and mosquito control must be effected. Recovery from infection or vaccination affords a long immunity. Inactivated organisms are administered through intramuscular or subcutaneous route. The vacuous given usually to young birds. However, vaccination against spirochetosis is not very common practice and the disease is usually prevented by proper managemental procedures to control vectors.

6. Ulcerative enteritis

6.1. Etiology

Corynebacterium percidum, a gram positive, pleomorphic, aerobic, non-motile, sporeforming, rod-shaped organism.

6.2. Hosts

Quails or most susceptible natural hosts; it can also affect chicken, turkeys and other Avians.

6.3. Transmission

Mainly through droppings and the infection of contaminated feed, water or litter. The organism sporulates in litter and establishes itself in the premises. The birds can also act as carriers. The disease organism is very resistant to disinfectants and will persist under varying environmental conditions.

6.4. Symptoms

This is an acute disease and death occurs within 3d. There may not be any symptoms; but, watery-white droppings, listlessness, partial closure of eyes, dull and ruffled feathers are seen in quails frequently. If the disease prolongs, emaciation (pectoral muscles) may be seen.

Birds with the acute form may die suddenly while in good flesh, whereas more chronically-affected birds become listless, have ruffled feathers, whitish watery diarrhea, and develop a humped-up posture. Such birds usually die in an extremely emaciated condition. The dropping may be confused with those of birds with coccidiosis and the two diseases are often seen in the same bird. Droppings of birds with only ulcerative enteritis never contain blood.

6.5. Morbidity and mortality

Mortality may go up to 100% in quails whereas in chicken, it seldom exceeds to 2 to 10%.

6.6. Gross lesions

Depend on duration of the disease. Quails show characteristic hemorrhagic enteritis in the duodenum and small punctate hemorrhages in intestines. In recovered birds, necrosis and ulceration occur in any portion of intestines and ceca. Spleen may be enlarged, congested and hemorrhagic.

6.7. Diagnosis

The postmortem lesions are characteristic. The entire intestinal tract often has button-like ulcers but the lower portion is most often affected. These ulcers often perforate, resulting in local or generalized peritonitis; isolation identification of the organism; serological tests.

6.8. Treatment

6.8.1. Feed

Chloramphenicol 500 g/tonne, Streptomycin 60 g/tonne, Bacitracin 100 g/tonne.

6.8.2. Water

Streptomycin 0.25 g/l.

6.9. Prevention and control

Litter contamination must be minimized by proper litter management as done in case of coccidiosis. Raising birds on wire is an effective preventative measure. Specific drugs (bacitracin or penicillin) fed at low levels, are effective for controlling the disease in operations where the use of wire flooring is impractical.

7. Necrotic Enteritis

7.1. Etiology

Clostridium perfringens, a spore-forming, rod-shaped bacterium. Bacterial organisms and their toxins are the primary cause but coccidiosis may be a contributing factor. Most of the damage to the intestinal lining apparently is due to toxins produced by the bacterial organisms.

7.2. Hosts

All species of birds.

7.3. Transmission

Ingestion of droppings from infected birds.

7.4. Symptoms

Necrotic enteritis is an acute disease that produces a marked destruction of the intestinal lining of the digestive tract. Necrotic enteritis appears suddenly in the affected flock. Apparently healthy birds may become acutely depressed and die within hours.

7.5. Mortality and morbidity

Mortality is usually between two and ten percent, but may be as high as thirty percent in severe outbreaks. Losses due to reduced growth and feed conversion may be more costly than flock mortality.

7.6. Lesions

Lesions of the disease usually involve the lower half of the small intestine, but in some instances the entire length of the tract is involved. The intestine is dilated, contains dark offensive fluid and a diphtheritic cauliflower-like membrane that involves the mucosa. The lining of the intestine will have a coarse Turkish-towel appearance and portions of the lining may slough off and pass out with the intestinal contents.

7.7. Diagnosis

Diagnosis in based upon history, symptoms and findings of the characteristic lesions.

7.8. Treatment

Bacitracin or virginiamycin is an effective treatment administered in the feed. Bacitracin can also be given in the drinking water. Supportive vitamin treatment may enhance the effectiveness of the treatment.

7.9. Prevention and control

Preventive medication may be of value on premises where prior infections have been observed. Since coccidiosis may be a contributing factor, attention must be given to an effective coccidiosis control program.

8. Botulism

8.1. Etiology

Botulism is a disease caused by the ingestion of a toxin produced by the *Clostridium botulinum* bacterium. Botulism is not a bacterial infection, but a condition produced by a byproduct of the bacteria's growth. The organism is common in nature and is widely dispersed in soils. Ingestion of the organism is not harmful. The organism is an obligatory anaerobe. Toxin types A and C cause the disease in birds.

8.2. Hosts

All domestic fowl and most wild birds are susceptible to the toxin's effects.

8.3. Symptoms

Weakness is generally the first sign of the illness and is followed by progressive flaccid paralysis of the legs, wings and neck. When neck muscles are affected the head hangs limp, thus causing a condition referred to as "limberneck". Affected birds may have a peculiar trembling, loose feathers that are pulled out easily and dull partly closed eyes. Some birds (turkey) do not develop loose feathers or limberneck symptoms. Because of the paralysis, birds are unable to swallow and mucous accumulates in the mouth. Fatally affected birds may lie in a profound coma appearing lifeless for several hours before death.

8.4. Lesions

Significant lesions are not usually observed in affected birds. Examining digestive contents may reveal insects, decomposed animal or vegetable material or other matter suggesting that the birds have consumed the toxin.

8.5. Diagnosis

A tentative diagnosis can be made from the history, symptoms and post-mortem findings.

8.6. Prevention and control

Prevention should be aimed at eliminating sources of toxin production and preventing access of birds to such materials. These practices include prompt removal of all dead animal from houses and pens, beak-trimming the birds, controlling fly and insect populations and avoiding access to decaying organic material. Contaminated water supplies are particularly dangerous.

9. Erysipelas

9.1. Etiology

Caused by *Erysipelothrix rhusiopathiae* (*Corynebacteriaceae* family).

9.2. Hosts

It affects mainly turkeys but can also infect chicken, pheasants and ducks.

9.3. Transmission

It is an acute fulminating infection of individuals within a flock. It is likely to be transmitted through contaminated material entering through breaks in mucous membranes and skin. Infected carrier birds and cannibalism on affected birds can also transmit the disease.

9.4. Symptoms

The symptoms include droopiness, sudden deaths and in some cases gradual emaciation, anemia and ultimately death due to endocarditis. Such a loss of hens after AI with peritonitis perineal congestion and skin discoloration. Weakness, depression, diarrhoea and sudden death are also noticed. Layers show conspicuous decrease in egg production.

9.5. Morbidity and mortality

Mortality may go up to 25 to 50% within a pen although adjacent pen may not be affected. Immunized and/or medicated birds show lower mortality.

9.6. Lesions

Lesions that can be noticed are congestion and intramuscular and subpleural ecchymotic hemorrhages. Turgid, reddish-purple colored snood in toms is pathognomonic. Most blood vessels appear injected, kidney and gonads congested with petichiae or ecchymoses under serous surface. Rose-colored hemorrhagic spots on pancreas, hydropericardium withpetichiae and ecchymotic hemorrhages on pericardium and epicardium. In chicken, septicaemic lesions, degeneration of fat on the anterior edge of the thigh, degeneration and hemorrhage in pericardial fat, hemorrhages in heart, friable liver, spleen and kidney. Recovered turkeys have high degree of immunity.

9.7. Treatment

Bacterins can be used under field conditions to prevent disease especially along with antibiotics like penicillin (22,000 units/kg i/m).

9.8. Prevention and control

Two or more doses of bacterin at 2-4-week intervals gives good protection with immunity declining 4-5 weeks post-vaccination equipments can be disinfected with the 1-2% NaOH before reuse. For breeders, two doses of bacterin with less than 4 weeks interval prior to onset of production.

10. Chlamydiasis (Psittacosis, Parrot fever)

10.1. Etiology

Chlamydia psittaci, are gram negative, spherical, (0.4-0.6 micron diameter), intracellular parasites; they use ATP produced by the host cell, hence, referred to as "energy parasites".

10.2. Hosts

An infection of turkeys, ducks, psittacines, pigeons, man, rarely chickens

10.3. Transmission

Transmission of this organism from one host to another is primarily through the air. The bacteria are shed from an infected bird in the nasal and or ocular secretions, fecal material, and feather dust. It is also transmitted by contact and wild bird carriers, especially pigeons and robins. Egg transmission does not occur but vertical transmission through the egg has been shown in domestic ducks..

Elementary bodies are highly resistant and can survive in dried faeces for many months.

10.4. Symptoms

Incubation periods in caged birds vary from days to weeks and longer. Most commonly this period is approximately 3 to 10 days. Latent infections are common and active disease may occur several years after exposure. In young birds clinical sings can include rough plumage, low body temperature, tremor, lethargy, conjunctivitis, dyspnea, emaciation, sinusitis, yellow to greenish droppings or grayish watery droppings may also be displayed. Adult birds may develop one or many of the symptoms such as tremors, lethargy, ruffled feathers, progressive weight loss, greenish diarrhea, occasional conjunctivitis, and high levels of urates in droppings. In some cases, birds may be asymptomatic carriers.

10.5. Morbidity and mortality

50-80 and 5-40%, respectively

10.6. Diagnosis

Fecal analysis, blood analysis, immunoflourescent testing, as well as PCR and nested PCR testing with whole blood sample in conjunction with a cloacal and/or throat swab wherever possible. Positive the bird has to be immediately quarantined and treated.

10.7. Treatment

Tetracycline and its derivatives (Vibramycin, Doxycycline, Oxytetracycline) by intravenous or intramuscular injections. Antibiotics can also be given orally or mixed with palatable food. Treatment periods generally last about 45 days varying slightly depending on the drug. Calcium should be withheld because tetracycline binds to calcium. Citric acid in the bird's drinking water can increase the levels of antibiotics in the blood.

10.8. Prevention and control

Preventing the organism from entering the facility is the best method of prevention. All new birds must be quarantined and tested; bird should not be purchased in marts and bird fares. Proper hygiene - removal of fecal material, and quality air circulation.

A major concern with *C. psittaci* is the zoonotic potential of the organism; it is related to *Chlamydia trachomatis*, the most common human STD, and Chlamydia pneumonia, a cause of human pneumonia. It is also one of the major causes of infectious abortion in sheep and cattle.

11. Rhinotracheitis

11.1. Etilogy

Ornithobacter rhinotracheale

11.2. Hosts

Turkeys and Chickens

11.3. Transmission

Probably by direct contact with affected birds or equipment contaminated with infective aerosols from the respiratory tract or via contaminated water systems.

11.4. Symptoms

Broilers - at 3-4 weeks of age show nasal discharge, facial swelling and sneezing. Increased condemnation rates due to "cheesy" air sacculitis. Broiler breeders - Usually affects flocks around peak production causing increased mortality and mild respiratory signs. Production and egg size are depressed although fertility and hatchability are usually unaffected. Turkeys - at 14 weeks or older although younger poults may be affected between 2 and 8 weeks of age. Coughing, sneezing, nasal discharge and sinusitis; leg problems and joint infections may also be seen. Turkey breeders - may cause reduced egg production and depress the number of settable eggs.

11.5. Morbidity and mortality

Mortality can range from 2-10%.

11.6. Lesions

Pneumonia, lung consolidation and creamy air sac exudates

11.7. Diagnosis

Lesions and confirmation is by isolation of the organism and ELISA.

11.8. Treatment, prevention and control

Antibiotic therapy can control clinical outbreaks. Specific autogenous vaccines may be effective in prevention of disease and commercial vaccines are becoming available in some countries. Biosecurity is important in preventing introduction of infection onto sites. Effective sanitization and cleaning of water systems to remove biofilms which may harbor the organism are considered important.

12. Turkey coryza

12.1. Etiology

Bordetella avium (previously designated *Alkaligenes faecalis*).

12.2. Hosts

Natural host is turkey

12.3. Transmission

By direct contact through litter and water; premise remains infective for a long period.

12.4. Symptoms

Incubation period is 7-9 days. The infected birds show snicking, excessive respiratory mucus, foamy ocular discharge, conjunctivitis, inanition and weight loss, dyspnoea and death. Secondly infection by *E.Coli* and ND is expectable. Loss of voice and nasal discharge are also seen occasionally. Usually causes mild respiratory disease in only a proportion of the flocks, and can be synergistic with other respiratory pathogens such as ND virus, *Ornithobacterium*, *Pasteurella* spp, *Mycoplasma* spp. and pneumoviruses.

12.5. Morbidity and mortality

Mortality ranges from 5-75% with a morbidity of 100%.

12.6. Lesions

The lesions noticed are collapse of cartilage of trachea, mucous in nasal turbinates and trachea, and submaxillary edema in uncomplicated cases. If secondary infection has occurred by *E. coli* infection, pneumonia, airsacculitis, perihepatitis and pericarditis.

12.7. Diagnosis

Clinical signs, isolation of agent.

12.8. Treatment

Antibiotics are not very effective, stress factors to be minimized, chlorination of drinking water may be of benefit.

12.9. Prevention and control

All-in all-out production. Good drinking water hygiene. No satisfactory vaccine available.

13. Gangrenous Dermatitis, Necrotic Dermatitis

13.1. Etiology

Clostridium septicum, occasionally *Staphylococcus aureus*, rarely *Clostridium noyvi / oedematiens*. It occurs due to invasion of 'normal' wounds by 'normal' bacteria in immunosuppressed birds. Immunosuppression is therefore a predisposing factor, especially following congenital Chick Anemia Virus infection or early Infectious Bursal Disease Virus infection (Gumboro Disease). The spores of Clostridial bacteria are highly resistant in the environment.

13.2. Hosts

Chickens

13.3. Symptoms

Loss of appetite, gangrenous skin, severe cellulitis especially of thighs, wings, wattles, sudden mortality.

13.4. Morbidity and mortality

Morbidity may be up to 50% and mortality is high.

13.5. Lesions

Patches of gangrenous skin with underlying emphysematous and/or sanguinous cellulitis, usually over wings and breast, sometimes thighs and other parts. Swelling and infarction of the liver, spleen. Foci in liver.

13.6. Diagnosis

Clinical signs and/or lesions. Isolation of organism.

13.7. Treatment

Sulphaquinoxaline, penicillin or amoxycillin.

13.8. Prevention and control

Good hygiene and management. Control of skin trauma and immunosuppression (congenital CAV infection, early Infectious bursal disease virus infection).

14. Arizona infection, Arizonosis

14.1. Etiology

Arizona hinshawii, renamed *Salmonella Arizonae.*

14.2. Hosts

It affects turkeys (mainly in North America).

14.3. Transmission

Vertical, transovarian, and also horizontal, through fecal contamination of environment, feed etc, from long-term intestinal carriers, rodents, reptiles; older birds are asymptomatic carriers.

14.4. Morbidity and mortality

Mortality is 10–50% in young birds

14.5. Symptoms

Inappetance, diarrhea, vent-pasting, nervous signs, paralysis, blindness, cloudiness in eye, huddling near heat.

14.6. Lesions

Enlarged mottled liver, unabsorbed yolk sac, congestion of duodenum, cheesy plugs in intestine or cecum, foci in lungs, salpingitis, ophthalmitis, pericarditis, and perihepatitis.

14.7. Diagnosis

Isolation and identification, methods as per *Salmonella* spp. (Since lesions noticed are similar to those of *E.*coli infection)

14.8. Treatment

Injection of streptomycin, spectinomycin, or gentamycin at the hatchery is used in some countries. Formerly in-feed medication with nitrofurans was also used. Generally, not satisfactory.

14.9. Prevention and control

Carriers must be removed from breeder population, fumigation of hatching eggs, good nest and hatchery hygiene, injection of eggs or poults with antibiotics; antibiotic sensitivity of the organism must be monitored regularly.

15. New duck disease (Infectious serositis)

Caused by *Moraxella anatipestifer*. This is similar to CRD in chicken. In addition, birds lose balance and fall on sides and backs; death due to dehydration. The disease is curable with Sulfonamides and other antibiotics.

16. Campylobacter infection

(The Merck Veterinary Manual, 2003)

16.1. Etiology

Campylobacter jejuni is the predominant species associated with food-borne infection derived from poultry. *Campylobacter coli* and *C lari* are occasionally recovered from the intestinal tract of poultry, and both have been implicated in food-borne infection. They are generally nonpathogenic in mature poultry. Toxigenic and invasive strains of *C jejuni* can cause enteritis and death in newly hatched chicks and poults.

16.2. Reservoirs

Commercial poultry and free-living birds are natural reservoirs of the thermophilic campylobacters (*C jejuni* , *C coli* , and *C lari*). It is estimated that over half of all commercial broiler and turkey flocks harbor *C jejuni*. The organism has been isolated from numerous birds, including Columbae and domestic and free-living Galliformes and Anseriformes.

16.3. Transmission

Environmental contamination is regarded as the source of infection for poults, chicks, and ducklings. Litter can remain infective for long periods, subject to at least a 10% moisture level and neutral pH. Infected chicks and poults become carriers. Contaminated water, rats, wild birds, and house flies and equipment and footwear contaminated with feces from an infected source. Once introduced into the environment, rapid transmission within the flock occurs. *Campylobacter jejuni* is not transmitted vertically, either on the surface of eggs or by transovarian transmission.

16.4. Symptoms

Infecting day-old chicks with a pathogenic and invasive strain of *C jejuni* results in a rapid onset of diarrhea that may persist for up to 4 days. Highly pathogenic isolates derived from people with enterocolitis may induce up to 32% mortality in chicks. Chicks >7 days old are refractory to infection.

16.5. Lesions

Gross lesions of *Campylobacter* infection in neonates include distention of the jejunum, disseminated hemorrhagic enteritis, and in some cases, focal hepatic necrosis.

16.6. Diagnosis

Fecal specimens used for isolation and identification

16.7. Control and prevention

Because *C jejuni* does not occur as a specific pathogen under commercial conditions, treatment of poultry flocks is not a consideration. If *C jejuni* is considered a problem in companion bird aviaries or in exotic species, antibiotics such as erythromycin can be administered in drinking water. Galliformes should receive a dose of 10-30 mg/kg for 4 consecutive days, and Psittaci and exotics should be medicated at 30-40 mg/kg.

Chlorination of drinking water to 2 ppm and operating farms on a strict all-in/all-out basis can reduce the prevalence of infection. Withholding feed from broilers and turkeys for at least 12 hr before slaughter and thorough decontamination of transport coops and modules reduce fecal contamination and lower the level of *C jejuni* introduced into processing plants.

16.8. Zoonotic risk

Campylobacter jejuni contamination can be reduced by improved washing of carcasses, the use of counter-flow scalding, elimination of immersion chillers, and reduction in manual handling by installation of advanced automated equipment. Chemical disinfectants, such as glutaraldehyde (0.125%) and succinic acid (3%), and organic compounds, such as lactic and acetic acids, may be used to destroy *C jejuni* when used as a spray or an immersion treatment.

Gamma irradiation at levels ranging from 1 to 3 kGy effectively eliminates *C jejuni* from poultry carcasses and products.

The only current measure to reduce the risk of *C jejuni* infection to consumers is thorough cooking of poultry to achieve a core temperature of 74°C for 1 min. This ensures destruction of *C jejuni*. Concurrent hygienic storage, handling, and preparation are necessary to prevent contamination of prepared foods, work surfaces, and utensils by raw poultry and other meats.

17. Amyloidosis

(ThePoultrySite.com)

17.1. Etiology

Chronic bacterial infection e.g. enterococcal or staphylococcal arthritis and perhaps *Mycoplasma synoviae.*

17.2. Hosts

A disease of ducks and chickens caused by prolonged stimulation of the immune system common in adult parent ducks; it also occurs in laying- and meat-type chickens.

17.3. Transmission

The disease itself is non-transmissible, though the predisposing primary infection may be.

17.4. Symptoms

Wasting, lameness and swelling of foot and leg joints

17.5. Morbidity and mortality

Morbidity is usually low; the course of the disease chronic and mortality is high.

17.6. Lesions

Enlarged liver, parboiled in appearance. Spleen enlarged, hemorrhagic or ruptured. Orange-colored deposits in joints. Primary inflammatory lesions (arthritis, endocarditis etc) are seen.

17.7. Diagnosis

Lesions, histopathology.

17.8. Treatment and control

None. Prevention is possible by control of predisposing infection.

18. Anatipestifer Disease

(New Duck Syndrome, Duck Septicemia)

18.1. Etiology

An acute or chronic septicaemic disease caused by *Riemerella anatipestifer*, Pasteurella, or *Moraxella*

18.2. Hosts

It affects ducks of any age, sometimes turkeys, and may also be isolated from chickens, game birds and wild waterfowl.

18.3. Transmission

Transmission is mainly direct, bird-to-bird, via toenail scratches, especially of the duckling foot, or through respiratory epithelium during respiratory disease. It can also be by fecal contamination of feed, water or the environment where survival of the infectious agent may be prolonged. Any kind of stress is a predisposing factor

18.4. Symptoms

Weakness, neck tucked in, head/neck tremor, ataxia, disinclined to walk, incoordination, dyspnoea, ocular and/or nasal discharge and hyperexcitability

18.5. Morbidity and mortality

Mortality is 2–75% in young ducks.

18.6. Lesions

Perihepatitis without much smell or liver damage, pericarditis, airsacculitis, enlarged liver and spleen, occasionally fibrinous meningitis, salpingitis, purulent synovitis, chronic arthritis, sometimes with erosions of the joint cartilage.

18.7. Diagnosis

Lesions, isolation and identification of organism

18.8. Treatment

Sulphonamides and potentiated sulphonamides are the products most commonly recommended for drinking water application. Subcutaneous injections of penicillin + dihydrostreptomycin, or streptomycin + dihydrostreptomycin are also highly effective.

18.9. Prevention and control

Good husbandry and hygiene, rigid depopulation and disinfection, adequate protection, 'hardening off', correct house relative humidity, sulphonamides in feed. Inactivated and attenuated vaccines available in some countries. Autogenous bacterins sometimes used.

19. Avian mycoplasmosis

(PPLO infection, chronic respiratory disease, Infectious sinusitis)

About 30 serotypes of *Mycoplasma* spp have been isolated from avian hosts; *M gallisepticum* (MG), *M iowae* (MI), *M meleagridis* (MM), and *M synoviae* (MS) are the most important. Mycoplasma is weakly Gram-negative and coccoid/hexagonal fastidious bacteria 0.3-0.8 μm in diameter; they lack a cell wall and require a rich growth medium containing serum. They do not survive for more than a few hours or days outside the host and are vulnerable to common disinfectants. Each has distinctive epidemiological and pathological characteristics.

19.1. MG infection

19.1.1. Hosts

Common in chicken and turkeys. Infection may also occur in pheasants, chukar partridges, and peafowl. Infection can occur in pigeons, quail, ducks, geese, and psittacine birds. Passerine type birds are quite resistant. The disease is ubiquitous. Its effects are most severe in large commercial operations during winter.

19.1.2. Transmission

By contact, air-borne, dust or droplets, contaminated equipment. It is frequently transmitted through eggs. Incubation period varies from 6 to 21 days.

19.1.3. Symptoms

19.1.3.1. Adults

The characteristic symptoms are tracheal rales, nasal discharge and coughing. Loss of body weight and reduced food intake and egg production. Usually the secondary invaders complicate the infection and result in Chronic Respiratory Disease (CRD). Severe air-sac infection frequently designated as complicated CRD or air-sac disease is most common.

19.1.3.2. Males, Younger birds (broilers)

More pronounced than in adults.

19.1.3.3. Turkeys

Nasal discharge, foaming of eye secretions, swelling of paranasal sinuses due to sinusitis, emaciation, tracheal rales, coughing and dyspnea may be seen.

19.1.3.4. Laying turkeys

Drop in egg production.

19.1.4. Morbidity and mortality

Morbidity in chicken is usually high during winter months and usually the secondary evaders produce CRD or complicated CRD. Mortality in chicken is highly variable; it may go up to 30% in complicated cases. In turkeys also, high morbidity and losses in the farm of carcass condemnation due to air-sacculitis.

19.1.5. Gross lesions

Catarrhal exudate in basal and paranasal passages, trachea, bronchi and air-sacs. Sinusitis is most common in turkeys. Air sacs frequently contain caseous exudate. Low degree of pneumonia and severe air-sac infection, fibrinous are fibrino-purulent perihepatitis and pericarditis along with air-sacculitis is seen. The organism can also produce salphingitis. The recovered birds act as carriers and spread the disease by contact or through egg.

19.1.6. Diagnosis

Isolation and identification; serological test; history and signs.

19.1.7. Treatment

Several antibiotics have been tried. Streptomycin, Oxytetracycline and Chlortetracycline (200 g/tonne of feed), Erythromycin, Magnamycin and Tylosin (7 to 11 mg/kg body weight subcutaneously or 0.4 to 0.6g/l in water for 2 to 4 days. MG can develop resistance to antibiotics.

19.1.8. Prevention and control

Proper sanitation and serological testing and elimination of carriers are very important. Immersion of warmed (37.8°C) hatching eggs in cold (1.67 to 4.44°C) antibiotic solution (especially Tylosin or Erythromycin at 400 to 1000 ppm) for 15 to 20 minutes before incubation has been found to result in absorption of sufficient antibiotic to eliminate egg transmission. Another procedure has been to heat the hatching eggs (at room temperature) to an internal temperature of 46.1°C in a forced-air incubator during a 12 to 14h period so that the MG and MS organisms are inactivated. Concomitant reduction in hatchability to a tune of 8 to 12% is expected due to heat treatment.

The use of birds free of *M gallisepticum* is desirable, but infection in multiple-age commercial egg farms where depopulation is not feasible is a problem. An inactivated, oil-emulsion bacterin is available in most countries; it prevents egg production losses but not infection. A live vaccine has been licensed in the USA for use in infected, multiple-age layer flocks but may be used only with permission of the state veterinarian. The vaccine consists of a mild strain of *M gallisepticum* (F-strain) and is usually given at ~10-14 wk of age. F-strain is of low pathogenecity for chickens but is fully virulent for turkeys. Vaccinated birds remain carriers, and immunity lasts through the laying season. Recently, two nonpathogenic live vaccine strains (6/85 and ts-11) have been introduced; these strains offer the advantage of improved safety (The Merck Veterinary Manual, 2003).

19.2. MI infection

Mycoplasma iowae was originally thought to be of low pathogenecity in producing air sac lesions in chickens and turkeys, but it is a potentially important cause of reduced hatchability in turkeys.

19.2.1. Hosts

Mainly turkey and a relatively uncommon infection of chickens.

19.2.2. Transmission

Mycoplasma iowae is egg transmitted, but little is known of other aspects of its epidemiology.

19.2.3. Symptoms, lesions and diagnosis

Affected turkey breeder flocks show no clinical signs other than reduced hatchability (usually 2-5%). In many flocks, the hatchability returns to normal after 1-2 mo. Most embryos die during the later stages of incubation. Dead turkey embryos are edematous, congested, and stunted; they may have clubbed down.

19.2.4. Diagnosis

Isolation and identification of the causative agent.

19.2.5. Treatment and control

The best method of control is to maintain flocks free of *M iowae* ; however, because serology is unreliable, this may be difficult. Dipping hatching eggs in solutions of enrofloxacin has significantly reduced losses in hatchability.

19.3. MM infection

19.3.1. Hosts

It is a specific pathogen of turkeys

19.3.2. Transmission

Primarily through eggs. Direct transmission by air-borne route and indirect transmission during sexing, artificial insemination and vaccination can also occur.

19.3.3. Symptoms

Air-sacculitis deficiency syndrome (TS-65) characterized by skeletal abnormalities and abnormal feathering has been observed. Air-sacculitis is also common but without respiratory signs.

19.3.4. Morbidity and mortality

Mortality is mainly due to cannibalism of affected birds. Usually, egg production and fertility are unaffected. Reduction in growth and increased carcass condemnations reduce the profits.

19.3.5. Gross lesions

19.3.5.1. Day-old poults

Thickening of walls of the air-sacs with or without yellow exudate and flecks of caseous material in the lumen.

19.3.5.2. Growing turkeys

Skeletal lesions, air-sacculitis and sinusitis.

19.3.6. Diagnosis

Isolation and identification; serological test.

19.3.7. Treatment

Tylosin (1000 to 3000 ppm) or Gentamicin (400 to 1000 ppm) along with disinfectants (250 ppm of a quaternary ammonium compound) is added into the egg-dipping solution. The solution is maintained at 2 to 8°C in which hatching eggs, warmed to 35 to 37°C, are immersed for 5 to 20 min.

19.3.8. Prevention and control

Since this is strictly a venereal infection, strict sanitary measures have to be followed during collection and insemination of semen to avoid cross-contamination. Serological testing and elimination of carriers also is an important step in control of MM infection.

19.4. MS infection

19.4.1. Hosts

Chicken, turkeys and guinea fowl

19.4.2. Transmission

Primarily through eggs. Direct transmission also helps in rapid spread of the disease

19.4.3. Symptoms

Pale or bluish-red comb, lameness and regarded growth, ruffled feathers, swellings around the joints (hock-joint and foot-pads usually), and breast blisters are commonly seen. Greenish discoloration of droppings with large amounts of uric acid or urates is also a common symptom. If infected through respiratory route, tracheal rales may be seen.

19.4.4. Morbidity and mortality

1 to 20% and up to 20%, respectively

19.4.5. Gross lesions

Viscous, creamy to grey or caseous exudate in sinovial membranes of joints, splenomegaly, swelling and mottling of kidneys and liver also are frequently seen. Air-sacculitis with mucus in trachea are seen occasionally, especially during winter.

19.4.6. Diagnosis

Isolation and identification; serological tests in conjunction with symptoms and lesions.

19.4.7. Treatment

Chlortetracycline (50 to 100 g/tonne for chicken and 200 g/tonne for turkeys) continuously (chicken) or for 5 to 7-day consecutively (turkeys) in feed provides protection against MS infection.

19.4.8. Prevention and control

Similar to MG and MM; control of egg-transmission is most important. Identification and culling of carriers has to be stringently followed. Treatment of hatching eggs as described for MG and MM infections is suitable for MS infection also.

Chapter **63**

Fungal Diseases

1. Aspergillosis

1.1. Etiology

Aspergillus fumigatus produces toxin which is hemotoxic, neuro toxic and histotoxic. It forms conidiophores.

1.2. Hosts

Chicken, turkeys and other birds.

1.3. Transmission

The fungus is likely to penetrate the egg shell during incubation, but the chances are very low. Infected incubator and hatcher can be a source of contamination. In most cases, the transmission of spores is by aerosol route. The bird comes in contact with the organisms through contaminated feed, litter or premises. The disease is not contagious and does not spread from one bird to another. Most healthy birds can withstand repeated exposure to these organisms. Inhalation of large amounts of the infectious form of the mold or reduced resistance of the bird apparently results in infection. In adult turkeys, the disease more often affects the male.

1.4. Symptoms

1.4.1. Acute form

In young birds, main symptoms dyspnea, gasping and polypnea usually without any sound until and unless the disease is complicated by infectious bronchitis are laryngotracheitis. Somnolence, inappetance, emaciation, increase thirst and pyrexia are also seen. Occasionally the organism invades the brain, causing paralysis or other forms of nervous symptoms.

1.4.2. Chronic form

The more chronic form in older birds usually results in loss of appetite, gasping or coughing and a rapid loss of body weight. Mortality is usually low and only a few birds are affected at one time.

1.5. Morbidity and mortality

Morbidity is very high and mortality may go up to 50%.

1.6. Gross lesions

Depends on site of infection – which can either be localized or generalized. Localized lesions in syrinx, air-sacs etc. may be seen. Lungs commonly show miliary to larger nodules with the without hepatization. Air passages and bronchi may show visible mycelia. Nodules of various sizes in lungs, thoracic and abdominal cavities may be seen.

1.7. Diagnosis

Typical masses of fungal growth in the air passages and lungs. Microscopic demonstration of the mycelia.

1.8. Treatment

Generally not useful; it is better to kill the infected birds.

1.9. Prevention and control

Avoid moldy litter or feed. Water contamination must be avoided to control the spread of disease. Proper sanitation and disinfection procedures must be followed.

2. Mycotoxicosis

2.1. Etiology

Details of mycotoxins in poultry feed are given in Chapter "Toxins in poultry feed". Mycotoxicosis is caused by ingestion of toxic substances produced by molds growing on feed, feed ingredients and possibly litter. Several types of fungi produce toxins that may cause problems in poultry, but the most commonly encountered mycotoxin is produced by *Aspergillus flavus* popularly known as Brazilian groundnut poisoning. It elaborates toxins B_1, G_1, B_2, G_2; the latter two being more potent than the former ones.

2.2. Hosts

Young poults, ducklings, pheasants and chicken are susceptible. However, among chickens, New Hampshire breed is more sensitive whereas White and Barred Plymouth Rocks, White Leghorns and Rhode Island Reds are less susceptible.

2.3. Tolerance limits

Maximum level of tolerance (chicken) up to 10 to 12 weeks of age appears to be 0.25 ppm whereas after 10 to 12 weeks of age, it can go up to 0.75 ppm in diets.

2.4. Symptoms

Mold toxins cause a wide variety of signs, many difficult to recognize. The aflatoxins under certain conditions cause death, reduced growth, reduced egg production, reduced hatchability, signs associated with "physiological stress" and impaired ability to develop immunity to infectious agents. Diagnosis is difficult because characteristic lesions usually are not present, and detection of the toxin is not conclusive.

2.4.1. Ducklings

Ducklings are most susceptible than other species. Lameness, purple discoloration of legs and feet (due to subcutaneous hemorrhage), ataxia, convulsions, opisthotonus and death.

2.4.2. Turkeys

Inappetance, lethargy, wing-droop, ruffled feathers and occasionally, nervous symptoms. Mortality may go up to 90%.

2.4.3. Young chicken

Depression of growth, inappetance and mortality

2.4.4. Adult chicken

Drop in egg production, egg weight and hatchability of eggs. Evidences indicate that aflatoxicosis impairs immune system subjecting the birds to secondary infections.

2.5. Lesions

2.5.1. Turkeys

Congestion and enlargement of liver, distension of gall bladder, congestion and swelling the kidneys, congestion of myocardium, distension of pericardial sac with amber-colored fluid and catarrhal inflammation of duodenum.

2.5.2. Ducklings

Pale reticulated network followed by atrophy and cirrhosis of liver, hydropericardium and ascites, petichiae in kidneys in pancreas.

2.5.3. Older ducklings

Cirrhosis and nodular hyperplasia of liver, petichiae in pancreas and kidney, swelling of kidneys and, a pathognomonic, severe subcutaneous hemorrhage in the shank and web of foot causing purplish discoloration are seen.

2.5.4. Chicken

Enlarged liver with reticulated network with peticheal hemorrhages; this will be followed by reduction in size and increased firmness and nodular lesions on the surface. Enlargement of kidneys with a fine network of urate deposits also can be seen. In broilers – liver, comb, shank and bone-marrow were pale, liver, spleen and pancreas were enlarged along with fatty infiltration of liver and regression of bursa of Fabricius. In certain cases, gallbladder is elongated and distended with bile, excessive fat deposited around the gizzard, in severe cases, liver distended, friable, and soft with diffused hemorrhage areas. Reticuloendothelial system also may be affected. There will be reduction in size of bursa of Fabricius and thymus.

2.6. Diagnosis

History and characteristic symptoms and lesions are helpful in making the tentative diagnosis. The feed samples have to be analyzed for aflatoxin residues for confirmation.

2.7. Treatment

There seems to be no satisfactory treatment. However, all species of poultry recover rapidly at once the toxin is removed from the diet. Use of mold inhibitors like propionate and Gentian violet can reduce toxin levels between mixing and consumption. Contaminated feed ingredients can be treated with chloramine or ammonia or extracted with isopropanol; but these methods are not commercially practicable. Hydrated sodium calcium aluminosilicate can bind the toxin and reverse the toxic effects. Several detoxification methods have been tried with variable success; details given in Chapter "Toxins in poultry feed".

3. Moniliasis (Thrush, Crop necrosis)

3.1. Etiology

Yeast-like fungus (*Candida albicans*)

3.2. Hosts

Poultry of all ages are susceptible to the effects of this organism. Chickens, turkeys, pigeons, pheasants, quail and grouse are species most commonly affected as well as other domestic animals and humans.

3.3. Transmission

Moniliasis is transmitted by ingestion of the causative organism in infected feed, water or environment. Unsanitary and unclean water troughs are an excellent reservoir of the Candida organism. The disease does not however, spread directly from bird to bird. The organism grows especially well on corn, so infection can be introduced by feeding moldy feed.

3.4. Symptoms

No specific symptoms. Young birds become listless, pale, show ruffled feathers and appear unthrifty. Affected caged layer hens become obese and anemic. Some birds exhibit a vent inflammation that resembles a diarrhea-induced condition having whitish incrustations of the feathers and skin around the area. Feed consumption may increase by ten to twenty percent.

3.5. Lesions

Gross lesions are mostly confined to the crop, proventriculus and gizzard. The lesions are found in crop consisting of thickening of mucosa with whitish, circular raised ulcers, often described as having a "turkish towel" appearance. Pseudomembrane formation and necrotic spots are also frequently seen in crop mucosa. These lesions may extend to proventriculus also. Erosion of the lining of the proventriculus and gizzard is commonly observed, as well as an inflammation of the intestines.

3.6. Diagnosis

History and typical lesions; the lesions are of high diagnostic value. Confirmation by isolation and identification of the organism.

3.7. Treatment

Treatment of the flock with an antimycotic drug will control the infection. Many broad spectrum antibiotics will enhance this disease; therefore they should not be used until after control of this condition is completed. Addition of Nystatin (100 g/Ton) or copper sulfate (1 kg/Ton) to the feed for seven to ten days should control moniliasis.

3.8. Prevention and control

Once introduced into the flock, moniliasis is perpetuated by suboptimal management conditions. Preventative measures include the continual use of mold inhibitors in the feed, proper feed handling and storage, daily cleaning and sanitizing of the watering system and periodic stirring and/or replacement of wet litter areas to prevent caking. An inexpensive, yet effective, water treatment is the continuous addition of household chlorine bleach to the drinking water at the rate of five parts per million (ppm).

4. Favus

4.1. Etiology

Trichophyton gallinae

4.2. Hosts

Chickens and turkeys; not common in commercial poultry production.

4.3. Transmission

Horizontally by contact

4.4. Symptoms and lesions

White, powdery spots and wrinkled crusts and scab on comb and wattles, feather loss, 'Honeycomb' skin, thick crusty skin, loss of condition.

4.5. Lesions

There may be some lesions in the mucosa of upper respiratory tract consisting of necrotic foci.

4.6. Diagnosis

Primarily by symptoms; confirmation by isolation.

4.7. Treatment

Various ointments are suggested – iodine and glycerine, green soap and 5% phenol, bi- chloride mercury and formalin in petroleum jelly etc.

4.8. Prevention and control

Good hygiene of facilities, culling affected birds.

Chapter **64**

Viral Diseases

1. Neoplastic diseases

Virus-induced neoplastic diseases are mesodermal in origin and are transferable. They are classified into:

1. Marek's disease
2. Leukosis, sarcoma and related neoplasms
3. Reticuloendotheliosis virus group
4. Lymphoproliferative disease in turkeys and
5. Tumors of unknown etiology.

Of these, the former two are of relatively higher importance than the latter three.

1.1. Marek's disease (Visceral leucosis)

1.1.1. Etiology

This the most common of the lymphoproliferative diseases of chicken reported by Marek (1907). It is caused by a Herpes (Group B) DNA virus. Of the three serotypes, serotype 1 is pathogenic.

1.1.2. Hosts

Chicken is the most important host and other species are not that important.

1.1.3. Transmission

Direct or indirect contact between birds, air-borne, affected feather-follicles, poultry house dust, and carriers. Beetles and mosquitoes are also suspected carriers of infection. Incubation period, generally, is 3 to 4 weeks.

1.1.4. Symptoms

Asymmetric progressive paralysis initially and finally complete paralysis of one or more of the extremities. The symptoms vary with the nerve involved - wing involvement characterized by drooping of wings; if nerves of the neck are involved, paralysis and dilation of crop and/or gasping; if sciatic nerve is involved, paralysis of the leg resulting in "sportsman's posture" characterized by one leg stretched forward and other backwards (unilateral paresis). In acute cases, severe depression, ataxia in few birds, unilateral or bilateral paralysis of extremities.

Many birds exhibit dehydration and emaciation, and fall comatose. When eyes (Iris) is involved, concentric spotty depigmentation and opacity of the iris is seen which causes blindness. Sometimes, feather-follicles are involved causing swelling of feather-follicles and sloughing of feathers.

1.1.5. Morbidity and mortality

Most of the mortality is due to inability of the birds to reach either food and/or water. Hence, morbidity and mortality are nearly the same. Morbidity depends on strains of virus, dosage, route of exposure and genetic resistance, age, immune-status and sex of the birds.

1.1.6. Gross lesions

Nerve-lesions are most consistent – they include, usually, one or more peripheral nerves but not the brain. The most commonly-affected nerves and plexuses are celiac, cranial, mesenteric, brachial and sciatic; nerve of Remak and greater splanchnic nerve. The affected nerve and/or plexus shows loss of cross-striations, grey or yellow discoloration and sometimes, edematous appearance. The affected portion will be 2 to 3 times the normal size and usually unilateral which is highly helpful for comparison.

Lymphoid tumors may occur in many organs – gonads (ovary frequently) are affected commonly and other organs that can be involved are lungs, heart, mesentry, kidney, liver, spleen, adrenal gland, pancreas, proventriculus, intestine, iris, skeletal muscle and skin. In more acute forms, visceral tumors are common. These tumors or not distinguishable with those occurring in lymphoid leukosis excepting in case of bursa of Fabricius which we usually atrophic but sometimes shows tumors that appear as a diffuse thickening because of inter-follicular distribution of tumor cells. Liver involvement – coarse granular appearance; non-producing and mature ovary when involved shows tumors; mature ovaries when severely involved show cauliflower-like appearance. Proventriculus becomes thickened and firm, lungs show solidification, heart becomes pale and nodular.

1.1.7. Diagnosis

By symptoms, gross lesions, virus isolation, antigen (MD tumor associated surface antigen, MATSA) and/or antibody demonstration.

1.1.8. Prevention and control

Although three types of vaccines were developed, the most extensively used is turkey Herpes virus (HVT) since it is more economical to produce and cell-free virus extracted from infected cells may be lyophilized for easy storage and handling. The other vaccines are – 1. Attenuated serotype 1 and 2. Naturally avirulant isolates of serotype 2. Although HVT is shed to some extent by the vaccinated chicks during the second week after vaccination, all chicks are vaccinated because horizontal transmission is usually relatively ineffective. Cell-associated with HVT vaccine affords 80 to 100% protection and is also thought to improve egg production and size. The vaccine is given by parenteral route (subcutaneous, intraperitoneal or intramuscular). HVT is antigenically related to Marek's disease virus and therefore, produces immunity in chicken. Suitable selection program can be practiced to develop chicken resistant to Marek's disease.It has been found that resistance to Marek's disease is independent of genetic factors controlling production traits.

Isolation and sanitation greatly help in preventing the reintroduction and spread of Marek's disease.

1.2. Leukosis/Sarcoma group

1.2.1. Etiology

Avian type C Oncovirus. Those which occur in chicken are divided into subgroups A and B - common exogenous viruses in the field; subgroups C and D – less frequent viruses in the field; subgroup E – ubiquitous endogenous leukosis viruses of low pathogenecity.

1.2.2. Hosts

Chicken is the natural host and these viruses have not been isolated from other avian species; however, Rous sarcoma virus can infect several avian species.

1.2.3. Transmission

The lymphoid leukosis viruses (LLVs) are transmitted through hatching eggs and also horizontally, bird to bird, by direct or, to a lesser extent, by in direct contact. Both vertical and·horizontal transmissions seem to be important in maintaining the infection from one generation to another. Birds with viremia and without antibodies (V+A-) transmit leukosis to a high proportion of their progeny whereas very few hens (less than 10%) without viremia but with antibodies (V-A+), which are genetically susceptible in a flock, do so. Birds from infection-free flock and genetically resistant birds in a susceptible flock will not have viremia as well as antibodies (V-A-). On the other hand, most of the birds showing no viremia but have antibodies (V-A+) are genetically susceptible in an infected flock.

Infection of cock does not influence the rate of vertical transmission since virus budding was not noticed in germinal cells of the male; but it acts as a carrier and a source of venereal infection to other birds. The infected hens shed the virus into the egg albumen via the albumen secreting glands (magnum). Not all eggs containing the virus give rise to infected embryos/chicks; only about 25 to 50% of embryos were infected from eggs with virus in the albumen. This may be due to neutralization of the viruses by antibodies in the yolk and loss because of thermal inactivation (incubation).

Under natural conditions, most chicks become infected by exogenous leukosis virus (from penmates or surroundings), after a transient viremia, develop a high titer of antibodies, which can give a lifelong immunity. The same is passed on to yolk to give the chicks maternal immunity for the first 4 weeks of life.

1.2.4. Symptoms and lesions

Depending on the virus involved, various conditions are encountered:

1.2.4.1. Lymphoid leukosis

Incubation period is 14 to 30 weeks; therefore, clinical disease seldom occurs at < 14 weeks of age; incidence is usually highest at about sexual maturity. No specific clinical sign is noticed although pale and cyanotic combs, emaciation and inappetance may be noticed. Abdomen is often enlarged and on palpation reveals enlargement of liver, bursa of Fabricius, and/or kidneys. The most characteristic gross lesion includes visible tumors especially in liver and also in spleen, bursa of Fabricius, kidney, liver, spleen, heart, bone-marrow and mesentery. The tumors are soft, smooth and glistening without necrotic areas. The growth may be nodular, miliary or diffused.

1.2.4.2. Erythroblastosis

Incubation period varies from 21 to 100 d. General weakness, paleness or cyanotic combs, emaciation and diarrhoea. There may be profuse hemorrhage from one or more feather follicles followed by anemia and paleness of comb. Gross lesions include general anemia and petichiae on various organs like muscles, subcutis and viscera. Thrombosis, infarction and rupture of liver or spleen may be seen. The most characteristic lesions are cheery-red to dark-mahogany colored, soft and friable liver and spleen (sometimes kidneys also) due to diffused enlargement. Due to severe anemia, spleen may undergo atrophy. Erythroblasts can be demonstrated in bone-marrow smears.

1.2.4.3. Myeloblastosis

Incubation period can be as low as 10d. Symptoms are similar to erythroblastosis; however, the gross lesions indicate general anemia resulting in pale red-colored liver and whitish bone-marrow (unlike erythroblastosis). The typical myeloblasts can be demonstrated in liver and bone-marrow smears. The myeloblasts are smaller than erythroblasts or lymphoblasts, and their cytoplasm is more acidophilic and is polygonal or angular.

1.2.4.4. Myelocytomatosis

Incubation period is 3 to 11 weeks. General symptoms are similar to myeloblastosis but unlike the latter, skeletal growths resulting in protuberances of head, thorax and shanks are also seen.

1.2.4.5. Hemangiomas

Incubation period is 3 to 16 weeks. Usually occur singly in skin and quite frequently with primary multiplicity. When the wall ruptures, it leads to profuse bleeding making the feather near the tumor blood-stained. Birds may die due to bleeding. Gross lesions include "blood-blisters" on skin,and other surface of visceral organs (gizzard mesentery etc.). Cannibalism usually develops in the flock due to bleeding from the affected skin surfaces.

1.2.4.6. Nephroma and Nephroblastoma

These are tumors of kidney induced by leukosis virus. The incubation period varies from 8 to 24 weeks. The symptoms appear as the tumors, which being initially small, start growing, resulting in emaciation and debility; paralysis may be seen if the tumors exert pressure on sciatic nerve. The gross lesions vary from small, pinkish-grey nodules embedded in kidney parenchyma to large, yellowish-grey lobulated masses that replaced most of the kidney tissue.

1.2.4.7. Hepatocarcinoma

Incubation period is 18 to 100 d. They cause primary tumors of liver which will be elevated above the surface of the liver. But, this has not been recorded under natural conditions.

1.2.4.8. Other epithelial tumors

These include thecomas and granulosa cell growths of the ovary, seminoma in testes, adenocarcinoma of pancreas and squamous cell carcinomas. However, there are very rare to detect in field cases.

1.2.4.9. Osteopetrosis

Incubation period is usually at least 4 weeks. Usually seen in long bones of limbs and is characterized by uniform or irregular thickening of the diaphyseal or metaphyseal regions which can be detected by inspection and/or palpation. In advanced cases, typical "boot-like shanks" are seen. The birds are stunted pale, and exhibit stilted-gait or lameness.

1.2.4.10. Connective tissue tumors

Most of these are caused by multipotent virus strains responsible for erythroblastosis, lymphoid leukosis and myeloblastosis. The connective tissue tumors include fibrosarcoma and fibroma, myxosarcoma and myxoma, histiocytic sarcoma, osteoma, osteogenic sarcoma and chondrosarcoma. The benign tumors grow slowly and do not invade the surrounding tissue whereas, Management tumors grow fast, infiltrate and metastasise. The incubation period for connective tissue tumors ranges from 8 to 12 months. Most of the tumors cause malfunctions only when they are extensive and metastasized. Some of the tumors of visceral organs are palpable. Mortality might occur especially in malignant forms of tumors due to secondary bacterial infection, toxemia, hemorrhage or dysfunction of an organ affected by the tumor.

1.2.5. Diagnosis

By isolation and testing for resistance inducing factor (RIF) and/or subjecting for complement fixation test for Avian leukosis (COFAL test), enzyme-linked immunosorbent assay (ELISA), fluorescent antibody (FA) tests and by tests based on phenotypic mixing of viruses. Various other tests are also evaluated but not commonly practiced. *In-vitro* tests by inoculating day-old chicks or chicken-embryos are also commonly practiced.

1.2.6. Prevention and control

Eradication of lymphoid leukosis virus infections depends on breaking of vertical transmission. Hatching eggs must be selected from hens negative in vaginal swab or egg-albumen test, avoiding manual vent-sexing and vaccination of the common needle and to prevent mechanical spread in a flock. Biological assay can be undertaken to identify the carriers and to cull them subsequently. The lymphoid leukosis virus-free flocks must be veiled in isolation on wire-floored cages. In the selection for birds resistant to lymphoid leukosis also has been tried but with varied success mainly because several serotypes as well as mutations of the virus itself. Removal of bursa of Fabricius affords protection against lymphoid leukosis but has several drawbacks including increased susceptibility to other diseases, lowered body weight, masculinization of birds affecting egg production, practicability of the procedure under commercial conditions etc.; and therefore, it is not practiced. Attempts to produce vaccine(s) against lymphoid leukosis have failed.

Table 64.1 Marek's disease *Vs* Lymphoid leukosis

	Lymphoid leukosis	Marek's disease
Age of onset	More than 16 weeks	less than 4 weeks
Mortality	Usually between 24 and 40 weeks of age	Usually between 10 and 20 weeks of age
Paralysis / Paresis	Absent	Usually present
Peripheral nerves / ganglia	Not involved	Usually involved
Bursa of Fabricius	Nodular tumors Intra-follicular infiltration of uniform blast cells	Atrophy or diffused enlargement Inter-follicular infiltration of pleomorphic mature and immature cells
MATSA antigen	Absent	Present in 5 to 40% of cells
		Source : Hofsted *et al.* 1984

2. Infectious bronchitis (IB)

2.1. Etiology

Corona virus (RNA virus); occurs in several strains.

2.2. Hosts

Chicken is the only host.

2.3. Transmission

Mainly through aerosol route.

2.4. Symptoms

Incubation period is 18 to 36h.

2.4.1. Young birds

Respiratory symptoms predominate – gasping, coughing, tracheal rales and nasal discharge. Chicks huddle under the brooder. Sinuses may be involved. Severe reduction in growth and feed consumption may occur.

2.4.2. Adults

Respiratory symptoms, reduced egg production, especially in those which are in the latter part of egg production period. In addition, shell-quality as well as internal quality of eggs deteriorate. The eggs maybe soft-shelled, misshapen or rough-shelled. The contents may show watery albumen or yolk separating from albumen or albumen sticking to the shell membrane.

2.5. Morbidity and mortality

Morbidity 100% whereas mortality is high only in young birds (25%)

2.6. Gross lesions

2.6.1. Young birds

Serous, catarrhal or caseous exudate in trachea, nasal passages and sinuses. Air-sacs may be cloudy with or without a yellowish caseous exudate.

2.6.2. Adults

Fluid yolk material in the abdominal cavity, reduction in length and weight of oviduct (isthmus and magnum are most affected) and necrosis (when the outbreak ease with the T strain) can be seen.

2.7. Diagnosis

By isolation of virus or by serological methods

2.8. Prevention and control

By proper management and vaccination. Efficiency of vaccination practices is greatly offset by the multiplicity of the serotypes. However, two strains, Massachusetts and Connecticut strains are commonly used for vaccination to vaccine can be given by spraying, dusting or in drinking water, but the easiest way is through drinking water. Maternal antibodies do not influence much on the immune response to live-virus vaccine. The common procedure is to vaccinate at 4 to 5d and again at 4 weeks. However, optimum immune response is obtained in chicken 6 weeks or older as they become immunologically mature. Replacement flocks should be vaccinated at 2 to 4 months of age. Vaccine against Newcastle disease and infectious bronchitis can be administered in combination.

3. Infectious laryngotracheitis (ILT)

3.1. Etiology

Herpes virus (DNA virus); occurs in single serotype.

3.2. Hosts

Chicken is the primary host and it causes little or no viremia. Virus multiplication is limited to respiratory tissues.

3.3. Transmission

Air-borne, intra-ocular, ingestion, carrier birds and the mechanical means (equipment and litter).

3.4. Symptoms

Incubation period is 6 to 12d.

3.4.1. Acute infection

Nasal discharge, moist rales, coughing and gasping. Expectoration of blood-stained mucus is characteristic in severe forms.

3.4.2. Mild infection

Unthriftyness, reduced egg production, watery eyes, conjunctivitis, nasal discharge and hemorrhagic conjunctivitis.

3.5. Morbidity and mortality

Morbidity 100% whereas mortality is between 10 and 20%; in mild infections, morbidity may be as low as 5% most of the adult chicken recovered from the mild form in 10 to 14d.

3.6. Gross lesions

Mucoid to hemorrhagic tracheitis which extend up to bronchi.

3.7. Diagnosis

In acute infections, typical coughing with expulsion of blood (which may appear on the walls) and high mortality rate are of diagnostic value. Virus isolation and serological testing are necessary to confirm the disease.

3.8. Prevention and control

General management procedures are valid for this disease also. However, vaccination gives a satisfactory protection.

4. Newcastle (Ranikhet) disease (ND, RD)

4.1. Etiology

Paramyxovirus (RNA). Depending on virulence, there are 3 strains – Velogenic (very virulent), Mesogenic (moderately virulent) and Lentogenic (less virulent). The most severe strain is called viscerotropic velogenic Newcastle disease (VVND) and is often referred to as "Exotic Newcastle Disease"

4.2. Hosts

Although the most common host is chicken, other species like turkeys, peafowls, guinea fowls, pheasants, quails, pigeons, ducks and geese are also affected.

4.3. Transmission

The most important route is aerosol. Wild birds and flies act as reservoirs of the virus. Contamination of food and water also helps disease spread.

4.4. Symptoms

4.4.1. Doyle's form

This is an acute disease and sometimes birds will be found dead without any symptoms. More often, listlessness, polypnea, weakness and death within 4 to 8d is seen. Greenish diarrhoea (sometimes blood-stained), muscular tremors, torticollis, opisthotonus are also frequently observed. There may be other nerve symptoms including paralysis of legs and wings. This is caused by velogenic farm of the virus and mortality is usually 90%.

4.4.2. Beach's form

This is also an acute form spreading very rapidly with characteristic respiratory symptoms like dyspnea, coughing and gasping, inappetance and reduced egg production. Diarrhoea is not usually observed. Paralysis of legs and wings, and torticollis may also be seen. Mortality is high amount immature birds (90%) and less in case of adults (maximum 50% and on average 10%). This form is caused by velogenic form of the virus.

4.4.3. Beaudette's form

This is an acute respiratory disease of adult chicken characterized by coughing but rarely, gasping. Appetite and egg production drop and internal quality of eggs get affected. Adults usually do not die of this form of disease but mortality is high in younger birds. This form is caused by mesogenic virus.

4.4.4. Hitchner's form

This is a mild form of disease usually goes unnoticed in adult birds. There may be respiratory symptoms detectable on close observation. Only when complicated by secondary infections, the mortality may go up to 30%. There will be air-sacculitis, resulting in higher carcass condemnations. This form is caused by lentogenic strains.

4.4.5. Asymptomatic enteric form

No clinical signs but detectable only by virus isolation from the gut or feces, and the presence of acidic antibodies. This is caused by lentogenic strains.

4.5. Gross lesions

Highly variable between birds, flocks, location etc. No truly pathognomonic lesions recognized for Newcastle disease. However, few gross lesions of diagnostic value are as follows:

4.5.1. Doyle's form

Dark-red or purple-red hemorrhagic lesions associated with some necrosis in the intestinal wall especially in the posterior half of the duodenum, jejunum and ileum. Diphtheroid necrotic inflammation of intestinal mucosa with button ulcers is highly suggestive of Newcastle disease. In older chicken, hemorrhagic and necrotic ovaries, and ruptured yolks in the peritoneal cavity are frequently seen.

4.5.2. Other forms

The primary lesions are in the respiratory tract. A serous catarrhal exudate in nasal passages, larynx and trachea (sometimes with hemorrhages), thickening of membranes of one or more air-sacs which may contain catarrhal or caseous exudate are seen.

4.6. Diagnosis

Isolation and identification of the disease is the method of chance. There are other serological tests useful in the diagnosis of this disease. An intracerebral pathogenicity index (ICPI) of at least 0.7 is necessary to define the disease as ND (OIE, 2000).

4.7. Prevention and control

A combination of sanitary measures and vaccination is important in controlling this disease. The lentogenic strains (B1, LaSota and F) are commonly used for birds of all ages for intranasal and intra-ocular instillation, admixture with drinking water or strain. When vaccine is given in water, skimmed milk powder may be used as stabilizer. Aerosol (spray) route produces the most severe reaction. The mesogenic strains (Roakin, Mukteshwar, H, Haifa) had been commonly used for wing-web (intra-dermal), intramuscular or feather follicle vaccination of stock more than 4 weeks of age. Chicks less than 3 weeks of age posses maternal immunity but since the passive immunity wanes off by 2nd or 3rd week, the common practice is to administer vaccines from lentogenic strains when the birds are 8 to 10 d old (or after 1 to 2 weeks old if the passive immunity is expected to be high). The same vaccine may be repeated after 4 to 5 weeks of the first vaccination; after this second vaccination, the flock may be vaccinated with a mesogenic strain during 8 to 10 weeks of age and repeated every 2 to 3 months for the birds destined for breeding.

5. Avian encephalomyelitis (AE)

5.1. Etiology

Picorna virus (RNA); occurs as a single serotype; enterotropic

5.2. Hosts

Affects only chicken, presents, quails and turkeys.

5.3. Transmission

Egg-borne, direct and indirect contact, oral and carriers.

5.4. Symptoms

Incubation period is 1 to 11d.

5.4.1. Chicks

Typical symptoms are shown during 1 to 2 weeks of age like dullness, progressive ataxia, sitting on hocks, uncoordinated gait, fine tremors of the head and neck, and inability of the chick to reach for food and water. Most of the deaths are due to inanition. Few of the survived birds might show opacity or bluish discoloration of lens in one or both the eyes and few of the progenies of the matured birds also show similar changes in the eye.

5.4.2. Adults

Only symptom is dropping egg production but no nervous symptoms are exhibited.

5.5. Morbidity and mortality

Usually, morbidity in younger birds goes up to 40 to 60% and mortality normally averages 25% but sometimes may go up to about 50%.

5.6. Gross lesions

Whitish areas in the muscularis of the ventriculus in young birds.

5.7. Diagnosis

By symptoms and confirmation by isolation and identification of the virus as well as by serological tests.

5.8. Prevention and control

Mainly by vaccination of the breeding flock to avoid egg-transmission. Since chicks from immunized breeding flock may retain passive immunity up to 8 weeks, there are vaccinated at or after 10 weeks of age by commercial vaccines.

6. Egg drop syndrome (EDS 1976)

6.1. Etiology

Adeno virus (DNA); occurs as a single serotype; enterotropic

6.2. Hosts

Affects usually chicken; it is possible that the virus is a duck or goose adeno virus and that chicken is not a natural host. Chicks can be experimentally infected by ducks only when they are kept in close contact, which under commercial conditions, is very rare.

6.3. Transmission

Mainly through embryonated eggs; horizontal transmission is slow and intermittent, especially in caged birds; spread appears to depend on birds coming into contact with infected feces. When the viremic birds are bled or inoculated and the same equipment is used on healthy birds, the disease spreads.

6.4. Symptoms

When the virus has been transmitted horizontally, the first sign is loss of color in pigmented eggs followed quickly by production of thin- and soft-shelled eggs or shell-less eggs. The thin-shelled eggs often have a rubbed, sand paper-like texture or a granular roughening of the shell at one end. There will be a sudden drop in egg production if infection has occurred in late production. If affected eggs are removed, incubation performance is unaffected. If the disease is due to reactivation of virus, small eggs and eggs with poor interior quality or produced. If birds have acquired antibodies before the latent virus is unmasked, then the birds do not attain the expected peak in egg production and there may be delayed sexual maturity.

6.5. Morbidity and mortality

Not defined

6.6. Gross lesions

In natural outbreaks, the only finding is that some birds may have inactive ovaries and atrophied oviducts

6.7. Diagnosis

By isolation and identification of the virus as well as by serological tests.

7. Prevention and control

Since the disease is egg-borne, birds must be obtained from unaffected flocks. Ducks should not be reared in close proximity with chicken. For immunization, and oil-adjuvant inactivated vaccine is widely used and the birds are vaccinated within 14 and 16 weeks of age. The vaccine gives good protection against clinical disease and reduces the amount of virus excreted.

8. Pox

8.1. Etiology

Pox viruses (DNA); turkey and fowl pox viruses can cause infection to chicken both when introduced by cutaneous and intravenous routes whereas, canary and pigeon pox can infect only through cutaneous route.

8.2. Hosts

The most important host is chicken.

8.3. Transmission

Mainly mechanical transmission of the virus to the injured or lacerated skin; Mosquitoes (Tulips and Aedes) are capable of extrinsic transmission (no multiplication of virus in the vector). In breeder turkeys, it may be transmitted through artificial insemination.

8.4. Symptoms

8.4.1. Cutaneous form

Cutaneous lesions of the head region and on featherless areas such as legs, feet and went.

8.4.2. Diphtheritic form

Diphtheritic lesions of the mouth region and infection of nasal chambers followed by coryza-like signs. In some cases, in broilers, lesions are seen primarily on feet.

8.5. Morbidity and mortality

Morbidity is usually high whereas mortality, especially in the cutaneous form, is quite low. Mortality in severe cases of diphtheritic form may go up as high as 50%.

8.6. Gross lesions

8.6.1. Chicken

8.6.1.1. Cutaneous form

Skin lesions appear as local epithelial hyperplasia followed by nodule formation. The adjoining nodules coalesce and become rough and grey or dark brown. This is followed by scab formation and desquamation of the epithelial layer. When the scab falls naturally, a smooth scar may be present.

8.6.1.2. Diphtheritic form

White opaque nodules on the mucous membranes which enlarge, coalesce and form a pseudo-diphtheritic or diphtheritic membrane. This may extend to sinuses and pharynx as well as larynx and esophagus.

8.6.2. Turkeys

Minute yellowish eruptions on dewlap, snood and other parts of the head. Corners of the mouth, eyelids and oral membranes are commonly affected. Lesions enlarge and coalesce and become covered with a dry scab or form wart-like mass. In young poults, head, legs and feet may be completely covered with pustules and extend to feathered portions also.

8.7. Diagnosis

By typical symptoms and lesions; confirmed by isolation identification of virus and serological tests.

8.8. Prevention and control

Two types of live-vaccines are used :

1. Fowl pox vaccine – highly virulent for chicks and hence used for birds more than 4 weeks of age and to birds 1 to 2 months before egg production. The vaccine is applied by stabbing the vaccinated needle dipped in vaccine to the wing-web region. The "takes" are observed after 7 to 10d for estimating the success of vaccination.
2. Pigeon pox vaccine – this is less pathogenic to chicken and turkeys; is applied by swabbing the vaccine to denuded feather-follicles on the thigh region at 8 to 10d of age. However, this vaccine is not being used and only fowl pox vaccine is used commercially.

9. Infectious bursal disease (IBD, Gumboro disease)

9.1. Etiology

Reo virus (RNA), Birnaviridae family; has 2 serotypes

9.2. Hosts

The primary host is chicken.

9.3. Transmission

Highly contagious; Water, feed and droppings from infected pens are highly infectious. There seems to be no carrier birds but it is likely to be arthropod-borne (mosquitoes, worms).

9.4. Symptoms

Incubation period is only 2 to 3d. Mortality begins 3rd day post-infection and will peak and recede in a period of 5 to 7d. This is referred to as "spiking death curve". The birds show a peculiar tendency of pecking at their own vents. There may be whitish or watery diarrhoea, anorexia, depression, dehydration, ruffled feathers, trembling, shrill cry while defecating and prostration before the birds actually die. Morbidity and mortality : Morbidity quickly approaches 100% and mortality seldom exceeds 20 to 30%.

9.5. Gross lesions

Darkened discoloration of pectoral muscles due to dehydration, frequently hemorrhages are seen on thigh and pectoral muscles. A sequel of changes are seen in bursa of Fabricius – a) 3rd d post-infection – increase in size and weight because of edema and hyperemia. There will be gelatinous yellow transudate in the bursa. Longitudinal striations on the surface become prominent and color changes from normal white to cream. b) 4th d post-infection – doubling of weight and subsequently reductions in size. c) 5th d post-infection : normal weight and continues to atrophy. d) 8th d post-infection – reduces to 1/3rds its original weight. The bursa may show necrotic foci, petichiae or ecchymoses, or extensive hemorrhage. Spleen is frequently enlarged with grayish foci uniformly distributed. There may be hemorrhage in the mucosa at the junction of proventriculus and gizzard.

9.6. Diagnosis

By typical symptoms and lesions; confirmed by isolation identification of wires and serological tests.

9.7. Prevention and control

Vaccination is the principal method of controlling the disease. There are 3 types of live-vaccines : 1. relatively virulent – causes disease (immunosuppression) in susceptible chicken. 2. Attenuated avirulant strains – do not cause lesions are immunosuppression. 3. Intermediate strains – meant what maternally-immune chicken. Commercially, it is a general practice to give intermediate-strain vaccine at 14d of age to be repeated after another 14 d; both times the vaccine is administered through oculonasal route or through drinking water. Broiler breeder – flocks may be vaccinated with live-attenuated vaccine at 1d and/or at 2 to 3 weeks of age followed by an oil-emulsion killed-virus vaccine at 16 to 18 weeks. Live-vaccine may be repeated at 10 to 14 weeks of age to get a better response for the killed-virus vaccine.

10. Avian influenza (AI, Fowl plague)

Highly pathogenic AI (fowl plague) was first documented in Italy more than 100 years ago. Highly pathogenic AI first occurred in the United States in 1924-1925 and again in 1929 but was eradicated both times. The epidemic HPAI in the northeastern United States in 1983-1984 required over 2 years to eradicate.

The United States has been HPAI-free since 1986, although LPAI viruses are present and some have caused significant losses in poultry. It was in 1964 that influenza viruses of low to moderate pathogenecity were first detected in poultry and they have been detected every year since 1964. The National Veterinary Service Laboratory in Ames, Iowa, can identify influenza virus serotypes and pathotypes (Halvorson, 2005).

10.1. Etiology

Influenza type A virus; Orthomyxovirus. The virus is further classified into subtypes based on:

1. The glycoproteins hemagglutinin (H) and neuraminidase (N) present on the virus – 15 H subtypes (H1-H15) and 9 N subtypes (N1-N9)(OIE, 2000); hence, it is obvious that as many as 135 combinations possible; of these, H5N2, H7N1 and H7N7 have been reported to be highly pathogenic causing high mortality in chicken.
2. The ability of the virus to cause disease – Highly Pathogenic Avian Influenza (HPAI) - Can cause up to100% mortality and Low Pathogenic Avian Influenza (LPAI) – Causes a mild disease with lower mortality; LPAI subtypes, especially H5 and H7, have the ability to mutate to HPAI after circulation in a poultry population.

HPAI is an OIE List A disease. Avian influenza is classified as HPAI if it conforms to these criteria:

1. Any influenza virus that is lethal for six, seven or eight of eight 4–8-week-old susceptible chickens within 10 days following intravenous inoculation with 0.2 ml of a 1/10 dilution of a bacteria-free, infective allantoic fluid.
2. The following additional test is required if the isolate kills from one to five chickens but is not of the H5 or H7subtype: growth of the virus in cell culture1 with cytopathic effect or plaque formation in the absence of trypsin. If no growth is observed, the isolate is not considered to be an HPAI isolate.
3. For all H5 and H7 viruses of low pathogenecity and for other influenza viruses, if growth is observed in cell culture without trypsin, the amino acid sequence of the connecting peptide of the hemagglutinin must be determined. If the sequence is similar to that observed for other HPAI isolates, the isolate being tested will be considered to be highly pathogenic.

10.1.1. EU Definition of HPAI (June 2003)

The definition of HPAI as stated in EU Directive 92/40/EEC, uses the intravenous pathogenecity index (IVPI) as a measure of virulence: "an infection of poultry caused by an influenza A virus that has an IVPI in 6-week old chickens >1.2 or any infection with influenza A viruses of H5 or H7 subtype for which nucleotide sequencing has demonstrated the presence of multiple basic amino acids at the cleavage site of the hemagglutinin".

Influenza type A virus can infect Avians, humans, swine, horses and occasionally, minks, seals and whales. The viral genome has eight segments which allows reassortment which gives it the property of antigenic shift. Strain that is pathogenic to one species need not be pathogenic to others; ducks have maximum number of strains of virus and turkeys, most outbreaks.

Influenza viruses vary widely in pathogenecity and ability to spread among birds. Two pathotypes are recognized: LPAI and HPAI. These pathotype designations are derived from laboratory inoculation of 8 susceptible chickens; LPAI isolates cause death in 0 to 5 of 8 chickens and HPAI isolates cause death in 6 or more. Although most H5 and H7 isolates are low path viruses, so far all HPAI outbreaks have been due to H5 or H7 viruses. All influenza viruses hemagglutinate chicken red blood cells. Most grow readily in embryonating chicken eggs and tissue culture. They are susceptible to detergents, disinfectants and heat (Halvorson, 2005).

10.2. Transmission

Migratory water birds, especially wild ducks, are the natural hosts of influenza A. They do not show clinical disease. The virus colonizes the intestinal tract and is spread in the faeces. They act as a reservoir for the infection of other species.

Wild birds while migrating can spread influenza viruses in the following ways (Raj, 2005):

1. Domestic poultry being free to mix with wild birds gives an opportunity for transmission through direct contact
2. Wild birds may defecate while flying and their droppings can fall in the areas where domestic poultry is reared
3. Water contaminated by droppings of the wild birds being used for domestic poultry
4. Poultry feed contaminated by droppings from infected wild birds; this is particularly true for birds given access to outdoor facilities (organic poultry farming)
5. Humans, vehicles and objects such as farm implements can get contaminated through contact with droppings from the wild birds
6. Illegal trade of wild and caged birds: this is particularly true in case of hawk eagles smuggled into Belgium from Thailand, the latter being one of the countries where the initial outbreaks of bird flu occurred during 2003. Similarly, illegal trade of meat and meat products, feathers etc. also to be potent sources of bird flu infection.

Infection of domestic poultry within a farm avian influenza is spread by direct contact, contact with contaminated equipment and farm staff. AI is spread between farms by the transfer of infected faeces, staff members, contaminated equipment and delivery trucks; air transmission over short distances. Live bird markets also play an important role in the spread of epidemics. Commercial swine facilities are also potential reservoirs.

Infected birds excrete the virus through respiratory tract, conjunctiva and feces; therefore, transmission occurs through direct and indirect contact, especially aerosol and fomites.

According to World Health Organization (WHO), the main route of human infection is through direct contact with infected poultry or surfaces and objects contaminated by their droppings. Risk of exposure to the virus is considered highest during slaughter, defeathering, butchering and preparation of poultry for cooking. There is no evidence that properly cooked poultry or poultry products can be a source of infection.

10.3. Symptoms

(The Merck Veterinary Manual, 2003)

Clinical signs are dependent on the pathogenecity of the infecting virus and the species infected. Infected ducks may show no clinical symptoms. Turkeys infected with LPAI may show severe clinical signs such as those seen with HPAI infection in chickens.

Birds are infected via the upper respiratory tract, the conjunctiva or the digestive tract. Viral replication takes place in the epithelium. The incubation period is very variable, a few hours to 18 days.

10.3.1. HPAI in chicken

Depression, decreased appetite, decreased egg production, nervous signs, swollen blue combs and wattles, coughing, sneezing, diarrhea, sudden death

10.3.2. LPAI in chicken

Mild respiratory disease, depression, decreased egg production; chicken infected with LPAI and other viruses may show severe clinical signs.

10.4. Morbidity and mortality

In outbreaks of Highly Pathogenic Avian Influenza (HPAI) mortality can be up to 100%. Low Pathogenic Avian Influenza (LPAI) in chickens may even go unnoticed.

10.5. Gross lesions

Variable.

10.5.1. Mild infection

In chicken and turkeys few striking lesions are noticeable because disease is mild; lesions in sinuses (catarrhal, fibrinous, sero- fibrinous, mucopurulant are caseous inflammation), sometimes edema of tracheal mucosa with serous to caseous exudate can be seen. The air-sacs may be thickened, catarrhal to fibrinous peritonitis, egg peritonitis may be seen. Catarrhal to fibrinous enteritis in ceca and/or intestines (especially in turkeys), exudates in oviduct of hens are also recorded.

10.5.2. Severe infection

No prominent lesions because of quick death. Congestive, hemorrhagic, transudative and necrobiotic changes may be seen. Necrotic foci in liver, spleen, kidney, lungs of chicken, swollen sinuses, cyanotic and congested combs and wattles, peritoneal edema, combs may show vesicles, edema, ecchymoses and frank necrosis are the other lesions. Petichiae on various serosal and mucosal surfaces (especially junction of proventriculus and gizzard), red areas along the length of pancreas, necrosis of lymphoid tissue (mottled appearance of spleen) may be seen. Many of the lesions are likely to be due to secondary bacterial infection(s).

Ducks infected with HPAI may not show any clinical signs or lesions

10.6. Diagnosis

Influenza must be differentiated from other poultry diseases including Newcastle disease, other Paramyxovirus infections, mycoplasmosis, Chlamydia infections, and fowl cholera. Highly pathogenic AI should be differentiated from velogenic viscerotropic Newcastle disease (Halvorson, 2005).

10.6.1. Samples

10.6.1.1. Live birds

Tracheal swabs and cloacal swabs or faeces; dead birds – organs and faeces

10.6.1.2. Serology

Clotted blood samples or serum

10.6.2. Procedures

10.6.2.1. Identification of the Agent

Inoculation of 9-11-day-old embryonated chicken eggs followed by Hemagglutination immunodiffusion test to confirm the presence of influenza A virus

10.6.2.1.1. Subtype determination with monospecific antisera

Strain virulence evaluation: by intravenous pathogenecity index (IVPI) in 4-8-week-old chickens.

The IVPI test is carried out as follows (OIE, 2000):

1. Fresh infective allantoic fluid with an HA titer of > 2^4 (> $\log_2 4$ or > 1/16) is diluted 1/10 in sterile isotonic solution.

2. 0.1 ml of the diluted virus is injected i/v into each of ten 6-week-old SPF chickens.
3. Birds are examined at 24-hr intervals for 10 d. At each observation, each bird is scored (subjectively) 0 if normal, 1 if sick (showing one of the signs), 2 if very sick (showing > 1 signs) and 3 if dead (at each of the remaining daily observations as well); the signs are – respiratory involvement, depression, diarrhea, cyanosis of the exposed skin and wattles, edema of the face and/or head, nervous signs.
4. The IVPI is the mean score per bird per observation over the 10 d period. Obviously, an index of 3.00 means all birds died within 24 hr and IVPI of 0.00 means no bird showed any clinical sign over the 10 d period.

10.6.2.1.2. Serology

ELISA / AGID (Agar Gel Immunodiffusion test): Detects antibodies to all AI virus, does not distinguish subtypes. Only suitable for testing chicken and turkey serum; Within 1 week of infection, antibodies are detected in more than half the specimens.

10.6.2.1.3. HI (Hemagglutination Inhibition test)

Serotype specific test; Test available for each H subtype. HI titers are positive a few days later than ELISA or AGID, titers persist long after infection. Standard test for all avian species

10.6.2.1.4. IFT (Immunofluoresence test)

Able to detect antibodies to specific N-subtype; can be used to detect infection in vaccinated birds if a heterologous vaccine is used.

10.6.2.1.5. RT-PCR (Reverse-transcriptase polymerase chain reaction)

This method, being one of the most frequently used, can detect influenza virus at very low levels. The presence of subtype H5 or H7 can be confirmed by using H5 or H7 specific oligonucleotide primers.

10.7. Prevention and control

Isolation of affected birds. Wild birds should not be allowed to come in contact with domestic birds. Usual biosecurity measures must be ensured to control spread of disease. Recovered birds act as carriers and hence they should not come in contact with healthy birds. Pigs should not be reared in proximity to poultry farms. Vaccination against most pathogenic strains is being attempted to. Since the disease is transmissible to human beings and recently it did cause serious concerns especially in countries like China, Philippines and other Asian countries, lot of research work is being carried out on control of Avian influenza. See Chapter "Biosecurity" for details of control measures.

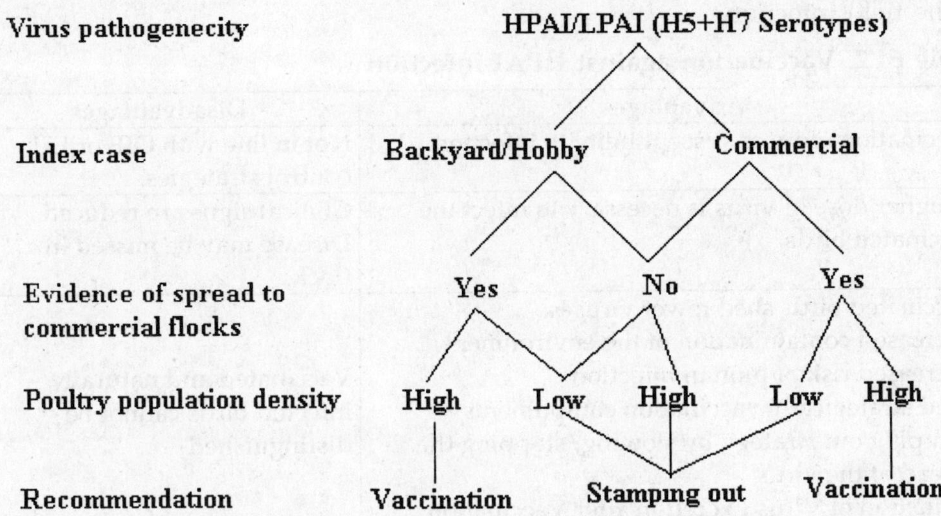

Fig.64.1 Control strategy for AI (Adopted from Capua and Marangon, 2003)

10.7.1. Vaccination

Immunity is hemagglutinin subtype specific. Because birds are susceptible to all 15 hemagglutinins preventive vaccination is not practical. Once an outbreak occurs and the subtype is identified, however, vaccination is a tool that may be used to help bring the infection under control. Influenza viruses are unstable so no live vaccines have been developed for poultry. Only killed, injectable vaccines are available and currently USDA prohibits use of H5 or H7 vaccines (Halvorson, 2005).The following types of vaccines against avian influenza are available:

10.7.1.1. Conventional vaccines

Inactivated oil emulsion vaccines that are used worldwide. May be homologous or heterologous. Infection in a vaccinated flock can be monitored through the use of sentinel birds or serology (heterologous vaccines).

10.7.1.2. Recombinant vaccines

Vectors used are ILT or Fowl Pox. Not licensed in the EU. Still used in Mexico. Have the potential to be used as "marker" vaccines

10.7.1.3. Heterologous Avian Influenza Vaccines

The influenza A vaccine strain in a heterologous vaccine contains the same hemagglutinin (H) sub-type as the field virus but a different neuraminidase (N).

10.7.1.4. Protection by H sub-type

Experimental data has shown that vaccinated birds are protected from avian influenza infection provided that the vaccine strain H subtype is the same as that of the field infection.

Table 64.2. Vaccination against HPAI infection

Advantages	Disadvantages
Vaccination reduces susceptibility to infection.	Not in line with OIE or EU control strategies.
A higher dose of virus is necessary to infect the vaccinated birds.	Clinical signs are reduced. Disease may be missed in a flock.
Vaccinated birds shed fewer viruses. Decreased contamination of the environment. Decreased risk of human infection Used strategically vaccination compliments a stamping out strategy by slowing/stopping the spread of the virus Reduction of Virus Excretion after Vaccination	Vaccinated and naturally infected birds cannot be distinguished

10.8. Bird flu (H5N1) outbreak during 2003 and after

The first known case of avian influenza (H5N1) was found in 1996 in a goose in China. The strain of avian influenza found in China is different from and older than the strain found in Thailand and Vietnam. Bird flu has infected more than 100 people in Thailand and Vietnam in the past two years, killing 60. Turkey confirmed an outbreak of the disease on a farm in northwestern Turkey, triggering a virtually pan-European ban on imports of all Turkish live birds and feathers. The Romanian government said on Saturday that a third place was affected by bird flu. Austria, Italy, Germany, Greece and Bulgaria have become the latest countries to report bird flu as the virus appears to spread across Europe. Russia has reported bird flu for the second time.

The outbreaks began in 2003 and have led to death and/or slaughter of about 140 million domestic birds, especially in Asian countries resulting in an estimated loss of about $10 billion; in addition, it has claimed the lives of 76 people.

The outbreak was earlier restricted to Indonesia, Vietnam, Thailand, Laos, Cambodia and China; but, during May to July this year (2005), the disease has appeared in Russia and Kazakhstan followed by Mongolia during August. Recently, bird flu has been confirmed in Turkey and Romania. It is assumed that the recent spread of disease is due to migratory birds which are carriers of the deadly H5N1 influenza virus. The UK has also recorded a case of AI in a parrot died at quarantine.

Twelve swans have also tested positive for bird flu in a second cluster in Romania. The European Commission has also ordered urgent tests on dead birds found in

Croatia (ThePoultrySite.com, 2005).

At this stage, neither of the new, outbreaks in Greece or Romania has been confirmed as the lethal H5N1 form of the bird flu virus. However, tests are continuing, and the Greek outbreak is known to involve the H5 strain, of which the deadly form H5N1 is a member.

On 18th October, 2005, the EU has declared the spread of bird flu from Asia into the EU a "Global threat" requiring international cooperation. Westward move of the virus is thought to be caused by migrating wild fowl. There is no human vaccine for H5N1 and, as per WHO report, the neuraminidase inhibiting drugs Oseltamivir phosphate (Tamiflu), Zanamivir (Relenza) are expected to help humans to fight the infection.

10.8.1. Bird flu in India

Please see Appendix - 1

11. Infectious stunting syndrome (ISS)

11.1. Etiology

Many viruses/virus-like particles are also isolated from/observed in intestinal fact/feces of affected birds including Calcivirus, Enterovirus, Parvovirus, Reovirus, atypical Rota virus and Toga virus-like particles. However, the syndrome is viral in origin.

11.2. Synonyms

Infectious runting, Runting and stunting syndrome, Malabsorption syndrome, Helicopter disease, Pale-bird syndrome, Osteoporosis or Brittle-bone disease and Femoral-head disease. However, runting in replacement pullets and breeders, guinea fowl and turkeys associated with proventriculitis (proventricular dilation syndrome) is unrelated to ISS. Economic losses are considerable in this disease although most of the affected birds recover partiallly.

11.3. Transmission

Vertical transmission is common in younger breeder flock whereas horizontal transmission is equally rapid and important. Poor housing and management accelerates transmission.

11.4. Symptoms

Small chicken detectable at 4 to 6d of age and remain weighing 100 to 200 g till 6 to 8 weeks of age (about 1 to 5% of birds). Another 10 to 15% of the flock may grow but poorly and stunting is obvious by 2 weeks of age. The "stunts" will have retarded feathering ("yellow-head"), pendulous abdomen and voracious appetite. Severely stunted birds are most active and spend much of their time near the feeders and they do not appear ill at all. Sometimes, abnormal feathering causes displacement of primary wing feathers (helicopter disease). Decreased pigmentation of carcass due to poor absorption (malabsorption syndrome) of carotenoid and related substances result in pale-bird syndrome and poor absorption of Vitamin E can cause encephalomalacia. Lameness is often seen during 2 to 4 weeks of age with tibiotarsus of affected birds being too brittle (brittle-bone disease, Femoral head disease) indicating the disorder in bone growth. Diarrhoea and/or pasting of vents is also a common symptom.

11.5. Gross lesions

Atrophy of lymphoid organs especially thymus and bursa of Fabricius although this is not specific to ISS. Tibial dyschondroplasia, retained embryonic cartilage in the metaphyses of tibiotarsi is rather characteristic in some of the affected birds. Enlarged parathyroids and pancreatic degeneration due to occlusion of pancreatic ducts may also be seen. Mucoid enteritis is also recorded. Poorly digested feed in the intestines is more commonly seen during necropsy examination with relative elongation of intestines in comparison to normal.

11.6. Pathogenesis

In those where pancreatic degeneration has occurred, stunting is expected. But, intestines are thought to be the target organ in ISS. Increased intestinal alkaline phosphatase activity, mucoid enteritis and diarrhoea in affected birds clearly indicate the intestinal disease. Histological changes in intestinal villi result in poor absorption which predisposes to stunting and intestinal elongation is the adoptive response of the bird. Activity of intestinal brush border also is reduced accentuating the impaired absorption.

11.7. Control

Proper cleaning and disinfection of the affected premises with formaldehyde fumigation, all in all-out system of rearing, proper housing, ventilation, litter management, temperature regulation.

Vertical transmission can be avoided by identification and culling of breeding birds. Antibiotics may only help reduce secondary infection(s). Low concentration of disinfectants (iodophores, quaternary ammonium compounds and organic acids) in water may inactivate the virus in intestines although such treatment may not be very helpful. Reduced absorption of fat-soluble nutrients may further add to the stunting of birds. Hence, parenteral administration of fat-soluble vitamins may help; but not practicable.

12. Hydropericardium syndrome

12.1. Etiology

Suspected to be a virus - may belong to Adenovirus, Reovirus or Parvo virus family

12.2. Transmission

Thought to be vertically transmitted.

12.3. Symptoms

Appear in about 2 weeks. The birds die and show hydropericardium.

12.4. Gross lesions

Hydropericardium, edema of heart and lungs

12.5. Diagnosis

By symptoms

12.6. Prevention and control

Obtaining chicks from disease-free breeders, proper management to reduce stress.

13. Hemorrhagic Enteritis (Turkey hemorrhagic enteritis)

13.1. Etiology

Adenovirus; the virus is immunosuppressive and may help precipitate intercurrent disease or secondary coli septicemia

13.2. Hosts

Turkeys usually seen between 6 and 12 weeks of age

13.3. Transmission

Not totally understood; likely to be vertically transmitted with significant horizontal spread through contact with infected faeces. Virus excretion occurs after maternal antibody has disappeared; hence, disease is rare in younger birds.

13.4. Morbidity and mortality

Mortality rates are usually low (1-2%) but may reach 50% in exceptional cases

13.5. Symptoms

Variable. dull and depressed, dark bloody droppings which frequently stain tail feathers and legs. In more acute outbreaks, birds may or may not show bloody droppings; birds in good body condition die suddenly with losses occurring over a short time period of 5-10 days.

13.6. Lesions

Pale carcasses and intestines , especially duodenal loop, contains fresh blood. Blood may even be found throughout the intestinal tract, often with dark red tarry cecal contents. The spleen is usually enlarged and mottled.

13.7. Diagnosis

Lesions and confirmation is by demonstration of antigen or antibody in spleen from affected turkeys by gel precipitation test or typical adenoviral, intranuclear inclusion bodies in enlarged spleens.

13.8. Prevention and control

Vaccines are available in some countries. Otherwise, limiting physical spread of infection via infected faeces, litter and fomites, and reducing intercurrent disease or environmental stressors is important.

14. Turkey rhinotracheitis

14.1. Etiology

Avian pneumovirus subgroup A (Paramyxoviridae family)

14.2. Hosts

Turkeys of all age groups

14.3. Transmission

Airborne with susceptible birds infected via the respiratory tract. Intercurrent viral or bacterial disease or mycoplasma infections exacerbate disease and prolong viral excretion and persistence. Limited evidence of fomite spread.

14.4. Symptoms

In meat and broiler turkeys, upper respiratory tract infection characterized by sudden onset, respiratory distress, swollen heads, coughing and conjunctivitis. Breeding turkeys - cough, prolapses, and drops in production and hatchability. Broiler breeders - swollen head syndrome, star gazing, peritonitis and drops in production. Layers - upper respiratory tract disease, persistent peritonitis, occasional nervous signs as opisthotonus. Drop in egg production and loss of shell quality and color.

14.5. Morbidity and mortality

High morbidity and variable mortality (up to 50%) depending on environmental conditions, ventilation and secondary bacterial infection.

14.6. Diagnosis

Clinical signs, ELISA or demonstration of virus in respiratory tract material.

14.7. Prevention and control

Antimicrobial medication reduces the effects of secondary bacterial disease. The main control measure is vaccination using an attenuated live vaccine.

15. Avian rhinotracheitis (Swollen head syndrome)

15.1. Etiology

Avian pneumovirus subgroup B (Paramyxoviridae family)

15.2. Hosts

Chickens, turkeys, guinea fowl and possibly pheasants

15.3. Transmission

Mainly by aerosol; vertical transmission is uncertain. Fomites can be important in moving infection between farms.

15.4. Symptoms

The incubation period is 5–7 days. Decreased appetite, weight gain and feed efficiency, facial and head swelling (though this can occur in other conditions), loss of voice, ocular and nasal discharge, conjunctivitis, snick, dyspnea, sinusitis.

15.5. Morbidity and mortality

Morbidity is 10–100% and mortality can be 1–10%.

15.6. Lesions

Serous rhinitis and tracheitis, sometimes pus in bronchi. If secondary invasion by *E. coli* then, pneumonia, airsacculitis and perihepatitis. Congestion, edema and pus in the air space of the skull occurs in a proportion of affected birds due to secondary bacterial infections.

15.7. Diagnosis

Clinical signs, serology, isolation of ciliostatic agent. Serology – Elisa normally used. To be differentiated from Infectious Bronchitis, Lentogenic Newcastle disease, low virulence avian influenza, *Ornithobacterium rhinotracheale*.

15.8. Treatment

Antibiotic not very effective. Control of respiratory stressors important, chlorination of drinking water and multivitamins might help combat secondary infections.

15.9. Prevention and control

All-in all-out production. Live vaccines can reduce clinical signs and adverse effects, inactivated vaccines may be used in breeders prior to lay.

16. Chicken anemia

(Chick anemia agent (CAA), Blue wing disease, anemia-dermatitis syndrome)

16.1. Etiology

Circovirus. Very resistant virus that can be removed from sites only with disinfectants with broad spectrum activity. The virus is dramatically immunosuppressive following *in ovo* infection.

16.2. Hosts

Appears to be unique to chicken

16.3. Transmission

Vertical transmission leads to destruction of bone marrow and atrophy of thymus, spleen and bursa. Infection met for the first time during lay leads to vertical transmission and disease in progeny. Horizontally acquired infection of broilers can have a significant adverse effect on performance. Chicks develop age resistance to disease but not infection, by 7-14 days of age.

16.4. Symptoms

Chicks show mortality from 10 days of age. Signs are associated with secondary bacterial or fungal infection such as gangrenous dermatitis, skin hemorrhages, fungal pneumonia and bacterial septicemia.

16.5. Morbidity and mortality

High mortality (up to 70%) seen in the progeny of breeder flocks infected whilst in lay.

16.6. Lesions

Atrophy of lymphoid organs, aplasia of bone marrow and clinical anemia.

16.7. Diagnosis

Lesions with Clinical history.

16.8. Control

Live and inactivated vaccines are available in different countries. Prevention of horizontal infection of young broilers requires effective cleansing and disinfection with products of proven efficacy against the virus.

17. Viral Arthritis

17.1. Etiology

Reovirus

17.2. Morbidity and mortality

Morbidity is high but mortality is usually low.

17.3. Transmission

Vertical and lateral through fecal contamination, carriers.

17.4. Symptoms

Lameness, low mobility, poor growth, inflammation at hock, swelling of tendon sheaths, unthriftyness, rupture of gastrocnemius tendons.

17.5. Lesions

Swelling and inflammation of digital flexor and metatarsal extensor tendon sheaths, foot pad swelling, articular cartilages may be ulcerated, hemorrhage in tissues; fibrosis in chronic cases.

17.6. Diagnosis

History, lesions, and isolation of the virus. Serology may be by FAT or Elisa. The disease has to be differentiated from mycoplasmosis, salmonellosis, Marek's, *Pasteurella*, erysipelas.

17.7. Treatment

No treatment is effective

17.8. Prevention

A live vaccine followed by an inactivated vaccine prior to coming into lay. All-in all-out rearing system advised.

18. Duck virus hepatitis

18.1. Etiology

RNA (Picorna) virus

18.2. Transmission

Highly contagious, recovered birds are carriers for about 8 weeks post-infection, wild birds may act as carriers

18.3. Symptoms

Symptoms include rapid onset and spread and mortality within 3 to 4 days, ducklings squat with closed eyes, fall sidewards, kick spasmodically and die with head drawn back.

18.4. Morbidity and mortality

Mortality is very high (95%) in very young stock (less than 7d) and about 50% in 1-3-week old ducklings. Ducklings over 4 weeks of age seldom die.

18.5. Lesions

Gross lesions include enlarged liver and/or spleen with punctate or ecchymotic hemorrhages, mottling of liver surface, kidneys may be swollen with congestion of renal vessels.

18.6. Prevention and control

Prevention is by immunization of the breeding stock to ensure high maternal antibodies with chicken-embryo attenuated vaccine, 0.5 ml undiluted egg-propagated virus i/m 2-4 weeks prior to collection of hatching eggs. Two to three doses six weeks apart recommended. Duck embryo adapted virus, which is better than that of chicken, is also available. Strict isolation during first 4-5 weeks of age is highly essential. For ducklings, chicken-embryo attenuated strain can be given through oral/aerosol/intranasal/foot-web stick/intramuscular route; but its effectiveness is highly variable depending on maternal antibodies, time and severity of exposure to the virulent virus.

19. Duck plague (Duck virus enteritis)

19.1. Etiology

Herpes virus

19.2. Transmission

Direct contact (aquatic environment) and indirect contact with contaminated environment.

19.3. Symptoms

In young ducks (7 weeks) the symptoms are dehydration, weight loss, blue beaks and often a blood-stained vent. In breeder ducks the symptoms include sudden high and persistent mortality, prolapse of penis (in dead males), drop in egg production, photophobia associated with half closed pasted eyelids, inanition, thirst, droopiness, ataxia, nasal discharge and diarrhoea. Affected birds cannot move and if forced to do so exhibit tremors of the head, neck and body. Birds fall with drooping head and outstretched wings.

19.4. Morbidity and mortality

Morbidity and mortality range between 5 and 100%, especially in young ducklings.

19.5. Lesions

Primary gross lesions are of vascular damage i.e., tissue hemorrhages (petichiae and ecchymoses or extravasations on or in myocardium and other visceral organs and supporting structures like mesentry and serosal investments) and presence of free blood in body cavities. Pericardium, especially in breeders, may show a paint-brushed appearance because of hemorrhages. Hemorrhages from ovary may fill the abdominal cavity. Intestines and gizzard are often filled with blood. Mucosal eruptions are seen in digestive tract and patchy diphtheritic membrane may also be formed in esophagus. All lymphoid organs are affected and show petichiae.

In ducklings, Bursa of Fabricius is intensely reddened with pinpoint yellow areas on the reddened surface upon opening; later on, it'll be filled with white coagulated exudate. At the beginning, liver appears pale copper color with speckled appearance and later on turns dark bronze or bile-stained without hemorrhages but with white foci on the surface.

In geese, the intestinal lymphoid discs become hemorrhagic giving button-like ulcers.

19.6. Diagnosis

Diagnosis is very easy based on lesions.

19.7. Prevention and control

The disease can be controlled by chicken-embryo-adapted vaccine 0.5ml s/c or i/m given to ducks over 2 weeks of age; the vaccinal virus does not spread by contact and it is not excreted by the vaccinated birds to a great degree.

20. Blue comb disease (Corona viral enteritis)

20.1. Etiology

Corona virus

20.2. Transmission

Transmitted by personnel, equipment and vehicles. Free-flying birds at this mechanical carriers; recovered birds excrete the virus in feces for several months.

20.3. Symptoms

Young poults show sudden symptoms like hypothermia, anorexia, frothy and watery droppings, chirping and tendency to seek heat. Growing turkeys suddenly show drop in feed and water intake, darkening ahead and skin with tucking of skin under the crop, watery droppings and hypothermia. Breeders exhibit similar symptoms along with drop in egg production with chalky egg shells.

20.4. Morbidity and mortality

Morbidity is nearly 100% while mortality ranges from 5 to 50% depending on age and provision of supplemental heat.

20.5. Lesions

Important lesions are contents of duodenum, jejunum and ceca being watery and geseous, ceca distended with watery, yellowish-brown contents having fetid odor. Small petichiae may be seen in the intestinal mucosa, breast muscle usually dehydrated in the carcass generally emaciated. Spleen may be subnormal and urate deposits are seen in kidneys and ureters. Pancreas may appear chalky with whitish areas. Supplemental heat in brooder house along with antibiotics will reduce mortality from secondary infection.

20.6. Treatment, prevention and control

A mixture of calf milk replacer (2.5 kg/100 lit), KCl (100 g/100 lit) and antibiotic (20 g/100 lit) for 4-5 d followed by untreated water and repeat medication for an additional 4-5 d. No vaccine is available and depopulation and disinfection before repopulation after a few weeks is recommended.

21. Turkey viral hepatitis

This is an Adenovirus infection which is antigenically related to duck hepatitis virus. The infection is confined to turkeys and is transmitted by directand indirect contact; trans-ovarian route is also suspected.

The infection goes mostly subclinical and becomes apparent under stress with sudden death in normal birds, reduced production, fertility and hatchability in breeders stock.

Mortality is low but can reach 25%. The disease is normally recorded in birds less than six weeks of age.

Gross lesions are focal, grey, sometimes depressed lesions on liver which often coalesce with vascular congestion and focal hemorrhage. If pancreas is involved, grey-pink circular areas often extending across a lobe are seen. These lesions are pathognomonic.

The disease can be controlled by proper management to minimize stress.

22. Turkey meningoencephalitis

This is a Group B Arbovirus infection of turkeys which is thought to be transmitted through mosquitoes and ornithophilic arthropods.

The symptoms are seen in turkeys taste above 10 weeks showing progressive paresis and paralysis of limbs and neck thereby making them reluctant to move.

Mortality may reach 50% and the disease can be controlled by a live-attenuated vaccine in endemic areas.

Chapter **65**

Protozoan Diseases

1. Coccidiosis

This is the most important protozoan disease causing greatest financial loss, especially in poultry. Very mild infections causing weight-loss, reduced food consumption and alterations in hematological parameters is called coccidiasis. Coccidiosis indicates a flock with severe economic losses.

1.1. Genera of coccidia

Oocysts of various genera consist of a thickened outer wall and a rounded mass of nucleated protoplasm. Just before sporulation, the protoplasm differentiates into sporocysts which, in turn, contain sporozoites. Depending on the number of sporocysts and sporozoites, the various genera of coccidia can be differentiated (Table 65.1). Most of the genera of coccidia are host-specific whereas *Tyzzeria* and *Cryptosporidium* seem to exhibit little or no host-specificity. The most important genus affecting various species of poultry is *Eimeria*, all members of which are highly host-specific.

Table 65.1. Number of sporocysts and sporozoites

Genus	Number of sporocysts	Numbers sporozoites per sporocysts
Eimeria	4	2
Isospora	2	4
Wenyonella	4	4
Tyzzeria	----	8 (free)
Cryptosporidium	----	4 (free)

Oocysts to become infective, must be shed in the feces and sporulate. Sporulation involves nuclear and cellular division to develop sporozoites. Time required for these changes is referred to as sporulation time. Time lapse between ingestion of oocysts and appearance of new generation of oocysts in the feces is called prepatent period. Very small proportion of millions of oocysts produced by a bird becomes infective since several environmental conditionsare essential for their sporulation. They include sufficient moisture and oxygen, optimum temperature (28°C) and the like.

The seven-day life-cycle of *Eimeria tenella* is considered as typical for all members of the genus *Eimeria* (Table 65.2).

Table 65.2. Life cycle of *E.tenella*.

Day	Developments
1st day	Mechanical action in gizzard and enzyme activity results in destruction of wall of the ingested oocyst as well as the sporocyst contained in the oocyst. Therefore, sporozoites are released. This is facilitated by the action of trypsin and the bile in the intestines.
2nd day	Sporozoites reach ceca along with the digestive contents and penetrate surface epithelium into the *lamina propria* and assume a round shape. Within the epithelial cells, the sporozoites grow causing hypertrophy of the host-cells. The nucleus of the parasite divides by multiple fission (referred to as schizogony or merogony) to form schizont or meront (I generation). The meronts elongate to form merozoites and arrange in layers (like a section of an orange) inside the schizont.
3rd day	The schizont ruptures releasing the merozoites (I generation) which penetrate into epithelial cells.
4th day	Penetration of the merozoites into the epithelial cells and formation of II generation schizont and merozoites.
5th day	Release of II generation merozoites causing severe denudation of the epithelium and subsequent bleeding. Bleeding begins by the end of 4th day itself and maximum bleeding occurs during the 5th day.
6th day	Beginning of sexual cycle – merozoites differentiate into microgametes (male) and macrogametes (female) and they unite to form oocyst (microgamete penetrates the epithelial cell containing macrogamete to form an oocyst inside the epithelial cell).
7th day	Oocysts are shed in the feces and bleeding nearly stops since there is not much damage to the epithelial cells.

1.2. Transmission

Ingestion of sporulated oocysts is the only route of transmission. Mechanically, shoes, hands and clothing serve to transfer the oocysts.

1.3. Coccidiosis of chicken

Table 65.3. Coccidiosis of chicken

Species	Zone of intestine affected	Lesions	Remarks
E.acervulina	Duodenum and upper jejunum	Whitish round lesions, sometimes in ladder-like streaks. In severe cases, plaques coalescing, thickening of intestinal wall*	

Species	Zone of intestine affected	Lesions	Remarks
E.brunetti	Lower small intestine, rectum and proximal area of ceca	Coagulative necrosis, mucoid bloody enteritis *. Blockade of digestive system may occur. Tips of villi most affected	Mortality 10 to 30%
E.hagani	Upper intestines	Catarrhal inflammation and hemorrhagic spots	Species is of doubtful validity
E.maxima	Middle of intestine above and below Mackle's diverticulum	Hemorrhagic enteritis, thickening and some ballooning	Oocysts have distinctive brownish walls *
E.mitis	Lower half of intestines and tubular portion of ceca	No distinctive lesions, mucoid exudate *	Subspherical, small oocyst; subclinical coccidiosis
E.mivati	Duodenal loop to rectum and ceca	Light infection - rounded plaques of oocysts *, heavy infection – thickened walls, coalescing plaques	
E.necatrix	Mid-intestines above and below Mackle's diverticulum	Massive ballooning *, white spots, petichiae, mucoid blood-stained exudate *, parasitic granuloma can be seen in lamina propria *	Oocysts develop in ceca, produces fewest oocysts
E.praecox	Upper third of the intestines	No lesions, mucoid exudate*, petichiae on the mucosa	High morbidity, lowest prepatent period (83 h) *
E.tenella	Ceca and adjacent areas of the digestive tract	Beginning – hemorrhages into the lumen. Later – thickening, whitish mucosa, cores of clotted blood *	Mortality highest (20% or more). Characteristic odor

* Pathognomonic, of diagnostic value — *Adopted from* : Saif, 2003

1.3.1. Symptoms and lesions

The birds will be pale, dull, keep away from the flock in a corner, pass bloody droppings, sleep while standing, show ruffled feathers, retract their head and finally show emaciation. Although the *Eimeria* species are highly host-specific, *E.tenella* has been reported from other species of birds. Day-old chicks are susceptible but may develop minimal infections because oocysts are not excysted in very young chickens. Usually, birds less than 11 d of age do not suffer from coccidiosis.

1.3.2. Diagnosis

Typical symptoms and gross lesions are sufficiently helpful in making a good diagnosis. However, microscopic examination for histopathological changes in the intestinal epithelium and also for different stages of light-cycle of the parasite and fecal examination for presence of oocysts are highly useful for confirmation.

1.3.3. Prevention and control

The life-cycle of coccidia may be intercepted a three stages - destruction of oocysts (at 1st and 7th d or by medication on 5th and 6th d), building immunity (when the merozoites are released, that means, during 2nd to 6th d) and by using anticoccidial drugs (from 1st to 6th d). Nine different approaches have been suggested which can be used singly or in suitable combinations depending on the situations:

1. Genetically resistant birds - Although it is possible to develop birds with genetic resistance, it is not economically-feasible.
2. Nutrition - During coccidiosis, hypovitaminosis A whereas after an outbreak of coccidiosis, vitamin A supplementation hastens recovery. Reduced protein levels in feed (5 to 10% of the ration) decreases *E.tenella* outbreaks presumably by decreasing trypsin activity thereby hindering release of sporozoites. Supplementation of selenium and Vitamin E decreased mortality and enhanced immune response. Birds fed on diets with corn or corn-fermentation solubles were more susceptible to coccidiosis.
3. Quarantine and depopulation - Since immunological methods to detect carrier-birds are not available, isolation, quarantine and sometimes depopulation may be essential to break the cycle.
4. Sanitation - Aimed at killing of oocysts, interrupting the life-cycle and preventing an outbreak. However, it is practically impossible to eliminate the oocysts. Notwithstanding these, litter management to avoid damp litter is extremely important in preventing outbreaks. Litter moisture must not exceed 30%.
5. Reducing exposure - Cage-rearing can nearly eliminate exposure of birds to the oocysts. But, initial cost, breast-blisters and other considerations have prevented the large-scale cage-rearing of birds.
6. Immunity a) Natural - Birds after obtaining few oocysts from litter develop a coccidiasis-like disease and localized tissue immunity. But, the duration of protective immunity is variable. However, as long as there are viable oocysts in litter, there will be reinforcement of immunity. b) Planned - Most extensively used program was feeding of smaller number of 8 species of oocysts, through water or feed, during 1st week of age. Attenuated oocysts (oocysts cultivated in egg embryos) are used more effectively for planned exposure in Western countries.

7. Chemotherapy – Treatment of the disease after the symptoms have appeared in the flock has severe limitations because most of the coccidicidals act very early in life-cycle of the parasite (that means before symptoms precipitate) but the medication would have started too late. Hence, soon it was recognized that chemotherapeutics had better be preventive measures.
8. Treatment – Some coccidicidals which are effective as late as 5th and 6th d of life-cycle of the parasites are still used to reduce mortality and morbidity (Table 65.4). For example – Sulfa drugs preferably given in water.
9. Prevention – Birds, especially broilers, reared on litter are universally kept on a preventive anticoccidial drug program; the drugs are commonly mixed in feed. However, drug residues in tissues posed certain problems and now, definite information on withdrawal period is available on most of the drugs used (Table 65.5)

1.3.4. Treatment

Table 65.4. Anticoccidial drugs – effective late in life-cycle (5th and 6th d)

Drug	Route	Content and duration	Pre-slaughter withdrawal period (d)
Sulfamethazine	Water	0.10%, 2d; 0.05%, 4d	10
Sulfaquinoxaline	Feed	0.10%, 2-3 d on; 3d off followed by 0.05% 2d on; 3d off and so on	10
	Water	0.04% 2d on; 3d off; 2d on and so on	10
Nitrofurazone	Feed	0.11%, 5d	5
	Water	0.00802%, 5d	5
2,4-diamino 5-(*p*-chlorophenyl) 6- ethyl-pyramidine + Sulfaquinoxaline	Water	0.0015% and 0.05%, respectively; 2-3 d on, 3d off and so on	5
Chlortetracycline	Feed	0.022%+ 0.8% calcium; not more than 3d	----
Oxytetracycline	Feed	0.02% + 0.18 to 0.55% calcium; not more than 5d	3
Amprolium	Water	0.012 to 0.024% for 3 to 5d; 0.006% for 1 to 2 weeks	----
Sodium Sulfachloropyrazine monohydrate	Water	0.03%, 3d	4
Sulfadimethoxine	Water	0.05%, 6d	5

1.3.5. Prevention

Table 65.5. Preventive coccidicidals

Drug	Approved level in feed (%)	Pre-slaughter withdrawal period (d)
Sulfaquinoxaline	0.015 to 0.025	10
Nitrofurazone	0.0055	5
Arsanilic acid (Sodium arsanilate)	0.04 for 8d	5
Butynorate	0.0375 (turkeys)	28
Nicarbazine	0.0125	4
Furazolidone	0.0055 to 0.011	5
Nitromide	0.025	5
Oxytetracycline	0.022	3
Amprolium	0.0125 to 0.025	-- --
Chlortetracycline	0.022	?
Zoalene	0.004to 0.0125	5
Amprolium with Ethopabate	0.0125 and 0.0004 to 0.004, respectively	-- --
Buquinolate	0.00825	-- --
Decoquinate	0.003	-- --
Sulfadimethoxine	0.0125	5
Monensin	0.01 to 0.0125	3
Robenidine	0.0033	5
Losalocid	0.0075 to 0.0125	3
Salinomycin	0.004to 0.0066	-- --

However, the following two aspects of use of anti-coccidials in poultry deserve greater attention:

1. Resistance to anti-coccidials and
2. Residues of anticoccidials in poultry products

1.3.5.1. Resistance to anti-coccidials

Can develop in one step as in case of quinolines (Deccox) or step-by-step over a period of time (generations) as in case of ionophores (Monensin, Salinomycin) and nicarbazine (Nicarzin). Resistance in a farm depends on various factors and in most cases, the process occurs in the following way:

1. In the beginning, almost no cases of subclinical coccidiosis
2. Subsequently, regular subclinical, chronic coccidiosis without serious effect on FCR or growth
3. High incidence of subclinical, chronic coccidiosis with significant effects on FCR and growth; and finally

4. None of the anti-coccidial drugs can control the disease and regular and severe losses occur due to the disease.

Interrupted use of anti-coccidials is very important. Some anti-coccidial drugs are preferred to be used during certain seasons (Table 65.6); for instance, ionophores, which has the water-sparing effect, is better to be used when wet litter is likely (such as during summer months).

1.3.5.2. Residues in poultry products

Some anti-coccidials depressed growth; such as nicarbazine. They have to be withdrawn well in advance to allow for compensatory growth. Withdrawal details are given in Chapter "Composition and nutritive value of eggs" and "Composition and nutritive value of poultry meat".

Table 65.6. Suggested anti-coccidial use for chicken

Summer period (6 to 8 months)	Winter period (4 to 6 months)
Annual change over of ionophore	Chemical products used in either a) a shuttle programs with nicarbazine or b) a rotational program
Salinomycin	Coyden
Avatec	Cycostat
Monensin	Procox, Stenerol

1.4. Coccidiosis of turkeys

Table 65.7. Coccidiosis of turkeys

Species	Zone of intestine parasitized	Symptoms and lesions	Shape index (length ÷ width)	Minimum prepatent period (h)	Minimum sporulation time (h)	Remarks
E. adenoiedes	mainly ceca and also lower small intestine and rectum	Loose whitish cecal cores *. Feces whitish to orange colored	1.54 *	103	24	Highly pathogenic*
E. Dispersa	Midgut area and sometimes any part from duodenum to rectum including ceca	Mild diarrhoea, reduced weight gain. Cream-colored changes in serosa of duodenum	1.24 *	120 *	36	Mildly pathogenic; more pathogenic to Bobwhite quail

Species	Zone of intestine parasitized	Symptoms and lesions	Shape index (length ÷ width)	Minimum prepatent period (h)	Minimum sporulation time (h)	Remarks
E. gallopavonis	Area posterior to Mackle's diverticulum; most severe in rectum	Edema and ulceration of mucosa of ileum, blood in feces	1.52 *	105 *	15 *	Highly pathogenic*
E. innocua	Area between duodenum and lower ileum	No gross lesions, slight reduction in body weight	1.07 *	114 *	< 45 *	Not pathogenic
E.meleagridis	Ceca	Cream colored caseous cecal cores	1.34 *	110	24	Not pathogenic*
E.meleagrimitis	Duodenum and upper jejunum	Enlargement and congestion of duodenum. Petichiae in other areas. Hemorrhages rarely	1.17 *	103	18 *	Most pathogenic than all species *
E. subrotunda	Area anterior to Mackle's diverticulum	No gross lesions	1.10 #	95 #	48 #	Not pathogenic

1.4.1. Treatment

Amprolium (0.012 to 0.024%), Sulfaquinoxaline (0.025%), Sulfadimethoxine (0.05%), Sulfachloropyrazine (0.005%), Sulfaquinoxaline (0.005%) + Pyrimethamine (0.0015%) + Sulfamethazine (0.1%). Withdrawal periods are similar to those with respect to chicken.

1.4.2. Prevention and control

Sulfaquinoxaline (0.0175%, withdrawal period 10d), Amprolium (0.015 to 0.25%, no withdrawal period), Butynorate (0.0375%, withdrawal period 28d), Zoalene (0.0125 to 0.01875%, no withdrawal period).

Combination of drugs to control of both coccidiosis and histomoniasis having recently been approved by the Federal Drug Administration (FDA). They are - Zoalene 0.0125% + Carbarsone 0.025 to 0.037%; Sulfadimethoxine 0.00625% + Ormetoprim 0.00375%+ Ipronidazole 0.00625%.

1.5. Coccidiosis in other species

1.5.1. Ducks

E.perniciosa, T. perniciosa and *W. philiplevinei*

1.5.2. Pigeon

E. labbeana

1.5.3. Geese

E. truncata (renal coccidiosis), *E. anseris* and *E. nocens* (causing intestinal coccidiosis).

1.6. Atoxoplasmosis

(The Merck Veterinary Manual, 2003)

1.6.1. Etiology

Atoxoplasma spp are host-specific coccidian protozoa with a direct life cycle that includes an extraintestinal, systemic phase.

1.6.2. Hosts

Passerine birds are affected, especially canaries, finches, sparrows, and species of the Sturnidae family (starlings, mynahs).

1.6.3. Transmission

Fecal-oral via ingestion of oocysts in droppings from infected birds. Infected canaries can shed oocysts for at least 2 yr.

1.6.4. Symptoms

Include listlessness, diarrhea, and anorexia.

1.6.5. Morbidity and mortality

Mortality can be high (up to 80%) in young birds.

1.6.6. Lesions

The enlarged liver is often visible through the abdominal wall, especially when moistened with alcohol. In acutely affected young birds, there is marked hepatomegaly and splenomegaly, often with multifocal necrosis.

1.6.7. Diagnosis

Diagnosis is difficult in chronically infected older birds. Only very few parasites are present in blood and tissues, and oocysts are shed intermittently, although sometimes in high numbers. Typical, nearly spherical oocysts averaging 19 × 21 μm are present in droppings of affected canaries. Buffy coat and organ smears are preferred, but negative findings should not be interpreted as uninfected. Hepatic and splenic enlargement persists because of infiltrations with high numbers of host mononuclear cells.

1.6.8. Treatment and control

There is no known effective treatment. Anticoccidial drugs do not affect parasites in the tissues. Use of antimalarial drugs has been suggested. Good management procedures such as isolation of age groups and scrupulous cleanliness, particularly daily cleaning before oocysts sporulate, help control the disease. Disinfectants have little effect on oocysts.

2. Histomoniasis (Black head)

2.1. Etiology

Histomonas meleagridis; can exist in non-amoeboid state or in an amoeboid state (as pleomorphic state) it has a single stout flagella.

2.2. Hosts

Turkey is the most susceptible host since most of the infected turkeys die. Chickens show a mild form of disease; many other species are also infected.

2.3. Transmission

Chicken which harbor the cecal worms can protect the heterakids and hence act as reservoirs or vectors of histomoniasis in turkeys especially when turkeys have access to droppings chicken. The intermediate host is cecal worm (*Heterakis gallinarum*) and several species of earthworms. Cecal-worm eggs contained the protozoan which is transmitted to the next generation. The mode of transmission of histomonads into the cecal worm is still inconclusive. Earthworms served as a means for collection and concentration of heterakid eggs from the poultry yard environment.

2.4. Symptoms

The incubation period is 7 to 12d. In turkeys, the disease is characterized by sulfur-colored feces, drowsiness, drooping of wings, stilted gait, closed eyes, heads bent close to the body and anorexia. Sometimes, head may be cyanotic (hence the name black head). In chicken, the only sign may be bloody feces resembling cecal coccidiosis.

2.5. Morbidity and mortality

Turkey confined to areas contaminated by chickens had a morbidity of 89% and mortality of 70%.

2.6. Gross lesions

Characteristic lesions are seen in turkeys. Ceca are filled with serous and hemorrhagic exudate which may turn caseous and cheesy as the disease progresses. Ulceration and perforation of ceca followed by generalized peritonitis may occur.

Liver shows depressed area of necrosis circumscribed by a raised ring; this is often the diagnostic lesion. White, rounded, necrotic areas may be seen in lungs, kidney, spleen and mesentry. At later stages, the necrotic area on liver may be replaced by scars.

2.7. Diagnosis

Gross lesions are of high diagnostic value. Microscopic demonstration of histomonads in specimens taken from freshly-filled birds.

2.8. Prevention and control

Turkey and chicken must be reared separately. Reducing access to earthworms is also useful in reducing incidence of the disease.

2.9. Treatment

Carbarsone (0.025 to 0.037%), Dimetridazole (0.015 to 0.02% or 0.06 to 0.08%) and Nitarsone (0.01875%) can be used in feed. However, the drug has to be withdrawn 4 to 5d before slaughter.

3. Trichomoniasis

3.1. Etiology

Trichomonas gallinae; a flagellated (4 anterior flagellae), pear-shaped parasites.

3.2. Hosts

Main host is pigeon.

3.3. Transmission

From parents to squabs or by contamination of feed and water by the infected material (oral fluid) in case of domestic fowl.

3.4. Symptoms

Depression, ruffled feathers, inanition, emaciation and death.

3.5. Gross lesions

Circumscribed caseous areas on the mucosal surface of buccal cavity, sinuses, pharynx, esophagus and crop (occasionally, proventriculus and conjunctiva). Solid (initially caseous), spherical to circular whitish to yellow masses in liver. The digestive tract below proventriculus is not invaded.

3.6. Diagnosis

Gross lesions are of high diagnostic value. Microscopic demonstration of the organisms in wet-smears from the mouth or crop to confirm the disease.

3.7. Treatment

Dimetridazole (0.05% in water for pigeons, 0.015 to 0.02% in turkey feed or 0.06 to 0.08% in water for turkeys). The drug has to be withdrawn 4 to 5d before slaughter.

3.8. Prevention and control

The infected birds must be removed from the flock.

4. Hexamitiasis

4.1. Etiology

Hexamita meleagridis (Spironucleus meleagridis)

4.2. Hosts

Main host is turkeys.

4.3. Transmission

Contamination of feed and water by the infected material.

4.4. Symptoms

Listlessness, huddling, yellowish watery diarrhoea; at later stages, conversions and death.

4.5. Gross lesions

Catarrhal enteritis, watery intestinal contents and occasionally, yellowish discoloration of liver.

4.6. Diagnosis

Symptoms and lesions along with microscopic demonstration of the organisms in fresh duodenal contents.

4.7. Treatment

Butynorate (0.0375%), Chlortetracycline (0.0055%) and Furazolidone (0.022%) alone or in combination with Oxytetracycline (0.044%), in feed.

4.8. Prevention and control

Removal of carrier birds and hygienic management.

5. Leucocytozoonosis

5.1. Etiology

Leucocytozoa are blood-parasites.

5.2. Hosts

Chicken (*L. caulleryi, L. subrezi* and *L.schoutedini*), turkeys (*L. smithi*) and ducks and geese (*L.simondi*).

5.3. Transmission

Primarily by insects (black-fly); the life-cycle consists of reproduction by schizogony (merogony) in tissue cells, gametogony in RBCs or WBCs and sporogony in insects.

5.4. Symptoms

More severe in ducks and geese. General symptoms include inappetance, dullness, ruffled feathers, dyspnea and anemia; there may be leucocytosis.

5.5. Gross lesions

There may be splenomegaly and liver degeneration and hypertrophy.

5.6. Diagnosis

Microscopic demonstration of intracellular parasite in blood smears.

5.7. Treatment

Pyrimethamine (1 ppm), Sulfadimethoxine (10 ppm) and Clopidol (0.0125 to 0.0250%).

5.8. Prevention and control

Black-fly must be eradicated.

6. Cryptosporidiosis

6.1. Etiology

Cryptosporidia are related to the coccidia, but much smaller poorly host specific when compared with coccidian. Bird strains do not infect mammals very well, and *vice versa*. *Cryptosporidium baileyi* can cause respiratory disease in chickens and turkeys and infections of the hindgut and cloacal bursa in chickens, turkeys, and ducks. *C. meleagridis* infects both chicken and turkeys.

6.2. Transmission

The oocysts are excreted already sporulated in the faeces and infection occurs by inhalation and ingestion.

6.3. Symptoms

Snick, cough, swollen sinuses, body weight loss and diarrhoea.

6.4. Lesions

Sinusitis, airsacculitis and pneumonia.

6.5. Diagnosis

Identification of the parasites attached to the epithelium by microscopic examination (smears histopathology acid-fast staining).

6.6. Treatment

Unfortunately there is currently no known effective treatment in poultry. If other disease processes are complicating the situation (e.g. coli-septicemia) there may be benefit in medicating for these.

6.7. Prevention

The oocysts of cryptosporidia are extremely resistant to chemicals. Steam cleaning is effective in reducing infection as oocysts are inactivated above about 65°C.

7. Trichomoniasis, Canker, Frounce

7.1. Etiology

Trichomonas gallinae

7.2. Hosts

Pigeons and doves (canker), raptors (frounce), turkeys and chickens.

7.3. Transmission

Transmission is via oral secretions in feed and water, and crop milk.

7.4. Symptoms

Mouth open, drooling and repeated swallowing movements, loss of condition, watery eyes in some birds and nervous symptoms (rare).

7.5. Morbidity and mortality

It has variable pathogenecity. Morbidity is high and mortality varies but may be high.

7.6. Lesions

Yellow plaques and raised cheesy masses in mouth, pharynx, esophagus, crop and proventriculus. Raptors may have liver lesions.

7.7. Diagnosis

Lesions, large numbers of protozoa in secretions. The disease has to be differentiated from Pox and candidiasis.

7.8. Prevention

Eliminate stagnant water, eliminate known carriers, and add no new birds.

7.9. Treatment

Previously approved products included dimetridazole, nithiazide, and enheptin. Organic arsenical compounds (where approved) and some herbal products may be of some benefit in managing this problem in food animals.

8. Hexamitiasis

8.1. Etilogy

H meleagridis is spindle-shaped protozoan, having six anterior and two posterior flagella.

8.2. Hosts

Turkeys, pheasants, quail, chukar partridges, and peafowl. Natural infection has not been observed in chickens.

8.3. Transmission

It is transmitted directly by ingestion of infected feces. Encysted hexamitids may be more important in transmission than free flagellates. Many survivors become carriers and shed parasites in their droppings.

8.4. Symptoms

Hexamitiasis produces acute, catarrhal enteritis. The nonspecific signs include watery diarrhea, dry unkempt feathers, listlessness, and rapid weight loss despite the fact that the birds continue to eat. Birds may die in convulsions.

8.5. Morbidity and mortality

Highest mortality occurs in birds 1-9 wk old.

8.6. Lesions

Bulbous dilatations of the small intestine (especially duodenum and upper jejunum) filled with watery contents are characteristic.

8.7. Diagnosis

Microscopic examination of scrapings of the duodenal and jejunal mucosa for flagellate protozoa that move with a rapid, darting motion (in contrast to the jerky motion of trichomonads).

8.8. Prevention and control

Because many birds remain carriers, breeder turkeys and poults should be raised on separate premises if possible, preferably with separate attendants. Wire platforms should be used under feeders and waterers. Pheasants and quail may also be carriers.

Furazolidone (0.0055% in the feed continuously), oxytetracycline (0.22% in the feed for 2 wk), chlortetracycline (0.022-0.044% in the feed for 2 wk), and butynorate have been used to prevent and treat hexamitiasis. Treatment does not substitute for adequate sanitation and management programs.

9. Aegyptianellosis

9.1. Etiology

Aegyptianella spp, a rickettsia in the family Anaplasmataceae. Organisms appear as round, "signet-ring" (0.5-4 μm) or irregular oval bodies in RBC often lateral to the nucleus.

9.2. Hosts

A variety of avian species, including chickens, turkeys, G. fowl, quail, pigeons, crows, waterfowl, ratites, passerines, and psittacines.

9.3. Transmission

Ticks, especially *Argas* spp, transmit the organism; infection can also be reproduced by blood inoculation.

9.4. Symptoms

Aegyptianellosis is an acute, tick-borne, febrile disease.

In endemic areas, infection is mild or asymptomatic. Ruffled feathers, anorexia, droopiness, diarrhea, fever, jaundice, and high mortality in younger birds occur in introduced or otherwise susceptible birds. Anemia, enlargement of the liver and spleen, enlarged discolored kidneys, and pinpoint serosal hemorrhages are seen. Infestation with larval argasid ticks and occurrence of borreliosis (Spirochetosis) may accompany the disease.

9.5. Treatment

Tetracyclines, especially doxycycline, are effective in controlling the disease and eliminating the organism from chronically infected birds. Tick control is an important adjunct to treatment.

10. Plasmodium infection

10.1. Etiology

Plasmodium spp,

10.2. Hosts

Plasmodium spp are not host-specific, infect a wide variety of domestic and wild birds in most areas of the world and can cause high losses.

10.3. Transmission

Asymptomatic infections in endemic or introduced birds can be spread via mosquitoes. Invertebrate hosts are ornithophilic mosquitoes, usually *Culex*, *Culiseta*, or *Aedes* spp.

10.4. Symptoms

Infection with *Plasmodium* spp may be nonclinical or cause illness characterized by weakness, lassitude, dyspnea, anemia, abdominal distension, ocular hemorrhage, and death. Death results from blockage of capillaries in the brain or other vital organs by exoerythrocytic schizonts in endothelial cells.

10.5. Lesions

Liver and spleen are markedly enlarged and often discolored (dark brown to black).

10.6. Mictoscopic examination of blood smears

Pigmented parasites including schizonts occur in RBC, and both immature and mature RBC can be infected. In birds that die acutely, organisms may be sparse or absent in blood, but numerous schizonts can be found in capillaries by examining squash or impression smears of brain, lung, liver, and spleen.

10.7. Treatment and Control

Chemotherapy is variably effective in treating infected birds or flocks. Chloroquine (5-10 mg/kg) potentiated with primaquine (0.3 mg/kg), or chloroquine in drinking water (250 mg/120 l) has been used. Grape or orange juice can disguise chloroquine's bitterness. Other antimalarial drugs have been used with success. Relapse may occur after treatment.

Chapter 66

Parasitic Diseases

Members of Classes *Insecta* and *Arachnida* under Phylum *Arthropoda* are the common external parasites. The important parasites belonging to Class *Insecta* are lice, flies, bugs and fleas whereas under Class *Arachnida* are ticks and mites.

It is convenient to divide the parasites into three broad groups with subgroups. These groups are ectoparasites, hemoparasites and endoparasites. The list only includes those species which have pathogenic and economic importance. Most of the external parasites are identified by their characteristic morphology and standard pictorial keys. Hemoparasites are identified by the microscopic examination of blood of the affected birds. Endoparasites are generally identified by identifying their eggs in excreta and/or the identification of the parasite itself in the excreta or while conducting necropsy examination.

1. Ectoparasites

Poultry are infested with a variety of insects and mites that live on the skin and feed on skin debris, feathers, and blood. This activity can lower growth rates, reduce egg production, and, if the infestation is heavy, cause debilitation and death of the birds. In addition, poultry houses can be a breeding source for a variety of flies that may be vectors of poultry diseases or, at the least, be a nuisance which may bring about litigation from neighbors.

The most common parasitic arthropods of poultry are ticks, lice and mites, fleas, bedbugs and chiggers (Table 66.1).

1.1. Fowl Ticks

Argas persicus is found worldwide in tropical and subtropical countries and is the vector of *Borrelia anserina* (Spirochetosis; see Chapter "Bacterial diseases"). In the USA, the *A persicus* complex has been divided to include *A sanchezi* and *A radiatus* in addition to *A persicus* (The Merck Veterinary Manual, 2003).

1.1.1. Features

These ticks are particularly active in poultry houses during warm, dry weather. All stages of life-cycle may be found hiding in cracks and crevices during the day. Larvae can be found on the birds because they remain attached and feed for 2-7 days. Nymphs and adults feed at night in ~15-30 min. Nymphs feed and molt several times before reaching the adult stage. Adults feed repeatedly, and the females lay 50-100 eggs after each feeding. Adult females may live ≥ 4 yr without a blood meal.

1.1.2. Symptoms

In addition to being vectors of some poultry diseases (spirochetosis, aegyptianellosis), fowl ticks produce anemia (most important), weight loss, depression, toxemia, and paralysis. Egg production decreases. Red spots can be seen on the skin where the ticks have fed. Because the ticks are nocturnal, the birds may show some uneasiness when roosting. Death losses are rare, but production may be severely depressed.

1.1.3. Prevention and control

After houses are cleaned, walls, ceilings, cracks, and crevices should be treated thoroughly (using a high-pressure sprayer) with carbaryl, coumaphos, malathion, stirofos, or a mixture of stirofos and dichlorvos. Cracks and crevices should be filled in.

1.2. Poultry Lice

Poultry lice are small (less than 0.4 cm) wingless insects with chewing mouthparts. The most common are brown chicken lice and chicken body lice. Less important are large chicken lice, shaft lice, chicken head lice, fluff lice, and several other species which are rarely present.

Poultry lice chew dry skin scales and feathers, but do not suck blood. Irritation from louse feeding and movement on birds causes appetite loss, weakened condition and susceptibility to diseases. Egg production is reduced, and heavily infested birds refuse to eat and gradually lose weight. Lice can be observed moving on the skin when feathers are parted, especially around the vent, head and under wings.

Chicken body lice (*Menacanthus stramineus*), can reduce egg production in caged layer hens. The skin of infested birds becomes irritated and red, with formation of localized scabs and blood clots. In addition to feeding on skin fragments, feathers, and debris, lice can attack young quill feathers, feeding on blood. Although naturally infected with the eastern encephalomyelitis virus, it is not considered an important vector. Adult chicken lice are flat-bodied, yellowish colored, and 0.2 cm long with chewing mouthparts.

1.3. Poultry Mites

Mites vary in size (0.2 to 0.4 cm long) and structure, have eight legs, and have mouthparts on the anterior of the body. Usually there are no clearly defined body divisions. Chicken mites are small and grey or yellow in color, but darken after filling with blood. Northern fowl mites remain on poultry. They are small and red or brown. Several kinds of mites attack poultry. The most common are chicken mites and northern fowl mites. Occasionally scaly-leg mites are a problem.

Table 66.1. Ectoparasites of importance in poultry production

Parasite	Hosts	Predilection site
Ticks		
Argas persicus	Chickens, Turkeys, Pigeons, Ducks, Geese	Skin
Mites		
Dermanyssus gallinae	Chickens, Turkeys, Ducks, wild birds	Skin
Ornithonyssus sylviarum		
Ornithonyssus bursa		
Cnemidocoptes mutans	Chickens, Turkeys	under the skin on legs, occasionally on comb and wattles
Flea		
Echidnophaga gallinacean	Chickens and other birds	Head
		Source : Permin and Hansen, 1998

Chicken mites feed at night. During the day they remain in cracks around roosts and interior portions of poultry houses. At night, they feed on the birds as they roost or nest. Chicken mites and northern fowl mites suck blood, resulting in emaciation and lowered egg production. Continued heavy infestations can kill the birds. Feathers of birds infested by Northern fowl mites are discolored by mite excrement and eggs, and the skin is scabby.

Scaly-leg mites burrow under the skin, especially on the lower legs and feet. Legs become scaly, swollen, and exude lymph fluid. Severely infested birds may be crippled or unable to walk.

1.3.1. Northern fowl mite (Feather mite)

Ornithonyssus sylviarum, is a very important external parasite of poultry with heavy populations capable of reducing egg production 10 to 15 percent. Mites can also annoy egg handlers and other persons.

Mites are often first noticed on eggs. Presence of mites is checked first on the vent, then tail, back and legs of layers. Feathers become soiled from mite eggs, cast skins, dried blood from feeding, and excrement.

The entire life cycle is completed on the bird and consists of the egg, larva, nymphal stages, and adult. The eight-legged adult is about 0.1 cm long and dark red to black in color. The entire life cycle can be completed under ideal conditions within a week.

With early detection, only some of the caged layers may need to be treated. Monitoring weekly at least 10 randomly selected birds from each cage row in the entire house will greatly assist control feather mites. Mite population will increase in cooler weather.

1.3.2. Chicken Mite

Dermanyssus gallinae, sucks blood from poultry at night and remains secluded during the day in cracks and crevices. When mites are numerous, weight gains and egg production can be reduced. These red and gray mites are difficult to see without a magnifying glass. The life cycle may be completed in 7 to 10 days during warm weather with inactivity during cold weather.

1.3.3. Scaly-leg Mite

The female *Knemidocoptes mutans* is small with a round body and short, stubby legs. These mites must be magnified to be seen, because they are only < 0.05 cm long. Young mites are at first six-legged, then metamorphose through two eight-legged nymphal stages.

The scaly-leg mite is distributed widely throughout the world. This mite attacks poultry, commonly chickens and turkeys. The scaly-leg mite also has been reported on pheasants, partridges, bullfinches, gold finches, and many passerine birds. Researchers suspect that wild birds transmit the mites to domestic flocks.

Females burrow under scales on the feet and legs of poultry and deposit eggs. They begin laying a short time after they burrow under the skin and continue to oviposit for about 2 months. Eggs hatch in about 5 days into six-legged larvae that soon molt into nymphs. Nymphs develop into mature males and immature females. The immature female transforms into a mature egg-laying female shortly after she is fertilized. The cycle to egg-laying female probably requires 10 to 14 days.

Burrowing beneath the scales (on feet and legs) results in formation of a powdery material which accumulates and binds into a scab of serum discharge. Affected feet and legs usually have red blotches. Glands in the mouthparts of mites may secrete an irritating fluid that causes the discharge and blotches. Eventually, the feet and legs may be covered with these crusts or scabs. Mites remain beneath the crusts in small oval vesicles. Irritation from mite infestation causes poultry to pick at the crusty formations. As formations extend over the feet and legs, they interfere with joint flexion and cause lameness. Severe infestations may cause loss of toes, loss of appetite, lowered egg production, emaciation, and death.

1.3.4. Depluming Mite

This mite, *Neocnemidocoptes laevis gallinae var. gallinae*, is similar to the scaly-leg mite, but is smaller and more oval.

Infestation occurs ubiquitous. Hosts include pigeons, pheasants, geese, canaries, and chickens. Many wild birds have been infested with this species or with closely related, unidentified species.

Development stages include egg, larva, nymph, male adult, and immature and mature female adult. The fertilized female begins depositing eggs within a few hours after starting to burrow and continues at 2- or 3-day intervals for about 2 months. Eggs hatch in about 5 d. The cycle from egg to egg-laying female requires 10 to 14 d. Less than 10 percent of the eggs mature into adults.

The depluming mite burrows into skin at the base of the feathers on the back, on the top of the wings, around the vent and on the breast and thighs. It causes intensive itching, often resulting in feather-pulling. Fowls may lose feathers over large areas of the body. Infestations, especially noticeable in spring and summer, may disappear in autumn.

1.4. Bedbugs

They are flat, wingless, blood-sucking insects that are about ½ cm long when fully grown. The common bedbug and several other closely related insects feed on poultry. They have a very distinctive pungent odor when crushed. Bedbugs feed at night, hiding and laying eggs behind insulation, in wall cracks, loose boards, nests and other dark areas during the day. At night they move to sleeping birds and suck their blood.

No specific symptoms except lowered performance. Small, dark fecal spots around cracks, roosts, and on chicken eggs frequently are observed. Bedbugs can be carried into poultry houses by other birds or into human dwellings where they become a pest of humans.

Cimex lectularius L., occasionally attacks poultry. It hides in cracks within the housing during the day and feeds mostly at night on blood while the host is asleep, causing small, hard, swollen, white welts that become inflamed and itch severely. It is rarely seen on poultry during daylight hours.

An infestation can sometimes be recognized by blood stains and dark spots of excreta. The adult is reddish-brown, oval-shaped, flattened and about 0.5 to 1.0 cm long. There may be three or more generations per year. There is no evidence that they spread disease.

1.5. Beetles

1.5.1. Darkling Beetle (Lesser meal worm, litter beetles)

Alphitobius diaperinus, is rapidly becoming more of a nuisance in the poultry operation. Large population of beetles sometimes migrate into nearby residence areas, especially during litter clean-out time. Although beetles can fly up to a mile, most crawl at night from litter disposed in neighboring fields and homes.

Beetles are frequently associated with poultry feed, preferring grain and cereal products that are damp, moldy, and slightly out of condition. Both adults and larvae consume poultry feed in amounts costly to the producer. Larvae are known as lesser mealworms. Increased importance has been placed on control of this beetle.

Because the beetles feed on poultry carcasses and because poultry may feed on them, litter beetles are mechanical vectors of several diseases; both adult beetles and larvae act as reservoirs for many poultry pathogens and parasites. Diseases spread by Darkling beetle include Marek's disease, avian influenza, salmonella, fowl pox, coccidiosis, botulism, and Newcastle disease. They also act as vectors of cecal worms and avian tapeworms. In addition, they cause considerable damage to insulation in poultry facilities.

In the poultry house, the beetle can lay up to 800 eggs in litter during a 42-day period. Eggs develop into larvae in 4 to 7 days. The life cycle requires about 42 to 97 days, depending on temperature. Beetles live up to 3 months to a year.

Adults are black or very dark reddish-brown, and about 0.65 cm long. Larvae are yellowish-brown (wireworm-like), up to 2.0 cm long, and accumulate in dark corners of manure or litter, especially under sacks, in bins, or in places where feed is stored. Pupation occurs in the litter, soil, and side walls of poultry houses. They migrate frequently throughout the litter, generally coming in soil contact. Late instar larvae may migrate upward and pupate in the insulation of the building. The result is extensive damage to all types of insulation.

Adult chickens and chicks are more likely than poults or turkeys to eat beetles and their larvae. Consumption of beetles and larvae, rather than providing "extra protein" in the diet, actually has a negative effect on feed conversion and rate of gain.

1.5.2. Hide Beetle

Mature larvae of the Hide beetle, *Dermestesmaculatus*, have the habit of boring into various hard surfaces to pupate, usually preferring softwoods. Some may climb 7 to 11 m and bore into wood posts, studs, and rafters, seriously weakening and "honey-combing" these structures. Larvae are especially troublesome in poultry houses, damaging yellow pine, foam insulation, Styrofoam air baffle boards, paneling, drywall, and even chemically treated wood, in some cases.

Larvae emerge from the litter, climb the walls, and bore into soft building material, often escaping cannibalism during the pupation period.

Hide beetles are larger than Darkling beetles, about 0.8 cm long, dark brown on top, with a mostly white undersurface (belly). Each female lays about 135 eggs, which hatch in 12 or more days. The life cycle requires 40 to 50 days. Larvae are thickly covered with long, brown hairs, grow to about 1.25 cm long, and have two spines on top near the tail end, which curve forward.

Reasonable control has been achieved by applying tetrachlorvinphos 50% WP in the dry form to building walls. Make treatments with an electrostatic duster to negatively charge the particles, providing better adhesion to the wall surface. Hide beetles can be killed with a 1.35% pyrethrin residual emulsion concentrate.

1.6. Gnats and Flea

Several kinds of gnats attack poultry, including black flies, buffalo gnats, and turkey gnats. The most common is the turkey gnat, *Simulium* spp., a vector of leucocytozoan parasites that cause a malaria-like disease in turkeys and ducks.

Eggs are deposited on objects on the surface of, or in, flowing water, usually at the edge. The eggs must be kept wet or submerged to hatch into larvae in 2 to 12 days. Larvae develop in water 1 to 6 weeks before transforming into pupae. Adults emerge after a 4- to 15-day pupal period.

Southern buffalo gnats appear during the first warm period of late winter or early spring. The turkey gnat usually appears later in spring.

Occasionally the flea is found in the poultry house. It is usually first noticed in the litter where a wide range of hosts are attacked, including rats, mice, chickens, humans, etc. Bites annoying egg handlers occur primarily on the ankles and legs, causing a raised (swollen) itching spot. The adult flea, an excellent jumper, passes through a complete life cycle consisting of egg, larva, pupa and adult. The life cycle varies from 2 weeks to 8 months depending on temperature, humidity, food and species.

1.6.1. Sticktight or Southern Chicken Flea

This flea, *Echidnophaga gallinacea*, attacks poultry, cats, dogs, horses, and humans. Adult males and females are found on the heads of fowl. Females remain attached by their mouthparts in the same spot as long as 2 or 3 weeks and the life cycle may be completed in 1 to 2 months. The flea is unique among poultry fleas in that the adults become sessile parasites and usually remain attached to the skin of the head for days or weeks. The adult females forcibly eject their eggs so that they reach surrounding litter. The larvae develop best in sandy, well-drained litter.

Hosts of the adult flea include chickens, turkeys, pigeons, pheasants, quail, man, and many other mammals. Irritation and blood loss may cause anemia and death, particularly in young birds. This pest thrives in dry, cool weather, and under these conditions adults may live several months. Sometimes they embed themselves in clusters about the face, eyes, ear lobes, comb, and wattles of poultry so that they cannot be brushed off. Young fowls are often killed; egg production and growth are reduced by loss of blood and irritation caused by bites.

1.6.2. Other fleas

The Western hen flea, *Ceratophyllus Niger* , seems to be confined to the Pacific coast area (USA). This flea actually breeds in the droppings and only feeds on birds occasionally. The European chick flea, *C gallinae* , is widespread in the USA. It breeds in nests and litter and is on the birds only to feed. It attacks many other birds besides chickens.

1.7. House flies

House flies are the most persistent and common fly pest, although other species such as blow flies and little house flies also are present. House flies do not bite poultry, but are a severe nuisance, and can spread some poultry diseases. House flies are present because of poultry manure and exposed wet feed, which are ideal feeding and breeding materials.

Details of control of house flies are given in Chapter "Pest management"

1.8. Other pests

Other blood-feeding insects or mites that may occasionally be pests of poultry include: chigger mites, biting midges, and black flies (turkey gnats).

1.8.1. Chiggers (red bugs, jiggers, and harvest mites)

Chiggers may be a problem where turkeys are kept on open range.

1.8.1.1. General morphology

Chiggers (adults) often are covered with dense, feathered hairs that give them a velvety appearance. They are often bright red with a figure eight-shaped body about 1 mm long. The parasitic larvae are about 0.2 mm long, reddish or straw-colored, and not as densely covered with feathered hairs as the adult. Larvae are barely visible to the naked eye. More than 700 species are known. Under laboratory conditions, as many as 4,764 eggs are deposited from a single female within 23 days.

The larval chigger usually feeds only once. It most often completes feeding in 1 to 4 days, but in some instances may require up to a month. When feeding is complete, larvae drop to the ground, burrow into upper layers of the soil and become quiescent and later on life cycle proceeds.

1.8.1.2. Activity

Chiggers feed in clusters on the thighs, breast, underside of the wings, and around the vent. Feeding by the chigger (larvae) creates scabby, reddish lesions that require two to three weeks to heal after the engorged mites leave the bird.

Neoschongastia americana americana is the most abundant external parasite on turkeys. Chiggers feed in clusters on the thigh, breast, and underside of wings, and around the vent. These clusters produce scabby lesions that require about 3 weeks to heal after engorged chiggers leave the host. These lesions and scabs result in downgrading of turkeys.

Chiggers normally do not burrow into the skin or suck blood. When the chigger is firmly attached, it injects a digestive enzyme into the wound that liquefies host tissue and is probably responsible for the severe irritation. It sucks up the partially digested, liquefied host tissue. The larval stage is the only parasitic stage in the chigger life cycle.

1.8.2. Biting midges and Black flies

1.8.2.1. General morphology

These are small (0.15 to 0.30 cm long), hump-backed flies hovering around the heads of birds and feeding on blood by piercing the skin. Biting midges and black flies develop as larvae in swiftly flowing water. When adults emerge, they quickly seek warm-blooded hosts on which to feed.

1.8.2.2. Activity

The feeding is painful and, in addition, leucocytozoan parasites that cause a malaria-like disease in birds may be transmitted during feeding.

2. Hemoparasites

Important hemoparasites of poultry are listed in Table 66.2; they are discussed in Chapter "Protozoan diseases"

Table 66.2. Hemoparasites of importance in poultry production

Parasite	Hosts	Predilection site
Leucocytozoan spp.	Chickens, Ducks, Geese, Turkeys	WBCs, RBCs
Plasmodium spp.	Chickens, Turkeys	RBCs
Hemoproteus spp.	Ducks, Geese, Chicken	RBCs
Aegyptinella spp.	Chicken, Turkeys, Dicks and Geese	
		Source : Permin and Hansen, 1998

3. Endoparasites

Modern confinement rearing of poultry has reduced the frequency and variety of endoparasite infections. However, severe parasitism still may occur in floor-reared layers, breeders, or pen-reared game birds; contributing factors may be the use of built-up litter, and/or the resistance of the parasites to therapeutic drugs. Some of the nematodes such as *Heterakis gallinarum* and *Syngamus trachea* may increase due to seasonal or climatic abundance of specific invertebrate hosts.

3.1. Nematodes

3.1.1. General

Nematodes (roundworms) are the most significant in number of species and in economic impact. Many field studies show that poultry maintained under free-range conditions may be heavily parasitized.

Table 66.3. Nematodes and Cestodes of poultry

Parasite	Host	Intermediate Host or Life Cycle	Organ Infected	Pathogenecity
		NEMATODES		
Amidostomum anseris	Duck, goose, pigeon	Direct	Gizzard	Severe
Ascaridia dissimilis	Turkey	Direct	Small intestine	Moderate
Ascaridia galli	Chicken turkey, duck, quail	Direct	Small intestine	Moderate
Capillaria caudinflata (columbae)	Chicken, turkey, duck, game birds, pigeon	Earthworms	Small intestine	Moderate to severe
Capillaria contorta (annulata)	Chicken, turkey, duck, game birds	None or earthworms	Mouth, esophagus, crop	Severe
Capillaria obsignata	Chicken, turkey, goose, pigeon, quail	Direct	Small intestine, ceca	Severe
Cheilospirura hamulosa	Chicken, turkey, game birds	Grasshoppers, beetles	Gizzard	Moderate
Cyathostoma bronchialis	Turkey, duck	Direct or earthworm	Trachea	Severe
Cyrnea colini	Turkey, game birds	Cockroaches	Proventric ulus	Mild
Dispharynx nasuta	Chicken, turkey, game birds, pigeon	Sowbugs	Proventric ulus	Moderate to severe
Gongylonema ingluvicola	Chicken, game birds	Beetles, cockroaches	Crop, esophagus, proventriculus	Mild
Heterakis gallinarum	Chicken, turkey, duck, game birds	Direct	Ceca	Mild, but transmits agent of histomoniasis
Heterakis isolonche	Quail, duck, pheasant	Direct	Ceca	Severe
Ornitho- strongylus quadriradiatus	Pigeon, dove	Direct	Small intestine	Severe
Oxyspirura mansoni	Chicken, turkey, guinea fowl, quail	Cockroach	Eye	Moderate

Parasite	Host	Intermediate Host or Life Cycle	Organ Infected	Pathogenecity
Strongyloides avium	Chicken, turkey, quail, goose	Direct	Cecum	Moderate
Subulura brumpti	Chicken, turkey, duck, game birds	Earwigs, grasshoppers, beetles, cockroaches	Ceca	Mild
Syngamus trachea	Chicken, turkey, pheasant, quail	None or earthworm	Trachea	Severe
Tetrameres americana	Chicken, turkey, duck, game birds, pigeon	Grasshoppers, cockroaches	Proventriculus	Moderate to severe
Trichostrongylus tenuis	Chicken, turkey, duck, game birds, pigeon	Direct	Ceca	Severe
		CESTODES		
Choanotaenia infundibulum	Chicken	House flies	Upper intestine	Moderate
Davainea proglottina	Chicken	Slugs, snails	Duodenum	Severe
Metroliasthes lucida	Turkey	Grasshoppers	Intestine	Unknown
Raillietina cesticillus	Chicken	Beetles	Duodenum, jejunum	Mild
Raillietina echinobothrida	Chicken	Ants	Lower intestine	Severe, nodules
Raillietina tetragona	Chicken	Ants	Lower intestine	Severe
		TREMATODES*		
Echinostoma revolutum	Ducks, Geese		Rectum, ceca	
Prosthogonimus Sp.	Chicken, Ducks, Geese		Bursa of Fabricius, oviduct, cloaca, rectum	
		PROTOZOA*		
Eimeria Sp.	Chicken, Turkeys, Ducks		Small intestine	
Histomonas meleagridis	Turkeys, Chicken		Ceca, liver	
* Permin and Hansen, 1998			*Source* : The Merck Veterinary Manual, 2003	

Generally, nematodes have separate sexes with morphologic differences. The size and shape of nematode species vary widely. Nematodes have either a species-specific, direct life cycle with bird-to-bird transmission by ingestion of infective eggs or larvae, or an indirect cycle that requires an intermediate host (insects, snails, or slugs). Eggs of many nematode species are resistant to low temperatures and disinfectants.

3.1.2. Symptoms and lesions

Ascaridia, *Heterakis*, and *Capillaria* spp are widely distributed and cause such nonspecific signs as general unthriftyness, inactivity, depressed appetite, and retarded growth; death may result. A mere few ascarids may depress weight, and larger numbers may block the intestinal tract. Ascarids may migrate up the oviduct (via the cloaca) to become enshelled later within the egg (an aesthetic, but not a public health problem).

Heterakis gallinarum , a mild pathogen, in large numbers may cause thickening, inflammation, or nodulation in the cecal walls producing cecal and hepatic granuloma. *Heterakis isolonche* , highly pathogenic in pheasants, may cause 50% mortality. *Heterakis gallinarum* carries *Histomonas meleagridis* , the protozoa that causes blackhead.

Capillaria contorta in the mucosa of the crop and esophagus, and *C obsignata* in the wall of the small intestine, cause marked thickening and inflammation of the organs. Birds harboring large numbers of these thread-like worms become weak and emaciated and may die.

Young birds are the most severely affected by gapeworms. Sudden death and verminous pneumonia characterize early outbreaks. Signs of gasping, choking, shaking of the head, inanition, emaciation, and suffocation may follow. Necropsy reveals adult gapeworms obstructing the lumina of the trachea, bronchi, and lungs. Respiratory inflammation may be present.

Oxyspirura mansoni is a slender nematode, 12-18 mm, found beneath the nictitating membrane of chickens and other fowl in tropical and subtropical regions. The parasite causes various degrees of inflammation, lacrimation, corneal opacity, and disturbed vision.

Among other nematodes, *Amidostomum anseris* attacks the gizzard lining of ducks and geese and causes dark discoloration, necrosis, and sloughing at the parasitic loci. *Dispharynx nasuta* causes ulceration, thickening, and maceration of the proventriculus; heavily infected birds may die. *Tetrameres americana* , a bright red worm discernible through the proventricular wall, causes diarrhea, emaciation, and with heavy infection, death. *Trichostrongylus tenuis* causes inflamed ceca, weight loss, anemia, and death, especially in young birds. *Ornithostrongylus quadriradiatus*, a blood-sucking parasite, causes pigeons to regurgitate bile-stained fluid mixed with food; greenish mucoid diarrhea from hemorrhagic intestines, emaciation, and death follow.

3.2. Cestodes

Table 66.4. Location and pathogenecity of Cestodes

Cestode	Location	Degree of pathogenecity
Amoebotaenia cuneata	Duodenum	Mild
Davainea proglottina	Small intestine	Severe
Hymenolepis contaneana	Small intestine	Relatively nonpathogenic
Railletina cesticillus	Duodenum and jejunum	Mild or harmless
Railletina tetragona	Lower half of intestine	Moderate to severe
Railletina echinobothrida	Small intestine	Moderate to severe #
# Nodular disease; common in India		Hofstad *et al* 1984

3.2.1. General

Cestodes (tapeworms) vary in size. The proglottids of individual tapeworms are hermaphroditic. Tapeworms have been recovered in the thousands from individual chickens and turkeys.

Cestodes require an intermediate host (insects, crustaceans, earthworms, or snails). Floor layers, breeders, and broilers are infected with *Raillietina cesticillus* by ingestion of the intermediate host, small beetles that breed in contaminated litter. Cage layers in unscreened houses may become infected with *Choanotaenia infundibulum* by eating its intermediate host, the housefly.

3.2.2. Symptoms

Most pathogenic tapeworms are found in the small intestine; the scolex, usually buried in the mucosa, generally causes mild lesions. *Davainea proglottina* may cause weight loss. *Raillietina tetragona* causes weight loss and decreased egg production; *R echinobothrida* produces granuloma at its attachment sites ("nodular disease").

3.3. Trematodes

Modern poultry production methods have diminished the incidence of fluke infections, although the parasites persist in poultry allowed contact with snails or other hosts, and in some wild birds

Prosthogonimus macrorchis , the oviduct fluke of poultry, infects birds after they consume infective metacercariae in developing or mature dragonflies, the secondary host. The piriform fluke matures in ~2 wk in the bursa of Fabricius or, in gallinaceous birds without a functional bursa (chickens, turkeys, pheasants), in the oviduct.

Light infections without clinical signs appear in ducks and other birds with a functional bursa. In gallinaceous birds, heavy infections in the oviduct cause inappetance, droopiness, weight loss, calcareous cloacal discharge, and depressed egg production (the eggs being soft-shelled). Lesions range from mild inflammation to distention or rupture of the oviduct due to exudate and egg components; death may result. Diagnosis by fecal examination is unreliable because fluke eggs are not consistently present. Adult flukes may appear in the bird's eggs or be found in the oviduct on necropsy.

To prevent fluke transmission, birds must be kept from feeding on dragonflies. There is no effective treatment approved for use in poultry. Carbon tetrachloride, is highly toxic to birds, especially chickens.

Table 66.5. Location and pathogenecity of Trematodes

Trematode	Location	Lesions and Pathogenecity
Hypoderaeum conoideum	Small intestine	Enteritis, moderate to severe
Ribeirola ondatrae	Proventriculus	Proventriculitis, moderate to severe
Amphimerus elongatus	Liver, pancreas, bile and pancreatic ducts	Thickening of duct walls, distension and stenosis of ducts, may be severe
Prosthogominus macrorchis	Oviduct, bursa of Fabricius	Reduced egg production, adhesive peritonitis, hyperemia of oviduct and intestines
		Hofstad *et al* 1984

Collyriclum faba appear as subcutaneous cysts 4-6 mm in diameter (usually containing 2 adults) anywhere on the body but more frequently near the vent in turkeys, chickens, and other birds. The cysts ooze exudate, which attracts flies and predisposes to bacterial infection. Signs in young birds include locomotor difficulty and inappetance; death may result in heavy infections. The parasites can be removed surgically. The life cycle is unknown but probably involves snails and insects such as dragonflies or mayflies. Prevention of infection requires restricting birds from areas frequented by aquatic insects.

4. Strategy for control of poultry parasites

The prevalence of most parasitic diseases in poultry seems to have been reduced significantly in commercial indoor poultry production systems due to improved housing, hygiene and management. However, parasitic diseases continue to be of great importance in deep-litter and free-range commercial systems. In traditional systems throughout the world a number of parasites are widely distributed and contribute significantly to the low productivity.

The most commonly mentioned parasites are *Eimeria* spp., *Ascaridia galli* and *Heterakis gallinarum* which is mainly due to the many studies carried-out on these parasites.

4.1. Diagnosis

Diagnosis is the main prerequisite for developing a control program. To diagnose parasitic infections in poultry is a time-consuming job, but it can be done with relatively simple tools. Diagnostic techniques include identification and quantification of parasitic infections through clinical and *post mortem* inspections, and the examination of fecal material and blood smears. The more sophisticated immunological and molecular biology techniques have not yet been developed for diagnosis of poultry parasites.

4.1.1. Clinical examination

Clinical examinations of individual animals may be of limited value. But when investigating a disease problem in a poultry flock it is often beneficial to examine a few animals. Before the examination is commenced, information related to the flock and the anamnesis (disease history) should be recorded, e.g. the flock size, management practices, vaccination procedures, feeding practices, trading habits, other poultry in the flock, symptoms before death, clinical signs, number of animals dead, time of death etc. Live birds have to be examined as follows:

The general attitude of live birds and all abnormal conditions should be carefully noted. Any indication of depression, anorexia, respiratory signs, blindness, incoordination, tremors, paralytic conditions, abnormal gait and leg weakness etc. in the flock has to be noted.

It may be useful to observe the waste disposal area (e.g., any dead animals been thrown there by the farmer).

Individual birds can be examined for ectoparasites on the head, under the wings and thighs (lice and mites can be found on the bird, whereas other species leave the animals during daytime).

Examinations should be made for diarrhoea, respiratory discharges, conjunctival exudates, nasal discharges, feather and comb conditions, skin changes, body condition, dehydration, abscesses, signs of cannibalism, tumors etc..

If necessary, blood samples and blood smears are made, swabs are made, fecal samples are collected etc.

If *post mortem* examinations are required, a number of chickens should be sacrificed and examined according to recommended procedure.

4.1.2. Fecal examination

Identification of helminth eggs in faeces is an easy and cheap way to diagnose many helminth infections and to get an impression of the infection level at individual as well as population level. All helminths must find a way or their eggs to become available for a new host. In poultry, the eggs simply pass with the faeces. Newly-deposited faeces may contain unembryonated eggs or eggs with well-developed embryos, e.g., *Tetrameres* spp., *Acuaria hamulosa*, *Allodapa suctoria* and *Gongylonema ingluvicola*. Hatched larvae are not seen in poultry faeces. The reader is advised to look into standard Parasitology books for pictorial charts for identification of parasitic eggs.

Table 66.6. Age group to sample most prevalent parasite species

Chicks (< 3 months)	Growers (3 - 7 months)	Adults (> 7 months)
Eimeria spp., A. galli, Heterakis spp., Capillaria spp.	Eimeria spp., A. galli, Tetrameres spp., Raillietina spp.	Raillietina spp., Heterakis, Tetrameres spp.

4.1.3. Blood examination for diagnosis of hemoparasites

Protozoans are often identified by microscopic examination of wet mounts, buffy coat, or blood smears; appropriate culturing techniques; or subinoculation of blood into susceptible birds. Microscopically, some are within blood cells (*Plasmodium*, *Haemoproteus*, *Leucocytozoon*, *Atoxoplasma*, *Babesia*, *Aegyptianella*), while others are free in the plasma. None lives exclusively in the blood; most are found in tissues but are present in blood during part of the disease process. Some, such as *Plasmodium*, may have a periodicity when numbers or stages of parasites are present at different times. In such cases, examining multiple smears at intervals will increase the likelihood of obtaining a diagnosis. Routine examination for blood-borne organisms should be included in the clinical and diagnostic procedures for any ill bird.

Thin blood smears should be made with blood directly from the bird if possible. Anticoagulants, storage, and cooling of the blood can distort protozoan morphology. A small drop of blood can be collected using a syringe and needle, or by selecting a small vessel in the wing web and, after cleaning the site thoroughly with alcohol and letting it dry, puncturing the vessel with a lancet so that a small drop of blood wells up from the wound. The drop should be picked up without touching the skin and spread on a clean glass slide at a 30° angle to make a thin smear. A good quality Romanowsky-type stain that gives good polychromatic coloration (Giemsa stain often preferred) should be used. At least 200 oil-immersion fields (~20,000 RBC) for single smears or 100 for multiple smears from the same bird should be examined. *Leucocytozoans* occur around the periphery of smears and can be easily seen on low-power magnification.

Blood-borne organisms in plasma or WBC are concentrated in the buffy coat. Stained buffy coat smears are recommended for detecting bacteremia and chronic *Leucocytozoan* or *Atoxoplasma* infections. For motile organisms such as spirochetes, the buffy coat and all of the plasma should be expressed, and a coverslip placed on the buffy coat and depressed slightly to spread the buffy coat. The buffy coat/ plasma interface should be examined with darkfield or reduced light microscopy to detect motile organisms. The latter is an excellent technique for identifying spirochetes and microfilariae when present in low numbers.

Plasmodium infections can be determined by subinoculation. Ideally, birds of the same or a known susceptible avian species should be used, but this is often not practical. In general, canaries are used for detecting passerine infections, and turkeys are susceptible to most plasmodia infecting gallinaceous birds. Parasites remain viable in blood stored at 0°C (in ice) for at least 7 days. Intravenous inoculation is preferred and will result in earlier parasitemia, but any parenteral route can be used. Recipients should be examined twice weekly for a minimum of 4 wk if exposed i/v; longer times are needed if other routes of inoculation are used.

4.1.4. Sampling

Most helminth infections in poultry are subclinical (no clinical signs are seen), therefore samples from diseased animals should never replace the sampling from a representative number of randomly-selected animals. Random sampling is necessary for a valid estimation of the flock/population problem. It should be noted that the full truth is only found when all animals are sampled, but when the sample size is calculated and the animals are sampled randomly a valid estimation of the flock/population problem is achieved.

Table 66.7. Suggested sample sizes

Number of animals in the age group	Number of animals to sample
1-10	1 - 9
11-25	10 - 20
26-100	21 - 49
101-200	49 - 65
201-500	65 - 81
500-1000	81 - 96
> 1000	96
Prevalence 50%, precision desired 10% and confidence level 95%	

5. General control and prevention of parasitic diseases

5.1. Ectoparasites

General control measures include judicious use of organophosphorus, carbamate and pyrethroid insecticides; the latter are most effective on flies.

The insecticides may be applied either by dusting or by spraying or by misting. Chlorinated hydrocarbons are banned from use on poultry in all the Western countries.

5.1.1. Ticks

The nymphal and adult stages of fowl ticks feed on their hosts for a limited period. The control of ticks therefore requires treatment of the environment (indoor and outdoor) in which the poultry are. After mechanical cleaning of the pen the entire building (walls, ceilings, cracks and crevices) is sprayed with a high pressure sprayer using carbaryl, coumaphos, Malathion or Stirofos. The outdoor facilities (feed troughs, woodpiles, tree trunks etc.) may be treated using approved insecticides, but this is not recommended due to environmental concerns.

5.1.2. Lice

Control of poultry lice requires treating the birds since lice remain on the bird throughout its life. Treat by dipping, dusting, or spraying the birds, and be careful to avoid contaminating eggs, feed and water. Treatment is easiest at night when birds are quiet. For best results, split treatments with half of the recommended amount of insecticide applied initially, and the second half applied soon after the first, since the wet feathers retain more active ingredient. Applying liquid sprays to dry feathers often results in loss of some of the insecticide due to runoff.

5.1.3. Mites

Infections with the chicken mite, tropical fowl mite and northern fowl mite may be controlled by using acaricides such as carbaryl, coumaphos, malathion, stirofos or a pyrethroid. Each bird should be sprayed as well as the stable. Especially all hiding places for the mites should be treated.

Prophylactic measures such as cleaning and disinfection of the house are advisable. Spraying of all surfaces with insecticides followed by lime wash is recommended. Furthermore, it should be avoided to insert new animals in a house straight after cleaning and disinfection.

Animals should not be introduced into an existing flock without examination first. Control of scaly leg mites should begin by isolation or culling of affected birds and thereafter the house should becleaned as recommended for other mites. Individual animals are treated by dipping the affected legs in kerosene, linseed or mineral oil or coated with Vaseline. Treatments should be given twice with a 10 day interval. Infections may also be treated using acaricides.

5.1.4. Beetles

Control efforts for the litter beetles generally are only partially successful. Applying insecticides after the building is emptied and thoroughly cleaned will further reduce beetle numbers. Residual insecticides should be applied to the soil after the litter is removed and in areas where migration occurs.

Disinfectants and insecticides should not be applied at the same time because each may destroy the toxic property of the other; a gap of 10-14 d is advised. Insecticides should be rotated periodically, i.e. a carbamate could be rotated with a phosphate and then with a pyrethroid which will delay or inhibit the onset of resistance (both to the beetles and to filth flies).

5.1.5. Fleas

Infections with fleas are best controlled by removing litter followed by treatment of the house with permethrin. Control may not be possible on non-confined birds. Confined facilities should be screened with mesh small enough to prevent these flies from entering.

Table 66.8. Some insecticides for control of ectoparasites on poultry

Parasite	Prophylaxis
Fowl ticks	Malathion 0.3% spray, Tetrachlorvinphos 1.0% spray (not direct), Carbaryl 0.2% spray
Mites, lice and fleas	Malathion (4 - 5%) dust, Permethrin (0.25%) dust, Rabon dust, Carbaryl (5%) dust, Malathion 0.5% spray, Permethrin 0.05% spray, tetrachlorvinphos 0.5% spray, Stirofos (23%) with dichlorfos (5.7%) 0.5% spray

5.1.6. Chiggers

Chiggers are most efficiently controlled by treating the range with insecticides before releasing the turkeys, however, repeated treatments may be necessary if the turkeys remain on the range more than a few weeks.

5.1.7. Insecticide application

Orthoboric acid provides long residual control up to 9 to 12 months or longer. Both adult beetles and larvae are killed by contact or ingestion.

Usually a quick-kill insecticide is used first to kill beetles away from the litter. For poultry houses where birds are grown on litter, birds are removed before applying the bait uniformly to the floor or to old litter by at the rate of 50 to 100g per m^2, in bands along feeder lines. Fresh litter at least 10 cm is spread uniformly over all treated areas (floor or old litter) and then birds are brought in. Repeated after every batch, if needed.

For poultry houses in which birds are grown in cages (layer or high-rise "pit type" houses), birds do not have to be removed prior to application of bait.

For control of beetle adults and larvae in poultry houses, birds are removed before dry and wet applications. For dry application, 50 to 100 g/m^2 of treated surface is the normal dose for dusting side walls, top plates, posts and framing. For wet application, dust is mixed at the rate of 40 to 80 g/l of water to apply over treated surface.

5.1.8. Insecticide resistance

House fly resistance is genetic in nature, developing more quickly under heavy doses of pesticide or very frequent application. Insects resistant to one insecticide can be cross-resistant to other insecticides of the same class or even having a similar mode of action. The only proven solution to resistance problems is to rotate the use of different classes of insecticides.

5.2. Hemoparasites

Hemoparasites are transmitted by mosquitoes, flies, biting midges etc. Control of the arthropods is thus of crucial importance. Screening of the poultry houses may avoid transmission of the hemoparasites. Also insecticides may be used to minimize the vectors

Table 66.9. Control of flies (vectors) in poultry houses

Name	Formulation
Bomyl (1%)	Bait
Dibrom	Bait
Dibrom (0.25%)	Spray
Dichlorvos (0.5%)	Spray
Dichlorvos (1%)	Bait
Methomyl (1%)	Bait
Permetrin (0.25%)	Spray
Pyrethrin (0.5%) and Piperonylbutoxide (3.75%)	Spray
Stirofos (1%)	Spray
Tetrachlorvinphos (1%)	Spray
Trichlorfon (1%)	Bait

A number of drugs are available for treatment against Avian malaria, Leucocytozoonosis and *Aegyptinella* spp., although treatment against Leucocytozoonosis is not effective.

Table 66.10. Some common drugs for use against hemoparasites

Drug and mode of application	Effective against
Chloroquine (1 mg/kg intra muscular (i/m) for 5 days)	Avian malaria
Chloroquine (2000 mg/l in water for 1 day)	Avian malaria
Quinacrine (1.6 mg/kg i/m. for 5 days)	Avian malaria
Primaquine (100 mg/kg per oral for 1 day)	Avian malaria
Sulfonamides + trimethoprim (i/m)	Avian malaria
Pyrimethamine (1 ppm) + sulfadimethoxine (10 ppm) oral	Prevention of Leucocytozoonosis
Pyrimethamine (25 ppm oral)	Cure of Leucocytozoonosis
Tetracycline (15 - 30 mg/kg oral)	Aegyptinella spp.

5.3. Endoparasites - General

In commercial indoor production systems management practices largely determine the extent of parasitosis: e.g., total enclosure principles, improvement of cleaning, disinfection procedures and production according to the "all in - all out" principle have apparently decreased the significance of parasitic infections. With the ban on battery cages in a number of countries, new intensive free-range systems have developed in which the prevention of parasitic infections has proven to be difficult. The use of out-door areas, where parasite eggs may persist in the environment for years have increased the risk of infections. The contact with wild birds also increases the chances of attracting parasitic diseases.

In backyard systems the birds are in permanent contact with soil, a range of intermediate hosts and wild birds. Parasitic diseases are thus difficult to avoid in such systems, but may be controlled by the use of management and treatment strategies.

The purpose of a parasitic control strategy is to keep the parasitic challenge (especially in young birds) at a minimum rate to avoid clinical symptoms and production losses. Total eradication from a geographical region is unlikely for most parasites due to the enormous numbers of eggs passed with the faeces and the high persistence of the infective stages in the environment.

5.3.1. Stocking rate

The density of birds (stocking rate) in any poultry production system should not be too high. Overcrowding will force the birds to come in a closer contact with material contaminated by faeces and may result in the consumption of a higher number of infective parasitic eggs.

5.3.2. Flock structure

Older animals are likely to be carriers of a range of parasitic diseases without showing clinical signs, e.g., *Eimeria* spp., *Ascaridia* spp, *Tetrameres* spp. etc. Therefore, it may be beneficial to separate different age groups *Vis-à-vis* the "all in - all out" principle.

5.3.3. Alternate use of pens

As poultry have few parasites in common with other livestock, management may include mixed use of pens (i.e., poultry scavenging together with other livestock) or alternate use of pens (poultry alternating with other livestock in the same pen). However, this opens up other risks in terms of Avian influenza as well as inter-relationship with life-cycle of certain intermediate hosts of parasites of either species involved. Hence, this alternative has found few takers.

5.3.4. Hygiene of pens

The floor should be kept as dry as possible because external stages of all parasites require nearly 100% relative humidity to develop. The draining capacity, and thus the dry microclimate at floor level, may be the main reason why slatted floors in intensive systems seem to be rather effective in reducing parasitic transmission indoors.

After mechanical removal of the litter and disinfection, lime-wash should be applied and allowed to dry. The effects achieved by this procedure are: 1) The drying effect of lime decreases the survival of parasite eggs, and 2) the pH - level exceeds 8, which also decreases the survival of parasite eggs. After application of lime, the house should be left empty for 2 - 4 weeks before new animals are introduced.

5.3.5. Dose and move

Before animals are moved to safe areas (outdoor, indoor), they may be dosed with an anthelmintic to remove any worms present in order to keep the environment free of contamination for as long as possible (dose and move). This principle has been shown to be rather effective, although, unfortunately, it also increases the risk of development of anthelmintic resistance.

5.3.6. Routine deworming

Routine deworming programs often appeal to farmers for reasons of convenience, and as a result worm treatments are generally the only control measure carried out. They are relevant in the control of nematodes in most management systems. Several programs for routine deworming of poultry have been worked out, and most are adjusted to the age or the reproduction cycle of the poultry. The standard procedure is treatment of hens shortly before the commencement of laying, followed by a move to a clean laying unit. The objective is to eliminate the worms from the hens thus reducing production losses and to prevent contamination of the environment.

The choice of drug should partly depend on the worm species present. Some drugs have a broader spectrum of activity than others and some nematodes may be controlled only by certain drugs; some drugs are more expensive than others etc.. Furthermore, it is important to alternate between drugs with different modes of action in order to reduce the risk of developing anthelmintic resistance, and to avoid drugs against which resistance has already developed.

5.3.7. Adequate nutritional level

The overall effect of helminth infections may be reduced by ensuring an adequate level of nutrition (especially proteins), although this should be no substitute for a sound parasite-control program.

5.3.8. Genetic resistance

Little is known about genetic resistance to parasitic infections in poultry, although a difference in infection levels between two breeds has been described.

5.4. Anthelmintic

5.4.1. Definition

An anthelmintic is a compound which destroys helminths or causes them to be removed from the gastro-intestinal tract or other organs and tissues they may occupy in their hosts.

Currently a series of safe anthelmintic is available, some with broad spectrum activity and others with activity against specific helminth infections. Many modern anthelmintic are effective against both adults and larval stages, including dormant larvae.

Due to their cost, their tendency to delay or interfere with natural host-immunity mechanisms, and not least the rapidly increasing prevalence of anthelmintic resistance, they should not be used indiscriminately.

5.4.2. Characteristics of an ideal drug

1. An ideal drug should have a broad spectrum activity against adult and larval helminth parasites. It is ideal if it can act on larval as well as adult stages of a number of parasite species simultaneously because an individual bird often harbors several different helminth species with different stages of each of the parasite.
2. The ideal drug should also be metabolized rapidly in order to avoid metabolic residues in animals slaughtered for human consumption, and thus reduce the slaughter withdrawal period.
3. A good drug should have low toxicity to the host, and the ratio of the therapeutic dose to the maximum tolerated dose of birds should be as large as possible.
4. There should be no unpleasant side-effects to the birds, the operator or to the environment.
5. The selected drugs should be competitively priced and ready to use in an easy way.
6. They should be stable and not lose activity on exposure to normal ranges of temperature, light and humidity.

5.4.3. Dosing methods

Oral dosing through feed or water is by far the most common and easiest way of administration of anthelmintic to birds. However, since the birds are often group-treated, sometimes a sub-therapeutic dose may reach some of the birds if the drug is administered only once. Hence, it is advisable to administer the drugs over several days.

Furthermore, the efficiency of a drug (especially Class I and III drugs) may be considerably increased by low dosing for several days. Anthelmintic are, so far, not available in a formulation for external application or as injections to birds.

5.4.4. Anthelmintic classes

On the basis of their mode of action, anthelmintic drugs can be subdivided into 5 classes:

5.4.4.1. Class I anthelmintic

Benzimidazoles and pro-benzimidazoles. These drugs exert their action on the intracellular polymerization of the tubulin molecules to microtubules. As the cellular functions are disrupted, the worms die. Examples of Class I compounds are albendazole, thiabendazole, fenbendazole, parbendazole, flubendazole, febantel, and thiophanate.

5.4.4.2. Class II anthelmintic

Imidazothiazoles and tetrahydropyrimidines. These drugs act on the acetylcholine receptor in the neuromuscular system of the worms causing a persistent depolarization of muscle cells and a spastic paralysis of the worms, which are then removed by gut motility. Examples of Class IIdrugs are levamisole, pyrantel, and morantel.

5.4.4.3. Class III anthelmintic

Avermectins and milbymicins. The compounds act on the nervous system of the worms, causing flaccid paralysis and removal by gut motility. Class III consists of two distinct types of drugs, i.e. the piperazines and the avermectins (ivermectin, doramectin, moxidectin), the latter having effects against some ectoparasites, e.g. mange mites.

5.4.4.4. Class IV anthelmintic

Salicylanilids and substituted nitrophenols. These drugs are typically used against bloodsucking parasites.

5.4.4.5. Class V anthelmintic

Acetylcholine esterase antagonists. These are organophosphorous compounds, which are only used to a limited extent. Examples are dichlorvos and neguvon.

Piperazines have previously been classified as Class III anthelmintic. These drugs act on the GABA receptors causing flaccid paralysis of the worms. However, their mode of action appears to be different from that of avermectins and milbymicins, and cross resistance has not been documented.

5.4.5. Selection of anthelmintic for administration

It is important first to identify the nature of the parasitic problem in order to select the appropriate drug. Generally, under commercial conditions, broad-spectrum drugs are appropriate.

Piperazines are relatively nontoxic and widely used against ascariasis. Several piperazine salts are available; because only the piperazine moiety is efficacious, doses should be calculated based on mg of active piperazine/bird. Piperazine should be consumed by birds within a few hours because only relatively high concentrations of the drug eliminate worms. It may be given to chickens as a single dose, 50-100 mg/bird, or at 0.2-0.4% in the feed or at 0.1-0.2% in the drinking water; it may be administered to turkeys at 100 mg/bird <12 wk old, 100-400 mg/bird =12 wk old, or in feed or water concentrations as for chickens. There is increasing evidence of piperazine resistance.

Phenothiazine controls cecal worms in chickens at 0.5 g/bird, in turkeys at 1 g/

bird, given in 1 day. Combined in drinking water, as a 1-day treatment, phenothiazine (0.5-0.56%) and piperazine (0.11%) remove both heterakids and ascarids.

Thiabendazole at 0.05% in the feed continuously for 2 wk effectively eliminates gapeworms from pheasants, and when given continuously for ≥ 4 days is said to help prevent and control infections.

Hygromycin B, 0.00088-0.00132% in feed, controls ascarids, cecal worms, and capillarids. Coumaphos, 0.004% in feed for10-14 days for replacements, or 0.003% in feed for 14 days for layers, is used more commonly against capillarids.

As a treatment for Manson's eyeworm, a local anesthetic is applied to the eye, and the worms in the lacrimal sac are exposed by lifting the nictitating membrane. A 5% cresol solution (1-2 drops) placed in the lacrimal sac kills the worms immediately. The eye should be irrigated with sterile water immediately to wash out the debris and excess solution. The eyes improve within 48-72 hr and gradually become clear if the destructive process caused by the parasite is not too far advanced.

Table 66.11. Common drugs for anthelmintic treatment of poultry

Active ingredient	Administration	Indication
Fenbendazole	In feed	Ascaridia galli, Capillaria spp., Syngamus trachea
Flubendazole	In feed	Intestinal nematodes, Cestodes in chickens
Hygromycin	In feed	Intestinal nematodes (mainly *Ascaridia galli*)
Levamisole	In feed	Ascaridia galli, Heterakis gallinarum, Capillaria spp., Syngamus trachea
Mebendazole	In feed	Intestinal nematodes, *Syngamus trachea*, Cestodes
Piperazine	In feed or drinking water	Ascaridia galli, (Heterakis gallinarum), Tetrameres spp.
Thiabendazole	In feed	Syngamus trachea (pheasants)

5.5. Endoparasites - Nematodes

5.5.1. Treatment

Several compounds are effective against nematode infections butare not currently approved for use. In chickens, *A galli* , *H gallinarum* , and *C obsignata* were effectively removed by tetramisole at 40 mg/kg. Pyrantel tartrate was highly effective against *A galli* and somewhat effective against *Capillaria* . Fenbendazole effectively removed experimental infections with *A dissimilis* , *H gallinarum* , and *C obsignata* in turkeys at a level of 45 ppm for 6 days. Levels as low as 30 ppm for 3 days were effective against *A dissimilis* . Levamisole at a level of 25-30 mg/kg is also effective against the same three species; it can also be given in the drinking water at 30-60 ppm. Injection SC of 1 ml of 10% methyridine in the pectoral region or leg of pigeons removed *Capillaria* but must be handled with care. Coumaphos removes *Capillaria* in quail. Haloxon at 25 and 50 mg/kg, or at 750 ppm in the feed for 5-7 days, has good activity against *Capillaria* in chickens and quail.

Fenbendazole at 20 mg/kg for 3-4 days is effective for removing gapeworms in pheasants. Tetramisole at 3.6 mg/kg for 3 consecutive days in the drinking water removes gapeworms. Poultry treated while larvae are migrating in the body develop immunity to gapeworms, even though therapy may abort larval migration. Levamisole fed at a level of 40 ppm for 2 days or at 2 g/gal. drinking water for 1 day each month has proved to be an effective control in game birds. Mebendazole fed prophylactically at 64 ppm or curatively at 125 ppm is effective in turkey poults. Cambendazole gave control when given in three treatments of 50 mg/kg for chickens and 20 mg/kg for turkeys.

There have been some reports of experimental drug treatment for other nematodes. Cambendazole (60 mg/kg), pyrantel (100 mg/kg), citran (40 mg/kg), mebendazole (10 mg/kg for 3 days), and fenbendazole have been reported to be effective against *Amidostomum anseris* . *Trichostrongylus tenuis* is controlled by cambendazole (30 mg/kg), pyrantel (50 mg/kg), thiabendazole (75 mg/kg), mebendazole (10 mg/kg for 3 days), and citrin (40 mg/kg). At recommended levels for chickens, mebendazole has some effect against *Dispharynx nasuta* , tetramisole against *Subulura brumpti* and *Strongyloides avium* , and piperazine against *Tetrameres* .

Table 66.12. Treatment for endoparasites - Nematodes

Nematode	Location	Chemotherapeutic agents
Ascaridia galli and other Ascarids	Small intestine	Hygromycin B (0.00088-0.00132%) in feed. Piperazine 0.2-0.4% in feed; 0.1-0.2% in water; 50-100 mg/bird single dose. Coumaphos for birds more than 10-12 weeks only at 0.003-0.004% in feed.
Capillarids	Small intestine, ceca, cloaca	Hygromycin B and Coumaphos in doses similar to Ascarids
Heterakis gallinarum	Ceca	Phenothiazine 0.5 g/bird, Hygromycin B and Coumaphos in doses similar to Ascarids
Syngamus trachea and other Syngamids	Trachea	No drug is approved by FDA for commercial use; Thiabendazole 0.5% in feed, Mebendazole 0.0064-0.0125% in feed, Levamisole 0.04%
		Hofstad *et al* 1984

5.5.2. Control

Proper sanitation to break the life-cycle of parasites, elimination of intermediate hosts and proper deworming program greatly assists in control of nematode infestation.

5.6. Endoparasites – Cestodes

5.6.1. Treatment of Cestodes

The only drug approved by FDA is dibutylin dilaurate (Butynorate). It can be tried in combination with Piperazine and Phenothiazine. Other drugs like Hexachlorophene, Niclosamide and Mebendazole have also been effective under experimental conditions.

5.7. Endoparasites – Trematodes

5.7.1. Treatment

For Trematodes of digestive system, carbon tetra chloride in doses of 1-5 ml/kg body weight can be tried.

5.7.2. Prevention and control

On the same lines as in the cases of Nematodes and Cestodes.

For additional information on poultry parasites and their control, the following sources can be read:

1. The Merck Veterinary Manual, 9th Edition
2. University of Nebraska website
3. Permin, A. and Hansen, J.W. 1998. An FAO handbook on "The epidemiology, diagnosis and control of poultry parasites"

Chapter **67**

Metabolic Diseases

1. Caged-layer fatigue

This condition is characterized by inability of birds to stand and marked fragility of bones. Thinning of long bones occurs and hence they break very easily. This also increases condemnations since there will be more formation of bone-splinters during handling of birds before and during processing. The ailing birds, when allowed on deep litter for 4-7 d, recover. Differences observed between strains and seasons. Several courses have been suggested – calcium deficiency, phosphorus deficiency etc. The condition is yet to be completely understood.

2. Fatty liver syndrome

This is, by far, the most important metabolic disorder in poultry. Two conditions are described under this syndrome:

1. Fatty liver kidney syndrome
2. Fatty liver hemorrhagic syndrome.

2.1. Fatty liver kidney syndrome

Incidence of this condition may vary from 1% in broiler flocks to 10% in broiler breeding flocks.

2.1.1. Etiology

Diet, sex and age of birds, diet of breeding flock, environmental temperature and other stress factors. Of these, diet is most consistently a predisposing factor. Wheat-based diets, low in fat or protein, produce high mortality due to FLKS.

Nutrient of crucial importance is biotin; its availability being the prime factor. However, birds with FLKS may not show biotin deficiency symptoms and the birds fed biotin-deficient diets (otherwise adequate) may not show any growth depression. But, increased fat and protein reduces FLKS but increases biotin deficiency symptoms; in contrast, reduced fat & protein reduced the appearance of biotin deficiency symptoms.

2.1.2. Symptoms

Lethargy is the most predominant symptom, severely of which may vary from slight or moderate dropping of head or inability to even stand with head sinking progressively lower. Symptoms may develop rapidly with birds moribund and dead within few hours.

2.1.3. Necropsy findings

Enlargement of liver and kidneys, characteristic massive accumulation of lipid in the liver and kidneys.

2.1.4. Treatment

Supplementation of diets with biotin, or increase in fat or protein content can cause significant reductions in incidence of FLKS.

2.2. Fatty liver hemorrhagic syndrome (FLHS)

2.2.1. Etiology

FLHS cannot be produced artificially unlike FLKS. Thus, factors which increase that the position in liver may predispose the layer to FLHS.

This condition is recognized in birds shifted from floor to cage and it is characterized by fat deposits, fatty livers and drop in egg production. Cause of this syndrome is not completely understood. In any case, FLHS could be caused by a combination of:

2.2.1.1. Physiological factors

1. A very high fatty acid synthesis in the liver and
2. A defective transport mechanism which is necessary for the discharge of these fatty acids via the blood to the various organs.
3. A high potential egg production does make the layer physiologically more sensitive to FLS. A higher intensity of lay is connected with a high estrogen activity which also has a stimulating effect on the manufacture of fat in the liver.

2.2.1.2. Housing conditions

1. The housing conditions of the birds also contribute towards the occurrence of FLS. This metabolic disease usually affects battery hens. The lack of space in which to move about (exercise), combined with high feed intake can stimulate fatty degeneration.
2. In caged birds, there is no access for droppings; hence, shortages of B vitamins are also a likely cause.

2.2.1.3. Environmental factors

Surrounding temperature also plays an important role. FLS mainly occurs when ambient temperatures are high. Heat stress can overtax the metabolism so much that imbalance occurs even sooner.

2.2.1.4. Feed factors

1. Over consumption of ME. However, excessive feed intake is not usually

recorded although it is suggested that energy requirements are reduced as a result of minimal exercise and/or higher temperature without concomitant reduction in food intake. Hence, the condition may be due to excessive calorie intake and/or decreased energy use, or due to stress of high egg production etc.

2. Higher levels of unsaturated fatty acids, especially linoleic acid reduce the incidence. Therefore, Soya oil, sunflower oil and corn oil are preferable fat sources
3. Toxins arising from moulds (aflatoxin in particular) and erucic acid which comes from rape seed products can cause degeneration of the liver
4. Lipotropic factors in feed, choline and inositol, in particular protect against fatty degeneration of liver.

2.2.2. Symptoms

Similar to FLKS, sudden death. The birds may be overweight, unduly nervous with reduced egg production. Mortality, normally, will be less than 5% but birds in good health suddenly die.

2.2.3. Lesions

Most important lesions are ruptured liver leading to massive hemorrhage (may be the cause of death), fatty liver, large fat deposits may be seen lining abdomen and covering the intestines. Pale and swollen kidneys. Affected birds when killed and examined show fatty liver with multiple small hemorrhages, pale comb, wattle and musculature. Large blood clots may be seen on the surface of the liver and the body cavity. Liver will be yellow, greasy and mush-like inconsistency.

2.2.4. Prevention and control

1. Decreasing the daily ME intake by means of feed restriction and/or reducing the ME value of the feed can be considered. If the carbohydrate content is decreased by means of increasing the amount of fat which is rich in linoleic acid, the ration will become less sensitive to FLHS because such high fat diets retard hepatic lipogenesis.
2. Increasing certain vitamins and lipotropc factors (lecithin, choline, inositol, betaine and methionine) in feed.
3. Lecithin in an unknown way decreases appetite and hence has a special value in controlling FLHS.
4. Vitamin B12, methionine and betaine are involved in choline synthesis; similarly, vitamin B1 and biotin are vital in carbohydrate metabolism. Hence, special attention has to be paid in ensuring proper levels of these nutrients in controlling FLKS
5. Treatment suggested is to add 100 g Choline chloride, 10,000 IU Vitamin E, 12 μg Vitamin B12 and 896 g Ionositol per tonne of mash in addition to the levels normally recommended.

3. Leg disorders

Improved growth rate in broilers induced by genetic maneuvers has resulted in greater strain on legs (tibia), therefore, in the beginning, due to very high specific-growth rate in broilers, there will be some sort of bending of tibia which, later on, gets corrected by itself when the specific-growth rate falls. However, many diseases and/or disorders manifest with impaired gait or reluctance to walk. Etiology of each of the leg disorders is not very easy to be defined. However, some of them which could be due to nutrition are:

3.1. Chondrodystrophy, enlarged hock disorders, turkey-syndrome '65

Most rapidly growing bones are most affected and the bones become shortened, thickened and more frequently, secondary bowed. Hock joints become enlarged and knobby with the birds becoming cow-hocked or bow-legged with or without slipping of gastrocnemius tendon.

Deficiency of manganese, choline, niacin, folic acid, zinc and pyridoxine also can cause similar condition. However, with modern-day balanced diets, it is rare to expect deficiency conditions. In any case, in turkeys, deficiency of niacin can cause these leg disorders of a considerable degree.

3.2. Rickets and osteomalacia

Bone maturation by mineralization occurs within first fortnight of age. Thus, defective mineralization can cause rubbery bones which, in growing birds, is called rickets and in adults, osteomalacia. Thus, the condition is most frequent during 2 to 3 weeks of age and detected early can be treated with water-soluble Vitamin D3 preparations. Cage layer fatigue is the commonest form of osteomalacia image there is no osteoporosis (organic matrix deficiency). Calcium and/or phosphorus deficiency also causes osteomalacia and rickets. In calcium deficiency, severe standing will always been noticed before skeletal changes and most of the problems of rickets and osteomalacia is due to deficiency of Vitamin D3 and/or available phosphorus. Mycotoxicosis (aflatoxins and ochratoxins) too can predispose for skeletal deformities similar to rickets by interfering with mineral absorption.

3.3. Twisted, bent and bowed legs and rotated tibias

Seen in broilers with an incidence not exceeding 5%, normally in those flocks which are grown to older ages. The syndrome is higher in caged birds than on floor. It could be due to autosomal recessive gene. High levels of thyroid-blocking agents (like thiouracil), or 5% propylene glycol, high levels of rape-seed meal, certain types of sorghum, fungal toxins, manganese deficiency etc. There is no reduction in skeletal strength.

3.4. Tibial dyschondroplasia (focal osteodystrophy)

This is quite common in all meat-type birds although it does not cause leg weakness. It is characterized by uncalcified plugs of avascular, hypertrophic cartilage in the proximal metaphyses of the tibiotarsi or, sometimes, tarsometatarsi. The etiology is not yet clear but high chloride diet, excessive pressure due to primary skeletal or postural abnormality resulting from very rapid early growth has been suspected. If the plugs bulge outwards into bone cortex, locomotor problems can appear following bowing and/or fracture of proximal end of tibia.

3.5. Spondylolisthesis (Kinky Back)

Seen only in broilers at the level of 6th thoracic vertebra during 3 to 6 weeks of age. In severe cases, leg weakness or paralysis may follow. This condition is probably genetic and severe restriction of early growth can prevent this condition.

3.6. Aseptic necrosis of femoral head

This condition is seen usually in turkeys; response to dietary supplementation of 0.5 to 2.5 ppm of molybdenum (as sodium molybdate).

3.7. Plantar pododermatitis

Characterized by lesions on ventral foot-pads in meat-type chicken and turkeys. It may be due to some corrosive substance in the litter, possibly arising from droppings. The initial change is necrotic degeneration of the epithelial cells followed by ulceration and inflammation. This is more common in overweight birds and the lesions are seen on pressure points such as keel and backs of hocks.

4. Altitude and gravitational force

High altitude of 3800 m elevation induced hypoxia in adult chicken with related changes like hypertrophy of heart, increased hematocrit (27%), erythrocyte number (33%), hemoglobin levels (36%), reduced number of shell pores and the like. These changes adequately protect birds against effects or altitude although reductions in egg production (by 5 to 6%) and hatchability are unavoidable with more morphogenic defects and reduced rate of embryogenesis.

Variation in gravitational force is not a natural phenomenon. However, changes due to alterations in gravitational force simulated those in birds acclimatized to high altitude (3800 m). Egg production reduces mainly due to the fact that avian oviduct prolapses during positive gravity achievement. Weightlessness *per se*, on the other hand, had little, if any, effect on hatchability probably because virtually all prenatal life in higher vertebrates is spent in a suspended or weightless state.

5. Over-production

Although the increasing egg production is a commercial proposition, considering that 16 kg of calcium has to be mobilized by a hen in one year, the limitation of available bone calcium itself is sufficient precludes the idea of producing super-producers. Multiple or super-ovulations are not practical because they cannot be calcified in shell gland and ME and other nutrient demand required force the animal into debility and finally the very process becomes counterproductive. The metabolic effort of the laying hen is extraordinary by considering that in one year, while producing 250 to 275 eggs, she deposits 2.0 kg protein, 1.7 kg fat, 0.6 kg calcium and more than 10 kg water. Therefore, it is extremely essential to provide ideal conditions in terms of management encompassing feed and water, temperature and humidity, disease control, floor, feeder and waterer space etc. to realize ideal/optimum production by the hens.

6. Hemorrhage syndrome

(Aplastic anemia, Hemorrhagic anemia, Hemorrhagic disease)

This is a blood dyscariasis of chicken characterized by hemorrhages in muscles, internal organs and aplastic bone-marrow.

6.1. Etiology

1. Chemicals – Vitamin K deficiency.
2. Coccidiostats – Sulfonamide toxicity, especially Sulfaquinoxaline.
3. Mycotoxins – doubtful.
4. Virus – Adenovirus.

6.2. Symptoms

This disease is primarily encountered in chicken between 3-15 weeks (mostly 5-9 weeks) of age. The symptoms are paleness or icteric discoloration of tissues about the head, ruffled feathers, huddling and sometimes diarrhea and hemorrhages in the anterior chamber of the eye. Morbidity highly variable whereas, mortality usually averages 5-10%.

6.3. Gross lesions

Hemorrhages in skin, musculature and viscera. Pale and fatty bone-marrow is of high diagnostic value. Hemorrhages may be seen in mucous membranes of various parts of the digestive system, kidney, spleen and myocardium

6.4. Diagnosis

Primarily by lesions and demonstration of etiological agent wherever applicable.

6.5. Treatment

No specific treatment. It has been controlled by

1. Judicious use of chemicals and Coccidiostats
2. By minimizing or avoiding mycotoxins and
3. If virus-induced, by taking proper measures for the eradication of the virus. However, the virus-induced anemia is not yet completely understood.

7. Sudden-death syndrome (SDS, Angara disease)

This is characterized by pulmonary edema, hydropericardium, full digestive tract and abnormally high mortality (up to 70%) in broilers during 10-15 d of its course.

7.1. Etiology

Etiology is unknown; but thought to be due to instant stress, cytotoxic contaminants, virus, *and Aegiptionella* infestation.

7.2. Lesions

Marked lesion is pulmonary edema leading to hypertension, hydropericardium and ventricular failure.

7.3. Control

The only method to alleviate this condition is to reduce the growth rate by restricting the feed intake. Supplementation of either corn or wheat-based broiler rations with meat-meal reduce the incidence of SDS whereas supplementation of either feed or water with taurine, a component of meat-meal, failed to protect the birds against SDS. It is hypothesized that heart-failure may be caused by pulmonary congestion associated with increased pressure in the pulmonary artery which conveys blood from the right ventricle to lungs. It is possible that SDS is related to ascites.

8. Urolithiasis

This condition is common in older laying chickens (analogous to gout). Brittle, white, staghorn calcium urate calculi form in one or both ureters. Most cases are due to feeding high-calcium laying feed to hens not in egg production, infection with infectious bronchitis virus, or severe vitamin A deficiency. If blockage is complete, acute post-renal failure develops, and birds die with acute urate deposition on visceral surfaces and less commonly in joint spaces. If blockage is incomplete or unilateral, chickens survive in compensated renal failure, and chronic urate deposits form in joint spaces.

Chapter **68**

Miscellaneous Diseases and Conditions

1. Miscellaneous diseases

1.1. Ascites syndrome

(Water-belly, Right ventricular failure, pulmonary hypertension syndrome)
(The Merck's Veterinary Manual, 2003)

Ascites is an accumulation of non-inflammatory transudate in one or more of the peritoneal cavities or potential spaces. The fluid, which accumulates most frequently in the two ventral hepatic, peritoneal, or pericardial spaces, may contain yellow protein clots. Ascites may result from increased vascular hydraulic pressure, vascular damage, increased tissue oncotic pressure, decreased vascular oncotic (usually colloidal) pressure, or blockage of lymph drainage.

1.1.1. Etiology

Varied; includes chronic hypoxia, toxicity of *Crotalaria spectabilis*, Furazolidone and coal-tar disinfectants; phosphorus, Se and vitamin E deficiencies, salt toxicity (through water or feed).

The most common cause of ascites is increased vascular hydraulic pressure in the venous system, which is most commonly caused by right ventricular failure (RVF) or hepatic fibrosis.

In poultry, RVF is usually secondary to valvular insufficiency and may result from inflammatory (myocarditis, valvular endocarditis) or degenerative disease of the myocardium or valves or from congenital heart disease. In turkeys, spontaneous cardiomyopathy is a common cause of ascites.

However, the most common cause of ascites in meat-type chickens is RVF in response to increased pulmonary arterial resistance. Pulmonary hypertension occurs frequently in chickens secondary to the hypoxia of altitude with the resultant polycythemia and increased blood viscosity. It also occurs frequently secondary to the red blood cell rigidity of sodium toxicity and less frequently from lung pathology.

When ascites occurs at low altitudes in meat-type chickens, which have a high metabolic oxygen requirement, it is usually caused by primary or spontaneous pulmonary hypertension because of insufficient capacity of the pulmonary capillaries.

In poultry, liver damage may be caused by aflatoxin or by toxins from plants such as *Crotalaria*. In broiler chickens, obstructive cholangiohepatitis (caused by *Clostridium perfringens* infection) is the most common cause of the liver damage, which results in ascites. In both meat-type ducks and breeders, amyloidosis of the liver frequently causes ascites.

1.1.2. Causes of hypoxia

Winter heating by gas heaters with windows closed can result in depletion of oxygen and formation of carbon monoxide, ammonia, water vapor (relative humidity) and dust. This predisposes to ascites and pulmonary damage. Nodules may develop in lungs limiting the respiratory efficiency thereby contributing to hypoxia. Pellet feeding during very fast growth further increases the oxygen demand in comparison to match-feeding. Incidence of ascites is higher in males (because of higher growth rate) and in certain breed-crosses. Nitrofurans and high energy diets also predispose birds for ascites. Ascites is seen in young and fast growing broilers. Hypoxia has been now thought to be the major factor and thus, ascites is recorded commonly at high altitude.

1.1.3. Predisposing factors

Predisposing factors for ascites are increased oxygen demand (e.g. cold), reduce oxygen-carrying capacity of the blood (e.g. acidosis, carbon monoxide), increase blood volume (e.g. sodium), or interfere with blood flow through the lung (e.g. by lung pathology that narrows or occludes capillaries, by increased RBC rigidity, or by polycythemia with increased blood viscosity) may result in flock outbreaks of PHS (Pulmonary Hypertension Syndrome) with or without ascites.

1.1.4. Pathogenesis

Bird lungs are rigid and fixed in the thoracic cavity. The small capillaries can expand very little to accommodate increased blood flow. Lung size in proportion to body weight, and particularly to muscle mass, decreases as meat-type chickens grow. The predicted sequalae are:

Hypoxia → Exceeds the ability of heart/lungs of broilers to send enough oxygen to the tissues → Tachycardia → Enlargement of heart, especially right side → Enlargement so great, it allows the blood to flow back into it after every beat (right-sided heart-failure) → Blood backs-up into the body → Hypertension within small blood vessels and congestion of all abdominal organs → Seepage of fluid into the abdomen and around the heart → Ascites and hydropericardium with accumulation of straw-colored serum-like fluid. In addition, constriction of capillaries may result which impedes the flow of erythrocytes compounding the effects of hypoxia.

1.1.5. Symptoms

Clinical signs are not seen until RVF (Fight Ventricular Failure) occurs and ascites develops. Clinically affected broilers are cyanotic, the abdominal skin may be red, and peripheral vessels congested. Because growth stops as RVF develops, affected broilers are smaller than their pen mates. The ascites increases the respiratory rate and reduces exercise tolerance. Affected broilers frequently die on their back; not all broilers that die from PHS have ascites. Death may occur suddenly before clinical signs are seen.

1.1.6. Lesions

Most lesions are due to increased venous hydraulic pressure secondary to RVF. There is a variable amount of clear yellow fluid and clots of fibrin in the hepatoperitoneal spaces. The liver may be swollen and congested, or firm and irregular with edema, and have clotted protein adherent to the surface. It may be nodular or shrunken; it may be white with subcapsular edema and a thickened capsule, or have large or small blebs of fluid between the capsule and the visceral peritoneum. Hydropericardium is mild to marked, and occasionally there is pericarditis with adhesions.

There is right ventricular dilatation and mild to marked hypertrophy of the right ventricular wall. The right atrium and vena cava are dilated. Occasionally, there is thinning of the left ventricle. The lungs are extremely congested and edematous. The intestine may or may not be empty.

1.1.7. Pathology

Elevated hematocrit, hemoglobin and erythrocyte count to counter hypoxia. This leads to hypertrophy of myocardium and, in turn, congestion of all organs in abdomen.

1.1.8. Control

Ascites caused by PHS can be prevented by reducing the birds' oxygen requirement; slowing growth or reducing feed lowers the metabolic oxygen requirement. Environmental temperature, humidity, and air movement should be controlled to prevent excessive loss of body heat. Ascites caused by other factors (e.g. sodium, lung damage, liver damage, etc) can be prevented by avoiding the etiologic agents involved. Altitudes >3000 ft (900 m) are unsatisfactory for meat-type chickens, and growth must be slowed to prevent mortality. More care to prevent chilling is also necessary at higher altitudes.

Adequate ventilation has to be ascertained. Disease(s) of air-sac/lungs must be checked. Floor space can be increased to provide more ventilation to the birds. Early feed restriction by light-restriction or reducing energy and/or protein is also efficient in controlling ascites.

1.2. Hock-burn and related conditions

(Breast-blisters, Pododermatitis, Scabby-hip syndrome, Contact dermatitis)

Hock-burn is a pressure-induced necrosis of the skin; related problems are breast-blisters, pododermatitis and scabby-hip syndrome; and a collectively referred to as contact dermatitis.

1.2.1. Etiology

Cause is mainly poor litter condition. Initially, they appear at brown-black erosions, later on, the skin is damaged followed by microbial invasion. The incident is variable between 15-20%. The economic importance is due to carcass downgrading. Higher moisture content of litter is the primary cause and birds recover if wet-litter is avoided/replaced.

Causes of wet-litter are many but primarily due to diarrhoea with the results in increased Moisture elimination through feces. Many nutritional factors lead to diarrhoea like high sodium, chlorine and potassium in diet, high soyabean meal (rich in potassium) in ration, use of tapioca and barley in the diet, and high-fat diets (which result in sticky feces). Diseases causing enteritis also result in diarrhoea.

High temperature and ammonia, poor ventilation and overcrowding also can predispose for wet-litter. Improper placement of sulfur and leaky sulfur can invariably pose problems of wet-litter. Therefore, proper litter-management augmented with balanced diet is the only way to avoid contact dermatitis.

1.3. Keel disease

Seen in young ducklings causing dehydration and emaciation; is a managemental disease. Proper sanitation of hatchers, warm brooding facilities, fresh water and good feed can prevent the disease.

1.4. Bumble foot

This is localized infection in the foot pad causing bulbous swelling of the foot-pad and surrounding tissues. This may be uni- or bi-lateral. In most cases, it is due to injury to the ball of the foot. At later stages, the birds become lame and show inappetance and stop laying. At early stages, surgery and antibiotic therapy can alleviate the condition.

1.5. Gout

Gout is the accumulation of urate deposits in joints (articular gout) or viscera (visceral gout).

1.5.1. Etiology

Kidney failure/insufficiency/damage.

Acute urate deposition occurs after rapidly progressing renal failure, or as a terminal event with acute decomposition of chronic renal disease.

1.5.2. Lesions

(The Merck's Veterinary Manual, 2003)

In acute cases, deposits develop most commonly on the pericardium, peritoneum, and liver capsule, and rarely on synovial surfaces of joints and tendons. They are usually present for too short a time to induce significant inflammation. Most clinical cases of acute renal failure and urate deposition in commercial poultry are due to dehydration, ingestion of feed containing >3% calcium by non-laying chickens, renal infection by nephrogenic strains of infectious bronchitis virus, or infection with avian nephritis virus. Other avian species commonly develop visceral deposits secondary to nephrotoxin exposure, most commonly aminoglycoside antibiotics or heavy metals.

Chronic urate deposition is less common and occurs after long-term increases in serum levels of uric acid. Deposits develop on synovial membranes in the toes and wing joints and incite a chronic granulomatous reaction to urate crystals (tophus). Chronic urate deposition is most frequently seen in chickens that have hereditary defects in uric acid metabolism or that are fed excessive protein.

1.6. Pullet disease

(Blue comb, Avian Monocytosis)

1.6.1. Etiology

It is associated with hot weather, water deprivation, toxin and possibly a virus infection. Affected birds have an increase in circulating monocytes in their blood. The disease is not much prevalent now.

1.6.2. Symptoms

A sudden onset condition of chickens early in lay with depression, loss of appetite, crop distension, dark comb and wattles, watery diarrhoea, dehydration and drop in egg production.

1.6.3. Morbidity and mortality

High morbidity and a mortality of 0–50%.

1.6.4. Lesions

Dehydration, skin cyanosis of head, skeletal muscle necrosis, crop distended, catarrhal enteritis, mucoid casts in intestine, liver swollen with foci, kidneys swollen and pancreas chalky.

1.6.5. Diagnosis

History, lesions, elimination of other causes.

1.6.6. Treatment

Adequate water supply, antibiotics, molasses, multivitamins in water.

1.6.7. Prevention and control

Good management, hygiene, diet and water supply.

2. Miscellaneous conditions

2.1. Autointoxication

This is due to absorption of waste-products of metabolism or of decomposition within intestines. For example, high fiber diets result in obstruction of digestive tract which will be followed by decomposition and autointoxication. The signs include anorexia, increased water intake, depression, weakness and finally nervous symptoms shortly before death.

2.2. Chemotherapeutic drugs

2.2.1. Sulfonamides

The toxic levels cause blood dyscariasis, kidney and liver dysfunction, delayed clotting time and super infection. Gross symptoms are ruffled feathers, depression, paleness, icterus, poor weight gain and hemorrhagic syndrome. The toxic level depends on the Sulfa drug in question.

2.2.2. Nitrofurans

Symptoms of toxicity are depression in some chicks and hyperexcitability in others. Recommended level is 0.01% in feed.

2.2.3. Nicarbazine

Reduced egg production and egg size, mottling of yolk, reduction in shell quality and hatchability are the predominant symptoms of nicarbazine toxicity. It is not advisable to exceed a level of 0.01% nicarbazine in feed.

2.2.4. Zoalene

Toxic levels cause stiff-neck, staggering, vertigo and tumbling-over when excited. The recommended level is 0.025 to 0.05% in feed.

2.2.5. Nitrophenide

Recommended level is 0.025% in feed. Toxicity is characterized by abnormal locomotion, retarded growth, ataxia and death.

2.2.6. Monensin

Toxic level depends on species. The symptoms are drowsiness, thirst and panting, flaccid paralysis and mortality.

2.3. Disinfectants

2.3.1. Cresol

Symptoms of toxicity are depression, weakness, huddling, rales, gasping, wheezing and extension of head and neck.

2.3.2. Potassium permanganate

In water, a concentration of 1: 500 or below seems to be non-toxic. There will be no clinical signs of toxicity. Gross lesions include severe cauterization of crop wall and hemorrhages and other tissues coming in contact with the chemical.

2.3.3. Quaternary ammonium compounds

The toxic levels depend on the chemical in question. The symptoms of toxicity are reluctance to drink, restlessness, persistent swallowing, spitting, foamy ocular discharge, nasal discharge, facial swelling and gasping, ataxia, conversions and death.

2.4. Chemicals

2.4.1. Copper sulfate

Toxic level depends on the species of poultry. Toxic symptoms are simulation followed by severe depression, weakness, conversions and death.

2.4.2. Mercury compounds

Oxides and chlorides of mercury are toxic. Symptoms of toxicity are incoordination and leg weakness followed by complete prostration.

2.4.3. Nitrates

Nitrates of potassium and sodium are toxic. Toxicity causes thirst, anorexia, vomition, diarrhoea, hypothermia and cyanosis of comb, wattle and skin.

2.4.4. Sodium bicarbonate

It is non-toxic at levels less than 0.1%. Toxicity characterized by weakness, increased water intake and watery droppings. Lesions produced are nephritis and visceral gout.

2.4.5. Sodium chloride

Salt poisoning can occur in most species of poultry. The level of salt to induce toxicity not only depends on species but also on its age, physiological status and environmental conditions. The minimum lethal dose for chickens seems to be 4 g/kg body weight. Symptoms of salt toxicity are inappetance, somnolence, thirst, dyspnoea, opisthotonus, convulsions and inability to stand. Lesions chicks are nephritis, myocardial hemorrhage, ascites and subcutaneous edema.

2.4.6. Calcium

Excess calcium causes diarrhoea, intense thirst and analysis with tetanic convulsions or opisthotonus.

2.5. Insecticides and pesticides

2.5.1. Chlorinated hydrocarbons

The most common of this group contaminating poultry feed and products is dichlorodiphenyl trichloroethane (DDT). A drop in egg production ranging from 20 to 70% over a period of one month has been recorded with a dietary DDT levels of 0.031 to 0.25%. Hatchability reduced to zero at 0.125% DDT. In 4 to 16d-old young chicks, 0.05% DDT caused 100% mortality. DDT toxicity is manifested as loss of weight, molting, cessation of lay, tremors, ataxia, walking on hocks and hyperexcitability. Concentration DDT in body fat is of public importance and DDT residues are seen in meat and eggs. Therefore, it is advisable not to use DDT in any form as an insecticide or for any other purpose in poultry production.

2.5.2. Organophosphorus compounds

The most commonly-used organophosphorus insecticides are Malathion and parathion. The maximum Malathion level for dusting chicken is 4%. The toxic symptoms include drowsiness, ataxia, resting on hocks, paralysis and death. Stringy mucus hanging from the beak, cyanosis of head and blood-tinged diarrhoea may be seen. At no time, birds are hyper-excitable, which aid in diagnosis. Parathion is the most toxic of the organophosphorus insecticides and his must be used with caution. The toxicity results in increased salivation and lacrimation followed by miosis, intermittent swallowing and gasping, and finally, clonic muscular convulsions, ataxia, tonic convulsions and death (Table 67.1).

2.5.3. Rodenticides

Zinc phosphide baits are sometimes used to control rodents in poultry sheds. Birds may get access to the bait due mainly to careless handling and placement of the bait. The lethal dose for birds has been estimated at 7 to 15 mg/kg body weight. Within an hour, birds become depressed and show ruffled feathers. This will quickly be followed by diarrhoea, weakness, nervous symptoms and death. Sublethally doses cause depression and greenish diarrhoea. High doses resulting in mortality without premonitory signs. Hydropericardium, ascites and enteritis (in the duodenum) are the lesions recorded.

Table 68.1. Tolerance and residues of common insecticides and pesticides

Insecticide	Tolerance (ppm)	Pre-slaughter withdrawal Period (d)	Concentration for dusting (%)	Concentration for spraying (%)
Carbaryl (Sevin)	5 in meat and fat	7	5	0.5
Coumaphos (Coral)	1 in meat and by-products	-- --	1	0.25
Malathion	4 in meat and by-products; 0.1 in eggs	-- --	5	0.5
Propoxyl (Baygon)	Zero degree	Not established	5	0.5

2.5.4. Others

Fumigants and gases (Carbon monoxide, Naphthalene, and Nicotine sulfate), fungicides (arasan and captan), several insecticides other than DDT, malathion and parathion (like Chlordane, Dieldrin, Heptachlor, Lindane, Chlorinated biphenyls, Diazinon and Selvin), metals (Lead, Selenium and Phosphorus), mycotoxins other than aflatoxins (Ergot, Oosporein and Fusariotoxin), phytotoxins in several plant issues (including cotton seed, rape seed and certain algae), rodenticides other than zinc phosphide (α-napthyl thiourea, sodium mono fluoracetate, arsenicals and strychnine) and others (like coal tar, ethylene glycol, toxic fats and certain insects) can cause physiological disturbances; but their incidence is very rare.

2.6. Prolapse

This is evagination of oviduct (vagina) and rectal region through vent; that is invariably followed by pecking leading to irreversible damage. This condition can be seen under the following situations:

2.6.1. Etiology

1. More prevalent in chicks hatch in spring coming to lay during June to August – because of increasing day lengths they are subjected to.
2. Early-maturing and heavier birds are more susceptible.
3. High producing birds subjected to improper management.
4. The duration of light if increased, birds mature earlier without proper physical development predisposing such birds to prolapse; particularly, the lower strength of abdominal muscle expectable in early-maturing birds.
5. If sufficient nests are not provided, birds lay eggs on floor, get infection onto the oviduct during oviposition.
6. If very big egg has to be laid, prolapse may occur.
7. Overcrowding of birds leads to competition for space, feed and water.
8. Birds which have enteritis are alsoproned to prolapse due to excessive bowel movements.
9. Nutritional deficiency of calcium can also cause prolapse.

2.6.2. Precautions to avoid prolapse

1. Beak-trimming of all birds 2. Smaller birds may be removed and reared separately 3. Proper floor space depending on age of birds 4. Birds have to be weighed at least once a week or fortnight to calculate food requirements and 5. Balanced poultry feed has been offered to birds.

2.7. Deformities, injuries, vices

2.7.1. Blue back

A permanent dark discoloration of skin on the back and sometimes of side and breast; seemed commonly in turkeys with dark plumage; likely to be due to recessive hereditary factor or due to direct sunshine on damaged developing feathers. Can be minimized by prevention of feathers picking and mating (injury).

2.7.2. Breast blisters and breast buttons

In chickens and turkeys, a bursa lined with synovial membrane normally exists over the anterior projection of the keel bone. When this bursa becomes inflamed by trauma or infection, fluid or exudate accumulates and appears as a fluid-filled blister 1-3 cm in diameter. Common in toms and those with no dimple on the breast. Lowers market grade.

Causes of trauma to this bursa include poor feathering, hard flooring, and leg weakness, which is associated with increased recumbence. Some young turkeys have pointed keels, which can lead to increased bursal trauma. Infectious causes of sternal bursitis include *Mycoplasma synoviae*, *Staphylococcus*, and *Pasteurella* spp.

Breast buttons are lesions in a similar location that have a hard crust on the surface and a core of dead skin and granulomatous reaction extending into the subjacent subcutis. Unlike breast blisters, they are likely to be chemical burns due to prolonged contact of poorly feathered skin with wet litter containing ammonia or toxins.

2.7.3. Pendulous crop (baggy/sour/dropped crop)

Incidence of pendulous crop is low in flocks of chickens and turkeys; in case of latter seen in breeds which are not broad-breasted. Crop and supporting tissue weakens – food stagnates – damage to crop lining – fermentation and souring and foul smell. Digestion is impaired, and afflicted birds become thin or emaciated. If these birds survive, they often are condemned at processing. Treatment is ineffective and is particularly seen

The etiology is not known, but a hereditary predisposition has been suggested in turkeys. Incidence may increase when the predisposed birds are exposed to hot weather forcing them to drink too much liquid and experimentally, by feeding rations that contain cerelose as a substitute for starch. There is no known efficacious treatment.

2.7.4. Cannibalism

Same as described under Chapter "Beak-trimming".

Chapter **69**

Diagnosis of Poultry Diseases

Disease diagnosis is extremely important in any animal enterprise because it is the prerequisite not only for control of the disease at the facility in question but also its spread to other facilities. Development of preventionand control strategies depend primarily on identification of the disease. Several techniques and procedures are available for diagnosis which also depends on factors like type of disease, nature of pathogen, site of prediliction of pathogen etc. Collection of information and material for diagnosis must be done in a most orderly fashion so that none of the possible clues will be overlooked. Whether essential are not, it is preferable to collect blood and tissue samples which can be discarded later, if found superfluous.

Most often it is not advisable to wait for the final diagnosis after all possible procedures because, by the time all the tests are conducted, if the flock is left without preliminary treatment and/prevention based on tentative diagnosis made depending on clinical symptoms and experience of the producer, probably no flock will be left at the facility!

It is not possible to give an exhaustive account of all the diagnostic procedures. Only brief notes on some of the important diagnostic procedures are provided below as a quick reference; however, it is necessary that each of the procedures described may have to be modified depending on the disease being investigated for. The reader is advised to refer to many publications in the fields of Veterinary microbiology, Veterinary clinical medicine, Veterinary clinical pathology, Veterinary parasitology etc. for further details. The OIE Manual of standards is also a very useful publication for details on epizootic diseases and their diagnosis. Several websites also provide excellent information; for instance, Univ. of Pennsylvania website, The Merck Veterinary manual, Ohio State Univ. website etc.

1. History

See Chapter "Necropsy examination" for details.

2. External examination

See Chapter "Necropsy examination" for details.

3. Collection of blood samples

See Chapter "Necropsy examination" for details.

4. Killing of ailing birds

See Chapter "Necropsy examination" for details.

5. Necropsy examination

See Chapter "Necropsy examination" for details.

6. Care of specimens

See Chapter "Necropsy examination" for details.

7. Tentative diagnosis

7.1. Clinical symptoms

The producer at times will be able to make a tentative diagnosis depending on the symptoms observed. This may be particularly true if she/he has earlier encountered similar situation or when the symptoms themselves are pathognomonic that corrective measures can be initiated at once. For instance, a disease like coccidiosis can be detected with sufficient accuracy without waiting for a laboratory report; at best, microscopic examination of the droppings with or without necropsy examination should be sufficient.

7.2. Microscopic examination

Many large-scale poultry producers are able to perform microscopic examination of droppings for detecting worm infestation and coccidiosis. In addition, routine microscopic examination of the droppings is also performed preceding deworming. However, detailed microscopic examination of blood and other material is best done at the disease diagnostic facility.

7.3. Necropsy examination

Necropsy examination of dead and/or ailing birds is a very important step in disease diagnosis and it should be performed by a qualified veterinarian/poultry specialist. Material collected during necropsy examination must be properly handled and sent to a diagnostic laboratory. For details see Chapter "Necropsy examination".

8. Specimen disposal

See Chapter "Necropsy examination" for details.

9. Material to be sent for a diagnostic laboratory

See Chapter "Necropsy examination" for details.

10. Laboratory diagnosis

A poultry disease-diagnosing laboratory is similar to most disease-diagnosing laboratories in the methodology toward the diagnosis. The procedure usually includes:

1. Scrutiny of the history obtained and recording relevant points about clinical symptoms and necropsy findings.
2. Ascertaining that the requisite samples from the desired number of birds have been sent following the prescribed procedure for each of the specimens received.
3. A laboratory will use primary and secondary tests in making a diagnosis. Primary tests include bacterial cultures, bird inoculations, direct microscopy of tissues and body fluids, and serum tests.
4. Secondary tests might include virus isolations, chemical analysis and microscopic examination of prepared tissues.

Results of all tests conducted would be considered in making the final report. Time required to get the final report might range from one day to six weeks. The longer period would be necessary for culture growth and identification of some types of bacteria such as avian tuberculosis and some fungi.

Many of the laboratories will give a preliminary report at or soon after the first examination of the birds. Then it is often necessary to change this tentative diagnosis as more information is accumulated. The objective is to get a true identification of the source of the disease and then to make sound recommendations that will help return the birds to good health with the least loss.

10.1. Microscopic examination

Microscopic examination of blood, droppings, skin/feather material, if obtained at the laboratory, is performed as per standard procedures. Details on preparation and staining of blood smears, fecal examination, and examination of skin/feather material can be obtained from many standard publications available on each of those topics.

10.2. Histopathological examination

The tissues obtained will be subjected to microscopic examination to detect histopathological changes; histopathology helps differential diagnosis of diseases especially when similar clinical symptoms are noticed. For instance, respiratory symptoms are expected in several bacterial, viral and fungal diseases.

10.3. Culturing of organisms

When it is suspected that the disease is caused by bacteria, fungi or viruses, a diagnostic laboratory has to culture the organisms for identification. Standard methods are available in many publications.

Table 69.1. List of tests for international trade*

Disease	Alternative tests
Highly pathogenic Avian Influenza**	AGID, HI
Newcastle disease**	HI
Infectious bursal disease	AGID, ELISA
Marek's disease	AGID
MG infection	Agg., HI
Avian chlamydiosis	---
Fowl typhoid and Pullorum disease	Agg., Agent id.
Avian infectious bronchitis	VN, HI, ELISA
Avian infectious laryngotracheitis	AGID, VN, ELISA
Avian tuberculosis	Tuberculin test, Agent id.
Duck virus hepatitis	---
Duck virus enteritis	---
Fowl cholera	---
Fowl pox***	---

* None of the Avian diseases has "Prescribed" test(s)

** Diseases under List A; others under List B

*** Not covered in the *Code*

Agent id. Agent identification; Agg., Agglutination test; AGID Agar Gel Immunodiffusion; ELISA Enzyme-Linked Immuno-Sorbent Assay; HI Hemagglutination Inhibition; VN Virus neutralization; --- No test designated yet

Adopted from : OIE Manual, 2000

10.4. Assessment of pathogenicity

10.4.1. Intravenous pathogenicity Index (IVPI)

The IVPI test is carried out as follows:

5. Fresh infective allantoic fluid with a HA titer of > 2^4 (> $\log_2 4$ or > 1/16) is diluted 1/10 in sterile isotonic solution.
6. 0.1 ml of the diluted virus is injected i/v into each of ten 6-week-old SPF chickens.
7. Birds are examined at 24-hr intervals for 10 d. At each observation, each bird is scored (subjectively) 0 if normal, 1 if sick (showing one of the signs), 2 if very sick (showing > 1 signs) and 3 if dead (at each of the remaining daily observations as well); the signs have to be listed specifically for the disease in question.
8. The IVPI is the mean score per bird per observation over the 10 d period. Obviously, an index of 3.00 means all birds died within 24 hr and IVPI of 0.00 means no bird showed any clinical sign over the 10 d period.

10.4.2. Mean Death Time (MDT) in eggs

1. Fresh infective allantoic fluid is diluted in sterile saline to give a tenfold dilution series between 10^{-6} and 10^{-9}.
2. For each dilution, 0.1 ml is inoculated into the allantoic cavity of each of five 9-10-d-old embryonated SPF fowl eggs, which are then incubated at 37°C.
3. The remaining virus dilutions are retained at 4°C and another five eggs are inoculated with 0.1 ml of each dilution 8 hr later and left at 37°C.
4. Each egg is examined twice daily for 7 d and the times of any embryo deaths are recorded.
5. The minimum lethal dose is the highest virus dilution that causes all the embryos inoculated with that dilution to die.
6. The MDT is the mean time in hr for the minimum lethal dose to kill all the inoculated embryos.

10.4.3. Intracerebral Pathogenicity Index (ICPI)

1. Fresh infective allantoic fluid with a HA titer of > 2^4 (> $\log_2 4$ or > 1/16) is diluted 1/10 in sterile isotonic saline with no additives, such as antibiotics.
2. 0.05 ml of the diluted virus is injected Intracerebral into each of ten chicks hatched from eggs from SPF flock. These chicks must be > 24 hr and < 40 hr old at the time of inoculation.
3. The birds are examined every 24 hr for 8 d.
4. At each observation, each bird is scored (subjectively) 0 if normal, 1 if sick, and 2 if dead (at each of the remaining daily observations as well).
5. ICPI is the mean score per bird per observation over the 8-d period.

10.5. Serological tests

Serology is used to detect antibodies in blood serum. Antibody detection can be qualitative or quantitative. In either case, sensitivity and specificity of the test are important considerations.

Sensitivity refers to the ability to detect minimal quantities of antibodies in a sample and specificity refers to the ability to detect antibodies against a certain/specific antigen (without any cross reaction). Statistically, sensitivity is the % of infected animals that are found positive and specificity is the % of uninfected animals found negative in the test (Cardoso, 2005).

High sensitivity minimizes false negative results whereas, high specificity tests minimize false positive results. Most often, more specific tests are not so sensitive and hence, require higher concentrations of antibodies to give a positive reaction. In some cases, like indirect immunofluorescence test, the specificity and sensitivity are high but the test itself is very costly (Cardoso, 2005).

The qualitative tests detect the presence/absence of antibodies and hence, the infection; results reported as positive/negative, respectively. Examples of qualitative tests are: rapid serum agglutination test (RSA), slow agglutination test (SAT), agar gel precipitation test etc. It is recommended to heat the sera samples from day-old chicks to be heated to 56°C for 30 min. for better results (Morrow and Hammond, 2003)

In qualitative/quantitative tests antibody titer is measured and expressed as the maximum dilution at which the antibodies were detected; but, not as an absolute number. Usually, the titer is expressed as dilution factor itself (1/4, 1/8, 1/16 etc.) or in corresponding logarithms (to base 10 or base 2; 4 $\log_{10}$ to mean 10^4 or 8 $\log_2$ to mean 2^8). Examples in this category are HI and serum neutralization tests (Cardoso, 2005).

In quantitative tests, the titers are expressed in absolute numbers depending on the working dilution used. These numbers are given in groups of titers and the values appear similar to those of qualitative/quantitative tests. ELISA falls into this category of tests (Cardoso, 2005).

10.5.1. Plate agglutination test

10.5.1.1. Purpose:

To detect the presence of serum antibody (e.g. *Mycoplasma gallisepticum* Infection (MG), *Mycoplasma synoviae* (MS) and *Mycoplasma meleagridis* (MM), pullorum/typhoid)

10.5.1.2. Method

1. Put one drop of sera on plate
2. Add one drop of antigen
3. Mix and incubate for two minutes
4. A positive reaction shows a clumping effect and indicates exposure to an infectious agent. It detects the presence of antibodies
5. Always run a positive and a negative control.

10.5.2. Agar gel immunodiffusion (AGID)

10.5.2.1. Purpose

To detect the presence of serum antibody (e.g. Avian Influenza (AI), Infectious Bursal Disease (IBD), Adeno and Reo)

10.5.2.2. Method

The test can be run in agar wells, on slides or petri dishes.

1. Antigen is placed in the center well. Positive control serum (antisera) is placed in wells on each side of the sample tested. A total of three samples can be tested in each pattern. The sample to be tested may be serum, plasma or yolks.
2. Dishes or slides are incubated at room temperature in a closed chamber for

a minimum of 24 hours

3. A white precipitant line will form between the antigen well and a positive sample.

10.5.3. Hemagglutination (HA) test

10.5.3.1. Purpose

To identify certain viruses (antigens) which hemagglutinate chicken RBCs (e.g. NDV or Avian Influenza). However, this is not a serological test but generally used as the basis for the Hemagglutination inhibition test; it detects antigen not antibody.

10.5.3.2. Method

1. Allantoic cavity fluid (possibly containing virus) from an embryonating chicken egg that had been inoculated with suspect material is removed
2. 0.1 ml of this fluid is placed on a glass plate and 0.1 ml of a 5.0% suspension of chicken RBCs is added
3. If a hemagglutinating agent is present, the RBCs will clump. It is determined by tilting the plate and observing the presence or absence of tear-shaped streaming of RBCs.
4. Highest dilution giving complete HA (no streaming) represents 1 HA unit (HAU)

10.5.4. Hemagglutination inhibition (HI) test

10.5.4.1. Purpose

To quantitate serum antibody to a specific avian antigen. Avian viruses which can agglutinate avian RBCs (hemagglutination) include NDV, Influenza, and others. In addition to viruses, Mycoplasmas are also capable of hemagglutinating avian RBCs. Antibodies directed against these organisms will inhibit hemagglutination.

10.5.4.2. Procedure

1. A constant amount of hemagglutinating (HA) antigen is added to each well in a microtiter plate
2. The test serum is then placed in the first well and serially diluted
3. The plates are incubated for one hour and then chicken RBCs are added to each well. If antibody is present in the test serum the RBCs will not agglutinate with the HA antigen
4. High negative wells will have a diffuse sheet of agglutinated RBCs covering the bottom
5. High positive wells will have a well circumscribed button of unagglutinated RBCs
6. HI titer is the highest dilution of the serum causing complete inhibition of 4 HAU of antigen. Only those wells in which the RBCs stream at the same rate

as the control wells (containing 0.025 ml RBCs and 0.05 ml PBS only) are considered to show HI

7. The validity of the result is assessed against a negative control serum, which should not give a titer > ¼ or > 2^2 or > $\log_2 2$, and a positive control serum for which the titer should be within one dilution of the known titer.

10.5.5. Enzyme-linked immunosorbent assay (ELISA)

10.5.5.1. Purpose

To quantitate serum antibody to a specific avian pathogen

10.5.5.2. Method

1. Viral antigen is precoated onto a microtiter plate
2. Serum is added to the coated wells. If antibody is present, it will bind specifically to the viral antigen. Unbound material is removed by a washing procedure
3. An anti-chicken antibody conjugated with horseradish peroxidase enzyme is next added to the viral antigen-antibody complex. This reagent binds to any chicken antibody that remains in the plate

Table 69.2. Commonly used serological tests

Antigen	Test	Remarks
ND	HI	HI antibody closely related to the VN antibody
	ELISA	Practical and good correlation with HI
IB	HI	Good test; but antigen production is complicated
	ELISA	Very variable result; difficulty in interpretation. High sensitivity
	SN	More specific and better for interpretation especially for differentiating serotypes. More costly and complex
	AGP	Specific; good for interpreting outbreaks in chicken. Low cost; but, not commonly employed
IBD	ELISA	Correlation with VN is good on the higher range and *vice versa*
	SN	More expensive and complex
	AGP	Used for diagnosis and initial evaluation of serotypes
EDS'76	HI	Good results
	ELISA	Good results
AE	ELISA	Variable results depending on vaccine used. More consistent results for outbreaks
Reo virus	ELISA	Good results but variable to some extent
	AGP	Mainly diagnostic. It is more specific
Salmonella	RPA	Sensitive test; used for screening
	SAT	Used for first serological confirmation; more specific
Mycoplasma	RPA	Sensitive test; used for screening
	HI	Confirmation or serological monitoring
	ELISA	

Source : Cardoso, 2005

4. Finally, O-phenylenediamine.2HCl (OPD) and hydrogen peroxide substrate are added to the wells of the plate. When antibody is present, the enzymatic reaction results in a color change which can be read visually or by a spectrophotometer.

Color development and intensity can be directly related to the amount of antibody in the original serum sample. Commercial kits are available for many of the poultry diseases (e.g. ND, IB, AE, AI, MG, IBD and others). Some laboratories have automated systems and computers which analyze results and generate a graphic and numerical report.

10.5.6. Virus neutralization (VN), Serum neutralization (SN)

10.5.6.1. Purpose

To quantitate antibody to specific viral avian pathogens.

10.5.6.2. Method

The test has two parts:

1. Neutralization - The virus and the serum are mixed in a test tube and incubated for a specified time period. If antibody to the specific virus being tested is present, it will neutralize the virus and
2. Assay - The unneutralized virus is assayed in a cell culture system or in embryonating chicken eggs. The amount of unneutralized virus depends on the amount of antibody present in the test serum.

This is a well established diagnostic and research tool but, compared to the HI and ELISA tests, is much more expensive and time-consuming.

10.6. Virus isolation

10.6.1. Purpose

To isolate and identify common avian viruses

10.6.2. Methods

10.6.2.1. Method A

Embryonating chicken eggs - Specific Pathogen Free embryonating chicken eggs provide an excellent medium in which most avian viruses will readily propagate.

1. Infected tissues or swabs of tissues may be used. The material, suspended in a medium (isotonic phosphate buffered saline, PBS) which contains antibiotics, is filtered through a millipore filter. The antibiotics and filtration are both used to eliminate bacterial contamination. The antibiotics used depend on the local conditions; generally, penicillin and mycostatin (2000 units/ml each), streptomycin (2mg/ml) and getamicin (50 ppm) are used. In case of fecal and cloacal swabs, 5 times the above concentrations are advisable.

2. Embryonating chicken eggs are inoculated with the prepared sample. The three most common routes of embryo inoculation are: allantoic sac (9-11 day embryos), chorioallantoic membrane (9-11 day embryos) and yolk sac (5-7 day embryos)
3. The route of inoculation depends on the suspected virus; allantoic sac - IB, AI and NDV, chorioallantoic membrane - ILT and Pox and yolk sac - AE and Reo

Some of these viruses produce lesions in the embryos (e.g. Infectious Bronchitis - dwarfing and stunting and excessive urates in kidneys) others require further testing such as ether sensitivity, hemagglutination.

10.6.2.2. Method B (Cell culture)

Can be used as an alternative to embryonating chicken eggs. Various types of tissues may be used (e.g. liver, kidney, fibroblasts). To prepare these cell cultures, organs are removed from specific pathogen-free embryonating eggs. Tissues are prepared as in Method A. Some viruses produce characteristic cytopathic effects (CPE) while others don't and must be examined using fluorescent antibodies (FA) or special dyes.

10.7. Polymerase chain reaction (PCR)

See Chapter "Fundamentals of genetic engineering"

Chapter **70**

Biosecurity

Biosecurity is a practice designed to prevent the spread of disease into and within a farm. It is accomplished by maintaining the facility in such a way that there is minimal movement of biological organisms (viruses, bacteria, rodents, etc.) across its borders. As successive flocks are raised on a farm, the concentration of organisms increases and so are the risks of disease, odor etc. Biosecurity is the cheapest, most effective means of disease control available and it is mandatory component of every disease prevention program.

An analysis of the disease (hazard) risk is a compulsory prerequisite to ensure not only production and economy from the facility but also to provide optimum welfare conditions for the animals. Of the several hazard analysis methods, Hazard Analysis of Critical Control Points (HACCP) is most popular.

1. Principles of HACCP

(Rushing and Ward on internet and Noordhuizen *et al.* 2003)

There are 7 HACCP principles (steps); the first one for risk assessment, the last one for documentation and the rest for risk management. HACCP principles are enumerated below taking the case of a hatchery:

1.1. Principle 1 Hazard Analysis

To identify hazards, both microbiological and physical, at eachstep in the process, from receiving through to delivery. Each section of the program is a step. Hazards must be assessed for severity and probability; the same may be quantified wherever possible.

1.2. Principle 2 Critical Control Points (CCP's)

At CCP's action can be taken to reduce or eliminate microbiological hazards. Attention must be paid to personal hygiene throughout the process, with the use of protective clothing, hand hygiene, foot-dips and showering in and out where possible. Application of a HACCP decision tree.

1.3. Principle 3 Critical Limits

The limits to which the hazard must be reduced. Cleaning and disinfecting will ensure that microbiologicalhazards meet suggested limits (for example Salmonella - 99.999% reduction of the number of organisms in the environment). In other words, establishment of target levels and tolerances for each CCP.

1.4. Principle 4 Monitoring

Observation and measurement of cleaning and disinfecting to ensure the critical limits are met at each step (establishment of monitoring system).

1.5. Principle 5 Correction

Action must be taken if the critical limits are not met at each step. A review of the application procedure is mandatory to ensure that the corrective actions are established.

1.6. Principle 6 Verification

Tests and procedures to ensure the HACCP system are working properly. Often performed by an outside person or organization.

1.7. Principle 7 Recording (Documentation)

Records must be kept that the biosecurity program is in place and implemented correctly and continuously. Records of products used, critical limits, cleaning schedules and any corrective action provide documentation for control and for monitoring. A complete set of records are important for legal action.

HACCP principles can be extended for any facility. Biosecurity for poultry flocks, hatchery and an egg-processing plant is explained below:

2. Biosecurity of poultry flocks

(www.ucdavis.edu)

Biosecurity has three major components:

1. Isolation
2. Traffic Control
3. Sanitation

Isolation refers to the confinement of animals within a controlled environment. A fence keeps birds in while keeping other animals out. Isolation also applies to the practice of separating birds by age group. In all-in-all-out management styles simultaneous depopulation offers time for periodic clean-up and disinfection to break the cycle of disease.

Traffic Control includes both the traffic onto the farm and the traffic patterns within the farm.

Sanitation addresses the disinfection of materials, people and equipment entering the farm and the cleanliness of the personnel on the farm. In order to assess how much biosecurity is practical the following factors are considered:

1. Economics
2. Common Sense
3. Relative Risk

Of all the possible breakdowns in biosecurity, the introduction of new birds and traffic pose the greatest risk to bird health. Properly managing these two factors should be a top priority in any farm.

2.1. Isolation

New birds represent a great risk to biosecurity because their disease status is unknown. They may have an infection or be susceptible to an infection that is already present in birds that appear normal (healthy carriers) on farm. New birds, including day-old chicks should not be put in contact with droppings, feathers, dust and debris left over from previous flocks.

While all-in-all-out management is not feasible for many breeding farms or farms raising exotic fowl or game-birds, it is possible to maintain a separate pen or place to isolate and quarantine all new, in-coming stock from the resident population. Isolation pens should be as far from the resident birds as possible. At least 2 weeks (4 weeks, if possible) of quarantine is suggested. Birds are observed for any signs of illness. Diagnostic blood tests for infectious diseases can also be performed at this time.

Infectious diseases can be spread from farm to farm by:

1. Introduction of diseased birds
2. Introduction of healthy birds who have recovered from disease but are now carriers
3. Shoes and clothing of visitors or caretakers who move from flock to flock
4. Contact with inanimate objects (fomites) that are contaminated with disease organisms
5. Carcasses of dead birds that have not been disposed of properly
6. Impure water, such as surface drainage water
7. Rodents, wild animals and free-flying birds
8. Insects
9. Contaminated feed and feed bags
10. Contaminated delivery trucks, rendering trucks, live hauling trucks
11. Contaminated premises through soil or old litter
12. Air-borne fomites
13. Egg transmission

2.2. Control of traffic

Biosecurity generally requires human traffic control, such as locking the doors and banning all visitors, or allowing entry to certain authorized and necessary personnel only after they have put on properly sanitized footwear, coveralls, and headgear. Human hands may spread infection and should be sanitized before anyone enters a poultry building or leaves the farm.

Traffic control is needed not only for humans, but for animals such as rats, mice, wild birds, and predators.

Openings in buildings should be screened with a properly sized protective wiring to keep out flying or walking predators. Protection from predators is almost impossible to achieve when birds are kept on a range, but steps can be taken to minimize the dangers. Grass and bushes can be kept trimmed for a few feet on each side of the fence. This will allow detection of burrowing rodents as well as force the predators to approach the fence without the benefit of cover.

Game bird pens are usually covered with netting. The mesh size of those nets determines the size of the wild birds kept out. It is ideal to keep all birds out except those which are the livelihood of the operation.

Only clean plastic coops have to be used for transfer of poultry. Wooden coops are difficult to clean and have been responsible for distributing poultry diseases over long distances.

Table 70.1. Longevity of disease causing organisms

Disease	Lifespan away from birds
Infectious Bursal Disease	Months
Coccidiosis	Months
Duck Plague	Days
Fowl Cholera	Weeks
Coryza	Hours to days
Marek's Disease	Months to years
Newcastle Disease	Days to weeks
Mycoplasmosis (MG, MS)	Hours to days
Salmonellosis (Pullorum)	Weeks
Avian Tuberculosis	Years

Flow of on-farm traffic must be directed from the youngest to the oldest birds and from resident to the isolation area. A "clear zone" free of vegetation around buildings is ideal to discourage rodent and insect traffic into the buildings or pens. A different pair of foot-wear while working in the isolation area and in the resident bird area is necessary to prevent the mechanical transfer of disease organisms on footwear. Footwear should be disinfected at each site. Disinfectant footbaths may help to decrease the dose of organisms on boots. Alternatively, it is a good idea to have a supply of cleanable rubber boots or strong-soled plastic boots for visitors.

Washing hands thoroughly after handling birds in isolation or birds of different groups is mandatory. Waterers and feeders have to be disinfected on a regular basis (daily). Periodic clean-out, clean-up and disinfection of houses and equipment have to be planned, at least once a year. Simulataneously, rodent and pest control procedures can be taken up when the houses are empty.

Dead birds must be disposed of promptly by rendering, burning, burying, composting or sending them to a sanitary landfill (See Chapter "Carcass disposal" for details).

2.2.1. Access restrictions

A record should be kept of all visitors to poultry farms. It should include the dates of visits, the names of the individuals, and the nature of their business. Addresses, if available, should also be recorded.

Once a disease has been identified, the producer can check the incubation period and identify the carrier by going over the visitor list. Different diseases have different incubation periods. For example, infectious bronchitis has an incubation period of 24 to 48 hours while infectious laryngotracheitis needs about 12 days to appear.

All visitors should be listed, even if they have not actually gone into a building or onto the range. Like other visitors, those delivering goods or making repairs, as well as all inspectors, should be considered a potential threat to a flock's health. Even visitors who claim they have not made a previous visit to other poultry facilities (including slaughter houses and processing plants) should be required to wear sanitized boots, coveralls, and caps when visiting or inspecting poultry facilities. Flock owners will have more control over security if they furnish the protective clothing used by visitors. Each farm should have several sets of clothing for visitors who must enter the building or the range. These items should be sanitized before the next use because their interiors may have been contaminated by the clothing of the visitor. Boots should be sanitized both inside and outside because they are worn to prevent visitors from contaminating the premises with their street shoes.

Considering the investment in a large poultry complex, security precautions are a small price to pay to keep disease away from a flock. On farms where top security management is practiced, visitors must disrobe, shower, wear the furnished apparel, shower again when finished with a call, and only then put their own clothes back.

All doors to poultry houses should be kept locked at all times, and the keys should not be easy for visitors to find. Some traffic in a poultry operation is necessary, such as feed delivery. However, there is no reason for the vehicle driver to enter any building. If the doors are locked and the keys kept out of reach, entry can be restricted. Feed delivery slips can be deposited in a glass or plastic container attached to the feed bin or any other nearby structure.

When chicks, poults, or young breeder stocks are delivered, bus or truck drivers, as well as the entire crew, must observe strict sanitary practices for two important reasons. First, the delivery personnel are entering a building that has just been cleaned and disinfected. Second, they may have made another delivery on their way. They should wear different protective clothing at each stop.

No responsible poultry grower should allow coops soiled with bird droppings or unclean trucks near the farm. This practice should be enforced even if the entire flock is removed from the premises.

2.2.2. Security involving service personnel

Flock supervisors and veterinarians who visit flocks on a routine basis should do all in their power to protect each flock from contamination. Hands and feet can move pathogens from one location to another and should be sanitized before service personnel leave a farm and again before they enter a premise where birds are kept. Boots should be sanitized inside and outside before being used on another farm. To avoid having to wear wet boots, it is advisable to have several pairs available.

Attendants should not ride in vehicles going between buildings or between farms, including feed trucks, chick buses, and egg trucks. Each vehicle should have a sturdy bucket, a container with a good brand of disinfectant (preferably phenolic acids), a stiff brush, and a supply of water.

After arriving at the farm, the flock supervisor or veterinarian should mix water and disinfectant in the bucket. Even if boots have already been sanitized, wearers should step into the bucket, brush their boots, and wash their hands before entering the premises. Upon leaving, service personnel should wash their boots again, inside and outside, and then place them in the bag with the used coverall. If necropsy was performed, that area should be washed. Finally, hands should be sanitized.

2.3. Sanitation

Primarily involves vaccination of birds on farm, cleaning and disinfection profiles adopted at the farm. See Chapters "Vaccination" and "Sanitation and disinfection" for further details.

3. HACCP at hatchery

The following four key areas in the production of quality chicks are identified for control of contamination:

1. The egg
2. Surfaces and equipment which can contaminate the egg.
3. Airborne contaminates, including the ventilation system.
4. Moveable equipment and personnel.

3.1. General

Modern hatchery design ensures that the hygiene of the egg is easy to maintain. By the use of material finishes on floor, walls, and equipment that are smooth and durable, and equipment including ventilation that can be dismantled for easy cleaning, biosecurity at hatchery has become all the more efficient and easy. However, not all facilities have the advantages of these features, and in these situations it is attentions to detail through training and motivation that will produce the best result; particularly to minimize contamination of *Salmonella* and *Aspergillus* which comes with infected eggs and chick fluff, air, personnel (both staff and visitors) and equipment. The general guidelines include the following:

All personnel must

1. Wear suitable protective clothing.
2. Wash their hands when entering or leaving the production area, and before chick and egg handling operations (traying up, setting, transfer, candling, take-off, sexing, vaccinating and packing).
3. Use antibacterial soap for hand hygiene.
4. Use a foot-dip before entering the production area.

The footdips have to be filled with a sanitizer at an appropriate dilution and renewed at least weekly.

3.2. Egg transport

Clean trays and trolleys have to be returned to the hatchery. Egg transport vehicles should be cleaned between collection runs, if used for chick delivery between deliveries and collections.

Disinfectant should be made available at the farm entrance or on the vehicle. Wheels should be disinfected by washing or spraying when arriving and leaving each collection point.

3.3. Egg reception and storage (Traying room and egg store)

If eggs have been disinfected on the farm they can be moved immediately to setter trays for storage. If eggs come from different sources or are of unknown hygiene status, all eggs must be disinfected before setting.

The use of formaldehyde gas for fumigation chambers with adequate safety and extraction facilities is well accepted. However, the hazard to staff due to acute or chronic exposure must be borne in mind. Repeated fumigation of eggs may lower hatchability too.

3.4. Eggs in setter

In multi-stage setter the risk of contamination is higher and cleaning and disinfection is more difficult than in single stage machines. It is therefore recommended to fog a disinfectant into the setter after new eggs have been set or transferred.

If eggs explode in the setter (bangers or rots) the level of contamination in the setter will rise dramatically unless a rapid disinfection is administered. Whether in single or multi-stage setters, fogging directly onto the affected area, followed by daily fogging of the setter until transfer will reduce this problem.

When setters are emptied either at the end of each set or periodically, all the accessible interior surfaces should be cleaned and disinfected by: removing all dust and debris; scrubbing all wall and floor surfaces with a suitable disinfectant and allowed to dry-up. Machine left to warm up. All surfaces sprayed with a disinfectant.

3.5. Setter rooms, candling and transfer area

(Walls, floors, ceilings, windows, fans, ducting, machine tops/fronts)

After egg transfer, all dust and debris are removed and disposed of.

All wall and floor surfaces scrubbed with an appropriate disinfectant and left to dry. Machine left to warm up. All surfaces sprayed with a disinfectant.

3.6. Hatchers

The hatcher is the main source of contamination for the whole hatchery. Debris, dead chicks and chick fluff are ideal breeding sites for bacteria and fungi in the warm and humid hatchery environment and this contamination is easily spread by air currents, personnel and equipment. Attention to ensure cleaning, disinfection and handling procedures are most important.

After transfer of hatcher trolleys with chicks to the take-off room, all equipments are taken out for cleaning. The equipment is scraped to remove maximum amount of debris out into containers for disposal. All interior surfaces are cleaned with a disinfectant. Walls and floor are also scrubbed to remove stubborn soiling.

Using a pressure washer, all interior surfaces are washed, walls and floors scrubbed, left soaked; then, washed with clean water. Excess water drained and all equipment replaced into the respective machinery. Machine left to warm up. All surfaces sprayed with a disinfectant.

3.7. Hatcher rooms, take off area, sexing and holding areas

(Walls, floors, ceilings, windows, fans, ducting, machine tops/fronts)

Daily after egg transfer or hatch all dust and debris are removed and disposed of. All wall and floor surfaces washed with a disinfectant and scrubbed where necessary to remove stubborn soiling; rinsed with clean water. Excess water removed and all surfaces sprayed with a disinfectant.

3.8. Wash room

After every setting up, transfer and hatch all walls and floors and equipment surfaces are washed with an appropriate disinfectant; rinsed with clean water. Excess water removed and all surfaces sprayed with a disinfectant.

Egg-trays, setter trays and chick crates must be thoroughly cleaned and disinfected before reuse or return to farm. Pressure washer is preferred with a disinfectant to soak all crates before washing off with clean hot water.

3.9. Other areas

All areas should be kept free from debris. Disinfectant is sprayed on all surrounding areas daily. Clean water is necessary to maintain humidity; if water becomes contaminated, suitable sanitizer is used until quality is restored.

Within the humidification and ventilation systems, water evaporation can leave deposits of calcium salts on fans, pipe-work etc. This build-up of scale is an ideal site for build-up of bacterial contamination. As required according to the rate of scale deposition, descaling should be taken up.

4. Biosecurity for an egg-processing plant

(Curtis *et al.* on internet)

In the context of biosecurity, egg-processing rooms and coolers are considered dirty areas because processing plants handle eggs from all flocks on the farm and may receive eggs from other farms. Processing plants receive eggs and equipment (pallets, racks and flats) from multiple sources, all of which could potentially bring infectious disease agents into one central location. Any and all traffic, direct or indirect, between the processing plant and chicken flocks should be considered a significant hazard and should be either avoided, or interrupted by thorough and effective cleaning and disinfection.

4.1. Parking

Ideally, processing plant visitors, inspectors and employees should park their vehicles off site in order to minimize the risk of carrying disease agents onto the premises. Parking areas that are paved and have gravel or some other surface that does not accumulate water, will minimize pathogen survival and, therefore, limit risk.

4.2. Plant Visitors

The soles of shoes should be disinfected upon entering the processing plant and covered with impermeable foot covering. In addition, hands should be sanitized with soap and hot water and clean protective outerwear should be donned, including a head covering. All protective clothing should be provided by the processing plant and not transported by the visitor. Under ideal conditions, a second shoe sanitation should be done before actually entering the processing area. When leaving the plant, the visitor should reverse the entry procedures, and leave the protective clothing at the plant.

It is always a good idea to require all visitors to log in and out including the date and time of the visit. No one should be allowed to visit more than one plant per day; therefore, all visitors should be asked whether they have been to another plant or production facility that day.

4.3. Egg inspectors

Egg inspectors that frequently visit the plant should be provided a place (locker or something similar) to leave their protective clothing for re-use and also the equipment they use (clipboards, pencils, forms, etc.). Clipboards, pencils and forms should not be used in or transported to other processing plants. Scales and other equipment that must be transported between processing plants by inspectors should be wiped down with sanitary cloths between uses and transported in clean plastic bags or containers that can be disinfected.

Egg plant visitors and inspectors should never go from the processing side of the farm to the production side because of the risk of transmitting pathogens to the chicken flock. If it is necessary for a visitor to go to both production and processing, they should visit the production side first before entering the processing plant.

4.4. Egg filler flats

Egg flats are a potential source of bacterial and viral pathogens and therefore should be sanitized between uses. Plastic flats should be washed in hot water with a detergent to remove any organic material, and then disinfected with a sanitizing agent. Most flat washers do not operate at high enough temperatures for a long enough time to kill all pathogens, such as the Avian Influenza virus; therefore the use of chlorine or some other suitable disinfectant is essential for sanitation. It is important not to over-fill the flat washer with flats so that adequate water penetration between all flats is achieved. All sanitizing agents are neutralized by organic material, making frequent changing of the rinse water an essential component of flat sanitation. During periods of high risk, when a pathogen is known to exist, a second wash cycle should be considered.

After the flats have been washed and sanitized, they should be stored in a clean environment to avoid recontamination. Stacking clean egg flats on racks or pallets that have not been sanitized should be avoided to prevent cross contamination. It is best that the sanitized egg flats be returned to their original source and not co-mingled with flats from other sources. If the processing plant receives eggs from multiple farms and or companies, using colored flats can help to keep flats from multiple sources separate.

4.5. Egg racks, trolleys and pallets

Except for in-line collection systems, eggs are transported to the processing plant on racks, trolleys or pallets, all of which have had contact with the production facility. This equipment must be considered contaminated with whatever pathogens exist on the farm and therefore should be cleaned and disinfected between uses.

The sanitation of racks, trolleys and pallets should include at least two steps, cleaning and disinfection. High pressure washing with a detergent capable of removing adherent organic material should always precede the final step of disinfection. All organic material must be removed paying particular attention to wheels and under surfaces of the equipment. If available, steam cleaning can be very effective at removing organic material such as dried egg yolk. If steam cleaning is used, the application of high temperature grease may be necessary to keep wheels well lubricated.

Following cleaning, the equipment should be disinfected thoroughly and then allowed to air dry before equipment is returned to the production facility. As with egg flats, the sanitized equipment should be stored in a clean environment away from the processing plant and farm traffic to prevent re-contamination. Racks, trolleys and pallets should be clearly identified (paint or other means) in order to minimize the likelihood of co-mingling of equipment from multiple sources. All equipment should be returned to the original source farm, and should not be shared with other farms.

4.6. Incoming Cooler Management

Many processing plants bring eggs in from multiple sources. Unprocessed eggs can carry dust, feathers and feces on their surfaces and can be a significant source of pathogens. If the processing plant is located on a farm that has all or part of its production off-line, this presents a significant challenge to prevent the transmission of a pathogen from off farm sources to resident birds. The attendant who brings the off-line eggs into the receiving cooler will go back and forth between production and processing. Therefore, it is essential that the attendant not come in contact with eggs from other farms.

The best way to prevent the attendant from becoming contaminated is through the separation of equipment (racks, trolleysor pallets) and traffic patterns. Ideally, there should be two separate entrances to the cooler, one for off-farm eggs and one for on-farm eggs, and the eggs should be stored in separate locations within the cooler.

5. General issues

5.1. Rodent and insect control

See Chapter "Pest management"

5.2. Dust control

See Chapter "Environmental implications of poultry production"

5.3. Disposing of dead birds and litter

See Chapter "Environmental implications of poultry production"

Chapter 71

Sanitation and Disinfection

Diseases and infections have always been a major concern to the poultry industry. Fortunately, microbial contamination can be prevented and controlled using proper management practices and modern health products.

Sanitation refers to a state wherein pathogenic organisms are present but are not a threat to the birds' health; disinfection indicates destruction of all vegetative forms of microorganisms whereas spores are not destroyed and sterilization means destruction of all infective and reproductive forms of all microorganisms (bacteria, fungi, virus, and the like).

Having an effective sanitation (cleaning) and disinfection program is a crucial step in every poultry-biosecurity program. A cleaning and disinfection program should be instituted after a poultry building has been depopulated and before restocking occurs on the farm. The main purpose of a cleaning and disinfection program is to reduce the number of pathogens (disease-causing agents) in the environment so as to reduce the potential for diseases to occur in subsequent flocks (Gordon and Morishita on internet). It is important to identify the pathogens to eliminate, as certain disinfectants are ineffective against certain disease agents.

1. Sanitation - general

Cleaning is the physical removal of organic material (i.e., manure, blood, feed, and carcasses). It is important to remove these organic materials before the disinfection process begins because disease agents are often protected in these materials and can survive the disinfection process. Hence, it is important to thoroughly clean a building before the disinfection process. The cleaning process can include a dry-cleaning and a wet-cleaning step.

Dry cleaning involves the physical removal of organic material, such as the removal of feed, litter, and manure. The process of dry-cleaning physically removes the organic material before the actual wet cleaning can occur. Wet cleaning, as its name implies, involves the use of water (Gordon and Morishita on internet).

There are four basic steps in the wet cleaning process — soaking, washing, rinsing, and drying. Although not necessary, detergents (wetting agents) can be used in the wet cleaning process. However, it is more important to have pressure washers with the proper pressure (35-55 kg/cm^2) to ensure all the organic materials are removed from the facilities (Gordon and Morishita on internet). In any case, poultry houses should be sprayed at a pressure of at least 14 to 21 kg/cm^2 or more.

The final step of ensuring a proper clean-up is having the wet areas of the building dried quickly. If the building is not dried properly, the excess moisture can result in bacteria multiplying to higher levels than seen before cleaning. Therefore, an improper cleaning can actually do more harm than good.

2. Housecleaning procedures
(www.ucdavis.com)

Housecleaning is the most arduous phase of biosecurity. After the flock has been depopulated and the litter removed from the poultry houses, it does little good to clean and disinfect the houses if chunks of manure remain around the buildings and in the driveways. Large chunks should be picked up and the small ones broken down with a water spray. Dilution and sunlight will reduce the danger from pathogens. Equipment such as rakes, shovels, trucks, manure spreaders, and bucket loaders should be cleaned and disinfected after use.

2.1. General recommendation for floor houses

A complete cleanout of houses and placement of new litter between each brood is most desirable. If cleaning that often is not possible, broiler houses should be cleaned completely at least once a year.

2.1.1. Total cleaning

The following procedures are recommended for a complete housecleaning of floor houses:

1. Litter removed from the poultry houses and shifted as far away from the houses as possible, or a minimum of 100 m.
2. House swept thoroughly to clean all floors, lighting fixtures, fan blades, and louvers. Burnt-out light bulbs replaced and all other bulbs cleaned.
3. All permanently installed waterers, feeders, and any other equipment scraped, scrubbed, and cleaned. Miscellaneous equipment (brooder guards, jugs, hand feeders) removed from the house to permit a thorough job of disinfection. Brooder guards (wire), hand feeders, and jugs that will be used for the next flock must be soaked, then hand scrubbed and disinfected.
4. The sills of the house scraped and cleaned. All material from the house interior, and clean up litter, trash, and debris from the outside of the house removed because residual dirt, debris, stains, and organic matter neutralize disinfectants.
5. The ceiling, curtains, walls, partitions, slats, feeders, waterers and other equipment thoroughly disinfected with a good disinfectant used at the rate recommended on the label.

6. The curtains are kept up and fully extended when cleaning and spraying. Both sides of the curtains and wire screens completely and thoroughly covered with spray to remove dirt, dust, and down. When dry, curtains should be dropped and the house allowed dry-out completely. All spray should be applied at a minimum of 14 kg/cm^2 for good penetration. However, care has to be exercised in case of aged material or thin covers. Extreme care should be taken about not getting any spray inside electric motors. Duct tape can be used to cover the slots in the motor housing; but to be removed when the work is completed. Start at the back and work toward the front of the building, spraying the ceiling first, then the walls, and finally the floor.
7. When the floor is dry, 10 cm of dry, absorbent litter is spread.
8. An approved insecticide is used on top of the new litter if insects are a problem. Insecticides and disinfectants should not be mixed for application.

2.1.2. Partial cleaning

The following procedures are recommended for a partial housecleaning of floor houses:

1. All fans and louvers (if installed) thoroughly cleaned.
2. The house swept clean.
3. Caked litter from the house, if any, removed.
4. Brooder guards (wire), hand feeders, and jugs that will be used for the next flock scrubbed and disinfected.

2.2. General recommendations for cage houses

2.2.1. Total cleaning for cage houses

A complete cleanout and disinfection between each brood of pullets is highly recommended. Cage layer houses and equipment must be cleaned thoroughly and disinfected after each flock is moved from the premises and before a new flock is started.

The following procedures are recommended for a complete housecleaning of cage houses:

1. Manure moved from the poultry house and as far away as possible or a minimum of 90 m.
2. House swept thoroughly to clean all floors, lighting fixtures, fan blades, and louvers. Burnt-out light bulbs replaced and all other bulbs cleaned.
3. The ceiling, curtains, walls, partitions, cages, feeders, waterers, and other equipment thoroughly disinfected. Extreme care should be taken about not getting any spray into electric motors. Duct tape can be used to cover the slots in the motor housing; but, should be removed when the work is completed.

4. If curtains are used, cover them completely and thoroughly with spray to remove dirt, dust, and down keeping them fully extended when cleaning and spraying. When dry, the curtains should be dropped and the house allowed to air out completely.
5. Between flocks, house repairs, if any, should be identified and completed.

2.2.2. Partial cleaning

No procedure is given for partial clean-out. In fact, cage layer houses and equipment must be cleaned thoroughly and disinfected after flock is removed from the premises.

3. Cleaning the feed system

In addition to the cleaning and disinfection practices mentioned above, all producers should implement the followingfeeding-system clean-up between each flock of birds, even if the cleaning is a partial one. The purpose of these procedures is to reduce the incidence of moldy feed and mold toxin problems. These are especially relevant where automatic feeding system is operative.

3.1. Feed bin, boot, and auger

1. All feed from the bin, boot, and auger removed as soon as possible after the flock is sold.
2. The boot disassembled to ensure thorough cleaning.
3. Insides of the feed bin, boot, and auger washed with a 5 percent solution of sodium hypochlorite (household bleach).

3.2. Feed hopper and feed cart

1. Feed hopper and feed lines emptied.
2. All feed from boxes removed and scraped or brushed away; all material such as caked feed from corners, seams, or lips inside and outside of the feed boxes must be removed.
3. All surfaces washed with a 5 percent solution of sodium hypochlorite.

3.3. Line feeder

1. All feed from the feed trough removed, including corners.
2. Line feeders removed and corners cleaned.
3. Sprayed with a 5 percent solution of sodium hypochlorite.

3.4. Line feeders with feed pans

1. All feed removed from feeding system.
2. Any mold ring present in the feeders scraped and removed.
3. The pan is removed and sprayed inside of the cone with a 5 percent solution of sodium hypochlorite.

4. Cleaning and sanitizing waterlines

If water in the well is used only for the flock and all birds have been removed from the farm, the well can be treated with sodium hypochlorite (household bleach) dumped directly into the well. Water is then run through all the lines until chlorine can be smelled at the end of each line. The water is allowed to stay in the lines a minimum of 24 hours.

If the well cannot be treated, the waterlines in the empty building can be treated through the medicator. When no birds are in the house, toxic levels of chlorine are not a concern, but undiluted bleach is very corrosive and should not be allowed to stand in any equipment. At 5 percent concentration, the bleach will destroy some synthetic sponges if they stand in the solution a few minutes.

The best way to distribute the chlorine solution quickly throughout the drinking system is to remove the drain plug at the end of each line or to temporarily disconnect the last drinker. Some drinkers do not lend themselves to easy cleaning. Cups in cages must be removed, soaked, and scrubbed before being returned to the cages.

A weekly treatment with baking soda (sodium bicarbonate), made through the medicator while the flock is in the building, helps reduce slime from drinkers, valves, and pipes. A baking soda treatment should follow immediately any treatment of water with antibiotics or vitamin packs.

5. Restarting after cleanout

The following checklist of procedures should put the house and its equipment back in running order after a cleanout.

1. Make sure that all electrical circuits operate properly.
2. Fill all light sockets with a clean, operative light bulb.
3. Check and adjust all time clocks (if installed). Remove coverings and tape used to protect electrical circuits and motors against entry of water and spray solutions.
4. Run all motors and lubricate them if necessary.
5. Check for proper tension on feed chains, winch cables, and fan belts.
6. Check all water lines for wear, cracks, and leaks.
7. Inspect thermometers and thermostats to ensure they are clean, operative, and accurate.
8. Adjust gas brooder stoves so they light up when needed and the flame is even and blue.
9. Check the feeding equipment for proper assembly.
10. Operate all equipment to ensure that it runs smoothly and without abnormal noises.

6. Hatchery cleaning

6.1. Cleaners

Any multi-purpose biocide suitable for use in the hatchery or egg handling environment should be selected.

6.1.1. General Properties

1. Broad spectrum biocidal activity.
2. Superior activity in the presence of organic challenge and at low temperatures
3. Independently proven bactericidal and fungicidal activity in Salmonella and Aspergillus.
4. Independently proven biocidal activity when sprayed on eggs
5. Can be used as a foam, spray or egg dip.
6. Non-staining and non-tainting.
7. Excellent foaming properties due to inclusion of foam booster.
8. Biodegradable.
   No residue after application

6.2. Hatching egg sanitation

(Earnst, 1975)

Hatching eggs can be sanitized by using a solution containing 250 ppm quaternary ammonium compounds and 10 ppm EDTA; the solution is made alkaline by adding sodium carbonate to increase the pH to 8.0. Other disinfectants which can be used on hatching eggs include: Iodophors (200 ppm), chlorine dioxide (100 ppm), synthetic phenols (800 ppm), chlorohexadiene (200 ppm), phenol (250 ppm) + quaternary ammonium compounds (200 ppm) in presence of formaldehyde (2500ppm).

7. Disinfection

Disinfection involves the use of a disinfectant that will reduce or kill the pathogens. Disinfectants are more effective at warmer temperatures. There are several types of disinfectants, and the one chosen should be effective against the disease agent(s).

7.1. Types of disinfectants

(Gordon and Morishita on internet)

1. Aldehydes (i.e., formalin, formaldehyde, and glutaraldehyde)
2. Chlorine-releasing agents (*i.e.*, sodium hypochlorite, chlorine dioxide, sodium dichloroisocyanurate, and chloramine-T)

3. Iodophors (*i.e.*, povidone-iodine and poloxamer-iodine)
4. Phenols and bis-phenols (*i.e.*, triclosan and hexachlorophene)
5. Quaternary ammonium compounds
6. Peroxygens (*i.e.*, hydrogen peroxide and peracetic acid). These, even at the recommended dilution for their use, are caustic and dangerous; therefore, not generally used in poultry facilities.

7.2. Properties of disinfectants

Disinfectants should not to be applied to animals directly, unless labeled for such use. Similarly, it must be certified fit for using around feeders and in animal quarters. A general recommendation is to rinse disinfectants off after the appropriate amount of contact time if animals will have contact with the disinfected surfaces. Label directions should be strictly followed, and different classes of disinfectants should not be mixed (Bowman and Shulaw on internet).

Disinfectants are effective against bacteria, viruses, and fungi. Disinfectants were not designed to be effective against parasites. In general, the descending order of resistance of disease agents is (Gordon and Morishita on internet):

1. Spores (*i.e.*, clostridial diseases like botulism) and acid-fast bacteria (*i.e.*, mycobacteria like *Mycobacterium avium* [avian tuberculosis])
2. Gram-negative bacteria (i.e., Pseudomonas, E. coli, Salmonella)
3. Fungi (*i.e.*, *Candida* and *Aspergillus*)
4. Non-enveloped viruses (*i.e.*, enteroviruses and adenoviruses)
5. Gram-positive bacteria (*i.e.*, *Staphylococcus aureus*)
6. Lipid enveloped viruses (*i.e.*, avian influenza virus).

Thus, spore-forming bacteria are harder to destroy by disinfectants than viruses. Avian parasites (*i.e.*, lice, mites, and endoparasites) are best treated using insecticides or by means of parasiticides. However, it has been reported that ammonia and phenolic disinfectants have been effective in reducing the numbers of coccidial oocysts.

Table 71.1. Properties and uses of some disinfectants

Properties	Chlorine	Iodine	Phenol	Quaternary Ammonium	Formaldehyde	Cresols	Lyes
Bactericidal	+	+	+	+	+	+	+
Bacteriostatic	-	-	+	+	+	+	+
Bacterial Spores	±	-	±	-	-	+	+
Fungicidal	±	+	+	±	+	+	+
Viricidal	±	+	+	±	+	+	+

Properties	Chlorine	Iodine	Phenol	Quarternary Ammonium	Formaldehyde	Cresols	Lyes
Insects	-	-	-	-	+	+	-
Worm eggs	-	-	±	-	-	+	±
Toxicity	+	±	+	-	+	+	+
Activity with organic matter*	++++	++	+	+++	++	±	±
Personnel	+	+	-	+	-	-	-
Egg "Washing"	+	-	-	+	+	-	-
Foot Baths	-	+	+	+	-	+	-
Rooms	±	+	±	+	+	-	+
Residual Effects	±	±	++	+	±	+++	±
Water sanitation	+	+	-	+	-	-	-
Hatchery	+	+	+	+	+	-	±
Poultry House	±	+	+	+	+	+	±
* Number of + indicates degree of affinity for organic material and corresponding loss of disinfecting action (negative property). + Positive property - Negative property ± Limited activity for specific property							

7.3. Advantages and disadvantages of disinfectants

Table 71.2. Commonly used disinfectants

Disinfectant	Advantages	Disadvantages
Iodine or Iodophors or organic iodine As disinfectant: 50 to 75 ppm As drinking water sanitizer: 25 ppm Most commonly used disinfectant	Effective against various pathogens; most gram positive and gram negative are killed. Also effective against fungi and some viruses if used in sufficient strength. They kill organisms by reacting with nucleic acids and are not affected by hard water or other chemicals such as detergents and acids which are often used in combination with disinfectants. They are effective in both acid and basic solutions, but are much more effective in the acid range. These are speedy disinfectants and good in combinations. They are good foot baths as they have a color indicator when used up. They turn a brownish orange when not effective.	They are quickly and easily tied up by organic material. This renders the iodophors useless for disinfection. Therefore, poultry facilities have to be cleaned thoroughly before iodophors can do their disinfecting. They leave no residual effect. Once applied to a clean surface and the iodophor has dried, the disinfecting properties are lost.

Disinfectant	Advantages	Disadvantages
Chlorines As Disinfectant: 200 ppm As sanitizer: 50 ppm	Effective against most poultry bacteria and fungi. Hypochlorite attacks both the outer coat and nucleic acids of viruses. Chlorine disinfectants kill very rapidly and are not affected by hard water and detergents. They are most effective in the acid range.	They are very easily tied up by organic matter and lose their disinfecting capabilities. Poultry houses have to be thoroughly cleaned before using chlorines as a disinfectant. They have very little residual effect after they have dried. They have odor when used in high concentrations. Face masks are needed to keep fumes from nose and eyes.
Quarternary Ammonias (n-alkyl dimethyl benzyl ammonium chloride) As disinfectant 400-500 ppm As drinking water sanitizer 200 ppm	Odorless, clear, generally non-irritating, has a deodorizing and detergent action. They kill pathogens, against which they are effective, very rapidly. These compounds leave a residual effect, after being applied to a surface. They continue to have a killing effect for a few hours or even a day after they have dried.	Their disinfectant range is not as wide as iodine or chlorine. They are effective under alkaline pH against gram-positive bacteria, moderately effective against the gram-negative. But, large number of bacteria encountered in a poultry house are gram-negative. They are not very effective against fungi and viruses. Marek's or adeno viruses are not killed They are less effective when tied up by organic matter.
Cresols and Cresylic Acid Disinfectant: 1 l in 4-5 l oil or 1 l in 30 l water Best disinfectant; but most expensive	Are effective against all bacteria, most fungi and most viruses. They are also effective against insects, eggs from insects and worms. They are corrosives and as such destroy ascarid (worm) eggs. They are the most complete disinfectants and insecticides. They kill quickly, penetrate organic matter and maintain their effectiveness after being in contact with organic matter, for some time. Residual disinfecting capability is very good. They retard bacterial, viral and fungal growth even having been on a surface for a month; especially true when used in combination with an oil base; however, they can also be used in water solutions and are effective. Coal tar derivatives are most effective in an acid range.	They have a strong odor and is probably effective as long as the odor can be detected. It is also very irritating to skin, and lungs if inhaled. Persons using this disinfectant must be protected by oil skins and facial breathing masks with air filters. This material is extremely combustible when mixed with oil; therefore, it can only be used when the power is off in the building and no smoking. They are expensive and hence, are excellent terminal disinfectants.

Disinfectant	Advantages	Disadvantages
Synthetic Phenols (Orthophenyl phenols) Disinfection: 1000 ppm Widely used and very effective	They kill organisms quickly and are not easily tied up by organic matter. This group leaves a residual effect and are most effective in the acid range. These disinfectants are effective against both gram positive and gram negative bacteria. They are effective against some bacterial spores. They are quite effective against fungi and for the most part, are effective against viruses. They are effective against Marek's Virus.	They are expensive when compared to iodines or chlorines. Worm eggs are destroyed by these disinfectants only when used in high concentration. They have no color indicator to tell the user that the disinfectant is tied up by organic matter. Their residual effect is not as long lasting as the creosols. High levels needed to disinfect.
Caustic Sodas (Lyes) Not widely used	These disinfectants are effective against all types of bacteria, bacterial spores, fungi and most viruses. Their corrosive properties destroy worm eggs.	The disinfectant is corrosive. Protective clothing (oil skins) and face masks are a must when using these disinfectants. They are usually used on floors.
Formaldehyde Gas Mainly used in hatcheries for fumigation	Formaldehyde gas kills bacteria fungi, viruses and insect life if used properly.	Formaldehyde gas is noxious to humans and animals. The building has to be airtight and fan inlets and windows should be sealed with polyethylene. During the winter period it can be difficult to keep the temperature above 24°C. A dry summer heat will make it difficult to keep the relative humidity above 75%. The formaldehyde gas does not penetrate layers of organic matter (dust, feces, etc.), so the building to be disinfected has to be thoroughly cleaned.

7.4. Selection of a disinfectant

Selection of disinfectant depends on:

1. The type of surface being treated.
2. The cleanliness of the surface.
3. The type of organisms being treated.
4. The durability of the equipment/surface material.
5. Time limitations on treatment duration.
6. Residual activity requirements.

If the surface is free of organic matter and residual activity is not required, quaternary ammonium compounds or halogen compounds can be used effectively. However, if surfaces are difficult to clean, residual activity is required, or the contaminating organisms are difficult to destroy, then multiple phenols or coal tar distillates may be needed.

7.5. Effectiveness of a disinfectant

Effectiveness of a disinfectant depends on:

1. The organism(s); all disinfectants are not effective on all organisms
2. Cleaning and disinfection if done separately adds to effectiveness.
3. Disinfectant solutions are more effective when applied as warm solutions rather than cold solutions. Hot solutions can reduce disinfectant efficiency. It is estimated that for each 4.4°C drop below 18.3°C, the effectiveness of most disinfectants is reduced by one half.
4. Few disinfectants are effective instantaneously; however, enough contact time should be given (usually 30 minutes is sufficient).
5. Embryos are very sensitive and severely affected by chemical vapors. Hence, disinfectants having least effect on embryo development must be selected.
6. All surfaces have to be allowed to dry thoroughly before reuse. Dryness reduces reproduction and spread and transport of germs.
7. Improper use of disinfectants can damage or hinder the function of equipment. Some disinfectants are corrosive or clog spray nozzles of water systems.

7.6. Recommendations for use of disinfectants

Table 71.3. Recommended uses and considerations for disinfectants

Disinfectant	Recommended Use	Considerations
Alcohols	Small utensils	Poor residual activity, fire hazard, expensive
Halogens	Water systems, foot baths	Corrosive, poor residual activity, ineffective in presence of organic material
Quaternary ammonias	Incubation equipment, feeding systems	Non-corrosive, non-irritating, limited residual activity and effectiveness with organic matter
Phenols	General house use	Slightly irritating, good residual activity, effective with organic matter
Coal tar distillates	General house use	Corrosive, irritating, good residual activity, effective with organic matter
Aldehydes	Fumigating incubators/eggs	Highly toxic, slight residual activity, sporicidal, fungicidal
Oxidizing agents	Small utensils	Poor residual activity, corrosive, ineffective in presence of organic material

Source : Smith on internet

8. Hatchery disinfection

Disinfection is of primary importance at every stage. The vehicles entering into the farm/hatchery will have to cross through a disinfectant pool so that their tyres are completely washed in the disinfectant. Personnel will have a disinfectant spray, especially on the underside of their shoes, and wherever practicable, change their dress and take bath.

8.1. Formaldehyde fumigation

Fumigation is by far the most common method of disinfection in a hatchery facility. Hatching eggs on receipt, in the setter, chicks in hatcher and all hatchery equipment (after thorough cleaning) are fumigated. Fumigation is also employed for poultry houses, equipment and many other related material on a regular basis.

8.1.1. Requirement for Proper Fumigation

There are four requirements that are essential in order to obtain the maximum germicidal activity from formaldehyde. If one or more of these requirements is ignored, the germicidal effect of the gas is reduced. Requirements are: temperature: 24 to 38°C; relative humidity: = 75%: Time: depending on temperature, humidity, concentration and area to be disinfected and concentration which depends on the material to be disinfected.

Formaldehyde (HCHO) is a gas sold commercially as a 40% solution (37% by weight) in water as formalin. It is also available in powder form called paraformaldehyde containing 91% of formaldehyde which is heated to 232°C liberate the gas.

8.1.2. Method

The most popular method of formaldehyde fumigation is to mix the 40% formalin onto potassium permanganate ($KMnO_4$) to liberate the gas. The reaction is exothermic and, in fact, the heat generated is useful for the release of the gas. Therefore, the $KMnO_4$ crystals are kept in an earthen pot deep enough to hold the volume several times that of the combined chemicals to avoid the spillage of contents during bubbling and splattering that take place in the process. For the same reason (excess heat generated), $KMnO_4$ must not be added to formalin, but the crystals must be kept first in the earthen pot and formalin is poured onto it. The door of the cabin must be closed immediately since the formaldehyde gas instantaneously released is harmful to the eyes of the operator.

8.1.2.1. Proportion of chemicals

Two parts by volume of formalin to one part by weight of $KMnO_4$. Usually, 40 cc of 40% formalin and 20 g of $KMnO_4$ for every 2.83 m³ is known as 1X concentration. For the same area, the quantities of the chemicals are doubled, it is called 2X and so on. If, after the reaction, the residue is purple, the quantity of formalin added is more; when proper quantities are used, a dry, brown powder will be left.

8.1.2.2. Neutralization

In the event that the fumigation must be stopped after a period of time, it can be effected, most of the times, by opening the air intakes and exhausts. However, it can be expedited by sprinkling the fumigated floor area with ammonium hydroxide (NH_4OH) prepared to contain 26-29% ammonia. The quantity of NH_4OH required is equal to one-half of the quantity of formalin used for fumigation (or 4 ml/g of paraformaldehyde).

8.1.2.3. Drawbacks of formaldehyde

It is highly volatile, has penetrating odor and caustic action, and it has a tendency to harden the skin. It is extremely irritant to conjunctiva and mucous membranes.

However, its advantages clearly outweigh the disadvantages – for instance, it can be used as a gas or vapor, it is relatively non-toxic to the animals including birds, it is efficient even in presence of organic matter and it does not corrode the utensils or spraying equipment. The maximum atmospheric concentration permitted in work area is 2 ppm. Suitable gas masks are also available.

8.1.2.4. Efficiency and concentrations required

Formaldehyde gas fumigation is widely used in poultry enterprise for disinfection; but fumigation of poultry house is not completely satisfactory since it is difficult to make the poultry house air-tight. For more efficiency, a fan to circulate warm humid air and fumigant, and a source of humidifying moisture are essential especially because formaldehyde, being soluble in water, acts better when humidity is higher (75% or more) and temperature at or above 21.1°C (ideal 24 to 38°C).

It is not advisable to fumigate eggs in the incubator, especially during the 1st - 4d, because embryos are highly sensitive. Similarly, embryos while pipping must not be fumigated. However, with multiple settings, the eggs may be fumigated as soon as they are incubated (with 2X concentration for 20 min) and chicks in hatcher can be fumigated (after the hatching is completed) with 1X concentration for 3 min and the gas is neutralized with NH_4OH. Similarly, trucks after being fumigated for 20 min with 5X concentration must be neutralized.

Table 71.4. Concentration and duration of fumigation

	Concentration	Time (min)
Hatching eggs, soon after lay	3X	20
Eggs in incubator, 1st day only	2X	20
Chicks in hatcher	1X	3
Incubator room	1X, 2X	30
Hatcher room, Chick room between hatches	3X	30
Wash room	3X	30
Chick boxes, pads, farm equipments	3X	30
Trucks	5X	20
		Source : North, 1984

Chapter **72**

Vaccination – Principles and Practices

Vaccines are an important part of disease prevention and control. However, vaccines have to be purchased and administered which not only adds to cost of poultry operations but also involves labor. Notwithstanding these, vaccines create immune response which needs metabolic adjustments many times leading to potential performance loss; vaccines (when injected) do produce localized tissue damage from vaccine injections. The increased metabolic demand can cause the animal to look depressed and therefore may be confused with illness. This is sometimes referred to as "vaccine sweat." If the risk of a particular disease is low, the insurance afforded by vaccination may not be required and the benefit provided might not be cost effective. If the risk of disease is high, the insurance afforded by vaccination may be very cost-effective even if it does not completely prevent sickness (Griffin *et al.* 2002).

Vaccines should not be expected to eliminate all disease problems. Many additional management procedures such as animal density, nutrition, environmental control, movement of animals, levels of stress, cleanliness of the environment, cleanliness and availability of drinking water, and the number of different disease-causing organisms in the environment can all influence how well a vaccine will work. Therefore, it is important to view vaccines only as an aid to health management and not the foundation of animal health (Griffin *et al.* 2002)

1. Characteristics of a vaccine

(Griffin *et al.* 2002)

1. Vaccines should stimulate the immune system similar to a natural infection. Vaccines may not be able to completely prevent a natural infection but the vaccine should reduce the severity or limit the frequency of the disease. This is especially true of many bacterial diseases such as those that cause respiratory or intestinal disease.

2. Immunity produced by a vaccine should last long enough to protect the animal through the typical time the animal will be exposed to the natural disease.

3. Vaccines should have minimal side-effects and if side effects are present, they must be acceptable.

4. The cost of vaccination should be less than the economic loss incurred by the naturally-occurring disease.

2. Types of vaccines

(Griffin *et al.* 2002)

2.1. Killed vaccines and toxoids

Inactivated vaccines are formulated with a high antigenic mass of bacterial or viral origin (killed microorganisms, organism components or organism by-products) formulated in a suitable adjuvants such as aluminum hydroxide or oil to produce a sufficient immune response. Adjuvants enhance the immune response by increasing the stability of the vaccine in the body, which then stimulates the immune system for a longer time. Vaccination results in a humoral response, the magnitude of which is directly related to the concentration of antigen administered per dose.

Inactivated vaccines must be administered by injection; success of vaccination is thus dependant on the skills of the vaccine administrator in administering a full dosage of vaccine to each bird. (Breytenbach, 2005 on internet). The major advantages of killed vaccines are safety and stability of the product.

2.2. Subunit vaccines

Subunit vaccines are a type of killed vaccine that contains only part of the virus or other microorganism. These vaccines were developed to either isolate or engineer the most important part of the microorganism needed to produce a proper immune response and eliminate the part(s) of the microorganism that caused adverse vaccine reactions or interfered with a proper immune response.

2.3. Autogenous bacterins

Autogenous bacterial vaccines (Autogenous bacterins) are produced from disease-causing organisms isolated from sick animals. The disease-causing bacteria are grown in culture, killed and mixed with an adjuvant. These vaccines frequently contain high levels of endotoxin and other by-products found in the culture (debris). Because autogenous vaccines frequently contain high levels of endotoxin and other by-products found in the culture (debris), they should be used with caution.

2.4. Modified live vaccines

Live poultry vaccines are generally attenuated bacterial or viral strains, which replicate in the vaccinated bird inducing a cellular and humeral immune response. Due to the ability to self-propagate, most live vaccines are suitable candidates for mass application by spray or via the drinking water. (Breytenbach, 2005 on internet).

The immunity produced by modified live vaccines typically lasts longer than the immunity produced by killed vaccines. Handling and storage of modified live vaccine products is critical. Exposure to high temperatures, sunlight, freezing temperatures, disinfectants or soaps can damage or destroy a modified live vaccine product. Modified live vaccine products require rehydration of lyophilized vaccine cake with provided diluent and should be used within an hour of mixing. The major advantage of modified live vaccines is a broader scope and duration of protection because the animal is exposed to all stages of the replicating virus or bacteria.

2.5. Chemically-altered vaccines

Chemically-altered vaccines contain modified live organisms that have been grown in a media containing adjusted levels of certain chemicals that trigger and amplify mutation of the organism, changing the organism's metabolism in such a way as to alter the ability to cause disease. Temperature-sensitive vaccine organisms are examples of an organism produced by this process. The temperature-sensitive organism loses the ability to grow at the animal's normal body temperature but can grow at the temperatures present in ocular or nasal mucosa. Chemically-altered organisms are considered safer than typical modified live organisms, but when given by a route other than direct contact with the mucus membranes stimulate little or no local immunity. The immune response produced is similar to modified live vaccine products but the duration of immunity is not considered to be as long. The major advantage of chemically-altered vaccines is they are safe because there is no systemic replication of the vaccine organism.

Vaccines used in the poultry industry are either live attenuated viral/bacterial strains (modified live vaccines) or inactivated viruses/bacteria (killed vaccine) formulated with a suitable adjuvant. Reputable vaccine manufacturers subject all batches of vaccines to stringent quality control tests, ensuring the end-user is supplied with a product containing sufficient antigen to initiate an immune response in the target bird. Thus assuming bird health, vaccine administered to the target bird in the prescribed dosage will induce immunity to the specific disease. Furthermore, a well-planned vaccination schedule will result in maximum immunity at the time when poultry are most susceptible to infection or when disease would have the highest negative impact on economic performance (Breytenbach, 2005 on internet).

Table 72.1 Advantages and disadvantages of major types of vaccines

Advantages	Disadvantages
Killed vaccines and toxoids	
Available for a wide variety of diseases. No risk of reverting to virulent form. No risk of vaccine organism spreading between animals. More stable in storage. No on-farm mixing, therefore less risk of contamination. Excellent stimulant of passive antibodies in yolk.	More likely to cause allergic reactions and post-vaccination lumps. Two initial doses required at least 10 days apart. Slower onset of immunity. May not produce as strong or as long-lasting immunity as modified live vaccine products. Produce a narrower spectrum of protection than modified live vaccine products. Tend to be more expensive than modified live vaccine products.
Modified live vaccines	
One initial dose is usually sufficient but additional booster doses may be required. More rapid protection than killed vaccine products. Produces a wider spectrum of protection than killed vaccine products. Less likely to cause allergic reactions or post-vaccination lumps than killed vaccine products. Less susceptible to passive antibody vaccine block than killed vaccine products. Tend to be less expensive.	Potential to mutate to a virulent form. Could exacerbate disease in immunosuppressed animals. Potential for excessive immune response. Must be handled and mixed with additional care
Chemically-altered vaccines	
Share many characteristics of modified live vaccine products. Safety similar to killed vaccine products. More rapid protection than killed vaccine products. No risk of reverting to virulent form.	Protection not as rapid as modified live vaccine products. Two initial doses required. May not produce as strong or as long-lasting immunity. Unless given on a mucus membrane, stimulates little or no mucosal immunity. Must be handled and mixed more carefully. Tend to be more expensive than modified live vaccines
	Source: Griffin *et al.* 2002

3. Vaccine Selection

(Griffin *et al.* 2002)

A vaccine that will protect at a cost that will be less than the financial loss due to the naturally-occurring disease is ideal; but, it may not be practically obtainable. Therefore, a vaccine that will protect the operation from catastrophic financial loss is selected, keeping in mind the cost of vaccination is like an insurance premium.

Some combination vaccines are available for convenience. However, they may not always be appropriate. For instance, ND and IB vaccines are available as bivalent product; but, when IB is not endemic, the additional cost involved is not justified.

Vaccines are production management tools and are not a substitute for proper animal husbandry. Therefore appropriate biosecurity measures should be followed. However, if there is a breakdown in production management and a disease outbreak occurs, the vaccination program needs to be adequate and effective to limit resulting losses. Thoroughly understanding causes of vaccine failure will help prevent future problems.

4. Methods of poultry vaccination

4.1. Individual bird application

4.1.1. Subcutaneous injection

Marek's Disease vaccine in the hatchery. Inactivated vaccines such as Newcastle Disease Virus, Infectious Bronchitis Virus, Reovirus and Infectious Bursal Disease Virus may be given to breeder hens at housing by subcutaneous injection into the back of the mid-neck region.

4.1.2. Conjunctival sac installation (eyedrop)

Newcastle Disease, Infectious Bronchitis, Infectious Laryngotracheitis vaccines. Head is held by one hand, thumb pressing the lower eyelid down; with the other hand, one drop of vaccine is deposited into the eye by use of a dropper. Popularly referred to as "Ocular" method. Simultaneously, some producers place a drop on a nostril and wait for few seconds for the vaccine to enter inside the nose (transnasal drop); however, ocular or transnasal drop is sufficient. When the vaccine is given through both eye and nose, it is referred to as "Oculonasal" method.

4.1.3. Wing-web puncture

This method is employed for vaccination against fowl pox (pigeon pox), and avian encephalomyelitis, fowl cholera in chickens. A double-prong sewing machine needle (supplied by vaccine manufacturer) is dipped into vaccine before each stab. One of the wings is spread to expose the underside (up). The vaccine-laden needle is stabbed through, without touching feather with needles and avoiding blood vessels. For fowl pox, examination of birds for "take" is done after 6-10 d; swelling followed by scab formation indicates "take". Non-reactors in the flock are revaccinated. In case of turkeys, the loose skin between the thigh and abdomen is punctured.

4.1.4. Feather follicle inoculation

Used for fowl pox vaccination in turkeys. Few (2-3) feathers are removed over the thigh and a vaccine-dipped brush (supplied by the manufacturer) is rubbed against the opening of the follicle or the vaccine may be sprayed on the area with a sprayer holding the tip 5.0 to 7.5 cm away. Birds are examined for "take" after 6-10 d.

4.1.5. Intramuscular injection

Vaccines against Infectious Bursal Disease, Newcastle Disease, *Mycoplasma gallisepticum* Fowl Cholera, Infectious Bronchitis, Reovirus alone or in combination are used in breeder birds usually just before housing. IM injection is in the pectoral muscles, using continuous flow, automatic syringe.

4.1.6. Embryo Injection (*in ovo* technique)

Vaccines against Marek's Disease, Infectious Bursal Disease, Newcastle Disease or Infectious Bronchitis are injected into 18-day-old embryonating eggs with an automated machine. More than 30 countries are practicing this technique, especially for administering MD vaccine (HVT). The following areas can be accessed for depositing MD vaccine: air-cell, allantoic sac, amniotic fluid, embryo itself and yolk sac. It was found that embryo or amniotic fluid was suitable for MD vaccination resulting in > 90% protection; the allantoic sac route was not satisfactory

4.2. Flock application

4.2.1. Aerosol

IB, ND and ILT vaccines can be given through this method. Day or night, house is closed sufficiently to prevent cross drafts. Sprayer as recommended by vaccine manufacturer is used for a minimum period of 3-4 min; house is kept closed for another 15 min. Workers must wear goggles and face mask. This method is also used at hatchery for administering ND and IBD vaccines in a spray cabinet. However, now-a-days, only Marek's disease vaccine is given at the hatchery.

4.2.2. Water Administration

ND and/or IB, AE, IBD and ILT vaccines can be administered through water. Infectious Bursal Disease and Infectious Laryngotracheitis vaccines. Waterer with plastic bottom (glass container being best), free of sanitizer in water or container is employed. Water is withheld for one hour in hot weather (longer in cold weather); vaccine is mixed in cold water (with 0.1% skim-milk powder as stabilizer) and offered. Sufficient waterer space must be ensured so that 2/3 of the birds of the flock should be able to drink at one time. All vaccine should be consumed within 45 minutes.

5. Vaccination schedule

(Adopted from Univ. of Pennsylvania website)

5.1. Broilers

Table 72.2. Vaccination schedule – Broilers

Age (d)	Disease	Route
Hatchery	MD	Subcutaneous back of neck
9-14	ND-IB*	Ocular or Nasal or Oculonasal or Water or Spray
8-12	IBD	
14-21	ILT**	Water
21	ND-IB*	Water or Spray
28	IBD	
* IB vaccine combined only in endemic areas		** Given only in endemic areas

5.2. Layers (Pullets)

Table 72.3. Vaccination schedule – Layers (Pullets)

Age	Disease	Route
Hatcher	MD	s/c back of neck
9-14 d	ND-IB*	Ocular or Nasal or Oculonasal or Water or Spray
14 d	IBD	
28 d	IBD	Water or Spray
4 weeks	ND-IB*	
8 weeks	ILT** Fowl Pox AE	Eye drop Wing web stab Wing web stab
13-14 weeks	ND-IB*	Water or Spray
16 weeks	ILT**	Spray or eye drop
Post-housing or 16 weeks	ND-IB* ND-IB*	Every 3 months in some operations - Spray or drinking water Injected i/m or s/c ***
* IB vaccine combined only in endemic areas *** Inactivated vaccine used		** Given only in endemic areas

5.3. Broiler breeders

Most broiler birds will be vaccinated for MD and ND/IB. Age of vaccination for ND or IB will vary. Other vaccines such as IBD and ILT are used in endemic areas.

Table 72.4. Vaccination schedule – Broiler breeders

Age	Disease	Route
1 d	MD	s/c back of neck
7-10 d	Coccivax**	Spray on feed
9-14 d	ND-IB*	Ocular or Nasal or Oculonasal or Water or Spray
4 weeks	ND-IB*	Water or Spray
8 weeks	ILT**	Eye Drop
	Fowl Pox – AE	Wing web stab
10 weeks	IBD	Drinking water
13-15 weeks	ND-IB*	Water or spray
	AE	Drinking water
	Cholera (killed)**	Injected i/m
16 weeks	ILT**	spray or eye drop
18 weeks	Inactivated IBD, NDV, IB*, Reo virus and Cholera**	Injected i/m
Post Housing	ND or ND-IB*	Every 3 months in some operations Spray or drinking water
* IB vaccine combined only in endemic areas		** Given only in endemic areas

5.4. Layer breeders

Table 72.5. Vaccination schedule – Layer breeders

Age	Disease	Route
1 d	MD IB*	s/c back of neck Spray
10 d	ND	Drinking water
11 d	Coccivax	
12 d	Coccivax	
17 d	IBD	
6 weeks	ND-IB*	
7 weeks	IBD	
8 weeks	Fowl Pox, AE ILT**	Wing web stab Wing web stab Eyedrop
14 weeks	ND-IB*	Drinking water
18 weeks	ILT**	Spray
19 weeks	ND, IB*, Reo virus, IBD	i/m
* IB vaccine combined only in endemic areas		** Given only in endemic areas

5.5. Commercial turkeys

Table 72.6. Vaccination schedule – Commercial turkeys

Age	Disease/Medicine	Route
1 d	Antibiotics	s/c injection in neck
10 d	Coryza* ND	Drinking water
14 d	Coryza (if endemic)	
23-24 d	Hemorrhagic enteritis	
6 Weeks	ND	
7 Weeks	Cholera (M9)	
9 Weeks	Cholera (varying serotypes)	
14 Weeks	Cholera (varying sere-types)	
* Given only in endemic areas		

5.6. Breeding turkeys

Table 72.7. Vaccination schedule – Turkey breeders

Age	Disease	Route
2 Weeks	ND	Drinking water
5 Weeks	Cholera	
10 Weeks	Cholera	s/c injection in neck
12 Weeks	ND	Drinking water
14 Weeks	Cholera	s/c injection in neck
18 Weeks	Cholera	
20 Weeks	Cholera ND	Drinking water
22 Weeks	AE IBD	
24 Weeks	Parainfluenza (PMV-3)*, Arizona, ND Cholera Pox	s/c injection in neck s/c injection in neck Wing web
28 Weeks	Cholera Swine Influenza*	s/c injection in neck
40 Weeks	Parainfluenza (PMV-3)* Swine influenza*	
* Not given for toms		

5.7. Duration of immunity

Table 72.8. Duration of immunity

Disease	Vaccine	Duration of immunity
Marek's	HVT (Live)	16 weeks
Newcastle	LaSota / B_1 / F (mild, live)	6 weeks
Fowl Pox	Pigeon Pox (mild, live)	8 weeks
Infectious bursal disease	Chick embryo adapted	8 weeks
Infectious bronchitis	Massachusetts strain (chick embryo adapted)	8 weeks
Fowl Pox	Chick embryo adapted	Lifelong
Infectious laryngotracheitis	Modified	8 weeks
Newcastle	R_2B (live, attenuated)	Lifelong
Avian encephalomyelitis	Chick embryo adapted	8 weeks
Fowl cholera	Lyophilized (live)	16 weeks

6. Poultry vaccination schedules in India

Table 72.9. Vaccination schedule - General

Disease	Vaccine	Age	Route	Broilers	Layers	Breeders
MD	HVT	Hatch	s/c or i/m	±	+	+
ND	F/LaSota	7-10 d	Ocular / Water	+	+	+
CocciVac	Irradiated	10 d	Water	±	±	±
IBD	Intermediate	14 d	Ocular / Water	+	+	+
IB	Attenuated	21 d	Ocular / Water	±	±	+
IBD	Intermediate	28 d	Ocular / Water	+	+	+
ND	F/Lasota	42 d	Ocular / Water	NA	+	+
Pox	Attenuated	49 d	Wing-web	NA	+	+
ND	R_2B	70 d	s/c	NA	+	+
ILT	Attenuated	70 d	Ocular / Water	NA	±	+
Cholera	Killed	12-13 Weeks	s/c	NA	±	+
AE	Attenuated	14-16 Weeks	Wing-web or Water	NA	±	+
ND, IB, IBD, EDS	Combined	16-18 Weeks	s/c	NA	+	+
ILT	Attenuated	18 Weeks	Ocular / Water	NA	±	+

± Given only in endemic areas s/c Subcutaneous i/m Intramuscular

Table 72.10. Vaccination schedule – Commercial broilers

Disease	Vaccine	Age (d)	Route
MD	HVT	At hatch (if endemic)	s/c or i/m
ND	F/B_1/LaSota	5 – 7	Ocular/Water
IBD	Intermediate plus	14 – 15	Ocular/Water
ND	LaSota	21 – 22	Water

Table 72.11. Vaccination schedule – Commercial layers

Disease	Vaccine	Age	Route
MD	HVT	At hatch	s/c or i/m
ND	F/B_1/LaSota	6 – 7 d	Ocular/Water
IBD	Intermediate plus	15 d	Ocular/Water
IBD	LaSota	22 d	Ocular/Water
ND and IB	LaSota and H_{120}/M_{41}	28 d	Ocular/Water
Coryza	A, B, C Killed	7th Week	s/c
Pox	Attenuated	8th Week	Wing-web
ND	K/R_2B	10th Week	i/m
Coryza	A, B, C Killed	12th Week	s/c
IB	Ma5	13th Week	Water
ND	K/Killed LaSota	16th Week	s/c or i/m
s/c Subcutaneous i/m Intramuscular			
Based on the level of antibody titer against ND, flock vaccination program should be planned during the laying period. Vaccination with LaSota (32nd week), killed vaccine (40th week), R_2B/K (48th week) and LaSota (56th week) can also be practiced. For IB, vaccination with Ma5 at 8 weeks with repeat vaccination every 8 weeks is advisable in endemic areas.			

7. Precautions for vaccination

7.1. General precautions

Although specific precautions during vaccination through different routes are listed separately, certain general precautions are given below which suit most of the commercial facilities:

1. Should be purchased from a reliable source. Batch number, date of manufacture and Supplier's details have to be recorded.
2. Expired vaccines should neither be purchased nor used.
3. If premixed vaccine has to be transported, it must be done in a flask filled with ice water and used as quickly as possible; preferably, within two to three hours.

4. Vaccines are normally stored at the temperature of 2-8°C; they should neither be frozen not kept above 8°C. While transportation also, the same temperature range is ideal.
5. Vaccine should not be mixed in or exposed to sunshine.
6. Vaccination be carried out during cooler part of the day; temperature above 30°C may affect vaccine potency.
7. Do not vaccinate sick birds.
8. Vaccines administered via drinking water should be opened underwater into which it is mixed; because, vaccines are normally vacuum-sealed and if opened in the air, they may draw contaminated air into the container.
9. Tap-water with chlorine or other disinfectants must not be used because they lower the potency of the vaccine.
10. Metal containers must be avoided because reactions of the surface of the metal, if any, can produce chemicals that affect potency of the vaccine.
11. Antibiotics must be removed from water before and after administering bacterial vaccines, in particular.
12. Careless spillage of vaccine must be avoided especially with live vaccines which can cause the disease.
13. While administering water-based vaccines, water should be withdrawn for 2-3 h before administering the vaccine. During hot-weather, the withdrawal period can be reduced.
14. Adequate waterer space has to be provided.
15. Vaccine should be administered so as to enable the vaccine to break through the maternal antibodies. Hence suitable information regarding vaccination of the breeding flock has to be obtained from the chick supplier.
16. Quantity of water to be used for reconstitution should be properly determined depending on type and age of birds, temperature etc.
17. Vaccine should be administered depending on manufacturer's recommendations.
18. When a vial is opened, complete contents must be used. The vaccine should be opened and mixed just before administering the vaccine.
19. The vaccinating equipment must be sterile and chemical disinfectants must not be used for sterilization.
20. Reconstituted vaccine must be kept in an ice-bath during the period of vaccination. It is advisable to use the reconstituted vaccine within 2h of its preparation.
21. Empty vials, left-over vaccines and other material containing the vaccine must be burnt and destroyed.

22. The birds must be provided with adequate heat, feed, water and ventilation depending on the requirements.
23. The vaccine and/or the diluent must be properly stored as per the manufacturer's recommendations and administered through the route specified at an appropriate age.
24. All birds must be vaccinated.

7.2. Drinking Water Administration

(Breytenbach, 2005 on internet)

The goal of a drinking water vaccination is for each chicken to drink a minimum of one dose of vaccine solution. It appears to be the simplest method of live poultry vaccine mass administration; however a drinking water application done correctly is time-consuming.

7.2.1. Water availability

1. Vaccine should be reconstituted in a sufficient volume of water to cater for a two-hour period.
2. Sufficient drinker space is required to allow free access to the vaccine solution.
3. Water consumption during the first three weeks of a chick's life is erratic, thus walking through the house during the vaccination chasing up inactive chicks, especially along the sides of the house, is recommended.
4. To further stimulate drinking birds should be deprived of water prior to the vaccination. Usually 1 to 2 hours of water deprivation is sufficient to stimulate thirst, however this is dependant on environmental conditions.
5. The water system must be drained prior to administering the vaccine solution, this to avoid the first chickens quenching their thirst on standard water.

7.2.2. Vaccine viability

1. The cold chain must be maintained during the transportation of vaccine to the site and on site prior to use.
2. Avoid exposure of vaccine vials and vaccine solution to direct sunlight.
3. To prevent neutralization of the vaccine virus/bacteria drinking water must have a neutral pH and be free of disinfectants (including chlorine), detergents and heavy metals.
4. The addition of skimmed milk or skimmed milk powder (2 g/l or 2 l/100 l water) to the water 20 - 30 min prior to adding the vaccine virus/bacteria is recommended as a stabilizer.
5. Vaccine has to be reconstituted in a smaller quantity of water (5-10 liters) before adding to the drinking water. This ensures a homogenous vaccine

solution.

6. Vaccine solution must be consumed within 2 - 3 hours of reconstitution.

7.2.3. Assessing water intake

To evaluate the uniformity of water intake a blue dye is added to the vaccine solution. The crops and tongues of birds that have consumed vaccine solution will be temporarily stained blue. The intensity of the coloring is an indication of the volume of water consumed. A correct vaccination procedure results in at least 90% of chickens coloring blue.

7.3. Spray Administration

(Breytenbach, 2005 on internet)

The spray application technique is especially suitable for the administration of respiratory type vaccines such as ND and IB, as the vaccine strain is deposited directly onto the target cells; respiratory mucosa.

Coarse Spray is intended to deliver large droplets of vaccine to the upper respiratory tract and eye. In addition the feathers of the bird are moistened stimulating preening which further increases the chance of vaccine uptake. Coarse spray is the preferred method of vaccine administration for young unprimed chicks.

Fine Spray or mist is intended to hang in the environment of the birds and is inhaled into the respiratory tract. The smaller droplet size allows for a deeper penetration into the lower respiratory tract of the bird. This method of vaccine administration is suitable for older birds that have previously been primed to a specific respiratory antigen.

7.3.1. Vaccine viability

1. The cold chain must be maintained during the transportation of vaccine to the site and on site prior to use.
2. Vaccine vials and vaccine solution should not be exposed to direct sunlight.
3. As smaller volumes of water are used with spray applications (200 - 1000 ml per 1000 doses) it is recommended to preferably use distilled or de-ionized water.
4. Where distilled or de-ionized water is not available care must be taken that water used has a neutral pH and is free of disinfectants (including chlorine), detergents and heavy metals. The addition of skimmed milk or skimmed milk powder (2 g/l or 2 l/100 l water) to the water 20 - 30 min prior to adding the vaccine virus is recommended as a stabilizer.

7.3.2. Equipment

1. The correct equipment must be selected for the desired application technique. In practical terms a spray is considered coarse when droplets gener-

ated fall to the ground as would be seen with a shower of rain. A fine spray hangs in the air forming a mist above the birds.

2. Equipment must be free of disinfectants and detergents, and should preferably be dedicated for the purpose of vaccine application.
3. Cleanliness of equipment is paramount especially where equipment is being shared among different sites.

7.3.3. Vaccine distribution in the poultry house

1. Uniform vaccination requires a uniform distribution of the vaccine spray in the poultry house at bird eye level.
2. Vaccine must be reconstituted in a sufficient volume of water to distribute the required number of dosages over the entire population.
3. Birds on floor or open-type housing must be crowded for an effective vaccination. Crowding is achieved using barriers or by herding birds to the sides of the house forming a bird-free passage down the center.
4. Birds in cages are sprayed cage by cage.
5. During vaccination, ventilation in windowless houses must be turned off and open-sided houses must have curtains raised to keep out cross winds. This is especially important when fine spray or mist is the method of application. In excessively hot climates minimum ventilation may be required, in which case application should be limited to a coarse spray.
6. Spray distribution and droplet size can be monitored using water sensitive paper, a special paper which changes color when exposed to moisture. Strips of water sensitive paper are placed at the back of cages or against the wall to test whether vaccine spray is being distributed to all corners of the house/cage or not.

7.4. Administration by Injection

(Breytenbach, 2005 on internet)

All inactivated as well as a few live poultry vaccines are administered by s/c or i/m injection. This method of application requires accurate deposition of the full dosage of vaccine to each individual bird. Technically this is the most demanding vaccination technique as both speed and accuracy are needed while vaccinating thousands of birds. The doses of vaccine used must be tallied with the number of birds vaccinated; overdosing is costly, under-dosing results in poor immunity.

However, the key to vaccine administration by injection is: "Quality is more important than speed". The immune response to inactivated vaccines is dose related, thus a large percentage of birds receiving a partial dose or being missed will result in a poor immune response for the flock. The importance of inactivated vaccine application cannot be over emphasized as inactivated vaccines administered to breeder type birds often have a direct impact on the performance of their offspring.

7.4.1. Equipment

1. Multi-dose syringe selected must be compatible with mineral oils often used as adjuvants.
2. Syringes must be calibrated prior to use and at regular intervals during vaccination procedure. Use of an accurately calibrated test tube is recommended for this purpose.
3. Syringes and all ancillary equipment must be sterilized prior to use.
4. The correct size needle must be selected according to age of birds being vaccinated, site of injection and type of vaccine being administered.
5. Needles should be regularly replaced, at least once every thousand birds.
6. Blunt needles and needles with burs must be replaced immediately.

7.4.2. Vaccination technique

The birds vaccinated must be physically checked for:

1. Deposition of vaccine at the correct site, wet feathers indicating that vaccine was poorly administered; either full dose or partial dose in feathers due to premature or delayed expulsion of vaccine from the syringe.
2. A blue dye may be used when administering vaccines reconstituted in a phosphate buffered saline type diluent, such as MD vaccines. This improves the visibility of the vaccine once administered. Use of a blue dye in mineral oil-based vaccines is however not recommended as the blue dye may persist for an extended length of time resulting in condemnations at slaughter.

8. Evaluation of immune response by serology

(Breytenbach, 2005 on internet)

Serology is a helpful tool in evaluating the efficacy of an individual operator. Checking the serological response to a specific antigen administered by the operator will indicate the accuracy of vaccination. A poor % CV (higher numerical value) in combination with a low mean titer could indicate a large percentage of birds are being missed or are not receiving the full vaccine dose.

Serology provides a useful profile on previous stimulation of the bird's immune system, but serology by no means provides the full picture. Serology only measures circulating antibody levels (Ig G and Ig M), a function of humeral immunity, and fails to give us insight into cellular and local immunity. Despite this shortcoming, used in the correct context, serology can provide valuable clues on the success or failure of a vaccination.

8.1. Sampling

The key to reliable serology results start on the farm with sample collection. For sampling to be statistically valid there are two basic conditions that have to be met:

8.1.1. Random selection of birds for sampling

This implies that every bird in the flock should have an equal chance of being selected for sampling. A protocol should be agreed upon whereby samples are collected throughout the house.

8.1.2. Proper sample size

The number of blood samples taken from a flock has a direct impact on the reliability of the results. The fewer the number of samples collected the higher the risk of calculating an inaccurate mean flock titer. Twenty-three samples is the minimum recommended number to be collected for a meaningful appraisal of flock immunity.

8.2. Sample handling and storage

Sample handling and storage is important to ensure a good quality serum sample is delivered to the laboratory. Samples with excessive hemolysis, bacterial or fungal contaminants or samples that have started to decompose will not deliver reliable serology results

8.3. Interpretation of Serology following vaccination

Following vaccination it generally takes 4 - 6 weeks for the development of significant antibody levels. Earlier sampling is possible, but peak antibody levels will not yet have been achieved especially when evaluating the serological response to inactivated vaccines. To interpret seroconversion mean, coefficient of variation (CV %) and standard deviation of the titers from the sample of birds selected are calculated.

A low mean titer could indicate samples were collected too soon after vaccination, or a poor vaccine application, or in the case of inactivated vaccines poor priming prior to vaccination. For most diseases the % CV following a correctly applied vaccination should be less than 40%. If the % CV is above 60 %, there is definite room for improvement of the vaccine application techniques.

9. Vaccination failure

(Griffin *et al.* 2002)

Modern poultry farming has resulted in the development of high-density poultry areas, which bring with them an increased risk of disease spread. The industry manages this risk by routine vaccination against known poultry pathogens of specific economic importance. The efficacy of a vaccination schedule however requires proper vaccine administration, a challenging feat when thousands of birds need to be vaccinated at one time (Breytenbach, 2005 on internet).

A vaccination fails when, following vaccine administration, the animals do not develop an immune response sufficient to protect the animal from disease exposure. When animals get sick following vaccination, the natural inclination is to blame the vaccine. Although this is certainly an important consideration, other factors must be evaluated to determine the cause of the failure. The factors concerned with vaccination failure are:

9.1. Stress

Stress can lead to immune suppression and may reduce the animal's ability to mount an immune response. Stress could include environmental extremes, handling, inadequate nutrition, parasitism, and other diseases. While it is common to vaccinate stressed animals, these animals are more susceptible to adverse vaccine reactions and frequently do not develop an adequate immune response. Immune stressed animals develop limited protection from vaccination. Frequently high stressed animals may be incubating disease at the time of vaccination. Although the subsequent sickness and death loss mimics a vaccine reaction, the underlying disease is to blame. Unless using the vaccine to stimulate immediate partial immunity, delaying vaccination of stressed animals is advisable. Similarly, properly vaccinated animals may still contract a disease if they are later stressed.

9.2. Passive antibody interference

Existing antibodies to the antigen(s) contained in the vaccine can block vaccine response. Antibodies acquired through yolk are critical to the day-old chick, but only last 2-3 weeks and while present can block the immune response to a vaccine. This is referred to as Passive Antibody Interference or Maternal Interference.

9.3. Improper timing

This is a common cause of vaccine failure. If an animal is incubating disease at the time of vaccination, the underlying disease may not be prevented by the vaccine. The clinical signs may mimic an adverse vaccine reaction although the symptoms are really due to a previous infection. An animal can succumb to a disease if exposed to the natural infection following vaccination; the protection afforded by vaccination cannot be expected until the immune response is complete.

9.4. Type of vaccine used

After the first exposure to a modified live vaccine product, interferon levels rise within hours followed by antibodies detectable in approximately four to five days. Killed vaccine products require training the immune system by two vaccinations given not less than 10 days apart. With killed vaccine products the immune system is not ready to protect the animal until after a rise in antibodies following the second vaccination.

9.5. Duration of protection offered by vaccine

An animal can succumb to a disease if exposed to the natural infection following vaccination if the length of time between vaccination and exposure has been longer than the protection afforded by the vaccine. Most vaccines do not ensure lifelong immunity from one or two injections.

9.6. Strain or serotype present in vaccine

The protection afforded by a vaccine may also be incomplete if the vaccine does not contain the proper strains or serotypes of organism required to stimulate protective immunity; particularly if the animals are stressed or exposed to an overwhelming level of infectious agents.

9.7. Quality of vaccine

Vaccine may be of poor quality (low vaccine titer, contaminated, etc.). Vaccine failure due to faulty vaccine manufacture is rare. In some situations, one type of vaccine may be better than another. Choosing the correct vaccine and using it properly is an important part of preventative animal health.

Chapter **73**

Inheritance of Qualitative Traits

1. Basic concepts

All living matter is made up of cells and in higher organisms; the cells form tissues and organs; and organs, the organism. The various organs are responsible for the appearance, structure, physiology and behavior, which are collectively referred to as "phenotype" of the organism.

The phenotype of the organism, therefore, must be determined from within the cell itself by its interaction with the environment and it is referred to as "genotype" which, in general terms, constitutes the phenomenon of life.

1.1. Types of cells

There are principally two types of cells namely, somatic cells which form most part of the body and reproductive cells or gametes which are concerned with propagation of the species. The sexual cell from male parent unites with its female counterpart and later on, develops into a new organism. Therefore, information about the phenotype of the new organism must be carried by the tiny sexual cells and this transfer information from parent to offsprings is called heredity.

1.2. Chromosomes

The genotype of the individual is determined by what is known as "genes", the smallest biological units of inheritance, carried on chromosomes. The number of chromosomes is constant for a species.

The chromosomes are of two types; autosomes which are similar in both the sexes and sex chromosomes (heterosomes) which differ between the two sexes. All species of bisexual animals contain a pair of chromosomes and known number of pairs of autosomes.

1.3. Chromosome number

The somatic cells contain the known number of pairs of chromosomes but the gametes get one chromosome from each pair of homologous chromosomes so that when gametes from male and female unite, they form the same number of pairs of chromosomes as in either of its parents. This reduction in number of chromosomes in the gametes is effected by a reduction division of the cell called "meiosis". The somatic cells, on the other hand, multiply by an equation division (in which the chromosome number is not halved) called "mitosis". As the gametes contain half the complement of chromosomes from each pair, they are haploid (N), whereas, the somatic cells are diploid (2N) in nature.

1.4. Sex-chromosomes (heterosomes)

In dioecious plants and most animals the sex chromosomes are designated as XX in case of female, and XY in case of male. Since autosomes are identical in both male and female within a species, it can be seen that the gametes produced in case of females are identical, whereas, those in males are dissimilar. Therefore, the females are "homogametic" and males "heterogametic". However, in case of birds and few other animals, the female is heterogametic and the male, homogametic.

It is also clear from the above discussion that the sex of the offspring is determined by the heterogametic parent because, the homogametic parent producers similar gametes. Obviously, the sex of the offsprings in birds is determined by the female. Hence, in birds, male is normally designated ZZ (instead of XX) and female ZW (instead of XY). The Y or W chromosome is smaller and thought to carry little or no information; hence, some of the use "Z-" to designate the female in case of birds.

In case of poultry, 2N = 78 to 82, depending on species. The karyotype (chromosomal appearance) consists of 6 to 8 relatively large, easily identifiable pairs and others are smaller (micro-) chromosomes. Z chromosome is the 4th or 5th of the larger pairs and W is among the six larger pairs of micro-chromosomes. There is a great similarity among species as far as karyotype is concerned which indicates extreme evolutionary conservation of chromosome structure in Aves. Number of chromosomes (2N) in Chicken (*Gallus domesticus*), Turkey (*Meleagris gallipavo*), Japanese quail (*Coturnix coturnix japonica*), Duck (*Anas platyrhynchos*) and Goose (*Anser anser*) is 78, 80±, 78±, 78± and 80±, respectively (Fechheimer, 1985). Muscovy and Mallard ducks have 78 and 80 chromosomes, respectively.

2. Nucleic acids

Table 73.1. DNA *Vis-à-vis* RNA

	DNA	RNA
Number of chains	Two	Usually one
Sugar	2-deoxyribose	Ribose
Pyrimidine bases	Cytosine (C) and Thymine (T)	Cytosine (C) and Uracil (U)

2.1. Structure

Chemical analysis has revealed that chromosome is a nucleoprotein. The nucleic acid moiety can be either deoxyribose nucleic acid (DNA which is present in nucleus) or ribose nucleic acid (RNA which is present mostly in cytoplasm). The DNA and RNA are different in many ways (Table 73.1).

Both DNA and RNA contain phosphoric acid and purines (Adenine, A and Guanine, G). The nitrogen bases (purines and pyramidines) are attached to the C1 of the sugar and the phosphoric acid attached to the C2 (RNA) or C3 (DNA) of the preceding and C5 of the succeeding sugar moiety resulting in a phosphodiester bridge.

The double-helical structure of the DNA with nitrogen bases towards the axis has been enumerated. Phosphoric acid-deoxyribose acts as backbone on which the nitrogen bases are located. There will be hydrogen bonds between A and T as well as between G and C (base pairing) so that the diameter of the helix remains uniform throughout (20Å). In any DNA molecule, A+G = T+C and (A+T) ÷ (G+C) is constant for a particular species.

The eukaryotic chromatin complex consists of equal quantities of DNA, histones (basic proteins) and non-basic chromosomal protein. Due to specificity of base pairing, DNA strands are complementary to each other and if the sequence of one strand is known, that of the other can be written. The strands run in opposite directions (3′-5′ and 5′-3′) and hence, referred to as "antiparallel".

2.2. Properties and functions

2.2.1. Duplication

Hydrogen bonds between the nitrogen bases weaken and disappear. This is followed by formation of complementary strand for each of the separated strands, with the latter as template. The interesting feature is that the daughter molecule consists of one parent molecule and the other, the newly synthesized one. Therefore, the duplication is "semi-conservative" which has been thought to occur in all forms of life; thus, is universal.

2.2.2. Transcription

The DNA-duplex unfurls during transcription and synthesis of messenger-RNA (mRNA) on thc DNA-template occurs in a 5′-3′ direction in presence of DNA-dependent RNA polymerase. The original DNA is preserved during mRNA synthesis and the latter therefore is a complementary poly- nucleotide to DNA-template. In eukaryotic cells, three types of RNA-polymerases, one each of ribosomal RNA (rRNA), transfer RNA (tRNA) and mRNA, all synthesized on a DNA template are identified.

2.2.3. Translation

Translation of mRNA is through nucleotide-amino-acid dictionary. With 4 nitrogen bases, considering two bases as a combination, only 4^2 combinations or 16 "codes" are possible. Similarly, considering three bases as a combination, 4^3 combinations or 64 codes are possible. Finally, considering four bases as the combination, only four permutations are possible. Keeping in view that there are 22 amino-acids to be coded, it is generally agreed that three consecutive nitrogen bases form a "triplet codon" and several successive codons may form a "gene".

It has also been known experimentally that the codons are "nonoverlapping"; for instance, if overlapping was to have been existing, single (point) mutation should bring more than one change in the polypeptide chain concerned but, no such effect has been reported. Similarly, overlapping invariably avoids certain amino-acids from lying side by side in a protein chain; which again, has not been proved.

With 64 codons available for 22 amino-acids, it is imperative that more than one code is available for some or many amino-acids; property is referred to as "degenerate" codons. In addition, if the dictionary of codons is carefully observed, it can be noticed that the third base in each of the codes has little or no meaning on most occasions; hence, the third base is often referred to as "wobble" base.

The degenerate nature and the presence of wobble bases confer a tremendous immunity to the DNA molecule against mutations. In addition, it is generally agreed that the genetic codes are common for all living matter indicating that the genetic codes are also "universal" in nature. Further, 'AUG', coding for methionine, is the chain-initiating and 'UAA, UAG and UGA' are the chain-terminating codons; this is also considered 'universal'.

2.2.4. Protein synthesis

The details of protein synthesis are beyond the scope of this book; the reader is advised to refer books related to detailed protein synthesis.

2.2.5. Mutation

Mutations are changes in genes which are the sources of hereditary variation on which evolutionary forces or selective action of the breeders act. Rate of mutation should neither be too low nor too high because, more often than not, mutations are deleterious and high mutation engineers the survival of the species. On the contrary, too low mutation fails to provide the new variability required by the species to adapt play changing environment.

It is also important note that any mutagen, which can cause mutation, can only produce a pro-mutation because the result of the mutation is (are) seen only in the progeny of the individual but not in the individual itself. Pro-mutations, therefore, are reparable but not mutations. In fact, the cell itself has an enzymatic system to reverse the pro-mutations.

If a codon specific for one amino-acid changes to specify a different amino-acid, the resultant protein will have an amino-acid replacement. This eventually may cause changes in the biological activity of the protein; a classical example is hemoglobin concerning sickle-cell anemia in human beings. The mutation brings about replacement of Glu by Val in two of the four polypeptides. Individuals homozygote to such hemoglobin condition will have abnormally shaped RBCs, reduced oxygen-carrying capacity of the blood, anemia, severe pain and reduced life-expectancy. However, individuals heterozygous to this condition will have enhanced resistance to falciform malaria because the cell membrane of the changed RBCs is less susceptible to attachment by the malaria parasites. On the same lines, it can be expected that the gene causing impaired digestion/absorptionof nutrients can affect growth and production of the animal concerned.

Mutation can also lead to a change from nonsense codon to a codon of an amino-acid. Such a mutation leads to an abnormally long polypeptide chain; on the other hand, if a codon for an amino-acid mutates to become a nonsense codon, the protein formed will be shorter; in either case, the resultant protein is likely to be biologically inactive.

Mutation in the nucleotide sequence of tRNA and rRNA can lead to alteration in amino-acid specificity and reading accuracy, respectively.

3. Gene structure

3.1. Split-genes

It is interesting to note that the concept of exons and introns and that the eukaryotic DNA is split was developed while working on ovalbumin gene in the oviduct of laying hens.

Number of genes appears to be split into structural sequences called "exons" interspersed by intervening sequences called "introns"; the function of the latter is unknown. During transcription, entire DNA, including both exons and introns, is transcribed; later on, the nonsense sequences (corresponding to introns) are spliced out to produce the "mature" mRNA which is ultimately conveyed to cytoplasm. For instance, chicken ovalbumin gene has more than 7000 base pairs with eight exons separated by seven introns. The mature mRNA produced on this gene has 1895 base pairs whereas the spliced mRNA sequences corresponding to the introns remain in nucleus.

Introns have been thought to have no function other than own preservation within the genome ('selfish DNA'!)

3.2. Antibody genes (Ab genes)

Ab genes are multi-gene families consisting of genes corresponding to ê and ë light (L)-chain as well as heavy (H)-chain. These are located on different autosomes. They have a gene structure with their variable (V) and constant (C) region-linked genes. One out of several (10%) 'V' gene joins with one 'C' gene to form a genetic unit which is finally transcribed as a single mRNA so that only one L- and one H-chain genes are expressed in a given cell.

3.3. Redundant genes

Concentrated DNA near centromere (whosefunction is not yet known), ribosomal DNA, transfer DNA, histone DNA coding for corresponding RNAs are present at several locations but with identical base sequence; these are referred to as "redundant genes".

4. Mendelian inheritance (Inheritance of qualitative traits)

Certain characteristics like comb shape, plumage etc. cannot be measured on a scale and hence are called qualitative traits. The term locus is used to indicate the location of a gene on a chromosome map. It also designates the unit whose variants act as alleles and it is usually named after the first variant allele found. Genes at different loci can modify, influence or alter the expression of gene/genes at other locus/loci. This sum total of all the gene-actions explains the great variability found between members of the same species. The genes in the same locus of homologous chromatids and having only two alternative expressions are referred to as alleles and hence the alleles represent the variants of the gene. When two members of the given pair of alleles are alike, then the individual is said to be homozygote, otherwise heterozygote, with respect to that allele. However, many and possibly, all genes, can change in different ways giving rise to several alternative states or variants of the gene; these genes are referred to as multiple alleles.

4.1. Types of gene action

4.1.1. Dominance

This is the ability of the gene to block-out or cover the expression of its allele. This is the interaction of alleles at the locus. The heterozygotes are phenotypically similar to the character in question.

4.1.2. Incomplete dominance

The heterozygotes are phenotypically different from either of the homozygote since one of the alleles is incompletely dominant.

4.1.3. Epistasis

When genes are different loci interact or when genes that one locus can modify the expression of genes at the different locus, the gene action is referred to as epistasis.

4.1.4. Pleotropy

When a gene influences more than one trait, it is referred to as a pleotropic gene.

4.1.5. Linkage

During meiosis, certain genes at different loci, but in close proximity, tend to go together. This tendency is referred to as linkage of genes. Genetic correlation between two or more characteristics can occur due to pleotropy or linkage.

4.2. Mendel's laws

Particulate theory of inheritance proposed by John Gregor Mendel (1865) rediscovered by Correns, de Vries and Tschermark (1900) has been a marvel of scientific discovery which has opened a wonderful field of genetics for application by animal breeders. Conspicuously, the general rules proposed by Mendel are valid even now barring certain explainable exceptions.

4.2.1. Mendel's I Law (or gametic purity or Law of segregation)

This states that the characteristics of individual are determined by pairs of genes; but the gametes contain only one gene from each pair. That means several genes come together in an individual for certain period of time and then part without changing each other at all. Therefore, genes that have a clear-cut phenotypic classification are transmitted without any modification by genes at other loci.

Table 73.2. Law of segregation - Classical

Trait: Shank feathering, Fs Nature: Autosomal, dominant to clean shanks, fs					
Case: Both parents being heterozygous					
	Male		Female		
Parent's genotype	Fs fs		Fs fs		
Gametes produced	Fs	fs	Fs	fs	
Probability of gametes	½	½	½	½	
Genotype of offsprings	Fs Fs		Fs fs		fs fs
Therefore, P(FsFs) = P(Fs)m X P(Fs)f = ½ x ½ = ¼ *, P(fsfs) = P(fs)m X P(fs)f = ½ x ½ = ¼ * and P(Fsfs) = P(Fs)m X P(fs)f + P(fs)m X P(Fs)f = ¼+ ¼ = ½ ** or 1-P(homozygotes) = 1 - ½ = ½.					
Genotype of the offsprings	FsFs		Fsfs		fsfs
Genotypic probability of offsprings	¼		½		¼
Phenotypic probability of offsprings	¾				¼
Genotypic ratio	1 : 2 : 1				
Phenotypic (Monohybrid) ratio	3 : 1				
* Since the events are independent ,multiplicative rule of probability applied ** Both multiplicative and addition rules of probability applicable					

Table 73.3. Law of segregation - deviation

Trait: Shank feathering, Fs Nature: Autosomal, dominant to clean shanks, fs					
One of the parents being heterozygous, the other homozygote recessive					
	Male		Female		
Parent's genotype	Fs fs		fs fs		
Gametes produced	Fs	fs	Fs	fs	
Probability of gametes	½	½	0	1	
Genotype of offsprings	Fs fs		fs fs		
Therefore, P(FsFs) = P(Fs)m X P(Fs)f = ½ x 0 = 0 *, P(fsfs) = P(fs)m X P(fs)f = ½ x 1 = ½ * and P(Fsfs) = P(Fs)m X P(fs)f + P(fs)m X P(Fs)f = ½ + 0 = ½ ** or 1-P(homozygotes) = 1 - ½ = ½.					

Genotype of the offsprings	FsFs	Fsfs	fsfs
Genotypic probability of offsprings	0	½	½
Phenotypic probability of offsprings	0	1	1
Genotypic ratio	1 : 1		
Phenotypic ratio	1 : 1		
* Since the events are independent, multiplicative rule of probability applied ** Both multiplicative and addition rules of probability applicable			

In other words, considering one locus, only one of the alleles is transmitted to the gamete (without being contaminated) with equal frequency. Monohybrid genotype refers to an individual which can produce two types of gametes with respect to one locus or which is heterozygous to the locus in question (Tables 73.2 and 73.3).

Table 73.4. Law of independent assortment - Classical

Loci : A and B Nature (both loci): Autosomal, dominant				
	Male		Female	
Parent's genotype	Aa Bb		Aa Bb	
Gametes produced				
'A' locus	A	a	A	a
Probability	½	½	½	½
'B' locus	B	b	B	b
Probability	½	½	½	½
Gametic assortment (as per I Law)				
'A' locus				
AA	½ x ½ = ¼			
Aa	½ x ½ + ½ x ½ = ½			
aa	½ x ½ = ¼			
'B' locus				
BB	½ x ½ = ¼			
Bb	½ x ½ + ½ x ½ = ½			
bb	½ x ½ = ¼			

Therefore, by the multiplicative rule of probability, the genotypic frequency of AABB. AABb, AAbb, AaBB, AaBb, Aabb, aaBB, aaBb and aabb will be 1/16, 2/16, 1/16, 2/16, 4/16, 2/16, 1/16, 2/16 and 1/16
Therefore, P(A-) = P(AA) + P(Aa) = ¾, P(aa) = ¼ and
P(B-) = P(BB) + P(Bb) = ¾, P(bb) = ¼
Since the gametes assort independently, by applying multiplicative rule of probability, the following phenotypic frequencies are obtained:

Phenotypic class	Probability
A-B-	9/16
A-bb	3/16
aaB-	3/16
aabb	1/16
Genotypic ratio	1 : 2 : 1 : 2 : 4 : 2 : 1 : 2 : 1
Phenotypic (Dihybrid) ratio	9 : 3 : 3 : 1

Exceptions to law of segregation: Deviation from the law of segregation can be expected when:

1. Chromosomes fail to separate at anaphase i.e., non-disjunction
2. Mutation
3. Change in penetrance of a gene i.e., certain proportion of offsprings not expressing the phenotype in relation to their genotype.
4. Pleotropism i.e., one gene controlling more than one trait.

Table 73.5. Law of independent assortment - deviation

Traits : Naked neck (N) and White skin (W) Nature : Autosomal, dominant to neck feathering (n) and yellow skin (w), respectively				
	Male		Female	
Parent's genotype	Nn Ww		Nn ww	
Gametes produced				
'N' locus	N	n	N	n
Probability	½	½	½	½
'W' locus	W	w	W	w
Probability	½	½	0	1
Gametic assortment (as per I Law)				
'N' locus				
NN	½ x ½ = ¼			
Nn	½ x ½ + ½ x ½ = ½			
nn	½ x ½ = ¼			
'W' locus				
WW	½ x 0 = 0			
Ww	½ x 1 + ½ x 0 = ½			
ww	½ x 1 = ½			

Therefore, P(N-) = P(NN) + P(Nn) = ¾ and P(W-) = P(WW) + P(Ww) = ½

Since the gametes assort independently, by applying multiplicative rule of probability, the following phenotypic frequencies are obtained:

Phenotypes	Genotypes	Probability
Naked neck, white skin	N-W-	P(N-) X P(W-) = 3/8 or 6/16
Naked neck, yellow skin	N-ww	P(N-) X P(ww) = 3/8 or 6/16
Normal neck, white skin	nnW-	P(nn) X P(W-) = 1/8 or 2/16
Normal neck, yellow skin	nnww	P(nn) X P(ww) = 1/8 or 2/16
Phenotypic ratio		6 : 6 : 2 : 2

4.2.2. Mendel's II Law (law of independent assortment)

This states that characters are inherited independently of one another. In other words, during the formation of gametes, the members of the individual allelic pairs segregated independently of each other. Following this, one pair of alternative characters is inherited quite independently of each other and probability of the composite genotype when more than one locus is considered is equal to the product of probabilities of its constituents. If an individual is heterozygous to both the loci and hence can produce 4 different types of gametes, it is called a dihybrid genotype.

4.3. Epistasis

The concept shown in Table 73.3 can be extended to many loci and also for calculating frequencies with epistatic effects (Table 73.6).

Phenotypic ratios are altered when there is epistasis depending on type of association between the two loci. There are several types of epistasis and not all are recorded in poultry; however, the most important epistasis in poultry is shown in Table 73.6. However, on the basis of the Classical dihybrid ratio explained in Table 73.3, different types of epistatis are listed in Table 73.7.

Table 73.6. Epistasis – Inhibitory gene

Nature : Autosomal, epistatic, I locus: 'II' completely and 'Ii' incompletely masking the expression C locus for color 'C-'				
	Male		Female	
Parent's genotype	Ii Cc		Ii Cc	
Gametes produced				
'I' locus	I	i	I	i
Probability	½	½	½	½
'C' locus	C	c	C	c
Probability	½	½	½	½
Gametic assortment (as per I Law)				
'I' locus				
II	½ x ½ = ¼			
Ii	½ x ½ + ½ x ½ = ½			
ii	½ x ½ = ¼			
'C' locus				
CC	½ x ½ = ¼			
Cc	½ x ½ + ½ x ½ = ½			
cc	½ x ½ = ¼			

Therefore, P(C-) = P(CC) + P(Cc) = ¾, P(cc) = ¼ and P(--) = 1

Since the gametes assort independently, by applying multiplicative rule of probability, the following phenotypic frequencies are obtained:

Phenotypes	Genotypes	Probability
White birds	II--, --cc	P(II) X P(--) + P(--) X P(cc) – P(IIcc)* = 7/16
Speckled birds	IiC-	P(Ii) X P(C-) = 3/8 or 6/16
Colored birds	iiC-	P(ii) X P(C-) = 3/16
		*Already accounted for under P(II--)
Phenotypic ratio		7 : 6 : 3

4.3.1. Types of epistasis

Table 73.7. Types of epistasis

Epistasis	Effect	Phenotypic Classes	Phenotypic Ratio	Example
Recessive	aa or bb (not both) dominant on other locus. Hence, considering aa as dominant, all aa– will have same phenotype	A - B - A – b b a a - -	¾ x ¾ = 9/16 ¾ x ¼ = 3/16 ¼ x 1 = 4/16	Coat color in mice 9 Agouti: 3 Black: 4 Albino
Duplicate recessive	Either aa or bb will produce a phenotype similar to aabb i.e., aab- and A-bb will be similar to aabb	A – B – Others	¾ x ¾ = 9/16 1 – 9/16 = 7/16	Color of Sweet pea flower 9 Purple : 7 White
Polymeric gene action	Same as duplicate recessive epistasis excepting that aabb is phenotypically distinguishable from aaB- and A-bb	A – B – a a b b Others	¾ x ¾ = 9/16 ¼ x ¼ = 1/16 1-10/16 = 6/16	Length of awns in barley 9 long awn : 6 Medium awn : 1 Awnless
Dominant	A- or B- (not both) dominant on other locus and phenotypically similar to aabb. Hence, considering A- as dominant, all A--- will have saved phenotype as that of aabb	a a b – Others	¼ x ¾ = 3/16 1-3/16 = 13/16	Inhibitory gene or dominant white in poultry when speckling is ignored and considered white 13 White : 3 Colored (See also Inhibitory gene)
Masking gene action	Same as dominant epistasis excepting that aabb is distinguishable from A---	A - - - a a B – a a b b	¾ x 1 = 12/16 ¼ x ¾ = 3/16 ¼ x ¼ = 1/16	Fruit color in Summer squash 12 White : 3 Yellow : 1 Green
Duplicate dominant	Presence of at least one dominant gene in either the loci, produces a common phenotype. Hence, aabb forms one type of phenotype and all others another type	a a b b Others	¼ x ¼ = 1/16 1–1/16 = 15/16	Shank feathering in poultry 15 Feathered : 1 Clean
Dominance modification of duplicate genes	At least one dominant gene being present at both the loci or any one of the loci being homozygous dominant, the phenotypes will be similar	A-B- + AAbb + aaBB aaBb + Aabb + aabb	¾ x ¾ + ¼ x ¼ +¼ x ¼= 11/16 ¼ x ½ + ½ x ¼ + ¼ x ¼= 5/16	Pigment glands in cotton plant 11 Glandular : 5 Glandless

Epistasis	Effect	Phenotypic Classes	Phenotypic Ratio	Example
Complete dominance at one locus and partial dominance at the other	Considering A locus as dominant, aa is epistatic to B locus. In the absence of aa condition, B locus modifies the effect of A locus	A–BB + aa-- A – B b A – b b	¾ x ¼ + ¼ x 1 = 7/16 ¾ x ½ = 6/16 ¾ x ¼ = 3/16	Hair direction in guinea pigs 7 Smooth : 6 Partly smooth : 3 Rough
Inhibitory gene with partial dominance at inhibitory locus	Dominant homozygosity in A locus completely masks whereas heterozygosity partially suppresses B locus. Homozygous recessive in B locus indisputable from the former group	A a B – a a B – Others	½ x ¾ = 6/16 ¼ x ¾ = 3/16 1-9/16 = 7/16	Inhibitory gene or dominant white in poultry 7 White : 6 Speckled : 3 Colored
Dominance is complete at one locus and partial at the other locus	Locus A completely dominate: presence of aa precludes the effect of B-, and aabb is similar to A-bb	A – B b A – B B a a b – Others	¾ x ½ = 6/16 ¾ x ¼ = 3/16 ¼ x ¾ = 3/16 1-12/16 = 4/16	Body color in floor beetle 6 Sooty : 3 Red 3 Jet : 4 Black
Complete dominance at one locus and partial dominance at the other	Same as above excepting phenotypic appearances	A-Bb + aaBB -- b b A – b b a a B b	¾ x ½ + ¼ x ¼ = 7/16 1 x ¼ = 4/16 ¾ x ¼ = 3/16 ¼ x ½ = 2/16	Body color in floor beetle 7 Sooty : 4 Black : 3 Red : 2 Dark sooty
Partial dominance at both genes; additive effects for each partially dominant gene	Same as classical dihybrid genotypes with additive effects resulting in 9 phenotypic classes. Value of 3 and 2 are assumed for each of A and B, respectively Ex : Flower color in beans	A A B B A A B b A a B B A A b b A a B b a a B B A a b b a a B b a a b b	1/16 (value 10) 2/16 (value 8) 2/16 (value 7) 1/16 (value 6) 4/16 (value 5) 1/16 (value 4) 2/16 (value 3) 2/16 (value 2) 1/16 (value 0)	Purple shade Purple shade Purple shade Purple shade Purple shade Purple shade Purple shade Purple shade White

5. Autosomal inheritance

(http://marsa_sellers.tripod.com/geneticspapers)

Inheritance of characters determined by the genes located on the autosomal chromosomes consists of several traits; some of them are discussed below (Note: + superscript indicates wild-type gene):

5.1. Ear lobe color

Most breeds have red ear lobes. The red color is due to the blood of the bird and is visible because the skin of the ear lobes, comb and wattles has a rich blood supply that is not masked in any way. These skin areas are so highly vascularized that squeezing a comb between thumb and forefinger is more than likely result in blood on fingers.

Breeds of the Mediterranean Class (Leghorn, Minorca and Spanish) have 'white' ear lobes. The white ear lobe is due to the purine pigment which is controlled by a number of genes. The trait is said to be polygenic. The red ear lobe is due to the lack of the genes that invoke the purine pigmentation. Sometimes the white ear lobe can have a greenish or yellowish tinge. The number and location of the genes responsible for white ear lobes is not presently known.

5.2. Eggshell color

Brown eggshell color is a complex trait and as many as 13 genes have been proposed to account for the range in eggshell color. The white eggshell color is due to an absence of blue and brown, and perhaps some modifying factors (genes), since there are different shades of white. The blue eggshell gene, O, expresses if it is present which is why it is considered to be dominant. The gene symbol for the recessive, wild-type gene is o or o^+. The locations of the brown eggshell genes are not known and it is not known how many brown modifying genes there are or where they are in relationship to the genes of known locations. Brown may itself be just an array of white modifiers. For example, there is a recessive sex-linked gene, pr that inhibits the expression of brown eggshell genes and can be used to help remove the brown tint from white eggs.

The eggshell color genes interact in the following way: The effect of the blue gene is dominant over white. The effect of the brown gene is dominant over white. When blue and brown genes are both present, both genes contribute to the eggshell color making the eggs appear green. In this case, the inside surface of the eggshell will be significantly less green and more blue than the outside surface, which is where most of the brown pigment is.

Since the blue and brown eggshell color genes should be at different locations, at least two pairs of genes to describe the genotypes of the blue, white, green and brown layers exist. If Br indicates a brown eggshell color gene and "br", its complementary recessive gene; and O indicates blue shell color o its complementary gene, then the genotype of a blue eggshell layer will be (O, O) with (br, br). Blue and white genes, (O, o) with (br, br) also yields a blue egg, but perhaps a lighter blue. The pair of eggshell color genes, (O, O) with (Br, Br), are the genes for producing a green egg, (o, o) with (Br, Br) produces a brown egg and (o, o) with (br, br) yields a white egg. Females having one blue gene and one or more brown genes will lay eggs having a greenish color. However, genetics picture of eggshell color is more complicated than a mere dihybrid cross because, there is certainly more than one gene for brown eggshell color. There must be a relatively large number of eggshell color modifying genes that are not yet known. Most people accept a rule of thumb to the effect that a daughter will lay eggs that are a color between that of the parent lines.

Surprisingly, daughters of a cross between green egg layer (faux-Araucanas or Easter Egg Chicken; genotype OO with BrBr) and a white egg layer (Leghorn;

genotype oo with brbr) will all have one gene for blue eggshell color and one gene for brown; still, all lay green eggs! This clearly demonstrates the complexity of inheritance of egg shell color.

5.3. Comb shape

Comb type in chickens is due to two genes on different chromosomes, the rose comb gene, R, and the pea comb gene, P; r and p are the genes that replace R and P when they are not present. At each locus complete dominance exists. Four types of combs namely, Walnut (R-P-), Rose (R-pp), Pea (rrP-), and Single (rrpp) are expressed by the dihybrid combinations; a Classical dihybrid ratio of 9:3:3:1 is expectable. Single comb is recessive to both rose and pea combs.

5.4. Shank/feet color

The shank/feet color is controlled by genes that affect the skin at different depths. The visible color is due to the combined effect of the different colors of the dermis and the epidermis. So, the shank/feet colors are a combination of upper skin and deeper skin pigmentations. The shank/feet colors result from the major gene combinations (the bird has two copies of each gene); but, other genes can modify shank and foot color. For example, the sex-linked barring gene, B, is a potent inhibitor of dermal melanin. The Barred Plymouth Rocks, for example, would not have light shanks and feet if it were not for the fact that they have sex-linked barring. The female Barred Rocks tend to have darker shanks due to the dose effect of the barring gene (Table 73.8).

Table 73.8. Some basic shank/feet color genetics

Shank/Foot Color	Homozygous loci
Near black with white soles	W^+, Id, E
White shanks and feet	W^+, Id, e^+
Black shanks, white soles	W^+, id^+, E
Blue shanks, white soles	W^+, id^+, e^+
Near black with yellow soles	w, Id, E
Yellow shanks and feet	w, Id, e^+
Black shanks with yellow soles	w, id^+, E
Green shanks with yellow soles	w, id^+, e^+
W Carotenoid locus, Id Dermal melanin locus, E Epidermal melanin locus	
Source : Crawford, 1990	

5.5. Dark skin color

The hypermelanic condition of some breeds, such as the Silkie breed, is due to a pigment cell activator (fibromelanosis) that causes pigmentation of connective tissue. The inheritance of the dark skin phenotype involves the fibromelanosis gene, Fm, as well as dermal melanin inhibitors, such as the sex-linked Id dermal melanin inhibiting mutation. The fowl with Fm and wild-type dermal melanin, id^+, will have darkly pigmented skin and connective tissue. The combination of

Fm and Id gives a bird that has little or no observable skin pigmentation. There are other dermal melanin inhibitors that may have an influence on the degree of melanization due to Fm (or the degree of expression of Fm). Some genes influencing plumage color have an effect on dermal melanin, such as the E-locus alleles, which may influence the expression of Fm. However, fibromelanotic Silkies exist with black, white, blue and partridge patterns.

5.6. Feather color

White is actually all the colors combined and black is the lack of reflection of light in the visible range, so one might argue that black and white are not really 'colors' technically. However, including black and white as colors, chickens have only three basic colors: black, white and red (gold).

The colors of chickens are achieved by diluting and enhancing or masking black and red (gold). For example, Rhode Island Reds have the gold gene with the dominant mahogany (red enhancing) gene. A blue chicken is a black bird that has the blue gene which dilutes black. Two copies of the blue gene give a splash effect. A white chicken can be achieved in a number of ways by inhibiting black and red pigmentation with combinations of genes (dominant white, recessive white, silver, Columbian, sex-linked 'Cuckoo' barring).

Some perceived colors of feathers are due to the structure of the feather and not any pigmentation. The purple and the 'beetle' green sheen that can be seen in some poultry is due to the way the feather structure reflects light rather than the presence of a pigment.

In poultry, there are primary and secondary color patterns. A secondary pattern is a pattern that appears on individual feathers. These are patterns like single and double lace, mottle, and so on. Primary patterns are color patterns that involve the entire body of the bird. An example is the silver Columbian pattern. In the Columbian bird, black is restricted to the hackles, wing bow and tail. The silver Columbian is a white bird with some black in the neck, wing and tail areas. Because this pattern is not manifest on individual feathers, it is a primary pattern.

To 'construct' a chicken having a particular color scheme, one begins with the 'background' or the E-locus gene(s). The other color and (secondary) pattern genes essentially modify this 'background'. Some of these are: E, extended black or nigrum; E^R, birchen; e^{Wh}, dominant wheaten; e^+, wild type; e^b, brown; e^s, speckled; e^{bc}, buttercup; and e^y, recessive wheaten. These genes cause recognizable chick down color and influence the adult feather color, sometimes male and female feather colors are influenced differently.

As an example, let's 'build' a white chicken. We can start with wild-type background, e^+, and require our bird to have two copies of this gene. We can suppress the red in the chicken by adding the silver gene, S, which has the effect of changing red to white. Black is suppressed (changed to white) by the dominant white gene, I; however this gene is 'leaky' and allows black specks through. A good 'helper' gene in this situation is the Columbian gene, Co, since it is a restrictor of black. Although this set of genes (e^+, S, I, Co) is not the only set that will yield a white chicken, it is one of the ways a white chicken can be obtained.

5.6.1. The influence of one versus two genes for a color trait

Feather color genes often display a 'dosing' effect: For example, since the locus of the sex-linked barring gene is on the Z sex chromosome, females that have 'Cuckoo' or sex-linked barring can have only one barring gene and have barring that is less well defined than the barring of males that have two barring genes. Also, Sil-Go-Link males that have only one silver gene often have some red color on their wings. So, in the Sil-Go-Link male, the one silver gene does not completely inhibit the red pigment. The silver gene is dominant but still some red is visible when only one silver gene is present. This 'dose effect' in which two genes for a trait reinforce or strengthen the expression of a trait is common in poultry.

Table 73.9. Autosomal genes causing white/essentially white plumage

Gene	Symbol	Inheritance	Phenotype of homozygotes	
			Plumage	Eye
Dominant white	I	Incompletely dominant	White	Pigmented***
Dun	i^D	Incompletely dominant	Whitish	Pigmented
Recessive white	c	Recessive to C^+	White	Pigmented****
Red-eyed white	c^{re}	Recessive to C^+, c	White	Dark red
Recessive albino	c^a	Recessive to C^+, c, c^{re}	White	Pink
Imperfect albino*	s^{al}	Recessive to S, s^+	Partially pigmented**	Light red

* Sex-linked mutation allelic to Silver and gold ** Black plumage becomes grayish tan and red plumage becomes light red *** Bay eye, iridial and retinal pigment epithelium melanized **** Usually bay-eyed; but, can be part/all brown with appropriate genotypes.

Source : Crawford, 1990

Table 73.10. Chicken plumage patterns

Pattern	Pg	Ml	Co	Db	E-locus	Hf
Partridge (or pencilled)	+	-	-	-	e^b	-
Salmon-breasted pencilled (female)	+	-	-	-	e^+	-
Autosomal barring - Hamburg	+	-	-	+	e^b	-
Autosomal barring - Fayoumi	+	-	+	+	E^R	-
Autosomal barring - Buttercup	+	-	-	+	e^{bc}	-
Single Lace (as in Wyandotte)	+	+	+	-	e^b	-
Single Lace (as in Polish)	+	+	+	+	E^R (or E)	-
Single Lace (as in Sebright)	+	+	+	+	E^R (or E)	+
Single Lace (as in Andalusian)	+	+	+	-	E	-
Double Lace (as in Dark Cornish)	+	+	-	-	e^{Wh},e^b,e^y	-
Spangling (as in Hamburg)	+	+	-	+	E^R (or E)	±

5.7. Patterns

New research has indicated that several major patterns in chickens are not due to independent genes as previous believed. The single lace, double lace, autosomal barring (horizontal penciling) and spangling are now thought to be due to interactions with the Pg gene (pattern gene). Single lace pattern can be obtained with the genes: Ml (melanotic), Pg (pattern gene) and Co (Columbian) on either e^b, e^y or e^{Wh} background genes. Double lace pattern is obtained from the single lace pattern by removing the Co gene (Table 73.9). Spangling is obtained with Db (dark brown), Ml (melanotic), Pg (pattern gene) on either the E or e^b background. Penciling is obtained with the Pg gene on the e^b or e^+ background. Autosomal barring (sometimes called horizontal penciling) is obtained with the Db and Pg genes.

5.8. Eye color

The wild-type eye is characterized by the Light Brown Leghorn. Eye color is a result of pigmentation of a number of structures within the eye (iris, retina, uveal tract, ciliary).

The bay-color eye (various shades of reddish brown) is due to carotenoid pigments and the blood supply of the iris. Brown eyes are increasingly melanized with the darkest eye color due to the fibromelanotic gene characterized by heavy eumelanin deposits throughout the eye. Little is known about pearl eye and it is speculated that it has the same eumelanin distributionas the bay but without the carotenoids.

Eye color is modified by a number of genes that are known to be associated with shank and plumage color. The sex-linked dermal melanin genes, id^+ and id^M enhance dermal shank and eye pigmentation. The inhibitor of shank dermal melanin, Id, also inhibits eye pigmentation. The id^M gene together with extended black, E, is likely to be responsible for dark brown eyes. id^M also darkens the eye on the e^+ background.

Λ dominant sex-linked inhibitor of eye pigmentation is known, Br. This trait is not useful for developing sexable day-old chicks because chickens do not get their final eye color until they reach sexual maturity.

In the absence of other melanin inhibitors, the E-locus alleles, E (extended black) and E^R, birchen, result in a brownish eye with the E allele making the darker eye. Sex-linked barring, B, and eumelanin inhibitors at the E-locus, like e^{Wh} have an effect on eye color. Recessive white seems to have no effect on eye color and dominant white, I, has a strong ability to inhibit eye pigmentation.

5.9. Chick down color

Some additional genes that affect chick down color are: Dominant and recessive white with extended black at the E-locus gives yellow chick down. If black spots leak through it is usually because the chick is heterozygous for dominant white.

Dark brown, Db, makes the black down of extended black to be a reddish brown. Blue or grey chicks can be extended black and Bl (blue) heterozygotes. These chicks can also look black. Chicks that are homozygous for recessive lavender are blue / grey.

Table 73.11. Some basic chick down colors

Chick Down Genotype	Phenotypic Remarks
E Extended Black	Basically black down. Variation includes gray and cream bellies. The cream can invade the head and face. Sex-linked barring puts the cream head spot on the black down. Recessive mottling (mo) makes these chicks look like penguins
e^{Wh} Dominant Wheaten	Without columbian (Co) Wheaten down is a light cream for both sex-linked silver and gold. Heterozygotes can have varying amounts of striping. Some New Hampshires have light reddish stripes on their backs at hatch. This may be due to heterozygosity or some other modifiers in these lines. There is a high degree of difference between silver and gold wheaten down when columbian is present. This is why commercial white-tailed reds have dominant wheaten and columbian.
e^{+} Wild-type	The dark eye stripe is characteristic of wild-type. The buff color of the face and back stripes are affected by sex-linked silver and gold. There are probably modifiers that affect the gold color of the pure line chicks.
e^{b} Brown	Can range from a solid dark mink brown to light brown with stripes. They have brown heads with no stripes on the head like e^{bc}.
e^{bc} Buttercup	More yellow than e^{+}. This dilution may be due to e^{bc} or Db that seems to be in all the crosses and pictures involving this allele. Adult females are like eb females and do not have salmon breasts.
e^{y} Recessive Wheaten	Sometimes pictured as being yellow but more brown than e^{Wh}. Both are said to be cream in color. Recessive wheatens are often called dark wheatens because the adult females have more stippling on their backs than dominant wheaten females.
e^{b},Co	Chicks having sex-linked silver in addition to the above genes are cream colored with varying amounts of gray on their backs. Their backs can be nearly black. Sex-linked gold chicks show buff on their flanks and faces and cream bellies with the same varying amount of gray on their backs. The less gray the more buff on the backs of sex-linked gold chicks.
e^{b}, Co, Db	Mostly light (yellow) body with brown head and back stripe. Buff breeds may have wheaten, Co, and Db.

6. Genotypes of some common breeds

(http://marsa_sellers.tripod.com/geneticspapers)

Partial genotypes for common breeds are given in Table 73.12. If a gene does not appear in the table, it is intended that wild-type be assumed although sometimes wild-type genes are listed for emphasis.

Table 73.12. Partial genotypes of common breeds

Breed	Autosomal Genes	Sex-Linked Genes	Comments
Australorp, Black	W/W, E/E, co^+/co^+, db^+/db^+, (Ml/Ml), i^+/i^+,(Pg)	S, Id	Evidence of unknown black enhancers and the pattern gene have been observed in Black Australorps.
Silver Spangled Hamburg	E^R/E^R, Co/Co, Db/Db, Ml/Ml, pg^+/pg^+	S, id^+	The combination of dark brown and melanotic may be responsible for the white undercolor.
Silver Laced Wyandotte	e^b/e^b, Co/Co, db^+/db^+, Ml/Ml, Pg/Pg	S,Id	Yellow legged blacks are usually based on the e^b E-locus allele.
New Hampshire	e^{Wh}/e^{Wh}, Co/Co, Mh/Mh, w/w	s^+, Id	With polygenes for red ear lobe and brown eggshell color. A primary difference between New Hampshire and Rhode Island Red is the wheaten allele at the E locus in the New Hampshires.
Rhode Island Red	e^y/e^y, Co/Co, Mh/Mh, w/w, Db?	s^+, Id	With polygenes for red ear lobe and brown eggshell color.
Barred Plymouth Rock	E/E, Co/Co, w/w	K, S, B, Id	With polygenes for red ear lobe and brown eggshell color. The slow feathering gene, K, is believed to aid in obtaining a cleaner barring. Barred Rocks have yellow shanks because of the dermal melanin inhibiting property of Cuckoo barring. Without this, it would have near black shanks with yellow soles.
White Leghorn	Homozygous for either E, E^R or e^+, I/I, w/w, o/o	B, S, Id	Lines of Leghorns have been found with different alleles at the E locus.
Delaware	e^b/e^b, Co/Co, w/w	B, S, Id	With polygenes for red ear lobe and brown eggshell color. Delawares have Barred Rock and New Hampshire genetics.

7. Chicken genes of common interest

(http://marsa_sellers.tripod.com/geneticspapers)

Chicken genes of common interest are given in Table 73.13. Allelic genes (genes that have the same location or locus on a chromosome, which are also genes that substitute for one another) are grouped together.

Table 73.13. Chicken genes of common interest – Sex-linked genes

Alleles	Symbol	Comments
Sex-linked barring	B	Barring, cuckoo barring. Dominant. Causes white barring pattern in red and black, sometimes used as a black inhibitor, most notably in Leghorns. Cuckoo barring is also an inhibitor of tissue pigmentation and is responsible for the yellow shanks of Barred Rocks. Shanks of females can be darker. Barring shows a distinct dosage effect. B/B gives wider bars than heterozygotes have. Incorporation of the slow feathering gene results in a cleaner, more sharply defined barring.
	b^+	Recessive wild-type gene. An allele of the sex-linked barring locus. Lack of barring.
Sex-linked dilution	B^{Sd}	Females that are hemizygous for B^{Sd} (having one B^{Sd} gene) have light blue and barred plumage as do the heterozygous males, however, homozygous males show a dosage effect and are essentially white. These homozygous males resemble dominant whites but differ in that they are epistatic to pheomelanin while dominant white is not.
Comments	Sex-linked barring, B, sex-linked dilution, B^{Sd} and the wild-type, b^+ are alleles of the same locus. The order of dominance is $B^{Sd} > B > b^+$.	
Brown eye	br	Not much is known about this gene and there may be a dominant inhibitor of brown eye. Many of the melanin-influencing genes have an effect on eye color.
Dwarf	dw	Recessive. Males are reduced in size by about 43%, females by 26-32%. Multiple alleles have been proposed. dw is responsible for some beneficial effects. dw homozygotes are more resistant to Marek's Disease and spirochetosis, fewer laying accidents, more aggressive immune response. Abnormal eggs are suppressed (soft-shelled, double yolks). Dwarfism, dw, does not affect mortality but does postpone the onset of lay in pullets up to two weeks. Although egg number and mass are slightly decreased by dw, feed efficiency (feed consumption per egg laid) in laying stocks is usually increased 13-25%.
	dw^B	Recessive but shows a dose effect; 'bantam' gene. Females reduced in size by 5-11% and males by about 5% in heterozygotes and 14% in homozygotes. Allelic with dw.
	dw^M	MacDonald dwarf. Reduces body weight by 13.5% and shank length by 9%. Allelic with dw.
	Dw^+	Wild-type gene. Lack of dwarfing alleles. Allows 'normal' size to develop.

Alleles	Symbol	Comments
Silver and Red-Gold	S	This gene is called 'silver'. Inhibits red pigment, pheomelanin. The expression of silver is sometimes affected by hormonal levels and is considered to be incompletely dominant and highly influenced by modifying genes.
	s^+	This gene is sometimes called 'gold'. Wild-type, recessive. Invokes red pigment.
Foot Color	Id	Light foot color. Dominant. Inhibits dermal melanin. Reported to have little influence on shank/foot color in birds with dark shanks due to E/E
	id^c	Recessive. This gene allows beak and sometimes plumage pigmentation in dominant white homozygotes.
	id^a	Allows green spots on shanks - this gene is not widely accepted and the effect of this gene may be due to the interaction of modifiers not allelic to this locus.
	id^M	Massachusetts mutation. Recessive. Unlike other alleles that belong to this locus, dermal melanin is present in shanks of day-old chicks. Other alleles take more time to express. The darkest shanks are produced in conjunction with E and i^+. The combination of id^M, E and I produces a pale blue or green color by about three months of age.
	id^+	Wild-type dermal melanin. Lack of dermal melanin inhibitors.
Sex-linked white skin	y	Recessive, causes white skin. Recessive sex-linked white skin causes yolks to be lighter in color and reduces xanthophyll levels in blood plasma. This is generally considered to be an inferior trait particularly since the autosomal white skin does not have these side-effects on yolk color.
	Y^+	Wild-type gene. Lack of recessive white skin mutation.
Feathering Rate Genes	k^+	Sometimes called rapid feathering. Recessive.
	K	Late feathering gene
	K^s	Slow feathering gene
	K^n	Very slow feathering or 'delayed' feathering gene. The order of dominance among the genes allelic to this locus is $K^n>K^s>K>k^+$. The slow feathering gene is believed to be associated with a bald patch on the back of the adolescent bird. The feathers do come in given enough time. Since this is likely due to a dose effect of the slow feathering gene, the homozygous males should be the most likely to exhibit the trait. In my personal flocks, I have both males and females exhibiting this. Many novice poultry-keepers wrongly attribute the bald back phenotype with a picking problem.
Brown eggshell color inhibitor	pr	This recessive gene results in a lack of protoporphyrin pigment (the brown eggshell pigment) even in hens with polygenic brown eggshell color. It can be employed to remove undesirable tints from eggs of white shelled strains.

Table 73.14. Autosomal linkage Group 1 genes

Genes	Symbol	Comments
Creeper	Cp	Short legged condition. Lethal in homozygous state. Dominant.
	cp^+	Recessive, wild-type gene. Lack of creeper trait.
Rose comb	R	Associated with poor fertility in some homozygous breeds. Dominant.
	r^+	Wild-type gene. Recessive. Lack of rose comb trait.
Lavender	lav	Recessive lavender has been associated with poor feather quality and even lack of feathers in some breeds. Lavender dilutes both black and red; changes black to grey and red to cream. Blue fowls termed "self blue" are normally lavender homozygotes. A mating of two lav homozygotes (blue fowls) will produce blue offspring. Lavender causes dilution by inhibiting the transfer of pigment granules from melanocytes, which produce them, to the feather structure. Lavender expression in homozygotes is present in chicks and adults.
	Lav^+	Dominant, wild-type gene. Lack of lavender trait.

Table 73.15. Autosomal linkage Group 2 genes

Genes	Symbol	Comments
Crest	Cr	Crest feathers are similar in shape and texture to hackle feathers. There may be more than one allele. Incompletely dominant.
	cr^+	Wild-type gene. Lack of crest.
Pied / Mottle	mo (pi)	The pied pattern is recessive black and white as in Exchequer Leghorn. Research has shown that the pied and mottle patterns are due to the mottle gene. Mottle causes a white tip at the distal end (end farthest from the skin) of the feather. Chicks with extended black and mottle (E/E mo/mo) as in the Exchequer Leghorn will often have black restricted from the belly and sometimes the head.
	Mo^+	Wild-type gene. Dominant. Lack of mottling.
Dominant white	I	Incompletely dominant. Influences eye pigment. Inhibits black pigment, eumalanin. This gene is 'leaky' and will allow black specks through. Generally not as efficient at producing a solid white bird as are two copies of recessive white. Heterozygotes of dominant white, I/i^+ are often grey with the grey color visible in the chick down. Dominant white dilutes, but does not eliminate, epidermal melanin.
Smoky	I^S	The smoky gene is an allele belonging to the dominant white locus. Smoky is dominant to dominant white in both chick down and adult plumage in that extended black with I/I^S (E/E I/I^S)results in grey chick down and adult plumage. Research to date indicates that $i+/I^S$ heterozygotes express more the wild-type phenotype with respect to this gene indicating a recessive character with respect to the wild-type. Smoky is dominant on the chick down of $I^S/i+$ heterozygotes in that down that should be black is grey. The melanosomes resulting from the expression of smoky resemble those resulting from Andalusian Blue. Smoky dilutes black much more than red/gold. An important difference between Smoky and Andalusian Blue is that Smoky in the homozygote state produces a grey/blue bird while Andalusian Blue homozygotes are splash. Therefore, Smoky fowl will breed true.
	I^D	This gene is often called 'Dun'. Incompletely dominant, off-white. Allelic with dominant white.
	i^+	Wild-type gene. Lack of dominant white.
Frizzle	F	Incompletely dominant. The action of the frizzle gene is localized in the feather follicle. It causes a structure abnormality in the feather and abnormalities of internal organs (enlarged heart, spleen, gizzard and alimentary canal) are common.
	f^+	Recessive, wild-type gene. Lack of frizzle.

Table 73.16. Autosomal linkage Group 3 genes

Genes	Symbol	Comments
Skin color	w	Yellow skin color. Recessive.
	W+	Dominant wild-type gene. Autosomal white skin gene. Prevents the transfer of xanthophyll into the skin, beak and shanks but does not effect the eye iris, egg yolk or blood serum. It is present in the Jungle Fowl.
Blue eggshell	O	The action of the blue eggshell gene is dominant to the action of the white eggshell gene, o. Blue and brown eggshell genes present simultaneously give a shade of green on the exterior of the egg. The blue eggshell color permeates the shell while brown is primarily an exterior coating.
	o	Recessive wild-type gene. Lack of blue eggshell color gene. Causes white eggshells in the absence of brown eggshell color genes.
Pea comb	P	Dominant. Sometimes referred to as triple comb. Heterozygotes often display a prominent central ridge with much smaller lateral points.
	p+	Wild-type gene. Recessive. Lack of pea comb.
Naked neck	Na	Incompletely dominant. Turkens. Causes bare skin on the neck which becomes reddish toward sexual maturity. Heterozygotes show a small tuft of feathers on the neck above the crop, which is almost missing in the homozygote. The Na allele is associated with increased tolerance for heat, which is probably due to the 30% reduction in overall plumage for heterozygotes and 40% for homozygotes. Na is also associated with a small increase in meat yield and lower body fat content. An increase in embryonic mortality of up to 10% is attributed to Na.
	na+	Recessive, wild-type gene. Lack of naked neck. Allows full feathering.
Silkie	h	Recessive. The barbs of the feathers are highly modified giving the silkie a 'woolly appearance.
	H+	Dominant, wild-type gene. Lack of silkie trait. Allows normal feather structure.
Melanotic	Ml	Dominant. Black intensifier, one of the genes which, in concert with Pg and other genes, is responsible for plumage patterns. There may be more than one eumelanin intensifying gene similar to Ml and non-allelic.
	ml+	Recessive, wild-type gene. Lack of melanotic eumelanin enhancing gene.
Pattern gene	Pg	Dominant. This is the pattern gene which, together with other genes is responsible for the patterns of plumage. The pattern gene doesn't seem to express in the absence of Ml in combination with some of the E locus alleles. The pattern gene with the Db and Co Columbian-like restrictors is believed to be responsible for autosomal barring.
	pg+	Recessive. Wild-type gene. Lack of pattern gene.
Dark brown	Db	Incompletely dominant. Changes black down of E, E^R to reddish-brown. Adults males exhibit a Columbian-type pattern of black, modifies red to orange-tan. Db is a better restrictor of black in males than females.
	db+	Wild-type gene. Recessive. Lack of dark brown-type Columbian restriction.

Table 73.17. Autosomal linkage Group 4 genes

Genes	Symbol	Comments
Duplex comb	D^v, D^c	Dominant alleles. The superscripts 'v' and 'c' indicate the 'V' and 'cup' shaped phenotypes and are considered to be separate genes.
	d^+	Recessive, wild-type gene. Lack of duplex trait.
Multiple spurs	M	Dominant. Causes more than one spur per shank on males.
	m^+	Wild-type gene. Recessive. Lack of multiple spur trait.
Polydactyly	Po	Dominant. Having too many toes. The fifth toe develops on top of the first toe and is longer than the first toe. There are several degrees of expression of this gene.
	Po^d	Duplicate polydactyly. Dominant to the wild-type allele. An extra toe is present as well as an elongation and splitting of the original first toe. Extreme expression can lead to duplication of the entire foot.
	po^+	Wild-type, recessive. Allows normal foot.

Table 73.18. Other autosomal genes

Genes	Symbol	Comments
Autosomal barring	Ab	Non-sex-linked barring. Sometimes called 'parallel pencilling'. This is not a real gene, rather autosomal barring is due to combinations of Pg, Co, Db with e^b, E^R, and e^{bc}. See text.
Breda combless	bd	Recessive. Birds with this gene are almost completely lacking comb and wattles. Females are considered to be completely combless and males have a tiny comb.
	Bd^+	Dominant, wild-type gene. Lack of breda combless trait. It is believed that this gene is necessary for chickens to produce a comb.
Blue	Bl	Incompletely dominant. Andalusian blue-dilutes black: blue pigment is a modified black. Two nigrum genes, E, and one Bl give a blue chicken; two Bl genes give splash.
	bl^+	Wild-type gene. Recessive. Lack of blue eumelanin dilution gene, Bl.
Brachydactyly	By	Dominant. Abnormally short digits (toes).
Recessive white genes	c	Thought to give a cleaner white than dominant white. Varieties of White Plymouth Rock, Wyandotte, Minorca, Orpington, Jersey Giant, Dorking, Langshan, Silky and others often carry recessive white genes. Many varieties carry both dominant and recessive white. Allows dark eyes. Pigmentation in chick down varies.
	c^{re}	Recessive white allele that allows red eyes.
	c^a	Autosomal albinism. Alellic with the recessive white genes. Evident via lack of eye pigment. Some melanin present in chick down.
	C^+	Wild-type gene. Dominant. Lack of recessive white mutations.

Genes	Symbol	Comments
Comments about the C locus	The order of dominance among the recessive white alleles is: $C^{+}>c>c^{re}>c^{a}$. The presence of other pigment inhibiting or enhancing genes will influence the chick down color. Some adults have a grey color.	
Champagne blond	Cb	Dominant. Inhibits pheomelanin (red / gold). The presence of the gene is not observable on the wild-type down
	cb^{+}	Wild-type gene. Recessive. Lack of champagne blond dilution.
Columbian	Co	Incompletely dominant. Confines black to hackle and tail in both sexes (called Columbian restriction). Thought to cause a gradient in color from head to tail. Modifies Wheaten to Buff Columbian. Has no effect on extended black, E.
	co^{+}	Wild-type gene. Lack of Columbian restriction. Recessive.
Red diluter	Di	Dominant. Dilutes red, changes red to buff.
	di^{+}	Wild-type gene. Lack of red diluter. Recessive.
Dark	Dk	A proposed gene of an allelemorphic series that darkens the shade of red. Pheomelanin enhancer(s). Dk^{l} was proposed for the dark brown Leghorn and dk^{+} for the wild-type allele.
Ginger	Gr	This gene may be Columbian, Co, or closely related. This may not be a distinct gene.
Grey	no symbol	Recessive, dilutes black to brown/grey.
The E-locus alleles	E	Often called 'extended black', 'nigrum' or 'self black'. Extends black, changes red to black, red inhibitor.
	E^{R}	Birchen. Resembles extended black, E, but with non-black breaks on head and hackle. Body is black with some stippling (flecks/dots) of other color. Used as red inhibitor in Leghorn.
	e^{b}	Partridge (brown). Sometimes represented as e^{p}, females have non-salmon breast with stippling. Males are wild-type.
	e^{Wh}	Dominant wheaten. Female body varies from light salmon to wheat color, some black may be present. Males are wild-type.
	e^{+}	Wild-type. Female: breast is salmon brown and devoid of stippling, body is black and brown in stippled pattern. Males: black breast and abdomen; non-black hackle, saddle and wings.
	e^{s}	Speckled. Resembles e^{b} but with less pronounced stippling. Males are wild-type.
	e^{y}	Recessive wheaten. Female: resembles dominant wheaten with more coarse black stippling on breast and back. Males are wild type.
	e^{bc}	Buttercup allele. Resembles the e^{b} phenotype.
Comments on E-locus alleles	The order of dominance among the generally accepted E-locus alleles is: $E>E^{R}>e^{+}>e^{b}>e^{s}>e^{bc}>e^{y}$. The birchen allele is incompletely dominant to dominant wheaten and the wild-type alleles. Additional alleles have been proposed for the E-locus but research to verify these as separate alleles has not been done. As of this writing, the buttercup allele has been sequenced and has been found to be the same sequence as the e^{b} allele. The buttercup phenotype then is due to modifiers or interactions with other genes. Every E-locus allele influences adult female phenotype. However, all the adult male phenotypes are the same as wild-type except for extended black and birchen.	

Genes	Symbol	Comments
Ear tuft	Et	Dominant. Lethal in homozygous state. Thought to be associated with birth defects, particularly in the ear structures.
	et^+	Wild-type gene. Recessive. Lack of ear tufts.
Erminette	no symbol	Black spots and flecks, variable black and white feathers, similar to pied.
Fibromelanosis	Fm	Dominant. Responsible for the deep skin pigmentation of silkie. Fm is strongly influenced by dermal melanin inhibitors such as the sex-linked Id mutation.
	fm^+	Wild-type gene. Recessive. Lack of fibromelanosis.
Long tail	Gt, mt	The Gt gene (dominant) allows continual growth of tail and saddle feathers. The mt gene allows certain tail and saddle feathers to be nonmolting.
Henny plumage	Hf	Dominant. The term comes from 'hen feathering' in which male plumage is indistinguishable from female plumage.
	hf^+	Wild-type gene. Recessive. Lack of henny feathering.
Cream	ig	Dilutes red (inhibitor of gold). Recessive. A major pheomelanin dilution gene.
	Ig^+	Wild-type gene. Dominant. Lack of cream dilution.
Lace	Lg	This is not a real gene. Partridge Rock, Silver Pencilled Rock.
Beard-Muff	Mb	Incompletely dominant. Characteristic of Ameraucanas, Easter Egg Chickens (faux-Araucanas)
	mb^+	Wild-type gene. Recessive. Lack of beard-muff.
Frizzle modifier	mf	Recessive. Reduces/modifies the effect or extreme expression of the frizzle gene. This gene can modify frizzle heterozygote expression to the point that they are almost indistinguishable from the wild type.
	Mf^+	Wild-type gene (uncertain). Dominant. Lack of frizzle modifier.
Recessive melanotic	mi	Enhances black, (helps) change red to black. E + mi gives a black chicken.
	Mi^+	Wild-type gene. Dominant. Lack of recessive melanotic enhancing.
Mottle	mo	Recessive. Makes a white tip on end of feather. Changes a black bird to Mottled and a Buff Columbian to a Mille Fleur. Dilutes epidermal melanin. There may be several alleles corresponding to this locus or non-allelic modifying genes.
	Mo^+	Wild-type gene. Dominant. Lack of mottling.
Mahogany	Mh	Dominant. Mahogany restricts eumelanin and enhances the color of red. Restricts black in the back and wing of both males and females. Down color seems to be unaffected by mahogany. Example RIR.
	mh^+	Wild-type gene. Recessive. Lack of mahogany.
Pink-eye dilute	p^K	Dilutes both feathers and eye color. Recessive.

Genes	Symbol	Comments
Recessive polydactyly	po-2	Recessive. A number of extra toes can be present even ascending the shank. Associated with leg deformities, significant decrease in hatchability and much higher post-natal mortality.
Feathered legs	Pti-1, Pti-2, Pti-1B, Pti-1L	Dominant. Two different feathered leg loci with perhaps four alleles for the Pti-1 locus (Pti-1, Pti-1B, Pti-1L and pti-1⁺). Pti-1 and Pti-2 genes are most likely not allelic. When both Pti-1 and Pti-2 alleles are present, heavy feathering as in Cochin, Sultan, Belgian d'Uccle results. If only one is present, the feathering is weaker as in Langshan, Faverolles, and Breda. These genes demonstrate a dose effect. Regarding the Pti-1B and Pti-1L genes, the Langshan and Brahma breeds were both shown to possess the same single shank-feathering locus, but because of their differences in phenotype and penetrance in the genetic crosses it was suggested that they possessed different alleles at this locus. This locus was designated as Pti-1, with Pti-1L being the Langshan allele and Pti-1B the Brahma allele. The Brahma allele was shown to be dominant over the Langshan allele. Both the Sultan and Cochin breeds were shown to possess two shank-feathering loci, one of the loci in the Sultan contained the Pti-1L allele. It is hypothesized that the comparable allele in the Cochin breed was Pti-1B. It is proposed that the second locus in both of these breeds is similar, and the symbol Pti-2 is suggested.
Recessive feathered legs	pti-3	The recessive leg feathering gene was identified in a Russian breed referred to as the Pavlov breed.
Dominant Rumplessness	Rp	No coccyx (tail vertebra), reduces hatchability.
	rp⁺	Wild-type gene. Recessive. Lack of dominant rumplessness.
Recessive Rumplessness	rp-2	A skeletal mutation commonly called 'roachback'.
Red splash white	rs	Recessive. Two copies of this gene give a white bird with splashes of red and black. Chicks are white with a red head spot; may be extinct
Recessive black	sg	Not much is known about this gene. Eumelanin intensifier. There may be a number of genes that play this role.
Spurlessness	sl	Recessive. Fowls have no spurs.
Snow-white down	sw	Recessive. The chick down is white rather than yellow.
Vulture hocks	v	Recessive. Long and stiff feathers on the posterior area of the tibia. Characteristic of Belgian Bearded d'Uccle, Breda, Sultan.
Dorking white	wh	Recessive.
Woolly	wo	Recessive.

8. Sex-linked inheritance

Genes carried on the sex chromosomes determine the sex-linked characters. The W chromosome being shorter than Z chromosome, cannot carry all the alleles corresponding to Z chromosome complement. Hence, a single recessive gene on the extra portion of Z chromosome not homologous to W chromosome (and thus not subjected to crossing over) can express in females whereas, a pair of such recessive genes is essential for expression in males. The function of W chromosome in poultry is thought to be in differentiation of early embryonic gonads into ovarian tissue triggered by its histocompatibility W antigen (H-W), the product of a W-linked structural gene.

In poultry, if 'n' sex-linked alleles present at a locus, total genotypes possible is n(n+1)/2 in males, of which, 'n' or homozygous and n(n-1)/2 are heterozygous. In females, only 'n' genotypes are possible, each having only one gene (referred to as hemizygous and question of dominance or recessiveness does not arise. Hence, type of gametes produced is dissimilar between sexes as far as sex-linked characters are concerned and hence, it does not follow Mendelian laws of inheritance.

8.1. Consequences of sex-linkage

1. Frequency of sex-linked recessive defects are always higher in females than males because given a single recessive gene can express in the former whereas males have to receive the recessive genes one each from both its parents which is very rare since its mother hardly survives with the recessive gene.
2. Response to selection for sex-linked genes is more in heterogametic sex (female) because effect of both dominant and recessive gene substitution is manifested; whereas in males, the dominant genes mask the recessive alleles and make identification of heterozygote difficult thereby including selection errors.
3. Decreased heterozygosity or inbreeding can be measured for sex-linked genes only in males (homogametic sex).
4. In case of quantitative traits, sex linkage alters the magnitude of resemblance between offspring and each parent. The female receives its sex chromosome only from its father and therefore the relationship between daughter-dam is zero while that between daughter-sire is 1. On the contrary, as the male receives sex chromosome from both its parents, it relationship with both sire and dam is ½.

8.2. Autosexing

That the females receive sex chromosome only from their father, is exploited for identification of chicks at hatch without the actual observation of the male papilla.

The principle is to mate the females sex-linked dominant gene to males homozygous to the recessive gene so as to obtain progenies wherein females express the recessive gene and the males the dominant gene.

8.3. Dwarf gene

This is a recessive gene designated dw. Hence, the males are dwdw and females dw-. At hatch, they weigh as much as normal chicks; but, their adult weights are reduced by 30% (females) to 40% (males). The most important effects of dw gene are (Nicholas, 1987):

1. Drastic reduction in feed consumption due to reduced basal metabolism per unit body weight and
2. Reduced floor space requirement and thus, the housing costs.

However, egg production is likely to be reduced by = 10% in lighter strains; but not in heavier strains. Egg weight is reduced in absolute terms although, as a % of body weight, it is actually increased. Effect on sexual maturity is variable and hatchability is improved (Nicholas, 1987).

In spite of the above, dw gene is not employed in commercial layer industry due mainly to the reduced egg production and egg size and delayed sexual maturity. But, in broiler breeding, this gene is exploited to reduce the cost of production of chicks (Nicholas, 1987).

When this gene is introduced to the female line, the body size may reduce the extent of 30%, whereas the reduction in egg size may be only 5-8%. Hence the space required as well as the feed consumed by the dwarf hens will be greatly reduced. These ultimately reduce the cost of production of hatching eggs and, in turn, of the day-old chicks. The resultant broilers have been found to have the same FCR and meat yield although body weight is reduced only by 3 to 4 %.

Therefore, this sex-linked (X-linked) dwarf gene is introduced to the broiler breeder dams (hens) to economize the broiler chick production. However, this necessitates the maintenance of homozygous dwarf males which have to be mated to normal hens to obtain dwarf breeder hens as their progeny. Reports indicating better feed efficiency on egg mass basis (egg number x egg weight) by dwarf dams (2.7 Vs 3.2 kg/kg of egg weight) is suggestive that egg weight as well as egg production are not severely affected by the dw gene (See Table 73.13).

8.4. Sex-influenced traits

These are controlled by autosomal genes whose expression depends on sex; mainly due to sex hormones. For example, males are heavier and grow faster than females. Similarly, baldness in humans is common in males than in females. In all these cases, the same genotype is producing different phenotype in the two sexes. Therefore, it is reasonable to state that the influence of sex on phenotype is an environmental source of variation.

Table 73.19. Autosexing crosses

Cock	Hen	Male Chick	Female Chick	Breeds
Early feathering (kk)	Late feathering (K-)	Late feathering At hatch: Primaries and secondaries are of same length and the former do not project out. At 10d: No tail feathers and slight growth of wing feathers	Early feathering At hatch: Primary and secondary feathers project well beyond the down. Primaries much longer than secondaries. At 10 d: Tail feathers up to 2.5 cm length and wing feathers up to or beyond the tail	Leghorn, Minorca, Ancona
Gold (ss)	Silver (S-)	Silver, at hatch appear white, silvery grey, black stippling, marbling or other pattern on a silvery background	Gold, at hatch color of down will be buff, red, chocolate, black stippling, marbling or other pattern on a golden background.	Gold-and silver-laced Campines, Golden and Silver Polish and Golden and Silver Sebrights
Non-Barred (bb)	Barred (B-)	Barred, at hatch, black or brown spot on the head	Non-barred, at hatch no spot on the head.	Barred Plymouth Rock with other breeds and varieties

8.5. Sex-limited traits

These are also controlled by autosomal genes but the expression is hormonally restricted to only one sex. For example, egg production in females, spur in males, anatomy and physiology of reproductivesystem, secondary sexual characteristics.

9. Lethal genes

9.1. Dominant lethals

Birds homozygous to this type of genes usually do not survive. They express even in the heterozygotes and hence can be identified and their frequency gets reduced. Example : Creeper.

9.2. Recessive autosomal lethals

9.2.1. Lethal to homozygote in the late embryonic life

These genes can be identified by observation of the dead embryo; Example : Stickiness, Congenital loco

9.2.2. Lethal to homozygote in early embryonic life

These are usually linked to or identical with some color or morphological character and therefore can be identified easily; Example : lethal with Recessive White

9.2.3. Lethal only in certain matings

They can be identified by abnormally high mortality in certain meetings in comparison to controls. Fowl does not seem to have any gene under this group.

9.3. Recessive sex-linked lethals

9.3.1. Lethal in early development

They can be detected by a deficiency of hatched females or by high mortality at one period of incubation; Example : Naked

9.3.2. Lethal in late development

Visible abnormalities and/or high mortality can be seen, first only in females. Example : Sex-linked albinism in turkeys.

Usually, most lethal effects occur before hatching or within a month after hatching depending on the nature of the lethal gene and homozygosity of the embryo. Those lethal genes which cause of death of the individual regardless of the environment are referred to as obligate lethals whereas those lethal genes which cannot cause mortality if the individual is given an extremely favorable environment are called facultative lethals.

Table 73.20. Important obligate lethal genes of poultry

Genes	Symbol
Creeper	Cp
Chondrodystrophy	ch
Cornish lethal	Cl
Abnormal maxillae	mx
Missing mandible	md
Wingless	wg
Diplopodia	dp
Micromelia	-
Stickiness	sy
Talpid	ta
Congenital loco	lo
Lethal with recessive white	l
Crooked-neck dwarf	cn
Congenital tremor	-
Bilateral microphthalmia	mi
Characters (facultative lethals)	Symbol
Short upper beak	su
Apterylosis	Ap
Naked	n
Dwarfism	td
Blindness	-
Short mandible	sm

Chapter **74**

Inheritance of Quantitative Traits

1. Quantitative traits

Characters such as egg production, egg weight, body weight etc. can be measured on a definite scale and hence are called as quantitative traits or metric traits. A comparison of characteristics of quantitative and qualitative traits is given in Table 74.1.

Table 74.1. Quantitative traits *Vs* Qualitative traits

Quantitative traits	Qualitative traits
Traits are measured rather than counted	Traits are more often qualitative; at best they can be counted
Traits show a continuous variation unlike qualitative traits which are usually of all-or-none type.	Traits are all-or-none type; no variability
Traits are controlled by polygenes; each having a minor effect; minor genes	Traits are controlled by one or few genes each having a major effect; major genes
Expression of the trait is often influenced by the environment	Expression of the trait is independent of the environment
Phenotypic classes do not reflect the genotypic classes	Phenotypic classes, on most occasions, are good indicators of the genotypes

1.1. Nicking

Since the quantitative traits are controlled by polygenes, it is possible to develop lines which are homozygous dominant for several but different factors. When these lines are crossed, the offsprings often perform better than either the parents. This property of the parents to combine and produced offsprings superior to themselves is called nicking. An efficient breeder can develop suitable lines which can nick and produced commercial chicks of high productivity.

1.2. Threshold characters

Certain polygenic traits do not show continuous variability phenotypically. A classical example for this being hatchability of eggs - until the total genetic, environmental and interaction influences, if any, which contributes/represents the phenotypic potential, is enough for the chick to hatch, the hatching doesn't occur. That means, there is a threshold value below which the chick fails to hatch and above the value, it hatches. Genotypically, there may be a distribution of threshold values of each of the embryos in the population but, phenotypically, the expression is binomial. In other words, the cause is a continuous variable whereas the effect is discrete. Such polygenic traits referred to as "threshold or quasi-continuous or all-or-none traits". Phenotypic potential of such traits can not be measured in an individual, but if coded suitably, it can be estimated from a group of individuals by arithmetic mean of the phenotypes.

1.3. Component characters

Many quantitative and threshold traits are mathematically and/or biologically determined by two or more contributing traits. For example - egg production is largely determined by clutch size, pause duration, age at sexual maturity and rate of lay. Similarly, body weight of broilers is determined by FCR, rate of gain, body conformation and others. Each of these contributing traits is a polygenic trait and therefore study of these contributing traits help in understanding the traits in question better and also to estimate the relative contribution of each of the traits to the component trait under consideration. It may also be possible that the mere manipulation of one or few of the contributing trait(s) can bring about desirable changes in the competent treat itself and more response per generation can be possible by developing most appropriate breeding program after ascertaining the role of each of the contributing traits.

2. Idealized population

Most often, animal breeder is primarily interested in improving the productive efficiency of domestic animals and hence is dealing with quantitative traits. Therefore it is a pre-requisite that a population in which no forces are operating to alter its genetic structure (referred to as idealized population) is studied so that the actual effects of manipulations by the animal breeder can be quantitatively predicted.

2.1. Gene and genotypic frequency

Considering 'n' number of alleles at a locus, number of possible genotypes is n (n+1)/2 and proportion of individuals belonging to each of the genotypes is referred to as "genotypic frequency" whereas proportion of each allele out of all the genes at that locus gives the "gene frequency".

2.1.1. Calculation of gene and genotypic frequencies (one locus, two alleles)

Total individuals in the population: 5000; Trait: Autosomal, Frizzled plumage (F) incompletely dominant to normal plumage (f); Number of individuals under each genotype:

Table 74.2. Calculation of gene and genotypic frequencies

Population considered: Homozygous frizzle (FF): 3042; Heterozygous weakly frizzle (Ff): 1716 and Normal (ff) : 242; Total: 5000	
Genotypic frequencies	Gene frequencies
FF = 3042/5000 = 0.6084 = 0.78^2 Ff = 1716/5000 = 0.3432 = 2 x 0.78 x 0.22 ff = 242/5000 = 0.0484 = 0.22^2 Total 1.0000 It can be observed that genotypic frequencies can be calculated by knowing the gene frequencies and *vice versa*. For instance, with two alleles at a locus with gene frequencies p (for dominant gene) and q (for recessive gene), the genotypic frequencies are p^2 for homozygous dominant, 2pq for heterozygotes and q^2 for the recessive homozygotes; in other words, the genotypic frequencies are represented by the binominal expansion of $(p+q)^2$	Total number of genes in the population = 5000 x 2 = 10000 Total number of 'F' genes = 3042 x 2 + 1716 = 7800 Total number of 'f' genes = 242 x 2 + 1716 = 2200 Frequency of 'F' gene = 7800/10000 = 0.78 (p) = $\sqrt{0.6084}$ Frequency of 'f' gene = 2200/10000 = 0.22 (q) = $\sqrt{0.484}$ Total = (0.78 + 0.22) = (p + q) = 1.00 p = √(frequency of 'FF' genotype) = $\sqrt{0.6084}$ = 0.78 q = √ (frequency of 'ff' genotype) = $\sqrt{0.0484}$ = 0.22

3. Hardy-Weinberg equilibrium

Assuming that the population with a genotypic frequency of 0.6084, 0.3432 and 0.0484 with respect to FF, Ff and ff; and gene frequency of 0.78 and 0.22 with respect to F and f underwent a random mating, allowing all matings, the resultant gene and genotypic frequencies will be as follows (Table 74.3):

Table 74.3. Calculation of gene and genotypic frequencies –I Generation

Gene and genotypic frequencies	
Base population	Reproduced population
Gene frequencies p = 0.78 q = 0.22 Hence, genotypic frequencies are: Homozygous dominant = p^2 = $(0.78)^2$ = 0.6084 or 60.84% of individuals, Heterozygotes = 2pq = 0.3432 or 34.32% of individuals and Homozygous recessive = q^2 = 0.0484 or 4.84% of individuals; Total of all genotypic frequencies = 0.6084+ 0.3432+ 0.0484 = 1.0000.	Gene frequencies: Since p+q = 1, $(p+q)^2$ $=p^2+2pq+q^2 = 1$; p= (p^2+½ of 2pq) ÷ ($p^2+2pq+q^2$). Since the denominator is $(p+q)^2 = 1^2$ Frequency of p = p^2+pq = 0.6084 + ½(0.3432) = 0.78; on the same lines, Frequency of q = (q^2+½ of 2pq) ÷ ($p^2+2pq+q^2$) = q^2+pq = 0.0484 + (0.3432/2) = 0.22

Therefore, it is interesting to note that in a random mating idealized population, in the absence of mutation, migration and selection, the gene and genotypic frequencies remain unaltered; this all important observation is named after the scientists who first discovered it. This can be checked by considering genotypic frequencies as well with one more generation of random mating (Table 74.4).

Table 74.4. Calculation of gene and genotypic frequencies – II Generation

Type of mating			Frequency of offsprings		
Male	Female	Frequency of mating	FF	Ff	ff
FF	FF	$p^2 \times p^2 = p^4 = (0.78)^4$	0.3702	-	-
Ff FF	FF Ff	$2(p^2 \times 2pq) = 4\,p^3q$ $= 4\,(0.78)^3(0.22)$	0.2088	0.2088	-
FF ff	Ff FF	$2(p^2q^2) = 2p^2q^2$ $= 2\,(0.78)^2(0.22)^2$	-	0.0589	-
Ff	Ff	$2pq \times 2pq = 4\,p^2q^2$ $= 4\,(0.78)^2(0.22)^2$	0.0294	0.0589	0.0294
Ff ff	ff Ff	$2(pq \times q^2) = 4\,pq^3$ $=4\,(0.78)\,(0.22)^3$	-	0.0166	0.0166
ff	ff	$q^2 \times q^2 = q^4 = (0.22)^4$	-	-	0.0024
Total		$p^4+4p^3q+6p^2q^2+4pq^3+q^4$ $= (p^2+2pq+q^2)^2 = (p+q)^4 = 1$	0.6084	0.3432	0.0484

Again, it can be seen that in an idealized population, gene and genotypic frequencies follow Hardy-Weinberg equilibrium. Conversely, if a random mating population continues to produce same gene and genotypic frequencies, then the population is said to be in Hardy-Weinberg equilibrium and it remains so, only as long as the Idealized conditions are met.

3.1. Factors affecting Hardy-Weinberg equilibrium

Improvement of livestock is not possible without selection and choice of mating system. Selection involves differential reproduction among individuals of different genotypes and therefore it invariably changes gene frequency favoring the selected one. Natural selection also works mainly by reducing the reproductive fitness of unfavorable genotypes.

Mating system regardless of type, alter only the genotypic frequency (not the gene frequency) to make desirable phenotypes more frequent.

Mutation and migration too can change gene frequency similar to selection. Practical example of migration is grading-up which is quite commonly employed in cattle breeding. Mutation, though can cause changes in gene frequency, has a very low occurrence in populations to cause quick deviations from equilibrium.

However, in the absence of mutation, migration and selection,sampling or chance factor can also alter the gene frequency in small (finite) populations; this is referred to as random-drift which forms a very important componentin animal production (inbreeding).

3.2. Sex-linked genes

In case of sex-linked genes, equilibrium is not achieved in one random mating but at each generation, difference in the gene frequency between the sexes is halved. In other words, if an allele is more frequent in female at the beginning, in the subsequent generation, it will be more abundant in males. That means, if base generation $(p_f - p_m) = x$, then, in Generation I $\rightarrow (p_f - p_m) = (-)\ ½\ x$, in Generation II $\rightarrow (p_f - p_m) = ¼\ x$ and in Generation III $\rightarrow (p_f - p_m) = (-)\ 1/8\ x$ and so on. Therefore, after about seven generations, $(p_f - p_m)$ will be approximately zero indicating the reaching of equilibrium frequency.

Another important difference in case of sex-linked genes is that the genotypic frequencies p^2, $2pq$ and q^2 applied only to the homogametic sex (male, in case of poultry) whereas, for heterogametic (hemizygous) sex (female in poultry), genotypic frequency is the same as equilibrium frequency of each allele.

4. Gene and genotypic frequencies of quantitative traits

Assuming a locus A, with alleles 'A' and 'a' with complete dominance, whose frequencies are p and q, respectively, in an idealized large population, genotypic frequencies are :

$(pA + qa)^2 = p^2AA + 2pq\ Aa + q^2\ aa$; and the phenotypic are

Dominant allele, $A = p^2 + 2pq$ and recessive allele, $a = q^2$.

In case of quantitative traits, the alleles will have equal effects and assuming $p = q = ½$ and that gene action is additive with the addition of 'A' gene to the phenotype causing one unit increase in phenotype whereas that of 'a' gene causing no change in the phenotypic value, the phenotypic values will be : AA = 2, Aa = 1 and aa = 0 whose frequencies in an idealized population will be $p^2 = ¼$, $2pq = ½$ and $q^2 = ¼$, respectively.

Extending the same to two unlinked, non-interacting loci (A and B) with equal gene frequencies (pA = qA = pB = qB = ½) and equal additive effects (a = b = 0 and A = B = 1), the phenotypic values will be 4, 3, 2, 1 and 0 with phenotypic frequency of 0.0625, 0.2500, 0.3750, 0.2500 and 0.0625, respectively (Table 74.5).

From the results obtained for two loci, it can be noticed that with 'n' number of unlinked, non-interacting loci, with equal allelic frequency and equal additive effects at each locus, 3^n distinct genotypes and (2n+1) distinct phenotypes are possible. Further, the phenotypic array can be found by expanding the trinomial $(¼+ ½+ ¼)^n$ and the frequency of individuals whose phenotype will be exactly 'k' units (f(k)) will be given by

$$f(k) = \frac{(2n)!}{k!(2n-k)!}\left(\frac{1}{2}\right)^{2n} \text{ or } {}^{2n}C_k\left(\frac{1}{2}\right)^{2n}$$

Table 74.5. Gene and genotypic frequencies – Quantitative trait

Genotype	Phenotypic value	Frequency	Phenotypic Class	Phenotypic Freq
AABB	4	$p^2A.p^2B = (½)^4 = 0.0625$	4	0.0625
AABb	3	$p^2A.2pBqB = (½)^2.(½) = 0.1250$	3	0.2500
AaBB	3	$2pAqA.p^2B = (½).(½)^2 = 0.1250$		
AAbb	2	$p^2A.q^2B = (½)^2.(½)^2 = 0.0625$	2	0.3750
AaBb	2	$2pA.qA.2pBqB = (½).(½) = 0.2500$		
aaBB	2	$q^2A.p^2B = (½)^2.(½)^2 = 0.0625$		
Aabb	1	$2pAqA.q^2B = (½).(½)^2 = 0.1250$	1	0.2500
aaBb	1	$q^2A.2pBqB = (½)^2.(½) = 0.1250$		
aabb	0	$q^2A.q^2B = (½)^4 = 0.0625$	0	0.0625

A close perusal of frequencies of different phenotypic classes indicates that:

1. As 'n' increases, number of distinct phenotypes increases arithmetically whereas that the genotypes increases geometrically.
2. At intermediate gene frequencies (p = q = ½), the frequency of extreme phenotypes are relatively rare.
3. Individuals can differ genotypically but still they can have the same phenotypes which, in fact, is unique to quantitative traits.

In case of complete dominance in all the 'n' loci, 3^n distinct genotypes and (n+1) distinct phenotypes are possible. Assuming that phenotypic value of aa = bb = cc = dd = …..nn = 0 and that of AA = Aa = BB = Bb = ……NN = Nn = 2, the phenotypic array can be obtained by expanding the bi-nominal $(¾ + ¼)^n$. Further, frequency of individuals whose phenotype will be exactly '2k' units with k lying between 0 and n can be obtained by

$$f(2k) = \frac{n!}{(n-k)!}\left(\frac{3}{4}\right)^k\left(\frac{1}{4}\right)^{n-k} \text{ or } {}^nP_k\left(\frac{3}{4}\right)^k\left(\frac{1}{4}\right)^{n-k}$$

Even with dominance, as 'n' increases, the number of phenotypic classes increases arithmetically whereas, frequency of one extreme class is high whereas that of the other is low.

In any case, it is noticeable that as 'n' increases, the phenotypic expression shows a continuous distribution. Since dominance is a rare phenomenon in polygenes, the frequency of the extreme classes will be low and that of the intermediate phenotypes will be high. Thus, when plotted on a graph, the frequencies follow, on most occasions, a symmetrical or Normal distribution.

Under realistic situations, most of the quantitative traits are controlled by more than two pairs of alleles but exact number cannot be known. Neither equal allelic frequency at each locus need be always true nor their effects equal and additive. In addition, when 'n' is large, linkage cannot be ruled out and environmental effects, more often than not, cannot be disregarded.

However, generalizations drawn on idealized populations are the best available as far as quantitative traits are concerned. Therefore, it is generally assumed and/or agreed that quantitative traits are continuously distributed with intermediate phenotypes being most common and extreme phenotypes least common.

5. Phenotype of an individual

Considering a polygenic trait, phenotype of individual (P) is determined at the moment of conception when its genotype (G) at all loci affecting the trait is fixed by union of gametes produced by its parents at random. The influence of G on P depends on effects of and interaction among the alleles each locus concerning the trait. The P is also influenced by the environmental factors (E) ever since the conception itself till the time the trait is actually measured. The effect of E is synonymous to the net environmental effect assigned the individual at random from all possible environmental effects that could have influenced the trait. The interaction between G and E (GE) also influences of the P of the individual. Therefore, $P = G + E + GE$.

Assuming that G and E are not correlated ($GE = 0$), $P = G + E$ and the P values of individuals in the population are continuously distributed. The distribution of the frequency of different phenotypic classes may appear differently for different traits. For instance, many of the quantitative traits show a skewed distribution whereas parasitic eggs/g of feces is exponentially distributed; many traits under practical situations could as well be irregularly distributed.

However, Normal distribution is generally assumed to suit many of the quantitative traits and statistical treatment is mainly based on this assumption. Wherever it is well-known that the trait self is not normally distributed, appropriate transformations have to be madeto make the statistic calculated more accurate and meaningful. For instance, logarithmic transformation (for counts), arcsine or angle values (for fertility, hatchability etc.), square root transformation (for small average proportions) etc. For retransformation of *arcsin* mean and standard deviation (SD) values to original scale see Appendix - 2.

Of the several measures of central tendency, "Average" or "arithmetic mean", which is the ratio of sum of all the determinations to the number of determinations is by far the most widely used. The best measure of dispersion has been the variance (V), the mean squared deviation or its squared root, the standard deviation (SD). Variance, by definition, is the average of the squared deviations of individual observations from the mean value.

6. Average effect of gene

Considering recessive dwarf gene (dw) in a population with the frequency of the dominant allele (DW) which does not depress body weight, p, as 0.7 and that of dw gene, q, as 0.3; and the day-old body weight of normal (DWDW and DWdw) as 40 g and that of dwarf chicks (dwdw) as 28 g, it is understandable that the day-old body weight is dependent on number of DW genes.

Therefore, a regression equation can be fit with day-old weight as the dependent variable (y) on number of DW genes (x). As per the Hardy-Weinberg law, frequency of homozygous dominants, heterozygotes and homozygous recessives will be 0.49 (i.e. 0.7^2), 0.42 (i.e. 2 x 0.7 x 0.3) and 0.09 (i.e. 0.3^2), respectively. Keeping these in view, a linear regression equation is fit (Table 74.6 and Fig. 74.1).

Population mean of DW genes = 1.40 and population mean of body weight = 38.92 g. The y values can be considered to be the phenotypic averages from large populations in nullify the net environmental effect so that the values measured represent those without error and thus reflect the genotypic values directly.

Table 74.6. Chick weight as a function of number of DW genes

No. of DW genes (x_i)	Chick weight (g) (y_i)	f_i	f_ix_i	f_iy_i	$f_ix_i^2$	$f_iy_i^2$	$f_ix_iy_i$
0	28	0.09	0.00	2.52	0.00	70.56	0.00
1	40	0.42	0.42	16.80	0.42	672.00	16.80
2	40	0.49	0.98	19.60	1.96	784.00	39.20
Σ 3	108	1.00	1.40	38.92	2.38	1526.56	56.00

Corrected sum of cross products = 56 – {(1.40 x 38.92) ÷ 1.00} = 1.512
Corrected sum of squares of x = 2.38 – (1.40^2 ÷ 1.00) = 0.42
Regression coefficient, b_{yx} = 1.512 ÷ 0.42 = 3.60 g/DW gene
Intercept of the line = {38.02 – (3.60 x 1.40)} ÷ 1.00 = 33.88 g
Regression equation is y = 33.88 + 3.60 x; predicted values of x when y assumes a value of 0, 1 and 2 are 33.88, 37.48 and 41.08 g, respectively as against corresponding actual values of 28, 40 and 40 g.
Corrected sum of squares of y = 1526.56 – (38.92^2 ÷ 1.00) = 11.7936

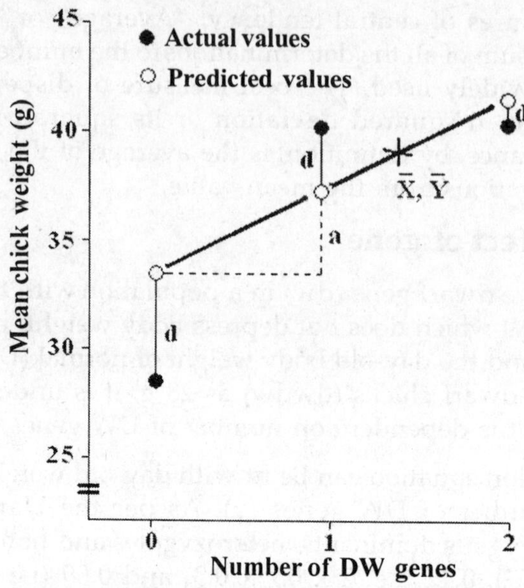

Fig. 74.1 Average effect of gene and breeding value

The population mean body weight (38.92 g) obtained is closer to normal birth weight (38 g). This is because the normal chicks represent $p^2 + 2pq = 0.91$ i.e., 91% of the population. The actual mean chick weight was 38.96 g which was 0.04 g more than the calculated mean weight which may be due to errors in the determination of the chick weight itself.

Regression coefficient, b_{yx}, indicates changing mean value of y per unit change in x; that is, change in the mean chick weight per single substitution of dw gene with DW gene; this is referred to as "Average effect of gene substitution, designated 'a', which, in this hypothetical example, is 3.60 g. The average of the gene or gene substitution is defined as deviation of the phenotypic mean of the progeny from that of population mean, where the progeny obtained one gene from its parents and the other is coming at random from the population; 'a' is the property of the gene frequency alone.

7. Breeding value (BV)

BV of an individual is the value of the individual judged by the mean value of its progeny. An individual contributes 50% of genes to its progenies produced over several matings and if each gene has an average effect of 'a', the total effect contributed is Σ a. However, as only half the genes of its progeny is contributed by the individual, the BV = 2 Σ a; and hence, BV is the property of both gene frequency as well as the population. In other words, BV is a transmitting ability of an individual and it is twice the phenotypic value (expressed as a deviation from the population mean) of individual inheriting one allele from a parent of that genotype, the other allele coming at random from population. The points on the regression line (Fig.74.1) represent the BV of individuals with 0, 1 and 2 DW genes.

7.1. Calculation of breeding values

Calculation of BV for each of the parents with 0, 1 and 2 DW genes is shown in Table 74.7. This conspicuous that the BV of the parents in actual values is nothing but the points of the regression line; hence, it is imperative to conclude that although, due to complete dominance, DWDW and DWdw are phenotypically indistinguishable, their BVs are not the same. Therefore, there is a difference between the actual mean value (genotypic value) and the BV (the corresponding points on the regression line) which is designated 'd' in Fig. 74.1. The deviation between genotypic value and BV is referred to as "dominance deviation" for the respective genotype.

The deviation of actual values from those of predicted ones from the regression line can also be due to the fact that the population is in Hardy-Weinberg equilibrium.

Table 74.7. Calculation of Breeding values

	Parents						
	dwdw		DWdw			DWDW	
Alleles transmitted	All dw		Half each DW and dw			All DW	
Frequency of alleles from parents	DW=0 dw=1		DW=½ dw=½			DW=1 dw=0	
Frequency of alleles from Population	DW=0.7 dw=0.3		DW=0.7* dw=0.3			DW=0.7 dw=0.3	
Type of progeny	DWdw	dwdw	DWDW	DWdw	dwdw	DWDW	DWdw
Frequency	0.70	0.30	0.35	0.50	0.15	0.70	0.30
Chick weight (g)	40	28	40	40	28	40	40
Mean Chick weight of progeny (g)	0.7x40+ 0.3x28= 36.4		0.35x40+ 0.5x40+ 0.15x28 = 38.2			0.7x40+ 0.3x40 = 40.0	
Mean chick weight as a deviation from Population mean (g); i.e. 'a'	36.4- 38.92 =(-) 2.52		38.2- 38.92 =(-) 0.72			40.0- 38.92 =1.08	
Breeding value (2? a), g	(-) 5.04		(-) 1.44			2.16	
Breeding value in actual units (Population mean +2? a), g	38.92- 5.04 = 32.88		38.92- 1.44 = 37.48			38.92+ 2.16 = 41.08	
Breeding value predicted from regression line, g	33.88		37.48			41.08	

* Half the frequency of the gametes combines with DW and the other half with dw

7.2. Calculation of additive, genotypic and phenotypic variances

It has been shown above that BV refers to transmitting ability of an individual and hence refers to additive genetic variance. If A, G and P* refer to BV, genotypic value and Population mean, the calculation of V(G), V(A) and V(P) are shown in Table 74.8 with an assumption that variance due to environment (V(E)) is zero and hence, V(P) = V(G).

Table 74.8. Calculation of additive, genotypic and phenotypic variances

	Genotypes of individuals		
Population mean, P* = 38.92 g	dwdw	DWdw	DWDW
Frequency (f)	0.09	0.42	0.49
Genotypic value (G), g	28.00	40.00	40.00
Breeding value (A), g	33.88	37.48	41.08
(A – P*), g	(-) 5.04	(-) 1.44	2.16
$f(A - P^*)^2$	2.286144	0.870912	2.286144
$\Sigma f(A - P^*)^2 = V(A)$		5.4432**	
(G – A)	(-) 5.88	2.52	(-) 1.08
$f(G - A)^2$	3.111696	2.667168	0.571536
$\Sigma f(G - A)^2 = V(D)$		6.3509	
(G – P*) i.e. row 4 + row 6	(-) 10.92	1.08	1.08
$f(G - P^*)^2$	10.732176	0.489888	0.571536
$\Sigma f(G - P^*)^2 = V(G)$ which is same as Corrected $\Sigma f_i y_i^2$ from Table 74.6.		11.7936	

** Can also be calculated as variance of breeding values i.e. V(A):
$\Sigma f_i A_i^2 - (\Sigma f_i y_i)^2 \div \Sigma f_i = 1520.2096 - (38.92)^2 \div 1.00 = 5.4432$

It can be seen from Table 74.8 that V(G) = V(A) + V(D). If there is no dominance, that means V(D) = 0, then it is easy to conclude that magnitude of V(A) is dependent on deviation of BV of each genotype from Population average and the frequency of each of the genotypes in the population. Similarly, V(D) is dependent on deviation of genotypic value from BV and genotypic frequency. Hence, changes in genotypic frequency can alter both V(A) and V(D), and in turn V(G). When many loci are involved, variance due to epistasis (V(EP)) is also a possibility. Statistically, it is possible to estimate V(A) and the deviation V(P) – V(A); the latter accounting for the sum of V(D), V(EP) and V(E).

In the above hypothetical example, with V(E) = 0,

$$h^2 = V(A) \div \{V(A)+V(D)\} \text{ or } V(A) \div V(G) = 5.443 \div 11.7936 = 0.4615$$

8. Heritability

Considering a trait X with the mean value of X* and variance V(X), the V(X) refers to the observed variance or phenotypic variance, V(P). But, as discussed earlier P

= G + E and hence, it is essential to know what proportion of V(P) or V(X) is due to genetic effects are genotypic variance, V(G). It is equally important to know the proportion of differences between two individuals that can be transmitted to the next generation. In other words, if an individual witha phenotypic superiority of 'a' units is selected, what proportion of this superiority is expectable in its offsprings? In still simpler terms, how reliable is the phenotype of the individual in predicting its genetic worth?

Extending the basic equation P = G + E to the individual X, P(X) = G(X) + E(X) + GE(X); the GE(X) representing the interaction between genotype and environment. It is reasonable to assume that when individuals are drawn independently and randomly from a large population, GE(X) = 0. Hence, P(X) = G(X) + E(X). Starting with this equation, the concept of heritability can be explained as follows:

Taking summation over all the n individuals in the population,

$$\sum_{i=1}^{n} P(X) = \sum_{i=1}^{n} G(X) + \sum_{i=1}^{n} E(X); \text{dividing both sides by n}$$

$$\bar{P} = \bar{G} + \bar{E}; \text{subtractin g this from } P = G + E$$

$$(P - \bar{P}) = (G - \bar{G}) + (E - \bar{E}); \text{this equation estimates the individual}$$

as a deviation from Population mean; therefore, for the entire population

$$\sum_{i=1}^{n} (P - \bar{P}) = \sum_{i=1}^{n} (G - \bar{G}) + \sum_{i=1}^{n} (E - \bar{E}); \text{squaring both the sides}$$

$$\sum_{i=1}^{n} (P - \bar{P})^2 = \sum_{i=1}^{n} (G - \bar{G})^2 + \sum_{i=1}^{n} (E - \bar{E})^2 + 2\left(\sum_{i=1}^{n} (G - \bar{G})\right)\left(\sum_{i=1}^{n} (E - \bar{E})\right)$$

Dividing both sides by (n – 1), the degrees of freedom,

$$V(P) = V(G) + V(E) + 2Cov(GE); \text{since G and E are assumed uncorrelat ed,}$$

$$cov(GE) = 0 \text{ and hence, } 2Cov(GE) = 0. \text{ Therefore, it follows that}$$

$$V(P) = V(G) + V(E); \text{dividing both sides by } V(P)$$

$$\frac{V(G)}{V(P)} + \frac{V(E)}{V(P)} = 1$$

It follows therefore that the ratio of genetic and environmental variances to that of phenotypic variance accounts for all the observed variance; of these, the first component, the ratio of genetic variance to phenotypic variance i.e. V(G) to V(P), is referred to as "heritability in broad sense". However, this cannot quantitate the reliability of P in predicting G; besides, it also cannot indicate the proportions of differences between individuals that can be transmitted to next generation.

Genetic component of variance, V(G), comprises of additive genetic variance, V(A), variance due to dominance deviation, V(D), and variance due to epistatic deviations, V(EP). In other words,

V(G)=V(A)+V(D)+V(EP) and V(P)=V(A)+V(D)+V(EP)+V(E). Statistically, it is possible to partition V(G) from V(E) and V(A) from V(D) and V(EP).

Hence, if V(E′) = V(D)+V(EP)+V(E), then V(P) = V(A) + V(E′); as per the earlier derivations, the following is also true:

$$(P-\bar{P}) = (A-\bar{A}) + (E'-\bar{E}')$$

Since additive effects cannot be measured directly, it can be estimated by regressing it on the easily measurable P i.e. by calculating b_{AP}.

$$\text{By definition, } b_{AP} = \frac{\text{Covariance AP}}{\text{Variance P}} = \frac{\sum(A-\bar{A})(P-\bar{P})\Big/(n-1)}{\sum(P-\bar{P})^2\Big/(n-1)}$$

$$\text{or } b_{AP} = \frac{\sum(A-\bar{A})(P-\bar{P})}{\sum(P-\bar{P})^2}; \text{substituting } (P-\bar{P}) = (A-\bar{A}) + (E'-\bar{E}')$$

$$b_{AP} = \frac{\sum(A-\bar{A})\left[(A-\bar{A}) + (E'-\bar{E}')\right]}{\sum(P-\bar{P})^2} = \frac{\sum(A-\bar{A})^2 + \left[(A-\bar{A})(E'-\bar{E}')\right]}{\sum(P-\bar{P})^2}$$

Since BV and environment are not correlated, $(A-\bar{A})(E'-\bar{E}') = 0$; and in a large population, the environmental effects will have a mean Zero; therefore,

$$b_{AP} = \frac{\sum(A-\bar{P})^2}{\sum(P-\bar{P})^2}; \text{dividing both numerator and denominator by degrees of freedom.}$$

$$b_{AP} = \frac{V(A)}{V(P)} \text{ which is also the heritability in narrow sense.}$$

Heritability narrow sense, designated h^2, is the ratio of additive genetic variances, V(A), to V(P). In fact, this the most useful estimate for animal breeders. In other words, h^2 indicates the proportion of V(P) attributable to V(A) or is the proportion of differences among individuals expected, on average of the population, to be transmitted to their progeny. It follows therefore that h^2 is the proportion of superiority of the selected parents that is expected to be realized in their progeny. Estimation of h^2 is also possible as the regression of BV on phenotypic value, V(P); the coefficient of determination (R^2) is the proportion of variability in BV that is attributable to its linear association with V(P); in other words, $R^2_{AP} = h^2$. If h^2 is large, it indicates that an animal which is above average phenotypically is likely to be correspondingly superior genetically and vice versa. That means, high h^2 suggests that phenotype is a reliable indicator of genotype and *vice versa.*

8.1. Characteristics of h^2

1. Since it is a ratio of variances, is always = 0. Further, as the animal breeder is not interested in traits whose V(A) = 0, h^2 is always > 0. Similarly, biological traits will seldom have V(P) = 0, and hence, h^2 is seldom infinity or indeterminate; i.e. h^2 = 0
2. Numerator is a part of the denominator in the formula for calculating h^2; therefore, h^2 is always = 1 or 0 = h^2 = 1.
3. h^2 is not a biological constant. The numerator V(A) is influenced by number of loci, type of gene action, average effect of gene and gene frequency. The denominator is determined by the magnitude of variances due to dominance, epistasis and environment.
4. h^2 for a given trait is not constant between populations because of differences in the variance components.
5. Within a population, h^2 can change over a period of time due to changes in any or many of the components of both the numerator and the denominator.

8.2. Estimation of heritability

8.2.1. Heritability in broad sense

Although, this estimate has little practical utility, its estimation necessitates a genetically uniform Population in which V(G) is likely to be zero so that V(P) of such population estimates V(E). However, in livestock, severe loss in fitness traits at high inbreeding coefficient precludes this method of estimation. But, monozygotic twins as may be available in case of cattle, sheep etc. (not in poultry) can be considered genetically uniform.

Variance calculated from a normal genetically variable population estimates V(P). Then, by definition, heritability in broad sense is the ratio of V(G) to V(P); the numerator can be estimated as V(P) – V(E). Therefore, heritability in broad sense −1 – V(E)÷V(P); in other words, it is equal to 1 – the ratio of V(P) in genetically uniform population to that in genetically variable population.

8.3. Methods of estimation of h^2 (narrow sense) :

8.3.1. Response to selection

By definition, h^2 is the superiority of selected parents transmitted to the next generation. If mean of the selected individuals was 'S' units superior to that of the base population from which they are selected (Selection differential), and the mean of the reproduced population was 'R' units superior to that of the base population (response), then $h^2 = R \div S$. In livestock population, including poultry, as the males form a smaller proportion, S is calculated separately for males and females, and averaged. h^2 calculated by this method is referred to as "realized h^2".

In livestock populations, the generations overlap and this would make the determination of R difficult. In reading depression can also offered to realized h^2; but this can be taken care of in poultry breeding by avoiding mating between close relatives and by increasing sample size. However, a serious drawback of realized h^2 is that it is based on the assumption that environmental influences are constant over the generations which is not true. Attempts have been made to estimate the environmental effects by a "control population". Similarly, bi-directional selection and estimation of differences in R between upward and downward selection which estimates environmental influences can be practiced; but it is a time-consuming and costly exercise. Alternatively, repeat matings of full-sibs which are genetically homogenous can also help estimates environmental effects. However, on most occasions, assumption that environment has uniform effect seems to distort realized h^2 to a very limited extent to warrant a costly corrective procedure. If S and R over several generations of known, h^2 can be estimated as regression of R on S.

Table 74.9. Calculation of realized h^2 (mean body weight, kg)

Population	Males	Females
Base population	1.80	1.50
Selected population	2.20	1.70
Selection differential	0.40	0.20
Mean Selection differential	0.30	
Reproduced population	1.90	1.60
Response	0.10	0.10
Mean response	0.10	
Realized h^2	$0.10 \div 0.30 = 0.33$	

8.3.1.1. Selection plateau

When response, R, has ceased, the population is said to be at the selection limit or selection plateau. This can occur when there is no more V(A) left out to be inherited; that is h^2 tending towards zero, which, in turn, minimize is the magnitude of selection differential (S)

8.3.2. Resemblance among relatives :

Any relationship can be used provided it is amenable to the following steps:

1. Quantification by correlation (r) or regression (b) coefficient
2. Explanation as to how much a relationship can cause more resemblance than those chosen strictly at random and
3. Equating the observed resemblance to theoretical causes so that h^2 can be estimated with minimum bias

8.3.2.1. Direct relationship

8.3.2.1.1. Offspring with one of the parents

The same trait is estimated at the same stage of life-cycle in both offspring and parent. Normally, offspring (O) –dam (D) relationship is commonly used because number of sires used in livestock breeding is low. If V(P) among dam and offsprings are homogeneous, either r_{OD} or b_{OD} can be calculated. But, as dams are subjected to selection, V(P) reduces among them. Therefore, if the relationship is linear, b_{OD} is a better statistic. If sire is known, numerator and denominator for calculation of b_{OD} can be obtained under each sire and then pooled to get intra-sire b_{OD}. On the same lines, if the data is stratified as per treatments or locations etc., "intra-class" r_{OD} or b_{OD} can be computed.

Sire and dam contribute 50% of genes to their offspring; in other words, covariance of offspring with any of the parent estimates ½ V(A) + other genes effects like variance due to interaction, maternal influence (parent is dam) etc. Since b_{OD} is the ratio of covariance_{OD} to variance among the dams, it estimates only ½ h^2; in other words, $h^2 = 2\ b_{OD}$

8.3.2.1.2. Offspring with mid-parent

Mid-parent values are not always practicable because of less number of sires in animal breeding. If mid-parent values are used, pooled mean (P) = ½ of mean of dams (D) + ½ of mean of sires (S); since each of them contribute ½ of genes (V(P)) to the next generation and they are unrelated, V(P) = ¼ V(D) + ¼ V(S). Further, assuming variances are homogeneous, V(D) = V(S) = V(P), it is clear that V(P) = ½ V(D) or ½ V(S) or ½ V(P) itself. Since it is known that covariance estimates ½ V(A), the regression coefficient b_{OD} = covariance ÷ variance = ½ V(A) ÷ ½ V(P) which is nothing but V(A) ÷ V(P) = h^2. Therefore, regression coefficient calculated by using mid-parent values is itself the value of h^2; use of mid-parent values is preferred because it is less influenced by maternal effects although interaction effects cannot be ruled out.

8.3.2.1.2.1. Calculation of h^2 by regression of offspring value on dam values

If x_i and y_i are the values of a trait recorded on i^{th} dam and offspring, respectively, where i takes values from 1 to n, then, the following are calculated:

$\sum_{i=1}^{n} x_i$, $\sum_{i=1}^{n} x_i^2$, $\sum_{i=1}^{n} y_i$ and $\sum_{i=1}^{n} x_i y_i$. They are designated A, B, C and D,

respectively. Then $b_{OD} = \dfrac{D-(A)(C)/n}{B-A^2/n}$ and $h^2 = 2b_{OD}$

8.3.2.2. Collateral relationship

8.3.2.2.1. Paternal half-sibs

Resemblance of half-sibs to that among individuals related by an average amount for the population is the criterion. This is estimated by analysis of variance (ANOVA) method (Table 74.10). That in the form of maternal half-sibs is available at least in poultry. The following calculations have to be made:

If X_{ij}'s are the observations on the jth half-sib (j=1,2,3,4...n) under i[th] sire (i=1,2,3,4...s), n_i is the number of observations under i[th] sire, $X_{i.}$ is the total of ith sire, n. is the total number of observations and X.. is the grand total, by using standard statistically procedures, the following are obtained:

$\sum_i^s \sum_j^{n_i} X_{ij}^2$ is the total uncorrected sum of squares

Sum of squares due to sires, $SS_S = \sum \frac{(X_{i.})^2}{n_i} - \frac{(X_{..})^2}{n_.}$

Total sum of squares, $TSS = \sum_i^s \sum_j^{n_i} X_{ij}^2 - \frac{(X_{..})^2}{n_.}$ and

Sum of squares within sires (Error), $SS_W = TSS - SS_S = \sum_i^s \sum_j^{n_i} X_{ij}^2 - \sum \frac{(X_{i.})^2}{n_i}$

number of half sib progenies per sire, $k = \frac{1}{s-1}\left(n_. - \sum_i^s \frac{n_i^2}{n_.}\right)$

V(S) can be calculated as $(MS_S - MS_w) \div k$, and V(P) = V(S) + V(E) and intra-class correlation, r_I = V(S) ÷ V(P).

Probability that to half-sibs inherit copy of the particular gene from their sire is ½ and therefore, the problem is that both of them will inherit the same gene is ½ x ½ = ¼ because the events are independent. It follows therefore that V(S) = ¼ V(A) or V(A) = 4 V(S); hence, h^2 = 4 V(S) ÷ V(P) or $h^2 = 4\ r_I$.

Table 74.10. ANOVA of half-sib relationship

Source	df*	Sum of squares (SS)	Mean sum of squares (MSS)**	Expected mean sum of squares (EMS)
Between sires	s-1	SS_S	$SS_S \div (s-1) = MS_S$	V(E)+kV(S)
Within sires	n.-s	SS_W	$SS_W \div (n.-s) = MS_W$	V(E)
* df indicates degrees of freedom		** Calculated as (SS ÷ df)		

8.3.2.2.2. Full-sibs

Litter-mates, progeny from repeat-mates or single-sire pedigree mating (which is very popular in poultry) can yield data of full-sibs. Full-sibs get 50% of genes each from sire and dam and hence their resemblance estimates 50% of V(A) of which 25% is from sire and the other 25% is from dam component. In other words, V(S) and variance (dams), V(D) account for ¼ V(A) each. They can be estimated by ANOVA (Table 74.11).

The preliminary calculations for obtaining ANOVA table are as follows:

Assuming that several sires (i = 1,2,3.....s) where each mated to several dams (j = 1,2,3....d) and each dam produced several progenies (k = 1,2,3....p). The value of each progeny for the trait in question represented by X_{ijk} is tabulated as per sire and dam.

Let $X_{ij.}$ be the total of the jth dam under i[th] sire (designated A), $X_{i..}$ be the total of ith sire (designated B), $X_{...}$ the grand total (designated C), n_{ij} the number of observations under j[th] dam under i[th] sire, $n_{i.}$ be the number of observations under i[th] sire and $n_{..}$ be the total number of observations.

The following are calculated from the data:

$\sum_i^s \sum_j^d \sum_k^p X_{ijk}^2$, the total uncorrected sum of squares (T)

$\sum_i^s \frac{B^2}{n_{i.}} - \frac{C^2}{n_{..}}$, the Sum of squares due to sires (SS_S)

$\sum_i^s \sum_j^d \frac{A^2}{n_{ij}} - \sum_i^s \frac{B^2}{n_{i.}}$, the Sum of squares due to dams within sires ($SS_{D/S}$)

$T - (SS_S + SS_{D/S}) = T - \sum_i^s \sum \frac{A^2}{n_{ij}}$, S = No. of sires, D=No. of dams/sire,

P = No. of progenies within dam within sire

the sum of squares between progenies within dams within sires ($SS_{P/D/S}$)

$$E = \sum_i^s \frac{\sum_j^d n_{ij}^2}{n_{i.}}, \quad F = \sum_i^s \sum_j^d n_{ij}^2, \quad G = \sum_i^s n_{i.}^2$$

$$K_1 = \frac{(n_{..} - E)}{(D - S)}, \quad K_2 = \frac{\left(E - \frac{F^2}{n_{..}}\right)}{(S-1)} \text{ and } K_3 = \frac{\left(n_{..} - \frac{G^2}{n_{..}}\right)}{(S-1)}$$

Table 74.11. ANOVA of full-sib relationship

Source	df*	SS	MSS**	EMS
Sires	S-1	SS_S	MS_S	$V(E)+K_2V(Dam)+K_3V(S)$
Dams within sires	S(D-1)	$SS_{D/S}$	$MS_{D/S}$	$V(E)+K_1V(Dam)$
Progenies within dams within sires	SD(P-1)	$SS_{P/D/S}$	$MD_{P/D/S}$	V(E)
* df indicates degrees of freedom	** Calculated as (SS ÷ df)			

From EMS, $V(E) = MS_{P/D/S}$, $V(Dam) = (MS_{D/S} - MS_{P/D/S}) \div K_1$

$V(S) = \{MS_S - MS_{P/D/S} - (K_2 \div K_1)(MS_{D/S} - MS_{P/D/S})\} \div K_3$

If $K_1 = K_2$, $V(S) = (MS_S - MS_{D/S}) \div K_3$

Note: K_1, K_2 and K_3 represent number of progenies per dam, per dam within sire and per sire, respectively.

After obtaining the variance components, V(P)=V(S)+V(Dam)+V(E); since V(S) and V(Dam) each estimates ¼ V(A),

4 V(S) = V(A) and 4 V(Dam) = V(A); hence

h^2(sire component), $h^2_S = 4\,\{V(S) \div V(P)\}$,

h^2(dam component), $h^2_D = 4\,\{V(Dam) \div V(P)\}$ and

h^2(sire + dam components), $h^2_{S+D} = \{2\,(V(S)+V(Dam)) \div V(P)\}$

Errors in the computing correlation of relationship is estimated by multiplying the calculated correlation with the reciprocalof the theoretical genetic relationship among the individualsconsidered. Value of reciprocal increases as the denominator reduces and therefore, as theoretical genetic relationship among the relatives considered reduces, error in estimation is magnified. Half-and full-sibs are, by far, the most popularly used relationships for estimating h^2.

8.3.3. h^2 of threshold characters

As discussed earlier, phenotypic expression of threshold characters is discrete. Therefore, h^2 can be estimated by assuming different numerical values to the alternate expression and carrying out ANOVA with data on full- and half-sibs. Alternatively, realized h^2 can be calculated with some modifications based on the assumption that the underlying distribution is a standard normal deviate with mean = 0 and SD = 1.

8.3.3.1. An example

Incidence of disease in the affected population (population mean = 0 and SD = 1) was 7%; this corresponds to 1.48 SD units (Z-table). The mean incidence was 3.5% corresponding to 1.92 SD units. In a related population the corresponding values were 12% (1.18 SD units) and 6% (1.67 SD units). Keeping these in view,

S = (Mean of the affected population – Mean of the base population)

= 1.92- 0.00 = 1.92 SD units.

R = (Mean of the related population – Mean of the base population); since both are 0,

R = (Incidence in base population – Incidence in related population)

= 1.48 – 1.18 = 0.30 SD units.

R/S = 0.156. The genetic relationship between parent and offspring is one half and therefore $h^2 = 2(R/S) = 0.312$

Another method has also been suggested. If Pi values are the survival of paternal half-sib families, and P* is the average survival, it is easy to visualize that, if the trait is not heritable, each of the Pi's will be the same as P* and if it is heritable, families differ from P*. The difference can be tested by ÷². Subsequently, if S is the number of paternal half-sib families each having K (effective number) of individuals, h^2 is calculated as:

$$h^2 = \frac{1}{\text{Genetic relationship}}\left(\chi^2 - \frac{(S-1)}{K(S-1)}\right)$$

8.4. Uses of h^2

Heritability in broad sense is less useful because the numerator includes dominance (which may be lost during segregation) and epistasis (which may be lost due to crossing over and independent assortment) during meiosis. The uses of h^2 are as follows :

1. Predicting BV of individuals; if an individual is K units superior to the population average, it is expected to be Kh^2 units superior in its BV.
2. Predicting response to selection ($R = h^2S$)
3. Developing suitable methods of selection to improve economic traits.

Table 74.12. Estimated h^2 values (%)

	Character (s)	h^2
Layers	Fertility, Chick livability	5
	Adult livability, Hatchability (FES)	10
	Egg production, Blood spots, Broodiness	15
	Keel length,	20
	Age at sexual maturity, Body depth, Shell texture, Albumin quality	25
	Adult body weight, Egg weight	55
	Egg shape	60
Broilers	Breast fleshing	10
	7-week feed conversion	35
	7-week broiler weight, Dressing percentage	45
	Fat deposition	50
	Total feed consumption	70
	Source : North, 1984	

9. Repeatability

Certain traits are expressed more than once in a lifetime; for example, egg weight. For such traits, V(E) has two components namely permanent and temporary sources of environmental variation designated $V(E)_{PMT}$ and $V(E)_{TMP}$, respectively. That means, $V(E) = V(E)_{PMT} + V(E)_{TMP}$. Those environmental causes that affect the trait every time the trait is expressed constitute $V(E)_{PMT}$; for instance, a permanent damage in yolk formation results in a permanent reduction in egg weight. On the other hand, factors that do not affect the trait every time form $V(E)_{TMP}$; for instance, effect of high temperature during a particular phase of egg production. Therefore, for repeatable traits, $V(P) = V(G) + V(E)_{PMT} + V(E)_{TMP} = V(A) + V(D) + V(EP) + V(E)_{PMT} + V(E)_{TMP}$.

Repeatability, designated R, is defined as a proportion of V(P) for a trait attributable to permanent differences among individuals. Since V(G) and $V(E)_{PMT}$ constitute permanent differences, R is the ratio of $\{V(G) + V(E)_{PMT}\}$ to V(P). Traits which are repeated may be influenced by the same genes each time they are expressed and so-called permanent environmental influence. $V(E)_{PMT}$ may not be always permanent; in spite of these, R is a useful tool to the animal breeders.

9.1. Characteristics of repeatability

1. Repeatability is not a constant value – it varies with the trait, for the same trait in different populations and the same population at different times (similar to h^2).
2. Since the numerator of repeatability includes V(G) and $V(E)_{PMT}$, $R > h^2$ in broad sense $> h^2$. In fact, most often, R is considered upper limit of h2, although, rarely, and in particular circumstances wherein effect of $V(E)_{PMT}$ is negative, R can even be less than h^2.

9.2. Estimation of repeatability

9.2.1. With two records per individual

If only two records per individual are available, repeatability is simply the product-moment correlation between the two records. In other words, repeatability is the ratio of covariance between the two records to the geometric mean of the variances of the two records. If X_{i1} and X_{i2} refer to two records of ith individual among 'n' individuals, then

$$R = \frac{\sum_{i=1}^{n} X_{i1}X_{i2} - \frac{\left(\sum_{i=1}^{n} X_{i1}\right)\left(\sum_{i=1}^{n} X_{i2}\right)}{n}}{\left[\sqrt{\sum_{i=1}^{n} X_{i1}^2 - \frac{\left(\sum_{i=1}^{n} X_{i1}\right)^2}{n}}\right]\left[\sqrt{\sum_{i=1}^{n} X_{i2}^2 - \frac{\left(\sum_{i=1}^{n} X_{i2}\right)^2}{n}}\right]}$$

9.2.2. With more than two records per individual

1. Sum of squares and cross products are calculated for all possible combinations within a sub-class (i.e. individual) and pooled to finally obtain repeatability value by the formula given above.

2. With many records per animal, repeatability can be calculated by the method of half-sib analysis (See Table 74.10) with only changes in terminologies; V(S) is termed variance between sub-class or individuals (V(B)) and V(E) is referred to as variance within subclass or individuals (V(W)), The intraclass correlation (r_I) gives the estimate of repeatability directly; $r_I = R = V(B) \div (V(B) + V(W))$.

9.3. Uses of repeatability

1. In estimating lifetime productivity of individuals; this is particularly employed in cattle breeding.

2. As upper limit of h^2 because it is very easy to compute unlike h^2.

10. Correlations

In practical situations, many quantitative traits are associated with each other. For example, body weight with egg weight, age at sexual maturity with egg production etc.. These associations can be due to genetic and/or environmental causes.

10.1. Types of correlations

10.1.1. Phenotypic correlation (r_p)

This measures linear association between two traits. In other words, it predicts deviation from population mean in one trait of an individual as a function of its deviation from the population mean of the other when both the measured in terms of their standard deviation units. As in the case of V(P), r_p also can be partitioned into genetic and environmental components.

10.1.2. Genetic correlation (r_g)

This measures the extent to which the same genes, are grossly linked genes, cause simultaneous variation in two different traits or the extent to which individuals genetically above average in one trait are genetically above, equal or below average for a second trait. r_g results primarily due to pleotropy of genes and also due to linkage. The latter appears to be of minor importance because, if it were to have been the main cause, r_g would reduce over many generations of random mating which break-up linkages by crossing over as the population approaches linkage equilibrium; further, r_g should have differed between populations because populations differ conspicuously in their genetic make-up. On the contrary, rg contributed by pleotropy is immuned to random mating or genetic structure of populations (as long as the gene(s) exist(s)). In any case, r_g between two traits in any population is certainly the net effect of pleotropy and linkage from as many segregating loci and linked loci affecting the traits, respectively. Analogous to h^2, r_g can be of two types

10.1.2.1. Genotypic correlation

Similar to h^2 (broad sense) – theoretically, it is the correlation that would exist if environmental conditions affecting the two traits are exactly the same ever since the conception till the two traits are measured; but it is unrealistic to accept that environment can cause no variation in either of the traits. Hence, this is of less practical value and also requires specialized populations for computation.

10.1.2.2. Additive genetic correlation

This is analogous to h^2 (narrow sense) and is very useful to estimate genetic merit of individuals and predict correlated response to selection. This is designated by r_g and it refers to correlation between individuals for the two traits considered.

10.1.3. Phenotypic correlation (r_p)

This measures linear association between two traits. In other words, it predicts deviation from population mean in one trait of an individual as a function of its deviation from the population mean of the other when both the measured in terms of their standard deviation units. As in the case of V(P), r_p also can be partitioned into genetic and environmental components.

10.1.4. Environment correlation (r_e)

If the same environmental effect causes simultaneous variation in both traits, a correlation results which is referred to as r_e. For example, increased light during growing period causes early sexual maturity, increased egg production and reduced egg weight. In real sense, r_e is not purely environmental but also includes dominance and epistatic effects because it is calculated as a deviation from r_g (which takes into account additive effects and a small portion of epistatic effects). Therefore, r_e is the net effect of environmental and non-additive genetic factors causing the two traits to vary simultaneously in either positive or negative direction.

Although r_g and r_e are parts of r_p; r_p is neither the sum nor the average of the former two; but their contribution to r_p is a function of relative V(P) contributed by them in each trait. It can be mathematically shown that $r_p = r_g h_x h_y + r_e e_x e_y$, as follows:

Designating P for phenotype, A for additive gene action, E′ for non-additive gene action + environmental effects and SD for standard deviation,

$P = A + E'$ and hence, for trait X $P_X = A_X + E'_X$ and $V(P)_X = V(A)_X + V(E')_X$;
dividing both sides by $V(P)_X$, $1 = \frac{V(A)_X}{V(P)_X} + \frac{V(E')_X}{V(P)_X}$,
$1 = h_X^2 + e_X^2$; and it follows by the formula of h^2 that
$h_X = \frac{SD(A)_X}{SD(P)_Y}$ and $e_X = \frac{SD(E')_X}{SD(P)_Y}$

Considering two traits X and Y, $P_X = A_X + E'_X$ and
$P_Y = A_Y + E'_Y$; therefore,
$CoV(P_X P_Y) = CoV(A_X A_Y) + CoV(A_X E'_Y) + CoV(E'_X A_Y) + CoV(E'_X E'_Y)$
Since A_X and E'_Y as well as E'_X and A_Y are not correlated,
$CoV(P_X P_Y) = CoV(A_X A_Y) + CoV(E'_X E'_Y)$;
dividing both sides by $SD(P_X)SD(P_Y)$

$$\frac{CoV(P_X P_Y)}{SD(P_X)SD(P_Y)} = \frac{CoV(A_X A_Y)}{SD(P_X)SD(P_Y)} + \frac{CoV(E'_X E'_Y)}{SD(P_X)SD(P_Y)}$$; after further manipulatons

$$r_P = \left(\frac{CoV(A_X A_Y)}{SD(P_X)SD(P_Y)}\right)\left(\frac{SD(A_X)SD(A_Y)}{SD(A_X)SD(A_Y)}\right) + \left(\frac{CoV(E'_X E'_Y)}{SD(P_X)SD(P_Y)}\right)\left(\frac{SD(E'_X)SD(E'_Y)}{SD(E'_X)SD(E'_Y)}\right)$$

By rearrangemet, $r_P = r_g h_X h_Y + r_e e_X e_Y$; Note : h_X, h_Y, e_X and e_Y are square roots of their corresponding squared values.

10.2. Estimation of correlations

10.2.1. Phenotypic correlation

Can be calculated as simple product-moment correlation. However, sums of cross products and sums of squares can be calculated within various groups and later on, pooled to get a more reliable estimate of r_p.

10.2.2. Environmental correlation

By using the formula $r_p = r_g h_x h_y + r_e e_x e_y$, r_e can be estimated. e_x and e_y can be estimated by using the formula $h^2_X + e^2_X = 1$ and $h^2_Y + e^2_Y = 1$.

10.2.3. Genetic correlation

10.2.3.1. Through correlated response to selection

It has been shown that $R = h^2S$ and it is more relevant to convert in terms of the corresponding SD to obtain "standardized selection differential" or "intensity of selection" designated 'i'. Hence, $i = S \div SD$ or $S = iSD$. Keeping this in view, r_g can be calculated from correlated response (CR) to selection. Let h be v h². Considering a trait X, $R_X = h^2_X S_X = h^2_X i_X SD(P_X)$; considering trait Y, $R_Y = h^2_Y S_Y = h^2_Y i_Y SD(P_Y)$

If the traits are genetically correlated, selection in trait X should bring a change in V(A) component of trait Y. Therefore, CR_Y = Regression of additive genetic component of trait on the phenotype of trait X times S_X. In other words, $CR_Y = b_{AyPx}S_X$; Further, b_{AyPx} is the product of change in A_X per unit change in P_X and change in A_Y per unit change in A_X. That means, $b_{AyPx} = b_{AxPx} \cdot b_{AyAx}$. Therefore, $CR_Y = b_{AxPx} \cdot b_{AyAx} S_X$.

By definition, $b_{YX} = r\left(\frac{SD_Y}{SD_X}\right)$, and substituting this

$$CR_Y = \left(r_{AxPx}\frac{SD_{Ax}}{SD_{Px}}\right)\left(r_{AyAx}\frac{SD_{Ay}}{SD_{Ax}}\right)(S_X);\ \text{However, } r_{AyAx} = r_g,$$

$r^2_{AxPx} = h^2_X$, and hence, $h_X = r_{AxPx}$; substituting these

$$CR_Y = h_X\left(\frac{1}{SD(P_X)}\right)r_g\,SD(A_Y)S_X;\ \text{since } h_y = \frac{SD(A_Y)}{SD(P_Y)}\ \text{and}$$

$$SD(A_Y) = h_Y SD(P_Y),\ CR_Y = \frac{h_X r_g h_Y SD(P_Y)S_X}{SD(P_X)};\ \text{but, } \frac{S_X}{SD(P_X)} = i_X$$

$$\text{Therefore, } CR_Y = h_X r_g h_Y SD(P_Y) i_X \text{ and } r_g = \frac{CR_Y}{h_X h_Y SD(P_Y) i_X}$$

Alternatively, if CR_Y and CR_X due to direct selection in trait X and trait Y, respectively are estimated simultaneously in different selection lines from the same base population, and substituting suitably for CR's and R's

$$\left(\frac{CR_X}{R_X}\right)\left(\frac{CR_Y}{R_Y}\right) = \left(\frac{h_X r_g h_Y SD(P_X) i_Y}{h^2_X i_X SD(P_X)}\right)\left(\frac{h_X r_g h_Y SD(P_Y) i_X}{h^2_Y i_Y SD(P_Y)}\right);\ \text{this reduces to}$$

$$\left(\frac{CR_X}{R_X}\right)\left(\frac{CR_Y}{R_Y}\right) = r^2_g \text{ and therefore, } r_g = \sqrt{\left(\frac{CR_X}{R_X}\right)\left(\frac{CR_Y}{R_Y}\right)}$$

If CR_X and R_X, and CR_Y and R_Y differ in sign, the negative square root and if they have positive signs, positive square root gives r_g. However, in practice, CR of only one trait is commonly available; in such cases

$$\left(\frac{CR_Y}{R_Y}\right) = \left(\frac{h_X r_g h_Y SD(P_Y) i_X}{h^2_Y i_Y SD(P_Y)}\right);\ \text{assuming that } i_X = i_Y, \text{ this reduces to}$$

$$\left(\frac{CR_Y}{R_Y}\right) = r_g\frac{h_X}{h_Y} \text{ and therefore, } r_g = \left(\frac{CR_Y h_Y}{R_Y h_X}\right)$$

10.2.3.2. From covariance among relatives

10.2.3.2.1.1. Parent-offspring relationship

Usually dam-offspring data are used because of limited number of sires. Relationship between dam and daughter is ½ i.e. $CoV(A_XA_Y)$ = ½. With traits X and Y between dams (P) and daughter (O), the possible covariances are $CoV(P_XO_X)$, $CoV(A_XA_Y)$, $CoV(P_YO_X)$ and $CoV(P_YO_Y)$. Ignoring dominance, epistasis and maternal influence, and assuming $CoV(P_XO_Y) = CoV(P_YO_X) = CoV(A_XA_Y) \div 2$ and $CoV(P_XO_Y) = V(A_X) \div 2$ and $CoV(P_YO_Y) = V(A_Y) \div 2$

$$r_g^2 = \frac{(CoV(D_XO_Y))(CoV(D_YO_X))}{(CoV(D_XO_X))(CoV(D_YO_Y))} = \frac{\left(\frac{1}{2}CoV(A_XA_Y)\right)\left(\frac{1}{2}CoV(A_XA_Y)\right)}{\left(\frac{1}{2}V(A_X)\right)\left(\frac{1}{2}V(A_Y)\right)}$$

$$\text{Therefore, } r_g^2 = \frac{(CoV(A_XA_Y))^2}{(V(A_X))(V(A_Y))} \quad \text{or} \quad r_g = \frac{CoV(A_XA_Y)}{(SD(A_X))(SD(A_Y))}$$

If in finite populations with r_g nearing zero, if $CoV(D_XO_Y)$ and

$$CoV(D_YO_X) \text{ differ in sign, then } r_g = \frac{\left(\frac{CoV(D_XO_Y)+CoV(D_YO_X)}{2}\right)}{\sqrt{(CoV(D_XO_X))(CoV(D_YO_Y))}}$$

$$\text{This condenses to } r_g = \frac{CoV(A_XA_Y)}{\sqrt{(SD(A_X))(SD(A_Y))}}$$

In the above formulae, regression equations can also be used.

10.2.3.2.1.2. Paternal half-sibs

Data on two traits X and Y are utilized to perform analysis of covariance (Table 74.13) analogous to ANOVA for such data while calculating r_I and h^2. Since half-sibs share ¼ of common additive genes, intra-class correlation obtained by covariance components ($r_{I(XY)}$) is multiplied by 4 to estimate r_g i.e. $r_g = 4\, r_{I(XY)}$.

Let X_{ij} and Y_{ij} represent j^{th} value under i^{th} sire (total sires = s) in respect of traits X and Y, respectively with number of half-sibs per sire being n_i and total number of parental half-sibs being n.; X.. and Y.. be the grand totals.

$\sum_i^s \sum_j^{n_i} X_{ij} Y_{ij}$ represents total covariance (designated A)

$\sum_i^s \frac{(X_{i.})(Y_{i.})}{n_i}$ represents total sum of cross products due to sires and is designated B.

Then, sum of cross products between sires $(SCP_S) = B - \frac{(X_{..})(Y_{..})}{n}$

and sum of cross products within sires $(SCP_W) = A - B$

number of half sib progenies per sire, $k = \frac{1}{s-1}\left(n_. - \sum_i^s \frac{n_i^2}{n_.}\right)$

Table 74.13. Analysis of covariance (ANACOVA) for half-sib relationship

Source	df*	Sum of cross products	Mean sum of cross products**	Estimated mean sum of cross products
Between sires	S-1	SCP_S	MCP_S	CoV_W+kCoV_S
Within sires	n.-S	SCP_W	MCP_W	CoV_W
* df indicates degrees of freedom	** Calculated as (SS ÷ df)			

$CoV_S = (MCP_S - MCP_W) \div k$, $r_{IXY} = CoV_S \div (CoV_S + CoV_W)$ and finally $r_g = 4\ r_{IXY}$.

10.2.3.2.1.3. Full-sibs

Data on full-sibs are frequently available from poultry breeding experiments. In such cases, ANOVA is performed separately for traits X and Y as described under estimation of h^2 by utilizing relationship between full-sibs (Table 74.11) and variance components $V(S)_X$, $V(Dam)_X$, $V(E)_X$ and $V(S)_Y$, $V(Dam)_Y$, $V(E)_Y$ as well as K_1, K_2 and K_3 values are obtained.

Covariance components are calculated (ANACOVA) analogous to variance component analysis performed by calculating h^2 by utilizing relationship between full-sibs. Finally, rg is computed as the ratio of covariance to geometric mean of SD of traits X and Y. The following are assumed:

1. Xijk and Yijk are the kth observation under j[th] dam under i[th] sire (P progenies per dam within sire)
2. Xij. and Yij. are the dam totals (D dams per sire)
3. Xi.. and Yi.. are the sire totals (total sires S)
4. X... and Y... are the grand totals
5. nij is the number of observations under j[th] dam under i[th] sire
6. ni. is the number of observations under i[th] sire and
7. n.. is the total number of observations

The following are calculated:

$$E = \sum_i^s \frac{\sum_j^d n_{ij}^2}{n_{i.}}, \quad F = \sum_i^s \sum_j^d n_{ij}^2, \quad G = \sum_i^s n_{i.}^2$$

$$K_1 = \frac{(n_{..} - E)}{(D - S)}, \quad K_2 = \frac{\left(E - \frac{F^2}{n_{..}}\right)}{(S-1)} \quad \text{and} \quad K_3 = \frac{\left(n_{..} - \frac{G^2}{n_{..}}\right)}{(S-1)}$$

The following are calculated from the data :

$\sum_i^s \frac{(X_{i..})(Y_{i..})}{n_{i.}}$ is designated as A and $\sum_i^s \sum_j^d \frac{(X_{ij.})(Y_{ij.})}{n_{ij}}$ is designated as B

$\sum_{i}^{s}\sum_{j}^{d}\sum_{k}^{p} X_{ijk}Y_{ijk}$, the total uncorrected sum of cross products (C)

Sum of cross products between sires $(SCP_S) = A - \frac{(X_{...})(Y_{...})}{n_{..}}$

Sum of cross products between dams within sire $(SCP_{D/S}) = B - A$

Sum of cross products between progenies within dam within sire

$(SCP_{P/D/S}) = C - B$

From EMCPs, $CoV(E) = MCP_{P/D/S}$, $CoV\ (Dam) = (MCP_{D/S} - MCP_{P/D/S}) \div K_1$

$CoV\ (S) = \{MCP_S - MCP_{P/D/S} - (K_2 \div K_1)(MCP_{D/S} - MCP_{P/D/S})\} \div K_3$

If $K_1 = K_2$, $V(S) = (MCP_S - MCP_{D/S}) \div K_3$

Note: K_1, K_2 and K_3 represent number of progenies per dam, per dam within sire and per sire, respectively (See Table 74.13).

Table 74.14. ANACOVA for full-sib relationship

Source	df*	Sum of cross products	Mean sum of cross products**	Estimated mean sum of cross products***
Between sires	S-1	SCP_S	MCP_S	$CoV_E+k_2CoV_D+k_3CoV_S$
Between dams within sire	S(D-1)	$SCP_{D/S}$	$MCP_{D/S}$	$CoV_E+k_1CoV_D$
Between progenies within dam within sire	SD(P-1)	$SCP_{P/D/S}$	$MCP_{P/D/S}$	CoV_E

* df indicates degrees of freedom ** Calculated as (SCP ÷ df) *** Designated EMCP

Utilizing the variance components of traits X and Y and covariance components between them, various correlation coefficients are calculated as (Table 74.14). On the same lines for h², rg can be computed utilizing any relationship provided the genetic relationship can be mathematically explained and quantitated.

Calculation of genetic, phenotypic and environmental correlations

Covariance components	Variance components Trait X	Trait Y	Correlation coefficients
CoV_S	$V(S)_X$	$V(S)_Y$	$r_{g(s)} = CoV_S / \sqrt{(V(S)_X)(V(S)_Y)}$
CoV_D	$V(D)_X$	$V(D)_Y$	$r_{g(s)} = CoV_D / \sqrt{(V(D)_X)(V(D)_Y)}$
CoV_E	$V(E)_X$	$V(E)_Y$	$r_{g(s)} = CoV_E / \sqrt{(V(E)_X)(V(E)_Y)}$
CoV_P	$V(P)_X$	$V(P)_Y$	$r_{g(s)} = CoV_P / \sqrt{(V(P)_X)(V(P)_Y)}$

10.3. Uses of correlations

1. To predict correlated response.
2. To estimate BV of individuals.
3. To formulate suitable selection indices.
4. For traits which are sex-linked or which can be measured only after slaughter or which in the measured only late in life, BV of individuals can be estimated by phenotypic values for genetically correlated trait.

Chapter 75

Applied Poultry Breeding

This chapter is mainly aimed to give description of practical aspects of poultry breeding developed based on the theory of inheritance of both qualitative and quantitative characters. It is common sense that to improve productivity in animals all animals can not be allowed to reproduce the next generation for the simple reason that, as per Hardy-Weinberg law, one such random mating is sufficient to bring back the gene and genotypic frequencies back to the ones from where the improvements began.

Therefore, only some of the animals will be chosen (selected) to reproduce and they will be mated by a specific system and method depending on the trait(s) and the requirement at the poultry facility. The principles and methods of selection and different systems and methods of mating are discussed below:

1. Selection

Selection is a process in which certain individuals in the population are referred to others for the production of next generation. Therefore, new gene(s) is(are) not introduced in the selection process, but frequency of the more desirable genes is increased.

1.1. Factors affecting efficiency of selection methods

1. The ability of the breeder to find superior breeding individuals.
2. Amount of selection differential (S) applied
3. Heritability of the trait
4. Generation interval: Progress made per unit of time is usually more important than progress for generation. Generation interval is defined as the interval of time between corresponding stages of life-cycle in successive generations; and is most commonly reckoned as the average age of parents when the offsprings are born that are destined to become parents in the next generation.
5. Genetic correlation among the traits

The smallest unit of selection is the individual and it can be evaluated for suitability for breeding purpose depending on:

1. Individual's own phenotype.

2. Average of all individuals in a family.
3. Individual's pedigree; that is evaluation of phenotype of the ancestors of an individual. The information from these criteria used as accurate estimate of breeding values of animals available for selection. The different selection methods deal with different exploitation of the phenotypic information (at a desired level) which is used for selection.

1.2. Methods of selection

Selection can be effected among individuals within a population (intra – population selection) or individuals from different populations (inter-population selection).

1.2.1. Intrapopulation selection methods

1.2.1.1. For single trait (See Table 75.1)

Table 75.1. Intrapopulation selection methods for single trait

Selection method	Criterion	Optimum h^2	Examples of traits
Individual/mass selection	Individual's own phenotype	≥ 0.5	Body weight, Egg weight, Body confirmation
Family selection	Family average. All members of a family are either selected or discarded depending on the average of the phenotypic values of all the individuals in the family in comparison to population mean	≤ 0.2	Egg production, Fertility, Hatchability, Viability
Sire-family (half-sib) Selection	The complete sire-family is either selected or discarded depending on the mean of the sire family. Particularly suitable when reproductive rate is low	0 - 0.1	Fertility, Viability
Dam-family selection	The complete dam-family is either selected or discarded depending on the mean of the sire family. Particularly suitable when reproductive rate is high	0.1 – 0.2	Egg Production, Hatchability
Sib selection	Individuals are selected depending on the performance of their brothers and sisters (Sex-limited traits)	-----	Selection of cockerels for egg production
Combined selection	To improve a single trait, information from various sources like the individual itself, its sire, dam etc. Suitable for low to medium heritable traits like egg production, viability, hatchability etc. A most popular combined selection method is Osborne index which is calculated as follows: $I=b_1(P-P^*)+b_2(FD-P^*)+b_3(FS-P^*)$ where I is the Osborne index, P is the individual's record, P* is the flock (Population) mean, FD and FS are the dam and sire family averages, respectively and b_1, b_2 and b_3 are relative weights attached to the three components, respectively. Osborne index (I) is calculated for each hen, arranged in a descending order and the top ones selected. Used for selecting birds for improving egg production		

1.2.1.2. For multiple traits (See Table 75.2)

Table 75.2. Intrapopulation selection methods for multiple traits

Method	Criterion and remarks
Independent culling level (Tandem Selection)	Level is set for each trait at which the birds are either selected or culled. The performance in one trait is considered independent of the performance in other traits. The method is easy to practice and birds can be selected at different ages. But, if the individual is inferior in any one of the traits, it is rejected regardless of its records in other traits.
Index selection (Total Score method)	Relative weightages are given to different traits and an index is constructed for each individual. Subsequently, the individuals are arranged in a descending order and the top ones are selected. If b_1, b_2, b_3 ....b_n are the relative weightages of the traits x_1, x_2, x_3.... X_n, respectively, then the index (I) is calculated as follows (See text for details): $I = b_1x_1 + b_2x_2 + b_3x_3 + ... + b_nx_n$

1.2.1.3. Selection index

(Ronningen and Vleck, 1985)

Selection index can be constructed for the following:

1. For a single trait by using information on the individual and some of relatives.
2. For more than two traits by using information from the individual.
3. For more than two traits by using information on the individual and some of relatives.
4. For selection of line crosses using information on the individual and some of relatives and also from data on specific crosses.

Several animals each with information X_1, X_2, X_3.....X_N where X_i's are records on various traits or on relatives. This information is used to predict True value (T) which consists mostly of additive genetic merit. The index, $I = \Sigma b_i X_i$ (i=1,2,4.....N) where bi are weighting factors chosen to optimize certain properties of the index, I and X_i's are expressed as deviation of population averages i.e. $X_i = (X'_i - \mu_i)$; μ_i represents Population mean. If Population mean is not available, an artificial value usually, least-squares mean, is used.

Properties desirable of Selection index, I which is a predictor of T are:

1. r_{TI} maximized.
2. $E(I - T)^2$ is minimal; expected or average squared difference.
3. The probability of selecting one of the largest sample values of T is maximal by selecting the largest value of the index criteria; in other words, r_{TI} is high

and approaching a value of 1.0.

4. The property of selecting the higher merit of any two individuals is maximal.
5. Genetic progressing any one-round of selection is maximal

Of the above properties, those at 3, 4 and 5 require multivariate normal distribution. The selection index theory is briefly outlined below by considering two traits.

The two traits are X_1 and X_2 and the factors involved in selection index I are shown in fig. 75.1. Ki's are partial regression coefficients of T (not I) on X_i ($b_i = \sigma x_i \div \sigma_T$); G_i's are measures of genetic source of variation in X_i's; E_i's, G_i's and X_i's are used to represent deviation from their respective means; e_i's and g_i's are standardized influence/path coefficients/standard partial regression coefficients; r_{e1e2} is the environmental and r_{g1g2}, the genetic correlation. d_i's are standardized partial regression coefficients and are usually function of relative economic values (V_i's) so that $d_i = V_i \sigma_{gi} \div \sigma_T = V_i g_i \sigma_{xi} \div \sigma_T$.

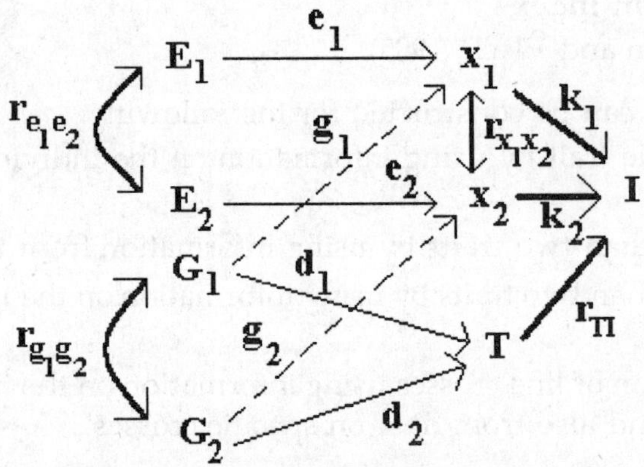

Fig. 75.1. Path diagram for selection index

V_i's are weighting factors for G_i's to establish the theoretical aggregate genetic value (T); that is $T = V_1G_1 + V_2G_2$.

b_i's can be obtained by solving the following simultaneous equations:

$k_1 + k_2 r_{x_1x_2} = r_{x_1T}$ (Eqn. 1)

$k_1 r_{x_1x_2} + k_2 = r_{x_2T}$ (Eqn. 2)

In terms of path coefficients,

$r_{x_1T} = g_1 (d_1 + r_{g_1g_2} d_2) = g_1(V_1 g_1 \sigma_{x_1} + V_2 g_2 \sigma_{x_2}) \div \sigma_T$

By substituting $k_i = b_i\sigma_{Xi} \div \sigma_T$ and multiplying both sides by σ_{X1} and σ_T in Eqn. 1 and σ_{X2} and σ_T in Eqn. 2, the simultaneous equations can also be written as

$$b_1\sigma^2X_1 + b_2\sigma_{X_1X_2} = \sigma_{X1T}$$

$$b_1\sigma_{X_1X_2} + b_2\sigma^2X_2 = \sigma_{X_2T}$$

By solving the above simultaneous equations, b_i's for the index I are obtained. The simultaneous equations can also be extended to any number of variables as follows:

$$b_1\sigma^2_{X_1} + b_2\sigma_{X_1X_2} + b_3\sigma_{X_1X_3} + b_4\sigma_{X_1X_4} \ldots\ldots + b_n\sigma_{X_1X_n} = C\sigma_{X_1T}$$

$$b_1\sigma_{X_2X_1} + b_2\sigma^2_{X_1} + b_3\sigma_{X_2X_3} + b_4\sigma_{X_2X_4} \ldots\ldots + b_n\sigma_{X_2X_n} = C\sigma_{X_2T}$$

....

....

$$b_n\sigma_{X_nX_1} + b_2\sigma_{X_nX_n} + b_3\sigma_{X_nX_3} + b_4\sigma_{X_nX_4} \ldots\ldots + b_n\sigma^2_{X_n} = C\sigma_{X_nT}$$

where $C = \sigma^2_I/\sigma_{TI}$ and $\sigma_{X_iX_j}$ are covariances between X_i and X_j. The above equations are solved by matrix inversion.

The r_{TI} can also be calculated as $r_{TI} = \sqrt{\dfrac{\sigma^2_T}{\sigma^2_T\sigma^2_I}}$; since $\sigma^2_I = \sigma_T$

$$r_{TI} = \sqrt{\frac{\sigma_{TI}}{\sigma^2_T}} = \sqrt{\frac{\sum_i b_i\sigma_{X_iT}}{\sigma^2_T}}$$

1.2.1.3.1.1. Restricted selection index

Restrictions can be included similar to linear programming for feed formulation; i.e. optimizing selection for a given genotype, subject to the conditions that no genetic change will be expected to occur in one or more of the traits. Best linear unbiased prediction (BLUP) procedure is normally used.

An exhaustive description of selection methods is beyond the scope of this book. However, the details can be obtained from several publications exclusively on selection methods, in general, and selection index, in particular. Selection index has become a very popular method especially for improving characters known for low heritability.

1.2.2. Interpopulation selection methods

1.2.2.1. Reciprocal recurrent Selection (RRS)

This method is practiced in two segregating populations, and each of them will be selected with respect to the other as follows (See Table 75.3):

Table 75.3. Reciprocal recurrent selection

Step	Breeding	Remarks
Step I	Cross-breeding	A males X B females and B males X A females; depending on the crossbred performance, superior males and females of A and B are selected; crossbreds sold as commercial birds.
Step II	Pure-breeding	A males X A females and B males X B females; A generation of improved populations of A and B are obtained.
Step III	Same as Step I	
Step IV	Same as Step II	
This process is continued till no more improvement in crossbred progeny is recorded. The chicks obtained by crossing strains or breeds or populations for testing will be utilized for commercial purposes.		

1.2.2.2. Recurrent selection (RS)

This method is similar to RRS excepting that one of the populations is an inbred tester line (reference line), with respect to which, the other population, which is open-bred or synthetic will be developed (improved). As the case of RRS, the crossbred chicks will be placed for commercial purposes.

2. Systems of mating

2.1. Random mating

Mating of individuals without any selection. This is used in developing a control population which is required to compare and measure the effects of other breeding systems. Control population also helps to estimate the effects of environment which, in turn, helps to estimate the true genetic gain through any breeding method.

2.2. Breeding for increased Homozygosity :

2.2.1. Inbreeding

Inbreeding is defined as mating between animals which are more closely related to each other than the average relationship between all individuals in a population. Inbreeding can be consistently carried out for several generations; the three distinct methods:

1. Close inbreeding – Mating between sibs or parents and progeny (incest). Full-Sib mating and back-crossing of the progeny to the younger of the parents are often practiced.
2. Strain formation – Development a small group of animals within a breed and variety with a special character in view. This is a mild form of inbreeding. Example : Babcock strain of SCWL developed to lay heavier eggs.

3. Line-breeding – This is inbreeding within an ancestral line and is the most intensive form of back-crossing. Line-breeding is back-crossing to the same parent for several generations in succession.

Mating based on phenotypic resemblance – It is mating of "like the like" without any weightage to the degree of relationship. In this case, not identical genes, but genes with similar effect(s) are brought together. This can be practiced when heritability of the trait(s) in question is(are) high.

2.3. Breeding for increased heterozygosity

2.3.1. Out-breeding

This is opposite of inbreeding in the sense that the relationship of the individuals which are mated is less close than the average relationship within the population. Mating between strains or inbred lines are the forms of out-breeding. The methods of out-breeding (cross-breeding) are

2.3.1.1. Single 2-way cross

Two different populations (inbred lines, strains are breeds) are crossed to produce a first filial (F_1) generation which is purely for commercial purpose but not for breeding. F_1 here usually exhibits hybrid vigor especially when inbred lines are involved. When two inbred lines of the same breed are crossed, the progeny is said to be in-crossbred.

$$A \times B \rightarrow AB$$

2.3.1.2. 3-way cross

In this, the F_1 crossbred females are mated to males of a third line, strain or breed so as to utilize hybrid vigor of the dams.

$$A \times B \rightarrow AB \times C \rightarrow ABC$$

2.3.1.3. 4-way cross or Double cross

Two different single crosses (AB, CD etc.) are crossed to obtain ABCD. This is usually practiced in poultry breeding for crosses between inbred lines of low viability since only relatively small number of animals of the lines A, B, C and D need be maintained. $A \times B \rightarrow AB$; $C \times D \rightarrow CD$; $AB \times CD \rightarrow ABCD$.

2.3.1.4. Back-cross

Crossing of F1 females to the males of one parental population. This is especially useful when F1 females are better mothers than mothers of the parental population (maternal influence). This commonly practiced in pig breeding.

2.3.1.5. Rotational cross-breeding

Cross-bred females mated with males from either of the populations; the males are alternated in each generation. This is usually practiced in pig breeding.

2.3.1.6. Top crossing

Mating of inbred males to females of non-inbred populations.

2.3.1.7. Crossing RRS

This is similar to RRS.

2.3.1.8. Grading-up

This is concerned only with increasing heterozygosity. Grading-up involves back-crossing of the same superior breed for several generations so that a population of mixed breed can be changed to "pure breed". This is of great value in grading-up local non-descript cows wait Jersey, Holstein- Friezian and other breeds.

2.3.1.9. Crossing for production of a new breed

Similar to grading-up, this also aims only at increasing heterozygosity. Different breed types have been crossed to produce the model-day breeds of farm animals so as to combine desirable traits from many sources. These foundations crosses have to be subjected to inbreeding combined with selection to consolidate them into true breeding populations (breeds). Example - Cornishdeveloped from Aseel, Malay and English Game breeds.

3. Methods of mating

3.1. Pen mating

A single male is segregated with the group of females in a pen during the breeding season. This requires marginally more labor. Number of females that can be allowed with a male (mating ratio) is 10-12 in case of Leghorns (egg-type birds) and 6-8 in case of meat-type birds. Pedigreeing is possible both on sire's as well as on dam's side. However, if Pedigreeing is done only on dam's side, multiple male matings can be employed in a larger pen.

3.2. Stud mating

The males are usually confined at all times too small individual pens (studs) within the large laying pen. The female is held in the stud for a known period of time (till mated) after which it is removed and another is added. Two matings per week or at least once every 5 days is desirable for optimum fertility. The method requires more labor and this method is seldom practiced nowadays.

3.3. Artificial insemination (AI)

3.3.1. Advantages

1. Allows unlimited number of single male matings without requiring extensive breeding equipment.
2. Preferential matings avoided.
3. Accurate pedigreeing possible.

4. If, due to some reason, a male of superior qualities cannot mate, it can still contribute to the next generation.
5. If the males are too heavy or two old precluding natural mating (as in the case of Broad-Breasted White turkeys), AI is the method of choice.
6. Hybridization between two different species possible; for instance, chicken-quail hybrids.
7. In caged layers also, fertile eggs can be obtained only by AI.
8. Problems of trap-nesting are avoided and
9. Incidences of sexually-transmitted disease(s) is/are minimized/avoided.

3.3.2. Collection of semen

Semen is obtained by massaging the male's back by moving the hand in the direction of bird's vent and supporting the bird by keeping its legs between the operator's legs or held by the assistant. The fingers of the hand are outspread at the start of the stroke so as to converge on the vent. It trained male would evert the vent by the strokings and the operator must quickly move his hands one above and other below so as to squeeze the protruded vent by the thumb and index finger inward and downward at a point just above the vent to force the yield of semen. Sustained pressure at this point will cause some males to yield more semen. Unrestrained and relaxed males are more semen than restrained and excited ones. The semen may be collected in a funnel (which is plugged in the bottom by paraffin wax) on small sugar tubes or any suitable container.

3.3.3. Characteristics of chicken semen (See Table 75.4)

3.3.4. Semen dilution

On most occasions, Semen dilution is not practiced; however, whenever dilution has to be made, the guidelines are : Temperature : 1-2°C, Dilution ratio : 1:2-1:4; Insemination dose - 0.03-0.05 ml or 20-25 million spermatozoa.

Table 75.4. Characteristics of poultry semen

	Heavy breed	Light breed
Volume, ml	0.75-1.0	0.4-0.6
Density	6.1×10^9 / mm^3	6.8×10^9 /mm^3
pH	7.0-7.2	7.0-7.2
Age at which semen can be collected	22-24 weeks	20-22 weeks
Appearance	Pearly white	Pearly White

3.3.5. Frequency of AI

Once in 5-7 days. However, whenever semen is diluted, it is seldom done for long-term usage but is done only to increase the volume and number of females that can be inseminated.

3.3.6. Transfer of semen into the oviduct

A hypodermic or tuberculin syringe (needle removed) or a medicine dropper can be used to introduce semen into the vagina. Abdomen of the female is pressed a degree the vent region is forced to evert exposing the opening from the cloaca into the vagina; the constricted opening of the latter will be seen at the left side of the vent (hen's left). The inseminating equipment is introduced about 2.5 cm into the vaginal opening before actually depositing the semen. The dose of insemination sufficient to effect of good fertility is 0.02 ml, although, in practice, 0.05 ml is deposited.

3.3.7. Time and frequency of insemination

Insemination of chicken must be done during the afternoon hours by which time most of the birds are expected to have laid eggs rendering their oviduct empty. Frequency of insemination preferable is twice a week or, at least, once every five days.

3.3.8. Disadvantages

1. Requires more labor
2. Chances of cross contamination of birds through the inseminating equipment especially of paratyphoid infections.
3. Handling of birds

3.4. Shift system of mating

It is desirable to obtain about 100-125 chicks to evaluate a male. This requires a total time which is inversely proportional to the mating ratio. If a male has 6, 12 and 16 dams, the approximate time required to obtain the desired number of progeny is 33, 17 and 13d, respectively. To reduce this time, the males can be shifted from one pen to another so that it will have more mates through which it can be evaluated. Similarly, the dams also have more mates making the comparison between dams more critical and accurate.

The major drawback of the system is the overlapping of the paternity when the males are changed since the viability of spermatozoa ranges from 7-21 d. It has been found that this problem of overlapping of pedigree can be avoided greatly by using AI because, the sperms held in the oviduct cannot compete with those in the fresher semen. Therefore, the eggs laid on the second day after AI is attributable to the new sire. Similar results of possible with natural mating; but the chances of preferential mating or time taken before the new male mates with the female concerned restricts the accuracy under natural mating. The other problems include the difference in vigor, virility and fertility of the successive cocks, difficulty in record keeping and more labor involvement etc.

The shift normally followed is as follows

1. 1st day – First shift males in breeding pens.
2. 2nd day – Start of collection of fertile eggs.
3. 15th day – Removal of first shift males.
4. 20th day – AI using the semen of second shift males and the males are allowed into the pen.
5. 22nd day – Eggs laid from 8th to 21st d (both days inclusive) are pooled and designated to have been fertilized by first shift males. Start of collection of fertile eggs on 22nd d and onwards.
6. 29th day – Second shift males removed.
7. 35th day – AI using the semen from third shift males and the males allowed into the pen. Eggs collected from 22nd to 34th d are pooled and designated to have been fertilized by second shift males; and so on...

However, pen mating or AI is the commonly practiced methods of mating in most of the poultry breeding farms.

4. Breeding for broiler production

All broilers are crossbreds and usually strains of Cornish breed are used as the male line and New Hampshire and/or Plymouth Rock (White variety usually) as the female line in a 2-way, 3-way or a multi-way cross. Cornish selected as the male line because of superior growth rate, feed efficiency and livability. However, due to the higher body weight, they consume more feed and require more space. In addition to these, they are poor egg producers (since body weight and egg production are inversely related), and hence the cost of production of hatching eggs (in turn, the day-old chicks) will be high. To reduce the cost of production of hatching eggs, the female line must be good in all the characters listed for the male line and also must be a moderately good egg producer. Therefore, the females are actually from New Hampshire and/or Plymouth Rock breeds.

The component traits of broiler production are:

1. Body weight (Rate of growth)
2. Feed efficiency
3. Viability
4. Dressing percentage
5. Body conformation

All the above traits are moderate to highly heritable (except viability) and hence, the males are selected by individual selection. The females have to be selected for reproductive traits also (egg number, fertility, hatchability etc.) and therefore, they are selected on an index or by other methods to improve egg production.

5. Breeding for egg production

Usually breeds belonging to Mediterranean Class, especially Leghorns are involved.

The component traits of egg production are:

1. Rate of lay
2. Age at sexual maturity
3. Clutch size
4. Pause duration and
5. Broodiness.

Other factors like egg number, feed efficiency and laying house survivability also decide the productive efficiency of layers.

Majority of the above traits are low to moderately heritable and therefore, egg production can be improved by combined selection (Osborne index) and sire-family selection. Inbreeding and hybridization also produces more response for this trait.

Inbred lines are produced by mating closely related individuals. Those birds with survive at an inbreeding coefficient (F) of at least 0.5 in the population are said to be inbred lines. Inbreeding brings about homozygosity of both favorable and unfavorable genes, and those with unfavorable genes cannot survive when F approaches or exceeds 0.5. Therefore, cross-breeding of two such inbred lines with favorable genes may bring about significant improvement in egg production.

The inter-population selection methods (RRS and RS) are also efficient in developing both broilers as well as laying-type birds.

6. Breeding for resistance to heat stress

(Gowe and Fairfull, 1995)

6.1. Heat-resistance

It is well-documented that temperature is one of the most important factors affecting poultry performance (See Chapters Thermoregulation and Principles of ventilation). It is also well-known that there is a very wide variety of climate depending on the geographical location of the particular place (See Chapter Climate and climatic indices). Therefore, it is natural to predict that Breeds and Varieties evolved and/or developed at the particular location may have higher adaptability than those with the exotic to the location. For instance, White leghorn has been shown to have a greater tolerance to high temperatures than heavier Breeds such as Rhode Island Reds, Barred Plymouth Rocks etc. The underlying reasons for the tolerance can be many; they include, a possible genetic basis which can be exploited. Heritability of survival of White Leghorns under heat stress has been reported to be high, varying from 0.30 to 0.45.

Some major genes have also been found to be associated with resistance to heat stress; one of them being the incompletely dominant autosomal gene for naked neck, designated Na. This gene offers resistance to heat stress by directly reducing feather cover facilitating heat loss by the bird. Another gene of interest is sex-linked recessive gene, dw, which not only reduces body size and also reduces metabolic heat output; the former effect facilitates heat transfer due to increased surface area per unit body weight (metabolic body size). Notwithstanding these two genes, there are other genes affecting heat tolerance in chicken.

6.1.1. Autosomal genes

6.1.1.1. Naked neck gene (Na)

This autosomal gene is incompletely dominant and therefore, in case of homozygotes it reduces feather covering by about 40% and in case of heterozygotes, the reduction in feather cover will be about 30%. The homozygotes are slightly superior to heterozygotes for body weight and FCR. However, both the groups are superior to normally feathered birds. It has been recorded that Na gene produces more positive effects in heavy- and/or medium-weight birds than in light-weight birds.

6.1.1.2. Frizzle gene (F)

This also is an incompletely dominant autosomal gene which causes the contour feathers to curve away from the body; the severity is extreme in case of homozygotes (no feather will have flat vane). Due to this effect, the gene reduces the insulating properties of the feather cover and also reduces feather weight facilitating radiation of heat from the body.

6.1.2. Sex-linked genes

6.1.2.1. Dwarf gene

See Chapter Inheritance of qualitative traits. In addition to the benefits in production characteristics, the reduced body weight (size) leads to increase surface area per unit body weight and reduced basal metabolic rate reduces heat produced; both facilitating resistance to thermal stress.

6.1.2.2. Slow feathering gene (K)

This is primarily used as an autosexing gene. In addition, K gene has been thought to produce indirect effects such as reduced protein requirement, reduced fat deposition during juvenile life and increase heat loss during early growth; all these factors promote resistance to heat stress.

6.1.3. Other genes

The recessives genes for silky (h), which affects barbules on the feathers may help dissipate heat. The dominant gene for pea comb (P) also reduces width of feather tracts and comb size while changing skin structure. All these may help to dissipate heat. The recessive sex-linked multiple allelic loci for dermal melanin (id) may also improve radiation from skin.

6.1.4. Interactions among major genes

It has been reported that the three genes Na, F and dw interact such a way that the combined effects of one or two genes is lower than the sum of their individual and additive effects still higher than the individual gene effects. The number of feathers on six-week-old broilers was in the descending order of the genotypes 'ff nana', 'ff Nana' and 'Ff Nana'. Similarly, association between K and Na genes has been recorded. However, further research is essential before any conclusions be drawn.

7. Development of heat-resistant strains

(Gowe and Fairfull, 1995)

Most of the research work on heat resistance has been on selection of birds for heat tolerance after subjecting them to heat stress; that means use of polygenes. The results have been, no doubt, encouraging. Even selecting birds for heat-shock proteins has been attempted. Selection for either feed efficiency and/or leanness has also benefited in terms of heat tolerance.

Notwithstanding the above, introduction of major genes such as Na and F into high producing lines (both meat and egg type) may have to be explored in detail. It may be possible that major genes in either of the parents may result in heat tolerance with little or no loss in other performance traits. Back-crossing for 6 to 8 generations into high-performing sire lines is the preferred method.

Of all the genes listed above, Na gene appears to be the most promising followed by F and dw genes. The Na gene can be incorporated by back-crossing into high-performing meat or egg stocks. F gene and/or dw gene can be used in combination with Na gene to develop stocks specifically for hot-humid tropics. The economic advantages expected by introducing dw gene is likely to be much more useful than F gene along with Na gene; this is particularly true because improvement in egg size produced by Na gene can counter the opposite effect of dw gene. In any case, further research is warranted both in the laboratory as well as under commercial conditions before drawing definite conclusion(s).

Indigenous village poultry native to hot climate may be mated to males from improved strains are likely to have not only higher productivity also a view major genes for heat tolerance. Such an improvement in performance at backyard was a long way to help the village poor. This is analogous to inseminating local cattle with exotic semen.

8. Breeding for disease resistance

Susceptibility to MD reduced drastically within 4 generations of selection in Cornell control-bred strain of chickens (Table 75.5). Further investigations revealed that the effect was associated with frequency of two alleles at B locus, the major histocompatibility locus in chicken. The resistant line had a high frequency of B^{21}allele and B^{19} allele appeared to confer susceptibility. Further comparisons between different B genotypes within families and mating results confirmed that indeed the B21 locus was responsible for disease resistance (Table 75.6).

However, commercial exploitation has not been done due mainly to difficulty in identifying blood group (a multiple allelic trait) in chicken which, in turn, has probably made it economically unsuitable; in addition, control of MD is quite effective by a cheaper vaccination.

Table 75.5. Results of selection for MD resistance

Incidence of MD			Frequency of B blood group alleles	
Before selection	Line	After selection	B^{21}	B^{19}
51%	Resistant	7%	1.00	0.00
	Susceptible	94%	0.00	0.97
				Source : Nicholas, 1987

Table 75.6. Results of matings among unselected chicken

Genotype	Number of chicken				Genotype with respect to B^{21}
	Exposed	Dead	% Dead*	% Dead**	
B^{21} B^{21}	65	0	0	0	B^{21} B^{21}
B^{21}-	76	2	3	4	B^{21} -
B^{21} B^{19}	60	4	7		
B^{19} -	50	19	38		
B^{19} B^{19}	13	6	46	56	- -
- -	93	63	68		
As per * genotype ** phenotype					*Source* : Nicholas, 1987

Chapter **76**

Biotechnology

At the outset it is in the fitness of things to emphasize that this publication is mainly intended to give basic and preliminary ideas about biotechnology, genetic engineering and recombinant DNA technology with special reference to domestic animal production, in general and poultry production, in particular. Therefore, it is neither intended nor possible to provide all the information pertaining to vast field of "Biotechnology". Hence, the reader requiring specific details had better go through dedicated publication(s).

In India, over the past 6 decades, there have been concerted efforts to make genetic improvements in both laying and broiler chicken through quantitative genetics approach. The results have been extremely beneficial to all concerned in the poultry industry. However, it now appears that genetic gain per generation and/or unit of time is continuously reducing, for obvious reasons of selection plateau. Therefore, reducing production cost has become the only parameter left for a breeder. Hence, it is imperative that molecular approach for improvement of economic traits has to be taken up. This requires detailed account of DNA (genes) and technology to further manipulate the genes for benefitof the industry.

1. Genetic engineering

(Fechheimer, 1985)

Genetic engineering, as applied to animal improvement schemes, refers to application of new methods of biology of reproduction and molecular biology to the task of enhancing the productive efficiency of livestock. Efficiency of genetic engineering methods depends on:

1. Extent to which the performance test and information on relatives are correlated with genotypic merit.
2. Amount of genetic variability in the population.
3. Proportion of population used for breeding
4. Generation interval

1.1. Gamete phenotype, selection and manipulation

Some RNAs are produced by nuclear genes in spermatids; therefore, future research may be required in the following fields:

1. Separation of Z and W (X and Y) chromosomes; but this is not easy because F body (fluorescing segment upon quinacrine staining of Y chromosome) is present only in man and gorilla.
2. Haplophase gene activity; may not be important for economic traits
3. Genotype analysis of sperms; this may be particularly important in domestic fowls because sperms undergo mitosis after penetrating oocyte in about 6% of embryos in some broiler flocks. In addition, many sperms enter and remain inside although only one of them fertilizes the ovum. Therefore, proper selection to increase such frequency will help facilitate study of genotype of sperms and also a possible culturing of haploid cells directly from embryo.
4. Super-ovulation and *in vitro* fertilization, although possible, are not of relevance in poultry production.

1.2. Egg and zygote manipulations

This approach is primarily for mammalian eggs and includes culture, storage and transplantation of eggs and zygotes; induction of monozygotic twins; nuclear removal and transplantation; and parthenogenesis.

Parthenogenesis is recorded in turkeys which can be increased by selection and vaccination of hens. Avian parthenogenomes develop from secondary oocytes and therefore are homozygous at loci that undergo pre-reduction (separation at I meiotic division) but remain heterozygous for loci that undergo post-reduction (separation at II meiotic division). But, gonosomic complement WW is lethal and hence all embryos are ZZ, males.

1.3. Chromosomal alterations

1.3.1. Tetraploidy

By suppressing cytokines (by cytochalasin B) of first cleavage division of the fertilized egg, tetraploidy can be induced. But, in higher animals, the embryo does not survive. However, if successful in future, it can bring in doubling of variability (and hence heterosis) and increase in size of cells or organs.

1.3.2. Duplication or trisomy

Duplication refers to addition of a chromosome segment whereas addition of a whole chromosome results in trisomy. Similar to tetraploidy, in animals duplication or trisomy is also lethal. But, if the animal can survive several practical implications are possible; for instance, trisomy increases B^{21} alleles which can confer higher resistance to MD (See Chapter "Population genetics"). If birds can be produced with three alleles, each of which can give greater resistance to different diseases, the prospects will be extremely useful.

1.4. Gene exchange

Some genes can be isolated, synthesized, attached to vectors and introduced into host cells derived from the same or different species. Host cells are then introduced into embryos.

1.4.1. Somatic cell fusion

Intraspecific/interspecific cell fusion can be induced with either inactivated Sendai virus or polyethylene glycol. The hybrid cells proliferate in a selective medium with careful monitoring. Cell clones containing complete genome of one species and single chromosome of other species are possible. This can be used to assign genes to a particular chromosome which later on can be useful in identifying and mapping of genes.

1.4.2. Chromosome-mediated transfer

The cell division of host cells in a culture medium is arrested at metaphase. Into the culture medium, extracteddonor chromosomes , separated by size and density gradient are added. Host cells take-up isolated chromosome fragments.

By this method, short segment of chromosomes with desired genes may be transferred from one breed are species to another; it avoids time, expenditure etc. required by regular breeding methods. In addition, this method also helps study gene linkage.

1.4.3. Gene transfer

To be successful in transferring genes depends on:

1. The capacity to identify the location of the particular gene required.
2. Removal of the gene and its controlling elements from the DNA molecule.
3. Insertion of the gene into a carrier DNA (vector)
4. Transfer of vector into the host cell and integrating it into host genome (See Restriction enzymes and Recombinant DNA outlined below)
5. Non-viral gene transfer (Table 76.1) – the most popular methods are calcium phosphate coprecipitation, DEAE dextran transfection and transfer of nucleic acids by physically opening the cells (microinjection, laser poring and electroporation)

2. Transgenic animals

(Puhler, 1993)

2.1. Transgenes and hemizygous animals

The term "Transgenic animal" refers to an animal whose genome has been altered to contain transferred genes (foreign DNA). Methods of gene transfer have been listed above. Since the nucleotides and codons are universal, ay DNA fragment (chemically synthesized or cloned or fragments of chromosomes, preferable linear molecule) can be micro-injected into the pronuclei of zygotes and/or embryos, will be taken up and integrated into the host genome with more or less the same frequency as that of the host genome itself. Mechanism of integration of foreign DNA with the host genome is still incompletely understood.

Table 76.1. Non-viral gene transfer methods

Methods	Remarks
Calcium phosphate coprecipitation	Large amounts of DNA can be delivered; most frequently used for adherent cells
DEAE-dextran	Large amounts of DNA can be delivered; most frequently used for suspension cells
Lipofection	Efficient method; RNA and proteins are also delivered; lipids have limited toxicity
Electroporation	Low amount of DNA or single copy integrates are transferred; used for lymphoid cells
Protoplast fusion	Frequently used for lymphoid cells; use of total protoplast may be a disadvantage
Fusion with erythrocyte ghosts	Nucleic acids and proteins can be delivered
Laser poring	Highly efficient; but expensive
Microinjection	Most efficient (100%); nucleus and/or cytoplasm can be targeted; but expensive
	Source : Puhler, 1993

Those animals which have stably integrated the transferred gene(s), transmit the same to their offspring; the inheritance being Mendelian (See Chapter "Inheritance of qualitative characters"). However, integration of foreign DNA occurs at a single site without any counterpart (allele) on the homologous chromosome and therefore, neither the term homozygous nor heterozygous is applicable; hence, the term "hemizygous" is used. Probability of offsprings inheriting a hemizygous condition is ½. In any case, the inheritance of transgenes over generations is not stable because of possible differences those may arise at integration of the transgene(s) during gametogenesis over all generations.

2.2. Inheritance of transgenes

Probability of offsprings inheriting at least one transgenic site increases with number of integration sites in the parent(s). For instance, from an animal with two integration sites, 75% of offsprings will inherit at least one transgene (25% inheriting one of the two sites and another 25% inheriting both the sites).

2.3. Transgenic chicken

Mainly adopted for improving disease resistance, FCR and growth. Gene transfer is primarily through vectors (retrovirus). The other methods are severely restricted because embryo will be at gastrula stage when the egg is laid. The retrovirus injection is given and the shell is sealed at the point of injection and incubated. However, safety and possible adverse effects on the embryo itself should be borne in mind. Use of replication-deficient retroviruses (suicide vectors) has reduced the probability of unwanted recombinations.

Chicken containing alv6 insert were resistant to infection by pathogenic subgroup of A of the avian leucosis complex. Replication-competentand replication-deficient retroviruses have been utilized in producing transgenic chicken (layers and broilers).

Irradiated sperms with marker genes have been used through AI. Similarly, transfected sperm cells are used as vectors. About 30 to 60% of the embryos showed transferred genes.

It has also been shown possible that DNA can be introduced into blastodermal cells; ad the latter from one embryo to another where they produce tissues (including germ cells).

2.4. Limitations of transgenic poultry

1. Progress may be slower due to higher generation interval; but, this is true for all methods, including conventional selection and breeding
2. Ethical aspects involved in hybridizing DNA by a foreign source as well as consumer aversion towards such transgenic animal production.

But, there are no reasons why these shortcomings can not be overcome and the technology becomes more useful to all concerned.

3. Recombiant DNA

3.1. Restriction enzymes (RE)

Certain bacteria produce enzymes which can degrade foreign DNA (which enters bacterial cell) so that the foreign DNA is 'restricted' from its activity; such enzymes are called "restriction enzymes". Now, more than200 REs are known. The activity of RE is through binding to specific sequences of nucleotides called "Recognition sequences" and each RE cuts the DNA at a specific "Cleavage site"; the latter, on most occasions is present within the recognition sequence. It is interesting to note that, in most cases, recognition sequences are palindromes; that is, the sequence of bases in both strands of DNA considered together is the same when read from right to left or *vice versa* (Table 76.2).

Table 76.2. Common restriction enzymes

Source	Enzyme	Recognition sequence
Arthrobacter luteus	*AluI*	3′ –T-C*-G-A- 5′
		5′ –A-G*-C-T- 3′
Bacillus amyloliquefaciens H	*BamHI*	3′ -C-C-T-A-G*-G- 5′
		5′ -G*-G-A-T-C-C- 3′
Escherichia coli	*EcoRI*	3′ -C-T-T-A-A*-G- 5′
		5′ -G*-A-A-T-T-C- 3′
Hemophilus influenzae R_d	*HindIII*	3′ -T-T-C-G-A*-A- 5′
		5′ –A*-A-G-C-T-T 3′
Haemophilus aegyptius	*HaeIII*	3′ -C-C*-G-G- 5′
		5′ -G-G*-C-C- 3′
Providencia stuartii	*PstI*	3′ -G*-A-C-G-T-C- 5′
		5′ -C-T-G-C-A-*G- 3′
Serratia marcescens	*SmaI*	3′ -G-G-G*-C-C-C- 5′
		5′ -C-C-C*-G-G-G- 3′
* Indicates cleavage site		*Source* : Nicholas, 1987

It can be observed from Table 75.1 that following action of *AluI* or *HaeIII* or *SmaI*, the DNA strands cleaved will have no unpaired nucleotides because the cleavage occurs at the same place in both the strands; hence referred to as "blunt ends". On the contrary, cleavage by *BamHI* or *EcoRI* or *HindIII* or *PstI* results in formation of DNA with unpaired nucleotides; hence, referred to as "sticky ends". The sticky ends are preferred because, such strands produced by the same RE on target DNA and the vector will have unpaired nucleotides which are complementary to each other facilitating their ligation (fusion). The DNA undergoing cleavage is said to have been "digested" into number of fragments.

It is not likely that the cleavage sites for any of the REs will be regularly positioned in any DNA. Therefore, it is obvious that following cleavage by an RE, DNA is broken into fragments of different lengths (or molecular weights); these can be separated into separate bands by electrophoresis wherein the fragments migrate at a rate inversely proportional to the logarithm of their molecular weights. In other words, the relative position of the bands indicates molecular weight. Extending the same, it is possible to draw a "restriction map" of the DNA by comparing the results of digestions by different enzymes.

3.2. Complementary DNA

Certain viruses (for instance certain tumor-producing viruses) do not contain DNA but still they propagate. This is possible only if their RNA is transcribed 'backwards' into DNA so that the resultant DNA duplicates and produces more RNA; hence, the term "reverse transcription" and "reverse transcriptase" for the enzyme which catalyzes the process. This enzyme is important in producing DNA from RNA in the laboratory. For example, a large quantity of ovalbumin mRNA can be easily isolated from chicken oviducal cells; then by supplying free nucleotides in the presence of reverse transcriptase enzyme a complementary strand of DNA (copy DNA)called cDNA can be synthesized. This can be followed by addition of DNA polymerase enzyme with additional nucleotides to produce double stranded cDNA. On the other hand isolation and purification of DNA is not only difficult but also the quantity realized will be small in comparison to production of double stranded cDNA.

3.3. Recombinant DNA and DNA cloning

The process of production of unlimited number of copies of a particular segment of DNA is called as "gene cloning" or "DNA cloning". This can be achieved by joining the DNA to be cloned (called 'foreign DNA') to a "vector" which is capable of replication within a particular "host". This joining is called "ligation or splicing" and it requires an enzyme called DNA ligase. The DNA molecule consisting of foreign DNA attached to vector is called "recombinant DNA".

The most commonly used vectors are plasmids, with the circles of double stranded DNA that exist in certain bacterial cells (hosts) independently of the main bacterial chromosome. The efficiency of transferring foreign DNA into a host cell can be facilitated by adding cohesive (cos) sites from ë phage DNA to a plasmid to form a new type of vector called "cosmid". These cosmids can be packaged *in vitro* into phage particles which then naturally attack a bacterial cell and transmit their DNA (now changed into a cosmid) into the bacterial cell. Through this process, larger pieces of foreign DNA can be easily introduced into the host cell. The host cell produces millions of copies of the foreign DNA during its natural process of the multiplication.

Recently, YAC (shuttle) vector (yeast artificial chromosome) and BAC (bacterial artificial chromosome) vectors are also identified. The description of these is beyond the scope of this book.

3.3.1. Production of polypeptide

Introduction of foreign DNA into the host cell does not automatically initiate production of polypeptide chain corresponding to the DNA introduced. For production of polypeptide chain the DNA has to be transcribed and then translated; the former occurring in the nucleus and the latter occurring with the help of tRNA and ribosomes in cytosol.

It is generally agreed that in most of the organisms have got a common control mechanism which is as follows:

1. The promoter region precedes transcription initiation site and it contains specific sequences that a highly 'conserved'; that means, the same sequence occurs at the same site in most, if not all genes. Such highly conserved sequences are called "boxes".

2. The promoter region of prokaryote genes contains the TATAAT box and TTG box which are thought to be sites for recognition and binding of RNA polymerase. In eukaryotes, promoters are likely to be TATA box and possible CCAAT box.
3. Adjacent to the promoter is the "leader" sequence which is transcribed but not translated. It contains a transcription initiation site (the CAT box), the ribosome-binding site and possibly, other regulatory sites.
4. The structural gene, which is split into exons and introns in case of eukaryotes (See Chapter "Inheritance of qualitative traits"), lies between the start and stop codons. In other words, the eukaryotes have split genes whereas prokaryotes do not; hence, prokaryotes do not require elaborate splicing mechanism for converting a primary mRNA into a mature mRNA. Therefore, eukaryotic genes consisting only of exons can be transcribed and translated by prokaryotes.
5. The transcription terminator is located after the stop codon.

Therefore, if mature mRNA is available, cDNA can be produced from it; otherwise, if the amino-acid sequence of the desired polypeptide is known, the nucleotides sequence can be deduced from dictionary of triplet codons and appropriate DNA molecule, at least of short sequences, can be synthesized by utilizing the required nucleotides through microprocessor controlled 'gene machines'.

Generally, only structural genes are used as foreign DNA and it is introduced immediately next to promoter and a leader sequence in the vector DNA so that the host cell cannot identify its presence and proceeds producing a hybrid polypeptide chain (especially because the genetic code is universal) from which the 'foreign polypeptide' can be released by appropriate chemical treatment. However, there are some differences between species for certain codons in case of amino acids having more than one codon. Hence it follows that if the foreign DNA consists of codons most favored by the host, the polypeptide production will be the most efficient.

3.3.2. Gene banks or libraries

The basic process involved in cloning the entire DNA from a species is as follows:
1. The genome is treated with REs to produce many fragments
2. The fragments of the combined with DNAs of suitable vectors
3. The vectors enter and multiply in a host cell, say *E.coli*; each host cell receiving only one vector (one fragment)
4. The host cell plated on nutrient substrate
5. The host cell forms colonies containing millions of copies of the fragment.
6. A set of colonies produced from different fragments should contain the entire DNA present in the organism and hence, a "gene bank" or a "gene library" is obtained.

It is very important to note that the process explained above is rather highly simplified. For mammalian gene banks, *HaeIII* and *AluI* REs are often used its give rise to fragments that the predominantly around 20 kilo base pairs (kbp) that is 20,000 base pairs long. The haploid genome of mammals contains as many as 3×10^9 nucleotide pairs (or 3×10^6 kbp) per cell. Therefore, digestion of mammalian genome yields as many as $3 \times 10^6 \div 20 = 1.5 \times 10^5$ or 150,000 host colonies to have all segments of the DNA in the gene bank; this requires about host 3.0×10^5 to 8.0×10^5 colonies.

3.3.3. Benefits of recombinant DNA technology

1. Production of any polypeptide provided the relevant fragment of DNA is either isolated and/or constructed. For instance, growth hormone, insulin, epidermal growth factor, interferon etc. have been produced in large quantities. Antigenic portion of many disease-causing organisms has also been produced which will be extremely useful for production of vaccines.
2. Detection of viruses and other disease organisms. If radioactively-labeled cloned DNA or RNA of the pathogen is mixed with DNA or RNA extracted from tissue of the individual being tested, DNA probes corresponding to the genes can identify individuals having particular genes. This procedure may help detect individuals carrying defective genes and/or disease-resistance genes, and in long run, help the animal breeders develop disease-resistant strains/varieties and improve productivity of animals.
3. It may be possible, in future, that foreign genes can be introduced into plants and animals to correct genetic differences and/or to enhance performance.

3.3.4. Limitations of recombinant DNA

1. It is possible, to produce harmful strains of microorganisms inadvertently which might lead to uncontrollable diseases/pandemics of both animals and human beings, regardless of its probability.
2. It is also possible that some major genes affected which might result in gross anatomical abnormalities in animals.
3. Lack of knowledge about eukaryotic genome: in case of chicken of the 40 given linkage loci, only two are assigned to a specific chromosome one of which to the Z chromosome.
4. In addition, there are still many technical and basic scientific problems to be solved before the technology can become more useful and productive.

4. Poultry genome mapping

Sequencing of DNA of *Gallus gallus* (Red jungle fowl) ancestor of most of the domestic chicken has indicated that humans and chicken share 60% of their genes which only confirms that the nature has used the same genes over and over gain, but in subtly different ways.

Chicken genome is estimated to contain 1.2 x 10^6 kilo base pairs (kbp) involved in coding about 6 x 10^4 genes. Like in other higher animals, chicken genome also has introns and exons (See Chapter "Inheritance of qualitative characters"). With such a huge number of genes, it is reasonable to expect that any two birds from an out-bred population differ at numerous loci; conversely, they are similar at few loci.

4.1. Micro-satellites

(Singh and Singh, 2002)

The loci controlling quantitative traits can be referred to as "quantitative trait loci (QTL)" and they can be located if they can be associated with a suitable marker because it is easy to monitor the latter. Commonly used markers are micro-satellites – a tandomly repeated sequence of usually two or a small number of nucleotides. It is definitely possible that the same tandom sequence is present at innumerable number of locations in the entire genome; but, the probability of being flanked on one/either side by the same DNA sequence (gene) is extremely low. Therefore, a polymerase chain reaction (PCR) with DNA primers that recognize the flanking sequences can easily make the marker unique for the locus. High polymorphicnature of micro-satellite markers further makes them amenable to identify alternative alleles as well.

Micro-satellite markers can be used to locate limited number of known genes (candidate gene approach) or to scan the entire genome for loci associated with variation in the traits of interest (Genome scan).

Several studies are being conducted in broilers and layers to identify QTL; for instance, QTL have been mapped for body weight, muscling and body composition, resistance to salmonellosis and susceptibility to MD. There QTL are available for exploitation by marker-assisted selection (MAS). Selection for a marker allele (which can be easily identified by blood samples from day-old chicks themselves) associated with a QTL will increase the frequency of that allele and consequently, improves the trait in question. It is therefore obvious that the prediction of genotype with regards to traits like egg production is possible by blood analysis at the day-old stage. In addition, sex-limited traits also can be assessed in the sex which does not express the trait (for e.g. egg production in males).

5. Polymerase chain reaction (PCR)

(Miesfeld, 1999)

5.1. Logic of PCR

The basic components of a PCR are:

1. One or more molecules of target DNA – which can be purified DNA or nucleic acid sample obtained from lysis of cells or any such source. The size of the target DNA can vary widely between 0.1 to 20.0 kilo bases (kb)

2. Oligonucleotide primers
3. Thermostable DNA polymerase enzyme
4. Deoxyribose-nitrogen base triphosphatases (dNTPs i.e. dATP, dTTP, dGTP and dCTP)

This mixture is repeatedly heated and cooled to 95 (template denaturation), 55 (sometimes 55-65; primer annealing) and 72°C (DNA polymerization), in that order, 25-35 times (cycles) to produce a > 10^6-fold amplification of the target DNA.

The DNA being double-stranded with nitrogen bases in the core, it is necessary that the target DNA molecule is denatured (by heat) to provide single strand regions for primer annealing. Annealing will be a rapid process because of overwhelming number of primer molecules surrounding few target DNA molecules and also reduced temperature from 95 to about 55°C. Subsequently temperature is increased to 72°C which is the thermal optimum for the enzyme, DNA polymerase.

Theoretically, number of molecules of target DNA replicated should double and hence, the total number of molecules synthesized should increase exponentially to yield 2^n molecules after n cycles. But, the efficiency of the reaction is not 100% because of suboptimal DNA polymerase activity, poor primer annealing and incomplete denaturation of target DNA itself. Therefore, the yield after n cycles can be calculated by using a formula similar to that of compound interest by substituting % efficiency to rate of interest and number of cycles to years as follows:

$$\text{PCR product yield} = \text{Amount of target DNA}\left(1+\frac{\%\ \text{Efficiency}}{100}\right)^{\text{No. of cycles}}$$

For example, if 1 pg (10^{-12} g) was subjected to 26 PCR cycles, each of 70% efficiency, there will be $(1.70)^{26}$ = 981006.66 ($\approx 10^6$) fold increase in the quantity of the initial product; therefore, $10^{-12} \times 10^6 = 10^{-6}$ g (1 µg) of PCR product is obtained.

Automated instrumentation for PCR assay has made the method faster, reliable and practically more useful. Duration of the processes in automated and water-bath based PCR assays is given below:

Table 76.3. Duration per cycle (sec) for PCR assay (Becker *et al.* 1990)

Equipment	Denaturation	Annealing	Polymerization
Automatic	30	30	60
Water-bath based	90	120	180

5.1.1. DNA polymerase

DNA polymerase for PCR should be heat-stable; it can be from the thermophilic *Thermus aquaticus* (*Taq* Pol), or *Thermococcus litoralis* (*Tli* Pol) or *Pyrococcus furiosus*

(*Pfu* Pol) or *Thermus thermophilus* (*Tth* Pol) or a recombinant form of the latter (*rTth* Pol) which can catalyze high-temperature reverse transcription of RNA.

5.1.2. PCR primers

Primers anneal to complementary sequences on the DNA template (target DNA) and hence determine the boundaries of the amplified product. The target DNA can be general DNA region whose location is flexible (as in case of genomic target) or specific (as in case of cloning, *in vitro* mutagenesis etc.).

5.1.2.1. General rules for primer design

After selection of the target DNA, the following have to be considered:

1. Primer length – The probability of finding the same sequence (specificity) is inversely proportional to the length of primers; therefore, primers that anneal to 18-24 nucleotide long complementary sequences are preferred to ensure specificity, the chance of finding the same sequence (which is about 4^{-20}) and at the same time to provide sufficient base-pairing for duplex formation
2. Duplex stability – both primers selected should have similar melting temperature which is prerequisite for similar hybridization kinetics during template annealing phase. Melting point of the primer can be calculated depending on nucleotide composition and sequence; generally, primers will have 45-55% G+C content.
3. Non-complementary nature of primers – if the primers themselves are complementary to each other, they combine to form a short hybrid (a dimer) which will compete with the target DNA for amplification.
4. Palindromic sequences should be avoided – because such sequences stabilize the structure leading to obstruction during annealing.
5. Inter-primer distance – most ideally, the distance between the opposing primers should be 150-500 bp; but, it does vary depending on the target DNA.

Computer programs are available to develop/predict PCR primer pairs most suitable for the nucleotide sequences in the target DNA.

5.1.3. Optimizing PCR assay

Important parameters to be optimized are:

1. Primers annealing temperature
2. DNA polymerase – concentration and type of DNA polymerase
3. Magnesium concentration – Low concentration of Mg ($MgCl_2$) increases specificity and high concentration stabilizes primer annealing with variable effects on specificity
4. Cycle parameters – denaturation temperature (increase temperature increases sensitivity), duration for primer extension (longer time increases sensitivity) and cycle number (depends upon initial concentration of target DNA)

Among them, the most important is to determine optimal primer annealing temperature. The most common method used to analyze PCR products is mini-agarose gel electrophoresis stained with ethidium bromide.

5.1.4. Amplification of cDNA

PCR can be used to amplify cDNA as well and is referred to as "Reverse transcriptase-mediated PCR or RT-PCR".

5.2. Applications of PCR

1. Detection of pathogens in tissue samples (disease diagnosis)
2. Identification of genetic mutations – if a locus of interest (may be connected with disease-resistance/disease susceptibility or quantitative traits of interest or any major gene) is identified through marker analysis, all genes in the region can be quickly scanned for presence of similar nucleotide sequences (breeding for disease-resistance).
3. Pedigree and forensic genotyping
4. Molecular genetics – for sub-cloning target DNA and *in vitro* mutagenesis
5. To detect transgenic genes (transgenes)/embryos.

6. Monoclonal antibodies

(Primrose, 2001)

On entry of antigen, antibodies are produced by the body and lymphocytes are associated with these antibodies. The antibodies which are synthesized are found in globulin fraction of the proteins that circulate in the blood and hence are referred to as "immunoglobulins". Although all antibodies have the same basic structure, they differ in amino acid sequence and spatial arrangement. Therefore, antibodies are highly specific to the antigen.

An antigen has many sites (epitopes) which can stimulate antibody production by the recipient (infected) animal. Therefore, following an infection, the serum of an animal will contain a variety of antibodies and hence the serum is referred to as "polyclonal antiserum" and such a serum can protect the animals against the antigen which is actually triggered the production of such a polyclonal antiserum. If serum with only one type of antibody can be produced, such antibodies are referred to as "monoclonal antibodies (MABs)".

Antibodies are produced by lymphocytes known as B-cells or plasma cells. If lymphocytes can be isolated and cultured *in vitro*, those which are producing the desired antibodies can be cloned and used as readymade antibody source against specific diseases. But, it is not been possible to culture lymphocytes in a laboratory. In addition, it is necessary that such a culture should be perpetual (immortal) so that an unlimited supply of antibodies would be possible.

Surprisingly, immortal monoclonal antibody-producing cells exist in nature; for instance, in multiple myeloma, a cancer of the B-lymphocytes, in which the resulting tumor produces homogeneous immunoglobulin through a neoplastic transformation of a single plasma cell. These cells can be easily cloned and grown *in vitro*. The myeloma protein is unique to an individual and therefore, different clones producing different MABs can be obtained.

Cells in culture fuse spontaneously to form heterokaryons containing both the nuclei of the cells (undergoing fusion) separate which later on fuse to generate stable hybrids expressing both parental genomes. The frequency of fusion can be increased by the addition of polyethylene glycol. Such a fusion between myeloma cells and lymphocytes produces the so-called "hybridoma" cells.

6.1. Properties of myeloma cells

1. It should not secrete antibodies by itself – otherwise, the very idea of MABs is defeated since instead of a single type of antibodies, a mixture of antibodies will be obtained
2. The cells are specially selected by growing them in the presence of 8-azaguanine so that most of the cells are killed and only the resistant once are left; these resistant cells have a defect in the enzyme hypoxanthine phosphoribosyl transferase (HPRT). This property will help survival of only hybridoma cells when grown in a mixture of hypoxanthine, aminopterin and thymidine (popularly called as HAT medium), because the lymphocytes with which they have fused have the ability to produce HPRT. In any case, even the lymphocytes will survive only for a few days.

6.2. Production of hybridoma cells

The steps involved in production of hybridoma Cells are as follows:

1. Immunization of an animal; for instance a mouse, with an appropriate antigen injected, usually, s/c or into the peritoneal cavity along with an adjuvant to stimulate immune system; several injections may be required.
2. 3 days after the final dose, the animal is sacrificed to harvest lymphocytes and RBCs from spleen. The lymphocytes are separated and mixed with HPRT negative myeloma cell line and mixed with the fusion-promoting agent (polyethylene glycol) for a few minutes.
3. The mixture of cells is grown on HAT medium to isolate hybridoma cells; the resultant mixture is made up of cells each of which is capable of producing unique MAB.
4. The mixture of cells is diluted to such an extent that individual sample has only one cell.
5. Each of the cells so separated is grown in a fresh medium and confirmed that it produces the desired MAB.

6.3. Limitations

1. Normally, only 5% of splenocytes will be producing antibodies; of these, only a small proportion will be responding to the antigen administered. Hence, the success rate is neither high nor guaranteed
2. The process is expensive and involves sacrifice of experimental animals which may be objected to by animal welfare activists
3. Hybridoma cells are likely to be contaminated with viruses, especially retroviruses among others; however, most MABs of purified before use.

6.4. Applications

(McCullough and Spier, 1990)

1. In studies of antigens and epitopes – Influenza A virus, ND virus etc.; antigenic drift in case of viruses like Influenza can be easily detected (antigen fingerprinting)
2. Identification of antigenic variations – Influenza A virus, parasites etc.
3. Leukocyte markers – in other species and human beings.
4. Identification of lymphocyte helper and suppressor factors
5. In cancer research.
6. Prophylaxis and therapeutic use – especially in case of human solid tumor diseases
7. Drug targeting – is an antibiotic is tagged with MABs, they can take them specifically to the antigen. Therefore, concentration of the antibiotic required is greatly reduced; these tagged antibiotics are also referred to as "magic bullets"
8. Other uses – for study of enzymes and hormones; for instance, lysozyme c in chicken egg white

7. Other related topics

7.1. Sub-unit vaccines

These vaccines refer to use of only that part (sub unit) of the antigen instead of the entire organism (as in case of live, attenuated or killed vaccines) which attracts immune reaction for production of vaccine. Such vaccines are not in vogue in case of poultry.

7.2. DNA probing

A suitably labeled random oligonucleotide (17 to 25 bp) is allowed to anneal at random to the denatured DNA molecule and subjected for synthesis. Then, the DNA can be probed for further studies based on the label(s) tagged onto it. This is used, though not extensively, for disease diagnosis, for transgenic animal production etc.

7.3. Embryo transfer technology

Useful for mammalian (viviparous) species.

8. Biotechnology in poultry industry

(McKay on internet)

8.1. Vaccines

Advances in microbial biotechnology have led to the manufacture of more effective vaccines, enzymes and dietary ingredients. For example, recombinant DNA technology is being used to produce novel vaccines that are multivalent – so are effective against multiple pathogens or several strains of the same pathogen. Multivalent vaccines should have less risk of reversion and they are commonly accepted as being more effective than natural variants.

In ovo biotechnology is advanced, it is widely accepted and practiced. The technology has already replaced the conventional administration of vaccines by injection of newly hatched broiler chicks. In ovo administration provides a precise dose of vaccine resulting in early protection to disease challenges in the field, significant welfare benefits for the chicks and management benefits for growers. In addition, vaccine compatibility continues to improve enabling multiple applications in a single dose - making vaccination administration less invasive and resulting in benefits in terms of cost effectiveness and animal welfare. Development continues to improve the vaccine product range available to be administered in ovo.

8.2. Nutrition

The same technology delivers key benefits through the manufacture of a new and wider range of enzymes such as Xylanases or Proteases. These enzymes have improved poultry gut function allowing the broiler to extract more nutrients, including nitrogen and phosphorus, from feed thus opening up the possibility of a more cost effective diet without having to compromise on quality (See Chapter "Performance enhancers in poultry feeds").

Transgenic or GM technology has been used to introduce herbicide tolerance or resistance to pests or diseases. Plant biotechnology may in the future lead to advances in the development of vaccine varieties more appropriate as animal feed raw materials. For example, plant varieties may be developed with improved vitamin or mineral content or an amino acid balance closer to the requirements of growing broilers.

8.3. Embryo sexing

A new technology is being developed to determine the sex of an embryo before hatch. Gender-sorting using in ovo technology may be a reality within the next two years. This technology will enable supply gender-sorted flocks that are handled significantly less and as a result will receive less stress in the vital early days of life.

8.4. Marker assisted selection

More than 2,500 DNA markers have been identified within the poultry genome map – enough to detect major genes affecting traits of interest (Quantitative Trait Loci (QTL)) that make Marker Assisted Selection (MAS) a reality. For example, Chromosome 2 contains regions with large effects on growth, fat content and white meat yield. By selecting for a combination of physical measurements and the presence of particular DNA markers we can be more accurate and efficient. Advances in this technology will facilitate development of products, to best meet changing needs of customers and markets worldwide.

However, the discovery of QTL does not identify the genes involved and it is only of limited value for changing those traits that are difficult to measure. Candidate gene MAS supports research into specific genes of interest and is significantly more efficient in identifying traits that are hard to measure or have low heritability, for example disease resistance or immune system function. Continued research in this area will reap significant genetic gain and provide a product that is more robust and consistent in its performance.

Functional genomics go further allowing the identification of specific genes affecting traits of interest. With such identification come a number of benefits, for example, the selection for certain traits or product development to suit specific local environments.

Continuing advances in biotechnology make product development more accurate across a growing number of traits; general fitness, skeletal quality, cardiovascular function and disease resistance resulting in a more robust bird that can perform predictably in a wide range of environments - a product that has improved growth rate, feed efficiency, yield and robustness.

8.5. Transgenics and cloning

In the area of poultry biotechnology two other techniques – transgenesis and cloning – have been slower to develop and currently, the benefits remain less clear. Unlike the application of genomics through MAS, these techniques require invasive measures. Access to single celled embryos is only possible through surgery and the male and female pronuclei are not easily visible or accessible. This makes genetic manipulation and cloning in poultry more technically challenging than in farm mammals. Recent progress in establishing embryonic stem cells in poultry will make both technologies more practical.

There is a widening range of acceptable technologies, notably microbial, plant and poultry technologies, which already deliver benefits to the flocks and to the broiler grower. Advances continue to lower the cost of production and improve the utilization of raw materials that, in turn, reduces the impact on the environment. In short, the field of biotechnology continues to make significant contributions to bird health and welfare and the success of broiler production.

Chapter **77**

Fundamentals of Economics

Economics has been defined in several ways; for instance, economics is defined as the study of the administration of scarce resources and of the determinants of income and employment. In other words, economics is a social science concerned with the proper use and allocation of resources for the achievement in maintenance of growth and stability or economics is a social science concerned chiefly with the way the society chooses to employ its limited resources, which have alternative uses, to provide goods and services four per cent in future consumption. However, the most suitable definition seems to be that the economics is a study of how, in a civilized society, one of against the share of what other people have produced and of how the total product of society changes and is determined (Dewett and Chand, 1996).

1. Theory of demand

(Dewett and Chand, 1996)

Theory of demand establishes relationship between the quantity of commodity demanded and its price; it also gives the reasons for variations in demand. The marginal utility analysis is the most accepted and it states that as price falls, demand is extended and *vice versa*. The word "demand" means that the various quantities of a given commodity or service which consumers would buy in one market in a given period of time at various prices, or at various incomes, or at various prices of related goods.

1.1. Types of demand

1.1.1. Price demand

Price demand refers to the various quantities of the commodity or service that a consumer would purchase at a given time in a market at various hypothetical prices when other parameters such as consumer's income, tastes and prices of integrated goods remained unchanged. The demand of the individual consumer is called "individual demand"; the total demand of all the consumers combined for the commodity or service is called "industry demand"; the total demand for the product of an individual firm at various prices is called "individual seller's demand".

1.1.2. Income demand

This refers to various quantities of goods and services purchased by consumers at various levels of incomes when other parameters remain unchanged.

1.1.3. Cross demand

Cross demand refers to the purchase of a commodity or service with reference to the change in the price of other interrelated goods which could be complementary or substitution goods.

Poultry products, eggs and meat, have all the above three types of demand; especially the price and cross demands and the poultry products also adhere to the law of demand which states that a rise in the price of the commodity or service is followed by a reduction in demand and *vice versa* if conditions of demand remain constant.

1.2. Changes in demand

Several factors cause changes in demand of a commodity, and poultry products are no exception. In addition, perishability of poultry products has restricted their elasticity. The major factors changing the demand for poultry products are:

1. Changes in tastes and preferences - between eggs and poultry meat, the latter is more influenced than the former because of variety/alternate commodity available.
2. Climate or weather changes - demand for eggs is severely reduced during summer months.
3. Changes in the size and composition of population - the population comprising of vegetarians in majority can produce a low demand for poultry products.
4. Changes in money supply - inflation, although provides an additional money, the prices also rise. Therefore, the demand for various goods will be readjusted by the consumer.
5. Change in price of the commodity itself - generally, all commodities for into the law of demands when the prices change.
6. Change in real income - real income indicates the quantity of goods and services the consumer can buy with the amount of money available; in other words, the purchasing power of money. Obviously, higher the purchasing power of money greater will be the demand.
7. Change the level and distribution of income - generally, larger the average household income, greater is the demand for the commodities they consume.
8. Prices of related goods - this can affect demand for poultry products, in general, and that of poultry meat, in particular because several alternate commodities are available for poultry meat.

9. Other factors - changes in savings, asset preferences, conditions of trade and other factors also have a minor influence on demand for poultry products.

(Note: Poultry products are "perishable goods" both with the producer and with the consumer and hence, they are consumed only once after which their services are not available for quantification; conversely, equipment and buildings in a poultry enterprise are "durable goods" which can not be consumed as such but consumed indirectly in the form of services they render.)

1.3. Elasticity of demand

Demand does not always change at a constant rate to the change in prices; if the rate of change the substantial, the demand is said to be "elastic"; otherwise, "inelastic". The following are the possible cases of elasticity: Graphically, price can be plotted as a function of demand to obtain "demand curve". Depending on the nature of demand curve, elasticity of demand can be a) perfectly elastic - price remains constant regardless of the demand; demand curve parallel to X-axis (slope = 0°) b) perfectly inelastic - demand remains constant regardless of price; demand curve parallel to Y-axis (slope 90°) c) relatively elastic; demand curve having a slope between 0 and 90° d) relatively inelastic; demand curve having a slope between 0 and 90° and e) unitary elastic - the total amount spent on purchase of a commodity at different prices is same; the demand curve will be a rectangular hyperbola (Fig. 77.1).

1.3.1.1. Price elasticity

As the name indicates, price elasticity indicates the responsiveness of consumers to changes in price. Poultry products or moderate to high in price elasticity. This is, by far, the most important criterion.

1.3.1.2. Income elasticity

This indicates changes in demand depending on changes in income when the price is assumed constant. Poultry products, especially eggs, appeared to be sufficiently inelastic to income because consumers generally cut down expenditure on other commodities to keep their egg consumption in spite of reduction in income. Income elasticity is calculated as the ratio of proportionate change in the quantity purchased to proportionate change in income; the income elasticity for eggs will be = 1.

1.3.1.3. Cross elasticity

Cross elasticity is the ratio of proportionate change in purchase of one commodity to that in price of another (interrelated) commodity. For instance, the purchase of poultry meat when the prices of other types of meats change. Poultry products are subject to cross elasticity; in particular, poultry meat.

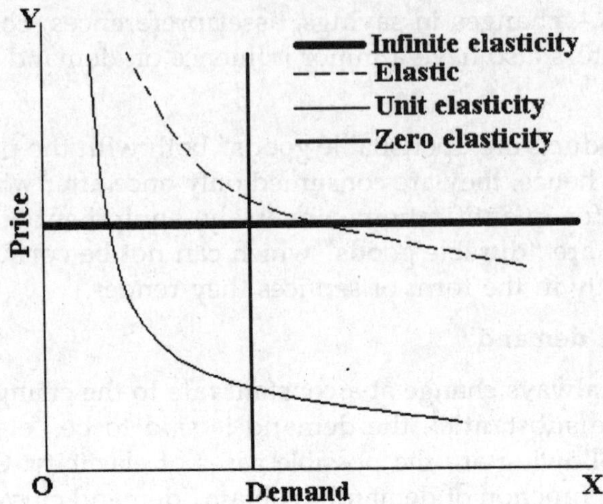

Fig. 77.1. Elasticity of demand

1.3.1.4. Substitution elasticity

Substitution elasticity indicates to what extent one commodity can be substituted for another without making any change in the total satisfaction derived by the consumer; in other words, the measures the ease with which one commodity can be substituted for another. Substitution elasticity can be infinite if the goods are perfect substitutes or zero if there is no substitution at all.

1.3.2. Factors affecting price elasticity

1. Necessaries and conventional necessaries – commodities which are absolutely essential, like salt, usually have inelastic demand whereas poultry products are not absolutely essential and hence, demand for poultry products is reasonably elastic
2. Nature of item – demand for luxuries is elastic; poultry products, especially meat, may be luxury for low income consumers. In addition, for vegetarians, poultry products may not be essential at all.
3. Proportion of total expenditure – if the commodity requires a higher proportion of total expenditure, its demand become is elastic and *vice versa.*
4. Substitutes – alternate commodities always introduce elasticity to go the commodities concerned.
5. Commodities having multiple uses – demand is generally elastic
6. Joint demand – for example demand for equipment and prices of eggs; the price of the former does not change depending on prices of eggs.
7. Items whose purchase can be postponed – generally, consumers wait till the prices of such commodities reduce and therefore demand for such items is elastic.

8. Level of prices – if the item is either very expensive or very cheap, a fall in price will not increase the demand substantially. This does not apply to poultry products.

2. Production

(Dewett and Chand, 1996)

Production, in simple terms, is the transformation of inputs (one set of goods) into outputs (another set of goods) or creation (or addition) of value. The goods can be transformed physically (form utility) are transported to the place of use (place utility) or stored till required (time utility).

2.1. Factors of production

A factor of production can be of use for only one purpose (specific factor) or it can be put to any (several) purposes (versatile factor). The factors of production include land, Labor, capital and organization.

2.1.1. Land

Land stands for all natural resources which yield an income of which have exchange value. It represents those natural resources which a useful and scarce, actually or potentially. It is generally accepted that the land has no supply price; in other words, the price of land prevailing in the market cannot affect its supply. Land is permanent and lacks mobility in the geographical sense.

2.1.2. Labor

Labor can be defined as any exertion of mind or body undergone partly or wholly with a view to some good other than the pleasure derived directly from the work. Labor is inseparable from the laborer and therefore it has to be sold in person. Labor is perishable and hence, has a very weak bargaining power (has to accept the wage offered). Rapid adjustment of the supply of labor is not possible. A fall in price (wage) below a certain point may increase supply under certain circumstances; but, sometimes, when alternate demand is available, supply will reduce. In addition to all that above, social factors has to be considered whenever an economic problem has to be solved concerning labor.

Several factors determining efficiency of labor:

1. Physical fitness – largely determined by heredity.
2. Climatic factors – tropical climate is more physically strenuous than temperate climate
3. Education level – both general and technical education has a direct bearing on the efficiency
4. Organization and equipment – organized establishments with proper equipment greatly enhance efficiency of labor; for instance, automatic egg collection in layer houses

5. Working environment - proper ventilation, sanitary surroundings etc. are conducive to efficient labor
6. Working hours - long hours impair labor efficiency
7. Fair and prompt wage payment - improves efficiency
8. Other factors. - social security at the workplace, organization of the establishment except

2.1.3. Capital

Capital refers to that part of the man's wealth which is used in producing further wealth or which yields an income and it is generally used for capital goods like plant and machinery, tools and accessories, stocks of raw materials, goods in process and fuel. The raw material is used up and the money spent on them is fully recovered than the finished goods of holding the market; but the investment on plant and machinery is permanent. Land is not considered the capital because it is a) not produced b) not perishable c) not mobile d) fixed and limited and e) a variable source of income in the sense that the rent varies.

2.1.4. Organization

Organization means the entrepreneur whose functions are as follows:

1. To initiate a business enterprise by mobilizing and harnessing the necessary productive resources
2. Taking the final responsibility of the business enterprise; in other words, risk-taking and uncertainty-bearing.
3. To provide innovation, wherever possible

2.2. Production possibility curve

On most occasions, it is possible to produce more than one product by utilizing the same resources and techniques; may be, with a few more inputs. The production possibility curve shows the maximum output of anyone commodity together with the prescribed quantities of other commodities produced and the resources utilized. The number of commodities that can be produced depends on the entrepreneur, demand, supply and other functions. For illustration, to commodities, broilers (X) and layers (Y) are considered:

Table 77.1. Production of broilers and layers (in thousands)

Possibilities	X	Y
A	0	15
B	1	14
C	2	12
D	3	9
E	4	5
F	5	0

Assuming, all the resources are made available only for production of X, only 5000 products are possible to be produced (Case F). Alternatively, if all the resources are allotted to produce only Y, 15,000 products will be produced (Case A). If both X and Y has to be produced, a part in both has to be sacrificed.

To the calculations reveal the following: when 1000 of X has to be produced, it requires the sacrifice of 1000 of Y with the total product being the same as 15,000. In other words, by this combination, 1000 of Y is transformed into 1000 of X; in economic terms, 100% marginal rate of transformation. It means that the marginal rate of transformation (MRT) is the amount of Y which has to be sacrificed for production of X. It can be calculated and observed that the MRT continuously declines as the number of alternate commodity to be produced increases; in the above example, for production of 3000 X, 6000 Y has to be sacrificed i.e. MRT 50%. In statistical terms, the MRT at any point on the production possibility curve is given by the slope of the curve at that point. The production possibility curve (Fig. 77.2) in this example is obtained by plotting Y as a function of X.

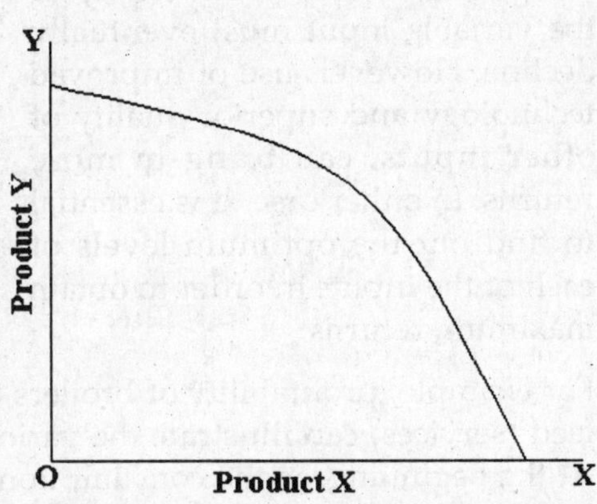

Fig. 77.2. Production possibility curve

2.3. Iso-revenue line

After developing details about MRT, it is necessary to know the exact quantities of X and Y to be produced. However, this is decided by the entrepreneur depending on the most profitable combination. Therefore, the price factor comes into picture and hence a line representing the quantity of products X and Y to be produced to realize the same revenue has to be drawn; such a line is called as "Iso-revenue line". These lines will be tangents to the production possibility curves at the different levels. A line from origin passing through the points of contact of the tangents is called as "output expansion path" (Fig. 77.3). The points of contact of this line with the iso-revenue line/production possibility curve, indicate the maximum revenue combination.

Iso-revenue lines are useful in planning combination of resources for optimum revenue and also to divert, if necessary, resources to capital goods like machinery to obtain higher production.

3. Returns

(Dewett and Chand, 1996)

Returns refer to output (product) from an enterprise. As per the law of variable proportions or law of proportionality or the law of eventually diminishing marginal physical productivity, as the quantity of any one input is increased which is combined with a fixed quantity of other inputs, the marginal physical productivity of the variable input must eventually decline. However, use of improved technology and superior quality of other inputs, can bring in more returns. In either case, it is essential to find out the optimum levels of each of the inputs in order to obtain maximum returns.

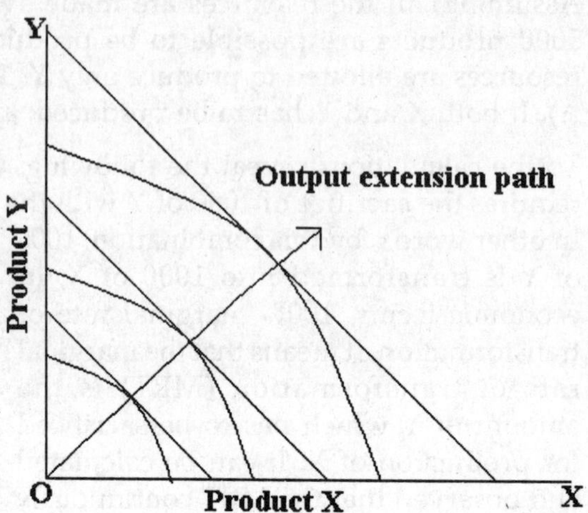

Fig. 77.3. Output extension path

For example, profitability of broilers (returns) as a function of protein content in feed (services) can illustrate the various stages of returns as a function of inputs. At the beginning, as the consumption of service increases the product (total body weight) also increases and reaches a maximum level after which any amount of increase in services does not give a proportionate increase in returns; finally, further increases in services will only be counter-productive resulting in reduced returns. The same thing is applicable for engaging labor (Table 77.2).

Table 77.2. Components of Law of proportionality

Number of laborers	Returns		
	Total	Marginal	Average
1	80	80	80
2	170	(170-80) = 90	(170÷2) = 85
3	270	(270-170) = 100	(270÷3) = 90
4	368	(368-270) = 98	(368÷4) = 92
5	430	(430-368) = 62	(430÷5) = 86
6	480	(480-430) = 50	(480÷6) = 80
7	504	(504-480) = 24	(504÷7) = 72
8	504	(504-504) = 0	(504÷8) = 63
9	495	(495-504) = - 9	(495÷9) = 55
10	440	(440- 495) = - 55	(440÷10) = 44
			Source : Dewett and Chand, 1996

It is clear from the Table that:

1. Total returns start reducing after the 9th worker – Law of total diminishing returns

2. Returns increased due to engaging of one additional labor (marginal return) increased up to 3rd worker, reduced thereafter, and the 9th and 10th workers actually obstructed return – Law of diminishing marginal returns
3. Average return per worker reached a maximum at the 4th worker and declined thereafter – Law of diminishing average returns.

It can also be noticed that both marginal and average returns equalized somewhere between 4 and 5 workers. In other words, ideally 4 workers on regular basis and one part-time are suitable for the enterprising question.

4. Cost of production

(Dewett and Chand, 1996)

4.1. Nominal cost

Nominal cost is otherwise called as money cost of production or in simple terms, expenses of production. This is extremely important for the producer because, on a long run, he must be able to obtain all the expenses back with the normal profit by selling the produce at a price fixed at the market; otherwise, he will run into the red and cannot afford to stay in business.

4.2. Real cost

Real cost includes the qualitative aspects such as the pains and sacrifices of labor; hence, it is very difficult to estimate and will be different from the nominal cost. Therefore, it is also referred to as "social cost".

4.3. Opportunity (Displacement) cost

This refers to the real cost of production of a commodity in terms of next best alternative sacrificed to obtain that commodity.

4.4. Economic costs

Economic costs mean those payments which must be resumed the resource owners in order to ensure that they will continue to supply them in the process of production. Economic cost includes normal profit.

4.5. Implicit and explicit costs

4.5.1. Implicit costs

These include costs of self-owned and self-employed resources such as returns on the entrepreneur's own investment, entrepreneur's salary (which is not paid) etc. Frequently, implicit costs are ignored while arriving at cost of production.

4.5.2. Explicit costs

These are paid-out costs like payments made for productive resources purchased (chicks, feed, medicines, vaccines) or hired (labor, transport etc.).

4.6. Full costs

Full cost of a firm is the total of money expenses, alternative or opportunity costs and normal profits; a producer can stay in business only with normal profits which have to be computed and therefore, normal profit is also considered as a cost item.

4.7. Entrepreneur's cost of production

Entrepreneur's cost of production includes a) wages of labor b) interest on capital c) rent or royalties paid to the owners of land or other property used d) cost of raw materials e) replacement and repairing charges of machinery f) depreciation of capital goods and g) normal profits to keep the entrepreneur in business.

4.8. Iso-quant curves

Analogous to the iso-revenue curve, iso-quant curve represents combinations in which the output will be equal (equal product curves). Entrepreneur will be squandering between the combinations and hence referred to as "product-indifference curves". Obviously, there will be separate curve for each of the output levels and the curves will be parallel to each other.

Assuming X and Y are the factors for production of a product, marginal rate of technical substitution (MRTS) of X for Y is the number of units of factor Y which can be replaced by one unit of factor X, with quantity of output kept unchanged. A hypothetical example is given below:

Table 77.3. Marginal rate of technical substitution (MRTS)

Combination	Factor		MRTS
	X	Y	X for Y*
A	1	12	
B	2	8	(12-8):(2-1) = 4:1
C	3	5	(8-5):(3-2) = 3:1
D	4	3	(5-3):(4-3) = 2:1
E	5	2	(3-2):(5-4) = 1:1

* MRTS = $\Delta X \div \Delta Y$ *Source* : Dewett and Chand, 1996

Incidentally, $\Delta X \div \Delta Y$ is also defined as the slope of the curve (regression coefficient), and hence, MRTS is also equal to the tangent of the angle made on the X-axis by a tangent drawn on the iso-quant curve.

The iso-quant curves are very useful in substituting machinery, raw material etc. and also in labor employment and deployment.

4.9. Iso-cost line

This is analogous to iso-revenue line drawn as a tangent to production possibility curve. The line connecting the points of contact of tangents and the origin is the "scale line" or the "expansion path line" (similar to "output expansion path") (Fig. 77.4).

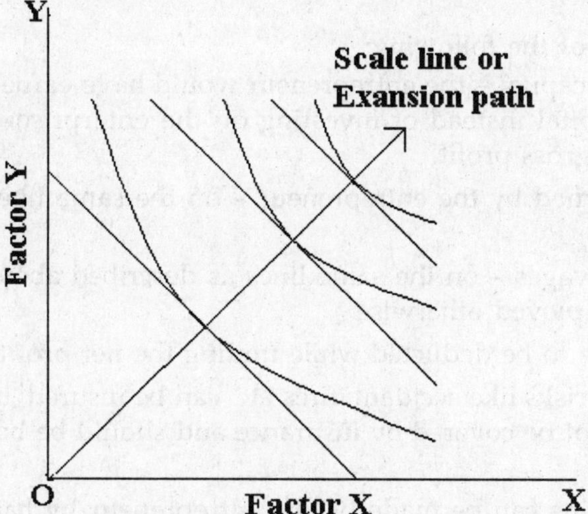

Fig 77.4. Scale line or Expansion path line

Iso-cost line and the expansion path line help the entrepreneur to plan production at the cheapest cost with the available variables at the disposal.

5. Supply

(Dewett and Chand, 1996)

Supply means the amount offered for sale at a given price. The law pertaining to supply states that "other things remaining same, as the price of the commodity rises, its supply is extended and as the price falls, its supply is contracted". In other words, supply and price are positively correlated.

5.1. Elasticity of supply

Elasticity of supply is a measure of ease with which the supply changes when the price is changed. Obviously, in the supply does not change regardless of the price it is termed "inelastic" and conversely, if the price does not change regardless of the supply, the supply is "infinitely elastic". Depending on the nature of commodity, it falls between these two extremes.

6. Equilibrium of the enterprise

6.1. Profit

Profit is the reward for the entrepreneur (not entrepreneurial functions). Profit differs from returns in the following aspects:

1. Profit is a residual income and not contractual are certain income
2. There are more fluctuations in profit

3. Profits may be negative

6.1.1. Gross profit

Gross profit consists of the following:

1. Interest on own capital – the entrepreneur would have earned interest if he had lent the capital instead of investing on the enterprise; therefore, this forms a part of gross profit.
2. Rent of land owned by the entrepreneur – on the same lines as described above
3. Entrepreneur's wages – on the same lines as described above, if the entrepreneur was employed otherwise.

The above three have to be deducted while finding the net profit.

4. Risk coverage – risks like accident, fires etc. can be insured; however, many other risks cannot be covered by insurance and should be borne by the entrepreneur.
5. Bargaining – gains can be made by the entrepreneur by bargaining while engaging labor, fixing of wages, purchase of inputs like feed, medicines and pharmaceuticals etc.
6. Monopoly gains – in case the poultry unit has an established in a remote area wherein there are no other competitors creating an artificial monopoly status, the entrepreneur can easily price his commodities higher and pay lower rewards for the inputs thereby increasing the gross profit.
7. Windfall profits – these are due to favorable circumstances due to pure luck; for instance, producers of essential commodities during periods of drought.
8. Efficiency of the entrepreneur – this includes not only the business efficiency but also the technical efficiency concerning the enterprise; for instance, production of new value-added products utilizing poultry meat and eggs.

6.1.2. Net profit

Net profit is defined as the amount that accrues to the entrepreneur for assuming the risk inseparable from all business under the system of production in anticipation of demand; in other words, it is a payment made exclusively for bearing risk. In simple terms, the net profit is a reward to the entrepreneur for assuming final responsibility, a responsibility that cannot be shifted on anyone else.

Net profit = Gross profit – (Interest on own capital + Rent of land owned by the entrepreneur + Entrepreneur's wages).

6.2. Break-even point

When revenue and cost are plotted against output (Fig. 77.5), the first intersection point of the lines indicates the minimum level of output when total cost and total

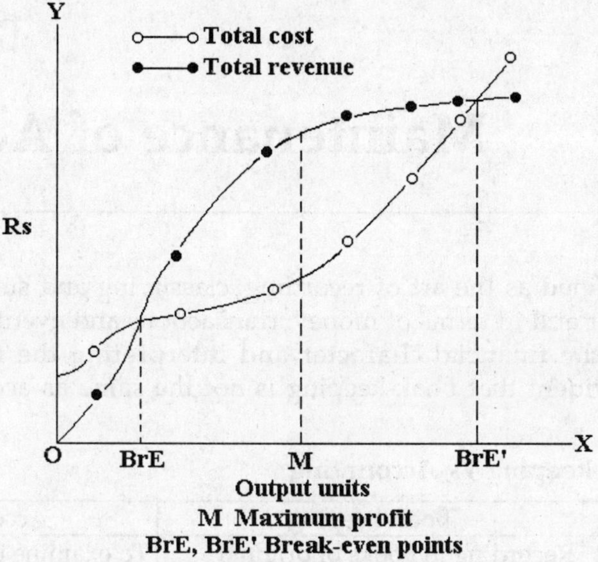

Fig. 77.5. Equilibrium analysis of an enterprise

revenue are equal (any other intersection between the lines also are break-even points; but at higher levels of production); the point is referred to as "break-even point". The output at which level there is maximum divergence (vertical distance) between revenue and cost lines, indicates the level of production for maximum profit.

Alternatively, if marginal cost and marginal returns are plotted against output, the point of intersection between the lines indicates the level of production for maximum profit.

Chapter **78**

Maintenance of Accounts

Accounting is defined as the art of recording, classifying and summarizing in a significant manner and in terms of money, transactions and events, which are, in part at least, of the financial character and interpreting the results thereof. Therefore, it is evident that book-keeping is not the same as accounting (Table 78.1)

Table 78.1. Book-keeping *Vs* Accounting

Parameters	Book-keeping	Accounting
Transactions	Recording in books of original entry	To examine the recorded transactions to find out their accuracy
Posting	To post in ledger	To examine the posting in order to ascertain its accuracy
Total land balances	To make total of the amount in journal and accounts of ledger. To ascertain balances in all accounts	To prepare trial balance with the help of balances of ledger accounts
Income statement and balance sheet	Preparation of trading, profit and loss account; and balance sheet is not included	Preparation of trading, profit and loss account; and balance sheet is included
Rectification of errors	Not included	Included
Special skill and knowledge	Not required	Required
Liability	A book-keeper is not liable for accountancy work	An accountant is liable for the work of book-keeper
		Source : Murthy, 2001

1. Basic concepts

1.1. Terminology

1.1.1. Account

Account is a statement of the various dealings which occur between a customer and the firm; it is also a clear and concise record of the transaction relating to a person or a firm or a property (assets) or a liability or an expense or an income.

1.1.2. Asset

Any physical thing or right owned that has the money value is an asset; in other words, it is that expenditure which results in acquiring of some property or benefits of a lasting nature.

1.1.3. Capital

Capital is an amount (money or assets) which the entrepreneur has invested in the firm or can claim from the firm. It is also called "owner's equity" or "net worth". Owner's equity means owner's claim against the assets and hence, Capital = Assets – Liabilities

1.1.4. Creditor

A creditor is a person to whom the firm owes money.

1.1.5. Debtor

A debtor is a person who owes money to the firm mostly on account of credit sales of goods.

1.1.6. Discount

Discount refers to any type of deduction in the prices of goods allowed by the businessman; discount allowed on the basis of sales of the items is called trade discount; discount allowed to debtors for quick payment is called cash discount.

1.1.7. Drawings

It easy amount of money or the value of goods which the proprietor takes for his domestic or personal use; hence, it is usually subtracted from capital.

1.1.8. Expenditure

In expenditure occurs when an asset or service is acquired. Purchase of goods is in expenditure whereas the cost of goods sold is an expense.

1.1.9. Expense

Expense refers to amount incurred in the process of earning revenue. If the benefit of an expenditure is limited to one year, it is treated as an expense as in case of salaries, rent etc. (also referred to as "revenue expenditure").

1.1.10. Goods

Goods are the articles in which the business is involved. Only those articles which are bought for resale for profits constitute goods.

1.1.11. Insolvent

A person whose liabilities are more than the realizable values of his assets is called any insolvent.

1.1.12. Invoice

An invoice is a statement made by the seller giving the particulars of quantity, price per unit, total amount payable, any deductions made and net amount payable by the buyer.

1.1.13. Liability

Liability means the amount which the firm owes to outsiders (excepting the proprietors).

1.1.14. Losses

In the strict sense, loss means something against which the firm receives no benefit or it represents money given up without any returns. For instance, losses due to fire, theft etc.

1.1.15. Proprietor

Proprietor is a person who makes the investment and bears all the risks connected with the business.

1.1.16. Purchases

Buying of goods by the trader for selling them to his customers is called purchases; purchase is the main function of trade. Purchase can be a cash purchase or credit purchase depending on the immediate payment in cash or postponed payment, respectively.

1.1.17. Revenue

Revenue is the amount added to the capital in lieu of operations; it is defined as the inflow of assets which results in an increase in the owner's equity. Obviously, it includes all incomes like sales, interest, commission, brokerage etc. However, receipts of capital nature (addition to capital, sale of assets etc.) do not form a part of revenue.

1.1.18. Sales

When the possession and ownership rights over the goods are transferred to the buyer, it is called as sales (business turnover or sales proceeds); as in case of purchase, it can be cash sales or credit sales.

1.1.19. Solvent

A person who has assets with realizable values exceeding his liabilities is solvent.

1.1.20. Stock

Stock refers to unsold goods kept with the trader till they are sold. Stock and in the accounting year is called closing stock which will be the opening stock for the subsequent year.

1.1.21. Transaction

It is an event which gives rise to an entry in accounting records thereby changes the balance sheet equation. In simple terms, it is an exchange of money or moneys worth from one account to another account.

Cash transaction refers to cash payment/receipt whereas credit transaction involves giving or receiving something, giving rise to debtor-creditor relationship. Non-cash transaction does not involve receipt/payment of cash but this involved in depreciation, return of goods etc.

1.1.22. Voucher

A voucher is a written document in proof of a transaction for the value stated in the voucher; it is essential for audit of accounts.

1.2. Advantages of accounting

1. Helps keep a systematic record of all business transactions for immediate after the future use
2. Gives the results of operations - profit or loss in a business during a particular period. It also helps compare the current year's profit, sales, expenses etc. with those of the previous years
3. Provides financial position of the business - availability of cash, position of assets and liabilities etc. at any time.
4. Gives the liquidity position - the financial report provides information about sources of capital, transactions with the capital, cash dividends etc.
5. It provides the knowledge of up to date information about the business so that the business can be easily protected against any mischief by anyone
6. It is a prerequisite for rational decision making - a complete knowledge about the business helps the entrepreneur to take proper decisions.
7. To satisfy the legal aspects of business - the business should be established as for the law of land; accounting is the fundamental requirement in every country.

1.3. Limitations of accounting

1. It does not reflect the current financial position or worth of a business
2. Transactions of non-monetary nature cannot be taken into account; the qualitative elements such as employee morale, reputation of the firm etc. cannot be accounted for
3. Financial statements are left the judgments of accountant or management. Therefore, valuation of inventory, provision for doubtful debts etc. are not uniform for all establishments.
4. Alternative accounting procedures are equally in vogue; hence, not all the accounting statements are comparable

5. Cost concept is found in accounting. Price changes are not considered. Money value is bound to change from time to time.
6. Impact of inflation is not included in the accounting statement
7. Accounting statements are equally indifferent towards net asset values

1.4. Branches of accounting

1.4.1. Financial accounting

This is concerned only with the financial state of affairs and financial results of operation; it is the woods will form of accounting. It is mainly concerned with the progression of financial statements for the use of outsiders (creditors, investors etc.)

1.4.2. Cost accounting

This is concerned with the cost of unit produced or sold or the services rendered by the business unit so as to control costs. It generally relates to the future costs which may have to be incurred and depends on the data provided by the financial accounting.

1.4.3. Management accounting

Management accounting, as the name indicates, provides information to the management for discharging its functions and helps in development of policy and the day-to-day operations of establishment.

2. Methods of accounting

2.1. Single entry system

In this method, cash book, personal accounts of debtors and creditors, sales day book, stock registers or maintained. Accounts regarding assets, nominal accounts for expenses and incomes, all of which are impersonal in nature, are ignored. Each transaction is entered (recorded) only once and hence is liable to fraud and misappropriations. Preparation of trading account is not possible because purchases and sales accounts are not available. For the same reason, profit and loss account as well as balance sheet cannot be prepared.

Therefore, this system is incomplete and not reliable making it unsuitable for are registered company to adopt.

2.2. Double entry system

In this system, every transaction is debited in one account and credited in another. In other words, according to this system, for every debit there is a credit and *vice versa.*

Debit (Latin *Debitum* meaning "due for that", usually abbreviated 'Dr') is the benefit-receiving aspect of a transaction, conventionally entered on the left-hand side of an account; conversely, credit (Latin *Creder* meaning "due to that", usually abbreviated 'Cr') is the benefit-giving aspect of a transaction, conventionally entered on the right-hand side of an account.

2.2.1. Advantages of double entry system

1. Since both debit and credit are recorded simultaneously for each of the transactions, arithmetical accuracy of accounts is automatically provided
2. Full and authentic information about all transactions and their results in the form of assets, liabilities, gains, losses etc. can be obtained
3. It is easier to compute profit and/or loss to guide the activities of the management accordingly.
4. Augmenting the above, the exact reasons for profit and/or loss can be analyzed for taking final decision
5. Chances of errors are very less so are the probabilities of frauds
6. The balance sheet can be prepared which gives the financial position of the enterprise; it will be a vital guide for the investors
7. Comparison of performances of the previous year(s) with the current year's is easily possible which also helps alter the strategies, if necessary
8. The accuracy of accounting adds to the efficiency of decisions taken by the management depending on the account statement.
9. Information regarding debtors helps collection of dues in time and conversely, helps prompt payment to the creditors; both these contribute to the health of the establishment.

3. The accounting cycle

From the very start of the business, the records of day-to-day transactions are recorded in the book of original entry called as "Journal". From the journal, the transactions move further to "Ledger" where accounts are written. From the ledger, the combined effect of debit and credit transactions pertaining to each account is arrived at in the form of "Balances", to prove the accuracy of the work done. These balances moves to a statement called "trial balance". By using the accounts, trading and profit and loss accounts are developed (Fig. 77.6).

4. Recording and posting of transactions

4.1. Journal

A journal (French *Jour* meaning "daily record") is the book of original entry in which all transactions are recorded in the order in which they take place (including those which are not provided for in any other register). The process of recording transactions in the journal is called as "journalizing". A specimen ruling of a journal is as follows:

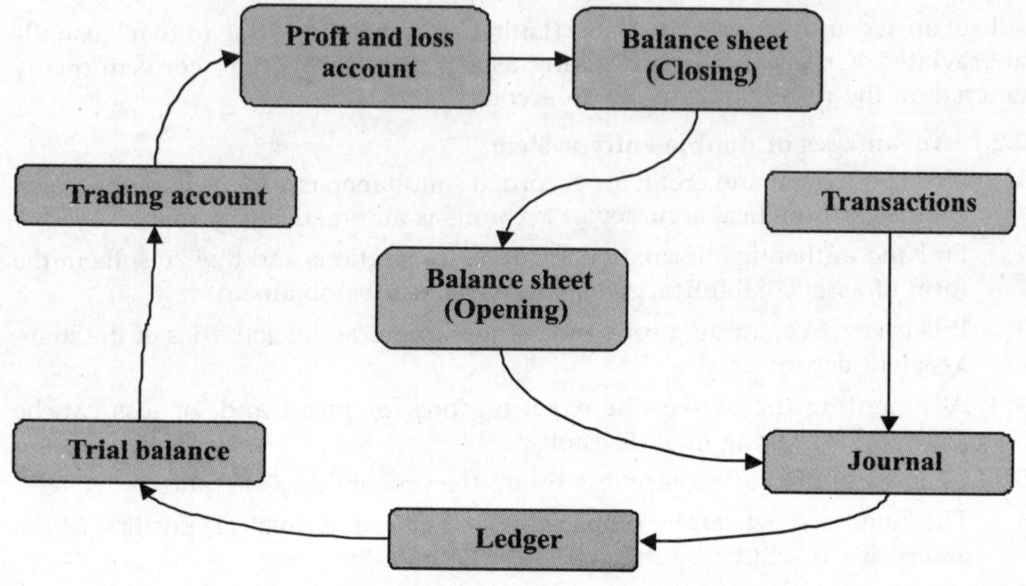

Fig. 78.1. Accounting cycle

Journal of …….				
Date (1)	Particulars (2)	L.F (3)	Debit (4) Rs p	Credit (5) Rs p

4.1.1. Guidelines for making entries in journal

1. Identify whether the account belongs to real/nominal/personal
2. Identify whether it is a debit or credit transaction
3. Money paid towards rent, salary etc. is a debit transaction for the firm
4. Expense towards acquiring fixed assets like machinery, building etc. comprises assets cost
5. Identify cash/credit transaction
6. Trade discount should be reduced from the list price of sales or purchases and only net amount should be entered
7. For goods purchased, "purchase account" is used and further goods sold "sales account" is used.
8. When goods are returned by the customers "sales returned or sales inward" is used and conversely, when goods are returned to the supplier "purchase returns or returns outward" is used.
9. If more than one debit/credit have to be entered (referred to as compound journal entry), all the debits have to be entered before the credits; the aggre-

gate amount of debits should be equal to aggregate amount of credits

10. When any amount is drawn by the proprietor or when among from the business cash is spent to meet personal expenses of the proprietor, it is debited to his personal account (drawings account)
11. When any expense becomes due, the expense account should be debited and since it is not yet paid, the expenses payable account should be credited
12. When any income is receivable, the amount is credited to income account and since it is not yet received, it is debited to income receivable account
13. When any amount/debt due from a person becomes irrecoverable, it becomes a loss (bad debt, a nominal account); it is debited from the nominal account and is credited to the debtor
14. Account of previous transaction(s) with a person or on account of balance payable or receivable, as the case may be, is referred to as "On account".
15. Loss by fire/theft/pilferage etc. are abnormal losses and credited to trading account and debited to respective loss accounts
16. If payment is made by a cheque (after allowing discount) is dishonored, the discount allowed also should be cancelled and credited to discount allowed account
17. If goods purchased are used for some other purpose, they have to be recorded accordingly. For instance, if proprietor takes goods, drawings account is debited and purchasers account is credited; if the goods are distributed as samples, advertisement account is debited and purchasers account credited; the goods are given a charity, charity (loss) account is debited and purchasers account credited
18. Opening entry: In the beginning of every financial year, the new books are opened and balances of assets, liabilities and capital appearing in previous year's balance sheet are brought forward as opening entries. A compound journal entry (opening entry) is passed in which all asset accounts are debited, liability accounts and capital account of owner(s) are credited.

4.1.2. Advantages of journal

1. Complete information about a transaction is available in a chronological order which can be easily located
2. Every transaction is suitably explained for clear understanding whenever the entry is scrutinized
3. The errors are minimized because both debits and credits are written side by side and hence, can be easily checked whether they are equal are not after every entry.

4.2. Ledger

The ledger is the principal book of accounts where similar transactions relating to particular person or thing are recorded. Therefore, a ledger will have several subdivisions which are generally as follows:

1. The general ledger – contains all accounts other than debtors and creditors; it includes accounts of properties, expenses, incomes, capital, drawings etc.
2. The debtor's ledger (Sold ledger or Customer's ledger) – contains the accounts of credit showing goods sold, cash received etc.
3. The creditor's ledger (Bought ledger or Supplier's ledger) – contains the account of those people from home the business purchases goods on credit showing goods purchased, cash paid etc.

A specimen of a ledger page is shown below:

Name of the account							
Date	Particulars	Journal folio	Amount Rs	Date	Particulars	Journal folio	Amount. Rs
	To …….				By ……		

It can be noticed that the ledger has two parts; the left half of the page is the debit side, usually designated 'Dr' and the right half of the page is a credit side, usually designated 'Cr'.

4.2.1. Guidelines for posting journal entries into ledger

1. All transactions related to a particular account should be recorded in the account already opened and no duplicate account allowed.
2. Posting will be made on the debit side of the account which is being debited in the journal and *vice versa*.
3. Every game three on the debit side of the account begins with the word "To" in the 'Particulars' column; on the credit side, by "By".
4. In the folio column, journal folio number where the entry was recorded in the journal is entered; page number of the ledger account will be entered in the 'L.F (Ledger Folio)' column in the journal.
5. Journal entries should be posted in the ledger chronologically.
6. Every ledger account should be balanced periodically – it is done by taking totals on both debit and credit sides; both should be equal. If the debit side exceeds (debit balance), the difference is entered on the credit side in the name of "By balance c/d"; on the same lines, if the credit side exceeds (credit balance), the differences entered on the debit side in the name of "To balance c/d" (c/d means "carried down").
7. Debit balance is brought down on the debit side by writing "To balance b/d" in the 'Particulars' column and similarly, credit balance is brought down

on the credit side by writing "By balance b/d" in the 'Particulars' column (b/d means "brought down").

4.3. Subsidiary books

Entering all transactions in journal and later on shifting to ledgers is time-consuming. In addition, in a large poultry farm, the journal becomes very bulky and it will be very difficult to obtain details of monthly sales, purchases, cash receipts or payments etc. It will be difficult even for tax transactions. Notwithstanding the above, when a single person handles such enormous transaction details, chances of fraud increase. Therefore, it is better to have various subsidiary books to help save time, increase efficiency and accuracy. In other words, the subsidiary books are simply the subdivisions of the journal wherein each of them is a special journal and a book of original or prime entry; they replace journal entries and accordingly, the ledger entries have to be modified.

The subsidy the books, their number and name of account depend on the nature of enterprise; for instance, in a poultry farm, the subsidiary books may be maintained for purchases, sales (for eggs, live poultry, dressed poultry, day-old chicks etc. separately), cash book etc. In any case, they will have the general format of the regular journal.

The subsidiary books are maintained by the information provided by:

1. Inward invoice - sent by suppliers of goods giving details of goods sent, price, value, discount etc. These are entered in purchases book.
2. Outward invoice - sent by the firm to the customers giving details of goods supplied, price, value, discount etc. These are entered in sales book.
3. Debit note - given by a person to another showing the amount debited to the account of the latter along with a brief explanation. This is issued by the trader regarding the purchase returns to claim abatement of dues to other party. These are serially numbered and treated as invoices.
4. Credit note - given by a person to another showing the amount credited to the account of the latter along with a brief explanation. This is issued regarding the sales returns to claim abatement of claims to other party. These are serially numbered and treated as invoices.

   Generally, when a customer returns goods to the supplier, he issues a debit note for the value of the goods returned and the supplier who receives the return goods issues a credit note.
5. Cash receipts and vouchers - these are the vouchers and receipts for cash received and paid, respectively. These are entered in the cash book; vouchers are extremely useful during audit.

General format of the various subsidiary books is given below:

Purchases book of					
Date	Particulars	L.F	Inward invoice No	Details Rs p	Amount. Rs p

Sales book of					
Date	Particulars	L.F	Inward invoice No	Details	Amount. Rs p

Purchase returns book of					
Date	Name of the supplier	L.F	Debit note No	Details	Amount. Rs p

Sales returns book of					
Date	Name of the customer	L.F	Credit note No	Details	Amount. Rs p

Simple or single column cash book of									
Date	Particulars	R.N	L.F	Amount	Date	Particulars	V.N	L.F	Amount

R.N = Receipt No V.N = Voucher No

Double column cash book (cash book discount and cash columns) of											
Date	Particulars	R.N	L.F	Discount allowed	Amount Rs	Date	Particulars	V.N	L.F	Discount received	Amount Rs

Three-column cash book of													
Date	Particulars	R.N	L.F	Discount allowed	Cash Rs	Bank Rs	Date	Particulars	V.N	L.F	Discount received	Cash Rs	Bank Rs

Analytical or Columnar petty cash book								
Amount received	Date	Particulars	V.N	Total amount paid	Cartage	Postage	Conveyance	miscellaneous

Bills receivable book										
Sl No	Date of receipt	From	Name of acceptor	Where payable	Date of bill	Term	When due	L.F	Amount	Remarks

Bills payable book								
Sl No	Date of bill	Drawers	Payee	Term	When due	L.F	Amount. Rs	Remarks

5. Preparation of trial balance

Trial balance is a statement containing the balances of all ledger accounts, as at any given date, arranged in the form of debit and credit columns placed side by side and prepared with the object of checking the arithmetical accuracy of ledger postings.

5.1. Objectives of preparing trial balance.

1. To obtain idea of balance of any desired account.
2. To check accuracy of posting of transactions into debit and credit accounts.
3. To facilitate preparation of profit and loss account and the balance sheet.

5.2. Methods of preparing trial balance

5.2.1. Total method

Debit and credit total of each account of ledger are recorded. The proforma for this method is as follows:

Trial balance as on …..				
Sl No	Name of account	L.F	Debit total Rs	Credit total Rs

5.2.2. Balance method

Debit and credit balances, as the case may be, from each ledger are recorded; this method is very popular. The proforma for this method is as follows:

Trial balance as on …..				
Sl No	Name of account	L.F	Debit balance Rs	Credit balance Rs
Note: Accounts of assets, expenses, losses and drawings are debit balances. Accounts of incomes, gains, liabilities and capital are credit balances.				

If the trial balance tallies, it is an indication that the accounts are correctly written up; but is not a conclusive proof. If the trial balance does not tally, when the difference amount is generally placed in "suspense account". A hypothetical example to demonstrate the various steps in trial balance is given below:

5.3. An example

A farmer has a capital of Rs 100,000 in cash and started a broiler farm with 1000 birds. He purchased equipment by paying Rs 10,000 and feed from M/s 'X' on credit worth to the 60,000. The following table gives the details of sales and return of credit:

Table 78.2. Hypothetical example of transactions

Week	Sales		Return of credit
	Number of birds	Amount (Rs)	
V	250	12,500	15,000
VI	250	16,250	15,000
VII	250	20,000	15,000
VIII	250	23,750	15,000

He also obtained Rs 1000 from sale of manure and Rs 600 from sale of gunny bags. These transactions will be entered in journal and suitable ledgers to describe a method of drawing trial balance.

Journal of Mr ....

Date	Particulars	L.F	Debit	Credit
	Cash A/c of Mr ...		100,000	
	To capital being cash introduced into the business			100,000
	Equipments A/c of Mr ...		10,000	
	To cash, being cash purchase			10,000
	Feed A/c Dr		60,000	
	To 'X' A/c, being credit purchase from 'X'			60,000
	Chicks A/c Dr		12,000	
	To cash, being cash purchase			12,000
	Cash A/c		12,500	
	To cash sales			12,500
	Cash A/c		16,250	
	To cash sales			16,250
	Cash A/c		20,000	
	To cash sales			20,000
	Cash A/c		23,750	
	To cash sales			23,750
	Cash A/c		1,000	
	To cash sales			1,000
	Cash A/c		600	
	To miscellaneous			600
	'X' A/c Dr		60,000	
	To cash			60,000

LEDGERS A/C

Capital A/c

Date	Particulars	J.F	Amount	Date	Particulars	J.F	Amount
	To balance, c/d		100,000		By cash		100,000
			100,000				100,000
					By balance, b/d		
							100,000

Cash A/c

Date	Particulars	J.F	Amount	Date	Particulars	J.F	Amount
	To capital		100,000		By equipment		
	To sales		12,500		By chicks		10,000
	To sales		16,250		By 'X'		12,000
	To sales		20,000		By balance, c/d		60,000
	To sales		23,750				92,100
	To sales		1000				174,100
	To miscellaneous		600				
			174,100				
	To balance, b/d						
			92,100				

			Equipment A/c				
	To cash		10,000		By balance, c/d		10,000
			10,000				
	To balance, b/d		10,000				
			Chicks A/c				
	To cash		12,000		By balance, c/d		12,000
			12,000				
	To balance, b/d		12,000				
			Feed A/c				
	To 'X'		60,000		By balance, c/d		60,000
			60,000				
	To balance, b/d		60,000				
			X's A/c				
	To cash		60,000		By feed		60,000
			60,000				60,000
					To balance, b/d		NIL
			Miscellaneous A/c				
	To balance, c/d		600		By cash		600
			600				600
					By balance		600
			Sales A/c				
	By balance, c/d		73,500		By cash		12,500
					By cash		16,250
					By cash		20,000
			73,500		By cash		23,750
					By cash		1,000
							73,500
					By balance, b/d		73,500

Trial balance as on				
Sl No	L.F	Name of the accounts	Debit	Credit
1		Capital A/c		100,000
2		Cash A/c	92,100	
3		Equipment A/c	10,000	
4		Chicks A/c	12,000	
5		Feed A/c	60,000	
6		Sales A/c		73,500
7		Miscellaneous A/c		600
		TOTAL	174,100	174,100

Production A/c (Expenditure)		
Sl No	Particulars	Amount
1	Chicks A/c	12,000
2	Feed A/c	60,000
3	Vaccine and medicine	---
4	Labor cost	---

Trading A/c			
Particulars	Amount	Particulars	Amount
To opening stock		By sales	73,500
To production	72,000		
		Closing stock	-----
To gross profit	1,500		
Transferred to profit and loss A/c	73,500		73,500

Profit and loss A/c			
Particulars	Amount	Particulars	Amount
Indirect expenses	Nil	By gross profit	1,500
Salaries, stationary etc.	2,100	By miscellaneous	600
Net profit	2,100		2,100

Balance sheet			
Liabilities	Amount	Assets	Amount
Capital	100,000	Cash in hand	92,100
Net profit	2,100	Equipment	10,000
	102,100		102,100

For further details on accounting and other applications like marginal costing and break-even analysis, the reader is advised to look in many publications available on "Accountancy".

Chapter **79**

Agricultural Marketing

Agricultural marketing can be defined as comprising of all activities involved in supply of farm inputs to the farmers and movement of agricultural products from the farms to the consumers; it includes the assessment of demand for farm inputs and their supply, post-harvest handling of farm products, performance of various activities required in transferring farm products from farm gate to processing industries and/or ultimate consumers, assessment of demand for farm products and public policies and programs relating to the pricing, handling and purchase sale of farm inputs and agricultural products (Acharya and Agarwal, 1999).

1. Differences between agricultural and manufactured goods

Table 79.1. Agricultural goods *Vs* Manufactured goods

Characteristic	Agricultural goods	Manufactured goods
Perishability	Highly variable; poultry products are highly perishable	Very low perishability
Seasonality of production	Many are seasonal; but poultry products are produced all through the year	No seasonality
Bulkiness of products	Generally bulky and hence transportation storage difficult and expensive	Extremely easy for storage and transportation
Variation in quality	Large variation	Narrow variation
Supply	Generally regular; depends on several factors	Can be decided accurately by the producer
Size of production centers	Usually small to medium in size and scattered over a large area	Both size and location can be controlled and determined
Processing	Most products require further processing	Generally does not require processing

Adopted from : Acharya and Agarwal, 1999

2. Market and prices

Market (*Latin: Marcatus* meaning merchandise or trade or a place where businesses conducted) is defined as any region in which buyers and sellers are in such a free interaction with one another that the price of the same goods tends to equality easily and quickly. Therefore, a market consists of:

1. A commodity which has to be sold
2. Presence of buyers and sellers
3. A specific place
4. Interaction between buyers and sellers so that only one price prevails for the same commodity at the same time (Dewett and Chand, 1996).

A market is said to be perfect if the same price for the same commodity at the same time prevails; otherwise, the market is said to be imperfect. On the same lines, competition in a market is said to be perfect (pure) when it has large number of buyers and sellers and a homogeneous product; otherwise, any imperfect competition. Any imperfect competition can take several forms depending on the number of firms as monopoly (one firm), duopoly (two firms) and oligopoly (many firms). Nature of commodity also varies with the type of market in competition (Table 78.2).

Market for poultry products falls into the category imperfect oligopoly, imperfect competition with a homogeneous commodity in case the eggs and differentiated in case of poultry meat.

Table 79.2. Markets and competition

Market	Number of firms	Nature of commodity
	Perfect competition	
Perfect	Infinite	Homogeneous
	Imperfect competition	
Monopolistic	Many	Differentiated
Perfect oligopoly	A few	Homogeneous
Imperfect oligopoly	A few	Differentiated
	Pure or absolute monopoly	
Absolute monopoly	One	Homogeneous
		Source : Dewett and Chand, 1996

2.1. Classification of markets
(Acharya and Agarwal, 1999)

2.1.1. Basis: Location

1. Village markets – located in a small village conducting major transactions among buyers and sellers of a village
2. Primary markets – located in big towns near centers of production. Producers themselves bring in the commodities and transact with traders directly.
3. Secondary wholesale markets – generally located at district headquarters or important trade centers; major transactions occur between village traders

and wholesalers

4. Terminal markets – produce is finally disposed off to the consumers of processors or assembled for export
5. Seaboard markets – located near the sea shore and meant mainly for import and/or export of goods.

2.1.2. Basis: Area/coverage.

1. Local or village markets – confined to village are nearby villages usually for perishable commodities in small lots
2. Regional markets – buyers and sellers are from a larger area; in India, regional markets exist for food grains
3. National markets – buyers and sellers or the national level; usually for durable goods like jute, tea etc.
4. World market – the participants are from the entire world for commodities having world-wide demand and/or supply; for instance, machinery, bullion etc.

2.1.3. Basis: Time span

1. Short-period markets – markets held only for day or few hours; usually highly perishable products like fish, fresh vegetables, liquid milk, poultry products are sold
2. Long-period markets – the commodities traded are usually of less perishable nature like food grains, oilseeds etc.
3. Secular markets – permanent markets to trade durable commodities which can be stored for years like machinery, manufactured goods etc.

2.1.4. Basis: Volumes of transactions

1. Wholesale markets – commodities are brought and sold in large volumes. These markets are located proximal to towns and cities. Primary wholesale markets located nearer to production centers comprise of traders assembling the commodities and bringing it to the secondary wholesale markets enroute terminal wholesale markets where the final transactions occur between primary wholesalers and traders.
2. Retail markets – commodities are bought and sold to the consumers as per their demand; transactions or between retailers and consumers. Retailers buy goods from wholesale market and sell in small lots to the consumers. These markets of the very near to the consumers.

2.1.5. Basis: Nature of transactions

1. Spot or cash markets – commodities are exchanged for money on the spot
2. Forward markets – a commodity is sold at the market but it is actually exchanged on some specified date; sometimes, the commodity may not be ex-

changed but only the difference in price is paid

2.1.6. Basis: Number of commodities

1. General markets – all types of commodities like food grains, livestock products etc. are sold
2. Specialized markets – only one or two commodities are sold like vegetable market, fish market etc.

2.1.7. Basis: Degree of competition

1. Perfect markets – large number of buyers and sellers having a perfect knowledge of demand, supply and prices and price is uniform at any one time over a geographical area and uniform at any one place over periods of time and uniform for different forms of the product; in all cases ± cost of getting the supply/storage/converting the product into another form, as the case may be.
2. Imperfect markets – markets wherein the characteristics of perfect markets does not exist. Depending on the number of sellers, the imperfect market or for the classified into a) monopoly market (one seller), b) duopoly market (two sellers), c) oligopoly market (more than two but a few sellers) and d) monopolistic competition where a large number of sellers deal in heterogeneous and differentiated form of a commodity.

2.1.8. Basis: Nature of commodities

1. Commodity markets – goods and raw materials are marketed
2. Capital markets – bonds, shares and securities are exchanged; money markets and share markets

2.1.9. Basis: Stage of marketing

1. Producing market – assemble commodities for further distribution to other markets; like the markets present in producing areas
2. Consuming markets – collect the produce for final disposal to consuming population; usually located in areas where production is inadequate as in case of thickly populated urban centers

2.1.10. Basis: Extent of public intervention

1. Regulated markets – markets at a working in accordance with statutory rules and regulations; marketing costs standardized and practices regulated
2. Unregulated markets – markets where traders frame the rules

2.1.11. Basis: Type of population served

1. Urban market
2. Rural market.

2.1.12. Basis: Accrual of marketing margins

Included in this category are co-operatives and other organizations handling marketing of various products. A classicalexample is the cooperativemilk unions.

3. Demand for and supply of farm products

(Acharya and Agarwal, 1999)

The aggregate demand for farm products (D) at the national level (Macro level) is determined by the rate of growth of the population (P) and rate of growth of per capita income (Y); $D = P + eY$ where e refers to income elasticity of demand for the farm product in question. It must be remembered that e varies between rural and urban population and also under certain conditions.

Domestic supply of farm products depends on several factors like whether, technology, input availability etc. Domestic supply can be supplemented, whenever necessary, by imports.

4. Price determination

(Acharya and Agarwal, 1999)

Assuming that demand and supply are influenced only by the price, an equilibrium price and equilibrium quantity can be determined either graphically by plotting demand and supply against price; the point of intersection of the lines indicates equilibrium price of the commodity. It can be solved by simultaneous equation method also.

The pricing system is considered efficient when it fulfils the following:

1. The prices through space (at different places) should vary of the basis of cost of transportation from one point to another.
2. The prices through time (on different days, periods) should not vary more than the cost of storage from one period to another.
3. The prices of different forms of products (dressed chicken, tandoori chicken etc.) should vary no more than the differences in the cost of processing

5. Characteristics of a developed market

(Acharya and Agarwal, 1999)

1. Commodities required by the consumers and for which they are prepared to pay should be available.
2. A variety of products should be made available so that the consumers can select a product of his choice
3. No harmful product should be offered for sale
4. Information about the goods available along with their relative merits must be divided for the consumers.

5. Consumers must be free to purchase without any pressure to buy a particular product, especially from traders.
6. Retailing service must also be available for small consumers
7. Prices must be fair and uniform for all categories of products and for all categories of consumers.
8. There should not be any wastage in the market.
9. Producers should be able to sell his surplus quickly at a consistent price in consonance with the demand and supply at the time of sale.

6. Characteristics of a good marketing system

(Acharya and Agarwal, 1999)

1. There should not been any government interference (in the form of restriction on commodity movements, quantity to be processed, construction of processing plant, price supports etc.) in free-market transactions
2. The marketing system should operate of the basis of independent, systematic and orderly decisions
3. It must be capable of developing into an intricate and far-flung marketing system to cope up with the rapid development of a non-industrial economy.
4. The marketing system should bring in both demand and supply; and establish equilibrium between the two.
5. It should be able to generate employment by supporting development of market-related industries

7. Marketing of farm products

(Acharya and Agarwal, 1999)

7.1. Marketing principles

It is advisable that the farmers follow the following principles in order to facilitate not only marketing but also to obtain better prices and easy sales:

1. Produce must be cleaned before setting for sale.
2. Different qualities of the products must be sold separately.
3. It is preferable to grade the product and sell.
4. The seller should be aware of day-to-day market information; especially of prevailing prices to facilitate smooth and appropriate sales
5. It is convenient if the product is made available in packs of standard weights
6. If co-operative marketing facilities available for the product, it had better be utilized
7. The product should be preferably sold at regulated markets
8. The product should have minimum chemical/toxic/other residues which are harmful to the consumers

7.2. Marketing functions

These include all the activities performed in carrying a product from the point of production to the ultimate consumer; obviously, it varies with the product, markets, level of economic development and the final form in which the product is consumed. For poultry products, marketing functions may be classified as follows:

1. Physical functions - comprises of assembly, grading (if practiced), processing (a converted into a different product), storage and transportation.
2. Exchange functions - involves buying and selling
3. Facilitating functions - includes standardization of grades, financing, risk-taking and dissemination of information about the product

7.3. Packaging

Packaging is extremely important for all the products, in general, and poultry products, the particular. Packing means the wrapping and crating a goods before they are transported and packaging (is part and parcel of packing) means placing the goods in small packages like filler flats, tin containers etc. so that they can be directly sold to the ultimate consumers.

7.3.1. Advantages of packing and packaging

1. Protection of goods against breakage or spoilage during their movement from production to the consumption point
2. Reduction in bulk - as in case of egg powder, dehydrated meat etc.
3. Facilitates handling - it is easier to handle eggs in filler flats at the time of packaging and large cartons at the time of packing and transportation
4. Facilitates quality identification, product differentiation, branding and advertisement.
5. Reduces marketing costs on handling and, in turn, reduces retailing costs.
6. Helps checking adulteration.
7. Helps maintain cleanliness of the product
8. Helps labeling the product giving details of composition of the product and instructions for use - as in case of egg powder, processed meat etc.
9. Enhances shelf life of the product - poultry products, processed or otherwise, will have a higher shelf life is properly packaged and packed in the marketing channel

7.3.2. Packing material

A volume of information is available on various packing materials with their advantages, disadvantages, procedure for usage etc. In general, the characteristics of good packing material include the following:

A good backing material should
1. have protective strength to prevent breakage or spoilage of the produce.
2. be attractive to draw attention of the consumers.
3. be convenient to the consumers.
4. be economical, bio-degradable and, if possible, reusable
5. be free from chemical reactions with the commodity itself and also conform to the safety standards prescribed from time to time

Commonly used packing material for poultry products are:
1. Filler flats – for eggs
2. Polyethylene covers for dressed chicken; usually vacuum sealed
3. Tin containers – for egg powder, canned chicken
4. Glass containers – for pickled eggs (quail)
5. Aluminium foils – for wrapping ready-to-eat poultry products for short storage

7.4. Transportation

It is not possible to produce poultry products at all points where they are demanded. Therefore, it is necessary that the products are transported from the points of production to the market and/or consumers and/or traders. Hence, transportation has become an indispensable part of the marketing system.

7.4.1. Advantages of transportation

1. Poultry products can reach a wider consumers (market)
2. Stabilization of prices over a larger area
3. Creation of employment
4. Processing for production of value-added poultry products like sausages, frankfurters etc. can be taken-up at any convenient place for such an activity and then transported to the desired market
5. Supports industrial growth by helping mobilize finished product from far and wide places to the market
6. Helps mobilization of factors of production to the centers of production

7.4.2. Cost of transportation

Cost of transportation depends on various factors like distance, quantity of the product, more transportation, condition of the road (in case of transport vehicles on road), nature of products, availability of return journey consignment and the risk associated. Nature of products includes perishability, bulkiness, fragility, inflammability and special type requiring entirely different type is transportation facility (as in case of livestock and poultry). Poultry products are perishable and lot number of eggs in cartons can be bulky.

8. Grading and standardization

Acharya and Agarwal, 1999)

Grading means the sorting of the unlike lots of the produce into different lots depending on the quality specifications laid down so that each lot has substantially the same characteristics in so far as quality is concerned. Grading is as a sub-function of standardization.

Standardization is the determination of the basic limits on grades or the establishment of model processes and methods of producing, handling and selling goods and services. Standards are established on the basis of certain characteristics such as weight, size, color, appearance etc.

8.1. Types of grading

1. Fixed (mandatory) grading – grading of goods according to size, quality and other characteristics which are of fixed standards which do not vary over time and space.
2. Permissive (variable) grading – the standards set for grading vary over time; this type of grading is not permitted in India
3. Centralized (decentralized) grading – this is based on the degree of supervision exercised by the government agencies on grading of various farm products. Centralized grading system involves the authorized packer two set-up a laboratory to test the products under the supervision of centralized authority and/or then get it approved by the concerned authorities. Decentralized grading system is followed for those commodities which do not require elaborate testing arrangements for quality assessment under the supervision and guidance of the Directorate of Marketing and Inspection; as in the case of eggs. For these commodities, grade is on the basis of physical characteristics. Grading standards for eggs and poultry meat set by BIS. These standards are compulsory for export of eggs and poultry meat.

8.2. Grading of poultry products

See chapters "Egg quality" and "Poultry meat quality".

8.3. Inspection and quality control

Inspection involves testing of the graded goods to determine whether they conform to the prescribed standards are not. Therefore, inspection ensures quality control. For purposes of inspection, samples of the product may be taken at various stages; in case of eggs and meat, usually they are obtained at the market. The inspection is carried out by the inspectors appointed by the government. Government of India has Central Agmark Laboratory at Nagpur which is the apex appellate laboratory; it has under its control several regional laboratories and commodity-specific laboratories at various places of the country.

The graded products, according to the standards by the Agricultural Marketing Advisor, Govrrnment of India, bear the label "AGMARK" as prescribed by the Advisor.

9. Buying and selling

(Acharya and Agarwal, 1999)

9.1. Methods

These are the primary activities in a market. Buying involves planning the purchase of goods, determining the sources of supply and establishing contacts with them, negotiation of price and terms and conditions of buying and finally deciding on a final agreement and transfer of goods. Similarly, selling also involves product planning and development, identifying the potential buyers and establishing contacts with them, creating demand for the produce and finally, negotiating the prices and terms and conditions of sales. In either of the activities, the most important is the negotiation of prices which can be done, under Indian markets in any one of the following methods:

1. Under cover of a cloth (Hatha) - not practiced for poultry products
2. Private negotiations - prices are fixed by mutual agreement; this is common in unregulated markets or village markets. Eggs and live poultry are still sold to the traders (middlemen) in this method at the village level.
3. Quotations on samples taken by commission agent - this is generally practiced for food grains
4. Dara sale methods - produce of different lots is mixed and then sold as one lot; not practicable for poultry products
5. Moghum sale methods - the sale is basically on a verbal understanding between the buyer and the seller without any pre-settlement of the price
6. Open auction method - practicable for food grains and not for poultry products because the latter cannot be put in an open yard for auction.
7. Close tender system - this is similar to auction but the bids are invited in the form of a close tender rather than by open announcement. Some regulated markets follow this method.

For poultry products, however, the negotiation of price depends on the commodity at stake. For live poultry and eggs, private negotiations are most popular. For other poultry products like value-added products, private negotiations as well as close tender system are practiced. In any case, small poultry farmers are the mercy of middlemen to sell their produce.

9.2. Demand creation

This is a sub-function of selling andin the present day, this is extremely important to move any product in the market. Demand creation can be effected by following methods:

1. Personal solicitation by salesman - this is not suitable for poultry products and is suitable mainly for manufactured goods
2. Advertisement - various media are used in this non-personal method to broadcast message about the product to the buyers and consumers. It could be in the form of printed material, pictures and diagrams, speeches, announcements, radio and television applications etc.
3. Other methods - include the display of goods in stores (particularly important for poultry products), arrangement of trade fairs and exhibitions to display and present the goods, distribution of free samples among the prospective buyers and offering credit facility etc.

10. Financing

In case of poultry production, sufficient time it involved before the investment made into the enterprise gives dividends. Therefore, poultry businesses possible only the financial support for both producers and traders (wholesalers, retailers and processors).

10.1. Factors affecting capital requirements
(Acharya and Agarwal, 1999)

1. Nature and volume of business.
2. Amount of stocks to be accumulated - in case of poultry, accumulating stocks is generally not advisable; excepting probably storage of eggs for about the week's time.
3. Continuity of business - poultry business runs all through the year and hence, provides continuity of business regardless of seasons; however, volumes may change within a limited range over seasons.
4. Time required between production and sale - in case of broilers, it is about 6 weeks and in case of layers, it is about 20 weeks. However, it can be reduced by procurement of batches at frequent intervals. For value-added products, the time it is sufficiently short.
5. Fluctuations in prices - this is serious problem with eggs during summer.
6. Risk-taking capacity - a trader with low risk-taking capacity of resorts to hedging and hence requires less finance
7. General conditions in the economy - like during recession; the products need to be held on stock for longer appeared necessitating higher finance.

Government of India has taken the following steps to improve the financial position of the farmers and to strengthen their holding capacity:

1. Nationalization of commercial banks during 1969
2. Development of co-operative system of financing and marketing in the agricultural sector.

3. Development of warehousing and cold-storage facilities – however, the cold-storag facilities need to be increased in capacity (during 1996, 3253 cold storage units were in operation with the capacity of 8.73 mt which is far too low for the volume of agricultural products, in general, and poultry products, in particular)

10.2. NABARD and agricultural marketing finance

The National bank for agriculture in rural development (NABARD) is the apex organization for rural finance. It re-finances the financial institutions (banks, co-operatives etc.) for the following:

1. Marketing of produce – refinance to state co-operatives banks by short-term credit limits
2. Construction of godowns and storage facilities – including cold storages
3. Construction of markets yards
4. Development of transportation facilities.
5. Establishment of processing units – like egg powder manufacture, canning to meat etc.

11. Marketing agencies

(Acharya and Agarwal, 1999)

11.1. Producers

Farmers, small or big, are producers. Processors who produce value-added products are also producers.

11.2. Middlemen

Middlemen, in case of poultry products, are involved in transfer of goods from the producer to the market and on some other occasions, big farmers may act as middlemen in assembling birds from small farmers

11.2.1. Merchant middlemen

These individuals buy and sell goods on their own and take the responsibility of profit of loss themselves.

11.2.1.1. Wholesalers

These are merchant middlemen who buy and sell in large quantities. They may buy directly from the producers or from other wholesalers. They sometimes have their own storage facilities for poultry products.

11.2.1.2. Retailers

These are also merchant middlemen who buy goods from wholesalers and then sell to the consumers in small quantities.

11.2.1.3. Itinerant traders and village merchants

These merchant middlemen move from village to village and directly purchase the produce. Such activity appears to be limited in case of poultry products.

11.2.2. Agent middlemen

These middlemen negotiate the purchase and/or sale but do not own the produce (sell services); e.g. commission agents, brokers etc. In case of poultry products, there appears to be very limited role of agent middlemen.

11.2.3. Speculative middlemen

These middlemen take title of the produce with a view to making profits at a later date; these are common for food grains but not for poultry products because storage of poultry product itself obviates such ventures.

11.2.4. Facilitative middlemen

Some middlemen assist the marketing process but do not buy or sell any produce. This group includes laborers, graders, transport agency, communication agency, advertising agency, auctioneers etc. These middlemen also play important role in case of poultry products.

11.3. Marketing institutions

These are big business organizations operating the marketing machinery; included in this category are individuals, corporate, co-operative and government institutions performing marketing functions. Some of the important marketing institutions are:

1. The State Trading Corporation (STC), The Food Corporation of India (FCI), The National Agricultural Co-operative Marketing Federation (NAFED), National Diary Development Board (NDDB), Agricultural Processed Products and Export Development Agency (APEDA) etc.
2. The Directorate of Marketing and Inspection, Government of India, State-level Agricultural Marketing Departments and Agricultural Marketing Boards.
3. State and lower level co-operative marketing societies, fair price shops, consumer's co-operative stores, milk unions.

12. Marketing channels

(Acharya and Agarwal, 1999)

Marketing channels are routes through which agricultural products move from producers to consumers; the length of the channel varies from commodity to commodity depending on the quantity to be moved, the form of consumer demand and degree of regional specialization in production.

12.1. Marketing channels for eggs

1. Producer to consumer
2. Producer to retailer to consumer
3. Producer to wholesalers to retailer to consumer
4. Producer to co-operative marketing society to wholesalers to retailer to consumer
5. Producer to egg powder factory.

Sometimes, the wholesaling and retailing functions are performed by the same firm in the channel.

12.2. Marketing channels for live poultry

As per a study conducted in the state of Punjab, the following are the marketing channels for live poultry:

1. Producer to consumer
2. Producer to itinerant trader to consumer
3. Producer to itinerant trader to retailer to consumer
4. Producer to wholesalers to retailer to consumer
5. Producer to co-operative marketing society to retailer to consumer
6. Producer to co-operative marketing society to consumer
7. Producer to hotels and institutions to consumer

13. Market integration

(Acharya and Agarwal, 1999)

Market integration can be defined as the process which refers to the expansion of firms by consolidating additional marketing functions and activities under a single management. Market integration is becoming a commonplace in poultry industry.

13.1. Types of market integration

13.1.1. Horizontal integration

A single firm or agency takes control of other firms or agencies performing similar marketing functions at the same level in the marketing sequence. Sellers combining to form a union to reduce their effective number and the extent of actual competition in the market are an example. Horizontal integration is advantageous for the members who join the group. Similarly, farmers joining to form cooperatives is also an example of horizontal integration. In general, horizontal integration does not benefit the consumers or buyers. In poultry industry, horizontal integration is not popular.

13.1.1.1. Advantages

1. Reduces competition.

2. Larger share of market is gained and therefore higher profits are expected.
3. Economics of scale can be attained.
4. Specialization in the trade is possible

13.1.2. Vertical integration

When a firm performs more than one activity in the sequence of the marketing process then, vertical integration occurs. Under a single ownership, two or more functions in the marketing process are taken up. For instance, a single firm producing day-old chicks, supplying them to needy farmers, supplying feed and other recurrent needs to the farmers, giving health coverage to the birds at the farmer's door-step and finally, assembling the produce at the farmer's door-step for further marketing is a classical example of vertical integration. Venkateshwara hatcheries and Suguna hatcheries are examples for vertical integration in poultry industry.

13.1.2.1. Advantages

1. Considerable control over both quantity and quality of the product from the beginning of the production process till it reaches the consumer
2. Reduction in number of middlemen in the marketing channel and hence more profits
3. Producers, especially smaller ones, have the facility of both supply of raw materials and assurance of the sale of their produce which will definitely stimulate more producers to enter into such an understanding.
4. Risk reduction because of improved market coordination.
5. Improvement in bargaining power and hence, can have more influence on prices.
6. Lowering cost of production by achieving operational efficiency

13.1.2.2. Disadvantages

1. The firm involved in vertical integration may become a monopoly over a period of time and dictate both buying and selling price to the disadvantage of all concerned.
2. The producer can expect only predetermined income for the produce.
3. Even if the commodity is commanding higher prices, the producer is forced to sell only at agreed prices.
4. The producers are reduced to the level of hired production units.

More often, vertical integration will be accompanied by the horizontal integration; in other words, all the activities concerning poultry production in a specific area, both at the same level as well as at different levels of production and marketing, are under the control of a single firm.

13.1.3. Conglomeration

A combination of agencies or activities not directly related to each other working under a unified management is called a conglomeration. For instance, one management taking-up different activities like textiles, poultry products, electronic goods, etc. the management is referred to as a conglomerate. The main advantage in conglomeration is the diversification which acts of the buffer for losses and hence reduces risk. NAFED is an example of a conglomerate.

14. Price spread

(Acharya and Agarwal, 1999)

The difference between the price paid the consumer and the price received by the producer for the same quantity of the same produce is called as far-retail spread or price spread; is also referred to as marketing margin.

14.1. Sources for price spread

1. The cost involved in moving the product from the point of production to the point of consumption - cost of marketing functions.
2. Profits of various marketing functionaries involved in the above process.

Therefore, marketing margin varies from channel to channel, market to market and time to time for the same produce.

14.2. Estimation of price spread

14.2.1. Computation of price spread

14.2.1.1. Lot method

A specific consignment is selected and changed through the marketing system until it reaches the ultimate consumer. During the process, cost and margin involved at each stage is assessed. However, this method is extremely difficult to identify the lot all through; in addition, there is no guarantee that the lot is representative sample of the entire produce. This method is not practiced for poultry products.

14.2.1.2. Sum of average gross margins method

Average gross margin obtained by all the intermediaries in the marketing channel are added to get the sum of average gross margins. If S_i is the sale value of a product, P_i is the purchase value paid and Q_i is the quantity of the product handled by the ith firm, the total marketing margin (M_T) is obtained by the formula $M_T = \Sigma\{(S_i - P_i) \div Q_i\}$ where i = 1,2,3... n number of firms in the marketing channel.

This method also involves considerable effort and examination of all the records by the intermediaries. In addition, the traders may not allow access to their account books and quantity purchased need not have to coincide with the quantities sold because there will be some wastage; this method needs to be adjusted for such wastages.

14.2.1.3. Comparison of prices are successive levels of marketing

Prices at the producer's, wholesaler's and retailer's are compared and the differences are taken as gross margin. This method gives an indication of movements of marketing costs and margins in relation to prices and cost indices. However, this method also has difficulties in the form of obtained series of prices for the same quality of the product and adjustment for the quality of the product as well as wastage and spoilage in processing and handling is difficult. In addition, the time lag between performances of various marketing operations is not properly accounted for by the method.

Notwithstanding the above, for perishable farm products like milk, meat and eggs wherein the time lag between the commodity entering the marketing system and the time of final consumption is very small, chasing a lot are consignment method is preferred.

14.2.2. Obtaining various measures of price spread

The following are the abbreviations used:

P_F = producer's price, P_A = wholesale price in the primary assembling market, C_F = marketing cost incurred by the farmer, P_s = producer's share in the consumer's Rs, P_r = retail price, A_{mi} = absolute margin of the ith middleman, P_{Ri} = total value of receipts per unit (sale price), P_{pi} = purchase value of goods per unit (purchase price), C_{mi} = cost incurred on marketing, per unit, P_{mi} = percentage margin of the ith middleman, M_i = percentage mark-up of ith middleman and C = total cost of marketing the commodity

14.2.2.1. Producer's price

This is the net price received by the farmer at the time of first sale and it is calculated as: $P_F = P_A - C_F$

14.2.2.2. Producer's share in the consumer's Rs

It is the price received by the farmer expressed as a percentage of the retail price; it is calculated as: $P_s = (P_F \div P_r) \times 100$

14.2.2.3. Marketing margin of a middleman

This is the difference between the total payments (cost + purchase price) and receipts (sale price of the middleman (i^{th} agency). Three alternative measures can be used:

1. $A_{mi} = P_{Ri} - (P_{pi} + C_{mi})$
2. $P_{mi} = [\{P_{Ri} - (P_{pi} + C_{mi})\} \div P_{Ri}] \times 100$
3. $M_i = [\{P_{Ri} - (P_{pi} + C_{mi})\} \div P_{pi}] \times 100$

14.2.3. Total cost of marketing

This includes all the costs either in cash or in kind by the producer-seller and the various intermediaries involved in the sale and purchase of the commodity to the commodity reaches the ultimate consumer; it is calculated as: $C = C_F + C_{m1} + C_{m2} + C_{m3} + \ldots\ldots + C_{mi} + \ldots\ldots C_{mn}$

In India, the general observations made regarding the marketing costs and margins indicate the following:

1. There is a wide variation in the producer's share in the consumer's Rs, and in the marketing costs and margins of the middlemen in different regions
2. The marketing margins or the middleman's share is relatively large and farmer's share is small for those commodities which undergo elaborate processing operations such as a case of food grains, oilseeds etc.
3. For highly perishable commodities which includes poultry products among other things, the middleman's share is much greater than the farmer's share; other things remaining the same

14.3. Factors affecting the cost of marketing

1. Perishability of the product - positively correlated
2. Extent of loss in storage and transportation - positively correlated
3. Volume of the product handled - negatively correlated
4. Regularity in the supply of the product - if regularly supplied, costs reduce
5. Extent of packaging - positively correlated.
6. Adoption to grading - cost of marketing of graded products is lesser than that of ungraded products.
7. Dependence on demand creation - if the product requires elaborate advertising, marketing costs go up
8. Bulkiness of the product - positively correlated
9. Necessity for retailing - adds to the marketing costs
10. Necessity for storage - adds to the marketing costs
11. Extent of risk - positively correlated; the risk may be in the form of price fluctuations, unfair practices in the market, business failure etc.
12. Facilities extended by dealers to the consumers - adds to the marketing costs.

The above factors clearly indicate the methods which have the adopted to reduce the marketing costs in any product, in general, and poultry products, in particular.

Chapter **80**

Economics of Poultry Operations

Like any other animal enterprise, the ultimate aim of the poultry farmer is to maximize or optimize the returns. Several factors contribute to efficiency; they include among other things, the quality of the stock procured, their maintenance, feeding and management, and the marketing efficiency. However, there are certain technical considerations which vary only a limited extent and other variables which have to be considered before venturing into commercial poultry farming. The poultry rearing has assumed an "industry" status and hence, the financial institutions are coming forward to finance poultry business if the proposal is made with all the details as per defined norms.

Economics of rearing broilers and layers

Before an attempt is made to approach for finance, it is necessary to be aware of an ideal economics of rearing broilers and layers. It is important to note that financial support is for various aspects of poultry business like, broiler production, layer production, hatchery, feed mixing unit, processing plant, pharmaceuticals etc. However, in this Chapter, economics of broiler and layer production along with basics of project report is given. The reader can get other information from various other publications. However, it isalso possible to developdifferent project proposals based on the information given in this publication as well by extension of the principles enunciated.

1. Inputs

Investment in poultry business is mainly classified into:

1. Non-recurring – which comprises of such expenditure which is incurred in the beginning and includes land, buildings and equipment. This expenditure does not repeat unless there is an expansion in the volume of output from the poultry farm.
2. Recurring – this constitutes the expenditure which repeats every time a new batch of chicks arrives. This includes cost of chicks, feed, medication, vacci-nation, labor etc.

Non-recurring and the recurring expenditures depend on type of house, birds and management, and number of birds and frequency of batches. Obviously, innumerable combinations are possible and hence it is not possible to give details for all possibilities. However, few of the constraints involved in both broiler and layer forming is outlined below.

1.1. Non-recurring expenditure

Comprises of such expenditure which is incurred in the beginning and includes land, buildings and equipment. This constitutes 60-70% of the total expenditure depending on the type of house, birds and management, and number of birds and frequency of batches (Table 80.1).

1.1.1. Housing

Major consideration in the construction of poultry house is floor space required. Usually, broilers are reared on deep litter. However, if there are grown in cages, floor space required is 60-70% that on deep litter. In case of layers, usually the birds are grown up to 16 weeks of age on deep litter and later on shifted to cages. The floor space required will be a 50% of that on deep litter. The floor space suggested in cages includes the space left for passage. On deep litter system, land required and building area are higher than for cage system (See Chapter "Management requirements and specifications").

Table 80.1. Non-recurring expenditure

Building	Cost of construction - Rs 1750 per m²	
	Deep litter	Cage system
	Floor space (cm² per bird)	
Broilers	900	540-630 (600)
Laying type		
Up to 8 weeks	700	350
9-16 weeks	1350	675
after 16 weeks	1800	900
	Cost of building	
Broilers	157.5	105
Laying type	551.25	275.63
	Equipment (Rs per bird)	
Broilers	10-12 (10)	30-35 (30)
Laying type	20-25 (25)	100-110 (100)
	Miscellaneous*	
Broilers	2	2
Laying type	10	10
	TOTAL (Rs per bird)	
Broilers	170	137
Layers **	586	386

Value in parentheses indicates the one which is considered for computation * Crates, egg filler-flats etc. ** A brood-grow house (cages) and a layer house (cage layer house) are constructed

Additional expenditure has to be made on equipment like feeders and waterers in case of deep litter system, whereas, in cage system, in addition to the cost of construction of building, investment has been made on cages for birds. Regardless of the land cost, considering the cost of construction of the building as same for both systems, cage system is cheaper than deep litter system; however, if the shed is already available, then deep litter system requires a lower initial investment than cage system.

Cost of construction of poultry sheds is variable depending on locality, material used and type of construction. Reasonable estimate, at present rates is Rs 1750 per m^2 of construction (See Chapter "Shelter engineering").

1.1.2. Equipment

Equipment cost for broilers on deep litter is also highly variable depending on the material costs. However, the current costs ranges between Rs 16-18 per bird. The equipment are regularly cleaned and disinfected and used for many subsequent batches stress to minimize investment on this account. In the case of cage system, cost of battery or single tier cages ranges between Rs 30-40 per bird. For layers, on deep litter system, in addition to feeders and waterers, nests have to be provided. In cage system, additional investment on cages ranges between 90 and 100 per bird. Total equipment cost in case of deep litter system will be Rs 35-40 per bird whereas in cage system since the feeders and waterers are built-in, only the egg-filler flats have to be purchased which might cost Rs 6 per bird.

1.1.3. Water and electricity

Power supply must be ensured to the poultry sheds, especially the brooder and layer houses. Expenditure on power supply depends on number of power points and total Wattage permitted. Wherever the size of the farm is high (10,000 birds or more), it is justifiable to install a generator of suitable output, especially to the brooder house. Cost of the generator depends on its make and output. To ensure continuous water supply, pump can be fixed to a well/bore-well.

1.2. Recurring expenditure

Although birds consume 3-4% less feed and grow (2-3%) and produce (1-2%) better in cages than on deep litter, for calculation of economic returns, such differences are not usually taken into account (Table 80.2).

1.2.1. Day-old chicks

Cost of day-old broiler as well as layer chicks is highly variable; the present cost averages Rs 15 and Rs 19, respectively. The hatchery supplies 2% extra chicks without additional cost. Packing and transportation costs are the farmer's responsibility.

1.2.2. Feed

Feed cost constitutes 60-70% of the recurring expenditure in both broiler and layer enterprise.

1.2.2.1. Broilers

Under commercial conditions, FCR ranges between 1.8 and 2.2; hence, it is reasonable to assume an FCR of 2.0 at 8 weeks of age. Many factors affect the feed efficiency at the farm level; for instance, a farmer who can sell all his broilers before 6 weeks of age itself can definitely expect a better feed efficiency; on the contrary, if the broilers cannot be sold evenafter 8 weeks of age, FCR will increase and drastically reduces returns. Similarly, average body weight that can be expected when the birds are sold is highly variable depending mainly on age, genetic potential, feed and other managemental conditions. The broilers at 6 weeks are likely to weigh between 1.8 and 2.0 kg whereas, at 8 weeks, the range expected is 2.4 to 2.8 kg. Under farmers' conditions, an average body weight of 1.80 kg at 6 weeks of age appears reasonable. Therefore, total feed consumed per broiler will be 3.60 kg (1.80 kg x 2.0). Cost of the broiler feed is also highly variable and present rate is Rs 12,000 per tonne.

1.2.2.2. Layers

Feed consumption up to 8 weeks of age (Chick starter) is around 1.5-1.75 kg per bird, whereas during 9-18 weeks and during the year they consume 5.0-6.0 kg and 38-40 kg of grower and layer ration, respectively. Cost of chick starter, grower and layer ration is also highly variable and the present rate is Rs 10,000, Rs 8000 and Rs 9000 per tonne, respectively. Present-day layers are likely to produce 300 eggs in a laying year.

1.2.3. Medication and administration

Cost of medication for broilers and layers is estimated at Rs 6 and Rs 15 per bird, respectively.

Table 80.2. Recurring expenditure

	Broilers		Layers	
	Quantity	Cost	Quantity	Cost
Cost per chick		Rs 15		Rs 19
		(per tonne)		(per tonne)
Feed, per bird	3.60 kg	Rs 12,000	2.0 kg starter	Rs 10,000
			5.0 kg grower	Rs 9,000
			38.0 kg layer	Rs 9,500
Medication, per bird	Rs 6		Rs 15	
TOTAL, per bird	Rs 64		Rs 460	

2. Income (Output)

2.1. Broilers

2.1.1. Sale of birds

In general, mortality never exceeds 5% and in most occasions, only extra chicks (2%) supplied by the hatchery will account for mortality. However, for calculation purposes, a total mortality of 3% can be considered. The average body weight at 6 weeks of age can be taken as 1.80 kg at a current rate of Rs 40 per kg live weight.

2.1.2. Manure

About 300 broilers produc fetches a tonne of manure which fetches of Rs 300 i.e., about Re 1.00 per bird.

2.1.3. Gunny bags

One gunny bag is expectable from 16 broilers and each gunny bag costs Rs 8 i.e., about Re 0.50 per bird.

2.2. Layers

Up to the beginning of egg production, mortality does not usually exceed 5% and during the laying period, a mortality rate of 1% per month is expected. Hence for purposes of calculation of eggs produced, mortality of 10% is considered whereas, for calculation of number of spent hens, mortality accounted for the 15% (considering 2% extra chicks).

2.2.1. Sale of eggs

It is reasonable to expect an overall rate of 300 eggs during the laying year; that means around 270 eggs per hen per annum after giving weightage to mortality (10%). The current price of eggs averages Rs 150 per 100 eggs. That means, Rs 405 per bird.

2.2.2. Sale of spent-hens

Average body weight will be 1.5 kg and selling price currently is Rs 20 per kg. Giving weightage to mortality (15%), Rs 25.50 per bird

2.2.3. Manure

150 birds up to 18 weeks of age as well as 40 hens in a laying year produce a tonne of manure (in other words, about 30 birds from day-old to the end of laying year produce one tonne of manure) which fetches between Rs 300 per tonne. Therefore, about Rs 10 per bird can be the realized from sale of manure.

2.2.4. Gunny bags

3 laying-type birds consume about 2 gunny bags feed and hence @ Rs 8 per bag,

Rs 5.33 per bird will be the income from this source.

Table 80.3. Income estimation (Rs per bird)

	Broilers	Layers
Sale of eggs	NA	405.00
Sale of birds/spent-hens	72.00	25.50
Manure	1.00	10.00
Gunny bags	0.50	5.33
TOTAL	73.50 (≈ 74)	445.83 (≈ 446)

3. Interest on capital and depreciation

Certain annual maintenance costs of buildings and equipments are expected; this includes white-washing of buildings, or replacement of feeders, sulfur and egg flats etc. During the first 5 years, depreciation on new building and equipments does not usually exceed 2.5% of the cost the building and 5% of the cost of equipment. Subsequently, the same may be 5% and 10%, respectively. Many commercial banks advance loans for poultry farming. Usually, the bank's share will be 75% of the non-recurring expenditure + recurring expenditure till the first batch of broilers is sold or first batch of layers have produced for 6 months, as the case may be. The annual simple interest charged, at present, is 12% and the actual loan amount has to be returned within a maximum of five years.

4. Profit estimation

Gross profit = Total income - Recurring expenditure and

Net profit = Gross profit - (Loan premiums + Interest on loan + Depreciation on building and equip)

Table 80.4. Profit estimation in broiler and layer farming (Rs / bird)

	Broilers		Layers	
	Deep-litter	Cages	Deep-litter	Cages
Non-recurring				
Building	158	105	551.25	275.63
Equipment	10	30	25	100
Miscellaneous	2	2	10	10
Total	170	137	586	386
Chicks	15	15	19	19
Feed	43	43	426	426
Medication etc.	6	6	15	15
Total recurring	64		460	
Capital required	234	201	1046	846
Bank's share (75%)	176	151	785	635
Eggs	NA		405	

	Broilers		Layers	
	Deep-litter	Cages	Deep-litter	Cages
Birds (Meat)	72		26	
Manure	1		10	
Gunny bags	0.50		5	
Total income	74		446	
Gross profit	+ 10		(-) 16	
Interest on capital*	3.3	2.83	176.6	142.8
Depreciation*				
Building, 2.5% pa	0.5	0.33	20.6	10.3
Equipment, 5% pa	0.06	0.19	1.88	7.5
Total deductions from gross profit	3.86	3.35	199.08	160.6
NET PROFIT	6.14	6.65	- 215.08	-176.6
* Calculated for 1½ months (broilers) and 18 months (layers) @ 15% pa				

It can be seen (Table 80.4) that in case of layers, the net profit is negative. This is the reflection of the fact that the building and equipment are not used to the fullest capacity and hence the birds start laying eggs, the brood-grow house and equipment therein are kept idle for another 12 months. Hence, from only one batch of layers, returns cannot meet the interest and depreciation on the non-recurring expenditure for 18 months along with the recurring costs. Therefore, to utilize the non-recurring expenditure efficiently, the chicks have to be procured in batches once every 4-5 months. Even in case of broilers, the profit shown is for all in all-out system and hence, the net profit is likely to increase when the chicks are procured in weekly, bi-weekly, fortnightly or monthly batches. Accordingly, suitable modifications have to be made while working out the gross and net profits, total capital required and cash-flow as per the requirement of the lending agency. Since the flock schedule is highly variable, a general model is given based on which any type of project can be developed by extending the underlying principles.

Preparation of project reports

A project report will contain the complete details of the farmer planning to venture into business, location and size of the farm, arrangements for inputs, sale of products and capital, expected expenditure under different heads, expected income and gross profit and finally, the cash flow showing the repayment of loan and interest. A feasibility report is the prerequisite to evaluate technical, economic and social feasibility of the proposed poultry business.

1. Feasibility report

Starting of a poultry farm should first be feasible technically, economically and socially. As far as possible, it must be environment-friendly also. In other words, all the internal and external constraints must be thoroughly studied and analyzed before imple nenting the idea into practice.

1.1. Technical feasibility

Technical feasibility involves a study to know whether or not:

1. The project proposal is technically sound.
2. All technical inputs are available.
3. The choice of technology is suitable.
4. The sequence of technical events is worked-out in a proper sequence.
5. Proper time has been allocated for each of the events.
6. Detailed work plan is thrashed out.
7. Technical expertise is available nearby.

1.2. Economic feasibility

Economic feasibility depends on whether or not:

1. Project cost, project operation cost and project fund requirements are estimated.
2. There is demand potential for the product produced.
3. The inputs are available easily and at reasonable price.
4. A reasonable price is expectable for the product.
5. Marketing of the product conveniently is possible.
6. Cost-benefit analysis has been worked out indicating viability of the project.

1.3. Social feasibility

Social feasibility includes whether or not:

1. The local public accepts the product without any stigma.
2. The location of the farm will not cause social disturbances and/or interferences.

It can be clearly visualized that starting of poultry farms will not have any feasibility problems as long as they are not in the midst of the tea populated areas. In addition, unlike ruminants, poultry do not produce CFCs and hence do not cause environment problems. Consumption of poultry does not have any social and/or religious stigma and therefore, poultry business will not be objected to, under normal circumstances.

Along with the feasibility reports, a financial statement giving details of proposed expenditure, income, profit and repayment of loan is necessary. The economics of poultry operations shown at the beginning of this Chapter is the details of the financial implications. For convenience, an algebraic model is given in Table 80.5.

Table 80.5. Economics of rearing commercial poultry – algebraic (N birds)

Head of account	Item	Description	Amount
Non-recurring expenditure	Building	@ Rs A/m² for B m²/bird	ABN
	Equipment	@ Rs C/bird	CN
	Miscellaneous	@ Rs D/bird	DN
	TOTAL		N(AB+C+D)
Recurring expenditure	Chicks	@ Rs E/chick	EN
	Feed	F kg/bird @ Rs G/kg	FGN
	Medication etc.	@ Rs H/bird	HN
	TOTAL		N(E+FG+H)
Income	Eggs (Layers)	Mortality 5%, I eggs/bird, Rs J/100 eggs	0.0095 IJN
	Culled layers	Mortality 10%, 1.5 kg/bird, @ Rs K/kg	1.35 KN
	Broilers	Mortality 2%, I kg/bird @ Rs J/kg	0.98 IJN
	Gunny bags	@ Rs L/bag	FLN ÷ 75
	Manure	M tons/100 birds @ Rs P/tonne	MPN ÷ 100
	TOTAL (Broilers)	{N (2.94 IJ + 4 FL + 3 MP)} ÷ 300	
	TOTAL (Layers)	{N (2.85J + 405K + 4F + 3MP)} ÷ 300	

Chapter **81**

Ethology of Poultry

An animal's "behavior" is the product of its genetic composition, the environment in which the animal functions, and the animal's experience (i.e., what it has learned given its previous genetic × environment interaction); the signs of observation and detailed description of behavior with the object of finding out how biological mechanisms function is termed "Ethology".

1. Terminology

(The Merck Veterinary Manual, 2003)

1.1. Aggression

Aggression can be defined in a narrow sense (attack) or in a broader sense as a specific example of agonistic behavior. In the latter case, aggression is an appropriate or inappropriate, in-context or out-of-context, inter- or intra-specific threat, challenge, or contest that results in deference or in combat and resolution.

1.2. Anxiety

Anxiety is the apprehensive anticipation of future danger or misfortune accompanied by somatic signs of tension (vigilance and scanning, autonomic hyperactivity, increased motor activity and tension). The focus of the anxiety can be internal or external.

1.3. Conflict

In this motivational state, tendencies to perform more than one type of activity are simultaneously present.

1.4. Displacement activity

This type of activity is performed out-of-context, or is "displaced," because the animal is "frustrated" in its attempt to execute another activity or otherwise occupy itself.

1.5. Dominance

Dominance refers to competitive control over a resource in a limited circumstance and to the ability of a higher-ranking animal to displace a lower-ranking one from that resource. Dominance is not interchangeable with a hierarchical rank. A "dominant" animal is *not* the one engaged in most fighting and combat. High-ranking animals are usually identified by character and frequency of deferential behaviors of others in their group.

1.6. Fear

This feeling of apprehension associated with the presence or proximity of an object, individual, or social situation is part of normal behavior and can be an adaptive response. Normal and abnormal fears are usually manifest as graded responses, with the intensity of the response proportional to the proximity (or the perception of the proximity) of the stimulus.

1.7. Frustration

This motivational state arises when an animal is engaged in a sequence of behaviors that it is unable to complete because of physical or psychological obstacles in the environment.

1.8. Phobia

Phobias are defined as profound and quickly developed fear reactions that do not diminish either with gradual exposure to the object or without exposure (as fears will) over time. Phobias involve sudden, all-or-nothing, profound, abnormal responses that result in extremely fearful behaviors.

1.9. Redirected Activity

These activities are directed away from the principal target and toward another, less appropriate target.

1.10. Stereotypic Behaviors

These behaviors involve a repetitious, relatively unvaried sequence of movements that has no obvious purpose or function but is usually derived from contextually normal maintenance behaviors (e.g. eating, walking).

1.11. Vacuum Activity

Such activity involves an instinctive, unconscious, or response behavior in the absence of the stimulus that would elicit that behavior. The activity seemingly has no apparent, contextual, useful purpose.

2. Behavioral domains

2.1. Embryonic behavior development

Embryos show behavior development in spite of the fact that their immediate problem is the very survival itself. Anatomical development of embryo is closely associated with behavioral development. Changes in behavior during embryonic development are outlined in Table 81.1.

Table 81.1. Pre-hatch behavior development

Age (d)	Development
4	First head and neck movements
5	First reaction to touch
6	First trunk movements
13	First reaction to sound, beak clapping begins
17	First reaction to light; first coordinated body movements
18	First calls
20	Reaction to parental calls
	Adopted from : Fraser and Broom, 1997

2.2. Juvenile behavior

In case of chicken and turkeys, the clutch of chicks with a hen maintains a close association with her and as indicated above, recognizes the physical characteristics and calls of the mother. After the development of feathers and disposal by the mother, the clutch gets dispersed and integrated into the peck order of the flock. A stable flock of about 40 chicken maintains a stable peck order.

2.3. Exploration behavior

With intensive system of rearing, birds have little opportunity for exploratory behavior in the form of food-searching, dust-bathing etc. given an opportunity, they often seek out nesting places, dust-bathing places etc.

2.4. Spacing behavior

Site attachment has been observed in domestic fowl in flocks of 200-400 hens. Crowing of the cock is considered a territorial pronouncement which gets limited with adequate territory is available. Therefore, cockerels crowded together in battery systems resort to excessive crowing. High-ranking males appear to move over smaller areas than low-ranking males.

2.5. Association behavior (Mother-offspring behavior)

Maternal behavior or broodiness has been selected out of commercial laying strains so it is not important in intensive poultry husbandry systems. For details thermoregulatory behavior see Chapter "Thermoregulation"

In a broody hen with chicks (in case of backyard rearing, especially under village conditions in India), a bond is formed and the chicks learn to respond to the maternal feeding call, distress call and to the hen's 'purring' sound as she settles down. Repeated exposure to her, accompanied by food, guidance and protection, strengthen the filial bond. Being precocial, birds are self-sufficient after hatching, but parents serve an important protective function while also teaching the chicks about edible and inedible foods. Precocial chicks imprint on their parents in the first few days of life. Imprinted chicks remain close to the imprinted object, which is normally a parent, but under laboratory conditions, may be a variety of different objects.

Newly hatched chicks show an excellent social response to adult calls and are attached to the hen by warmth, contact, clucking and body movements. Under natural conditions, this behavior greatly helps the newly born to learn to eat, drink, roost and protect itself from enemies. The intensity of this behavior starts developing even before hatch and is most intense at hatch. The chicks also recognize the voice and appearance of the mother. When the down is replaced by normal feathers, the mother pecks them away to break the association.

2.6. Social behavior

2.6.1. Peck order

After separating from the association of the mother, the chicks (birds) develop separate peck orders as per sex. Peck order among males is less stable than that among females mainly because of greater aggressiveness among the cocks. Dominant hens mate less frequently whereas dominant males mate frequently. Subordinate hens fail to obtain sufficient feed and hence lay fewer eggs.

The social organization differs in different systems of rearing but peck orders emerge in cages and breeder sheds. This has not been shown in meat chickens. In cages, there is a definite hierarchy established by pecking and threatening when the hens are placed in the cage, usually a few weeks before laying commences at six months. The social order in broiler flocks is relatively unimportant as they are generally processed at an age when the establishment of social stratification is just beginning. Laying hens have complex interrelationships involvingsocial rank, aggression, feeding behavior and egg production.

In large groups kept together for some months, subgroups form and become restricted to an area. This means that birds can recognize their own group members and those of an overlapping territory. It was suggested that this territorial behavior is important in large flocks as it reduces the numbers of conflicts when strangers meet. It has also been shown that individuals are more dominant in the area where they spend most time. Thus in larger flocks, hens tend to live in neighborhoods where they are well-acquainted.

Peck orders are regarded as highly stable once established, and in mixed groups, males and females have their own peck order. Agonistic pecking begins to occur within a few weeks after hatching, stable dominance and subordinate relationships usually do not become established until 6–8 weeks of age in cockerels and 8–10 weeks in pullets. A potential problem in the industry, depending on spacing and the strain of poultry, is the frequency and severity of agonistic acts. Social interaction rate increased as space decreased then suddenly falls off as space decreased further. It has been shown that individuals behave less aggressively towards subordinates in the near presence of dominant flock-mates. This 'third-party-effect' is associated with a reduction in agonistic behavior; it may be due to the lack of space for threat displays.

Selection for productivity traits may cause behavioral changes. Increased aggressiveness and social dominance, prior to full maturity has accompanied the selection for early onset of egg production in several genetic stocks studied. Higher-ranking hens may have better egg production than the lowest ranked bird in a cage, possibly because the higher ranked birds have greater access to feed.

2.6.2. Feeding

Laying hens choose to feed close to each other when given a choice of feeding locations, which demonstrates the importance of social attraction. Hens that are in the same cage and in neighboring cages synchronize their feeding. Chickens show socially facilitated feeding; in particular, they peck more at feed when they have company than when alone.

Most aggression is seen at the feed trough, where there is some competition among the chickens. Aggression in cages is relatively low, as the small group size in the cages allows the hens to establish a stable dominance hierarchy. Once a social group becomes organized, the incidence of agonistic interactions decreases.

2.6.3. Movement

It has also been shown that hens do not move randomly in normal intensive housing conditions – they maintain their heads at regular patterns of spacing and orientate them to avoid the frontal aspects of other birds. However, they turn, probably in defense, to face approaching birds. In cages that are too low for the chickens to raise their heads in a threat, aggression is provoked by an approaching bird rather than by a bird that is in continuous close proximity.

2.6.4. Recognition

Recognition of each other is based on features of the head, the comb being the most important cue. Hens can distinguish between breeds that are dissimilar but are unable to distinguish between individuals of such breeds. The ability of flock mates to recognize and remember one another becomes very difficult under commercial poultry husbandry conditions where group sizes are very large. Dim or colored lighting can affect a chicken's ability to discriminate between other birds.

2.6.5. Other related facts

Mortality, production and behavioral problems are all worse in large groups of hens, which imply the formation of unstable social groups, so this is particularly a problem in barn/aviary egg-production systems.

In case of turkeys, the most common social interaction is a simple threat; if the bird threatened does not submit, both birds warily circle each other with wing feathers spread, tails fanned and producing high-pictured trill. Both will leap and attempt to claw one another. The encounter lasts for a few minutes till the victor pushes or pulls or presses down the head of the other. The losing bird may be injured in which case proper care has to be taken to isolate and treat such birds. Both parents usually look after the poults at least until eight weeks of age.

2.7. Mating behavior

Treading is an important mechanical feature of mating in poultry. Waltzing by the cock is the main pre-copulatory behavior. In waltzing, the cocks adopt a stilted walk around the hen, tilting the body to one side. High frequency of treading is observed in freely mating poultry and most often result in crouching is by the females.

Sexual preferences between particular pairs of birds are also noted. In case of ducks, mating may take place on water or land and sometimes the mating drake is assisted by another male pressing the female's neck onto the ground when the mating drake is mounting.

A series of displays occurs before mating, based on a stimulus-response sequence initiated by the male. Male courtship displays are generally elaborate, involving vocalizations and noises, postures, spreading of the feathers to increase apparent size and emphasize plumage characteristics. Sexual approach by the male → female has 3 alternatives viz. escape or avoid or crouch; if it crouches → male mounts and treads → female moves tail to one side and everts cloaca → male spreads tail and everts cloaca → vents meet → male ejaculates, dismounts and goes off → female stands, shakes and moves off.

Sexual behavior and dominance relationships are important in the management of mating. Because the female must crouch to elicit courting behavior in the male and this is, for the female, a submissive behavior; hence, high-status females are often difficult to mate. Although it is never done commercially, research suggests that to overcome this, chickens may be sub-flocked and this reduces the number of individuals each may dominate or be submissive towards. When high-ranking hens are isolated from hens lower in the peck order, they crouch more often than when in the larger flock, and hens in the middle and lower thirds of the peck order crouched less often.

3. Behavioral needs

All animals have strong instincts to carry out behavior that is important to their species. In the case of hens, this behavior includes scratching and pecking the ground, laying eggs in a nest, dust-bathing, and normal exercise such as wing flapping.

Modern hens are descended from the Red Jungle fowl, and they still have the instincts of this ancestor. Their behavioral needs have not been substantially altered by domestication and generations of selective breeding.

3.1. Space and opportunity for exercise

One of the Five Freedoms (see Chapter "Poultry welfare") is the freedom to express normal behavior. In the case of hens, this means they must have enough space for normal movements and comfort behaviors, a nest in which to lay eggs, material for dust-bathing, and material to scratch and peck. These needs are acknowledged by the European Union in their recent directive, and form the basis of the new laws applying after 2012, when battery cages will be banned in member countries (see Chapter "Rearing systems").

Researchers have videoed hens and calculated the space they occupied during various normal behavioral activities as follows:

Table 81.2. Space required for some of the normal behavioral activities

	Space (cm^2/brd)	
Behavior	Average	Range
Stand	475	428-592
Ground scratch	856	655-1217
Turn	1272	978-1626
Wing stretch	829	660-1476
Wing flap	1876	1085-2606
Feather ruffle	873	609-1362
Preen	1150	800-1977

Comparing the above with the fact that 3 or more hens in cages have only 450 cm^2 of space, gives the idea about the chronic behavioral stress they are subjected to. Lack of exercise to the legs the wings leads to weak bones that break more easily. It has been recorded that even before hens are taken out of cages around 1 in 6 has broken bones. A further 14% suffer broken bones when they are taken from cages very carefully. When they pulled from cages by one leg in the normal rough manner, 24% break bones. By the time they are hung up by the legs in the killing chain at the slaughterhouse, around 1 in every 3 hens has at least 1 broken bone. In another study, 31% of battery hens compared to 14% of free range hens had recently broken bones before they were killed.

3.2. Nests for laying eggs

One of the strongest instincts of a hen is to lay her egg in a nest. Modern hens still have this instinct, as shown when they are released into a natural environment. They become very active in the hours before laying and look for a nest site. They scrape a rough nest and sit in it until they lay their egg. This behavior is triggered by hormones, and still exists in caged hens. In cages, hens become extremely frustrated when they can't find a nest. They show this frustration by pacing and other restless behavior.

Pacing up and down has been accepted as a sign of frustration since the study where hungry hens were shown food they couldn't reach. These hens were definitely frustrated, and they showed their frustration by pacing. Similarly, they show their frustration at not having a laying nest by pacing. Undoubtedly, a laying nest is very important to hens, but in a cage they have to lay their egg on the wire floor. Hence, the enriched cages are being made mandatory in several countries (For details of furnished cages etc. see Chapter "Rearing systems").

3.3. Material for dust-bathing

Hens regularly take dust-baths to keep their feathers in good condition - without dust-bathing material their feathers become oilier and less fluffy. In addition, dust-bathing is also a behavioral need in hens, so much so that they try to dust-bathe on the wire in cages but, only in further damaging their feathers.

3.4. Material to scratch and peck

Red Jungle fowl, the ancestors of modern hens, spent nearly 2/3 of the active part of their day pecking the ground, and 1/3 time scratching the ground. When hens are given their food either easily accessible in a tray or mixed with litter, they still spend the same amount of time feeding. In other words, all hens forage, even when they don't need to in order to get food; and hence, pecking in association with food is so much part of natural feeding behavior of hens that the animals seem to 'need' to peck and scratch even if they are not hungry at all.

Hens raised in cages are still attracted to litter. Not surprisingly, hens prefer large cages to small cages when given a choice. However, they choose to enter the smallest cages if they have litter on the floor and the larger cages do not.

Apart from the behavioral importance of litter, allowing hens to express a strong instinct and giving them something to occupy their time, litter to scratch also keeps the feet and claws in better condition. Hens on litter have less foot damage and fewer overgrown or broken claws than hens on wire.

3.5. Environmental preferences

3.5.1. Floor

The hens prefer fine hexagonal mesh over coarse rectangular mesh and over perforated steel sheet possibly because hexagonal mesh supported the bird's foot at more points than the other two floors. In comparing wire and litter floors, it was found that previous experience with either wire or litter floors affected the choice: birds reared on litter spent more time on litter than those raised on wire.

3.5.2. Cage *Vs* free run

See Chapter "Rearing systems"

3.5.3. Nest

Before egg laying, hens will work to gain access to nest sites. The demand for this resource is inelastic. Hens in cages without nests often show abnormal activities during prelaying, such as increased pacing, reduced sitting and displacement behaviors. So, if cages contain a nesting box, there is an opportunity for more normal behavior.

3.5.4. Rearing system

Hens used to living outside in the garden all preferred the run. Hens previously used to living in cages tended to choose the cage on first trial, although subsequently they came to choosing the run. So choice is strongly influenced by previous experience. The fact that the hens prefer an outside run to a cage is not indicative of suffering in a cage. Preference in itself is no indication of suffering.

Cages showed overall advantages in economy and hygiene; there was significantly less social conflict among the birds with a lesser number and intensity of threats, agonistic pecks and associated vocalizations, and fewer deaths from cannibalism than in the floor systems. However, when compared to floor-housed birds, those in cages showed more behavior indicative of conflict and frustration during nest selection and egg laying behavior. It has been recommended that further research into completely new husbandry systems or modifications of existing systems be carried out to provide optimal conditions in all aspects of egg production

4. Abnormal behavior

1. Sometimes males will hound other males, which can be a problem.
2. Caged birds may exhibit some abnormal behavior such as head flicks and feather pecking, i.e., pecking and pulling the feathers of other birds. Feather-pecking may be a form of redirected ground pecking. Experience in early life with ground pecking may influence pecking behavior in later life.
3. In some housing systems, cannibalism can be a problem.
4. Pseudo-mating occurs most frequently between high-ranking males and low-ranking males, who are pursued and trodden and indicates that dominance relationships are important. The same situation may occur in flocks of hens.

Chapter **82**

Poultry Welfare

The welfare of an individual is its state as regards its attempts to cope with its environment; therefore, welfare refers to the state of the animal and not to any human care for the animal. Human attitudes to poultry often differ from those to sheep, cattle, goats or pigs. The chicken is less often thought of as an individual and few people would ascribe much intellectual ability to it. These attitudes are partly a consequence of the very large numbers of these animals which are kept in one place, partly to the fact that they are birds and therefore harder for a person to identify with than are the larger mammals, and only slightly because of any real difference in behavioral ability (Fraser and Broom, 1997).

1. The Five Freedoms

Table 82.1. The five freedoms

Freedom	Provided by
Freedom from hunger and thirst	Ready access to fresh water and a diet to maintain full health and vigor
Freedom from discomfort	An appropriate environment including shelter and a comfortable resting area
Freedom from pain injury or disease	Prevention or rapid diagnosis and treatment
Freedom to express normal behavior	Sufficient space, proper facilities and company of the animals' own kind
Freedom from fear and distress	Ensuring conditions and treatment to avoid mental suffering.

Source : Webster, 1994

Welfare concerns regarding animals are slowly gaining momentum in India and it is expected that in near future, welfare laws may as well be imposed similar to other Western countries. The Chapter therefore gives the welfare concerns and laws in developed countries like UK, USA etc.

2. Poultry welfare guidelines

2.1. General

The welfare guidelines given below apply to all animals, in general and poultry, in particular. They are excerpts from directions in vogue in UK and for further details, the reader is advised to visit www.defra.gov.uk/animalh/welfare/farmed/

Appropriate issues specific to some species of poultry are also given; otherwise, all guidelines are applicable to all poultry species. Similarly, stocking density, feeder and drinker space along with ventilation requirements are given in Chapter "Management requirements and specifications" and hence, they are not repeated in this Chapter.

2.1.1. Care of birds

Animals shall be cared for by a sufficient number of staff who possesses the appropriate ability, knowledge and professional competence. It is essential that sufficient well-motivated and competent personnel are employed to carry out all necessary tasks. Staff should be well managed and supervised, fully conversant with the tasks they will be required to undertake and competent in the use of any equipment.

A good flock-keeper will have a compassionate and humane attitude, will be able to anticipate and avoid many potential welfare problems, and have the ability to identify those that do occur and respond to them promptly.

2.1.2. Housing

Materials used for the construction of accommodation, and, in particular for the construction of pens, cages, stalls and equipment with which the animals may come into contact, shall not be harmful to them and shall be capable of being thoroughly cleaned and disinfected.

A well designed house will incorporate insulation and heaters, ventilation fans and vents, effective light-proofing, and a lighting system providing controllable light levels with uniform distribution.

Accommodation and fittings for securing animals shall be constructed and maintained so that there are no sharp edges or protrusions likely to cause injury to them. It is important to ensure that the design of housing and equipment is suitable for the intended use. The incorporation of facilities for raising drinkers and feeders to aid access for handling equipment should be considered. Consideration should also be given to the incorporation of weighing, handling and loading facilities.

2.1.3. Freedom of movement *Vis-à-vis* stocking density

The details of different systems of rearing are given in Chapter "Rearing systems".

On most occasions, birds for meat are grown on deep-litter which takes care of most of the welfare requirements. The only constraint is the stocking density for which stipulations are already in vogue in many developed countries (Fraser and Broom, 1997).

The freedom of movement of animals, having regard to their species and in accordance with established experience and scientific knowledge, shall not be restricted in such a way as to cause them unnecessary suffering or injury. Where animals are continuously or regularly confined, they shall be given the space appropriate to their physiological and ethological needs in accordance with established experience and scientific knowledge.

The maximum stocking density for chickens kept to produce meat for the table should be 34 kg/m^2, which should not be exceeded at any time during the growing period. This stocking density is satisfactory for chickens reared to the usual slaughter weights (1.8 - 3.0 kg) but it should be reduced for birds being reared to significantly lower slaughter weights. Stocking density for breeding birds should not exceed 25 kg/m^2 calculated by dividing the total weight of all the birds (males and females) in the house by total area available to the birds.

Irrespective of the type of system, all chickens should have sufficient freedom of movement to be able, without difficulty, to stand normally, turn around and stretch their wings. They should also have sufficientspace to be able to sit without interference from other birds.

2.1.3.1. Birds allowed outdoor

2.1.3.1.1. Management

Enclosed range areas should be used in rotation, and flocks should be moved before the land becomes contaminated with organisms that can cause or carry disease to an extent which could seriously prejudice the health of the birds. The time taken for land to become heavily contaminated depends on the type of land and the density of stocking. Portable houses should be moved regularly to avoid continuously muddy conditions. Drinking facilities should be moved every one or two days to avoid the immediate vicinity becoming contaminated.

Shade and shelter from extreme weather conditions should always be available. Windbreaks should be provided on exposed land. Water sprinklers may be useful in very hot weather.

2.1.3.1.2. Housing

When birds are transferred to range houses, precautions should be taken to avoid crowding and suffocation, particularly during the first few nights. Cannibalism is a danger under this system and birds should not be confined for too long during hours of daylight or subjected to direct sunlight during confinement.

2.1.4. Protection

Animals not kept in buildings shall, where necessary and possible, be given protection from adverse weather conditions, predators and risks to their health and shall, at all times, have access to a well drained lying area.

Land on which range birds are kept for prolonged periods may become 'fowl sick', i.e. contaminated with organisms which cause or carry disease to an extent which could seriously prejudice the health of the birds on the land. Land should be frequently monitored for worm burden.

Sufficient housing should be available to the birds at all times and it may be necessary to exclude birds from the range in bad weather if there is a clear danger that their welfare will be compromised.

Birds should be encouraged to use the outdoor area by provision of adequate suitable, properly managed vegetation, a fresh supply of water and overhead cover, all sufficiently far from the house to encourage the birds to range.

2.1.5. Ventilation and thermal comfort

Air circulation, dust levels, temperature, relative air humidity and gas concentrations shall be kept within limits which are not harmful to the animals. Ventilation rates and house conditions should at all times be adequate to provide sufficient fresh air for the birds and keep the litter dry and friable. Accumulations of ammonia, hydrogen sulfide, carbon dioxide, carbon monoxide and dust should be avoided. In particular, the concentration of ammonia should not exceed 20ppm of air measured at bird height level.

Extremes of temperature should be avoided. Maximum and minimum temperatures should be monitored and recorded daily to assist management. Birds should be protected from cold draughts. Efforts should be made to ensure that the ventilation systems do not result in large differences in air speed across the house.

Chicks should be placed in the brooding area when they arrive in the house and their behavior monitored carefully.

Birds should not be exposed to strong, direct sunlight or hot, humid conditions long enough to cause heat stress as indicated by prolonged panting.

2.1.6. Feed and water

Animals shall be fed a wholesome diet which is appropriate to their age and species and which is fed to them in sufficient quantity to maintain them in good health, to satisfy their nutritional needs and to promote a positive state of well-being.

No animals shall be provided with food or liquid in a manner, nor shall such food or liquid contain any substance, which may cause them unnecessary suffering or injury. To enrich the environment, insoluble grit should be offered (spread on the litter) from about 6 weeks of age. This will also help the gizzard to break down any litter or feathers which may have been consumed, and encourage scratching.

All animals shall have access to feed at intervals appropriate to their physiological needs (and, in any case, at least once a day) except where a veterinary surgeon acting in the exercise of his profession otherwise directs.

All animals shall either have access to a suitable water supply and be provided with an adequate supply of fresh drinking water each day or be able to satisfy their fluid intake needs by other means.

Feeding and watering equipment shall be designed, constructed, placed and maintained so that contamination of food and water and the harmful effects of competition between animals are minimized.

No other substance, with the exception of those given for therapeutic or prophylactic purposes or for the purpose of treatment shall be administered to animals unless it has been demonstrated by scientific studies of animal welfare or established experience that the effect of that substance is not detrimental to the health or welfare of the animals.

All birds should have daily access to feed. When introducing birds to a new environment, the flock-keeper should ensure that the birds can find feed and water. To prevent birds having access to stale or contaminated feed or water these should be replaced on a regular basis. Provision must be made for supplying water in freezing conditions. In intensively housed systems, the maximum distance which any bird should have to travel in a house to reach feed and water should not be more than 4 m. However, in some situations, such as some outdoor production systems, it may be necessary for the birds to travel more than 4 m; in these situations, all birds must be adequately cared for in terms of stocking density, feeding and drinking space to allow for such movements.

Sudden changes in the type, quantity and make-up of feed should be avoided. Any changes in diet should be introduced gradually. Compounded feeds which have been prepared for other species should be avoided as certain substances can be toxic to birds.

For meat chickens, feed should not be withheld for more than 12 hours before the birds are slaughtered or delivered to a new farm. This period of 12 hours must include the catching, loading, transport lairaging and unloading time prior to slaughter.

Daily access to water throughout the period of lighting and a sufficient number of drinkers, well distributed and correctly adjusted, should be provided.

2.1.6.1. Breeding birds – feed restriction

Feed offered to breeding chickens is a fine balance between offering too much feed (because birds fed to demand would become obese, fail to survive through the laying period and breeding would be severely impaired) and causing suffering due to hunger and starvation (See Chapter "Restricted feeding" for details. The weight of present evidence is that the overall welfare of the bird is better if feed is restricted. However it is particularly important that the effects on the individual bird are carefully monitored by skilled staff.

In no circumstances should breeding birds be induced to molt by withholding feed and water. Skip a day regimes are not acceptable.

Birds should not be fed on the day of transportation as they travel more comfortably with an empty crop. Increased feed should be given to breeding birds on the day before travel and water should be made available up to the time of catching.

During the first 6 weeks of life for breeding birds, feed levels should be adequate to ensure good skeletal development. The level of feed intake throughout rearing should be managed to achieve a steady growth, not less than 7% week-on-week, and the desired weight and condition at point-of-lay. Feed should be offered to the birds at least daily throughout the production cycle with the exception of the day before depopulation, when a more generous allocation should be made in anticipation of starvation of the birds before slaughter.

In addition to routine daily checks, the body weight and condition of the birds should be systematically monitored and recorded on a weekly basis. As the amount of feed offered to the birds is so small its nutritional quality must be carefully

The equipment used to prevent cockerels taking feed intended for hens should be carefully adjusted to ensure that access for hens is maintained and cockerels are not injured.

Breeding birds should be reared in houses in which temperature, humidity, ventilation rates, light levels and photoperiods are carefully regulated.

2.1.7. Equipment

All equipment and services, including feed hoppers, feed chain and delivery systems, drinkers, ventilating fans, heating and lighting units, fire extinguishers and alarm systems, should be cleaned and inspected regularly and kept in good working order.

Ventilation, heating, lighting, feeding, watering and all other equipment or electrical installation should be designed, sited and installed so as to avoid risk of injuring the birds.

All equipment should be constructed and maintained in such a way as to avoid subjecting the birds to excessive noise.

2.1.8. Litter

Meat chickens and breeding chickens spend their lives in contact with litter and their health and welfare are linked to its quality. Conditions such as pododermatitis, hock burn, foot pad lesions and breast blisters are consequences of poor litter quality. Well-designed equipment and high standards of management are important if good litter quality is to be maintained.

The ventilation capacity should be sufficient to avoid overheating and to remove excess moisture. The feed composition should be well balanced to avoid problems with wet or sticky droppings.

Litter should be kept loose and friable and measures should be taken to minimize the risk of mould and mite infestation. It should be inspected frequently for signs of deterioration and appropriate action should be taken to rectify any problem. Moldy litter should not be used.

Litter should also be inspected to ensure it does not become excessively wet or dry. A water system which minimizes water spillage should be used, such as water nipples with drip cups positioned at an appropriate height for all birds. Nipple drinkers without cups may be used if they are well managed and the water pressure is checked frequently.

Moldy litter should not be used. There should be frequent checks to ensure that litter does not become excessively wet or dry, or infested with mites or other harmful organisms.

Foraging behavior has the added advantage of improving litter quality in birds provided with outdoor facility. Suitable perches in the rearing house may provide a form of enrichment to aid the birds in performing another of their natural behaviors. Perches will also aid the birds' adaptation from litter to raised, perforated floors when they move to the laying house.

2.1.9. Light

Animals kept in buildings shall not be kept in permanent darkness. Where the natural light available in a building is insufficient to meet the physiological and ethological needs of any animals being kept in it then appropriate artificial lighting shall be provided. Animals kept in buildings shall not be kept without an appropriate period of rest from artificial lighting. Chickens should be housed at light levels which allow them to see clearly and which stimulate activity. This should be provided by lighting systems designed, maintained and operated to give a minimum light level of 10 lux at bird eye height. Illumination of the house to at least 20 lux will further encourage activity. Houses should have a uniform level of light.

Meat chickens which do not have access to daylight should be given at least 8 hours of artificial lighting each day. It is important for bird welfare to provide them with a period of darkness (not less than 30 minutes) in each 24- hour cycle.

Recommended minimum light intensities and photoperiods for breeding birds are: Up to 10 days - minimum of 60 lux at day old, reducing to 10 lux and an uninterrupted day length minimum of 8 hours by 10 days of age. Up to point of lay - minimum of 10 lux. Uninterrupted day-length minimum of 8 hours. Laying - minimum of 20 lux. Uninterrupted day-length increasing from 8 hours to a maximum of 18 hours.

If a behavioral problem such as cannibalism occurs, it may be necessary to dim the lights for a few days.

2.1.10. Slaughter

When an animal (bird) is routinely slaughtered or killed on farm, this must be done using a permitted method. The permitted methods of killing poultry include decapitation and neck dislocation. However, for commercial slaughter, the following are the guidelines:

Welfare aspects of poultry before and during slaughter are still a matter of concern. It is not possible to elaborate all aspects in this book and the reader is advised to obtain information on many sources both in print as well as on Internet (www.thestationeryoffice.com); only salient features have been given for ready reference.

The major problems involved in transportation are (Webster, 1994):

1. Fear and pain associated with handling and mixing.
2. Thermal or motion stresses during the journey.
3. Hunger, thirst and exhaustion.
4. Risks of infection.

At every stage, extreme caution and consideration are required to ensure highest level of welfare to the birds, which after all, are going to be slaughtered for human food.

2.1.10.1. Unloading the birds for slaughter

The problem for poultry is that life is cheap, and for spent hens, very cheap. Thus, the processor is prepared to accept a very low incidence of broiler birds found dead on arrival at the abattoir but it considerably higher incidence among spent hens. During unloading, the personnel involved have to handle such large numbers in a limited time that they often treat the birds with absolute indifference. Very high proportion of fractures of the leg, especially in spent hens, is a common problem which precipitates as problems at hanging the bird upside-down on shackles. Rough handling, in addition to the above, will increase the chances of bruising and condemnations in the carcass.

2.1.10.2. Stunning

A volume of information is available on various methods are stunning and the reader is advised to refer publications dedicated to the topic for further information; stunning methods generally used for poultry are outlined in Chapter

"Ready-to-cook chicken". The most popularly practiced method is the electrical stunning method and if practiced as per standard procedure, it shall ensure the least pain to the birds destined for slaughter.

2.1.11. Inspection

All animals kept in husbandry systems in which their welfare depends on frequent human attention shall be thoroughly inspected at least once a day to check that they are in a state of well-being. Animals kept in systems other than husbandry systems in which their welfare depends on frequent human attention shall be inspected at intervals sufficient to avoid any suffering.

Where animals are kept in a building adequate lighting (whether fixed or portable) shall be available to enable them to be thoroughly inspected at any time. In order to reduce the risk of welfare problems developing on meat chicken or breeding chicken units, it is recommended that a systematic inspection of all flocks should be undertaken at least twice each day at appropriate intervals. Young birds, in the first few days of life, should be inspected more frequently.

All automated or mechanical equipment essential for the health and well-being of the animals shall be inspected at least once a day to check that there is no defect in it. Where defects in automated or mechanical equipment of the type are discovered, these shall be rectified immediately, or if this is impossible, appropriate steps shall be taken to safeguard the health and well-being of the animals pending the rectification of such defects including the use of alternative methods of feeding and watering and methods of providing and maintaining a satisfactory environment.

Any animals which appear to be ill or injured shall be cared for appropriately without delay; and where they do not respond to such care, veterinary advice shall be obtained as soon as possible.

2.1.12. Disease control

Vaccinations, injections and similar procedures should be undertaken by competent, trained operators. Care should be taken to avoid injury and unnecessary disturbance of the birds. Measures to control diseases caused by external parasites should be taken by using the appropriate parasiticides.

Where necessary, sick or injured animals shall be isolated in suitable accommodation and where appropriate, with dry comfortable bedding.

Should the flock-keeper decide that there is a good chance of a sick bird recovering, it should be isolated in a hospital pen, providing it is able to eat, drink and stand unassisted. Birds should be examined frequently throughout the day.

However, if a bird is suffering and cannot be treated or if it fails to show significant improvement within 24 hours of being placed in the hospital pen it should be humanely killed without delay.

Record shall be maintained of any medicinal treatment given to animals, and the number of mortalities found on each inspection of animals the record shall be retained for a period of at least three years from the date on which the medical treatment was given, or the date of the inspection, as the case may be, and shall be made available to an authorized person when carrying out an inspection or when otherwise requested by such person.

All those in contact with birds should practice strict hygiene and disinfection procedures. Where possible the site should be managed so that all houses are empty simultaneously to facilitate effective cleaning, disinfection and disinfestations.

When houses are emptied and cleaned, old litter should be removed from the site before restocking so as to reduce the risk of the carry over of disease.

2.1.13. Contingency plans

Farmers should make advance plans for dealing with emergencies such as fire, flood, power or equipment failure, or disruption of supplies, and should ensure that all staff is familiar with the appropriate emergency action.

Contingency arrangements should be made to ensure that adequate supplies of water and suitable feed can be made available in emergencies. Efforts should be made to minimize the risk of drinking water freezing.

On artificial ventilation systems, (a) provision shall be made for an appropriate back-up system to guarantee sufficient air renewal to preserve the health and well-being of the animals in the event of failure of the system, and (b) an alarm system (which will operate even if the principal electricity supply to it has failed) shall be provided to give warning of any failure of the system.

The back-up system shall be thoroughly inspected and the alarm system shall be tested at least once every seven days in order to check that there is no defect in the system and, if any defect is found it shall be rectified immediately.

All automated equipment upon which the birds' welfare is dependent, must incorporate a fail safe and/or standby device and an alarm system to warn the flock-keeper of failure.

2.1.14. Transportation

Chicks for dispatch should be healthy and vigorous, and should be placed in suitably ventilated boxes without overcrowding. Care should be taken to ensure adequate ventilation of the boxes, particularly when they are stacked, and to protect the chicks, especially ducklings, from direct sunlight and cold draughts.

Among other requirements, that no person shall transport any animal in a way which causes or is likely to cause injury or unnecessary suffering to that animal.

No person shall transport any animal unless: (a) it is fit for its intended journey, and (b) suitable provision has been made for its care during the journey and on arrival at the place of destination.

For these purposes an animal shall not be considered fit for its intended journey if it is ill, injured, infirm or fatigued, unless it is only slightly ill, infirm or fatigued and the intended journey is not likely to cause it unnecessary suffering.

Any person shall ensure that the animals are transported without delay to their place of destination.

In the case of animals transported in a crate or a container, any person in charge of animals shall ensure that they are not caused injury or unnecessary suffering while they are in the crate or a container either waiting to be loaded on to the means of transport or after they have been unloaded.

Means of transport and crate or a container shall be constructed, maintained, operated and positioned to provide adequate ventilation and air space.

Crate or a container in which animals are carried shall be:

1. Constructed and maintained so that they allow for appropriate inspection and care of the animals.
2. Of such a size as to protect the animals from injury or unnecessary suffering during transport.
3. Constructed and maintained so that they prevent any protrusion of the heads, legs or wings from them. The catching and handling of birds without causing them injury or stress requires skill.

Catching and handling should be carried out quietly and confidently exercising care to avoid unnecessary struggling which could bruise or otherwise injure the birds.

Unless they are caught and carried around the body (using both hands to hold the wings against the body), birds should be caught and carried by both legs. No catcher should carry by the legs more than three chickens (or two adult breeding birds) in each hand. Birds must not be carried by the wings or by the neck. One possible way of avoiding the potential for damage to the birds is to collect the birds mechanically; only devices proven to be humane should be considered for use in gathering birds.

Journeys should be carefully planned so that birds are not left on the vehicle for long periods either at the start of the journey or at their destination. The provision of adequate ventilation and protection from adverse weather and extremes of temperature are essential during loading and transport. Animals (including birds) should be unloaded as soon as possible after arrival at a slaughterhouse. After unloading, animals must be protected from adverse weather conditions and be provided with adequate ventilation.

2.1.14.1. Layers

Growing layers in cage system makes them most vulnerable during assembly, loading and unloading for fractures in the leg. The stress is evident in the form of increased adrenal activity. Physical conditions of journey are extremely important in minimizing fractures and carcass condemnations (Fraser and Broom, 1997).

2.1.14.2. Broilers

Proper precautions have to be taken during assembly, crating, loading, transportation and unloading of broilers; they are particularly sensitive to temperature and crowding during transportation (Fraser and Broom, 1997).

2.2. Commercial layers

The most serious welfare problems for the caged layers are the frustration of normal behavior and the lack of sustained fitness. They are deprived of free movement (exploration), scratching, nest boxes, dust-bathing, perches and brooding (both the eggs and the chicks). Hence, the conventional battery cages are totally unsuitable for laying chicken; the European Union has stipulated that all the member countries replace the conventional cages with modified (furnished, enriched) cages or resort to alternate methods of rearing success free range or aviary system etc. (see Chapter "Rearing systems").

Table 82.2 Welfare of layers in battery cage *Vs* Free range

System	Battery cage	Free range
Hunger and thirst	Adequate	Adequate
Comfort, thermal	Good	Variable
Comfort, physical	Poor	Usually good
Ill-health, disease	Low	Parasitism?
Ill-health, pain	Feet and legs	Injury
Behavior	Very restricted	Cannibalism?
Fear and stress	Frustration	Agoraphobia
		Source : Webster, 1994

2.3. Broilers and meat turkeys
(Webster, 1994)

Commercially broilers are grown in large flocks with near continuous lighting (23L : 1D) to reach a slaughter weight of 1.8-2.0 kg as early as 35-40 d. Similarly, turkeys also reach a slaughter weight of 8-20 kg by 16-24 weeks of age. On most occasions, they are grown on litter, fed *ad libitum* on nutritionally adequate diets, given access to continuous water, most of the diseases controlled by vaccines and they are protected from extremes of weather by properly planned houses. The main limitation as far as production aspect is concerned is the stocking density. Quite often, they are overcrowded; no matter what the efficiency in management (ventilation, litter management etc.), there is some amount of compromise with respect to welfare. The recommended maximum stocking density for broiler chickens is 34 kg bird mass per m^2; in other words, 750 cm^2 for a 2 ½ kg bird. At this stocking density, birds continue to find feed and water and problems of aggression, feather pecking and cannibalism are significantly minimized. However in case of turkeys, fighting and cannibalism are common and hence, beak trimming is virtually mandatory.

Another aspect of serious concern is the leg problems the broilers are prone to. Although when grown on deep litter most of the five freedoms are ensured, increased frequency of birds showing the symptoms of leg weakness warrants a rethink. Notwithstanding that etiology of leg weakness is still incompletely understood, that a considerable proportion of animal have to suffer to serve human beings in spite of the pain, exposes the man's inhumanity to another sentient animal. Hence, concerted study to ameliorate to leg problems in heavy strains of broiler chicken and turkeys is an urgent necessity.

2.4. Turkey breeders

2.4.1. Beak trimming

When birds are kept in daylight conditions they can be vicious, and beak trimming is an essential aid to management. It is usual to trim beaks as a routine measure before birds leave the brooder or the rearing accommodation and normally it need be done once in the lifetime of the stock. When birds are kept in buildings with a light control system, beak trimming should be carried out only when it is clear that more suffering would be caused in the flock if it were not done. Beak trimming should be done by a skilled operator or under his supervision.

2.4.2. Desnooding

If desnooding is done, it should be as soon as possible after hatching. A veterinary surgeon must carry out the operation if it is performed after the first 21 days of life.

2.4.3. Saddling of hens

Before hens are mated they should be fitted with strong saddles, made for example of canvas, to prevent injury to the backs and sides by the males.

2.4.4. Toe cutting

To avoid injury to hens during mating, even when saddled, the last joint of the inside toes of the male breeding birds should be removed. This must be done within the first 72 hours of life. A veterinary surgeon must carry out the operation if it is performed after the first 72 hours of life.

2.5. Ducks

In addition to the guidelines mentioned above, the following are relevant to ducks:

All floors, particularly slatted or metal mesh ones, should be designed, fitted and maintained so as to avoid injury or distress to the birds. Remedial action should be taken if either of these occurs. The front and sides of raised pens for ducklings should be kept properly adjusted so that birds have access to feed and water but cannot escape and fall to the floor. Nest boxes and roosting areas should not be so high above floor level that birds have difficulty or risk injury in using them. Adequate litter should be provided on solid floors and in nest boxes.

A newly-hatched duckling has poor control of its body temperature. Environmental conditions during the early part of its life should therefore allow it to maintain its normal body temperature without difficulty.

2.5.1. Handling

Day-old and young ducklings should be picked up bodily in the palm of the hand. It may be necessary to catch older ducks by the neck and they should be supported either by taking the weight of the bird by a hand placed under its body, or by holding the bird with a hand on either side of its body with the wings in the closed position. Birds should never be carried by the legs.

2.5.2. Bill trimming

Bill trimming should be carried out only when it is clear that more suffering would be caused in the flock if it were not done. It should be done by a skilled operator or under his supervision. If practiced, only the rim at the front of the upper bill should be removed and before the birds leave the brooder or the rearing accommodation. Normally it need be done only once in the lifetime of the stock.

2.6. Prohibited operations

Mutilations can cause considerable pain and therefore constitute a major welfare insult to farm animals. Whenever absolutely necessary, mutilations should be carried out humanely, by trained, competent staff. High standards of hygiene are essential.

2.6.1. Beak-trimming

The operation of beak-trimming (sometimes known as debeaking) means the removal from a bird by means of a suitable instrument of (i) not more than a one-third part of its beak, measured from the tip towards the entrance of the nostrils, if carried out as a single operation; or (ii) not more than a one-third part of its upper beak only, measured in the same way; and the arrest of any subsequent hemorrhage from the beak by cauterization.

Beak trimming of birds reared for meat should not be necessary because they are normally slaughtered before reaching sexual maturity. Beak trimming of breeding chickens should be avoided if at all possible, and used only if veterinary advice is that the procedure is essential to prevent worse welfare problems of injurious feather pecking and cannibalism.

It is unnecessary to beak trim female breeding chicks and only the tip of the beak should be removed from male breeding chicks. This is best done at 5 to 10 days of age in order to allow the chicks to establish eating and pecking behaviors before the operation takes place.

Beak trimming of older birds should only be carried out when advised by a veterinary surgeon.

2.6.2. Wing mutilation

De-winging, pinioning, notching or tendon severing, which involves mutilation of wing tissues. When it is necessary to reduce the effects of flightiness, the flight feathers of one wing may be clipped.

2.6.3. Blinkers

The use of blinkers which pierce the nasal septum. Other forms of device fitted to bird's heads (such as spectacles, contact lenses and nasal bits) may also cause welfare problems and should not be used.

2.6.4. Castration and devoicing

Surgical castration and devoicing: Operations on birds (other than feather clipping) to impede their flight and the devoicing or surgical castration of male birds.

2.6.5. Dubbing

The removal of all, or part, of the male comb is known as dubbing. Removal of the comb offers few, if any, welfare advantages in comparison with the disturbance and pain likely to be caused and should be avoided. Where the operation occurs, it is usually performed when the chicks are one day old using sharp scissors and should only be undertaken by appropriately trained personnel. Once chicks are over 72 hours old, the procedure must only be carried out by a veterinary surgeon

2.6.6. Despurring

This is the removal, at day-old, of the spur bud on the back of the male's leg using a heated wire. If the spur grows to be very pronounced it may cause damage to females during mating. Selection of breeding male stock with the genotype of short, blunt spurs should be encouraged, so that routine despurring should not be necessary.

2.6.7. Claw removal

Some parts of the industry remove the dew and pivot claw from the feet of breeding males to prevent damage to females during natural mating. The procedure is usually carried out at day-old and must be carried out by a trained, competent person.

Toe removal (cutting) for purposes of identification is an unnecessary mutilation and should be avoided. Instead alternative methods of identification should be used that do not adversely affect the chicks' welfare.

2.7. Records

Records are an essential aid to management and those kept should include:

1. The number and sex of chicks placed.
2. Daily mortality and the number and average weight of birds removed for slaughter or when thinning the flock to reduce stocking density.
3. Number of culls with reason for cull to be recorded (leg culls to be specifically identified).
4. Where possible, feed consumed (daily and cumulative).
5. Body weight in relation to expected growth rates.
6. The internal floor area of the house.
7. Daily water consumption (water meters should be fitted in each house).
8. Testing and maintenance of automatic equipment, including alarms, fail safes, fire extinguishers and stand-by generators.
9. Daily maximum and minimum temperature.
10. The lighting regime - intensity and duration.
11. Dates of cleaning/disinfection and bacterial counts between placements.
12. Veterinary consultation, date and outcome.
13. Medicine and vaccine administration records.

3. Abnormal behavior

(Fraser and Broom, 1997)

Birds deprived of basic welfare needs frequently exhibit abnormal behavior. Identification of the abnormality and timely rectification will definitely be reflected in better performance of the entire flock. Some of the abnormalities are give below:

3.1. Stereotypies

A stereotypy is repetition of a sequence of movements several times with little or no variation. Stereotypies occur in situations where the individual lacks control of its environment, as in case of frustration. Pacing or route-racing by hens is an example when they are hungry or when there are expecting feed or before oviposition is no nest material is available. Similarly, head-shaking or-nodding is seen in fowls wherein they show rotary movement of the head with a series of rapid sided turns ending with a slight downward movement. The stereotypy has been found the increase in presence of an observer in certain strains of birds.

3.2. Self-or environment-directed

Some caged birds pull their own feathers; chicks and turkeys may over-eat the little material when they are not provided with sufficient feed leading to digestive problems; are some of the self-directed abnormal behavior recorded in poultry.

3.3. Addressed to another animal

Egg eating is one of the abnormal behaviors coming under this category. This is recorded in birds on litter as well as on wire floors. It begins with a bird pecking at an egg till it is broken and then developing a habit of consuming the contents. The other birds also quickly mimic and the vice spreads quickly to the entire flock. Even the egg shells may be eaten giving a false impression that the feed is deficient in calcium.

Feather-pecking, body-pecking and eating pecked matter is another condition found in various species of poultry falling under this category. Crowding of birds with the few objects for them to peck at has been thought to be one of the reasons for development of this condition. Feather pecking may stimulate cannibalism when the injury is sufficient to cause hemorrhage.

4. Euthanasia

(www.ucdavis.edu)

Euthanasia is a humane death that occurs with a minimum of pain, fear, and distress. In any poultry production system, it is inevitable that some birds will become ill, debilitated, or injured. If the bird is unlikely to respond favorably to treatment, or if treatment is not feasible because of economic or public health considerations, euthanasia may be the best option to prevent the bird from suffering. In addition, healthy spent hens may be killed on-farm for subsequent rendering because their low market value makes it impractical to send them to a processing facility.

4.1. Decision Making

Questions to be considered in deciding whether a sick, debilitated or injured bird should be euthanized include:

1. Is the bird experiencing pain or distress?
2. Is recovery likely?
3. Is the bird likely to transmit disease to other birds?
4. Is the bird able to access the feed and water?
5. Can the bird be treated?
6. Is the bird or its eggs suitable for human consumption, or will they be suitable for consumption after recovery or treatment?

General economic considerations may also play a role in deciding whether or not to euthanize a bird.

4.2. Considerations

4.2.1. Poultry Welfare

The method chosen should minimize the pain and distress experienced by the bird. However, the choice of techniques may be limited in certain environments. In all cases, proper restraint can help to decrease the bird's fear and distress. When possible, poultry should be held gently in an upright position with their wings closed to prevent flapping, not carried upside down by the legs. Covering the eyes with a hand or a piece of cloth exerts a calming effect, as does holding the bird in contact with the handler's body.

4.2.2. Human Safety

The method chosen should not pose undue risks to the individual performing the euthanasia. Some methods are more dangerous than others, and should only be used under controlled conditions with proper equipment or protection.

4.2.3. Skill

Appropriate training of personnel is important to ensure that poultry are euthanized appropriately. Untrained personnel in an emergency situation can use some methods, while others, like cervical dislocation, require skill and training to carry out correctly.

4.2.4. Aesthetics

Some methods may be objectionable to the person performing the procedure because of blood loss or involuntary reflex movements by the bird. Personnel that may euthanize birds should be trained to understand how birds respond to particular euthanasia methods.

4.2.5. Cost

Some methods are more costly than others are. Some have initial costs associated with the purchase of equipment, but are thereafter inexpensive.

4.2.6. Limitations

Some methods may be suitable for only certain ages or types of poultry. In addition, some methods involve administration of controlled drugs by a veterinarian.

4.3. Methods

4.3.1. Cervical Dislocation

If carried out near the head area, dislocation of the neck vertebrae from the cranium damages the lower brain region, causing rapid unconsciousness. In order to be humane, dislocation must cause severance of the brain from the spinal cord and carotid arteries. This is best achieved using a stretching motion rather than by crushing the vertebrae. Training of personnel is critical. Small birds can be dislocated by applying a rotational movement to the neck. Adult poultry should be held by the shanks with one hand, and the head grasped immediately behind the skull with the other hand. The neck is then extended and dislocated using a sharp downward and backward thrust.

The necks of larger or heavily muscled birds like broiler breeders, turkeys, geese, ratites, and waterfowl are extremely difficult to dislocate. It is therefore recommended that other methods like captive bolt or gas euthanasia be used for birds weighing more than 3 kg.

Flapping and other body movements may persist for several minutes after cervical dislocation, although if the vertebrae have been properly dislocated these are reflex reactions. Securing the bird's wings prior to performing the dislocation can prevent involuntary flapping. To ensure death, the bird's throat should be cut after cervical dislocation. If large numbers of birds are to be euthanized cervical dislocation is not an appropriate method because personnel performing the procedure rapidly become fatigued due to the physical effort required.

4.3.2. Argon

Argon gas is an acceptable method for killing all poultry species except waterfowl, and is not an irritant like CO_2. Exposure to argon causes hypoxia. A concentration of 90% argon in air, or a mixture of argon and CO_2, should be used to for euthanasia of newly hatched fowl chicks, ratites, and poults. Older birds should be euthanized using argon with less than 2% residual oxygen.

4.3.3. Carbon Dioxide (CO_2)

Carbon dioxide causes rapid onset of anesthesia with subsequent death due to respiratory arrest. Death occurs in 2-5 minutes depending on the species and concentration of CO_2 used. Poultry can be euthanized using carbon dioxide gas by being placed in containers that are sufficiently airtight to maintain CO_2 at desired level. Depending on how many birds are being euthanized, a circulation system may be necessary to ensure that the gas does not become stratified. Birds should be added to the chamber gradually so that proper CO_2 levels are maintained. CO_2 should always be delivered from vapor delivery cylinders or, if from a liquid delivery cylinder, vaporized first to prevent it from turning into dry ice. To meet the criteria for humane euthanasia, birds already in the chamber must be unconscious before being overlain by other birds loaded after them, and unconsciousness must be maintained until death occurs.

Domestic fowl chicks should be euthanized using a concentration of CO_2 of at least 80% in air; higher concentrations (at least 90%) are required for newly hatched turkey poults and ratite chicks. However, such high concentrations of CO_2 are repugnant to adult birds. Adult chickens should be killed using approximately

50% CO_2 in air. A mixture of 30% CO_2 and 60% argon or 90% argon (with less than 5% residual oxygen) is effective and less aversive to adult chickens than CO_2 alone. CO_2 is not an acceptable method for killing waterfowl.

It is especially important to confirm death when birds are euthanized using gas, since they can appear dead but then regain consciousness. Containers in which birds are euthanized should be clear or have a window through which the birds can be observed. When large numbers of poultry are to be killed, for instance during the depopulation of spent hen flocks, it is important that CO_2 be injected frequently into the chamber to maintain these levels.

4.3.4. Carbon Monoxide (CO)

Carbon monoxide is a relatively rapid and effective method of euthanasia for birds. Carbon monoxide combines with the hemoglobin in the red blood cells in preference to oxygen, causing hypoxia. Only a pure, commercially compressed source of CO should be used. Vehicle exhaust is not an acceptable source of CO for euthanasia because it is hot and contains contaminants. High levels of CO are deadly to humans, and chronic exposure of pregnant women to even low levels of CO can cause birth defects. Only well-trained personnel should therefore use carbon monoxide and then only under properly controlled circumstances. The gas should be delivered into tightly sealed containers and the area around the containers monitored for leakage. Depending on how many birds are being euthanized, a circulation system may be necessary to ensure that the gas does not become stratified. The gas being extremely toxic to humans forcing too many rigid restrictions has limited the use of the method to experimental situations only.

4.3.5. Gunshot

Larger birds like ratites can be euthanized by gunshot directly to the head, causing extensive damage to the brain. The gun must be correctly positioned to ensure that the brain is destroyed. Care must be taken to ensure human safety when using firearms. It is recommended that the carotid arteries and jugular veins be severed immediately afterwards to ensure death.

4.3.6. Captive Bolt

Captive bolt pistols designed for livestock can be used to euthanize larger poultry species like waterfowl and ratites. The pistol should be applied correctly. Because there is motion after use of the captive bolt, it is advisable to restrain the bird to prevent injury to personnel. It is recommended that the carotid arteries and jugular veins of the bird be severed immediately afterwards to ensure death.

4.3.7. Electrocution

Electrocution is a rapid and acceptable method of euthanasia provided that a sufficient current passes first through the brain to ensure unconsciousness, and then through the heart to induce cardiac arrest. Specialized equipment is required to ensure humaneness and personnel safety.

4.3.8. Exsanguination/Decapitation

Birds can be killed by severing the jugular veins, carotid arteries, and trachea. Full decapitation also results in a rapid decrease in blood pressure and brain stem trauma. However the blood vessels may seal after being severed, delaying the onset of unconsciousness, and brain responses do persist for a brief period of time after decapitation. For this reason, exsanguination or decapitation should only be used as sole methods of euthanasia in extreme emergencies involving animal suffering where alternative methods are not feasible because of lack of equipment or trained personnel. Exsanguination and decapitation are acceptable methods of euthanasia when the bird is first stunned or anesthetized. Hand-held electrical stunning knives are available for stunning and exsanguinating chickens and turkeys, although these do pose personnel dangers if used in an area where there are wet surfaces. Birds can also be stunned first by administering a blow to the head.

4.3.9. Maceration

Maceration in a high-speed grinder results in rapid death, and is considered a humane method for disposing of young chicks and embryonated eggs. Only grinders specifically designed for disposal of poultry, which have blades that turn at 5000 or more revolutions per minute, should be used for this purpose. The grinder should be properly maintained and must not be overloaded, since birds may be incompletely macerated under these circumstances.

4.3.10. Anesthetic Overdose

When properly administered by the intraperitoneal route, barbiturate overdose produces rapid unconsciousness and anesthesia followed by respiratory depression and cardiac arrest. Poultry euthanized using barbiturates must be properly disposed of in accordance with local regulations.

4.4. Confirmation of loss of consciousness and death

Confirmation of death is critical regardless of the method chosen. The cessation of reflexes in the head area can be used to confirm loss of consciousness:

1. Lack of response to a hard pinch delivered to the comb, wattles, or snood
2. Lack of blink reflex when the eye is touched
3. The following signs can be used to confirm death: Cessation of respiration, Cessation of heartbeat

5. Poultry as experimental animal

Most of the countries do have their own regulatory authorities to monitor the use of animals for experimentation. InIndia Committee for the Purpose of Control and Supervision of Experiments on Animals (CPCSEA) has been instituted during 1998. The committee has been able to create greater awareness among scientists about the need to treat animals humanely and also has the credit for improving the conditions of animal houses across the country. However, CPCSEA's real job is to facilitate scientific research within the regulatory framework balancing even apparently conflicting objectives between animal activism and scientific research. Scientists on their part should try to adopt the '3R' philosophy – Replace animals by *in vitro* methods wherever possible, Reduce the number of animals needed for experiment and Refine experimentation to lessen pain and distress, in letter and spirit.

Biotechnology and genetic engineering are being used to tinker with animals for profit without considering the animal welfare aspects. For instance, what will be the impact on the animal (poultry) as far as taste and suitability of the feed when animal feedstuffs are manipulated to increase nutrient yield and composition? There are so many such research projects being taken up having very doubtful safety norms as far as animal welfare is concerned. Animals are being isolated during the reader experimentation which in itself can cause stress and modify behavioral and physiological responses. However, poultry are primarily exposed to genetically modified feedstuffs rather than being directly employed as experimental animals for biotechnological research. It is indeed important to note that biotechnology has come to stay and all concerned have to make suitable modifications or adjustments on the principle of "give and take" keeping the welfare of both human beings and animals.

Appendix - 1

Bird flu in India

First case of bird flu has been reported from Navapur (Nandurbardistrict), Hated, Sawada, Salve and Marul (Jalgaon district,) of Maharashtra state. The disease has been confirmed by the High Security Animal Disease Laboratory (HSADL) at Bhopal. As per the reports till the end of 2nd week of March, 2006, no human case has been reported. The disease spread to Madhya Pradesh (Ichchapur in Burhanpur district; close to Jalgaon of Maharashtra) by the end of March.

However, as per the report of WHO, 185 human infections with 104 deaths has been recorded since the first case of AI during 2003. It is also predicted that 5 to 150 m human deaths is a possibility, if the infection spreads between humans due to some changes in the virus. Already, the disease has been detected in dogs and cats, the pets which are in close contact with humans. Notwithstanding these, it is generally thought that only cells in the deep respiratory system of humans contain receptors for the virus particles thereby necessitating a huge dose of virus infection and sustained proximity to the infected birds to actually cause human infection. This also explains the fact that the humans affected are those who were in close contact with affected poultry. It is hoped that the disease doesn't become a human to human pandemic. In any case, there appears to be no way to prevent the virus from becoming endemic in birds everywhere posing perennial problems in future. Meanwhile, the Asia Development Bank (ADB) has pledged $ 470 m to fight the disease.

All steps necessary to curtail the disease have been taken in India. For instance, approximately 400,000 birds have been culled and/or killed and disposed under technical supervision. Similarly, about 1,475,000 eggs have been destroyed. All poultry farms are rigorously practicing optimum sanitation and disinfection procedures. House-to-house surveys for identifying possible sick bird(s) and also to spread awareness regarding the disease have been taken-up on war footing. Strict vigil on vehicle movement from and to the infected area/state has been initiated. Rapid Response Team (RRT) is constituted. Protective gears have been supplied to all the personnel involved in culling and disposal of birds. In addition to the available stock of 100,000 doses of Tamiflu, additional 50,000 doses are being procured. Wherever feasible and vaccine available, birds are vaccinated with H5N2 vaccine. All the neighboring states and Health Department personnel are kept under red alert to obviate any contingency. Print and electronic media are put to service for spreading day-to-day updates about the disease in India.

In spite of all measures, there has been a sudden crash in prices of both eggs and poultry meat (the prices have fallen to a tune of 40 to 60%). Till now, about Rs 15 m has beer distributed as compensation to the farmers. Many farmers are going for distress sales and the small and marginal farmers are the worst hit. Print and electronic media are trying their level best to inform the public that cooking beyond 70°C makes the food absolutely safe. Poultry food *melas* are being conducted in worst-hit areas to make the people believe that properly cooked poultry products are totally safe. Notwithstanding all the measures, it is estimated that the industry is incurring a loss of about Rs 2000 m per day. As per the report of *Financial express* (24th April, 2006), the Poultry industry has suffered a loss of about Rs 90 billion.

As a relief measure, working capital has been converted as loan and a 4% reduction in interest rate for all loans has been ordered. Further, one year moratorium for refund of loan has been granted.

Appendix - 2

Retransformation of *arcsin* values to the original scale

Many traits in biological experiments are recorded as percentage values; the following are some of the examples pertaining to poultry:

1. Binomial traits like fertility, hatchability, survivability (mortality)
2. Proportions like albumen, yolk and shell as a percentage of egg weight, Cut-up parts as a percentage of live-weight and the like
3. Certain values as a percentage of the Control treatment like weight gain in a particular treatment group as a percentage of control

These percentage values do not follow normal distribution which is a prerequisite for applying routine statistical procedures like Analysis of Variance, Tests of significance etc., especially if the values are below 20 % or above 80 %.

Such data are recommended for an *arcsin* transformation as follows:

Let x be the percentage value and y, its arcsin value; then

$$y = \sin^{-1}\left(\sqrt{\frac{x}{100}}\right)$$

The statistical analyses will be performed replacing each of the x values with its corresponding y value calculated as above. In fact, ready-reference tables are available for obtaining *arcsin* values.

After the statistical analyses, arcsin Mean ($\bar{y}$) and *arcsin* Standard deviation (SD_{arc}) values are calculated and tests of significance performed for significant effects, if any.

Finally, when the results are tabulated, $\bar{y}$ is retransformed to the original scale as follows:

Let $\bar{x}$ be the mean in the original scale; then $\bar{x} = 100\left(\sin^2 \bar{y}\right)$

However, for retransformation of SD_{arc} values, no procedure appears to be available.

It is in the fitness of things to note that microbiological data are subjected to logarithmic transformation and as in *arcsin* transformation, the statistical analyses are performed utilizing the log values. The mean value on the logarithmic scale is retransformed to the original scale by taking the antilog of the log mean value.

For retransformation of log SD value, the following procedure is shown:

Let sd be the standard deviation in the original scale; then

$$sd = \left[\frac{\text{antilog}\left(\bar{X}+SD\right) - \text{antilog}\left(\bar{X}-SD\right)}{2} \right]$$

Where $\bar{X}$ = log mean and SD = log SD

On the same lines, SD_{arc} can be retransformed to original scale by

$$sd = \left[\frac{100\left\{\sin^2\left(\bar{y}+SD_{arc}\right)\right\} - 100\left\{\sin^2\left(\bar{y}-SD_{arc}\right)\right\}}{2} \right] \text{--------(1)}$$

Where $\bar{y}$ = *arcsin* mean. It is evident that retransformation of SD_{arc} values by the above formula is quite circuitous and hence time consuming. Therefore, it is necessary that by applying certain basic trigonometric rules, the equation (1) be simplified into a computational formula.; the same is enumerated below:

$$\text{Equation (1)} = 50\left[\sin^2\left(\bar{y}+SD_{arc}\right)\right] - \left[\sin^2\left(\bar{y}-SD_{arc}\right)\right] \text{-----(2)}$$

Let $A = \bar{y}+SD_{arc}$ and $B = \bar{y}-SD_{arc}$

$$\text{Then, } sd = 50\left(\sin^2 A\right) - \left(\sin^2 B\right) \text{-----------(3)}$$

$$\therefore\ sd = 50(\sin A + \sin B)(\sin A - \sin B) \text{----------(4)}$$

By applying trigonmometric identity we get

$$sd = 50\left[2\sin\left(\frac{A+B}{2}\right)\cos\left(\frac{A-B}{2}\right)\right]\left[2\sin\left(\frac{A-B}{2}\right)\cos\left(\frac{A+B}{2}\right)\right] \text{-------(5)}$$

By rearranging the terms, we get

$$sd = 50\left[2\sin\left(\frac{A+B}{2}\right)\cos\left(\frac{A+B}{2}\right)\right]\left[2\sin\left(\frac{A-B}{2}\right)\cos\left(\frac{A-B}{2}\right)\right] \text{---------(6)}$$

$$\text{Since } \left[2\sin\left(\frac{A+B}{2}\right)\cos\left(\frac{A+B}{2}\right)\right] = \sin(A+B)$$

$$\text{and } \left[2\sin\left(\frac{A-B}{2}\right)\cos\left(\frac{A-B}{2}\right)\right] = \sin(A-B),$$

by substituting in equation (6), we get

$$sd = 50[\sin(A+B)\sin(A-B)] \quad \text{----------}(7)$$

Now, by substituti ng $A = \bar{y} + SD_{arc}$ and $B = \bar{y} - SD_{arc}$ in equation (7),

we get $$sd = 50\,(\sin 2\,\bar{y})(\sin 2\,SD_{arc}) \quad \text{---------}(8)$$

and this is the computatio nal formula for retransfor ming the SD value in the arcsin scale to the original percentage scale.

That means, standard deviation in percentage scale is equal to 50 times the product of the sine of twice the *arcsin* mean and sine of twice the *arcsin* standard deviation (Sreenivasaiah, 1985).

This short-cut method is accurate, simple and time-saving.

A Sample data pertaining to hen-housed egg production (%) over a period of 20 days is shown below to demonstrate the use of the short-cut method for retransformation of mean and standard deviation on *arcsin* scale back to the percentage scale.

Hypothetical data	
% Hen-day egg prodn	*Arcsin* value
8	16.43
10	18.43
10	18.43
18	25.10
12	20.27
16	23.58
25	30.00
20	26.57
30	33.21
38	38.06
45	42.13
40	39.23
42	40.40
39	38.65
40	39.23
48	43.85
56	48.45
35	36.27
40	39.23
36	36.87

The result of calculations			
	Percentage scale	*Arcsin* scale	Retransformed (%)
Sum	608	654.39	NA
Sum of squares	22428	23173.6395	NA
Mean	30.40	32.72	29.22
Standard deviation	14.41	9.61	14.97

It is clear from the above that the retransformation proposed can be successfully used.

Abbreviations

a/α	Average effect of the gene
AA	Arachidonic acid
Ab	Antibody
A/c	Account
ADA	American Dietitics Association
ADP	Adenosine diphosphate
AE	Avian encephalomyelitis
Ag	Antigen
AGID	Agar gel immunodiffusion
AGP	Antibiotic growth promoters/Agar gel precipitation
AHA	American Heart Association
AI	Avian influenza / Artificial insemination
AME	Apparent metabolizable energy
ANACOVA	Analysis of covariance
ANOVA	Analysis of variance
Apo	Apoprotein
AI	Artificial insemination
ATPase	Adenosine tri-phosphatase
ATP	Adenosine tri-phosphate
ACTH	Adreno-corticotropic Hormone
AI	Artificial Insemination
BHC	Benzehexachloride
BHT	Butylated hydroxyl toluene
BIS	Bureau of Indian Standards
BMI	Body mass index
BMR	Basal metabolic rate
b/d	Brought down
bp	Base pairs
BSE	Bovine spongiform encephalopathy
BV	Breeding value
CAA	Chicken anemia agent
CaBP	Calcium Binding Protein
CAM	Chorio-allantoic membrane
c/d	Carried down
cDNA	Copy deoxyribose nucleic acid
CDS	Cluster designation surface
CE	Competitive exclusion
cfu	Colony forming units
CHD	Coronary heart disease
CHO	Carbohydrate
CLA	Conjugated linoleic acid
cm	Centimeter(s)

CM	Chylomicrons
CMI	Cell-mediated immunity
COFAL	Comlement fixation for avian leucosis
COV(D/S)	Covariance (dams/sires)
COV(S)	Covariance (sires)
COV(W)	Covariance (progenies/dams/sires or error)
CP	Crude protein / Calorie-Protein
CPCSEA	Committee for purpose of control and supervision of experiments on animals
CPE	Cytopathic effects
Cr	Credits
CRD	Chronic respiratory disease
CRP	C-reactive protein
Cf	Unlike
CoA/CoASH/SCoA	Coenzyme A
CSF	Cerebro-spinal fluid
CT	Connective tissue
Da	Daltons
d	Day(s)
°	Degree(s)
°C	Degree Celsius
DAG	Diacylglycerol
dATP	Deoxyribose adenosine triphosphate
DCP	Digestible crude protein
dCTP	Deoxyribose cytosine triphosphate
DDGS	Distillers dried grain with solubles
DDE	Dichloro diphenyl dichloroethane
DDS	Distillers dried solubles
DDT	Dichloro diphenyl trichloroethane
DE	Digestible energy
dGTP	Deoxyribose guanidine triphosphate
DHA	Docosahexaenoic acid
DLA	Dihomo γ-linoleic acid
DM	Dry matter
DNA	Deoxyribose nucleic acid
dNTPs	Deoxyribose nucleotide triphosphates
DON	Deoxynivalenol
DPA	Docosapentaenoic acid
Dr	Debits
dTTP	Deoxyribose thymine triphosphate
EDS	Egg drop syndrome
EDTA	Ethylene diamine tetra acetic Acid
ELISA	Enzyme-linked immuno-sorbent assay
EMS	Expected mean square
EMSCP	Expected mean sum of cross products
EPA	Eicosapentaenoic acid
EU	European union

Abbreviations

FA	Fatty acid
FABP	Fatty acid binding protein
FAD/$FADH_2$	Flavin adenine dinucleotide/reduced form
FAO	Food and Agricultural Organization
FAS	Foreign Agricultural Service
FAT	Fluorescent antibody technique
FCR	Food Conversion Ratio
FDA	Federal drug administration
FE	Fecal energy
FEN	Fenvalerate
FFA	Free fatty acid
FLHS	Fatty liver hemorrhagic syndrome
FLKS	Fatty liver kidney syndrome
FLS	Fatty liver syndrome
FMN	Flavin mononucleotide
FSH	Follicle Stimulating Hormone
g	Gram(s)
GE	Gross energy/Genotype-environment interaction
GIT	Gastro-intestinal tract
GLN	γ-linolenic acid
GMO	Genetically modified organism
GNC	Groundnut (Peanut)cake
GnRH	Gonadotropin Releasor Hormone
GXP	Good manufacturing practice
h^2	Heritability
HA	Hemagglutination
HACCP	Hazard analysis of critical control points
HAT	Hypoxanthine, aminopterin and thymine
HAU	Hemagglutination unit
HCB	Hexachlorobenzene
HCH	Hexachlorohexane
HDL	High density lipoprotein
HI	Hemagglutination inhibition
HPAI	Highly pathogenic avian Influenza
HPE	Histo-patholgoical examination
HPRT	Hypoxanthine phosphoribosyl transferase
hr	Hour(s)
HSCAS	Hydrated sodium calcium aluminosilicate
HUFA	Highly unsaturated fatty acid
HVT	Herpes virus of turkeys
Hz	Hertz
I	Index
IB	Infectious bronchitis
IBD	Infectious bursal disease
ICPI	Intracerebral pathogenicity index
ICU	International chick units
IDL	Intermediate density lipoprotein

IFN	Interferon
Ig	Immunoglobulin
ILT	Infectious laryngotracheitis
In	Inch(es)
IP	Identity preserved
ISS	Infectious stunting syndrome
IU	International unit
IVPI	Intravenous pathogenicity index
J	Joule
J.F	Journal folio
kbp	Kilo base pairs
kg	Kilogram(s)
kJ	kilo joule
km/h	Kilometers per hour
kPa	Kilo Pascals
kWh	Kilo Watt hour
L/l	Liter
LA	Linoleic acid
LD_{50}	Lethal dose 50
LDL	Low density lipoprotein
L.F	Ledger folio
LHS	Left hand side
LLV	Lymphoid leucosis viruses
LNA	Linolenic acid
LP	Lipoprotein
LPAI	Low pathogenic avian influenza
LPP	Linear programming problem
LT	Leukotrienes
LX	Lipoxins
lb	Pound(s)
$<$	Less than
$\leq$	Less than or equal to
LH	Leutinizing Hormone
m	Meter(s)
MABs	Monoclonal antibodies
MAG	Monoacylglycerol
MAS	Marker-assisted selection
MATSA	Marek' disease tumor associated surface antigen
MD	Marek's disease
MDPM	Mechanically deboned poultry meat
MDT	Mean death time
ME	Metabolizable energy
MG	*Mycoplasma gallisepticum*
MI	Myocardial infarction/*Mycoplasma iowae*
M.I.U	Moisture, impurities and unsaponifiables
MJ	Mega Joule
m^3/min	Cubic meter per minute

Abbreviations

MM	*Mycoplasma meleagridis*
mmol	milli mole
MOS	Mannan oligosaccharide
MPA	Methyl parathion
MRT	Margial rate of transformation
MRTS	Marginal rate of technical substitution
ms	millisecond
MSS	Mean sum of squares
MCP_S	Mean sum of cross products (sires)
$MCP_{D/S}$	Mean sum of cross products (dams/sires)
$MCP_{P/D/S}$	Mean sum of cross products (progenies/dams/sires)
MCP_W	Mean sum of cross products (progenies/dams/sires or Error)
mRNA	Messenger ribose nucleic acid
MSS_E	Mean sum of squares (Error)
MSS_S	Mean sum of squares (sires)
$MSS_{D/S}$	Mean sum of squares (dams/sires)
$MSS_{P/D/S}$	Mean sum of squares (progenies/dams/sires)
MS_W	Mean sum of squares (progenies/dams/sires or Error)
mt	Metric tonne(s)
MUFA	Monounsaturated fatty acid
μm	Micrometer
mm	Millimeter(s)
min	Minute(s)
>	More than
≥	More than or equal to
NABARD	National bank for agriculture and rural development
NAD/NADH/NADH+H^+	Nicotinamide adenine dinucleotide/reduced for
NADP	Nicotinamide adenine dinucleotide phosphate
ND	Newcastle disease
NDF	Non-digestible fiber
NDV	Newcastle disease virus
NE	Net energy / Necrotic enteritis
NE_M	Net energy (maintenance)
NE_P	Net energy (production)
NRC	National research council
NSP	Non-starch polysaccharides
OA, OB, OM, ON	Ochratoxin forms
OC	Organochlorine
OP	Organophosphate
OPD	*O*-phenylenediamine
OPIDN	Organophosphate-induced delayed neurotoxicity
OR	Odds ratio
Pa	Pascals
PA	Pre-albumin
PAH	Polycyclic aromatic hydrocarbon
PBS	Phosphate buffer solution
PCB	Polychlorinated biphenyl

PCR	Polymerase chain reaction
PDI	Pellet durability index
PHS	Pulmonary hypertension syndrome
PPLO	Pleuropneumonia like organisms
PPP	Prepatent period
PTC	Plasma thromboplastin components
PTH	Parathyroid Hormone
%	Percent
PC	Prostacyclin
PG	Prostaglandin
PL	Phospholipid(s)
ppm	parts per million
ppb	parts per billion
PUFA	Polyunsaturated fatty acids
QAC	Quaternary ammonium compounds
QTL	Quantitative trait loci
r	Correlation coefficient (product-moment)
r_e	Environmental correlation
r_g	Genetic correlation
r_p	Phenotypic correlation
r_I	Intraclass correlation
R	Repeatability/Intraclass correlation
RBC	Red blood corpuscle(s)
RBP	Retinyl-binding protein
RD	Ranikhet disease
RDA	Recommended daily allowance
RE	Retinyl esters/Restriction enzyme
RGR	Relative Growth Rate
RHS	Right hand side
RIF	Resistance-inducing factor
R.N	Receipt number
RNA	Ribose nucleic acid
ROS	Reactive oxygen species
RPA	Rapid plate agglutination
rpm	rotations per minute
rRNA	Ribosomal ribose nucleic acid
RRS	Reciprocal recurrent selection
RS	Recurrent selection
RSA	Rapid serum agglutination
RT-PCR	Reverse transcriptase polymerase chain reaction
RVF	Right ventricular failure
SAT	Slow agglutination test
Sec	Second(s)
SFA	Saturated fatty acids
SI	Le Systeme International d'Unites
SN	Serum neutralization
SS	Sum of squares

Abbreviations

SCP_S	Sum of cross products (sires)
$SCP_{D/S}$	Sum of cross products (dams/sires)
$SCP_{P/D/S}$	Sum of cross products (progenies/dams/sires)
SCP_W	Sum of cross products (progenies/dams/sires or Error)
SD	Standard deviation
$SS_{D/S}$	Sum of squares (dams/sires)
SS_E	Sum of squares (Error)
$SS_{P/D/S}$	Sum of squares (progenies/dams/sires)
SS_S	Sum of squares (sires)
SS_W	Sum of squares (progenies/dams/sires or Error)
T	True value
TCR	T-cell receptor
Th	T-helper
Thpp	Thiamine pyrophosphate
TNF	Tumor necrosis factor
tRNA	Transfer ribose nucleic acid
TS '65	Turkey syndrome'65
TSH	Thyroid stimulating hormone
TX	Thromboxanes
UFA	Unsaturated fatty acids
VN	Virus neutralization
USDA	United States Department of Agriculture
UV	Ultraviolet
VFA	Volatile fatty acids
VLDL	Very Low Density Lipoprotein(s)
V.N	Voucher number
V(A)	Variance (Additive)
V(D)	Variance (Dominance)/Variance (Dams)
V(E)	Variance (Environment)
$V(E)_{PMT}$	Variance (Environment) permanent
$V(E)_{TMP}$	Variance (Environment) temporary
V(Ep)	Variance (Epistasis)
V(P)	Variance (Phenotypic)
V(S)	Variance (Sires)
Vs	Versus
VVND	Very virulent Newcastle disease
W	Watts
WBC	White blood corpuscle(s)
WHO	World health organization
WSNSPS	Water-soluble non-starch olysaccharides
yr	Year(s)

Index to Tables

Sl.	No.	Title	Page
1.	1.1	Development of global poultry industry	2
2.	1.2	Growth of poultry industry	3
3.	1.3	Number of laying hens and eggs produced	4
4.	1.4	Broiler production	4
5.	1.5	Economics of broiler and egg production	6
6.	1.6	Indian broiler industry – global competitiveness	7
7.	2.1	Mandatory requirements for sustained flight	12
8.	2.2	Characteristic features of Avians	16
9.	3.1	Sequence of domestication	19
10.	3.2	Description of wild jungle fowls	22
11.	3.3	Body size, reproductive traits and other features of *Gallus* species	23
12.	3.4	Diffusion of chicken in Indus Valley	24
13.	3.5	Zoological classification of birds (Class : *Aves*)	29
14.	4.1	Nomenclature to denote vaious species of poultry	48
15.	5.1	Economic or Type classification	49
16.	5.2	Breeds and varieties of American Class	51
17.	5.3	Breeds and varieties of Asiatic Class	52
18.	5.4	Breeds and varieties of English Class	52
19.	5.5	Breeds and varieties of Mediterranean Class	53
20.	5.6	Breeds and varieties of Continental Class (North European)	53
21.	5.7	Breeds and varieties of Continental Class (Polish)	54
22.	5.8	Breeds and varieties of Continental Class (French)	54
23.	5.9	Breeds and varieties of Game Class	54
24.	5.10	Breeds and varieties of Oriental Class	55
25.	5.11	Breeds and varieties of Miscellaneous Class	56
26.	5.12	Breeds and varieties of Ducks : Heavy weight Class	57
27.	5.13	Breeds and varieties of Ducks : Medium weight Class	58
28.	5.14	Breeds and varieties of Ducks : Light weight Class	58
29.	5.15	Breeds and varieties of Ducks : Bantam weight Class	60
30.	5.16	Plumage pattern of Ducks	61
31.	5.17	Breeds and varieties of Geese : Heavy Class	66
32.	5.18	Breeds and varieties of Geese : Medium Class	66
33.	5.19	Breeds and varieties of Geese : Light Class	67
34.	5.20	Breeds and varieties of Geese : Plumage patterns	67
35.	5.21	Varieties of Turkeys	69
36.	5.22	Breeds of Poultry	73
37.	5.23	Breeds of Ducks and Geese	74
38.	5.24	Sitters and Non-Sitters	74
39.	6.1	Scale of points : Chicken	92
40.	6.2	Scale of points : Ducks	93
41.	6.3	Scale of points : Geese	93

Index to Tables

42.	6.4	Scale of points : Turkeys	94
43.	7.1	Photo-schedule *Vs* time of ovulation / oviposition	109
44.	7.2	Carotenoids in alfalfa, yellow maize and algal meal	119
45.	7.3	Total carotenoid content (ppm) of feed ingredients	120
46.	7.4	Summary of egg formation in oviduct	121
47.	7.5	Gross composition of eggs from different species	129
48.	7.6	Surface area and volume of eggs from different species (cm^2)	130
49.	7.7	Chemical composition of egg shell calcium	131
50.	7.8	Major proteins in albumen	133
51.	8.1	Formation and differentiation during incubation	134
52.	8.2	Volume and concentration of semen	138
53.	8.3	Number of sperms *Vs* number of ejaculations	139
54.	8.4	Motility of sperms at different regions	139
55.	8.5	Composition of seminal plasma *Vs* blood plasma	141
56.	9.1	Gross composition of eggs from different species	149
57.	9.2	Gross composition of eggs, maize, rice and wheat	150
58.	9.3	Composition of albumen and yolk : Gross	151
59.	9.4	Composition of albumen and yolk : Amino-acids, mg	152
60.	9.5	Composition of albumen and yolk : Lipids, g	153
61.	9.6	Composition of albumen and yolk : Major minerals	154
62.	9.7	Composition of albumen and yolk : Trace minerals	154
63.	9.8	Composition of albumen and yolk : Vitamins	155
64.	9.9	Major proteins of albumen	156
65.	9.10	RDA of ME (in MJ) for moderately active adult	158
66.	9.11	RDA of amino-acids supplied by one egg	159
67.	9.12	RDA of minerals met by one egg	160
68.	9.13	RDA of vitamins met by one egg	160
69.	9.14	Composition of selected egg products, g/100g	161
70.	9.15	Composition of frozen, dried and cooked whole-eggs	161
71.	9.16	Tolerance of organochloride insecticides in poultry meat, fat and eggs	162
72.	10.1	Structure and functions of lipoproteins (LPs)	164
73.	10.2	Apolipoproteins and their functions	165
74.	10.3	Nomenclature of unsaturated fatty acids (UFA)	167
75.	10.4	Some important risk factors of CHD	170
76.	10.5	Fatty acid composition and content of standard hen's egg	175
77.	10.6	Fatty acid composition of lipid fractions of yolk	175
78.	10.7	Composition of low density lipoproteins of egg	176
79.	10.8	Ingredients/substances having health functions	179
80.	10.9	Composition of NEE *Vs* Commercial egg	180
81.	10.10	Nutritive value of enriched and conventional eggs	181
82.	10.11	Fatty acid profile of ω3-enriched eggs	181
83.	10.12	Change in plasma of humans after consuming two ω3 PUFA-enriched eggs with their habitual diets further period of three weeks (%)	182
84.	10.13	Fatty acid composition of breast-milk as influenced by consumption of designer eggs for six weeks	182

85.	10.14	Enrichment products to fortify hen's egg with ω3 FA	184
86.	10.15	Percentage of specific types of fat in common oils and fats	191
87.	10.1	Comprehensive cardiovascular risk profile	192
88.	11.1	Methods of scalding	201
89.	11.2	*Post-mortem* examination guide	206
90.	11.3	Minimum chilling (ageing) time (in ice and water)	207
91.	11.4	Yields (%) of poultry species (raw)	207
92.	11.5	Proportion of meat, skin and bone in chicken fryer parts (%)	207
93.	11.6	Proportion of various cuts too eviscerated weight in broilers	208
94.	11.7	BIS Grading characteristics for dressed chicken	209
95.	12.1	Composition of meat – Gross (g / 100 g edible portion)	211
96.	12.2	Composition of meat – Vitamins (per 100 g edible portion)	211
97.	12.3	Composition of meat – Minerals (mg / 100 g edible portion)	212
98.	12.4	Composition of meat – lipids (g / 100 g edible portion)	212
99.	12.5	Amino-acid composition of chicken and turkey meat (% of protein)	213
100.	12.6	Amino-acid composition of poultry meat (mg/100 g)	213
101.	12.7	Fatty acid composition of poultry tissues (% of lipid)	214
102.	12.8	Fatty acid composition of different species of poultry (% of lipid)	214
103.	12.9	Minerals and vitamins supplied by 100 g broiler meat	214
104.	12.10	Proportion of meat, skin and bone in chicken fryer parts (%)	216
105.	12.11	Proximate composition of light and dark meat (per 100 g edible portion)	217
106.	12.12	Proximate composition of broiler parts (100 g edible portion)	217
107.	12.13	Effect of processing on nutritive value of raw poultry meat	218
108.	12.14	Effect of cooking methods on lipid content of chicken meat	220
109.	12.15	Effect of cooking methods on retention of B vitamins in chicken meat (%)	221
110.	12.16	Mineral content of cooked whole carcasses (with giblets)	221
111.	12.17	Nutrient retention in cooked chicken products (%)	222
112.	12.18	Commonly used antibacterials and effects of their residues	225
113.	13.1	Dietary intake of various nutrients and fecal excretion of certain metabolites in low- and high-risk individuals	231
114.	14.1	USDA quality and grade specifications	239
115.	14.2	Modified constants for calculation of HUS for eggs of different species	246
116.	14.3	HUS calculated by the formula for chicken eggs *Vs* the modified formulae	246
117.	14.4	USDA quality based on HUS	247
118.	14.5	Off-odors and off-flavors in eggs	251
119.	14.6	Influence of time, temperature and humidity on egg quality	252
120.	14.7	Grade designations and quality of table eggs (as per BIS)	254
121.	14.8	Summary of US Standards for quality of individual shell-eggs	255
122.	14.9	Summary of UK Standards for quality of individual shell-eggs	256
123.	15.1	Color changes in a bruise over time for broiler muscle	258
124.	15.2	Pigments in fresh, cured and cooked meat	259
125.	16.1	Perishability of foods	265
126.	16.2	Water in foods	266

Index to Tables

127.	16.3	a_w *Vs* Microbial inhibition	267
128.	16.4	Minimum a_w *Vs* microbial growth and toxin production	267
129.	18.1	Rate of freezing *Vs* changes in histology of muscle (meat)	286
130.	18.2	Suggested periods for storage of poultry meat	288
131.	18.3	Reactions of nitrous acid with normal constituents of meat	292
132.	18.4	Components of smoke	293
133.	19.1	Minimum Pasteurization requirement of liquid egg products in the US	301
134.	19.2	Cooking time *Vs* degree of doneness	310
135.	20.1	Checking doneness	329
136.	20.2	Barbecuing *Vs* Grilling	338
137.	21.1	Types of microorganisms on the shell and in rotten eggs	347
138.	21.2	Egg rots produced by pure cultures	350
139.	22.1	Common contaminants during production of ready-to-cook chicken	354
140.	22.2	Spoilage of meat by *Clostridium welchii*	355
141.	22.3	Superficially recognizable symptoms of spoilage	356
142.	22.4	Temperature (range) for growth of food poisoning microorganisms (°C)	356
143.	23.1	Non-food applications of eggs	360
144.	23.2	Handling, storage and transport of slaughter-house by-products	363
145.	24.1	Incubation period of different Avian Species	370
146.	24.2	Physical conditions for chicken eggs	371
147.	24.3	Effect of temperature on growth of chicken embryo	372
148.	24.4	Dry- and wet-bulb readings (°C) for incubating chicken eggs	373
149.	24.5	Physical conditions for incubating eggs of different species	379
150.	24.6	Summary of embryo development	389
151.	24.7	Composition of albumen during incubation	393
152.	24.8	Composition of yolk during incubation	394
153.	24.9	Gross O_2 uptake and CO_2 production by chicken embryo	396
154.	24.10	Moisture content of shell and shell membranes (%)	397
155.	24.11	Embryo mortality pattern	399
156.	24.12	Amino-acid requirements of chicken embryo	400
157.	24.13	Functions of nutrients to the embryo	400
158.	24.14	Hypothetical incubation results	401
159.	24.15	Suggested mating ratios	405
160.	24.16	Nutrition *Vs* incubation performance	410
161.	24.17	Causes of poor hatchability	411
162.	24.18	Operating Temperature (°C) for still-air incubators	414
163.	25.1	Koeppen-Geiger Classification: I Level	439
164.	25.2	Koeppen-Geiger Classification: II Level	429
165.	25.3	Koeppen-Geiger Classification: III Level	430
166.	25.4	Characteristics of Tropical climate	430
167.	26.1	Behavioral and functional differences between Mammals and Avians	435
168.	26.2	Deep body temperature of selected species of birds	436
169.	26.3	Equations for calculating SMR (in W) and Total evaporative water loss (TEWL) at 25°C (in ml /d) in relation to body mass (M, kg)	

		(for all birds)	444
170.	26.4	Total evaporative heat loss (E) of birds as percentage of metabolic heat production (H) at high ambient temperatures	445
171.	26.5	Effect of temperature on heat output	448
172.	26.6	Body weight, thermoneutral zone (TNZ) and oxygen consumption in Rhode Island Red birds	452
173.	26.7	SMR ($W/kg^{0.75}$) in young *Vs* adults	452
174.	27.1	Effect of temperature on laying birds	464
175.	27.2	Adverse effects of temperature on birds	465
176.	27.3	Effect of temperature on White Leghorn layers	465
177.	27.4	Heat production of fowls ($kJ/W^{0.75}/d$)	466
178.	27.5	Effect of temperature on egg production	468
179.	27.6	Factors affecting water intake other than temperature	472
180.	27.7	Dew-point temperatures (°C)	474
181.	27.8	Heat, moisture and fecal production by broilers at 21.1°C	476
182.	27.9	Heat and moisture production by chicken at 21.1°C	477
183.	27.10	Heat and moisture production by layers (per bird/h)	477
184.	29.1	Minimum winter ventilation requirements for broilers	495
185.	29.2	Typical air flow requirement at 30-60% relative humidity	499
186.	29.3	Summer ventilation guidelines	507
187.	29.4	Poultry ventilation guidelines	507
188.	29.5	Heat stress index number *Vs* comfort of birds	508
189.	29.6	Recommended Ventilation Rates (m^3/min for 1000 birds)	511
190.	29.7	Sample performance data for exhaust fans	515
191.	29.8	Fan capacity with and without louvers for a 18.75 cm diameter blade fan, 50 W motor turning at 3400 rpm	515
192.	29.9	Fan capacity of 35 cm diameter blade with 93 W motor turning at the four indicated rpm	515
193.	29.10	Performance and efficiencies of most and least efficient of the 90 cm blade fans	516
194.	29.11	Wind effect on static pressure	517
195.	29.12	Typical indoor set point temperatures and ventilating rates for turkeys	519
196.	29.13	Heat and moisture production estimates for turkeys	520
197.	29.14	Minimum ventilation requirement of straight-run chicken during cold weather	521
198.	29.15	Suggested ventilation rate for chicken and turkeys	526
199.	29.16	Airflow through hatchery rooms (m^3/min)	527
200.	29.17	Capacities of exhaust fans	527
201.	30.1	Bulk density, thermal conductivity and vapor resistivity of some common building material	534
202.	30.2	Thermal conductivity (k) of brickwork and concrete as a function of moisture content and bulk density	534
203.	30.3	Solar absorptivity, long-wave emissivity over the temperature range of 0-100°C for some common building material	536
204.	30.4	R-values recommended for different climates	536
205.	30.5	R-values of common building material	537

206.	30.6	Suggested design R values for tropical housing	537
207.	30.7	U-values required to overcome condensation	540
208.	30.8	Effect of temperature (X, in °C) on various parameters	542
209.	31.1	*Pro forma* for calculation of quantities	549
210.	31.2	*Pro forma* for abstract of estimated cost	550
211.	31.3	Standards for selected items	553
212.	31.4	Current rates (Finished) of certain works and material	553
213.	31.5	Measurements and estimation of quantities	556
214.	31.6	Abstract of estimated cost	559
215.	33.1	Moisture holding capacity of litter material (g/100 g)	580
216.	33.2	The European directive - a summary	587
217.	33.3	Minimum standards for laying hens	598
218.	33.4	Comparative performance - Cage *Vs* Aviary system	599
219.	34.1	Brooding temperature (°C)	602
220.	34.2	Space requirements during brooding (per bird)	610
221.	35.1	Temperature and duration of cauterization	630
222.	36.1	Space requirements of broilers and roasters	637
223.	36.2	Ventilation requirements of broilers (m^3/min/1000 birds)	638
224.	36.3	Uniformity of flock	644
225.	36.4	Rearing broilers in cages	652
226.	36.5	Space allowances - Capons (per bird)	655
227.	37.1	Standard body weights - egg-type growing pullets	658
228.	37.2	Body weight recommendations for breeders	660
229.	38.1	Effect of color of light on poultry	665
230.	38.2	Effect of cleanliness of bulbs and reflectors	666
231.	38.3	Height for fixing bulbs in a poultry house (m)	666
232.	38.4	Intermittent lighting program for layers	673
233.	38.5	Ahemeral lighting	674
234.	38.6	Lighting scheme - Commercial laying-type birds	675
235.	38.7	Lighting scheme - Broiler and Commercial laying-type breeders	675
236.	38.8	Lighting scheme - Broilers and Roasters	676
237.	38.9	Lighting scheme - Turkeys	676
238.	39.1	Judging present production	679
239.	39.2	Order of bleaching of pigments	684
240.	40.1	Identifying sexing errors	687
241.	40.2	Sample size *Vs* Flock size	689
242.	40.3	Score card for degree of uniformity	690
243.	40.4	Sensitivity of balance *Vs* uniformity	690
244.	40.5	Suggested mating ratios (Males per 100 Females)	694
245.	40.6	Minimum egg weight for settable eggs (g/egg)	695
246.	40.7	Flock size *Vs* egg cooler requirements	696
247.	40.8	Flock Uniformity Guideline	697
248.	43.1	Weight-loss targets (%)	724
249.	43.2	Conventional molting program for layers	726
250.	43.3	California molting program (with water *ad libitum*)	726
251.	43.4	Minimum nutrient nevels of recovery diets	728
252.	44.1	Nutrient excretion (% of intake)	729

253.	44.2	Systems used in high-density, large volume operations	730
254.	44.3	Gross composition of poultry manure (%)	730
255.	44.4	Nutrient composition of manure (Average)	731
256.	44.5	Mineral composition of manure (Average)	732
257.	44.6	Amino acid composition of manure (Average)	733
258.	44.7	Manure production by chicken	745
259.	44.8	Manure storage volumes required	746
260.	45.1	Residual sprays for fly control	763
261.	46.1	Odor threshold concentrations of selected compounds from a rendering plant	787
262.	46.2	Common biogenic amines and their precursors	789
263.	46.3	Advantages and disadvantages of fermentation process	789
264.	46.4	Advantages and disadvantages of anaerobic digestion.	792
265.	46.5	Use and toxicity of various types of disinfectants	796
266.	46.6	Background information on six major disinfectant groups	796
267.	47.1	Gases in poultry houses (%)	802
268.	47.2	Maximum gas concentration (ppm) for air quality and odor control in poultry buildings.	809
269.	48.1	Broiler carcass yield	819
270.	48.2	F values for calculation of uniformity in weights	820
271.	48.3	Calculation of hen-housed and hen-day egg production	822
272.	48.4	Weight gain, feed consumption and mortality in a broiler farm	823
273.	48.5	Calculation of FCR	823
274.	48.6	Simplified method of calculating bird days	824
275.	50.1	Space requirements - Major species	835
276.	50.2	Space requirements - Other species	836
277.	50.3	Floor, feeder and waterer space requirements of chicken (per bird) (All litter system)	836
278.	50.4	Floor, feeder and waterer space - Cage system	837
279.	50.5	Floor space (m^2/bird) for chicken under different systems of rearing	837
280.	50.6	Cage dimensions (cm)	838
281.	50.7	Feed and water intake by Standard Leghorn pullets (21.1°C)	838
282.	50.8	Growth and feed consumption of broilers and roasters (21.1°C)	839
283.	50.9	Feed and water intake by Straight-run broilers (21.1°C)	840
284.	50.10	Standard body weights - egg-type growing pullets	840
285.	50.11	Body weight recommendations for breeders	841
286.	50.12	Incubation, hatching and brooding specifications - Major species	842
287.	50.13	Incubation, hatching and brooding specifications - Other species	842
288.	50.14	Specifications for light and natural mating - Major species	843
289.	50.15	Specifications for light and natural mating - Other species	843
290.	50.16	Lighting scheme for broilers	844
291.	50.17	Lighting program for broiler breeders in a black-out house	844
292.	50.18	Lighting program according to housing type	845
293.	50.19	Suggested mating ratio	845
294.	50.20	Heat and moisture production by broilers at 21.1°C	846
295.	50.21	Poultry ventilation rates	846

296.	50.22	Sample performance data for exhaust fans	847
297.	50.23	Running current requirements of electric motors	847
298.	51.1	Essential inorganic elements for poultry	853
299.	51.2	Estimate of dry matter intake by laying hens	855
300.	51.3	Estimate of dry matter intake by broilers	856
301.	51.4	Summary of actions of proteolytic enzymes	860
302.	51.5	Crude protein (CP) and digestibility of common feedstuffs	866
303.	51.6	True digestibility coefficients (%, average) of selected amino-acids in poultry feedstuffs	867
304.	51.7	Digestibility (%) and metabolizable energy (MJ/kg) of feed-grade fats	868
305.	51.8	Content and availability of vitamins	870
306.	52.1	Average GE and ME of nutrients	881
307.	52.2	Gross energy of some compounds	882
308.	52.3	Vitamins in foods and their bioavailability	885
309.	52.4	Vitamin and mineral deficiency in chicken embryo	903
310.	52.5	Symptoms associated with nutrient deficiencies	904
311.	53.1	Nutrient requirement of broilers	906
312.	53.2	Nutrient requirements of laying-type chicken	907
313.	53.3	Nutrient requirements of broilers	908
314.	53.4	Nutrient requirements of laying-type chicken	909
315.	53.5	Nutrient requirements of brown-egg laying chicken	911
316.	53.6	Nutrient requirements of meat-type breeding hens	912
317.	53.7	Nutrient requirements of meat-type breeding males	913
318.	53.8	ME requirement (kcal/hen/d) of layers *Vs* egg production and body weight	913
319.	53.9	Nutrient requirements of immature turkeys	913
320.	53.10	Nutrient requirements of breeding turkeys	915
321.	53.11	Nutrient requirements of Japanese quails	916
322.	53.12	Nutrient requirements of Bobwhite quails	917
323.	53.13	Nutrient requirements of ducks	918
324.	53.14	Nutrient requirements of Geese	919
325.	53.15	Nutrient requirements of Ring-necked Pheasants	919
326.	53.16	Guidelines for suitability of drinking water	920
327.	53.17	Maximum acceptable levels of contaminants in water	920
328.	53.18	Water intake by chicken and turkeys (ml/bird/week, 20-25°C)	921
329.	53.19	Growth and feed consumption by broilers	922
330.	53.20	Body weight (g) and feed intake (g/week) of immature layers	922
331.	53.21	Growth and feed consumption by turkeys	923
332.	53.22	Body weight and feed intake of large-type turkeys	924
333.	53.23	Growth and feed consumption by Pekin ducks	924
334.	53.24	Body weight and feed intake of Geese	924
335.	53.25	Toxicity of inorganic elements and compounds	925
336.	53.26	Stability of vitamins and losses during storage	927
337.	54.1	Composition and nutrient content of molasses products	933
338.	54.2	Trace minerals and vitamins in molasses	933
339.	54.3	Rates of inclusion of fishmeal for optimum benefit	936

340.	54.4	Limiting amino-acids in selected protein sources	939
341.	54.5	Anti-nutritional factors in poultry feedstuffs	941
342.	54.6	Mineral sources in poultry feeds	942
343.	54.7	Rates of inclusion of common feed ingredients	943
344.	54.8	Composition of feed ingredients – Gross, on as-fed basis	944
345.	54.9	Composition of feed ingredients – Amino-acids (%), on as-fed basis	945
346.	54.10	Composition of feed ingredients – Inorganic elements, on as-fed basis	946
347.	54.11	Composition of feed ingredients – Vitamins (ppm), on as-fed basis	946
348.	55.1	American Feed Manufacturer's Association sieves and pore sizes	954
349.	55.2	Protein, amino-acid and mineral requirements of hens in different phases of egg production	968
350.	56.1	Some enzymes that are or can be used in the feed industry	979
351.	56.2	Nomenclature and action of β-glucan-degrading enzymes	980
352.	56.3	Benefits obtained from the addition of enzymes to poultry feeds	981
353.	57.1	Variables for iteration	986
354.	57.2	Tentative formula – I	987
355.	57.3	Tentative formula - II	987
356.	57.4	Final formula	988
357.	57.5	Hypothetical data for LPP	997
358.	57.6	Vitamin-mineral premix for broilers	1002
359.	58.1	Routes of contamination by natural toxicants	1003
360.	58.2	Toxicants in feedstuffs	1003
361.	58.3	Endogenous toxic factors in feeds of plant origin	1004
362.	58.4	Some phytoestrogen-containing plants	1008
363.	58.5	Important mycotoxins	1011
364.	58.6	Partial list of known mycotoxins	1012
365.	58.7	Tolerance limits (ppm) for mycotoxin contamination of mixed feed and general symptoms of toxicity	1013
366.	58.8	Tolerance levels of dietary aflatoxin	1015
367.	58.9	Additives for protection against aflatoxins in broiler diets	1018
368.	58.10	Methods of detoxificaton of aflatoxin	1019
369.	59.1	Concentration of immunoglobulins in serum of adult fowl	1034
370.	60.1	Principal routes of transmission of diseases	1055
371.	60.2	Factors influencing occurrence of an infectious disease	1057
372.	60.3	Risk analysis for salmonellosis in broiler breeder flocks	1061
373.	61.1	General observations during necropsy examination of birds	1071
374.	61.2	Material to be sent to diagnostic laboratory	1073
375.	64.1	Marek's disease *Vs* Lymphoid leukosis	1116
376.	64.2	Vaccination against HPAI infection	1132
377.	65.1	Number of sporocysts and sporozoites	1143
378.	65.2	Life cycle of *E.tenella*	1144
379.	65.3	Coccidiosis of chicken	1144
380.	65.4	Anticoccidial drugs – effective late in life-cycle (5th and 6th d)	1147

381.	65.5	Preventive coccidicidals	1148
382.	65.6	Suggested anti-coccidial use for chicken	1149
383.	65.7	Coccidiosis of turkeys	1149
384.	66.1	Ectoparasites of importance in poultry production	1162
385.	66.2	Hemoparasites of importance in poultry production	1168
386.	66.3	Nematodes and Cestodes of poultry	1169
387.	66.4	Location and pathogenecity of Cestodes	1172
388.	66.5	Location and pathogenecity of Trematodes	1173
389.	66.6	Age group to sample most prevalent parasite species	1175
390.	66.7	Suggested sample sizes	1176
391.	66.8	Some insecticides for control of ectoparasites on poultry	1178
392.	66.9	Control of flies (vectors) in poultry houses	1179
393.	66.10	Some common drugs for use against hemoparasites	1179
394.	66.11	Common drugs for anthelmintic treatment of poultry	1184
395.	66.12	Treatment for endoparasites - Nematodes	1185
396.	68.1	Tolerance and residues of common insecticides and pesticides	1201
397.	69.1	List of tests for international trade	1207
398.	69.2	Commonly used serological tests	1211
399.	70.1	Longevity of disease causing organisms	1217
400.	71.1	Properties and uses of some disinfectants	1231
401.	71.2	Commonly used disinfectants	1232
402.	71.3	Recommended uses and considerations for disinfectants	1235
403.	71.4	Concentration and duration of fumigation	1237
404.	72.1	Advantages and disadvantages of major types of vaccines	1241
405.	72.2	Vaccination schedule – Broilers	1244
406.	72.3	Vaccination schedule – Layers (Pullets)	1244
407.	72.4	Vaccination schedule – Broiler breeders	1245
408.	72.5	Vaccination schedule – Layer breeders	1245
409.	72.6	Vaccination schedule – Commercial turkeys	1246
410.	72.7	Vaccination schedule – Turkey breeders	1246
411.	72.8	Duration of immunity	1247
412.	72.9	Vaccination schedule - General	1247
413.	72.10	Vaccination schedule – Commercial broilers	1248
414.	72.11	Vaccination schedule – Commercial layers	1248
415.	73.1	DNA *Vis-à-vis* RNA	1258
416.	73.2	Law of segregation -Classical	1263
417.	73.3	Law of segregation - deviation	1263
418.	73.4	Law of independent assortment - Classical	1264
419.	73.5	Law of independent assortment – deviation	1265
420.	73.6	Epistasis – Inhibitory gene	1266
421.	73.7	Types of epistasis	1267
422.	73.8	Some basic shank/feet color genetics	1270
423.	73.9	Autosomal genes causing white/essentially white plumage	1272
424.	73.10	Chicken plumage patterns	1272
425.	73.11	Some basic chick down colors	1274
426.	73.12	Partial genotypes of common breeds	1275
427.	73.13	Chicken genes of common interest – Sex-linked genes	1276

428.	73.14	Autosomal linkage Group 1 genes	1278
429.	73.15	Autosomal linkage Group 2 genes	1278
430.	73.16	Autosomal linkage Group 3 genes	1279
431.	73.17	Autosomal linkage Group 4 genes	1280
432.	73.18	Other autosomal genes	1280
433.	73.19	Autosexing crosses	1286
434.	73.20	Important obligate lethal genes of poultry	1287
435.	74.1	Quantitative traits *Vs* Qualitative traits	1288
436.	74.2	Calculation of gene and genotypic frequencies	1290
437.	74.3	Calculation of gene and genotypic frequencies –I Generation	1290
438.	74.4	Calculation of gene and genotypic frequencies –II Generation	1291
439.	74.5	Gene and genotypic frequencies – Quantitative trait	1293
440.	74.6	Chick weight as a function of number of DW genes	1295
441.	74.7	Calculation of Breeding values	1297
442.	74.8	Calculation of additive, genotypic and phenotypic variances	1298
443.	74.9	Calculation of realized h^2 (mean body weight, kg)	1302
444.	74.10	ANOVA of half-sib relationship	1304
445.	74.11	ANOVA of full-sib relationship	1305
446.	74.12	Estimated h^2 values (%)	1307
447.	74.13	Analysis of covariance (ANACOVA) for half-sib relationship	1314
448.	74.14	ANACOVA for full-sib relationship	1315
449.	75.1	Intrapopulation selection methods for single trait	1318
450.	75.2	Intrapopulation selection methods for multiple traits	1319
451.	75.3	Reciprocal recurrent selection	1322
452.	75.4	Characteristics of poultry semen	1325
453.	75.5	Results of selection for MD resistance	1331
454.	75.6	Results of matings among unselected chicken	1331
455.	76.1	Non-viral gene transfer methods	1335
456.	76.2	Common restriction enzymes	1337
457.	76.3	Duration per cycle (sec) for PCR assay	1342
458.	77.1	Production of broilers and layers (in thousands)	1354
459.	77.2	Components of Law of proportionality	1356
460.	77.3	Marginal rate of technical substitution (MRTS)	1358
461.	78.1	Book-keeping *Vs* Accounting	1362
462.	78.2	Hypothetical example of transactions	1373
463.	79.1	Agricultural goods *Vs* Manufactured goods	1377
464.	79.2	Markets and competition	1378
465.	80.1	Non-recurring expenditure	1396
466.	80.2	Recurring expenditure	1398
467.	80.3	Income estimation (Rs per bird)	1400
468.	80.4	Profit estimation in broiler and layer farming (Rs / bird)	1400
469.	80.5	Economics of rearing commercial poultry – algebraic	1403
470.	81.1	Pre-hatch behavior development	1406
471.	81.2	Space required for some of the normal behavioral activities	1410
472.	82.1	The five freedoms	1413
473.	82.2	Welfare of layers in battery cage *Vs* Free range	1424

Index to Figures

Sl.No.	Number	Title	Page
1	4.1	Nomenclature - Male	44
2	4.2	Nomenclature - Female	44
3	4.3	Skeleton of fowl	45
4	4.4	Single comb	45
5	4.5	Rose comb	45
6	4.6	Pea comb	46
7	4.7	Cushion comb	46
8	4.8	V-shaped comb	46
9	4.9	Strawberry comb	46
10	4.10	Buttercup comb	46
11	4.11	Nomenclature of shanks and toes - Male	46
12	4.12	Nomenclature of shanks and toes - Female	46
13	4.13	Fifth toe	47
14	4.14	Toe feathering	47
15	4.15	Measuring tail angle	47
16	4.16	Duck head	47
17	4.17	Parts of wing	47
18	4.18	Penciled feather	47
19	4.19	Laced feather	47
20	4.20	Barred feather	47
21	4.21	Striped feather	47
22	4.22	Spangled feather	48
23	4.23	Spangle-tipped feather	48
24	4.24	Mottled feather	48
25	4.25	Stippled feather	48
26	6.1	Crow head	94
27	6.2	Lopped Rose comb	94
28	6.3	Lopped Single comb	94
29	6.4	Coarse comb	94
30	6.5	Thumb marks	94
31	6.6	Twisted comb	94
32	6.7	Side strips	95
33	6.8	Split-comb	95
34	6.9	Split-wing	95
35	6.10	Stripped wing and twisted feathers	95
36	6.11	Mossy feather	95
37	6.12	Frosty feather	95
38	6.13	Splashed feather	95
39	6.14	Shafting in feather	95
40	6.15	Mealy feather	95
41	6.16	Vulture hocks	95

42	6.17	Wry tail	95
43	6.18	Scoop bill	95
44	6.19	Duck foot	95
45	7.1	Hormones in female reproduction	105
46	7.2	Reproductive system of a hen	113
47	7.3	Vertical section of egg	113
48	7.4	Biochemistry of shell formation	118
49	7.5	Germinal disc on surface of yolk in a fertile and an infertile egg	124
50	7.6	Geometry of chicken egg	129
51	9.1	Composition of yolk	157
52	11.1	Flow-chart for conversion of live poultry into poultry meat	194
53	12.1	Environmental contaminants in meat and meat products	223
54	12.2	Contaminants in meat and meat products from animal production	224
55	12.3	Contaminants in meat and meat products from processing and packaging	226
56	12.4	Bio-contaminants in meat in meat products	226
57	15.1	Factors affecting poultry meat quality	257
58	15.2	Components of poultry meat flavor	262
59	17.1	Methods of preservation of eggs	276
60	18.1	Methods of preservation of poultry meat	284
61	18.2	Color of cured and cooked-cured meat	291
62	18.3	Formation and general structure of nitrosamines	292
63	24.1	Air-cell size during incubation	374
64	24.2	Turned *Vs* unturned eggs	376
65	24.3	Fertility testing during incubation	382
66	24.4	Stages of embryo development	389
67	26.1	Ambient temperature *Vs* heat loss and body temperature	437
68	26.2	Major routes of heat production and heat loss	443
69	27.1	Variables affecting water balance in chicken	473
70	28.1	Natural ventilation in a gable-roof building	482
71	28.2	Ridge ventilators with adjustable openings	486
72	28.3	Ridge ventilators with 'upstands'	486
73	28.4	Sidewall air inlets for summer ventilation	488
74	28.5	Butterfly doors with bottom-mounted cables (open position)	492
75	28.6	Desired air flow patterns in a modified open-front mono-slope-roof building	492
76	28.7	Cross partitions to prevent draft.	493
77	29.1	Propeller, Tube-axial and Vane-axial fans	510
78	29.2	Triangle of ventilation	518
79	31.1	Modular brick with frog	544
80	31.2	Foundation	547
81	31.3	Steel truss with roofing material	547
82	31.4	Calculation of pitch	549
83	31.5	Individual-wall method – Floor diagram	552
84	31.6	Center-line method	552
85	31.7	Steps	555
86	31.8	Cross-section of a house for estimation of quantities	555

87	31.9	Roofing truss details	559
88	32.1	Cross-section of a poultry house - Deep-litter system	565
89	32.2	Cross-section of a poultry house - Conventional cage system	566
90	32.3	Floor diagram - Conventional cage-system	566
91	32.4	High-rise house (2L, 1M)	568
92	32.5	High-rise house (2Ms)	568
93	34.1	Brooding temperature *Vs* comfort of chicks	605
94	34.2	Placement of equipment during brooding	606
95	34.3	Battery brooder	608
96	35.1	Severity of beak-trimming	630
97	35.2	Beak-trimming procedure	632
98	36.1	Growth rate of broilers	644
99	38.1	Bulb-reflector assembly	667
100	38.2	Arrangement of bulbs - Cage system	667
101	38.3	Distribution of 60W bulbs - Deep-litter house	668
102	38.4	Lighting programs for growing pullets	670
103	39.1	Molting *Vs* persistency of production	683
104	39.2	Order of bleaching of pigments	684
105	40.1	Nests for layers	692
106	42.1	Conventional and reverse cages	711
107	42.2	Arrangement of laying cages	713
108	43.1	Feathers on the wing	718
109	50.1	Comparison of linear and circular feeders/drinkers	849
110	51.1	Digestive system of chicken	856
111	51.2	Process of digestion - Mouth, crop and proventriculus	857
112	51.3	Process of digeston - Gizzard, duodenum and intestines	858
113	51.4	Formation of active proteolytic enzymes	859
114	51.5	Autoenzymatic digestion in birds - Carbohydrates and lipds	861
115	51.6	Autoenzymatic digestion in birds - Proteins	861
116	51.7	Digestibilit (%) and metabolizable energy (MJ/kg) of feed-grade fats	876
117	52.1	Inter-relationship between variables affecting water balance	878
118	52.2	Vitamin A and optic cycle	887
119	52.3	Hormone-like properties of Vitamin D_3	889
120	55.1	Egg production parameters during a laying cycle	967
121	57.1	Pearson's Square method	988
122	64.1	Control strategy for AI	1131
123	74.1	Average effect of gene and breeding value	1296
124	75.1	Path diagram for selection index	1320
125	77.1	Elasticity of demand	1352
126	77.2	Production possibility curve	1355
127	77.3	Output extension path	1356
128	77.4	Scale line or Expansion path line	1359
129	77.5	Equilibrium analysis of an enterprise	1361
130	78.1	Accounting cycle	1368

Index to Plates

Sl.No.	Number	Title	Page
1	5.1	New Hampshire - Cock	70
2	5.2	New Hampshire - Hen	70
3	5.3	White Plymouth Rock - Cock	70
4	5.4	White Plymouth Rock - Hen	70
5	5.5	Barred Plymouth Rock - Cock	70
6	5.6	Barred Plymouth Rock - Hen	70
7	5.7	Rhode Island Red - Cock	70
8	5.8	Rhode Island Red - Hen	70
9	5.9	Buff Cochins - Cock	70
10	5.10	Buff Cochins - Hen	70
11	5.11	Dark Cornish - Cock	70
12	5.12	Dark Cornish - Hen	70
13	5.13	White Cornish - Cock	70
14	5.14	White Cornish - Hen	70
15	5.15	Black Australorp - Cock	70
16	5.16	Black Australorp - Hen	70
17	5.17	White Leghorn - Cock	71
18	5.18	White Leghorn - Hen	71
19	5.19	Black Minorca - Cock	71
20	5.20	Black Minorca - Hen	71
21	5.21	Black Araucanas - Cock	71
22	5.22	Black Araucanas - Hen	71
23	5.23	Black-breasted Aseel - Cock	71
24	5.24	Black-breasted Aseel - Hen	71
25	5.25	Red Naked-neck - Cock	71
26	5.26	Red Naked-neck - Hen	71
27	5.27	Black Sumatra - Cock	71
28	5.28	Black Sumatra - Hen	71
29	5.29	White Sultan - Cock	71
30	5.30	White Sultan - Hen	71
31	5.31	Black Muscovy - Drake	71
32	5.32	Black Muscovy - Duck	71
33	5.33	Khaki Campbell - Drake	72
34	5.34	Khaki Campbell - Duck	72
35	5.35	Fawn & White Runner - Drake	72
36	5.36	Fawn & White Runner - Duck	72
37	5.37	White Sebastopol - Drake	72
38	5.38	White Sebastopol - Duck	72
39	5.39	Pilgrim - Drake	72
40	5.40	Pilgrim - Duck	72
41	5.41	Beltsville Small White - Tom	72

Plates

42	5.42	Beltsville Small White - Hen	72
43	5.43	Bronze - Tom	72
44	5.44	Bronze - Hen	72
45	5.45	White Holland - Tom	72
46	5.46	White Holland - Hen	72
47	5.47	Black - Tom	72
48	5.48	Black - Hen	72
49	11.1	Trussing	210
50	11.2	Trussing	210
51	11.3	Trussing	210
52	14.1	Highly porous shell	238
53	14.2	Rotten egg	241
54	14.3	Blood ring	242
55	14.4	Blood spot	242
56	14.5	Blood spot	242
57	14.6	Blood spot	242
58	24.1	Chick embryo development	391
59	41.1	Conditions related to poor litter management	707

Index to Sites on Internet

Topic	Site
Alternatives to antibiotics for growth promotion	Doyle, E.M. 2001. University of Wisconsin-Madison website Revington, B. 2002. on internet
Aflatoxins	Herrman, T. 2002 www.oznet.ksu.edu/grsiext
Avian Influenza – FAO global strategies	http://www.fao.org/ag/againfo/resources/documents /empres/AI_globalstrategy.pdf
Biosecurity – egg processing plant	wwwz.ncsu.edu
Biosecurity – poultry farm	www.ucdavis.edu
Biotechnology	McKay, J www.aviagen.com
Carcass disposal	Kansas State University website and www.cps.gov.ca
	www.texaserc.tamu.edu
Circulatory physiology	Riddel, C. at http://caltest.vet.upenn.edu/poultry/syllabus_home.htm
Disease transmission	http://caltest.vet.upenn.edu/poultry/syllabus
Dubbing	chickenchronicles@groups.msu.com
Dust control	Atia, A. 2004. www1.agric.gov.ab.ca. Paulson, S.P. 1999. www.agf.gov.bc.ca/poultry
Egg quality evaluation	EggTester.com
Epidemiology	Brachman, P.S. http://gsbs.utmb.edu/microbook/ch009.htm
The European Directive on minimum standards for protection of laying hens	http://europa.eu.int/eur-lex/en/index.html
Euthanasia	www.ucdavis.edu
Fan performance	Arnold and Veenhuizen, ohioline.ag.ohio-state.edu
Feed formulation	www.spesfeed.com
Feed restriction	www.animallaw.info; 2003
Fish meal	www.gafta.com/FIN.html
HACCP principles	1. J.E. Rushing and D.R. Ward 2. Curtis, P.A., Anderson, K.E. and Jones, F.T. Carolina Univ. website
House site	Berry, J. http://www.osuextra.com

Index to Sites on Internet

Topic	Site
Industry statistics	www.wattpoultry.com
Judicious Use of Antimicrobials for poultry veterinarians	The American Veterinary Poultry association website
Inheritance of qualitative traits	http://marsa_sellers.tripod.com/geneticspapers
Lighting poultry houses	Winchell, W. 2001. www.cps.gov.on.ca.
Linear programming	http://www-fp.mcs.anl.gov/otc/Guide/OptWeb/continuous/constrained/linearprog/section2_1_1.html. or http://www.ece.nwu.edu/OTC/ Beasley, J.E at http://people.brunel.ac.uk/~mastjjb/jeb/or/lp.html Waner, S. and Costenoble, S.R. 2001 at http://people.hofstra.edu/faculty/Stefan_Waner/RealWorld/Summary4.html#top
Litter management	www.defra.gov.uk/animalh/welfare/farmed/meatchks
Management specifications	H.R. Wilson, F. B. Mather, and J.P. Jacob, http://edis.fas.ufl.edu/PSO42
Meat products	www.hormel.com
Molting	Elli, M.R. Agriculture Western Australia
	North Carolina University website
Muscovy ducks	Gourley, J., Univ. of California, Davis website
Odor in broiler farms	www.rirdc.gov.au
Odor control	Chastain, J.P. 2000. Iowa State Univ. website
Pest management	Hoelscher, C.E. The Texas A&M Univ Website
Santation and disinfection	Gordon, J.C. and Morishita, T.Y. http://ohioline.osu.edu/ Bowman, G.L. and Shulaw, W.P. http://ohioline.osu.edu/ Smith, T.W http://msucares.com/pubs/infosheets/
Trussing	www.hormel.com
Vaccination	Griffin, D., Ensley, S., Smith, D. and Dewell, G. Univ. of Nebraska website; NebGuide; http://ianrpubs.unl.edu/pubs/
	Breytenbach, J.H. 2005. Auditing vaccine application procedures in poultry. ThePoultrySite.com
	NS Department of Agriculture and Fisheries
Ventilation	Jacobson, L.D. and Chastain, J.P. Univ. of Minnesota Extension Service
	Arnold, G.J. and Veenhuizen, M.A. Ohio State Univ; ohioline.ag.ohio-state.edu
Welfare during slaughter	www.thestationeryoffice.com

Literature Cited

1. Abril, J.R., Barclay, W.R. and Abril. P.G. 2000. Usage of microalgae (DHA GOLD™) in laying hen feed for the production of DHA-enriched eggs. *In* Egg nutrition and biotechnology. 2000. Sim, J.S.. Nakai, S. and Guenter, W. (Editors). CABI Publishing.
2. Acharya, S.S. and Agarwal, N.L. 1999. Agricultural marketing in India. (III Ed). Oxford and IBH Publishing Co. Pvt. Ltd., New Delhi
3. Agricultural Marketing Service. 1995. Poultry Grade Yield Report, Poultry Grading Branch, United States Department of Agriculture, Washington, D.C.
4. Alchalabi, D. 2001a. Two-stage air cooling for very hot environments. Poultry Int. 40(11) : 28-32
5. Alchalabi, D. 2001b. Triangle of ventilation. Poultry Int. 40(13) : 36-41
6. Alchalabi, D. 2002. Correlation between litter pH and airflow pattern. Poultry Int. 41(5): 42-46
7. Alchalabi, D. 2004. Help is at hand to solve temperature distribution and other ventilation problems in the broiler house. Poultry Int. 43(3) : 28-33
8. Aldrigde, D. and Tahourdin, C. 1998. *In* "Natural toxicants in food" by Watson, D.H. (Ed), CRC Press LLC, USA and Canada
9. All India poultry business directory (year-book). 2003-04. Second edition. Sadana publishes and distributors, Delhi
10. American Poultry Association, Inc. 1998 Edition. The American standard of perfection.
11. Arora, M.P. 1992. Embryology. Himalaya publishing house, Bombay
12. Austic, R.E. and Nesheim, M.C. 1990. Poultry production, 13th edition. Lea & Febiger
13. Australian egg corporation, Ltd. 2003. Focus on research – iron-enriched eggs. *In* Poult. Int. 42(3) : 43
14. Baerdemaeker, J.D., Ketelaere, B.D., Bamelis, P., Kemps, B. and Decuypere, B.D. 2005. New practical systems to measure egg shell quality. Poult. Int. 44(3) : 28-35
15. Bains, B.S. 1994. The working mechanism of the Avian immune system. World poultry 10(5): 51-53
16. Banks, S. 1984. The complete handbook of poultry keeping. Wardlock Ltd., London
17. Barnwell, R. 2003. Maximizing performance during hot weather. Int. Poultry Prodn 11(1) : 11-17
18. Baron, R.B. 2005. Lipid abnormalities. *In* Current medical diagnosis and treatment. By Tierney, L.M.Jr., McPhee, S.J. and Papadakis, M.A. 2005. Lange medical books/ McGraw-Hill.
19. Becker, J.M., Caldwell, G.A. and Zachgo, E.A. 1996. Biotechnology – laboratory course (II Ed). Academic press
20. Belyavin, 1991. Water-I: A significant output. Poultry Int. 30(14): 24-30
21. Bietz, D.C. and Hansen, R.E. 1982. Animal products in human nutrition. Academic press, New York, London, Paris, San Diego, San Francisco, Sao Paulo, Sydney, Tokyo, Toronto
22. Bird, N.A and Munroe J.A. 1996. Manure handling from barn to storage for poultry farms. Agriculture and Agric-Food Canada; Ottawa, Ontario
23. Boon, N.A., Fox, K.A.A. and Bloomfield, P. 1999. *In* Davidson's principles and practice of medicine, 18th edition (1999) by Haslett, C., Chilvers, E.R., Hunter, J.A.A. and Boon, N.A.

(Editors), Churchill, Livingstone, Edinburgh.

24. Butcher, G.D. and Miles, R.D. 2002. Interrelationship of nutrition and immunity. Univ. of Florida IFAS Extension Bulletin No. VM139
25. Capua, I. and Marangon, S. 2003. Vaccination in the control of Avian Influenza in the EU. The Vet. Rec., March 1, 2003
26. Cardoso, B. Understanding serology. Int. Poultry Prodn. 13(5): 30
27. Carey, J. 2002. Evolution of bird flight linked to parental care. The Hindu, Feb 7, 2002
28. Chapman, A.B. 1985. General and Quantitative genetics. (Ed). World animal science series, A-4, Elsevier Pub.
29. Charles, D.R. 1981. *In* Environmental aspects of housing for animal production by Clark, J.A. (Editor). Butterworths
30. Charles, D.R. 1994. *In* Livestock housing by Wathes, C.M. and Charles, D.R. (Editors), CAB International
31. Charles, D.R., Eson, H.A. and Haywood, M.P.S. 1994. *In* Livestock housing by Wathes, C.M. and Charles, D.R. (Editors), CAB International
32. Chuong, C-M. 2002. Uncovering secrets of feather formation. The Hindu, Nov 7, 2002
33. Coker, R.D. 1998. *In* "Natural toxicants in food" by Watson, D.H. (Ed), CRC Press LLC, USA and Canada
34. Collins, E. 1996. Poultry litter management and carcass disposal. Publication No.442-910, Virginia Cooperative Extension bulletin
35. Crafer, R., Deeming, D.C. and Eady, P.E. 2005. The allometric relationship between egg mass and incubation period. Int. Hatchery Pract. 19(3) : 17
36. Crawford, R.W. 1990. Poultry breeding and Genetics. Elsevier Science Publishers B.V.,
37. Cruikshank, E.M. 1934. Studies in the fat metabolism in the fowl. I. The composition of the egg fat and depot fat of the fowl as affected by the ingestion of large amounts of different fats. Biochemical journal 28 : 965-977. *Quoted from* Watkins *et al.*, 2000.
38. Curtin, L.V. 1983. Molasses – general considerations. National feed ingredients association, Iowa
39. Czarick, M. and Lacy, M. 1997. Controlling litter moisture. The University of Georgia Cooperative Extension Service
40. Daghir, N.J. 1995. Poultry production in hot climates. Daghir, N.J. 1995. CAB International
41. Davis, R.H., Hassan, O.E.M. and Sykes, A.H. 1972. *cited by* Kampen, 1981
42. Dawson, W.R. and Whittow, G.C. 2000. Regulation of body temperature. *In* Sturkie's Avian physiology. 2000. Whittow, G.C. (Editor), Academic press.
43. Deeming, C. 2005. Incubation technology and 21st century: are we close to replacing the hatchery manager? Poultry Int. 44(7) : 18-19
44. Degraeve, P.Ir. 2004. Technical innovations in incubation. Int. Hatch. Pract. 18(8) : 21-23
45. Dewett, K.K. and Chand, A. 1996. Modern economic theory. (XXI Ed). Shyam Lal Charitable trust, New Delhi
46. Dozier, W.A. 2001. Cost-effective pellet quality for meat birds. Feed management, 52(2)
47. Duckworth, R.B. 1975. Water relations of foods. Academic Press, London, New York, San Francisco.
48. Editorial. 2001. Impact of globalization of trade on livestock, poultry and their products. Indian Vet. J. 78(5) : 13
49. Elibol, O and Brake, J. 2004. Identification of critical periods for turning broiler hatching eggs during incubation. British Poultry Sci., 45 : 631-637

50. Elson, A. 2002. Half a century of egg production – the industry's development from 1962 to 2012. Poult. Int. 41(5) : 8-16
51. Ensmingı , M.E. 1993. Poultry Science (3rd Ed). International Book Distributing Co., Lucknow, India
52. Erf, G.F. 2004. Cell-mediated immunity in poultry. Poultry Sci., 83: 580-590
53. Ernst, R.A. 1975. Hatchery and hatching egg sanitation. Univ. of California "Planed disease prevention" series publication
54. Ernst, R.A. and Coates, W.S. 1977. Raising geese. Leaflet No. 2225, Univ. of California, Davis
55. Ernst, R.A. *In* Poultry production in hot climates. Daghir, N.J. 1995. CAB International
56. Esmail, S.H.M. 2003. How nutrition affects egg quality. Poult. Int. 42(3) : 32-34
57. Etches, R.J., John, T.M. and Verrinder Gibbins, A.M. 1995. *In* Poultry production in hot climates. Daghir, N.J. 1995. CAB International
58. Etches, R.J. 1996. Reproduction poultry. CAB International
59. Evans, T. 2003. Challenge to educate consumers about the good egg news. Poult. Int. 42(1) : 38-39
60. Evans, T. 2005. Marketing the nutritional value of eggs. Poult. Int. 43(1) : 18-19
61. Evans, T. 2005. Massive potential for egg products in Asia. Poult. Int. 44(1) : 30-31
62. Farrell, D.J. 2000. Not all ω-3-enriched eggs are the same. *In* Egg nutrition and biotechnology. 2000. Sim, J.S.. Nakai, S. and Guenter, W. (Editors). CABI Publishing.
63. Fechheimer, N.S. 1985. Prospects of genetic engineering in domestic animals. *In* General and Quantitative genetics. Chapman, A.B. (Ed). World animal sciemce series, A-4, Elsevier Pub.
64. Firth, C.A. 1977. The normal lymphatic system of the domestic fowl. The Vet. Bull. 47(3): 167-179
65. Forbes, K. 2002. Management and egg quality. Int. Poultry prodn. 10(5) : 18-19
66. Frazer, A.F. and Broom, D.M. 1997. Farm animal behavior and welfare. (III Ed.). CAB International
67. Frier, B.M., Truswell, A.S., Shepherd, J., DeLooy, A. and Jung, R. 1999. Diabetes mellitus and nutritional and metabolic disorders. *In* Davidson's principles and practice of medicine, 18th edition (1999) by Haslett, C., Chilvers, E.R., Hunter, J.A.A. and Boon, N.A. (Editors), Churchill, Livingstone, Edinburgh.
68. Froning, G. 2004. New ways in which eggs are good for you. Poult. Int. 43(10) : 50
69. Gerrits, A.R. and Dijk, D.J. 1991. How hatching chicks react when inhaling formalin. Misset World Poultry 7(10) : 17
70. Gill, C. 2001. Formulation for nutraceutical eggs. Feed Int. 22(12) : 16-19
71. Glatz, P.C. 2000. Review of beak trimming methods. A report for the Rural Industries Research and Development Corporation, South Africa.
72. Gleaves, E.W. 1997. Cannibalism – Cause and prevention in poultry. Univ. of Nebraska Extension bulletin No. G84-78-A
73. Gopal, R., Sachdev, A.K. and Singh, R. 2001. Study on distribution and expectation of Indian poultry industry. Indian J. Poult. Sci., 36(3) : 280-283
74. Gowe, R.S. and Fairfull, R.W. 1995. Breeding for resistance to heat stress. *In* Poultry production in hot climates. Daghir, N.J. 1995. CAB International
75. Gregory, N.G. 1992. Catching damage. Broiler Industry 55:14-16.
76. Griffin, D., Ensley, S., Smith, D. and Dewell, G. 2002. Understanding vaccines. Univ. of

Nebraska Extension bulletin. F-12

77. Gwinner, E. and Han, N. *In* Sturkie's Avian physiology. 2000. Whittow, G.C. (Editor), Academic press.
78. Halvorson, D. 2005. Overview of Avian Influenza. University of Minnesota Extension Service - May 2005
79. Hastings, M.M. 2003 The Dollar Hen. Norton Creek Press
80. Haugh, R.R. 1937. The Haugh unit for measuring egg quality. U.S. Egg Poultry Magazine, No. 43, pages 552-555 and 572-573.
81. Heber, A.J., Zimmerman, N.J. and Linton, R.H. 1995. Ventilation of poultry slaughtering and processing plants. Purdue University Extension Bulletin
82. Herrman, T. 2001. Sampling: procedures for feed. Feed manufacturing, July 2001, Kansas state Univ.
83. Herrman, T. 2002. Mycotoxins in feed grains and ingredients. Kansas State University Extension Bulletin No. M.F.2061
84. Hill, D. 2004a. Performance losses in incubation and brooding: selected critical management points. Part 1. Poultry Int. 43(8) : 19-23
85. Hill, D. 2004b. Performance losses in incubation and brooding: selected critical management points. Part 1. Poultry Int. 43(13) : 48-52
86. Hofstad, M.S., Barner, H.J., Calnek, B.W., Reid, W.M. and Yoder, H.M.Jr. 1984. Diseases of poultry. The Iowa state Univ. press, Ames
87. Horrocks, L.A. and Yeo, Y.K. 2000. Docosahexaenoic acid-enriched foods : production of eggs and health benefits. *In* Egg nutrition and biotechnology. 2000. Sim, J.S.. Nakai, S. and Guenter, W. (Editors). CABI Publishing.
88. Howell, W.H. 2000. Food cholesterol and plasma lipid and lipoprotein response : Is food cholesterol still a problem or overstated ?. *In* Egg nutrition and biotechnology. 2000. Sim, J.S.. Nakai, S. and Guenter, W. (Editors). CABI Publishing.
89. Hunton, P. 1987. Factors affecting egg yolk pigmentation. Poultry Int. 26(13): 64
90. Jacobson, L.D. and Janni, K.A. 2003. Mechanical ventilation for poultry. University of Minnesota Extension Publication.
91. Janni, K.A. and Jacobson, L.D. 2003. Natural ventilation for poultry. University of Minnesota Extension Publication
92. Jeffrey, J.S. 1999. Use of competitive exclusion products for poultry. Poultry fact sheet No. 30. Univ of California
93. Jiang, J.K. and Sands, J.R. 2000. Odor and ammonia emission from broiler farms – a report for the Rural Industrial Research and Development Corporation, Australia, Publication No. 00/2
94. Johnson, A.L. *In* Sturkie's Avian physiology. 2000. Whittow, G.C. (Editor), Academic press.
95. Jones, D.D., Friday, W.H., Sherwood, S. and DeForest, P.E. 1980. Natural ventilation for livestock housing. Purdue Extension Publication AE-97
96. Jones, F.T., Genter, M.B., Hagler, W.M., Hansen, J.A., Mowrey, B.A., Poore, M.H. and Whitlow, L.w. 1994. Understanding and coping with effects of mycotoxins in livestock feed and forage. Electronic Publication Number DRO-29, North Carolina State University.
97. Kaminski, M.V.Jr. 2000. Giving eggs a bad rap : what physicians should know and tell their patients about cholesterol. *In* Egg nutrition and biotechnology. 2000. Sim, J.S.. Nakai, S. and Guenter, W. (Editors). CABI Publishing.
98. Kampen, V.M. 1974. *cited by* Kampen, 1981

99. Kampen, V.M. 1981. *In* Environmental aspects of housing for animal production by Clark, J.A. (Editor). Butterworths
100. Kan, C.A. 1991 Residues in poultry products. Poultry Int. 30(9):18-22
101. Kirby, J.D. and Froman, D.P. *In* Sturkie's Avian physiology. 2000. Whittow, G.C. (Editor), Academic press.
102. Klasing, D. C., B. J. Johnstone, and B. N. Benson, 1991. Implications of an immune response on growth and nutrient requirements of chicks. *Cited from* Butcher and Miles, 2002.
103. Klasing, K.C. 1998. Comparative Avian nutrition. CAB International
104. Koelkebeck, K.W. 2005. What is Egg Quality and Conserving It. Illini PoultryNet
105. Kramer, A., 1951. What is quality and how it can be measured: From a food technology point of view. *In* Market Demand and Product Quality. Mktg. Res. Workshop Rept., Michigan State College
106. Kritchevsky, D. 2000. Dietary fat and disease : What do we know and where do we stand. *In* Egg nutrition and biotechnology. 2000. Sim, J.S.. Nakai, S. and Guenter, W. (Editors). CABI Publishing.
107. Lands, W.E.M. 2000. Lipid mediators in a paradigm shift : balance between ù-6 and ù-3. *In* Egg nutrition and biotechnology. 2000. Sim, J.S.. Nakai, S. and Guenter, W. (Editors). CABI Publishing.
108. Lawless, H. 1991. The sense of smell in food quality and sensory evaluation. J. Food Quality 14:33-60.
109. Lawrie, R.A. 1979. Meat science. Pergamon press; Oxford, New York, Toronto, Sydney, Paris, Frankfurt.
110. Leaf, A., Kang, J.X. and Xiao, Y.-F. 2000. Antiarrhythmic effects of n-3 polyunsaturated fatty acids. *In* Egg nutrition and biotechnology. 2000. Sim, J.S.. Nakai, S. and Guenter, W. (Editors). CABI Publishing.
111. Leeson, S. 2001. Feeding program for laying hens. American Soybean Asociation Technical Bulletin No. P049-2001
112. Leeson, S. 2003. Breeder age has unanticipated influences on the performance of their progeny. Poultry Int. 42(10) : 36-44
113. Leeson, S. and Summers, J.D. 2000. Broiler breeder production. International book distributing company, Lucknow
114. Leeson, S. and Zubair, A.K. 2004. Digestion in Poultry I: Proteins and fats. Novus International Inc.,
115. Leeson, S. and Zubair, A.K. 2004. Digestion in Poultry I: Carbohydrates, minerals and vitamins. Novus International Inc.,
116. Lever, S.H. 2000. Wright's Encyclopedia of poultry science. Biotech books, Delhi
117. Marquardt, R.R. 2005. Enzyme enhancement of the nutritional value of cereals: role of viscous, water-soluble, non-starch polysaccharides in chick performance. Development international, Canada
118. Massie, S.M. and Granger, C.B. 2005. *In* Current medical diagnosis and treatment. By Tierney, L.M.Jr., McPhee, S.J. and Papadakis, M.A. 2005. Lange medical books/ McGraw-Hill.
119. McAdam, J. 2004. Balanced breeding for a better performance. Intern. Poult. Prodn. 12(3) : 7-8
120. McArthur, A.J. 1981. *In* Environmental aspects of housing for animal production by Clark, J.A. (Editor). Butterworths
121. McCullough, K.C. and Spier, R.E. 1990. Monoclonal antibodies in biotechnology – theo-

retical and practical aspects. Cambridge Univ. press

122. McNamara, D.J. 2000. Eggs, dietary cholesterol and heart disease risk : an international perspective. *In* Egg nutrition and biotechnology. 2000. Sim, J.S.. Nakai, S. and Guenter, W. (Editors). CABI Publishing.
123. Menezes, R.M., Satyanarayana Reddy, P.V.V., Venkataramaiah, V., Sudhakara Reddy, P. and Rama Prasad, J. 2001. Studies on incubation, hatching and early growth rates of emus. Indian J. Poult. Sci. 36(3): 268-270
124. Miesfeld, R.L. 1999. Applied molecular genetics. Wiley-Liss, John Wiley & Sons., publication
125. Mountney, G.J. 1995. Poultry products technology. (3rd Edition). Food products press, NY, London
126. Moreng, R.E. and Avens, J.S. 1985. Poultry science and production. Reston Pub. Co., Inc. Reston, Virginia
127. Mossel, D.A.A., 1975. Water and micro-organisms in foods – a synthesis. In *Water relations of foods* by Duckworth, 1975
128. Murthy, A. 2001. Accounting and financial management. S Viswanathan printers and publishers Pvt. Ltd., Chennai, India
129. Nalbandov, A.V. 1970. Reproductive physiology. II Edition (Indian), D.B.Taraporevala sons & Co. Private Ltd., Bombay
130. Narahari, D. 2001. Nutritionally enriched eggs. Poult. Int. 40(10) : 22-30
131. Narahari, D. 2003. Health-promoting and therapeutic uses of egg. Poult. Int. 42(9) : 45-47
132. Narahari, D. 2004. Feeds and feedstuffs. Pixie publications India (P), Ltd. Karnal, India
133. National Research Council. 1994. Nutrient requirements of poultry. NRC, Washington. D.C.
134. Neshiem, M.C., Austic, R.E. and Card, L.E. 1988. Poultry production. Lea and Febiger, Philadelphia
135. Nicholas, F.W. 1987. Veterinary genetics. Clarendon press, Oxford.
136. Nicholson, D. 2003. The art of male management. Poultry Int. 42(11) : 24-26
137. Noble, R and Penny, p. 2002. Egg fat – a fair hearing. Poult. Int. 41(5) : 18-22
138. Noordhuizen, J.P.T.M., Frankena, K., Thrusfield, M.V. and Graat, E.A.M. 2003. Application of quantitative methods in Veterinary epidemiology. Int. Book Distrib. Co., Lucknow, India
139. North, M.O. 1972. Commercial chicken production manual. The AVI publishing company, Inc.
140. North, M.O. 1984. Commercial chicken production manual. The AVI publishing company, Inc.
141. North, M.O. and Bell, D.D. 1990. Commercial chicken production manual. The AVI publishing company, Inc.
142. Northcutt, J.K. 1997. Factors Affecting Poultry Meat Quality. Bulletin 1157/June, 1997, The University of Georgia College of Agricultural and Environmental Sciences and the U.S. Department of Agriculture cooperating.
143. Nutsch, A. and Spire, M. 2004. Carcass disposal: a comprehensive review. National Agricultural Biosecurity Center, Kansas State Univ.
144. Ockerman, H.W. 1978. Reference book for food scientists. The AVI publishing Co. Inc.
145. Oderkirk, A. 2001a. Modern cold weather ventilation management techniques in poultry barns. NS Department of Agriculture and Fisheries, Canada

146. Oderkirk, A. 2001b. Air inlets: air inlet design and control. NS Department of agriculture and fisheries, Canada
147. Office International des Epizooties – Manual of standards for diagnostic tests and vaccines. 2000. 4th Ed.
148. O'Neill, S.J.B., Balnave, D. and Jackson, N. 1971. *cited by* Kampen, 1981
149. Osuga, , D.T. and Fenny, R.E. 1974. Avian egg whites. *In* Toxic constituents in animal food-stuffs by Liener (Ed). Academic press
150. Parsai, G. 2005. FAO warning on bird flu in Asia. The Hindu, Feb 2, 2005
151. Pearl, G.G. 2002. The future of animal proteins in poultry diets. Fats and proteins research foundation Inc.,
152. Permin, A. and Hansen, W. 1998. The epidemiology, diagnosis and control of poultry parasites. An FAO Handbook
153. Plattel, Ed. 2003. Improving egg safety through ultraviolet. Poult. Int. 42(3) : 12-14
154. Poczopko, P. 1981. *In* Environmental aspects of housing for animal production by Clark, J.A. (Editor). Butterworths
155. Posudin Yu.I., Tsarenko P.P., Tsyganjuk O.V.(1992), Eggs quality evaluation with optical and laser methods, Agr. Biology, 4 : 69-74.
156. Poultry times of India. 1996. Some interesting food facts about India. Feb 1996, Pp.6 Quoted from The Hindu, Jan 31, 1996
157. Pousga, S., Boly, H. and Ogle, B. 2005. Choice feeding of poultry – a review. Livestock research for rural development 17(4)
158. Primrose, S.B. 2001. Molecular biotechnology. (II Ed, II Indian Reprint). Panima publishing corporation, New Delhi, Bangalore
159. Press release. 2003. APPF president submits memorandum to the government. Poult. Planner 4(8) : 22-23
160. Puhler, A. 1993. Genetic engineering of animals. (Ed) VCH publication
161. Raj, N.G. 2005. Bird fly: how far and fast can it spread? The Hindu 128(246): 11 (October 18, 2005)
162. Rao, P.V. 1982. Aflatoxins. Avian Research, Dec 1982 (Spl Number on Aflatoxins)
163. Reddy, B.S. 1982. Influence of dietary fat, protein and fiber on colon cancer. *In* Animal products in human nutrition. 1982. Bietz and Hansen.
164. Richey, E.J. 2003. The immune system. Univ. of Florida, IFAS Extension Bulletin No. VM 101
165. Riemann, H. and Bryan, F.L., 1979. Food-borne intoxications (Eds). Academic Press, New York.
166. Robertshaw, D. 1981. *In* Environmental aspects of housing for animal production by Clark, J.A. (Editor). Butterworths
167. Ronningen, K. and Vleck, D.V. 1985. Selection index theory with practical applications. *In* General and Quantitative genetics. Chapman, A.B. (Ed). World animal sciemce series, A-4, Elsevier Pub.
168. Ruszler, P.L. 1997. The keys to successful induced molting of leghorn-type hens. Publication Number 408-026, Virginia Tech
169. Saif, Y.M. 2003. Diseases of poultry. 11th Ed (Editor-in-chief). The Iowa state Univ. press, Ames
170. Sainsbury, D. and Sainsbury, P. 1988. Livestock health and housing. ELBS., Bailliere, Tindall, London

171. Salazar, A. 2001. The effects of high altitude on hatchability. Poultry Int. 40(12) : 22-24
172. Salazar, A. 2003. All you willing to pay for the cost of high-quality chicks? Or do you prefer the more expensive alternative? Poultry Int 42(2) : 16-19
173. Savory, J. 2003. Trying to make sense of poultry welfare issues in the EU. Poultry Int. 42(8) : 16-18
174. Scudamore, K.A. 1998. *In* "Natural toxicants in food" by Watson, D.H. (Ed), CRC Press LLC, USA and Canada
175. Sharma, J.M. 1997. The structure and function of the avian immune system. Acta Veterinaria Hungarica 45 (3): 229–238
176. Sim, J.S. 2000. Designer egg concept : perfecting egg through diet enrichment with ù-3 PUFA and cholesterol stability. *In* Egg nutrition and biotechnology. 2000. Sim, J.S.. Nakai, S. and Guenter, W. (Editors). CABI Publishing.
177. Sim, J.S., Nakai, S. and Guenter, W. 2000. Egg nutrition and biotechnology. CABI publishing.
178. Singh, B.P. and Singh, S.P. 2002. Integrating molecular approaches with quantitative genetics in breeding. Poultry Int. 41(4): 26-31
179. Smith, T.W.Jr. 2000. Poultry disease diagnosis. The Mississippi State Univ. Extension publication No. 1276
180. Smith, W.K. 1981. *In* Environmental aspects of housing for animal production by Clark, J.A. (Editor). Butterworths
181. Sreenivasaiah, P.V. 1977. Studies on the growth rate, egg production, fertility and hatchability of Japanese quails (*Coturnix coturnix japonica*) hatched in different seasons. M.V.Sc Thesis, Rohilkhand Univ., Uttar Pradesh, India
182. Sreenivasaiah, P.V. 1985. Studies on the use of sodium tri-polyphosphate (STPP) in chilling medium on selected quality characteristics of spent-hen meat during storage. Ph.D Thesis, Andhra Pradesh Agricultural Univ., Hyderabad, India
183. Sreenivasaiah, P.V. 1987. Scientific poultry production, I Edition. IBH Prakashana, Bangalore, India
184. Sreenivasaiah, P.V. 1998. Scientific poultry production, II Edition. IBH Prakashana, Bangalore, India
185. Stadelman, W.J. and Cotterill, C.J. 1977. Egg science and technology (Ed), IV Edition, Food products press, NY, London
186. Stadelman, W.J., Olson, V.M., Shemwell, G.A. and Pasch, S. 1988. Egg and poultry meat processing. Ellis Harwood Ltd. and VCH
187. Stanislaw, W. and Katarzyna, C-B. 2002. Waterfowl research in the spotlight. Poult. Int. 41(3) : 32-36
188. Starr, J.R. 1981. *In* Environmental aspects of housing for animal production by Clark, J.A. (Editor). Butterworths
189. Steele, R. 1989. Food research news (newsletter of the Food Research Laboratory, CSIRO, Australia), issue no. 5 February 1989
190. Stull, C. 1998. Animal care series: Broiler care practices. (Ed) Poultry workgroup series, Univ. of California
191. Taylor, L.W. 2003. Fertility and hatchability of chicken and turkey eggs. Greenworld publishers
192. Tazawa, H. and Whittow, G.C. 2000. Incubation physiology. *In* Sturkie's Avian physiology. 2000. Whittow, G.C. (Editor), Academic press.
193. The Merck's Veterinary Manual. 2003. 8th Ed.

194. Thomison, P.R. and Loux, M.M. 2001. Commonly used method for detecting GMOs in grain crops. Ohio State Univ. Fact-sheet, AGF-149-01
195. Thornton, G. 2004. Future Competitiveness of the US poultry industry in the world broiler and turkey trading arenas Poultry Int. 43(1) : 20-26
196. Trevidy, J-J. 2005. Mash or pellet? The question of feed presentation, feed factory technology and manufacturing cost. Hubbard technical bulletin, Jan 2005
197. Tzschentke, B. 2005. Development of thermoregulation and impact of environmental factors. Int. Hatchery Pract. 19(5) : 17
198. United States Department of Agriculture, Agricultural Marketing Service Agricultural Handbook Number 75, Egg-Grading Manual
199. Varela-Alvarez, H. 1995. Diet formulation. *In* "Basic animal nutrition and feeding" 4th Ed, by Pond, W.G., Church, D.C. and Pond, K.R. John Wiley & Sons, New York, Chichester, Brisbane, Toronto, Singapore
200. Ven, L van de. 2004. Storage of hatching eggs in the production process. Int. Hatch. Pract. 18(8) : 27-31
201. Verkerk, R., Dekker, M. amd Jongen, W.M.F. 1998. *In* "Natural toxicants in food" by Watson, D.H. (Ed), CRC Press LLC, USA and Canada
202. Vest, L.R. 1997. Broiler breeder and replacement pullet management. Extension bulletin No. 861. Univ. of Georgia.
203. Wathes, C.M. 1981. *In* Environmental aspects of housing for animal production by Clark, J.A. (Editor). Butterworths
204. Watkins, B.A., Devitt, A.A., Yu, L. and Latour, M.A. 2000. Biological activities of conjugated linoleic acids and designer eggs. *In* Egg nutrition and biotechnology. 2000. Sim, J.S.. Nakai, S. and Guenter, W. (Editors). CABI Publishing.
205. Watson, D. 1998. Natural toxicants in food. (Ed) CRC Press LLC, USA, Canada
206. Webster, J. 1994. Animal welfare: a cool eye towards Eden. Blackwell Science Ltd., Oxford, London, Edinburgh, Cambridge, Massachusetts, Victoria
207. Wells, R.G. and Belyavin, C.G. 1987. Egg quality – current problems and recent advances. Butterworths and Co., (Pub) Ltd.
208. West, B. 1996. Odor control for livestock facilities. Canada Plan service. M 10704 96:08
209. H.R. Wilson, H.R., Mather, F.B. and Jacob, J.P. 1997. Poultry management specifications. IFAS Extension bulletin, Univ. of Florida
210. Winchell, W. 2001. Lighting for poultry housing. Canada Plan Scheme No. 5602, 2001:04
211. Woodard, A.E. 1982. Raising Chukar Partridges. Leaflet No. 21321, Univ. of California, Davis
212. Woodard, A.E., Abplanalp, H., Wilson, W.O. and Vohra, P. 1973. Japanese quail husbandry in the laboratory. Booklet, Univ. of California, Davis
213. Woolford, R. 2003. Feed restriction is today's best answer for broiler breeder fertility. Poultry Int. 42(10) : 61
214. Wright, C. 2005. Special focus of attention in Hanover: alternative layer housing systems. Poultry Int. 44(2) : 10-15
215. Zeidler, G. 2000. Egg products around the world : today and tomorrow. *In* Egg nutrition and biotechnology. 2000. Sim, J.S.. Nakai, S. and Guenter, W. (Editors). CABI Publishing.

Subject Index

A

Abdominal massage method 143
Abnormal behavior1 412
Acute pain 618
Acute septicemia 1083
Adaptation for flight 12
Aegyptianellosis 1157
Aerobic Lagoons 735
Aerobic Processing 735
Aerosol 1243
Aerosols 811
Aflatoxins 227, 1013
Agar gel immunodiffusion 1209
Agent middlemen 1389
Aggression 1404
Agricultural Marketing 1377
Ahemeral Lighting 674
Air inlets 522
Air leaks 507
Air movement 576
Air pollution 778
Air pressure 407
Air velocity` 462
Air-cell 238
Air-curtain incineration 778
Albumen 132
Albumen index 244
Albumen quality 244
Albumen utilization3 92
Alfalfa meal 939
Alkaline hydrolysis 789
All-slat house 572
Alternaria toxins .. 1023
Alternate day feeding 688
Alternative Husbandry Practices 984
Altricial animals 435
Amino acids 851, 983
Amino-acid deficiency 883
Ammonia 1017
Amyloidosis 1098
Anabolics 224
Anaerobic Digesters 734
Anaerobic digestion 791
Anaerobic Lagoons 733
Anaerobic Processing 732
Anatipestifer Disease 1098
Anesthetic Overdose 1433
Angara disease 1193
Animal Odors 805
Animal protein controversy 934
Anthelmintic 1181
Antibacterial agents 223
Antibiotics 969
Antibody genes 1261
Antigen 1051
Antivitamins 1010
Anxiety 1404
Applied Poultry Breeding 1317
Aqueous phase 156
Arabinoxylans 981
Arboreal theory 10
Archaeopteryx 13
Argemone contamination 1027
Arginine vasotocin 440
Argon 1431
Arizona infection 1095
Artificial insemination 1324
Artificial inseminaton 142
Ascites syndrome 1194
Ascorbic acid 897
Asiatic Class 52
Aspergillosis 1105
Asset 1363
Association behavior 1406
Autoclaving 752
Automatic Egg Turning 415
Autosexing 1284
Autosomal genes 1329
Autosomal inheritance 1268
Avian encephalomyelitis 1121
Avian influenza 1125
Avian mycoplasmosis 1099
Avian Pasteurellosis 1083
Avian rhinotracheitis 1136
Avian salmonellosis 1076
Aviary 589
Avidin 151
Avoiding CHD 188

B

Baby Chick 697
Back-cross 1323
Bacteria 814, 1051
Bacterial Diseases 1076
Bacteriophages 977
Baited jug trap 759
Baits 764
Baked custard 319
Bambara groundnut meal 940
Bantams 56
Barbecuing 337
Barley 929, 931
Barn 588
Basal Heat Production 537
Basal metabolic rate 443
Basal NST 453
Battery brooding 606
Beach's form 1119
Beak trimming 612
Beak-trimming 615
Beating egg white 306
Beaudette's form 1119
Beet Molasses 932
Beetles 1177
Behavioral needs 1409
Behavioral thermoregulation 448
Betaine 983
Bio beak-trimming 625
Bio-contaminants 226
Biofilters 780
Biosecurity 783, 1214
Biosolids 791
Biotechnology 1332
Biotin 895
BIS recommendations 271
Black head 1152
Blind chickens 633
Blood meal 936
Blood sample 1064
Bloody egg 242
Blow flies 757
Blue back 1202
Blue comb disease 1141
Bobwhite quail 832, 917
Body weight recommendations 840
Body-checked egg 240
Bone development 887
Botulism 1089
Break-even point 1360
Breast-blisters 649
Breeding value 1296
Broiler breeders 696
Broiler Diets 951
Broiler problems 646
Broiler-Fryer 321
Broilers 469, 819
Bronchitis 810
Brooder guard 604
Brooder house 602
Brooding 601, 640
Brooding cages 708
Brooding temperature 601
Building location 484
Building material 543
Bulking agents 780
Bumble foot 1197
Bursa 1030
Buying and selling 1386

C

C-reactive protein 174
Cage dimensions 838
Cage Management 708
Cage system 581
Caged-layer fatigue1 187
Calcium 898, 1200
California molting program 725
Campylobacter infection 1096
Candling 237, 281
Cane Molasses 932
Canker 1156
Cannibalism 619
Canopy brooding 604
Capon production 653
Capons 321
Captive Bolt 1432
Carbohydrates 851, 856, 879
Carcass Disposal 772
Care of birds 1414
Cash-book 818
Cassava root meal 940
Castor seed cake 939
Catecholamines 441
CE products 978
Cell-Mediated Immunity 1042
Cement Concrete 545
Cement mortar 544

Cement plaster 545
Cereal 231
Cereal by-products 930
Cereal grains 928
Cerebral spinal fluid 887
Chemical beak-trimming 624
Chemical composition 131
Chemically-altered vaccines 1240
Chemotherapeutic drugs 1199
Chick down color 1273
Chicken anemia 1137
Chicken bouillon 344
Chicken genes 1275
Chicken ham 344
Chicken parts 321
Chicken sausage 343
Chicken steaks 343
Chiggers 1178
Chlamydiasis 1091
Chlorine 900
Cholesterol 171
Choline 897
Chromosomal alterations 1333
Chromosome number 1257
Chromosome-mediated transfer 1334
Chronic farmer lung disease 811
Chronic pain 618
Citrinin 1021
Citrus Molasses 932
Cleavage 123
Climate classification 429
Climatic Indices 429
Coagulation 303
Coat resistance 460
Coated eggs 241
Cobalt 903
Coccidiosis 1143
Cockerels 727
Coconut meal 939
Coddled egg 315
Cohort study 1059
Cold blade 624
Cold mash moisture9 54
Cold Stress 422
Cold vasodilatation 444
Coligranuloma 1083
Collection of semen 1325
Colombian 89
Colon Cancer 228
Color of light 665
Color of shell 118
Color of yolk 119
Comb shape 1270
Comfort Zone 496, 521
Commercial egg products 297
Commercial Layers 686
Complementary DNA 1337
Complex carbohydrates 187
Conalbumin 153
Conflict 1404
Confuciusornis sanctus 14
Conjugated linoleic acid 976
Conjunctival sac installation 1242
Contaminants 220
Contamination from animal production 223
Continental Class 53, 54
Conventional cage system 571
Cook-loss 219
Cooler capacity 504
Cooling pad efficiency 502
Cooling the poultry house 501
Copper 901
Copper sulfate 1027
Coronary Heart Disease 164
Corrugated eggs 240
Corticosterone 441
Cost of construction 548
Cost of production 1357
Cotton seed cake 938
Courtship 141
Crack detection 250
Creditor 1363
Cresol 1200
Cross Contamination 351
Cross demand 1350
Cross elasticity 1351
Crossing RRS 1324
Cryptosporidiosis 1155
Cursorial theory 10
Custards 319
Cut-up parts 208
Cutaneous evaporative heat loss 446
Cuticle 129
Cyanocobalamin 896
Cyanogens 1007
Cytokines 976

D

Dark meat 216

Dark skin color 1270
Data processing 818
Dead germs 382
Debilling 826
Debtor 1363
Deep-litter house 571
Deep-pit drying system 738
Defeathering 202
Dehydrated meat 357
Deli chicken roast 343
Depopulation 700
Deterioration of eggs 273
Dewinging 613
DF value 295
Diabetes mellitus 171
Diced chicken 342
Die maintenance 954
Diet eggs 177
Digestion 850, 856
Disease control 1421
Disinfectants 225
Disinfection 1225, 1230
Displacement activity 1404
Disqualifications 83
Distribution and marketing 7
DM value 295
DNA polymerase 1342
DNA probing 1346
Domestic turkeys 25
Domestication 19
Dominance 1404
Doolittle method 991
Dosage assessment 294
Double entry system 1366
Double Pearson's Square 989
Double-tube bag-triers 964
Double-yolked eggs 241
Doyle's form 1119, 1120
Draft Control 493
Drawings 1363
Dried Egg Products 298
Dried poultry manure 750
Drinking Water Administration 1250
Dry Manure Management 759
Dry-rot 545
Dubbing 613, 636
Duck plague 1140
Duck virus hepatitis 1139
Ducks 56, 826, 918
Dust 809
Dust bath 595
Dust control 812, 1224
Dwarf gene 1285, 1329

E

E.coli infections 1082
Ear lobe color 1268
Eave Vent Design 487
Economics 1349
Ectoparasites 1160, 1176
Egg 96
Egg Consumption 164
Egg drop syndrome1 122
Egg filler flats 1223
Egg formation 111
Egg handling 712
Egg inspectors 1223
Egg omelette 317
Egg production 467
Egg Products 297
Egg Quality 235
Egg racks 1223
Egg transport 1220
Egg weight 468
Egg-eating vice 622
Egg-processing plant 1222
Egg-type pullets 657
Eggshell color 1269
Eicosanoids 167
Elasticity of demand 1351
Electric Soldering Iron 623
Electrocution 1432
Embryo 449
Embryo Injection 1243
Embryo inoculation 1072
Embryo mortality pattern 398
Emu 834
Endogenous toxins 1003
Endoparasites 1168, 1179
Energy deficiency 882
Energy Efficiencies 515
Energy sources 928
Energy utilization 392
English Class 52
Ensiled poultry manure 751
Environment-controlled poultry houses 567
Environmental contaminants 220
Environmental correlation 1311
Environmental resistance 459

Enzymes 978
Eoalulavis hoyasi 14
Equipment Cleanliness 1025
Equivalent temperature 434
Erysipelas 1090
Essential vitamins 852
Estrogens 1008
Euthanasia 1429
Evisceration 202
Evolution of Birds 10
Excretion 875
Exhaust fan efficiency 847
Exogenous toxicants 1010
Expenditure 1363
Expense 1363
Experimental animal 1433
Explicit costs 1357
Exploration behavior 1406
Eye color 1273

F

Fabric Swatch 804
Facilitative middlemen 1389
Faking 77
Fan efficiency 512
Fan Ratings 513
Fan selection 509
Farm buildings 561
Farm Data – Recording 816
Fat application 954
Fats 884
Fatty acids 166, 851
Fatty liver syndrome 1187
Favus 1109
Fear 1405
Feasibility report 1401
Feather color 1271
Feather cover 440
Feather follicle inoculation 1243
Feather meal 937
Feather-pecking 621
Feather-pulling 621
Fecal moisture 538
Feed Additive 765
Feed conversion ratio 645, 823
Feed efficiency 822
Feed factory technology 961
Feed wastage 622
Feedstuffs for Poultry 928
Fiber Components 231
Fibrinogen 173
Filled omelette 318
Financing 1387
Fish meal 934
Five Freedoms 1413
Fixed-facility incineration 777
Fleas 1177
Flies 756
Flock application 1243
Fluorescent light 664
Fly Belts 764
Fly-parasites 761
Fly-predators 761
Foam Formation 302
Folic acid 896
Follicular maturation 110
Food Preservation 265
Foot-candle 664
Forced-air 456
Forced-draft 380
Foreign investments 8
Formation of feathers 15
Fowl cholera 1083
Fowl typhoid 1078
Fragile bones 647
Free radicals 174
Free-choice (Cafeteria) system 954
Freeze-drying method 626
French toast 319
Fried egg 316
Frittata 319
Frizzle gene 1329
Frounce 1156
Frozen egg products 298
Frozen meat 357
Frustration 1405
Frying Chicken 333
Full costs 1358
Fumigation 366
Fumonisin 1022
Fungal Diseases 1105
Fungi 1052
Furnished cages 593
Future competitiveness 8

G

Game Class 54
Gangrenous Dermatitis 1094
Gas beak-trimming 623
Gas Chromatography 804

Gas Detectors 805
Gaseous exchange 395
Geese 66, 829
Gene exchange 1334
Gene structure 1261
Gene transfer 1334
Genetic correlation 1311
Genetic engineering 1332
Global poultry 1
Glucosides 1006
Glucosinolates 1006
Glucostatic theory 853
Goitrogens 1007
Goods 1363
Goose 29
Gossypol 1008
Gout 1197
Grading of eggs 252
Grading-up 1324
Granulocytes 1038
Gravity convection 456
Greasy Capped Litter 704
Green leafy products 939
Grinder grills 961
Ground chicken 342
Ground nut cake 937
Ground Turkey 325
Growing cages 709
Guar meal 938
Guinea fowl 27, 832
Gular flutter 446
Gunny bags 1399
Gunshot 1432

H

Halal method 199
Hammer and grill wear 962
Hardy-Weinberg equilibrium 1290
Hatchers 1221
Hatchery cleaning 1230
Hatchery disinfection 1236
Hatchery ventilation 526
Hatching 364
Hatching Egg Care 699
Hatching eggs 695
Hatching maturity 388
Haugh Unit Score 245
HCHO 801
Head picking 621
Heart-puncture 1064
Heat balance 442
Heat balance equation 457
Heat conservation 447
Heat loss 445
Heat spot 242
Heat Stress 422
Heat stress 438
Heat stress index number 507
Heat-processed meat 358
Heat-resistant strains 1330
Heaters 484
Heavy metals 162, 220, 226
Hemicellulose extract 932
Hemoparasites 1168, 1178
Hemorrhage syndrome 1192
Hemorrhagic Enteritis 1135
Hen depreciation 722
Herbicide bioassay 947
Heritability 1298
Hexamitiasis 1154, 1156
High ambient temperature 465
High-rise houses 571
Histomoniasis 1152
Hitchner's form 1119
Homocysteine 172
Homogenization 282
Hot blade 622
House flies 756
House heating 603
House Mouse 766
Housecleaning procedures 1226
Housing density 534
Human Safety 1430
Humane slaughter 198
Humoral Immunity1 041
Hydropericardium syndrome 1134
Hypervitaminosis 894
Hypervitaminosis D 890

I

Iberomesornis romerali 14
Idealized population 1289
Immune System 1029
Implicit costs 1357
Incandescent light 664
Incineration 744
Income demand 1350
Income elasticity 1351

Incoming Cooler Management 1224
Incubation 364
Incubation periods 369
Indian Broiler industry 7
Indian Poultry Industry 1
Inedible viscera 204
Infectious bronchitis 1116
Infectious bursal disease 1124
Infectious coryza 1085
Infectious laryngotracheitis 1118
Infectious stunting syndrome 1133
Infundibulum 112
Ingestion 853
Inorganic elements 852
Insecticides 162
Insemination dose 144
Insoluble Fiber 232
Insolvent 1363
Insulating animal sheds 535
Insulation 487, 532
Intake Control 496
Interferon system 1039
Intermittent lighting 669
Intestinal Health 698
Intramuscular injection 1243
Invoice 1364
Iodine 902
Ionising radiation 294
Ionostatic theory 854
Iron 901
Irradiated meat 358
Iso-cost line 1358
Iso-quant curves 1358
Iso-revenue line 1355
Isthmus 114

J

Japanese quail 26
Japanese quails 830, 916
Jojoba meal 940
Judging breeding males 685
Judging Pullets 677
Juiciness 263
Jutka method 199
Juvenile behavior 1406

K

Keel disease 1197
Killing of birds 1065
Kosher 325
Kosher's method 199

L

Lactic acid fermentation 787
Laser beak-trimming 625
Layer diets 951
Layers 478
Laying cages 710
Laying hens 855
Lectins 1004
Leg disorders 1190
Lethal genes 1286
Leucocytozoonosis 1154
Liability 1364
Liaoningornis 14
Lice 1177
Light intensity 663, 672
Light-proof houses 674
Lighting Management 663
Lime 544
Linear drinkers 848
Linear feeders 848
Lipids 875
Lipoproteins 164
Lipostatic theory 854
Liquid Spray 765
Liquor 792
Litter Management 702
Litter Moisture 704
Litter moisture 524
Live-hang 198
Losses 1364
Low blood cholesterol 186
Low-pressure fogging system 569
LP metabolism 165
Lung tissue 1072
Lysine 957
Lysozyme 152

M

Maceration 1433
Magnesium 900
Magnum 112
Maintenance of Accounts 1362
Maize 928, 930
Male Management 699
Male Reproductive System 134
Mallard Ducks 28

Management during lay 691
Management Requirements 835
Manganese 900
Mannan oligisaccharides 976
Manual Egg Turning 415
Manure 798
Manure Gases 799
Manure sales register 818
Marker assisted selection 1348
Market integration 1390
Marketing 649
Marketing agencies 1388
Marketing channels 1389
Marketing institutions 1389
Mash feed 950
Masonry 550
Mass burial 775
Mating 141
Mating behavior 1409
Mating ratio 693, 845
Meat scrap 936
Meat spots 241
Meat-cum-bone meal 936
Meat-type breeding hens 912
Mechanical Ventilation 494, 809
Mediterranean Class 53
Mendelian inheritance 1262
Mendel's laws 1262
Merchant middlemen 1388
Mercury vapor 665
Mesotocin 440
Metabolic Diseases 1187
Metabolizable energy 881
Methane 792
Methionine 957
Micro-satellites 1341
Microbial load 345
Microbiology of Poultry Meat 353
Microwave 314
Middlemen 1388
Millets 929
Mineral sources 942
Minerals 874, 898, 983
Miscellaneous Diseases 1194
Miscellaneous mycotoxins 1023
Mites 1177
Moisture condensation 472
Moisture removal 539
Moisture retarders 535
Molasses 931
Molting 680, 717
Molybdenum 902
Monensin 1199
Moniliasis 1108
Moniliformin 1022
Monoclonal antibodies 1344
Mononuclear phagocytes 1039
Mortars 544
Mottled shell 240
Mouse Baiting 770
Movement of air 498
Mucosal Immunity 1042
Multiple mycotoxins 1023
Muscovy Ducks 28
Muscovy ducks 827
Mustard oil cake 938
Mycotoxicosis 1106
Mycotoxins 227

N

Naked neck gene 1329
Napalm 794
Natural mating 141
Natural molting 717
Natural Ventilation 479, 809
Necropsy Examination 1063
Necrotic Enteritis 1088
Neem seed cake 939
Neoplastic diseases 1111
Nest-boxes 594
Neuromas 618
New duck disease 1096
Newcastle (Ranikhet) disease 1119
Niacin 893
Nicarbazine 1199
Nicking 1288
Nitrofurans 1199
Nitrogen 754
Nitrogen in the litter 705
Nitrophenide 1199
Nitrosamines 225
Non-food Applications 360
Non-recurring expenditure 1396
Non-starch polysaccharides 1009
Norway Rat 766
Novel Pyrolysis 795
Nucleic acids 1258
Nutrient requirements 906
Nutrients 850

Nutritive Value of Eggs 147

O

Oats 929
Occupational asthma 811
Ocean Disposal 794
Ochratoxin A 227, 1018
Odor control 806
Oily bird syndrome 648
Open-air burning 777
Open-handled spiral probe 963
Open-handles grain probe 963
Opened-out eggs 275
Operative temperature 434
OR potential 357
Organic acids 977, 982
Organic dust toxic syndrome 811
Organic poultry production 596
Organoleptic measurement 804
Organophosphorus compounds 1201
Oriental Class 55
Oscillation response 250
Osteomyelitis 648
Ostrich 834
Over-production 1191
Oviposition 122
Ovoflavoprotein 153
Ovoinhibitor 152
Ovomucoid 152
Ovotransferrin 153
Ovulation 105
Ovulatory cycle 108
Ovulatory sequence 109
Oxidation Ditches 736

P

P-Band 520
Packing material 1383
Pad and fan system 501, 569
Palm kernel meal 940
Panting 462
Pantothenic acid 895
Parasites 814
Parasitic Diseases 1160
Parasiticides 224
Paratyphoid infections 1080
Parental-care theory 11
Partition of energy 880
Partridge 88
Partridges 833
Passive immunity, 1044
Pasteurization 283
Paternal half-sibs 1313
Pathogen reduction 973
Patulin 1022
Peafowl 834
Peanut 937
Pearson's square method 988
Peck order 1407
Pelican grain sampler 964
Pellet quality 951
Pelleted feed 951
Pen mating 1324
Penicillic acid 227
Perches 594
Performance Enhancers 969
Pest Management 756
Pesticides 221
Pests 814
pH of albumen 247
pH of egg 275
Phase feeding 967
Pheasants 832
Phenols 1008
Phenotypic correlation 1311
Phobia 1405
Phosphorus 754, 898
Photo-periodic response 101
Photo-periodism 98
Photo-refractoriness 100
Photoperiod 668
Physical thermoregulation 451
Phytates 1010
Phytohaemagglutinins 1004
Phytotoxins 227
Pickled eggs 312
Pigeons 833
Pigmentation 682
Pimpling 241
Placement of semen 145
Plain omelette 318
Plant Visitors 1222
Plantar pododermatitis 647
Plasmodium infection 1158
Plastering 551
Plastic vent plugs 416
Plate test 948
Plinth-filling 550
Plinth-wall 546

Poaching 339
Polychlorinated biphenyls 222
Polycyclic aromatic hydrocarbons 224
Polymerase chain reaction 1341
Portland Cement 544
Potassium 754, 899
Potassium permanganate 1200
Poultry by-product meal 937
Poultry Farming 798
Poultry feeds 950
Poultry flocks 1215
Poultry genome mapping 1340
Poultry house ventilation 498
Poultry Housing 560
Poultry Manure 729
Poultry Meat 211
Poultry Meat Consumption 228
Poultry Meat Products 320
Poultry Meat Quality 257
Poultry Operations 1395
Poultry Welfare 1413
Poultry welfare guidelines 1413
Pox 1123
Precautions for vaccination 1248
Precocial animals 435
Preservation of Eggs 270
Preservation of Poultry Meat - 284
Pressure-cooking 219
Price demand 1349
Price determination1 381
Price elasticity 1351
Price spread 1392
Principles of Ventilation 455
Processed meat 216
Processed poultry manure 752
Processing waterfowl 210
Producers 1388
Production possibility curve 1354
Profiling A Disease 1051
Profit estimation 1400
Proprietor 1364
Protease inhibitors 1004
Protein sources 934
Proteins 851, 872, 883
Proteins of albumen 149
Protozoa 1052
Protozoan Diseases 1143
Public health benefits 936
Pullet disease 1198
Pyridine Equivalent 804
Pyridoxine 894

Q

Quantitative traits 1288
Quasi-static compression 250

R

Radiation death time 294
Radionuclides 223
Rancidity 876
Random mating 1322
Rat Baiting 769
Ration Formulation 985
Raw eggs 150
Raw meat 211
Re-feeding 794
Ready-to-cook Chicken 194
Rearing Systems 578
Recombiant DNA 1336
Recurrent selection 1322
Redhocks 648
Redirected Activity 1405
Redundant genes 1261
Reflectors 665
Reinforced Cement Concrete 545
Release of LH 108
Removal of moisture 491
Resin Strips 764
Respiratory diseases 810
Respiratory moisture 539
Resting metabolic rates 443
Restricted Feeding 656
Restricted selection index 1321
Restriction enzymes 1336
Retailers 1388
Reuse of old litter 702
Revenue 1364
Rhinotracheitis 1092
Riboflavin 892
Rice 929, 931
Ridge ventilators 486
Ring-necked Pheasant 27, 919
Risk of contamination 788
Roasted egg 317
Roaster 321
Roasting 331
Robotic beak-trimming 624
Rodent Biology 767
Rodenticides 768, 1201

Rodents 766
Roof slope 487
Roof-coverings 548
Roofing material 546
Rotary drum dryers 743
Rotational cross-breeding 1323
Routine deworming 1181
Routine feeding 610
Rubratoxin B 1021
Rule of thumb 610

S

Safflower meal 939
Sal seed meal 940
Sales 1364
Salt 942
Sanitation 1225
Saponins 1007
Sautéing 340
Scalding 200
Scale of points 92
Scotch eggs 312
Scrambled egg 315
Seasonal demands 6
Secondary sex glands 139
Segmentation 123
Selection index 1319
Selenium 902
Semen 137
Semen evaluation 144
Sensible heat loss 439
Sesame meal 938
Sex-chromosomes 1258
Sex-influenced traits 1285
Sex-linked genes 1329
Sex-linked inheritance 1284
Shackling 651
Shape index 244
Shed and equipment5 88
Shell formation 117
Shell gland 114
Shell membranes 132
Shell porosity 251
Shell quality 238, 248
Shell structure 130
Shelter Engineering 543
Shift system of mating 1326
Shirred-egg 315
Shivering thermogenesis 453
Shredded chicken 342
Sidewall height 487
Sidewall Vent Design 487
Sight Culling 677
Silver penciled 87
Simplex method 999
Single entry system 1366
Single-cell protein 940
Sino-sauropteryx 14
Skeleton photo-periods 672
Slack space 986
Slat-and-litter system 583
Slaughter 1420
Slotted grain probe 963
Slow feathering gene 1329
Smoked meat 358
Soap Formation 869
Social behavior 1407
Sodium 899
Sodium bicarbonate 1200
Sodium chloride 1200
Soft-shelled egg 240
Soil injection 749
Soluble Fiber 232
Solvent 1364
Somatic cell fusion 1334
Sorghums 930
Source of microorganisms 354
Soyabean meal 937
Spacing behavior 1406
Spanish omelette 318
Specialty Egg Products 299
Specialty Turkeys 325
Speck fly count 758
Speculative middlemen 1389
Sperm storage 120
Spermatozoa 140
Spirochaetosis 1086
Spleen 1032
Split-genes 1261
Spondylolisthesis 647
Spot-heating 603
Spray Administration 1251
Sprinkler irrigation 749
Squab broilers 652
Stability of vitamins 927
Stack effect 489
Standard classification 50
Standard metabolic rate 443
Staphylococcal arthritis 647

Starch Molasses 932
Static Pressure 512
Stationary Building Atomizers 764
Steaming 339
Stereotypic Behaviors 1405
Stereotypies 1428
Sterigmatocystin 1022
Stewing 219
Stewing Chickens 321
Sticky fly tapes 758
Still Air Model Incubator 412
Still-air 380
Stir-frying 341
Stirred custard 319
Stock 1364
Stocking Density 706
Strip test 948
Structural constraints 562
Structure of egg 128
Stud mating 1324
Stuffed chicken breasts 342
Stuffed egg 312
Stunning 198, 1420
Sub-droplets 155
Sub-embryonic fluid 395
Subabul meal 940
Subcutaneous injection 1242
Subsidiary books 1371
Sudden-death syndrome 1193
Sulfonamides 1199
Summer ventilation 482, 490, 505
Summit metabolism4 53
Sunflower seed meal 938
Surface irrigation 748
Swans 833
Synovitis 647
Systems of mating 1322

T

T-2 toxin 227, 1020
Tamarind seeds 939
Tandoori chicken 219
Tannins 1009
Tapered bag-triers 964
Temperature requirements 477
Temperature-humidity index 434
Temporary trimming 624
Testes 135
Tetraploidy 1333
Texture 260
Thawing 326
Thawing Turkey 327
The European Directive 585
Theory of demand 1349
Thermal buoyancy 479
Thermal depolymerization 793
Thermal influences on poultry 463
Thermal tachypnea 446
Thermoregulation 435
Thermostatic theory 854
Thin-shelled egg 240
Threshold Determination Level 804
Thymus 1031
Thyreostatics 224
Tibial dyschondroplasia 647
Ticks 1176
Tiles 543
Timing of AI 145
Toe-clipping 613
Toe-picking 621
Top crossing 1324
Total thermal resistance 459
Toxicity of inorganic elements 925
Toxins in Poultry Feed 1003
Trace Mineral Requirements 1049
Tranquillizers 224
Trans Fats 190
Transaction 1365
Transfer and Storage 747
Transgenic animals 1334
Transgenic chicken 1336
Transgenics 1348
Tray dryers 742
Tremortirn A 1021
Trench burial 774
Triacylglycerol 172
Trial balance 1372
Triangle of ventilation 518
Trichomoniasis 1153, 1156
Trisomy 1333
Triticale 930
Tropical zone 432
Trussing 204
Turkey coryza 1093
Turkey meningoencephalitis 1142
Turkey rhinotracheitis 1136
Turkey viral hepatitis 1142
Turkeys 69, 829
Turning 374, 384

Types of light 664

U

Ulcerative enteritis 1087
Unnamed fossils 13
Unsaturated fatty acids 187
Urolithiasis 1193
USDA chart 247
USDA recommendations 271

V

Vaccination 1238
Vaccination failure 1254
Vaccination program 1043
Vaccine Selection 1241
Vaccines 976, 1347
Vacuum Activity 1405
Van Waggenen Chart 247
Vascular heat exchange 444
Vegetable Fiber 231
Venipuncture 1064
Vent pecking 620
Ventilation 376, 846
Ventilation Air 806
Ventilation Capacity 511
Ventilation systems 456
Vertical integration 1391
Veterinary drugs 162
Vibration analysis 250
Viral Arthritis 1138
Viral Diseases 1111
Virus isolation 1212
Viruses 814, 1051
Vitamin C 1047
Vitamin-mineral premix 1001
Vitamins 852, 884
Vomitoxin 1021
Voucher 1365

W

Wash room 1221
Washing of eggs 270
Wastewater 787
Water 850
Water activity 266
Water Administration 1243
Water balance 375
Water consumption 921
Water deficiency 878
Water in foods 266
Water intake 470
Water loss during incubation 423
Water quality 920
Water-glass method 277
Wet Bulb Globe Temperature 433
Wheat 930, 931
Whole Chickens 322
Wholesalers 1388
Wild jungle fowls 22
Wild turkey species 25
Wind 480
Wind-chill factor 433
Windowless houses 639
Wing Molt 681
Wing-web puncture1 242
Winter ventilation 482

X

Xenobiotics 221

Y

Yolk 133
Yolk index 248
Yolk infrastructure 155
Yolk membranes 132
Yolk quality 248
Yolk utilization 393
Yolk-less eggs 242

Z

Zearalenone 227, 1022
Zinc 901
Zoalene 1199
Zone samplers 964